OXIDATIVE STRESS IN AQUATIC ECOSYSTEMS

OXIDATIVE STRESS IN AQUATIC ECOSYSTEMS

Edited by Doris Abele, José Pablo Vázquez-Medina, and Tania Zenteno-Savín

WILEY-BLACKWELL
A John Wiley & Sons, Ltd., Publication

Wiley-Blackwell is an imprint of John Wiley & Sons, formed by the merger of Wiley's global Scientific, Technical and Medical business with Blackwell Publishing.

Registered office: John Wiley & Sons Ltd, The Atrium, Southern Gate, Chichester, West Sussex, PO19 8SQ, UK

Editorial offices: 9600 Garsington Road, Oxford, OX4 2DQ, UK
350 Main Street, Malden, MA 02148-5020, USA
2121 State Avenue, Ames, Iowa 50014-8300, USA
111 River Street, Hoboken, NJ 07030-5774, USA

For details of our global editorial offices, for customer services and for information about how to apply for permission to reuse the copyright material in this book please see our website at www.wiley.com/wiley-blackwell

Library of Congress Cataloguing-in-Publication Data

Oxidative stress in aquatic ecosystems/[edited by] Doris Abele, José Pablo Vázquez-Medina, Tania Zenteno-Savín. – 1
p. cm.
Includes index.
ISBN 978-1-4443-3548-4 (hardback)
1. Aquatic ecology. 2. Aquatic biodiversity. 3. Oxidative stress. 4. Oxidation, Physiological. I. Abele, Doris. II. Vázquez-Medina, Jose Pablo. III. Zenteno-Savín, Tania.
QH541.5.W3O95 2012
577.6′14--dc23

2011014027

A catalogue record for this book is available from the British Library.

This book is published in the following electronic formats: epdf 9781444345957; Wiley Online Library 9781444345988; epub 9781444345964; MobiPocket 9781444345971

Set in 9/11pt Photina by Laserwords Private Limited, Chennai, India
Printed and bound in Malaysia by Vivar Printing Sdn Bhd

1 2012

CONTENTS

CONTRIBUTORS

DORIS ABELE *Alfred Wegener Institute for Polar and Marine Research, Department of Functional Ecology, Am Handelshafen 12, 27570 Bremerhaven, Germany*

JOSÉ AGUILERA *Photobiology Laboratory, Medical Research Center, Department of Dermatology, Faculty of Medicine, University of Málaga, 29071-Málaga, Spain*

EDUARDO ALVES DE ALMEIDA *Department of Chemistry and Environmental Sciences, IBILCE-UNESP, São José do Rio Preto, SP, Brazil*

LÍLIAN L. AMADO *Biological Sciences Institute, Federal University of Rio Grande- FURG, Rio Grande, RS, Brazil; and Post-graduation Program in Physiological Sciences, Comparative Animal Physiology, FURG, Rio Grande, RS, Brazil*

AFONSO CELSO DIAS BAINY *Department of Biochemistry, Federal University of Santa Catarina, UFSC, Florianópolis, Santa Catarina, Brazil*

MAURA BENEDETTI *Department of Biochemistry, Biology and Genetics, Polytechnic University of Marches, Ancona, Italy*

MARÍA BELÉN AGUIAR *Physical Chemistry-PRALIB, School of Pharmacy and Biochemistry, University of Buenos Aires, Junín 956, C1113AAD, Buenos Aires, Argentina*

RAFFAELLA BOCCHETTI *Department of Biochemistry, Biology and Genetics, Polytechnic University of Marches, Ancona, Italy*

ANNA BOGDANOVA *Institute of Veterinary Physiology, Vetsuisse Faculty and Zurich Center for Integrative Human Physiology (ZIHP), University of Zurich, Winterthurerstrasse 260, CH-8057 Zurich, Switzerland*

KATJA BROEG *Alfred Wegener Institute for Polar and Marine Research, Am Handelshafen 12, 27570 Bremerhaven, Germany*

NADIEZHDA CANTÚ-MEDELLÍN *Northwestern Center for Biological Research, La Paz, Baja California Sur, 23090, Mexico; and University of Alabama Birmingham, Department of Pathology, Birmingham, AL, 35294, USA*

BETUL CATALGOL *Institute of Nutrition, Friedrich Schiller University, Jena, Germany; Institute of Biological Chemistry and Nutrition, University Hohenheim, Stuttgart, Germany; Department of Biochemistry, Faculty of Medicine, Marmara University, 34668 Haydarpasa, Istanbul, Turkey*

PAOLO DI MASCIO *Department of Biochemistry, Institute of Chemistry, University of Sao Paulo, São Paulo, SP, Brazil*

ROBERT ELSNER *University of Alaska Fairbanks, School for Fisheries and Ocean Science, Institute of Marine Science, Fairbanks, AK, 99775-7220, USA*

MARÍA LUISA FANJUL-MOLES *National Autonomous University of Mexico, School of Sciences, C.P. 04510, México, D. F.*

LAURA A. FLORES-RAMÍREZ *University of Colima, Aquaculture and Biotechnology, FACIMAR, Manzanillo, Colima, México*

JOSÉ PEDRO FRIEDMANN ANGELI *Department of Biochemistry, Institute of Chemistry, Sao Paulo University, São Paulo, SP, Brazil*

CAROLINA A. FREIRE *Department of Physiology, Section of Biological Sciences, Federal University of Parana, Curitiba, PR, Brazil*

FLORÊNCIO PORTO FREITAS *Department of Biochemistry, Institute of Chemistry, University of Sao Paulo, São Paulo, Brazil*

MIRIAM FURNÉ *Department of Animal Biology, University of Granada, Campus Universitario Fuentenueva s/n, 18071 Granada, Spain*

CAMILA CARRIÃO MACHADO GARCIA *Department of Biochemistry, Institute of Chemistry, University of Sao Paulo, São Paulo, SP, Brazil*

MAX GASSMANN *Institute of Veterinary Physiology, Vetsuisse Faculty and Zurich Center for Integrative Human Physiology (ZIHP), University of Zurich, Winterthurerstrasse 260 CH-8057 Zurich, Switzerland*

MARISA HELENA GENNARI DE MEDEIROS *Department of Biochemistry, Institute of Chemistry, University of Sao Paulo, São Paulo, SP, Brazil*

OSMAR FRANCISCO GOMES *Department of Biochemistry, Institute of Chemistry, University of Sao Paulo, São Paulo, Brazil*

JAIME GÓMEZ-GUTIÉRREZ *Interdisciplinary Center for Marine Sciences, Department of Plankton and Marine Ecology, La Paz, Baja California Sur 23096, Mexico*

MARÍA E. GONSEBATT *Institute for Biomedical Research, National Autonomous University of Mexico, A.P. 70-228 Ciudad Universitaria, C.P. 04510, México, D.F.*

PAULA MARIELA GONZÁLEZ *Physical Chemistry-PRALIB, School of Pharmacy and Biochemistry, University of Buenos Aires, Junin 956, 1113 Buenos Aires, Argentina*

STEFANIA GORBI *Department of Biochemistry, Biology and Genetic, Polytechnic University of Marches, Ancona, Italy*

STEFANIE GRIMM *Institute of Nutrition, Friedrich Schiller University, Jena, Germany; and Institute of Biological Chemistry and Nutrition, University Hohenheim, Stuttgart, Germany*

MATTHEW B. GRISHAM *Immunology and Inflammation Research Group, Department of Molecular and Cellular Physiology, Louisiana State University Health Sciences Center, Shreveport, LA, USA*

TILMAN GRUNE *Institute of Nutrition, Friedrich Schiller University, Jena, Germany*

HELGA GUDERLEY *Department of Biology, University of Laval, Québec, Québec, Canada G1K 7P4*

NORBERT HÄUBNER *Department of Ecology and Evolution, Uppsala University, Villavägen 14, SE-752 36 Uppsala, Sweden*

MARCELO HERMES-LIMA *Laboratory for Free Radicals, Department of Cellular Biology, University of Brasilia, Brasilia, DF, Brazil*

JOANNA JOYNER-MATOS *Department of Biology, Eastern Washington University, Cheney, WA, USA*

DAVID JULIAN *Department of Biology, University of Florida, Gainesville, FL, USA*

ANDREAS KRELL *Alfred Wegener Institute for Polar and Marine Research, 27570 Bremerhaven, Germany*

GISELA LANNIG *Alfred Wegener Institute for Polar and Marine Research in the Hermann von Helmholtz Association of National Research Centres e.V., Integrative Ecophysiology, 27570 Bremerhaven, Germany*

MICHAEL P. LESSER *Department of Molecular, Cellular and Biomedical Sciences, University of New Hampshire, Durham, NH 03824, USA*

RAFAELA ELIAS LETTS *Post-graduation program in Physiological Sciences, Comparative Animal Physiology, Federal University of Rio Grande – FURG, Rio Grande, RS, Brazil; Department of Biochemistry, Biology and Genetic, Polytechnic University of Marches, Ancona, Italy*

MARCO A. LIÑÁN-CABELLO *University of Colima, Aquaculture and Biotechnology, FACIMAR, Manzanillo, Colima, México*

SIMONE LIPINSKI *Institute of Clinical Molecular Biology, Cell Biology Department, Christian-Albrechts University Kiel, Schittenhelmstrasse 12, 24105 Kiel, Germany*

ROBERTO I. LÓPEZ-CRUZ *Laboratory of Biochemisty Research, Graduate School in Molecular Biomedicine, National School of Medicine and Homeopathy (ENMyH-IPN), México D.F.*

ALCIR LUIZ DAFRE *Department of Physiological Sciencies, Center of Biological Sciences, Federal University of Santa Catarina, 88040-900 Florianópolis, SC, Brazil*

OLEH V. LUSHCHAK *Department of Biochemistry and Biotechnology, Vassyl Stefanyk Precarpathian National University, 57 Shevchenko Str., 76025, Ivano-Frankivsk, Ukraine*

VOLODYMYR I. LUSHCHAK *Department of Biochemistry and Biotechnology, Vassyl Stefanyk Precarpathian National University, 57 Shevchenko Str., 76025, Ivano-Frankivsk, Ukraine*

ALFONSO N. MAEDA-MARTÍNEZ *Northwestern Center for Biological Research (CIBNOR), La Paz, Baja California Sur 23090, Mexico*

GABRIELA MALANGA *Physical Chemistry-PRALIB, School of Pharmacy and Biochemistry, University of Buenos Aires, Junin 956, C1113AAD, Buenos Aires, Argentina*

GLAUCIA REGINA MARTINEZ *Molecular Biology, Section for Biological Sciences, Federal University of Parana, Curitiba, PR, Brazil*

J. C. MASSABUAU *University of Bordeaux 1, CNRS, UMR 5805 EPOC, Place du Dr Peyneau, 33120, Arcachon, France*

SAYURI MIYAMOTO *Department of Biochemistry, Institute of Chemistry, University of Sao Paulo, São Paulo, SP, Brazil*

JOSÉ MONSERRAT *Biological Sciences Institute, Federal University of Rio Grande- FURG, Rio Grande, RS, Brazil; Post-graduation program in Physiological Sciences, Comparative Animal Physiology, Federal University of Rio Grande – FURG, Rio Grande, RS, Brazil*

AMALIA E. MORALES *Department of Animal Biology, University of Granada, Campus Universitario Fuentenueva s/n, 18071 Granada, Spain*

FLÁVIA DANIELA MOTTA *Department of Biochemistry, Institute of Chemistry, Universiy of Sao Paulo, São Paulo, Brazil*

ETSUO NIKI *National Institute of Advanced Industrial Science and Technology, Health Research Institute, Osaka 563–8577, Japan*

MIKKO NIKINMAA *Department of Biology, University of Turku, FI-20014, Turku, Finland*

LÍLIAN NOGUEIRA *Departament of Chemistry and Environmental Sciences, IBILCE-UNESP, São José do Rio Preto, SP, Brazil*

AMALIA PÉREZ-JiMÉNEZ *Department of Animal Biology, University of Granada, Campus Universitario Fuentenueva s/n, 18071 Granada, Spain*

EVA E. R. PHILIPP *Institute of Clinical Molecular Biology, Christian-Albrechts-University Kiel, Schittenhelmstrasse 12, 24105 Kiel, Germany*

REBECCA POGNI *Chemistry Department, University of Siena, Italy*

PAUL J. PONGANIS *Center for Marine Biotechnology and Biomedicine, Scripps Institution of Oceanography, University of California San Diego, La Jolla, CA 92093-0204, USA*

SUSANA PUNTARULO *Physical Chemistry-PRALIB, School of Pharmacy and Biochemistry, University of Buenos Aires, Junin 956, 1113 Buenos Aires, Argentina*

JONATHAN RAST *Sunnybrook Health Sciences Centre, Department of Medical Biophysics and Department of Immunology, University of Toronto, 2075 Bayview Avenue, Rm. S126B, Toronto, ON M4N 3M5, Canada*

RALF RAUTENBERGER *Institute for Polar Ecology, Christian Albrechts University of Kiel, Wischhofstraße*

1–3, 24148 Kiel, Germany; Department of Marine Botany, University of Bremen, Faculty of Biology and Chemistry, 28359 Bremen, Germany; and Department of Botany, University of Otago, Dunedin, 9016, New Zealand

FRANCESCO REGOLI *Department of Biochemistry, Biology and Genetics, Polytechnic University of Marches, Ancona, Italy*

HÉCTOR REYES-BONILLA *Autonomous University of Baja California Sur, La Paz, Baja California Sur, México*

JOSENCLER L. RIBAS FERREIRA *Post-graduation program in Physiological Sciences, Comparative Animal Physiology, Federal University of Rio Grande – FURG, Rio Grande, RS, Brazil*

ALESSANDRA M. ROCHA *Post-graduation program in Physiological Sciences, Comparative Animal Physiology, Federal University of Rio Grande – FURG, Rio Grande, RS, Brazil*

GUSTAVO RODRÍGUEZ-ALONSO *Department of Molecular Medicine and Bioprocesses, Institute of Biotechnology, Autonomous University of Mexico, México; Faculty of Sciencies, Autonomous University of the State of Morelos, México*

GRAZIELLA ELIZA RONSEIN *Department of Biochemistry, Institute of Chemistry, University of Sao Paulo, São Paulo, SP, Brazil*

PHILIP ROSENSTIEL *Institute of Clinical Molecular Biology, Cell Biology Department, Christian-Albrechts University Kiel, Schittenhelmstrasse 12, 24105 Kiel, Germany*

HALYNA M. SEMCHYSHYN *Department of Biochemistry and Biotechnology, Vassyl Stefanyk Precarpathian National University, 57 Shevchenko Str., 76025, Ivano-Frankivsk, Ukraine*

DANILO GRUNIG HUMBERTO SILVA *Department of Chemistry and Environmental Sciences, IBILCE-UNESP, São José do Rio Preto, São Paulo, Brazil*

PAULINE SNOEIJS *Department of Systems Ecology, Stockholm University, Svante Arrhenius väg 21A, SE-106 91 Stockholm, Sweden*

INNA M. SOKOLOVA *Department of Biology, University of North Carolina at Charlotte, Charlotte, NC, USA*

DIETER STEINHAGEN *University of Veterinary Medicine Hannover, Centre for Infection Medicine, Fish Disease Research Unit, Buenteweg 17, 30559 Hannover, Germany*

JANET M. STOREY *Institute of Biochemistry, Carleton University, Ottawa, ON, Canada*

KENNETH B. STOREY *Institute of Biochemistry, Carleton University, Ottawa, ON, Canada*

JULIA STRAHL *Alfred Wegner Institute for Polar and Marine Research, Department of Functional Ecology, Am Handelshafen 12, 2570 Bremerhaven, Germany*

ALEXEY A. SUKHOTIN *White Sea Biological Station, Zoological Institute of Russian Academy of Sciences, 199034 St. Petersburg, Russia*

PETER SYLVANDER *Department of Systems Ecology, Stockholm University, Svante Arrhenius väg 21A, SE-106 91 Stockholm, Sweden*

NELLY TREMBLAY *Interdisciplinary Center for Marine Sciences, Department of Plankton and Marine Ecology, Av. IPN, Col. Palo de Santa Rita s/n, La Paz, Baja California Sur, 23096, Mexico*

BRENDA VALDERRAMA *Department of Molecular Medicine and Bioprocesses, Institute of Biotechnology, Autonomous University of Mexico, México*

JOSÉ PABLO VÁZQUEZ-MEDINA *Northwestern Center for Biological Research (CIBNOR). Mar Bermejo 195; Playa Palo Santa Rita, La Paz, Baja California Sur, 23090, Mexico; School of Natural Sciences, University of California Merced, Merced, CA 95343, USA*

JULIANE VENTURA-LIMA *Biological Sciences Institute, Federal University of Rio Grande- FURG, Rio Grande, RS, Brazil; Post-graduation program in Physiological Sciences, Comparative Animal Physiology, Federal University of Rio Grande – FURG, Rio Grande, RS, Brazil*

ALEXIS F. WELKER *Laboratory for Free Radicals, Department of Cell Biology, University of Brasilia,*

Brasilia, DF, Brazil; Faculty of Ceilandia, University of Brasilia, DF, Brazil

DANILO WILHELM FILHO *Department of Ecology and Zoology, Center for Biological Sciences, Federal University of Santa Catarina, 88040-900 Florianópolis, SC, Brazil*

DOROTHEE WILHELMS-DICK *University of Bremen, Geoscience Department, Klagenfurter Straße, 28359 Bremen, Germany*

XIANG-PING NIE *Department of Systems Ecology, Stockholm University, Svante Arrhenius väg 21A, SE-106 91 Stockholm, Sweden; Department of Ecology, Jian University, 601 West Huangpu Street, Guangzhou 510632, China*

YASUKAZU YOSHIDA *National Institute of Advanced Industrial Science and Technology, Health Research Institute, Osaka 563–8577, Japan*

TANIA ZENTENO-SAVÍN *Northwestern Center for Biological Research (CIBNOR). Mar Bermejo 195; Playa Palo Santa Rita, La Paz, Baja California Sur, 23090, Mexico*

ACKNOWLEDGMENTS

Michael P. Lesser (Chapter 1) acknowledges the support from various funding agencies including NOAA and NSF for research on coral reef bleaching. Additionally, the Coral Reef Targeted Research (CRTR) Program, a partnership between the Global Environmental Facility and the World Bank provided funding and a challenging environment to explore the underpinnings and ramifications of global climate change on coral reefs around the world.

Research in the Storey laboratory (Carolina A. Freire, Alexis F. Welker, Janet M. Storey, Kenneth B. Storey, and Marcelo Hermes-Lima, Chapter 3) is supported by NSERC Canada; Kenneth B. Storey holds the Canada Research Chair in Molecular Physiology. Brazilian authors thank CNPq, INCT-CNPq Redoxoma, and DAAD (German Academic Exchange Service).

The work presented by Pauline Snoeijs, Peter Sylvander, and Norbert Häubner in Chapter 5 and by Pauline Snoeijs, Norbert Häubner, Peter Sylvander and Xiang-Ping Nie in Chapter 29 was supported by the research grants Formas 21.9/2003-1033, Formas 21.0/2004-0313, and EU Stukturstöd FiV Dnr 231-0692-04.

Nelly Tremblay, Tania Zenteno-Savin, Jaime Gómez-Gutiérrez, and Alfonso N. Maeda-Martínez (Chapter 6) wish to thank C.J. Robinson, the crews of the R/V *El Puma* and R/V *Francisco de Ulloa*, and the graduate students and researchers at ICMyL-UNAM, UABCS, and CICIMAR-IPN for recording hydroacoustic, environmental information, and collecting zooplankton samples; N.O. Olguín-Monroy for technical help in the biochemical analyses, O.Calvario M. for training in the use of the Oxymat2000, and S. Martínez-Gómez, O. Angulo-Campillo, J. R. Morales, H. Urias-Leyva, and J. Cruz for helping to sort out the krill specimens from the zooplankton samples. Nelly was supported by graduate student grants Programa Institucional de Formación de Investigadores (PIFI-IPN) and Secretaría de Relaciones Exteriores. This research was supported by CICIMAR-IPN, CONACYT-FOSEMARNAT (2004-01-144), CONACYT- SAGARPA (S007-2005-1-11717), CIBNOR (PC2.0, PC2.5, PC2.6), and ICMyL-UNAM (IN219502, IN210622).

Marco A. Liñán-Cabello, Michael P. Lesser, Laura A. Flores-Ramírez, Tania Zenteno-Savín, and Hector Reyes-Bonilla (Chapter 9) would like to thank everyone directly and indirectly involved in the research included in this chapter, too many to be enumerated individually. Research was supported by the Alvarez-Buylla de Aldana Foundation, Universidad de Colima, PROMEP of the Secretaría de Educación Pública, Mexico (Marco), CIBNOR (Tania), and UABCS (Hector).

Work presented in Chapter 11 by Roberto I. López-Cruz, Alcir Luiz Dafre, and Danilo Wilhelm Filho was funded by grants from CONACYT, CIBNOR, and a fellowship from Programa de Estudios de Posgrado (CIBNOR) (to Roberto). The authors wish to thank Marco Antonio Salazar Bermúdez (UABCS) for the artwork.

Mikko Nikinmaa (Chapter 12) is supported by the Centre of Excellence grants from the Academy of Finland and the University of Turku, Max Gassmann and Anna Bogdanova (Chapter 12) are supported by the Swiss National Science Foundation (# 320030-125013, #310030-124970 and #310030-124970/1) and by the Zurich Center for Integrative Human Physiology.

Research presented by Tania Zenteno-Savín, José Pablo Vázquez-Medina, Nadiezhda Cantú-Medellín, Paul J. Ponganis, and Robert Elsner (Chapter 13) was funded by grants from ONR, SEMARNAT-CONACYT, CIBNOR (to Tania), OPP 0944220 (to Paul), and fellowships from Programa de Estudios de Posgrado at CIBNOR (to José Pablo and Nadia). José Pablo is currently supported by UC-Mexus, CONACYT and Secretaría de Educación Pública fellowships.

While working on their manuscript (Chapter 14), Alexey A. Sukhotin was supported by Russian Foundation for Basic Research (grant #10-04-00316), Julia Strahl by the German Science foundation (DFG), grant numbers AB124/10-1 and DR262/10-1, and Eva E.R. Philipp by the DFG Cluster of Excellence "The Future Ocean."

The research presented in Chapter 15 by María Luisa Fanjul-Moles and María E. Gonsebatt was partially supported by PAPIIT IN-207008 (María Luisa) and by PAPIIT IN-207408 (María Eugenia). María Luisa and María Eugenia thank Julio Prieto-Sagredo for his help with the figures.

Brenda Valderrama, Gustavo Rodríguez-Alonso, and Rebecca Pogni (Chapter 16) were funded by the Executive Program of Scientific and Technological Cooperation Mexico-Italy 2006–2009. Brenda and Gustavo received additional support from the SNI-STUDENTS fund. Brenda and Gustavo acknowledge financial support from CONACYT.

While working on the manuscript Eva Phillip and Philip Rosentiel (Chapter 17) were supported the by the German Science foundation (DFG) Cluster of Excellence "The Future Ocean," Eva Phillip and Simone Lipinski by the Cluster of Excellence "Inflammation at Interfaces," and Simone Lipinsky by the DFG grant RO2994/5-1 "Reactive oxygen species as modulators and effectors of epithelial defense: A role for Nod-like receptors."

Inna M. Sokolova, Alexey A. Sukhotin, and Gisela Lannig (Chapter 19) the work of Amalia E. Morales, Amalia Pérez-Jiménez, Miriam Furné, and Helga Guderley (Chapter 20) was primarily supported by NSERC of Canada, as well as by DFO, with strong collaborative support from Jean-Denis Dutil of DFO gratefully acknowledge the following programs and organizations for support during the work on this manuscript: NSF awards IOS-0921367 and IBN-0347238 to Inna, Russian Foundation for Basic Research Grant #10-04-00316 to Alexey and the PACES research program of the Alfred Wegener Institute to Gisela.

Volodymyr I. Lushchak (Chapters 21 and 32), Halyna M. Semchyshyn, and Oleh V. Lushchak (Chapter 32) wish to thank Nadia Semchuk who helped with figures and artwork.

José María Monserrat, Rafaela Elias Letts, Josencler L. Ribas Ferreira, Juliane Ventura-Lima, Lílian L. Amado, Alessandra M. Rocha, Stefania Gorbi, Raffaella Bocchetti, Maura Benedetti, and Francesco Regoli (Chapter 23) were supported by funds from the Brazilian agency CNPq (Productivity Research Fellowship) to José and by a grant from the LASPAU/Fincyt Peruvian Research Fund to Rafaela. Josencler and Alessandra are graduate fellows from CNPq and CAPES, respectively. Juliane receives a post-doctoral fellowship from the Brazilian Agency CAPES. The support from CAPES (PROCAD Program, Proc. 089/2007) is acknowledged by Lílian and José.

Matthew B. Grisham (Chapter 24) wishes to thank all current and former students and post-doctoral fellows who contributed greatly to our understanding of the role of reactive oxygen and nitrogen species in acute and chronic inflammation.

Work presented in Chapters 25 (Graziella Eliza Ronsein, Glaucia Regina Martinez, Eduardo Alves de Almeida, Sayuri Miyamoto, Marisa Helena Gennari de Medeiros, and Paolo Di Mascio), 28 (Eduardo Alves de Almeida, Danilo Grunig Humberto Silva, Afonso Celso Dias Bainy, Florêncio Porto Freitas, Flávia Daniela Motta, Osmar Francisco Gomes, Marisa Helena Gennari de Medeiros and Paolo Di Mascio), 30 (Eduardo Alves de Almeida, Glaucia Regina Martinez, and Paolo Di Mascio), 34 (Sayuri Miyamoto, Eduardo Alves de Almeida, Lílian Nogueira, Marisa Helena Gennari de Medeiros, and Paolo Di Mascio), 37 (José Pedro Friedmann Angeli, Glaucia Regina Martinez, Flávia Daniela Motta, Eduardo Alves de Almeida, Marisa Helena Gennari de Medeiros, and Paolo Di Mascio) and 38 (Camila Carrião Machado Garcia, José Pedro Friedmann Angeli, Eduardo Alves de Almeida, Marisa Helena Gennari de Medeiros, and Paolo Di Mascio) was supported by the Brazilian research funding institutions FAPESP (Fundação de Amparo à Pesquisa do Estado de São Paulo), CNPq (Conselho Nacional para o Desenvolvimento Científico e Tecnológico), CAPES (Coordenação de Aperfeiçoamento de Pessoal de Nível Superior), Pró-Reitoria de Pesquisa USP, Instituto do Milênio: Redoxoma and INCT de Processos Redox em Biomedicina – Redoxoma. The authors also thank L'ORÉAL-UNESCO for Women in Science (Sayuri) and The John Simon Memorial Guggenheim Foundation (Paolo) for the fellowships provided.

Betul Catalgol, Stefanie Grimm, and Tilman Grune (Chapter 33) thank COST (B35 and CM1001) for support.

Doris Abele thanks the Alfred-Wegener Institute of Polar and Marine Research in Bremerhaven for supporting her oxidative stress working group and its scientific works in the cold South for many years; and to her colleagues at the AWI, especially Thomas Brey, Christian Wiencke, Fritz Buchholz, and Victor Smetacek for a creative and inspiring working atmosphere. Doris gratefully acknowledges support by German Science Foundation (DFG) throughout her career track, during the time when we worked on the book it was DFGAb124/10-1.

We would like to thank all co-authors for joining this venture and the effort each and every one of you invested in making this book reality. We appreciate your straightforward cooperation and quick responses to our (sometimes manic) e-mails. We are especially proud of the participation of young authors such as Julia Strahl, Nelly Tremblay, Dorothee Dick, Paula Gonzalez, Laura Flores-Ramírez, Gustavo Rodríguez-Alonso, Josencler Ribas Ferreira, Alessandra Rocha, Juliane Ventura-Lima, Sayuri Miyamoto, Danilo Grunig Humberto Silva, Flávia Daniela Motta, Roberto López-Cruz, and Nadia Cantú-Medellín, who wrote brilliant text passages. Tania and José Pablo wish to thank the so-called *Secta de Estrés Oxidativo* for their patience, support and dedication.

Citlali Guerra, Stefanie Meyer, Michiel Rutgers van der Loeff, and Gerhard Diekmann from the Alfred-Wegener Institute, Bremerhaven, and Kai Bischoff from the University of Bremen, Norma Olguín-Monroy, Patricia Parrilla-Taylor, Paola Tenorio-Rodríguez, Marcela Vélez-Alavez, Vanessa Labrada-Martagón, Ramón Gaxiola-Robles, and Orlando Lugo-Lugo from Centro de Investigaciones Biológicas del Noroeste (CIBNOR), kindly took the time to review and improve chapters of the book. The photograph in the front cover was taken by Rigoberto Moreno (www.rigobertomoreno.com) in the coast of Nayarit, Mexico. We enormously appreciate Rigo's kind contribution of his artwork for our book.

We dedicate this book to all our students and post-doctoral fellows, past, present and future, who are a continuous source of inspiration, and who make all our efforts worth the while.

As editors, Doris, José Pablo, and Tania acknowledge the support of our home institutions AWI, DFG, CIBNOR, UC-Merced, as well as the funding provided by DAAD, CONACYT, and Secretaría de Educación Pública for our research. The cover photograph was taken by Rigoberto Moreno on the coast of Nayarit, Mexico. The editors appreciate Rigo's kind contribution of his artwork for our book.

LIST OF ABBREVIATIONS

A•: ascorbyl radical
aaMF: alternatively activated macrophages
AA: arachidonic acid
ABC: ATP-binding cassette
AChE: acetylcholinesterase
AH•: ascorbate
AMPK: adenosine monophosphate kinase
AMT: aminotriazole
AOX: total antioxidant capacity
APAF: poptosis protease-activating factor
APx: ascorbate peroxidase
AhR: aryl hydrocarbon receptor
ALAD: δ-aminolevulinic acid dehydratase
ARNT: AhR nuclear translocator
ARE: antioxidant response element
ATP: adenosine triphosphate
B[*a*]P: benzo[*a*]pyrene
BHT: butylated hydroxytoluene
BKD: bacterial kidney disease
β-OHBDH: β hydroxybutyrate dehydrogenase
caMF: classically activated macrophages
cAMP: cyclic adenosine monophosphate
CAR: constitutively active receptor
CAT: catalase
CDK: cyclin-dependent kinases
CDOM: colored dissolved organic matter
CFCs: chlorofluorocarbons
CGD: chronic granulomatous disease
cGMP: cyclic nucleotide guanosine monophosphate
CHH: crustacean hyperglycemic hormone
COX: cyclooxygenase
CPD: cyclobutane pyrimidine dimers
CS: citrate synthase
CSF: colony-stimulating factor
Cu,Zn-SOD: copper and zinc-dependent superoxide dismutase
Cyt *c*: cytochrome *c*
DBNBS: 3,5-dibromo-4-nitrosobenzensulfonate
DCFH-DA: 2′, 7′-dichlorofluorescin-diacetate (DCFH-DA)
DDC: diethyldithiocarbamate
DEB: dynamic energy budget
DHAR: dehydroascorbate reductase
DMPO: 5,5-dimethyl-1-pyrroline-N-oxide
DMSP: dimethylsulphoniopropionate
DMT1: divalent metal cation transporter 1
DNA: deoxyribonucleic acid
DNIC: dinitrosyl iron complex
DNPH: 2,4-dinitrophenylhydrazine
DOM: dissolved organic matter
DPX: DNA-protein cross-links
ECM: extracellular matrix
EC-SOD: extracellular superoxide dismutase
EDC: endocrine disrupting chemicals
ELISA: enzyme linked immunosorbent assay
EPR: electron paramagnetic resonance
EST: expressed sequence tag
ETC: electron transport chain
FAD: flavin-adenine dinucleotide
FAP: fluorescent age pigments
Fe-SOD: iron-containing superoxide dismutase
FOC: ferrous oxidation-xylenol orange
FOXO: forkhead box class O
G6PDH: glucose-6-phosphate dehydrogenase
GCL glutamate-cysteine ligase
GGT: γ-glutamyl transpeptidase
GIH: gonad inhibiting hormone
GCL: glutamate-cysteine ligase
GR: glutathione disulphide reductase
GPx: glutathione peroxidase
Grx: glutaredoxin
GSH: glutathione
GSSG: glutathione disulphide
GST: glutathione *S*-transferase
GTP: guanosine triphosphate
H_2O_2: hydrogen peroxide

H_2S: hydrogen sulfide
HCB: hexachlorobenzene
HGPRT: hypoxanthine guanine phosphoribosyl transferase
HIF-1: hypoxia inducible factor 1
HNE: 4-hydroxy-2-nonenal
HO: heme oxygenase
$HO^{\bullet}$: hydroxyl radical
HO^-: hydroxyide anion
$HO_2^{\bullet}$: hydroperoxyl radical
HOAD: hepatic b-hydroxyacyl CoA dehydrogenase
HOCl: hypochlorous acid
HOG: high-osmolarity glycerol
HRE: hypoxia-response element
HS^-: hydrosulphide anion
HSF: heat shock inducing factor
HSP: heat shock proteins
HSR: heat shock response
HX: hypoxanthine
IMP: inosine monophophate
IRP: iron-regulatory proteins
LDH: lactate dehydrogenase
LOX: lipoxygenase
LPO: lipid peroxidation
LPS: lipopolysaccharide
$LOO^{\bullet}$: peroxyl radical
LOOH: lipid peroxide
LRR: leucine-rich repeats
MAA: mycosporine-like amino acid
MAMP: microorganism-associated molecular pattern
MAP: Mehler-ascorbate pathway
MAPK: mitogen-activated protein kinase
MDA: malondialdehyde
MDAR: monodehydroascorbate reductase
MF: methyl farnesoate
MHC: major histocompatibility complex
MIH: molt inhibiting hormone
MLH: multilocus heterozygosity
Mn-SOD: manganese-dependent superoxide dismutase
MNIC: mononitrosyl iron tris(thiolate) complex
MNP: 2-methyl-2-nitroso propane
MPO: myeloperoxidase
MT: metallothionein
MV: methyl viologen
MXR: multixenobiotic resistance proteins
NADH: nicotinamide adenine dinucleotide hydrogen
NADPH: nicotinamide adenine dinucleotide phosphate hydrogen
NCBI: National Center for Biotechnology Information
NF-κB: nuclear factor κB
$NO^{\bullet}$: nitric oxide, nitrogen monoxide
NLR: NOD-like receptor
NOD: nucleotide-binding oligomerization domain
NOS: nitric oxide synthase
NOX: NADPH oxidase
NQO1: NADPH-quinone reductase 1
Nrf2: nuclear factor erythroid 2–related factor 8-oxodGuo: 8-oxo-7, 8-dihydro-2′-deoxyguanosine
$O_2^{\bullet -}$: superoxide radical
$^1O_2$: singlet oxygen
$ONOO^-$: peroxinitrite anion
ONOOH: peroxinitrous acid
OP: organophosphate
PAH: polycyclic aromatic hydrocarbons
PAMP: pathogen-associated molecular pattern
PAR: photosynthetic active radiation
PB: phenobarbitol
PBL: peripheral blood lymphocytes
PBN: phenyl-t-butyl-nitrone
PCBs: polychlorinated biphenyls
PCN: pregnenolone-16α-carbonitrile
PCR: polymerase chain reaction
PGG_2: prostaglandin endoperoxide G_2
PGH_2: prostaglandin endoperoxide H_2
PGHS: prostaglandin H synthase
PGN: peptidoglycan
PHD: prolyl hydroxylases
P_{O_2}: oxygen partial pressure
POBN: α(4-pyridyl-1-oxide)-N-t-butyl nitrone
POM: particulate organic matter
PPAR: peroxisomal proliferator activated receptor
PPRE: peroxisome proliferator responsive elements
proPO: prophenoloxidase system
PRR: pattern recognition receptor
Prx: peroxiredoxin
PS I: photosystem I
PS II: photosystem II
PSSG: protein-glutathione mixed disulfide
PSU: practical salinity units
PUFA: polyunsaturated fatty acid
pVHL: von Hippel-Lindau protein
PXR: pregnane X receptor
$R^{\bullet}$: alkyl radical
RBC: red blood cell
RNA: ribonucleic acid
RNS: reactive nitrogen species

ROM: reactive oxygen metabolism
$RO^{\bullet}$: alkoxy radical
$ROO^{\bullet}$: peroxyl radical
ROOH: lipid hydroperoxide
ROS: reactive oxygen species
RPP: reversible protein phosphorylation
RUBISCO: ribulose-1,5-carboxylase/oxygenase
RXR: retinoid X receptor
S^{2-}: sulphide anion
SMR: standard metabolic rate
SOD: superoxide dismutase
SRCR: scavenger receptor cysteine-rich
ST: sulfotransferases
TBARS: thiobarbituric acid reactive substances
TCDD: 2,3,7,8-tetrachlorodibenzo-p-dioxin
TF: transcription factor
TfR: transferrin receptor
TLR: Toll-like receptor
TNF-α:tumor necrosis factor alpha
TOSC: total oxyradical scavenging capacity
Trx: thioredoxin
TSE: transmissible spongiform encephalopathies
UDP-GT: UDP-glucoronosyl transferases
UTR: untranslated region
UVR: ultraviolet radiation
VEGF: vascular endothelial growth factor
VHSV: viral haemorrhagic septicemia virus
VSH: vitellogenesis-stimulating hormone
VTG: vitellogenin
WSSV: white spot syndrome virus
WWC: water–water cycle
XDH: xanthine dehydrogenase
XO: xanthine oxidase

INTRODUCTION TO OXIDATIVE STRESS IN AQUATIC ECOSYSTEMS

Doris Abele[1], *José Pablo Vázquez-Medina*[2,3], *and Tania Zenteno-Savín*[2]

[1] Alfred Wegener Institute for Polar and Marine Research, Bremerhaven, Germany
[2] Centro de Investigaciones Biológicas del Noroeste, S.C. (CIBNOR), La Paz, Baja California Sur, Mexico
[3] School of Natural Sciences, University of California Merced, Merced, CA, USA

Aquatic ecosystems house a large biosphere of marine and freshwater organisms, highly diverse in their tolerance of fluctuations in P_{O_2} and temperature, two major modulators of metabolism. Often, both factors act in concert, and some of the most hypoxia-tolerant fish and molluskan species are indeed from cold-water environments. Other marine invertebrate and fish specialists thrive in the mixed waters at hydrothermal vent sites, underwater volcanic outflows where warm and hydrogen-sulfide-enriched, deoxygenated vent waters mix with colder and oxygenated oceanic waters, and temperatures and oxygen concentrations are extremely variable. Many vent species can even deal with toxic hydrogen sulfide that threatens to inhibit their mitochondrial electron transporters. More than 700 Myr of aquatic evolution have fostered a huge variety of ectothermic life-forms that can deal with the most extreme and fluctuating environmental conditions. The discovery of many fascinating underwater biota has raised an interest in the respiratory capacities of aquatic organisms and in how they deal with, from our air breathing perspective, way too little or way too much and fluctuant oxygen concentrations. As long ago as 1982, James Dykens and Malcolm Shick (*Nature* 297, 579–580) discovered that high oxygen concentrations, produced by endosymbiontic microalgae, represent a toxic assault which induces antioxidant activities in the cnidarian host cells. In 1984, Janice Blum and Irvin Fridovich investigated the activities of superoxide dismutases (Cu,Zn-, Mn- and Fe-SOD) in tissues of the hydrothermal vent tube worm *Riftia pachyptila* and the bivalve *Calyptogena magnifica* (*Archives of Biochemistry and Biophysics*, 228(2), 617–620). Superoxide dismutases detoxify superoxide anions ($O_2^{\bullet-}$) by adding another electron and converting $O_2^{\bullet-}$ to the less reactive, and therefore less toxic reactive oxygen species (ROS) hydrogen peroxide (H_2O_2). Both vent species rely largely on energy production by endosymbiontic sulfide-oxidizing bacteria but are still endowed with considerable SOD activity, just as are their sulfide-metabolizing endosymbionts, which feature a special procaryotic Fe-SOD isoform. The central message of Blum and Fridovich's paper is that cellular antioxidants are ubiquitous and therefore not

Oxidative Stress in Aquatic Ecosystems, First Edition. Edited by Doris Abele, José Pablo Vázquez-Medina, and Tania Zenteno-Savín.

only present in organisms relying primarily on aerobic energy production. Indeed, SOD enzyme forms developed early in evolution when oxygen started to accumulate: a toxicant in a primarily anoxic world. Together these two seminal papers started a whole new field of research, relating oxidative stress and antioxidant parameters in marine and freshwater organisms to the conditions prevailing in different aquatic habitats and microhabitats, such as the host cell environments of endosymbionts.

In 2010, a Google Scholar search for "oxidative stress" and "marine" yielded 50,000 publication hits ("oxidative stress" and "aquatic" 25,000 hits). This is indicative of the enormous interest and intensive research in this field, which prompted us to initiate this book project. There is also a growing interest in aquatic organisms as models for clinical and aging studies, which is expected to boost comparative research. A great number of diseases in animals and humans involve oxidative stress phenomena, and many aquatic organisms tolerate extreme states, which are pathological in humans (e.g. ischemia/reperfusion). Finally, global change and pollution massively threaten and change the Earth's ecosystems and, as over 70% of our planet's surface area is covered by water, aquatic species have become important sentinels and indicators of change. Since most forms of environmental and pollution stress eventually cause an imbalance between oxygen radical-producing and -scavenging processes, oxidative stress parameters are broadly employed in marine and terrestrial impact studies.

In preparing the concept for this book, it seemed fundamental to determine how climate effects in tropical versus polar habitats and natural scenarios in extreme environments shape the basic levels of oxidative stress parameters in aquatic ectotherms (Part I, Climate Regions and Special Habitats). Individual chapters focus on life strategies in special habitats in terms of oxygen availability, such as the sulfidic sedimentary and hydrothermal vent environments, the oxygen minimum layer of the ocean, or the cnidarian host cell of zooxanthellate endosymbionts. Fluctuations of abiotic parameters during tidal cycles confer stress hardening on intertidal species and populations; Chapter 3 delves into the effect of these fluctuations on antioxidant concentrations and enzyme activities in animals and plants from the higher littoral zone. Furthermore, long-term seasonal and climate related fluctuations modulate oxidative stress parameters in aquatic ecosystems, and Chapters 4 and 5 have a special focus on the expected consequences for primary producers at the base of aquatic food chains.

Part II of this book addresses the specific features of oxidative stress parameters with respect to respiration in water- and air-breathing aquatic animals. The respiratory medium water contains 30 times less oxygen per liter than air, and water-breathing organisms are generally adapted to perform at these lower oxygen concentrations. What this means for animal respiratory performance, including active swimmers such as sharks, and how cellular oxygen sensing mechanisms have evolved under aquatic conditions is explored in Part II (Aquatic Respiration and Oxygen Sensing). Furthermore, aquatic animals are increasingly discussed and tested as model organisms for aging and disease. The longest lived of all noncolonial organisms so far known is the hard clam *Arctica islandica*. Several authors have summarized what is new in the field of aging in marine ectotherms, a recent hot topic in aging research. Aquatic models for human diseases, including fish and invertebrate immune function and cellular signaling pathways, where ROS play different roles in development of cancer, are reviewed in Part III (Marine Animal Models for Aging, Development, and Disease). Many current papers on oxidative stress in aquatic organisms lack information about gender, reproductive or molting state, and age distribution in the experimental animals. While we know that in many cases it is still difficult to supply these data, we strongly encourage choosing model species that help us to understand the relevance of life-history-related physiological change on oxidative stress parameters in aquatic ectotherms.

Part IV (Marine Animal Stress Response and Biomonitoring) delves into the general stress response in aquatic fauna and the applicability of oxidative stress markers as indicators of environmental stress and pollution in biomonitoring studies. One important take-home message in many chapters, especially in this Part, is that it does not suffice for stress assessment to compare only the levels of antioxidants, or measure the rates of radical production alone. A stress response should be characterized by measurements of different oxidative damage markers and antioxidants, ideally complemented by a confirmation of higher radical production under stress. On one hand, the mere increase in antioxidant activity of animal tissues is not a confirmation of a physiological stress condition and, much to the contrary, can indicate the activation of antioxidant defense systems in control or anticipation of increased ROS production. On the other hand, different toxicants can interfere with each other, and a decline in antioxidant defense systems or the absence of a stress signaling (e.g. for immune stimulation) are, in many cases, the result of toxicant cross-effects, often worsening the situation.

The last and most comprehensive part of the book (Methods of Oxidative Stress Detection) presents an evaluation of classic and modern methods for the assessment of oxidative stress in aquatic animals and plant material. We asked experts in different analytical fields to describe the relevant methods and their analytical background. Many of our colleagues not only provide detailed measurement protocols but also suggest where to start troubleshooting. Importantly, the authors of the method chapters make suggestions concerning the applicability of different methods. Indeed, the classic methods to assess lipid or protein oxidation are widely used and applicable in environmental studies, in spite of known constraints with respect to accuracy and specificity. More accurate techniques are now available, including those for direct analysis of various radical species or oxidative damage parameters, such as DNA adducts. Often these require complex and costly analytical equipment, such as an EPR (electron paramagnetic resonance spectrometry) or chromatography with mass spectrometric detection. The authors share their expertise and at the same time evaluate the usefulness of alternative methods for different problems in aquatic oxidative stress research.

New tools are also coming into reach for genetic and genomic stress research, which promise a rapid advance in the understanding of molecular pathways in the response of aquatic organisms to different stressors and stress scenarios. At present, measurements of transcript levels can be compared to the antioxidant enzyme activities in most aquatic organisms, as a growing amount of partial or full sequences become available in gene banks. Antibodies for measuring antioxidant protein levels are less available, perhaps because for many questions the catalytic activity seems more functionally important than the amount of enzyme subunits present in a sample. However, antibodies that tag regulatory proteins and transcription factors in aquatic species are urgently needed for the mechanistic assessment of stress response capacities in different species. Further work is needed to verify the applicability of mammalian cell stress research kits designed to detect activity of cellular processes, such as apoptosis and autophagy, in aquatic invertebrates, often genetically distant from the originally targeted model system.

In future research it will also be important to establish closely related model species or single species with wide geographical distribution (migrating species) for functional studies of animal adaptation and effects of climate change in marine and freshwater systems. Cultures of different cell types, such as hemocytes or liver cells of aquatic species, need to be established as test systems and for intercalibration of methods among laboratories. These mechanistic model systems and the enormous advances in organic environmental chemistry, especially with respect to identification and elucidation of chemical compound structures, can be instrumental in the assessment of pollution and anthropogenic disturbance in aquatic habitats and, within a short time, will allow chemists to identify sources of pollution in the globally interconnected oceanic environments.

An important motivation for us as editors of this book was the great enthusiasm of our fellow authors. The readiness with which many young authors engaged with this project was inspiring. We are especially proud of the fact that several chapters were co-authored or have been reviewed by graduate students from different laboratories, who have greatly contributed to improve the understandability of the text and the completeness of the experimental protocols.

We hope that this book can further stimulate research in the exciting field of oxygen toxicity, stress and molecular signaling in marine and freshwater organisms.

SUGGESTED READINGS

Abele, D., Strahl, J., Brey, T., Philipp, E.E.R. (2008) Imperceptible senescence: Ageing in the ocean quahog *Arctica islandica*. *Free Radical Research* 42, 474–480.

Aldini, G., Yeum, K-J., Niki, E., Russell, R.M. (eds) (2010) *Biomarkers for Antioxidant Defense and Oxidative Damage. Principles and Practical Applications*. Wiley-Blackwell.

Antezana, T. (2009) Species-specific patterns of diel migration into the oxygen minimum zone by euphausiids in the Humboldt Current Ecosystem. *Progress in Oceanography* 83, 228–236.

Austad, S.N. (2009) Is there a role for new invertebrate models for aging research? *The Journals of Gerontology Series A* 64, 192–194.

Bagarinao, T. (1992) Sulfide as an environmental factor and toxicant: tolerance and adaptations in aquatic organisms. *Aquatic Toxicology* 24, 21–62.

Bailey-Serres, J., Mittler, R. (2006) *Reactive oxygen species. Plant Physiology* (Special Issue) 141(2).

Banaszak, A.T., Lesser, M.P. (2009) Effects of ultraviolet radiation on coral reef organisms. *Photochemical and Photobiological Sciences* 8, 1276–1294.

Bayne, B.L. (1985) Responses to environmental stress: tolerance, resistance and adaptation. *In* Gray J.S., Christiansen, M.E. (eds). *Marine Biology of Polar Regions and Effect of Stress on Marine Organisms*. John Wiley & Sons, pp. 331–349.

Beauchamp Jr, R.O., Bus, J.S., Popp, J.A., Boreiko, C.J., Andjelkovich, D.A. (1984) A critical review of the literature on hydrogen sulfide toxicity. *CRC Critical Reviews in Toxicology* 13, 25–97.

Bouverot, P. (1985) *Adaptation to Altitude-hypoxia in Vertebrates*. Springer-Verlag, Berlin, Heidelberg, New-York, Tokyo.

Calabrese, E.J., Baldwin, L.A. (2003) Hormesis: The dose-response revolution. *Annual Review of Pharmacology and Toxicology* 43, 175–197.

Cavanaugh, C.M. (1983) Symbiotic chemoautotrophic bacteria in marine invertebrates from sulphide-rich habitats. *Nature* 302, 58–61.

Cooper, T.F., Gilmour, J.P., Fabricius, K.E. (2009) Bioindicators of changes in water quality on coral reefs: review and recommendations for monitoring programmes. *Coral Reefs* 28, 589–606.

Corliss, J.B., Dymond, J., Gordon, L.I., Edmond, J.M., von Herzen, R.P., Ballard, R.D., Green, K., Williams, D., Bainbridge, A., Crane, K., van Andel, T.H. (1979) Submarine thermal springs on the Galápagos Rift. *Science* 203, 1073–1083.

Császár, N.B.M., Ralph, P.J., Frankham, R., Berkelmans R., van Oppen M.J.H. (2010) Estimating the potential for adaptation of corals to climate warming. *PLoS ONE* 5(3), e9751. doi:10.1371/journal.pone.0009751.

Dahlhoff, E.P. (2004) Biochemical indicators of stress and metabolism: Applications for marine ecological studies. *Annual Review of Physiology* 66, 183–207.

Dejours, P. (1981) *Principles of Comparative Respiratory Physiology*, 2nd edn. Elsevier/North-Holland, Amsterdam.

Ekau, W., Auel, H., Pörtner, H.-O., Gilbert, D. (2010) Impacts of hypoxia on the structure and processes in pelagic communities (zooplankton, macro-invertebrates and fish). *Biogeosciences* 7, 1669–1699.

Escribano, R., Hidalgo, P., Krautz, C. (2009) Zooplankton associated with the oxygen minimum zone system in the northern upwelling region of Chile during March 2000. *Deep-Sea Research Part II* 56, 1083–1094.

Felbeck, H. (1981) Chemoautotrophic potential of the hydrothermal vent tube worm, *Riftia pachyptila* Jones (Vestimentifera). *Science* 213, 336–338.

Felbeck, H., Childress, J.J., Somero, G.N. (1981) Calvin-Benson cycle and sulphide oxidation enzymes in animals from sulphide-rich habitats. *Nature* 293, 291–293.

Fenchel, T.M., Riedl, R.J. (1970) The sulfide system: a new biotic community underneath the oxidized layer of marine sand bottoms. *Marine Biology* 7, 255–268.

Gerhard, G.S. (2007) Small laboratory fish as models for aging research. *Ageing Research Reviews* 6, 64–72.

Giles, G.I., Tasker, K.M., Jacob, C. (2001) Hypothesis: The role of reactive sulfur species in oxidative stress. *Free Radical Biology and Medicine* 31, 1279–1283.

Grieshaber, M.K., Völkel, S. (1998) Animal adaptations for tolerance and exploitation of poisonous sulfide. *Annual Review of Physiology* 60, 33–53.

Halliwell, B., Gutteridge, J. (2007) *Free Radicals in Biology and Medicine*, 4th edn. Oxford University Press, Oxford.

Hermes-Lima, M. (2004) Oxygen in biology and biochemistry: role of free radicals. *In* Storey, K.B. (ed.). *Functional Metabolism: Regulation and Adaptation*. John Wiley & Sons, New York pp. 319–368.

Hochachka P.W., Guppy M. (1987) *Metabolic Arrest and the Control of Biological Time*. Harvard University Press.

Hochachka P.W., Somero G.N. (2002) *Biochemical Adaptation: Mechanism and Process in Physiological Evolution*. Oxford University Press.

Hulbert, A.J., Else, P.L. (2000) Mechanisms underlying the cost of living in animals. *Annual Review of Physiology* 62, 207–235.

Ignarro, L.J., (ed.) (2010) *Nitric Oxide: Biology and Pathobiology*, 2nd edn. Academic Press.

Kassahn K.S., Crozier R.H., Pörtner H.O., Caley M.J. (2009) Animal performance and stress: responses and tolerance limits at different levels of biological organization. *Biological Reviews of the Cambridge Philosophical Society* 84, 277–292.

Kimura, H., Nagai, Y., Umemura, K., Kimura, Y. (2005) Physiological roles of hydrogen sulfide: Synaptic modulation, neuroprotection, and smooth muscle relaxation. *Antioxidants and Redox Signaling* 7, 795–803.

Kobayashi, M., Li, L., Iwamoto, N., Nakajima-Takagi, Y., Kaneko, H., Nakayama, Y., Eguchi, M., Wada, Y., Kumagai, Y., Yamamoto, M. (2009) The antioxidant defense system Keap1-Nrf2 comprises a multiple sensing mechanism for responding to a wide range of chemical compounds. *Molecular Cell Biology* 29, 493–502.

Lesser, M.P. (2006) Oxidative stress in marine environments: Biochemistry and physiological ecology. *Annual Review of Physiology* 68, 253–278.

Lesser, M.P. (2007) Coral reefs bleaching and global climate change: Can corals survive the next century? *Proceedings of the National Academy of Sciences* 104, 5259–5260.

Martchecknez-Álvarez, R.M., Morales, A.E., Sanz, A. (2005) Antioxidant defenses in fish: Biotic and abiotic factors. *Reviews in Fish Biology and Fisheries* 15, 75–88.

McClanahan, T.R. Weil, E., Cortés, J., Baird, A.H., Ateweberhan, M. (2009) Consequences of coral bleaching for sessile reef organisms. *Ecological Studies* 205, 121–138.

Navarro, I., Gutiérrez, J. (1995) Fasting and starvation. *In* Hochachka, P.W., Mommsen, T.P. (eds), *Biochemistry and Molecular Biology of Fishes*. Elsevier, Amsterdam, pp. 393–434.

Nilsson, G.E. (ed.) (2010) *Respiratory Physiology of Vertebrates.* Cambridge University Press.

Patnaik, B.K., Mahapatro, N., Jena B.S. (1994) Ageing in fishes. *Gerontology* 40, 113–132.

Pauly, D. (2010) *Gasping Fish and Panting Squids: Oxygen, Temperature and the Growth of Water-breathing Animals.* International Ecology Institute, Oldendorf/Luhe.

Pörtner, H.O. (2010) Oxygen- and capacity-limitation of thermal tolerance: a matrix for integrating climate-related stressor effects in marine ecosystems. *Journal of Experimental Biology* 213, 881–893.

Powell, M.A., Somero, G.N. (1986) Hydrogen sulfide oxidation is coupled to oxidative phosphorylation in mitochondria of *Solemya reidi. Science* 233, 563–566.

Reznick, D. (1993) New model systems for studying the evolutionary biology of aging: Crustacea. *Genetica* 91, 79–88.

Richards J.G., Farrell A.P., and Brauner C.J. (eds) (2010) *Fish Physiology*, Vol. 27, *Hypoxia*. Academic Press/Elsevier, San Diego.

Saltzman, J., Wishner, K.F. (1997) Zooplankton ecology in the eastern tropical Pacific oxygen minimum zone above a seamount: 1. General trends. *Deep-Sea Research Part I* 44, 907–930.

Sheehy, M.R.J., Greenwood, J.G., Fielder, D.R. (1995) Lipofuscin as a record of "rate of living" in an aquatic poikilotherm. *Journal of Gerontology: Biological Sciences B* 50, 327–336.

Shick, J.M., Lesser, M.P. Jokiel, P. (1996) Effects of ultraviolet radiation on corals and other coral reef organisms. *Global Change Biology* 2, 527–545.

Sokolova, I.M., Lannig, G. (2008) Interactive effects of metal pollution and temperature on metabolism in aquatic ectotherms: implications of global climate change. *Climate Research* 37, 181–201.

Stambler, N. (2010) Coral symbiosis under stress. *Cellular Origin, Life in Extreme Habitats and Astrobiology* 17 (3), 197–224.

Storey, K.B. (ed.) (2001) *Molecular Mechanisms of Metabolic Arrest: Life in Limbo.* Society for Experimental Biology/Garland Science.

Storey, K.B., Storey, J.M. (2004) Oxygen limitation and metabolic rate depression. *In* Storey, K.B. (ed.). *Functional Metabolism: Regulation and Adaptation.* Wiley-Liss, Hoboken, NJ, pp. 415–442.

Thompson, D.M., van Woesik, R. (2009) Corals escape bleaching in regions that recently and historically experienced frequent thermal stress. *Proceedings of the Royal Society – Biological Sciences* 276(1669), 2893–2901.

Tremblay, N., Gómez-Gutiérrez, J., Zenteno-Savcheckn, T., Robinson C.J., Sánchez-Velasco, L. (2010) Role of oxidative stress in seasonal and daily vertical migration of three species of krill in the Gulf of California. *Limnology and Oceanography* 55(6), 2570–2584.

Tunnicliffe, V. (1991) The biology of hydrothermal vents - ecology and evolution. *Oceanography and Marine Biology: an Annual Review* 29, 319–407.

Ungvari, Z., Philipp, E.E.R. (2010) Comparative gerontology – from mussels to man. *The Journals of Gerontology Series A: Biological Sciences and Medical Sciences.* doi:10.1093/gerona/glq198.

Vidal-Dupiol, J., Adjeroud, M., Roger, E., Foure, L., Duval, D., Mone, Y., Ferrier-Pages, C., Tambutte, E., Tambutte, S., Zoccola, D., Allemand, D., Mitta, G. (2009) Coral bleaching under thermal stress: putative involvement of host/symbiont recognition mechanisms. *BMC Physiology* 9, 14.

Vismann, B. (1991) Sulfide tolerance: physiological mechanisms and ecological implications. *Ophelia* 34, 1–27.

Wang, T., Hung, C.C.Y., Randall, D.J. (2006) The comparative physiology of food deprivation: From feast to famine. *Annual Review of Physiology* 68, 223–251.

Warner, M.E., Lesser, M.P., P. Ralph. (2010) Chlorophyll fluorescence in reef building corals. *In* Suggett, D., Prasil, O., Borowitzka, M. (eds). *Chlorophyll a Fluorescence in Aquatic Sciences: Methods and Applications.* Springer-Verlag, Berlin, pp. 209–222.

Weis, V.M. (2010) The susceptibility and resilience of corals to thermal stress: adaptation, acclimatization or both? *Molecular Ecology* 19, 1515–1517.

Woodhead, A.D. (1998) Aging, the fishy side: An appreciation of Alex Comfort's studies. *Experimental Gerontology* 33, 39–51.

Yakovleva, I.M., Baird, A.H., Yamamoto, H.H., Bhagooli, R., Nonaka, M., Hidaka, M. (2009) Algal symbionts increase oxidative damage and death in coral larvae at high temperatures. *Marine Ecology Progress Series* 378, 105–112.

Part I

Climate Regions and Special Habitats

Chapter 1

OXIDATIVE STRESS IN TROPICAL MARINE ECOSYSTEMS

Michael P. Lesser

Department of Molecular, Cellular and Biomedical Sciences, University of New Hampshire, USA

The accumulation of oxygen in Earth's atmosphere has had profound effects on the geochemistry, physiology, and evolution of life on the planet. However, most organisms must also contend with the negative aspects of living in a world with oxygen. Reactive oxygen species (ROS) production is prevalent in the world's oceans and oxidative stress is an important component of the stress response in marine organisms exposed to a variety of environmental stressors such as thermal stress, which is now becoming more prevalent because of climate change. In tropical environments exposure to high irradiances of visible and ultraviolet radiation (UVR) contributes significantly, through both direct and indirect processes, to ROS production in the water as well as in many tropical marine taxa of plants and animals. The negative effects of ROS must also be balanced by their role in signal transduction that facilitates processes such as apoptosis, autophagy and necrosis. Because of the high irradiances of solar radiation and exposure to high air and seawater temperatures, oxidative stress in tropical marine environments is ubiquitous and is normally kept in check by a suite of antioxidants, both enzymatic and nonenzymatic, in diverse tropical marine taxa in order to survive, grow and reproduce.

HISTORY AND CHEMISTRY OF OXYGEN ON EARTH

Life on Earth began in the Archean at least 3.5 Gyr, and possibly as far back as 3.8 Gyr (Nisbet and Sleeo 2001). The early atmosphere of the Earth was highly reduced and dominated by microbes (Kasting and Siefert 2002), with additional evidence for the presence of biogenic structures that supported an oxidizing environment as far back as 3.5 Gyr (Nisbet and Sleeo 2001). By the mid- to early–Archean, cyanobacteria had evolved and were carrying out oxygenic photosynthesis (Nisbet and Sleeo 2001; Kasting and Siefert 2002); and with ample amounts of CO_2, water as a reductant, and solar radiation, oxygenic cyanobacteria flourished and evolved into other taxa by multiple endosymbiotic events (Falkowski et al. 2004). The end result of this was that molecular oxygen, or dioxygen (O_2), accumulated in significant amounts in the Earth's atmosphere ~2.5 Gyr, and in the upper atmosphere it formed O_3 which filtered out the shortest wavelengths of harmful UVR (<290 mn), and changed the course of biological evolution.

The accumulation of oxygen changed terrestrial and shallow oceanic habitats from a reduced state to an oxidized state and provided strong selective pressures on

Oxidative Stress in Aquatic Ecosystems, First Edition. Edited by Doris Abele, José Pablo Vázquez-Medina, and Tania Zenteno-Savín.

anaerobic life-forms existing at the end of the Archean. The evolution of aerobic respiration with its greater efficiency and higher yields of energy was critical to the development of complex multicellular eukaryotic organisms but not without having to solve additional problems associated with gas and nutrient transport. The percentage of oxygen in the Earth's atmosphere is now $\sim 21\%$. This makes oxygen the second most abundant element in the atmosphere, behind nitrogen at $\sim 78\%$.

Oxygen is a stable, odorless, tasteless, and colorless gas at room temperature that was isolated and characterized in the 1770s. While Joseph Priestly (USA) and Antoine Lavoisier (France) are generally given credit for the discovery and naming of oxygen, it is now widely accepted that Carl Scheele (Sweden) discovered it in 1771. Oxygen has a low solubility coefficient in water that decreases with increasing temperature and affects its availability for a wide range of taxa in both aquatic and marine habitats. Normoxic air dissolved in water contains a higher percentage of oxygen (34%) than does ambient air (21%) because, despite its low solubility, it is more soluble in water than nitrogen. These differences in solubility have important implications for availability and transport of oxygen for oxidative metabolism in aquatic and marine organisms.

Oxygen Can Be Toxic!

In a world where the presence of oxygen in the Earth's atmosphere is taken for granted many biologists still do not comprehend its potential toxicity. Reactive oxygen species are responsible for the toxic effects of oxygen, and this is because in its ground-state oxygen is distinctive among the elements as it is a biradical and has two unpaired electrons in its outer orbit (Asada and Takahashi 1987; Cadenas 1989; Fridovich 1998; Halliwell and Gutteridge 1999). Therefore, oxygen is usually non-reactive with organic molecules that have paired electrons with opposite spins. This spin restriction means that the most common mechanisms of oxygen reduction in biochemical reactions are those involving the transfer of a single, or univalent reduction, electron (Asada and Takahashi 1987; Cadenas 1989; Fridovich 1998; Halliwell and Gutteridge 1999).

The univalent reduction of molecular oxygen produces reactive intermediates such as $O_2^{\bullet-}$, $^1O_2$, H_2O_2, $HO^{\bullet}$, and finally water (Halliwell and Gutteridge 1999). H_2O_2 is not technically a free radical because all of its electrons are paired but is usually included in the definition of ROS. All photosynthetic and respiring organisms produce ROS via the univalent pathway, and subsequently H_2O_2 with the continued reduction of $O_2^{\bullet-}$, and eventually $HO^{\bullet}$, which is then reduced to HO^- and water as a consequence of exposure to, and use of, molecular oxygen (Halliwell and Gutteridge 1999).

The production of ROS is directly, and positively, related to the concentration or P_{O_2} of oxygen (Jamieson et al. 1986) and since most photosynthetic organisms are hyperoxic while photosynthesizing, the production of ROS that occurs requires robust antioxidant defenses or oxidative stress will occur (Asada and Takahashi 1987; Cadenas 1989; Fridovich 1998; Halliwell and Gutteridge 1999). Oxidative stress, the production of ROS beyond the capacity of an organism to quench these reactive species, can cause damage to lipids, proteins, and DNA (Halliwell and Gutteridge 1999). The primary purpose of antioxidant defenses in biological systems is to quench $^1O_2$ at the site of production, and quench or reduce the flux of other ROS such as $O_2^{\bullet-}$ and H_2O_2 to ultimately prevent the production of $HO^{\bullet}$ the most damaging of the ROS (Asada and Takahashi 1987; Cadenas 1989; Fridovich 1998; Halliwell and Gutteridge 1999).

H_2O_2 causes significant damage because it's diffusion within the cell from its point of synthesis is less restrictive than other forms of ROS and it can enter into numerous other reactions. Exposure to H_2O_2 can damage many cellular constituents directly, such as DNA and enzymes involved in carbon fixation (Asada and Takahashi 1987; Cadenas 1989; Fridovich 1998; Halliwell and Gutteridge 1999), but H_2O_2 is also involved in pathways such as programmed cell death or apoptosis (Halliwell and Gutteridge 1999). If H_2O_2 is further reduced, it can produce $HO^{\bullet}$. One source of electrons for that reduction in biological systems is transition metals, such as Fe, in what is known as Fenton chemistry (Halliwell and Gutteridge 1999).

Most organisms have developed mechanisms to sequester and transport essential metal ions in order to reduce the occurrence of Fenton chemistry. Proteins such as transferrin and lactoferrin afford significant Fe-binding capacity, and can reduce the availability of free Fe to zero, while intracellular Fe is stored in ferritin (Halliwell and Gutteridge 1999). Many bacteria produce Fe-chelating proteins, known as siderophores, which are an important indicator of pathogenic potential in clinical settings and essential for the survival of bacteria in natural habitats where free-Fe availability is also low. Exposure of organisms to Fe or Cu beyond their capacity to chelate or bind these metal ions potentially exposes them to the production of $HO^\bullet$ in the presence of other ROS.

While the Haber–Weiss and Fenton reactions have dominated the theoretical underpinnings of most studies on oxidative stress, the discovery that many cells produce $NO^\bullet$, a molecule initially believed to be primarily involved in signal transduction (e.g., neurotransmission), is now known to be involved in processes involving oxidative stress (Fang 2004). $NO^\bullet$ is a colorless gas and a weak reducing agent with moderate solubility in water but by itself is relatively nonreactive despite the fact that it contains an unpaired electron making it paramagnetic and a free radical (Halliwell and Gutteridge 1999). It is now known that the microbicidal activity of white blood cells is primarily a function of the inducible enzyme NOS, which produces $NO^\bullet$ that reacts with $O_2^{\bullet-}$, to form $ONOO^-$, a potent oxidant (Fridovich 1986). Because the solubility of $NO^\bullet$ is similar to water it can readily diffuse across biological membranes, where it reacts at near diffusion-limited rates with free radicals, especially $O_2^{\bullet-}$, to form $ONOO^-$, which can diffuse across biological membranes at rates 400 times greater than $O_2^{\bullet-}$ (Marla et al. 1997; Denicola et al. 1998). It has been suggested that the high concentrations of $NO^\bullet$ in a wide range of taxonomically diverse organisms creates significant competition between $NO^\bullet$ and superoxide dismutase (SOD) for $O_2^{\bullet-}$. This competition for $O_2^{\bullet-}$ may be a major determinant of oxidative stress in many organisms. Many investigators are now re-evaluating the role of $O_2^{\bullet-}$ in oxidative stress because of these new insights, and because many of the observed *in vitro* effects ascribed to $O_2^{\bullet-}$ may in fact be mediated by $ONOO^-$ (Halliwell and Gutteridge 1999).

Principal Cellular Sites of ROS Production

The maximum quantum efficiency of photosynthesis occurs when photosynthesis increases in a linear fashion with the irradiance of visible, or photosynthetically active radiation (PAR: 400–700 nm). Additional increases in PAR cause this relationship to become nonlinear and can be attributed to changes in electron transport and light-harvesting processes (Asada and Takahashi 1987; Asada 1994; Falkowski and Raven 1997). These processes lead to a decrease in photosystem II (PS II) photochemistry, and additional increases in PAR exceed the capacity to protect PS II by dissipating the excess absorbed excitation energy, and will increase the probability of damage to the primary quinone acceptor (e.g., D1 protein) from the production of ROS in the chloroplast (Asada and Takahashi 1987; Richter et al. 1990; Falkowski and Raven 1997; Lupínkov a and Komenda 2004). Oxidative stress in the chloroplasts not only has direct effects on PS II but also inhibits the repair of damage to PS II (Nishiyama et al. 2001). In addition to the production of $^1O_2$ within PS II (Macpherson et al. 1993) it has been shown that $O_2^{\bullet-}$ and $HO^\bullet$ are also produced in the PS II reaction center (Liu et al. 2004).

There are several sites within the chloroplast that can reduce oxygen. Photosensitized chlorophyll is in a singlet-excited state and normally transfers its excitation energy to the photosynthetic reaction centers, but under high irradiance conditions the long-lived chlorophyll triplet state occurs and can interact with oxygen to form $^1O_2$. The reducing side of PS I can reduce oxygen to $O_2^{\bullet-}$ by the Mehler reaction and is the most significant site of $O_2^{\bullet-}$ production in the chloroplast (Asada and Takahashi 1987; Asada 1994, 1999). The Mehler reaction is often described as an alternative sink for electrons when sink limitation (e.g., carbon or nitrogen limitation) occurs. Under normal circumstances the $O_2^{\bullet-}$ produced is rapidly dismutated to H_2O_2 by SOD and the H_2O_2 converted to water by ascorbate peroxidase (APx; Asada and Takahashi 1987; Asada 1994, 1999). The production of $O_2^{\bullet-}$ is exacerbated under stressful conditions such as exposure to xenobiotics or pollutants, high PAR irradiances, exposure to UVR, and exposure to thermal stress that can overwhelm antioxidant defenses to produce damage to both PS II and carbon fixation (Asada and Takahashi 1987; Asada 1994). Molecular oxygen can also be reduced during photorespiration, the oxygenase or C_2 pathway for ribulose 1, 5-bisphosphate decarboxylase/oxygenase (Rubisco) under conditions where the ratio of CO_2 to oxygen is low, forming glycolate. Although this does not generate ROS in the chloroplast, the subsequent conversion of glycolate to glyoxylate in glyoxisomes does generate H_2O_2 via the divalent reduction of oxygen by glycolate oxidase (Asada and Takahashi 1987; Asada 1994, 1999).

In nonphotosynthetic eukaryotic cells the mitochondrion is the primary site of ROS production. Within this organelle there are two primary sites of $O_2^{\bullet-}$ generation in the inner mitochondrial membrane: NADH dehydrogenase at complex I, and the interface between ubiquinone and complex III (Brookes 2005). $O_2^{\bullet -}$ produced at these sites is then converted by spontaneous dismutation or by SOD to H_2O_2 (Brookes 2005). The inner membrane is also permeable to protons (H^+). There are apparent benefits to H^+ leakage despite the loss of energy as it is associated with a decrease in the production of ROS. Other important sites of ROS production in cells include the endoplasmic reticulum of animals, plants, and some bacteria that contain several cytochromes, collectively known as cytochrome P-450, that can form $O_2^{\bullet -}$ (Halliwell and Gutteridge 1999). The substrate for these reactions is commonly an organic xenobiotic, such as herbicides, alcohols, insecticides, and a long list of hydrocarbons (Halliwell and Gutteridge 1999). The wide occurrence of these compounds in nature, their detoxification, and the subsequent $O_2^{\bullet -}$ production, can contribute significantly to the oxidative load of organisms.

Microbodies are cellular organelles that include peroxisomes and glyoxysomes that contain enzymes involved in the β-oxidation of fatty acids and photorespiration. The H_2O_2 synthesized by these microbodies is produced by a two-electron transfer and by the dismutation of $O_2^{\bullet -}$ to H_2O_2 via SOD in both peroxisomes and glyoxisomes (Sandalio and del Rio 1988).

REACTIVE OXYGEN SPECIES ARE BOTH GOOD AND BAD

Oxidative Damage

Breathing oxygen is hazardous to your health! The consequence of respiring oxygen is the production of ROS. Oxidative stress can also represent a significant energetic drain on an organism. For example, recent work has shown an increased susceptibility to oxidative stress during periods of reproduction when energy is diverted away from the costs of antioxidant defenses (Alonso-Alvarez et al. 2004).

The reaction of ROS with lipids, especially membrane-associated lipids, is considered one of the most prevalent mechanisms of cellular injury (Halliwell and Gutteridge 1999). The peroxidation of lipids involves three well-defined steps: initiation, propagation and termination (Yu 1994; Halliwell and Gutteridge 1999) ultimately forming $ROO^{\bullet}$ (peroxyl radical) which then participates in a chain reaction of lipid peroxidation. The lipid hydroperoxide (ROOH) formed during these reactions is unstable in the presence of Fe or other metal catalysts because ROOH will participate in a Fenton reaction leading to the formation of $RO^{\bullet}$ (alkoxy radical). Therefore, in the presence of Fe, the chain reactions are not only propagated but also amplified. Among the degradation products of ROOH are aldehydes such as malondialdehyde (MDA), and hydrocarbons such as ethane and ethylene all of which are commonly measured end-products of lipid peroxidation (Freeman and Crapo 1982; Gutteridge and Halliwell 1990). Lipid peroxidation in mitochondria is particularly cytotoxic, with multiple effects on enzyme activity and ATP production, as well as the initiation of apoptosis (Bindoli 1988; Green and Reed 1998).

Oxidative attack on proteins results in site-specific amino acid modifications, fragmentation of the peptide chain, aggregation of cross-linked reaction products, altered electrical charge, and increased susceptibility to removal and degradation. The primary, secondary, and tertiary structure of proteins determines the susceptibility of each amino acid to attack by ROS. Certain amino acids, such as the sulfur-containing amino acids tryptophan, tyrosine, phenylalanine, histidine, methionine, and cysteine, are particularly susceptible. The structure of proteins can limit the exposure of some of the most susceptible amino acids to attack by ROS (Freeman and Carpo 1982; Halliwell and Gutteridge 1999; Yu 1994). For many enzymes the oxidation of Fe-sulfur centers by $O_2^{\bullet -}$ inactivates enzyme function (Freeman and Crapo 1982; Hyslop et al. 1988), and other amino acids, such as histidine, lysine, proline, arginine, and serine, form carbonyl groups when oxidized (Stadtman 1986). The oxidative degradation of any protein is enhanced in the presence of transition metals that are capable of redox cycling, such as Fe (Stadtman 1986; Davies 1987).

The most sensitive of cellular constituents to damage by ROS is nuclear and organelle DNA. Reactive oxygen species generation can induce numerous lesions, including base degradation, single strand breakage, and cross-linking to proteins (Imlay and Linn 1988; Imlay et al. 1988; Imlay 2003), which result in base deletions, mutations, and lethal genetic effects. The main cause of single strand breaks is oxidation of the sugar moiety by $HO^{\bullet}$. *In vitro*, H_2O_2 or $O_2^{\bullet -}$ cannot by themselves cause strand breaks under normal physiological conditions, and therefore, their toxicity *in vivo* is most likely the result of Fenton reactions (Imlay and Linn 1988; Imlay et al. 1988; Imlay 2003). When a bound transition metal is reduced by a small diffusible molecule, such as ascorbate or $O_2^{\bullet -}$, it will react with H_2O_2 to form $HO^{\bullet}$ radicals (Imlay and Linn 1988).

The short-lived $HO^{\bullet}$ then oxidizes an adjacent sugar or base, causing DNA chain breakage.

DNA damage is also the least tolerable of all damage caused by ROS. DNA is effective in binding metals that are involved in Fenton reactions, and is susceptible to many conditions that are conducive to oxidative damage. Because of this susceptibility both prokaryotic and eukaryotic cells have a number of DNA repair enzymes (Beyer et al. 1991). It has been suggested that one reason why eukaryotic organisms have compartmentalized DNA in the nucleus, away from sites of redox cycling that are high in reductants, has been the selective pressure to avoid oxidative damage.

Signal Transduction

All cells produce ROS and, while some of ROS synthesis is the result of insults to the integrity of the cell, ROS are also produced by ubiquitous enzyme systems for specific purposes (Schrek and Baeuerle 1991). It has been proposed that the antioxidant systems of cells regulate the intracellular levels of ROS, and the small size and general diffusibility of ROS make them ideal candidates as second messengers (Schrek and Baeuerle 1991). Reactive oxygen species production does, in fact, result in a downstream cascade of gene expression for many important pathways under normal physiological conditions. This has been shown for the expression of several transcription factors and other signal transduction molecules, such as heat shock inducing factor (HSF1), nuclear factor (NF-κB), *p53*, mitogen-activated protein kinase (MAPK), and *oxy*R gene products where expression was preceded by ROS accumulation (Zheng et al. 1998; Finkel and Holbrook 2000; Martindale and Holbrook 2002). Many of these pathways have the dual role of regulating normal cellular functions and responding to a variety of stressors besides increased ROS production levels.

Oxidative stress is also known to play a role in apoptosis through several cell cycle genes (Johnson et al. 1996) subsequent to DNA damage. Two apoptotic pathways have been described and are known as the death-receptor pathway and the mitochondrial pathway. The mitochondrial pathway is commonly associated with DNA damage and up-regulation or activation of the cell cycle gene *p53* (Hengartner 2000). Both the death receptor and mitochondrial pathways converge at the mitochondria and the Bcl-2 family of genes. Bcl-2 can also be directly down-regulated, and therefore promote apoptosis, by exposure to ROS (Hildeman et al. 2003). In mitochondria, release of proapoptotic effectors (e.g., cytochrome *c*, ROS, caspase 9) subsequently leads to the assembly of the apoptosome, which activates caspase-dependent DNase (Green and Reed 1998; Rich et al. 2000). Both caspases and Bcl-2 genes are present in sea anemones and are involved in apoptosis (Cikala et al. 1999; Dunn et al. 2006).

Exposure to ROS and subsequent apoptosis is also a common feature of higher plants and many caspases, homologous to animal caspases, have been identified and also appear to control apoptosis (Korthout et al. 2000; Lam and del Pozo 2000). Reactive oxygen species are also an important component of plant defense systems against pathogens (Mehdy 1994), and $O_2^{\bullet-}$ is directly involved in the apoptotic hypersensitive reaction of higher plants against pathogens (Jabs et al. 1996). Caspases have also been identified in unicellular photoautotrophic eukaryotes (i.e., phytoplankton) and they are regulated in a similar fashion when compared to higher plant and metazoan caspases during experimentally induced apoptosis (Segovia et al. 2003).

TROPICAL MARINE ENVIRONMENTS AND OXIDATIVE STRESS

Reactive Oxygen Production in Seawater

The absorption of solar radiation, especially UVR (290–400 nm), by dissolved organic matter (DOM) in seawater leads to the photochemical production of a variety of reactive transients, including ROS (Mopper and Kieber 2000). Many of the refractory components of DOM are degraded by ROS and potentially enhance biological productivity by releasing small molecular weight organic compounds that can be taken up by bacteria and other microorganisms (Mopper and Kieber 2000). Reactive oxygen species produced in seawater could also have deleterious effects on bacteria and phytoplankton by affecting cell membranes and potentially inhibiting photosynthesis. H_2O_2 has the longest lifetime in seawater, the highest steady-state concentrations (10^{-7} M), and can readily pass through biological membranes (Halliwell and Gutteridge 1999; Mopper and Kieber 2000).

On tropical coral reefs the most comprehensive studies describing the underwater light field of coral reefs as it relates to UVR, and its potential effects, come from work in the Florida Keys (Lesser 2000; Zepp et al. 2008). Lesser (2000) examined the vertical attenuation of UVR down to 30 m depth, while Zepp et al. (2008) examined many reef sites over a wide geographic area in the Florida Keys using both shipboard and continuous *in situ* measurements over a seven year period. The waters of the Florida Keys have a spatially and temporally dynamic UVR environment, primarily as a result of changes in the concentration of colored dissolved organic matter (CDOM) (Zepp et al. 2008). When CDOM is highest, the K_d for UVR is also high, which decreased the amount of biologically effective UVR incident on a shallow (3 m) coral reef (Zepp et al. 2008). Concentrations of CDOM also show large spatial and temporal variability due to changes in the source of the principal absorbing compound (of terrigenous or oceanic origin), as has been reported for several islands in French Polynesia (Maritorena and Guillocheau 1996), or with ebb and flood tides, as has been described in the Bahamas (Otis et al. 2004) and the Lower and Middle Keys and Dry Tortugas (Zepp et al. 2008). The effects of ocean acidification on the composition of CDOM and particulate organic matter (POM) have not been investigated, but could be another source of spatial and temporal variability of these constituents in the waters over coral reefs. Studies on the effects of ocean acidification on the photochemical production of ROS should be an important area of study in the future.

Oxidative Stress in Tropical Marine Organisms

Marine organisms are exposed to wide variety of environmental factors on varying temporal and spatial scales. For tropical ecosystems this means that their exposure to the high extremes of abiotic factors, such as temperature and UVR, requires robust metabolic adjustments in order to maintain homeostasis, grow, and reproduce. It should be no surprise that the potential for ROS production is high in tropical marine organisms, and antioxidant defenses are required to maintain the steady-state concentration of ROS at low levels in order to prevent oxidative stress and subsequent oxidative damage.

The production of ROS is a consistent feature of photoautotrophs, and marine algae are no exception. In unicellular eukaryotic algae, especially dinoflagellates, all three SOD metalloproteins have been identified (Lesser and Shick 1989; Matta and Trench 1991; Hollnagel et al. 1996; Okamoto et al. 2001). Many of these cells exhibit a daily cycling of maximum SOD activities, and other antioxidant enzymes, associated with peak midday irradiances and ROS production (Hollnagel et al. 1996; Okamoto et al. 2001). This daily rhythm may be under transcriptional control and new SOD proteins produced on a daily basis (Okamoto et al. 2001). Distinct seasonal regulation of antioxidant enzymes is based on total daily irradiance in addition to the daily rhythms (Butow et al. 1997). Some species of phytoplankton are responsible for red tides and produce neurotoxins, but others elicit direct cytotoxic effects on bacteria and fish that are caused by enhanced extracellular ROS production (Oda et al. 1992; Yang et al. 1995).

Macrophytes are conspicuous components of many marine ecosystems, but are most commonly studied from the temperate rocky intertidal area where these algae withstand some of the harshest environmental conditions known, including freezing, desiccation, carbon limitation, and heat stress. These environmental extremes lead to ROS formation and contribute to the photoinhibition of photosynthesis (Collén and Davison 1999, 2001). Studies on tropical marine macrophytes include a survey of the antioxidant potential of a taxonomically diverse group of marine macrophytes that showed an enhanced protective capacity against ROS production (Zubia et al. 2007). In an environment where temperature stress and exposure to high UVR occur, this level of protection has distinct advantages but is also likely to be energetically costly. Exposure of the chlorophyte *Ulva fasciata* to UVR, specifically UVB (290–320 nm), caused an increase in H_2O_2 production, subsequent lipid peroxidation, and increased antioxidant enzyme activity (Shiu and Lee 2005).

Reactive oxygen species production is also a common feature of many marine invertebrates (Abele and Puntarulo 2004). In bivalve mollusks ROS are produced in response to exposure to xenobiotics (Winston et al. 1990; Mitchelmore et al. 1998), and changes in temperature and especially heat stress (Abele et al. 2002). Exposure of bivalves to pollutants in tropical environments also results in an increase in oxidative stress (Angel et al. 1999). Sponges (Phylum: Porifera) are a common tropical marine taxon and many are photoautotrophic with symbiotic cyanobacteria. These sponges should experience hyperoxia due to elevated P_{O_2} in their tissues from photosynthetically produced oxygen, as do corals symbiotic with zooxanthellae (Lesser 2006), and are exposed to significant pro-oxidant pressure (Regoli et al. 2000, 2004). Any additional increase in oxygen or exposure to another abiotic factor, such as UVR or thermal stress, can lead to elevated ROS production. In fact, for the Mediterranean sponge, *Petrosia ficiformis*, exposure to summer highs in seawater temperature resulted in the highest values of total oxidative scavenging capacity (TOSC) and catalase (CAT) activity, which was attributed to the increased production of H_2O_2 (Regoli et al. 2000, 2004). Sponges are sensitive to elevated temperature in a similar way to corals (Webster et al. 2008) and can up-regulate the expression of critical stress proteins, such as the protein chaperon HSP70 (López-Legentil et al. 2008). The end result of thermal stress is often a similar process of "bleaching", which involves the loss of sponge bacterial symbionts and in particular their cyanobacterial symbionts (Vicente 1990; López-Legentil et al. 2008). If thermal stress continues sponges undergo apoptosis or programmed cell death (Wagner et al. 1998) with the expression of signature, highly conserved, apoptotic markers such as Bcl-2 and caspases (Weins et al. 2003; Weins and Müller 2006), which are also expressed by cnidarians, worms, mice, flies, and humans (Lesser 2006; Weis 2008).

The surface waters of nearshore coastal habitats often contain the planktonic life-history phases of many species of fish, macrophytes, and benthic invertebrates. Many species of echinoderms are important broadcast spawning members of benthic temperate marine communities. Exposure to UVR potentially affects fertilization success, the timing of cleavage, and development time for embryos that survive, as observed in laboratory studies on the sea urchin, *Strongylocentrotus droebachiensis* (Adams and Shick 1996, 2001; Lesser et al. 2003, 2006). Exposure to UVR induces oxidative stress, DNA damage, and apoptosis (Adams and Shick 1996, 2001; Lesser et al. 2003, 2006). Field exposures of temperate embryos at fixed depths reveal similar results down to 5 m depth, where UVR attenuation mitigates the physiological effects; for developing embryos, oxidative stress and DNA damage does occur shallower than 5 m (Lesser 2010). The developing stages of tropical species of sea urchin (e.g., *Diadema savigny*), despite being exposed to greater UVR than either temperate or polar species, exhibit decreased amounts of DNA damage and enhanced activities of DNA repair enzymes such as photolyase (Lamare et al. 2006, 2007). These repair rates are greater than either temperate or polar species and are presumably the result of evolutionary exposure to higher UVR irradiances. It would be interesting to also assess whether these tropical species express higher activities and quantities of antioxidant enzymes as well, and whether there are fitness costs associated with mounting these defenses in an environment of constant exposure to UVR and ROS.

One of the best understood marine invertebrate systems, as it relates to oxidative stress, are cnidarians (sea anemones, corals, jellyfish) with symbiotic dinoflagellates (*Symbiodinium* sp.), more commonly known as zooxanthellae (Fig. 1.1). Scleractinian corals in particular are an important component of coral reefs around the world (Fig. 1.2). Global climate change, caused by the emission of greenhouse gases (e.g., CO_2, CH_4), has resulted in elevated seawater temperatures in tropical ecosystems and is the primary cause of "coral bleaching" events observed around the world (Hoegh-Guldberg 1999; Lesser 2004; Hoegh-Guldberg et al. 2007). Bleaching, a generalized stress response characterized by the visual loss of coloration (Fig. 1.3), is caused by the loss of zooxanthellae, the end point of a series of molecular and cellular events (Hoegh-Guldberg 1999; Lesser 2004; Hoegh-Guldberg et al. 2007). Both field and laboratory studies on

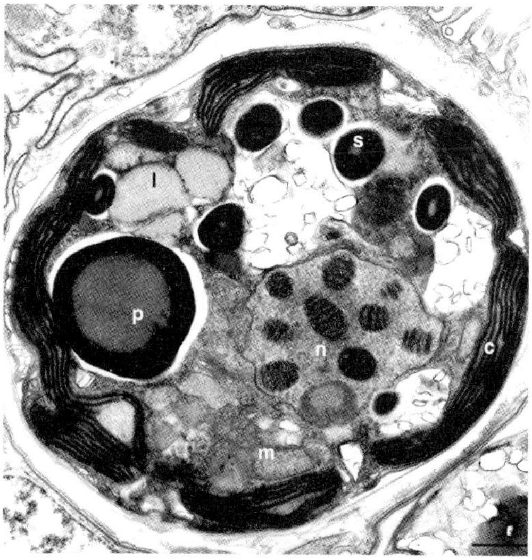

Fig. 1.1 Transmission electron micrograph of symbiotic dinoflagellate or zooxanthella (*Symbiodinium* sp.) from the sea anemone, *Aiptasia pallida*. Scale bar = 1 μm; n = nucleus; c = chloroplast; m = mitochondrion; l = lipid droplet; s = starch granule; p = pyrenoid.

Fig. 1.2 Underwater photograph of shallow coral reef in the Indonesian archipelago near Komodo Island in the year 2000. (Photograph by Michael Lesser.) See plate section for a color version of this image.

Fig. 1.3 Bleached coral (*Montastraea faveolata*) from the shallow coral reefs of Puerto Rico during 2005 bleaching event. (Photograph by Ernesto Weil.) See plate section for a color version of this image.

bleaching in corals and other symbiotic cnidarians have established a causal link between temperature stress and bleaching (Hoegh-Guldberg 1999; Lesser 2004; Hoegh-Guldberg et al. 2007), and the extent of bleaching, subsequent mortality, and the underlying mechanism(s) are related to the magnitude of temperature stress, exposure to other interacting abiotic factors, and the duration of exposure to the stressor(s).

Strong morphological evidence exists that apoptosis, cell necrosis, and autophagy occur in both host and algal cells of thermally stressed symbiotic sea anemones (Dunn et al. 2002, 2007), and that these processes are preceded by the accumulation of ROS, leading to oxidative damage of critical cellular components (i.e., DNA). Which underlying process occurs may be a function of the time frame of exposure to the stressor, and whether or not simultaneous exposure to secondary abiotic factors, such as high irradiances of solar radiation, occurs because of their important contribution to ROS production (Lesser 2006).

CONCLUSIONS AND FUTURE DIRECTIONS

As we continue to grapple with the multiple threats associated with global climate change on natural ecosystems at all trophic levels, oxidative stress continues to emerge as a common feature of the organismal response to these stressors. Several studies have already examined the effects of ocean acidification on calcification for tropical marine organisms (Hoegh-Guldberg et al. 2007), but future studies should include the physiological effects of hypercapnia on cellular pH, and various metabolic pathways including ROS production and apoptosis. Investigation will continue to depend on integrative studies using molecular genetics, microarrays, proteomics, and biochemical activities of key genes and pathways well into the future to re-visit old, and address new, questions.

REFERENCES

Abele, D., Puntarulo, S. (2004) Formation of reactive species and induction of antioxidant defence systems in polar and temperate marine invertebrates and fish. *Comparative Biochemistry and Physiology A* 138, 405–415.

Abele, D., Heise, K., Pörtner, H., et al. (2002) Temperature-dependence of mitochondrial function and production of reactive oxygen species in the intertidal mud clam *Mya arenaria*. *Journal of Experimental Biology* 205, 1831–1841.

Adams, N., Shick, J.M. (1996) Mycosporine-like amino acids provide protection against ultraviolet radiation in eggs of the green sea urchin *Strongylocentrotus droebachiensis*. *Photochemistry and Photobiology* 64, 149–158.

Adams, N., Shick, J.M. (2001) Mycosporine-like amino acids prevent UVB induced abnormalities during early development of the green sea urchin *Strongylocentrotus droebachiensis*. *Marine Biology* 138, 267–280.

Alonso-Alvarez, C., Bertrand, S., Devevey, G. et al. (2004) Increased susceptibility to oxidative stress as a proximate cost of reproduction. *Ecology Letters* 7, 363–368.

Angel, D.L., Fiedler, U., Eden, N., Kress, N. (1999) Catalase activity in macro-and microorganisms as an indicator of biotic stress in coastal waters of the easterm Mediterranean Sea. *Helgoland Marine Research* 53, 209–218.

Asada, K. (1994) Mechanisms for scavenging reactive molecules generated in chloroplasts under light stress. *In* Baker, N.R.; Bowyer, J.R. (eds). *Photoinhibition of Photosynthesis: From Molecular Mechanisms to the Field.* Bios Scientific Publishers, Oxford, pp. 129–142.

Asada, K. (1999) The water-water cycle in chloroplasts: scavenging of active oxygens and dissipation of excess photons. *Annual Review of Plant Physiology and Plant Molecular Biology* 50, 601–639.

Asada, K., Takahashi, M. (1987) Production and scavenging of active oxygen in photosynthesis. *In* Kyle, D.J., Osmond, C.B., Arntzen, C.J. (eds). *Photoinhibition.* Elsevier, Amsterdam, pp. 228–287.

Beyer, W., Imlay, J., Fridovich, I. (1991) Superoxide dismutases. *Progress in Nucleic Acid Research* 40, 221–253.

Bindoli, A. (1988) Lipid peroxidation in mitochondria. *Free Radical Biology and Medicine* 5, 247–261.

Brookes, P.S. (2005) Mitochondrial H^+ leak and ROS generation: An odd couple. *Free Radical Biology and Medicine* 38, 12–23.

Butow, B.J., Wynne, D., Tel-Or, E. (1997) Superoxide dismutase activity in *Peridinium gatunense* in Lake Kinneret: effect of light regime and carbon dioxide concentration. *Journal of Phycology* 33, 787–793.

Cadenas, E. (1989) Biochemistry of oxygen toxicity. *Annual Review of Biochemistry* 58: 79–110.

Cikala, M., Wilm, B., Hobmayer, E., Böttger, A., David, C.N. (1999) Identification of caspases and apoptosis in the simple metazoan *Hydra. Current Biology* 9, 959–962.

Collén, J., Davison, I.R. (1999) Stress tolerance and reactive oxygen metabolism in the intertidal red seaweeds *Mastocarpus stellatus* and *Chondrus crispus. Plant Cell and Environment* 22, 1143–1151.

Collén, J., Davison, I.R. (2001) Seasonal and thermal acclimation of reactive oxygen metabolism in *Fucus vesiculosus* (Phaeophyceae). *Journal of Phycology* 37, 474–481.

Davies, K.J.A. (1987) Protein damage and degradation by oxygen radicals. *Journal of Biological Chemistry* 262, 9895–9901.

Denicola, A., Souza, J.M., Radi, R. (1998) Diffusion of peroxynitrite across erythrocyte membranes. *Proceedings of the National Academy of Sciences USA* 95, 3566–3571.

Dunn, S.R., Bythell, J.C., Le Tessier, D.A. et al. (2002) Programmed cell death and necrosis activity during hyperthermic stress-induced bleaching of the symbiotic sea anemone *Aiptasia* sp. *Journal of Experimental Marine Biology and Ecology* 272, 29–53.

Dunn, S.R., Phillips, W.S., Spatafora, J.W. et al. (2006) Highly conserved caspase and Bcl–2 homologues from the sea anemone *Aiptasia pallida*: lower metazoans as models for the study of apoptosis evolution. *Molecular Evolution* 63, 95–107.

Dunn, S.R., Schnitzler, C.E., Weis, V.M. (2007) Apoptosis and autophagy as mechanisms of dinoflagellate symbiont release during cnidarian bleaching: every which way you lose. *Proceedings of the Royal Society of London. Series B, Biological Sciences* 274, 3079–3085.

Falkowski, P.G., Raven, J.A. (1997) *Aquatic Photosynthesis.* Blackwell Science, Malden, MA.

Falkowski, P.G., Katz, M.E., Knoll, A.H. et al. (2004) The evolution of modern eukaryotic phytoplanbkton. *Science* 305, 354–360.

Fang, F.C. (2004) Antimicrobial reactive oxygen and nitrogen species: concepts and controversies. *Nature Reviews Microbiology* 2, 820–832.

Finkel, T., Holbrook, N.J. (2000) Oxidants, oxidative stress and the biology of ageing. *Nature* 408, 239–247.

Freeman, B.A., Crapo, J.D. (1982) Biology of disease, free radicals and tissue injury. *Laboratory Investigation* 47, 412–426.

Fridovich, I. (1986) Biological effects of the superoxide radical. *Archives of Biochemistry and Biophysics* 247, 1–11.

Fridovich, I. (1998) Oxygen toxicity: a radical explanation. *Journal of Experimental Biology* 201, 1203–1209.

Green, D.R., Reed, J.C. (1998) Mitochondria and apoptosis. *Science* 281, 1309–1312.

Gutteridge, J.M.C, Halliwell, B. (1990) The measurement and mechanism of lipid peroxidation in biological systems. *Trends in Biochemical Science* 15, 129–135.

Halliwell, B., Gutteridge, J.M.C. (1999) *Free Radicals In Biology and Medicine.* Oxford University Press, New York.

Hengartner, M.O. (2000) The biochemistry of apoptosis. *Nature* 407, 770–776.

Hildeman, D.A., Mitchell, T., Aronow, B. et al. (2003) Control of Bcl–2 expression by reactive oxygen species. *Proceedings of the National Academy of Sciences USA* 100, 15035–15040.

Hoegh-Guldberg, O. (1999) Climate change, coral bleaching and the future of the world's coral reefs. *Marine and Freshwater Research* 50, 839–866.

Hoegh-Guldberg O., Mumby, P.J., Hooten, A.J. et al. (2007) Coral reefs under rapid climate change and ocean acidification. *Science* 318, 1737–1742.

Hollnagel, H.C., di Mascio, P., Asano, C.S. et al. (1996) The effect of light on the biosynthesis of β-carotene and superoxide dismutase activity in the photosynthetic alga *Gonyaulax polydra.* Brazil. *Journal of Medicine and Biomedical Research* 29, 105–110.

Hyslop, P.A., Hinshaw, D.B., Halsey Jr., W.A. et al. (1988) Mechanisms of oxidant-mediated cell injury. *Journal of Biological Chemistry* 263, 1665–1675.

Imlay, J.A. (2003) Pathways of oxidative damage. *Annual Review of Microbiology* 57, 395–418.

Imlay, J.A., Linn, S. (1988) DNA damage and oxygen radical toxicity. *Science* 240, 1302–1309.

Imlay, J.A., Chin, S.M., Linn, S. (1988) Toxic DNA damage by hydrogen peroxide through the Fenton reaction *in vivo* and *in vitro*. *Science* 240, 640–642.

Jabs, T., Dietrich, R.A., Dang, J.L. (1996) Initiation of runaway cell death in an *Arabidopsis* mutant by extracellular superoxide. *Science* 273, 1853–1856.

Jamieson, D., Chance, B., Cadenas, E. et al. (1986) The relation of free radical production to hyperoxia. *Annual Review of Physiology* 48, 703–719.

Johnson, T.M., Yu, Z., Ferrans, V.J. et al. (1996) Reactive oxygen species are downstream mediators of p53-dependent apoptosis. *Proceedings of the National Academy of Sciences USA* 93, 11848–11852.

Kasting, J.F., Siefert, J.L. (2002) Life and the evolution of Earth's atmosphere. *Science* 296, 1066–1068.

Korthout, H.A.A.J., Berecki, G., Bruin, W. et al. (2000) The presence and subcellular localization of caspase 3-like proteinases in plant cells. *FEBS Letters* 475, 139–144.

Lam, E., del Pozo, O. (2000) Caspase-like protease involvement in the control of plant cell death. *Plant Molecular Biology* 44, 417–428.

Lamare, M.D., Barker, M.F., Lesser, M.P. et al. (2006) DNA photorepair in echinoid embryos: effects of temperature on repair rate in Antarctic and non-Antarctic species. *Journal of Experimental Biology* 209, 5017–5028.

Lamare, M.D., Barker, M.F., Lesser, M.P. (2007) *In situ* rates of DNA damage and abnormal development in Antarctic and non-Antarctic sea urchin embryos. *Aquatic Biology* 1, 21–32.

Lesser, M.P. (2000) Depth-dependent effects of ultraviolet radiation on photosynthesis in the Caribbean coral, *Montastraea faveolata*. *Marine Ecology Progress Series* 192, 137–151.

Lesser, M.P. (2004) Experimental coral reef biology. *Journal of Experimental Marine Biology and Ecology* 300, 217–252.

Lesser, M.P. (2006) Oxidative stress in marine environments: biochemistry and physiological ecology. *Annual Review of Physiology* 68, 253–278.

Lesser, M.P. (2010) Survivorship, oxidative stress, and DNA damage of sea urchin (*Strongylocentrotus droebachiensis*) embryos and larvae exposed to ultraviolet radiation (290–400 nm) in the Gulf of Maine. *Photochemistry and Photobiology* 86, 382–388.

Lesser, M.P., Shick, J.M. (1989) Effects of irradiance and ultraviolet radiation on photoadaptation in the zooxanthellae of *Aiptasia pallida*: primary production, photoinhibition, and enzymic defenses against oxygen toxicity. *Marine Biology* 102, 243–255.

Lesser, M.P., Kruse, V.A., Barry, T.M. (2003) Exposure to ultraviolet radiation causes apoptosis in developing sea urchin embryos. *Journal of Experimental Biology* 206, 4097–4103.

Lesser, M.P., Barry, T.M., Lamare, M.D. et al. (2006) Biological weighting functions for DNA damage in sea urchin embryos exposed to ultraviolet radiation. *Journal of Experimental Marine Biology and Ecology* 328, 10–21.

Liu, K., Sun, J., Song, Y. et al. (2004) Superoxide, hydrogen peroxide, and hydroxyl radical in D1/D2/cytochrome *b*–559 photosystem II reaction center complex. *Photosynthesis Research* 81, 41–47.

López-Legentil, S., Song, B., McMurray, S.E. et al. (2008) Bleaching and stress in coral reef ecosystems: *hsp70* expression by the giant barrel sponge *Xestospongia muta*. *Molecular Ecology* 17, 1840–1849.

Lupínková, L., Komenda, J. (2004) Oxidative modifications of the photosystem II D1 protein by reactive oxygen species: From isolated protein to cyanobacterial cells. *Photochemistry and Photobiology* 79, 152–162.

Macpherson, A.N., Telfer, A., Barber, J. et al. (1993) Direct detection of singlet oxygen from isolated photosystem II reaction centers. *Biochimica et Biophysica Acta* 1143, 301–309.

Maritorena, S., Guillocheau, N. (1996) Optical properties of the water and spectral light absorption by living and non-living particles and by yellow substances in coral reef waters of French Polynesia. *Marine Ecology Progress Series* 131, 245–255.

Marla, S.S., Lee, J., Groves, J.T. (1997) Peroxynitrite rapidly permeates phopholipid membranes. *Proceedings of the National Academy of Sciences USA* 94, 14243–14248.

Matta, J.L., Govind, N.S., Trench, R.K. (1991) Polyclonal antibodies against iron-superoxide dismutase from *Escherchia coli* cross-react with superoxide dismutases from *Symbiodinium microadriaticum* (Dinophyaeae). *Journal of Phycology* 28, 343–346.

Martindale, J.L., Holbrook, N.J. (2002) Cellular response to oxidative stress: signaling for suicide and survival. *Journal of Cellular Physiology* 192, 1–15.

Mehdy, M.C. (1994) Active oxygen species in plant defense against pathogens. *Plant Physiology* 105, 467–472.

Mitchelmore, C.L., Birmelin, C., Chipman, J.K. et al. (1998) Evidence for cytochrome P–450 catalysis and free radical involvement in the production of DNA strand breaks by benzo[a]pyrene and nitroaromatics in mussel (*Mytilus edulis* L.) digestive gland cells. *Aquatic Toxicology* 41, 193–212.

Mopper, K., Kieber, D.J. (2000) Marine photochemistry and its impact on carbon cycling. *In* De Mora, S., Demers, S., Vernet, M. (eds). *The Effects of UV Radiation in the Marine Environment*. University Press, Cambridge, pp. 101–130.

Nisbet, E.G., Sleeo, N.H. (2001) The habitat and nature of early life. *Nature* 409, 1083–1091.

Nishiyama, Y., Yamamoto, H., Allakhverdiev, S.I. et al. (2001) Oxidative stress inhibits the repair of photodamage to the photosynthetic machinery. *EMBO Journal* 20, 5587–5594.

Oda, T., Ishimatsu, A., Shimada, M. et al. (1992) Oxygen-radical-mediated toxic effects of the red tide flagellate *Chattonella marina* on *Vibrio alginolyticus*. *Marine Biology* 112, 505–509.

Okamoto, O.K., Robertson, D.L., Fagan, T.F. et al. (2001) Different regulatory mechanisms modulate the expression of a dinoflagellate iron-superoxide dismutase. *Journal of Biological Chemistry* 276, 19989–19993.

Otis, D.B., Carder, K.L., English, D.C. et al. (2004) CDOM transport from the Bahamas Banks. *Coral Reefs* 23, 152–160.

Regoli, F., Cerrano, C., Chierici, E. et al. (2000) Susceptibility to oxidative stress of the Mediterranean demosponge *Petrosia ficiformis*: role of endosymbionts and solar irradiance. *Marine Biology* 137, 453–461.

Regoli, F., Cerrano, C., Chierici, E. et al. (2004) Seasonal variability of prooxidant pressure and antioxidant adaptation to symbiosis in the Mediterranean demosponge *Petrosia ficiformis*. *Marine Ecology Progress Series* 275, 129–137.

Rich, T., Allen, R.L., Wyllie, A.H. (2000) Defying death after DNA damage. *Nature* 407, 777–783.

Richter, M., Rüle, W., Wild, A. (1990) Studies on the mechanism of photosystem II photoinhibition II. The involvement of toxic oxygen species. *Photosynthesis Research* 24, 237–243.

Sandalio, L.M., del Rio, L.A. (1988) Intraorganellar distribution of superoxide dismutase in plant peroisomes (glyoxisomes and leaf peroxisomes). *Plant Physiology* 88, 1215–1218.

Schrek, R., Baeuerle, P.A. (1991) A role for oxygen radicals as second messengers. *Trends in Cell Biology* 1, 39–42.

Segovia, M., Haramaty, L., Berges, J.A. et al. (2003) Cell death in the unicellular chlorophyte *Dunaliella tertiolecta*. A hypothesis on the evolution of apoptosis in higher plants and metazoans. *Plant Physiology* 132, 99–105.

Shiu, C.-T., Lee, T.-M. (2005) Ultraviolet-B-induced oxidative stress and responses of the ascorbate-glutahione cycle in a marine macroalga *Ulva fasciata*. *Journal of Experimental Botany* 56, 2851–2865.

Stadtman, E.R. (1986) Oxidation of proteins by mixed-function oxidation systems: implication in protein turnover, aging and neutrophil function. *Trends in Biochemical Science* 11, 11–12.

Vicente, V.P. (1990) Response of sponges with autotrophic endosymbionts during the coral-bleaching episode in Puerto Rico. *Coral Reefs* 8, 199–202.

Wagner, C., Steffen, R., Koziol, C. et al. (1998) Apoptosis in marine sponges: a biomarker for environmental stress (cadmium and bacteria). *Marine Biology* 131, 411–421.

Webster, N.S., Cobb, R.E., Negri, A. (2008) Temperature thresholds for bacterial symbiosis with a sponge. *THE ISME JOURNAL* 2, 830–842.

Weis, V.M. (2008) Cellular mechanisms of cnidarian bleaching: stress causes the collapse of symbiosis. *Journal of Experimental Biology* 211, 3059–3066.

Weins, M., Müller, W.E.G. (2006) Cell death in Porifera: molecular players in the game of apoptotic cell death in living fossils. *Canadian Journal of Zoology* 84, 307–321.

Weins, M., Krasko, A., Perovic, S. et al. (2003) Caspase-mediated apoptosis in sponges: cloning and function of the phylogenetic oldest apoptotic proteases from metazoa. *Biochimica et Biophysica Acta* 1593, 179–189.

Winston, G.W., Livingstone, D.R., Lips, F. (1990) Oxygen reduction metabolism by the digestive gland of the common marine mussel, *Mytilus edulis* L. *Journal of Experimental Zoology* 255, 296–308.

Yakoleva, I., Bhagooli, R., Takemura, A. et al. (2004) Differential susceptibility to oxidative stress of two scleractinian corals: antioxidant functioning of mycosporine-glycine. *Comparative Biochemistry and Physiology B* 139, 721–730.

Yang, C.Z., Albright, L.J., Yousif, A.N. (1995) Oxygen-radical-mediated effects of the toxic phytoplankter *Heterosigma carterae* on juvenile rainbow trout *Oncorhynchus mykiss*. *Marine Ecology Progress Series* 23, 101–108.

Yu, B.P. (1994) Cellular defenses against damage from reactive oxygen species. *Physiology Review* 74, 139–162.

Zepp, R.G., Shank, G.C., Stabenau, E. et al. (2008) Spatial and temporal variability of solar ultraviolet exposure of coral assemblages in the Florida Keys: Importance of colored dissolved organic matter. *Limnology and Oceanography* 53, 1909–1922.

Zheng, M., Å slund, F., Storz, G. (1998) Activation of the OxyR transcription factor by reversible disulfide bond formation. *Science* 279, 1718–1721.

Zubia, M., Robledo, D., Freile-Pelegrin, Y. (2007) Antioxidant activities in tropical marine macroalgae from the Yucatan Peninsula, Mexico. *Journal of Applied Phycology* 19, 449–458.

Chapter 2

OXIDATIVE CHALLENGES IN POLAR SEAS

Francesco Regoli[1], *Maura Benedetti*[1], *Andreas Krell*[2], *and Doris Abele*[2]

[1]Dipartimento di Biochimica, Biologia e Genetica, Università Politecnica delle Marche, Ancona, Italy

[2]Alfred Wegener Institute for Polar and Marine Research, Bremerhaven, Germany

OXYGEN RADICALS IN ICY WATERS

Reactive oxygen species (ROS) are produced in open oceanic and coastal surface waters through interactions of sunlight with photoreactive dissolved organic matter (DOM). Photons in the visible (VIS) and ultraviolet (UV) light range are absorbed by molecules containing chromophores, organic structures carrying delocalized π-electrons in aromatic rings and conjugated double-bond systems. Electrons that absorb the energy of a photon are elevated from the energetic ground state (S_o) to a higher, excited energy state (singlet state: S_1). Excited electrons can either return to S_o by releasing energy as fluorescence light and heat (so that the emitted fluorescence wavelength is energetically lower than the absorbed wavelength/photon energy). In the aquatic environment, photoactivated electrons are absorbed by an electrophilic acceptor (A), or persist shortly as freely dissolved aquatic electrons (e_{aq}).

Oxygen and transition metals (Me^+) such as Fe^{3+} are readily available acceptors for electrons released from photoactivated organic chromophores in surface waters. Molecular oxygen, especially, with a redox potential ($E^{o\prime}$) of +0.82 V undergoes univalent reduction to $O_2^{\bullet -}$ (Fig. 2.1).

$$O_2 + DOM\text{-}e^* \Rightarrow DOM^{\bullet +} + O_2^{\bullet -} \quad (2.1)$$

Due to its high reactivity, $O_2^{\bullet -}$ has a half-life of milliseconds and either donates the radical electron to other molecules or accepts another electron from a second $O_2^{\bullet -}$ and disproportionates to oxygen and the more persistent H_2O_2.

$$2O_2^{\bullet -} + 2H^+ \Rightarrow O_2 + H_2O_2 \quad (2.2)$$

H_2O_2 is relatively stable in natural, slightly alkaline seawater. Its half-life is a matter of hours and depends mainly on the presence and activity of H_2O_2 decomposing enzymes, catalase (CAT), and peroxidases in the water. H_2O_2 is membrane-permeable and its degradation is vastly accelerated by bacteria, whereas H_2O_2 breakdown by marine phytoplankton is less efficient but still accelerated over the rates in sterile filtered (0.2 μm pore size) seawater. Once released into the water column as cellular debris and organism exudates, cell-free enzymes (especially CAT and peroxidases) preserve their catalytic activity. Recorded half-life of H_2O_2 ranges from slightly over 1 h in coastal environments to >100 h in open oceanic waters (Petasne and Zika 1997).

Chemical reactions consuming H_2O_2 involve dissolved and complexed Fe and DOM in surface waters. Fe^{2+} from eroding rocks and hydrothermal vents is soluble at seawater pH (8.2), but usually dissolved Fe^{2+}

Oxidative Stress in Aquatic Ecosystems, First Edition. Edited by Doris Abele, José Pablo Vázquez-Medina, and Tania Zenteno-Savín.

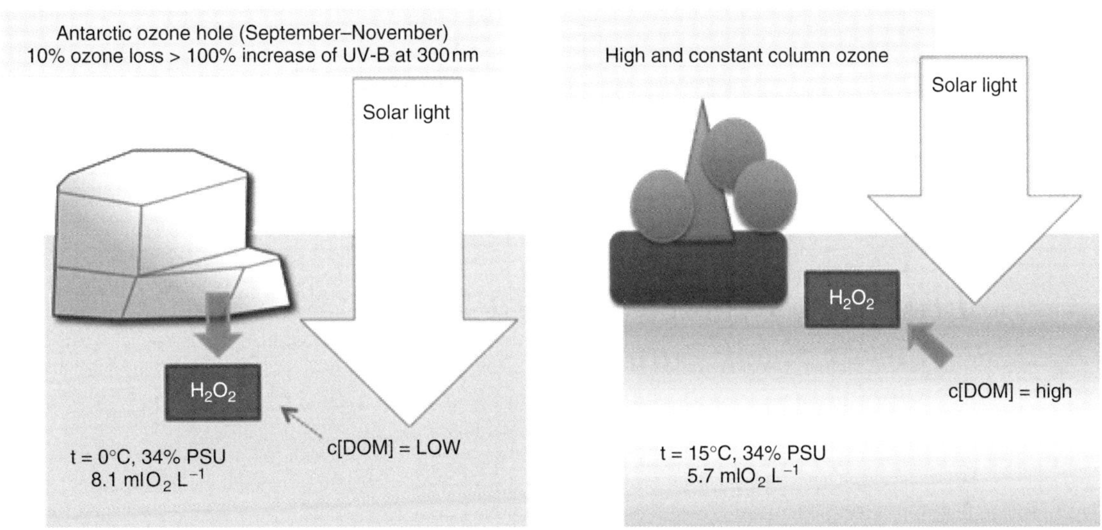

Fig. 2.1 Radiation scenario in polar (left) and temperate regions (right). In oligotrophic polar seas solar light and especially high energy UV-B radiation penetrates up to 60 m deep into the water column. Low DOM (dissolved organic matter) concentrations limit the photochemical formation of H_2O_2. In temperate waters DOM is high and limits light penetration, but also intensifies photochemical H_2O_2 formation in surface waters. In polar regions especially, melting of accumulated snow during sea-ice break-up releases substantial amounts of H_2O_2 into the water column. See plate section for a color version of this image.

is oxidized to ferric iron Fe^{3+} by oxygen and H_2O_2. Ninety-nine percent of Fe in oxygen-rich oceanic water is present as oxidized, insoluble Fe^{3+} in the colloidal or particulate matter fraction. The Fenton reaction (2.3) involves H_2O_2 instead of oxygen and generates highly reactive and very hazardous $HO^•$.

$$Fe^{2+} + H_2O_2 \Rightarrow Fe^{3+} + HO^- + HO^• \quad (2.3)$$

When formed in open waters, $HO^•$ are of little danger to organisms because they immediately interact with either DOM or other seawater ions (e.g. Br^-, Cl^-, and CO_3^{2-}). If Fe concentrations are high enough, $HO^•$ can also oxidize Fe^{2+} to Fe^{3+}.

$$Fe^{2+} + HO^• \Rightarrow Fe^{3+} + OH^- \quad (2.4)$$

If it were not for the effect of sunlight, all Fe released into the water column would rapidly become oxidized. However, UV and VIS (< 560 nm) driven photoreduction reconverts Fe^{3+} to bioavailable Fe^{2+} either directly or via the intermediate formation of $O_2^{•-}$ (2.5).

$$Fe^{3+} + O_2^{•-} \Rightarrow Fe^{2+} + O_2 \quad (2.5)$$

Of the solar surface irradiance spectrum, UV-B photons are most efficiently absorbed by DOM in the range of 300 nm, which makes them powerful drivers of H_2O_2 formation and of Fe^{3+} photoreduction in the water column (Rijkenberg et al. 2005). UV-B photons have a 10-fold stronger yield in H_2O_2 photoformation than UV-A photons in natural seawater (Abele-Oeschger et al. 1997). Iron photoreduction yields with UV-B radiation are 7 to 10-times higher than under UV-A light.

The fraction of UV-B in solar surface irradiance and water surface mixing, together with the concentrations of oxygen and DOM (quantified as light absorbance at 300 nm) can therefore be regarded as the major determinants of H_2O_2 photoproduction in sea- and freshwater. It results that H_2O_2 steady-state concentrations in surface waters follow a diurnal cycle, with maxima usually around noon and a pre-dawn minimum (Zika et al. 1983). This is most clearly observed in the photic zone of stratified water bodies with little mixing above the pycnocline. Typical H_2O_2 surface maxima in open waters range between 10 nmol L^{-1} in oligotrophic and 300 nmol L^{-1} in more eutrophic coastal areas (Zika et al. 1983; Cooper and Lean, 1989). On sunny days water in shallow tidal pools can even reach concentrations of more than 4 μmol L^{-1} (Abele-Oeschger et al. 1997).

The capacity for photochemical H_2O_2 formation of Antarctic waters is limited by lower concentrations of DOM compared to temperate seas (Table 2.1). When exposed to natural irradiance levels, water in coastal areas of the Antarctic Peninsula yields only one-third of the H_2O_2 formation rate (110 nmol H_2O_2 L^{-1} h^{-1}) measured in water from the North Sea (300 nmol H_2O_2 L^{-1} h^{-1}) under the same experimental conditions. This may, however, change in the course of climate warming in the Antarctic Peninsula region, where coastal vegetation, both terrestrial and intertidal, is beginning to sprout, adding to the phytogenic DOM pool in Antarctic coastal waters. At present, steady-state H_2O_2 concentrations recorded for photochemical generation in Antarctic waters are still much lower than in temperate areas. Recorded values amount to 220 ± 40 nmol L^{-1} in intertidal pools at King George Island (between October and November; Abele et al. 1999), and to only up to 30 nmol L^{-1} in Southern Ocean surface waters (Weller and Schrems 1993; Resing et al. 1993).

Photoactivation of DOM is a temperature-independent process, whereas the electron transfer to oxygen and the dismutation of $O_2^{\bullet -}$ are accelerated at higher temperature (Scymczak and Waite 1988). Further, the enzyme catalyzed decomposition of H_2O_2 and its interaction with Fe^{2+} are temperature-dependent and, together, these processes determine the steady state H_2O_2 concentration in sunlight-exposed natural waters. Dissolved organic matter is the rate-limiting component for ROS photoformation in cold polar waters, and the high levels of dissolved oxygen are of minor significance. In contrast, in temperate tide-water pools with high DOM concentrations, oxygen can be completely depleted by heterotrophic respiration and chemical oxidation processes and can become rate-limiting for H_2O_2 formation.

Table 2.1 Examples of DOM concentrations measured in Antarctica and North Atlantic waters.

Location	DOM (μmol C L^{-1})	Source
South Shetlands, Antarctica, coastal	150–400	Abele et al. (1999)
Bransfield and Gerlache Straits, Antarctica (upper mixed layer)	60–50	Doval et al. (2002)
German Wadden Sea, intertidal	6000–7000	Abele-Oeschger et al. (1997)
North Atlantic, open	100–150	Kepkay and Wells (1992)

Real concentrations of H_2O_2 in surface waters, tidal pools, or sea-ice can be higher than predicted from the yields of DOM photo-oxidation alone, as additional H_2O_2 originates from atmospheric wet deposition of snow and rain. As H_2O_2 is produced photochemically in the atmosphere, seasonal differences of H_2O_2 precipitation can vary by a factor of five between winter and summer snow samples in Antarctica (Sigg and Neftel 1988). Eicken et al. (1994) reported H_2O_2 concentrations between 53 and 5000 nmol L^{-1} in 65 snow samples they collected from the Weddell Sea ice. At Queen Maud Island, Antarctica, Kamiyama et al. (1996) recorded seasonal peak concentrations of 6700 and 12,000 nmol H_2O_2 L^{-1} in snow deposits in October 1991 (spring) and only maximum values of 1700 nmol L^{-1} in July/August (winter). Concentrations of 10,000 and 13,600 nmol L^{-1} in freshly fallen snow are reported from the Antarctic Peninsula region during the ozone minimum (October–November; Abele et al. 1999). Although chemical and biological decomposition of H_2O_2 in snow is rather quick, precipitation produces summer and winter bands with large concentration differences in ice cores (Sigg and Neftel 1988; Eicken et al. 1994), and H_2O_2 has been proposed as a tracer for ice-core dating. Coastal surface waters in spring show the signal of the melting sea-ice layer and its snow cover with salinity decreasing from 34 to below 20 PSU (practical salinity units), and H_2O_2 reaching peak concentrations of 1500 nmol L^{-1} (Abele et al. 1999). Thus, Antarctic marine organisms must be able to cope with elevated ROS levels although the H_2O_2 photoformation in the water is limited by the lower DOM levels in polar regions (Table 2.1).

Adaptation of polar marine organisms to oxidative stress is often insufficient because the light climate in both polar regions is characteristically low. Owing to a large solar zenith angle, solar light travels long distances to the poles, and much of the biologically damaging UV radiation is attenuated before reaching the Earth's surface. Light is also the most important seasonal variable in polar regions, where the long polar nights in winter months mean an absence of light, which together with shading by sea-ice reduces H_2O_2 photoformation to practically zero for three months each year. As a consequence, the pelagic microalgae and shallow-water benthic micro- and macroalgae in Antarctica suffer stress from the rapid increase of radiation during the ice break-up in the austral spring and thus from an elevated photoformation of H_2O_2 in conjunction with an increased release of H_2O_2 from melting sea-ice (Janknegt et al. 2008; Karsten et al. 2009). Similar fluctuations of radiation-induced oxidative stress occur during advective transport of phytoplankton between ice-covered and ice-free zones in coastal waters and at the sea-ice edge (Janknegt et al. 2008; Karentz and Bosch 2001).

As a consequence of the anthropogenic release of ozone-destroying chlorofluorocarbons (CFCs), a reduction of total column ozone, the "ozone hole", develops in September and dissipates in November over the Antarctic continent and the southern ocean (Fig. 2.2). This implies a vast increase of the biologically active UV-B radiation (Frederick and Snell 1988; Villafañe et al. 2001; Hernandez et al. 2002) during the spring ice-opening period and affects organisms in surface waters around Antarctica and Southern Patagonia. As stratospheric ozone filters mostly UV-B radiation below 320 nm, the ozone hole specifically enhances the small percentage (1%) of biologically damaging radiation of the solar spectrum. A diminishment by 60% of column ozone over the Antarctic causes 2.2-fold enhancement of irradiance at 315 nm and a 14-fold enhancement at 305 nm (Frederick and Snell 1988), increasing the risk of DNA mutations and photochemical oxidation of UV-absorbing biomolecules in UV transparent organisms, such as amphipod crustaceans. After the measures taken to regulate usage and release of CFCs in the Montreal and Kyoto Protocols, predictions are that CFCs concentration will start to decline in the atmosphere during the coming years, but recovery of the ozone layer is not expected until 2050 or later (http://www.ozonelayer.noaa.gov). At present nobody can foresee what the final effects will be of a 100-year duration ozone-hole condition affecting large parts of the Antarctic continent and surrounding waters.

THE SEA-ICE ENVIRONMENT

Sea-ice, a habitat for a multitude of cryoresistant marine organisms, differs considerably from other environments, notably for two reasons. Sea-ice organisms have to deal with a multitude of extreme abiotic conditions, especially with respect to temperature, salinity, and the light regime (Fig. 2.3). Second, sea-ice represents a semi-enclosed system where the flux of nutrients for photosynthetic growth is restricted, and the exchange of gases is limited. In spite of such harsh conditions, sea-ice can be densely populated with organisms ranging from viruses to small metazoans (Lizotte 2003), with major representatives being diatoms (Bacillariophyceae) and prymnesiophytes (e.g. *Phaeocystis* spp.) (Bartsch 1989; Gleitz et al. 1994; Lizotte 2001).

With the onset of freezing, a developing ice cover restricts gas exchange between ocean and atmosphere. It was believed that this layer forms an impermeable barrier to the exchange of gases, and global climate

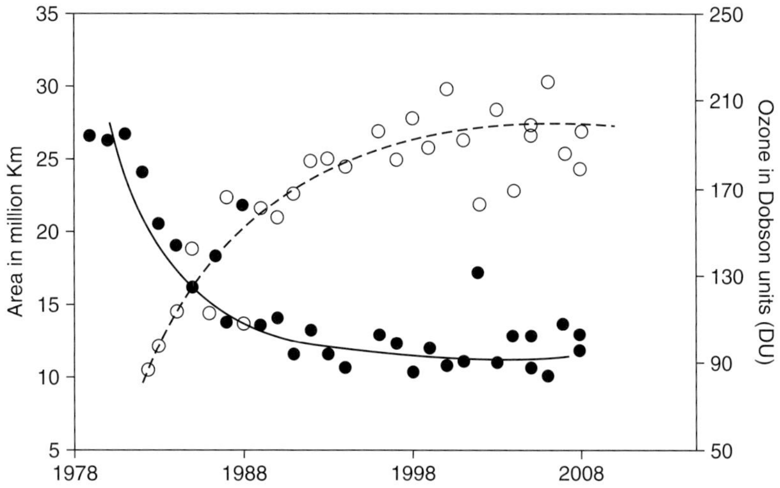

Fig. 2.2 Development of column ozone concentrations in Dobson units (filled circle) and areal extent of the South Pole ozone hole in million square kilometers (open circle) (Data from NASA and the National Oceanic and Atmospheric Administration Center (NOAA), courtesy of R. McPeters).

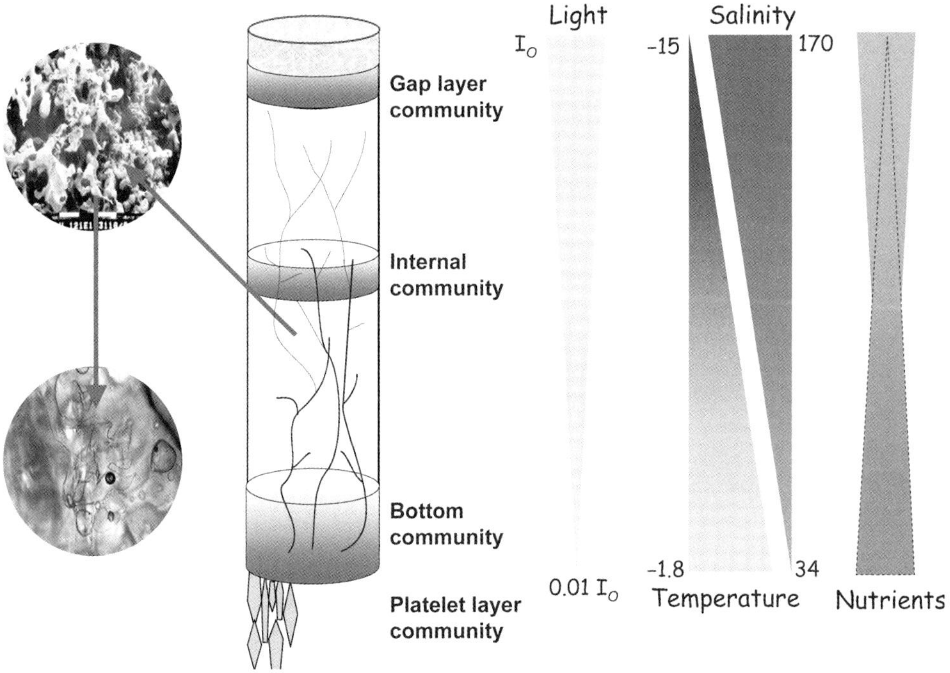

Fig. 2.3 Schematic drawing of a sea-ice column with the different communities that can be encountered. The two close-ups depict the brine-channel system and a single brine pocket with diatoms (courtesy C. Krembs). Gradients in abiotic factors are shown on the right. In cases of flooding events nutrients are equally supplied via the top of the ice sheet.

models did not include CO_2 and O_2 exchanges between sea ice and the atmosphere. Recent findings suggest, however, that sea ice is permeable for gases, and gas migration through sea ice might very well be an important factor in winter ocean-atmosphere exchange at sea-ice surface temperatures above $-10^{\circ}C$ (Delille et al. 2007).

Seawater with a salinity of 34 commences to freeze at a temperature of $-1.86^{\circ}C$. During this process only water molecules form a solid crystal matrix, whereas other ions and dissolved constituents accumulate within a system of interconnected channels and isolated pockets (Weissenberger et al. 1992). Therefore, besides gas inclusions and small amounts of precipitated salts, the interior of sea-ice consists primarily of two phases: solid ice and brine (hypersaline water inclusions) (Weeks and Ackley 1982). Brine volume and salinity are strictly temperature-dependent and decreasing temperatures lead to a decrease in brine volume and an increase in salinity (Cox and Weeks 1983). Brine salinity can attain values of 70 to 144 at -4 to $-10^{\circ}C$, respectively. Without the interference of organisms, oxygen, like other solutes, behaves conservatively during freezing, i.e. it increases proportionally to salinity and can attain concentrations twice or three times higher than in the underlying seawater. From a physical perspective, this is brought about by two contrasting effects influencing the solubility of oxygen. The solubility of oxygen in water at constant atmospheric pressure and water vapour pressure is temperature-dependent and can be described using the Bunsen absorption coefficient $\alpha(\theta)$ and P_{O_2}. With decreasing temperature, the solubility of oxygen in water increases. However, the Bunsen absorption coefficient $\alpha(\theta)$ is equally dependent on salinity. Higher salinities result in a decreased solubility of oxygen (Fig. 2.4), a phenomenon characteristic for the solubility of many nonelectrolytes, known as the salting-out effect. This leads to the occurrence of high oxygen concentrations, not harmful in themselves, but potentially driving ROS formation.

Whereas underlying polar waters can often be undersaturated in O_2, with saturation levels dropping

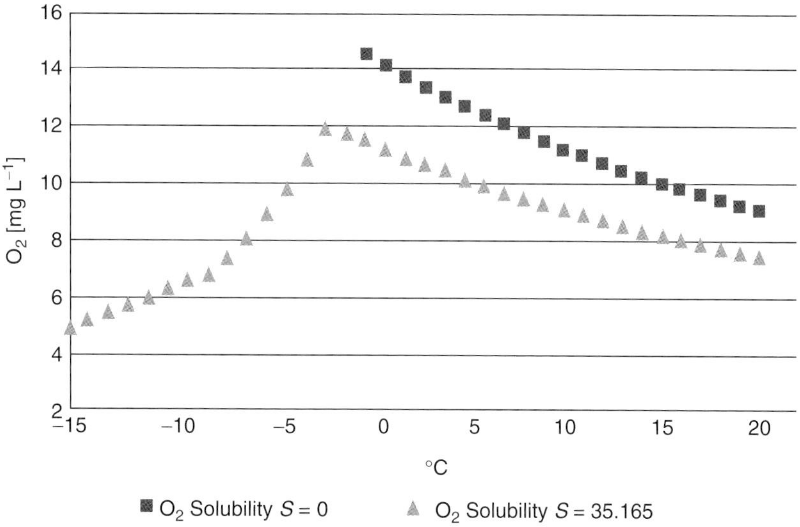

Fig. 2.4 Solubility of oxygen in freshwater ($S = 0$) and standard seawater ($S = 35.165$) versus temperature at 100% air saturation and normal pressure. Note the general decreased solubility in seawater and the strong decline with the onset of freezing.

as low as 82% of normoxic levels (Hoppema et al. 1995; Delille et al. 2007), the opposite can be true for the sea-ice system. The oxygen saturation can vary considerably over the sea ice season depending firstly on the physical conditions of ice growth or melt. During freezing, oxygen concentrations increase in the liquid phase by the above described mechanism which ultimately leads to the degassing of brine and formation of oxygen bubbles as the ice gets colder, whereas the melting water released during thawing is undersaturated and thus hypoxic (Glud et al. 2002). The situation is even more complex when considering these abiotic processes and the state of the prevailing biological community. In communities with predominantly heterotrophic activity, oxygen consumption can even cause anoxic conditions (Rysgaard et al. 2008). In contrast, in dense sea-ice algal communities (Fig. 2.5) under favourable light conditions and consequently high photoautothrophic activity, oxygen concentrations are elevated and supersaturation can reach 160% (Gleitz et al. 1995; Delille et al. 2007). Because of the heterogeneity and the compartmentalization of the sea-ice system, microsites of high and low O_2 concentrations may occur in close proximity.

The distribution pattern of DOM in sea-ice is also highly heterogeneous and does not always correlate with biomass (Thomas et al. 2001). However, DOM in bottom ice was found to co-vary with the colonization

Fig. 2.5 Bottom section of a sea-ice core is colored brownish due to the high number of microalgae (courtesy M. Nicolaus). See plate section for a color version of this image.

by ice microalgae and was enriched relative to the interior of the ice as well as to the underlying seawater (Table 2.2), suggesting autochthonous sources of DOM in sea-ice.

A high oxygen concentration in itself negatively affects the growth of photosynthetic organisms since oxygen is a competitive inhibitor of the carboxylase activity at the site of ribulose-1,5-carboxylase/oxygenase (RUBISCO). The combination of high oxygen and DOM concentrations and, at times, high irradiance levels, especially in the UV range, results

Table 2.2 Levels of DOC measured in brine and bulk ice from different locations and seasonal periods.

Location	DOC (μmol C L^{-1})	Season	Source
Amundsen and Bellingshausen Sea	134–23284 (brine) 16–556 (bulk sea ice)	Jan/Feb, multi-year ice	Thomas et al. (2001)
Weddell Sea	147–866 (brine)	Apr/May, first-year ice	Thomas et al. (2001)
	124–18451 (brine) 16–1842 (bulk sea ice)	Jan–Mar, multi-year ice	
	106–701 (bulk sea ice)	Nov–Jan, first-year ice	Dumont et al. (2009)
East Antarctica	80–717 (bulk sea ice)	Sept/Nov, first-year ice	Dumont et al. (2009)
Chukchi Sea, Arctic	50–258 (bulk sea ice)	Mar, first-year ice	Krembs et al. (2002)
Baltic Sea	99–152 (bulk sea ice)	Mar, first-year ice	Granskog et al. (2005)

in an elevated production of ROS, which exert considerable oxidative stress on sea-ice organisms and drastically inhibits growth (McMinn et al. 2005; Buma et al. 2006; Janknegt et al. 2008). Thus, there is a strong necessity to ameliorate the potentially toxic effects of ROS. However, especially in sea-ice, in relation to other aquatic systems this is hampered through the plethora of extreme abiotic conditions affecting the cellular integrity of sea-ice organisms.

Although the general metabolism of sea-ice-associated microalgae does not vary from those living in lower latitudes, physiological adaptations to enhance survival in psychrophilic microalgae are evident. A number of metabolites have been discerned to play a key role in the adaptation potential of sea-ice algae. One of the substances put forward as a potent ROS scavenger in marine algae is dimethylsulfoniopropionate (DMSP), which at the same time can act as an active osmolyte to counterbalance osmotic stress and serves as a cryoprotectant (Dickson and Kirst 1986). DMSP and its breakdown products (DMS, acrylate, dimethylsulfoxide and methane sulfinic acid) are able to readily scavenge $HO^{\bullet}$ and other ROS (Sunda et al. 2002). High concentrations of DMSP and DMS are frequently encountered in sea-ice, attaining values of DMS and associated compounds three orders of magnitude higher than those observed in the water column (Trevena and Jones 2006; Delille et al. 2007).

Another powerful scavenger of ROS in sea-ice algae is proline (Schriek 2000; Rodriguez and Redman 2005). Proline can alleviate the negative effects of various abiotic stresses such as UV light, heat, and H_2O_2 toxicity, and in doing so mitigates cellular stress levels and prevents apoptosis (Chen and Dickman 2005). Besides its function as a ROS scavenger, proline also acts as the major organic osmolyte in a number of sea-ice microalgae (Plettner 2002). The pivotal role of proline is illustrated by the fact that in the psycrophilic diatom *Fragilariopsis cylindrus* anabolic and catabolic pathways under salt stress substantially differ from those in higher plants (Krell et al. 2007).

Ultraviolet B radiation promotes $O_2^{\bullet-}$ production. Microalgae exposed to enhanced UV radiation at higher latitudes are able to counteract oxidative stress with elevated levels of superoxide dismutase (SOD) (Martínez 2007; Janknegt et al. 2008). High salinities even lead to a higher SOD activity (Roncarati et al. 2008), suggesting that microalgae living in the sea-ice brine channel system with elevated salt concentrations might be protected by the salt effect.

Although all the above-mentioned compounds, and many more (ascorbate peroxidase, glutathione disulphide reductase (GR), and glutathione (GSH) redox status), play a major role in the oxidative stress response, recent investigations suggest that the antioxidant response in microalgae is highly species-specific and presumably not related to either geographic origin or phylogenetic position (Janknegt et al. 2009).

PRO-OXIDANT CHALLENGE AND ANTIOXIDANT DEFENSES IN POLAR ORGANISMS

Adaptation to a permanently cold climate in polar marine organisms can modulate their susceptibility to oxidative stress. However, the effect is not as straightforward as previously thought, and the concept that the high oxygen levels in cold seawater cause increased oxidative stress in marine ectotherms has been outdated in the most recent literature (see also Chapter 11). Whereas some features of cold-adapted invertebrates and fish may exacerbate oxidative stress, low temperatures *a priori* reduce metabolic rates in ectotherms thereby limiting the routine levels of mitochondrial ROS output. This could be one reason why several ectotherms, especially in the Arctic Ocean, attain centenarian life spans (Abele et al. 2009). Neither the compensatory increase of mitochondrial density, as seen in red and white muscle of Antarctic fish (Johnston et al. 1998) and in temperate fish upon acclimation to cold winter temperatures (Egginton et al. 2000), nor the adjustment of respiratory enzyme activities per mitochondrial unit in the cold (Guderly 2004), can reasonably support claims of elevated basal ROS formation in polar fish species. Among Antarctic mollusks, slower moving, crawling snails and burrowing clams do not compensate the mitochondrial densities but rather submit to the cold temperatures and adjust mobility and burrowing activity at slower pace (Morley et al. 2009). These animals are not likely to produce high levels of ROS in their cold-adapted cells under nonstressful conditions.

The major trait that enhances oxidative stress sensitivity in cold-adapted species is the elevated unsaturation level in membrane lipids. A higher percentage of the polyunsaturated fatty acid (PUFA) represents an important mechanism by which polar ectotherms maintain biological membrane structural and functional properties against low temperature (Hazel 1985; Hochachka and Somero 2002; Logue et al. 2000). Both animals and plants change the relative phospholipid content and increase the PUFA proportion in membrane lipids and depot fat. A higher ratio of the bilayer-destabilizing phosphatidylethanolamine to the bilayer-stabilizing phosphatidylcholine is characteristic of polar species and is probably associated with the regulation of dynamic phase behavior of the bilayer (Hazel 1985; Pruitt 1988; Acierno et al. 1996). The higher proportions of unsaturated fatty acids in cold-adapted species has been directly associated with a decrease in membrane viscosity. Investigations of artificial monolayers showed that the presence of the first double bond in the acyl chain greatly increases the molecular area and reduces the phospholipid packing, whereas subsequent double-bond insertions cause only minor additional modifications (Coolbear et al. 1983; Acierno et al. 1996). Since PUFA is highly susceptible to ROS attack (Fig. 2.6), and high unsaturation levels enhance the velocity and propagation of lipid radical chain reactions (Halliwell and Gutteridge 2007), the difference in fatty acid composition increases the vulnerability of cold-adapted organisms to oxidative injuries.

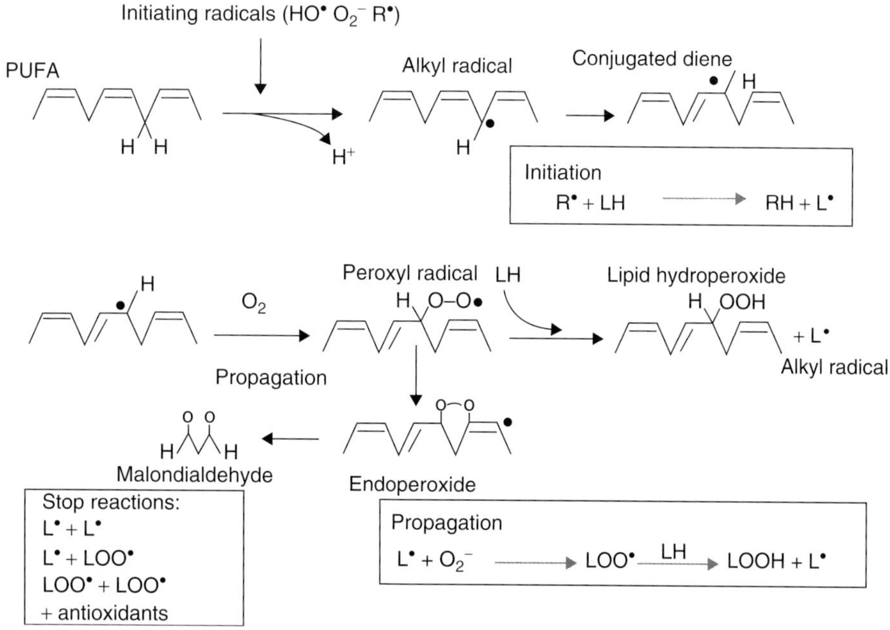

Fig. 2.6 Schematic overview of initiation, propagation, and stop reactions during membrane lipid peroxidation.

Adaptation to cold environments can also influence protein biosynthesis and turnover. Contrary to the assumption that low temperature causes lower protein synthesis rates, both RNA and protein synthesis are temperature compensated in polar scallops and fish, and equal those measured in temperate species (Marsh et al. 2001; Fraser et al. 2002; Storch and Pörtner 2003). Elevated RNA:protein ratios in polar ectotherms were often interpreted as an adaptation to a low translational efficiency (Fraser et al. 2002) but may rather reflect an enhanced RNA stability resulting from low RNA turnover rate in polar organisms (Storch and Pörtner 2003). Although ROS production levels are low at cold temperatures, the cellular longevity of individual RNA molecules theoretically prolongs the period of ROS exposure to these important macromolecules (Camus et al. 2005). Also, the capability to repair DNA damage caused by UV radiation is not temperature compensated and has been shown to range significantly lower in Antarctic echinoids in comparison to temperate or tropical species, another suggestive trait of higher susceptibility of polar species to oxidative stress (Lamare et al. 2006).

Comparisons of antioxidant activities without a record of physiological or life-history state (age, reproductive state) can be very misleading when used as an isolated parameter to compare the general background of oxidative stress in different climatic regions. To establish a clear picture of oxidative stress in permanently cold areas, an integrated approach should include quantification of ROS formation, susceptibility of membrane lipids to ROS attack, analyses of oxidative damages to different cellular compartments, and antioxidant capacity.

OXIDATIVE STRESS IN POLAR VS. TEMPERATE SCALLOPS

Compared with the Mediterranean scallop *Pecten jacobaeus*, membrane fluidity in the Antarctic scallop *Adamussium colbecki* (Fig. 2.7) is maintained through a higher cholesterol/phospholipid ratio, a higher amount of short-chained saturated fatty acids, and in particular, by a higher amount of unusual branched and saturated fatty acids. Thus, *A. colbecki* departs from the most common assumption that cold adaptation of ectothermal membranes comprises an increase in the levels of unsaturated fatty acids (Viarengo et al. 1994).

Adamussium colbecki is endowed with a more efficient antioxidant system than the Arctic *Chlamys islandicus* and the Mediterranean *Pecten jacobaeus* (Viarengo et al. 1995; Regoli et al. 1997, 2000). Particularly high values of CAT activity are reported in gills and especially in digestive tissues of the Antarctic scallop, which may represent an important mechanism to limit $HO^{\bullet}$ formation by decomposing one of the main cellular precursors, H_2O_2. Both tissues of polar scallops further contain higher basal activities of glutathione peroxidase (GPx) (Viarengo et al. 1995). However, this does not equally apply to other antioxidant enzymes and, for example, SOD activity differs little between the digestive gland and gill of Antarctic and Mediterranean scallops (Viarengo et al. 1995; Regoli et al. 1997, 2000), or between mantle tissues of the Antarctic *A. colbecki* and the temperate queen scallop (*Aequipecten opercularis*) (Philipp et al. 2006). As a corollary, differences in most antioxidant enzymes were inconspicuous and within the natural range of seasonal (Viarengo et al. 1991) or age-related variations seen in temperate mollusks (Philipp et al. 2006).

The levels of total GSH are quite similar in Antarctic, Arctic, and Mediterranean scallops, and also overlap with values reported for scallops and other bivalves from temperate latitudes (Table 2.3). When plotting GSH concentration over age, higher levels of total GSH are found in the longer lived Antarctic *A. colbecki* (>30 yr of life span) than in the short-lived *A. opercularis* from the Irish Sea (<10 yr of life span; see Philipp et al. 2006), which presumably relates more to life expectancy than temperature. Although, polar scallops exhibit significantly higher GR activities, indicating a greater capability to maintain GSH cellular redox status (Regoli et al. 2000), the glutathione disulphide (GSSG):GSH ratio was significantly more oxidized in the polar than the temperate scallop (Philipp et al. 2006), especially when plotting the data over age. A greater capacity to reconvert GSSG to GSH renders these species more tolerant to glutathione depletion, especially during exposure to oxidative insult (Regoli and Principato 1995).

Efficiency of ROS protection *in A. colbecki* has been evaluated by integrating the analyses of individual antioxidants with the measurement of total oxyradical scavenging capacity (TOSC, see Chapter 26). The TOSC toward both $ROO^{\bullet}$ and $HO^{\bullet}$ is more elevated in *A. colbecki* than in temperate *P. jacobaeus*, confirming the general enhancement of antioxidant defenses in

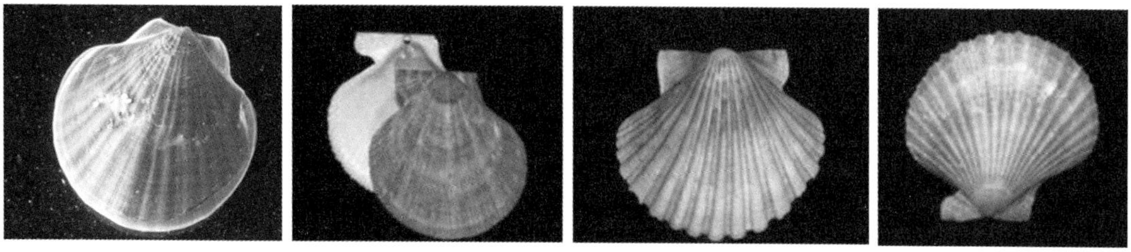

Fig. 2.7 Polar and temperate scallop species. See plate section for a color version of this image.

Table 2.3 Levels of total glutathione (as GSH equivalents) in tissues of scallops and bivalves from polar and temperate regions. Data normalized to μmol g^{-1} FW or nmol mg^{-1} protein.

	Gills	Digestive gland	Mantle tissue
A. colbecki (Antarctic)	0.18 ± 0.02[1]	1.21 ± 0.14[1]	2.00–0.43 from age 3 to 14 yr[2] (with GSSG:GSH ratio 0.20–0.06)
A. opercularis (Temperate)			0.90–0.30 from age 1 to 5 yr[2] (with GSSG:GSH ratio 0.1–0.02)
P. jacobaeus (Temperate)	0.27 ± 0.02[1]	1.89 ± 0.26[1]	
M. galloprovincialis (Temperate)	0.13 ± 0.03[4]	0.73 ± 0.07[4] 1.28 ± 0.07[5]	
M. edulis (Temperate)	0.19 ± 0.006[6]	0.61 ± 0.06[6]	
A. colbecki (Antarctic)		12.2 ± 2.33[3]	
C. islandicus (Arctic)		12.9 ± 2.17[3]	
P. jacobaeus (Temperate)		13.1 ± 1.52[3]	
M. galloprovincialis (Temperate)		7.48 ± 0.95[5]	

Data from: [1]Viarengo et al. 1995; [2]Philipp et al. 2006 ; [3]Regoli et al. 2000 ; [4]Viarengo et al. 1990; [5]Regoli and Principato 1995; [6]Canesi and Viarengo 1997.

this polar bivalve (Viarengo et al. 1995; Regoli et al. 1997). In particular, elevated TOSC values towards $HO^•$ agree with somewhat higher CAT and GPx activities, indicating an elevated resistance of the Antarctic scallop to the effects of pro-oxidant factors that can generate H_2O_2, $HO^•$, and organic peroxides (Regoli et al. 1997). The high efficiency of antioxidant defenses in *A. colbecki* is reflected by the reduced accumulation of oxidative stress products such as protein carbonyls and the fluorescent age pigment lipofuscin (Philipp et al. 2006). Also, elevated stability of lysosomal membranes, typical targets of ROS attack, was reported in long-lived polar scallops (Regoli et al. 1998a). The oxidative challenge and antioxidant protection in *A. colbecki* are significantly modified by the seasonality of food availability in Antarctica. The TOSC values towards $HO_2^•$ and $HO^•$ indicate highest ROS protection during the phytoplankton bloom. Since feeding activities also increase aerobic metabolic rates in polar ectotherms, bloom conditions create a need to adjust defenses against ROS (Regoli et al. 2002).

Overall, elevated levels of antioxidant defenses and their maintenance over age, low natural ROS formation rates, seasonal caloric restriction, and the presumably efficient elimination of ROS damaged macromolecules, seem to constitute a complex and powerful oxidative damage prevention system in polar mollusks, which stimulates the hypothesis of increased longevity in polar compared to temperate marine ectotherms (Philipp et al. 2006; Abele et al. 2009).

ANTIOXIDANT LEVELS IN POLAR VS. TEMPERATE FISH

Antioxidant enzyme activities in gills, liver, heart, muscle, and blood of Antarctic fish do not differ much from the levels in ecologically comparable temperate species (Witas et al. 1984; Cassini et al. 1993; Ansaldo et al. 2000; Benedetti et al. 2010). SOD activity did not differ in Antarctic and North Sea eelpout, *Pachycara brachycephalum* and *Zoarces viviparous*, both belonging to the zoarcid family, when measurements were carried out at the environmental temperature: 0°C for the Antarctic and 6°C for the North Sea species (Heise et al. 2007). The GPx hepatic activities were identical in both fish (at 20°C assay temperature). In the North Sea eelpout, higher enzyme activities occurred during summer because of higher food intake and higher metabolic rates during the warmer season. Other comparisons of SOD activities were conducted between tissues of red blooded, white blooded, and temperate fish from different families, but interpretations relating these activities to different blood oxygen levels remain speculative (Witas et al. 1984; Cassini et al. 1993; Ansaldo et al. 2000). Arctic fish have similar or even slightly reduced antioxidant enzyme activities compared to temperate congeners or to temperate species of similar ecotype (Speers-Roesch and Ballantyne 2005).

Despite their inconspicuous antioxidant enzyme activities, notothenioids retain an elevated capability to neutralize both $ROO^{\bullet}$ and $HO^{\bullet}$ seemingly by non-enzymatic antioxidants (Benedetti et al. 2010). Relatively high levels of GSH, ascorbic acid, α-tocopherols, and carotenoids have been measured in liver and plasma of various Antarctic fish, whereas more variable antioxidant concentrations were recorded in muscle tissues (Tables 2.4 and 2.5). The polar eelpout (*P. brachycephalum*) has twice as much GSH and α-tocopherol in liver tissue than the temperate North Sea eelpout *Z. viviparus* (Table 2.4). However, GSH redox status was twice more oxidized in Antarctic eelpout liver (Heise et al. 2007), indicating that at thermally depressed metabolic rates the amount of reduced nicotinamide adenine dinucleotide phosphate (NADPH) invested into enzymatic GSH reduction by GR may be limited by more fundamental metabolic requirements.

Gieseg et al. (2000) compared vitamin E (α-tocopherol) and vitamin C (ascorbic acid) plasma levels in Antarctic notothenoids to New Zealand blue cod and banded wrasse (Table 2.5). The comparison is difficult as these species are in no way related to each other, and also the authors compared pelagic temperate fish to demersal Antarctic fish. Five to six times higher vitamin E concentrations were found in Antarctic than temperate fish. Higher levels of vitamin C were found in the plasma of *Trematomus bernacchii*. The elevated

Table 2.4 Levels of low molecular weight scavengers in liver of various polar and temperate fish species.

Species	Liver content		
	Vitamin E (μmol g^{-1} FW)	GSH (μmol g^{-1} FW)	2GSSG/GSH
Pachycara brachycephalum 0°C (Antarctic eelpout)	300 ± 200[1]	2.1 ± 0.5[1]	2.7 ± 0.9[1]
Pachycara brachycephalum 5°C (Antarctic eelpout)	500 ± 100[1]	2.4 ± 0.3[1]	3.0 ± 1.0[1]
Zoarces viviparous 6°C winter (North Sea eelpout)	93 – [1]	1.0 ± 0.4[1]	1.8 ± 0.9[1]
Zoarces viviparous 12°C summer (North Sea eelpout)	240 ± 80[1]	0.9 ± 0.3[1]	0.36 ± 0.04[1]
Trematomus bernacchii (Antarctic red-blooded)		1.07 ± 0.16[2]	
Trematomus hansoni (Antarctic red-blooded)		0.53 ± 0.05[2]	
Trematomus newnesi (Antarctic red-blooded)		0.54 ± 0.17[2]	

Data from: [1]Heise et al. 2007; [2]Benedetti et al. 2010.

Table 2.5 Levels of low molecular weight scavengers in plasma and muscle of various polar and temperate fish species.

Species	Plasma		Muscle
	Vitamin E (μM)	**Vitamin C (μM)**	**Vitamin E (ng mg^{-1} protein)**
Pagothenia borchgrevinki Antarctic red-blooded	116 ± 41[1]	12.3 ± 6.5[1]	
Trematomus bernacchii Antarctic red-blooded	106 ± 57[1]	30.3 ± 16.3[1]	27.6 ± 5.4[2]
Parapercis colias Temperate (blue cod)	14 ± 4.5[1]	8.4 ± 4.1[1]	
Notolabrus fucicola Temperate (banded wrasse)	18.9 ± 5.9[1] (ng mg protein^{-1})	8.1 ± 4.8[1] (ng mg protein^{-1})	
Notothenia coriiceps Antarctic red-blooded	142.0 ± 8.3[2]		73.2 ± 9.3[2]
Notothenia gibberifrons Antarctic red-blooded	40.0 ± 6.1[2]		
Champsocephalus gunnari Antarctic white-blooded	17.9 ± 3.3[2]		
Chionodraco hamatus Antarctic white blooded			86.0 ± 9.1[2]
Chaenocephalus aceratus Antarctic white blooded			34.6 ± 6.3[2]
Oncorhynchus mykiss Temperate	11.0 ± 2.1[2]		
Mugil cephalus Temperate			32.2 ± 6.3[2]

Data from: [1]Gieseg et al. 2000; [2]Colella et al. 2000.

vitamin E levels presumably reflect a difference in nutritional status in both regions (Ansaldo et al. 2000; Gieseg et al. 2000). Higher levels of plasma vitamin E were reported in Antarctic red-blooded compared to both white-blooded and temperate fish, and also muscle of Antarctic fish contained more vitamin E (Table 2.5).

EXAMPLES OF OXIDATIVE STRESS IN CHANGEABLE POLAR HABITAT CONDITIONS

Several species in polar marine ecosystems are exposed to more than the basal pro-oxidant challenge at low and constant temperatures. Situations bearing the risk of increased oxidative stress are associated with special scenarios and environmental niches and often with marked seasonal fluctuations of environmental parameters. Sea-ice partially protects the associated fauna from direct irradiation. Upon ice break-up these species are exposed to marked increase of solar radiation and release of H_2O_2 from melting sea-ice.

The ice-associated (sympagic) Arctic amphipod *Gammarus wilkitzkii* grazes sea-ice algae and copepods, mobilizing nutrients that fertilize the underlying seawater (Werner 2000; Scott et al. 2001). During summer this amphipod has much stronger scavenging capacity towards ROO$^\bullet$ and HO$^\bullet$ than during winter (Krapp et al. 2009), suggesting that antioxidant protection is adjusted to meet seasonal metabolic requirements and possibly also higher oxidative stress levels under the melting ice. Levels of thiobarbituric acid reactive substances (TBARS) were elevated during the summer in amphipod extracts, indicating higher rates of lipid peroxidation when the sea-ice cover is at its minimum (Krapp et al. 2009). Laboratory experiments simulating Arctic summer sea-surface (0.7 W m^{-2}) and under-ice UV radiation levels (0.2 W m^{-2} at 2 m water depth) resulted in significant depletion of ROS scavenging capacity in amphipods directly exposed to surface UV radiation intensity (Krapp et al. 2009). Obermüller et al. (2005) recorded the immediate response to changing surface UV-B and UV-A intensities on the metabolic rate of Antarctic shallow-water amphipods under experimental conditions. The effect can be produced by both UV ranges or limited to UV-B, depending on the species, and it consists in a significant increase of locomotory activity and oxygen consumption in addition to the detrimental effect of direct exposure to hazardous UV light. Differential susceptibility to various UV ranges relates to the fact that amphipods incorporate sunscreen compounds such as mycosporine-like amino acids and carotenoids, not only into their tissues, but also into their chitinous exoskeleton. What sunscreens they incorporate depends on their algal diet.

One of the strongest effects caused by even mildly elevated experimental UV radiation to the half maximal natural dose consisted in the inactivation of CAT activity. This effect is caused by UV-A light, where CAT's chromophoric group is known to absorb (k_{max} at 405 nm; Gantchev and van Lier 1995). The consequences of this effect are still unclear since no altered levels of protein carbonylation were found in illuminated Antarctic amphipods; whether or not DNA damage or lipid peroxidation occurred following experimental radiation exposure remains unknown.

In its adult stage the Antarctic silverfish *Pleuragramma antarcticum* is the only pelagic fish widely distributed and abundant in shelf waters around the continent, where it constitutes a major component of the Antarctic food chain. At the final stage of their embryonic development, eggs with fully developed yolk-sac embryos accumulate in huge numbers in the platelet ice below the sea-ice layer (Vacchi et al. 2004). The structure of the irregularly disk-shaped ice platelets provide a large surface area for the growth of algal and microzooplankton communities, which represent an important food source, but also a favorable environment which offers protection from predation for the early life stages of several under-ice species (Gutt 2002). At the beginning of the austral spring, the rapid growth of sea-ice algal communities, the massive release of nutrients and the photoactivation of dissolved organic carbon and nitrates, represent important sources for ROS formation (Fig. 2.8). All these processes happen in a period of 3–6 weeks before the lower layer of sea-ice loses its biological richness. *Pleuragramma antarcticum* embryos are associated with platelet ice and exhibit a marked temporal enhancement of antioxidant defenses, possibly representing an adaptive strategy to meet the increasing environmental oxidative challenge in the beginning of Antarctic spring. Other enzymes not directly involved in ROS defense, such as cytochrome P450, do not exhibit a significant variation in this period. Despite the rapid antioxidant response, especially within the GSH metabolism, the capability to neutralize $HO^{\bullet}$ was reduced, and lipid peroxidation occurred in *P. antarcticum* tissues. Exposure to a strong pro-oxidant chemical, benzo[a]pyrene, produced only minor oxidative stress effects in these fish, indicating that exposure to elevated natural pro-oxidant scenarios, such as the sea-ice environment, may indeed reduce the susceptibility of the fish with respect to other pro-oxidant stressors (Regoli et al. 2005a).

ANTARCTIC SYMBIOSES

In symbioses of Antarctic sponges with diatoms, photosynthetic products secreted by the symbionts represent an additional food source for the host, while benefits for microalgae include the use of nutrients released by the sponge cells and a protective habitat. At the same time, increased levels of photosynthetically produced oxygen and ROS form and induce an adaptive response in the host tissues (see Chapter 1). Contrary to tropical symbioses, Antarctic symbioses are characterized by strong seasonality (Dunlap and Shick 1998; Cerrano et al. 2004). In the demosponge *Haliclona dancoi*, both, chlorophaeopigments and frustules are absent in early November at the beginning of summer, indicating that the diatoms do not overwinter inside the sponge but are newly incorporated during the phytoplankton bloom in each year (Cerrano et al. 2004). A marked enhancement of antioxidant capacity in sponge tissues is correlated with the period of maximum symbiotic association when the highest quantities of diatoms are present. These results indicate the plasticity of antioxidant defenses as an important feature for *H. dancoi* to counteract seasonal pro-oxidant pressure due to elevated photosynthetic activity of its diatom symbionts (Regoli et al. 2004).

COASTAL GEOCHEMISTRY AND OXIDATIVE STRESS

Elevated basal levels of pro-oxidant trace metals are typical in many Antarctic regions characterized by both geochemical anomalies and natural inputs of these elements (Bargagli 2005). At the Antarctic Peninsula,

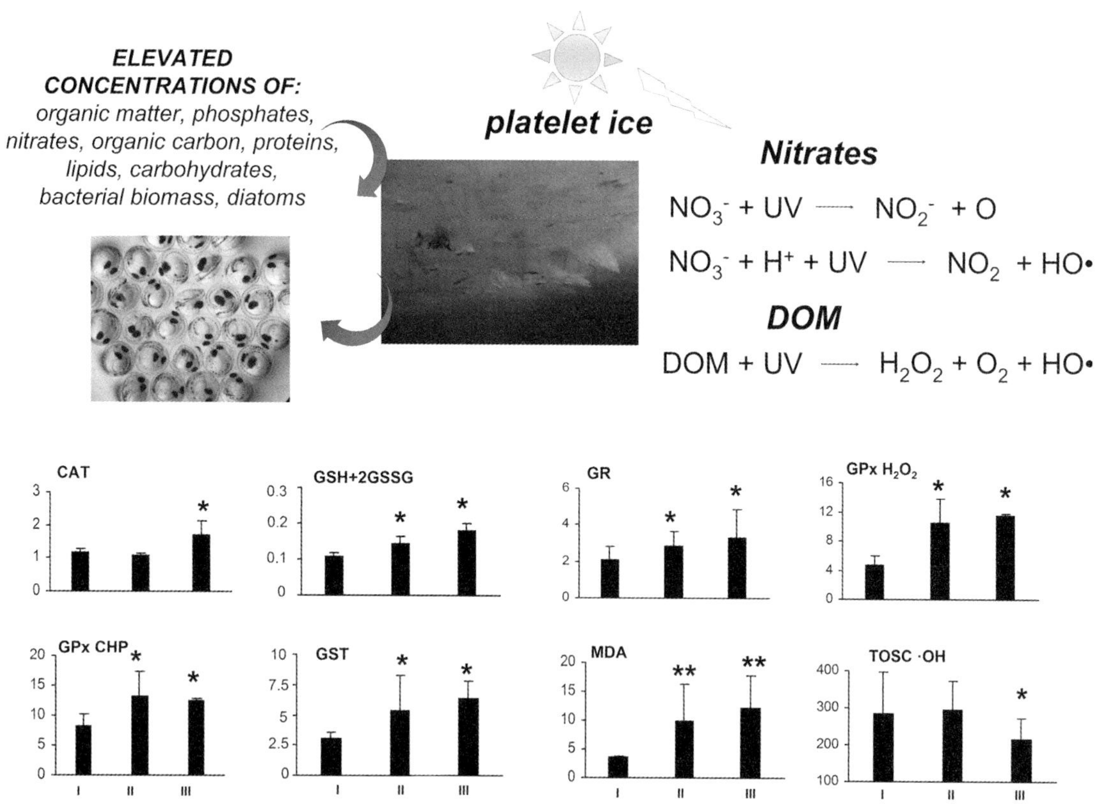

Fig. 2.8 Embryonated eggs of *P. antarcticum* are exposed to increasing pro-oxidant challenge in platelet ice and consequently enhance their antioxidant defenses: CAT (catalase, μmol/min/mg prot), GSH+2GSSG (total glutathione, μmol/g tiss), GR (glutathione reductase, nmol/min/mg prot), GP× H_2O_2 (Se-dependent glutathione peroxidases, nmol/min/mg prot), GP×CHP (Se-dependent and Se-independent glutathione peroxidases, nmol/min/mg prot), GST (glutathione S-transferases, nmol/min/mg prot), MDA (malondialdheyde, nmol/g tiss), TOSC •OH (total oxyradical scavenging capacity, UTOSC/mg prot). I, II, III: sampling dates on 2, 9, 16 November respectively. See plate section for a color version of this image.

sediment ablation from glacier melting transports terrigenous particles into coastal waters, causing marked enrichment of Fe, Al, Cu and Zn concentrations (Ahn et al. 1996; Dierssen et al. 2002). Volcanic rocks at King George Island contain 5 to >7% Fe (Tatur et al. 1999), which upon erosion is washed into the sea in turbid meltwater streams. Most of the Fe in coastal seawater is bound to suspended sediment particles with concentrations ranging between 0.2 and 318 μg Fe L^{-1} (Ahn et al. 2004). Once ingested by sedimentary grazers or benthic filter feeders, these metals can exacerbate the risk for oxidative stress. The Antarctic limpet *Nacella concinna* splits into permanently sublittoral and one seasonally intertidal, migratory subpopulation. Significantly higher Fe, Al, Zn concentrations are measured in sublittoral limpets, which graze surface sediments and ingest particles charged with terrigenous trace elements. Elevated levels of these elements increase ROS generation, and sublittoral limpets have higher SOD activities in the digestive gland to ward off metal-induced oxidative stress (Weihe et al. 2010).

Cadmium is of special interest in organisms from Terra Nova Bay (TNB, Ross Sea), as this element is naturally enriched because of regional upwelling (Kurtz and Bromwich 1985). In November before the sea-ice melts and phytoplankton blooms occur, Cd concentrations in the Ross Sea are ~0.7 nmol L^{-1}, whereas nonmelt-influenced surface waters typically contain <0.02 nmol L^{-1} (Scarponi et al. 2000; Bargagli 2005).

Cadmium levels in TNB biota are 10–20-fold higher than in similar temperate species (Nigro et al. 1997; Bargagli 2005). Although the natural enrichment of Cd has no direct adverse effect on local organisms, some indirect oxidative effects have been detected, which interfere with the metabolism of other chemicals, i.e. limited biotransformation capability of polycyclic aromatic hydrocarbons in fish (Regoli et al. 2005b; Benedetti et al. 2007). Most of these interactions appear modulated by the oxidative response and in particular by GSH and GSH-dependent enzymes (see below). Cadmium bioavailability increases during phytoplankton blooms, and algae efficiently remove Cd from the water column. This, in turn, increases Cd flux to the benthic compartment, and liver concentrations in the fish *T. bernacchii* almost double between November and December, but without triggering an antioxidant response (Canapa et al. 2007; Benedetti et al. 2010). These results suggest that polar organisms naturally exposed to high fluctuations of potentially toxic elements might be less sensitive and more apt to deal with increased metal bioaccumulation. But it does not imply that they are not highly sensitive to the multitude of chemical stressors from anthropogenic activities that are presently increasing the pollution in the Antarctic environment.

POLAR ECOTOXICOLOGY AND OXIDATIVE STRESS RESPONSES AS BIOMARKERS OF CHEMICAL CONTAMINATION

Studies on susceptibility to oxidative stress in polar organisms are of great importance for monitoring human impact in these remote areas. Geologists believe that one-third of the planet's crude oil is located under the Arctic sea-bed (USGS 2000) and exploitation of these resources, presently limited by technological constraints, is coming into reach as fast regional warming of the Arctic causes rapid shrinking of ice sheets. This enhances the potential of mining pollution and accidental oil spills in Arctic ecosystems. Higher temperatures alter both the chemical bioavailability of some toxic compounds and the responsiveness of species to toxic insult. In Antarctica, anthropogenic pressure is still more restricted, but contamination arising from around scientific bases and from cruise ships and tourism presently affects terrestrial and marine coastal ecosystems locally (Curtosi et al. 2007; Bargagli 2008). Although Antarctic oceanic and atmospheric circulation are natural "barriers" for lower latitude water and air masses, data on concentrations of metals, pesticides, and other persistent pollutants in air, snow, mosses, lichens, and marine organisms show that persistent contaminants from the industrialized regions of Earth are infiltrating pristine Antarctic environments.

Biological adaptations to extreme environmental conditions and the different behavior of contaminants at low temperature may affect ROS metabolism and antioxidant responses of polar species. These organisms concentrate reproduction, feeding, and storage of energy reserves to the short summer season: a contamination event in this period of highest anthropogenic activity might be difficult to repair (Regoli et al. 2002). Variations of antioxidant responses in sentinel organisms have been investigated as potential biomarkers for a rapid detection of both anthropogenic and natural impact.

Arctic populations of the clam *Macoma balthica* revealed significant variations of the main antioxidant components in five estuaries in the White Sea and Pechora Sea differently impacted by metals, supporting the utility of these biomarkers even in mildly polluted Arctic regions (Regoli et al. 1998b). The impact of polycyclic aromatic hydrocarbons on ROS metabolism of Arctic invertebrates was measured in the clam *Mya truncata*, the spider crab *Hyas araneus*, and the scallop *Chlamys islandicus*, exposed to oils via injection and/or contaminated sediments (Camus et al. 2002a, b, 2003). Although the low temperatures reduce bioavailability and oxidative biotransformation of polycyclic aromatic hydrocarbons (PAHs), these chemicals increase ROS generation, and antioxidants are important for mitigating the resulting cellular toxicity.

Among vertebrates, the polar cod *Boreogadus saida* has recently been tested as a potential bioindicator for monitoring oil pollution in the Arctic (Nahrgang et al. 2009; Nahrgang et al. 2010a,b). Oxidative stress genes (catalase, glutathione peroxidases, Cu,Zn-SOD and Mn-SOD) were generally up-regulated following exposure (7 h), and expression levels remained elevated (up to 70-fold for catalase) until 1–2 weeks following exposure. Catalase transcripts showed a biphasic response, with a first peak reflecting the initial oxidative challenge by PAHs, and a second induction peak probably related to cytochrome-P450-mediated ROS formation. In this context, it is well known that oxidative biomarkers rarely respond in a clearly dose-dependent manner, which is unsurprising when considering the complexity of interactions and cascade effects between oxidant stressors and antioxidant mechanisms (Regoli et al. 2005b).

Contrasting results at transcriptional and catalytic (activity) levels are common for antioxidant responses, reflecting the effects of various oxidant stressors at different cellular levels, and the importance of post-transcriptional modifications on antioxidant enzymes (Benedetti et al. 2007, 2009). Antioxidant responses to pollutants have further been characterized in several Antarctic ectotherms in field and laboratory experiments. The scallop *A. colbecki* exhibited a decrease of antioxidant defenses when challenged with sublethal concentrations of various trace metals (Cu 20 μg L^{-1} or Hg 5 μg L^{-1}); this resulted in a marked enhancement of lipid peroxidation and damage of lysosomal membranes (Regoli et al. 1998a). The Antarctic limpet *N. concinna* was exposed to low doses of ship diesel in seawater (0.05%), which had no effect on antioxidant defense but resulted in a significant enhancement of protein oxidation and lipid peroxidation levels (Ansaldo et al. 2005). Limpets treated with a higher dose (0.1% diesel) featured significant induction of antioxidants that prevented the onset of oxidative damage. These results suggest that a threshold of ROS production upon exposure to pollutants has to be reached before antioxidant defenses are induced.

Elevated responsiveness and variability of oxidative biomarkers has recently been demonstrated in Antarctic rock cod *T. bernacchii* exposed to different chemicals and chemical mixtures (Regoli et al. 2005b; Benedetti et al. 2007, 2009). Exposure to dioxin

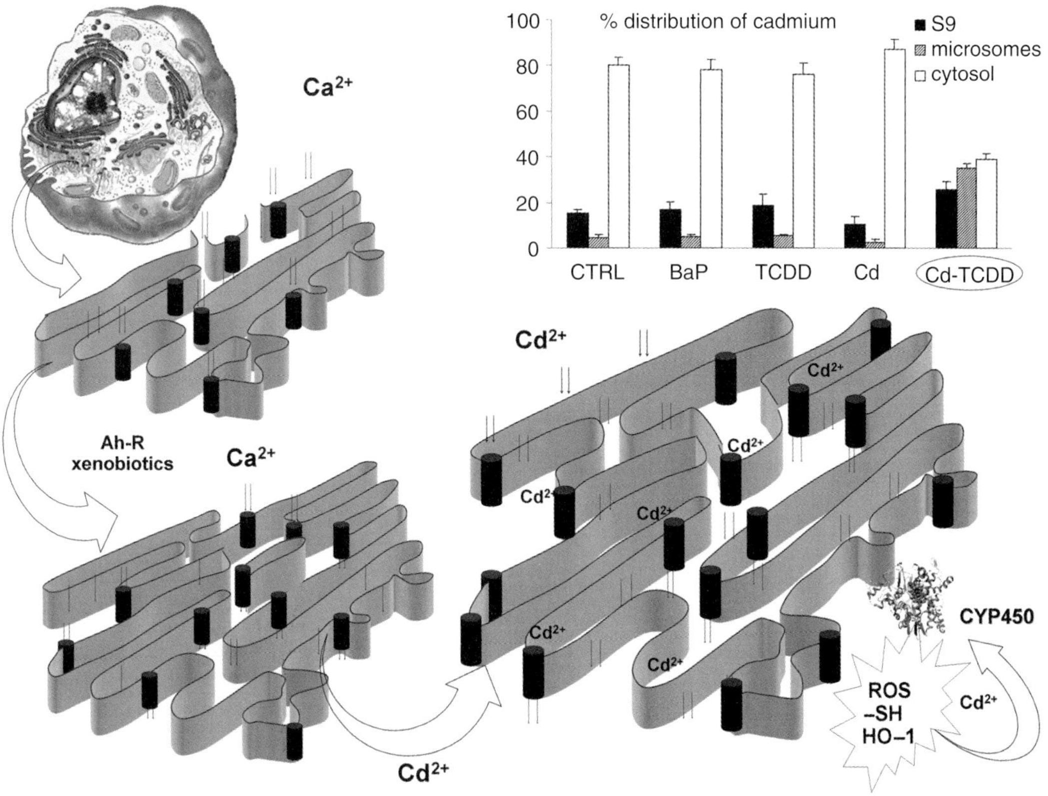

Fig. 2.9 Ah-R xenobiotics induce proliferation of endoplasmic reticulum where Cd can accumulate through Ca^{2+} channels, as confirmed by elevated content of the metal in the microsomal fraction of co-exposed fish. Through post-transcriptional mechanisms, Cd suppresses the induction of cytochrome P450 at protein and catalytic levels. See plate section for a color version of this image.

(2,3,7,8-tetrachlorodibenzo-p-dioxin, TCDD) caused a marked oxidative perturbation and loss of antioxidant efficiency in fish, whereas Cd exposure caused induction of several antioxidants (Regoli et al. 2005b). Reciprocal interactions between metabolism of dioxin and Cd were observed in fish co-exposed to both chemicals. Dioxin enhanced the accumulation of Cd, stored within proliferating endoplasmic reticulum, whereas Cd suppressed the induction of cytochrome P4501A at protein and catalytic levels (Fig. 2.9). Another oxidative effect is the induction by Cd of HO-1, the enzyme degrading heme to biliverdin, which has antioxidant properties. These results suggest Cd activates specific stress signals or induces antioxidant response elements in the presence of aromatic xenobiotics, modulating the oxidative toxicity of these chemicals. The sensitivity of specific antioxidants to Cd in *T. bernacchii* might represent a consequence of the elevated bioavailability of this element at TNB, which would thus affect the biotransformation capability of organic xenobiotics, and potentially the sensitivity of fish toward these chemicals.

Further investigations with different chemical mixtures confirmed that oxidative interactions can influence metabolism of metals and aromatic xenobiotics, with implications for monitoring both bioaccumulation and biological effects in key sentinel species. Different mechanisms can occur depending on the metals and aromatic xenobiotics, involving transcriptional, translational, and catalytic levels, with indirect or cascade effects difficult to predict for various chemicals or mixtures. The limited response to aromatic compounds in fish co-exposed to metals suggests that elevated natural levels of trace elements in many Antarctic areas might alter the susceptibility of these fish to bioaccumulation and to the toxic effects of organic chemicals that might come from local anthropogenic emitters (stations, vessels). Oxidative mechanisms and antioxidant responses of polar organisms are critical in mitigating the toxic effects of chemical mixtures released by accidental oil spills or transported by atmospheric circulation from the industrialized regions.

REFERENCES

Abele, D., Brey, T., Philipp, E. (2009) Bivalve models of aging and the determination of molluscan lifespans. *Experimental Gerontology* 44, 307–315.

Abele, D., Ferreyra, G.A., Schloss, I. (1999) H_2O_2 accumulation from photochemical production and atmospheric wet deposition in Antarctic coastal and off-shore waters of Potter Cove, King George Island, South Shetland Islands. *Antarctic Science* 11, 131–139.

Abele-Oeschger, D., Tüg, H., Röttgers, R. (1997) Dynamics of UV-driven hydrogen peroxide formation on an intertidal sandflat. *Limnology and Oceanography* 42, 1406–1415.

Acierno, R., Maffia, M., Sicuro, P. et al. (1996) Lipid and fatty acid composition of intestinal mucosa of two Antarctic teleosts. *Comparative Biochemistry and Physiology A* 115, 303–307.

Ahn, I.Y., Lee, S.H., Kim, K.T. et al. (1996) Baseline heavy metal concentrations in the Antarctic clam, *Laternula elliptica* in Maxwell Bay, King George Island Antarctica. *Marine Pollution Bulletin* 32, 592–598.

Ahn, I.Y., Chung, K.H., Choi, H.J. (2004) Influence of glacial runoff on baseline metal accumulation in the Antarctic limpet *Nacella* concinna from King George Island. *Marine Pollution Bulletin* 49, 119–141.

Ansaldo, M., Luquet, C.M., Evelson, P.A. et al. (2000) Antioxidant levels from different Antarctic fish caught around South Georgia Island and Shag Rocks. *Polar Biology* 23, 160–165.

Ansaldo, M., Najle, R., Luquet, C.M. (2005) Oxidative stress generated by diesel seawater contamination in the digestive gland of the Antarctic limpet Nacella concinna. *Marine Environmental Research* 59, 381–390.

Bargagli, R. (2005) *Antarctic Ecosystems: Environmental Contamination, Climate Change, and Human Impact*. Springer-Verlag, Berlin.

Bargagli, R. (2008) Environmental contamination in Antarctic ecosystems. *Science of the Total Environment* 400, 212–226.

Bartsch, A. (1989) Sea ice algae of the Weddel Sea (Antarctica): Species composition, biomass and ecophysiology of selected species. *Reports on Polar Research* 63, 110.

Benedetti, M., Martuccio, G., Fattorini, D. et al. (2007) Oxidative and modulatory effects of trace metals on metabolism of polycyclic aromatic hydrocarbons in the Antarctic fish Trematomus bernacchii. *Aquatic Toxicology* 85, 167–175.

Benedetti, M., Fattorini, D., Martuccio, G. et al. (2009) Interactions between trace metals (Cu, Hg, Ni, Pb) and 2,3,7,8-tetrachlorodibenzo-p-dioxin in the Antarctic fish Trematomus bernacchii: oxidative effects on biotransformation pathway. *Environmental Toxicology and Chemistry* 28, 818–825.

Benedetti, M., Nigro, M., Regoli, F. (2010) Characterization of antioxidant defences in three Antarctic Notothenioids species from Terra Nova Bay (Ross Sea). *Chemistry and Ecology* 26, 305–314.

Buma, A.G.J., Wright, S.W., Van Den Enden, R. et al. (2006) PAR acclimation and UVBR-induced DNA damage in

Antarctic marine microalgae. *Marine Ecology Progress Series* 315, 33–42.
Camus, L., Jones, M.B., Borseth, J.F. et al. (2002a) Total oxyradical scavenging capacity and cell membrane stability of haemocytes of the Arctic scallop, Chlamys islandicus, following benzo[a]pyrene exposure. *Marine Environmental Research* 54, 425–430.
Camus, L., Jones, M.B., Borseth, J.F. et al. (2002b) Heart rate respiration and total oxyradical scavenging capacity of the Arctic spider crab, *Hyas araneus*, following exposure to polycyclic aromatic compounds via sediment and injection. *Aquatic Toxicology* 61, 1–13.
Camus, L., Birkely, S.R., Jones, M.B. et al. (2003) Biomarker responses and PAH uptake in Mya truncata following exposure to oil-contaminated sediment in an Arctic fjord (Svalbard). *Science of the Total Environment* 308, 221–34.
Camus, L., Gulliksen, B., Depledge, M.H., Jones, M.B. (2005) Polar bivalves are characterized by high antioxidant defences. *Polar Research* 24, 111–118.
Canapa, A., Barucca, M., Gorbi, S. et al. (2007) Vitellogenin gene expression in males of the Antarctic fish *Trematomus bernacchii* from Terra Nova Bay (Ross Sea): a role for environmental cadmium? *Chemosphere* 66, 1270–1277.
Canesi, L., Viarengo, A. (1997) Age-related differences in glutathione metabolism in mussel tissues (*Mytilus edulis* L.). *Comparative Biochemistry and Physiology B* 116, 217–221.
Cassini, A., Favero, M., Albergoni, A. (1993) Comparative studies of antioxidant enzymes in red-blooded and white-blooded Antarctic teleost fish *Pagothenia bernacchii* and *Chionodraco hamatus*. *Comparative Biochemistry and Physiology C* 106, 333–336.
Cerrano, C., Calcinai, B., Cucchiari, E. et al. (2004) Are diatoms a food source for Antarctic sponges? *Chemistry and Ecology* 20, 57–64.
Chen, C., Dickman, M.B. (2005) Proline suppresses apoptosis in the fungal pathogen *Colletotrichum trifolii*. *Proceedings of the National Academy of Sciences USA* 102(9): 3459–3464.
Colella, A., Patamia, M., Galtieri, A., Giardina, B. (2000) Cold adaptation and oxidative metabolism of Antarctic fish. *Italian Journal of Zoology Supplement* 1, 33–36.
Coolbear, K.P., Berde, C.B., Keough, K.M.W. (1983) Gel to liquid-crystalline phase transitions of aqueous dispersion of polyunsaturated mixed-acid phosphatidylcholines. *Biochemistry* 22, 1466–1473.
Cooper, W.J., Lean, D.R.S. (1989) Hydrogen peroxide concentration in a northern lake: Photochemical formation and diurnal variability. *Environmental Science and Technology* 23, 1425–1428.
Cox, G.F.N., Weeks, W.F. (1983). Equations for determining the gas and brine volumes in sea-ice samples. *Journal of Glaciology* 29, 306–316.
Curtosi, A., Pelletier, E., Vodopivez, C.L., Mac Cormack, W.P. (2007) Polycyclic aromatic hydrocarbons in soil and surface marine sediment near Jubany Station (Antarctica). Role of permafrost as a low-permeability barrier. *Science of the Total Environment* 383, 193–204.
Delille, B., Jourdain, B. et al. (2007) Biogas (CO_2, O_2, dimethylsulfide) dynamics in spring Antarctic fast ice. *Limnology and Oceanography* 52(4), 1367–1379.
Dickson, D.M.J., Kirst, G.O. (1986) The role of β-dimethylsulphonio-propionate, glycine betaine and homarine in the osmoacclimation of *Platymonas subcordiformis*. *Planta* 167, 536–543.
Dierssen, H.M., Smith, R.C., Vernet, M. (2002) Glacial meltwater dynamics in coastal waters west of Antarctic peninsula. *Proceedings of the Natural Academy of Sciences USA* 99, 1790–1795.
Doval, M.D., Álvarez-Salgado, X.A., Castro, C.G., Pérez, F.F. (2002) Dissolved organic carbon distributions in the Bransfield and Gerlache Straits, Antarctica. *Deep-Sea Research Part II: Topical Studies in Oceanography* 49, 663–674.
Dumont, I., Schoemann, V., Lannuzel, D. et al. (2009) Distribution and characterization of dissolved and particulate organic matter in Antarctic pack ice. *Polar Biology* 32, 733–750.
Dunlap, W.C., Shick, J.M. (1998) Ultraviolet radiation-absorbing mycosporine-like amino acids in coral reef organisms: a biochemical and environmental perspective. *Journal of Phycology* 34, 418–430.
Egginton, S., Cordiner, S., Skilbeck, C. (2000) Thermal compensation of peripheral oxygen transport in skeletal muscle of seasonally acclimatized trout. *American Journal of Physiology* 279, R375–R388.
Eicken, H., Lange, M.A., Hubberten, H-W., Wadhams, P. (1994) Characteristics and distribution patterns of snow and meteoric ice in the Weddell Sea and their contribution to the mass balance of sea ice. *Annales Geophysicae* 12, 80–93.
Fraser, K.P.P., Clarke, A., Peck, L.S. (2002) Low-temperature protein metabolism: Seasonal changes in protein synthesis and RNA dynamics in the Antarctic limpet *Nacella concinna* Strebel 1908. *Journal of Experimental Biology* 205, 3077–3086.
Frederick, J.E., Snell, H.E. (1988) Ultraviolet radiation levels during the Antarctic spring. *Science* 241, 438–440.
Gantchev, T.G., van Lier, J.E. (1995) Catalase inactivation following photosensitization with tetrasulfonated metal-lophthalocyanines. *Photochemistry and Photobiology* 62, 123–134.
Gieseg, S.P., Cuddihy, S., Hill, J.V., Davison, W. (2000) A comparison of plasma vitamin C and levels in two Antarctic and two temperate water fish species. *Comparative Biochemistry and Physiology B* 125, 371–378.
Gleitz, M., Bathmann, U.V., Lochte, K. (1994) Build-up and decline of summer phytoplankton biomass in the eastern Weddell Sea, Antarctica. *Polar Biology* 14, 413–422.
Gleitz, M., van der Loeff, M.R., Thomas, D.N. et al. (1995) Comparison of summer and winter inorganic carbon,

oxygen and nutrient concentrations in Antarctic sea ice brine. *Marine Chemistry* 51, 81–91.

Glud, R.N., Rysgaard, S., Kühl, M. (2002) A laboratory study on O_2 dynamics and photosynthesis in ice algal communities: quantification by microsensors, O_2 exchange rates, ^{14}C incubations and a PAM fluorometer. *Aquatic Microbial Ecology* 27, 301–311.

Granskog, M.A., Kaartokallio, H., Thomas, D.N., Kuosa, H. (2005) Influence of freshwater inflow on the inorganic nutrient and DOM within coastal sea ice and underlying waters in the Gulf of Finland (Baltic Sea). *Estuarine, Coastal and Shelf Science* 65, 109–122.

Guderly, H. (2004) Metabolic responses to low temperature in fish muscle. *Biological Reviews* 79, 409–427.

Gutt, J. 2002. The Antarctic ice shelf: an extreme habitat for notothenioid fish. *Polar Biology* 25, 320–322.

Halliwell, B., Gutteridge, J.M.C. (2007) *Free Radicals in Biology and Medicine*, 4th edn. Oxford University Press, New York.

Hazel, J.R. (1985) Thermal adaptation in biological membranes: is homeoviscus adaptation the explanation? *Annual Review of Physiology* 57, 19–42.

Heise, K., Estevez, M.S., Puntarulo, S. et al. (2007) Effects of seasonal and latitudinal cold on oxidative stress parameters and activation of hypoxia inducible factor (HIF–1) in zoarcid fish. *Journal of Comparative Physiology B* 177, 765–777.

Hernandez, E., Ferreyra, G.A., MacCormack, W.P. (2002) Effect of solar radiation on two Antarctic marine bacterial strains. *Polar Biology* 25, 453–459.

Hochachka, P.W., Somero, G.N. (2002) *Biochemical Adaptation: Mechanism and Process in Physiological Evolution*. Oxford University Press, New York.

Hoppema, M., Fahrbach, E., Schröder, M. et al. (1995) Winter-summer differences of carbon dioxide and oxygen in the Weddell Sea surface layer. *Marine Chemistry* 51, 177–192.

Janknegt, P.J., van de Poll, W.H., Visser, R.J.W. et al. (2008) Oxidative stress responses in the marine Antarctic diatom *Chaetoceros brevis* (Bacillariophyceae) during photoacclimation. *Journal of Phycology* 44, 957–966.

Janknegt, P.J., de Graaff, C.M., van de Poll, W.H. et al. (2009) Short-term antioxidative responses of 15 microalgae exposed to excessive irradiance including ultraviolet radiation. European *Journal of Phycology* 44, 525–539.

Johnston, A., Calvo, J., Guderley, H. et al. (1998) Latitudinal variation in the abundance and oxidative capacities of muscle mitochondria in perciform fishes. *Journal of Experimental Biology* 201, 1–12.

Kamiyama, K., Motoyama, H., Fujii, Y., Watanabe, O. (1996) Distribution of hydrogen peroxide in surface snow over Antarctic ice sheet. *Atmospheric Environment* 30, 967–972.

Karentz, D., Bosch, I. (2001) Influence of ozone-related increases in ultraviolet radiation on Antarctic marine organisms. *American Zoologist* 41, 3–16.

Karsten, U., Wulff, A., Roleda, M.Y. et al. (2009) Physiological responses of polar benthic algae to ultraviolet radiation. *Botanica Marina* 52, 639–654.

Kepkay, P.E., Wells, M.L. (1992) Dissolved organic carbon in North Atlantic surface waters. *Marine Ecology and Progress Series* 80, 275–283.

Krapp, R.H., Baussant, T., Berge, J. et al. (2009) Antioxidant responses in the polar marine sea-ice amphipod *Gammarus wilkitzkii* to natural and experimentally increased UV levels. *Aquatic Toxicology* 94, 1–7.

Krell, A., Funck, D., Plettner, I. et al. (2007) Regulation of proline metabolism under salt stress in the psychrophilic diatom *Fragilariopsis cylindrus* (Bacillariophyceae). *Journal of Phycology* 43, 753–762.

Krembs, C., Eicken, H., Junge, K., Deming, J.W. (2002) High concentrations of exopolymeric substances in Arctic winter sea ice: implications for the polar ocean carbon cycle and cryoprotection of diatoms. *Deep-Sea Research Part* 49, 2163–2181.

Kurtz, D.D., Bromwich, D.H. (1985) A recurring, atmospherically forced polynya in Terra Nova Bay. *Oceanology of Antarctic Continental Shelf* 177–201.

Lamare, M.D., Barker, M.F., Lesser, M.P., Marshall, C. (2006) DNA photorepair in echinoid embryos: Effects of temperature on repair rate in Antarctic and non-Antarctic species. *Journal of Experimental Biology* 209, 5017–5028.

Lizotte, M.P. (2001) The contributions of sea ice algae to Antarctic marine primary production. *American Zoologist* 41, 57–73.

Lizotte, M.P. (2003) The microbiology of sea ice. *In* Thomas, D.N., Dieckmann, G.S. (eds). *Sea Ice*, 2nd edn. Wiley-Blackwell, Chichester, pp. 184–210.

Logue, J.A., de Vries, A.L., Fodor, E., Cossins, A.R. (2000) Lipid compositional correlates of temperature-adaptive interspecific differences in membrane physical structure. *Journal of Experimental Biology* 203, 2105–2115.

Marsh, A.G., Maxson R.E., Jr., Manahan, D.T. (2001) High macromolecular synthesis with low metabolic cost in Antarctic sea urchin embryos. *Science* 291, 1950–1952.

Martínez, R. (2007) Effects of ultraviolet radiation on protein content, respiratory electron transport system (ETS) activity and superoxide dismutase (SOD) activity of Antarctic plankton. *Polar Biology* 30, 1159–1172.

McMinn, A., Pankowski, A. Delfatti, T. (2005) Effect of hyperoxia on the growth and photosynthesis of polar sea ice microalgae. *Journal of Phycology* 41, 732–741.

Morley, S.A., Hirse, T., Pörtner, H.-O., Peck, L.S. (2009) Geographical variation in thermal tolerance within Southern Ocean marine ectotherms. *Comparative Biochemistry and Physiology A* 153, 154–161.

Nahrgang, J., Camus, L., Gonzalez, P. et al. (2009) PAH biomarker responses in polar cod (*Boreogadus saida*) exposed to benzo(a)pyrene. *Aquatic Toxicology* 94, 309–319.

Nahrgang, J., Camus, L., Carls, M.G. et al. (2010a) Biomarker responses in polar cod (*Boreogadus saida*) exposed to the water soluble fraction of crude oil. *Aquatic Toxicology* 97, 234–242.

Nahrgang, J., Camus, L., Carls, M.G. et al. (2010b) Biomarker responses in polar cod (*Boreogadus saida*) exposed to dietary crude oil. *Aquatic Toxicology* 96, 77–83.

Nigro, M., Regoli, F., Rocchi, R., Orlando, E. (1997) Heavy metals in Antarctic molluscs. *In* Battaglia, B., Valencia, J., Walton, D.W.H. (eds). *Antarctic Communities: Species, Structure and Survival*. Cambridge University Press, Cambridge, pp. 408–412.

Obermüller, B., Karsten, U., Abele, D. 2005 Response of oxidative stress parameters and sunscreening compounds in Arctic amphipods during experimental exposure to maximal natural UVB radiation. *Journal of Experimental Marine Biology and Ecology* 323, 100–117.

Petasne, R.G., Zika, R.G. (1997) Hydrogen peroxide lifetimes in South Florida coastal and offshore waters. *Marine Chemistry* 56, 215–225.

Philipp, E., Brey, T., Heilmayer, O. et al. (2006) Physiological ageing in a polar and a temperate swimming scallop. *Marine Ecology Progress Series* 307, 187–198.

Plettner, I. (2002) *Stress physiologie bei antarktischen Diatomeen – Ökophysiologische Untersuchungen zur Bedeutung von Prolin bei der Anpassung an hohe Salinitäten und tiefe Temperaturen*. Thesis, Universität Bremen.

Pruitt, N. L. (1988) Membrane lipid composition and overwintering strategy in thermally acclimated crayfish. *American Journal of Physiology* 254, 870–876.

Regoli, F., Principato, G. (1995) Glutathione, glutathione-dependent and antioxidant enzymes in mussels, *Mytilus galloprovincialis*, exposed to metals under field and laboratory conditions: implications for the use of biochemical biomarkers. *Aquatic Toxicology* 31, 143–164.

Regoli, F., Principato, G.B., Bertoli, E. et al. (1997) Biochemical characterization of the antioxidant system in the scallop *Adamussium colbecki*, a sentinel organism for monitoring the Antarctic environment. *Polar Biology* 17, 251–258.

Regoli, F., Nigro, M., Orlando, E. (1998a) Lysosomal and antioxidant responses to metals in the Antarctic scallop *Adamussium colbecki*. *Aquatic Toxicology* 40, 375–392.

Regoli, F., Hummel, H., Amiard-Triquet, C. et al. (1998b) Trace metals and variations of antioxidant enzymes in Arctic bivalve populations. *Archives of Environmental Contamination and Toxicology*. 35, 594–601.

Regoli, F., Nigro, M., Bompadre, S., Winston, G.W. (2000) Total oxidant scavenging capacity (TOSC) of microsomal and cytosolic fractions from Antarctic, Arctic and Mediterranean scallops: differentiation between three potent oxidants. *Aquatic Toxicology* 49, 13–25.

Regoli, F., Nigro, M., Chiantore, M., Winston, G.W. (2002). Seasonal variation of susceptibility to oxidative stress in *Adamussium colbecki*, a key bioindicator species for the Antarctic marine environment. *Science of the Total Environment* 289, 205–211.

Regoli, F., Nigro, M., Chierici, E. et al. (2004). Variations of antioxidant efficiency and presence of endosimbiontic diatoms in the Antarctic porifera *Haliclona dancoi*. *Marine Environmental Research* 58, 637–640.

Regoli, F., Nigro, M., Benedetti, M. et al. (2005a) Antioxidant efficiency in early life stages of the Antarctic silverfish, *Pleuragramma antarcticum*: responsiveness to pro-oxidant conditions of platelet ice and chemical exposure. *Aquatic Toxicology* 75, 43–52.

Regoli, F., Nigro, M., Benedetti, M. et al. (2005b) Interactions between metabolism of trace metals and xenobiotics agonists of the Ah receptor in the Antarctic fish *Trematomus bernacchii*: environmental perspectives. *Environmental Toxicology and Chemistry* 24, 1475–1482.

Resing, J., Tien, G., Letelie, R., Karl, D.M. (1993) Palmer LTER: hydrogen peroxide in the Palmer LTER region: II Water column distribution. *Antarctic Journal of the United States* 28, 227–228.

Rijkenberg, M. J. A., Fischer, A.C., Kroon, J.J. et al. (2005) The influence of UV irradiation on the photoreduction of iron in the Southern Ocean. *Marine Chemistry* 93, 119–129.

Rodriguez, R., Redman, R. (2005) Balancing the generation and elimination of reactive oxygen species. *Proceedings of the National Academy of Sciences USA* 102(9), 3175–3176.

Roncarati, F., Rijstenbil, J., Pistocchi, R. (2008) Photosynthetic performance, oxidative damage and antioxidants in *Cylindrotheca closterium* in response to high irradiance, UVB radiation and salinity. *Marine Biology* 153(5), 965–973.

Rysgaard, S., Glud, R., Sejr, M.K. et al. (2008) Denitrification activity and oxygen dynamics in Arctic sea ice. *Polar Biology* 31, 527–537.

Scarponi, G., Capodaglio, G., Barbante, C. et al. (2000) Concentration changes in cadmium and lead in Antarctic coastal seawater (Ross Sea) during the Austral summer and their relationship with the evolution of biological activity. *In* Faranda, F.M., Guglielmo, L., Ianora, A. (eds). *Ross Sea Ecology*. Springer-Verlag, Berlin, pp. 585–594.

Schriek, R. (2000) Effects of light and temperature on the enzymatic defense systems in the Antarctic ice diatom *Entemoneis kufferathii* (Manguin). *Reports on Polar Research* 349, pp 130.

Scott, C., Falk-Petersen, S., Gulliksen, B. et al. (2001) Lipid indicators of the diet of the sympagic amphipod *Gammarus wilkitzkii* in the marginal ice zone and in open waters of Svalbard (Arctic). *Polar Biology* 24, 572–576.

Scymczak, R., Waite, T.D. (1988) Generation and decay of hydrogen peroxide in estuarine waters. *Australian Journal of Freshwater Research* 39, 289–299.

Sigg, A., Neftel, A. (1988) Seasonal variations in hydrogen peroxide in polar ice cores. *Annals of Glaciology* 10, 157–162.

Speers-Roesch, B., Ballantyne, J.S. (2005) Activities of antioxidant enzymes and cytochrome c oxidase in liver of Arctic and temperate teleosts. *Comparative Biochemistry and Physiology A* 140, 487–494.

Storch, D., Pörtner, H.O. (2003) The protein synthesis machinery operates at the same expense in eurythermal and cold stenothermal pectinids. *Physiological and Biochemical Zoology* 76, 28–40.

Sunda, W., Kieber, D.J. Kiene, R.P., Huntsman, S. (2002) An antioxidant function for DMSP and DMS in marine algae. *Nature* 418, 317–320.

Tatur, A., Valle, R., Barczuk, A. (1999) Discussion on the uniform pattern of holocene tephrochronology in SouthShetland Island, Antarctica. In *Polish Polar Studies, Proceedings of XXVI Polar Symposium*, pp. 305–321.

Thomas, D.N., Kattner, G., Engbrodt, R. et al. (2001) Dissolved organic matter in Antarctic sea ice. *Annals of Glaciology* 33, 297–303.

Trevena, A.J., Jones, G.B. (2006) Dimethylsulphide and dimethylsulphoniopropionate in Antarctic sea ice and their release during sea ice melting. *Marine Chemistry* 98(2–4), 210–222.

USGS. (2000) *World Petroleum Assessment 2000 – Description and Results*. US Geological Survey, Reston. VA.

Vacchi, M., La Mesa, M., Dalu, M., Macdonald, J. (2004) Early life stages in the life cycle of Antarctic silverfish, *Pleuragramma antarcticum* in Terra Nova Bay, Ross Sea. *Antarctic Science* 16, 299–305.

Viarengo, A., Canesi, L., Pertica, M. et al. (1990) Heavy metal effects on lipid peroxidation in the tissues of *Mytilus galloprovincialis* Lam. *Comparative Biochemistry and Physiology C* 97, 37–42.

Viarengo, A., Canesi, L., Pertica, M., Livingstone, D.R. (1991) Seasonal variations in the antioxidant defence system and lipid peroxidation of the digestive gland of mussels. *Comparative Biochemistry and Physiology C* 100, 187–190.

Viarengo, A., Accomando, R., Roma, G. et al. (1994) Differences in lipid composition of cell membranes from Antarctic and Mediterranean scallops. *Comparative Biochemistry and Physiology B* 109, 579–584.

Viarengo, A., Canesi, L., Martinez, P.G. et al. (1995) Prooxidant processes and antioxidant defence systems in the tissue of the Antarctic scallop (*Adamussium colbecki*) compared with the Mediterranean scallop (*Pecten jacobaeus*). *Comparative Biochemistry and Physiology B* 111, 119–126.

Villafañe, V.E., Helbling, E.W., Zagarese, H.E. (2001) Solar ultraviolet radiation and its impact on aquatic systems of Patagonia, South America. *Ambio* 30, 112–117.

Weeks, W.F., Ackley, S.F. (1982) *The Growth, Structure and Properties of Sea Ice*. Monograph 82-1, Cold Regions Research and Engineering Laboratory, US Army, 130 pp.

Weihe, E., Kriews, M., Abele, D. (2010) Differences in heavy metal concentrations and in the response of the antioxidant system to hypoxia and air exposure in the Antarctic limpet Nacella concinna. *Marine Environmental Research* 69, 127–135.

Weissenberger, J., Dieckmann, G., Gradinger, R., Spindler, M. (1992) Sea ice: a cast technique to examine and analyze brine pockets and channel structure. *Limnology and Oceanography* 37, 179–183.

Weller, R., Schrems, O. (1993) H_2O_2 in the marine troposphere and seawater of the Atlantic Ocean (48°N–63°S). *Geophysical Research Letters* 20, 125–128.

Werner, I. (2000) Faecal pellet production by Arctic underice amphipods – transfer of organic matter through the ice/water interface. *Hydrobiologia* 426, 89–96.

Witas, H., Gabryelak, T., Matkovics, B. (1984) Comparative studies on superoxide dismutase and catalase activities in livers of fish and other Antarctic vertebrates. *Comparative Biochemistry and Physiology C* 77, 409–411.

Zika, R.G., Petasne, R.G., Coope, W.J. (1983) Photochemical formation of hydrogen peroxide in ground water exposed to sunlight. *National Meeting – American Chemical Society, Division of Environmental Chemistry* 23, 362–364.

Chapter 3

OXIDATIVE STRESS IN ESTUARINE AND INTERTIDAL ENVIRONMENTS (TEMPERATE AND TROPICAL)

Carolina A. Freire[1], *Alexis F. Welker*[2,3], *Janet M. Storey*[4], *Kenneth B. Storey*[4], *and Marcelo Hermes-Lima*[2]

[1]Departamento de Fisiologia, Setor de Ciências Biológicas, Universidade Federal do Paraná, Curitiba, PR, Brazil
[2]Laboratório de Radicais Livres, Departamento de Biologia Celular, Universidade de Brasília, DF, Brazil
[3]Faculdade da Ceilândia, Universidade de Brasília, DF, Brazil
[4]Institute of Biochemistry, Carleton University, Ottawa, ON, Canada

ESTUARINE/INTERTIDAL HABITATS

Coastal aquatic habitats are extremely unstable in terms of abiotic parameters, basically due to tidal influence. Tidal movement of ocean water masses results in a fluctuating sea level, caused by the gravitational attraction of the Moon and the Sun, and centrifugal forces from the rotation of the Earth (Nybakken, 1988). The vertical magnitude and horizontal extension of the fluctuation depends on the phase of the Moon, the time of year, the latitude, the geography, the slope of the beach, and the winds (Nybakken, 1988; Willmer et al. 2005).

Estuarine environments (boundary between marine and freshwater systems) and intertidal environments (boundary between marine and terrestrial systems) are characterized by cyclical fluctuations of physical characteristics, at the rhythm of the tides (Willmer et al. 2005; Gracey et al. 2008). The three most relevant physical parameters are temperature, dissolved

Oxidative Stress in Aquatic Ecosystems, First Edition. Edited by Doris Abele, José Pablo Vázquez-Medina, and Tania Zenteno-Savín.

oxygen, and salinity (Nybakken, 1988; Ross et al. 2001; Sagasti et al. 2001; Oliveira et al. 2005). Ultraviolet radiation (UVR), risk of desiccation upon air exposure, and strong wave action are also important in tidally exposed species (Willmer et al. 2005; Lesser 2006). Estuarine areas can be as different as coastal mudflats, swamps, salt marshes, and mangrove forests, whereas intertidal areas are mostly rocky or sandy shores. In any case, the very fast (within a few hours) fluctuation in environmental parameters will constantly challenge the inhabitants, e.g. in the vicinity of mangrove roots, or of an intertidal pool (Fig. 3.1).

During low tides, algae and animals that inhabit the upper intertidal zone spend more time uncovered by seawater, thereby exposed to wave action and desiccation, as well as to heavy rains or intense sunshine and variable levels of hazardous UVB radiation. Additionally, intertidal organisms experience high and low thermal extremes depending on climatic conditions, and can be exposed to hypoxic conditions and intermittent fasting. If not fully exposed, organisms may hide in rock crevices, or may be trapped in tidal pools in the upper littoral zone; alternately, infaunal species may burrow in sandy/muddy substrates. In fact, the intertidal zone is considered one of the most

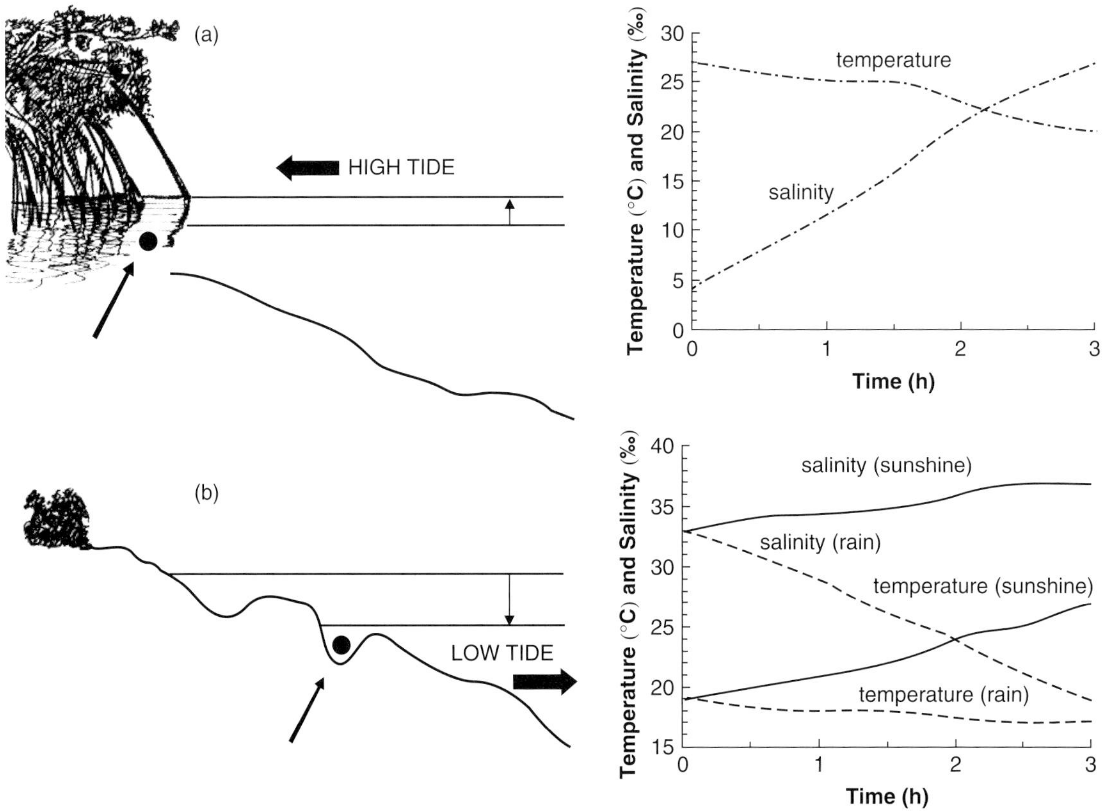

Fig. 3.1 Fast fluctuation in environmental parameters in estuarine (a) and rocky intertidal (b) habitats, upon seawater movements during high or low tides. Only temperature and salinity have been illustrated, for clarity. Arbitrary values of temperature and salinity, compatible with tropical environments, measured at a certain spot (black dot indicated by an arrow) below the low-tide water levels. The vertical thin arrow indicates the change in water level. In (a) a fast decrease in temperature and increase in salinity occur during incoming flood tide in an estuary. In (b) a fast increase in both temperature and salinity (solid lines) occur when the intertidal pool is exposed to intense sunshine during a daytime low tide, or a fast decrease in both temperature and salinity (dashed lines) occur when the intertidal pool is exposed to rain precipitation during low tides.

challenging environments on Earth (Halpin et al. 2004; Gracey et al. 2008).

For organisms that inhabit estuarine areas, the major challenge to cope with is salinity increase during flood tide and salinity decrease during ebb tide, albeit accompanied by relevant oscillations in temperature, dissolved oxygen, and pH. All of these effects are intensified in intertidal/estuarine areas during spring tides of greater amplitude, when the effects of the Sun and Moon are additive (e.g. the intertidal environment in Fig. 3.2), whereas during neap tides, when the Sun and Moon are in opposition, the effects of the tides are less strong (Nybakken 1988).

TEMPERATURE INCREASE AS STRESS INDUCER

Thermal stress often involves high temperatures as tides recede, but during winter in temperate and polar regions, low tides can also expose intertidal organisms to freezing. Plants and animals exhibit a heat shock response (HSR) involving synthesis of heat shock proteins (HSPs) that act as molecular chaperones (Halpin et al. 2004; Gracey et al. 2008). The expression of these HSPs is especially high in sessile species inhabiting the upper littoral areas of the rocky coasts, compared with individuals from low-intertidal or sublittoral zones (Downs et al. 2002; Nakano and Iwama 2002;

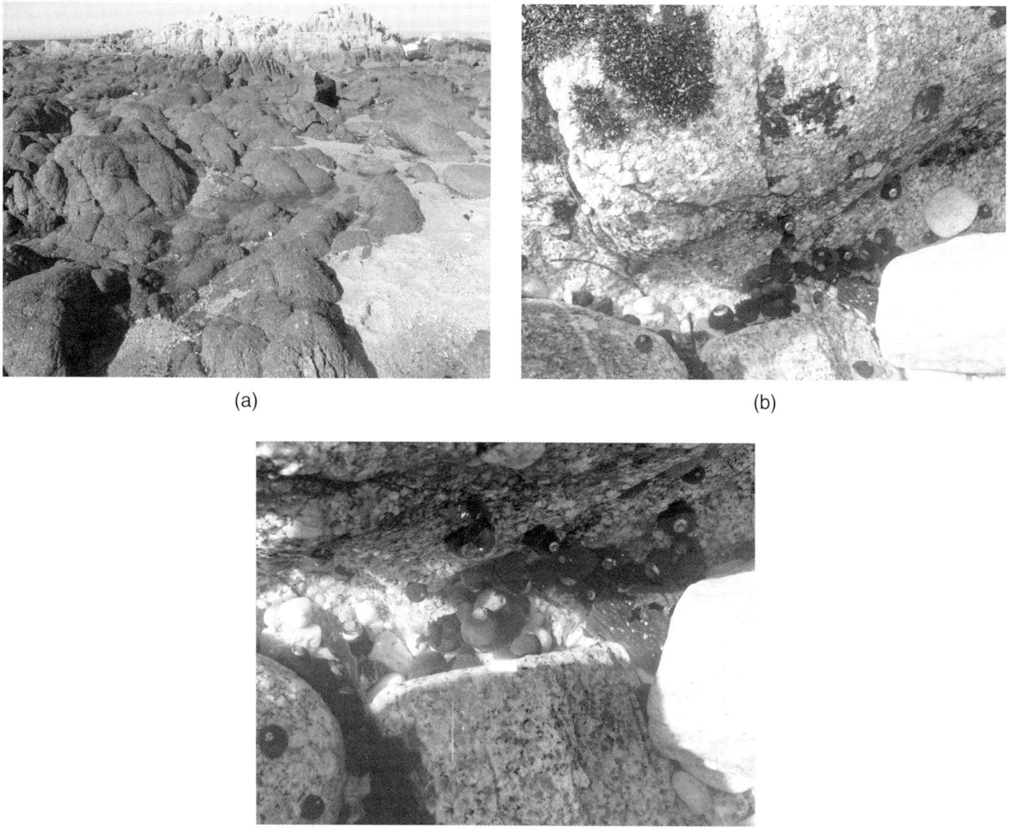

(a) (b) (c)

Fig. 3.2 Example of an intertidal environment at a rocky beach in Pacific Grove, California. Observe the water pools during a low tide (a). Many of these pools are rich with life, such as algae, bivalves, and marine gastropods (b and c). Progressive evaporation of these pools exposes marine invertebrates to increased salinity and temperature, as well as a decrease in dissolved oxygen. (Photographs by Marcelo Hermes-Lima, November 2010.) See plate section for a color version of this image.

Halpin et al. 2004; Lesser 2006; Gracey et al. 2008; Henkel et al. 2009). For example, in mussels (*Mytilus californianus*), the amount of the inducible HSP72 isoform was higher in individuals from the upper littoral zone than in those from the lower littoral zones (Halpin et al. 2004). In intertidal gastropods (genus *Tegula*), the ratio of expression of HSP72 (a strongly heat-induced isoform) to HSP74 is higher in intertidal species, those subjected to greater heat stress, than in their subtidal relatives (Tomanek and Somero 2002). Subtidal species of both invertebrates and fish display a lower temperature of activation for the HSR than their intertidal congeners (Nakano and Iwama 2002; Tomanek and Sanford 2003). Several intertidal invertebrates exhibit daily cycles of HSP expression with the tides, with variable patterns in different tissues, as observed in the chiton *Acanthopleura granulata* (Schill et al. 2002). Changes in HSP (and other proteins) expression can be detected within a few hours (1–6 h) in aquatic species (Sanders et al. 1991; Gracey et al. 2008), which is in the time frame of physiological events happening during tidal cycles.

Other stress markers responding during cyclic fluctuations in intertidal organisms include blood cortisol in fish (Yamashita et al. 2003), and the endogenous antioxidant system in corals, crustaceans, mollusks, and fish (Ross et al. 2001; Downs et al. 2002; Malanga et al. 2004; Cailleaud et al. 2007; Togni 2007). For example, comparison between the intertidal limpet *Nacella (Patinigera) magellanica* and its subtidal congener *N. (P.) deaurata* showed a more potent antioxidant system (e.g. catalase and superoxide dismutase (SOD) activities) in the intertidal species that faces daily variability on the shores of the Beagle Channel in southern Argentina (Malanga et al. 2004). In the intertidal seaweeds (*Fucus* spp.) a correlation was found between position in the intertidal zone and antioxidant protection (Colleén and Davison 1999). Clearly, in the intertidal zone, no individual stress factor acts alone (Gardeström et al. 2007; Helmuth 2009), and the cellular stress response combines protection from a suite of challenges. The HSR and the antioxidant defense system are members of this integrated stress response (Pritchard 2002; Scholz 2002; Gardeström et al. 2007). Sublethal stress renders organisms more resistant to subsequent more severe stress in a preconditioning pattern (Pritchard 2002; Scholz 2002; Halpin et al. 2004).

Temperature accelerates reactive oxygen species (ROS) formation in marine ectotherms simply by accelerating mitochondrial respiration. For example, isolated mitochondria of eurythermal bivalves (*Mya arenaria*) doubled the *in vitro* rate of ROS release when warmed from a habitat temperature of 15°C to heat stress at 25°C (Abele et al. 2002). Further, in several invertebrates, high temperature exposure was found to induce antioxidant enzyme activities (Abele et al. 1998b, 2002; Wilhelm Filho et al. 2001b; Downs et al. 2002; Heise et al. 2003) and, likewise, in a freshwater fish, the cichlid cará (*Geophagus brasiliensis*; Wilhelm Filho et al. 2001a). In addition to temperature, relatively fast changes (in a matter of hours) in environmental parameters such as dissolved oxygen and pH during tidal cycles induced changes in the activities of catalase (CAT) and SOD in body wall tissue of a sedimentary capitellid worm (Abele et al. 1998a).

Seasonal modulations of antioxidant defense (SOD, CAT, glutathione S-transferase (GST)) are triggered by higher temperatures and reproductive activity in early summer and early autumn, e.g. in intertidal mussels (*Perna perna*; Wilhelm Filho et al. 2001b) and the worms *Heteromastus filiformis* and *Nereis diversicolor* from North Sea intertidal mud flats (Abele et al. 1998a). It is important to note that the short-range (tidally related) environmental variation in estuaries and intertidal regions is also affected by seasons (e.g. winter/summer differences; Niyogi et al. 2001; Wilhelm Filho et al. 2001b; Lau et al. 2004; Ferreira et al. 2005), and last, but not least, that inhabitants of coastal estuarine areas are frequently exposed to pollution (Niyogi et al. 2001; Geracitano et al. 2004; Manduzio et al. 2004). All these conditions contribute synergistically to accelerate oxidative stress in intertidal marine organisms (Zaccaron da Silva et al. 2005).

Thus, seasonal patterns, overlapping with the effects of pollution on oxidative stress biomarkers, were observed in estuarine oysters (*Saccostrea cucullata*; Niyogi et al. 2001) and in estuarine fish (mullet and flounder; Ferreira et al. 2005). Activities of antioxidant enzymes or indicators of oxidative damage are very often employed as biomarkers for pollution in estuarine areas (Geracitano et al. 2004; Zaccaron da Silva et al. 2005). However, a full characterization of seasonal fluctuations to oxidative stress parameters in estuarine and intertidal habitats is required before any biomarker pattern can be related to pollution (Niyogi et al. 2001).

VARIATION OF ENVIRONMENTAL OXYGEN LEVELS, ESPECIALLY HYPOXIA–REOXYGENATION AS A STRESS INDUCER

A situation that prompts ROS generation is the transition between hypoxia/anoxia and reoxygenation. Tissues of hypoxia-intolerant species submitted to ischemia show tissue and cellular damage when they are reperfused with oxygenated solution/blood (Vetterlein et al. 2003; Walshe and D'Amore, 2008). Aquatic animals face variable oxygen availability, either in a circadian/circumtidal pattern, or seasonally in ponds susceptible to ice cover, when compared to terrestrial animals with aerial respiration. But the most stressful pattern of variation is again found in intertidal/estuarine environments (e.g. Nilsson and Renshaw 2004). Intertidal and estuarine organisms have been shown to display high expression levels of antioxidants (Ross et al. 2001; Brouwer et al. 2004; Malanga et al. 2004; Oliveira et al. 2005). Higher specific activities of antioxidant enzymes were found in fish caged in the estuary that displayed the widest amplitude in dissolved oxygen and the highest level of oxygen supersaturation among all estuaries studied (Ross et al. 2001).

Most hypoxia-tolerant animals enter metabolic depression during a period of oxygen restriction (which could last from a few hours to weeks, depending on the species), associated with suppression of protein synthesis (Storey and Storey 2004). In spite of this, various studies reported an activation of selected antioxidant enzymes (activity and/or expression) during hypometabolic states (Hermes-Lima and Zenteno-Savín, 2002; see also Table 3.1). These apparently contradictory observations serve to minimize oxidative stress arising when animals leave the hypometabolic state during arousal (Orr et al. 2009). An increase in levels of oxidative stress markers, such as lipid peroxidation, protein carbonyls or the glutathione (GSH): glutathione disulphide (GSSG) ratio has been observed in several animal models during reoxygenation after a hypoxic/anoxic event. This increase in the activity/expression of antioxidant enzymes during hypoxia/hypometabolism is considered a key phenomenon in the protection against post-hypoxic free radical damage and was coined "preparation for oxidative stress" (Hermes-Lima and Zenteno-Savín 2002).

The first study demonstrating increased oxidative stress during reoxygenation was undertaken with the nonintertidal anoxic-tolerant goldfish *Carassius auratus* (Lushchak et al. 2001). Animals held under 8 h of anoxia showed increased CAT activity in liver and glutathione peroxidase (GPx) activity in brain. After reoxygenation, lipid peroxidation was increased in liver and brain. Since goldfish survive cycles of anoxia/reoxygenation in nature, the authors concluded that this is a case of "physiological oxidative stress" (Hermes-Lima and Zenteno-Savín, 2002), and that damage would have been worse if these animals did not enhance the antioxidant capacity. In this case, the increase in the activity/expression of antioxidant enzymes under anoxia (where ROS production is halted – a state which can last for many hours) should minimize ROS-induced oxidative damage during reoxygenation. Other studies (see Table 3.1) demonstrated controlled oxidative stress associated with activation of the antioxidant system before reoxygenation events in several aquatic and intertidal species. For example, elevated activities of CAT and GPx were observed in brain of the freshwater fish *Cyprinus carpio* under hypoxia, whereas no changes in oxidative stress markers occurred in the brain after 1 h reoxygenation (Lushchak et al. 2005). Unfortunately, several other studies evaluating changes in antioxidants by hypoxia/anoxia in aquatic animals did not analyze oxidative stress parameters, such as lipid peroxidation (Table 3.1) – one example is the case of the estuarine fish *Leiostomus xanthurus*, where hypoxia led to increased SOD activity (Cooper et al. 2002).

Air exposure is another situation that causes gradual hypoxia in internal organs of aquatic animals and is a major stress for intertidal/estuarine species (Table 3.1). For example, crabs *Paralomis granulosa* (a marine species) exposed to air showed activation of antioxidant enzymes. Oxidative stress markers were also elevated during aerial exposure; however, this study did not analyze reoxygenation (Romero et al. 2007).

The responses of the antioxidant system to fluctuations in oxygen availability might be triggered by ROS (especially nitric oxide, superoxide radical, and H_2O_2), involved in cellular signaling (Hermes-Lima 2004). For example, in the North Sea eelpout (*Zoarces viviparus*), functional hypoxia derived from cold-exposure, which reduces blood circulation and oxygen transport, caused increased liver GSH levels (Heise et al. 2006). Animals also responded to cold-exposure with

Table 3.1 Aquatic animals showing signs of "preparation for oxidative stress."

Animal	Species	Exposure to:	Antioxidant alteration[1]	Reference
Freshwater fish	*Carassius auratus*	Anoxia	↑CAT and GPX	Lushchak et al. (2001)
	Oreochromis niloticus	Severe hypoxia	↑GPX (tendency: $P = 0.07$)	Welker (2009)
	Cyprinus carpio	Hypoxia	↑GPX and CAT	Lushchak et al. (2005)
	Cyprinus carpio	Hypoxia	↑SOD	Vig and Nemcsok (1989)
	Perccottus glenii	Hypoxia	↑SOD and GST ↓GR and CAT	Lushchak and Bagnyukova (2007)
Estuarine fish	*Leiostomus xanthurus*	Hypoxia	↑SOD	Cooper et al. (2002)
Crab	*Callinectes ornatus*	Air exposure	↑GPX and CAT	Togni (2007)
	Chasmagnathus granulata	Anoxia	↑CAT and GST ↓SOD	Oliveira et al. (2005)
	Paralomis granulosa	Air exposure	↑SOD, CAT and GST	Romero et al. (2007)
Freshwater snail	*Biomphalaria tenagophila*	Anoxia	↑GPX ↓CAT	Ferreira et al. (2003)
	Biomphalaria tenagophila	Aestivation	↑GPX ↓SOD	Ferreira et al. (2003)
Marine gastropod	*Littorina littorea*	Anoxia	↑GSH ↓several antioxidant enzymes	Pannunzio and Storey (1998)
Oyster	*Crassostrea gigas*	Severe hypoxia	↑expression of GPX	David et al. (2005)
Mussel	*Perna perna*	Air exposure	↑GST	Almeida et al. (2005)
	Perna perna	Air exposure	↑SOD	Almeida and Bainy (2006)
	Mytilus edulis	Air exposure	↓SOD-total ↑SOD (two isoforms, relative to total-SOD)	Letendre et al. (2008)
Marine polychaete	*Heteromastus filiformis*	Anoxia	↑CAT	Abele et al. (1998a)
Intertidal Polychaeta	*Nereis diversicolor*	Anoxia	↑SOD in mature animals ↓ SOD in immature animals	Abele-Oeschger et al. (1994)

[1]Note that in some cases, certain antioxidant defenses are activated while others are down.

activation of the transcription factor HIF-1 in the liver (considered a "master- switch" in the redox-regulated transcriptional response to low oxygen in vertebrates; Nikinmaa and Rees 2005; Chapter 12). Recovery from cold-exposure increased thiobarbituric acid reactive substance (TBARS) levels, similar to what happens during hypoxia–reoxygenation (Heise et al. 2006).

SALINITY VARIATION AS A STRESS INDUCER

Severe fluctuations in salinity add to the stress caused by the above-mentioned variability in temperature and dissolved oxygen in estuarine and intertidal habitats. Long-term effects of salinity variation on the antioxidant response have been studied in the mussel *Perna viridis* in the estuarine waters of Hong Kong (Lau et al. 2004), and the mangrove oyster *Crassostrea rhizophorae* in Brazilian southern estuaries (Zaccaron da Silva et al. 2005). However, there is an obvious lack of experimental investigation of the relationship between tidal salinity variation and oxidative stress in intertidal or estuarine species. Starting to fill this gap, an up-regulation of glutathione disulphide reductase (GR) expression has been reported in the intertidal copepod *Tigriopus japonicus* upon salinity increase (Seo et al. 2006). In another copepod (*Eurytemora affinis*, Calanoida) an experimental tide with salinities varying from 0 to 25‰ over 6 h resulted in a variable GST response, with a peak of activity at 5‰ salinity (Cailleaud et al. 2007). In this study the authors only stated that "salinity stress could interfere with the oxidative stress response in *E. affinis*" without further reasoning. In the estuarine mud crab *Scylla serrata*, after weeks of exposure to 10, 17, and 35‰ (17‰ being the acclimated control), lower metabolic rates and higher levels of oxidative damage (lipid peroxidation and protein carbonyls) were noted at 35‰ (Paital and Chainy 2010). In some circumstances, reduced metabolic rates can be associated with lower mitochondrial proton H^+ leak (Brookes 2005), which may increase ROS production (Murphy 2009). Furthermore, SOD activity – which is activated by increased ROS formation in many species – was found to protect the intertidal diatom *Cylindrotheca closterium* from oxidative damage at very high salinity (60‰; Roncarati et al. 2008).

A causal relationship between an osmotic challenge and oxidative stress was also observed in a freshwater fish (the sturgeon *Acipenser naccarii*) that tolerates the full range of marine salinity fluctuations and can be characterized as "euryhaline." After gradual acclimation and an additional 20 days in fully marine water (34‰), the sturgeons displayed increased activities of the antioxidant enzymes SOD and CAT in plasma, as well as SOD and GPx in red blood cells. This was accompanied by intensified lipid peroxidation in plasma, and decreased CAT activity in liver (Martínez-Álvarez et al. 2002). In another study, 2 h exposure of swimming crabs *Callinectes ornatus* to hypersaline seawater (40‰, controls at 33‰) led to increased GPx and CAT activities in three tissues, as well as increased lipid peroxidation (Togni 2007). Exposure to hyposaline water (10‰) for 2 h produced no effect in oxidant/antioxidant parameters in this species. On the other hand, in response to hypo-osmotic shock, the blue crab *Callinectes sapidus* showed a rise in metallothionein-like proteins in gills, which play a role in ROS scavenging (Martins and Bianchini 2009). Salinity reduction causes an increase in metabolic rate due to the demands for osmoregulation, which, in turn, could result in increased ROS production. In addition, stressful reduction of salinity from full-strength seawater down to 4‰ over several days resulted in a progressive increase in GPx and GST expression in the liver of the marine olive flounder (*Paralichthys olivaceus*); these marine fish were submitted to progressive seawater dilution (35, 17.5, 8.75, 4‰), and remained for 2 days in each experimental salinity (Choi et al. 2008a). Interestingly, two salinity treatments, i.e., salinity reduction from 30‰ (control) down to 5‰, or increase from 30 to 50‰ within 24 h both caused decreased activities of SOD and GPx in the shrimp *Litopenaeus vannamei* (Liu et al. 2007).

Recent studies showed that hypo-osmotic stress, from 35 to 10‰ (after 24 h), induced increases in both activity and expression of hepatic SOD, CAT, and GPx (as well as increase in H_2O_2 production and lipid peroxidation) in black porgy (*Acanthopagrus schlegeli*), a fish from coastal areas, including estuaries (An et al. 2010). In a study with ark shells (*Scapharca broughtonii*) An and Choi (2010) observed that exposure to either hypersaline (45‰) or hyposaline (25‰) waters caused oxidative stress, which was more extensive at the low salinity exposure. It appears that, depending on the aquatic animal species, either an increase or decrease in salinity may affect the redox balance by prompting ROS generation.

Future research to elucidate the molecular/cellular mechanisms that deal with fluctuating salinity regimes in intertidal animals will certainly profit from existing knowledge on the intracellular pathways involved in the detection of hyperosmolality by osmosensing genes in the mammalian kidney. Since the seminal work by Ferraris and collaborators (1999), major advances in the detection of osmotically regulated genes have been made, especially in the laboratory of M. Burg (Burg et al. 2007).

Although still not fully explored in intertidal environments (but likely a relevant source of stress to their inhabitants, and directly demonstrated in mammalian renal cell systems), there is an intimate direct link between high salinity (NaCl) and ROS generation (Zhou et al. 2005; Carlström et al. 2009). High NaCl leads to oxidative stress (Zhang et al. 2004; Zhou et al. 2005); hyperosmolality can lead not only to protein damage (Kwon et al. 2009), but also to damage of genetic material (Kültz and Chakravarty 2001). Furthermore, it has been shown that $O_2^{\bullet-}$ and other ROS increase the transactivation of the transcription factor tonicity-responsive enhancer binding protein/osmotic response element binding protein (TonEBP/OREBP) in human kidney cells (Zhou et al. 2005). This transcription factor is activated by hypertonic stress and then increases the transcription of osmoprotective genes, in turn leading to the accumulation of compatible organic osmolytes and up-regulation of protective HSPs in mammals (reviewed in Zhou et al. 2005; Kwon et al. 2009). Thus, a relationship between hypoxia, salt, oxidative stress, the HSR, and salt transport is increasingly evident. Furthermore, it is also apparent that research on intertidal/estuarine organisms will add valuable knowledge that makes reciprocal contributions to our understanding of mammalian salt-related hypertension.

Several mitogen-activated protein kinases (MAPKs) have been related to salt stress in fish gills (ERK1, SAPK1, and SAPK2 subgroups) and several other systems, such as "high-osmolarity glycerol" (HOG) in *Saccharomyces cerevisiae*, obviously based on a phylogenetically ancient signaling system with which eukaryotes respond to changes in external osmolality (Kültz and Ávila 2001). Thus, it seems reasonable to predict – in a rather simplistic way – that intertidal and estuarine species will display increased constitutive expression of osmoprotective transcription factors and antioxidant enzymes, similar to what has been shown for HSPs (Downs et al. 2002; Nakano and Iwama 2002), potentially regulated via MAPK signaling as predicted by Cowan and Storey (2003).

POST-TRANSCRIPTIONAL AND POST-TRANSLATIONAL CONTROLS ON ANTIOXIDANT ENZYMES

It is widely accepted that changes in the activities of antioxidant enzymes derive mainly from regulating the expression of antioxidant genes (discussed later) but there is now good evidence that post-transcriptional and post-translational controls must also be considered.

Post-transcriptional controls are well documented in the regulation of proteins of iron transport and storage. Although these are not normally classed as antioxidants, the management of free iron is intimately linked with antioxidant defense because of the central role of Fe^{2+} as a catalyst in producing $HO^{\bullet}$ via the Fenton reaction. Hence, as much as possible, free iron is minimized and instead iron is sequestered in association with storage proteins, transport proteins, or functional proteins (e.g. hemoglobin, cytochromes). Expression of the intracellular storage protein, ferritin, is mainly controlled at the post-transcriptional level by the action of iron-regulatory proteins (IRPs) that bind to specific sequences in the 5′ untranslated region (UTR) of mRNA to block the translation of ferritin heavy (H) and light (L) chains (Muckenthaler et al. 2008). Dissociation of IRPs from ferritin transcripts allows translation to go forward and is triggered not just by high Fe^{2+} but also by hypoxia (reversed by reoxygenation; Schneider and Leibold 2003), $NO^{\bullet}$, and ROS (Fillebeen and Pantopoulos 2002). Iron-regulatory proteins provide coordinated regulation of at least 10 other mRNA types, including the plasma membrane transferrin receptor (TfR) that imports iron–transferrin complexes from blood or hemolymph (Salahudeen and Bruick 2009). The ferritin H chain is one of the genes that is up-regulated by anoxia exposure in *L. littorea* (mRNA and protein levels rose two-fold within 24–72 h), suggesting a need to enhance iron storage either during anoxia or in preparation for oxidative stress during reoxygenation (Larade and Storey 2004). Furthermore, gene screening studies spanning multiple organisms and environmental challenges have repeatedly shown stress-responsive up-regulation of ferritin and TfR2, including under anoxia and/or freezing in freeze-tolerant frogs and turtles, and hibernation in bats (Storey and Storey 2007).

Another relevant mode of post-transcriptional regulation of mRNA is the differential distribution of transcripts between translationally active polysomes and inactive monosomes. In *L. littorea*, for example, ferritin H transcripts were preferentially maintained with the polysome fraction under anoxia (Larade and Storey 2002a,b, 2004). Indeed, the dissociation of polysomes, increase in monosomes, and sequestration of most mRNA transcripts into translationally inactive monosome or ribonuclear protein fractions – all actions that suppress protein synthesis – is a widespread component of anaerobiosis and other hypometabolic states across phylogeny (Storey and Storey 2004). Pytharopoulou et al. (2006) reported that polysome contents were also reduced in digestive gland of *Mytilus galloprovincialis* exposed to polluted waters in the Mediterranean Sea as compared with cleaner waters. However, a distinct seasonal pattern to polysome content was also identified that would limit the value of this parameter as a biomarker of pollution. Interestingly, polysome content in *M. galloprovincialis* was negatively correlated with metallothionein (MT) levels, which suggests that MT transcripts are also probably retained in the polysome fraction and preferentially translated under stress conditions that include metal ion challenge. Metallothionein binds copper, cadmium, zinc, and other metal ions, all of which cause oxidative stress (Hansen et al. 2007).

Indeed, paralleling ferritin, MT was up-regulated in response to both anoxia and freezing in tissues of *L. littorea* (English and Storey 2003).

A final form of post-transcriptional regulation of mRNA transcripts should be mentioned because it is the hot new topic in molecular biology that will definitely impact our understanding of gene regulation in comparative fields. This is microRNA (miRNA). These are short noncoding regulatory RNAs of about 20–23 nt in length. Differential miRNA expression has already been demonstrated in mammalian hibernation and frog freeze tolerance, both conditions that involve wide variation in oxygen availability (Morin et al. 2008; Biggar et al. 2009). Changes in miRNA patterns were also seen in the responses by zebrafish to osmotic stress (Flynt et al. 2009) and during muscle development at different water temperatures (Johnston et al. 2009). A new study shows that miR-1-1 increased by 62% in hepatopancreas of *L. littorea* in response to anoxia whereas miR-34a was unaffected (Biggar et al. 2011). Hence, miRNAs will undoubtedly prove to be intimately involved in animal responses to environmental stress in the intertidal and estuarine environments and future research in this area will be critical. Indeed, a recent study has developed methodology to detect and amplify miRNAs from species with unsequenced genomes and shown differential regulation of miR-1-1 and miR-34 in hepatopancreas from control versus 24 h anoxic L. littorea (Biggar et al. 2011).

Post-translational controls on proteins are also critical regulatory mechanisms. Reversible protein phosphorylation (RPP) is by far the best known because of its widespread role in the control of intermediary metabolism, particularly as a coordinating mechanism in stress-responsive metabolic rate depression (Storey and Storey 2004, 2007) but many other types of modifications exist (e.g. methylation, acetylation, ubiquitylation, SUMOylation, S-nitrosylation, *O*-GlcNAcylation, ADP-ribosylation, S-sulfhydration). It should be noted that the role of S-sulfhydration in metabolic regulation, mediated by hydrogen sulfide (H_2S) (Mustafa et al. 2009) could turn out to be very important for intertidal and aquatic organisms. Significantly, Budde and Roth (2010) have recently shown that H_2S can regulate the hypoxia inducible factor 1 (HIF-1) setting up the potential for reciprocal control of gene expression by oxygen and H_2S.

The role of post-translational modification in the control of antioxidant enzymes is wide open to exploration. Only a handful of studies have been undertaken to date and these suggest that there is much to be learned. For example, a recent study of the land snail *Otala lactea* was the first to show that RPP can control glucose-6-phosphate dehydrogenase (G6PDH) activity (Ramnanan and Storey 2006). Given that the pattern of glycolytic enzyme modification by RPP is virtually identical between land snails entering estivation and marine snails enduring anoxia (Brooks and Storey 1997), it is highly likely that comparable regulation of G6PDH by RPP occurs in marine mollusks.

Studies with mammalian cell systems are also beginning to show that mainline antioxidant enzymes are subject to post-translational controls and this may be relevant for intertidal/estuarine species. Peroxiredoxin (Prx) isoforms 1 and 2 were inactivated after phosphorylation by the cyclin-B-dependent kinase Cdc2 (Rhee et al. 2005), and Prx 6 activity is modulated by the mitogen-activated protein kinases ERK and p38 (Wu et al. 2009). CAT was phosphorylated by the nonreceptor tyrosine kinases, c-Abl and Arg, that are activated when cells are treated with H_2O_2 and this led to enhanced CAT activity by four- to five-fold (Cao et al. 2003a). GPx1, the ubiquitous cytosolic member of the GPx family, was also activated by these tyrosine kinases (Cao et al. 2003b).

OXIDATIVE STRESS AND TRANSCRIPTION FACTOR CONTROL OF GENE EXPRESSION

Transcription factors (TFs) are the central link between detection of a stress and activation of a gene expression response to ameliorate the stress. Transcription factors typically control a suite of genes so that a coordinated response can be made by genes/proteins serving different parts of the adaptive response. The best-known of the TFs that mediate cell responses to oxidative stress is the nuclear factor E2 p45-related factor 2 (Nrf2) that regulates expression of multiple genes for both constitutive and inducible antioxidant defenses by binding to the antioxidant response element in their gene promoters (Kensler et al. 2007). Examples of the many genes under Nrf2 control are GST isozymes, GPx, GR, Cu/Zn-SOD, ferritin, heme oxygenase, and glutathione synthesis enzymes. Other TFs of relevance to oxidative stress responses include nuclear factor kappa B (NFκB), the metal-responsive TF (that regulates MT), the forkhead box class O (FOXO) TFs, and the aryl hydrocarbon receptor (AhR) (Pahl 1999; Van der Horst and Burgering 2007; Mitchell and Elferink 2009). NFκB is a critical regulator of cellular response to stress that is activated by many kinds of stress and regulates at least 150 genes involved in cell functions, including immune response, stress response, antioxidant defense (e.g. SOD, ferritin H), and cell growth, differentiation and apoptosis (Pahl 1999). NFκB, like HIF-1, is an oxygen-sensitive TF and both are activated under hypoxic conditions as the consequence of inhibition of oxygen-dependent hydroxylases (Taylor and Cummins 2009). This may make NFκB a critical target for study with respect to organisms in intertidal and estuarine zones where cyclic exposures to hypoxia/anoxia (and variations in temperature, salinity, pH) occur on both daily and seasonal time frames. On the other hand, FOXO signaling, which can up-regulate Mn-SOD and CAT, is interesting because of its demonstrated role in mediating protective responses in quiescent cells (Kops et al. 2002). There is a vast literature on each of the above TFs but virtually nothing that is specific to their action in stress-responsiveness by intertidal and estuarine species. However, excellent array screening technologies are now available that allow researchers to quantify stress-responsive changes in the nuclear levels of dozens of TFs (TF presence in the nucleus being a requirement for gene expression; Storey 2008). Application of these tools to identify which TFs are activated under different stress conditions will lead to critical advances in our understanding the adaptive responses of intertidal and estuarine species. The gene targets of each individual TF are highly conserved across the animal kingdom and so the identification of which TFs respond to a particular environmental insult immediately provides the researcher with a list of probable genes/proteins that are likely to be involved in either adaptive (e.g. antioxidants, chaperones, etc.) or destructive (e.g. apoptosis, autophagy) responses to stress.

OXIDATIVE STRESS AND GENE SCREENING IN MARINE AND ESTUARINE ORGANISMS

Gene screening is a powerful tool in modern biology and for the comparative biologist has two major advantages: (1) the opportunity to gain a global overview of gene expression changes in response to any given environmental stress; (2) the ability to identify novel responses that were never previously considered to be a part of adaption to stress. Technical support for a gene screening approach to study stress responses by intertidal and estuarine species has increased massively in recent years. For example, continuing additions to the National Center for Biotechnology Information (NCBI) nucleotide and protein databases allow researchers to assemble multispecies sequence comparisons. Multiple genome sequencing projects are now complete or in progress for a number of invertebrate species that live in estuarine, intertidal, or shallow coastal waters (see www.genamics.com). Among these are the starlet sea anemone (*Nematostella vectensis*), an estuarine/brackish inhabitant that is often used as a sentinel species (Sullivan et al. 2008), the purple sea urchin (*Strongylocentrotus purpuratus*), the California sea hare (*Aplysia californica*), the mussel *Mytilus californianus*, and the Atlantic horseshoe crab (*Limulus polyphemus*). In addition, expressed sequence tag (EST) libraries are available for a variety of species including (among others) the bay scallop (*Argopecten irradians*), oysters (*Crassostrea virginica*, *C. gigas*), the shore crab (*Carcinus maenas*), mussels (*Mytilus edulis*, *M. galloprovincialis*) and the tunicate sea squirt (*Ciona intestinalis*), a shallow-water sessile species (Tanguy et al. 2008; Fleury et al. 2009; Nava et al. 2009; Venier et al. 2009). The mitochondrial genomes of numerous species have also been sequenced. Reitzel et al. (2008) used the sequenced genome of *N. vectensis* and the predicted protein sequences available in Stellabase in an interesting way to learn about the potential for this estuarine sea anemone to respond to environmental, pathogenic, and wounding stresses. By searching the database for known motifs associated with stress-responsive proteins, the authors were able to show the presence in sea anemones of multiple proteins/enzymes involved in responses to oxidative, oxygen, and metal stresses.

Current technology offers a variety of approaches to identify gene expression responses to oxidative stresses in marine and estuarine environments. At a "basic" level, researchers can use the polymerase chain reaction (PCR) to amplify and quantify mRNA transcript levels for selected genes. This can begin with primer design based on consensus sequences, leading to amplification of the species-specific sequence, possible redesign to produced species-specific primers, and then RT-PCR or Q-PCR to quantify stress-responsive changes in mRNA expression. For example, such an approach was used by Jo et al. (2008) to amplify transcripts from total RNA of oyster tissues (*Crassostra gigas*), and assess time and dose effects of cadmium on SOD, CAT, and GPX mRNA levels. Nava et al. (2009) took a different approach and used consensus sequences derived from the NCBI to identify 12 genes involved in GSH metabolism from the sequenced genome of the tunicate *C. intestinalis*. After confirming the identity and documenting substantial homology of these tunicate genes with those of mammals, PCR was used to assess gene expression in three tissues and the responses to tert-butylhydroquinone (a prooxidant) by genes involved in multiple aspects of GSH metabolism.

cDNA library screening is another approach. Commercial kits for cDNA library construction and screening are now quite easy to use, but the method itself still has a variety of limitations. Larade and Storey (2008) provide a detailed description of the use of this method for differential screening of cDNA libraries from hepatopancreas of aerobic versus anoxic *L. littorea* as well as the follow-up methods that can be used to characterize clones encoding novel proteins. Indeed, the ability to detect novel, stress-responsive genes (and their protein products) that are potentially unique to a given species is the critical advantage of this method. For example, differential screening of *L. littorea* hepatopancreas libraries comparing aerobic versus 24 h anoxic states identified the ferritin H chain as anoxia-responsive (Larade and Storey 2004) but also multiple novel genes including *kvn* that encoded a ferredoxin-like protein (Larade and Storey 2002b). A similar technology, suppression subtractive hybridization, was applied to assess the effects of thermal stress on oysters, identifying both up- and down-regulated genes in gill and mantle of oysters exposed to 25°C seawater (4°C higher than normal summer exposures) compared with 13°C controls (Meistertzheim et al. 2007). Polymerase chain reaction confirmed strong up-regulation of a number of transcripts in response to heat stress including Se-independent GPx, HSP70, and HSP23.

DNA arrays are now the most popular and widely used method of gene screening and a number of examples of their use with marine species are available for either heterologous (array constructed from another species) or homologous screening. With appropriate attention to optimization of hybridization conditions as well as controls to detect both false positives and false negatives, heterologous screening can be highly cost-effective for studying stress-responsive gene expression with animals that are not among the mainstream model species (Eddy and Storey 2008). This is an especially useful option now that commercial arrays are available for an increasing number of invertebrates including *Drosophila melanogaster* and *Caenorhabditis elegans*. Since many genes contain domains and motifs that are highly conserved across phylogeny, impressive results can be obtained from heterologous screening and these provide "lead generation" that can be followed up with PCR or immunoblotting. For example, a human 19,000 gene array was used to screen for anoxia-responsive genes in *L. littorea* hepatopancreas. Although cross-reactivity was low (only 18.4%) the screening still revealed over 300 genes that were putatively up-regulated under anoxia (Larade and Storey 2009). These genes fell into several classes including antioxidant enzymes and iron-binding proteins.

Despite the time and expense involved in their production, species-specific arrays are becoming more common and are effective tools for studying stress-responsive gene expression. Several arrays are now available for commercially important shellfish, increasing the options for both homologous and heterologous screening of intertidal and estuarine species. These include a low-density oligonucleotide microarray for *Mytilus* (Dondero et al. 2006), oyster and mussel cDNA microarrays (Venier et al. 2006; Jenny et al. 2007), and oligo arrays for the intertidal copepod *T. japonicus* (Lee and Raisuddin 2008; Ki et al. 2009). Brouwer et al. (2004) used homologous microarrays to screen for effects of hypoxia on the expression of 10 genes, including Mn-SOD and hemocyanin in blue crabs, *Callinectes sapidus*. In addition, a functional genomics initiative known as the Marine Genomics project (www.marinegenomics.org) has prepared and analyzed microarray data for 28 species databases to date (McKillen et al. 2005). By enabling cross-species data comparison and data mapping, users can refine and share their research within a marine-focused environment.

Major goals of research on gene/protein expression in intertidal and estuarine organisms are often the development of effective biomarkers of the health of populations and/or for toxicity testing to assess effects of marine pollutants including heavy metals, endocrine disrupting chemicals (EDCs) and others. Several recent studies have effectively employed gene screening of antioxidant and related enzymes for these purposes.

Green et al. (2009) selectively bred Sydney rock oysters (*Saccostrea glomerata*) for high disease resistance (R) and compared them with wild-caught controls (W) when both were exposed under field conditions to infection with disseminated neoplasia. The authors used suppression subtractive hybridization to retrieve 183 sequences from hemocytes that were differentially expressed between R and W oysters. Subsequent analysis via qRT-PCR showed, among other changes, that expression of extracellular SOD was increased in R oysters, whereas Prx 6 was reduced. Gene screening is also effective for toxicity evaluation and testing. For example, studies with the intertidal copepod *T. japonicus* assessed the expression of several antioxidant genes to different stressors including H_2O_2, metals, and EDCs using a 12K oligochip and identified GST-σ as a potential biomarker gene for oxidative stress and trace metal contamination (Lee and Raisuddin 2008). Ki et al. (2009) used a 6K oligochip for *T. japonicus* to demonstrate that copper exposure altered the expression of multiple genes including those involved in antioxidant functions. Various other studies have focused on Cd since it is one of the most toxic of heavy metals that pollute marine ecosystems and it can induce oxidative stress. Jo et al. (2008) used RT-PCR to demonstrate increases in SOD, CAT, and GPx mRNA levels in response to Cd exposure in gills of oysters (*C. gigas*). For example, expression of SOD at 0.1 ppm Cd were 90-fold greater after 7 days exposure as compared with controls. Metallothionein and Hsp 90 mRNA levels also soared in these animals, with MT expression increasing over 100-fold in gill and digestive gland while Hsp 90 increased 15–35 fold (Choi et al. 2008b). These data and others support the suggestion that MT would be a useful biomarker of heavy metal stress in marine environments (Boutet et al. 2002). However, several lines of evidence argue against the use of MT as a biomarker. First, the gene is also strongly up-regulated in response to bacterial challenge (Fasulo et al. 2008). Second, studies with *M. galloprovincialis* suggested that the many cysteine residues of MT (typically ~30% of total amino acids) may give the protein an antioxidant function that is perhaps more important than its metal binding role (Viarengo et al. 1999). Finally, MT was up-regulated in *L. littorea* in response to both anoxia and freezing (English and Storey 2003). Since periodic hypoxia/anoxia is very common in intertidal and estuarine environments as a consequence of the tidal cycle, as is freeze/thaw for various temperate and polar species during the winter months, the effects of these environment factors on MT could easily confound any attempt at correlating MT levels with heavy metal challenge.

CONCLUSION

In conclusion, the investigation of oxidative stress in estuarine and intertidal species has been increasingly addressed in recent years, with studies ranging from the eco-physiological/toxicological to the cellular/genomic/proteomic levels. These studies are not only relevant for developing a full appreciation and understanding of how coastal life works, but also illuminate the need for in-depth analysis of the impact of temperature, salt, and hypoxia on oxidative stress processes in vertebrate and invertebrate species. The study of how coastal species tolerate hypersalinity and/or hypoxia will certainly have biotechnological impact for aquaculture of commercial species and could illuminate fundamental principles that reach all the way to applications such as the treatment of cardiovascular disorders in humans, including hypertension and heart/brain ischemia.

Moreover, given the central role played by the endogenous antioxidant system of organisms that dwell in the especially challenging intertidal and estuarine habitats, such studies are highly valuable tools in the evaluation of potential impacts of climate change on coastal systems at all latitudes (Downs et al. 2002; Lesser 2006; Gracey et al. 2008; Helmuth 2009). This investigation is urgent, given that these coastal systems are already endangered by intensively destructive human action, through widespread chemical contamination and habitat destruction.

REFERENCES

Abele, D., Groβpietsch, H., Pörtner H.O. (1998a) Temporal fluctuations and spatial gradients of environmental P_{O_2}, temperature, H_2O_2 and H_2S in its intertidal habitat trigger enzymatic antioxidant protection in the capitellid worm *Heteromastus filiformis*. *Marine Ecology Progress Series* 163, 179–191.

Abele, D., Burlando, B., Viarengo, A., Pörtner, H.O. (1998b) Exposure to elevated temperatures and hydrogen peroxide elicits oxidative stress and antioxidant response in the

Antarctic intertidal limpet *Nacella concinna*. *Comparative Biochemistry and Physiology B* 120, 425–435.

Abele, D., Heise, K., Pörtner, H.O., Puntarulo, S. (2002) Temperature-dependence of mitochondrial function and production of reactive oxygen species in the intertidal mud clam *Mya arenaria*. *Journal of Experimental Biology* 205, 1831–1841.

Abele-Oeschger, D., Oeschger, R., Theede, H. (1994). Biochemical adaptations of *Nereis diversicolor* (Polychaeta) to temporarily increased hydrogen peroxide levels in intertidal sandflats. *Marine Ecology Progress Series* 106, 101–110.

Almeida, E.A., Bainy, A.C.D. (2006) Effects of aerial exposure on antioxidant defenses in the brown mussel *Perna perna*. *Brazilian Archives of Biology and Technology* 49, 225–229.

Almeida, E.A., Bainy, A.C.D., Dafre, A.L., Gomes, O.F., Medeiros, M.H.G., Di Mascio, P. (2005) Oxidative stress in digestive gland and gill of the brown mussel (*Perna perna*) exposed to air and re-submersed. *Journal of Experimental Marine Biology and Ecology* 318, 21–30.

An, M.I., Choi, C.Y. (2010) Activity of antioxidant enzymes and physiological responses in ark shell, *Scapharca broughtonii*, exposed to thermal and osmotic stress: effects on hemolymph and biochemical parameters. *Comparative Biochemistry Physiology B* 155, 34–42.

An, K.W., Kim, N.N., Shin, H.S., Kil, G.S., Choi, CY. (2010) Profiles of antioxidant gene expression and physiological changes by thermal and hypoosmotic stresses in black porgy (*Acanthopagrus schlegeli*). *Comparative Biochemistry Physiology A* 156, 262–268.

Biggar, K., Dubuc, A., Storey, K.B. (2009) MicroRNA regulation below zero: Differential expression of miRNA–21 and miRNA–16 during freezing in wood frogs. *Cryobiology* 59, 317–321.

Biggar, K.K., Kornfield, S., Storey, K.B. (2011) Amplification and sequencing of mature microRNAs in uncharacterized animal models using stem-loop RT-PCR. *Analytical Biochemistry* in press. DOI: 10.1016/j.ab.2011.05.015.

Boutet, I., Tanguy, A., Auffret, M., Riso, R., Moraga, D. (2002) Immunochemical quantification of metallothioneins in marine molluscs: characterization of a metal exposure bioindicator. *Environmental Toxicology and Chemistry* 21, 1009–1014.

Brooks, S.P.J., Storey, K.B. (1997) Glycolytic controls in estivation and anoxia: a comparison of metabolic arrest in land and marine molluscs. *Comparative Biochemistry and Physiology A* 118, 1103–1114.

Brookes, P.S. (2005) Mitochondrial H(+) leak and ROS generation: an odd couple. *Free Radical Biology and Medicine* 38, 12–23.

Brouwer, M., Larkin, P., Brown-Peterson, N., King, C., Manning, S., Denslow, N. (2004) Effects of hypoxia on gene and protein expression in the blue crab, *Callinectes sapidus*. *Marine Environmental Research* 58, 787–792.

Budde, M.W, Roth, M.B. (2010) Hydrogen sulfide increases hypoxia-inducible factor-1 activity independently of von Hippel–Lindau tumor suppressor-1 in *C. elegans*. *Molecular Biology of the Cell* 21, 212–217.

Burg, M.B, Ferraris, J.D., Dmitrieva, N.,I. (2007) Cellular response to hyperosmotic stresses. *Physiological Reviews* 87, 1441–1474.

Cailleaud, K., Maillet, G., Budzinski, H., Souissi, S., Forget-Leray, J. (2007) Effects of salinity and temperature on the expression of enzymatic biomarkers in *Eurytemora affinis* (Calanoida, Copepoda). *Comparative Biochemistry and Physiology A* 147, 841–849.

Cao, C., Leng, Y., Kufe, D. (2003a) Catalase activity is regulated by c-Abl and Arg in the oxidative stress response. *Journal of Biological Chemistry* 278, 29667–29675.

Cao, C., Leng, Y., Huang, W., Liu, X., Kufe, D. (2003b) Glutathione peroxidase 1 is regulated by the c-Abl and Arg tyrosine kinases. *Journal of Biological Chemistry* 278, 39609–39614.

Carlström, M., Brown, R.D., Sällström, J., et al. (2009) SOD1 deficiency causes salt sensitivity and aggravates hypertension in hydronephrosis. *American Journal of Physiology* 297, R82–R92.

Choi, C.Y., An, K.W., An, M.I. (2008a) Molecular characterization and mRNA expression of glutathione peroxidase and glutathione S-transferase during osmotic stress in olive flounder (*Paralichthys olivaceus*). *Comparative Biochemistry and Physiology A* 149, 330–337.

Choi, Y.K., Jo, P.G., Choi, C.Y. (2008b) Cadmium affects the expression of heat shock protein 90 and metallothionein mRNA in the Pacific oyster, *Crassostrea gigas*. *Comparative Biochemistry and Physiology C* 147, 286–292.

Collén, J., Davison, I.R. (1999) Reactive oxygen metabolism in intertidal *Fucus* spp. (Phaeophyceae). *Journal of Phycology* 35, 62–69.

Cooper, R.U., Clough, L.M., Farwell, M.A., West, T.L. (2002) Hypoxia-induced metabolic and antioxidat enzymatic activities in the estuarine fish *Leiostomus xanthurus*. *Journal of Experimental Marine Biology and Ecology* 279, 1–20.

Cowan, K.J., Storey, K.B. (2003) Mitogen-activated protein kinases: new signaling pathways functioning in cellular responses to environmental stress. *Journal of Experimental Biology*, 206, 1107–1115.

David, E., Tanguy, A., Pichavant, K., Moraga, D. (2005) Response of the Pacific oyster *Crassostrea gigas* to hypoxia exposure under experimental conditions. *FEBS Journal* 272, 5635–5652.

Dondero, F., Piacentini, L., Marsano, F., Rebelo, M., Vergani, L., Venier, P. (2006) Gene transcription profiling in pollutant exposed mussels (*Mytilus* spp.) using a new low-density oligonucleotide microarray. *Gene* 376, 24–36.

Downs, C.A., Fauth, J.E., Halas, J.C., Dustan, P., Bemiss, J., Woodley, C.M. (2002) Oxidative stress and seasonal coral bleaching. *Free Radical Biology and Medicine* 33, 533–543.

Eddy, S.F., Storey, K.B. (2008) Comparative molecular physiological genomics: heterologous probing of cDNA arrays. *Methods in Molecular Biology* 410, 81–110.

English, T.E., Storey, K.B. (2003) Freezing and anoxia stresses induce expression of metallothionein in the foot muscle and hepatopancreas of the marine gastropod, *Littorina littorea*. *Journal of Experimental Biology* 206, 2517–2524.

Fasulo, S., Mauceri, A., Giannetto, A., Maisano, M., Bianchi, N., Parrino, V. (2008) Expression of metallothionein mRNAs by in situ hybridization in the gills of *Mytilus galloprovincialis* from natural polluted environments. *Aquatic Toxicology* 88, 62–68.

Ferraris, J.D., Williams, C.K., Ohtaka, A., García-Pérez, A. (1999) Functional consensus for mammalian osmotic response elements. *American Journal of Physiology* 276, C667–C673.

Ferreira, M., Moradas-Ferreira, P., Reis-Henriques, M.A. (2005) Oxidative stress biomarkers in two resident species, mullet (*Mugil cephalus*) and flounder (*Platichthys flesus*), from a polluted site in River Douro Estuary, Portugal. *Aquatic Toxicology* 71, 39–48.

Ferreira, M.V.R., Alencastro, A.C.R., Hermes-Lima, M. (2003) Role of antioxidant defenses during estivation and anoxia exposure in the freshwater snail *Biomphalaria tenagophila* (Orbigny, 1835). *Canadian Journal of Zoology* 81, 1239–1248.

Fillebeen, C., Pantopoulos, K. (2002) Redox control of iron regulatory proteins. *Redox Report* 7, 15–22.

Fleury, E., Huvet, A., Lelong, C. et al. (2009). Generation and analysis of a 29,745 unique expressed sequence tags from the Pacific oyster (*Crassostrea gigas*) assembled into a publicly accessible database: the GigasDatabase. *BMC Genomics* 10, 341.

Flynt AS, Thatcher EJ, Burkewitz K, Li N, Liu Y, Patton JG. (2009) miR-8 microRNAs regulate the response to osmotic stress in zebrafish embryos. *Journal of Cell Biology* 185, 115–127.

Gardeström, J., Elfwing, T., Löf, M., Tedengren, M., Davenport, J.L., Davenport, J. (2007) The effect of thermal stress on protein composition in dogwhelks (*Nucella lapillus*) under normoxic and hyperoxic conditions. *Comparative Biochemistry and Physiology A* 148, 869–875.

Geracitano, L.A., Monserrat, J.M., Bianchini, A. (2004) Oxidative stress in *Laeonereis acuta* (Polychaeta, Nereididae): environmental and seasonal effects. *Marine Environmental Research* 58, 625–630.

Gracey, A.Y., Chaney, M.L., Boomhower, J.P., Tyburczy, W.R., Connor, K., Somero, G.N. (2008) Rhythms of gene expression in a fluctuating intertidal environment. *Current Biology* 18, 1501–1507.

Green, T.J., Dixon, T.J., Devic, E., Adlard, R.D., Barnes, A.C. (2009) Differential expression of genes encoding anti-oxidant enzymes in Sydney rock oysters, *Saccostrea glomerata* (Gould) selected for disease resistance. *Fish and Shellfish Immunology* 26, 799–810.

Halpin, P.M., Menge, B.A., Hofmann, G.E. (2004) Experimental demonstration of plasticity in the heat shock response of the intertidal mussel *Mytilus californianus*. *Marine Ecology Progress Series* 276, 137–145.

Hansen, B.H., Garmo, O.A., Olsvik, P.A., Andersen, R.A. (2007) Gill metal binding and stress gene transcription in brown trout (*Salmo trutta*) exposed to metal environments: the effect of pre-exposure in natural populations. *Environmental Toxicology and Chemistry* 26, 944–953.

Heise, K., Puntarulo, S., Pörtner, H.O., Abele, D. (2003) Production of reactive oxygen species by isolated mitochondria of the Antarctic bivalve *Laternula elliptica* (King and Broderip) under heat stress. *Comparative Biochemistry and Physiology C* 134, 79–90.

Heise, K., Puntarulo, S., Nikinmaa, M., Lucassen, M., Pörtner, H., Abele, D. (2006) Oxidative stress and HIF-1 DNA binding during stressful cold exposure and recovery in the North Sea eelpout (*Zoarces viviparus*). *Comparative Biochemistry and Physiology A* 143, 494–503.

Helmuth, B. (2009) From cells to coastlines: how can we use physiology to forecast the impacts of climate change? *Journal of Experimental Biology* 212, 753–760.

Henkel, S.K., Kawai, H., Hofmann, G.E. (2009) Interspecific and interhabitat variation in hsp70 gene expression in native and invasive kelp populations. *Marine Ecology Progress Series* 386, 1–13.

Hermes-Lima, M. (2004) Oxygen in biology and biochemistry: role of free radicals. *In* Storey, K.B. (ed.). *Functional Metabolism: Regulation and Adaptation*. John Wiley & Sons, New York, pp. 319–368.

Hermes-Lima, M., Zenteno-Savín, T. (2002) Animal response to drastic changes in oxygen availability and physiological oxidative stress. *Comparative Biochemistry and Physiology C* 133, 537–556.

Jenny, M.J., Chapman, R.W., Mancia, A., et al. (2007) A cDNA microarray for *Crassostrea virginica* and *C. gigas*. *Marine Biotechnology* 9, 577–591.

Jo, P.G., Choi, Y.K., Choi, C.Y. (2008) Cloning and mRNA expression of antioxidant enzymes in the Pacific oyster, *Crassostrea gigas* in response to cadmium exposure. *Comparative Biochemistry and Physiology C* 147, 460–469.

Johnston, I.A., Lee, H.T., Macqueen, D.J., et al. (2009) Embryonic temperature affects muscle fibre recruitment in adult zebrafish: genome-wide changes in gene and microRNA expression associated with the transition from hyperplastic to hypertrophic growth phenotypes. *Journal of Experimental Biology* 212, 1781–1793.

Kensler, T.W., Wakabayashi, N., Biswal, S. (2007) Cell survival responses to environmental stresses via the Keap1-Nrf2-ARE pathway. *Annual Review of Pharmacology and Toxicology* 47, 89–116.

Ki, J.S., Raisuddin, S., Lee, K.W., et al. (2009) Gene expression profiling of copper-induced responses in the intertidal copepod *Tigriopus japonicus* using a 6K oligochip microarray. *Aquatic Toxicology* 93, 177–187.

Kops, G.J., Dansen, T.B., Polderman, P.E., et al. (2002) Forkhead transcription factor FOXO3a protects quiescent cells from oxidative stress. *Nature* 419, 316–321.

Kültz, D., Ávila, K. (2001) Mitogen-activated protein kinases are *in vivo* transducers of osmosensory signals in fish gill cells. *Comparative Biochemistry and Physiology B* 129, 821–829.

Kültz, D., Chakravarty, D. (2001) Maintenance of genomic integrity in mammalian kidney cells exposed to hyperosmotic stress. *Comparative Biochemistry and Physiology A* 130, 421–428.

Kwon, M.S., Lim, S.W., Kwon, H.M. (2009) Hypertonic stress in the kidney: a necessary evil. *Physiology* 24, 186–191.

Larade, K., Storey, K.B. (2002a) Reversible suppression of protein synthesis in concert with polysome disaggregation during anoxia exposure in *Littorina littorea*. *Molecular and Cellular Biochemistry* 232, 121–127.

Larade, K., Storey, K.B. (2002b) Characterization of a novel gene up-regulated during anoxia exposure in the marine snail *Littorina littorea*. *Gene* 283, 145–154.

Larade, K., Storey, K.B. (2004) Accumulation and translation of ferritin heavy chain transcripts following anoxia exposure in a marine invertebrate. *Journal of Experimental Biology* 207, 1353–1360.

Larade, K.F., Storey, K.B. (2008) Constructing and screening a cDNA library: methods for identification and characterization of novel genes. *Methods in Molecular Biology* 410, 55–80.

Larade, K., Storey, K.B. (2009) Living without oxygen: anoxia-responsive gene expression and regulation. *Current Genomics* 10, 76–85.

Lau, P.S., Wong, H.L., Garrigues, Ph. (2004) Seasonal variation in antioxidative responses and acetylcholinesterase activity in *Perna viridis* in eastern oceanic and western estuarine waters of Hong Kong. *Continental Shelf Research* 24, 1969–1987.

Lee, J.-S., Raisuddin, S. (2008) Copepod modulation of expression of oxidative stress genes of the intertidal copepod *Tigriopus japonicus* after exposure to environmental chemicals. *In* Murakami, Y., Nakayama, K., Kitamura, S.-I., Iwata, H., Tanabe, S. (eds). *Interdisciplinary Studies on Environmental Chemistry – Biological Responses to Chemical Pollutants*. TERRAPUB, Tokyo, pp. 95–105.

Lesser, M.P. (2006) Oxidative stress in marine environments: biochemistry and physiological ecology. *Annual Review of Physiology* 68, 253–278.

Letendre, J., Chouquet, B., Rocher, B., Manduzio, H., Leboulenger, F., Durand, F. (2008) Differential pattern of Cu/Zn superoxide dismutase isoforms in relation to tidal spatio-temporal changes in the blue mussel *Mytilus edulis*. *Comparative Biochemistry and Physiology C* 148, 211–216.

Liu, Y., Wang, W.-N., Wang, A.-L., Wang, J.-M., Sun, R.-Y. (2007) Effects of dietary vitamin E supplementation on antioxidant enzyme activities in *Litopenaeus vannamei* (Boone, 1931) exposed to acute salinity changes. *Aquaculture* 265, 351–358.

Lushchak, V.I., Bagnyukova, T.V. (2007) Hypoxia induces oxidative stress in tissues of a goby, the rotan *Perccottus glenii*. *Comparative Biochemistry and Physiology B* 148, 390–397.

Lushchak, V.I., Lushchak, L.P., Mota, A.A., Hermes-Lima, M. (2001) Oxidative stress and antioxidant defenses in goldfish *Carassius auratus* during anoxia and reoxygenation. *American Journal of Physiology* 280, R100–R107.

Lushchak, V.I., Bagnyukova, T.V., Lushchak, O.V., Storey, J.M., Storey, K.B. (2005) Hypoxia and recovery perturb free radical processes and antioxidant potential in common carp (*Cyprinus carpio*) tissues. *International Journal of Biochemistry and Cell Biology* 37, 1319–1330.

Malanga, G., Estevez, M.S., Calvo, J., Puntarulo, S. (2004) Oxidative stress in limpets exposed to different environmental conditions in the Beagle Channel. *Aquatic Toxicology* 69, 299–309.

Manduzio, H., Monsinjon, T., Galap, C., Leboulenger, F., Rocher, B. (2004) Seasonal variations in antioxidant defences in blue mussels *Mytilus edulis* collected from a polluted area: major contributions in gills of an inducible isoform of Cu/Zn-superoxide dismutase and of glutathione S-transferase. *Aquatic Toxicology* 70, 83–93.

Martínez-álvarez, R.M., Hidalgo, M.C., Domezain, A., Morales, A.E., García-Gallego, M., Sanz, A. (2002) Physiological changes of sturgeon *Acipenser naccarii* caused by increasing environmental salinity. *Journal of Experimental Biology* 205, 3699–3706.

Martins, C.M.G., Bianchini, A. (2009) Metallothionein-like proteins in the blue crab *Callinectes sapidus*: effect of water salinity and ions. *Comparative Biochemistry and Physiology* 152A, 366–371.

McKillen, D.J., Chen, Y.A., Chen, C., Jenny, M.J., Trent, H.F. (2005). Marine genomics: a clearing-house for genomic and transcriptomic data of marine organisms. *BMC Genomics* 6, 34.

Meistertzheim, A.L., Tanguy, A., Moraga, D., Thebault, M.T. (2007) Identification of differentially expressed genes of the Pacific oyster *Crassostrea gigas* exposed to prolonged thermal stress. *FEBS Journal* 274, 6392–6402.

Mitchell, K.A., Elferink, C.J. (2009) Timing is everything: consequences of transient and sustained AhR activity. *Biochemical Pharmacology* 77, 947–956.

Morin, P., Dubuc, A., Storey, K.B. (2008). Differential expression of microRNA species in organs of hibernating ground squirrels: a role in translational suppression during torpor. *Biochimica et Biophysica Acta* 1779, 628–633.

Muckenthaler, M.U., Galy, B., Hentze, M.W. (2008) Systemic iron homeostasis and the iron-responsive element/iron-regulatory protein (IRE/IRP) regulatory network. *Annual Review of Nutrition* 28, 197–213.

Murphy, M.P. (2009) How mitochondria produce reactive oxygen species. *Biochemical Journal* 417, 1–13.

Mustafa, A.K., Gadalla, M.M., Sen, N., et al. (2009) H_2S signals through protein S-sulfhydration. *Science Signaling* 2, ra72.

Nakano, K., Iwama, G.K., 2002. The 70-kDa heat shock protein response in two intertidal sculpins, *Oligocottus maculosus* and *O. snyderi*: relationship of hsp70 and thermal tolerance. *Comparative Biochemistry and Physiology* 133A, 79–94.

Nava, G.M., Lee, D.Y., Ospina, J.H., Cai, S.-Y., Gaskins, H.R. (2009) Genomic analyses reveal a conserved glutathione homeostasis pathway in the invertebrate chordate *Ciona intestinalis*. *Physiological Genomics* 39, 183–194.

Nilsson, G.E., Renshaw, G.M.C. (2004) Hypoxic survival strategies in two fishes: extreme anoxia tolerance in the North European crucian carp and natural hypoxic preconditioning in a coral-reef shark. *Journal of Experimental Biology* 207, 3131–3139.

Niyogi, S., Biswas, S., Sarker, S., Datta, A.G. (2001) Antioxidant enzymes in brackish water oyster, *Saccostrea cucullata* as potential biomarkers of polyaromatic hydrocarbon pollution in Hooghly Estuary (India): seasonality and its consequences. *Science of the Total Environment* 281, 237–246.

Nybakken, J.W. (1988) *Marine Biology: An Ecological Approach*. Harper-Collins, New York.

Oliveira, U.O., Araújo, A.S.R., Belloó-Klein, A., da Silva, R.S.M., Kucharski, L.C. (2005). Effects of environmental anoxia and different periods of reoxygenation on oxidative balance in gills of the estuarine crab *Chasmagnathus granulata*. *Comparative Biochemistry and Physiology B* 140, 51–57.

Orr, A.L., Lohse, L.A., Drew, K.L., Hermes-Lima, M. (2009) Physiological oxidative stress after arousal from hibernation in Arctic ground squirrel. *Comparative Biochemistry and Physiology A* 153, 213–221.

Pahl, H.L. (1999) Activators and target genes of Rel/NF-kappaB transcription factors. *Oncogene* 18, 6853–6866.

Paital, B., Chainy, G.B.N. (2010) Antioxidant defenses and oxidative stress parameters in tissues of mud crab (*Scylla serrata*) with reference to changing salinity. *Comparative Biochemistry and Physiology C* 151, 142–151.

Pannunzio, T.M., Storey, K.B. (1998) Antioxidant defenses and lipid peroxidation during anoxia stress and aerobic recovery in the marine gastropod *Littorina littorea*. *Journal of Experimental Marine Biology and Ecology* 221, 277–292.

Pritchard, J.B. (2002) Comparative models and biological stress. *American Journal of Physiology* 283, R807–R809.

Pytharopoulou, S., Kouvela, E.C., Sazakli, E., Leotsinidis, M., Kalpaxis, D.L. (2006) Evaluation of the global protein synthesis in *Mytilus galloprovincialis* in marine pollution monitoring: seasonal variability and correlations with other biomarkers. *Aquatic Toxicology* 80, 33–41.

Ramnanan, C.J., Storey, K.B. (2006) Glucose–6-phosphate dehydrogenase regulation during hypometabolism. *Biochemistry and Biophysics Research Communications* 339, 7–16.

Reitzel, A.M., Sullivan, J.C., Traylor-Knowles, N. Finnerty, J.R. (2008) Genomic survey of candidate stress-response genes in the estuarine anemone *Nematostella vectensis*. *Biological Bulletin* 214, 233–254.

Rhee, S.G., Yang, K.-S., Kang, S.W., Woo, H.A., Chang, T.-S. (2005) Controlled elimination of intracellular H_2O_2: regulation of peroxiredoxin, catalase, and glutathione peroxidase via post-translational modification. *Antioxidants and Redox Signaling* 7, 619–626.

Romero, M.C., Ansaldo, M., Lovrich, G.A. (2007) Effect of aerial exposure on the antioxidant status in the subantarctic stone crab *Paralomis granulosa* (Decapoda: Anomura). *Comparative Biochemistry and Physiology C* 146, 54–59.

Roncarati, F., Rijstenbil, J.W., Pistocchi, R. (2008) Photosynthetic performance, oxidative damage and antioxidants in *Cylindrotheca closterium* in response to high irradiance, UVB radiation and salinity. *Marine Biology* 153, 965–973.

Ross, S.W., Dalton, D.A., Kramer, S., Christensen, B.L. (2001) Physiological (antioxidant) responses of estuarine fishes to variability in dissolved oxygen. *Comparative Biochemistry and Physiology C* 130, 289–303.

Sagasti, A., Schaffner, L.C., Duffy, J.E. (2001) Effects of periodic hypoxia on mortality, feeding and predation in an estuarine epifaunal community. *Journal of Experimental Marine Biology and Ecology* 258, 257–283.

Salahudeen, A.A., Bruick, R.K. (2009) Maintaining mammalian iron and oxygen homeostasis: sensors, regulation, and cross-talk. *Annals of the New York Academy of Sciences* 1177, 30–38.

Sanders, B.M., Hope, C., Pascoe, V.M., Martin, L.S. (1991) Characterization of the stress protein response in two species of *Colisella* limpets with different temperature tolerances. *Physiological Zoology* 64, 1471–1489.

Schill, R.O., Gayle, P.M.H., Köhler, H.-R. (2002) Daily stress protein (hsp70) cycle in chitons (*Acanthopleura granulata* Gmelin, 1791) which inhabit the rocky intertidal shoreline in a tropical ecosystem. *Comparative Biochemistry and Physiology C* 131, 253–258.

Schneider, B.D., Leibold, E.A. (2003) Effects of iron regulatory protein regulation on iron homeostasis during hypoxia. *Blood* 102, 3404–3411.

Scholz, H. (2002) Adaptational responses to hypoxia. *American Journal of Physiology* 282, R1541–R1543.

Seo, J.S., Lee, K.-W., Rhee, J.-S., Hwang, D.-S., Lee, Y.-M., Park, H.G., Ahn, I.-Y., Lee, J.-S. (2006) Environmental stressors (salinity, heavy metals, H_2O_2) modulate expression of glutathione reductase (GR) gene from the intertidal copepod *Tigriopus japonicus*. *Aquatic Toxicology* 80, 281–289.

Storey, K.B. (2008) Beyond gene chips: transcription factor profiling in freeze tolerance. *In* Lovegrove, B.G., McKechnie, A.E. (eds). *Hypometabolism in Animals: Hibernation, Torpor and Cryobiology*. University of KwaZulu-Natal, Pietermaritzburg, pp. 101–108.

Storey, K.B., Storey, J.M. (2004) Metabolic rate depression in animals: transcriptional and translational controls.

Biological Reviews of the Cambridge Philosophical Society 79, 207–233.

Storey, K.B., Storey, J.M. (2007) Putting life on 'pause' – molecular regulation of hypometabolism. *Journal of Experimental Biology* 210, 1700–1714.

Sullivan, J.C., Reitzel, A.M., Finnerty, J.R. (2008) Upgrades to *StellaBase* facilitate medical and genetic studies on the starlet sea anemone, *Nematostella vectensis*. *Nucleic Acids Research* 36, D607–D611.

Tanguy, A., Bierne, N., Saavedra, C., et al. (2008) Increasing genomic information in bivalves through new EST collections in four species: development of new genetic markers for environmental studies and genome evolution. *Gene* 408, 27–36.

Taylor, C.T., Cummins, E.P. (2009) The role of NF-κB in hypoxia-induced gene expression. *Annals of the New York Academy of Sciences* 1177, 178–184.

Togni, V.G. (2007) *Efeito da salinidade sobre a resposta do sistema antioxidante e expressão de hsp70 em siris (gênero Callinectes)*. PhD thesis, Universidade Federal do Paraná, Brazil.

Tomanek, L., Sanford, E. (2003) Heat-shock protein 70 (Hsp70) as a biochemical stress indicator: an experimental field test in two congeneric intertidal gastropods (genus: *Tegula*). *Biological Bulletin* 205, 276–284.

Tomanek, L., Somero, G.N. (2002) Interspecific- and acclimation-induced variation in levels of heat-shock proteins 70 (hsp70) and 90 (hsp90) and heat-shock transcription factor-1 (HSF1) in congeneric marine snails (genus *Tegula*): implications for regulation of *hsp* gene expression. *Journal of Experimental Biology* 205, 677–685.

Van der Horst, A., Burgering, B.M.T. (2007) Stressing the role of FoxO proteins in lifespan and disease. *Nature Reviews Molecular and Cell Biology* 8, 440–450.

Venier, P., De Pitta, C., Pallavicini, A., et al. (2006) Development of mussel mRNA profiling: can gene expression trends reveal coastal water pollution? *Mutation Research* 602, 121–134.

Venier, P., De Pittà, C., Bernante, F., et al. (2009) MytiBase: a knowledgebase of mussel (*M. galloprovincialis*) transcribed sequences. *BMC Genomics* 10, 72.

Vetterlein, F., Schrader, C., Volkmann, R., et al. (2003) Extent of damage in ischemic, nonreperfused, and reperfused myocardium of anesthetized rats. *American Journal of Physiology* 285, H755–H765.

Viarengo, A., Burlando, B., Cavaletto, M., Marchi, B., Ponzano, E., Blasco, J. (1999) Role of metallothionein against oxidative stress in the mussel *Mytilus galloprovincialis*. *American Journal of Physiology* 277, R1612–R1619.

Vig, E., Nemcsok, J. (1989) The effects of hypoxia and paraquat on the superoxide-dismutase activity in different organs of carp, *Cyprinus carpio* L. *Journal of Fish Biology* 35, 23–25.

Walshe, T.E., D'Amore, P.A. (2008) The role of hypoxia in vascular injury and repair. *Annual Review of Pathology* 3, 615–643.

Welker, A.F. (2009) *Efeitos da flutuação da disponibilidade de oxigênio e da privação alimentar sobre o metabolismo de radicais livres*. PhD thesis, Universidade de São Paulo, Brazil.

Wilhelm Filho, D., Torres, M.A., Tribess, T.B., Pedrosa, R.C., Soares, C.H.L. (2001a) Influence of season and pollution on the antioxidant defenses of the cichlid fish acará (*Geophagus brasiliensis*). *Brazilian Journal of Medical and Biological Research* 34, 719–726.

Wilhelm Filho, D., Tribess, T., Gáspari, C., Cláudio, F.D., Torres, M.A., Magalhães, A.R.M. (2001b) Seasonal changes in antioxidant defenses of the digestive gland of the brown mussel (*Perna perna*). *Aquaculture* 203, 149–158.

Willmer, P., Stone, G., Johnston, I. (2005) *Environmental Physiology of Animals*. Blackwell, Oxford.

Wu, Y., Feinstein, S.I., Manevich, Y., et al. (2009) Mitogen-activated protein kinase-mediated phosphorylation of peroxiredoxin 6 regulates its phospholipase A_2 activity. *Biochemical Journal* 419, 669–679.

Yamashita, Y., Tominaga, O., Takami, H., Yamada, H. (2003) Comparison of growth, feeding and cortisol level in *Platichthys bicoloratus* juveniles between estuarine and nearshore nursery grounds. *Journal of Fish Biology* 63, 617–630.

Zaccaron da Silva, A., Zanette, J., Ferreira, J.F., Guzenski, J., Marques, M.R.F., Bainy, A.C.D. (2005) Effects of salinity on biomarker responses in *Crassostrea rhizophorae* (Mollusca, Bivalvia) exposed to diesel oil. *Ecotoxicology and Environmental Safety* 62, 376–382.

Zhang, Z., Dmitrieva, N.I., Park, J.H., Levine, R.L., Burg, M.B. (2004) High urea and NaCl carbonylate proteins in renal cells in culture and *in vivo*, and high urea causes 8-oxoguanine lesions in their DNA. *Proceedings of the National Academy of Sciences USA* 101, 9491–9496.

Zhou, X., Ferraris, J.D., Cai, Q., Agarwal, A., Burg, M.B. (2005) Increased reactive oxygen species contribute to high NaCl-induced activation of the osmoregulatory transcription factor TonEBP/OREBP. *American Journal of Physiology* 289, F377–F385.

Chapter 4

OXIDATIVE STRESS TOLERANCE STRATEGIES OF INTERTIDAL MACROALGAE

José Aguilera[1] *and Ralf Rautenberger*[2,3,4]

[1]Photobiology Laboratory, Medical Research Center, Department of Dermatology, Faculty of Medicine, University of Málaga, Spain
[2]Institute for Polar Ecology, Christian Albrechts University of Kiel, Kiel, Germany
[3]Department of Marine Botany, University of Bremen, Faculty of Biology and Chemistry, Bremen, Germany
[4]Department of Botany, University of Otago, Dunedin, New Zealand

Macroalgae are sessile and, therefore, restricted to their growth sites in coastal ecosystems where they are exposed to considerable changes in environmental conditions. Macroalgae that occur in the upper part of the rocky shore must be exceptionally capable of coping with many forms of abiotic stress (especially with respect to overexposure to solar radiation) that affect algal physiology. Primary production is reduced due to inhibition and destruction of the photosynthetic apparatus, DNA, proteins, and membranes by reactive oxygen species (ROS) formation. In addition, the damaging effects of solar ultraviolet radiation (UVR) are exacerbated by other abiotic stress factors. The environmental tolerance of intertidal macroalgae is mediated by antioxidant enzymatic pathways and ROS-scavenging substances. Ultraviolet-absorbing compounds, dynamic distribution of photosynthetic pigments, transient inhibition of macroalgal photosynthesis and even morphological changes are successful protection strategies.

Benthic macroalgae are exposed to abiotic environmental factors that structure the coastal ecosystem. Their frequent and rapid changes cause different adaptation strategies to maintain algal survival and produce the vertical zonation pattern of benthic algae. Whereas physical stress determines the upper boundaries of macroalgae, biological stress controls their lower boundaries.

Light is the major factor controlling macroalgal zonation and has a strong impact from the highest intertidal zone down into the deep subtidal areas. Since photosynthetically active (PAR) and UVR regimes are high in the intertidal environment,

Oxidative Stress in Aquatic Ecosystems, First Edition. Edited by Doris Abele, José Pablo Vázquez-Medina, and Tania Zenteno-Savín.

macroalgae inhabiting this zone generally tolerate these light regimes better by developing specific adaptation and acclimation mechanisms. In contrast, subtidal algae are more sensitive to light since the PAR and UV irradiances are permanently low. Temperature regimes affecting macroalgae result from tidal fluctuations, wave actions, and air temperatures. Temperature tolerance determines algal distribution and is a factor for seasonality of natural populations together with solar irradiance. Intertidal algae are generally able to cope better with colder and warmer waters than species from greater depths. Moreover, tidal fluctuations, aerosols formed by wave action and air exposure change salt content in seawater. High dynamics in salinity are better tolerated by euryhaline species, mostly intertidal species, than by deeper-water species, which are usually more stenohaline (Kain and Norton 1990). Algal survival to desiccation is determined by the velocity of water loss off thalli and their metabolic adaptation (Williams and Dethier 2005). The maintenance of the photosynthetic activity at low water content and faster recovery of photosynthesis after emersion determine algal survival.

LIGHT STRESS AND PHOTOPROTECTIVE MECHANISMS

Intertidal macroalgae are simultaneously exposed to dynamic and high irradiances of PAR (400–700 nm), UV-A (315–400 nm), and UV-B (280–315 nm). In contrast, algae from the subtidal are mainly exposed to low PAR and (almost) no UVR. Strong sunlight, alone or in combination with other environmental factors, has a great impact on photosynthetic performance and growth (Davison and Pearson 1996). As sessile organisms, benthic algae are not able to escape from unfavorable light conditions. This requires effective adaptation and acclimation mechanisms in order to avoid cellular damage.

Intertidal macroalgae acclimate physiologically to high PAR by regulation of excitation energy and decreasing the photosynthetic activity to prevent chronic photoinhibition or photodamage (Franklin and Forster 1997). In Chlorophyta and Phaeophyceae, the xanthophyll cycle is the dominant photoprotective mechanism that dissipates excess excitation energy as heat. Due to the rapid enzymatic interconversion of violaxanthin to antheraxanthin and zeaxanthin by violaxanthin de-epoxidase, excess excitation energy is transferred from chlorophylls to zeaxanthin owing to the lower orbital level of the singlet excited state. When low-light-grown algae are transferred to high PAR, zeaxanthin accumulates due to interconversion of violaxanthin. Thus, the zeaxanthin content appears to be crucial to direct quanta to photosystem (PS) II under conditions of high light stress. In contrast to green and brown macroalgae, it seems that the xanthophyll cycle is lacking in red macroalgae. Although zeaxanthin is detectable, its particular role in photoprotection remains unclear.

Photoinhibition is measurable with a pulse-amplitude modulated chlorophyll fluorometer by the decrease in maximal PS-II-quantum yield under high radiation regimes and degree of recovery subsequently in dim light. Two kinds of photoinhibition were identified: dynamic and chronic photoinhibition (Osmond 1994). Generally, sun-adapted algae exhibit a dynamic photoinhibition, characterized by a reversible PS II down-regulation due to their increased xanthophyll cycle activity. The fast and complete recovery reflects an efficient repair of damaged photosynthetic proteins such as D1 protein in PS II. In contrast, when shade-adapted algae are transferred to high PAR, chronic photoinhibition is observed. This phenomenon is characterized by photodamage of PS II ascribed to the damage of D1 protein. Because this protein is integrally involved in charge separation and electron transport in PS II, the functioning of PS II is significantly impaired when D1 protein is damaged. The recovery from photodamage is slow since *de novo* synthesis and replacement of damaged D1 proteins to retrieve PS II activity is a time-consuming process.

In terms of the ecological relevance, dynamic photoinhibition of intertidal macroalgae is a strategy of photoprotection to cope with high light conditions (Hanelt 1996). Chronic photoinhibition of subtidal macroalgae indicates their lower tolerance to environmental stress. Macroalgal susceptibility to photoinhibition depends on the irradiance dose, season, water transparency, and the specific habitat.

Recent insights contribute to elucidate photodamage in PS II via a two-step model (Murata et al. 2007): (1) inactivation of the oxygen emitting complex by releasing manganese due to UVR and strong blue light results in (2) inactivation of the PS II reaction centers when high amounts of photons are absorbed. Presumably, ROS increase the extent of photoinhibition by inhibiting repair of PS II instead of attacking PS II reaction centers, as previously assumed.

ROM IN ALGAE LIVING IN THE INTERTIDAL ZONE

Reactive oxygen metabolism (ROM) refers to the entirety of reactions and biochemical pathways leading to the production and detoxification of ROS. Whereas the present chapter only concentrates on the biochemical pathways, Dring (2005) gives an excellent overview of the reactions between ROS and transition metals.

Fighting Against $^1O_2$

Singlet oxygen ($^1O_2$) is exclusively produced by deactivation of triplet excited chlorophylls ($^3Chl^\bullet$) in thylakoid membranes of chloroplasts under high PAR (Fig. 4.1). It is highly toxic to cells because it readily oxidizes all biological structures (Triantaphylidès and Havaux 2009). Therefore, algae have developed protective mechanisms to scavenge both $^3Chl^\bullet$ and $^1O_2$ directly on their formation site, the light antennae and in PS II. Carotenoids and antioxidants act as physical and chemical quenchers, respectively. β-Carotene is the major pigment that is able to scavenge these two highly reactive oxidizers efficiently through energy transfer (Fig. 4.2).

So far, research on formation and detoxification of $^1O_2$ in algae is underrepresented. *Mastocarpus stellatus* and *Chondrus crispus* incubated in 100 μM Rose Bengal produced large amounts of $^1O_2$ upon subsequent exposure to high PAR. Consequently, differential adverse effects on photochemistry in PS II were detected: whereas PS II activity remained unaffected in *M. stellatus*, excess $^1O_2$ significantly reduced PS II activity in *C. crispus* (Collén and Davison 1999c). The higher resistance of *M. stellatus* to $^1O_2$ was attributed to its higher content of β-carotene and α-tocopherol. In *Porphyra umbilicalis*, the increase in the total carotenoid content (mainly β-carotene) during desiccation stress rather than upon low PAR exposure supports the role of β-carotene as a $^1O_2$ scavenger (Sampath-Wiley et al. 2008).

Water–Water Cycle Detoxifies $O_2^{\bullet-}$ in PS I

Other types of ROS are formed by the reduction of oxygen with one (superoxide anion: $O_2^{\bullet-}$) and two electrons (hydrogen peroxide: H_2O_2) (Fig. 4.3). These ROS readily oxidize proteins, lipids, pigments and nucleic acids, leading to adverse effects on the cellular metabolism or even to cell death. Therefore, they have to be detoxified by the Mehler-ascorbate pathway (MAP), a part of the water–water cycle (WWC). A further reduction of H_2O_2 in the presence of transition metals (e.g. iron, copper) generates the hydroxyl radical ($HO^\bullet$), the most reactive oxygen radical. It oxidizes all biological molecules and initiates free-radical chain reactions.

Photoreduction of Oxygen (Mehler Reaction)

The WWC is a complex pathway that is not locally constrained, although the MAP is located at the "acceptor side" of PS I. It can be defined as the formation of water in PS I through multiple electron donation steps to oxygen, in which the electrons result from water splitting in PS II (Fig. 4.4). Consequently, the measurable net exchange of oxygen is zero. After the transfer of electrons released by photolysis in PS II through the photosynthetic electron transport chain, one electron can be donated to oxygen at the "acceptor side" of PS I to generate a $O_2^{\bullet-}$ (Fig. 4.5). This process is called photoreduction of oxygen or the Mehler reaction. Under nonstressful conditions, the rate of ROS production in PS I is low. When the algae are exposed to different sources of stress, however, the rate of ROS production increases significantly. This is favored when reduced nicotinamide adenine dinucleotide phosphate (NADPH) accumulates, for example, due to a decline in carbon assimilation (Asada 1999).

SOD is the First Line of Defense

The first step in the MAP is achieved by superoxide dismutase (SOD, EC 1.15.1.1). However, cells are not completely detoxified by SOD since it disproportionates $O_2^{\bullet-}$ to H_2O_2 and oxygen.

Green macroalgae such as *Ulva fasciata* feature only two isozymes of SOD (Wu et al. 2009). Whereas Fe-SOD and Mn-SOD are present in chloroplasts and mitochondria, respectively, CuZn-SOD is completely

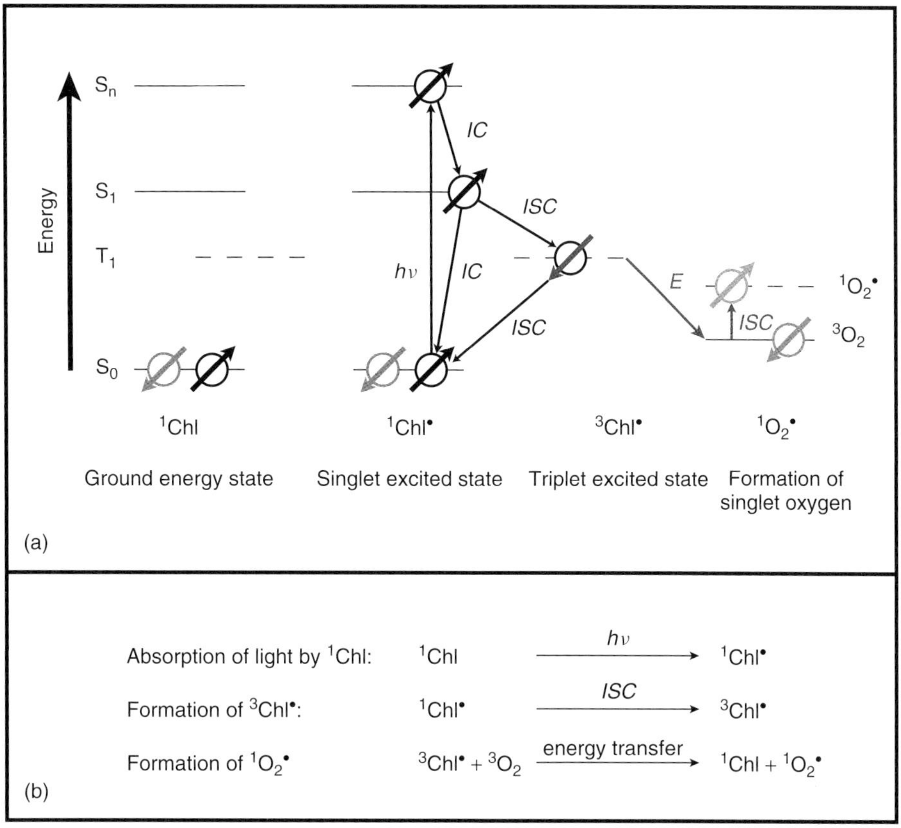

Fig. 4.1 (a) Jablonski diagram of the different electronic energy states of chlorophyll in association with the formation of singlet oxygen ($^1O_2$). In unexcited singlet chlorophylls (^{1}Chl), even-numbered electrons occupying the lowest orbital level exhibit antiparallel spin (indicated by the circles with opposite transverse arrows). In this ground energy state of Chl, S_0, electrons have their energetically most favorable configuration. Due to the absorption of light energy (*hν*), one electron from S_0 is elevated to a higher orbital level ($S_0 \rightarrow S_1 \rightarrow S_n$) without changing its spin. Now this chlorophyll has a singlet excited state (^{1}Chl$^•$). The elevation to a given orbital level corresponds to the amount of energy absorbed. During the relaxation process, the return of the elevated electron to a lower orbital level ($S_n \rightarrow S_1 \rightarrow S_0$), excitation energy is normally released either as fluorescence or heat to its surroundings. The de-excitation process between orbital levels with the same spin is called internal conversion (*IC*). Sometimes, however, the spin of the excited electron flips and, consequently, a triplet excited chlorophyll (^{3}Chl$^•$) arises. This de-excitation process between orbital states with different spins is called intersystem crossing (*ISC*). In the light antennae, ^{3}Chl$^•$ arises directly by *ISC*, whereas in photosystem II (PS II), the triplet excited states of P680$^•$ are formed (^{3}P680$^•$) when the primary electron acceptor in PS II, the quinone Q_A, is reduced during charge separation. Due to the longer lifetime of ^{3}Chl$^•$ ($\sim 10^{-3}$ s) compared to ^{1}Chl$^•$ ($\sim 10^{-8}$ s), there is a higher possibility of interaction between ^{3}Chl$^•$ and molecular (triplet) oxygen ($^3O_2$). By transferring excitation energy (*E*) from ^{3}Chl$^•$ to $^3O_2$, highly reactive $^1O_2$ is produced through changing the spin from a parallel ($^3O_2$) to an antiparallel state ($^1O_2$). The basal electron state of $^3O_2$ is achieved when the two electrons of the lowest orbital level exhibit parallel spin. (b) Reactions of chlorophyll and oxygen summed up in reaction equations.

absent (de Jesus et al. 1989). Thus, Fe-SOD is of major interest regarding photoprotection because light stress increases $O_2^{•-}$ in chloroplasts (Davison and Pearson 1996).

The addition of methyl viologen (MV) to plants is highly toxic because it generates high amounts of $O_2^{•-}$ by photoreduction of oxygen. Macroalgal tolerance to MV was ascribed to higher activities of antioxidant enzymes, larger pool sizes of non-enzymatic antioxidants, or higher regeneration rates of antioxidants. Resistance of *M. stellatus* and *C. crispus* to MV was attributed to their high activities of SOD. In contrast,

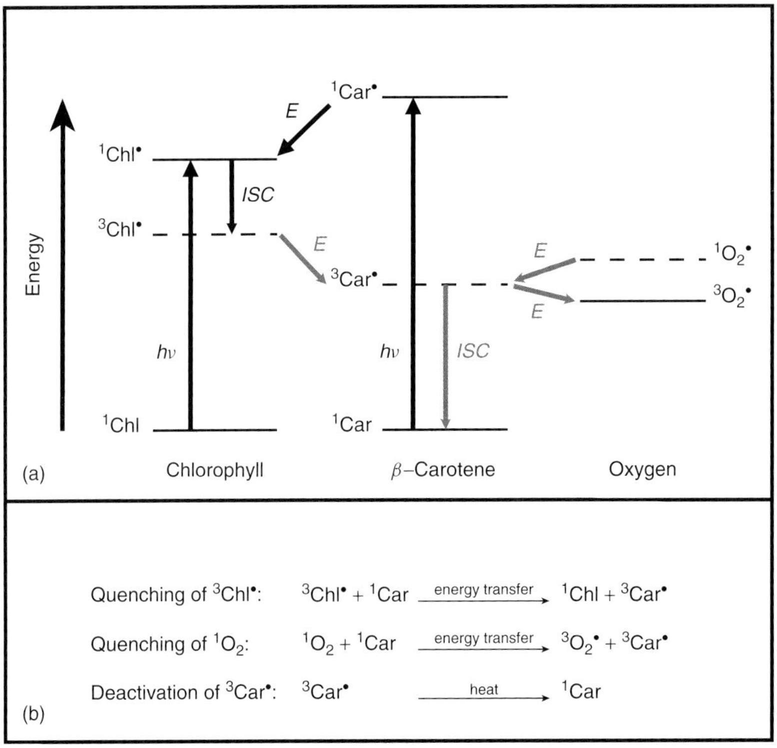

Fig. 4.2 (a) Jablonski diagrams of the protective mechanism by β-carotene against triplet excited chlorophylls (^{3}Chl$^•$) and singlet excited oxygen ($^1O_2$) in algal light antennae. Light energy (*hν*) that was absorbed by chlorophylls (Chl) and carotenoids (Car) such as β-carotene is directly transferred as excitation energy (*E*) to other chlorophylls (bold black arrows) towards the reaction centers of photosystem II. Deactivation of ^{3}Chl$^•$ and $^1O_2$ formed through intersystem crossing (*ISC*) and *E*, respectively, is achieved by β-carotene. Energy transfer from ^{3}Chl$^•$ and $^1O_2$ to β-carotene is energetically possible since the orbital level of triplet excited β-carotene (^{3}Car) is below those of both ^{3}Chl$^•$ and $^1O_2$. Thus, β-carotene acts as acceptor and donor of energy. The close distance and favorable orientation of β-carotene to the chlorophylls is a spatial prerequisite for energy transfer. β-Carotene dissipates excitation energy through *ISC* as heat. (b) Reaction equations of energy quenching and deactivation by β-carotene.

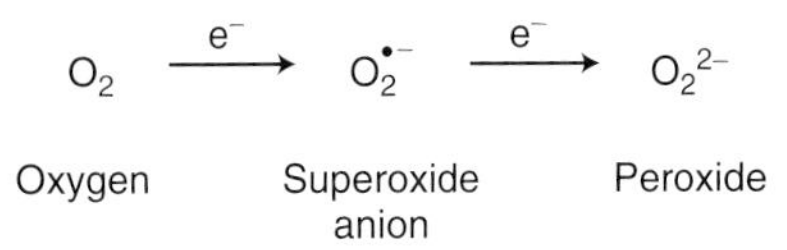

Fig. 4.3 Formation of superoxide and peroxide anions by electron transfer.

Fucus spp. exhibit 10 times lower SOD activity and are characterized by a high sensitivity to MV (Collén and Davison 1999a–c).

H_2O_2 is Detoxified by APx and Excretion

Removal of H_2O_2 is performed by ascorbate peroxidase (APx, EC 1.11.1.11). In green macroalgae, H_2O_2 is detoxified by APx rather than by catalase (CAT; Collén and Pedersén 1996). APx reduces H_2O_2 to H_2O consuming reduced ascorbate (AsA) in the neighborhood of SOD (Asada 1999). AsA donates electrons to APx and is regenerated in the ascorbate–glutathione cycle (AsA–GSH cycle).

In macroalgae, excretion of surplus H_2O_2 into the surrounding seawater represents an additional protective strategy because it easily diffuses through biological membranes (Collén and Davison 1999a). In wave-exposed systems, H_2O_2 in seawater does not sufficiently accumulate to cause substantial oxidative damage. However, in more protected habitats where the undulation is reduced (e. g. tide pools), H_2O_2 may accumulate in seawater up to concentrations that cause adverse effects in macroalgae and associated

Water-water cycle

Photolysis of water in PS II:	$2H_2O$	$\longrightarrow$	$O_2 + 4H^+ + 4e^-$
Photoreduction of O_2 in PS I:	$2O_2 + 2e^-$	$\longrightarrow$	$2O_2^{\bullet -}$
Disproportionation of O_2 by SOD in PS I:	$2O_2^{\bullet -} + 2H^+$	$\xrightarrow{SOD}$	$O_2 + H_2O_2$
Reduction of H_2O_2 to water by APX in PS I:	$H_2O_2 + 2AsA$	$\xrightarrow{APX}$	$2H_2O + 2MDA$
Sum reaction of the water–water cycle:	$2H_2O + O_2$	$\longrightarrow$	$O_2 + 2H_2O$

Regeneration of AsA from its oxidized forms by multiple pathways:

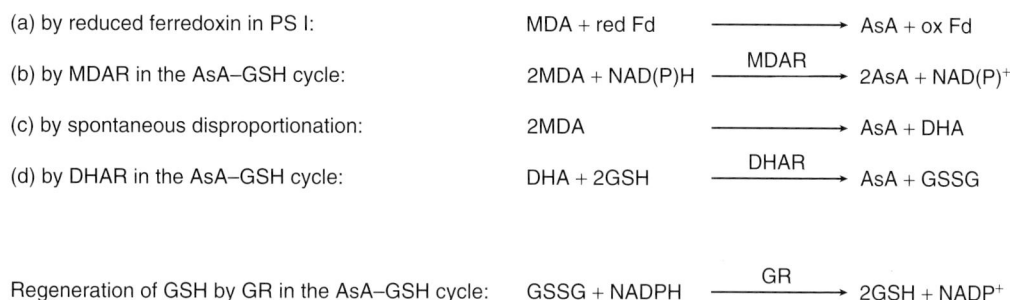

(a) by reduced ferredoxin in PS I:	MDA + red Fd	$\longrightarrow$	AsA + ox Fd
(b) by MDAR in the AsA–GSH cycle:	2MDA + NAD(P)H	$\xrightarrow{MDAR}$	2AsA + NAD(P)$^+$
(c) by spontaneous disproportionation:	2MDA	$\longrightarrow$	AsA + DHA
(d) by DHAR in the AsA–GSH cycle:	DHA + 2GSH	$\xrightarrow{DHAR}$	AsA + GSSG

Regeneration of GSH by GR in the AsA–GSH cycle:	GSSG + NADPH	$\xrightarrow{GR}$	2GSH + NADP$^+$

Fig. 4.4 Reactions of the photolysis of water in photosystem II (PS II) and the reactive oxygen metabolism in photosystem I (PS I). The sum reaction of the water–water cycle shows that equal amounts of water and O_2 were consumed (by photolysis and the photoreduction of O_2) and produced again. Thus, the measurable net exchange of O_2 is zero. Reduced ascorbate (AsA) is regenerated from its oxidized forms (MDA and DHA) enzymatically (by MDAR and DHAR), nonenzymatically by reduced ferredoxin (red Fd to oxidized Fd), and by spontaneous disproportionation. Reduced glutathione (GSH) is also recycled from oxidized glutathione (GSSG) in the ascorbate–glutathione (AsA–GSH) cycle by a NADPH-dependent glutathione reductase (GR).

organisms (Choo et al. 2004). Excretion rates depend on incident light intensities and seawater pH (Collén et al. 1995). In *Ulva rigida*, H_2O_2 excretion increases more than four times when seawater pH rises from 8.2 to 9.0 due to photosynthetic activity.

Catalase

Catalase (CAT, EC 1.11.1.6) is another enzyme that reduces H_2O_2 to water. Catalase is located in algal peroxisomes to scavenge H_2O_2 produced by oxidation of glycolate (by glycolate oxidase) to glyoxylate in the photorespiratory pathway (Gross 1993). In *U. rigida*, however, no CAT activity has been detected, which suggests that glycolate is reduced to glyoxylate by glycolate dehydrogenase in mitochondria rather than oxidized in peroxisomes. Thus, the Mehler reaction is the only source of H_2O_2 production in this *Ulva* species. In contrast, CAT activities measured in brown macroalgae indicate two sources of H_2O_2: the Mehler reaction and photorespiration. In *Fucus vesiculosus*, CAT is regulated independently of ROM components and responds primarily to light stress (Collén and Davison 2001). In *M. stellatus*, elevated CAT activities are responsible for the high scavenging ability of externally added H_2O_2 (Collén and Davison 1999c).

Regeneration of Antioxidants is Implemented in the AsA-GSH Cycle

The rapid regeneration of AsA from its two oxidized forms seems to be crucial for proper H_2O_2 scavenging, because AsA leads electrons over

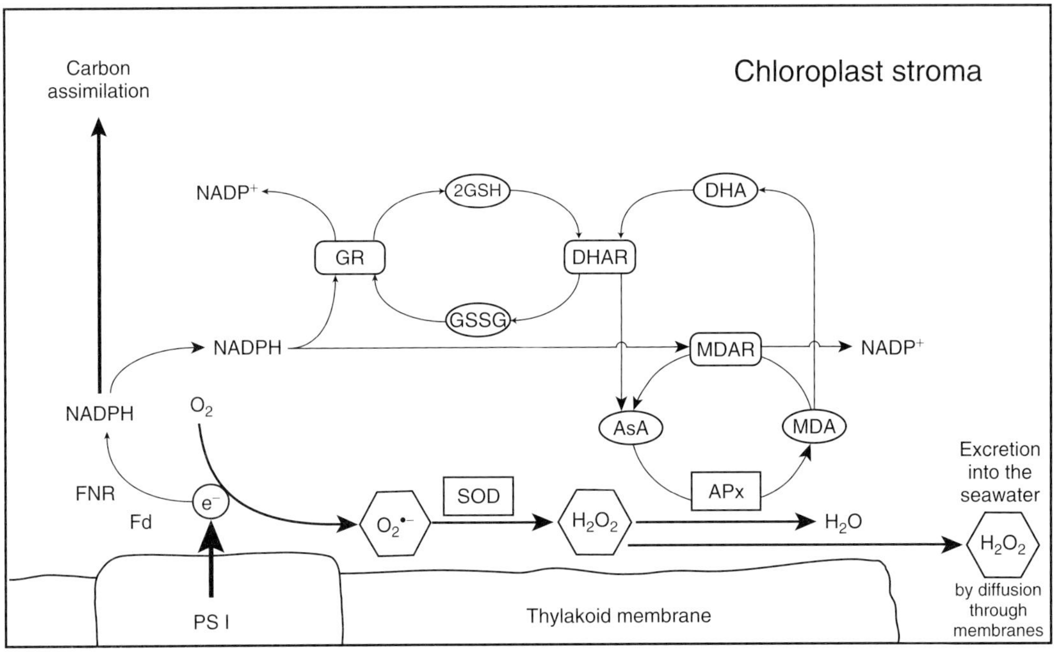

Fig. 4.5 Simplified model of ROS production and detoxification in the stroma of chloroplasts as investigated in marine algae so far. In this model, the importance of ferredoxin (Fd) and other stromal factors with respect to electron transfer in photosystem I (PS I) (particularly to O_2) as well as direct reduction of monodehydroascorbate (MDA) for recycling reduced ascorbate (AsA) is not represented. A more detailed model was given by Asada (1999). At the "acceptor side" of PS I, electrons (e^-) from the photosynthetic ETC can be transferred via Ferredoxin-NADP$^+$ oxidoreductase (FNR) from Fd to NADP$^+$. NADPH produced through this mechanism is mainly utilized by carbon assimilation in the Calvin–Benson cycle as well as other pathways. However, e^- from the photosynthetic ETC are increasingly transferred to oxygen (O_2) under environmental stress. Due to this process called photoreduction of oxygen, or the Mehler reaction, superoxide anions ($O_2^{\bullet -}$) arise. $O_2^{\bullet -}$ are detoxified by superoxide dismutase (SOD) to hydrogen peroxide (H_2O_2). Highly toxic H_2O_2 is removed quickly by ascorbate peroxidase (APx) which requires AsA for its proper functioning. Alternatively or additionally to the intracellular detoxification by APx, overproduced H_2O_2 is excreted by macroalgae into the surrounding seawater. Regeneration of AsA takes place in the ascorbate–glutathione (AsA-GSH) cycle. AsA is regenerated enzymatically from MDA by a NADPH-dependent monodehydroascorbate reductase (MDAR) and from dehydroascorbate (DHA) by dehydroascorbate reductase (DHAR). Regeneration of AsA from DHA by DHAR is coupled with recycling of reduced glutathione (GSH) from oxidized glutathione (GSSG) by a NADPH-dependent glutathione reductase (GR). NADPH utilized by these steps is derived from PS I. Key: hexagons, ROS produced in the MAP; rectangles with sharp edges, antioxidant enzymes involved in the MAP; rectangles with rounded edges, enzymes of the AsA–GSH cycle; ellipse, antioxidants of the AsA–GSH cycle.

to APx. AsA regeneration is mediated by both monodehydroascorbate reductase (MDAR, EC 1.6.5.4) and dehydroascorbate reductase (DHAR, EC 1.8.5.1). In the latter case, AsA regeneration strongly collaborates with glutathione recycling by glutathione reductase (GR, EC 1.6.4.2). Because of this interrelationship, the regenerating cycle is called the AsA-GSH cycle.

There is a mechanistic connection of GR and DHAR for AsA regeneration. Glutathione reductase is responsible for regeneration of glutathione (GSH) from its oxidized form (GSSG). If GR activity is broken down due to damage, inadequate provision of DHAR with GSH abrogates AsA regeneration. Consequently, H_2O_2 scavenging by APx is hampered because dehydroascorbate (DHA) accumulates in the chloroplast instead of being converted to AsA. However, a high MDAR activity seems to ensure H_2O_2 scavenging by providing sufficient AsA levels to APx (Shiu and Lee 2005). Monodehydroascorbate reductase catalyzes the reduction of monodehydroascorbate (MDA) to

AsA under consumption of NADPH. The rapid AsA regeneration from MDA and DHA is crucial for proper ROM functioning because all of the available ascorbate would be oxidized within seconds during photosynthesis (Sampath-Wiley et al. 2008). Otherwise, ROS increase to lethal levels due to inhibition of APx. AsA itself has ROS-scavenging functions too. It directly reacts with $HO^{\bullet}$, $O_2^{\bullet-}$, $^1O_2$, and H_2O_2 and is involved in the regeneration of α-tocopherol.

Significantly increased pool sizes of GSSG may entail an increase in GR activity as proposed for *Porphyra umbilicalis* under desiccation stress (Sampath-Wiley et al. 2008). Rapid GSH regeneration by GR ensures both DHAR-mediated AsA regeneration and nonenzymatic scavenging of $^1O_2$, $O_2^{\bullet-}$, and $HO^{\bullet}$ by GSH. In contrast, GSH depletion is attributed to rapid oxidation of GSH to GSSG and GR inhibition under environmental stress (Rijstenbil 2005). This is reflected by the decline in the redox ratio of GSH to the total pool size of glutathione, representing an important indicator of the intensity of oxidative stress: GSH/(GSH + 0.5 GSSG) in nonstressed algae is higher than in stressed cells. The nonstressed benthic, semi-pelagic diatom *Cylindrotheca closterium* and some planktonic diatoms exhibit redox ratios between 0.5 and 0.96. They decrease down to 0.34 and 0.53 for *C. closterium* and the planktonic diatoms, respectively, under UV-B-stress (Rijstenbil and Wijnholds 1996, Rijstenbil 2001).

ROM Efficiency

The efficiency of the ROM is represented by the removal rate of ROS due to increased levels of molecular antioxidants, activities of antioxidant enzymes, and enzymes of the AsA-GSH cycle. For example, *M. stellatus* is characterized by a more efficient ROM than *C. crispus* because of its slightly higher removal capacity for externally supplied H_2O_2 through increased CAT activity (Collén and Davison 1999c).

THE ZONATION OF MACROALGAE DEPENDS ON THEIR STRESS TOLERANCE

Coastal ecosystems exhibit characteristic vertical zonation patterns of macroalgae along the shore, which are controlled by the differential impact of environmental factors in intertidal and subtidal zones. Intertidal algae are exposed to frequently and rapidly changing environmental conditions predominantly caused by tidal fluctuations (Fig. 4.6). During low tide macroalgae are exposed to air for a certain period and are (also interactively) jeopardized by:

- desiccation stress
- high solar radiation (PAR and UV radiation)
- changing temperature regimes
- hypo- and hypersalinity.

Regarding the ecology of benthic algae, the general pattern of vertical zonation can be well explained by their differences in tolerance or sensitivity to environmental stress. Shallow-water macroalgae are more stress-tolerant than those from greater depths. Hence, they exhibit less pronounced stress effects than algae from the subtidal or even the lower intertidal. This seems to be related to the efficiency of the ROM which is more efficient in intertidal macroalgae than in subtidal species or conspecifics (Collén and Davison 1999a–c). ROM is a major factor for algal zonation. Generally, stress-tolerant intertidal algae feature:

- higher activities of antioxidant enzymes
- higher H_2O_2 excretion rates
- increased pool sizes of antioxidants
- faster regeneration of antioxidants

than stress-sensitive algae from greater depths. These responses can be either constitutive or induced by environmental stress. Thus, algal stress tolerance can be understood as their ability to withstand environmental stress successfully due to their efficient ROM. *Vice versa*, sensitivity of an algal species to environmental stress is a consequence of higher ROS production because of insufficient ROS detoxification. However, it must be pointed out that although stress tolerance is species-specific it may manifest differently within the same species (Aguilera et al. 2002b).

Decreasing ROS production by avoiding light stress is the main goal of algal zonation. For example, by growing underneath the canopies of larger macroalgal thalli, smaller macroalgae such as *Porphyra leucosticta* avoid direct contact with high PAR and UVR (Salles et al. 1996). Although growth of *P. leucosticta* is significantly reduced in the shade, the algae remain

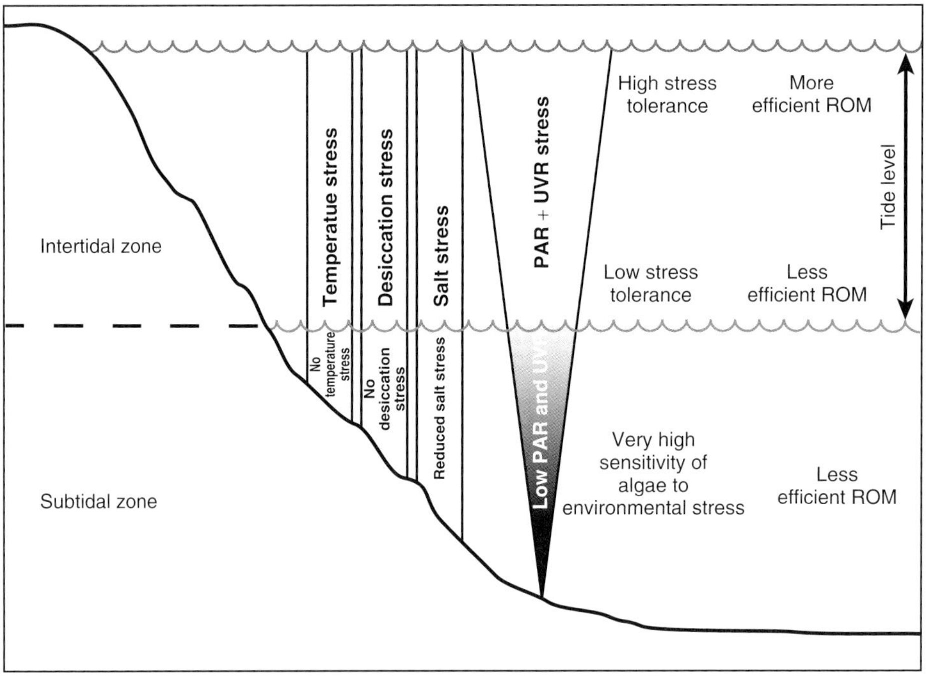

Fig. 4.6 Temperature, desiccation, salinity, and solar radiation (PAR and UVR) represent major abiotic environmental factors controlling vertical zonation of macroalgae. The regular decrease in the seawater level due to tidal fluctuations is predominantly responsible for their impact on benthic algae. The lack of the protecting water body during low tides may significantly increase their effects in the intertidal zone – either individually or in combination. The intensities of these environmental factors are differentially pronounced at given water depths as depicted by geometric forms. Temperature, desiccation, and salt stress have the greatest impact on macroalgae in the intertidal zone. The vertical light gradient is the major factor determining algal distribution since it may significantly affect algae in the intertidal down to the subtidal zones. Consequently, sessile benthic algae have to respond to these stressful conditions by protective mechanisms. Algae from highest positions on the shore characterized as highly tolerant to environmental stress generally exhibit low rates of ROS production and/or high rates of ROS detoxification. This represents an efficient reactive oxygen metabolism (ROM). In contrast, algae from deeper waters such as low intertidal and the subtidal are characterized by a higher sensitivity to environmental disturbances. Their fast and massive ROS production under stressful conditions is attributed to a less efficient ROM. Hence, these algae have to select more protected habitats (e.g. deeper water depths, canopies).

undamaged and their photosynthetic performance is unaffected by the low irradiances under the protective canopy. A shade-acclimation characteristic is the increase in the size of light antennae in order to enhance the possibility to absorb as much as possible of the incoming light quanta. Due to the changes in spectral composition of the available light, algae acclimate by adjusting the pigment composition according to the light qualities (Talarico 1996). Moreover, subcanopy individuals are characterized by a high UV sensitivity because UV protective mechanisms are not a priority (Bischof et al. 1998).

Light Stress

The increase in activities of MAP and AsA-GSH cycle enzymes and pool sizes of antioxidants represents a physiological acclimation mechanism to cope with rising ROS levels under high irradiances of solar radiation (PAR and UVR). An efficient ROM prevents photoinhibition and photo-oxidative damage by ROS. The xanthophyll cycle in light antennae deactivates excessively absorbed light energy to avoid $^1O_2$ formation and to diminish "excitation pressure" in PS II. Excessively produced $O_2^{\bullet -}$ and H_2O_2 upon high PAR exposure are removed by SOD and APx, respectively, as well as by excretion into the seawater (Collén and Pedersén 1996). However, in nature, PAR and UV-BR act synergistically because high UV-B irradiances inhibit photoprotective mechanisms (e.g. xanthophyll cycle), and high PAR exposure enhances ROS production (Bischof et al. 2002). Moreover, when the increase in SOD activity is insufficient to detoxify $O_2^{\bullet -}$, macroalgal survival under the high light regimes is significantly reduced due to increased photo-oxidative damage (lipid peroxidation). For example, in spite of increased SOD activities due to PAR-induced rising ROS levels, excessively produced ROS caused to a complete disintegration of *U. rotundata*'s thallus under high solar UV-BR (Bischof et al. 2002, 2003). Reduction of UV-B-induced oxidative stress by an efficient ROM allows the tropical ecotype of *U. fasciata* to exist even under high tropical UV-BR regimes in summer. Increased SOD and APx activities supported by antioxidant regeneration in the AsA-GSH cycle avoid lethal lipid peroxidation up to tropical maximum summer UV-B irradiances (Shiu and Lee 2005).

Generally, intertidal macroalgae exhibit higher activities of antioxidant enzymes and antioxidants upon UV exposure than species inhabiting greater depths. This represents a fast acclimation mechanism to cope with the fluctuating radiation regimes (Aguilera et al. 2002b). Green macroalgae dominating the intertidal Arctic waters on Spitsbergen exhibit higher SOD activities than brown and red macroalgae from greater depth. When *Monostroma arcticum* is suddenly exposed to high solar radiation due to the break-up of sea-ice in the Arctic spring, SOD and CAT activities increase significantly to avoid photoinhibition and photodestruction by ROS (Aguilera et al. 2002a).

In contrast to intertidal macroalgae, subtidal algae are generally characterized by a lack of photoprotective mechanisms, which is the prerequisite for quickly rising oxidative stress (Bischof et al. 2000). Being less photoprotected by physiological acclimation or adaptation mechanisms, these macroalgae have to colonize light-protected habitats or greater depths.

Desiccation Tolerance

Desiccation stress can be tolerated when ROS produced during dehydration and subsequent rehydration are efficiently removed by the ROM.

Ulva lactuca is characterized by its wide vertical distribution from the intertidal down to the subtidal area. Intertidal individuals exhibit lower rates of ROS production during desiccation than those from the subtidal zone because of their increased H_2O_2 detoxification capacity through higher APx and GR activities. Both SOD and CAT play a rather negligible role since their activities were identical in individuals from both zones. Thus, *U. lactuca's* ability to colonize the dynamic intertidal area is attributed to the phenotypic plasticity of APx and GR, representing an intraspecific variability of the ROM (Ross and Van Alstyne 2007).

In *Stictosiphonia arbuscula* the AsA–GSH cycle is crucial for withstanding periods of desiccation in the intertidal zone (Burritt et al. 2002). It generally acclimates to desiccation stress by the increase in GR, DHAR, and APx activities. Elevated GR and DHAR activities ensure the rapid AsA and GSH regeneration to support APx proper functioning. However, the remarkable difference in ROM efficiency between individuals (intraspecific variability) from the intertidal area is attributed to the differential increase in activities of these three enzymes. The more effective H_2O_2 detoxification in individuals of the high intertidal zone results from greater increases in APx, GR, and DHAR activities compared to conspecifics in the low intertidal zone. This minimizes cellular damage.

Changing Temperature Regimes – Freezing Stress

Production of ROS in macroalgae during freezing stress consists of two components: freezing and thawing. During freezing events at subzero temperatures, membranes may become structurally impaired, which stops all biological processes. Subsequently, when frozen thalli start to thaw at rising temperatures, ROS are increasingly produced due to electron leakage (e.g. by photoreduction of oxygen in PS I) at impaired membranes. Hence, ROS production during freezing stress is attributed to thawing rather than to freezing (Collén and Davison 1999b).

Intertidal macroalgae are characterized by enhanced freezing stress tolerance compared to subtidal macroalgae, resulting in a vertical pattern of freezing stress tolerance (Collén and Davison 1999b). Although *Fucus spiralis* and *Fucus evanescens* from the intertidal area exhibit similar rates of ROS production at −18°C, *F. spiralis* is more tolerant to freezing stress than *F. evanescens*. This is attributed to its faster H_2O_2 excretion resulting in less oxidative stress. Slower H_2O_2

excretion in *F. evanescens* generated a higher degree of oxidative stress. In contrast, subtidally positioned *Fucus distichus* was characterized as highly sensitive to freezing since its less efficient ROM was responsible for serious oxidative damage.

Salinity Stress

Benthic algae exhibit inter -or intraspecific patterns of tolerance to changing salinities: intertidal algae are generally more tolerant to salinity stress than those from the subtidal zone.

Ulva lactuca's ability to cope with hyposalinity in the intertidal zone is attributed to phenotypic plasticity of APx and GR (Ross and Van Alstyne 2007). Higher APx and GR activities in intertidal individuals might be responsible for their capability of withstanding hyposalinity. High APx and GR activities in intertidal individuals result in low H_2O_2 excretion rates under both full-strength (35‰) and moderate (17.5‰) hyposaline seawater. Even when salinity was 3.5‰ ROS were effectively detoxified, as indicated by moderate H_2O_2 excretion rates. In contrast, lower APx and GR activities in subtidal conspecifics resulted in faster H_2O_2 excretion under the same hyposaline conditions.

The tolerance of intertidal macroalgae to short-term changing salinities regarding their ROM was investigated in *U. fasciata* (Sung et al. 2009). $O_2^{\bullet-}$ and H_2O_2 increasingly produced under acute hypersalinity (90‰) were detoxified by enhanced activities of Fe-SOD, Mn-SOD, APx, and GR. These increased activities were induced by co-expression of SOD, APx, and GR genes. Thus, an elevated ROM providing a high level of ROS protection is crucial to *U. fasciata*'s survival to temporarily changing salinities in the intertidal zone. This survival ability, however, seems to be decreased after long-term exposure to hypo- and hypersaline conditions due to a generally reduced antioxidant defense capacity (Lu et al. 2006). To effectively counteract increased ROS production, pool sizes and regeneration rates of antioxidants as well as activities of antioxidant enzymes increased, depending on the respective salt content. This means that ROM is highly regulated to allow for adaptation to variable seawater salinities. Rising oxidative stress was related to the inhibition of the photosynthetic electron transport. Under moderate hypo- (15‰) and hypersalinities (60–90‰), APx, CAT, GR, and MDAR activities increased due to lipid peroxidation by elevated production of intracellular H_2O_2. Under extreme hypersalinities (120–150‰), the decreased oxidative damage resulted from reduced photoreduction of oxygen in PS I due to inhibition of the photosynthetic electron transport. Detoxification of intracellular H_2O_2 under moderate hyposalinity (15‰) is achieved by increased CAT, GR, and DHAR activities. At extremely low salinity (5‰), increased SOD and CAT activities as well as low intracellular and extracellular H_2O_2 levels suggest that both $O_2^{\bullet-}$ and H_2O_2 generated by the inhibition of the photosynthetic electron transport were efficiently removed. Increased activities of GR, MDAR, and DHAR promote GSH and AsA regeneration.

OTHER ECOLOGICAL AND PHYSIOLOGICAL MECHANISMS OF STRESS TOLERANCE

Accumulation of UV-absorbing Compounds

Production and accumulation of UV-absorbing compounds represent an adaptive strategy to protect macroalgae from UV-AR and UV-BR. Mycosporine-like amino acids (MAAs) are low-molecular-weight nitrogenous UV-absorbing compounds with high absorption bands between 310 and 360 nm. The MAAs are natural sunscreens predominantly found in red macroalgae from polar to tropical habitats (Bandaranayake 1998). Beside their role in UV screening, several MAAs also have antioxidant properties and act as compatible solutes and nitrogen reserves (Korbee et al. 2005).

In several intertidal brown and green macroalgae, UV protection is provided by the accumulation of large amounts of phenolic compounds such as phlorotannins absorbing between 195 and 265 nm (Pavia et al. 1997). The photoprotective effect of polyphenols, phlorotannins, and coumarins also results from their high antioxidant and radical-scavenging capacities (Shibata et al. 2003). Contents of polyphenols depend on season, light, desiccation, salinity, and nutrient availability (Pérez-Rodríguez et al. 2001).

Damage and Repair of DNA

Although mutagenic effects of UV-BR on macroalgae are well recognized, they are not yet well understood.

As a consequence of direct absorption of high-energy UV-BR, DNA is structurally modified. The two most common DNA lesions are cyclobutane pyrimidine dimers (CPDs), which account for 75%, and (6-4) photoproducts (Mitchell and Nairn 1989). These distortions are cytotoxic since they block DNA and RNA polymerases. This inhibits genome replication and gene expression (van de Poll et al. 2007). DNA-protein cross-links, DNA strand breaks, and deletions or insertions of base pairs can also be induced upon UV exposure (Smith 1989).

An adaptive strategy to cope with environmental stress is represented by an efficient repair of damaged DNA. There are two mechanisms of enzymatic DNA repair in macroalgae. Photoreactivation by photolyases leads rapidly to the monomerization of dimers (photorepair). This process requires light between 350 and 450 nm (Lois and Buchanan 1994). The second mechanism is called excision repair since damaged nucleic acids are excised from the DNA molecule. This light-independent multistep process is much slower than the photoreactivation process (Stapleton 1992). Furthermore, abiotic factors may affect DNA repair significantly. For example, DNA repair slows down when *Palmaria palmata* is cultivated at low temperatures. Especially, repair pathways requiring numerous gene expressions may be inhibited by unfavorable external conditions.

CONCLUSION

Rapid and fluctuating changes in major abiotic environmental factors (high PAR and UVR, desiccation, changes in salinity, and high or low seawater temperatures) may lead to enhanced ROS production in algae living in the intertidal zone. These ROS produced either directly by excess excitation energy ($^1O_2$) or by electron leakage from PS I ($O_2^{\bullet -}$, H_2O_2, $HO^\bullet$) have a great impact on cellular health and integrity due to oxidation of proteins, pigments, DNA, and peroxidation of lipids in biological membranes. In order to avoid and counteract these disadvantages special photoprotective mechanisms (e.g. the xanthophyll cycle) and biochemical pathways have evolved in algae. The MAP in combination with the AsA–GSH cycle detoxifies ROS produced by the Mehler reaction enzymatically and nonenzymatically with antioxidants. The efficiency of ROS detoxification is determined by (inducible) high activities of antioxidant enzymes, large pool sizes of antioxidants, as well as their rapid regeneration. Macroalgae additionally excrete H_2O_2 to protect themselves against this harmful oxidizer.

From an ecological point of view, the efficiency of the ROM is crucial for macroalgal survival. Because it is intra- and interspecies specific under each kind of stress, the efficiency of ROS detoxification is responsible for the vertical zonation of macroalgae. Generally, macroalgae from higher positions on the shore are regarded as stress-tolerant because of their more efficient ROM. In contrast, stress-sensitive algae preferentially occur in deeper waters or other protected habitats, since they are not able to efficiently counteract ROS.

REFERENCES

Aguilera, J., Bischof, K., Karsten, U., Hanelt, D., Wiencke, C. (2002a) Seasonal variation in ecophysiological patterns in macroalgae from an Arctic fjord. II. Pigment accumulation and biochemical defence systems against high light stress. *Marine Biology* 140, 1087–1095.

Aguilera, J., Dummermuth, A., Karsten, U., Schriek, R., Wiencke, C. (2002b) Enzymatic defences against photooxidative stress induced by ultraviolet radiation in arctic marine macroalgae. *Polar Biology* 25, 432–441.

Asada, K. (1999) The water-water cycle in chloroplasts: Scavenging of active oxygen and dissipation of excess photons. *Annual Review of Plant Physiology and Plant Molecular Biology* 50, 601–639.

Bandaranayake, W. (1998) Mycosporines: are they nature's sunscreens? *Natural Product Reports* 15, 159–172.

Bischof, K., Hanelt, D., Wiencke, C. (1998) UV-radiation can affect depth-zonation of Antarctic macroalgae. *Marine Biology* 131, 597–605.

Bischof, K., Hanelt, D., Wiencke, C. (2000) Effect of ultraviolet radiation on photosynthesis and related enzyme reactions of marine macroalgae. *Planta* 211, 555–562.

Bischof, K., Peralta, G., Kräbs, G., van de Poll, W. H., Pérez-Lloréns, J. L., Breeman, A. M. (2002) Effects of solar UV-B radiation on canopy structure of *Ulva* communities from southern Spain. *Journal of Experimental Botany* 53, 2411–2421.

Bischof, K., Janknegt, P. J., Buma, A. G. J., Rijstenbil, J. W., Peralta, G., Breeman, A. M. (2003) Oxidative stress and enzymatic scavenging of superoxide radicals induced by solar UV-B radiation in *Ulva* canopies from southern Spain. *Scientia Marina* 67, 353–359.

Burritt, D., Larkindale, J., Hurd, C. (2002) Antioxidant metabolism in the intertidal red seaweed *Stictosiphonia arbuscula* following desiccation. *Planta* 215, 829–838.

Choo, K.-S., Snoeijs, P., Pedersén, M. (2004) Oxidative stress tolerance in the filamentous green algae *Cladophora glomerata* and *Enteromorpha ahlneriana*. *Journal of Experimental Marine Biology and Ecology* 298, 111–123.

Collén, J., Davison, I. (1999a) Reactive oxygen metabolism in intertidal *Fucus* spp. (Phaeophyceae). *Journal of Phycology* 35, 62–69.

Collén, J., Davison, I. (1999b) Reactive oxygen production and damage in intertidal *Fucus* spp. (Phaeophyceae). *Journal of Phycology* 35, 54–61.

Collén, J., Davison, I. (1999c) Stress tolerance and reactive oxygen metabolism in the intertidal red seaweeds *Mastocarpus stellatus* and *Chondrus crispus*. *Plant, Cell and Environment* 22, 1143–1151.

Collén, J., Davison, I. (2001) Seasonality and thermal acclimation of reactive oxygen metabolism in *Fucus vesiculosus* (Phaeophyceae). *Journal of Phycology* 37, 474–481.

Collén, J., Pedersén, M. (1996) Production, scavenging and toxicity of hydrogen peroxide in the green seaweed *Ulva rigida*. *European Journal of Phycology* 31, 265–271.

Collén, J., Jimènez del Río, M., García, G., Pedersén, M. (1995) Photosynthetic production of hydrogen peroxide by *Ulva rigida* C. Ag. (Chlorophyta). *Planta* 196, 225–230.

Davison, I. R., Pearson, G. A. (1996) Stress tolerance in intertidal seaweeds. *Journal of Phycology* 32, 197–211.

De Jesus, M. D., Tabatabai, F., Chapman, D. J. (1989) Taxonomic distribution of copper-zinc superoxide dismutase in green algae and its phylogenetic importance. *Journal of Phycology* 25, 767–772.

Dring, M. (2005) Stress resistance and disease resistance in seaweeds: the role of reactive oxygen metabolism. *In* Callow, J. (ed.). *Advances in Botanical Research*, Vol. 43. Academic Press, San Diego, pp. 175–207.

Franklin, L. A., Forster, R. M. (1997) The changing irradiance environment: consequences for marine macrophyte physiology, productivity and ecology. *European Journal of Phycology* 32, 207–232.

Gross, W. (1993) Peroxisomes in algae: their distribution, biochemical function and phylogenic importance. *In* Round, F. E., Chapman, D. J. (eds). *Progress in Phycological Research*, Vol. 9. Elsevier Biomedical Press, Amsterdam, pp. 47–78.

Hanelt, D. (1996) Photoinhibition of photosynthesis in marine macroalgae. *Scientia Marina* 60, 243–248.

Kain J. M., Norton T. A. (1990) Marine ecology. *In* Cole, K. M., Sheath, R. G. (eds). *Biology of the Red Algae*. Cambridge University Press, Cambridge, pp. 377–408.

Korbee, N., Figueroa, F. L., Aguilera, J. (2005) Effect of light quality on the accumulation of photosynthetic pigments, proteins and mycosporine-like amino acids in the red alga *Porphyra leucosticta* (Bangiales, Rhodophyta). *Journal of Photochemistry and Photobiology B: Biology* 80, 71–78.

Lois, R., Buchanan, B. B. (1994) Severe sensitivity to ultraviolet radiation in an *Arabidopsis* mutant deficient in flavonoid accumulation. *Planta* 194, 504–509.

Lu, I.-F., Sung, M.-S., Lee, T.-M. (2006) Salinity stress and hydrogen peroxide regulation of antioxidant defense system in *Ulva fasciata*. *Marine Biology* 150, 1–15.

Mitchell, D. L., Nairn, R. S. (1989) The biology of the (6–4) photoproduct. *Photochemistry and Photobiology* 49, 805–19.

Murata, N., Takahashi, S., Nishiyama, Y., Allakhverdiev, S. I. (2007) Photoinhibition of photosystem II under environmental stress. *Biochimica et Biophysica Acta* 1767, 414–421.

Osmond, C. B. (1994) What is photoinhibition? Some insights from comparisons of shade and sun plants. *In* Baker, N. R., Bowyer, J. R. (eds). *Photoinhibition of Photosynthesis: From Molecular Mechanisms to the Field*. Bios Scientific Publishers, Oxford, pp. 1–24.

Pavia, H., Cervin, G., Lindgren, A., Aaberg, P. (1997) Effects of UV-B radiation and simulated herbivory on phlorotannins in the brown alga *Ascophyllum nodosum*. *Marine Ecology Progress Series* 157, 139–146.

Pérez-Rodríguez, E., Aguilera, J., Gómez, I., Figueroa, F. L. (2001) Accumulation and excretion of coumarins in response to environmental stress by the Mediterranean green alga *Dasycladus vermicularis*. *Marine Biology* 139, 633–639.

Rijstenbil, J. W. (2001) Effects of periodic, low UVA radiation on cell characteristics and oxidative stress in the marine planktonic diatom *Ditylum brightwellii*. *European Journal of Phycology* 36, 1–8.

Rijstenbil, J. W. (2005) UV- and salinity-induced oxidative effects in the marine diatom *Cylindrotheca closterium* during simulated emersion. *Marine Biology* 147, 1063–1073.

Rijstenbil, J. W., Wijnholds, J. A. (1996) HPLC analysis of non-protein thiols in planktonic diatoms: pool size, redox state and response to copper and cadmium exposure. *Marine Biology* 127, 45–54.

Ross, C., Van Alstyne, K. (2007) Intraspecific variation in stress-induced hydrogen peroxide scavenging by the ulvoid macroalga *Ulva lactuca*. *Journal of Phycology* 43, 466–474.

Salles, S., Aguilera, J., Figueroa, F. L. (1996) Light field in algal canopies: changes in spectral light ratios and growth of *Porphyra leucosticta*. *Scientia Marina* 60, 29–38.

Sampath-Wiley, P., Neefus, C., Jahnke, L. (2008) Seasonal effects of sun exposure and emersion on intertidal seaweed physiology: Fluctuations in antioxidant contents, photosynthetic pigments and photosynthetic efficiency in the red alga *Porphyra umbilicalis* Kützing (Rhodophyta, Bangiales). *Journal of Experimental Marine Biology and Ecology* 361, 83–91.

Shibata, T., Nagayama, K., Tanaka, R., Nakamura, T. (2003) Inhibitory effects of brown algal phlorotannins on phospholipase A2s, lipoxygenases and cyclooxygenases. *Journal of Applied Phycology*, 15, 61–66.

Shiu, C.-T., Lee, T.-M. (2005). Ultraviolet-B-induced oxidative stress and responses of the ascorbate-glutathione cycle in a marine macroalga *Ulva fasciata*. *Journal of Experimental Botany* 56, 2851–2865.

Smith, K. C. (1989) UV radiation effects: DNA repair and mutagenesis. *In* Smith, K. C. (ed.). *The Science of Photobiology*. Plenum Press, New York, 111–134.

Stapleton, A. E. (1992) Ultraviolet radiation and plants: Burning questions. *Plant Cell* 4, 1353–1358.

Sung, M.-S., Hsu, Y.-T., Hsu, Y.-T., Wu, T.-M., Lee, T.-M. (2009) Hypersalinity and hydrogen peroxide upregulation of gene expression of antioxidant enzymes in *Ulva fasciata* against oxidative stress. *Marine Biotechnology* 11, 199–209.

Talarico, L. (1996) Phycobiliproteins and phycobilisomes in red algae: adaptive response to light. *Scientia Marina* 60, 205–222.

Triantaphylidès, C., Havaux, M. (2009) Singlet oxygen in plants: production, detoxification and signaling. *Trends in Plant Science* 14, 219–228.

Van de Poll, W. H., Hanelt, D., Hoyer, K., Buma, A. G. J., Breeman, A. M. (2007) Ultraviolet-B-induced CPD formation and repair in arctic marine macrophytes. *Photochemistry and Photobiology* 76, 493–500.

Williams, S. L., Dethier, M. N. (2005) High and dry: Variation in net photosynthesis by the intertidal seaweed, *Fucus gardneri*. *Ecology* 86, 2373–2379.

Wu, T.-M., Hsu, Y.-T., Lee, T.-M. (2009) Effects of cadmium on the regulation of antioxidant enzyme activity, gene expression, and antioxidant defenses in the marine macroalga *Ulva fasciata*. *Botanical Studies* 50, 25–34.

Chapter 5

OXIDATIVE STRESS IN AQUATIC PRIMARY PRODUCERS AS A DRIVING FORCE FOR ECOSYSTEM RESPONSES TO LARGE-SCALE ENVIRONMENTAL CHANGES

Pauline Snoeijs[1], *Peter Sylvander*[1], *and Norbert Häubner*[2]

[1]Department of Systems Ecology, Stockholm University, Stockholm, Sweden
[2]Department of Ecology and Evolution, Uppsala University, Uppsala, Sweden

OXIDATIVE STRESS AND ECOSYSTEMS

In this chapter we review the protection systems of aquatic primary producers against excess reactive oxygen species (ROS) and the involvement of ROS in environmental and nutritional stress. We further explore how changes in oxidative balance and species composition may affect the nutritional quality of aquatic primary producers for higher trophic levels and the flows of essential nutrients, such as carotenoids and vitamins in aquatic food webs. In a wider perspective, the cellular process of oxidative stress emerges as a driving force for food-web and ecosystem responses to large-scale environmental changes. The Earth's ecosystems are being transformed with accelerating speed.

Oxidative Stress in Aquatic Ecosystems, First Edition. Edited by Doris Abele, José Pablo Vázquez-Medina, and Tania Zenteno-Savín.

Direct human interferences such as habitat disturbance, overexploitation of natural resources and introductions of nonindigenous species can radically change food-web structures and ecosystems. Human actions that produce large-scale environmental change, such as global warming, eutrophication, and contamination with hazardous substances, generate environmental stress and transform ecosystems more slowly. However, in an evolutionary perspective, these latter changes also occur extremely fast, and many species suffer from stress until they ultimately cannot survive or are outcompeted by other species, which are better adapted to the new environment.

ENVIRONMENTAL STRESS GENERATES OXIDATIVE STRESS

Ecological effects of environmental and nutritional stress on organisms are often mediated through the cellular process of oxidative stress (Monaghan et al. 2009). The oxidative reagents, ROS, are essential in the physiological control of cell function in biological systems. Oxidative stress occurs when the generation of ROS exceeds a cell's ability to neutralize and eliminate them. The imbalance can result from a lack of antioxidant capacity caused by a disturbance in the production, cellular distribution, or uptake of antioxidants. It can also result from an overproduction of ROS as a response to external stressors. If not regulated properly, the excess ROS can damage macromolecules such as membrane lipids, proteins, and DNA and so inhibit normal cell functioning. In recent years it has become clear that while organisms possess excessive mechanisms to combat elevated ROS levels during environmental stress situations, they also use ROS as signaling molecules to control various processes. The rates and cellular sites of ROS production during environmental stress are thought to play a central role in stress perception and protection through ROS levels and signals controlled by the ROS gene network (Suzuki and Mittler 2006). Recently, ROS have also been discussed as universal constraints in life-history evolution of animals (Dowling and Simmons 2009). In this context, ROS are suggested to be important mediators of the cost of reproduction, and of trade-offs between metabolic rate and life span, and between immunity, sexual ornamentation, and sperm quality.

OXIDATIVE STRESS AND TROPHIC INTERACTIONS

Most antioxidants or their precursors in aquatic ecosystems are produced by photoautotrophs and are essential in the diet of heterotrophs (Table 5.1). The major reason for this is that photosynthesis requires effective safety valves to eliminate hazardous excess energy from solar radiation and, thus, to prevent oxidative damage to the phototrophic cells (Fig. 5.1). Most of the antioxidant compounds that protect plant and algal cells also protect the cells of organisms at higher trophic levels (Demmig-Adams and Adams 2002). Therefore, enhancing the photosynthesizers' own protective systems may also improve the nutritional quality of foods for heterotrophic organisms, since fundamental cellular signaling processes and protective mechanisms are highly conserved (Bouvier et al. 1998; Suzuki and Mittler 2006). The quality of the aquatic primary producers as food for other organisms can both increase or decrease as a result of increased oxidative stress, through accumulation of antioxidants or through lower growth rates by oxidative damage, respectively. Continuously changing interactions between organisms and between organisms and their environment constantly change the oxidative balance in cells. This can have consequences for physiological processes and ecological interactions when normal environmental and/or seasonal fluctuations are replaced by directional trends such as large-scale environmental change.

Several methodological and conceptual breakthroughs – from ecological stoichiometry (Sterner and Elser 2002) to stable isotope measurement (Peterson and Fry 1987) – have increased our knowledge on the molecular currencies with which organisms interact with their environment and between each other (McGraw et al. 2010). Mechanisms of production and transfer of antioxidants are key features in trophic interactions, which have been largely neglected in aquatic food-web ecology, in spite of their quality and usefulness in explaining population declines and shifts in community composition in nature. Studies in this field are complicated by the heterogenic nature of antioxidant defenses and need multivariate approaches. Changes in the environment, such as temperature and salinity shifts or increased exposure to photosynthetic active radiation (PAR), ultraviolet radiation (UVR), toxic transition metals or organic pollutants, can cause

Table 5.1 Summary of antioxidants and vitamins in aquatic food webs.

Compound	Production and antioxidant function
Algal carotenoids	Fat-soluble pigments produced by photoautotrophs. Carotenoids provide protection of the photosynthetic apparatus by quenching triplet chlorophyll and singlet oxygen and by dissipating excess excitation energy as heat (Shimidzu et al. 1996).
Animal carotenoids	All animal carotenoids originate from photoautotrophs and occur in animals either in their original form or they are transformed to other carotenoids in the animal body (Miki 1991; Matsuno 2001).
Phycobiliproteins	Phycobiliproteins are water-soluble pigments produced by cyanobacteria and some eukaryotic algal groups. They are components of the light-harvesting antenna complexes and their antioxidant capacity has recently been described by, e.g., Benedetti et al. (2010) and Cano-Europa et al. (2010).
Vitamin A (Retinol)	Some carotenoids, e.g. α-carotene, β-carotene, γ-carotene and β-cryptoxanthin, act as "provitamin A". They are first transferred to the aldehyde retinal and then to retinol.
Vitamin B_1 (thiamine)	Vitamin B_1 plays a vital role as a coenzyme in carbohydrate metabolism, but can also act as an antioxidant. It inhibits lipid peroxidation (Lukienko et al. 2000) and possesses free radical scavenging activity (Gliszczyńska-Owigło 2006). Most cyanobacterial, red algal, brown algal, and diatom species can produce thiamine, but most cryptophyte, euglenophytes, and haptophyte species cannot. Most heterotrophs are unable to produce thiamine.
Vitamin C (ascorbic acid)	The major water-soluble antioxidant in cell plasma. The ascorbate ion is required in the cellular metabolism of all organisms. In most species it is a normal metabolite of glucose and not an essential dietary constituent.
Vitamin D (calciferol)	Fat-soluble prohormones that occur as ergocalciferol (vitamin D_2) and cholecalciferol (vitamin D_3) in phyto- and zooplankton (Boland 1986; Takeuchi et al. 1991; Rao and Raghuramulu 1996). Vitamin D in fish originates from algae, which synthesize it with the help of solar UV. In vertebrates only vitamin D_3 occurs. As antioxidants, vitamin D_2 and D_3 inhibit liposomal lipid peroxidation (Wiseman 1993) and vitamin D_3 decreases oxidative DNA damage (Fedirko et al. 2010).
Vitamin E (tocopherol)	The fat-soluble tocopherols are produced by photoautotrophs and serve as singlet oxygen scavengers under conditions of photoinhibition. α -tocopherol, the most abundant form of the tocopherols in nature is produced in the chloroplasts of plant cells and has the highest antioxidant activity *in vivo* among the tocopherols.
Glutathione (GSH)	A water-soluble tripeptide, which is crucial for oxidative stress management. It is an essential component of the glutathione–ascorbate cycle, which reduces excess H_2O_2. Glutathione is not an essential nutrient since it can be synthesized from several amino acids.

increased ROS production and thereby a larger need for antioxidant defenses (Fig. 5.1). Environmental change can also trigger a lower production of antioxidants in aquatic food webs, e.g. when large species shifts occur in the primary producers. If the new algal community composition has lower nutritional and/or antioxidant levels than the previous community, this can cause oxidative stress at higher trophic levels.

PROTECTION AGAINST OXIDATIVE STRESS IN PHOTOAUTOTROPHS

Oxidants and Antioxidant Defenses

The four major ROS in photoautotrophs are $^1O_2$, $O_2^{\bullet -}$, $HO^\bullet$, and H_2O_2. Cellular antioxidant defenses involve several strategies, both enzymatic and nonenzymatic. In the lipid phase, carotenoids, retinol (vitamin A),

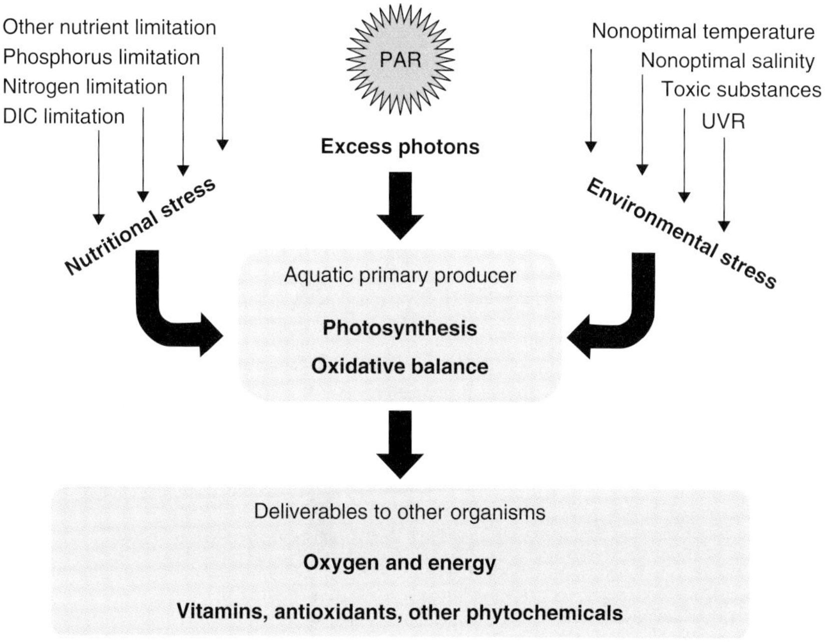

Fig. 5.1 Summary of external causes of ROS production in cells of aquatic primary producers. These factors affect the cell's photosynthetic rate and oxidative balance, which affect the quality of the deliverables from primary producers to other organisms in terms of oxygen, energy (food), and phytochemicals. The terms vitamins, antioxidants, and other phytochemicals are not mutually exclusive. DIC = dissolved inorganic carbon; PAR = photosynthetically active radiation; UVR = ultraviolet radiation.

Table 5.2 Summary of major antioxidant enzymes in aquatic primary producers.

Abbreviation	Full name	Antioxidant function
SOD	Superoxide dismutase	Catalyzes the dismutation of the superoxide radical ($O_2^{\bullet-}$) into O_2 and H_2O_2
CAT	Catalase	Catalyzes the decomposition of H_2O_2 into H_2O and O_2
APx	Ascorbate peroxidase	Catalyzes the reaction ascorbate + H_2O_2 = DHA + 2 H_2O
DHAR	Dehydroascorbate reductase (DHA reductase)	Catalyzes the reaction 2 glutathione + DHA ⇔ glutathione disulfide + ascorbate
MDHAR	Monodehydroascorbate reductase (MDA reductase)	Catalyzes the reaction NADH + H^+ + 2 MDA ⇔ NAD^+ + 2 ascorbate
GPx	Glutathione peroxidase (GSH peroxidase)	Catalyzes the reaction 2 glutathione + H_2O_2 ⇔ glutathione disulfide + 2 H_2O
GR	Glutathione reductase (GSH reductase)	Catalyzes the reaction glutathione disulfide + NADPH + H^+ ⇔ 2 glutathione + $NADP^+$

calciferol (vitamin D), and tocopherol (vitamin E), and in the aqueous phase phycobiliproteins, ascorbate (vitamin C), glutathione (GSH), and flavonoids are examples of antioxidant molecules (Table 5.1). Some of these molecules are substrates for antioxidant enzymes, notably superoxide dismutase (SOD), catalase (CAT), glutathione peroxidase (GPx), glutathione reductase (GR), ascorbate peroxidase (APx), dehydroascorbate reductase (DHAR), and monodehydroascorbate reductase (MDHAR) (Table 5.2).

The enzymatic defenses concentrate on $O_2^{\bullet-}$ and H_2O_2, whereas carotenoids and tocopherols are efficient quenchers of the highly reactive $^1O_2$ (Sies 1997; Ledford and Niyogi 2005). The local concentration of carotenoids in a phototrophic cell is decisive in determining the efficiency of $^1O_2$ quenching. The most reactive ROS is HO•, which is generated by the interaction of H_2O_2 with reductants such as reduced transition metal ions (Asada 1999), and the main strategy is to prevent the formation of HO• through enzymatic removal of H_2O_2 with the help of CAT, APx and GPx. Several other small-molecule antioxidant compounds are involved in cellular defense against ROS in photoautotrophs, e.g. the osmolytes mannitol and dimethylsulphoniopropionate (DMSP), and the enzymatic cleavage product of the latter, dimethylsulphide (DMS), can quench HO, (Shen et al. 1997, Sunda et al. 2002). Also some mycosporine-like amino acids (MAAs), which absorb UVR, have antioxidant activity (Dunlap and Yamamoto 1995). These MAAs are synthesized *de novo* by the shikimic acid pathway in photoautotrophs and are acquired by animals through their diet (Shick and Dunlap 2000).

Most antioxidant defenses are located in the chloroplasts and the mitochondria, where O_2 is most active during photosynthesis and respiration, respectively (Wise 1995; Lesser 2006). Mitochondria are the main source of ROS in heterotrophs (Abele 2002), but in photoautotrophs only a small proportion of the ROS are produced in mitochondria, because an alternative endoxidase offers an alternative pathway for electron flow through the electron transport chain to this final alternative reduction site for oxygen (Purvis 1997).

Protection of the Photosynthetic Apparatus

Photosynthetic organisms harvest light energy and transform it into chemical energy to be used mainly in cellular maintenance and growth. In the ideal case, all energy harvested by the antenna complexes in the chloroplasts would be either utilized in photosynthesis or emitted back to the surroundings. However, even under optimal conditions, the photosynthetic apparatus cannot utilize all energy absorbed and part of the excess energy is transferred to molecular oxygen and ROS formation (Falkowski and Raven 2007). When singlet excited states of chlorophyll (^{1}Chl•) are not readily processed by photochemistry they can convert to triplet excited states (^{3}Chl•), which can transfer their energy directly to oxygen. The resulting highly reactive excited form of oxygen ($^1O_2$) is believed to be the main oxygen species causing photoinhibition, i.e. the light-induced loss of photosystem (PS) II activity (Ledford and Niyogi 2005; Krieger-Liszkay et al. 2008). After two billion years of fine-tuning (Des Marais 2000), oxygenic photosynthetic organisms have developed a wide range of protective mechanisms, both to prevent the formation of ROS and to detoxify ROS once they have formed. Following Asada (1999), phototrophs use three major routes to dissipate energy from absorbed photons that cannot be utilized for CO_2 assimilation: the water–water cycle, down-regulation of PS II and photorespiration (Fig. 5.2).

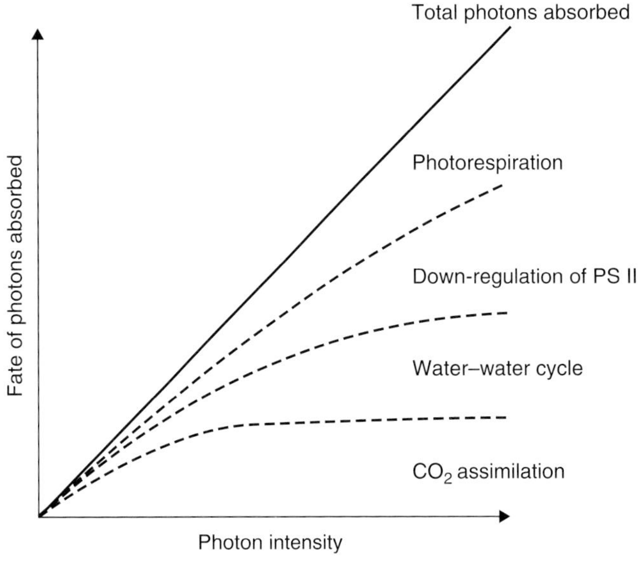

Fig. 5.2 The fate of photon energy absorbed by chlorophylls in chloroplasts under various photon intensities. When the photon intensity is in excess of CO_2 assimilation, excess photon energy is dissipated by the water–water cycle, down-regulation of PS II, and photorespiration. The photo-utilizing capacity for CO_2 assimilation is also affected by environmental and nutritional stress (see Fig. 5.1). (Figure adapted from Asada 1999.)

In the water–water cycle (Fig. 5.2), O_2 produced by reduction of H_2O in PS II is reduced in photosystem I (PS I), and $O_2^{\bullet -}$ is formed. $O_2^{\bullet -}$ is then disproportionated to H_2O_2 and O_2 with the aid of SOD (Wolfe-Simon et al. 2005). The H_2O_2 can further be reduced to H_2O with the aid of APx and its substrate ascorbate (Asada 1997). Through the water–water cycle adenosin triphosphate (ATP) is produced via a proton gradient across the thylakoid membrane, and the mechanism might therefore, even though it is not directly involved in carbon fixation, harvest some of the excess energy that the main photosynthetic pathway is unable to exploit. Since the electron flow from H_2O in PS II to H_2O in PS I occurs in this process, it was named the water–water cycle (Asada 1999). The most important function of this cycle is a rapid, immediate scavenging of $^1O_2$ and H_2O_2 at the site of its generation prior to their interaction with target molecules and the formation of $HO^\bullet$.

Down-regulation of PS II (Fig. 5.2) involves protonation and conformational change of the light harvesting complex as a result of the transthylakoid proton gradient of the water–water cycle, as well as the ability of photoprotective carotenoids like β-carotene, zeaxanthin, antheraxanthin, and diatoxanthin to quench singlet-excited chlorophylls ($^1Chl^\bullet$) and harmlessly dissipate excess excitation energy as heat. These nonphotochemical quenching (NPQ) processes occur in almost all photosynthetic eukaryotes (Asada 1999; Müller et al. 2001). Additional mechanisms that very rapidly mobilize photoprotective pigments are xanthophyll cycles. Xanthophyll cycles include the de-epoxidation of light harvesting into photoprotective xanthophylls (Demmig-Adams and Adams 1996). This reaction can occur rapidly when light conditions change and happens on the scale of minutes for microalgae and hours for higher plants (Falkowski and Raven 2007).

Photorespiration (Fig. 5.2) is a metabolic pathway that consumes O_2, releases CO_2, produces no ATP, and lowers the net photosynthetic output. Photorespiration occurs when the O_2 concentration rises relative to CO_2 in the plant cell and the enzyme ribulose-1,5-bisphosphate carboxylase oxygenase (RuBisCO) catalyzes the oxygenation of the sugar ribulose-1,5-bisphosphate (RuBP) instead of the carboxylation. Photorespiration dissipates excess harvested light energy and mitigates the potential effects of chronic photoinhibition when photosynthetic CO_2 assimilation is decreased (Asada 1999).

Responses of Primary Producers to Light

Fluctuating light conditions are natural events and cannot be classified as environmental stress conditions. The need to dissipate excess photon energy is everpresent for aquatic primary producers. Attached algae in intertidal zones strongly rely on their photoprotective mechanisms when they are exposed to high irradiation at low tide. Also free-floating planktonic algae need to be able to increase their photoprotection when they are maneuvered towards the water surface. In a mesocosm experiment, Barros et al. (2003) showed that low-biomass phytoplankton communities had higher lipid oxidative damage, higher SOD and CAT activities, and higher excretion of H_2O_2 to the surrounding water, compared with high-biomass phytoplankton communities which were less exposed to PAR through self-shading.

An increase of the cellular content of protective carotenoids is a common response to increased PAR, but it is not universal. Some antioxidants are destroyed in the process of oxidant quenching (Trebst 2003) and some antioxidants are themselves destroyed directly by intensive light (Young and Britton 1990). For the net change in antioxidant concentrations in a cell to be positive, up-regulation of the production must be larger than the consumption. Young and Britton (1990) found that the green microalga *Dunaliella* sp. exposed to extreme PAR (4000 μmol m^{-2} s^{-1}) for 2 h lost approximately 20% of its carotenoid content and, also, approximately 15% of its chlorophyll content. Contrary, zeaxanthin increased by approximately 200%. A general observation in this and later studies (Siems et al. 2002) is that carotenes (e.g. α- and β-carotene) are more sensitive to photobleaching than xanthophylls (oxygenated carotenoids, e.g. lutein, zeaxanthin, neoxanthin, violaxanthin, α- and β-cryptoxanthin), and that chlorophyll shows sensitivity to bleaching which ranges between carotenes and xanthophylls.

Closely related coexisting algal species can have different strategies for protection against oxidative stress. For example, the littoral green alga *Cladophora* sp. invests more in carotenoid protection against oxidative stress than the coexisting littoral green alga *Ulva* sp. With increasing PAR, *Cladophora* sp. increased its carotenoid/chlorophyll ratio through *de novo* synthesis of carotenoids and activated its violaxanthin xanthophyll cycle (Choo et al. 2005). *Cladophora* sp. also has high baseline activities of the H_2O_2 scavenging enzymes CAT and APx (Choo et al. 2004). *Ulva* sp., on the other hand, has no xanthophyll cycle and releases much more H_2O_2 to the seawater medium, while intracellular CAT and APx levels are low unless induced by additional oxidative stress (Abrahamsson et al. 2003; Choo et al. 2004). This renders *Ulva* sp. more vulnerable to oxidative stress when it is exposed to sudden UV-B radiation (which directly inhibits PS II; Schofield et al. 1995) than *Cladophora* sp. While the ecophysiological traits of *Cladophora* sp. seem to be directed toward persistence, those of *Ulva* sp. seem to be more engaged with short-term gains. This is in accordance with their different life strategies (Choo et al. 2005). Ursi et al. (2003) showed that *de novo* synthesis of carotenoids can also take place in red algae as a response to increased PAR. They found that zeaxanthin decreased proportionally to a decrease in violaxanthin, and that β-carotene increased through *de novo* synthesis in the red macroalga *Gracilaria* sp. (Fig. 5.3).

Different responses to light between coexisting species demonstrate that the antioxidant content of photoautotrophs as food for higher trophic levels in

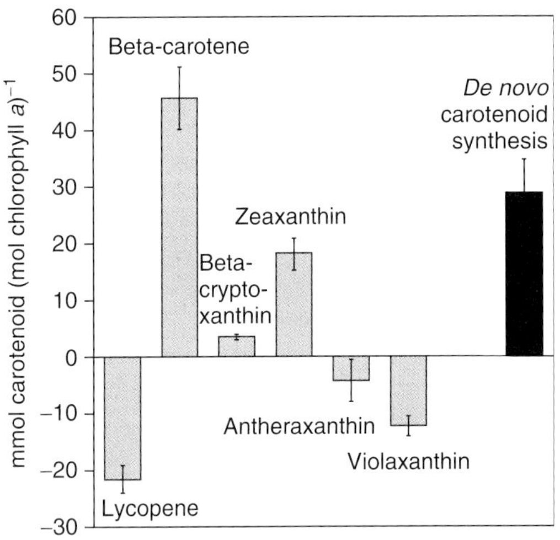

Fig. 5.3 Budget of carotenoid concentrations in six strains of the tropical red macroalga *Gracilaria birdiae*. Grey bars show the difference in carotenoid contents between light and dark incubations. In the light incubations the quenchers β-carotene and zeaxanthin increased and in the dark incubations lycopene and violaxanthin increased. The black bar shows the net *de novo* synthesis of the six carotenoids in the light incubations. Error bars indicate one standard error of the mean ($n = 6$ strains) (Figure adapted from Ursi et al. 2003).

natural waters is affected by species composition. The cyanobacterium *Anabaena* was shown to increase its β-carotene content upon exposure to high PAR (Han et al. 2003). Bhandari and Sharma (2006) found that β-carotene also increased in the cyanobacterium *Nostoc* sp. after exposure to high PAR, but it decreased in the cyanobacterium *Phormidium* sp., while chlorophyll *a*, phycobiliproteins, and α- and β-cryptoxanthin increased in both species. Depka et al. (1998) found an increased content of zeaxanthin in the green microalga *Chlamydomonas* sp. under light stress that could not be explained by a simultaneous decrease in violaxanthin alone, and they assumed that a part of β-carotene, which also decreased in concentration, had been rapidly converted to zeaxanthin.

Not only light intensity, but also the spectral composition of the light, affects algal pigment dynamics. For example, under red and green light, the carotenoid myxoxanthophyll decreased while β-carotene increased when compared to white light in the cyanobacterium *Spirulina* sp. (Olaizola and Duerr 1990). Green light stimulated the synthesis of α-carotene and lutein in the red alga *Halymenia* sp., whereas synthesis of the phycobiliproteins phycocyanin and phycoerythrin was stimulated by blue light (Godínez-Ortega et al. 2008). In the cyanobacterium *Fremyella* sp., green light stimulated the accumulation of phycoerythrin at the expense of phycocyanin and red light *vice versa* (Kehoe and Gutu 2006). The antioxidant capacity of phycobiliproteins was recently assessed by Benedetti et al. (2010) and Cano-Europa et al. (2010).

ENVIRONMENTAL STRESS AND OXIDATIVE STRESS IN PRIMARY PRODUCERS

Temperature Stress

The potential consequences of global warming on the biota in terrestrial and aquatic ecosystems represent one of the major questions in contemporary ecological research. In the face of observed and predicted future increases in global surface temperature (IPCC 2007), it is mandatory to know how temperature change affects organisms, food webs, and ecosystem processes. High-latitude aquatic ecosystems are considered to be especially sensitive to the changing climate. For example, the annual average surface water temperature in the subarctic Baltic Sea has already increased by 1.4°C during the past 100 yr, more than in any of the world's other 63 large marine ecosystems (MacKenzie and Schiedek 2007), and is expected to increase by 1–5°C during the coming 100 yr depending on choice of prediction model (BACC Author Team 2008) and global political decisions (IPCC 2007).

Temperature stress can have a devastating effect on the metabolism of primary producers by disrupting cellular homeostasis and uncoupling major physiological processes. Too low and too high temperatures limit the fixation of CO_2 and have been shown to accelerate photoinhibition (Takahashi and Murata 2008). Choo et al. (2004) found that lipid oxidative damage in two green algal species was higher in the combination of high irradiance and low temperature than in the combination of high irradiance and high temperature (Fig. 5.4). At lower temperatures the photosynthetic enzymatic processes (carbon fixation) slow down, whereas the photochemical processes (light reaction) remain relatively unaffected, and, thus, the excess excitation energy in PS II is larger. This imbalance of the kinetics in the photosynthetic reactions increases ROS production and can induce chilling-enhanced photo-oxidation manifested as the light- and oxygen-dependent bleaching of photosynthetic pigments (Wise 1995). Chemical reactions are generally slower at lower temperature and the chemical quenching of ROS can decrease with temperature. However, the responses of photosynthetic organisms to temperature are not uniform because each species has its optimal, suboptimal, and lethal temperature levels.

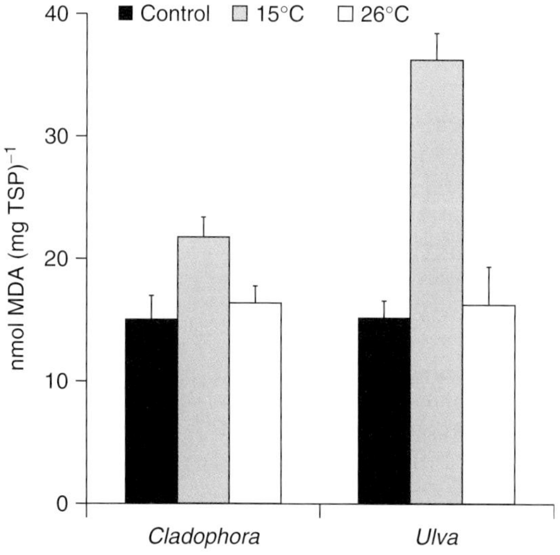

Fig. 5.4 Oxidative damage expressed as mean thiobarbituric acid-reactive substances (TBARS) in *Cladophora glomerata* and *Ulva procera*, measured as malondialdehyde (MDA) equivalents per unit total soluble protein (TSP). Control = field samples, 15 and 26°C = after 6 h incubation at 600 μmol photons PAR m^{-2} s^{-1} at 15 and 26°C, respectively. Error bars indicate one standard error of the mean ($n = 4$). For both species, the MDA levels for the Control and the incubation at 26°C were not significantly different and the MDA levels for the two incubations at 15°C were higher than for those at 26°C. The MDA level in *Ulva* at 15°C was higher than that in *Cladophora* (Figure adapted from Choo et al. 2004).

Many studies have shown that ROS-scavenging mechanisms have an important role in protecting photoautotrophs against temperature stress (e.g. Wise 1995; Yabuta et al. 2002). In response to increased ROS production by temperature stress, different types of antioxidant mechanisms are mobilized in algae. In the green macroalga *Ulva* sp., Choo et al. (2004) found increased activity of APx and increased release of H_2O_2 to the seawater medium in suboptimal temperature. In two other studies, the green microalga *Dunaliella* sp. exposed to suboptimal temperature increased its β-carotene and α-carotene levels, respectively (Ben-Amotz 1996; Orset and Young 1999). In both studies also a higher 9Z/all−*E* stereoisomer ratio was found in the β-carotene globules at suboptimal temperature. This can be interpreted as a protection mechanism in which the oily 9Z stereoisomer shields the crystallization of the all-*E* isomer at low temperature (Ben-Amotz 1996). This isomeric difference might be explained by the higher reactivity of *Z* compared to *E* bonds (Levin and Mokady 1994).

Not only carotene levels but also xanthophyll cycles can be involved in the defense against temperature-induced oxidative stress. Ursi et al. (2003) were able to manipulate the xanthophyll cycle in the red macroalga *Gracilaria* sp. by varying the temperature in a laboratory experiment. Higher concentrations of the light-harvesting carotenoid violaxanthin were found at 30°C (optimal temperature) and higher concentrations of the quencher zeaxanthin at 20°C (suboptimal temperature).

Salinity Stress

Although salinity changes are predicted to occur in the course of climate change in certain areas such as the brackish Baltic Sea (Meier et al. 2006), changes in salinity frequently occur in nature and are not a large-scale environmental problem. Especially algae growing in intertidal areas are exposed to strong environmental stress by extreme salinities in combination with desiccation and exposure to high PAR and UVR. Water stress, as that simulated by hypersaline conditions, is known to enhance $O_2^{\bullet -}$ production in chloroplasts (Asada and Takahashi 1987). Experimental salinity increase caused oxidative damage to proteins and lipids in the tidal-flat diatom *Cylindrotheca* sp., and increases in the β-carotene/chlorophyll *a* ratio as well as SOD and MDHAR activities (Rijstenbil 2003; Roncarati et al. 2008). APx and GR activities were unaffected by the salinity change. Jahnke and White (2003) showed that both hyper- and hypo-osmotic stress induced antioxidant responses in the salt-tolerant green microalga *Dunaliella* sp., but that the respective responses were different. MDHAR and APx activities increased under hyperosmotic stress, whereas hypo-osmotic stress caused major increases in glutathione and α-tocopherol and reduced ascorbate. SOD, CAT, DHAR, and GR activities were not different between normal and extreme salinities. Lu et al. (2006) studied salinity stress in the green macroalga *Ulva* sp. and found that increased thallus H_2O_2 concentrations were positively correlated to lipid oxidative damage (TBARS).

UV Stress

Depletion of stratospheric ozone that normally filters the UV-B fraction (280–320 nm) of sunlight has been linked to the use of volatile halogenated compounds, particularly chlorinated and brominated methanes and chlorofluorocarbons. Despite ozone depletion leveling off (McKenzie et al. 2003), the UV transparency of inland aquatic ecosystems remains highly variable and subject to increased UVR exposure due to climate change. UVR mainly affects terrestrial systems, but penetrates to ecologically significant depths in aquatic systems and can affect both marine and freshwater systems from phytoplankton to zooplankton and fish (Häder et al. 2007). Especially algae living at shallow depths in clear waters or in intertidal zones can be subjected to UVR. Exposure to UV-B radiation creates oxidative stress in aquatic photoautotrophs. For example, Shiu and Lee (2005) documented production of $O_2^{\bullet -}$ and H_2O_2 when the green macroalga *Ulva* sp. was exposed to UV-B radiation. For the cyanobacterium *Anabaena* sp., He et al. (2002) showed enhanced lipid peroxidation, increased DNA strand breaks, and elevated chlorophyll bleaching upon exposure to moderate UV-B radiation. However, *Anabaena* sp. fully adapted to the UV-B radiation level in two weeks. An efficient defense system presumably repaired the damaged photosynthetic apparatus and DNA, and induced *de novo* synthesis of proteins and lipids, allowing the organisms to adapt successfully to UV-B stress and survive as well as grow. Different responses of aquatic photoautotrophs to UV-B generated oxidative stress in primary producers have been recorded. For example, in the green alga *Monostroma* sp. and the red algae *Coccotylus* sp. and *Phycodry* sp., GR was stimulated by UV-B radiation (Aguilera et al. 2002). In the diatom *Thalassiosira* sp. both SOD and GR activities increased, but APx activity remained unaffected (Rijstenbil 2002). Results from laboratory cultures are related to growth phase. In log-phase cultures of the green microalga *Chlorella* sp., UV-B radiation caused 284% and 145% increases in ROS generation and lipid peroxidation, respectively, whereas stationary-phase cultures showed no significant changes in these parameters with UV-B radiation (Malanga and Puntarulo 1995; see Chapter 36).

Toxic Stress

ROS production is promoted by hazardous substances in the environment, both anthropogenic through chemical pollutants (Chapters 21 and 22) and natural, such as algal toxins. At high levels of hazardous substances, damage to algal cells occurs because ROS levels exceed the capacity of the cell to neutralize them. At lower chronic levels, algae can accumulate hazardous substances and pass them on to other trophic levels. ROS increase either through a general disturbance of metabolism and other cellular mechanisms or by acting directly on specified cellular reactions (Limón-Pacheco and Gonsebatt 2009). Oxidative biomarkers in aquatic organisms can, in combination with other types of biomarkers (hepatic, genotoxic, hemato-immunotoxic), be used in large-scale environmental monitoring programs (van der Oost et al. 2003; Valavanidis et al. 2006; Chapter 23).

The toxicity of transition metals is due either to the inactivation of functional proteins, including enzymes, caused by direct binding of the metal ion, or to oxidative damage caused by the accelerated generation of ROS (Stohs and Bagchi 1995; Pinto et al. 2003a). Excess free metal ions can cause increased $HO^{\bullet}$ production through the Fenton reaction. Some metals such as zinc and copper are also essential in cell function, and phytochelatins, peptides synthesized from glutathione, play important roles in the binding and detoxification of transition metals as well as maintenance of the homeostasis of intracellular levels of essential metal ions in eukaryotes and cyanobacteria (Hirata et al. 2005).

Microcystins (peptide hepatotoxins), which are produced by some strains of the cyanobacterial genera *Microcystis, Anabaena, Oscillatoria, Nostoc, Hapalosiphon* and *Anabaenopsis*, have been shown to generate oxidative stress in other organisms, such as laboratory mammals (van Apeldoorn et al. 2007), fish (Amado and Monserrat 2010), the invertebrate *Daphnia* sp. (Wiegand et al. 2002), and the cyanobacterium *Synechococcus* sp. (Hu et al. 2005). Another hepatotoxin nodularin, produced by the cyanobacterium *Nodularia* sp., can cause oxidative stress in a wide range of organisms, including algae (Pflugmacher et al. 2007), crustaceans, mammals (Bouaicha and Maatouk 2004), and fish (Persson et al. 2009). Nodularin can be transferred to higher trophic levels from cyanobacteria to planktivores through grazing copepods (Karjalainen et al. 2007). However, in contrast to the biomagnification seen in the trophic transfer of many persistent organic pollutants, nodularin is biodiluted in higher trophic levels, and this route of exposure is therefore not a major source of oxidative stress in top predators.

Combined Environmental Stresses

While single environmental stresses can be simulated in laboratory experiments, in nature they occur in many different combinations, with different intensities and with temporal variations. In addition, it is difficult to separate the role of each individual ROS, and also antioxidants have overlapping functions (Ledford and Niyogi 2005). UV and toxic stress are usually also combined with direct toxic effects, whereas temperature and salinity stress only affect the cellular redox balance and no direct toxicity is involved. Species are able to adapt to increased UVR (He et al. 2002) and increased metal concentrations (Contreras et al. 2005) to a certain limit, but some species have a higher basal stress tolerance than others, which is coupled to their ability to deal with oxidative stress. This correlation between reactive oxygen metabolism and general stress tolerance is best illustrated by examples from intertidal seaweeds, which at low tide are daily exposed to high levels of PAR and UVR, in combination with desiccation, high salinity, and sometimes even freezing. Collén and Davison (1999a–1999c) showed that red and brown algae growing higher up in the intertidal zone are more tolerant to environmental stress and that this is coupled to oxidative stress. The red alga *Mastocarpus* sp. (growing higher up in the intertidal zone) had higher levels of ascorbate and β-carotene and higher CAT and GR activities than *Chondrus* sp. (growing lower down in the intertidal zone), whereas tocopherol content, SOD and APx activities were similar in both species. The activities of ROS-scavenging enzymes increased with tidal height within the species *Mastocarpus* sp., which indicates adaptation to ambient stress conditions (Collén and Davison 1999a). Similar results were obtained for three *Fucus* species (brown algae) growing at different levels in the intertidal zone (Collén and Davison 1999b, 1999c). These latter studies also suggested that the ratio between ROS protection and ROS production is more important than the absolute content of antioxidants.

NUTRITIONAL STRESS AND OXIDATIVE STRESS IN PRIMARY PRODUCERS

Excess input of inorganic N and P to aquatic ecosystems is a global problem which causes predictable increases in the biomass of primary producers in aquatic ecosystems (Smith 2003). Nutrient availability and stoichiometry also have strong selective roles in determining species composition of photosynthetic organisms, and eutrophication is a major driver for shifts in community composition. With respect to oxidative stress the effect of eutrophication is multifaceted. On one hand, higher growth rates and biomass increase self-shading and decrease antioxidant production; on the other hand, one nutrient can become limiting with lower growth rates and higher antioxidant production as results (Turner et al. 2003). Pinto et al. (2003b) found that SOD levels in batch cultures of the diatom *Nitzschia* sp. were positively correlated with cell density, which seems contradictory to the results of Barros et al. (2003) that self-shading reduces light-induced oxidative stress and SOD levels in phytoplankton mesocosms. Most probably, however, nutrient depletion in the batch cultures increased ROS production more than it was decreased by the lower light conditions. The results from both experiments suggest that eutrophication generally decreases oxidative stress in algal communities by lowering nutrient and light stress.

Nutrient limitation can cause oxidative stress as a result of disturbed cellular metabolism. When phototrophic cells are starving, cellular components are catabolized to maintain photosynthesis rates (Grossman et al. 1993; Falkowski and Raven 2007). Thereby, key proteins of the PS II reaction centers are destroyed and excess absorbed light energy leads to ROS production. Multiple laboratory studies on oxidative stress and nutrient limitation in algae have been carried out with the cosmopolitan green rock-pool microalga *Haematococcus* sp. This algae experiences extreme temperature, salinity, and desiccation stress in its natural habitat and is able to form cysts with high concentrations of the carotenoid astaxanthin stored in cytosolic lipid bodies. In the laboratory, astaxanthin formation in *Haematococcus* can be induced by high light intensity, salinity stress, and N, P or S starvation (Boussiba 2000; Guerin et al. 2003).

IMPLICATIONS FOR FOOD-WEB INTERACTIONS

Oxidative Stress and Nutritional Quality of Aquatic Primary Producers

The above assessment of environmental and nutritional stress demonstrates that the cellular process of oxidative stress in primary producers is a universal response to different ecological stress factors. Photoautotrophs supply the aquatic food webs with antioxidants and constitute a crucial addition to the antioxidant systems of heterotrophs. Some of these antioxidants are essential in the food of heterotrophs, e.g. tocopherol, others are essential precursors, e.g. β-carotene for retinol and astaxanthin, or they can be produced by heterotrophs but are also supplied in their food. As an example of the latter, Hapette and Poulet (1990) found that short- and long-term variations of ascorbic acid in copepods reflects the dependence of vitamin incorporation from phytoplankton. In turn, copepods and their fecal pellets are substantial carriers of ascorbic acid, constituting a potential pathway from phytoplankton to consumers at higher trophic levels.

Somewhat like a paradox, increases of ROS in photoautotrophs stimulate the production of antioxidants and, thus, increase antioxidant levels in the heterotrophs consuming them so that ROS levels in the heterotrophs decrease. This implies that factors that stimulate growth and decrease the production and levels of antioxidants in primary producers (e.g. nutrient additions, higher temperature in cold areas) pose a threat to food-web and ecosystem functioning because they can induce deficiency syndromes at higher trophic levels. As in the case of toxic pollution or overexploitation of top predators, antioxidant deficiency has the potential to hamper ecosystem functioning through trophic cascades (Pace et al. 1999).

The oxidative stress responses of primary producers to eutrophication and climate change are not uniform. Eutrophication decreases antioxidant levels in photoautotrophs through higher biomass and self-shading (Barros et al. 2003; Pinto et al. 2003b), but it may also generate the opposite when one macronutrient becomes limiting (Fig. 5.5). Climate change can lower the production of antioxidants in photoautotrophs when higher temperature increases carbon fixation so that photon energy can be exploited to a larger extent in the main photosynthetic pathway, resulting in higher growth rates and lower need for antioxidant accumulation in the cells (Ursi et al. 2003; Choo et al. 2004). However, each species has its optimal, suboptimal, and lethal temperature levels and the results may also be the opposite (Fig. 5.5). Laboratory cultures, which are subjected to sudden temperature changes can produce ROS at both lower and higher temperatures than they were adapted to (Halfen and Francis 1972, Olaizola and Duerr 1990). Toxic agents, such as hazardous substances and UV-B radiation, generally increase the production of antioxidants in primary producers, but they also directly damage cells, resulting in lower growth rates (Fig. 5.5).

Community Composition and Nutritional Quality of Aquatic Primary Producers

Besides their direct effects on the nutritional quality of primary producers in terms of antioxidant production (Fig. 5.5), eutrophication and climate change have profound effects on species composition (Worm and Lotze 2006; Hillebrand et al. 2010). The different

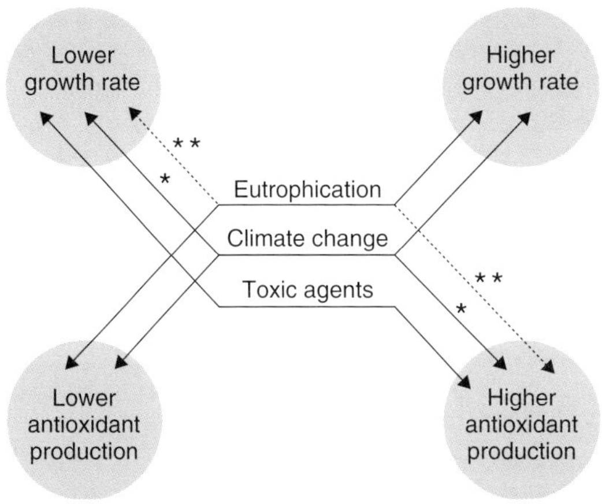

Fig. 5.5 Summary of general effects of large-scale environmental changes (eutrophication, climate change, and increased toxic agents) on growth rates and antioxidant content of photoautotrophic cells. Qualitative aspects, such as changes in species composition and changes in type of antioxidant defense, may interfere with these general effects in different ways. Note that higher oxidant production is not automatically reflected in measurements of cellular levels because antioxidants are also used to combat ROS. * = temperature stress when approaching lethal high or low temperature; ** = nutrient stress when one nutrient is limiting. Solid lines: most frequent reactions; dotted lines: reactions occurring under specific conditions.

species in a phytoplankton community can have very different species-specific antioxidant and vitamin contents. Andersson et al. (2003) and Van Nieuwerburgh et al. (2004) found that a varied diet dominated by diatoms yielded the highest astaxanthin production in zooplankton. Copepods fed with diets containing higher amounts of the astaxanthin precursors β-carotene and zeaxanthin produced high levels of astaxanthin and higher E/Z isomer ratios of astaxanthin (Rhodes 2007). In aquatic systems, the reddish-pink carotenoid astaxanthin is a major antioxidant and crustaceans are the most important producers, but they need to acquire the precursors from the photoautotrophs. Astaxanthin is involved in gonadal maturation, early development and growth, and acts as a precursor of retinol (Linán-Cabello et al. 2002). Through crustaceans, the astaxanthin is transferred to fish, which use it as an antioxidant in muscles and gonads. Zooplankton crustaceans are a major link in pelagic food webs; one important reason is the transport of vitamins and antioxidants to higher trophic levels.

Food-web Disruption and Deficiency Syndromes

Deficiency syndromes in top predators have been reported from different aquatic ecosystems, e.g. the Baltic Sea (Bengtsson et al. 1999) and the Great Lakes in North America (Fitzsimons et al. 1999). In the latter case this could be explained by food-web disruption through a nonindigenous clupeid prey-fish, the alewife, with high thiaminase levels. In the case of the Baltic Sea the explanation pattern seems more complicated and the phenomenon is still not fully understood. Atlantic salmon in the Baltic Sea show astaxanthin and thiamine (vitamin B_1) deficiencies in conjunction with a reproductive disturbance, known as the "M74 syndrome" (Bengtsson et al. 1999). Also seabirds in the Baltic Sea area die of thiamine deficiency (Balk et al. 2009). Salmon roe and yolk sac fry, i.e. life-stages that still rely on the yolk for nutrition, from astaxanthin and vitamin B_1 deficient females display imbalance in fatty acid composition and increased mortality rates (Amcoff et al. 1999; Pettersson and Lignell 1999; Pickova et al. 1999). In M74 fry, levels of α-tocopherol and ubiquinone are reduced, the cellular GSH/GSSG ratio is altered in favor of the oxidized form, and the activities of redox enzymes in the liver (GPx, GR, glutathione-S-transferase) are increased (Vuori and Nikinmaa 2007). Also adult Baltic salmon suffer from oxidative stress and the muscle have low astaxanthin levels compared to that of their Atlantic counterpart (Pettersson and Lignell 1998; Lundström et al. 1999). Recently it was found that astaxanthin is effective in improving mitochondrial function through retaining mitochondria in the reduced state (Wolf et al. 2010). Thiamine is active in the mitochondria as a coenzyme for α-ketoacid dehydrogenase and transketolases in the Krebs cycle and the oxidative pentose phosphate pathway. It can thus be hypothesized that astaxanthin helps to protect thiamine functions in salmon, but that the low astaxanthin content in the Baltic salmon populations suffering from M74 might not provide sufficient protection.

The deficiency syndrome in the Baltic Sea salmon could not be coupled to contaminants or genetic factors and the general assumption is that it is caused by large-scale environmental change and resulting food-web disruption (Bengtsson et al. 1999). During the past 50 years synchronous regime shifts have been reported for all trophic levels in the pelagic system of the Baltic Sea. These shifts are mainly attributed to climate change with higher temperature, less ice cover, and lower salinity, as well as to overfishing of the major predator fish, the cod (Alheit et al. 2005; Österblom et al. 2007; Casini et al. 2009). Major documented changes in the phytoplankton are decreases in diatoms and increases in (dino)flagellates and cyanobacteria (Wasmund and Uhlig 2003; Alheit et al. 2005; Suikkanen et al. 2007). In the zooplankton, dominance has changed from larger to smaller copepods and a community composition thriving in lower salinity and more eutrophic waters (Alheit et al. 2005; Möllmann et al. 2008). The Baltic herring, a major food item for the salmon, suffers from starvation displayed as loss of body fat (Cardinale and Arrhenius 2000). In the light of all these major changes, it may be hypothesized that deficiency syndromes in top predators such as the Baltic salmon reflect a problem much wider than that of a single species, namely that the fast changing food webs cannot keep up the balance of antioxidant and vitamin production in lower trophic levels so that oxidative stress in higher trophic levels increases.

REFERENCES

Abele, D. (2002) The radical life-giver. *Nature* 420, 27.

Abrahamsson, K., Choo, K.S., Pedersén, M., Johansson, G., Snoeijs, P. (2003) Effects of temperature on the production of hydrogen peroxide and volatile halocarbons by brackish-water algae. *Phytochemistry* 64, 725–734.

Aguilera, J., Dummermuth, A., Karsten, U., Schriek, R., Wiencke, C. (2002) Enzymatic defenses against photooxidative stress induced by ultraviolet radiation in Arctic marine macroalgae. *Polar Biology* 25, 432–441.

Alheit, J., Möllman, C., Dutz, J., Kornilovs, G., Loewe, P., Mohrholz, V., Wasmund, N. (2005) Synchronous ecological regime shifts in the central Baltic and the North Sea in the late 1980s. *ICES Journal of Marine Science* 62, 1205–1215.

Amado, L.L., Monserrat, J.M. (2010) Oxidative stress generation by microcystins in aquatic animals: why and how. *Environment International* 36, 226–235.

Amcoff, P., Börjesson, H., Landergren, P., Vallin, L., Norrgren, L. (1999) Thiamine (vitamin B_1) concentrations in salmon (*Salmo salar*) brown trout (*Salmo trutta*) and cod (*Gadus morhua*) from the Baltic Sea. *Ambio* 28, 48–54.

Andersson, M., Van Nieuwerburgh, L., Snoeijs, P. (2003) Pigment transfer from phytoplankton to zooplankton with emphasis on astaxanthin production in the Baltic Sea food web. *Marine Ecology Progress Series* 254, 213–224.

Asada, K. (1997) The role of ascorbate peroxidase and monodehydroascorbatereductase in H_2O_2 scavenging in plants. *In* Scandalios, J.G. (ed.). *Oxidative Stress and the Molecular Biology of Antioxidant Defense*. Cold Spring Harbor Laboratory Press, New York, pp. 715–735.

Asada, K. (1999) The water–water cycle in chloroplasts: scavenging of active oxygens and dissipation of excess photons. *Annual Review of Plant Physiology and Plant Molecular Biology* 50, 601–639.

Asada, K., Takahashi, M. (1987) Production and scavenging of active oxygen in photosynthesis. *In* Kyle, D.J., Osmond, C. B., Arntzen, C.J. (eds). *Photoinhibition*. Elsevier, Amsterdam, pp. 227–287.

BACC Author Team. (2008) *Assessment of Climate Change in the Baltic Sea Basin*, 2nd edn. Springer-Verlag, Heidelberg.

Balk, L., Hägerroth, P. Å., Åkerman, G., et al. (2009) Wild birds of declining European species are dying from a thiamine deficiency syndrome. *Proceedings of the National Academy of Sciences USA* 106, 12001–12006.

Barros, M.P., Pedersén, M., Colepicolo, P., Snoeijs, P. (2003) Self-shading protects phytoplankton communities against H_2O_2-induced oxidative damage. *Aquatic Microbial Ecology* 30, 275–282.

Ben-Amotz, A. (1996) Effect of low temperature on the stereoisomer composition of β-carotene in the halotolerant alga *Dunaliella bardawil* (Chlorophyta). *Journal of Phycology* 32, 272–275.

Benedetti, S., Benvenuti, F., Scoglio, S., Canestrari, F. (2010) Oxygen radical absorbance capacity of phycocyanin and phycocyanobilin from the food supplement *Aphanizomenon flos-aquae*. *Journal of Medicinal Food* 13, 223–227.

Bengtsson, B.E., Hill, C., Bergman, Å., et al. (1999) Reproductive disturbances in Baltic fish: a synopsis of the FiRe project. *Ambio* 28, 2–8.

Bhandari, R., Sharma, P.K. (2006) High-light-induced changes on photosynthesis, pigments, sugars, lipids and antioxidant enzymes in freshwater (*Nostoc spongiaeforme*) and marine (*Phormidium corium*) cyanobacteria. *Photochemistry and Photobiology* 82, 702–710.

Boland, R.L. (1986) Plants as a source of vitamin D_3 metabolites. *Nutrition Reviews* 44, 1–8.

Bouaicha, N., Maatouk, I. (2004) Microcystin-LR and nodularin induce intracellular glutathione alteration, reactive oxygen species production and lipid peroxidation in primary cultured rat hepatocytes. *Toxicology Letters* 148, 53–63.

Boussiba, S. (2000) Carotenogenesis in the green alga *Haematococcus pluvialis*: Cellular physiology and stress response. *Physiologia Plantarum* 108, 111–117.

Bouvier, F., Backhaus, R.A., Camara, B. (1998) Induction and control of chromoplast-specific carotenoid genes by oxidative stress. *Journal of Biological Chemistry* 273, 30651–30659.

Cano-Europa, E., Ortiz-Butrón, R., Gallardo-Casas, C.A., et al. (2010) Phycobiliproteins from *Pseudanabaena tenuis* rich in c-phycoerythrin protect against $HgCl_2$-caused oxidative

Chapter 6

MIGRATING TO THE OXYGEN MINIMUM LAYER: EUPHAUSIIDS

Nelly Tremblay[1], *Tania Zenteno-Savín*[2], *Jaime Gómez-Gutiérrez*[1], *and Alfonso N. Maeda-Martínez*[2]

[1]Centro Interdisciplinario de Ciencias Marinas, Departamento de Plancton y Ecología Marina, Av. IPN s/n, Col. Palo de Santa Rita, La Paz, Baja California Sur, Mexico
[2]Centro de Investigaciones Biológicas del Noroeste. La Paz, Baja California Sur, Mexico

EUPHAUSIID DAILY VERTICAL MIGRATIONS

The largest animal daily migration on the planet takes place in the water column of the oceans. Every day, massive numbers of zooplanktonic and micronektonic organisms move from deeper to surface waters during dusk and in the opposite direction during dawn, a process known as daily vertical migration (DVM). It means that animals must dynamically respond to multiple vertical gradients. Gradients in temperature, food, and oxygen concentrations are the most influential factors affecting animal behavior and physiology. Apparently, ecophysiological responses of euphausiids (Crustacea: Euphausiacea), commonly known as krill, during their DVM involve changes in the reactive oxygen species/total antioxidant capacity (ROS/AOX) balance. Euphausiids that perform the deepest migrations and cross the oxygen minimum layer (OML) exhibit the highest antioxidant enzyme activities at the surface.

Although there are only 86 euphausiid species worldwide, they are a key component of marine pelagic ecosystems. Some of them form massive aggregations or even schools that are voracious predators of phytoplankton, marine snow, micro- and mesozooplankton, and are, in turn, the main prey of carnivorous zooplankton, fish, squids, marine birds, and mammals. As part of their strategies to avoid predation, euphausiids display extensive DVM (Zaret and Sufferen 1976; Ohman 1984). When they migrate to deeper waters, euphausiids are exposed to low oxygen concentrations, sometimes penetrating into the OML (Brinton 1979; Fernández-Álamo and Färber-Lorda 2006) and decreasing their metabolic rates (McLaren 1963; Enright 1977). Thus, krill migrate daily through significant vertical oxygen gradients that imply fast metabolic and respiratory adjustments during descent or ascent migration (Mauchline and Fisher 1969; Simmard et al. 1986; Onsrud and Kaartvedt 1998). Most krill species measure less than 3 cm in length and the depth amplitude of their DVM varies among

Oxidative Stress in Aquatic Ecosystems, First Edition. Edited by Doris Abele, José Pablo Vázquez-Medina, and Tania Zenteno-Savín.

species. Considering their relatively small size, they can migrate between 100 and 600 m depth every day (Brinton 1979). Krill species of the genus *Stylocheiron* do usually not migrate and instead remain relatively close to the surface throughout their circadian cycle (Brinton et al. 2000). Other species, like the Antarctic krill *Euphausia superba*, recently reported to migrate down to abyssal depths (>3000 m; Clarke and Tyler 2008), perform large DVMs. Previous studies in the Gulf of California have shown that regional bathymetry, depth and intensity of thermocline, and the depth of the OMLs influence the interspecific and intraspecific variability of krill DVM (Brinton 1962, 1979; Lavaniegos 1996). Studying krill physiological responses associated with the DVM represents a significant logistical and technological challenge.

THE OXYGEN MINIMUM LAYER

Oxygen minimum layers typically occur at intermediate depth (300–2500 m) in most of the oceans (Emelyanov 2005). The larger, more pronounced, and shallower OMLs are located in the northern Indian Ocean, the eastern Atlantic off northwest Africa, and the eastern Tropical Pacific (Wyrtki 1962; Kamykowski and Zentara 1990; Olson et al. 1993). In the eastern Tropical Pacific, the upper limit of the OML, here defined as seawater strata with <1 mL O_2 L^{-1}, can be as shallow as 60 m depth at the entrance to the Gulf of California (Fiedler and Talley 2006). Among the mechanisms that cause OMLs are: (1) formation of a strong thermocline, which limits oxygen diffusion inside the OML; (2) a weak or nonexistent oceanic circulation; and/or (3) high local primary production, which contributes to a decrease in oxygen concentration (Wyrtki 1962; Olson et al. 1993). These low-oxygen pelagic environments provide stable conditions throughout the year allowing the partial or permanent establishment of highly adapted zooplanktonic assemblages (Childress and Seibel 1998). It was recently demonstrated that the OMLs of the eastern Tropical Pacific and the eastern Atlantic off northwest Africa have expanded into higher latitudes during the past 50 yr (Stramma et al. 2008), suggesting that zoogeographic patterns and regionalization of biomass production can change in latitude extension.

KRILL RESPIRATION RATES

Temperate and polar krill, such as *Euphausia pacifica* and *Euphausia superba*, which apparently do not inhabit regions with strong or shallow OMLs, typically decrease their respiration rates when exposed experimentally to depleted oxygen concentration (Small and Hebard 1967; Teal and Carey 1967; Childress 1975). Antezana (2002) measured respiration rates of *Euphausia mucronata*, an endemic species from the upwelling region off Peru–Chile. This region has high productivity and an extremely shallow OML, originating from the coastal Peru–Chile undercurrent that transports nutrient-rich anoxic waters of equatorial origin poleward. Respiration rates of *E. mucronata* did not change when animals were exposed to dissolved oxygen above and below the OML and the authors concluded that this is a species adapted to low oxygen concentrations (Antezana 2002). *Euphausia mucronata* has a larger gills/cephalothorax surface ratio compared with other species of the genus *Euphausia*, a feature interpreted as an adaptation to the hypoxic waters of the Humboldt Current System (Antezana 2002). Indeed, larger gills increase the surface area for oxygen uptake (Antezana 2002), supporting respiration during daily migration through the OML.

Nyctiphanes simplex and *Nematoscelis difficilis* are the most abundant euphausiid species in the northern and central regions of the Gulf of California; *N. simplex* is a subtropical and *N. difficilis* a temperate species. Under experimental conditions, *N. simplex* specimens do not survive when incubated at O_2 concentrations lower than 2.5 mL O_2 L^{-1} (exposure <15 min) (Fig. 6.1), indicating a significant lack of tolerance to low oxygen concentrations (Tremblay 2008). Unlike temperate and polar krill that decrease respiration rates in hypoxic waters (Small and Hebard 1967; Teal and Carey 1967; Childress 1975), *N. simplex* and *N. difficilis* increase their respiration rates when exposed to low oxygen conditions (Fig. 6.1), which suggests that they also increase their respiration rates when entering the OML (Tremblay 2008).

BIOCHEMICAL BIOMARKERS ASSOCIATED WITH POORLY OXYGENATED WATER

Anaerobic metabolism has been studied in several krill species. Under natural and laboratory conditions, lactate content in the temperate krill species *Meganyctiphanes norvegica* increased when oxygen concentration decreased (Spicer et al. 1999). *M. norvegica*, a species that does not normally distribute in zones with low oxygen content, did not survive

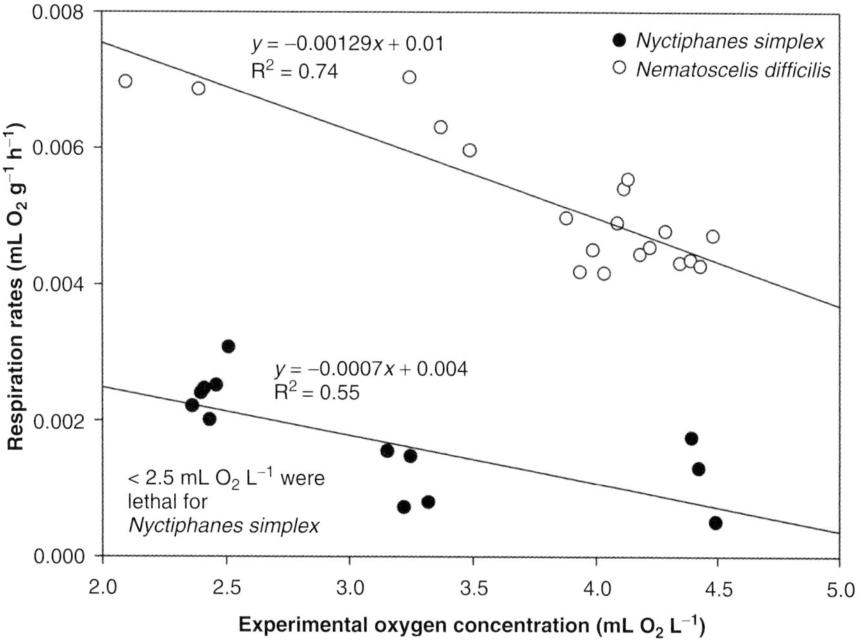

Fig. 6.1 *Nyctiphanes simplex* and *Nematoscelis difficilis* respiration rates (mL O_2 g^{-1} h^{-1}) measured as a function of experimental dissolved O_2 concentration (mL L^{-1}) at constant temperature (16± 1°C). (From Tremblay 2008.)

under prolonged hypoxic conditions (>12 h; Spicer et al. 1999). A comparative study with the euphausiid *E. mucronata*, which regularly enters the OML, and the copepod *Calanus chilensis*, which always distributes above the OML, found that *E. mucronata* had higher lactate dehydrogenase (LDH) activity than *C. chilensis* (Gonzalez and Quiñones 2002). These results were interpreted as a physiological adaptation of *E. mucronata* to cross and inhabit the OML. Thus, distinct metabolic capacities might enable different daily vertical distribution patterns in euphausiids and copepods, two distinct taxonomic groups of crustaceans with marked differences in morphology, behavior, DVM, metabolic rates, longevity, and size. Krill respire with external gills on the lower part of the cephalothorax (Mauchline and Fisher 1969), whereas copepods respire through the body surface (Mauchline 1998). Therefore, it is complicated to interpret the differences in LDH activity between *E. mucronata* and *C. chilensis*.

OXIDATIVE STRESS IN KRILL DURING DVM

Nyctiphanes simplex, *Nematoscelis difficilis*, and *Euphausia eximia* are the most abundant krill species in the northern and central regions of the Gulf of California. They are exposed to a seasonal shallower OML, showing interspecific differences in oxidative stress. *N. simplex* and *N. difficilis* are distributed on both sides of the Baja California peninsula, but they do not extend their distribution southwards into the eastern Tropical Pacific (Brinton 1962, 1979; Brinton et al. 2000) (Fig. 6.2). *Euphausia eximia* is a tropical endemic krill species that mostly proliferates along the cold margins of the eastern Tropical Pacific (Fig. 6.2; Brinton 1962; Brinton et al. 2000). Each species has a distinct DVM range: *N. simplex* migrates between the sea surface and about 200 m depth, typically forming dense surface swarms during winter and spring; *N. difficilis* has a DVM of about 400 m; and *E. eximia* can migrate even deeper than 400 m (Brinton 1962; Lavaniegos 1996; Tremblay et al. 2010). *Euphausia eximia* has a considerably larger gill/cephalothorax surface ratio, suggesting that this species is adapted to inhabit shallow hypoxic waters (Antezana 2002).

Vertical profiles of temperature and dissolved oxygen concentrations recorded under different seasonal conditions (January, July and October 2007) in the Gulf of California are shown in Fig. 6.3 (Tremblay et al. 2010). During the cold season in January, the water column is cold (<17°C) and completely mixed. The upper limit of the OML is located in the strata between 110 and 200 m depth (average of 150 m). In

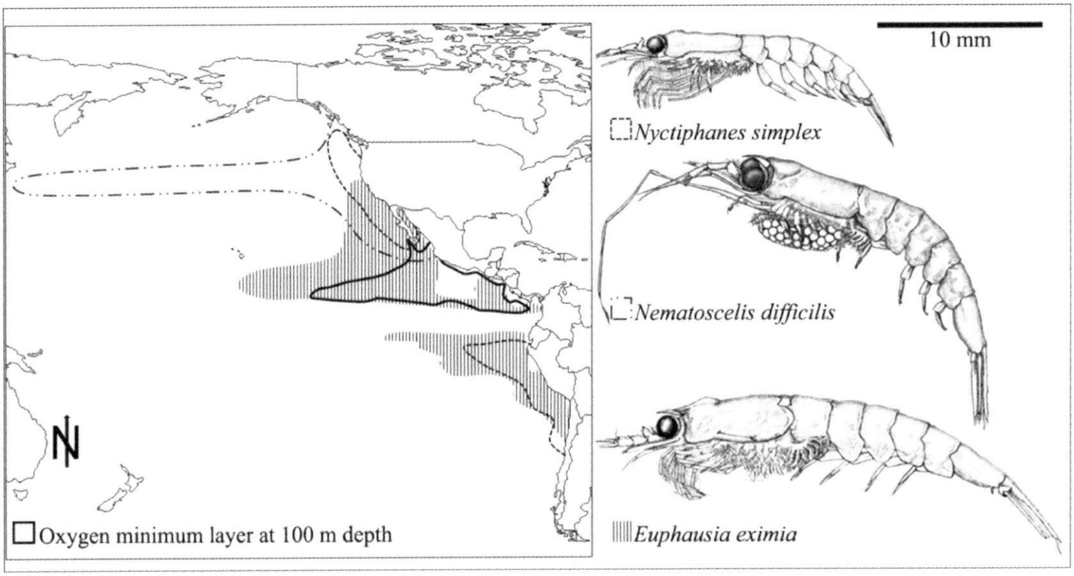

Fig. 6.2 *Nyctiphanes simplex*, *Nematoscelis difficilis*, and *Euphausia eximia* (modified after Brinton et al. 2000) zoogeographic distribution patterns (composite of data from Brinton 1962; Lavaniegos 1996; Brinton et al. 2000) and geographical extension of the eastern Tropical Pacific oxygen minimum layer at 100 m depth (solid line; defined as <1 mL O_2 L^{-1}). (After Fiedler and Talley 2006.)

July, the average sea-surface temperature is 27°C and the mean depth of the seasonal thermocline is 24 m. The upper boundary of the OML depth was detected between 160 and 200 m with an average depth of 183 m. In October, the mean depth of the seasonal thermocline and the OML is shallower (40 m) than during January (90 m), with an average sea-surface temperature of 29°C. Therefore, *N. simplex*, *N. difficilis* and *E. eximia* inhabit a region with pronounced variability of OML depth and a significant increase of temperature above the thermocline between July and October, considered the warm season.

The vertical distribution (integrated average abundance) by depth strata and sampling local time of the three species is shown in Fig. 6.4. During January, *N. simplex* and *E. eximia* are displaying a typical DVM pattern migrating deeper during daytime and staying close to the surface at night, whereas *N. difficilis* do not show a day–night vertical migration pattern, concentrating mostly between 100 and 200 m depth. *Nyctiphanes simplex* forms diurnal surface swarms during January and is never observed below the OML depth.

Oxidative stress indicators measured for each species at different sampling depths in different seasons were associated with their observed daily vertical distribution (Table 6.1) (Tremblay 2008). $O_2^{\bullet-}$ production and superoxide dismutase (SOD) activity in *N. simplex* are higher in deeper water than in shallow layers. *N. difficilis* shows a similar pattern. This indicates relatively similar physiological responses for these subtropical and temperate krill species when they migrate to deeper layers with low dissolved oxygen. These may be associated with the increase in respiration rates observed when both species were experimentally exposed to low dissolved oxygen concentrations. In theory, each live cell generates about 0.1% ROS from the oxygen consumed (Fridovich 2004). The observed $O_2^{\bullet-}$ production and the increased SOD activity in both species suggest that when krill has higher respiration rates, $O_2^{\bullet-}$ production is higher. A similar physiological response was detected in an Arctic benthic amphipod, *Eurythenes gryllus*, in which higher ROS production was associated with elevated respiration rates (Camus and Gulliksen 2005).

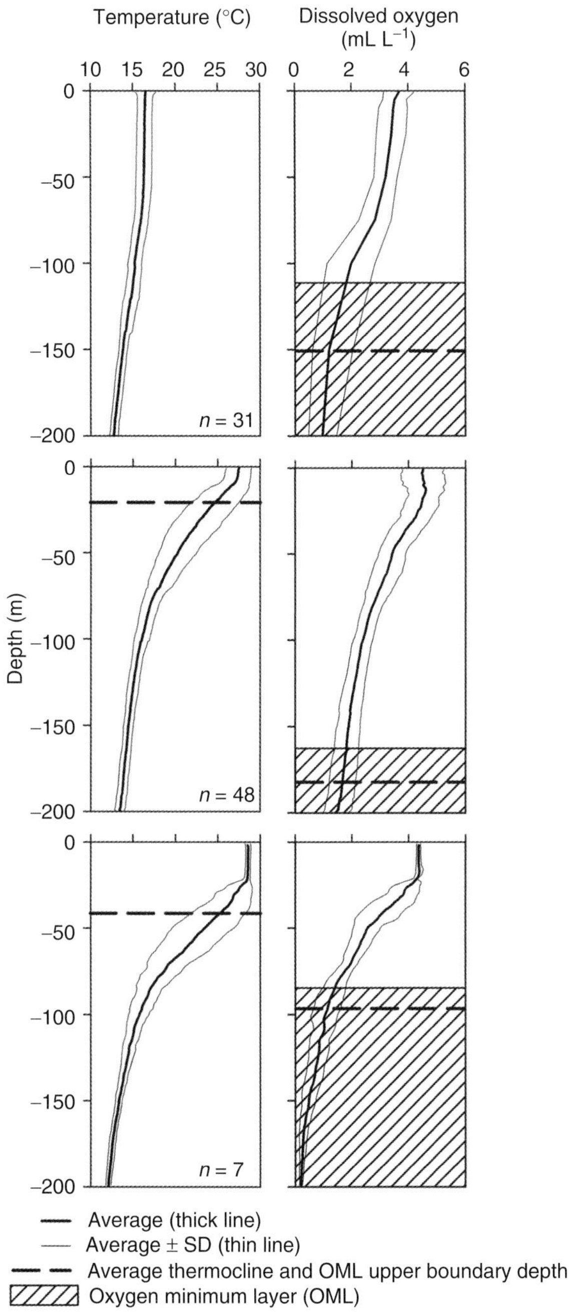

Fig. 6.3 Vertical profiles of temperature (°C) and dissolved O_2 concentrations (mL O_2 L^{-1}) during January, July, and October 2007 in the Gulf of California, Mexico. Dissolved O_2 concentrations was recorded from water collected with Niskin bottles using a Yellow Spring Instrument multi-sensor (YSI-1556) in January (cold season, left) and throughout the water column with a Seabird SB09 conductivity, temperature, depth (CTD) system during July and October (warm season, right). (From Tremblay et al. 2010.)

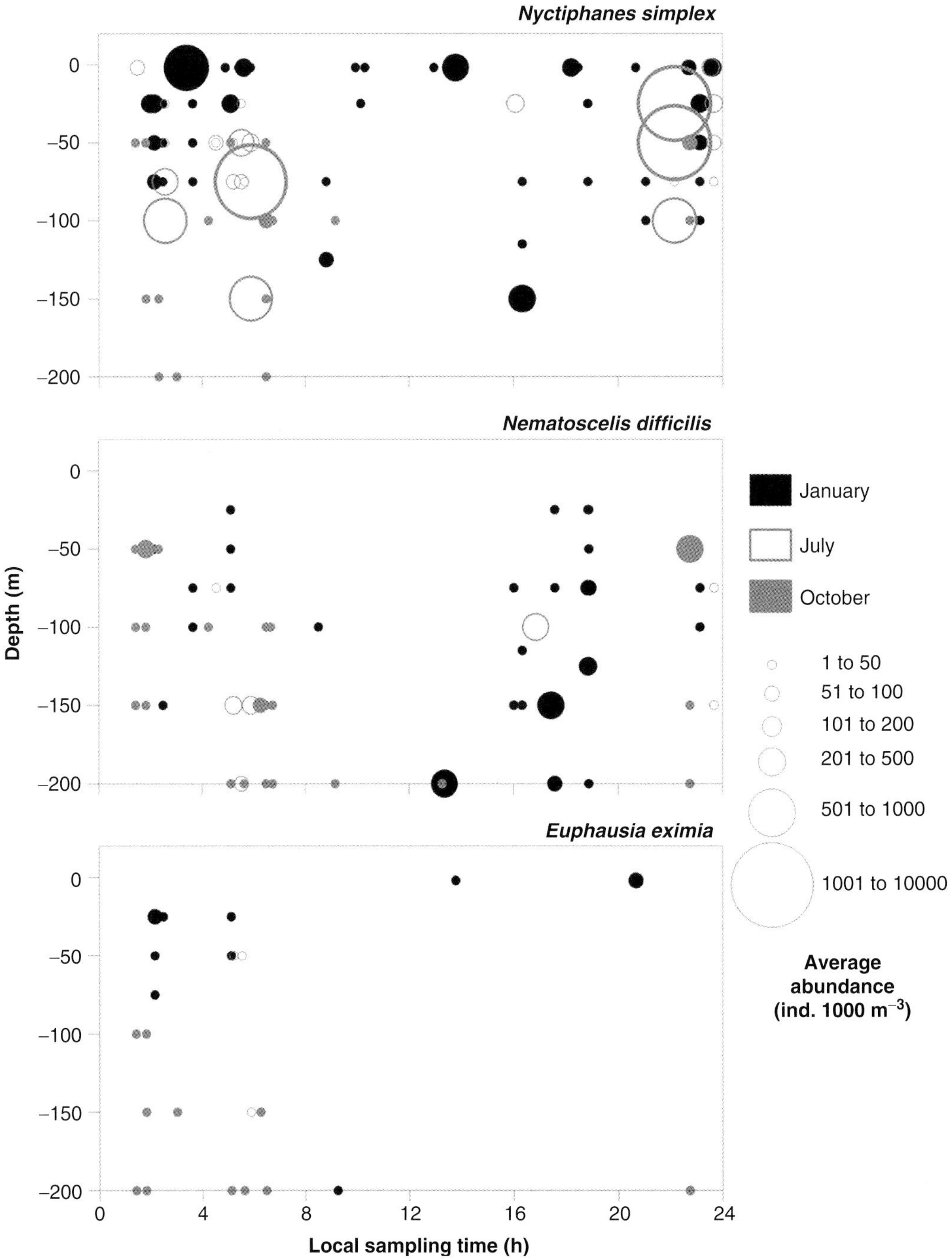

Fig. 6.4 Vertical profiles of integrated average abundance (number of individuals 1000 m^{-3}) of *Nyctiphanes simplex*, *Nematoscelis difficilis*, and *Euphausia eximia* as a function of local sampling time (h) and depth strata during January, July, and October 2007 in the northern and central Gulf of California, Mexico. (From Tremblay 2008.)

Table 6.1 Average and standard errors of oxidative stress indicators of *Nyctiphanes simplex*, *Nematoscelis difficilis*, and *Euphausia eximia* per sampling depth layer (m) and season during 2007 (January = cold season, July and October = warm season). The oxidative stress indicators are: ($O_2^{\bullet-}$) superoxide radical production (nmol $O_2^{\bullet-}$ min^{-1} mg^{-1} protein), (TBARS) lipid peroxidation (nmol TBARS mg^{-1} protein), activities of (SOD-total) total superoxide dismutase, (CAT) catalase, (GPx) glutathione peroxidase, (GR) glutathione reductase, and (GST) glutathione S-transferase (U mg^{-1} protein) (From Tremblay 2008.)

Species	Month	Depth	$O_2^{\bullet-}$	TBARS	SOD-total	CAT	GPx	GR	GST
N. simplex	January	0–50	$36e^{-5} \pm 6e^{-5}$ (16)	0.21 ± 0.03 (12)	6568 ± 888 (16)	0.30 ± 0.07 (16)	$34e^{-4} \pm 7e^{-4}$ (16)	$6e^{-3} \pm 2e^{-3}$ (16)	$8e^{-4} \pm 1e^{-4}$ (16)
		100–150	$8e^{-4} \pm 3e^{-4}$ (3)	0.16 ± 0.04 (3)	17 959 ± 9460 (4)	0.14 ± 0.08 (4)	$4e^{-3} \pm 4e^{-3}$ (3)	$4e^{-3} \pm 2e^{-3}$ (4)	$10e^{-4} \pm 2e^{-4}$ (4)
	July	50–100	n/d	12 ± 4 (5)	306 492 ± 128255 (8)	6 ± 3 (8)	0.07 ± 0.02 (8)	0.24 ± 0.09 (8)	$8e^{-3} \pm 4e^{-3}$ (8)
		100–150	n/d	18 ± 12 (5)	131 237 ± 46541 (10)	5 ± 2 (10)	0.14 ± 0.10 (10)	0.2 ± 0.1 (10)	$7e^{-3} \pm 1$ (10)
N. difficilis	Jan	0–50	$21e^{-5} \pm 6e^{-5}$ (3)	0.09 ± 0.06 (3)	14 325 ± 6521 (3)	0.14 ± 0.07 (3)	$6e^{-3} \pm 3e^{-3}$ (3)	$5e^{-3} \pm 1e^{-3}$ (3)	$3e^{-3} \pm 2e^{-3}$ (3)
		50–100	$23e^{-5} \pm 7e^{-5}$ (11)	0.35 ± 0.09 (13)	5964 ± 1267 (13)	0.11 ± 0.02 (13)	$5e^{-3} \pm 2e^{-3}$ (13)	$6e^{-3} \pm 2e^{-3}$ (13)	$17e^{-4} \pm 3e^{-4}$ (13)
		100–150	$20e^{-5} \pm 7e^{-5}$ (7)	0.16 ± 0.06 (7)	30 831 ± 9960 (7)	0.21 ± 0.07 (7)	$12e^{-3} \pm 6e^{-3}$ (7)	$10e^{-3} \pm 2e^{-3}$ (7)	$23e^{-4} \pm 8e^{-4}$ (7)
		150–200	$60e^{-5} \pm 8e^{-5}$ (8)	0.23 ± 0.06 (9)	33 641 ± 10849 (10)	0.27 ± 0.10 (10)	0.03 ± 0.01 (10)	$11e^{-3} \pm 3e^{-3}$ (10)	$3e^{-3} \pm 1e^{-3}$ (10)
N. difficilis	July and October	0–50	0.13 (1)	$5e^{-5} \pm 4e^{-5}$ (2)	9630 ± 7370 (2)	0.09 ± 0.04 (2)	$81e^{-4} \pm 4e^{-4}$ (2)	$8e^{-3} \pm 1e^{-3}$ (2)	$3e^{-3} \pm 1e^{-3}$ (2)
		50–100	0.15 ± 0.09 (3)	$7e^{-4} \pm 3e^{-4}$ (2)	17 461 ± 7419 (3)	1.0 ± 0.5 (3)	$7e^{-3} \pm 3e^{-3}$ (3)	$12e^{-3} \pm 3e^{-3}$ (3)	$5e^{-3} \pm 3e^{-3}$ (3)
		100–150	0.03 (1)	$6e^{-5} \pm 4e^{-5}$ (3)	13 638 ± 7151 (5)	1.0 ± 0.2 (4)	$8e^{-3} \pm 5e^{-3}$ (5)	$11e^{-3} \pm 3e^{-3}$ (5)	$3e^{-3} \pm 1e^{-3}$ (5)
		150–200	2.65 (1)	$8e^{-4}$ (1)	54,288 (1)	n/d	$3e^{-3}$ (1)	n/d	$7e^{-3}$ (1)
E. eximia	January	0–50	$8e^{-4} \pm 5e^{-4}$ (7)	0.4 ± 0.1(5)	29 214 ± 18 449 (7)	0.4 ± 0.1 (7)	$4e^{-3} \pm 1e^{-3}$ (7)	$10e^{-3} \pm 3e^{-3}$ (7)	$21e^{-4} \pm 9e^{-4}$ (7)
		50–100	$2e^{-4} \pm 1e^{-4}$ (2)	0.15 ± 0.07 (2)	9390 ± 6796 (2)	0.17 ± 0.06 (2)	$24e^{-4} \pm 9e^{-4}$ (2)	$5e^{-3} \pm 1e^{-3}$ (2)	$153e^{-5} \pm 5e^{-5}$ (2)
		100–200	$7e^{-5}$ (1)	0.49 (1)	11,765 (1)	0.53 (1)	$3e^{-4}$ (1)	$5e^{-3}$ (1)	$3e^{-3}$ (1)
	October	0–50	$3e^{-3} \pm 2e^{-3}$ (11)	1.3 ± 0.5 (7)	92 772 ± 65 210 (13)	1.0 ± 0.7 (12)	0.02 ± 0.01 (13)	0.05 ± 0.04 (9)	$10e^{-3} \pm 5e^{-3}$ (13)
		50–100	$10e^{-5} \pm 3e^{-5}$ (4)	1.6 ± 0.9 (4)	18 039 ± 6844 (7)	1.4 ± 1.0 (6)	0.02 ± 0.01 (7)	0.02 ± 0.01 (6)	$3e^{-3} \pm 2e^{-3}$ (6)
		100–200	$12e^{-5} \pm 9e^{-5}$ (5)	0.4 ± 0.2 (4)	5302 ± 2166 (7)	0.17 ± 0.05 (7)	$13e^{-3} \pm 9e^{-3}$ (7)	$10e^{-3} \pm 6e^{-3}$ (7)	$9e^{-4} \pm 4e^{-4}$ (7)

Average ± standard error; (*n*); n/d = no data available; significant differences.

In *Nematoscelis difficilis*, GR activity is also higher in animals collected at 100–150 m layer than at 50–100 m (Table 6.1; Tremblay 2008). High SOD and GR activities detected in *N. simplex* and *N. difficilis* in deeper than in shallower strata may prepare euphausiids for the reoxygenation derived from their earlier migrations to upper water layers, which have higher dissolved oxygen concentrations than the deep hypoxic waters. This mechanism, known as preparation for oxidative stress, has been observed in several invertebrates, amphibians, reptiles, and fish (Hermes-Lima et al. 1998; Hermes-Lima and Zenteno-Savín 2002; Lushchak et al. 2005). Given that thiobarbituric acid reactive substance (TBARS) levels are not significantly higher in *N. simplex* and *N. difficilis* than in *E. eximia* (Table 6.1), it seems that SOD and GR are efficiently balancing $O_2^{\bullet-}$ production during DVM in January (cold season) (Tremblay 2008). *Nyctiphanes simplex* and *N. difficilis* may minimize or even neutralize the effects of exposure to hypoxic deep waters during a short period.

During July and October (warm season), the DVM of the three krill species was completely different from the pattern observed during January. None of the three species seemed to migrate vertically (Fig. 6.4; Tremblay 2008). In July, *N. simplex* was found in large densities below the thermocline and above the OML, mainly during the first hours of the night, with no clear DVM pattern compared to January (Fig. 6.4). *Nematoscelis difficilis* concentrated between 75 and 125 m depth and *E. eximia* was encountered in low densities at 150 m depth. In October, *N. simplex* and *E. eximia* were collected just after sunset between 25 and 50 m water depth, likely during their ascent to the surface (Fig. 6.4). During the day, *N. simplex* is always at <75 m depth, above average OML. *Nematoscelis difficilis* is scarce in October, can be collected only in the first 100 m just before dawn, presumably already in descent. The higher SOD activity measured at the core abundance depth of *N. simplex* could be a response to a stressful environment, probably warm temperature and stratified environmental conditions, which induces a higher metabolic rate (Table 6.1). Similar physiological and abundance changes have been detected in other marine benthic invertebrates (Polychaeta and Decapoda) associated with *in situ* changes of temperature, photoperiod, and salinity (Buchner et al. 1996; Kong et al. 2008). SOD and CAT activities of the polychaete *Arenicola marina* in sandflats of the German Wadden Sea increased significantly in response to a temperature increase during summer and to a high H_2O_2 concentration on intertidal sandflats (Buchner et al. 1996). In the decapod *Scylla serrata*, glutathione peroxidase (GPx) activities and TBARS increased significantly during summer, demonstrating that GPx was not sufficient to avoid oxidative damage (Kong et al. 2008). *Nematoscelis difficilis* did not show differences in SOD activity or TBARS concentration, possibly because this deeper living species is less exposed to seasonal variability (Fig. 6.4 and Table 6.1). The distinct DVM pattern of *N. simplex* and *N. difficilis*, mostly distributed above the OML during the warm season, provides another piece of evidence that these two species are not adapted to penetrate and inhabit warmer and/or hypoxic waters for long periods (Brinton 1962, 1979; Lavaniegos 1996; Brinton et al. 2000; Fernández-Álamo and Färber-Lorda 2006). Because all *N. simplex* specimens died at low dissolved oxygen concentrations under laboratory conditions (Fig. 6.1) and had higher antioxidant enzyme activities in July than in January (Table 6.1), it is clear that this species does not tolerate hypoxic conditions. Thus, *N. simplex* simply avoids entering regions with low dissolved oxygen levels, has a relatively short DVM pattern, and is unable to survive under the tropical conditions prevailing in the eastern Tropical Pacific.

Euphausia eximia display higher SOD activity and TBARS levels at the surface (0–50 m) than in deep (150–200 m) layers (Table 6.1), probably as a result of exposure to a relatively high oxygen concentration (Tremblay 2008). No differences were detected in SOD activity and TBARS between seasons, and all the other antioxidant enzymes measured for *E. eximia* follow a pattern similar to that of SOD activity during October. In general, they have an inverse trend to that observed for the subtropical and temperate species. *Euphausia eximia* TBARS levels were positively associated with increases in all antioxidant enzyme activities as a function of sampling depth, probably in association with its large gill/cephalothorax ratio. Drastic reoxygenation after exposure to hypoxia in muscle and hepatopancreas of the white shrimp *Litopenaeus vannamei* induces an increase in lactate concentration and $O_2^{\bullet-}$ production while total antioxidant capacity falls (Zenteno-Savín et al. 2006). This can be explained by the external position of the gills in krill (Mauchline and Fisher 1969) and the exposure to the high oxygen concentration gradient.

CONCLUSIONS

Considering the currently known zoogeographic patterns (Fig. 6.2; Brinton et al. 2000), the location of the highest density peak of each species, and their contrasting DVM range under cold and warm environmental conditions in the north and central Gulf of California, two scenarios of antioxidant responses can be identified. Subtropical (*N. simplex*) and temperate (*N. difficilis*) krill show higher respiration rates, $O_2^{\bullet -}$ production, and antioxidant capacity when they migrate to deeper low-oxygen layers, while the tropical species (*E. eximia*) have higher oxidative damage and antioxidant activity when collected in surface layers (0–50 m depth). These interspecific differences in the physiological capabilities likely have been shaping the current zoogeographical patterns of these three krill species. *Euphausia eximia* is an endemic species of the eastern Tropical Pacific and has specific physiological patterns that can be explained by their larger gill/cephalothorax surface ratio in comparison to the rest of the species of the genus *Euphausia* (Antezana 2002). Apparently, the major role of this adaptation is to maintain efficient oxygen supply when the specimens are migrating to deeper hypoxic layers (Antezana 2002).

The physiological processes that allow krill, and likely other zooplankton crustaceans, to inhabit, or not inhabit, water masses with low dissolved oxygen concentrations (like in the OML) are strongly associated with their specific antioxidant system and respiration rates. It can be extrapolated conceptually that if the OML expands to higher latitudes (Stramma et al. 2008), or hypoxic conditions have longer seasonal or interannual persistence in the Gulf of California, *N. simplex*, and perhaps to a lesser extent *N. difficilis*, will shift their core abundance in response to the depth and horizontal distribution of the OML. *Nyctiphanes simplex* is the most abundant krill species with the highest biomass in the Gulf of California (Martínez-Gómez 2009). Any change in the zoogeographic distribution of these krill species and their associated zooplankton assemblages will modify food availability for krill predators with considerable consequences for the regional ecosystem.

REFERENCES

Antezana T. (2002) Adaptive behavior of *Euphausia mucronata* in relation to the oxygen minimum layer of the Humboldt Current. *In* Färber-Lorda, J. (ed.). *Oceanography of the Eastern Pacific*, Vol. II. Editorial CICESE, Ensenada, pp. 29–40.

Brinton, E. (1962) The distribution of Pacific euphausiids. *Bulletin of the Scripps Institute of Oceanography* 8, 51–270.

Brinton, E. (1979) Parameters relating to the distribution of planktonic organisms, especially euphausiids in the Eastern Tropical Pacific. *Progress in Oceanography* 8, 125–189.

Brinton, E., Ohman, M.D., Townsend, A.W., Knight, M.D., Bridgeman, A.L. (2000) *Euphausiids of the World Ocean*. Series: World Biodiversity Database CD-ROM, Series Windows Version 1.0, Expert Center for Taxonomic Identification, Amsterdam.

Buchner, T., Abele-Oeschger, D., Theede, H. (1996) Aspects of antioxidant status in the polychaete *Arenicola marina*: tissue and subcellular distribution, and reaction to environmental hydrogen peroxide and elevated temperatures. *Marine Ecology Progress Series* 143, 141–150.

Camus, L., Gulliksen, B. (2005) Antioxidant defense properties of Arctic amphipods: comparison between deep-, sublittoral and surface species. *Marine Biology* 146, 355–62.

Childress, J.J. (1975) The respiratory rates of midwater crustaceans as a function of depth occurrence and relation to the oxygen minimum layer off Southern California. *Comparative Biochemistry and Physiology A* 50, 787–799.

Childress, J.J., Seibel, B.A. (1998) Life at stable low oxygen levels: Adaptations of animals to oceanic oxygen minimum layers. *Journal of Experimental Biology* 201, 1223–1232.

Clarke, A., Tyler, P.A. (2008) Adult antarctic krill feeding at abyssal depths. *Current Biology* 18, 282–285.

Emelyanov, E.M. (ed.) (2005) *The Barrier Zones in the Ocean*. Springer-Verlag, New York.

Enright, J.T. (1977) Diurnal vertical migration: adaptative significance and timing. Part 1 – Selective advantage: a metabolic model. *Limnology and Oceanography* 22, 856–872.

Fernández-Álamo, M.A., Färber-Lorda, J. (2006) Zooplankton and the oceanography of the Eastern Tropical Pacific: A review. *Progress in Oceanography* 69, 318–359.

Fiedler, P.C., Talley, L.D. (2006) Hydrography of the Eastern Tropical Pacific: a review. *Progress in Oceanography* 69, 143–180.

Fridovich, I. (2004) Mitochondria: are they the seat of senescence? *Aging Cell* 3, 13–16.

Gonzalez R.R., Quiñones, R.A. (2002) Ldh activity in *Euphausia mucronata* and *Calanus chilensis*: implications for vertical migration behaviour. *Journal of Plankton Research* 24, 1349–1356.

Hermes-Lima, M., Storey, J.M., Storey, K.B. (1998) Antioxidant defenses and metabolic depression. The hypothesis of preparation for oxidative stress in land snails. *Comparative Biochemistry and Physiology B* 120, 437–448.

Hermes-Lima, M., Zenteno-Savin, T. (2002) Animal response to drastic changes in oxygen availability and physiological oxidative stress. *Comparative Biochemistry and Physiology C* 133, 537–556.

Kamykowski, D, Zentara, S.J. (1990) Hypoxia in the world ocean as recorded in the historical data set. *Deep-Sea Research* 37, 1861–1874.

Kong, X., Wang, G., Li, S. (2008) Seasonal variations of ATPase activity and antioxidant defenses in gills of the mud crab *Scylla serrata* (Crustacea, Decapoda). *Marine Biology* 154, 269–276.

Lavaniegos, B.E. (1996) Vertical distribution of euphausiid life stages in waters adjacent to Baja California. *Fishery Bulletin* 94, 300–312.

Lushchak, V.I., Bagnyukova, T.V., Lushchak, O.V., Storey, J.M., Storey, K.B. (2005) Hypoxia and recovery perturb free radical processes and antioxidant potential in common carp (*Cyprinus carpio*) tissues. *International Journal of Biochemistry and Cell Biology* 37, 1319–1330.

Martínez-Gómez, S. (2009) *Producción de biomasa del eufáusido Nyctiphanes simplex (Crustácea: Euphausiacea) en el Golfo de California, B.C.S., México.* Master of Science thesis, Centro Interdisciplinario de Ciencias Marinas, La Paz BCS, Mexico.

Mauchline, J. (1998) The biology of calanoid copepods. *In* Blaxter, J.H.S., Southward, A.J., Tyler, P.A. (eds). *Advances Marine Biology*, Academic Press, London, 710 pp.

Mauchline, J., Fisher, L.R. (1969) The biology of euphausiids. *In* Russel, F.S., Yonge, M. (eds). *Advances Marine Biology*. Academic Press, Plymouth, 454 pp.

McLaren, I.A. (1963) Effects of temperature on growth of zooplankton and the adaptive value of vertical migration. *Journal of the Fisheries Research Board of Canada* 20, 685–727.

Ohman, M.D. (1984) Omnivory by *Euphausia pacifica*: The role of copepod prey. *Marine Ecology Progress Series* 19, 125–131.

Olson, D.B., Hitchcock, G.L., Fine, R.A., Warren, B.A. (1993) Maintenance of the low-oxygen layer in the central Arabian Sea. *Deep-Sea Research II*. 40, 673–685.

Onsrud, M.S.R., Kaartvedt, S. (1998) Diel vertical migration of the krill *Meganyctiphanes norvegica* in relation to physical environment, food and predators. *Marine Ecology Progress Series* 171, 209–219.

Simmard, Y., de Ladurantaye, R., Therriault, J.C. (1986) Aggregations of euphausiids along a coastal shelf in an upwelling environment. *Marine Ecology Progress Series* 32, 203–215.

Small, L.F., Hebard, J.F. (1967) Respiration of a vertically migrating marine crustacean *Euphausia pacifica*Hansen. *Limnology and Oceanography* 12, 272–280.

Spicer, J.I., Thomasson, M.A., Stromberg, J-O. (1999) Possessing a poor anaerobic capacity does not prevent the diel vertical migration of Nordic krill *Meganyctiphanes norvegica* into hypoxic waters. *Marine Ecology Progress Series* 185, 181–187.

Stramma, L., Johnson, G.C., Sprintall, J., Mohrholz, V. (2008) Expanding oxygen-minimum zones in the tropical oceans. *Science* 320, 655–658.

Teal, J.M., Carey, F.G. (1967) Respiration of a euphausiid from the oxygen minimum layer. *Limnology and Oceanography* 12, 548–550.

Tremblay, N. (2008) *Variación estacional de los indicadores de estrés oxidativo asociada a la migración vertical de los eufáusidos subtropicales del Golfo de California, B.C.S., México.* Master of Science thesis, Centro Interdisciplinario de Ciencias Marinas, La Paz BCS, Mexico.

Tremblay, N., Gómez-Gutiérrez, J., Zenteno-Savín, T., Robinson, C.J., Sánchez-Velasco, L. (2010) Role of oxidative stress in seasonal and daily vertical migration of three krill species in the Gulf of California. *Limnology and Oceanography* 55, 2570–2584.

Wyrtki, K. (1962) The oxygen minima in relation to ocean circulation. *Deep-Sea Research* 9, 11–23.

Zaret, T.M., Sufferen, J.S. (1976) Vertical migration in zooplankton as a predator avoidance mechanism. *Limnology and Oceanography* 21, 804–813.

Zenteno-Savín, T., Saldierna-Martínez, R., Ahuejote-Sandoval, M. (2006) Superoxide radical production in response to environmental hypoxia in cultured shrimp. *Comparative Biochemistry and Physiology C* 142, 301–308.

Chapter 7

OXIDATIVE STRESS IN SULFIDIC HABITATS

Joanna Joyner-Matos[1] *and David Julian*[2]

[1]Department of Biology, Eastern Washington University, Cheney, WA, USA
[2]Department of Biology, University of Florida, Gainesville, FL, USA

Many coastal and deep-sea habitats contain the toxin hydrogen sulfide, which results from human activities, geological processes, or bacterial metabolism. In these habitats sulfide reaches millimolar concentrations, which typically are considered lethal for aerobic organisms because sulfide disrupts mitochondrial function and hemoglobin oxygen transport. Nonetheless, sulfidic habitats have rich species assemblages, including vestimentiferan tube worms, mollusks, annelids, and arthropods. Sulfide oxidation in seawater and in animal tissues generates oxygen-centered and sulfur-centered free radicals, and accumulating evidence suggests that sulfide exposure causes oxidative stress. Some sulfide-tolerant organisms up-regulate the cellular oxidative stress response, especially the antioxidants, when exposed to sulfide. Additionally, sulfide exposure can damage nucleic acids and disrupt mitochondrial and lysosomal structures in a manner consistent with oxidative stress. High levels of copper and iron in these habitats may interact with sulfide to increase reactive oxygen species (ROS) production. Paradoxically, sulfide has antioxidant properties in mammalian cells *in vitro*. Furthermore, sulfide is endogenously produced in vertebrates and invertebrates and at low concentrations has vasoregulatory and neuromodulatory roles. Consequently, sulfide may be toxic at high concentrations but beneficial at low concentrations. Whether animals in sulfidic habitats have specific adaptations to reduce oxidative stress from exposure to high sulfide levels remains unknown.

ENVIRONMENTAL CHEMISTRY OF HYDROGEN SULFIDE

Many aquatic habitats contain the toxin hydrogen sulfide (H_2S), which can be produced from bacterial metabolism, geological processes and human activities. As a gas, H_2S is colorless and has the odor of rotten eggs. Dissolved in aqueous solution, H_2S dissociates to hydrosulfide anion (HS^-) and then to sulfide anion (S^{2-}), depending upon the solution's temperature, salinity, sulfide:O_2 ratio, and pH (Cline and Richards 1969; Millero et al. 1987) (see Box 7.1). We refer to the equilibrium sum of these three ionization states as "sulfide." In solutions that contain metal catalysts, as is typical of marine habitats, sulfide oxidizes rapidly to sulfate, generating oxygen-centered and sulfur-centered free radicals (Chen and Morris 1972b; Tapley et al. 1999).

In coastal mudflats and marshes, microbially produced sulfide can accumulate in the interstitial water and, to a lesser extent, in the water column (Fenchel and Riedl 1970; Vismann 1991). In mangrove swamps, high rates of organic decomposition lead to millimolar sulfide concentrations (up to 20 mM; Lee et al. 2008) in the sediment and micromolar sulfide concentrations in the water column surrounding

Oxidative Stress in Aquatic Ecosystems, First Edition. Edited by Doris Abele, José Pablo Vázquez-Medina, and Tania Zenteno-Savín.

Box 7.1 Overview of sulfide chemistry*

Sulfide dissociates in aqueous solutions: $H_2S \leftrightharpoons HS^- + H^+ \leftrightharpoons S^{2-} + 2H^+$

Sulfide is oxidized in aqueous solutions in several steps, some of which have oxygen- and sulfur-centered free radical intermediates

Overall reactions: $2HS^- + 2O_2 \rightarrow S_2O_3^{2-} + H_2O$ (thiosulfate) or $HS^- + 2O_2 \rightarrow SO_4^{2-} + H^+$ (sulfate)

Potential free radical intermediates include: $HS^{\bullet}$, $O_2^{\bullet -}$, $HSO_2^{\bullet}$, $SO_2^{\bullet -}$, and $SO_3^{\bullet -}$

Sulfite (derived from thiosulfate) is enzymatically oxidized to sulfate with glutathione cycling between reduced (GSH) and oxidized (GSSG) forms: $SO_3^{2-} + 2\ GSH \rightarrow SO_4^{2-} + GSSG$

Similarly, sulfide scavenges free radicals such as superoxide: $2O_2^{\bullet -} + 2H_2S \rightarrow HS\text{-}SH + O_2 + 2\ OH^-$

Sulfide reacts with sulfur-containing free amino acids like hypotaurine: (hypotaurine) $^+NH_3\text{-}CH_2\text{-}CH_2\text{-}SO_2^- + HS \rightarrow {}^+NH_3\text{-}CH_2\text{-}CH_2\text{-}SO_2^-\text{-}SH$ (thiotaurine)

Notes on field measurements of sulfide

Sulfide concentrations in aqueous solutions are typically measured using a spectrophotometric method, in which sulfide reacts with ferric ions and N,N-dimethyl-p-phenylenediamine, to produce methylene blue (Pomeroy 1936; Budd and Bewick 1952; Cline 1969). Newer techniques utilizing electrodes (Berner 1963; Ma et al. 2007) or chromatography (Dattagupta et al. 2007) increase spatial resolution and allow integration over time. With the exception of techniques that utilize *in situ* electrodes, the rapid oxidation of suphide in aqueous solution necessitates that field sulfide measurements first "trap" sulfide with zinc acetate (forming stable zinc sulfide), after which the sulfide can be released by acidification.

**Information on sulfide chemistry was assembled from Cline and Richards (1969); Tapley et al. (1999); Pruski et al. (2000)*

decaying wood (Laurent et al. 2009). Sulfide is also present in the water column of limestone cave systems (e.g. Sarbu et al. 1996) and in anoxic basins like the Black Sea, Scandinavian fjords (for review, Morse et al. 1987; Diaz and Rosenberg 1995), and the Urania basin in the Mediterranean Sea (Borin et al. 2009).

In the deep sea, sulfide chemistry is particularly well-studied at hydrothermal vents, cold seeps, and large food falls (e.g. whale falls). Hydrothermal vents form along mid-ocean ridges where seawater contacts exposed basaltic lava, heating to $>350°C$ and leaching metals from the basalt, which are then precipitated into structures like black-smoker chimneys (for review, Van Dover 2000). Plumes emitted from these depositions tend to have high concentrations of sulfide (up to 7 mM) and transition metals. Deep-sea cold seeps take multiple forms, from sulfide-rich brine seeps to mud volcanoes (for review, Sibuet and Olu 1998). The water chemistry of cold seeps varies extensively, but seeps may have micromolar sulfide concentrations. At food falls, sulfide is produced by sulfate-reducing bacteria that metabolize lipids in the carcass (Smith et al. 1998; Treude et al. 2009).

Human activities increase sulfide concentrations in aquatic habitats, typically by enhancing microbial production. This is particularly true in the vicinity of industrial plants and sewage outfalls and is particularly well-studied along the southern California coast (Rittenberg et al. 1958; Swartz et al. 1986).

ENDOGENOUS PRODUCTION OF SULFIDE

Sulfide is endogenously produced from the amino acid L-cysteine in tissues of vertebrates (Kimura 2011) and invertebrates (Julian et al. 2002). It is proposed to have multiple tissue-specific functions in vertebrates (Kimura 2011). In aquatic invertebrates, sulfide alters muscle tone (Julian et al. 2005b) and gill muscle contraction (Gainey and Greenberg 2005).

PHYSIOLOGICAL ECOLOGY OF SULFIDE

Species Assemblages

Most sulfidic habitats are characterized by abundant and diverse communities, and in these habitats sulfide plays a complex ecological role, influencing recruitment, settlement, succession, and community structure. Species assemblages in coastal sulfidic habitats include cyanobacteria, spirochaetes, foraminifers, turbellarians, gastrotrichs, nematodes, platyhelminthes, polychaetes, bivalves, gastropods, crustaceans, echinoderms, saltmarsh plants like *Spartina* spp., and mangroves (Fenchel and Riedl 1970; King et al. 1982; Bagarinao 1992; Lee et al. 2008). For these organisms, laboratory measured sulfide tolerances are predictive of distribution patterns, even among congeners (Maricle et al. 2006) and conspecifics (e.g. Arp et al. 1992). Distribution is closely linked to dissolved oxygen availability, particularly for burrowing macrofauna (Vismann 1991; Atkinson and Taylor 2005). A common theme from these studies is that sulfide's complex role in coastal habitats is influenced by dose, tidal cycles, and seasonality.

Hydrothermal vent communities typically have high biomass and density (e.g. Gebruk et al. 2000) but vary in microbial (Jørgensen and Boetius 2007) and eukaryotic diversity (Van Dover 2000). Macrofauna include cnidarians, vestimentiferan tube worms, polychaetes, gastropods, bivalves, pycnogonids, crustaceans, and fishes. Sulfide acts as one of several settlement cues to new vents (Renninger et al. 1995) by signaling the availability of food and influences food-web dynamics on established vents (Levesque et al. 2006). Cold-seep communities also are highly variable in microbial (Jørgensen and Boetius 2007) and eukaryotic biomass and diversity (Sibuet and Olu 1998). They tend to have bivalves, gastropods, vestimentiferan tube worms, sponges, crustaceans, polychaetes, oligochaetes, nematodes, and foraminifers. In these habitats, sulfide influences recruitment (Levin et al. 2006), succession and faunal abundance (Cordes et al. 2009). Deep-sea food falls are similar to vents and seeps in both species assemblage and ecology (Smith et al. 1998; Treude et al. 2009).

In contrast, sulfidic caves are relatively depauperate in eukaryotes, although they have extensive microbial communities (e.g. Engel et al. 2004) and a small number of fish (e.g. Tobler et al. 2006) and invertebrate (e.g. Dattagupta et al. 2009) species. Sulfide appears to play a key role in the ecology of cave animals, influencing processes from phenotypic evolution (Tobler et al. 2008) to behavior (Plath et al. 2007).

TOXICOLOGY OF SULFIDE

Given the abundance of animals in sulfidic habitats, it is perhaps surprising that sulfide is considered lethal for aerobic organisms, particularly since the H_2S form diffuses freely across epithelial membranes (Powell et al. 1979, Julian and Arp 1992). The main cellular-level effects of sulfide involve interactions with metalloproteins and disulfide bridges, and these have been reviewed for mammals (Beauchamp et al. 1984), aquatic invertebrates and fish (e.g. Bagarinao 1992, Grieshaber and Völkel 1998). Several antioxidants are specifically susceptible to sulfide, including catalase (CAT), superoxide dismutase (SOD), and glutathione peroxidase (GPx), as will be discussed in detail below. Furthermore, sulfide's tendency to act as a strong reducing agent, disrupting protein disulfide bridges (Smith and Abbanat 1966), is particularly important for glutathione (GSH) function. These cellular-level effects lead to detectable changes in whole-organism function.

The whole-organism toxicity of sulfide was first described in the 1700s (for review, Roth 2004) (see Box 7.2). Sulfide is a broad-spectrum toxicant that has dose-dependent effects on the cardiovascular, respiratory, endocrine, gastrointestinal, immune, nervous, and olfactory systems (Beauchamp et al. 1984). In aquatic invertebrates, sublethal sulfide exposure reduces feeding activity (Oseid and Smith Jr. 1974), heart rate (Vetter et al. 1987), and swimming (Takagi et al. 2005), and induces panic behavior and paralysis (Vismann 1996). Sublethal sulfide doses in fish lead to damaged gill and liver tissue (Kiemer et al. 1995), lower overall condition (Tobler et al. 2006), and decreased ventilation, equilibrium, and mobility (Forgan and Forster 2010). In aquatic plants, sublethal sulfide exposure decreases nitrogen uptake (e.g. Bradley and Morris 1990), reduces growth and biomass (e.g. King et al. 1982), and is implicated in die-backs (e.g. Carlson et al. 1994).

Box 7.2 Chronology of sulfide biology milestones*

1700s	First reports of H_2S poisoning in European sewer workers.
1775	Scheele generates H_2S gas through two distinct chemical reactions.
1785	Investigative committee in Paris describes symptoms of poisoning in sewer workers but initially did not attribute them to sulfide. In the next five years sulfide gas is confirmed as the cause of accidents and deaths of sewer workers.
1800s	Reports of accidents and deaths in sewer and tunnel workers continue in France, England, and America.
1803	Chaussier conducts the first animal experimental study confirming that inhaled or injected sulfide is lethal.
1862	Victor Hugo uses sickening in sewer workers as a theme in *Les Miserables*.
1863	Hoppe-Seyler shows that blood exposed to sulfide contained a dark green pigment similar to the greenish hue of cadavers of sewer workers; first use of term "sulfmethemoglobin."
1870s–1930s	Intense experimental work on the effects of sulfide on blood, blood components, and respiration accompanies numerous reports of sulfide poisoning by inhalation in workers in multiple industries.
1917–1920	Multiple studies show bacterial synthesis of sulfide.
1949	Fogo and Popowsky publish methylene blue method of sulfide measurement.
1960s–1970s	Oxidation of sulfide to thiosulfate and sulfate is confirmed *in vitro* and *in vivo*; action of sulfide on cytochrome-containing enzymes is determined.
1969	Cline publishes optimized method for determining sulfide concentrations in solution; Theede and coworkers explore sulfide tolerance of invertebrates.
1970	Fenchel and Riedl describe the worldwide existence of the sulfide system in marine sediments.
1970s	Other sulfidic habitats, like anoxic basins, gain increasing attention.
1972	Chen and Morris show that sulfide oxidation can produce free radicals, this is confirmed in seawater in 1999 by Tapley and colleagues.
1979	Deep-sea hydrothermal vents discovered by Corliss and coworkers.
1979–1989	Geology, physiology, biochemistry, and ecology of sulfide-based chemoautotropy and symbioses at deep sea vents explored.
1984	Blum and Fridovich document activities of antioxidant enzymes in deep-sea hosts of chemoautotrophic symbionts.
1988	Morrill and colleagues establish a relationship between sulfide tolerance and antioxidant activity in meiofauna.
1989–1990	Several research groups measure endogenous sulfide levels in mammalian brains.
1990	Koch and coworkers document that sulfide slows wetland plant growth.
1996	Abe and Kimura hypothesize that sulfide may act as a neuromodulator.

**Information in this box was assembled from (Mitchell and Davenport 1979, Beauchamp et al. 1984, Roth 2004)*

Sulfide Tolerance Strategies

Organisms in sulfidic habitats employ a suite of strategies to detoxify sulfide and maintain aerobic metabolism. Perhaps the most widespread sulfide detoxification strategy in aquatic invertebrates (for review, Bagarinao 1992, Grieshaber and Völkel 1998) and plants (e.g. Lee et al. 1999) is sulfide oxidation (see Box 7.1), especially oxidation to thiosulfate ($S_2O_3^{2-}$). Sulfide oxidation can occur via metal-catalyzed reactions (Huxtable 1986) and enzymatic oxidation in "sulfide-oxidizing bodies" putatively derived from mitochondria or lysosomes (Powell and Somero 1985; Wohlgemuth et al. 2007). Furthermore, since sulfide is highly reduced, sulfide oxidation can be coupled to oxidative phosphorylation in mitochondria (e.g. Powell and Somero 1986; Bagarinao and Vetter 1990).

Vestimentiferan tube worms and some bivalves have formed endosymbiotic relationships with chemoautotrophic bacteria by housing the symbionts intracellularly (e.g. bivalves) or in specialized body compartments (e.g. tube worms). The invertebrate hosts provide reduced sulfur compounds and oxygen to the symbionts and rely so heavily on symbiont-produced carbohydrates that they have lost their digestive systems (e.g. Cavanaugh 1983; Felbeck 1983; Fisher 1990). Some hosts have hemoglobin that reversibly binds and transports sulfide without impairing oxygen affinity (for review, Weber and Vinogradov 2001). Sulfide-oxidizing bacteria also may form ectosymbiotic relationships with animals (e.g. Ott et al. 1998; Dattagupta et al. 2009) and plants (Joshi and Hollis 1977), forming a barrier between ambient sulfide and the sulfide-permeable host membranes.

Sulfur-containing free amino acids like hypotaurine and thiotaurine may play a role in sulfide detoxification, transport, and/or storage (see Box 7.1). This strategy may safely transport sulfide to the endosymbionts of deep-sea (e.g. Alberic and Boulegue 1990) and coastal (Joyner et al. 2003) invertebrate hosts, and may reduce sulfide toxicity in nonsymbiotic invertebrates (e.g. Rosenberg et al. 2006). Hypotaurine reacts with sulfide to produce thiotaurine, which may act as a stable sulfide storage compound (e.g. Rosenberg et al. 2006) and may be cytoprotective (Ortega et al. 2008). This method of sulfide detoxification with the thiol groups in free amino acids is similar to the detoxification and protective effects of glutathione disulphide (GSSG; see Box 7.1; Smith and Abbanat 1966; Truong et al. 2006).

Many organisms in sulfidic habitats employ some or all of the strategies described above to detoxify sulfide and/or transport sulfide through their tissues to endosymbionts. However, within some of the studies cited above are hints of what has become a new focus of ecophysiological studies in sulfide biology: the impact of sulfide on ROS metabolism.

SULFIDE AND OXIDATIVE STRESS

Evidence that the Oxidative Stress Response is Stimulated by Sulfide Exposure

Early studies explored the potential links between environmental sulfide and the steady-state presence or activity of antioxidants, the main components of the oxidative stress response (Table 7.1). For example, the deep-sea vent tube worm *Riftia pachyptila* and vent bivalve *Calyptogena magnifica*, both of which contain chemoautotrophic symbionts, had high activities of all three SOD forms (cytoplasmic or Cu,Zn-SOD; mitochondrial or Mn-SOD; and bacterial iron or Fe-SOD) and GPx, but no detectable CAT (Blum and Fridovich 1984). Sulfide-tolerant meiofauna had detectable CAT and total SOD activities (sum of Mn-SOD and Cu,Zn-SOD), and higher CAT activity than non-sulfide-tolerant species (Morrill et al. 1988). Similarly, in two populations of the vent mussel *Bathymodiolus azoricus*, environmental sulfide conditions were positively correlated with mitochondrial SOD and cytoplasmic SOD activities in gill and mantle tissue, and with mantle GPx activity, but were negatively correlated with gill CAT activity (Company et al. 2007). In contrast, environmental sulfide concentration did not correlate with total SOD activity in the gills of several coastal bivalve and polychaete species (Abele-Oeschger 1996). Inconsistencies in relationships between enzyme activities and environmental sulfide are in part complicated by tissue-specific differences (Table 7.1). Overall, there is no obvious correlation between environmental sulfide concentration and steady-state antioxidant activity in aquatic invertebrates.

The next logical step was to test whether sulfide exposure alters antioxidant activities in sulfide-tolerant organisms. However, the results were inconsistent, which may be attributed to tissue-specific differences and variance in the dissolved oxygen concentration during sulfide exposure (Table 7.1). Since the severity of sulfide exposure is tightly linked to dissolved oxygen availability, we will henceforth indicate whether treatments were normoxic (dissolved oxygen at atmospheric or saturating levels), hypoxic (dissolved oxygen at less than atmospheric levels), or anoxic (no dissolved oxygen). Normoxic exposure to 100 μM sulfide did not alter total SOD or CAT activities in the polychaete *Heteromastus filliformis* body wall (Abele et al. 1998), but exposure to higher sulfide concentrations

Table 7.1 Summary of studies exploring oxidative stress and hydrogen sulfide in aquatic organisms.

Organism	Sulfide tolerant?[1]	Sulfide treatment (O_2 level)[2]	Antioxidant activity or expression[3] (tissue)	ROS production[3,5]	Damage or repair markers[3]	Organelle integrity, cell viability	Reference
Riftia pachyptila	Yes	–	Cu,Zn-SOD (Ms), Mn-SOD (Ms), Fe-SOD (T), GPx (M, G) No CAT (Ms,T)				[1]
Calyptogena magnifica	Yes	–	Cu,Zn-SOD (Ms, G), Mn-SOD (Ms, G), Fe-SOD (T), GPx (Ms, G) No CAT (M, G)				
Gastrotrich and turbellarians	Yes and no	–	CAT α S SOD [4] not α S				[2]
Bathymodiolus azoricus	Yes	–	SOD (G, M) α S GPx (M) α S; (G) → CAT(G) inv α S; (M) →		LPO (G, M) →		[3]
Bivalves and polychaetes	Yes and no	–	SOD (G) α S				[4]
Heteromastus filliformis	Yes	100 μM (N)	SOD (Bw) → CAT (Bw) →				[5]
Astarte borealis	Yes	200 μM (N)	SOD (Hm) ↓ (G) →				[4]
Arenicola marina	Yes	200–500 μM (N)	SOD (C) ↓				
Solemya velum	Yes	100 μM (N)	Mn-SOD (G,F) ↑ SOD (G) ↑ GR (G,RT) ↑ CAT (G,F,RT) →				[6,7]
Yoldia limatula	Yes	100 μM (N)	SOD (RT) ↑ GR (G,F,RT) ↑ CAT (G,F,RT) →				
Arenicola marina	Yes	200 μM (H)	SOD (C,Bd) → CAT (C) ↓ GR (C) →	(Bd) ↑			[8]
Astarte borealis	Yes	200 μM (H)	SOD (Hm) ↓ (G) → CAT (G) ↓ GR (G) ↑	(Hm) ↑			

Nereis diversicolor	Yes (?)	100 μM (A)	CAT (At) ↑, (Ep) → SOD (At) ↑, (Ep) →				[9]
Halicryptus spinulosus	Yes	200 μM (A)	CAT (Bw) →				[10]
Heteromastus filliformis	Yes	100 μM (A)	SOD (Bw) ↓ CAT (Bw) ↓				[5]
Bathymodiolus azoricus	Yes	Transplant low to high S	Cu,Zn-SOD (G) ↑ GPx (G) ↑ CAT (G) ↓ Mn-SOD (M) →		LPO (G, M) ↑		[3]
Donax variabilis	No	100 μM (N)	Mn-SOD ↑ (exp) Cu,Zn-SOD ↑ (exp) GPx → (exp)		HNE ↓ OGG1-m ↑		[11]
Glycera dibranchiata	Yes	0–3 mM (N)		(Cl) ↑	8-oxodGuo (C, Bw) ↑ 8-oxoGuo (C, Bw) ↑		[12]
Glycera dibranchiata	Yes	0–2 mM (N)		(Cl) ↑		Mito depol (C) ↑	[13]
Tubificoides benedii	Yes	200 μM (H)				Mito structure (Epd, In) →	[14]
Halicryptus spinulosus	Yes	200 μM (A)				Rough ER structure (Epd) ↓ Mito swell (Epd, Ms) ↑ Mito structure (Ms) ↓	[15]

[1] May be based on sulfide tolerance tests in the laboratory and/or presence of sulfide in habitat.

[2] Abbreviations for oxygenation of sulfide dose: N = normoxia, H = hypoxia, A = anoxia; see references for specifics.

[3] Significant changes are indicated by up or down arrows, nonsignificant changes indicated by horizontal arrows. Where body parts are not noted, assays were conducted on whole individuals. Abbreviations kept consistent with references: Ms, muscle; T, trophosome; G, gill; M, mantle; α S, proportional to sulfide presence in habitat; inv α S, inversely proportional to sulfide presence in habitat; Hm, hemolymph; SMB, sulfur metabolizing body; Ept, epithelium; Bw, body wall tissue; C, chloragog; F, foot; RT, remaining tissues; Bd, blood; Ep, epitokous; At, atokous; (exp), expression levels, not enzyme activity levels; Cl, coelomocytes; Epd, epidermis; In, intestine.

[4] Indicates measurements of total SOD activity (assumed to be the sum of all SODs).

[5] For ROS production – see papers for specifics about which species measured.

References:[1] Blum and Fridovich 1984; [2] Morrill et al. 1988; [3] Company et al. 2007; [4] Abele-Oeschger 1996; [5] Abele et al. 1998; [6] Tapley 1993; [7] Tapley and Shick 1991; [8] Abele-Oeschger and Oeschger 1995; [9] Abele-Oeschger et al. 1994; [10] Oeschger and Vetter 1992; [11] Joyner-Matos et al. 2006; [12] Joyner-Matos et al. 2010; [13] Julian et al. 2005a; [14] Dubilier et al. 1997; [15] Janssen and Oeschger 1992.

(200–500 μM) decreased total SOD activities in the bivalve *Astarte borealis* hemolymph (but not gill) and the polychaete *A. marina* chloragog (= polychaete digestive gland tissue; Abele-Oeschger 1996). In contrast, antioxidant activities in the coastal symbiotic clam *Solemya velum* and the nonsymbiotic clam *Yoldia limatula* were stimulated by normoxic exposure to moderate sulfide (100 μM; Tapley and Shick 1991; Tapley 1993). In both species, the sulfide treatment increased total SOD and glutathione reductase activities but not CAT activities in most tissues, whereas Mn-SOD activity increased in tissues from *S. velum* but not *Y. limulata*. Hypoxic and anoxic sulfide exposures produced similarly conflicting results (Table 7.1) in an assortment of sulfide-tolerant polychaetes (Abele-Oeschger et al. 1994; Abele-Oeschger and Oeschger 1995), a bivalve (Abele-Oeschger and Oeschger 1995), and a priapulid (Oeschger and Vetter 1992). Glutathione reductase (GR) activity either increased or remained unchanged in the three species studied, reflecting a potential role for GR in the cycling of reactive sulfur species (Giles et al. 2001). While SOD activity showed no clear pattern across studies, CAT activity decreased following sulfide exposure in all species except *Nereis diversicolor*. In a field study of *B. azoricus* transplanted from vent sites with low-sulfide concentrations to sites with high sulfide concentrations, gill CAT activity decreased but GPx and Cu,Zn-SOD activities increased (Company et al. 2007). In many of these studies, seasonality (e.g. Abele et al. 1998) and reproductive cycle (e.g. Abele-Oeschger et al. 1994) also influence the effects of sulfide exposure.

Given that the oxidative stress response incorporates multiple metabolic pathways, it is perhaps not surprising that measuring a few antioxidants has not conclusively demonstrated that sulfide exposure alters ROS metabolism. One complication is that sulfide itself may directly interact with antioxidants, influencing the balance between ROS production and detoxification. This cycle can be illustrated by the effects of sulfide on CAT (a metalloprotein). Sulfide inhibits CAT activity *in vitro* (100–200 μM, normoxic; Khan et al. 1987) and *in vivo* (100 μM, normoxic; Carlsson et al. 1988) by interacting with the heme portion of the enzyme and preventing H_2O_2 binding (Keilin and Hartree 1936). Catalase inhibition can exacerbate free radical stress, as demonstrated in yeast that were exposed to sulfide and up to 50 μM H_2O_2; these treatments did not trigger an oxidative stress response but nonetheless resulted in a 100-fold increase in mutation frequency and in mortality (Carlsson et al. 1988). This suggests that studies exploring sulfide toxicity should not rely solely on whether the oxidative stress response is up-regulated.

A second potential complication is that the organisms may invest metabolic resources in preparing their cells for oxidative stress. Storey (1996) first proposed that animals that routinely encounter oxidative stresses may maintain high steady-state antioxidant or repair enzyme levels. Alternative strategies would be for animals to rapidly "turn on" antioxidant defenses once they are faced with oxidative stress, or to simply repair oxidative damage after it occurs. Storey and collaborators have found the most support for the strategies of maintaining high antioxidant levels and turning on antioxidant defenses, depending on whether an organism experiences the stresses frequently or rarely, respectively (Storey 1996; Hermes-Lima et al. 1998, 2001). Dong and coworkers (2008) termed the maintenance strategy (in the case of heat shock proteins) a "preparative defense." With this perspective, it is possible that the mixed results summarized in Table 7.1, particularly those studies in which sulfide exposure caused no significant changes, are in fact examples of preparative defense in sulfide-tolerant organisms.

These results suggest that our ability to detect evidence of oxidative stress from sulfide exposure may be limited by the traditional approach of selecting a study organism that tolerates extremes of the environmental factor of interest (Hoffmann and Parsons 1991; Spicer and Gaston 1999). We adopted an alternative approach of testing whether sulfide exposure alters ROS metabolism in the bivalve *Donax variabilis* (Joyner-Matos et al. 2006), which is not sulfide-tolerant. Sulfide exposure (100 μM, normoxic) caused elevated Mn-SOD and Cu,Zn-SOD protein expression, but no changes in GPx expression. Additional findings regarding oxidative damage and repair markers in this bivalve are discussed below.

Evidence that Sulfide Elevates ROS Production and/or Cellular Damage

A recent avenue of sulfide toxicity research has been to test whether sulfide alters steady-state ROS levels or causes oxidative damage, organelle damage, or cell death. In biomedical studies, potential links between sulfide and cardiovascular signaling, tissue damage, and disease have been reviewed multiple times (e.g. Beauchamp et al. 1984; Whiteman and Moore 2009).

In this section, we will review the studies conducted on aquatic organisms, with limited comparisons to biomedical research.

Sulfide oxidation in solution is accompanied by formation of reactive oxygen and sulfur species (Chen and Morris 1972a; Tapley et al. 1999) (see Box 7.1), reactions that also may occur in cells and tissues (Tapley 1993, Giles et al. 2001, Eghbal et al. 2004). For example, H_2O_2 concentrations significantly increased in the sulfide-tolerant bivalve *A. borealis* and the polychaete *A. marina* following hypoxic sulfide exposure (Abele-Oeschger and Oeschger 1995; see Table 7.1). Similarly, mitochondrial (Julian et al. 2005a) ROS production increased in the sulfide-tolerant polychaete *Glycera dibranchiata* following normoxic sulfide exposure, and sulfide exposure increased cellular oxidative stress (Eghbal et al. 2004; Julian et al. 2005a; Joyner-Matos et al. 2010). Finally, the presence of ROS can lead to the chemical formation of reactive sulfur species like thiyl radicals, particularly if metals are present (Abedinzadeh 2001; Giles et al. 2001), and sulfide may combine with nitric oxide to form a radical-containing compound (for review, Whiteman and Moore 2009). Although reactive sulfur species have not been directly measured in an aquatic organism, it is reasonable to assume that when ROS, metals, or nitric oxide are present, sulfur-centered radicals are present as well.

Oxidative damage can be detected when ROS production exceeds the capacity for detoxification or repair. The few studies of sulfide-induced oxidative damage in aquatic organisms have focused on lipid peroxidation (LPO) or nucleic acid damage (Table 7.1). Lipid peroxidation did not correlate with environmental sulfide exposure in two populations of *B. azoricus*, but transplantation of mussels from low-sulfide to high-sulfide sites significantly increased LPO in gill and mantle (Company et al. 2007). In contrast, abundance of 4-hydroxy−2*E*-nonenol-adducted protein (HNE) was significantly reduced by sulfide exposure in a nonsulfide-tolerant bivalve (Joyner-Matos et al. 2006), a seemingly contradictory finding that will be discussed in greater detail below. A more consistent pattern has been detected for sulfide-induced oxidative damage to nucleic acids. Sulfide exposure led to increased oxidation of RNA and DNA guanine bases in the body wall and coelomocytes of *G. dibranchiata* (Joyner-Matos et al. 2010). Similarly, sulfide exposure increased expression of the mitochondrial form of 8-oxoguanine DNA glycosylase (OGG1-m), an enzyme responsible for repairing guanine lesions, in *D. variabilis* (Joyner-Matos et al. 2006). This tight linkage between sulfide exposure and DNA damage or repair has been reported many times in mammalian cell lines or other preparations (Shi and Mao 1994; Attene-Ramos et al. 2006, 2007; Baskar et al. 2007; Yaegaki et al. 2008). Interestingly, this also suggests that sulfide could be mutagenic (Attene-Ramos et al. 2007; Joyner-Matos et al. 2010).

It is still unclear whether altered organelle structure or function indicate sulfide exposure or tolerance (e.g. Duffy III and Tyler 1984). For example, exposure to 200 μM sulfide (anoxic) led to disruption of the rough endoplasmic reticulum and swelling and disintegration of mitochondria in the sulfide-tolerant priapulid *Halicryptus spinulosus* (Janssen and Oeschger 1992). In contrast, 200 μM sulfide (hypoxic) caused no changes in mitochondrial structure in the sulfide-tolerant oligochaete *Tubificoides benedii* (Dubilier et al. 1997). Research in this area often has overlapped with descriptions of electron-dense organelles, which are inducible by sulfide exposure and may be linked to degradation of mitochondria via autophagy (Wohlgemuth et al. 2007).

In contrast to the ambiguous relationship between mitochondrial structure and sulfide, several studies have demonstrated that sulfide disrupts mitochondrial function. Sulfide reversibly inhibits cytochrome *c* oxidase (e.g. Nicholls 1975), halting adenosine triphosphate (ATP) production by the mitochondrial electron transport chain with an effectiveness several times greater than that of cyanide. An additional pathway for alteration of mitochondrial function involves ROS and GSH. The mitochondrial membrane potential is essential to oxidative phosphorylation and is generated by the proton gradient across the mitochondrial inner membrane. Disruption of that membrane potential may lead to permeabilization of the inner membrane, causing formation of the mitochondrial permeability transition pore (PTP) (Rasola and Bernardi 2007). Formation of PTP is increased by the presence of ROS (Zoratti and Szabó 1995) and by imbalances in the GSH pool (Costantini et al. 1996, Kowaltowski et al. 1998). PTP opening can also lead to the release of cytochrome *c*, ultimately causing apoptotic cell death. Sulfide exposure produced mitochondrial membrane depolarization in *G. dibranchiata* coelomocytes (Julian et al. 2005a), although it was not clear in that study if PTP opened. In similar studies of rat hepatocytes, sulfide exposure led to mitochondrial membrane depolarization (Eghbal et al. 2004) and PTP opening (Thompson et al. 2003).

Sulfide as an Antioxidant

Paradoxically, low doses of sulfide have antioxidant and cytoprotective properties in addition to the regulatory roles discussed above. Essentially all of this work is biomedical in nature, but it will be reviewed here in the hope that similarly constructed studies may be conducted in aquatic organisms. Sulfide and derived-sulfur-containing compounds interact with a variety of ROS and with several types of oxidative damage, relieving oxidative stress and reducing cell death (for review, Battin and Brumaghim 2009; Whiteman and Moore 2009). Sulfide has multiple "antioxidant" functions, including repair of lipid peroxidation products (Geng et al. 2004; Chang et al. 2008; Muellner et al. 2009) and protein carbonyls (Whiteman et al. 2004, 2005), direct scavenging of H_2O_2 and superoxide (see Box 7.1; Geng et al. 2004; Yan et al. 2006; Chang et al. 2008), and binding to nitric oxide or peroxynitrite (Whiteman et al. 2006). These protective effects typically result in increased cell viability (Whiteman et al. 2004, 2005; Yan et al. 2006).

Sulfide can also up-regulate antioxidant enzyme activities. In two plant studies, addition of sulfide exposure to copper exposure (Zhang et al. 2008) and osmotic stress (Zhang et al. 2009) protected cells by up-regulating activities of SOD, CAT, and ascorbate peroxidase (osmotic stress only). It is not known why sulfide might stimulate CAT in plant tissues but inhibit CAT in other organisms. The multiple reports of increases in total SOD activity following sulfide exposure (Zhang et al. 2008; Liu et al. 2009; Su et al. 2009; Zhang et al. 2009) may in fact reflect the ability of sulfide to directly bind to Cu,Zn-SOD and increase its activity (Searcy et al. 1995). Sulfide also may increase GPx activity by interacting with GPx itself or with GSH cycling in general (Szabó 2007).

A third potential mechanism by which sulfide could decrease oxidative stress is by altering cellular metabolism. Low doses of H_2S gas (up to 80 ppm) induced a reversible hypometabolic state in mice (Blackstone et al. 2005), a result that could not be directly replicated in sheep (Haouzi et al. 2008) or indirectly replicated in *Caenorhabditis elegans* (Miller and Roth 2007). Nonetheless, induction of hypometabolism via sulfide exposure (100 ppm H_2S) decreased apoptosis and tissue damage and increased whole-animal survival following a cycle of ischemia/reperfusion in mouse kidneys (Bos et al. 2009). Although the mechanism by which sulfide induces hypometabolism is unknown, these studies have prompted speculation that sulfide could be therapeutic for a number of diseases or conditions (Szabó 2007).

PERSPECTIVES ON SULFIDE BIOLOGY

The study of oxidative stress must incorporate not only biochemistry and physiology, but also ecology and evolution. For example, individual variation in ROS metabolism might underlie the link between genetic variation for fitness and the honesty of sexual ornamentation as indicators of fitness (von Schantz et al. 1999). Several aspects of ROS metabolism appear to be heritable (Olsson et al. 2008) and influence multiple life-history trade-offs (Constantini 2008; Dowling and Simmons 2009) and patterns in species distribution (e.g. Abele and Puntarulo 2004). This relationship is most apparent in aquatic habitats (Lesser 2006), in which many abiotic factors influence ROS metabolism. What is less clear is whether this is also true for sulfide. The multiple potential mechanisms of sulfide toxicity undoubtedly influence ecological and evolutionary processes, but to what degree are these toxic effects related to sulfide's ability to alter ROS metabolism? This remains an open and potentially rich area of investigation. Traditional sulfide ecophysiology studies have focused on mitochondrial structure and function and whole-animal respiration. A new direction integrating antioxidants with oxidative damage and apoptosis may be more successful. This new approach would allow us to determine whether animals in sulfidic habitats have adaptations to reduce oxidative stress from sulfide exposure, and this information would also be fundamental to the ongoing biomedical research.

An important consideration of any sulfide study is the question of dosage. The complex patterns of the cellular effects of sulfide may in part be indications of hormesis. In a hormetic response, low doses of a compound stimulate physiological responses while high doses inhibit them (Calabrese and Baldwin 2003). Although patterns in sulfide toxicity are complicated by the studied organism's sulfide tolerance and oxygen conditions, there does appear to be a general pattern: low doses of sulfide have antioxidant and cytoprotective effects and higher doses cause damage. If true, at what concentration does sulfide transition between beneficial and toxic effects, and how does this transition point vary across species? Does this explain the proposed widespread production and maintenance of low levels of sulfide in mammalian and invertebrate tissues?

We would argue that there is now sufficient evidence that sulfide exposure induces a detectable oxidative stress response, and that it causes oxidative damage and secondary alterations of mitochondrial function. However, it is not clear whether this has biological significance for organisms in sulfidic habitats. In *G. dibranchiata*, similar dosing regimens that caused ROS production (Julian et al. 2005a; Joyner-Matos et al. 2010), mitochondrial depolarization (Julian et al. 2005a), and oxidative damage to nucleic acids (Joyner-Matos et al. 2010) also caused significant cell death and decreased cell proliferation (Hance et al. 2008). There are numerous examples in the biomedical literature of studies linking sulfide exposure to cell death, some of which incorporate mitochondrial or lyosomal cell death pathways (e.g. Cheung et al. 2007). Sulfide-induced oxidative damage to DNA is accompanied by apoptosis in human fibroblasts (Baskar et al. 2007, Yaegaki et al. 2008), and enhanced ROS production and PTP opening coincides with apoptosis in sulfide-exposed human gingival epithelial cells (Calenic et al. 2009). However, it is unclear whether these pathways are attributed solely to free radicals or to ROS (e.g. Truong et al. 2006), and to what extent they may vary with degree of sulfide tolerance in aquatic organisms.

REFERENCES

Abedinzadeh, Z. (2001) Sulfur-centered reactive intermediates derived from the oxidation of sulfur compounds of biological interest. *Canadian Journal of Physiology and Pharmacology* 79, 166–170.

Abele-Oeschger, D. (1996) A comparative study of superoxide dismutase activity in marine benthic invertebrates with respect to environmental sulphide exposure. *Journal of Experimental Marine Biology and Ecology* 197, 39–49.

Abele-Oeschger, D., Oeschger, R. (1995) Hypoxia-induced autoxidation of haemoglobin in the benthic invertebrates *Arenicola marina* (Polychaeta) and *Astarte borealis* (Bivalvia) and the possible effects of sulphide. *Journal of Experimental Marine Biology and Ecology* 187, 63–80.

Abele, D., Puntarulo, S. (2004) Formation of reactive species and induction of antioxidant defence systems in polar and temperate marine invertebrates and fish. *Comparative Biochemistry and Physiology A* 138, 405–415.

Abele-Oeschger, D., Oeschger, R., Theede, H. (1994) Biochemical adaptations of *Nereis diversicolor* (Polychaeta) to temporarily increased hydrogen peroxide levels in intertidal sandflats. *Marine Ecology Progress Series* 106, 101–110.

Abele, D., Grosspietsch, H., Pörtner, H.O. (1998) Temporal fluctuations and spatial gradients of environmental P_{O_2}), temperature, H_2O_2 and H_2S in its intertidal habitat trigger enzymatic antioxidant protection in the capitellid worm *Heteromastus filiformis*. *Marine Ecology Progress Series* 163, 179–191.

Alberic, P., Boulegue, J. (1990) Unusual amino compounds in the tissues of *Calyptogena phaseoliformis* (Japan Trench): Possible link to symbiosis. *Progress in Oceanography* 24, 89–101.

Arp, A.J., Hansen, B.M., Julian, D. (1992) Burrow environment and coelomic fluid characteristics of the echiuran worm *Urechis caupo* from populations at three sites in northern California. *Marine Biology* 113, 613–623.

Atkinson, R.J.A., Taylor, A.C. (2005) Aspects of the physiology, biology and ecology of thalassinidean shrimps in relation to their burrow environment. *Oceanography and Marine Biology: an Annual Review* 43, 173–210.

Attene-Ramos, M.S., Wagner, E.D., Plewa, M.J., Gaskins, H.R. (2006) Evidence that hydrogen sulfide is a genotoxic agent. *Molecular Cancer Research* 4, 9–14.

Attene-Ramos, M.S., Wagner, E.D., Gaskins, H.R., Plewa, M.J. (2007) Hydrogen sulfide induces direct radical-associated DNA damage. *Molecular Cancer Research* 5, 455–459.

Bagarinao, T. (1992) Sulfide as an environmental factor and toxicant: tolerance and adaptations in aquatic organisms. *Aquatic Toxicology* 24, 21–62.

Bagarinao, T., Vetter, R.D. (1990) Oxidative detoxification of sulfide by mitochondria of the California killifish *Fundulus parvipinnis* and the speckled sanddab *Citharichthys stigmaeus*. *Journal of Comparative Physiology B* 160, 519–527.

Baskar, R., Li, L., Moore, P.K. (2007) Hydrogen sulfide induces DNA damage and changes in apoptotic gene expression in human lung fibroblast cells. *FASEB Journal* 21, 247–255.

Battin, E.E., Brumaghim, J.L. (2009) Antioxidant activity of sulfur and selenium: A review of reactive oxygen species scavenging, glutathione peroxidase, and metal-binding antioxidant mechanisms. *Cell Biochemistry and Biophysiology* 55, 1–23.

Beauchamp Jr, R.O., Bus, J.S., Popp, J.A., Boreiko, C.J., Andjelkovich, D.A. (1984) A critical review of the literature on hydrogen sulfide toxicity. *CRC Critical Reviews in Toxicology* 13, 25–97.

Berner, R.A. (1963) Electrode studies of hydrogen sulfide in marine sediments. *Geochimica et Cosmochimica Acta*, 27, 563–575.

Blackstone, E., Morrison, M., Roth, M.B. (2005) H_2S induces a suspended animation-like state in mice. *Science*, 308, 518.

Blum, J., Fridovich, I. (1984) Enzymatic defenses against oxygen toxicity in the hydrothermal vent animals *Riftia pachyptila* and *Calyptogena magnifica*. *Archives of Biochemistry and Biophysics* 228, 617–620.

Borin, S., Brusetti, L., Mapelli, F. et al. (2009) Sulfur cycling and methanogenesis primarily drive microbial colonization of the highly sulfidic Urania deep hypersaline basin. *Proceedings of the National Academy of Sciences USA* 106, 9151–9156.

Bos, E.M., Leuvenink, H.G.D., Snijder, P.M. et al. (2009) Hydrogen sulfide-induced hypometabolism prevents renal ischemia/reperfusion injury. *Journal of the American Society of Nephrology* 20, 1901–1905.

Bradley, P.M., Morris, J.T. (1990) Influence of oxygen and sulfide concentration on nitrogen uptake kinetics in *Spartina alterniflora*. *Ecology* 71, 282–287.

Budd, M.S., Bewick, H.A. (1952) Photometric determination of sulfide and reducible sulfur in alkalies. *Analytical Chemistry* 24, 1536–1540.

Calabrese, E.J., Baldwin, L.A. (2003) Hormesis: The dose-response revolution. *Annual Review of Pharmacology and Toxicology* 43, 175–197.

Calenic, B., Yaegaki, K., Murata, T. et al. (2009) Oral malodorous compound triggers mitochondrial-dependent apoptosis and causes genomic DNA damage in human gingival epithelial cells. *Journal of Periodontal Research* 45, 31–37.

Carlson Jr, P.R., Yarbro, L.A., Barber, T.R. (1994) Relationship of sediment sulfide to mortality of *Thalassia testudinum* in Florida bay. *Bulletin of Marine Science* 54, 733–746.

Carlsson, J., Berglin, E.H., Claesson, R., Maj-Britt, K.E., Persson, S. (1988) Catalase inhibition by sulfide and hydrogen peroxide-induced mutagenicity in *Salmonella typhimurium* strain TA102. *Mutation Research* 202, 59–64.

Cavanaugh, C.M. (1983) Symbiotic chemoautotrophic bacteria in marine invertebrates from sulphide-rich habitats. *Nature* 302, 58–61.

Chang, L., Geng, B., Yu, F., Zhao, J., Hongfeng, J., Du, J., Tang, C. (2008) Hydrogen sulfide inhibits myocardial injury induced by homocysteine in rats. *Amino Acids* 34, 573–585.

Chen, K.Y., Morris, J.C. (1972a) Kinetics of oxidation of aqueous sulfide by O_2. *Environmental Science and Technology* 6, 529–537.

Chen, K.Y., Morris, J.C. (1972b) Oxidation of sulfide by O_2: catalysis and inhibition. *Journal of the Sanitation Engineering* 98, 215–227.

Cheung, N.S., Peng, Z.F., Chen, M.J., Moore, P.K., Whiteman, M. (2007) Hydrogen sulfide induced neuronal death occurs via glutamate receptor and is associated with calpain activation and lysosomal rupture in mouse primary cortical neurons. *Neuropharmacology* 53, 505–514.

Cline, J.D. (1969) Spectrophotometric determination of hydrogen sulfide in natural waters. *Limnology and Oceanography* 14, 454–458.

Cline, J.D., Richards, F.A. (1969) Oxygenation of hydrogen sulfide in seawater at constant salinity, temperature, and pH. *Environmental Science and Technology* 3, 838–841.

Company, R., Serafim, A., Cosson, R.P., Fiala-Médioni, A., Dixon, D.R., Bebianno, M.J. (2007) Adaptation of the antioxidant defence system in hydrothermal-vent mussels (*Bathymodiolus azoricus*) transplanted between two Mid-Atlantic Ridge sites. *Marine Ecology* 28, 93–99.

Cordes, E.E., Bergquist, D.C., Fisher, C.R. (2009) Macroecology of Gulf of Mexico cold seeps. *Annual Review of Marine Science* 1, 143–168.

Constantini, D. (2008) Oxidative stress in ecology and evolution: lessons from avian studies. *Ecology Letters* 11, 1238–1251.

Costantini, P., Chernyak, B.V., Petronilli, V., Bernardi, P. (1996) Modulation of the mitochondrial permeability transistion pore by pyridine nucleotides and dithiol oxidation at two separate sites. *Journal of Biological Chemistry* 271, 6746–6751.

Dattagupta, S., Telesnicki, G., Luley, K., Predmore, B., Mcginley, M., Fisher, C.R. (2007) Submersible operated peepers for collecting porewater from deep-sea sediments. *Limnology and Oceanography Methods* 5, 263–268.

Dattagupta, S., Schaperdoth, I., Montanari, A. et al. (2009) A novel symbiosis between chemoautotrophic bacteria and a freshwater cave amphipod. *ISME Journal* 3, 935–943.

Diaz, R.J., Rosenberg, R. (1995) Marine benthic hypoxia: A review of its ecological effects and the behavioral responses of benthic macrofauna. *Oceanography and Marine Biology: an Annual Review* 33, 245–303.

Dong, Y., Miller, L.P., Sanders, J.G., Somero, G.N. (2008) Heat-shock protein 70 (Hsp70) expression in four limpets of the genus *Lottia*: Interspecific variation in constitutive and inducible synthesis correlates with *in situ* exposure to heat stress. *Biological Bulletin* 215, 173–181.

Dowling, D.K., Simmons, L.W. (2009) Reactive oxygen species as universal constraints in life-history evolution. *Proceedings of the Royal Society of London B* 276, 1737–1745.

Dubilier, N., Windoffer, R., Grieshaber, M.K., Giere, O. (1997) Ultrastructure and anaerobic metabolism of mitochondria in the marine oligochaete *Tubificoides benedii*: effects of hypoxia and sulfide. *Marine Biology* 127, 637–645.

Duffy III, J.E., Tyler, S. (1984) Quantitative differences in mitochondrial ultrastructure of a thiobiotic and an oxybiotic turbellarian. *Marine Biology* 83, 95–102.

Eghbal, M.A., Pennefather, P.S., O'brien, P.J. (2004) H_2S cytotoxicity mechanism involves reactive oxygen species formation and mitochondrial depolarisation. *Toxicology* 203, 69–76.

Engel, A.S., Porter, M.L., Stern, L.A., Quinlan, S., Bennett, P.C. (2004) Bacterial diversity and ecosystem function of filamentous microbial mats from aphotic (cave) sulfidic springs dominated by chemolithoautotrophic "Epsilonproteobacteria". *FEMS Microbiology Ecology* 51, 31–53.

Felbeck, H. (1983) Sulfide oxidation and carbon fixation by the gutless clam *Solemya reidi*: an animal-bacteria symbiosis. *Journal of Comparative Physiology B* 152, 3–11.

Fenchel, T.M., Riedl, R.J. (1970) The sulfide system: a new biotic community underneath the oxidized layer of marine sand bottoms. *Marine Biology* 7, 255–268.

Fisher, C.R. (1990) Chemoautotrophic and methanotrophic symbioses in marine invertebrates. *Critical Reviews in Aquatic Sciences* 2, 399–436.

Forgan, L.G., Forster, M.E. (2010) Oxygen consumption, ventilation frequency and cytochrome c oxidase activity in blue cod (*Parapercis colias*) exposed to hydrogen sulphide or isoeugenol. *Comparative Biochemistry and Physiology C* 151, 57–65.

Gainey Jr, L.F., Greenberg, M.J. (2005) Hydrogen sulfide is synthesized in the gills of the clam *Mercenaria mercenaria* and acts seasonally to modulate branchial muscle contraction. *Biological Bulletin* 209, 11–20.

Gebruk, A.V., Chevaldonné, P., Shank, T., Lutz, R.A., Vrijenhoek, R.C. (2000) Deep-sea hydrothermal vent communities of the Logatchev area (14°45'N, Mid-Atlantic Ridge): diverse biotopes and high biomass. *Journal of the Marine Biological Association of the UK* 80, 383–393.

Geng, B., Chang, L., Pan, C. et al. (2004) Endogenous hydrogen sulfide regulation of myocardial injury induced by isoproterenol. *Biochemical and Biophysical Research Communications* 318, 756–763.

Giles, G.I., Tasker, K.M., Jacob, C. (2001) The role of reactive sulfur species in oxidative stress. *Free Radical Biology and Medicine* 31, 1279–1283.

Grieshaber, M.K., Völkel, S. (1998) Animal adaptations for tolerance and exploitation of poisonous sulfide. *Annual Review of Physiology* 60, 33–53.

Hance, J.M., Andrzejewski, J.E., Predmore, B.L., Dunlap, K.J., Misiak, K.L., Julian, D. (2008) Cytoxicity from sulfide exposure in a sulfide-tolerant marine invertebrate. *Journal of Experimental Marine Biology and Ecology* 359, 102–109.

Haouzi, P., Notet, V., Chenuel, B., Chalon, B., Sponne, I., Ogier, V., Bihain, B. (2008) H_2S induced hypometabolism in mice is missing in sedated sheep. *Respiratory Physiology and Neurobiology* 160, 109–115.

Hermes-Lima, M., Storey, J.M., Storey, K.B. (1998) Antioxidant defenses and metabolic depression. The hypothesis of preparation for oxidative stress in land snails. *Comparative Biochemistry and Physiology B* 120, 437–448.

Hermes-Lima, M., Storey, J.M., Storey, K.B. (2001) Antioxidant defenses and animal adaptation to oxygen availability during environmental stress. *In* Storey, K.B., Storey, J.M. (eds). *Cell and Molecular Responses to Stress*. Elsevier, Amsterdam, pp. 263–287.

Hoffmann, A.A., Parsons, P.A. (1991) *Evolutionary Genetics and Environmental Stress*, Oxford University Press, New York.

Huxtable, R.J. (1986) *Biochemistry of Sulfur*. Plenum Press, New York.

Janssen, H.H., Oeschger, R. (1992) The body wall of *Halicryptus spinulosus* (Priapulida) – ultrastructure and changes induced by hydrogen sulfide. *Hydrobiologia* 230, 219–230.

Jørgensen, B.B., Boetius, A. (2007) Feast and famine – microbial life in the deep-sea bed. *Nature Reviews Microbiology* 5, 770–781.

Joshi, M.M., Hollis, J.P. (1977) Interaction of *Beggiatoa* and rice plant: detoxification of hydrogen sulfide in the rice rhizosphere. *Science* 195, 179–180.

Joyner, J.L., Peyer, S.M., Lee, R.W. (2003) Possible roles of sulfur-containing amino acids in a chemoautotrophic bacterium-mollusc symbiosis. *Biological Bulletin* 205, 331–338.

Joyner-Matos, J., Downs, C.A., Julian, D. (2006) Increased expression of stress proteins in the surf clam *Donax variabilis* following hydrogen sulfide exposure. *Comparative Biochemistry and Physiology A* 145, 245–257.

Joyner-Matos, J., Predmore, B.L., Stein, J.R., Leeuwenburgh, C., Julian, D. (2010) Hydrogen sulfide induces oxidative damage to RNA and DNA in a sulfide-tolerant marine invertebrate. *Physiological and Biochemical Zoology* 83, 356–365.

Julian, D., Arp, A.J. (1992) Sulfide permeability in the marine invertebrate *Urechis caupo*. *Journal of Comparative Physiology B* 162, 59–67.

Julian, D., Statile, J.L., Wohlgemuth, S.E., Arp, A.J. (2002) Enzymatic hydrogen sulfide production in marine invertebrate tissues. *Comparative Biochemistry and Physiology A* 133, 105–115.

Julian, D., April, K.L., Patel, S., Stein, J.R., Wohlgemuth, S.E. (2005a) Mitochondrial depolarization following hydrogen sulfide exposure in erythrocytes from a sulfide-tolerant marine invertebrate. *Journal of Experimental Biology* 208, 4109–4122.

Julian, D., Statile, J., Roepke, T.A., Arp, A.J. (2005b) Sodium nitroprusside potentiates hydrogen-sulfide-induced contractions in body wall muscle from a marine worm. *Biological Bulletin* 209, 6–10.

Keilin, D., Hartree, E.F. (1936) On some properties of catalase haematin. *Proceedings of the Royal Society of London B* 121, 173–191.

Khan, A.A., Schuler, M.M., Coppock, R.W. (1987) Inhibitory effects of various sulfur compounds on the activity of bovine erythrocyte enzymes. *Journal of Toxicology and Environmental Health* 22, 481–490.

Kiemer, M.C.B., Black, K.D., Lussot, D., Bullock, A.M., Ezzi, I. (1995) The effects of chronic and acute exposure to hydrogen sulfide on Atlantic salmon (*Salmo salar* L). *Aquaculture* 135, 311–327.

Kimura, H. (2011) Hydrogen sulfide: its production, release and functions. *Amino Acids* 41, 113–121.

King, G.M., Klug, M.J., Wiegert, R.G., Chalmers, A.G. (1982) Relation of soil water movement and sulfide concentration to *Spartina alterniflora* production in a Georgia salt marsh. *Science* 218, 61–63.

Kowaltowski, A.J., Netto, L.E., Vercesi, A.E. (1998) The thiol-specific antioxidant enzyme prevents mitochondrial permeability transition: Evidence for the participation of reactive oxygen species in this mechanism. *Journal of Biological Chemistry* 273, 12766–12769.

Laurent, M.C., Gros, O., Brulport, J.P., Gaill, F., Bris, N.L. (2009) Sunken wood habitat for thiotrophic symbiosis in mangrove swamps. *Marine Environmental Research* 67, 83–88.

Lee, R.W., Kraus, D.W., Doeller, J.E. (1999) Oxidation of sulfide by *Spartina alterniflora* roots. *Limnology and Oceanography* 44, 1155–1159.

Lee, R.Y., Porubsky, W.P., Feller, I.C., Mckee, K.L., Joye, S.B. (2008) Porewater biogeochemistry and soil metabolism in dwarf red mangrove habitats (Twin Cays, Belize). *Biogeochemistry* 87, 181–198.

Lesser, M.P. (2006) Oxidative stress in marine environments: Biochemistry and physiological ecology. *Annual Review of Physiology* 68, 253–278.

Levesque, C., Juniper, S.K., Limén, H. (2006) Spatial organization of food webs along habitat gradients at deep-sea hydrothermal vents on Axial Volcano, Northeast Pacific. *Deep-Sea Research I* 53, 726–739.

Levin, L.A., Ziebis, W., Mendoza, G.F., Growney-Cannon, V., Walther, S. (2006) Recruitment response of methane-seep macrofauna to sulfide-rich sediments: An *in situ* experiment. *Journal of Experimental Marine Biology and Ecology* 330, 132–150.

Liu, H., Bai, X.B., Shi, S., Cao, Y.X. (2009) Hydrogen sulfide protects from intestinal ischaemia-reperfusion injury in rats. *Journal of Pharmacy and Pharmacology* 61, 207–212.

Ma, S., Luther, G.W., Scarborough, R.W., Mensinger, M.G. (2007) Voltammetry: An *in situ* tool to monitor the health of ecosystems. *Electroanalysis* 19, 2051–2057.

Maricle, B.R., Crosier, J.J., Bussiere, B.C., Lee, R.W. (2006) Respiratory enzyme activities correlate with anoxia tolerance in salt marsh grasses. *Journal of Experimental Marine Biology and Ecology* 337, 30–37.

Miller, D.L., Roth, M.B. (2007) Hydrogen sulfide increases thermotolerance and lifespan in *Caenorhabditis elegans*. *Proceedings of the National Academy of Sciences USA* 104, 20618–20622.

Millero, F.J., Hubinger, S., Fernandez, M., Garnett, S. (1987) Oxidation of H_2S in seawater as a function of temperature, pH, and ionic strength. *Environmental Science and Technology* 21, 439–443.

Mitchell, C.W., Davenport, S.J. (1979) Appendix II: Hydrogen sulphide literature. *In* Council, N.R. (ed.) *Hydrogen Sulfide*. University Park Press, Baltimore, pp. 141–153.

Morrill, A.C., Powell, E.N., Bidigare, R.R., Shick, J.M. (1988) Adaptations to life in the sulfide system: a comparison of oxygen detoxifying enzymes in thiobiotic and oxybiotic meiofauna (and freshwater planarians). *Journal of Comparative Physiology B* 158, 335–344.

Morse, J.W., Millero, F.J., Cornwell, J.C., Rickard, D. (1987) The chemistry of the hydrogen sulfide and iron sulfide systems in natural waters. *Earth-Science Reviews* 24, 1–42.

Muellner, M.K., Schreier, S.M., Laggner, H. et al. (2009) Hydrogen sulfide destroys lipid hydroperoxides in oxidized LDL. *Biochemistry Journal* 420, 277–281.

Nicholls, P. (1975) The effect of sulphide on cytochrome *aa*3: isoteric and allosteric shifts of the reduced alpha-peak. *Biochimica et Biophysica Acta* 396, 24–35.

Oeschger, R., Vetter, R.D. (1992) Sulfide detoxification and tolerance in *Halicryptus spinulosus* (Priapulida): a multiple strategy. *Marine Ecology Progress Series* 86, 167–179.

Olsson, M., Wilson, M., Uller, T. et al. (2008) Free radicals run in lizard families. *Biology Letters* 4, 186–188.

Ortega, J.A., Ortega, J.M., Julian, D. (2008) Hypotaurine and sulfhydryl-containing antioxidants reduce H_2S toxicity in erythrocytes from a marine invertebrate. *Journal of Experimental Biology* 211, 3816–3825.

Oseid, D.M., Smith Jr., L.H. (1974) Factors influencing acute toxicity estimates of hydrogen sulfide to freshwater invertebrates. *Water Research* 8, 739–746.

Ott, J.A., Bright, M., Schiemer, F. (1998) The ecology of a novel symbiosis between a marine peritrich ciliate and chemoautotrophic bacteria. *PSZN: Marine Ecology* 19, 229–243.

Plath, M., Tobler, M., Riesch, R., De Leon, F.J.G., Giere, O., Schlupp, I. (2007) Survival in an extreme habitat: the roles of behaviour and energy limitation. *Naturwissenschaften* 94, 991–996.

Pomeroy, R. (1936) The determination of sulphides in sewage. *Sewage Works Journal* 8, 572–591.

Powell, E.N., Crenshaw, M.A., Rieger, R.M. (1979) Adaptations to sulfide in the meiofauna of the sulfide system.I. ^{35}S-sulfide accumulation and the presence of a sulfide detoxification system. *Journal of Experimental Marine Biology and Ecology* 37, 57–76.

Powell, M.A., Somero, G.N. (1985) Sulfide oxidation occurs in the animal tissue of the gutless clam, *Solemya reidi*. *Biological Bulletin* 169, 164–181.

Powell, M.A., Somero, G.N. (1986) Hydrogen sulfide oxidation is coupled to oxidative phosphorylation in mitochondria of *Solemya reidi*. *Science* 233, 563–566.

Pruski, A.M., Fiala-Médioni, A., Prodon, R., Colomines, J.-C. (2000) Thiotaurine is a biomarker of sulfide-based symbiosis in deep-sea bivalves. *Limnology and Oceanography* 45, 1860–1867.

Rasola, A., Bernardi, P. (2007) The mitochondrial permeability transition pore and its involvement in cell death and in disease pathogenesis. *Apoptosis* 12, 815–833.

Renninger, G.H., Kass, L., Gleeson, R.A. et al. (1995) Sulfide as a chemical stimulus for deep-sea hydrothermal vent shrimp. *Biological Bulletin* 189, 69–76.

Rittenberg, S.C., Mittwer, T., Ivler, D. (1958) Coliform bacteria in sediments around three marine sewage outfalls. *Limnology and Oceanography* 3, 101–108.

Rosenberg, N.K., Lee, R.W., Yancey, P.H. (2006) High contents of hypotaurine and thiotaurine in hydrothermal-vent gastropods without thiotrophic endosymbionts. *Journal of Experimental Zoology A* 305, 655–662.

Roth, S.H. (2004) Toxicological and environmental impacts of hydrogen sulfide. *In* Wang, R. (ed.). *Signal Transduction and the Gasotransmitters: NO, CO, H_2S in Biology and Medicine*. Humana Press, Totawa, NJ, pp. 293–313.

Sarbu, S.M., Kane, T.C., Kinkle, B.K. (1996) A chemoautotrophically based cave ecosystem. *Science* 272, 1953–1955.

Searcy, D.G., Whitehead, J.P., Maroney, M.J. (1995) Interaction of Cu,Zn superoxide dismutase with hydrogen sulfide. *Archives of Biochemistry and Biophysics* 318, 251–263.

Shi, X., Mao, Y. (1994) 8-Hydroxy-2'-deoxyguanosine formation and DNA damage induced by sulfur trioxide anion radicals. *Biochemical and Biophysical Research Communications* 205, 141–147.

Sibuet, M., Olu, K. (1998) Biogeography, biodiversity and fluid dependence of deep-sea cold-seep communities at active and passive margins. *Deep-Sea Research II* 45, 517–567.

Smith, C.R., Maybaum, H.L., Baco, A.R. et al. (1998) Sediment community structure around a whale skeleton in the deep Northeast Pacific: Macrofaunal, microbial and bioturbation effects. *Deep-Sea Research II* 45, 335–364.

Smith, R.P., Abbanat, R.A. (1966) Protective effect of oxidized glutathione in acute sulfide poisoning. *Toxicology and Applied Pharmacology* 9, 209–217.

Spicer, J.I., Gaston, K.J. (1999) *Physiological Diversity and its Ecological Implications*. Blackwell Science, Oxford.

Storey, K.B. (1996) Oxidative stress: animal adaptations in nature. *Brazilian Journal of Medicine and Biomedical Research* 29, 1715–1733.

Su, Y.-W., Liang, C., Jin, H.-F. et al. (2009) Hydrogen sulfide regulates cardiac function and structure in adriamycin-induced cardiomyopathy. *Circulation Journal* 73, 741–749.

Swartz, R.C., Cole, F.A., Schults, D.W., Deben, W.A. (1986) Ecological changes in the Southern California Bight near a large sewage outfall: benthic conditions in 1980 and 1983. *Marine Ecology Progress Series* 31, 1–13.

Szabó, C. (2007) Hydrogen sulphide and its therapeutic potential. *Nature Reviews. Drug Discovery* 6, 917–935.

Takagi, S., Kikuchi, E., Doi, H., Shikano, S. (2005) Swimming behavior of *Chironomus acerbiphilus* larvae in Lake Katanuma. *Hydrobiologia* 548, 153–165.

Tapley, D.W. (1993) *Sulfide-dependent oxidative stress in marine invertebrates, especially thiotrophic symbioses*. University of Maine, Orono.

Tapley, D.W., Shick, J.M. (1991) Antioxidant enzyme activities in the protobranch bivalves *Solemya velum* and *Yoldia limatula*: Responses to sulfide exposure. *American Zoologist* 31, 135A.

Tapley, D.W., Beuttner, G.R., Shick, J.M. (1999) Free radicals and chemiluminescence as products of the spontaneous oxidation of sulfide in seawater, and their biological implications. *Biological Bulletin* 196, 52–56.

Thompson, R.W., Valentine, H.L., Valentine, W.M. (2003) Cytotoxic mechanisms of hydrosulfide anion and cyanide anion in primary rat hepatocyte cultures. *Toxicology* 188, 149–159.

Tobler, M., Schlupp, I., Heubel, K.U. et al. (2006) Life on the edge: hydrogen sulfide and the fish communities of a Mexican cave and surrounding waters. *Extremophiles* 10, 577–585.

Tobler, M., Dewitt, T.J., Schlupp, I. et al. (2008) Toxic hydrogen sulfide and dark caves: phenotypic and genetic divergence across two abiotic environmental gradients in *Poecilia mexicana*. *Evolution* 62, 2643–2659.

Treude, T., Smith, C.R., Wenzhöfer, F., Carney, E., Barnardino, A.F., Hannides, A.K., Krüger, M., Boetius, A. (2009) Biogeochemistry of a deep-sea whale fall: sulfate reduction, sulfide efflux and methanogenesis. *Marine Ecology Progress Series* 382, 1–21.

Truong, D.H., Eghbal, M.A., Hindmarsh, W., Roth, S.H., O'brien, P.J. (2006) Molecular mechanisms of hydrogen sulfide toxicity. *Drug Metabolism Reviews* 38, 733–744.

Van Dover, C.L. (2000) *The Ecology of Deep-Sea Hydrothermal Vents*. Princeton University Press, Princeton, NJ.

Vetter, R.D., Wells, M.E., Kurtsman, A.L., Somero, G.N. (1987) Sulfide detoxification by the hydrothermal vent crab *Bythograea thermydron* and other decapod crustaceans. *Physiological Zoology* 60, 121–137.

Vismann, B. (1991) Sulfide tolerance: physiological mechanisms and ecological implications. *Ophelia* 34, 1–27.

Vismann, B. (1996) Sulfide species and total sulfide toxicity in the shrimp *Crangon crangon*. *Journal of Experimental Marine Biology and Ecology* 204, 141–154.

Von Schantz, T., Bensch, S., Grahn, M., Hasselquist, D., Wittzell, H. (1999) Good genes, oxidative stress and condition-dependent sexual signals. *Proceedings of the Royal Society of London B* 266, 1–12.

Weber, R.E., Vinogradov, S.N. (2001) Nonvertebrate hemoglobins: functions and molecular adaptations. *Physiology Review* 81, 569–628.

Whiteman, M., Moore, P.K. (2009) Hydrogen sulfide and the vasculature: a novel vasculoprotective entity and regulator of nitric oxide bioavailability? *Journal of Cellular and Molecular Medicine* 13, 488–507.

Whiteman, M., Armstrong, J.S., Chu, S.H. et al. (2004) The novel neuromodulator hydrogen sulfide:an endogenous peroxynitrite 'scavenger'? *Journal of Neurochemistry* 90, 765–768.

Whiteman, M., Cheung, N.S., Zhu, Y.-Z. et al. (2005) Hydrogen sulphide: a novel inhibitor of hypochlorous acid-mediated oxidative damage in the brain? *Biochemical and Biophysical Research Communications* 326, 794–798.

Whiteman, M., Li, L., Kostetski, I., Chu, S.H., Siau, J.L., Bhatia, M., Moore, P.K. (2006) Evidence for the formation of a novel nitrosothiol from the gaseous mediators nitric oxide and hydrogen sulphide. *Biochemical and Biophysical Research Communications* 343, 303–310.

Wohlgemuth, S.E., Arp, A.J., Bergquist, D.C., Julian, D. (2007) Rapid induction and disappearance of electron-dense organelles following sulfide exposure in the marine annelid *Branchioasychis americana*. *Invertebrate Biology* 126, 163–172.

Yaegaki, K., Qian, W., Murata, T., Imai, T., Sato, T., Tanaka, T., Kamoda, T. (2008) Oral malodorous compound causes apoptosis and genomic DNA damage in human gingival fibroblasts. *Journal of Periodontal Research* 43, 391–399.

Yan, S.-K., Chang, T., Wang, H., Wu, L., Wang, R., Meng, Q.H. (2006) Effects of hydrogen sulfide on homocysteine-induced oxidative stress in vascular smooth muscle cells. *Biochemical and Biophysical Research Communications* 351, 485–491.

Zhang, H., Hu, L.-Y., Hu, K.-D., He, Y.-D., Wang, S.-H., Luo, J.-P. (2008) Hydrogen sulfide promotes wheat seed germination and alleviates oxidative damage against copper stress. *Journal of Integrative Plant Biology* 50, 1518–1529.

Zhang, H., Ye, Y.-K., Wang, S.-H., Luo, J.-P., Tang, J., Ma, D.-F. (2009) Hydrogen sulfide counteracts chlorophyll loss in sweetpotato seedling leaves and alleviates oxidative damage against osmotic stress. *Plant Growth Regulation* 58, 243–250.

Zoratti, M., Szabó, I. (1995) The mitochondrial permeability transition. *Biochemica et Biophysica Acta* 1241, 139–176.

Chapter 8

IRON IN COASTAL MARINE ECOSYSTEMS: ROLE IN OXIDATIVE STRESS

Paula Mariela González[1], Dorothee Wilhelms-Dick[2], Doris Abele[3], and Susana Puntarulo[1]

[1]Physical Chemistry-PRALIB, School of Pharmacy and Biochemistry, University of Buenos Aires, Buenos Aires, Argentina
[2]University of Bremen, Geoscience Department, Bremen, Germany
[3]Alfred Wegener Institute for Polar and Marine Research, Department of Functional Ecology, Bremerhaven, Germany

Iron is an essential element for growth and well-being of heterotrophic and autotrophic marine organisms. It is often the central ion in molecules of diverse biological functions, especially in electron transport systems. Further, Fe is coordinated with different inorganic and protein ligands. In living cells Fe has access to a wide range of redox potentials that are involved in different electron transfer reactions, or stored in Fe protein complexes such as ferritin (Ft – FeIII; Galatro et al. 2007). In spite of being one of the abundant metals in the Earth's crust, biologically available Fe can be limiting primary production in well-aerated aquatic systems. At normal seawater pH 8.2 and under oxidizing conditions, Fe(II) is oxidized to insoluble Fe(III) in open ocean waters. Several physiological strategies have been developed by organisms to improve Fe uptake. These mechanisms include siderophores, small, high-affinity Fe-chelating compounds secreted by fungi, plants, and marine bacteria (Martinez et al. 2000), synthesis of species-specific Fe-chelating molecules, specific plasma membrane translocating systems, or uptake from food (Harrison and Arosio 1996).

IRON AVAILABILITY IN MARINE ENVIRONMENTS

Iron content in the upper Earth's crust is around 6% (Wedepohl 1995). The Fe concentration in sediments influences the Fe concentration in the associated surrounding seawater. However, the concentration of dissolved Fe (defined as Fe that can diffuse through a membrane of less than 0.45 μm) in open oceanic waters (<20 m) is extremely low. For all areas except the high-depositional regions the surface ion concentration is 0.25 ± 0.23 nM ($\pm 1\ \sigma$), and the mean in the high deposition regions is 0.76 ± 0.27 nM (the North

Oxidative Stress in Aquatic Ecosystems, First Edition. Edited by Doris Abele, José Pablo Vázquez-Medina, and Tania Zenteno-Savín.

Atlantic and North Indian basins are high deposition regions; Moore and Braucher 2008). Areas outside the high-deposition regions can be further subdivided between high-nutrient, low-chlorophyll (HNLC) zones (0.15 ± 0.16 nM) and the non-HNLC regions (0.27 ± 0.23 nM; Moore and Braucher 2008).

Natural parameters that augment Fe levels in coastal and central oceanic areas are: eolian deposition of dust, river discharge, washout of dust particles in the atmosphere by rainfall, ground water discharge, glacial melting, volcanic sediments, coastal erosion and upwelling of Fe-rich deep waters over hydrothermal vents (Watson 2001). Human activities also have a great impact on Fe levels, especially around coastal areas. Chemical and mining industries, disposal of waste metal, ports, eolian deposition of atmospheric dust from polluted areas, are some of the human activities bringing Fe and other metals to the marine ecosystem. Therefore waters from different regions may have different Fe concentrations.

Figure 8.1 shows a global Fe (oxyhydr)oxide cycle (Raiswell et al. 2006). The contribution of each different source to the global Fe pool can be found in Poulton and Raiswell (2002) and Raiswell and Anderson (2005). Sediments from glacial meltwaters contain (oxyhydr)oxide nanoparticles, ~ 5 nm in diameter, and potentially bioavailable (Raiswell et al. 2006). In the dissolved fraction, Fe is found in seawater as ferrous (Fe^{2+}) and ferric (Fe^{3+}) forms. Fe^{3+} is the thermodynamically stable form in oxygenated seawater and exists as insoluble oxyhydroxides or colloidal matter (Stum and Morgan 1981; Turner et al. 1981; Hudson and Morel 1990). Fe^{2+} is rapidly oxidized at seawater pH ~ 8. It is only a transient species with a half-life of 90 min in seawater (Croot and Laan 2002; Millero and Sotolongo 1989). The mechanisms proposed by King et al. (1995) for oxidation of Fe^{2+} are:

$$Fe^{2+} + O_2 \xrightarrow{k_1} Fe^{3+} + O_2^{\bullet -} \quad (8.1)$$

$$2H^+ + Fe^{2+} + O_2^{\bullet -} \xrightarrow{k_2} Fe^{3+} + H_2O_2 \quad (8.2)$$

$$Fe^{2+} + H_2O_2 \xrightarrow{k_3} Fe^{3+} + OH^- + OH^\bullet \quad (8.3a)$$

$$Fe^{2+} + H_2O_2 \xrightarrow{k_3} FeO^{2+} + H_2O \quad (8.3b)$$

$$Fe^{2+}OH^\bullet \xrightarrow{k_4} Fe^{3+} + OH^- \quad (8.4)$$

Equation (8.3a) is known as the Fenton reaction, but other studies (Kremer 1999; Pierre and Fontecave 1999; Dunford 2002) found evidence for the formation of the ferryl ion FeO^{2+} (equation 8.3b). Croot et al. (2005) measured Fe and H_2O_2 concentrations in the Southern Atlantic and postulated the major pathways of Fe cycling in seawater in the presence of algal cells (Fig. 8.2).

Today, methodological improvements with respect to trace-metal sampling enable even the extremely low-level trace-element concentrations in the deep ocean to be sampled (e.g. De Baar et al. 2008). Figure 8.3 (Moore and Braucher 2008; M. Klunder: Thesis in preparation) shows information about dissolved Fe in filtered ocean water. In the framework of the ongoing project GEOTRACES (www.geotraces.org) the available data set will increase rapidly over the next decade.

The situation is very different in some coastal areas receiving relatively high Fe concentrations. Glacial runoff has been found to increase Fe content locally in the water around the volcanic islands of the South

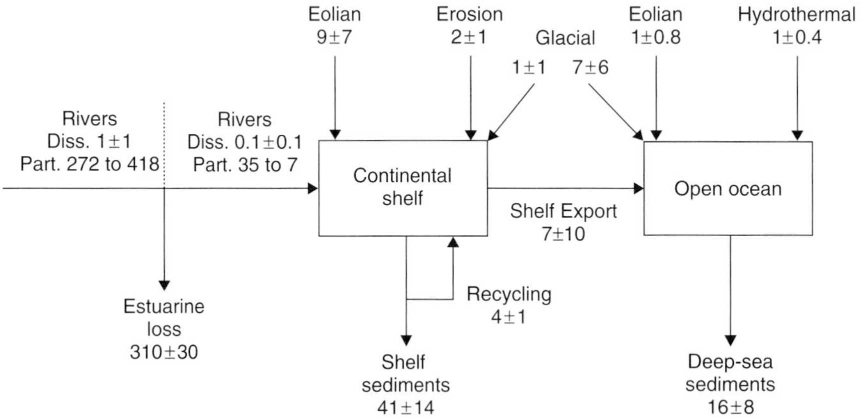

Fig. 8.1 The global Fe (oxyhydr)oxide cycle. Units in Tg yr^{-1}. (Reproduced with permission of Elsevier, Raiswell et al. 2006.)

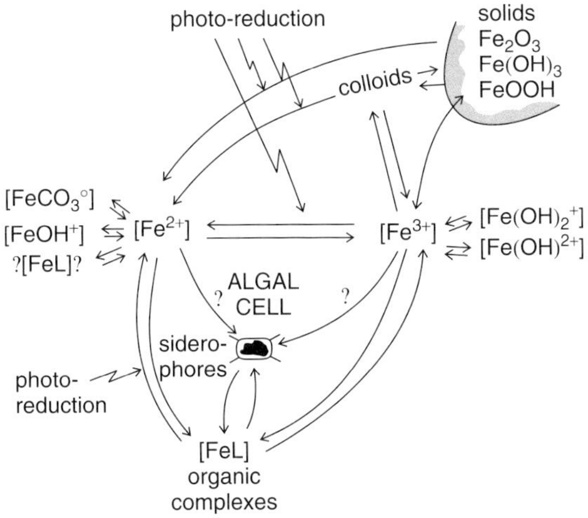

Fig. 8.2 Schematic representation of the speciation of Fe in natural seawater and the possible uptake pathways of iron by an algal cell. (Reproduced with permission of Elsevier, Gerringa et al. 2000.)

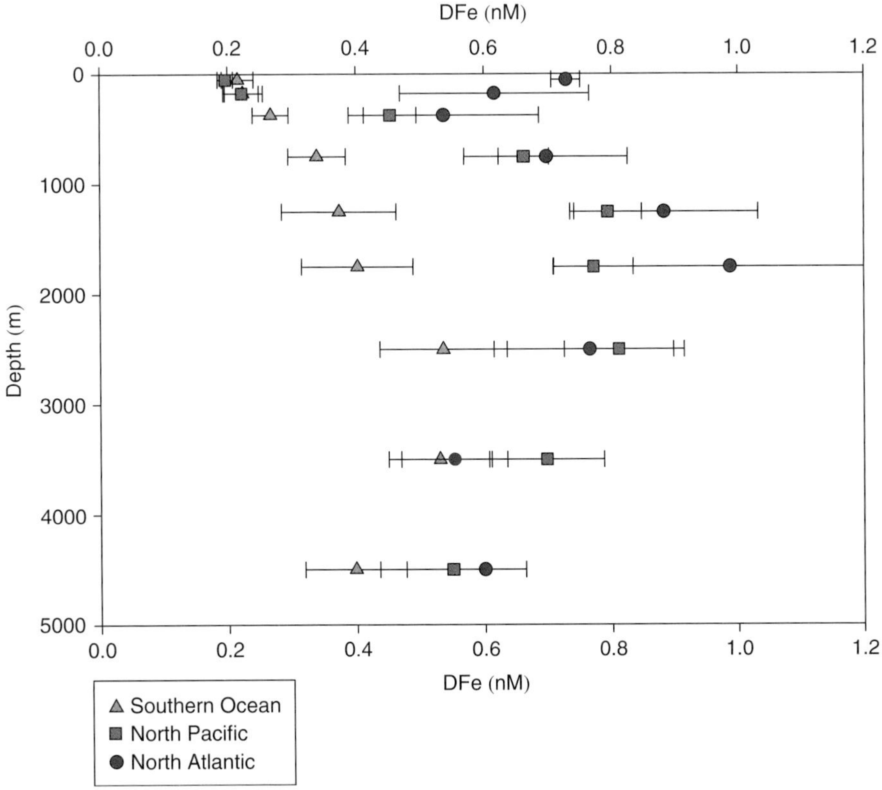

Fig. 8.3 Mean profiles of dissolved iron in the North Atlantic, North Pacific, and Southern Ocean averaged over depth intervals: 0–100 m, 100–250 m, 250–500 m, 500–1000 m, 1000–1500 m, 1500–2000 m, 2000–3000 m, 3000–4000 m, 4000–5000 m. Error bars reflect the 95% confidence interval. (Moore and Braucher 2008 with courtesy of M.B. Klunder.) See plate section for a color version of this image.

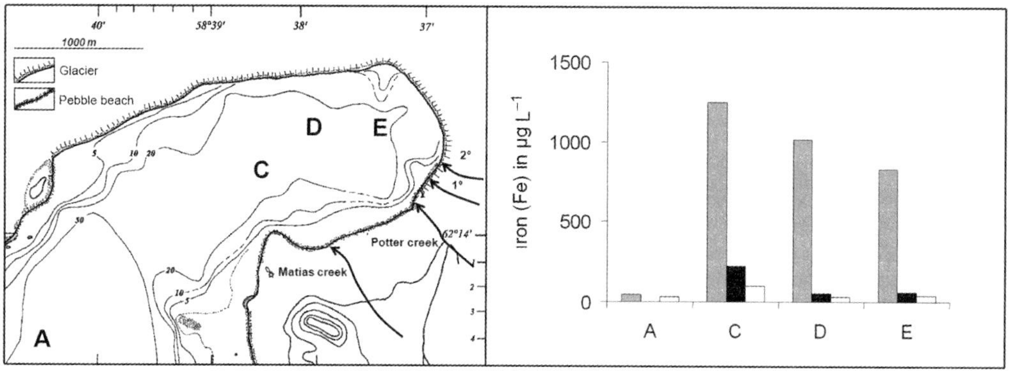

Fig. 8.4 Left. Potter Cove map showing sampling stations A, B, C, D in 2004. Creeks are indicated by arrows. (Courtesy of G. Veit-Köhler, DZMB; Wilhelmshaven, Germany.) Right. Water column depth profiles of particulate (>4 μm) Fe concentrations for the four-station transect between the inside (E) and the outside (A) of Potter Cove, on January 2004. Column colors indicate water depth: grey 0 m, black 15 m, white 30 m (Abele et al. 2008).

Shetland archipelago. Discharge of particulate Fe initially affects shallow coastal waters and the surface layer of the water column (Fig. 8.4; Abele et al. 2008). The particulate matter release is a local phenomenon and outside the cove from King George Island particulate matter content and, consequently, Fe levels dropped considerably. It is unknown how this affects dissolved Fe levels around the island and further transport into Bransfield Strait. About 24 mg Fe g^{-1} dry weight (DW) were found in the sediment (Ahn et al. 2004).

IRON METABOLISM IN MARINE INVERTEBRATES

Iron is an important micronutrient, and Fe deficiency may limit primary productivity in some ocean regions (Martin and Fitzwater 1988; Martin et al. 1990). Fe can control plankton blooms, which in turn have an effect on the biogeochemical cycles of C, N, Si, and S, and therefore influences the Earth climate system (Boyd et al. 2007).

Marine animals incorporate Fe bound to inorganic particles or to organic matter during food ingestion. Further, dissolved Fe is absorbed over the respiratory surfaces and mantle tissue in some marine mollusks, such as filter-feeding mollusks. The extrapallial water around these tissues is in constant exchange with the surrounding seawater. The cellular absorption of Fe^{2+} occurs, at least partially, by a divalent cation exchanger in lobster (Chavez-Crooker et al. 2001). Since Fe is a highly reactive and therefore potentially toxic element, its accumulation in cells and storage in tissues of marine animals is controlled by a network of transporting and binding proteins, in addition to the presence of a labile Fe pool.

Importantly, Fe is the active center of many enzymes and of heme-based oxygen carrier proteins. In vertebrates, Fe is essential for Hb, Mb, and catalase (CAT) formation. However, in invertebrates this metal plays a major role in biomineralization, i.e., formation of inorganic solids deposited in biological systems. For example chitons and limpets incorporate Fe, in magnetite form (Fe_3O_4), into mineralized deposits within the teeth of the radula to harden them (Lowenstam 1981; Webb and Macey 1983; St. Pierre et al. 1990; Lu et al. 1995).

The main Fe storage protein ferritin (Ft), binds Fe and, in so doing, keeps its reactive oxygen species (ROS) forming capacities under control and thus acts as an antioxidant within cells. Ferritin is also thought to play a role in the process of biomineralization. It has been identified in periwinkles (along with hemosiderin) (Taylor 1995), in limpets (*Nacella deaurata*, *Nacella magellanica*, and *Cellana toreuma*; González et al. 2008a; Lu et al. 1995), and in oysters and lobsters (Durand et al. 2002). In the chiton *Clavarizona hirtosa* hemolymph Ft functions as an Fe transporter to the mineralized front of the radula (Kim et al. 1986). Similar Fe-transport systems involving Ft seem to occur in limpets (Burford et al. 1986; Webb et al. 1986). Ferritin may be involved in Fe distribution in the outer side of the mantle, thus playing a role in shell biomineralization of the bivalve *Pinctada fucata* (Zhang et al. 2003). In the snails

Planorbarius corneus and *Lymnaea stagnalis*, Ft is the major protein involved in Fe transportation towards the developing oocyte (Bottke 1982).

The marine ecosystem can be seen as an integrative system with many factors that interact in their effect on the biota. Once metals become bioavailable they can enter the food web at different levels starting with the primary producers and also the heterotrophic organisms at the bottom of the marine food chain, such as benthic filter feeder. Metals follow a bioaccumulation process inside the animals, depending on the animal's detoxification capacities. Fe content in the biota depends as well on exogenous Fe availability. Fe concentration data of several marine mollusks from different regions are presented in Table 8.1.

Winston et al. (1996) reported the presence of Fe in the hemocytes of the marine mussel *Mytilus edulis*. Hemocytes remove metal ions from hemolymph plasma and transfer them to excretion sites. However, the major organ involved in sequestering and detoxifying metals in invertebrates appears to be the digestive gland. Total Fe concentration in *Mya arenaria* mantle represents 35% of the value in the digestive gland, suggesting that Fe is ingested via food particles and then transferred to the mantle tissue to be sequestered into the shell, probably by the hemolymph cells (González et al. 2008b). In *Mercenaria mercenaria* most elements are more concentrated on the exterior surfaces of the shell than within the main shell matrix; however, Fe concentration peaks are located near the surface (30 μm) and further (180 μm) into the shell matrix (Thorn et al. 1995). Highest incorporation of Fe and other elements are found in the growth bands of young *Laternula elliptica* (Figs 8.5 and 8.6; Dick et al. 2007). This might simply reflect higher specific metabolism and filtration rates, or a more efficient transfer of the investigated elements into the shell in young specimens. Similar age-dependent shell profiles of Fe deposition have been recorded in *M. arenaria* (Dick and Abele, unpublished results), and could represent a common phenomenon in bivalve development.

Kidneys and digestive gland of marine invertebrates can produce insoluble phosphate salts containing some detoxified metals, such as Fe, inside the cells, in the form of mineral granules (Nott and Nicolaidou 1996). These granules pass through the gut when the digestive cells break down and they also pass straight through the gut of an invertebrate predator that eats the digestive gland of its prey (Nott and Nicolaidou 1990). In gastropods, transition metals accumulate in residual lysosomes and in the intracellular phosphate concretions in the digestive gland, which are regularly released into the lumen of the gland, passed through the gut, and excreted as fecal pellets (Nott and Nicolaidou 1996). Fe present in fecal pellets and in granules from the digestive gland has been

Table 8.1 Total Fe content in different mollusk species.

Species	Tissue	Fe content (μmol g^{-1} FW)	Reference
Laternula elliptica (King George Is.)	DG	5.0 ± 1.0	Malanga et al. (2008)
Nacella concinna (subtidal) (King George Is.)	DG	3.0 ± 2.0	Malanga et al. (2008)
Nacella concinna (intertidal) (King George Is.)	DG	1.8 ± 0.5	Malanga et al. 2008
Nacella magellanica (Beagle Channel)	DG	1.0 ± 0.1	Malanga et al. 2004
Nacella deaurata (Beagle Channel)	DG	1.9 ± 0.3	Malanga et al. 2004
Mytilis edulis (Helgoland, North Sea)	Whole animal	0.5 ± 0.2	Malanga et al. 2004
Mactra corallina (Mediterranean)	Whole animal	0.9 ± 0.3	Herut et al. 1999
Donax sp. (Mediterranean)	Whole animal	0.5 ± 0.2	Herut et al. 1999
Mytilis edulis (New Zealand)	Whole animal	0.9	Nielsen and Nathan 1975
Perna canaliculus (New Zeland)	Whole animal	0.6	Nielsen and Nathan 1975
Aulacomya maoriana (New Zeland)	Whole animal	0.5	Nielsen and Nathan 1975
Modiolus neozelanicus (New Zeland)	Whole animal	2.6	Nielsen and Nathan 1975
Brachidontes pharaonis (Turkey)	Whole animal	0.05 ± 0.004	Göksu et al. 2005
Pinctada radiata (Turkey)	Whole animal	0.007 ± 0.001	Göksu et al. 2005
Ruditapes decussatus (Spain)	Whole animal	1.9 ± 1.2	Usero et al. 1997
Ruditapes philippinarum (Spain)	Whole animal	1.2 ± 0.8	Usero et al. 1997

DG stands for digestive gland.

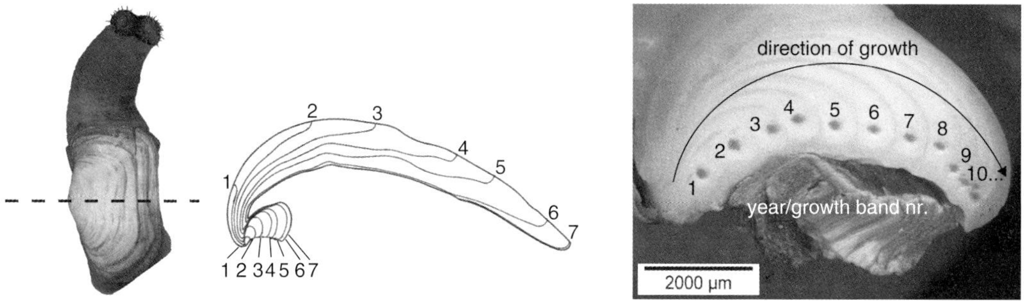

Fig. 8.5 Left. Photograph of the bivalve *L. elliptica*. Middle: Schematics of a cross section of the bivalve shell. Right. Cut and polished umbo of *L. elliptica* with clearly visible annual growth bands after laser ablation. (Reproduced with permission of Elsevier, Dick et al. 2007.)

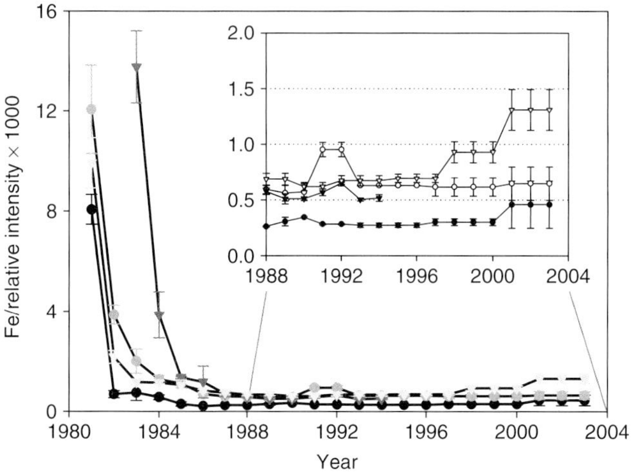

Fig. 8.6 Relative intensities ± SD (×1000) of Fe in annual growth bands of four different *L. elliptica* between 1980 and 2003. Small diagram shows a magnification of the time course from 1988 to 2003. (Reproduced with permission of Elsevier, Dick et al. 2007.)

found in the marine gastropods *Cerithium vulgatum* and *Monodonta mutabilis* (Nott and Nicolaidou 1996) and in fecal pellets of the crab *Pugeltia producta* (Boothe and Knauer 1972). These metal-containing granules are virtually present in every invertebrate phylum (Brown 1982; Mason and Simkiss 1982; Al-Mohanna and Nott 1985; Viarengo 1989). The Stegocephalidae, a family of amphipods, stores Fe obtained from their diet of cnidarians in large Ft crystals that are subsequently lost from the ventral ceca (Moore and Rainbow 1984, 1989).

Metal flux from hemolymph filtration through renal organs could be of great importance; however, there has been little investigation in invertebrates. At least four organs may be involved in regulating dietary and environmental metals in crustaceans: gut, gills, kidney, and integument (Ahearn et al. 2004). The exact mechanism and the contribution of each tissue to metal-eliminating activity are far from clear.

Hemocyanin is the most abundant protein in crustacean hemolymph; it has Cu in its oxygen-binding center and has the ability to bind Fe (Howard and Simkiss 1981; Robinson et al. 1984). An Fe-binding protein in the crab *Cancer magister* fulfills all the criteria of transferrin (Tf) (Huebers et al. 1982). Fe is also stored in lysosomes of marine mussels (Fowler et al. 1975; Janssen and Ertelt-Janssen 1983; Soto et al. 1996).

ROLE OF IRON IN THE CATALYSIS OF REACTIVE OXYGEN SPECIES REACTIONS

Iron excess is believed to generate oxidative stress (Abele and Puntarulo 2004) due to the Fe-dependent conversion of $O_2^{\bullet -}$ and H_2O_2 into the extremely reactive and toxic $HO^{\bullet}$ (Haber–Weiss reaction; equation 8.5).

$$O_2^{\bullet -} + H_2O_2 \rightarrow O_2 + OH^- + OH^{\bullet} \qquad (8.5)$$

A labile Fe pool is defined as a low-molecular-weight pool of weakly chelated Fe that passes through the cell. It likely contains both Fe^{2+} and Fe^{3+} associated with a variety of ligands with low affinity for Fe ions. The labile Fe pool represents a minor fraction of the total cellular Fe (Kruszewski 2004). It has been proposed that Fe is complexed by diverse low-molecular weight chelators, such as citrate and other organic ions, phosphate, carbohydrates and carboxylates, nucleotides and nucleosides, polypeptides, and phospholipids (Kakhlon and Cabantchik 2002; Petrat et al. 2002; Kruszewski 2003); however, the actual nature of the intracellular ligands participating in labile Fe pool formation remains obscure. The accessibility of cellular Fe to chelators, such as deferioxamine (DF), is commonly used as the criterion of "lability". In cell lysates, loosely bound Fe representing the labile Fe pool may be scavenged by weaker or nonpenetrating ligands, including ethylenediaminetetraacetic acid (EDTA), adenosin triphosphate (ATP), tris(hydroxymethyl)aminomethane (TRIS), and glycine. The choice of the Fe chelator changes the amount of Fe in the labile Fe pool (Petrák and Vyoral 2001). The affinity of a chelator for Fe, its ability to permeate different cell compartments, as well as the steric accessibility of specific metal binding sites on proteins could all influence the "labile" pool. Thus, it was suggested that the labile Fe pool actually represents Fe bound relatively weakly to prosthetic groups in functional sites of Fe-containing proteins, such as non-Tf and non-Ft proteins, the functions of which are not yet known (Petrák and Vyoral 2001). Despite the mysterious nature of the labile Fe pool, its source is better defined than that of its ligands. A continuous demand for Fe available for synthesis of Fe-containing proteins forces a permanent Fe flux from the extracellular milieu to the cytoplasm.

The presence of a labile Fe pool in the digestive gland of *M. arenaria* could be involved in the basal stress detected in the lipophilic and hydrophilic fractions of the tissue (González et al. 2008b). Although total Fe content is different between gills and mantle, the labile Fe pool and lipid peroxidation levels are identical (Table 8.2). Since mantle and gills have significantly less labile Fe pool and thiobarbituric acid reactive substances (TBARS) levels than digestive gland it seems that only the labile Fe pool (and not the total Fe) in these tissues is responsible for triggering lipid damage.

Molecules such as Ft are part of the mechanism involved in oxidative protection by sequestering Fe. The potential antioxidant properties of Fe binding are related to Fe concentrations and antioxidant levels in tissues of marine animals. Fe storage in Ft has been suggested as a crucial mechanism for preventing the occurrence of ROS-driven reactions catalyzed by Fe that generate lipid radicals ($LR^{\bullet}$) in digestive gland of the sub-Antarctic limpets *N. magellanica* and *N. deaurata* (González et al. 2008a). Estévez et al. (2002) showed the Fe-dependent formation of $LR^{\bullet}$ in the digestive gland of *L. elliptica*. Moreover, the authors reported higher CAT activity and α-tocopherol concentration in *L. elliptica* digestive gland compared to the bivalve *M. arenaria* characterized by a lower Fe content. However, higher antioxidant content may not suffice to prevent lipid peroxidation. Higher activity of superoxide dismutase (SOD) and CAT in *N. magellanica*, as compared to *N. deaurata*, digestive gland, contributes to decreased lipid peroxidation and to reduced $HO^{\bullet}$ formation by scavenging $O_2^{\bullet -}$ and H_2O_2 (Malanga et al. 2004, 2005). The sublitoral population of the Antarctic limpet *N. concinna* has higher Fe content and SOD activity in digestive gland than specimens collected in the Antarctic intertidal rocky shore, presumably to protect from the oxidative stress (Weihe et al. 2010).

Table 8.2 Fe content and TBARS in tissues of *M. arenaria* from the Wadden Sea.

Tissue	Total Fe (μmol g^{-1} FW)	Labile Fe Pool (nmol g^{-1} FW)	TBARS (nmol g^{-1} FW)
DG	1.89 ± 0.70^2	118 ± 9^2	57 ± 8^2
Mantle	$0.66 \pm 0.11^{1,2}$	59 ± 9^1	26 ± 1^1
Gills	$0.17 \pm 0.01^{1,2}$	52 ± 8^1	25 ± 1^1

[1] Significantly different from digestive gland (DG) ($p < 0.001$). ANOVA.
[2] Taken from González et al. (2008b).

The glutathione (GSH) status is known as a key determinant of the cellular redox environment (Weihe et al. 2010) and is frequently measured as an indicator of oxidative stress (Han et al. 2006). In *M. arenaria* digestive gland, the tissue with highest labile Fe pool content (Table 8.2), showed higher total GSH content than gills and mantle (Table 8.3). However, glutathione disulphide (GSSG) was also higher in digestive gland than in mantle or gills, yielding a lower GSH redox ratio in mantle as compared to the other tissues, suggesting a better antioxidant protection at the hydrophilic level.

The elimination of Fe can also be a strategy to minimize the hazardous effects of this potentially toxic transition metal. To be incorporated into Ft, Fe^{3+} is primarily reduced to Fe^{2+}, but it is stored as the inert form Fe^{3+}. By redox reaction with other cellular components (i.e. $O_2^{\bullet -}$), Fe^{3+} stored in Ft can be reduced again to Fe^{2+}, released to the cytoplasm and become a catalyst of Fenton-type reactions as part of the labile Fe pool. Furthermore, the role of other soluble and insoluble Fe-storage proteins, the formation and contribution of Fe-nitrosyl complexes (dinitrosyl iron complex, DNIC-Fe, and mononitrosyl iron tris(thiolate) complex, MNIC-Fe), GSH, etc., as possible candidates for Fe transport and storage under stress conditions needs to be considered in future work.

IRON OVERLOAD EFFECTS IN MARINE INVERTEBRATES

Toxicity depends on the concentration, bioavailability, and the organism's capabilities to deal with a pollutant by regulating/limiting its internal concentrations. The regulation can be physiological, and when this is not possible, some tissues have the ability to detoxify metals, or to sequester them inside the tissues in special cellular compartments or even in proteins, which apparently is a less harmful way to store metals. Vertebrates can regulate the concentration of some metals over a relatively wide concentration range. However, marine invertebrates are less tolerant of metal accumulation and can be affected at lower metal concentrations.

Fe supplementation and its effects on survival and physiological conditions have been studied in marine organisms such as algae, mollusks, and fish. Studies of Fe effects on wildlife often compare organisms living in polluted areas with others of the same species in non-polluted regions.

In the marine alga *Chlorella vulgaris* Fe concentrations higher than 11 mg L^{-1} lead to a drastic decrease in culture growth, while a linear dependency between growth and intracellular Fe occurs in the low concentration range; $LR^{\bullet}$ and ascorbyl radical (A) content increases significantly with increasing Fe concentration in the growth medium (Estévez et al. 2001). Fe supplementation in the mussel *Mytilus galloprovincialis* produces lipid peroxidation and Fe accumulation in digestive gland (Viarengo et al. 1999). Exposure to 500 µg Fe^{2+} L^{-1} up to 72 h increases glutathione peroxidase (GPx) activity but not malondialdehyde (MDA) levels in digestive gland of the bivalve *Perna perna* (Alves de Almeida et al. 2004). This negative correlation suggests that GPx protects tissues from lipid peroxidation. However, after 120 h of exposure both GPx activity and MDA concentrations increase (Alves de Almeida et al. 2004). Fe exposure generates ROS in the lysosomal compartment of *M. edulis* hemocytes; this response is related to the defense against endocytosed pathogens (Winston et al. 1996).

Most of the studies from polluted and less affected regions document metal accumulation, but few have studied the oxidative metabolism. For instance, Pempkowiak et al. (1999) showed that higher concentration of Fe in Baltic compared to Norwegian Sea sediments is reflected in increased Fe concentration in *M. arenaria* and *Astarte borealis*. Herut et al. (1999) also demonstrated how in less contaminated areas, e.g. around

Table 8.3 Glutathione content and redox potential (ΔE) in tissues of *M. arenaria* from the Wadden Sea.

Tissue	GSH (nmol/mg FW)	GSSG (nmol/mg FW)	Total glutathione (GSSG*2) + GSH	GSSG/GSH	ΔE (mV)
DG	0.60 ± 0.06	0.30 ± 0.02	1.19 ± 0.09	0.50 ± 0.06	−215 ± 3
Mantle	0.33 ± 0.03[1]	0.05 ± 0.01[1]	0.47 ± 0.05[1]	0.17 ± 0.03[1]	−250 ± 2[1]
Gills	0.24 ± 0.01[1]	0.18 ± 0.03[1]	0.53 ± 0.06[1]	0.79 ± 0.18	−224 ± 4[1]

[1]Significantly different to digestive gland (DG) ($p <$ 0,01). ANOVA.

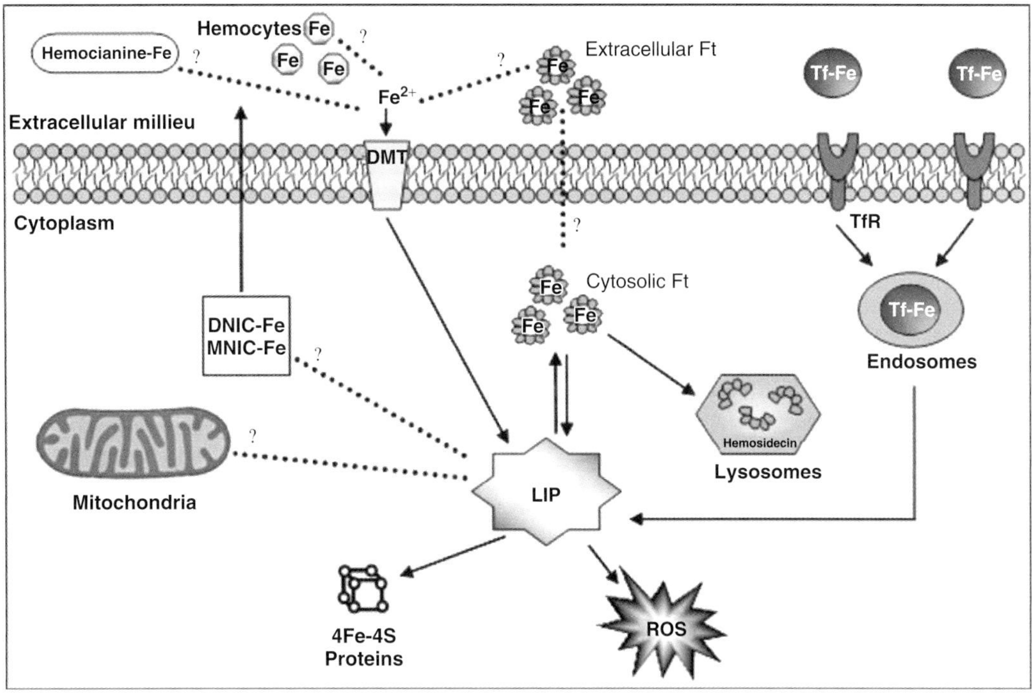

Fig. 8.7 Scheme summarizing Fe distribution in cells from marine invertebrates.

Helgoland in the North Sea, bivalves such as *M. edulis* have lower Fe content in the whole soft tissue than the same species in polluted areas, such as the Elbe river estuary and the North Sea. In *M. arenaria* collected in the St Lawrence estuary positive relationships between Fe-content and sex ratio, gonad DNA strand breaks, digestive gland monoamine oxidase, gill lipid peroxidation, digestive gland glutathione S-transferase (GST) activity, and gonad HO• production were found (Gagné et al. 2006). These observations suggest that Fe accumulation in tissues can produce oxidative stress and other harmful effects.

CONCLUSIONS AND PERSPECTIVES

Figure 8.7 briefly summarizes the cellular Fe origin and fate in marine invertebrates. Delivery pathways include the uptake of Fe via receptor-dependent endocytosis of Tf-bound Fe and via the low-affinity divalent metal cation transporter 1 (DMT1) with a broad substrate specificity (Gunshin et al. 1997). Little is known about the mechanism of Fe export from the cell but it has been proposed that Fe ions might leave the cells as a complex with low molecular weight thiols and/or nitric oxide (Kruszewski 2004). Ferritin plays a dual role in labile Fe pool homeostasis. Under Fe-enriched conditions it acts as Fe-sequestering protein, protecting cells against toxicity and at low Fe conditions it acts as a source of Fe ions necessary for Fe-containing protein synthesis. However, the physiological mechanism of Fe release from Ft, and the features of the extracellular Ft and possible existence of cytosolic or organelle Ft (as in lysosomes and mitochondria) remain unclear. Alterations in Fe metabolism should be carefully analyzed before evaluating cellular responses, since Fe serves as a micronutrient and as a catalyst of ROS reactions.

REFERENCES

Abele, D., Puntarulo, S. (2004) Formation of reactive species and induction of antioxidant defence systems in polar and temperate marine invertebrates and fish. *Comparative Biochemistry and Physiology A* 138, 405–415.

Abele, D., Atencio, A., Dick, D. et al. (2008) Iron, copper and manganese discharge from glacial melting into Potter

Cove and metal concentrations in *Laternula elliptica* shells. *In* Wiencke, C., Ferreyra, G.A., Abele, D., Marenssi, S. (eds). *The Antarctic ecosystem of Potter Cove, King-George Island (Isla 25 de Mayo)*. Alfred-Wegener-Institute for Polar and Marine Research, Bremerhaven, Germany, pp. 39–46.

Ahearn, G.A., Mandal, P.K., Mandal, A. (2004) Mechanisms of heavy-metal sequestration and detoxification in crustaceans: a review. *Journal of Comparative Physiology B* 174, 439–452.

Ahn, I.-Y., Chung, K.H., Choi, H.J. (2004) Influence of glacial runoff on baseline metal accumulation in the Antarctic limpet *Nacella concinna* from King George Island. *Marine Pollution Bulletin* 49, 119–141.

Al-Mohanna, S.Y., Nott, J.A. (1985) The accumulation of metals in the hepatopancreas of the shrimp *Penaeus semisulcatus* de Haan (Crustacea: Decapoda) during the moult cycle. *In* Halwagy, R., Clayton, D., Behbehani, M. (eds). *Marine Environmental Pollution*. Kuwait University, pp. 195–209.

Alves de Almeida, E., Miyamoto, S., Bainy, A.C.D. et al. (2004) Protective effects of phospholipid hydroperoxide glutathione peroxidase (PHGPx) against lipid peroxidation in mussels *Perna perna* exposed to different metals. *Marine Pollution Bulletin* 49(5–6), 386–392.

Boothe, P.N., Knauer, G.A. (1972) The possible importance of fecal material in the biological amplification of trace and heavy metals. *Limnology and Oceanography* 17, 270–274.

Bottke, W. (1982) Isolation and properties of vitellogenic ferritin from snails. *Journal of Cell Science* 58(1), 225–240.

Boyd, P.W., Jickells, T., Law, CS. et al. (2007) Mesoscale iron enrichment experiments 1993–2005, synthesis and future directions. *Science* 315, 612–617.

Brown, B.E. (1982) The form and function of metal-containing "granules" in invertebrate tissues. *Biological Reviews* 57, 621–667.

Burford, M.A., Macey, D.J., Webb, J. (1986) Hemolymph ferritin and radula structure in the limpets *Patelloida alticostata* and *Patella peronii* (Mollusca: Gastropoda) *Comparative Biochemistry and Physiology A* 83, 353–358.

Chavez-Crooker, P., Garrido, N., Ahearn, G.A. (2001) Copper transport by lobster hepatopancreatic epithelial cells separated by centrifugal elutriation: measurements with the fluorescent dye Phen Green. *Journal of Experimental Biology* 204, 1433–1444.

Croot, P.L., Laan, P. (2002) Continuous shipboard determination of Fe(II) in Polar waters using flow injection analysis with chemiluminescence detection. *Analytica Chimica Acta* 466, 261–273.

Croot, P.L., Laan, P., Nishioka, J., et al. (2005) Spatial and temporal distribution of Fe(II) and H_2O_2 during EisenEx, an open ocean mesoscale iron enrichment. *Marine Chemistry* 95, 65–88.

De Baar, H.J.W., Timmermanns, K.R., Laan, P. et al. (2008) Titan: A new facility for ultraclean sampling of trace elements and isotopes in the deep oceans in the international Geotraces Program. *Marine Chemistry* 111, 4–21.

Dick, D., Philipp, E., Kriews, M., Abele, D. (2007) Is the umbo matrix of bivalve shells (*Laternula elliptica*) a climate archive? *Aquatic Toxicology* 84, 450–456.

Dunford, H.B. (2002) Oxidation of Fe(II)/(III) by hydrogen peroxide: from aquo to enzyme. *Coordination Chemistry Reviews* 233/4, 311–318.

Durand, J.P., Goudard, F., Barbot, C. et al. (2002) Ferritin and hemocyanin: ^{210}Po molecular traps in marine fish, oyster and lobster. *Marine Ecology Progress Series* 233, 199–205.

Estévez, M.S., Malanga, G., Puntarulo, S. (2001) Iron-dependent oxidative stress in *Chlorella vulgaris*. *Plant Science* 161, 9–17.

Estévez, M.S., Abele, D., Puntarulo, S. (2002) Lipid radical generation in polar (*Laternula elliptica*) and temperate (*Mya arenaria*) bivalves. *Comparative Biochemistry and Physiology B* 132, 729–737.

Fowler, B.A., Wolte, D.A., Hettler, W.F. (1975) Mercury and iron uptake by cytosomes in mantle epithelial cells of quahog clams (*Mercenaria mercenaria*) exposed to mercury. *Journal of the Fisheries Research Board of Canada* 32, 1767–1775.

Gagné, F., Blaise, C., Pellerin, J. et al. (2006) Health status of *Mya arenaria* bivalves collected from contaminated sites in Canada (Saguenay Fjord) and Denmark (Odense Fjord) during their reproductive period. *Ecotoxicology and Environmental Safety* 64, 348–361.

Galatro, A., Rousseau, I., Puntarulo, S. (2007) Ferritin role in iron toxicity in animals and plants. *Research Trends and Current Topics in Toxicology* 4, 65–76.

Gerringa, L.J.A., De Baar, H.J.W., Timmermans, K.R. (2000) A comparison of iron limitation of phytoplankton in natural oceanic waters and laboratory media conditioned with EDTA. *Marine Chemistry* 68, 335–346.

Göksu, M.Z.L., Akar, M., Çevik, F., Findik, Ö. (2005) Bioaccumulation of some heavy metals (Cd, Fe, Zn, Cu) in two bivalvia species (*Pinctada radiata* Leach, 1814 and *Brachidontes pharaonis* Fisher, 1870). *Turkish Journal of Vetinary and Animal Science* 29, 89–93.

González, P.M., Malanga, G., Puntarulo, S. (2008a) Ferritin and labile iron pool in limpets from the Beagle Channel. *In* Svensson E.P. (ed.). *Aquatic Toxicology Research Focus*, Vol. 9. Nova Science Publishers, New York, pp. 177–188.

González, P.M., Abele, D., Puntarulo, S. (2008b) Iron and radical content in *Mya arenaria*. Possible sources of NO generation. *Aquatic Toxicology* 89, 122–128.

Gunshin, H., Mackenzie, B., Berger, U.V., et al. (1997) Cloning and characterization of a mammalian proton-coupled metal-ion transporter. *Nature* 388, 482–488.

Han, D., Hanawa, N., Saberi, B., Kaplowitz, N. (2006) Mechanisms of liver injury. III. Role of glutathione redox status in liver injury. *American Journal of Physiology* 291, G1–G7.

Harrison, P.M., Arosio, P. (1996) The ferritins: molecular properties, iron storage function and cellular regulation. *Biochimica et Biophysica Acta* 1275, 161–203.

Herut, B., Kress, N., Shefer, E. (1999) Trace element levels in mollusks from clean and polluted coastal marine sites in the Mediterranean, Red and North Seas. *Helgoland Marine Research* 53, 154–162.

Howard, B., Simkiss, K. (1981) Metal binding by *Helix aspersa* blood. *Comparative Biochemistry and Physiology A* 70, 559–561.

Hudson, R.J., Morel, F.M.M. (1990) Iron transport in marine-phytoplankton-kinetics of cellular and medium coordination reactions. *Limnology and Oceanography* 35, 1002–1020.

Huebers, H.A., Finch, C.A., Martin, A.W. (1982) Characterization of an invertebrate transferring from the crab Cancer magister (Arthropoda). *Journal of Comparative Physiology B* 148, 101–109.

Janssen, H.H., Ertelt-Janssen, U. (1983) Cytochemical demonstration of cadmium and iron in experimental blue mussels (*Mytilus edulis*). *Mikroskopie* 40, 329–340.

Kakhlon, O., Cabantchik, Z.I. (2002) The labile iron pool: characterization, measurement, and participation in cellular processes. *Free Radical Biology and Medicine* 33, 1037–1046.

Kim, K.-S., Webb, J., Macey, D.J. (1986) Properties and role of ferritin in the hemolymph of the chiton *Clavarizona hirtosa*. *Biochimica et Biophysica Acta*. 884, 387.

King, D.W., Lounsbury, H.A., Millero, F.J. (1995) Rates and mechanisms of Fe(II) oxidation at nanomolar total iron concentrations. *Environmental Science and Technology* 29, 818–824.

Kremer, M.L. (1999) Mechanisms of the Fenton reaction. Evidence for a new intermediate. *Physical Chemistry Chemical Physics* 1(15), 3595–3605.

Kruszewski, M. (2003) Labile iron pool: the main determinant of cellular response to oxidative stress. *Mutation Research* 531, 81–92.

Kruszewski, M. (2004) The role of labile iron pool in cardiovascular diseases. *Acta Biochimica Polonica* 51(2), 471–480.

Lowenstam, H.A. (1981) Minerals formed by organisms. *Science* 90, 1126–1130.

Lu, H.-K., Huang, C.–M., Li, C.-M. (1995) Translocation of ferritin and biomineralization of goethite in the radula of the limpet *Cellana toreuma* Reeve. *Experimental Cell Research* 219, 137–145.

Malanga, G., Estévez, M.S., Calvo, J., Puntarulo, S. (2004) Oxidative stress in limpets exposed to different environmental conditions in the Beagle Channel. *Aquatic Toxicology* 69, 299–309.

Malanga, G., Estévez, M.S., Calvo, J. et al. (2005) Oxidative stress in gills of limpets from the Beagle Channel: comparison with limpets from the Antarctic. *Scientia Marina* 69(2), 297–304.

Malanga, G., González, P.M., Estévez, M.S., et al. (2008) Oxidative stress in Antarctic algae and molluscs. *In* Wiencke, C., Ferreyra, G.A., Abele, D., Marenssi, S. (eds). *The Antarctic ecosystem of Potter Cove, King-George Island (Isla 25 de Mayo)*. Alfred-Wegener-Institute for Polar and Marine Research, Bremerhaven, Germany, pp. 208–215.

Martin, J.H., Fitzwater, S.E. (1988) Iron deficiency limits phytoplankton growth in the north-east Pacific subarctic. *Nature* 331, 341–343.

Martin, J.H., Gordon, R.M., Fitzwater, S.E. (1990) Iron in Antarctic waters. *Nature* 345, 156–158.

Martinez, J.S., Zhang, G.P., Holt, P.D. et al. (2000) Self-assembling amphiphilic siderophores from marine bacteria. *Science* 287, 1245–1247.

Mason, A.Z., Simkiss, K. (1982) Sites of mineral deposition in metal accumulating cells. *Experimental Cell Research* 139, 383–391.

Millero, F.J., Sotolongo, G. (1989) The oxidation of Fe(II) with H_2O_2 in seawater. *Geochimica et Cosmochimica Acta* 53, 1867–1873.

Moore, J.K., Braucher, O. (2008), Sedimentary and mineral dust sources of dissolved iron to the world ocean. *Biogeosciences* 5, 631–656.

Moore, P.G., Rainbow, P.S. (1984) Ferritin crystals in the gut caeca of *Stegocephaloides chrstianiensis* Boeck and other Stegocephalidae (Amphipoda: Gammaridea): a functional interpretation. *Philosophical Transactions of the Royal Society of London' Series B* 306, 219–245.

Moore, P.G., Rainbow, P.S. (1989) Feeding of the mesopelagic gammaridean amphipod *Parandania boecki* (Stebbing, 1888) (Crustacea: Amphipoda: Stengocephalidae) from the Atlantic Ocean. *Ophelia* 30, 1–9.

Nielsen, S.A., Nathan, A. (1975) Heavy metal levels in New Zeland molluscs. *N. Z. Marine and Freshwater Research* 9(4), 467–681.

Nott, J. A., Nicolaidou, A. (1990) Transfer of metal detoxification along marine food chains. *Journal of the Marine Biological Association of the UK* 70, 905–912.

Nott, J. A., Nicolaidou, A. (1996) Kinetics of metals in molluscan faecal pellets and mineralized granules, incubated in marine sediments. *Journal of Experimental Marine Biology and Ecology* 197, 203–218.

Pempkowiak, J., Sikora, A., Biernacka, E. (1999) Speciation of heavy metals in marine sediments vs their bioaccumulation by mussels. *Chemosphere* 39(2), 313–321.

Petrák, J., Vyoral, D. (2001) Detection of iron-containing proteins contributing to the cellular labile iron pool by a native electrophoresis metal blotting technique. *Journal of Inorganic Biochemistry* 86(4), 669–675.

Petrat, F., De Groot, H., Sustmann, R., Rauen, U. (2002) The chelatable iron pool in living cells: a methodically defined quantity. *Biological Chemistry* 383, 489–502.

Pierre, J. L., Fontecave, M. (1999) Iron and activated oxygen species in biology: the basic chemistry. *Biometals* 12(3), 195–199.

Poulton, S. W., Raiswell, R. (2002) The low-temperature geochemical cycle of iron: from continental fluxes to marine sediment deposition. *American Journal of Science* 302, 774–805.

Raiswell R., Anderson, T. F. (2005) Reactive iron enrichment in sediments deposited beneath euxinic bottom waters: constraints on supply by shelf recycling. *In* McDonald, I., Boyce, A. J., Butler, I., Herrington, R. J., Polya, D. (eds). *Mineral Deposits and Earth Evolution. Geological Society of London Special Publication* 218, 179–194.

Raiswell, R., Tranter, M., Benning, L. G. et al. (2006) Contributions from glacially derived sediment to the global iron (oxyhydr)oxide cycle: Implications for Iron delivery to the oceans. *Geochimica et Cosmochimica Acta* 70(11), 2765–2780.

Robinson, W. E., Ryan, D. K., Morse, M. P. (1984) Potencial role of Mercenaria mercenaria blood plasma in metal transport. *American Zoologist* 24, 70A.

Soto, M., Cajaraville, M. P., Marigómez, I. (1996) Tissue and cell distribution of cooper, zinc and cadmium in the mussel *Mytilus galloprovincialis* determined by autometallography. *Tissue Cell* 28, 557–568.

St. Pierre, T. G., Kim, K.-S., Webb, J., Mann, S., Dickson, D. P. E. (1990) Biomineralization of iron: Mössbauer spectroscopy and electron microspcopy of ferritin cores from the chiton *Acanthopleura hirtosa* and the limpet *Patella laticostata. Inorganic Chemistry* 29(10), 1870–1874.

Stumm, W., Morgan, J. J. (1981) *Aquatic Chemistry – An Introduction Emphasizing Chemical Equilibria in Natural Waters.* John Wiley and Sons, New York, 780 pp.

Taylor, M. G. (1995) Mechanisms of metal immobilization and transport in cells. *In* Cajaraville, M. P. (ed.). *Cell Biology in Environmental Toxicology*. University of the Basque Country Press, Bilbao, pp. 155–170.

Thorn, K., Cerrato, R. M., Rivers, M. L. (1995) Elemental distribution in marine bivalve shells as measured by synchrotron x-ray fluorescence. *Biological Bulletin* 188, 57–67.

Turner, D. R., Whitfield, M., Dickson, A. G. (1981) The equilibrium speciation of dissolved components in freshwater and seawater at 25°C al 1atm. preassure. *Geochimica et Cosmochimica Acta* 45, 855–882.

Usero, J., González-Regalado, E., Gracia, I. (1997) Trace metals in the bivalve molluscs *Ruditapes decussatus* and *Ruditapes philippinarum* from the Atlantic coast of Southern Spain. *Environment International* 23, 291–298.

Viarengo, A. (1989) Heavy metals in marine invertebrates: mechanisms of regulation and toxicity at the cellular level. *Reviews in Aquatic Science* 1, 295–317.

Viarengo, A., Brulando, B., Cavaletto, M., et al. (1999) Role of metallothionein against oxidative stress in the mussel *Mytilus galloprovincialis. American Journal of Physiology* 277, R1612–R1619.

Watson, A. J. (2001) Iron limitation in the oceans. *In* Turner, D. R., Hunter, K. A. (eds). *The Biogeochemistry of Iron in Seawater*. John Wiley & Sons, New York, pp. 85–121.

Webb, J., Macey, D. J. (1983) Plasma ferritin Polyplacophora and its possible role in the biomineralization of iron. *In* Westbroek, P., De Jong, E. W., (eds). *Biomineralization and Biological Metal Accumulation.* Reidel, Doredrecht, pp. 413–422.

Webb, J., Mann, S., Bannister, J. V., Williams, R. J. P. (1986) Biomineralization of iron: Isolation of ferritin from the hemolymph of the limpet *Patella vulgata. Inorganica Chimica Acta* 124, 37–40.

Wedepohl, K. H. (1995) The composition of the continental crust. *Geochimica et Cosmochimica Acta* 59, 1217–1232.

Weihe, E., Kriews, M., Abele, D. (2010): Differences in heavy metal concentrations and in the response of the antioxidant system to hypoxia and air exposure in the Antarctic limpet *Nacella concinna. Marine Environmental Research* 69, 127–135.

Winston, G. W., Moore, M. N., Kirchin, M. A., Soverchia, C. (1996) Production of reactive oxygen species by hemocytes from the marine mussel, *Mytilus edulis*: lysosomal localization and effect of xenobiotics. *Comparative Biochemistry and Physiology C* 113, 221–229.

Zhang, Y., Meng, Q., Jiang, T., et al. (2003) A novel ferritin subunit involved in shell formation from the pearl oyster (*Pinctada fucata*). *Comparative Biochemistry and Physiology B* 135, 43–54.

Chapter 9

OXIDATIVE STRESS IN CORAL-PHOTOBIONT COMMUNITIES

Marco A. Liñán-Cabello[1], *Michael P. Lesser*[2], *Laura A. Flores-Ramírez*[1], *Tania Zenteno-Savín*[3], *and Héctor Reyes-Bonilla*[4]

[1]Acuacultura/Biotecnología, FACIMAR, Universidad de Colima, Manzanillo, Colima, Mexico
[2]Department of Molecular, Cellular and Biomedical Sciences, University of New Hampshire, Durham, NH, USA
[3]Centro de Investigaciones Biológicas del Noroeste, S.C. (CIBNOR), La Paz, Baja California Sur, Mexico
[4]Universidad Autónoma de Baja California Sur, La Paz, Baja California Sur, Mexico

Coral reefs are among the largest, most widely distributed, diverse, and complex ecosystems. They provide areas of refuge, feeding, reproduction, and residence for a variety of invertebrates and fish. Although they represent only 0.2% of the ocean bottom, they offer a large array of ecosystem services and support the economy of entire countries with respect to tourism and fisheries (McClanahan 2006). Destruction or damage to coral reefs has serious impacts on the oceanic environment (Hallock 2005). Climate change is one of the main factors affecting the health of coral reefs due to the expected thermal stress which, combined with high solar irradiance, leads to metabolic and physiologic alterations such as immunosupression, starvation of the colonies, and eventual death of entire coral populations (Baker et al. 2008). The usual response to these effects is the loss of coloration caused by the expulsion or death of the symbiotic dinoflagellates (zooxanthellae), a phenomenon known as coral bleaching (Brown 1997; Douglas 2003). The biochemical mechanisms of coral bleaching involve impaired photosynthetic pathways, dysfunction of cnidarian-zooxanthellae interactions and/or disruption of the antioxidant defenses (Lesser and Schick 1989a; Lesser et al. 1990; Iglesias-Prieto et al. 1992).

Exposure to ultraviolet radiation (UVR) can induce the formation of reactive oxygen species (ROS), such as $^1O_2$, $O_2^{\bullet -}$, and H_2O_2. These ROS can disrupt protein synthesis and damage cell membranes and induce the formation of $HO^{\bullet}$, which may trigger oxidative damage and bleaching of coral reefs (Lesser 1996; Shick et al. 1999; Downs et al. 2002).

Oxidative Stress in Aquatic Ecosystems, First Edition. Edited by Doris Abele, José Pablo Vázquez-Medina, and Tania Zenteno-Savín.

CAUSES OF CORAL BLEACHING

While thermal stress is seen as the principal cause of coral bleaching (Lesser 2004, 2006; Hoegh-Guldberg et al. 2007), other environmental factors interact by effectively lowering the threshold temperature at which coral bleaching occurs. The primary abiotic factor that influences the severity of thermally induced coral bleaching is solar radiation (Banaszak and Lesser 2009). UVR has independent detrimental effects on photosynthesis and zooxanthellae and coral growth (Lesser 2004; Banaszak and Lesser 2009). For sessile corals, exposure to UVR in shallow tropical waters is unavoidable and is particularly important because corals experience hyperoxic conditions on a daily basis, simultaneously with UVR exposure, as a result of the photosynthetic activity of their symbiotic zooxanthellae (Dykens and Shick 1982; Kühl et al. 1995). UVR exposure also leads to photodynamic ROS production (Dykens and Shick 1982; Asada and Takahashi 1987; Halliwell and Gutteridge 1999) while coral response to UVR involves synthesis of UVR absorbing compounds (e.g., mycosporine-like amino acids) and antioxidant enzymes to protect both the host and symbiont from oxidative stress (Lesser 1996; Lesser and Farrell 2004).

Photoinhibition of photosynthesis in zooxanthellae occurs as a result of the reduction in photosynthetic electron transport, combined with the continued high absorption of excitation energy and ROS production (Lesser and Shick 1989a; Lesser 1996, 2006; Lesser and Farrell 2004). Additionally, there are genetic differences in the sensitivity of zooxanthellae to thermal stress and ROS production (Warner and Berry Lowe 2006; Reynolds et al. 2008; Suggett et al. 2008). Exposure to elevated temperatures alone (Iglesias-Prieto et al. 1992; Lesser 1997; Smith et al. 2005), UVR alone (Lesser and Shick 1989b), or in combination (Lesser 1996) can result in photoinhibition of photosynthesis in zooxanthellae. Exposure to elevated temperatures functionally lowers the set-point for light-induced photoinhibition, ROS production, and bleaching (Lesser 2006).

Oxidative stress has been proposed as a unifying mechanism for several environmental insults that cause bleaching (Lesser 1996, 2006; Lesser and Farrell 2004). Both the cnidarian host and zooxanthellae express a full suite of antioxidant enzymes to quench ROS production (Lesser and Shick 1989a,b; Lesser and Farrell 2004). The host experiences hyperoxia imposed by its photosynthetic symbionts and, therefore, the level of antioxidant defense occurs in proportion to the potential for photo-oxidative damage, which also includes the amount of solar radiation they are exposed to (Lesser 2006; Richier et al. 2008). Exposure of corals to thermal stress and high solar radiation results in high ROS fluxes in the host or zooxanthellae (Lesser and Shick 1989a; Matta and Trench 1991; Dykens et al. 1992; Lesser 1996; Nii and Muscatine 1997). Oxidative stress can lead to bleaching of zooxanthellae via exocytosis from coral host cells or apoptosis (Gates et al. 1992; Lesser 1997; Dunn et al. 2002; Franklin et al. 2004; Lesser and Farrell 2004). A cellular model of bleaching in symbiotic cnidarians includes oxidative stress, photosystem (PS) II damage, sink limitation, DNA damage, and apoptosis as underlying processes (Lesser and Farrell 2004; Lesser 2006; Weis 2008). This model is consistent with biomarker proteins expressed in corals during thermal stress and with the differential sensitivity of zooxanthellae thylakoid membranes to thermal stress (Downs et al. 2000, 2002; Tchernov et al. 2004), but is contrary to results for thylakoid and host membranes, however, do not support membranes as the primary target of thermal stress (Sawyer and Muscatine 2001; Hill et al. 2009; Díaz-Almeyda et al. 2011). The presence of nitric oxide synthase (NOS) activity (Morrall et al. 2000; Trapido-Rosenthal et al. 2005) means that within symbiotic cnidarians it is also possible for $NO^{\bullet}$ to react with $O_2^{\bullet-}$ and form $ONOO^-$; the diffusion of the latter through membranes is much greater than that of $O_2^{\bullet-}$.

High ROS production is a consistent feature of coral physiology, especially during exposure to thermal stress and UVR, in both the symbiont and host (Lesser 1996, 1997, 2006; Lesser and Farrell 2004). ROS, especially H_2O_2, inhibit Rubisco and cause damage to PS II (Asada and Takahashi 1987; Cadenas 1989; Fridovich 1998). The cascade of events that ultimately induces the expulsion of zooxanthellae from their host could include the enhanced production of ROS and damage to PS II, a decrease in the amount of translocated photosynthate, or both (Lesser 2004, 2006; Weis 2008). Significant DNA damage occurs in host tissues upon exposure to thermal stress combined with exposure to solar radiation (Lesser and Farrell 2004). DNA damage can lead to apoptosis if not repaired, and the expression of the cell cycle checkpoint gene *p53* in *Montastraea faveolata* after exposure to thermal stress and high irradiances of solar radiation is consistent with the observed pattern of DNA damage (Lesser and Farrell 2004).

ECOLOGICAL EFFECTS OF BLEACHING IN EASTERN PACIFIC CORAL REEFS

The eastern Pacific Ocean is among the areas in the world with less abundance of coral reefs. Coral reefs in the eastern Pacific can be found from the southern Gulf of California to Ecuador (Glynn et al. 2007; Reyes-Bonilla and López-Pérez 2009). They have relatively low reef-coral richness (Reyes-Bonilla 2002; Glynn et al. 2007) and reduced bottom cover (rarely over 30%; Glynn 2001). These communities are formed by a mixture of geologically young species (most dating from the Pleistocene) with western Pacific immigrants and small (usually $<$ 3 to 5 m in height) reef frameworks (Glynn and Ault 2000; Cortés 2003; López-Pérez 2005). The principal reason for the limited growth in these communities is that environmental conditions in the region are somewhat unfavorable (Veron 1995). The eastern Pacific has a narrow continental shelf, relatively low temperature, high nutrient concentrations and productivity, and is also one of the regions with lowest pH levels in the global ocean surface, a condition that promotes low omega aragonite concentrations in seawater and thus, low calcification rates (Manzello et al. 2008). Despite these environmental limitations, the west coast of the Americas, especially Panamá and Costa Rica, hosts some well-developed coral ecosystems with complex trophic structures and high richness (Glynn 2004; Guzmán and Cortés 2007; Robertson and Cramer 2009).

Another important area for coral growth in the eastern Pacific is Mexico, where reefs occur from Tangolunda Bay, Oaxaca (15°N) to San Gabriel Bay (24°N) in the southern Gulf of California (Reyes-Bonilla 2003). These reefs present a clear coral vertical zonation in which shallow areas (0–6 m) are mostly occupied by branching corals of the genus *Pocillopora*, whereas deeper zones (usually 6–12 m) are dominated by massive corals of the genus *Pavona* and *Porites* (Reyes-Bonilla 2003). The highest species richness is reported for the oceanic Revillagigedo Islands. This archipelago also functions as the key stepping stone for coral colonization from the west (Reyes-Bonilla and López-Pérez 1998; Pérez-Vivar et al. 2006), enriching the reefs and making them among the most topographically complex and biodiverse in Mexico (Carriquiry and Reyes-Bonilla 1997; Glynn 2004; Liñán-Cabello et al. 2008).

Coral bleaching has been a recurrent phenomenon in the eastern Pacific. The first event was reported in 1982–1983 as a result of the warming brought about by the El Niño–Southern Oscillation (ENSO). Its consequences were catastrophic: 90% coral mortality in Galápagos Islands, and over 75% in Panamá and Costa Rica (Glynn 1990). Subsequently, loss of coloration in corals was detected at Cabo Pulmo reef, Gulf of California, Mexico, in 1997, but there was no significant coral death (Reyes-Bonilla 1993). Moreover, Glynn and Leyte-Morales (1997) suggested that the size distribution of *Pavona gigantea* Verrill, 1869, as well as the presence of large dead or eroded framework areas at La Entrega reef in the Oaxaca reef tract, could reflect perturbations caused by ENSO events in the early 1980s.

Finally, there was another episode in 1997–1998, when the eastern Pacific suffered the strongest ENSO in the 20th century (Goreau and Hayes 2008). This time, the loss of coral cover in Central America was low (Guzmán and Cortés 2007), but in Mexico the scenario was completely different: scleractinian mortality was high, up to 80% in Banderas Bay and other areas of Nayarit, while in Oaxaca and the Gulf of California there was a decrease of less than 30% in coral cover (Reyes-Bonilla et al. 2002). Glynn et al. (1998) and Calderón-Aguilera and Reyes-Bonilla (2006) suggested that cooling of oceanic water caused by high frequency of hurricanes along the coast in 1997, and the presence of oceanographic fronts at the entrance of the gulf, buffered the ENSO effect at the sites referred to.

After the occurrence of these ENSO events, others of smaller scale have not caused any bleaching whatsoever in the eastern Pacific and the reefs have recovered in Central America (Guzmán and Cortés 2007; Alvarado et al. 2009). In sharp contrast, the status of this ecosystem in Mexico is poorer as other perturbations, such as a string of strong hurricanes, have directly affected the main reef areas. Among them we can cite "Olaf", "Pauline," and "Rick" striking the Oaxaca coast in 1997; "Juliette" and "Isis" in the Gulf of California (1995 and 1998, respectively), and "Kenna" in Banderas Bay (2002). For these reasons, Calderón-Aguilera and Reyes-Bonilla (2006) hypothesized that coral reefs in the Mexican Pacific are structured by environmental factors rather than biotic interactions, and that the disentangling of the effects of natural disturbances from those of human-induced perturbations will require further work and permanent monitoring programs.

PHYSIOLOGICAL RESPONSES TO BLEACHING

Several environmental factors have been described as causes of coral bleaching, including a variety of physical, chemical, biological, and environmental conditions (Jokiel 2004). These factors may cause oxidative stress, which in turn leads to changes at the cellular and biochemical levels, which are usually the first detectable responses to perturbations (Bierkens 2000). Bleaching is a normal stress response in corals, but it is recognized that temperature and radiation are the main causes of this phenomenon (Douglas 2003; Baker et al. 2008). In Mexico, Flores-Ramírez and Liñán-Cabello (2007) performed an experiment on temperature gradient and examined the behavior of the coral–zooxanthellae relationship *in vitro* by using *Pocillopora capitata* fragments collected at La Boquita reef (19°N). Results showed that specimens with normal appearance and those with partial discoloration contained higher levels of carotenoid pigments than of chlorophyll *a* (Chl *a*). The levels of both pigment increased with temperature from 22° to 28°C in normal samples, whereas partially discolored coral samples showed less cellular damage (measured as lipid peroxidation) at temperatures between 26° and 28°C, and had higher superoxide dismutase (SOD) activity at temperatures between 26° and 30°C. SOD activity increased with temperature, peaking at 30–32°C in normal and control corals, with the highest levels overall seen in partially discolored samples harvested at 28°C. This is consistent with the SOD activity levels observed by Yakovleva et al. (2004) in *Stylophora pistillata* corals exposed to 33°C for 6 h. From these experiments, it was concluded that temperature has a direct effect on the antagonistic relationship between temperature-induced damage and protective antioxidant mechanisms in this species (Flores-Ramírez and Liñán-Cabello 2007).

The sole relevant source of UVR is solar radiation, so for that reason the amount of sunlight usually plays a role in coral bleaching. Corals in shallow waters suffer from excess of UV light, especially in exposed parts of their colonies (Banaszak and Lesser 2009). Both photosynthetically active radiation (PAR, 400–700 nm) and UVR (280–400 nm) have been implied in bleaching events (Banaszak and Lesser 2009; Harrell and Barron 2010).

Corals have various defense mechanisms against photochemical damage derived from exposure to high temperatures, UVR or a combination of both (Lesser 1996, 1997, 2004; Lesser and Shick 1989b). These mechanisms include nucleotide excision repair and DNA recombination (Mitchell and Karentz 1993; Van de Poll et al. 2001), photoactivation of DNA (mediated by UV-A and PAR radiation), accumulation of lipid- and water-soluble antioxidants, and production of antioxidant enzymes (Cockell and Knowland 1999). They also have other biomolecules with antioxidant and sunscreen effect, which aid in preventing the effects of UVR. For instance, compounds such as carotenoids, phycobiliproteins, phenols, coumarin, and mycosporine-like amino acids (MAAs) have been found in corals (Korbee et al. 2006; Carignan et al. 2009). Anderson et al. (1997) used DNA damage as a marker of UVR exposure, whereas Lesser and Farrell (2004) investigated the effects of solar radiation and thermal stress on DNA damage. According to Britt (1996) UVR can cause DNA damage through the formation of cyclobutane pyrimidine dimers (CPDs), which induce genetic mutations and inhibit RNA and DNA polymerases. In a study on the algae *Porphyra yezoensis*, Misonou et al. (2003) suggested that MAAs blocked the production of both 6-4 pyrimidine–pyrimidone photoproducts and CPDs formation. Torregiani and Lesser (2007) showed that populations of the coral *M. verrucosa* exposed to UVR at greater depth had increased capacity for MAA synthesis and generate less CPDs compared to those at shallower depths. A close relationship between bleaching and cellular stress indicators was emphasized by Downs et al. (2005), who showed that protein carbonyls, lipid peroxidation, and DNA damage, all serve as valuable prognostic indicators of coral health.

Recent studies on *Pocillopora capitata* specimens from reefs in Colima, Mexico, were conducted in order to characterize the radiation-induced response in short-term exposure (Liñán-Cabello et al. 2010a). Corals exposed to UVR had lower levels of carotenoids and SOD activity than those exposed to PAR, as well as lower carotenoid pigments to Chl *a* ratios. Moreover, in corals exposed to PAR, SOD activity was directly correlated with Chl *a*, and inversely linked with expelled zooxanthellae, and glutathione S-transferase (GST) and glutathione peroxidase (GPx) activities were directly related with CPDs and expelled zooxanthellae. In these experiments the maximum concentrations of MAA occurred at 4 h and 7 h after exposure to UVR and PAR, respectively. Similar results were presented by

Carreto et al. (1990), who showed that exposure of the dinoflagellate *Alexandium excavantum* to high PAR resulted in the synthesis of MAA during the first 3–6 h of exposure. Although MAAs and carotenoid pigments are rapidly produced as nonenzymatic antioxidants in response to UVR in corals, these were not sufficient, even in the dark phase of the experiment, to mitigate oxidative damage. ROS caused breakdown of the symbiotic relationship between zooxanthellae and the host coral to an extent 33-fold greater than seen after PAR exposure (Carreto et al. 1990). Concomitant with the short-term adjustments observed, *P. capitata* exhibited (at the enzymatic level) a series of responses designed to resist the effects of ROS and, thus, to adapt to shallow oceanic environments that commonly exhibit high UVR levels (Liñán-Cabello et al. 2010a). A similar response to experimental time (exposure for 3 days) was reported by Torregiani and Lesser (2007), who conducted an *in vitro* experiment with *Montipora verrucosa* specimens that were collected from depths of 1.5–10 m and exposed to UVR of 290–400 nm. They also found that corals showed increased concentrations of MAAs as a response to reduced quantum yield, and the appearance of both CPDs and (6-4) pyrimidine–pyrimidone were indicators of UVR damage to DNA. The effects of radiation are not only at the molecular level. Torres et al. (2008) studied the effect of UVR exposure on the fertility of shallow *Acropora cervicornis* populations. Histological analyses showed that sudden increases in UVR can completely stop broadcasting ramose sexual reproduction in this coral species. This, in turn, can affect species dominance and thus composition and structure of shallow reef environments. The study of physiological responses of *Pocillopora capitata* to environmental conditions during February–March (winter) and June–July (summer) at La Boquita reef (Mexico) indicate that increasing temperature and irradiation associated with seasonal changes induce an increase in CAT and GST activities (Table 9.1) (Liñán-Cabello et al. 2010b). This effect might be attributable to the presence of lipid peroxidation products, epoxides, and organic hydroperoxides, as well as to increased H_2O_2 levels caused by an increased $O_2^{\bullet-}$ production observed in summer. During summer, MAAs were also increased; however, this was not sufficient to completely offset oxidative stress and oxidative damage to lipids, quantified as thiobarbituric acid-reactive substances. $O_2^{\bullet-}$ production was 7.6-fold higher in summer than in winter, which could be associated with greater exposure to PAR, UVR, higher temperature, and high concentrations of Chl *a* during this time (Liñán-Cabello et al. 2010b). Lesser (1996, 1997) suggested that increased ROS production and higher activities of antioxidant enzymes might be expected during summer. However, in our studies SOD activity was lower in summer than in winter (Table 9.1). Some authors have reported that SOD can be inactivated through direct absorbance of light (Streb et al. 1993; Schittenhelm et al. 1994). Scandalios et al. (1997) suggested that in CAT-expressing plants

Table 9.1 Average values of pro-oxidant ($O_2^{\bullet-}$), antioxidant (SOD, CAT, GST, GPx, GR, MAAs), oxidative damage (TBARS), and bleaching markers [(Chl *a*, CPDs), and zooxanthellae density (ZD)] in *Pocillopora capitata* during winter and summer. Data are expressed as means ± SD ($n = 12$).

	Parameters (units)	Winter	Summer	*p* level
Pro-oxidant marker	$O_2^{\bullet-}$ (pmol mg^{-1} protein)	1.25 ± 0.62	9.51 ± 2.80	0.004[1]
Oxidative damage	TBARS (nmol mg^{-1} protein)	0.22 ± 0.01	16.01 ± 4.91	0.001[1]
Antioxidants	SOD (U SOD mg^{-1} protein)	619.78 ± 195.93	267.17 ± 65.25	0.197
	CAT (U CAT mg^{-1} protein)	109.54 ± 57.63	334.27 ± 86.28	0.045[1]
	GST (U GST mg^{-1} protein)	0.06 ± 0.03	1.70 ± 0.51	0.001[1]
	GPx (U GPx mg^{-1} protein)	13.07 ± 2.96	24.01 ± 4.18	0.050
	GR (U GPx mg^{-1} protein)	2.36 ± 0.59	5.15 ± 1.82	0.108
	MAAs (pmol g^{-1} protein)	201.99 ± 46.15	433.73 ± 59.32	0.000[1]
Bleaching markers	CPDs (μg cm^{-2})	0.715 ± 0.174	0.3160 ± 0.204	0.001[1]
	Chl *a* (μg cm^{-2})	9.76 ± 2.81	26.24 ± 8.87	0.000[1]
	ZD (cells cm^{-2})	5.29x10^{6} ± 1.80	5.16 × 10^{6} ± 1.24	0.877

[1]Significant differences at $p < 0.05$.

induction of Cat2 can be mediated by photoreceptors in the UV-A/blue and UV-B range. These findings could explain why the highest CAT and the lowest SOD activities were observed in summer.

The photoprotective effect of oxocerbonyl-MAAs, including mycosporine-glycine and mycosporine-taurine, is related to their capacity to neutralize ROS (Yakovleva et al. 2004; Zhang et al. 2007; Carignan et al. 2009). Liñán-Cabello et al. (2006) suggest that the observed decrease in total carotenoids in shallow-water corals exposed to high levels of solar radiation could be due to the demand of these antioxidants as a consequence of increased ROS generation. The observed carotenoid decrease was associated with a breakdown in the relationship between cnidarian corals and their symbiotic dinoflagellates. This may represent a short-term response providing a specific biochemical advantage over enzymatic mechanisms.

Irradiance can cause a change in carotenoids to chlorophyll ratio. Such change may be caused by the susceptibility of chlorophyll to photo-oxidation and the consequent use of carotenoid-like photoprotective pigments in this oxidative process (de Carvalho et al. 2001; Stambler and Dubinsky 2004). It has been suggested that the carotenoids to chlorophyll ratio is a potential indicator of photo-oxidative damage (Hendry and Price 1993; de Carvalho et al. 2001).

POTENTIAL EFFECTS OF CLIMATE CHANGE

Guzmán and Cortés (2001) reported that populations of massive and ramose corals in Costa Rica were more tolerant to bleaching caused by severe thermal stress in 1997–1998 than they were in 1982–1983, even when in both cases the positive temperature anomaly was similar. The authors interpreted this finding as evidence of coral adaptation to warmer conditions. In contrast, Liñán-Cabello et al. (2010b) found that in reefs situated near Juluapan lagoon (19°06′48.03″ N, and 104°24′21.540″ W), low concentrations of Chl *a* and zooxanthellae density in the corals may indicate a stressful state arising from sediment load, as well as from the natural thermal and saline fluctuations in the environment. These observations suggest that despite the existence of specific antioxidant adaptive mechanisms in *Pocillopora*, their capacity might not be enough to counteract the effects caused by extreme environmental changes.

The Mexican Pacific is an area that has been severely affected by ENSO events and corals live in a state of chronic pressure due to hurricanes that break colonies and consequently decrease their reproductive output and even their survival rates (Reyes-Bonilla 2003). Furthermore, in this region economic activities are increasing, and consequently its human population and use of resources is continuously increasing causing deforestation, changes in agricultural practices, and the emission of compounds derived from the industry (Fig. 9.1) (Ortíz-Lozano et al. 2005).

Sedimentation, eutrophication and the presence of chemical agents can cause coral bleaching as corals are exposed to stressful situations that may lead to decreased antioxidant capacity (Fig. 9.1; Douglas 2003; Downs et al. 2005). There are cases in Latin America in which the situation is so severe that entire reefs have been obliterated (Cortés and Risk 1985). Even when coral reefs have the ability to stand physiological changes and show adaptive and acclimation responses to bleaching by increasing the synthesis and concentration of enzymatic and nonenzymatic antioxidants, climate change can potentially affect coral populations and thereby the entire coral reef ecosystem. Unpublished analyses by Reyes-Bonilla et al. indicate that by 2020 it is expected that corals in the Gulf of California may surpass their bleaching threshold on an annual basis, and by 2050 the effects would be seen more often. In fact, loss of zooxanthellae and coral mortality might occur so often that the future of the reefs in the region could be imperiled.

CONCLUSIONS

Coral populations in the Eastern Pacific are subject to both natural and anthropogenic insult, causing oxidative stress and bleaching. This is one of the most serious problems for coral reefs. We agree with other researchers (Downs et al. 2000, 2005; Fauth et al. 2003) that a cellular diagnostic approach measuring various biochemical indicators can help to correlate observed stress with the underlying causes, providing diagnostic and prognostic biomarkers of coral health. Rapid anthropogenic climate change has added another major stress to the combination of environmental stressors, which could cause irreversible damage to the reef's productivity, adaptability, and regeneration capacity.

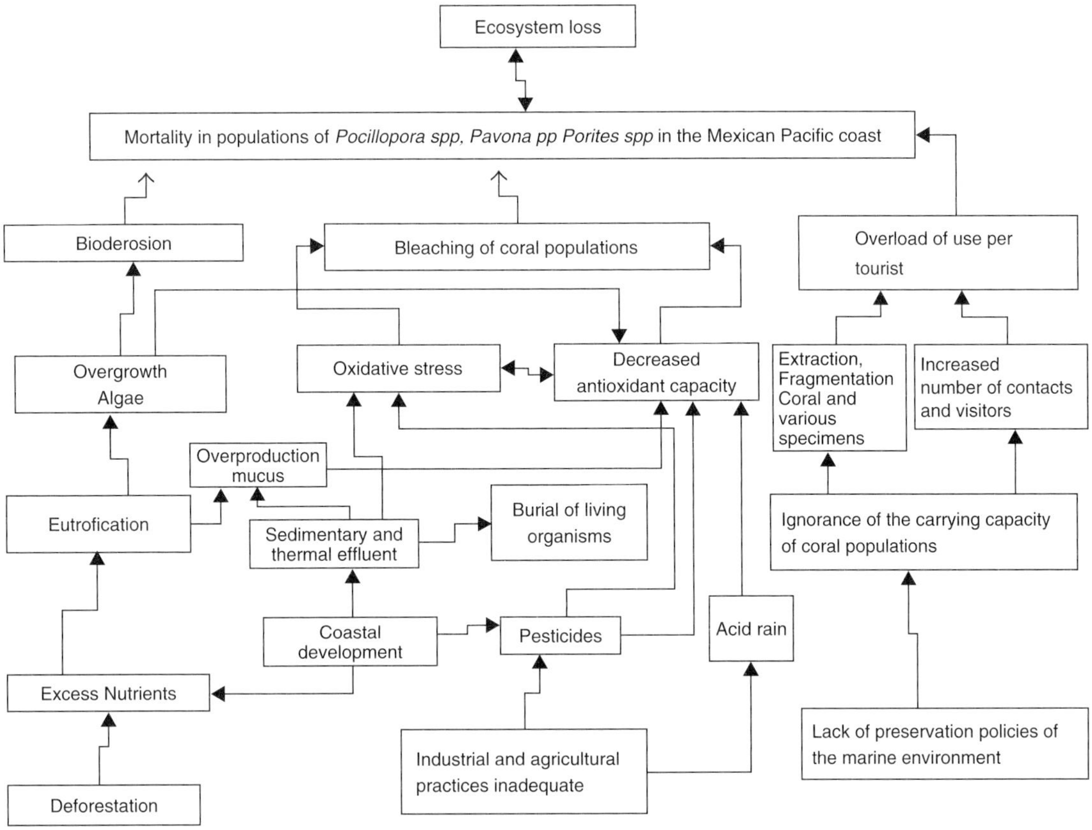

Fig. 9.1 Interaction of various anthropogenic effects with coral populations of the Pacific coast and the relationships of such interactions to oxidative stress.

REFERENCES

Alvarado, J.J., Reyes-Bonilla, H., Buitraco, F., Aguirre Rubí, J., Álvarez del Castillo-Cárdenas, P. A (2009) Coral reefs of the Pacific coast of Nicaragua. *Coral Reefs* 29, 201.

Asada, K., Takahashi, M. (1987) Production and scavenging of active oxygen in photosynthesis. *In* Kyle, D.J., Osmond, C.B., Arntzen, C.J. (eds). *Photoinhibition*. Elsevier, Amsterdam, pp. 228–287.

Anderson, J.M., Park, Y.I., Chow, W.S. (1997) Photoinactivation and photoprotection of photosystem II in nature. *Physiologia Plantarum* 100, 214–233.

Baker, A.C., Glynn, P.W., Riegl, B. (2008) Climate change and coral reef bleaching: an ecological assessment of long-term impacts, recovery trends and future outlook. *Estuarine, Coastal and Shelf Science* 80, 435–471.

Banaszak, A.T., Lesser, M.P. (2009) Effects of ultraviolet radiation on coral reef organisms. *Photochemistry and Photobiology Science* 8, 1276–1294.

Bierkens, J. (2000) Applications and pitfalls of stress-proteins in biomonitoring. *Toxicology* 153, 61–72.

Britt, A. (1996) DNA damage and repair in plants. *Annual Review of Plant Physiology and Plant Molecular Biology* 47, 75–100.

Brown, B.E. (1997) Coral bleaching: causes and consequences. *Coral Reefs* 16, S129–S138.

Cadenas, E. (1989) Biochemistry of oxygen toxicity. *Annual Review of Biochemistry* 58, 79–110.

Calderón-Aguilera, L.E., Reyes-Bonilla, H. (2006) Can local oceanographic conditions in the Mexican Pacific buffer the El Niño–Southern Oscillation effects on coral reefs?. *Proceedings of 10th International Coral Reef Symposium*, pp. 1138–1143.

Carignan, M.O., Cardozo, K.H.M., Oliveira Silva, D., Colepicolo, P., Carreto, J.L. (2009) Palythine–threonine, a major novel mycosporine-like amino acid (MAA) isolated from the hermatypic coral *Pocillopora capitata*. *Photochemistry and Photobiology* 94, 191–200.

Carreto, J.I., Lutz, V.A., De Marco, S. G & Carignan., M.O. (1990) Fluence and wavelength dependence of mycosporine-like amino acids synthesis in the dinoflagellate *Alexandrium excavatum*. *In* Granéli, E., Sunström, B., Edler L., Anderson, D.M. (eds). *Toxic Marine Phytoplankton*. Elsevier, Amsterdam, pp. 275–279.

Carriquiry, J.D., Reyes- Bonilla, H. (1997) Community structure and geographic distribution of coral reefs in Nayarit, western México. *Ciencias Marinas* 23, 223–248.

Cockell, C.S., Knowland, J. (1999) Ultraviolet radiation screening compounds. *Biological Reviews* 74, 311–345.

Cortés, J. (2003) Coral reefs of the Americas. *In* Cortés, J. (ed.). *Latin American Coral Reefs*. Elsevier, Amsterdam, pp. 1–8.

Cortés, J., Risk, M.J. (1985) A reef under siltation stress: Cahuita, Costa Rica. *Bulletin of Marine Science* 36, 339–356.

De Carvalho, J.F.G., Marenco, R.A., Viera, G. (2001) Concentration of photosynthetic pigment and chlorophyll fluorescence of mahogany and tonka bean under two light environments. *Revista Brasileira de Fisiologia Vegetal* 13, 149–157.

Díaz-Almeyda, E., Thome, P.E., Hafidi, M., Iglesias-Prieto, R. (2011) Differential stability of photosynthetic membranes and fatty acid composition at elevated temperature in *Symbiodinium*. *Coral Reefs* 3, 217–225.

Douglas, A.E. (2003) Coral bleaching – how and why? *Marine Pollution Bulletin* 46, 385–392.

Downs, C.A., Mueller, E., Philips, S., Fauth, J.E., Woodley, C. M. (2000) A molecular biomarker system for assessing the health of coral (*Montastraea favedata*) during heat stress. *Marine Biotechnology* 2, 533–544.

Downs, C.A., Fauth, J.E., Halas, J.C., Dustan, P., Bemiss, J., Woodley, C.M. (2002) Oxidative stress and seasonal coral bleaching. *Free Radical Biology and Medicine* 33, 533–543.

Downs, C.A., Fauth, J.E., Robinson, C.E., Curry, R. (2005) Cellular diagnostics and coral health: Declining coral health in the Florida Keys. *Marine Pollution Bulletin* 51, 558–569.

Dunn, S.R., Bythell, J.C., Le Tessier, D.A. et al. (2002) Programmed cell death and necrosis activity during hyperthermic stress-induced bleaching of the symbiotic sea anemone *Aiptasia sp. Journal of Experimental Marine Biology and Ecology* 272, 29–53.

Dykens, J.A., Shick, J.M. (1982) Oxygen production by endosymbiotic algae controls superoxide dismutase activity in their animal host. *Nature* 297, 579–580.

Dykens, J.A., Shick, J.M., Benoit, C., Buettner, G.R., Winston, G.N. (1992) Oxygen radical production in the sea anemone *Anthopleura elegantissima* and its endosymbiotic algae. *Journal of Experimental Biology* 168, 219–241.

Fauth, J.E., Downs, C.A., Halas, J.C., Dustan, P., Woodley, C.M. (2003) Mid-range prediction of coral bleaching: A molecular diagnostic system approach. *In* Valette-Silver, N., Scavia, D. (eds). *Ecological Forecasting: New Tools for Coastal and Ecosystem Management*. NOAA Technical Memorandum NOS NCCOS 1, pp. 5–12.

Flores-Ramírez, L., Liñán-Cabello, M.A. (2007) Relationships among thermal stress, bleaching and oxidative damage in the hermatypic coral, *Pocillopora capitata*. *Comparative Biochemistry and Physiology C* 146, 194–202.

Franklin, D.J., Hoegh-Guldberg, O., Jones, R.J., Bergesn J.A. (2004) Cell death and degeneration in the symbiotic dinoflagellates of the coral *Stylophora pistillata* during bleaching. *Marine Ecology Progress Series* 272, 117–130.

Fridovich, I. (1998) Oxygen toxicity: a radical explanation. *Journal of Experimental Biology* 201, 1203–1209.

Gates, R.D., Baghdasarian, G., Muscatine, L. (1992) Temperature stress causes host cell detachment in symbiotic cnidarians: implications for coral bleaching. *Biological Bulletin* 182, 324–332.

Glynn, P.W. (1990) Coral mortality and disturbances to coral reefs in the tropical eastern Pacific. *In* Glynn P.W. (ed.). *Global Ecological Consequences of the 1982–83 El Niño-Southern Oscillation*. Oceanography Series Vol. 52. Elsevier, Amsterdam, pp. 55–126,

Glynn, P.W. (2001) Eastern Pacific coral reef ecosystems. *In* Seeliger, U., Kjerfve, B. (eds). *Coastal Marine Ecosystems of Latin America*. Springer-Verlag, Berlin, pp. 281–305.

Glynn, P.W. (2004) High complexity food webs in low-diversity eastern Pacific coral reef communities. *Ecosystems* 7, 358–367.

Glynn, P.W., Ault, J.S. (2000) A biogeography analysis and review of the far eastern Pacific coral reef region. *Coral Reefs* 19, 1–23.

Glynn, P.W., Leyte-Morales, G.E. (1997) Coral reefs of Huatulco, west Mexico: Reef development in upwelling Gulf of Tehuantepec. *Revista de Biología Tropical* 45, 1033–1048.

Glynn, P.W., Lirman, D., Baker, A.C., Leyte-Morales, G.E. (1998) First documented hurricane strikes on eastern Pacific coral reefs reveal only slight damage. *Coral Reefs* 17, 368.

Glynn, P.W., Wellington, G.M., Rieg, B., Olson, D.B., Bornemann, E., Wieters, E. A (2007) Diversity and biogeography of the scleractinian coral fauna of Easter Island (Rapa Nui). *Pacific Science* 61, 67–90.

Goreau, T.J., Hayes, R.L. (2008) Effects of rising seawater temperature on coral reefs, in fisheries and aquaculture. *In* Safran, P. (ed.). *Encyclopedia of Life Support Systems (EOLSS)*. Developed under the Auspices of the UNESCO, EOLSS Publishers, Oxford.

Guzmán, H., Cortés, J. (2001) Changes in reef community structure after fifteen years of natural disturbances in the eastern pacific (Costa Rica). *Bulletin of Marine Science* 69, 133–149.

Guzmán, H.M., Cortés, J. (2007) Reef recovery 20 years after the 1982–1983 El Niño massive mortality. *Marine Biology* 151, 401–411.

Harrell, S.Y., Barron, M.G. (2010) Predicting coral bleaching in response to environmental stressors using 8 years of global-scale data. *Environmental Monitoring and Assessment* 161, 423–438.

Hallock, P. (2005) Global change and modern coral reefs: New opportunities to understand shallow-water carbonate depositional processes. *Sedimentary Geology* 175, 19–33.

Halliwell, B., Gutteridge, J.M.C. (1999) *Free Radicals in Biology and Medicine*. Oxford University Press, New York.

Hendry, G.A.F., Price, A.H. (1993) Stress indicators: Chlorophylls and carotenoids. *In* Hendry, G.A.F., Grime, J.P. (eds). *Methods in Comparative Plant Ecology*. Chapman & Hall, London, pp. 148–152.

Hill, R., Ulstruo, K.E., Ralph, P.J. (2009) Temperature induced changes in thylakoid membrane thermostability of cultured, freshly isolated, and expelled zooxanthellae from scleractinian corals. *Bulletin of Marine Science* 85, 223–244.

Hoegh-Guldberg O., Mumby, P.J., Hooten, A.J. et al. (2007) Coral reefs under rapid climate change and ocean acidification. *Science* 318, 1737–1742.

Iglesias-Prieto, R, Matta J.L., Robins W.A., Trench, R.K. (1992) Photsynthetic responses to elevated temperature in the symbiotic dinoflagellate *Symbiodinium microadriaticum* in culture. *Proceedings of the National Academy of Sciences USA* 89, 10302–10305.

Jokiel, P.L. (2004) Temperature stress and coral bleaching. *In* Rosenberg, E., Loya, Y. (eds). *Coral Health and Disease*. Springer-Verlag, Berlin, pp. 401–425.

Korbee N., Figueroa, F., Aguilera, J. (2006) Acumulación de aminoácidos tipo micosporina (MAAs): Biosíntesis, fotocontrol y funciones ecofisiológicas. *Revista Chilena de Historia Natural* 79, 119–132.

Kühl, M., Cohen, Y., Dalsgaard, T. et al. (1995) Microenvironment and photosynthesis of zooxanthellae in scleractinian corals studied with microsensors for O_2, pH, and light. *Marine Ecology Progress Series* 117, 159–172.

Lesser M.P. (1996) Elevated temperatures and ultraviolet radiation cause oxidative stress and inhibit photosynthesis in symbiotic dinoflagellates. *Limnology and Oceanography* 41, 271–283.

Lesser, M.P. (1997) Oxidative stress causes coral bleaching during exposure to elevated temperatures. *Coral Reefs* 16, 197–192.

Lesser, M.P. (2004) Experimental Biology of coral reef ecosystem. *Journal of Experimental Marine Biology and Ecology* 300, 217–252.

Lesser, M.P. (2006) Oxidative stress in marine environments: biochemistry and physiological ecology. *Annual Review of Physiology* 68, 253–278.

Lesser M.P., Farrel J.H. (2004) Exposure to solar radiation increases damage to both host tissues and algal symbionts of coral during thermal stress. *Coral Reefs* 23, 367–377.

Lesser, M.P., Shick, J.M. (1989a) Photoadaptation and defences against oxygen toxicity in zooxanthellae from natural populations of symbiotic cnidarians. *Journal of Experimental Marine Biology and Ecology* 134, 129–141.

Lesser M.P., Shick J.M. (1989b) Effects of irradiance and ultraviolet radiation on photoadaptation in zooxanthellae of *Aiptasia pallida*: primary production, photoinhibition, and enzyme defenses against oxygen toxicity. *Marine Biology* 102, 243–255.

Lesser M.P., Stochaj W.R., Tapley, D.W., Shick JM (1990) Bleaching in coral reef anthozoans: effects of irradiance, ultraviolet radiation, and temperature on the activities of protective enzymes against active oxygen. *Coral Reefs* 8, 225–232.

Liñán-Cabello M.A., Flores-Ramírez, L., Zacarías-Salinas, J.S., Hernández-Rovelo, O., Lezama-Cervantes, C. (2006) Correlation of chlorophyll *a* and total carotenoid concentrations with coral bleaching from locations on the Pacific coast of México. *Marine and Freshwater Behavioral Physiology* 39, 279–291.

Liñán-Cabello, M.A., Hernández-Medina, D., Florián-Álvarez, P., Mena-Herrera, A. 2008. Estado actual del arrecife coralino "La Boquita", Colima. *IRIDIA* 5, 16–27.

Liñán-Cabello, M.A., Flores-Ramírez, L.A., Cobo-Díaz, L.F. et al. (2010a) Response to short term ultraviolet stress in the reef-building coral *Pocillopora capitata* (Anthozoa: Scleractinia). *Revista de Biología Tropical* 58(1), 103–118.

Liñán-Cabello, M.A., Flores-Ramírez, L.A., Zenteno-Savin, T., Olguín-Monroy, N.O., Sosa-Avalos, R., Patiño-Barragan, M. (2010b) Seasonal changes of antioxidant and oxidative parameters in the coral *Pocillopora capitata* on the Pacific coast of Mexico. *Marine Ecology* 31, 407–417.

López-Pérez, R.A. (2005) The Cenozoic hermatypic corals in the eastern Pacific: a review. *Earth Science Reviews* 72, 67–87.

Manzello, D.P., Kleypas, J.A., Budd, D.A., Eakin, C.M., Glynn, P.W., Langdon, C (2008) Poorly cemented coral reefs of the eastern tropical Pacific: Possible insights into reef development in a high-CO_2 world. *Proceedings of the National Academy of Sciences USA* 29, 10450–10455.

Matta, J.L., Trench, R.K. (1991) The enzymatic response of the symbiotic dinoflagellate *Symbiodinium microadriaticum* (Freudenthal) to growth *in vitro* under varied oxygen tensions. *Symbiosis* 11, 31–45.

McClanahan, T. (2006) Challenges and accomplishments towards sustainable reef fisheries. *In* Coté, I.M., Reynolds, J.D. (eds). *Coral Reef Conservation*. Cambridge University Press, Cambridge, pp. 147–182.

Misonou, T., Saitoh, J., Oshiba, S., Tokitomo, Y., Maegawa, M., Inoue, Y., Hori, H., Sakurai, T. (2003) UV-absorbing substance in the red algae *Porphyra yezoensis* (Bangiales, Rhodophyta) block thymine dimer production. *Marine Biotechnology* 5, 194–200.

Mitchell, D.L., Karentz, D. (1993) The induction and repair of DNA photodamage in the environment. *In* Young, A.R.,

Björn, L.O., Moan, J., Nultsch, W. (eds) *Environmental UV Photobiology*. Springer-Verlag, New York, pp. 345–377.

Morrall, C.E., Galloway, T.S., Trapido-Rosenthal, H.G. et al. (2000) Characterization of nitric oxide synthase activity in the tropical sea anemone *Aiptasia pallida*. *Comparative Biochemistry and Physiology B* 125, 483–491.

Nii, C.M., Muscatine, L. (1997) Oxidative stress in the symbiotic sea anemone *Aiptasia pulchella* (Calgren, 1943): Contribution of the animal to superoxide ion production at elevated temperature. *Biological Bulletin* 192, 444–456.

Ortíz-Lozano, L., Granados-Barba, A., Solís-Weiss, V., García-Salgado, M.A. (2005) Environmental evaluation and development problems of the Mexican coastal zone. *Ocean and Coastal Management* 48, 161–176.

Pérez-Vivar, T.L., Reyes-Bonilla, H., Padilla, C. (2006) Stony corals (Scleractinia) from the Marías Islands, Mexican Pacific. *Ciencias Marinas* 32, 259–270.

Reyes-Bonilla, H. (1993) 1987 coral reef bleaching at Cabo Pulmo reef, Gulf of California, México. *Bulletin of Marine Science* 52, 832–837.

Reyes-Bonilla, H. (2002) Checklist of valid names and synonyms of stony corals (Anthozoa: Scleractinia) from the eastern Pacific. *Journal of Natural History* 36, 1–13.

Reyes-Bonilla, H. (2003) Coral reefs of the Pacific coast of Mexico. *In* Cortés, J. (ed.) *Latin America Coral Reefs*. Elsevier, Amsterdam, pp. 331–350.

Reyes-Bonilla, H., López-Pérez, R.A. (1998) Biogeography of stony corals (Scleractinia) from the Mexican Pacific Ocean. *Ciencias Marinas* 24, 211–224.

Reyes Bonilla, H., López-Pérez, R. A (2009) Corals and coral reef communities in the Gulf of California. *In* Johnson, M.E., Ledesma- Vásquez, J. (eds). *Atlas of Coastal Ecosystems of the Western Gulf of California*. University of Arizona Press, Tucson, pp. 43–55.

Reyes-Bonilla, H. Carriquiry, J.D., Leyte-Morales, G., Cupul, A.L. (2002) Effects of the El Niño-Southern Oscillation and the anti- El Niño event (1997–98) on coral reefs of the western coast of México. *Coral Reefs* 21, 368–372.

Reynolds, J.M., Brigitte, U., Bruns, B.U., Jitt, W.K., Schmidt, G.W., (2008) Enhanced photo protection pathways in symbiotic dinoflagellates of shallow-water coral and other unidarians. *Proceedings of the National Academy of Sciences USA*. 105, 13674–13678.

Richier, S., Furla, P., Plantivaux, A. et al. (2008) Symbiosis-induced adaptation to oxidative stress. *Journal of Experimental Biology* 208, 277–285.

Robertson, D.R., Cramer, K. (2009) Marine shore-fishes and biogeographic subdivisions of the Tropical Eastern Pacific. *Marine Ecology Progress Series* 380, 1–17.

Sawyer, S.J., Muscatine, L. (2001) Cellular mechanisms underlying temperature-induced bleaching in the tropical sea anemone *Aiptasia pulchella*. *Journal of Experimental Biology* 204, 3443–3456.

Scandalios J.G., Guan L., Polidoros A.N. (1997) Catalases in plants: Gene structuret properties, regulation, and expression. *In* Scandalios, J.G. (ed.). *Oxidative Stress and Molecular Biology of Antioxidant Defenses*. Cold Spring Harbor Laboratory Press, New York, pp 343–406.

Schittenhelm J.S., Toder S., Fath S., Westphal S., Wagner E. (1994) Photoactivation of catalase in needless of Norway spruce. *Physiologia Plantarum* 90, 600–606.

Shick, J.M., Romaine-Lioud, S., Ferrier-Pagès, C., Gattuso, J.P. (1999) Ultraviolet-B radiation stimulates shikimate pathway-dependent accumulation of mycosporine-like amino acids in the coral *Stylophora pistillata* despite decreases in its population of symbiotic dinoflagellates. *Limnology and Oceanography* 44, 1667–1682.

Streb P., Michael-Knauf A., Feierabend J. (1993) Preferential photoinactivation of catalase and photoinhibition of photosystem II are common early symptoms under various osmotic and chemical stress conditions. *Physiologia Plantarum* 88, 590–598.

Stambler, N., Dubinsky, Z. (2004) Stress effects on metabolism and photosynthesis of hermatypic corals. *In* Rosenberg, E., Loya, Y. (eds). *Coral Health and Disease*. Springer, Berlin, pp. 195–215.

Suggett, D.J., Warner, M.E., Smith, D.J. et al. (2008) Photosynthesis and production of hydrogen peroxide by *Symbiodinium* (Pyrrhophyta) phylotypes with different thermal tolerances. *Journal of Phycology* 44, 948–956.

Tchernov, D., Gorbunov, M.Y., de Vargas, C. et al. (2004) Membrane lipids of symbiotic algae are diagnostic of sensitivity to thermal bleaching in corals. *Proceedings of the National Academy of Sciences USA* 101, 13531–13535.

Torres, J.L., Armstrong, R.A., Weil, E. (2008) Enhanced ultraviolet radiation can terminate sexual reproduction in the broadcasting coral species *Acropora cervicornis* Lamarck. *Journal of Experimental Marine Biology and Ecology* 358, 39–45.

Trapido-Rosenthal, H., Zielke, S., Owen, R. et al. (2005) Increased zooxanthellae nitric oxide synthase activity is associated with coral bleaching. *Biological Bulletin* 208, 3–6.

Torregiani, J.H., Lesser, M.P. (2007) The effects of short-term exposures to ultraviolet radiation in the Hawaiian Coral *Montipora verrucosa*. *Journal of Experimental Marine Biology and Ecology* 340, 194–203.

Van de Poll, W.H., Eggert, A., Buma, A.G.J., Breeman, A.M. (2001) Effects of UV-B-induced DNA damage and photoinhibition on growth of six temperate marine red macrophytes: habitat-related differences. *Journal of Phycology* 37, 30–37.

Veron, J.E.N. (1995) *Corals in Space and Time*. Comstock-Cornell, Ithaca, 325 pp.

Warner, M.E., Berry-Lowe, S. (2006) Xanthophyll cycling and photochemical activity in symbiotic dinoflgellates in

multiple locations of three species of Caribbean coral. *Journal of Experimental Marine Biology and Ecology* 339, 86–95.
Weis, V.M. (2008) Cellular mechanisms of cnidarian bleaching: stress causes the collapse of symbiosis. *Journal of Experimental Biology* 211, 3059–3066.
Yakovleva, I., Bhagooli, R., Takemura, A., Hidaka, M. (2004) Differential susceptibility to oxidative stress of two scleractinian corals: antioxidant functioning of mycosporine–glycine. *Comparative Biochemistry and Physiology B* 139, 721–730.
Zhang L., Li, L., Wu, Q. (2007) Protective effects of mycosporine-like amino acids of *Synechocystis* sp. PCC 6803 and their partial characterization. *Photochemistry and Photobiology B* 86, 240–245.

Part II

Aquatic Respiration and Oxygen Sensing

Chapter 10

PRINCIPLES OF OXYGEN UPTAKE AND TISSUE OXYGENATION IN WATER-BREATHING ANIMALS

J. C. Massabuau[1] *and Doris Abele*[2]

[1]Université Bordeaux 1, Arcachon, France
[2]Alfred Wegener Institute for Polar and Marine Research, Department of Functional Ecology, Bremerhaven, Germany

Oxygen started to appear in very low amounts (<1% present atmospheric oxygen level, 1 kPa [today, in air or air-equilibrated water, $P_{O_2} \approx$ 21 kPa at sea level]) around 3.5 Gyr ago in a primitive ocean, and the first unicellular and pluricellular aquatic organisms obviously inhabited a mildly oxygenated and often anoxic environment. Everything suggests that the basic machinery of cellular life, such as DNA synthesis, gene transcription, translation and regulation should have become established, and largely irrevocably so, under these conditions (Segerer et al. 1985; Trevors 2003) thus explaining why cellular processes function preferentially under hypoxia. Then, a rapid rise in oxygen occurred approximately 2 Gyr ago, probably up to 3 kPa (Canfield 1998). At that time mitochondria became established as oxygen reducing energetic power plants in eukaryotic cells, which boosted the biological evolution of aerobic life on Earth, allowed the appearance of complex water-breathing animals, and culminated much later in the appearance of energetically optimized, but highly energy dependent, endothermic life-forms such as mammals and birds (Buttemer et al. 2010).

Arguably, oxygen fostered life but it is also one of the most hazardous cellular toxicants. Oxidative stress occurs when the production of reactive oxygen species (ROS) exceeds the antioxidant capacities. Too much free unbound oxygen in the cellular environment is deleterious, and one of the primary ways to prevent ROS from forming is to keep tissue oxygen activity (oxygen partial pressure, P_{O_2}) as low as possible. This is summarized as the "the low P_{O_2} strategy" (where "low P_{O_2}" stands for 1–3 kPa; Massabuau 2001, 2003), which certainly applies to most animals but was first described in water-breathers. It is based on three points: the first steps of the evolution of aerobic life occurred in a

Oxidative Stress in Aquatic Ecosystems, First Edition. Edited by Doris Abele, José Pablo Vázquez-Medina, and Tania Zenteno-Savín.

low oxygen environment during 1 Gyr (Barnabas et al. 1982; Hoogewijs et al. 2007); most water-breathers live at low blood and tissue P_{O_2}; mammalian tissue oxygenation is in the same "low P_{O_2}" range (Vanderkooi et al. 1991).

A key message in this chapter is that oxygen uptake in ectotherms is adjusted on morphological and behavioral levels, as well as through physiological adaptations, in a way that combines the most efficient usage of energetic resources with the highest possible maintenance of protective mechanisms against oxygen toxicity from the earliest animals onwards and over evolutionary time scales. The critical high cellular P_{O_2} when one electron oxygen reduction, and therewith the toxic side of tissue oxygenation becomes significant, varies between species with different mitochondrial terminal oxidases and antioxidant levels, but can be approximated to be 5 kPa. Onset of ROS formation was approximated at 20 μM cellular oxygen ($\approx$ 2 kPa) by Papa and Skulachev (1997).

BASIC PRINCIPLES OF OXYGEN DIFFUSION BETWEEN COMPARTMENTS

How does oxygen diffuse from extra- to intra-cellular compartments and, through the gill epithelium, from water to blood? Generally speaking, the movement of gas molecules from a compartment of higher to lower chemical activity (for oxygen, it can be characterized by the P_{O_2} that is also nearly synonymous with the fugacity of oxygen) is described by Fick's first law:

$$\dot{M}_{O_2} = A \cdot 1/E \cdot K_{O_2} \cdot \Delta P_{O_2} \qquad (10.1)$$

The symbol $\dot{M}_{O_2}$ designates the oxygen quantity flowing from one compartment to the other per unit of time; A and E are the surface area and the thickness of the membrane separating both compartments, respectively; K_{O_2} is the Krogh's constant of diffusion (often referred to as the permeation coefficient); and ΔP_{O_2} is the difference of oxygen partial pressure between the two compartments (Dejours 1981).

It is worth noting that the oxygen concentration does not appear in Fick's law of diffusion. This is particularly important as a common confusion when discussing limits of oxygen uptake in water-breathers involves the misuse of oxygen concentration (C_{O_2}) and P_{O_2}. Numerous authors use C_{O_2} and P_{O_2} as equivalent and inappropriately use absolute values of oxygen concentration (at constant P_{O_2}) to claim, for example, that normoxic (air-equilibrated) ecosystems at high temperature are "hypoxic" compared to colder water, and vice versa. The contrary is true: the oxygen concentration *per se* is basically never a key limiting factor for respiration in an animal that is living in a large enough water body and has access to an unlimited volume of water. Indeed, equation (10.1) shows that the oxygen driving force to diffuse from water to the animal is the difference in P_{O_2} and in no case the C_{O_2} difference (Foster 1964). This is further demonstrated in Fig. 10.1, which illustrates a simple experiment wherein we prepare a not very savoury French dressing by intimately mixing and shaking olive oil and water in a flask. Air is bubbled into the mixture to ensure an equal distribution of oxygen in the whole system. When bubbling and mixing stops, the light olive oil separates from the denser water, filling the upper part of the flask while the water stays at the bottom. Note two fundamental points here:

1. there is no membrane that separates both compartments, meaning that there is no diffusion barrier limiting oxygen exchange;
2. there is a difference of oxygen solubility between olive oil and water, with much higher solubility of oxygen in the oil than in water, which explains the unequal distribution of oxygen molecules between the two compartments.

Now, one can wait for days, months or even longer, and the C_{O_2} will stay five times higher in the oil compared to the water, but the P_{O_2} will remain identical in both liquids of the vinaigrette. This simple experiment demonstrates that differences in the C_{O_2} amounts do not govern oxygen distribution between compartments.

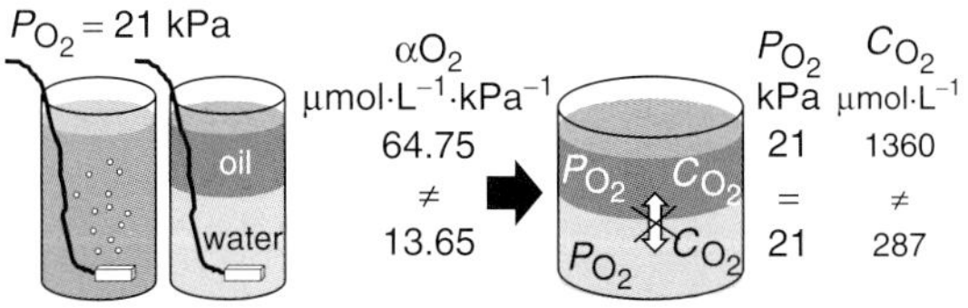

Fig. 10.1 Distribution of oxygen between compartments exhibiting different O_2 solubility. In a simple diffusion model, made of distilled water and olive oil, the distribution of oxygen between both compartments at rest is driven by the O_2 partial pressure gradient but not the O_2 concentration gradient. See text.

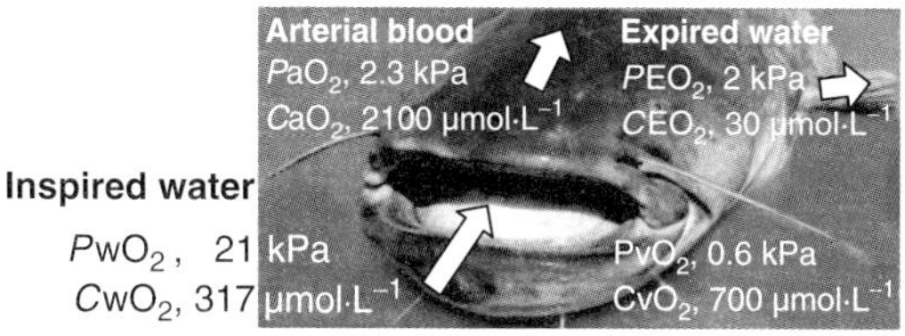

Fig. 10.2 The O_2 equilibrium in resting sheatfish *Silurus glanis* in normoxia. Note that O_2 enters the animal following its pressure gradient and against its concentration gradient; P_{O_2} in the fluids leaving the branchial cavity, the arterial P_{O_2}, and the expired water P_{O_2} are not significantly different. (Redrawn from Forgue et al. 1989.)

In an ecologically relevant context this means that when comparing life in polar waters to tropical waters, the C_{O_2} amounts in the water differ dramatically for the same P_{O_2} amounts. However, for the same water P_{O_2}, the driving forces for oxygen to enter the gill blood are the same, provided that the blood P_{O_2} amounts in both animals are the same. Thus, when both P_{O_2} amounts are the same, it is inappropriate to say that warm waters are "hypoxic" by comparison to cold waters and vice versa, as long as the animals live in open waters where chemical and biological oxygen demands do not change P_{O_2}. In closed systems – or confined systems such as poorly ventilated burrows or tide water pools at night with reduced water exchange, and only under these circumstances – the situation is different, of course, as the total amount of oxygen molecules available for the animals is finite and lower at high compared to low temperatures.

Conclusion 1

The driving force for oxygen diffusion between two compartments is the difference in P_{O_2}. It can operate against a concentration gradient as it does between air and blood in the lungs, water, and blood at the gill surface, and between extra- and intracellular compartment in tissues where myoglobin is present and alters the tissue-specific oxygen "solubility". An example is given in Fig. 10.2, which presents the recorded values of oxygen distribution between the inspired water, the expired water, and blood in a resting teleost fish, the sheatfish *Silurus glanis* (Forgue et al. 1989). The P_{O_2} gradient between the inspired water and the venous or the arterial blood favours the influx of oxygen against a concentration gradient. We will discuss below why P_{O_2} in the arterial blood and the expired water can be so close together (2.3 and 2.1 kPa, respectively). (For reference, 1 kPa = 7.5 mm Hg or torr and in a saline solution, $P_{O_2} = 1$ kPa corresponds to an O_2 fraction of $\approx 1\%$ of the total present atmospheric pressure at sea level.)

THE PRINCIPLES OF INTERNAL OXYGENATION IN WATER BREATHING ANIMALS

The low P_{O_2} strategy was first proposed for resting crayfish respiring at various oxygen levels and under steady-state conditions (Massabuau and Burtin 1984). Ventilation was found to be adjusted in such a way that the P_{O_2} in the blood that leaves the branchial cavities remains nearly constant over time against variable ambient oxygenation levels. This was deduced from the following facts: (i) the most frequently measured P_{O_2} in the fluids leaving the branchial cavities – the arterial blood and the expired water – remained within a narrow P_{O_2} range independent of the inspired water P_{O_2}; (ii) oxygen chemoreception is located in the branchial cavities and very likely in the gills (Massabuau et al. 1980; Massabuau and Burtin 1985; Ishii et al. 1989; Jonz and Nurse 2006). As long as the blood P_{O_2} and the arterio-venous C_{O_2} difference can be maintained through ventilatory adjustment, the oxygen consumption can be maintained without the need to change the blood flow rate via changes in cardiac activity (see equation (10.2), Fick's second law). This is evidently important for physiological performance.

To learn if this strategy is special only to crayfish or part of a more general pattern, regulation of arterial P_{O_2} was studied in a number of physiologically different species, representative of the most common groups (other crustaceans, mollusks, and fish) of water-breathers. It resulted that resting arterial P_{O_2} is adjusted to a low apparent set-point in the range of 1–3 kPa in diverse species and phyla. It is the same for marine or freshwater species differing in behavior (active vs. sluggish) and ecological niche, and for different seasons. It is independent of organization of the respiratory system (gill type, ventilatory pump, open vs. closed circulatory system) as well as presence, absence, concentration, and oxygen affinity of respiratory pigments. Figure 10.3 gives examples for crustaceans, mollusks, and fish. It shows that in normoxic water most of the arterial P_{O_2} values are

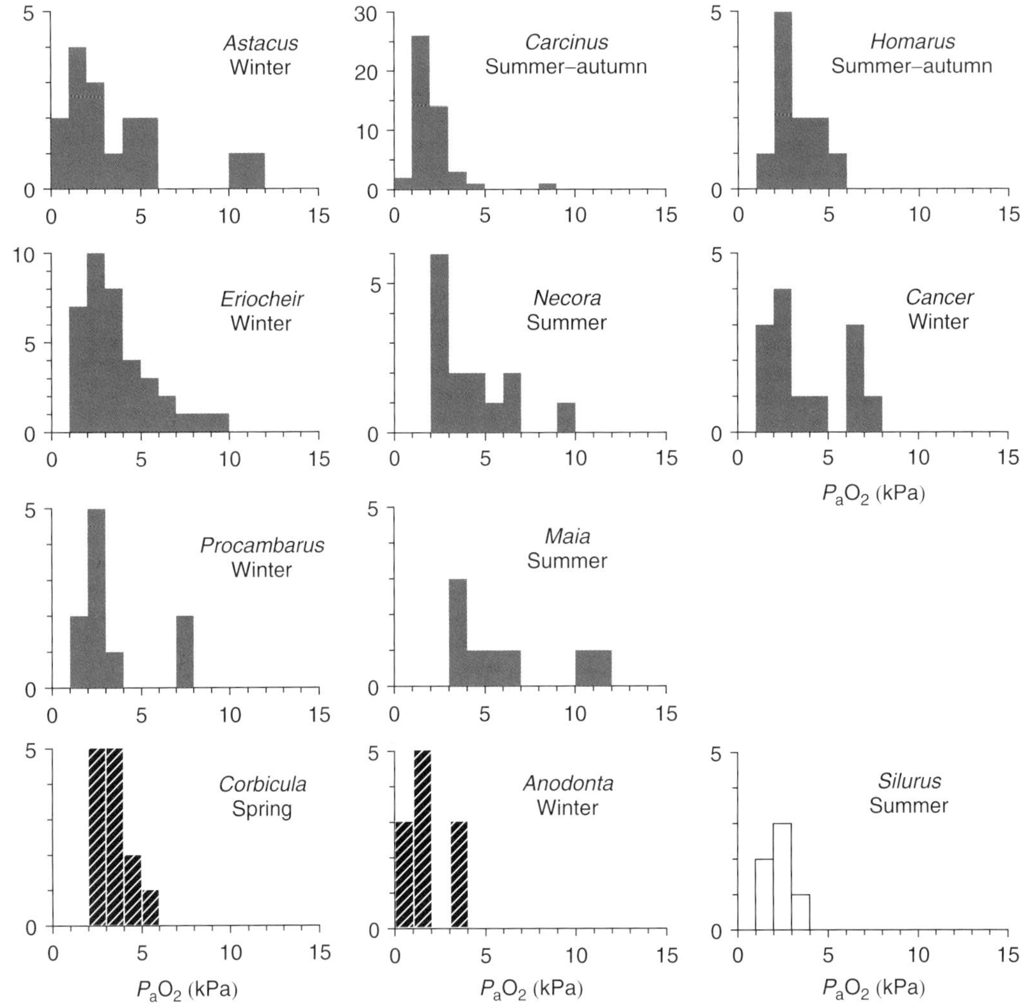

Fig. 10.3 Frequency distribution of arterial O_2 partial pressure, P_aO_2, in various water-breathers at rest and in normoxia (inspired $P_{O_2} = 20–21$ kPa, inspired $P_{CO_2} = 0.1$ kPa, T = 15°C, unfed). For all species, values were not normally distributed and most frequently measured P_aO_2 were in the range 1–3 kPa (note the low occurrence of exceptional values of 10–12 kPa). Filled histograms, crustaceans representative of most European species; hatched histograms, data from the Asiatic clam *C. fluminea* and the freshwater mussel *A. cygnea*, i.e. two mollusks; open histogram, data from the teleost fish *Silurus glanis*. (Reprinted from JEB 170: 257-264, 1992, with permission from The Journal of Experimental Biology.)

low and the data are not normally distributed. The most frequently measured arterial P_{O_2} values (i.e. the modes) are in the range 1–3 kPa. Note that higher arterial P_{O_2} values of up to 10–12 kPa are occasionally observed (Forgue et al. 1992a). Apparently, these transiently higher values are of outstanding importance as they participate in shaping the normal physiological repertoire of the animal life cycle (see below and Clemens et al. (2001) for a review in the field of neurobiology; see Forgue et al. (2001) for a discussion about resting vs. exercise state). The values in resting and unfed animals are just above arterial P_{O_2} at the anaerobic threshold, 0.5–0.7 kPa, as determined in the crabs *Eriocheir sinensis* and *Carcinus*

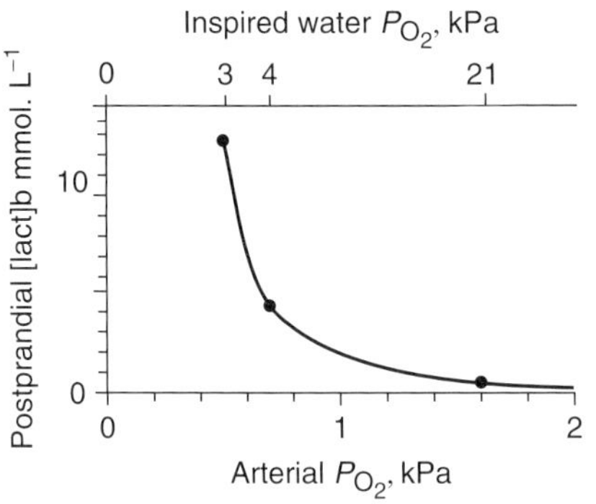

Fig. 10.4 Anaerobic threshold in green crab *Carcinus maenas*. Crabs were fed in normoxia with a single mussel and then transferred to hypoxic water (3 or 4 kPa) to experimentally limit their arterial P_{O_2} during digestion in order to induce a switch on anaerobiosis, and to study the impact on protein synthesis. Here, blood lactate was used as a marker of the initiation of anaerobic metabolism. It revealed that during postprandial change in O_2 consumption, the arterial P_{O_2} at the anaerobic threshold is in the 1 kPa range. T = 15°C. (Redrawn from Mente et al. 2003.)

maenas (Forgue et al. 1992b; Legeay and Massabuau 2000) and in trout *Oncorhynchus mykiss* (1.3 kPa; Thomas et al. 1988). In fed *Carcinus maenas*, Mente et al. (2003) found an arterial P_{O_2} at the anaerobic threshold of ≈ 1 kPa (Fig. 10.4). Consequently, the above data set shows that water-breathers applying the low blood P_{O_2} strategy most of the time feature a narrow range of tissue oxygenation (1–3 kPa), just above the levels where underoxygenation starts but sheltered from problems of oxidative stress. It is now largely appreciated that living at low blood and tissue P_{O_2} is part of the normal physiological features of water-breathers and a very basal strategy in the animal kingdom (Hetz and Bradley 2005; Lane 2005; Bradley 2006; Janvier et al. 2006; Lin et al. 2008; Ivanovic 2009).

WHICH ANIMALS FUNCTION AT LOW BLOOD OXYGENATION LEVELS? HOW DO THEY ACCOMPLISH IT?

The behavioural and physiological mechanisms to achieve low blood P_{O_2} may differ between animal models and different bauplans and ecotypes, but the basic principles remain the same as in crayfish: in normoxic and hyperoxic water, the animals establish a hypoxic and incidentally hypercapnic environment around their gas exchange organs. Ventilatory activity controls – and literally limits – oxygen influx into the branchial cavities, to restrict oxygen diffusion between water and blood. This is not a very classic view, but "normoxic" animals (respiring in normoxic water air equilibrated at 21 kPa) following the low P_{O_2} strategy perform a kind of "re-breathing." The branchial cavity must be viewed as an "antechamber" in which the water renewal is slowed compared to open waters. The consequence is that the blood flowing through the gills extracts oxygen from a relatively confined environment – the branchial cavities – where P_{O_2} is lowered against the outside medium, to achieve low oxygenated and slightly hypercapnic conditions in the arterial blood. Consequently, P_{O_2} is also very low in the expired water independently of the inspired water P_{O_2} (Fig. 10.5; see also the expired water P_{O_2} in normoxic sheatfish in Fig. 10.2). In public aquaria with well controlled and air-equilibrated water, one can observe quiet fish with extremely low ventilatory activity and virtual absence of bucco-opercular movements: they are practicing the low P_{O_2} strategy. Clearly, a tight control of the water renewal in their branchial cavity, antechamber bathing the gills, is the first adaptive step to maintain low blood and tissue oxygenation.

Figure 10.6 presents the ventilatory response of a trout facing rapid fluctuations of water oxygenation. It was experimentally exposed to short pulses of hyperoxygenated water (80 kPa during 5 s). The track record shows that within seconds the amplitude and frequency of the bucco-opercular movements goes down, corresponding to instantaneous reduction in ventilatory activity of reflex origin (Eclancher 1972; Bamford 1974; Shelton et al. 1986). Hyperoxic pulse treatment further caused mild tachycardia. What is the immediate adaptive value of this physiological response? A study was carried out with crayfish in which one can simultaneously record the ventilatory reflex response and the oxygenation status of the arterial blood as it leaves the gills (Massabuau et al. 1980; Massabuau and Burtin 1985). Figure 10.7 shows that, following a brief hyperoxic pulse supplied at the opening of the branchial cavities, the crayfish stops ventilating within seconds, which abrogates the rise of P_{O_2} in its arterial blood and therewith the intrusion of excess oxygen into its milieu intérieur. Ventilation is only resumed when the arterial P_{O_2} goes back towards its reference set

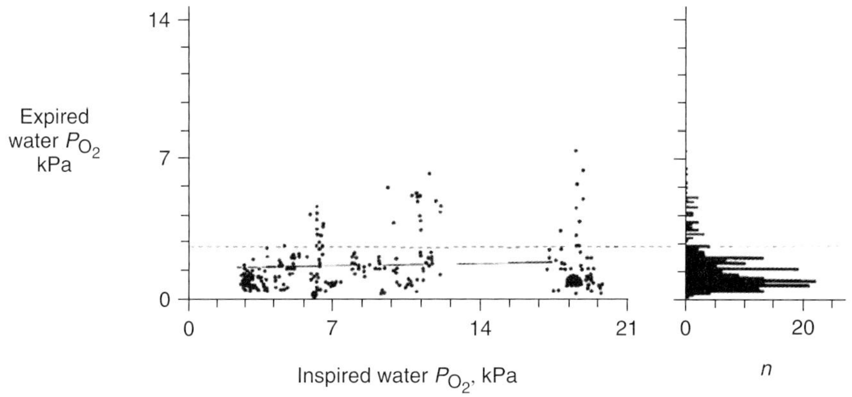

Fig. 10.5 (Left) Changes of P_{O_2} in the expired water of crayfish *Astacus leptodactylus* exposed to selected values of inspired water P_{O_2} (unfed). (Right) Distribution of the expired P_{O_2} values; most frequently measured value ≈ 1–2 kPa (T = 13°C; Massabuau and Burtin 1984). Same experiment as in Fig. 10.10 (left). (Reprinted from *Journal of Comparative Physiology B* 155: 43–49. Massabuau, J.-C. and B. Burtin: Regulation of the oxygen consumption in the crayfish *Astacus leptodactylus*: role of the peripheral O_2 chemoreception. With permission from Springer-Verlag.)

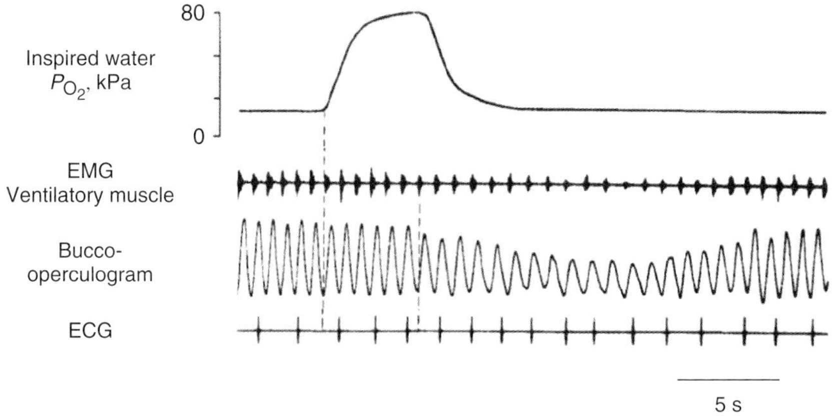

Fig. 10.6 Velocity of ventilatory and circulatory responses in trout *Onchorynchus mykiss* exposed to an abrupt change of water oxygenation from 20 to 80 kPa. The ventilatory activity slows down within 5 s and a transient tachycardia occurs. (Based on data from Eclancher 1980.)

point value. Note that arterial P_{O_2} recovers because there is a continuous cardiac activity (tachycardia in fish, Fig. 10.6) during the apnoeic period. It clearly demonstrates the limiting and protective role of ventilation in water-breathers facing oxygen surges.

Various other animals were also shown to reduce P_{O_2} in the confined environments of their shells (bivalves and limpets), or by seeking low P_{O_2} areas in the sediment or installing low P_{O_2} in their immediate surrounding (ostracod crustaceans). This is certainly true for microoxophilic meiofauna species found in hypoxic sediments. For them, normoxic conditions (21 kPa) can already be too oxidizing (Abele-Oeschger et al. 1998; Tschischka et al. 2000). The ability to establish hypoxic conditions and survive at reduced aerobic rates and/or switch to mitochondrial anaerobic energy pathways for long-term survival are good indicators for the capacity of species to resist adverse environmental conditions (Morley et al. 2007: seasonal food shortage in polar bivalves; Weihe and Abele 2008: limpets surviving air exposure on polar beaches; for review see Storey and Storey 2004).

Among the microoxophilic species we find primitive crustaceans that lack the ability for ventilatory

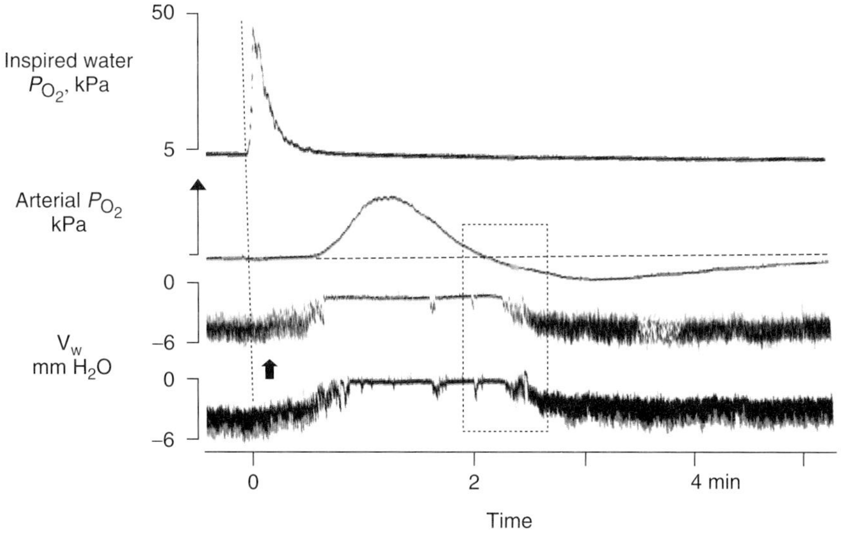

Fig. 10.7 Adaptive value of the ventilatory response to an abrupt water O_2 change in the crayfish *Astacus leptodactylus*. From top to bottom, inspired water P_{O_2} change measured at the entrance of the branchial cavities; P_{O_2} change in the arterial blood leaving the gills (heart level); V_W, ventilatory activity measured as hydrostatic pressure changes in the left and right branchial cavities. Following hyperoxic water inhalation the ventilatory activity decreases within seconds (start at arrow). The result is a rapid and efficient slow down of the blood O_2 enrichment followed by a transient undershoot. Dashed insert: the ventilation resumes to offset the arterial P_{O_2} decrease below its reference value (T = 13°C; Massabuau, unpublished data). See Massabuau et al. (1980) and Massabuau and Burtin (1985) for technical details.

adjustments and literally seek hypoxic sedimentary environments through migration in order to stabilize low tissue oxygenation. Corbari and co-authors studied minute ostracod crustaceans (Corbari et al. 2004, 2005a,b) that were already present 400–450 Myr ago and have not changed much morphologically since then. All different groups of ostracods possess ventilatory appendages, physiologically analogous to the scaphognathites of modern crustaceans. Among the groups of ancient crustaceans, different evolutionary states can be studied. For example, the podocopids, the largest ostracod group, lack a heart as well as gills. Myodocopid ostracods are further evolved from the morpho-functional perspective and possess a cardiovascular system with a well-differentiated heart, but several myodocopid species lack gills. Are they regulating their internal oxygenation status and, if so, how do they do it? The answer is: all of them regulate tissue oxygenation, but podocop ostracods adjust oxygenation of their internal medium by migrating along oxygen gradients in the sediment, systematically avoiding both oxygen-rich and -depleted regions. By contrast, myodocop ostracods build hypoxic nests on the bottom of the sea in which they perform collective "re-breathing" and produce a confined hypoxic environment. As, in addition to the fossil record dating back to the early Paleozoic, numerous aspects of the ostracod respiratory physiology can be taken as primitive, suggesting that 400–500 Myr ago, early crustaceans were already managing their tissue oxygenation levels when necessary, by avoiding oxygen enriched areas to keep out of the "death zone" of free radical attack. Interestingly, they used a behavioral rather than a physiological strategy, which was conserved over evolutionary times.

On the seashore, barnacles (sessile crustaceans) also maintain a hypoxic internal milieu in a normoxic environment. Davenport and Irvin (2003) concluded that their tissues are exposed to hypoxic conditions during most of their life. Barnacles are typical for the high intertidal and splash zones and tolerate extreme fluctuations of environmental oxygen during tidal cycles.

Bivalve and gastropod mollusks are renowned for their profound tolerance of hypoxia, based on their intrinsic capacity to reduce metabolic rate and switch on anaerobic metabolism to bridge times of oxygen deficiency (Brooks and Storey 1997; Guppy et al. 2000). Evolution of the molluskan clade dates back to the early Cambrian Period (550 Myr ago), with primitive soft-bodied and shell-carrying ancestors already abundant in the Vendian Period (550–800 Myr ago). Interestingly, the hemocyanin of mollusks (and the parallel evolved crustacean hemocyanin) is considered to have evolved from phenol oxidase enzymes that helped oxygen-sensitive early animals to detoxify photosynthetic oxygen. Modern hemocyanins have conserved some weak phenoloxidase activity as they took over the function of oxygen-binding and transport in modern mollusks (Decker and Terwilliger 2000; Van Holde et al. 2001).

Extant mollusks maintain mantle cavity water P_{O_2} inside their shell at levels below normoxia by ventilatory control (bivalves) and shell lifting movements (patellogastropods) (Abele et al. 2010). Again, in most bivalves the mantle cavity surrounding the gills can be considered an antechamber from which oxygen is taken up into the arterial blood. The authors recorded P_{O_2} frequencies in mantle cavity water of four bivalves and the limpet *Patella vulgata*, as representative of the patellogastropod family, using optical oxygen sensors. All animals were maintained in normoxic experimental conditions and at 10°C. It resulted that no species allows mantle cavity P_{O_2} to become permanently normoxic. The two scallops had especially high mantle cavity P_{O_2} median (6.2 kPa in *Aequipecten opercularis*, 8.3 kPa in *Pecten maximus*) associated with higher metabolic rate than mud clams, possibly to serve the energy requirements of swimming. Note that the mantle cavity P_{O_2} median is considerably lower in the more hypoxia-tolerant mud clams (0.4 kPa in *Mya arenaria*, 3.6 kPa in *Arctica islandica*) and the limpet (2.6 kPa in *Patella vulgata*) although the animals were maintained in air-equilibrated water. In contrast to the observed differences in shell water oxygenation, P_{O_2} in the arterial hemolymph sampled from the heart of two bivalves, the river pearl mussel *Anodonta cygnea* (Massabuau et al. 1991) and the Asiatic clam *Corbicula fluminea* (Tran et al. 2000), had the same low-frequency distribution previously observed in crustacean arterial blood. Further, in the lugworm *Arenicola marina*, a marine polychaete inhabiting sandy beaches and belonging to the fourth largest group of aquatic animals, blood sampled from the ventral and dorsal vessels was in the low range of 1 kPa in animals kept in normoxic water at 15–17°C (Toulmond 1973). Taken together, these results indicate that blood P_{O_2} is indeed kept low and protective across different phyla, despite obvious major physiological differences in lifestyle, motility, or metabolic rate.

Conclusion 2

Modern and primitive water-breathers maintain P_{O_2} in their internal milieu at levels very similar to those reported for the primitive atmosphere in which life started to evolve (below 3 kPa). The low P_{O_2} strategy applies to different ecotypes, independently of other physiological characteristics. Hypoxia-tolerant marine species are our window to the past, enabling understanding of the early features of metabolism and how early animals generally dealt with oxygen. The basic machinery of cellular life was established in a low oxygenated environment and the level of cellular oxygenation seems to have been conserved in a similarly low range during evolution. Still, in modern animals, too much oxygen disturbs cellular homeostasis and behavioral, morphological, as well as physiological adaptations evolved to reduce P_{O_2} levels and maintain it within tolerated limits (below 5 kPa). Maintaining a low *in vivo* P_{O_2} environment in animal tissues must be viewed as a first firewall protecting against the negative consequences of overoxygenation.

THE LOW P_{O_2} STRATEGY IS INDEPENDENT OF METABOLIC RATE

In water-breathers, living at low blood P_{O_2} does not mean living at low metabolic rate and succumbing to an oxygen-limited mode of living. When tissue oxygen consumption increases, oxygen must simply be transported faster to tissues and cells and readily delivered to the mitochondria or other oxygen-consuming cellular reactions. First key adaptations are to modulate the blood flow rate (cardiac activity) and to establish an appropriate peripheral vascular bed, i.e. adjust local micro-perfusion to the tissue-specific metabolic standards. Tissue-specific oxygen demand can be very different and, within a given species, the demand of different organs can be variable over time, e.g. with activity or maturation state of the animal. Frequent high organ-specific oxygen demand is met by high vascular density (fish red muscle, crustacean ventilatory

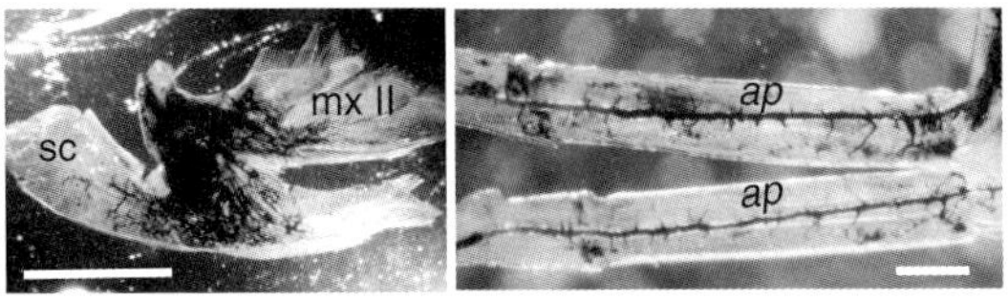

Fig. 10.8 Microcirculation in continuously or periodically active crustacean muscles. The microcirculation was traced by injection of India ink into the heart. (a) Scaphognathite, the ventilatory plate. (b) The meropodite of a limb. Dissected preparations were cleaned with methyl salicylate to reveal the finely stained microvascularization. Note the rich vascularisation in the continuously active ventilatory muscles of the scaphognathites (a) and a few microvessels in the locomotor muscles of the limb (b). sc, scaphognathite; ap: arteria pedalis; mx II, maxille 2. Scale bar: 1 mm. (Massabuau, unpublished data.)

muscles, see Fig. 10.8) whereas low local oxygen demand can be limited by a low local vascular density. Low perfusion is evidently a simple adaptation to limit oxidative stress by unconsumed oxygen.

Fick's second law shows the parameters which determine the oxygen-transport by convection from gills to cells (Dejours 1981):

$$\dot{M}_{O_2} = \dot{V}_b(C_aO_2 - C_vO_2) = \dot{V}_b \cdot ß_bO_2(P_aO_2 - P_vO_2) \quad (10.2)$$

where $\dot{M}_{O_2}$ is metabolic rate, $\dot{V}_b$ is the blood flow rate, C_aO_2 and C_vO_2 are the oxygen concentrations in the arterial and venous blood, respectively, $ß_bO_2$ is the blood oxygen affinity and P_aO_2 and P_vO_2 stand for P_{O_2} in arterial and venous blood.

All arterial P_{O_2} data reported in Fig. 10.3 were measured in resting animals. How is arterial P_{O_2} adjusted when metabolism of these resting poikilotherms increases as animals are acclimated or naturally exposed to higher temperatures within their thermal range of optimal physiological performance? Several reports demonstrate that it remains low and constant under steady-state conditions. In crayfish, *Astacus leptodactylus*, arterial P_{O_2} remained unchanged upon warming from 10 to 20°C (3.2 ± 0.1 and 3.7 ± 0.1 kPa, respectively; Angersbach and Decker 1978). The same stability of arterial P_{O_2} has been reported in carp, *Cyprinus carpio* (3.8 ± 2.1 and 3.3 ± 1.1 kPa at 10 and 25°C; Itazawa and Takeda 1978). Furthermore, in green crab *Carcinus maenas*, sampled at water temperatures between 10 and 20°C in the Bay of Arcachon, France, arterial P_{O_2} was invariably at 1–3 kPa (Massabuau and Forgue 1996). Only when animals are subject to stressfully rapid warming or cooling beyond the limits of their thermal optimum range, hemolymph or arterial P_{O_2} declines dramatically, eventually causing death through asphyxiation. The sea spider *Maya squinado* has a thermal optimum range between 7 and 17°C and rapid cooling by 1°C/h to 0°C or rapid warming to 40°C caused hemolymph P_{O_2} in the cardiac sinus to decrease as ventilation and circulation ceased in the cold. This was also observed under critical warming. Although experimental warming was performed very rapidly, hemolymph P_{O_2} in the sea spider was maintained constant within the thermal optimum range (Frederich and Pörtner 2000).

The crayfish, the carp, and the crabs are animals of moderate oxygen demand but similarly low arterial P_{O_2} values are also recorded in more active marine species such as peneid shrimps, which are comparable in terms of metabolic rate to trout, salmon, or even tuna. In spite of their exceptionally high oxygen demand, *Penaeus japonicus* and *Litopenaeus stylirostris* are working between ≈ 1 and 2 kPa in their arterial blood (Massabuau and Soyez 2004; Wabete et al. 2008).

In addition to blood flow rate and microcirculation adjustments, modulations of the blood oxygen affinity by changes in acid–base balance are important to adjust oxygen delivery to different organs. Numerous reviews and textbooks cover the topic of acid–base regulation in aquatic species (Truchot 1992; Heisler 1999; Perry and Gilmour, 2006; Brauner and Baker 2009). More often, the aim is a tight blood pH regulation at precise set-points that can eventually vary according, for example, to the circadian rhythm of activity (Sakakibara et al. 1987) or the feeding status (Legeay and Massabuau 1999). Low pH and high P_{CO_2} lower the blood oxygen affinity and facilitate on-site oxygen delivery at cellular level. Higher pH and lower P_{CO_2} increase blood oxygen affinity and facilitate oxygen loading in the gills. How these adjustments are performed is not simple, but to make a long story short, CO_2 excretion and acid–base regulation are linked through the reversible reactions of CO_2, H_2O, and intermediate acid–base equivalents (H^+, HCO_3^-, CO_3^{2-}). There are two routes through which blood acid–base balance can be regulated. Respiratory compensation: ventilation and blood flow rate can modulate the elimination rate of metabolic CO_2 and compensate to some extent for changes of P_{CO_2} in the ambient water. In metabolic compensation, rates of H^+, HCO_3^-, and eventually CO_3^{2-} exchange between blood, urine, and water are adjusted to regulate/compensate blood pH. Thus, adjustments of blood flow (general and/or local) and of oxygen binding respiratory protein properties (as well as concentrations) participate in the choreography of a tiered and flexible system of oxygen delivery finely tuned to the tissue-specific oxygen demand.

Conclusion 3

Setting arterial P_{O_2} in the low range over most of an animal's lifetime does not imply either low aerobic metabolism or sluggish behavior and performance. It is a large enough head pressure to satisfy a significant range of oxygen demands. The key points to ensure oxygen flux from gill to cells are the velocity of blood flow, appropriate local vascular beds corresponding to low or high tissue oxygen demand, and adjustment of blood oxygen-carrying capacity by acid–base balance regulation. Appropriate signal transduction must function in order to orchestrate the oxygen distribution on acute demand and, at the same time, avoid overoxygenation and ROS formation in metabolically less active tissues. However, we will now see that living most of the time at low P_{O_2} does not preclude the existence of transient arterial P_{O_2} rise.

LOW ARTERIAL P_{O_2} DOES NOT MEAN FIXED ARTERIAL P_{O_2}

Rapid changes of arterial P_{O_2}, corresponding to higher arterial P_{O_2} in the frequency distribution plots in Fig. 10.3 are produced by transient hyperventilatory bouts and short-term increase in blood flow rate (Angersbach and Decker 1978). These arterial P_{O_2} peaks are well described during specific actions, such as spontaneous or forced walking or feeding. In the crab *Callinectes sapidus* (Gannon and Wheatly 1995) and the crayfish *Astacus leptodactylus* (Forgue et al. 2001), arterial P_{O_2} increased from about 2 to 4–5 kPa between resting and walking. In lobster *Homarus gammarus*, Clemens et al. (1998) observed a transient increase of arterial P_{O_2} from 1–2 to 2–4 kPa two to five hours after feeding. A similar small rise in arterial blood P_{O_2} following feeding was also observed in the green crab *Carcinus maenas* (Fig. 10.9) and participates in setting the limit for postprandial protein synthesis rates

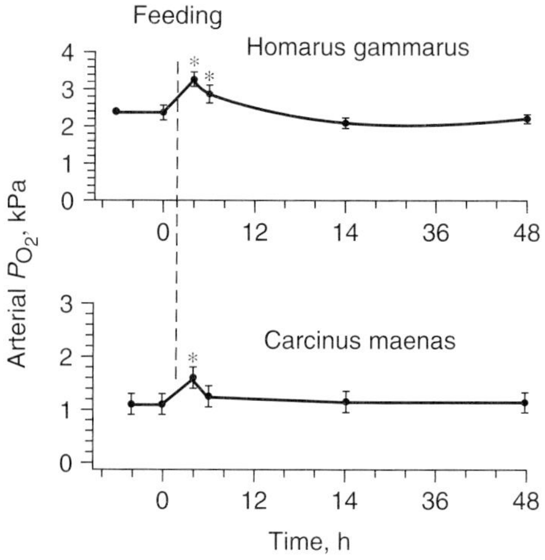

Fig. 10.9 Time course of arterial P_{O_2} changes in the lobster *Homarus gammarus* and the green crab *Carcinus maenas* under normoxic conditions at different times after food intake (vertical dashed line, one mussel). T = 15°C. (Based on data from Clemens et al. 1998 and Mente et al. 2003.)

(Mente et al. 2003). Note that these small changes in absolute values, as they are in the low range, can represent a doubling of the arterial head pressure at the tissue level.

In crustaceans a rise in arterial P_{O_2} accompanies the increased gastro-pyloric muscle activity during the first steps of the digestive process. It was shown that these P_{O_2} changes, acting specifically at the level of a particular ganglion, the stomatogastric ganglion, contribute to the multiple neuronal network modulation necessary for feeding-related behavior. Oxygen modulates the expression of oscillating neural networks and their interconnecting synaptic pathways in the central nervous system of the lobsters (Massabuau and Meyrand 1996; Clemens et al. 1998). Increasing evidence indicates that animal oxygen uptake and internal P_{O_2} regulation not only serve energy production and mitochondrial metabolism, but also modify neuronal networks and the organismal bioenergetic state in terms of growth and cell death (apoptosis). In 2001, Clemens et al. reviewed the role of tissue oxygenation, notably in controlling the operation of neuronal networks in the central nervous system during feeding and molting in lobster. They concluded that oxygen is able to act in a manner equivalent to a classic neuromodulator. Local P_{O_2} variations within the low, but physiological range of 1–6 kPa shape ongoing activity of neuronal networks and therewith the motor behaviours in which they are involved. Importantly, these P_{O_2} effects are not related to hypoxic depression, but are highly specific with respect to the network, neuron, and even the synapse targeted. Again, this highlights the necessity of well-tuned blood and tissue P_{O_2} status.

Conclusion 4

Although in resting water-breathers the most frequently observed blood P_{O_2} is in the low range between 1 and 3 kPa, bouts of higher arterial P_{O_2} are part of the normal respiratory response pattern, and allow the animals to perform specific activities associated to changes in activity and metabolic level through modulation of neuronal networks.

OXYGEN-DEPENDENT RESPIRATION AND ARTERIAL P_{O_2} AT VARIOUS WATER OXYGENATION LEVELS IN OXYREGULATING WATER-BREATHERS

We have seen above that fish and crustacean possess O_2 sensors in the gill chambers. Likewise, patellogastropods have chemosensory organs, so called osphradia, located above the head and also in the anterior pallial roof. These osphradia enable the animals to sense the physio-chemical properties of the water pumped through their shell, including the concentrations of organic and inorganic solutes, pheromones, and P_{O_2} (Kamardin et al. 2001; Lindberg and Ponder 2001). The control loop is connected to the visceral arch of their central nervous system to adjust ventilation. These chemosensors should be at the origin of the animals' ability to achieve stepwise reduction of P_{O_2} between the milieu extérieur and intérieur (seawater/mantle cavity–hemolymph–tissues). In other words, they form part of a loop that enables adjustment of ventilatory and circulatory performance patterns to changing environmental oxygen levels. Most water-breathers respond to lowered ambient P_{O_2} with an increase of ventilation frequency and/or amplitude. Only on prolonged hypoxia is blood oxygen-carrying capacity increased, either through release of more red blood cells from the spleen in fish or by increasing hemoglobin oxygen affinity, and hemoglobin concentration, to maintain constant oxygen delivery to the tissues.

Most fish and invertebrates with oxygen-binding pigments belong to the group of oxyregulators, which maintain constant respiration rates through increased ventilation against declining ambient P_{O_2} down to a critical P_{O_2} (P_{crit}). Below P_{crit} aerobic respiration rate declines as tissues and mitochondria become oxygen-limited, while anaerobic energy metabolism is initiated to support survival. The low critical P_{O_2} therefore marks the threshold of transition between oxyregulation and oxyconforming respiration, where respiration rates decrease with declining tissue/cellular P_{O_2}. Invertebrates are the classic "oxyconformers," mostly the infaunal organisms such as bivalves, polychaetes, or other primitive worms (sipunculids and priapulids), which have two critical P_{O_2} levels (high and low $P_{O_2 crit}$). The low $P_{O_2 crit}$, again, marks the onset of oxygen limitation within the tissues, as seen in oxyregulators. The high $P_{O_2 crit}$ is sometimes even in the normoxic range and marks the point where oxyconforming respiration becomes supersaturated.

Oxyconforming respiration in these animals is consistent between whole animal, tissue, and cellular level, as their mitochondria are endowed with alternative end-oxidases that have P_{50} values well above the P_{50} of cytochrome *c* oxidase (for review see Abele et al. 2007). Mitochondria of the infaunal bivalve *Arctica islandica* isolated from the animals' tissues and exposed to experimental hyperoxic P_{O_2} (up to 40 kPa) respire in an oxyconforming manner, meaning respiration increases nearly linearly with P_{O_2} (Tschischka et al. 2000). The low oxygen affinity alternative end oxidase is active when cytochrome oxidase becomes inhibited by sulfide or supersaturated with oxygen under hyperoxic conditions, and intracellular P_{O_2} rises. The alternative electron pathway branches off the classic mitochondrial electron transport chain (ETC) after respiratory complexes I and II and circumvents the ultimate ETC phosphorylation site (Abele et al. 2007). With this alternative end-oxidase, often a cytochrome *o* type oxidase, oxygen-sensitive species such as *Arctica islandica* reduce and eliminate excess oxygen in their tissues without disturbing cellular energetic balance by too high rates of phosphorylation. Oxyconforming species often feature diffusion limited tissues, low in vascularization and often, but not always, lack respiratory pigments, such as the bivalve *Arctica islandica*. As these species are mostly sensitive to overoxygenation, the low oxygen sedimentary environment is their refuge from the oxygenated world, as seen in the ostracod crustacean described earlier in this chapter or in the low-oxygen-adapted polychaete worm *Heteromastus filiformis* (Abele-Oeschger et al. 1998). Seeking environments of low ambient P_{O_2} they achieve low metabolic rates and avoid "hyper-active respiration" to prevent ROS formation. The same effect is produced in molluskan shellfish that maintain mantle cavity water P_{O_2} as low as possible and appropriate for their metabolic and lifestyle requirements by species-specific shell ventilation patterns (see Abele et al. 2010).

Next we will discuss the adaptive strategies of resting oxyregulators following the low P_{O_2} strategy when they are exposed to changes in environmental P_{O_2}. Figure 10.10 summarizes the data for the crayfish (Massabuau and Burtin 1984), the sheatfish (Forgue et al. 1989), and the Asiatic clam *Corbicula fluminea* (Tran et al. 2000). All experiments were performed during daytime and with resting animals (close to basal metabolism). The measured values correspond to physiological steady states following 24 h acclimation at experimentally fixed water P_{O_2} between 3 and 40 kPa. At this time point, the initial period of experimental disturbance and subsequent adaptation to the experimental conditions had been completed, and the animals were resting again and breathing steadily. This is based on the observation that metabolic rate was identical to its reference level at 21 kPa (normoxia). In all three examples the adaptation was primarily achieved by ventilatory adjustment: elevated ventilation frequency under hypoxia and decreased frequency under hyperoxia. No major change in blood flow rate was recorded. Another fundamental observation was that the arterial P_{O_2} was maintained low, above the anaerobic threshold ($\approx$ 1 kPa) but clearly below P_{O_2} at which oxygen toxicity might become a problem (< 5 kPa).

Again, note that blood acid–base changes are fundamental in adjusting respiratory pigment characteristics and favour oxygen-loading in the gills, as well as oxygen delivery at the cellular level. Changes in water P_{O_2} entail changes in blood partial pressure

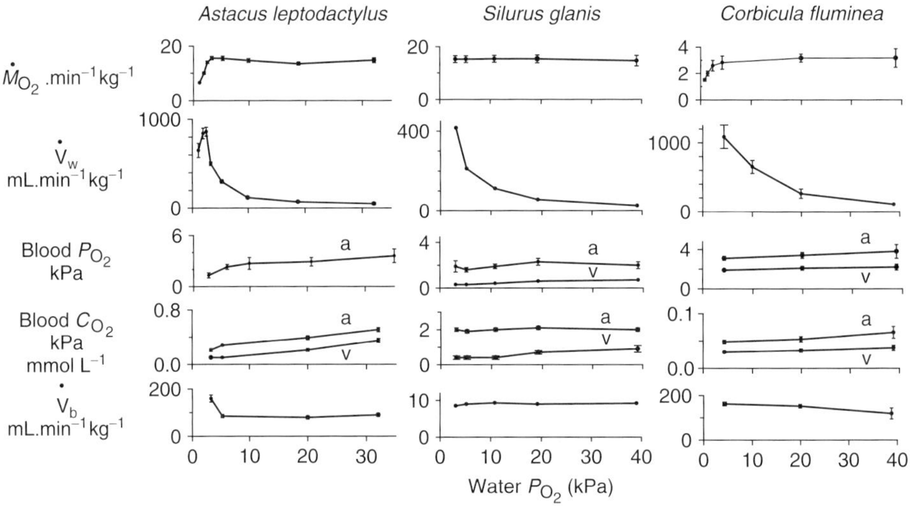

Fig. 10.10 Summary diagram of main respiratory adaptations in three physiologically different water-breathers (unfed) representative of three main groups of aquatic animals: the crustaceans (a crayfish, *Astacus leptodactylus*), the teleost fish (the sheatfish, *Silurus glanis*), and the mollusks (the Asiatic clam, *Corbicula fluminea*). In all animals O_2 consumption is maintained in a large water P_{O_2} range. The primary adaptation is a ventilatory adjustment inversely related to the water oxygenation level. The blood oxygenation status remains in a low and narrow range without a large change of blood flow rate except in *A. leptodactylus* at very low water P_{O_2}, 3 kPa. $\dot{M}_{O_2}$, O_2 consumption; $\dot{V}_w$, ventilatory flow rate; blood P_{O_2}, arterial (a) and venous (v) blood P_{O_2}; blood C_{O_2}, arterial (a) and venous (v) blood O_2 concentration; $\dot{V}_b$, circulatory blood flow rate. (Reprinted from *Respiratory Physiology* 128, 249–261. Massabuau: From a low arterial- to low tissue-oxygenation strategy, an evolutionary strategy. © 2002, with permission from Elsevier.)

of CO_2, P_{CO_2}, as a consequence of the ventilatory adjustments. In the examples shown in Fig. 10.10, hypoxic exposure caused a ventilatory-induced hypocapnic alkalosis whereas hyperoxia results in a ventilatory-induced hypercapnic acidosis. These changes are fully compensated within 1 day in *Astacus* and *Silurus* by metabolic means (for further explanations see for example the textbook of Dejours 1981), which is considered of adaptative value in terms of blood oxygen-affinity regulation. On the contrary, bivalves, which lack a respiratory pigment, do not compensate such ventilatory-induced blood pH disturbance (Heming et al. 1988; Massabuau et al. 1991). It is often speculated that absence of blood pH regulation and respiratory pigment are cause-related in bivalves. In Fig. 10.10, only in *Astacus* at water $P_{O_2} < 6$ kPa does the blood pH become more alkaline by 0.1 pH units compared to normoxia, which helps oxygen transport by inducing a Bohr effect that favours blood oxygen loading (Sakakibara et al. 1987). Finally, it must be mentioned that the very high oxygen-demanding peneid shrimps also follow the low P_{O_2} strategy as they maintain arterial P_{O_2} in the 1 kPa range for water P_{O_2} ranging from 3.5 to 40 kPa, despite their specifically high metabolic rate (Massabuau and Soyez 2004; Wabete et al. 2008).

Conclusion 5

Respiration is regulated on constant levels by ventilatory adjustments in oxyregulating water-breathers. In contrast, in evolutionary early oxyconforming species respiration rates of isolated tissues and mitochondria change as a function of P_{O_2}. In these species, whole animal respiration rates can be maintained by ventilation of the water surrounding their respiratory organs (gills or the whole body surface).

Low arterial P_{O_2} is maintained independent of water oxygenation in a large range of environmental P_{O_2}, independent of the animal group and whether oxyconformer or oxyregulator. There is no demonstration that all water-breathers are following the low blood P_{O_2} strategy, but animals functioning in this way are present in all groups of aquatic ectotherms. From a comparative point of view, all animals, including air-breathers and homeotherms, should be expected to follow the low tissue P_{O_2} strategy (Massabuau 2001, Massabuau 2003). This strategy limits excess formation of ROS and, moreover, enables highly sensitive oxygen-dependent modulations of systemic (neuronal networks) and cellular function.

GENERAL CONCLUSION

Mitochondrial ROS production becomes significant above 4–5 kPa P_{O_2} and overproduction of ROS associated with cellular overoxygenation results in oxidative stress, a mediator of cell damage and disease. Actually, ROS are well known today for playing a dual role as deleterious but also beneficial molecules (Valko et al. 2007). For example, ROS are involved in the immune response and have second messenger and even a neuromodulator-like function. Small deviations from precise arterial P_{O_2} set-points can control the operation of neuronal networks in the central nervous system. The low arterial P_{O_2} spontaneously set at various oxygen levels has also been reported to limit the rate of protein synthesis after a meal and the oxygen consumption of locomotor muscles. Thus, both the absolute blood and tissue P_{O_2} levels and the change in oxygen activity (P_{O_2}) and ROS formation have tremendous physiological effects. This highlights the necessity of a well-tuned management of oxygen exchange: the balance between oxygen supply and consumption must be perfectly controlled, free oxygen excess at cellular level must be either avoided or stay under perfect control.

This chapter has outlined the low P_{O_2} strategy (i) as part of the normal physiological repertoire in water-breathers, and (ii) as a first firewall protecting tissues against hypoxia on the one hand and from oxidative stress on the other (Massabuau 2003). We have argued that ventilation (strictly speaking: breathing) controls, and literally limits, oxygen flux into the branchial cavities, to restrict oxygen diffusion from water to blood. The fundamental idea is that oxygen deficiency as well as an oxygen excess, which produces oxidative stress, results from a mismatch between oxygen supply and consumption. Oxygen management is a key problem in aerobic organisms. The equilibrium must be perfect and the ventilatory regulation must definitively be taken as a first adaptive step protecting the tissues of water-breathers against oxidative stress. Clearly, oxidative stress is a significant factor underlying the susceptibility of any animal to a variety of health disorders. This must be taken into account when investigating aspects of respiratory control and oxidative stress in water-breathing animals.

REFERENCES

Abele, D., Philipp, E., Gonzalez, P., Puntarulo, S. (2007) Marine invertebrate mitochondria and oxidative stress. *Frontiers in Bioscience* 12, 933–946.

Abele, D., Kruppe M., Philipp, E.E.R., Brey, T. (2010) Mantle cavity water oxygen partial pressure (P_{O_2}) in marine molluscs aligns with lifestyle. Canadian Journal of Fisheries and Aquatic Science 53, 1–11.

Abele-Oeschger, D., Großpietsch, H., Pörtner, H.O. (1998) Temporal fluctuations and spatial gradients of environmental P_{O_2}, temperature, H_2O_2 and H_2S in its intertidal habitat trigger enzymatic antioxidant protection in the capitellide worm *Heteromastus filiformis*. *Marine Ecology Progress Series* 163, 179–191.

Angersbach, D., Decker, H. (1978) Oxygen transport in crayfish blood: effect of thermal acclimation, and short term fluctuations related to ventilation and cardiac performance. *Journal of Comparative Physiology B* 123, 105–112.

Bamford, O.S. (1974) Oxygen reception in the rainbow trout (*Salmo gairdneri*). *Comparative Biochemistry and Physiology A* 48, 69–76.

Barnabas, J., Schwartz, R.M., Dayhoff, M.O. (1982) Evolution of major metabolic innovations in the Precambrian. *Origins of Life* 12, 81–91.

Bradley, T.J. (2006) Discontinuous ventilation in insects: protecting tissues from O_2. *Respiratory Physiology and Neurobiology* 154, 30–36.

Brauner, C.J., Baker, D.W. (2009) Patterns of acid-base regulation in fish. *In* Glass, M.L., Wood, S.C. (eds). *Cardio-Respiratory Control in Vertebrates: Comparative and Evolutionary Aspects*. Springer-Verlag, pp. 43–63.

Brooks, S.P.J., Storey, K.B. (1997) Glycolytic controls in estivation and anoxia: a comparison of metabolic arrest in land and marine molluscs. *Comparative Biochemistry and Physiology A* 118, 1103–1114.

Buttemer, W.A., Abele, D., Costantini, D. (2010) From bivalves to birds: Oxidative stress and longevity. *Functional Ecology*, 24, 971–983.

Canfield, D.E. (1998) A new model for Proterozoic ocean chemistry. *Nature* 396, 450–453.

Clemens, S., Massabuau, J.C., Legeay, A., Meyrand, P., Simmers, J. (1998) *In Vivo* modulation of interacting central pattern generators in lobster stomatogastric ganglion: influence of feeding and partial pressure of oxygen. *Journal of Neuroscience* 18, 2788–2799.

Clemens, S., Massabuau, J.-C., Meyrand, P., Simmers, J. (2001) A new neuromodulatory-like role for oxygen in shaping the activity of neuronal networks in lobsters. *Respiratory Physiology* 128, 299–315.

Corbari, L., Carbonel P., Massabuau J.-C. (2004) How was the low tissue O_2 strategy maintained in early crustaceans? The example of the podocopid ostracods. *Journal of Experimental Biology* 207, 4415–4425.

Corbari, L., Carbonel P., Massabuau J.-C. (2005a) The early life history of tissue oxygenation in crustaceans: the strategy of the myodocopid ostracod *Cylindroleberis mariae*. *Journal of Experimental Biology* 208, 661–670.

Corbari, L., Mesmer-Dudons, N., Carbonel, P., Massabuau, J.-C. (2005b) Cytherella as a tool to reconstruct deep-sea paleo-oxygen levels: the respiratory physiology of the ostracod platycop *Cytherella* cf. *abyssorum*. *Marine Biology* 147, 1377–1386.

Davenport, J., Irvin, S. (2003) Hypoxic life of intertidal acorn barnacles. *Marine Biology* 143, 555–563.

Decker, H., Terwilliger, N. (2000) Cops and robbers: putative evolution of copper oxygen-binding proteins. *Journal of Experimental Biology* 203, 1777–1782.

Dejours, P. (1981) *Principles of Comparative Respiratory Physiology*, 2nd edn. Elsevier/North-Holland, Amsterdam.

Eclancher, B. (1972) Action des changements rapides de P_{O_2} de l'eau sur la ventilation de la truite et de la tanche. *Journal de Physiologie (Paris) A* 65, 397.

Eclancher, B. (1980) *Action des changements soudains de l'oxygénation du milieu ambiant sur les activités ventilatoire et cardiaque de quelques téléostéens d'eau douce. Arguments en faveur d'un contrôle chémoréflexe*. Thèse d'Etat. University Louis Pasteur, Strasbourg, France. pp. 264.

Forgue, J., Burtin, B., Massabuau, J.-C. (1989) Maintenance of oxygen consumption in resting teleost *Silurus glanis* at various levels of oxygenation. *Journal of Experimental Biology* 143, 305–319.

Forgue, J., Truchot, J.-P., Massabuau, J.-C. (1992a) Low arterial P_{O_2} in resting crustaceans is independent of blood O_2 affinity. *Journal of Experimental Biology* 170, 257–264.

Forgue, J., Massabuau, J.-C., Truchot, J.-P. (1992b) When are resting water-breathers lacking O_2? Arterial P_{O_2} at the anaerobic threshold in crab. *Respiratory Physiology* 88, 247–256.

Forgue, J., Legeay, A., Massabuau, J.-C (2001) Is the resting rate of oxygen consumption of locomotor muscles limited by the low blood oxygenation strategy in crustaceans? *Journal of Experimental Biology* 204, 933–940.

Foster, R.E. (1964) Diffusion of gases. *In* Fenn, W.O., Rahn, H. (eds). *Handbook of Physiology, Respiration*, Vol. 1. American Physiological Society, Washington, DC, pp. 839–872.

Frederich, M., Pörtner, H.O. (2000) Oxygen limitation of thermal tolerance defined by cardiac and ventilatory performance in the spider crab, *Maja squinado*. *American Journal of Physiology* 279, R1513–R1538.

Gannon, A.T., Wheatly, M.G. (1995) Physiological effects of a gill barnacle on host blue crabs during short-term exercise and recovery. *Marine Behaviour and Physiology* 24, 215–225.

Guppy, M., Reeves, D.C., Bishop, T., Withers, P., Buckingham, J., Brand, M. (2000) Intrinsic metabolic depression in cells isolated from the hepatopancreas of estivating snails. *FASEB Jornal* 14, 999–1004.

Heisler, N. (1999) Limiting factors for acid-base regulation in fish: branchial transfer capacity versus diffusive loss of acid-base relevant ions. *In* Egginton, S., Taylor, E.W., Raven, J. A. (eds). *Acid-Base Regulation of Acid-Base Status in Animals and Plants.* Society for Experimental Biology Seminar Series, Cambridge University Press, Cambridge, UK, pp. 125–154.

Heming, T.A., Vinogradov, G.A., Klerman, A.K., Komov, V.T. (1988) Acid-base regulation in the freshwater pearl mussel *Margaritifera margaritifera*: effects of emersion and low water pH. *Journal of Experimental Biology* 137, 501–511.

Hetz, S.K., Bradley, T.J. (2005) Insects breathe discontinuously to avoid oxygen toxicity. *Nature* 433, 516–519.

Hoogewijs D., Terwilliger N.B., Webster K.A. et al. (2007) From critters to cancer: bridging trajectories between comparative and clinical research of oxygen sensing, HIF signalling and adaptation towards hypoxia. *Integrative and Comparative Biology* 47, 552–577.

Ishii, K., Ishii, K., Massabuau, J.-C., Dejours, P. (1989) Oxygen-sensitive chemoreceptors in the branchio-cardiac veins of the crayfish *Astacus leptodactylus*. *Respiratory Physiology* 78, 73–81.

Itazawa, Y., Takeda, T. (1978) Gas exchange in the carp gills in normoxic and hypoxic conditions. *Respiratory Physiology* 35, 263–269.

Ivanovic, Z. (2009) Hypoxia or *in situ* normoxia: the stem cell paradigm. *Journal of Cellular Physiology* 219, 271–275.

Janvier, P., Desbiens, S., Willet, J.A., Arsenault, M. (2006) Lamprey-like gills in a gnathostome-related Devonian jawless vertebrate. *Nature* 440, 1183–1185.

Jonz, M.G., Nurse, C.A. (2006) Ontogenesis of oxygen chemoreception in aquatic vertebrates. *Respiratory Physiology and Neurobiology* 154, 139–152.

Kamardin, N.N., Shalanki, Y., Sh.-Rozha, K., Nosdrachev, A.D. (2001) Studies of chemoreceptor perception in mollusks. *Neuroscience and Behavioral Physiology* 31, 227–235.

Lane, N. (2005) *Power, Sex, Suicide: Mitochondria and the Meaning of Life.* Oxford University Press, Oxford.

Legeay, A., Massabuau, J.-C. (1999) Blood oxygen requirement in resting crab (*Carcinus maenas*) 24h after feeding. *Canadian Journal of Zoology* 77, 784–794.

Legeay, A. Massabuau, J.-C. (2000) The ability to feed in hypoxia follows a seasonally dependent pattern in shore crab *Carcinus maenas*. *Journal of Experimental Marine Biology and Ecology* 247, 113–129.

Lin, Q., Kim, Y., Alarcon, R.M., Yun, Z. (2008) Oxygen and cell fate decisions. *Gene Regulation and Systems Biology* 2, 43–51.

Lindberg, D.R., Ponder, W.F. (2001) The influence of classification on the evolutionary interpretation of structure a re-evaluation of the evolution of the pallial cavity of gastropod molluscs. *Organisms, Diversity and Evolution* 1, 273–299.

Massabuau, J.-C. (2001) From a low blood- to low tissue-oxygenation strategy, an evolutionary theory. *Respiratory Physiology* 128, 249–262.

Massabuau, J.-C. (2003) Primitive, and protective, our cellular oxygenation status? *Mechanisms of Ageing and Development* 124, 857–863.

Massabuau, J.-C., Burtin, B. (1984) Regulation of the oxygen consumption in the crayfish *Astacus leptodactylus*: role of the peripheral O_2 chemoreception. *Journal of Comparative Physiology B* 155, 43–49.

Massabuau, J.-C., Burtin, B. (1985) Ventilatory C_{O_2} reflex response in hypoxic crayfish *Astacus leptodactylus* acclimated to 20°C. *Journal of Comparative Physiology B* 156, 115–118.

Massabuau, J.-C., Forgue, J. (1996) A field *vs* laboratory study of blood O_2-status in normoxic crabs at different temperatures. *Canadian Journal of Zoology* 74, 423–430.

Massabuau, J.-C., Meyrand, P. (1996) Modulation of a neural network by physiological levels of oxygen in lobster stomatogastric ganglion. *Journal of Neurosciences* 16, 3950–3959.

Massabuau, J.-C., Soyez, C. (2004) La gestion du flux d'oxygène chez les animaux aquatiques. Application au cas des crevettes pénéides. *In* Goarant, C., Harache, Y., Herbland, A., Mugnier, C. (eds) *Styli 2004 – Trente ans de crevetticulture en Nouvelle-Calédonie. Actes du Colloque* 38, Edition Ifremer, pp. 66–74.

Massabuau, J.-C., B. Eclancher, Dejours, P. (1980) Ventilatory reflex response to hyperoxia in the crayfish, *Astacus pallipes*. *Journal of Comparative Physiology* 140, 193–198.

Massabuau, J.-C., Burtin, B., Wheatly, M. (1991) How is O_2-consumption maintained independent of ambient oxygen in mussel *Anodonta cygnea*. *Respiratory Physiology* 83, 103–114.

Mente, E., Legeay, A., Houlihan, D., Massabuau J.-C. (2003) Influence of oxygen partial pressures on protein synthesis in feeding crabs. *American Journal of Physiology* 284, R500–R510.

Morley, S.A., Peck, L.S., Miller, A.J., Pörtner, H.O. (2007) Hypoxia tolerance associated with activity reduction is a key adaptation for *Laternula elliptica* seasonal energetic. *Oecologia* 153, 29–36.

Papa, S., Skulachev, V.P. (1997) Reactive oxygen species, mitochondria, apoptosis and aging. *Molecular and Cellular Biochemistry* 174: 305–319.

Perry S.F., Gilmour K.M. (2006) Acid-base balance and CO_2 excretion in fish: unanswered questions and emerging models. *Respiratory Physiology and Neurobiology* 154, 199–215.

Sakakibara, Y., Burtin, B., Massabuau, J.-C. (1987) Circadian rhythm of extracellular pH in crayfish at different levels of oxygenation. *Respiratory Physiology* 69, 359–3677.

Segerer, A., Stetter, K.O., Klink, F. (1985) Two contrary modes of chemolithotrophy in the same archaebacterium. *Nature* 313, 787–789.

Shelton, G., Jones, D., Milsom, W. (1986) Control of breathing in ectothermic vertebrates. *In* Macklem, P.R., Mead, J. (eds). *Handbook of Physiology, Section 3: The Respiratory System, Control of Breathing*, Part 1, Vol. II. American Physiological Society, Bethesda, MD, pp. 857–909.

Storey, K.B., Storey, J. M (2004) Metabolic rate depression in animals: transcriptional and translational controls. *Biological Reviews* 79, 207–233.

Thomas, S., Fievet, B., Claireaux, G., Motais, R. (1988) Adaptive respiratory responses of trout to acute hypoxia. Effects of water ionic composition on blood acid–base status response and gill morphology. *Respiratory Physiology* 74, 77–90.

Toulmond, A. (1973) Tide-related changes of blood respiratory variables in the lugworm *Arenicola marina* (L.). *Respiratory Physiology* 19, 130–144.

Tran, D., Boudou, A., Massabuau, J.-C (2000) Mechanism of oxygen consumption maintenance under varying levels of oxygenation in the freshwater clam *Corbicula fluminea*. *Canadian Journal of Zoology* 78, 2027–2036.

Trevors, J.T. (2003) Early assembly of cellular life. *Progress in Biophysics and Molecular Biology* 81, 201–217.

Truchot, J.-P. (1992) Respiratory function of arthropod hemocyanin. *In* Mangum, C.P. (ed.). *Advances in Comparative and Environmental Physiology*, Vol. 13. Springer-Verlag, Berlin, pp. 377–410.

Tschischka, K., Abele, D., Pörtner, H.O. (2000) Mitochondrial oxyconformity and cold adaption in the polychaete *Nereis pelagic* and the bivalve *Arctica islandica* from the Baltic and the White Seas. *Journal of Experimental Biology* 203, 3355–3368.

Valko, M., Leibfritz, D., Moncol, J., Cronin, M.T., Mazur, M., Telser, J. (2007) Free radicals and antioxidants in normal physiological functions and human disease. *International Journal of Biochemistry and Cell Biology* 39, 44–84.

Vanderkooi, J.M., Erecinska, M., Silver, I.A. (1991) Oxygen in mammalian tissue: methods of measurement and affinities of various reactions. *American Journal of Physiology* 260, C1131–C1150.

Van Holde, K. Miller, K., Decker, H. (2001) Hemocyanins and invertebrate evolution. *Journal of Biological Chemistry* 276, 15563–15566.

Wabete, N., Chim, L., Lemaire, P., Massabuau, J.-C. (2008) Life on the edge: physiological problems in penaeid prawns *Litopenaeus stylirostris*, living on the low side of their thermopreferendum. *Marine Biology* 154, 403–412.

Weihe, E., Abele, D. (2008) Differences in the physiological response of inter- and subtidal Antarctic limpets (*Nacella concinna*) to aerial exposure. *Aquatic Biology* 4, 155–166.

Chapter 11

OXIDATIVE STRESS IN SHARKS AND RAYS

Roberto I. López-Cruz[1], *Alcir Luiz Dafre*[2], *and Danilo Wilhelm Filho*[3]

[1]Laboratorio de Investigación Bioquímica, Posgrado en Biomedicina Molecular, Escuela Nacional de Medicina y Homeopatía (ENMyH-IPN), México D.F.
[2]Departamento de Ciências Fisiológicas, Centro de Ciências Biológicas, Universidade Federal de Santa Catarina, SC, Brazil
[3]Departamento de Ecologia e Zoologia, Centro de Ciências Biológicas, Universidade Federal de Santa Catarina, SC, Brazil

Data on tissue oxygen consumption, reactive oxygen species (ROS) production, and antioxidant capacity in elasmobranchs (sharks, rays, skates, and chimaeras) are scanty. To our knowledge, there are no reports on skates and chimaeras. Physiological antioxidant adaptations of fish to ROS generation suggest that less active swimmers, such as rays, possess lower antioxidant levels compared to active swimmers, such as most shark species (Wilhelm Filho and Boveris 1993). This also applies to the comparison between elasmobranchs and marine teleosts (Wilhelm Filho and Boveris 1993; Wilhelm Filho et al. 1993). Therefore, the trade-off between metabolic rate/oxygen consumption and antioxidant defenses in elasmobranch, as well as in teleost fish enables them to cope with high ROS generation without facing oxidative stress (Wilhelm Filho and Boveris 1993; Rudneva 1997; Gorbi et al. 2004; Solé et al. 2009). Similarly, active shark species (e.g. endothermic sharks belonging to the Lamnidae family), which are exposed to ischemia/reperfusion processes associated with a high oxygen demand during an explosive burst-type swimming, display higher superoxide dismutase (SOD) and catalase (CAT) activities in their tissues than less active sharks. This seems to be correlated to higher oxygen consumption and therefore to inherent elevated ROS generation (López-Cruz et al. 2010). Intracellular reducing power depends on energy-producing systems in order to counteract the oxidizing processes of oxidative metabolism. Some abundant proteins are highly sensitive to the redox environment, among them the thiol-rich hemoglobin (Hb), present in some vertebrate groups may play an important antioxidant role (Reischl 1986; Torsoni and Ogo 2000; Dafre et al. 2007; Reischl et al. 2007).

GENERAL ASPECTS OF PHYSIOLOGICAL ANTIOXIDANT ADAPTATIONS IN ELASMOBRANCH FISH

Elasmobranch fish comprise an ancient group of specialized vertebrates that display some peculiarities compared to modern fish (teleosts). In elasmobranchs there is no bone formation except in the jaws (because

Oxidative Stress in Aquatic Ecosystems, First Edition. Edited by Doris Abele, José Pablo Vázquez-Medina, and Tania Zenteno-Savín.

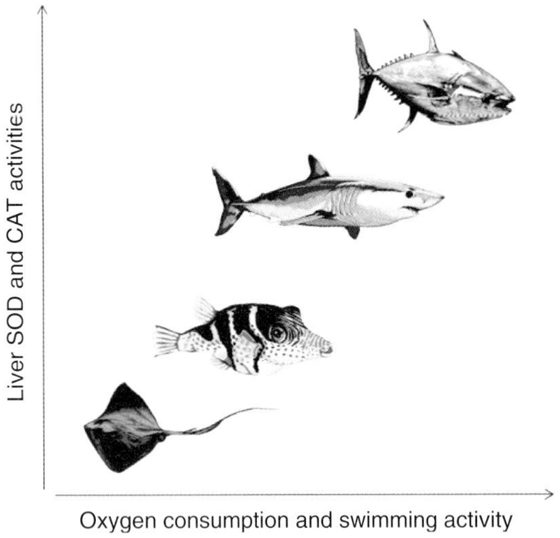

Fig. 11.1 Antioxidant *status* in fish based on the oxygen consumption rate and swimming activities (Wilhelm Filho et al. 1993; Wilhelm Filho and Boveris, 1993).

of this characteristic they are also known as cartilaginous fish), teeth are not fused to jaws, they lack swim bladder or lungs, have high urea and trimethylamine oxide concentrations in tissues, and are relative less diverse as compared with teleosts (Nelson 1994).

Comparative studies in marine fish haven shown that CAT and SOD activities in liver decrease in the following order: active teleosts, sharks, sluggish teleosts, and rays (Wilhelm Filho and Boveris 1993; Wilhelm Filho et al. 1993) (Fig. 11.1). Thus, the lower antioxidant enzyme activities generally found in sharks and rays, compared to marine teleosts suggest that elasmobranchs have an antioxidant system proportional to their oxygen consumption and lifestyle (Wilhelm Filho and Boveris 1993; Wilhelm Filho et al. 1993; Rudneva 1997; Gorbi et al. 2004; Solé et al. 2009). Likewise, the hammerhead shark *Sphyrna zygaena* showed an average ($n = 3$) liver oxygen consumption of $0.70 \pm 0.06\ \mu\text{mol min}^{-1}\ \text{g}^{-1}$, which is roughly twice the value recorded in the ray *Psammobatis scobina* ($0.33 \pm 0.09\ \mu\text{mol min}^{-1}\ \text{g}^{-1}$; $n = 4$), while the values in the marine teleost *Trichiurus lepturus* were 0.98 ± 0.07 and $2.83 \pm 0.10\ \mu\text{mol min}^{-1}\ \text{g}^{-1}$ for winter and summer respectively; ($n = 15$) (Wilhelm Filho et al. 1993; Wilhelm Filho and Boveris 1993). This correlation between antioxidant levels and oxygen consumption or metabolic rate does not necessarily apply to freshwater teleosts. Instead, more unstable and variable physicochemical characteristics of the different freshwater environments compared to generally stable marine environments seem to influence the levels of enzymatic and nonenzymatic antioxidant defenses in freshwater teleosts (Wilhelm Filho et al. 1993).

Sharks and rays seem to avoid oxidative damage in particular tissues through the ubiquitous antioxidant strategy, i.e. sustaining an efficient balance between ROS production and the main antioxidant enzymes, such as SOD, CAT, glutathione peroxidase (GPx), and glutathione reductase (GR), as well as by maintaining cellular redox balance through nonenzymatic antioxidants, such as glutathione (GSH) and vitamin E. In this regard, the hammerhead shark *S. zygaena* showed higher (39.12 ± 1.2 nmol mL^{-1}; $n = 3$) vitamin E content in plasma than in the liver (6.03 ± 0.5 nmol g^{-1}). In another hammerhead shark species, *Sphyrna lewini*, vitamin E contents in liver similar to those found in *S. zygaena* (4.07 ± 0.7 nmol g^{-1}; $n = 2$) were recorded.

Different types of antioxidant molecules seem to act synergistically to prevent tissue damage related to ROS overgeneration in elasmobranchs. Hypoxic preconditioning in the epaulette shark *Hemiscyllium ocellatum*, a species that inhabits shallow reef platforms that can become hypoxic, seems to decrease neuronal damage during hypoxic exposure and after reoxygenation (Mulvey and Renshaw 2000). Interestingly, elasmobranchs also seem to be protected against oxidative damage by their highly (physiological) urea concentrations. Wang et al. (1999) showed that isolated hearts from sharks and rats are capable of tolerating electrolysis-induced damage, but only sharks are able to face post ischemia/reperfusion-induced cardiac injury. The protection is attributable to the antioxidant properties of urea derivatives, such as hydroxyurea, dimethylurea, and thiourea.

Constitutive High GSH Levels in Elasmobranch Red Cells

Elasmobranch fish exhibit, in general, high blood GSH content and seem to tolerate high levels of metabolites derived from liver lipid peroxidation processes when compared to teleosts and other vertebrates (Wilhelm Filho and Boveris 1993;

Table 11.1 Intraerythrocytic glutathione (GSH) levels in rays. Last column: Intraerythrocytic values calculated as $A_{412}/14.1 \times 137.5$ (dilutions).

Ray species	Time (min)					GSH
	2	5	10	20	30	
Myliobatis goodie	0.005	0.009	0.014	0.019	0.021	2.88
Sympterygia acuta	0.007	0.011	0.018	0.028	0.031	4.28
Raja agassizi	0.009	0.016	0.021	0.040	0.043	5.86
Narcine brasiliensis	0.010	0.017	0.031	0.043	0.044	6.06
Sympterygia bonapartei	0.018	0.030	0.065	0.086	0.087	11.96

Wilhelm Filho et al. 2000). Thus, even considering that GSH is found in relatively high concentrations in different tissues from aerobic organisms including fungi, plants, invertebrates, and vertebrates (Wilhelm Filho et al. 2000), some rays display remarkably high intraerythrocytic levels of GSH (up to 12 mM!) (Table 11.1). To our knowledge, only two opossum species (*Didelphis albiventris* and *D. marsupialis*) from South America showed higher intraerythrocytic GSH levels than these rays (ca.30 mM!; Wilhelm Filho unpublished data). The remarkably low hematocrit values that characterize the elasmobranch taxon are responsible for the relatively high GSH content. However, the GSH : Hb ratio is still very high, surpassing the approximately equimolar ratio usually found in other vertebrates (Wilhelm Filho et al. 2000).

High Lipid Peroxidation in Sharks and Rays

Compared to other vertebrates, fish liver displays unusually high levels of lipid peroxidation (Wilhelm Filho and Boveris 1993; Wilhelm Filho et al. 1993, 2000). While mammals as a group showed thiobarbituric acid reactive substance (TBARS) contents of $0.24 \pm 0.02\,\mu mol\ g^{-1}$ ($n = 16$), teleost fish showed levels of $0.71 \pm 0.19\,\mu mol\ g^{-1}$ ($n = 22$) (Wilhelm Filho et al. 2000). Moreover, sharks and rays display even higher levels of peroxidation ($3.38 \pm 1.07\,\mu mol\ g^{-1}$; $n = 6$), almost five times higher than the values of teleost fish and ca. 14 times higher compared to those of mammals (Wilhelm Filho and Boveris 1993). Elasmobranchs possess high liver contents of squalene, which is a low-density polyunsaturated fatty acid (PUFA) that supports their neutral buoyancy. This lipid fuel might be also involved in the higher levels of lipid peroxidation that characterize this taxon (Wilhelm Filho and Boveris 1993).

Passive H_2O_2 Diffusion in Fish Gills and Elasmobranch SH-Rich Hemoglobin

Cytosolic H_2O_2 concentrations of 10^{-7} to 10^{-6} M are usually found in mammalian tissues that are relatively devoid of peroxisomes. H_2O_2 diffuses to the blood, which is loaded with erythrocytic CAT, which catabolizes the bulk of this ROS (Giulivi et al. 1993). However, several teleost and elasmobranch fish exhibit diminshed CAT activity or are even devoid of erythrocytic CAT, i.e., acatalassemia (Smith 1976; Wilhelm Filho et al. 1993; Rudneva 1997; Wilhelm Filho et al. 2000). Teleost fish usually release the excess H_2O_2 through gill diffusion (Wilhelm Filho et al. 1994; Wilhelm Filho et al. 2000). Considering the high diffusion capacity of H_2O_2 through biological membranes and the acatalassemia observed (Rudneva 1997), it is possible that elasmobranch fish also release excess H_2O_2 through the gills. Another antioxidant adaptation present in elasmobranchs might be the use of SH-rich Hb that characterizes sharks and rays (Reischl 1986).

ADAPTATIONS TO BURST-SWIMMING IN SHARKS

High Swimming Performance and Antioxidant Skeletal Muscle Adaptations in Sharks

Lamnid sharks display body temperatures above the surrounding water that enable them to sustain a high locomotory performance and, therefore, a fast swimming capacity (Bernal et al. 2001; Compagno 2001; Wilhelm Filho 2007). These sharks have a counter-current heat exchange mechanism that allows them to maintain high temperatures in eyes, brain, muscle, and viscera, and are, thus, considered to be regionally endothermic (Carey and Teal 1969; Block and Carey 1985; Bernal et al. 2001, Compagno 2001). A sharp increase in physical activity, such as burst-type swimming in sharks like *Isurus oxyrinchus*, would increase oxygen consumption and consequently ROS production (Boveris 1977; Wilhelm Filho et al. 2000). Skeletal muscle of *I. oxyrinchus* has higher GPx and glutathione S-transferase (GST) activities and lower TBARS levels as compared to the ectothermic *Carcharhinus falciformis* and *S. zygaena*. These findings suggest that an enhanced antioxidant capacity allows *I. oxyrinchus* to avoid oxidative damage associated to increased $O_2^{\bullet-}$ production resulting from burst-type swimming; while enzymatic antioxidant activities in ectothermic sharks are modulated by ambient temperature and metabolic rate (López-Cruz et al. 2010). *Sphyrna zygaena* is considered an active swimmer (Compagno 1999), although less active than other shark species (see Kohler et al. 1998). Relatively lower GPx and GST activities found in *S. zygaena* in comparison with other shark species reflect a lower capacity for burst-type swimming (López-Cruz et al. 2010).

Metabolic Adaptations to Different Swimming Activities or Lifestyles in Sharks

Evolutionary convergences between lamnid sharks and tunas include morphological and biochemical adaptations such as large gill-surface area, high myoglobin (Mb) concentrations and higher capillary density in red muscle (Bernal et al. 2001, 2003). *Isurus oxyrinchus* white muscle displays higher activities of some metabolism indicator enzymes, i.e. citrate synthase (CS; aerobic metabolism) and lactate dehydrogenase (LDH; anaerobic metabolism) in comparison with ectothermic sharks, rays, and some teleost fish (Dickson et al. 1993). Similar observations have been reported for white muscle of *Squalus acanthias* and of the deep-sea shark *Centroscyllium fabricii* (Treberg et al. 2003).

Mitochondrial density in red muscle of *I. oxyrinchus* is significantly higher than that of *Prionace glauca*, but not higher than that of *Triakis semifasciata*, while mitochondrial density in liver is similar in *I.oxyrinchus* and *P. glauca*, but is lower than in *T. semifasciata* (Duong et al. 2006). Apparently, there is not a general pattern between swimming activities and/or endothermy and mitochondrial densities. However, mitochondrial respiration rates (states 3 and 4) in red muscle from *I. oxyrinchus* are higher than those from ectothermic sharks, and are comparable to those from active teleosts, whereas the respiration rates in liver are similar in *I. oxrinchus* and *P. glauca*, but are lower in *T. semifasciata* (Duong et al. 2006).

This evidence supports the hypothesis of Wilhelm Filho and Boveris (1993) and confirms that not only antioxidant enzyme activities, but also metabolic enzyme activities and biochemical adaptations synergistically interact to protect against oxidative stress. Therefore, the most active species, such as *I. oxyrinchus*, display higher protection than less active species, such as *S. zygaena*.

Phylogenetic position of elasmobranchs as a group is recognized and they are considered as living fossils. Nucleotide substitutions in mitochondrial and nuclear shark genes seem to be one order of magnitude slower than in mammals (Martin 1999). Nam et al. (2006) showed that Cu,Zn-SOD from sharks (*Scyliorhinus torazame*) has considerable amino acid identities with SODs from teleosts (57–72%) and other advanced vertebrates (53–65%). As a consequence, this antioxidant enzyme from sharks possesses intermediate characteristics between teleosts and more advanced vertebrate SODs. For example, a valine residue in position 20 is shared only between sharks and teleosts, while other advanced vertebrates commonly have isoleucine in this position; valine 121 is shared by sharks and advanced vertebrates. Further research on this subject will provide information to better understand evolution of antioxidant defenses at molecular level in teleost and elasmobranch fish.

CELLULAR REDOX (GSH, PROTEIN THIOLS) ADAPTATIONS IN SHARKS

Cellular Redox Environment

The reductive intracellular compartment contrasts to the more oxidizing extracellular environment. The intracellular reducing power depends on energy-producing systems that counteract the oxidizing processes of a normal cell metabolism (Jensen et al. 2009). Biomolecules such as lipids, proteins and low molecular weight antioxidants are continuously oxidized. Protein thiols and the low molecular weight GSH are the main redox active elements. Due to counterbalancing oxidizing and reducing forces, cells maintain a certain ratio between glutathione disulfide (GSSG) and GSH, which is considered the main redox buffer (Gilbert 1990). Under prooxidant conditions the GSSG/GSH ratio increases and becomes an important marker of oxidative stress (Jensen 2009; Jensen et al. 2009; Novitch and Butler 2009; Paulsen and Carroll 2009). Thiol/disulfide exchange is a slow process, contrasting to the fast radical-mediated thiol oxidation. Regardless of the mechanism, GSH and protein thiols become oxidized under oxidative stress. Fast-reacting thiols in some proteins are the first elements to be oxidized and are associated to a physiological role (Nulton-Persson et al. 2003; Colombo et al. 2009; Paulsen and Carroll 2009).

Hemoglobin Function and Redox Metabolism

Certain abundant proteins are very sensitive to the redox environment, for instance actin (Chai et al. 1994a), carbonic anhydrase (Mallis et al. 2002; Zimmerman et al. 2004) and thiol-rich Hb present in some vertebrate groups (Reischl 1986; Torsoni and Ogo 2000; Dafre et al. 2007; Reischl et al. 2007). The large amount of reactive thiols in proteins can be viewed as a "sacrifice" power. Preferential oxidation of these thiols can alleviate the oxidative impact on other cell proteins extending the cellular function for a longer period. Reischl et al. (2007) identified groups of vertebrates that have a large number of thiols in their Hb, which are considered fast-reacting due to their position externally in the Hb molecule. Hb from elasmobranchs, amphibians, crocodylians and fresh water turtles have high number of surface thiols, indicating that they may be very reactive. Thiols in Hb molecules from aves and lepidosauria are partially exposed; thus, have lower reactivity, while Hb from most mammals and teleosts do not have reactive thiols. There is an average of 10 cysteinyl residues per tetrameric Hb position in the thiol rich-Hb group of elasmobranchs. Electropherograms of elasmobranch Hb are blurred, an indication of polymer formation during sample handling. This provides evidence for the presence of reactive thiols in elasmobranch Hb. In addition, around 50% of the alpha and beta chains so far analyzed present external or partially external cysteinyl residues, an indication of reactivity. Scalloped hammerhead shark (*S. lewini*), subjected to stress of capture and transportation, presented several electrophoretic components which were shown to be due to S-thiolation of its single Hb, and elevated levels of GSH in their erythrocytes (Dafre and Reischl 1990; Dafre and Reischl 1997). The three most abundant proteins in erythrocytes, Hb, carbonic anhydrase and peroxiredoxin, are redox active (Neumann et al. 2003). Combining highly reactive thiols in the Hb along with elevated non-protein thiols (GSH) can provide increased protection against oxidative stress to elasmobranch erythrocytes (Reischl et al. 2007; Jensen 2009).

Although it has not been reported for non-mammalian species yet, the ability of Hb-thiols to intercept thiol reagents could preserve GSH, as demonstrated *in vivo* and *in vitro* for rodent species (Giannerini et al. 2001; Colombo et al. 2009). Highly reactive thiols in Hb behave as an active barrier for thiol-reactive compounds, including xenobiotics and metal ions, possibly preventing them to gain access to other molecules and tissues (Reischl et al. 2007).

The $NO^{\bullet}$ transport function of Hb has been extensively discussed in the literature, and seems to be an important mechanism for regulation of $NO^{\bullet}$ delivery to tissues (Jia et al. 1996; Stamler et al. 2008; Allen et al. 2009). The existence of NOx molecules in the plasma of early and recent fish species, including the little skate *Raja erinacea* and the spiny dogfish *S. acanthias*, has been recently demonstrated (Williams et al. 2008). A vasoregulatory response to $NO^{\bullet}$ has been observed in the spiny dogfish under hypoxia, but not at normoxia (Swenson et al. 2005). Hypoxic water is a natural habitat for some shark species (Jorgensen et al. 2009), in which $NO^{\bullet}$ metabolism would be operating. The participation of Hb, either at the heme or at the thiol sites has not been addressed, and needs special attention to establish if elasmobranch Hb participates in $NO^{\bullet}$ metabolism.

Protein Thiol Oxidation as a Protective Mechanism Against Irreversible Oxidation

Protein oxidation, leading to protein-glutathione mixed disulfides (PSSG) formation, has been considered an important mechanism to avoid irreversible oxidation of enzymes, also known as S-thiolation (Grimm et al. 1985; Coan et al. 1992; Dalle-Donne et al. 2008). Under oxidative stress, formation of PSSG usually leads to enzymatic inactivation (Thomas et al. 1995; Mallis et al. 2002; Nulton-Persson et al. 2003; Jensen et al. 2009; Paulsen and Carroll 2009). Upon restoration of the redox status, mixed disulfides can be reduced back to their thiol forms in a process called dethiolation, recovering enzymatic function. For instance, rat carbonic anhydrase is irreversibly inactivated by oxidizing species in the absence of GSH, while in the presence of GSH the activity is preserved after reduction of the carbonic anhydrase-glutathione mixed disulfide (Mallis et al. 2002; Zimmerman et al. 2004). A similar phenomenon has been demonstrated for other proteins (Chai et al. 1994b; Dafre et al. 2007; Reischl et al. 2007). The presence of GSH bound to tiger shark (*Galeocerdo cuvieri*) carbonic anhydrase has been shown (Bergenhem et al. 1986), indicating that S-thiolation of this protein can be an important physiological mechanism in non-mammalian vertebrates.

Redox Regulation

An on/off switch from reduced/oxidized forms can result in the control of certain signaling pathways, termed redox regulation (Neumann et al. 2003; Rhee 2006; Novitch and Butler 2009). Trx, Grx and Prx are the main players in redox metabolism (Rhee 2006; Ahsan et al. 2009). Grx and Trx act as thiol/disulfide isomerases, reducing intra- or inter-molecular protein disulfides. Prx is a thiol peroxidase displaying an important catalytic activity towards H_2O_2 (Neumann et al. 2003), with rate kinetics similar to GPx (Peskin et al. 2007; Hall et al. 2009), one of the most efficient peroxidases (Tosatto et al. 2008). Under unstressed metabolic conditions Prx, along with other antioxidant defenses, scavenges H_2O_2, maintaining redox homeostasis. What is odd in Prx function is its inactivation upon oxidative stress, reinforcing the oxidative influence. During the catalytic cycle, a sulfenic acid is formed in the catalytic cystein that under excess H_2O_2 formation undergoes to sulfinic and sulfonic forms inactivating Prx catalytic activity. This Prx inhibition is taken as seminal process in the redox regulation of a series of cell function such as neuronal differentiation, cell cycle and apoptosis. Decreased catalytic activity of Prx leads to a series of oxidative modifications in proteins controlling transcription factors related to a response to the prevailing oxidative stress (Neumann et al. 2003; Rhee 2006; Ahsan et al. 2009; Hall et al. 2009).

Redox regulation seems to be a general regulatory phenomenon for a number of cellular functions that needs a close attention from the scientific community studying non-mammalian vertebrates. The lack of information on elasmobranch proteins and genes related to oxidative stress and redox regulation is evident in the literature (http://www.ncbi.nlm.nih.gov). A few publications on SOD from *P. glauca* and *S. torazame*, HIF−1α from *H. ocellatum* and *Mustelus canis*, and metallothionein from *S. torazame* are available. The limited data regarding redox regulation at the molecular level in elasmobranch species available precludes further inference.

REFERENCES

Ahsan, M.K., Lekli, I., Ray, D., Yodoi, J., Das, D.K. (2009) Redox regulation of cell survival by the thioredoxin superfamily: an implication of redox gene therapy in the heart. *Antioxidants and Redox Signaling* 11, 2741–2758.

Allen, B.W., Stamler, J.S., Piantadosi, C.A. (2009) Hemoglobin, nitric oxide and molecular mechanisms of hypoxic vasodilation. *Trends in Molecular Medicine* 15, 452–460.

Bergenhem, N., Carlsson, U., Strid, L. (1986) The existence of glutathione and cysteine disulfide-linked to erythrocyte carbonic anhydrase from tiger shark. *Biochimica et Biophysica Acta* 871, 55–60.

Bernal, D., Dickson, K.A., Shadwick, R.E., Graham, J.B. (2001) Analysis of the evolutionary convergence for high performance swimming in lamnid sharks and tunas. *Comparative Biochemistry and Physiology A* 129, 695–726.

Bernal, D., Sepulveda, C., Mathieu-Costello, O., Graham, J.B. (2003) Comparative studies of high performance swimming in sharks I. Red muscle morphometrics, vascularization and ultrastructure. *The Journal of Experimental Biology* 206, 2831–2843.

Block, B.A., Carey, F.G. (1985). Warm brain and eye temperatures in sharks. *Journal of Comparative Physiology B* 15, 229–236.

Boveris, A. (1977) Mitochondrial production of superoxide radical and hydrogen peroxide. *Advances in Experimental Biology and Medicine* 78, 67–82.

Carey, F.G., Teal, J.M. (1969) Mako and porbeagle: warm-bodied sharks. *Comparative Biochemistry and Physiology* 28, 199–204.

Chai, Y.C., Ashraf, S.S., Rokutan, K., Johnston, R.B., Thomas, J.A. (1994a) S-thiolation of individual human neutrophil proteins including actin by stimulation of the respiratory burst: evidence against a role for glutathione disulfide. *Archives of Biochemistry and Biophysics* 310, 273–281.

Chai, Y.C., Hendrich, S., Thomas, J.A. (1994b) Protein S-thiolation in hepatocytes stimulated by t-butyl hydroperoxide, menadione, and neutrophils. *Archives of Biochemistry and Biophysics* 310, 264–272.

Coan, C., Ji, J.Y., Hideg, K., Mehlhorn, R.J. (1992) Protein sulfhydryls are protected from irreversible oxidation by conversion to mixed disulfides. *Archives of Biochemistry and Biophysics* 295, 369–378.

Colombo, G., Dalle-Donne, I., Giustarini, D. et al. (2009) Cellular redox potential and hemoglobin S-glutathionylation in human and rat erythrocytes: A comparative study. *Blood Cells, Molecules, and Diseases* 44, 133–139.

Compagno, L.J.V. (1999) Systematics and body form. In: Hamlett, W.C. (Ed.), *Sharks, Skates and Rays: The Biology of Elasmobranch Fishes*. John Hopkins University Press, Baltimore, pp. 1–42.

Compagno, L.J.V. (2001) *Sharks of the world. An annotated and illustrated catalogue of shark species known to date. Volume 2: bull-head, mackerel and carpet sharks (Heterodontiformes, Lamniformes and Orectolobiformes)*. FAO, Rome.

Dafre, A.L., Reischl, E. (1990) High hemoglobin mixed disulfide content in hemolysates from stressed shark. *Comparative Biochemistry and Physiology B* 96, 215–219.

Dafre, A.L., Reischl, E. (1997) Asymmetric hemoglobins, their thiol content, and blood glutathione of the scalloped hammerhead shark, *Sphyrna lewini*. *Comparative Biochemistry and Physiology B* 116, 323–331.

Dafre, A.L., Brandao, T.A.S., Reischl, E. (2007) Involvement of vertebrate hemoglobin in antioxidant protection: chicken blood as a model. *Canadian Journal of Zoology* 85, 404–412.

Dalle-Donne, I., Milzani, A., Gagliano, N., Colombo, R., Giustarini, D., Rossi, R. (2008) Molecular mechanisms and potential clinical significance of S-glutathionylation. *Antioxidants and Redox Signaling* 10, 445–474.

Dickson, K.A., Gregorio, M.O., Gruber, S.J., Loefler, K.L., Tran, M., Terrell, C. (1993) Biochemical indices of aerobic and anaerobic capacity in muscle tissues of California elasmobranch fishes differing in typical activity level. *Marine Biology* 117, 185–193.

Duong, C.A., Sepulveda, C.A., Graham, J.B., Dickson, K.A. (2006) Mitochondrial proton leak rates in the slow, oxidative myotomal muscle and liver of the endothermic shortfin mako shark (*Isurus oxyrinchus*) and the ectothermic blue shark (*Prionace glauca*) and leopard shark (*Triakis semifasciata*). *The Journal of Experimental Biology* 209, 2678–2685.

Giannerini, F., Giustarini, D., Lusini, L., Rossi, R., Di Simplicio, P. (2001) Responses of thiols to an oxidant challenge: differences between blood and tissues in the rat. *Chemico-Biological Interactions* 134, 73–85.

Gilbert, H.F. (1990) Molecular and cellular aspects of thiol-disulfide exchange. *Advances in Enzymology and Related Areas of Molecular Biology* 63, 69–172.

Giulivi, C., Hochstein, P., Davies, K.J.A. (1993) Hydrogen peroxide production by red blood cells. *Free Radical Biology and Medicine* 16, 123–129.

Gorbi, S., Pellegrini, D., Tedesco, S., Regoli, F. (2004) Antioxidant efficiency and detoxification enzymes in spotted dogfish *Scyliorhinus canicula*. *Marine Environmental Research* 58, 293–297.

Grimm, L.M., Collison, M.W., Fisher, R.A., Thomas, J.A. (1985) Protein mixed-disulfides in cardiac cells. S-thiolation of soluble proteins in response to diamide. *Biochimica et Biophysica Acta* 844, 50–54.

Hall, A., Karplus, P.A., Poole, L.B. (2009) Typical 2-Cys peroxiredoxins–structures, mechanisms and functions. *The FEBS Journal* 276, 2469–2477.

Jensen, F.B. (2009) The role of nitrite in nitric oxide homeostasis: a comparative perspective. *Biochimica et Biophysica Acta* 1787, 841–848.

Jensen, K.S., Hansen, R.E., Winther, J.R. (2009) Kinetic and thermodynamic aspects of cellular thiol-disulfide redox regulation. *Antioxidants and Redox Signaling* 11, 1047–1058.

Jia, L., Bonaventura, C., Bonaventura, J., Stamler, J.S. (1996) S-nitrosohaemoglobin: a dynamic activity of blood involved in vascular control. *Nature* 380, 221–226.

Jorgensen, S.J., Klimley, A.P., Muhlia-Melo, A.F. (2009) Scalloped hammerhead shark *Sphyrna lewini*, utilizes deep-water, hypoxic zone in the Gulf of California. *Journal of Fish Biology* 74, 1682–1687.

Kohler, N.E., Casey, J.G., Turner, P.A. (1998) NMFS cooperative shark tagging program, 1962-1993: an atlas of shark tag and recapture data. *Marine Fisheries Review* 60, 1–87.

López-Cruz, R.I., Zenteno-Savín, T., Galván-Magaña, F. (2010). Superoxide production, oxidative damage and enzymatic antioxidant defenses in shark skeletal muscle. *Comparative Biochemistry and Physiology A* 56, 50–56.

Mallis, R.J., Hamann, M.J., Zhao, W., Zhang, T., Hendrich, S., Thomas, J.A. (2002) Irreversible thiol oxidation in carbonic anhydrase III: protection by S-glutathiolation and detection in aging rats. *Biological Chemistry* 383, 649–662.

Martin, A.P. (1999) Substitution rates of organelle and nuclear genes in sharks: implicating metabolic rate (again). *Molecular Biology and Evolution* 16, 996–1002.

Mulvey, J.M., Renshaw, G.M.C. (2000) Neuronal oxidative hypometabolism in the brainstem of the epaulette shark (*Hemiscyllium ocellatum*) in response to hypoxic pre-conditioning. *Neuroscience Letters* 290 (1), 1–4.

Nam, Y.K., Cho, Y.S., Kim, K.Y., et al. (2006) Characterization of copper, zinc superoxide dismutase from a cartilaginous shark species, *Scyliorhinus torazame* (Carcharhiniformes). *Fish Physiology and Biochemistry* 32, 305–315.

Nelson, J.S. (1994) *Fishes of the world*. 3rd edition. John Wiley and Sons. New York.

Neumann, C.A., Krause, D.S., Carman, C.V., et al. (2003). Essential role for the peroxiredoxin Prdx1 in erythrocyte antioxidant defence and tumour suppression. *Nature* 424, 561–565.

Novitch, B.G., Butler, S.J. (2009). Reducing the mystery of neuronal differentiation. *Cell* 138, 1062–1064.

Nulton-Persson, A.C., Starke, D.W., Mieyal, J.J., Szweda, L.I. (2003) Reversible inactivation of alpha-ketoglutarate dehydrogenase in response to alterations in the mitochondrial glutathione status. *Biochemistry* 42, 4235–4242.

Paulsen, C.E., Carroll, K.S. (2009) Orchestrating redox signaling networks through regulatory cysteine switches. *ACS Chemical Biology* 5, 47–62.

Peskin, A.V., Low, F.M., Paton, L.N., Maghzal, G.J., Hampton, M.B., Winterbourn, C.C. (2007) The high reactivity of peroxiredoxin 2 with H(2)O(2) is not reflected in its reaction with other oxidants and thiol reagents. *The Journal of Biological Chemistry* 282, 11885–11892.

Reischl, E. (1986) High sulfhydryl content in turtle erythrocytes: is there a relation with resistance to hypoxia? *Comparative Biochemistry and Physiology B* 85, 723–726.

Reischl, E., Dafre, A.L., Franco, J.L., Wilhelm Filho, D. (2007) Distribution, adaptation and physiological meaning of thiols from vertebrate hemoglobins. *Comparative Biochemistry and Physiology C* 146, 22–53.

Rhee, S.G. (2006) Cell signaling. H2O2, a necessary evil for cell signaling. *Science* 312, 1882–1883.

Rudneva, I.I. (1997) Blood antioxidant system of Black Sea elasmobranch and teleosts. *Comparative Biochemistry and Physiology C* 118, 255–260.

Smith, A.C. (1976) Catalase in fish red blood cells. *Comparative Biochemistry and Physiology B* 54, 331–332.

Solé, M., Rodríguez, S., Papiol, V., Maynou, F., Cartes, J.E. (2009) Xenobiotic metabolism markers in marine fish with different trophic strategies and their relationship to ecological variables. *Comparative Biochemistry and Physiology C* 149, 83–89.

Stamler, J.S., Singel, D.J., Piantadosi, C.A. (2008) SNO-hemoglobin and hypoxic vasodilation. *Nature Medicine* 14, 1008–1009.

Swenson, K.E., Eveland, R.L., Gladwin, M.T. and Swenson, E.R. (2005) Nitric oxide (NO) in normal and hypoxic vascular regulation of the spiny dogfish, *Squalus acanthias*. *Journal of Experimental Zoology A* 303, 154–60.

Thomas, J.A., Poland, B., Honzatko, R. (1995) Protein sulfhydryls and their role in the antioxidant function of protein S-thiolation. *Archives of Biochemistry and Biophysics* 319, 1–9.

Torsoni, M.A., Ogo, S.H. (2000) Hemoglobin-sulfhydryls from tortoise (*Geochelone carbonaria*) can reduce oxidative damage induced by organic hydroperoxide in erythrocyte membrane. *Comparative Biochemistry and Physiology i* 126, 571–577.

Tosatto, S.C., Bosello, V., Fogolari, F., et al. (2008) The catalytic site of glutathione peroxidases. *Antioxidants and Redox Signaling* 10, 1515–1526.

Treberg, J.R., Martin, R.A., Driedzic, W.R. (2003) Muscle enzyme activities in a deep-sea squaloid shark, *Centroscyllium fabricii*, compared with its shallow-living relative, *Squalus acanthias*. *Journal of Experimental Zoology A* 300, 133–139.

Wang, X., Lingyun, W., Aouffen, M., Mateescu, M.A., Nadeau, R., Wang, R. (1999) Novel cardiac protective effects of urea: From shark to rat. *British Journal of Pharmacology* 128, 1477–1484.

Williams, D.A., Flood, M.H., Lewis, D.A., Miller, V.M., Krause, W.J. (2008) Plasma levels of nitrite and nitrate in early and recent classes of fish. *Comparative Medicine* 58, 431–439.

Wilhelm Filho, D. (2007) Reactive oxygen species, antioxidants and fish mitochondria. *Frontiers in Bioscience* 12, 1229–1237.

Wilhelm Filho, D., Boveris, A. (1993) Antioxidant defences in marine fish-II. Elasmobranchs. *Comparative Biochemistry and Physiology C* 106, 415–418.

Wilhelm Filho, D., Giulivi, C., Boveris, A. (1993) Antioxidant defences in marine fish-I. Teleosts. *Comparative Biochemistry and Physiology C* 65, 409–413.

Wilhelm Filho, D., Gónzalez-Flecha, B., Boveris A. (1994) Gill diffusion as a physiological mechanism for hydrogen peroxide elimination by fish. *Brazilian Journal of Medical and Biological Research* 27, 2879–2882.

Wilhelm Filho, D., Torres, M.A., Marcon, J.L., Fraga, C.G., Boveris, A. (2000) Comparative antioxidant defenses in vertebrates - Emphasis on fish and mammals. *Trends in Comparative Biochemistry and Physiology* 7, 33–45.

Zimmerman, U.J., Wang, P., Zhang, X., Bogdanovich, S., Forster, R. (2004) Anti-oxidative response of carbonic anhydrase III in skeletal muscle. *IUBMB Life* 56, 343–347.

Chapter 12

OXYGEN SENSING: THE ROLE OF REACTIVE OXYGEN SPECIES

Mikko Nikinmaa[1], *Max Gassmann*[2], *and Anna Bogdanova*[2]

[1]Department of Biology, University of Turku, Turku, Finland
[2]Institute of Veterinary Physiology, Vetsuisse Faculty and Zurich Center for Integrative Human Physiology (ZIHP), University of Zurich, Zurich, Switzerland

In treating oxygen sensing we consider here all the molecules that initiate apparently oxygen-dependent responses (although in the strictest sense the phenomenon would be restricted to cases where molecular oxygen is the sensed species). Oxygen sensing with this definition has been reviewed, for example, by Ward (2008). We focus on examining the roles of reactive oxygen species (ROS) and reactive nitrogen species (RNS). In addition to their commonly accepted toxic effects, these molecules are important in normal physiological signaling.

Because the oxygen capacity of water (concentration change per unit change in the partial pressure) is much (about 30 times) smaller than that of air, hypoxic conditions commonly occur (for a review, see, e.g., Nikinmaa and Rees 2005). In addition to hypoxia, eutrophic waters are characterized by supra-atmospheric oxygen tensions (hyperoxia) during active photosynthesis of green plants. Owing to this wide variation in the environmental oxygen levels of water, it is no surprise that organisms respond to variations in environmental oxygen tension in multiple ways. Also, cells are exposed to different oxygen levels depending on their body location (Nikinmaa 2002). There is no such thing as a single hypoxic oxygen level in the body. Rather, what is hypoxia (i.e., inadequate oxygen level to support maximal metabolism) to a myocardial cell can be adequate oxygenation to a chondrocyte. The function of the cells studied differs at different oxygen tensions, as exemplified by results of a cell line derived from mouse macrophages: iNOS is activated differently by a change in oxygen tension at low (5 kPa) and at high (20 kPa) oxygen level (Otto and Baumgardner 2001). Thus, whenever oxygen sensing is discussed, one needs to consider the cell type and the oxygen tension experienced. Furthermore, one has to consider differences between species, as different species have markedly different oxygen requirements.

Generally, hypoxic conditions are reducing and hyperoxic oxidizing, which means that changes in the redox state of the cell may take place simultaneously with changes in oxygen tension. Consequently, redox sensing may be associated with apparently oxygen-dependent responses. In addition, any apparently oxygen-sensitive system may be modulated by temperature, pH, salinity, the presence of inflammation,

Oxidative Stress in Aquatic Ecosystems, First Edition. Edited by Doris Abele, José Pablo Vázquez-Medina, and Tania Zenteno-Savín.

Table 12.1 A list of molecules that may be sensed when apparently oxygen-dependent responses occur.

Molecular oxygen	O_2
ROS (reactive oxygen species)	H_2O_2
	HO
	$O_2^{\bullet-}$
Nitric oxide	$NO^{\bullet}$
Carbon monoxide	CO
Hydrogen sulphide	H_2S
Adenosine and its phosphates	Adenosine
	AMP
	ADP
	ATP

etc. One must also bear in mind that the distinction between oxygen-dependent and general stress responses remains unclear at present, and thus many of the responses to anoxic stress may have little to do with oxygen (Wenger and Gassmann 1996).

THE SENSED MOLECULES

Table 12.1 gives the major molecules that are sensed when oxygen-dependent responses take place. Figure 12.1 gives a representation of the pathways linking the different gaseous messengers to oxygen levels, and the temporary and irreversible protein modifications associated with changes in oxygen level. In the following, the sensed molecules are described in detail.

Molecular Oxygen

Molecular oxygen serves as a ligand of heme-containing molecules and prolyl/asparaginyl hydroxylases (Lahiri et al. 2006). Many of these oxygen-binding molecules also reduce oxygen and transform it to various forms of ROS. These molecules may initiate the apparently oxygen-sensitive signaling pathways. ROS may also influence sensing of molecular oxygen by affecting cellular redox state.

Reactive Oxygen and Nitrogen Species

In many apparently oxygen-sensitive systems, ROS are the molecules sensed by the downstream targets. Often the production of ROS is directly proportional to the amount of oxygen present, but in hypoxia a lack of terminal electron acceptor in the electron transport chain of mitochondria may lead to the generation of ROS. In some cell types hypoxia decreases ROS production, and in others increases. This indicates either that the oxygen affinities of mitochondrial proteins involved in ROS generation show marked cell-type-specific differences or that extramitochondrial ROS production can also be increased in hypoxia. ROS and RNS can couple redox- and oxygen-dependent signaling. Cysteine residues of proteins are the primary targets of at least H_2O_2 and $HO^{\bullet}$ (Michiels et al. 2002).

Reactive Oxygen Species

$O_2^{\bullet-}$ is produced by nicotinamide adenine dinucleotide phosphate hydrogen (NADPH) oxidases and nitric oxide synthases (NOS). It is also a subproduct of the electron transport chain in the mitochondria (Gonzalez et al. 2007). Most of the available data suggest that physiologically relevant hypoxic conditions are not severe enough to cause increased formation of $O_2^{\bullet-}$ in mitochondria, although suggestions about increased ROS production in hypoxic mitochondria have also been made (Guzy and Schumacker 2006). One possible reason to the apparently conflicting conclusions is that different cells have different oxygen dependencies of mitochondrial function. The $O_2^{\bullet-}$ may couple ROS- and nitric oxide ($NO^{\bullet}$) dependent signaling, since it readily interacts with $NO^{\bullet}$.

The H_2O_2 levels depend on its generation and conversion to molecular oxygen and water (see Fig. 12.1). The reported H_2O_2-related functions include involvement in oxidative biosynthesis in fertilized sea-urchin eggs (Stone and Yang 2006). It is thought that the major effects of H_2O_2 are caused by the molecule oxidizing cysteine (-SH-groups) in the catalytic centre of the tyrosine phosphatases (Gloire et al. 2006). Tyrosine phosphatases regulate protein phosphorylation, a major regulatory factor of cellular functions. Hydrogen peroxide, thus, can be important in coupling oxygen- and redox state-dependent responses.

$HO^{\bullet}$ may also be an important signaling ROS (Gloire et al. 2006). Whereas H_2O_2 is quite stable (Lesser 2006), $HO^{\bullet}$ have a short life time (10^{-7} s). Consequently, they have an effect radius of only 4–5 nm, and generate spatial resolution to radical effects

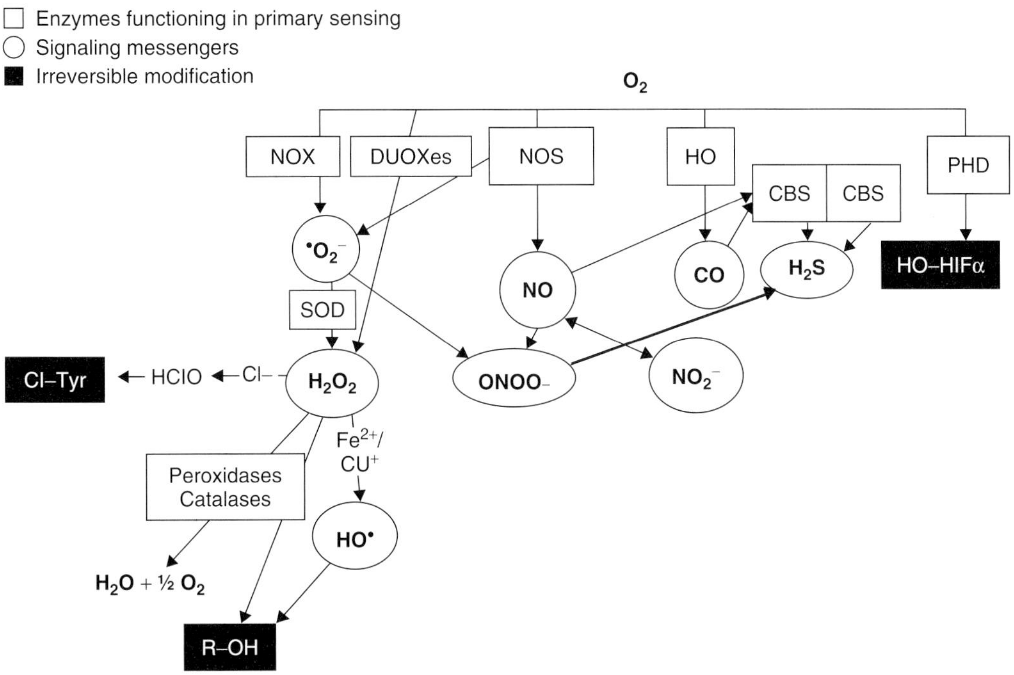

Fig. 12.1 Schematic representation of the pathways showing oxygen sensors, the messenger molecules involved in oxygen sensing, and the protein modifications that take place and are probably important in functionally oxygen-dependent responses. Apart from molecular oxygen, all the sensed molecules are produced via enzyme pathways. The molecules used in apparently oxygen-dependent signaling (superoxide anion, hydrogen peroxide, hydroxyl radical, nitric oxide, carbon monoxide, peroxynitrite) are enclosed in ovals. The enzymes involved in their metabolism (NADPH oxidases, NOX; NO synthetases, NOS; heme oxygenases, HO; superoxide dismutase, SOD; cystathionine β-synthase, CBS; cystathionine γ-lyase, CSE; and catalase), and the enzymes involved in hypoxia-inducible factor breakdown (prolyl hydroxylases, PHD) are enclosed in rectangles. Irreversible protein modifications caused by the interaction of the protein with the second messengers are shown in black boxes.

(Lesser 2006). $HO^{\bullet}$ reacts with virtually all molecules in its vicinity (Halliwell and Gutteridge 2007). It is very difficult to separate H_2O_2 and $HO^{\bullet}$ effects because H_2O_2 is readily converted to $HO^{\bullet}$ in the Fenton reaction (see Fig. 12.1).

Reactive Nitrogen Species

Nitric oxide ($NO^{\bullet}$) may be involved in apparently oxygen-dependent signaling because its production requires oxygen. To some extent, the oxygen dependency of its function stems from its interaction with $O_2^{\bullet -}$. The latter fact may lead to a paradoxical situation where the generally reducing hypoxic conditions (associated with decreased $NO^{\bullet}$ production) can result in oxidative stress. Since $O_2^{\bullet -}$ are produced continuously, and since $NO^{\bullet}$ is an $O_2^{\bullet -}$ scavenger, the reduction of $NO^{\bullet}$ availability will stimulate $O_2^{\bullet -}$ conversion to H_2O_2.

$NO^{\bullet}$ production in different tissues and conditions may vary depending on how the NOS isoform ratio is affected by oxygen tension. Notably, effects of hypoxia have been described both for NOS activity and the level of $NO^{\bullet}$ (Kumar et al. 1996; Lacza et al. 2001). Tissues may also differ in the effects of hypoxia on $NO^{\bullet}$–$O_2^{\bullet -}$ interaction.

The reaction between $NO^{\bullet}$ and $O_2^{\bullet -}$ produces peroxynitrite anion ($ONOO^-$; Fig. 12.1), which is a powerful membrane-permeant oxidant (Fridovich 1986) with a life time near 0.1 s. $ONOO^-$ interacts with thiols or amines forming nitrosothiols and nitrosoamines.

$NO^{\bullet}$ influences the oxygen affinity of mitochondrial function (Koivisto et al. 1997). $NO^{\bullet}$ reduces both oxidative phosphorylation and electron leak under limited oxygen supply (Erusalimsky and Moncada 2007). If the effects occur at physiologically relevant oxygen concentration, mitochondrial $NO^{\bullet}$ can be a molecule participating in apparent oxygen sensing. Nitrite can be an important source of $NO^{\bullet}$ in hypoxia (Jensen 2007). Fish have relatively high basal nitrite levels, and, therefore, nitrite metabolism plays a role in functional oxygen sensing in this group.

Carbon Monoxide and Hydrogen Sulfide

CO and H_2S may also play a role in oxygen-dependent signaling. Recent reviews on the possible roles of these molecules in signaling include those of Lowicka and Beltowski (2007) and Kaczorowski and Zuckerbraun (2007). Endogenously formed CO affects cellular respiration (D'Amico et al. 2006). CO-dependent signaling may at least partly proceed via a cGMP-dependent pathway (Morita et al. 1995).

H_2S level is typically elevated in anoxic reducing conditions. H_2S production is directly inhibited by $NO^{\bullet}$ and CO (Szabo 2007). In addition to affecting – SH groups, H_2S functions as $ONOO^-$ scavenger (Whiteman et al. 2004). The molecule appears to be involved in the oxygen-dependent regulation of vascular muscle tone, and may be involved in oxygen sensing in fish gills (Olson et al. 2006; 2008).

Adenosine and its Phosphates

Adenosine or adenosine phosphate concentrations commonly change in hypoxia. For example, hypoxia leads to a decrease in cellular adenosin triphosphate (ATP) concentration (Lutz and Nilsson 2004). Ecto-5′-nucleoside dephosphorylates (adenosine monophosphate – AMP) to adenosine in hypoxia (Adair 2005). The extracellular adenosine concentration remains elevated, since hypoxia both increases adenosine release from the cells (Conde and Monteiro 2004) and decreases its cellular re-uptake (Eltzschig et al. 2005). In addition, hypoxia appears to induce the formation of cellular adenosine receptors (Kong et al. 2006). Since many hypoxia responses in animals have been associated with adenosine (Nilsson 2001), it may be one molecule used in oxygen sensing. Similarly, since hypoxia causes a decrease in cellular ATP concentration, any mechanism detecting disturbances in the energy balance would be highly useful for maintaining cellular function in hypoxia. Changes in energy balance (AMP/ATP ratio) are sensed by adenosine monophosphate kinase (AMPK) (Hardie 2003).

THE SENSORS

Possible molecules involved in functional oxygen sensing and their locations in the cells are illustrated in Fig. 12.2. A major class of these proteins is heme-containing. Non-heme prolyl and asparaginyl hydroxylases bind molecular oxygen to a cysteine-rich cluster. Also molecules that respond indirectly to a change in oxygen tension, for example because oxygen affects the energy charge of cells, are considered here to be functional oxygen sensors, which adds AMPK, to the list.

Heme-Containing Proteins

Globins

All the different globins, the hemoglobin, myoglobin, cytoglobin, neuroglobin and globin X have a capacity to act as oxygen sensors. Oxygen-sensitive ion transport across erythrocyte membranes, where hemoglobin may play an important role, has recently been reviewed (Bogdanova et al. 2009).

The possibility that other globins would be involved in oxygen sensing has not been verified experimentally. Myoglobin plays a role in regulating NO function in mitochondria (Brunori 2001; Wittenberg and Wittenberg 2003), whereby it is possible that if myoglobin has a role in oxygen sensing it stems from the interaction of the myoglobin molecule with NO. At present, the functions of neuroglobin, cytoglobin, and globin X are poorly known (Hankcheckeln et al. 2005). It is possible that their functions are tissue- and species-specific. In addition to possible regulation by hypoxia, cytoglobin and neuroglobin can be regulated by redox state (Hamdane et al. 2003).

Cytochromes

NADPH Oxidases

NADPH oxidases appear at present as the most important non-mitochondrial cytochromes involved in oxygen sensing. Traditionally NADPH oxidases have been considered to be constituents of phagocytic

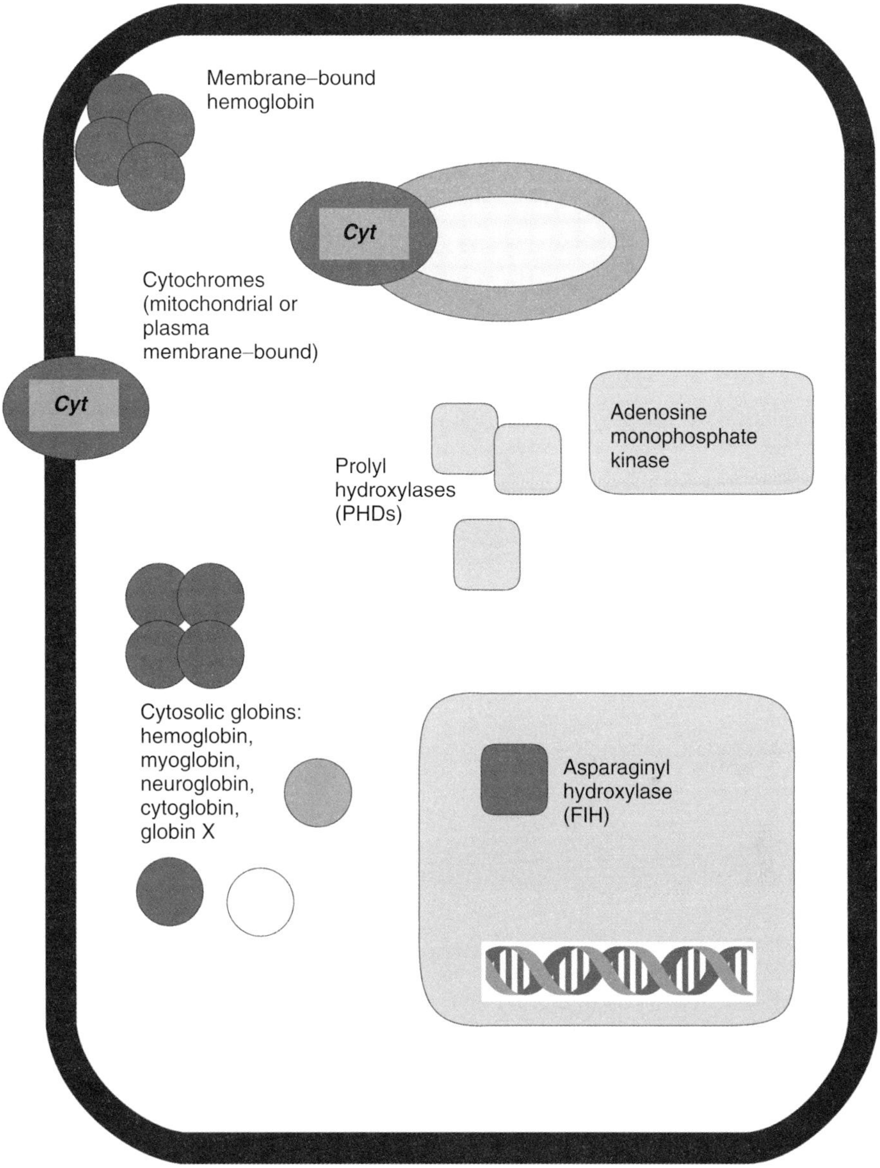

Fig. 12.2 The principal oxygen sensor molecules and their intracellular locations.

cells generating the respiratory burst (Decoursey and Ligeti 2005). However, recent studies have indicated that NADPH oxidases are also present in nonphagocytic cells (Bedard and Krause 2007). The signaling by NADPH oxidases, involved in oxygen sensing, is most likely mediated via oxygen-regulated ROS (mainly $O_2^{\bullet-}$) formation (Ehleben et al. 1998; Porwol et al. 2001). Although the $O_2^{\bullet-}$ are mainly released to the extracellular compartment, intracellular ROS generation also has been demonstrated for the isoforms present in nonphagocytic cells (Dinger et al. 2007).

Mitochondrial Cytochromes

Mitochondrial cytochromes in complexes IIV may sense oxygen. Inhibitors of these complexes affect the hypoxic response in the glomus cells of the carotid body (Lopez-Barneo et al. 2008). A response similar to that in hypoxia is seen in the glossopharyngial nerve preparation of the gill arch of rainbow trout as a result of cyanide treatment (Burleson and Milsom 1995). The exact mechanism of apparent oxygen sensing by mitochondrial cytochromes is not known. However, an increase in ROS production may play a role. Alternatively, a reduction of ATP levels or activation of AMPK can be important (Lopez-Barneo et al. 2008).

Nitric Oxide Synthases

Three types of NOS, i.e., neuronal (nNOS), inducible (iNOS), and endothelial NOS (eNOS), are currently known (Mungrue et al. 2003). Since the oxygen affinities of nNOS and iNOS are relatively low (with P_{50} values of 6–8 kPa), they are well suited for oxygen sensors (Lahiri et al. 2006), but conclusive proof of their role is lacking. All NOS require calcium/calmodulin. However, whereas nNOS and eNOS bind calmodulin reversibly, interaction of iNOS with calmodulin is irreversible (Li and Poulos 2005).

Heme Oxygenases

Heme oxygenases (HO-1 and HO-2) use oxygen and NADPH to convert heme to biliverdin, iron and CO. The HO-1 function is affected by hypoxia (Lee et al. 1997). CO donors mimic hypoxic response in carotid bodies, HO-2 co-immunoprecipitates with potassium channels that respond to changes in oxygenation. RNA interference of HO-2 reduces the oxygen sensitivity of these potassium channels in cells expressing both the channels and the heme oxygenase. However, although these findings suggest a role for HO-2 in oxygen sensing, oxygen chemoreception of carotid bodies in mice with HO-2 knockout remains unaltered (Ortega-Saenz et al. 2007), which does not agree with an oxygen sensing role for HO-2. Induction of HO-1 in hypoxia is protective in goldfish (*Carassius auratus*; Wang et al. 2008).

Hydroxylases and the Hypoxia-Inducible Factor Pathway

Oxygen-dependent regulation of transcription generally involves the hypoxia-inducible factor (HIF). Principles of HIF function are described in Fig. 12.3, and have been reviewed repeatedly (e.g., Semenza 2009). The transcriptionally active form of HIF is a dimer consisting of α and β subunits. The α subunits give the oxygen sensitivity while the function of β subunits appears to be oxygen-insensitive. There are at least three different classes of α subunits, denoted as HIF1α, HIF2α (also called EPAS1), and HIF3α. The recently described HIF4α (Law et al. 2006) of a teleost group belongs to class HIF3α on the basis of phylogenetic analyses. HIF1α is the most studied member of the HIFα family.

The oxygen-dependent effect of HIF1α on transcription is regulated mainly either by the stability of the protein or its DNA binding. The stability of the protein is influenced by prolyl hydroxylase (PHD) enzymes which hydroxylate conserved prolines (proline 402 and 564 in the human protein) in an oxygen-dependent fashion. The PHD function has been reviewed by, for example, Fandrey et al. (2006). Three oxygen-dependent PHDs (PHD1–3; EGLN1–3) have so far been described. PHD2 (EGLN1) appears to be the most important oxygen sensor (Berra et al. 2003). Prolyl hydroxylation takes place in normoxia. It enables the interaction of HIFα and the von-Hippel-Lindau protein (VHL). The complex can subsequently be ubiquitylated and degraded proteasomally. In hypoxia, prolyl hydroxylation does not occur, and HIF1α protein is stabilized. It is transported from cytoplasm to nucleus, where it forms a dimer with AhR nuclear translocator (ARNT; also named HIFß) and recruits the general transcriptional activator CBP/p300. The activator can only be recruited in hypoxia, since in normoxia a conserved asparagine residue (Asp803 in the human molecule – similar residues are found in all vertebrates studied to date; Bracken et al. 2003) is hydroxylated and not available for interaction. The oxygen affinity of asparaginyl hydroxylase is much higher than those of PHD. Consequently, it is possible that the two enzymes either regulate different genes or HIF function at different levels (Dayan et al. 2006, 2009). Prolyl and asparaginyl hydroxylases use molecular oxygen as substrate.

The heterodimeric HIF (HIFα+ARNT) binds to hypoxia-response elements (HREs) present in the promoter/enhancer region of the hypoxia-inducible genes stimulating transcription. HREs may also be present in the introns of oxygen-dependent genes (Rees et al. 2001). The minimal consensus HRE of most organisms is A/GCGTG (Camenisch et al. 2002),

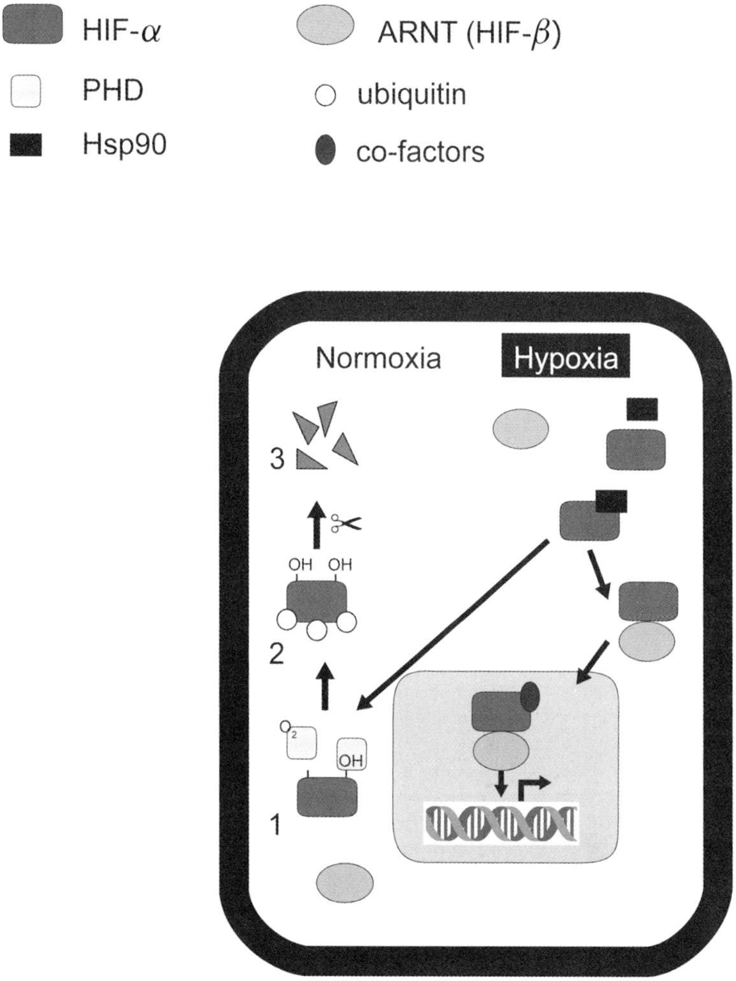

Fig. 12.3 Function of hypoxia-inducible factors. The molecule is produced continuously, but in normoxia prolyl hydroxylase tags the protein for ubiquitin mediated breakdown. In the cytoplasm the protein is somewhat stabilized via its interaction with HSP90. In hypoxia the molecule is not broken down and is transported to the nucleus where it forms a dimer with ARNT, binds to hypoxia response elements in the regulatory regions of genes, and induces the transcription of hypoxia-inducible genes.

although a novel sequence (GATGTG) has also been found in the vicinity of fish LDH gene (Rees et al. 2009). In some cases the presence of HREs alone is not sufficient for hypoxic induction of the genes (Firth et al. 1995), but additional elements such as binding sites for various molecules, for example AP1, ATF1/CREB1, HNF4 or Smad3, may be required (Bracken et al. 2003).

Both the DNA binding and the transcriptional activation by HIF appear to be under redox control (Lando et al. 2000). Serine-to-cysteine mutation at a specific residue in the DNA binding domain confers redox sensitivity of DNA binding (Lando et al. 2000). Rainbow trout HIF1α, which has cysteine in critical position, is redox sensitive (Nikinmaa et al. 2004). The nuclear redox regulator Ref1 potentiates the hypoxic induction of a reporter gene (Lando et al. 2000). There can be interaction between redox-sensitive transcription factors and HIF (Khomenko et al. 2004), and cellular redox state correlates with HIF induction (Heise et al. 2006).

The hydroxylase enzymes need α-ketoglutarate, ascorbate, and ferrous ions as cofactors (Kaelin 2005). These factors are redox sensitive. The need for α-ketoglutarate as a cofactor brings together the oxygen-dependent hydroxylase function and the citric acid cycle. The regulation of HIF-dependent gene expression by NO is largely due to the effects of $NO^{\bullet}$

on PHD activity (Berchner-Pfannschmidt et al. 2007). Also CO has a significant influence on HIF-function, but the regulatory mechanism is as yet unknown (Chin et al. 2007). In addition, calcium affects the function of HIF (Mottet et al. 2003).

It now appears that the transcription of HIFα also may play a role in the regulation of the HIF pathway in some, mainly hypoxia-tolerant animals (Shams et al. 2005; Law et al. 2006; Rissanen et al. 2006), although it was suggested earlier that transcriptional regulation would not occur. In addition to hypoxia, there are several conditions where HIFα is induced in normoxic conditions (Dery et al. 2005; Hirota and Semenza 2005). While the normoxic function of HIFs in water-breathing animals has been studied very little, interactions of temperature acclimation and HIF function are likely to occur (Treinin et al. 2003; Heise et al. 2006; Rissanen et al. 2006). A set of examples of known HIF-dependent genes and their functions is given in Table 12.2. More than 500 genes may be transcriptionally affected by hypoxia in fish (Ju et al. 2007). The emphasis in the studies has been HIF induction;

Table 12.2 Selected examples of genes under transcriptional regulation by the hypoxia-inducible factor (HIF) (for details see e.g. Gardner et al. 2001; Goda et al. 2003; Schnell et al. 2003; Fandrey 2004; Gorr et al. 2004; Semenza 2004; Greijer et al. 2005; Fukuda et al. 2007; Jelkmann 2007).

Oxygen transport	Iron metabolism	Ceruloplasmin
		Transferrin
		Transferrin receptor
	Red cell production	Erythropoietin
	Hemoglobin synthesis	Globin genes of *Daphnia*, for example
	Angiogenesis	VEGF (vascular endothelial growth factor)
		VEGF receptor 1
		Endothelin
Energy production	Glycolysis	Aldolase A
		Fructose-2,6-bisphoshatase 3 and 4
		Enolase
		Lactate dehydrogenase
	Substrate availability	Glucose transporter
	Mitochondrial effects	LON (mitochondrial protease involved in COX4 breakdown)
Hormonal regulation and cellular signaling		Leptin
		Atrial natriuretic peptide
NO production		Nitric oxide synthase-2
CO production		Heme oxygenase-1
Adrenergic signaling		α-adrenergic receptor
		Tyrosine hydroxylase (an enzyme needed in the biosynthesis of catecholamines–hydroxylating tyrosine)
Immunological responses		Thymopoietin (factor involved in T-cell development)
Cell cycle and apoptosis		p21
		p27 (both proteins are inhibitors of cyclin-dependent kinase)
		NIP3 (proapoptotic factor)
Cytoskeleton and extracellular matrix		Fibronectin
		Keratin components

the mechanisms of hypoxia-induced down-regulation of gene transcription have been little studied.

Adenosine Monophosphate Kinase

The function of AMP-activated protein kinase has been reviewed recently (Hardie et al. 2006). The enzyme can couple oxygen and energy sensing (Wyatt et al. 2007). It is composed of three subunits, the catalytic α subunit, and the regulatory β and γ subunits. The multisubunit enzyme remains inactive even in the presence of AMP, if not phosphorylated at a critical threonine residue (Hardie 2003). Since AMPK is involved in regulating cellular energy balance, its activation switches energy-consuming pathways off and energy-producing ones on. One of the major oxygen consuming processes in the cells involves mRNA translation to proteins. Translation also is inhibited by AMPK in hypoxia independently from HIF regulation (Liu et al. 2006), showing the importance of energy sensing in hypoxia regulation.

AN EXAMPLE OF OXYGEN-SENSITIVE INTEGRATIVE FUNCTION: GILL VENTILATION

Ventilation in water-breathers responds mainly to changes in ambient oxygen level (Dejours 1975). In addition, it can respond to changes in carbon dioxide tension and pH, the major factors affecting ventilation in air-breathers (Gilmour and Perry 2006). The oxygen-sensing neuroendocrine cells are located in all gill arches and characteristically contain serotonin (Milsom and Burleson 2007). Figure 12.4 gives a representation of gill neuroendocrine cell function, which has also been reviewed by Perry et al. (2002). Both external (responding to changes in the oxygenation of water) and internal oxygen receptors (responding to changes in the oxygenation of blood or some other body constituent) have been considered to be present (reviewed e.g., by Perry and Gilmour, 2002). Regardless of their location, the sensors must be neuroendocrine cells with neural connections to the central nervous system. Gill neuroendocrine cells were first described microscopically in the early 1980s in

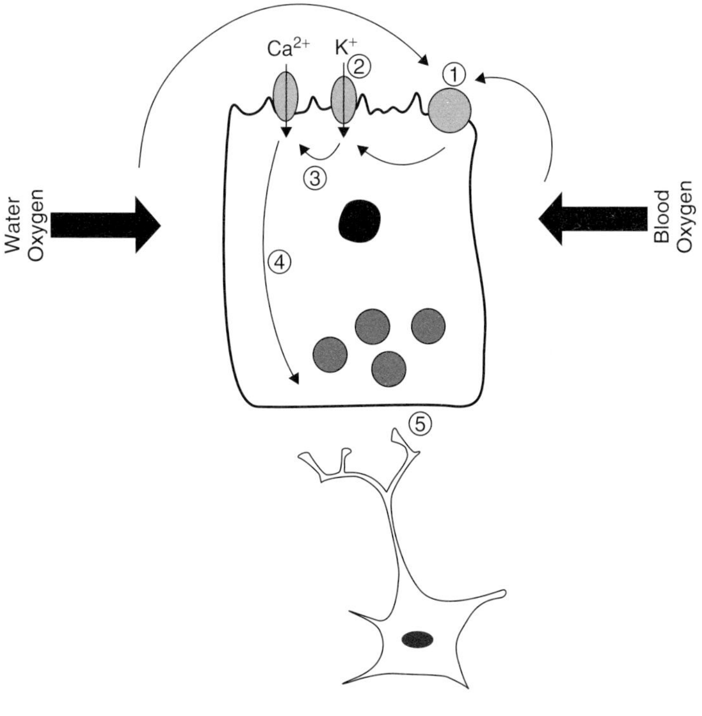

Fig. 12.4 Schematic representation of the function of gill neuroendocrine cells sensing water and blood oxygenation. Oxygen affects, via a primary oxygen sensor – the nature of which is currently unknown (1), the activity of the potassium channel (2). The efflux of potassium via the channel affects the membrane potential (3), which influences calcium flux via a calcium channel (4). The change in intracellular calcium level affects the neurotransmitter release (5), and influences neural activities. See plate section for a color version of this image.

trout (Dunel-Erb et al. 1982). However, these cells were characterized as oxygen-sensing cells in zebrafish first in the early 2000s (Jonz et al. 2004). Putative oxygen-sensing cells have also been characterized from the gills of the channel catfish (Burleson et al. 2006). Since the neuroendocrine cells are located in the lamellae, they can respond to changes in both the ambient and blood oxygen level. While the above discussion is based on fish, similar oxygen-sensing cells/structures are most likely present in the gills of water-breathing invertebrates.

REFERENCES

Adair, T.H. (2005) Growth regulation of the vascular system: an emerging role for adenosine. *American Journal of Physiology* 289, R283–R296.

Bedard, K., Krause, K.H. (2007) The NOX family of ROS-generating NADPH oxidases: Physiology and pathophysiology. *Physiological Review* 87, 245–313.

Berchner-Pfannschmidt, U., Yamac, H., Trinidad, B., Fandrey, J. (2007) Nitric oxide modulates oxygen sensing by hypoxia-inducible factor 1-dependent induction of prolyl hydroxylase 2. *Journal of Biological Chemistry* 282, 1788–1796.

Berra, E., Benizri, E., Ginouves, A., Volmat, V., Roux, D., Pouyssegur, J. (2003) HIF prolyl-hydroxylase 2 is the key oxygen sensor setting low steady-state levels of HIF-1 alpha in normoxia. *EMBO Journal* 22, 4082–4090.

Bogdanova, A., Berenbrink, M., Nikinmaa, M. (2009) Oxygen-dependent ion transport in erythrocytes. *Acta Physiologica* 195, 305–319.

Bracken, C.P., Whitelaw, M.L., Peet, D.J. (2003) The hypoxia-inducible factors: key transcriptional regulators of hypoxic responses. *Cellular and Molecular Life Sciences* 60, 1376–1393.

Brunori, M. (2001) Nitric oxide moves myoglobin centre stage. *Trends in Biochemical Sciences*, 26, 209–210.

Burleson, M.L., Milsom, W.K. (1995) Cardio-ventilatory control in rainbow trout: I. Pharmacology of branchial, oxygen-sensitive chemoreceptors. *Respiratory Physiology* 100, 231–238.

Burleson, M.L., Mercer, S.E., Wilk-Blaszczak, M.A. (2006) Isolation and characterization of putative O-2 chemoreceptor cells from the gills of channel catfish (*Ictalurus punctatus*). *Brain Research* 1092, 100–107.

Camenisch, G., Wenger, R.H., Gassmann, M. (2002) DNA-binding activity of hypoxia-inducible factors (HIFs). *Methods in Molecular Biology* 196, 117–129.

Chin, B.Y., Jiang, G., Wegiel, B. et al. (2007) Hypoxia-inducible factor 1 alpha stabilization by carbon monoxide results in cytoprotective preconditioning. *Proceedings of the National Academy of Sciences USA* 104, 5109–5114.

Conde, S.V., Monteiro, E.C. (2004) Hypoxia induces adenosine release from the rat carotid body. *Journal of Neurochemistry* 89, 1148–1156.

D'Amico, G., Lam, F., Hagen, T., Moncada, S. (2006) Inhibition of cellular respiration by endogenously produced carbon monoxide. *Journal of Cell Science* 119, 2291–2298.

Dayan, F., Roux, D., Brahimi-Horn, M.C., Pouyssegur, J., Mazure, N.M. (2006) The oxygen sensor factor-inhibiting hypoxia-inducible factor-1 controls expression of distinct genes through the bifunctional transcriptional character of hypoxia-inducible factor-alpha. *Cancer Research* 66, 3688–3698.

Dayan, F., Monticelli, M., Pouyssegur, J., Pecou, E. (2009) Gene regulation in response to graded hypoxia: The non-redundant roles of the oxygen sensors PHD and FIH in the HIF pathway. *Journal of Theoretical Biology* 259, 304–316.

Decoursey, T.E., Ligeti, E. (2005) Regulation and termination of NADPH oxidase activity. *Cellular and Molecular Life Sciences* 62, 2173–2193.

Dejours, P. (1975) *Principles of Comparative Respiratory Physiology*. North-Holland, Amsterdam.

Dery, M.A.C., Michaud, M.D., Richard, D.E. (2005) Hypoxia-inducible factor 1: regulation by hypoxic and non-hypoxic activators. *International Journal of Biochemistry and Cell Biology* 37, 535–540.

Dinger, B., He, L., Chen, J. et al. (2007) The role of NADPH oxidase in carotid body arterial chemoreceptors. *Respiratory Physiology and Neurobiology* 157, 45–54.

Dunel-Erb, S., Bailly, Y., Laurent, P. (1982) Neuroepithelial cells in fish gill primary lamellae. *Journal of Applied Physiology* 53, 1342–1353.

Ehleben, W., Bolling, B., Merten, E., Porwol, T., Strohmaier, A.R., Acker, H. (1998) Cytochromes and oxygen radicals as putative members of the oxygen sensing pathway. *Respiratory Physiology* 114, 25–36.

Eltzschig, H.K., Abdulla, P., Hoffman, E. et al. HIF-1-dependent repression of equilibrative nucleoside transporter (ENT) in hypoxia. *Journal of Experimental Medicine* 202, 1493–1505.

Erusalimsky, J.D., Moncada, S. (2007) Nitric oxide and mitochondrial signaling: from physiology to pathophysiology. *Arteriosclerosis, Thrombosis, and Vascular Biology* 27, 2524–2531.

Fandrey, J. (2004) Oxygen-dependent and tissue-specific regulation of erythropoietin gene expression. *American Journal of Physiology* 286, R977–R988.

Fandrey, J., Gorr, T.A., Gassmann, M. (2006) Regulating cellular oxygen sensing by hydroxylation. *Cardiovascular Respiration* 71, 642–651.

Firth, J.D., Ebert, B.L., Ratcliffe, P.J. (1995) Hypoxic regulation of lactate dehydrogenase A: Interaction between hypoxia-inducible factor 1 and cAMP response elements. *Journal of Biological Chemistry*, 270, 21021–21027.

Fridovich, I. (1986) Biological effects of the superoxide radical. *Archives of Biochemistry and Biophysics* 247, 1–11.

Fukuda, R., Zhang, H.F., Kim, J.W., Shimoda, L., Dang, C.V., Semenza, G.L. (2007) HIF-1 regulates cytochrome oxidase subunits to optimize efficiency of respiration in hypoxic cells. *Cell* 129, 111–122.

Gardner, L.B., Li, Q., Park, M.S., Flanagan, W.M., Semenza, G.L., Dang, C.V. (2001) Hypoxia inhibits G(1)/S transition through regulation of p27 expression. *Journal of Biological Chemistry* 276, 7919–7926.

Gilmour, K.M., Perry, S.F. (2006) "Branchial Chemoreceptor Regulation of Cardiorespiratory Function. *In* Hara, T.J., Zielinski, B.S. (eds). *Fish Physiology*, Vol. 25, *Sensory Systems Neuroscience*. Academic Press, New York, pp. 97–151.

Gloire, G., Legrand-Poels, S., Piette, J. (2006) NF-[kappa]B activation by reactive oxygen species: Fifteen years later. *Biochemistry and Pharmacology* 72, 1493–1505.

Goda, N., Ryan, H.E., Khadivi, B., McNulty, W., Rickert, R.C., Johnson, R.S. (2003) Hypoxia-inducible factor 1 alpha is essential for cell cycle arrest during hypoxia. *Molecular and Cellular Biology* 23, 359–369.

Gonzalez, C., Agapito, M.T., Rocher, A. et al. (2007) Chemoreception in the context of the general biology of ROS. *Respiratory Physiology and Neurobiology* 157, 30–44.

Gorr, T.A., Cahn, J.D., Yamagata, H., Bunn, H.F. (2004) Hypoxia-induced synthesis of hemoglobin in the crustacean Daphnia magna is hypoxia-inducible factor-dependent. *Journal of Biological Chemistry* 279, 36038–36047.

Greijer, A.E., van der Groep, P., Kemming, D. et al. (2005) Up-regulation of gene expression by hypoxia is mediated predominantly by hypoxia-inducible factor 1 (HIF-1). *Journal of Pathology* 206, 291–304.

Guzy, R.D., Schumacker, P.T. (2006) Oxygen sensing by mitochondria at complex III: the paradox of increased reactive oxygen species during hypoxia. *Experimental Physiology* 91, 807–819.

Halliwell, B., Gutteridge, J. (2007) *Free Radicals in Biology and Medicine*, 4th ed. Oxford University Press, Oxford.

Hamdane, D., Kiger, L., Dewilde, S. et al. (2003) The redox state of the cell regulates the ligand binding affinity of human neuroglobin and cytoglobin. *Journal of Biological Chemistry* 278, 51713–51721.

Hankeln, T., Ebner, B., Fuchs, C. et al. (2005) Neuroglobin and cytoglobin in search of their role in the vertebrate globin family. *Journal of Inorganic Biochemistry* 99, 110–119.

Hardie, D.G. (2003) Minireview: The AMP-activated protein kinase cascade: The key sensor of cellular energy status. *Endocrinology* 144, 5179–5183.

Hardie, D.G., Hawley, S.A., Scott, J. (2006) AMP-activated protein kinase – development of the energy sensor concept. *Journal of Physiology* 574, 7–15.

Heise, K., Puntarulo, S., Nikinmaa, M., Abele, D., Portner, H. O. (2006) Oxidative stress during stressful heat exposure and recovery in the North Sea eelpout *Zoarces viviparus* L. *Journal of Experimental Biology* 209, 353–363.

Hirota, K., Semenza, G.L. (2005) Regulation of hypoxia-inducible factor 1 by prolyl and asparaginyl hydroxylases. *Biochemical and Biophysical Research Communications* 338, 610–616.

Jelkmann, W. (2007) Erythropoietin after a century of research: younger than ever. *European Journal of Haematology* 78, 183–205.

Jensen, F.B. (2007) Nitric oxide formation from nitrite in zebrafish. *Journal of Experimental Biology* 210, 3387–3394.

Jonz, M.G., Fearon, I.M., Nurse, C.A. (2004) Neuroepithelial oxygen chemoreceptors of the zebrafish gill. *Journal of Physiology* 560, 737–752.

Ju, Z.L., Wells, M.C., Heater, S.J., Walter, R.B. (2007) Multiple tissue gene expression analyses in Japanese medaka (*Oryzias latipes*) exposed to hypoxia. *Comparative Biochemistry and Physiology C* 145, 134–144.

Kaczorowski, D.J., Zuckerbraun, B.S. (2007) Carbon monoxide: medicinal chemistry and biological effects. *Current Medicinal Chemistry* 14, 2720–2725.

Kaelin, W.G. (2005) Proline hydroxylation and gene expression. *Annual Review of Biochemistry* 74, 115–128.

Khomenko, T., Deng, X.M., Sandor, Z., Tarnawski, A.S., Szabo, S. (2004) Cysteamine alters redox state, HIF-1 alpha transcriptional interactions and reduces duodenal mucosal oxygenation: novel insight into the mechanisms of duodenal ulceration. *Biochemical and Biophysical Research Communications* 317, 121–127.

Koivisto, A., Matthias, A., Bronnikov, G., Nedergaard, J. (1997) Kinetics of the inhibition of mitochondrial respiration by NO. *FEBS Letters* 417, 75–80.

Kong, T.Q., Westerman, K.A., Faigle, M., Eltzschig, H.K., Colgan, S.P. (2006) HIF-dependent induction of adenosine A2B receptor in hypoxia. *FASEB Journal* 20, 2242–2250.

Kumar, M., Liu, G.J., Floyd, R.A., Grammas, P. (1996) Anoxic injury of endothelial cells increases production of nitric oxide and hydroxyl radicals. *Biochemical and Biophysical Research Communications* 219, 497–501.

Lacza, Z., Puskar, M., Figueroa, J.P., Zhang, J., Rajapakse, N., Busija, D.W. (2001) Mitochondrial nitric oxide synthase is constitutively active and is functionally upregulated in hypoxia. *Free Radical Biology and Medicine* 31, 1609–1615.

Lahiri, S., Roy, A., Baby, S.M., Hoshi, T., Semenza, G.L., Prabhakar, N.R. (2006) Oxygen sensing in the body. *Progress in Biophysics and Molecular Biology* 91, 249–286.

Lando, D., Pongratz, I., Poellinger, L., Whitelaw, M.L. (2000) A redox mechanism controls differential DNA binding activities of hypoxia-inducible factor (HIF) 1alpha and the HIF-like factor. *Journal of Biological Chemistry* 275, 4618–4627.

Law, S.H.W., Wu, R.S.S., Ng, P.K.S., Yu, R.M.K., Kong, R.Y.C. (2006) Cloning and expression analysis of two distinct HIF-alpha isoforms – gcHIF-1alpha and gcHIF-4alpha – from the hypoxia-tolerant grass carp, *Ctenopharyngodon idellus* *BMC Molecular Biology* 7, 15.

Lee, P.J., Jiang, B.H., Chin, B.Y. et al. (1997) Hypoxia-inducible factor-1 mediates transcriptional activation of the heme oxygenase-1 gene in response to hypoxia. *Journal of Biological Chemistry* 272, 5375–5381.

Lesser, M.P. (2006) Oxidative stress in marine environments: Biochemistry and physiological ecology. *Annual Review of Physiology* 68, 253–278.

Li, H., Poulos, T.L. (2005) Structure-function studies on nitric oxide synthases. *Journal of Inorganic Biochemistry* 99, 293–305.

Liu, L.P., Cash, T.P., Jones, R.G., Keith, B., Thompson, C.B., Simon, M.C. (2006) Hypoxia-induced energy stress regulates mRNA translation and cell growth. *Molecular Cell* 21, 521–531.

Lopez-Barneo, J., Ortega-Saenz, P., Pardal, R., Pascual, A., Piruat, J.I. (2008) Carotid body oxygen sensing. *European Respiratory Journal* 32, 1386–1398.

Lowicka, E., Beltowski, J. (2007) Hydrogen sulfide (H_2S) – the third gas of interest for pharmacologists. *Pharmacological Reports* 59, 4–24.

Lutz, P.L., Nilsson, G.E. (2004) Vertebrate brains at the pilot light. *Respiratory Physiology and Neurobiology* 141, 285–296.

Michiels, C., Minet, E., Mottet, D., Raes, M. (2002) Regulation of gene expression by oxygen: NF-kappa B and HIF-1, two extremes. *Free Radical Biology and Medicine* 33, 1231–1242.

Milsom, W.K., Burleson, M.L. (2007) Peripheral arterial chemoreceptors and the evolution of the carotid body. *Respiratory Physiology and Neurobiology* 157, 4–11.

Morita, T., Perrella, M.A., Lee, M.E., Kourembanas, S. (1995) Smooth muscle cell-derived carbon monoxide is a regulator of vascular cGMP. *Proceedings of the National Academy of Sciences USA* 92, 1475–1479.

Mottet, D., Michel, G., Renard, P., Ninane, N., Raes, M., Michiels, C. (2003) Role of ERK and calcium in the hypoxia-induced activation of HIF-1. *Journal of Cellular Physiology* 194, 30–44.

Mungrue, I.N., Bredt, D.S., Stewart, D.J., Husain, M. (2003) From molecules to mammals: what's NOS got to do with it? *Acta Physiologica Scandinavica* 179, 123–135.

Nikinmaa, M. (2002) Oxygen-dependent cellular functions – why fishes and their aquatic environment are a prime choice of study. *Comparative Biochemistry and Physiology A* 133, 1–16.

Nikinmaa, M., Rees, B.B. (2005) Oxygen-dependent gene expression in fishes. *American Journal of Physiology* 288, R1079–R1090.

Nikinmaa, M., Pursiheimo, S., Soitamo, A.J. (2004) Redox state regulates HIF-1 alpha and its DNA binding and phosphorylation in salmonid cells. *Journal of Cell Science* 117, 3201–3206.

Nilsson, G.E. (2001) Surviving anoxia with the brain turned on. *News in Physiological Science* 16, 217–221.

Olson, K.R., Dombkowski, R.A., Russell, M.J. et al. (2006) Hydrogen sulfide as an oxygen sensor/transducer in vertebrate hypoxic vasoconstriction and hypoxic vasodilation. *Journal of Experimental Biology* 209, 4011–4023.

Olson, K.R., Healy, M.J., Qin, Z. et al. (2008) Hydrogen sulfide as an oxygen sensor in trout gill chemoreceptors. *American Journal of Physiology: Regulatory, Integrative and Comparative Physiology* 295, R669–R680.

Ortega-Saenz, P., Pascual, A., Piruat, J.I., Lopez-Barneo, J. (2007) Mechanisms of acute oxygen sensing by the carotid body: Lessons from genetically modified animals. *Respiratory Physiology and Neurobiology* 157, 140–147.

Otto, C.M., Baumgardner, J.E. (2001) Effect of culture Po_2 on macrophage (RAW 264.7) nitric oxide production. *American Journal of Physiology: Cell Physiology* 280, C280–C287.

Perry, S.F., Gilmour, K.M. (2002) Sensing and transfer of respiratory gases at the fish gill. *Journal of Experimental Zoology* 293, 249–263.

Perry, S.F., Jonz, M.G., Gilmour, K.M. (2009) Oxygen sensing and the hypoxic ventilatory response. *In* Richards, J.G. (ed.). *Fish Physiology*, Vol. 27, *Hypoxia*. Academic Press, New York, pp. 193–253.

Porwol, T., Ehleben, W., Brand, V., Acker, H. (2001) Tissue oxygen sensor function of NADPH oxidase isoforms, an unusual cytochrome aa3 and reactive oxygen species. *Respiratory Physiology* 128, 331–348.

Rees, B.B., Bowman, J.A., Schulte, P.M. (2001) Structure and sequence conservation of a putative hypoxia response element in the lactate dehydrogenase-B gene of *Fundulus*. *Biological Bulletin* 200, 247–251.

Rees, B.B., Figueroa, Y.G., Wiese, T.E., Beckman, B.S., Schulte, P.M. (2009) A novel hypoxia-response element in the lactate dehydrogenase-B gene of the killifish *Fundulus heteroclitus*. *Comparative Biochemistry and Physiology A* 154, 70–77.

Rissanen, E., Tranberg, H.K., Sollid, J., Nilsson, G.E., Nikinmaa, M. (2006) Temperature regulates hypoxia-inducible factor-1 (HIF-1) in a poikilothermic vertebrate, crucian carp (*Carassius carassius*). *Journal of Experimental Biology* 209, 994–1003.

Schnell, P.O., Ignacak, M.L., Bauer, A.L., Striet, J.B., Paulding, W.R., Czyzyk-Krzeska, M.F. (2003) Regulation of tyrosine hydroxylase promoter activity by the von Hippel-Lindau tumor suppressor protein and hypoxia-inducible transcription factors. *Journal of Neurochemistry* 85, 483–491.

Semenza, G.L. (2004) Hydroxylation of HIF-1: Oxygen sensing at the molecular level. *Physiology*, 19, 176–182.

Semenza, G.L. (2009) Regulation of oxygen homeostasis by hypoxia-inducible factor 1. *Physiology* 24, 97–106.

Shams, I., Nevo, E., Avivi, A. (2005. Ontogenetic expression of erythropoietin and hypoxia-inducible factor-1 alpha genes in subterranean blind mole rats. *FASEB Journal* 19, 307–309.

Stone, J.R., Yang, S. (2006) Hydrogen peroxide: a signaling messenger. *Antioxidants and Redox Signaling* 8, 243–270.

Szabo, C. (2007) Hydrogen sulphide and its therapeutic potential. *Nature Reviews: Drug Discovery* 6, 917–935.

Treinin, M., Shliar, J., Jiang, H.Q., Powell-Coffman, J.A., Bromberg, Z., Horowitz, M. (2003) HIF-1 is required for heat acclimation in the nematode *Caenorhabditis elegans*. *Physiological Genomics* 14, 17–24.

Wang, D., Zhong, X.P., Qiao, Z.X., Gui, J.F. (2008) Inductive transcription and protective role of fish heme oxygenase-1 under hypoxic stress. *Journal of Experimental Biology* 211, 2700–2706.

Ward, J.P.T. (2008) Oxygen sensors in context. *Biochimica et Biophysica Acta – Bioenergetics* 1777, 1–14.

Wenger, R.H., Gassmann, M. (1996) Little difference. *Nature* 380, 100.

Whiteman, M., Armstrong, J.S., Chu, S.H., Jia-Ling, S., Wong, B.S., Cheung, N.S., Halliwell, B., Moore, P.K. (2004) The novel neuromodulator hydrogen sulfide: an endogenous peroxynitrite 'scavenger'? *Journal of Neurochemistry* 90, 765–768.

Wittenberg, J.B., Wittenberg, B.A. (2003) Myoglobin function reassessed. *Journal of Experimental Biology* 206, 2011–2020.

Wyatt, C.N., Mustard, K.J., Pearson, S.A. et al. (2007) AMP-activated protein kinase mediates carotid body excitation by hypoxia. *Journal of Biological Chemistry* 282, 8092–8098.

Chapter 13

ISCHEMIA/REPERFUSION IN DIVING BIRDS AND MAMMALS: HOW THEY AVOID OXIDATIVE DAMAGE

Tania Zenteno-Savín[1], *José Pablo Vázquez-Medina*[1,2], *Nadiezhda Cantú-Medellín*[1,3], *Paul J. Ponganis*[4], *and Robert Elsner*[5]

[1]Centro de Investigaciones Biológicas del Noroeste, S.C., La Paz, Baja California Sur, Mexico
[2]University of California Merced, School of Natural Sciences, Merced, CA, USA
[3]University of Alabama Birmingham Department of Pathology, Birmingham, AL, USA
[4]Center for Marine Biotechnology & Biomedicine, Scripps Institution of Oceanography, University of California San Diego, La Jolla, CA, USA
[5]University of Alaska Fairbanks, School for Fisheries and Ocean Science, Institute of Marine Science, Fairbanks, AK, USA

Seals, whales, and penguins must dive in order to feed and escape from surface-active predators. The diving capacity of marine birds and mammals varies according to species. Deep and long diving seals, such as northern elephant seals (*Mirounga anguistirostris*), hooded seals (*Cystophora cristata*), and Weddell seals (*Leptonychotes weddellii*) are capable of dives of nearly 1000 m depth and over 1 h duration (Kooyman 1966; Folkow and Blix 1999), whereas short and shallow divers, such as ringed seals (*Phoca hispida*) and California sea lions (*Zalophus californianus californianus*) exhibit routine dives of less than 120 m depth and 4–5 min duration (Kooyman 1989; Kelly and Wartzok 1996). The baleen whales, such as blue

Oxidative Stress in Aquatic Ecosystems, First Edition. Edited by Doris Abele, José Pablo Vázquez-Medina, and Tania Zenteno-Savín.

(*Balaenoptera musculus*) and fin whales (*B. physalus*), perform dives with maximum durations of approximately 25 min (Croll et al. 2001). Dolphins, in general, can dive for 2 to 10 min to depths of 4–700 m (Stewart 2002; Clarke 2003), whereas sperm whales (*Physeter macrocephalus*) have the capacity to dive for nearly 2 h to depths of 2000 m, while both dwarf and pygmy sperm whales (*Kogia sima* and *Kogia breviceps*, respectively) can perform dives that last up to 90 min with maximum depths of 1100 m (Davis et al. 1988; Cardona-Maldonado and Mignucci-Giannoni 1999; Kooyman 2002; Whitehead 2002). Although ducks, cormorants, pelicans, and other aquatic birds can swim underwater, they do so for a short time (on average 1 min). Penguins are the only avian species that are capable of extended diving. Adélie penguins (*Pygoscelis adeliae*) can dive to 27 m depths for 1–4 min, Macaroni penguins (*Eudyptes chtysolophus*) can dive to 80 m, rarely exceeding 2 min duration, Gentoo penguins (*Pygoscelis papua*) can dive to 100 m depths, and Emperor penguins (*Aptenodytes forsteri*) can dive to 500 m depths, and routinely dive for periods of 5 to 10 min (Croxall et al. 1988; Naito et al. 1990; Kooyman and Ponganis 1998).

Because they are air-breathing vertebrates, seals, whales, and penguins are routinely exposed to progressive asphyxia (the combination of hypoxia, hypercapnia, and acidosis) during dives (Elsner and Gooden 1983). Diving vertebrates have developed a number of mechanisms that allow them to tolerate prolonged apnea (breath-holding) periods and increase their diving capacity. These include elevated hemoglobin and myoglobin concentrations, high mitochondrial content in specific tissues, high hematocrits and elevated blood volumes (Kooyman and Ponganis 1998; Elsner 1999; Ramírez et al. 2007). Diving vertebrates also undergo a number of cardiovascular adjustments, such as bradycardia (lower heart rate) and peripheral vasoconstriction. This results in a redistribution of blood flow preferentially towards oxygen-sensitive tissues, namely the central nervous system, and in selective ischemia (restriction of blood flow) to the most hypoxia-tolerant tissues during diving (Bevan and Butler 1992; Kooyman and Ponganis 1998; Elsner 1999). After the dive, a prompt tachycardia (elevated heart rate) increases cardiac output and restores blood flow to tissues. These cyclic bouts of ischemia and reperfusion present the diving birds and mammals with the potential for the production of reactive oxygen species (ROS) and thus oxidative stress.

Ischemia/reperfusion promotes ROS production and oxidative stress, because adenosine triphosphate (ATP) depletion during ischemia results in purine nucleotide accumulation and proteolytic conversion of xanthine dehydrogenase (XDH) to xanthine oxidase (XO). During reperfusion, XO is able to reduce purine nucleotides in the presence of oxygen generating ROS such as $O_2^{\bullet -}$ and H_2O_2 (Fig. 13.1). ROS are able to react with proteins, lipids, and nucleic acids causing cellular injury (Kuppusamy and Zweier 1989; Thompson-Gorman and Zweier 1990; Halliwell and Gutteridge 1999). Penguins, seals, dolphins, and whales do not show the adverse effects associated with the ischemia/reperfusion-induced oxidative stress that have been reported in humans and other vertebrates (Halliwell and Gutteridge 1999).

DO DIVING VERTEBRATES PRODUCE REACTIVE OXYGEN SPECIES?

Early *ex vivo* studies on the effects of experimental ischemia on potential ROS production in diving and nondiving vertebrates showed that the heart and the kidney of the ringed seal (*Phoca hispida*) and the domestic pig (*Sus scrofa*) accumulate the ATP degradation product hypoxanthine (HX) after simulated ischemia, with kidney samples from both species accumulating more HX than heart samples (Elsner et al. 1995, 1998). These investigations also showed that both heart and kidney of ringed seals accumulate less HX than the heart and the kidney of domestic pigs. They suggest that the ringed seal tissues may counteract effects of diving-induced ischemia/reperfusion by the activation of mechanisms that ameliorate postischemic ROS production, such as HX salvage, by a pathway that utilizes it for resynthesis of ATP via inosine monophophate (IMP) catalyzed by the enzyme hypoxanthine guanine phosphoribosyl transferase (HGPRT) (Elsner et al. 1998). The differences between HX accumulation in heart and kidney of the ringed seal may be related to the differential perfusion to these tissues during diving. While coronary circulation may be intermittently reduced during dives, renal blood flow is often lowered or suspended entirely during extreme dives (Elsner et al. 1966, 1985; Kjekshus et al. 1982).

Further investigations demonstrated that the basal levels of $O_2^{\bullet -}$ production, measured with the cytochrome *c* assay (see Chapter 24), are higher in the heart, kidney, and muscle of ringed seals than in the heart,

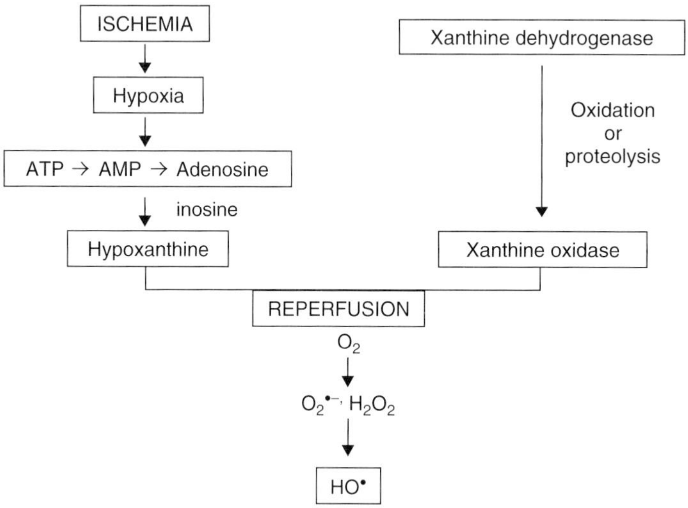

Fig. 13.1 Schematic representation of the molecular events leading to reactive oxygen species production during ischemia/reperfusion in mammals: O_2 = molecular oxygen; $O_2^{\bullet-}$ = superoxide radical; H_2O_2 = hydrogen peroxide; $HO^\bullet$ = hydroxyl radical. (Modified from Halliwell and Gutteridge 1999; after McCord 1987.)

kidney, and muscle of domestic pigs, with the muscle of the ringed seal having substantially higher $O_2^{\bullet-}$ production than heart or kidney (Zenteno-Savín and Elsner 1998, 2000). Furthermore, Zenteno-Savín et al. (2002) showed that the heart, kidney, and muscle of ringed seals have higher capacity to produce $O_2^{\bullet-}$ than the same tissues of domestic pigs. These results indicate that seals have higher capacity to produce ROS than nondiving mammals, presumably as a direct consequence of the cyclic bouts of ischemia/reperfusion to which they are routinely exposed during dives. When comparing $O_2^{\bullet-}$ production in a wide variety of marine birds and chicken (*Gallus domesticus*), Zenteno-Savín et al. (2010) found that the capacity to produce $O_2^{\bullet-}$ in pectoral muscle of emperor penguin is similar to that of pectoral muscle of nondiving birds and of sartorius muscle of chicken. This supports the concept that the tissues of diving birds and mammals can produce ROS in response to ischemia/reperfusion, and points to the potential role of antioxidants in avoiding reperfusion-derived oxidative damage and/or in modulating ROS production to activate ROS-mediated protective pathways (Halliwell and Gutteridge 1999). Recent studies (Vázquez-Medina et al. 2010, 2011a) also showed that ROS production increases with maturation in the muscle of elephant and hooded seals but that oxidative damage accumulation does not. These results, together with those showing maturation-related increases in antioxidant enzyme activities and protein content (Vázquez-Medina et al. 2010, 2011a, 2011b) support the idea that ROS may be playing a role in the activation of the antioxidant mechanisms that apparently allow seal and penguin tissues to counteract potential oxidative damage associated with diving-induced ischemia/reperfusion. Thus, present knowledge suggests that avoidance of ROS production is not the mechanism by which diving birds and mammals deal with the potential problems associated with ischemia/reperfusion.

Studies in our laboratories have also shown that $O_2^{\bullet-}$ production in brain, muscle, heart, lung, kidney, and liver of the deep-diving hooded seal is higher than in the same tissues of short-duration, shallow-diving harp seals and California sea lions (Vázquez-Medina et al. 2007b). Similar results were reported by Cantú-Medellín et al. (2011) when comparing tissues from bottlenose dolphins with tissues from deep-diving dwarf and pygmy sperm whales (kogiid whales). The same pattern of $O_2^{\bullet-}$ production among tissues has been found in almost all of the phocid and cetacean species analyzed by using the cytochrome *c* assay (see Chapter 24), with the liver being the tissue with highest $O_2^{\bullet-}$ production among the analyzed tissues, followed by muscle, which is exposed to progressive ischemia during apnea in seals (Ponganis et al. 2008). These results support the concept that the differences in ROS production among tissues of diving vertebrates are related to the degree of perfusion during dives, and that ROS in marine birds and mammals elicit protective pathways against acute progressive ischemia derived from extended diving, such as the potential activation of the HIF-1 pathway (Johnson et al. 2004, 2005).

ARE TISSUES OF DIVING ANIMALS SUSCEPTIBLE TO OXIDATIVE DAMAGE?

Remarkably, despite higher basal levels of $O_2^{\bullet-}$ production (Zenteno-Savín and Elsner 1998, 2000) and increased ROS production in heart, kidney, and muscle of the ringed seal than in the same tissues of the domestic pig (Zenteno-Savín et al. 2002), neither higher lipid peroxidation (thiobarbituric acid reactive substances (TBARS) content) nor higher protein oxidation (protein carbonyl levels) were found in the heart, kidney, liver, lung, or muscle of the ringed seal than in the domestic pig (Zenteno-Savín et al. 2002; Vázquez-Medina et al. 2007a). Similarly, TBARS content in red blood cell (RBC) lysates of a group of marine mammals (elephant seals, manatee, dwarf minke whale, striped and franciscana dolphins) is lower than in the RBC of a group of terrestrial mammals (deer, anteater, raccoon, ferret, and monkey) (Wilhelm-Filho et al. 2002). No increase in lipid peroxidation (TBARS) levels were found in the liver and pectoral muscle of the emperor penguins when compared to liver and muscle of a wide variety of marine birds and the chicken (Zenteno-Savín et al. 2010). Despite finding higher $O_2^{\bullet-}$ production in the skeletal muscle of adult hooded seals than in the skeletal muscle of newborn hooded seals, neither TBARS content, protein carbonyl levels, nor oxidatively modified DNA content (8-oxo-7, 8-dihydro-2′-deoxyguanosine) were found to be increased in muscle of adult hooded seals (Vázquez-Medina et al. 2011a). Moreover, the higher levels of $O_2^{\bullet-}$ production found in tissues of long deep divers, such as hooded seals and kogiid whales, than in tissues of short divers, such as California sea lions and dolphins, were not associated with higher levels of TBARS or protein carbonyls (Vázquez-Medina et al. 2007b; Cantú-Medellín et al. 2011). The combined results suggest the active participation of the antioxidant system in preventing oxidative damage accumulation, activation of effective protective pathways and/or enhanced repair systems in tissues of diving birds and mammals.

ARE ANTIOXIDANT ENZYME ACTIVITIES ELEVATED IN DIVING VERTEBRATES?

An early comparative study demonstrated that the basal activity of superoxide dismutase (SOD) was higher in the heart of the ringed seal than in the heart of the domestic pig (Elsner et al. 1998). Higher glutathione peroxidase (GPx), glutathione reductase (GR) and glucose-6-phosphate dehydrogenase (G6PDH) activities have also been found in the heart of the ringed seal than in the heart of the domestic pig (Vázquez-Medina et al. 2006, 2007a). These authors suggest that the intermittent coronary blood flow during diving may precondition seal's heart by activating its enzymatic antioxidant system.

Other tissues of the ringed seal also show constitutively higher antioxidant enzyme activities than tissues of the domestic pig. The lungs of the ringed seal have higher SOD, GPx, GR, and G6PDH activities than the lungs of the domestic pig; the skeletal muscle of the ringed seal has higher GPx, GR, and G6PDH activities than the muscle of the domestic pig; the liver of the ringed seal has higher catalase (CAT), GR, and G6PDH activities than the liver of the domestic pig, and the kidney of the ringed seal has higher GR and G6PDH activities than the kidney of the domestic pig (Vázquez-Medina et al. 2006, 2007a). Similarly, SOD, GPx, and GR activities are higher in the RBC of a group of diving mammals than in the RBC of a group of terrestrial mammals (Wilhelm-Filho et al. 2002). Elevated activity and/or protein content of Cu,Zn-SOD, Mn-SOD, CAT, GPx, GR, GST, G6PDH, glutamate-cysteine ligase (GCL), γ-glutamyl transpeptidase (GGT) and 1-cys peroxiredoxin (PrxVI) have also been found in the muscle and/or the RBC of young, post-weaning northern elephant seals immediately before the beginning of their diving lifestyle (Vázquez-Medina et al. 2010, 2011b). SOD activities and Mn-SOD protein content are higher in muscle of adult deep-diving hooded seals than in muscle of neonatal and weaned hooded seal pups (Vázquez-Medina et al. 2011a). These determinations suggest that the tissues of diving mammals have an acute enzymatic antioxidant system that reduces post-dive ROS production and avoids oxidative damage (Zenteno-Savín et al. 2002; Vázquez-Medina et al. 2006, 2007a, 2010, 2011a, 2011b). Of special interest are the results of the glutathione (GSH)-related antioxidant enzymes. GR, G6PDH, and/or GPx activities have been shown to be generally higher in marine mammal tissues than in those of their terrestrial counterparts. This suggests the importance of the combined roles of both the antioxidant enzyme system and the nonenzymatic antioxidant GSH.

Higher CAT and GST activities have also been found in the pectoral muscle of emperor penguins than in the pectoral muscle of a wide variety of nondiving marine birds, and the sartorious muscle of the chicken (Zenteno-Savín et al. 2010). Emperor penguin liver also has higher CAT, GPx, and GST activities than the liver of several marine birds and the liver of the chicken (Zenteno-Savín et al. 2010). These data suggest that the contribution of CAT, GPx, and GST protects against dive-related oxidative damage in emperor penguin liver and muscle. Phylogenetic differences, as well as differences in muscle fiber type and oxidative capacity, metabolic rates, mass-specific O_2 flow, and/or degree of ischemia during dives (Meir and Ponganis 2009; Meir et al. 2009) may be responsible for the differences in the antioxidant enzyme pattern observed between penguins and seals (Zenteno-Savín et al. 2010).

In a comparison of antioxidant enzyme profiles, similar antioxidant enzyme activities were found in the brain, heart, kidney, liver, or muscle of the long-diving hooded seal than in tissues of the brief-diving harp seal or the California sea lion (Vázquez-Medina et al. 2007b). Similar results were reported by Cantú-Medellín et al. (2011) when comparing antioxidant enzyme activities in tissues of deep-diving kogiid whales and shallow-diving cetacean species. These results suggest that enhanced basal antioxidant enzyme activities allow both shallow- and deep-diving birds and mammals to avoid oxidative damage induced by ischemia/reperfusion.

ARE NONENZYMATIC ANTIOXIDANTS HIGHER IN DIVING VERTEBRATES?

The low molecular weight tri-peptide thiol γ-L-glutamyl-L-cysteinyl glycine, known as GSH, is the most important nonenzymatic endogenous antioxidant in animal cells (Forman et al. 2009). GSH content is modulated by factors such as its rate of utilization in cellular oxidation/reduction reactions, the enzymatic reduction of hydroperoxides, xenobiotic metabolism, and regeneration of reduced forms of redox pairs (Kirlin et al. 1999; Schafer et al. 2000). The role of GSH as a potential protective mechanism against dive-induced ischemia reperfusion in marine birds and mammals has hardly been investigated (Murphy and Hochachka 1981; Wilhelm-Filho et al. 2002; Vázquez-Medina et al. 2007a).

Blood GSH content in Weddell seals was found to be almost twice that in humans (Murphy and Hochachka 1981). The same study showed that blood GSH content drops dramatically during forced submersion and rapidly increases above pre-dive levels upon recovery; suggesting that GSH may have a potential protective role against reoxygenation-derived ROS production. A more recent investigation supports the earlier study by showing that the mean GSH concentration in RBC lysates from a group of diving mammals (elephant seals, manatee, mink whale, striped dolphin, and franciscana dolphin) is about two times higher than the mean GSH content in the RBC of a group of terrestrial mammals (monkey, otter, ferret, raccoon, and anteater) (Wilhelm-Filho et al. 2002).

GSH content is also higher in the heart, kidney, lungs and muscle of ringed seal than in the heart, kidney, lungs, and muscle of domestic pig (Vázquez-Medina et al. 2007a). Remarkably, GSH content in the heart of the ringed seal is almost 20 times higher than in the heart of the pig, while the levels in the lungs and muscle are four times higher, and those in kidney two times higher (Vázquez-Medina et al. 2007a). Moreover, the latter is associated with increased GR and G6PDH activities in the heart, kidney, liver, lungs, and muscle of the ringed seal (Vázquez-Medina et al. 2007a). Both, GR and G6PDH are key players in the recycling of GSSG to GSH. GR restores GSH from GSSG at the expense of NADPH, which is recycled from $NADP^+$ by G6DPH maintaining the reduced intracellular GSH pool. GSH is synthesized from glutamate, cysteine, and glycine in a two-step enzymatic process catalyzed by the enzymes GCL and glutathione synthase. GSH homeostasis is partially maintained by GGT, which breaks down extracellular GSH providing cysteine, the rate-limiting substrate for intracellular GSH *de novo* synthesis (Kosower and Kosower 1978; Meister and Anderson 1983).

An elevated rate of GSH recycling in diving mammals may allow the maintenance of this tripeptide in order to support its functions as a redox homeostasis modulator, $HO^\bullet$ and $^1O_2$ scavenger, and co-substrate for the antioxidant enzymes GPx, GST, and PrxVI (Kosower and Kosower 1978; Meister and Anderson 1983; Schaefer et al. 2000). Supporting evidence for this is provided by the higher GPx and GST activities found in heart, lungs, and muscle of ringed seals than in the same tissues of domestic pigs (Vázquez-Medina et al. 2006). Higher GPx activity has been reported in the blood of diving mammals than in the blood of terrestrial mammals (Wilhelm-Filho et al. 2002), as well as in the liver and muscle of emperor penguin than in the liver and muscle of nondiving birds (Zenteno-Savín et al. 2010). Furthermore, increased GR, G6PDH, GCL,

GGT, GPx, and PrxVI protein expression and activity in muscle and RBC have been found in elephant seal pups immediately before the beginning of their diving lifestyle, suggesting the potential protective role of the GSH system against dive-induced ROS production (Vázquez-Medina et al. 2010, 2011b).

In terrestrial models, dietary supplementation of GSH protects the heart against ischemia/reperfusion (Ahmad et al. 2001; Ramires and Ji 2001). GSH levels also appear to be the cellular control mechanism that regulates GPx activity and H_2O_2 levels in response to changes in oxygen tension in coronary arteries (Mohazzab-H et al. 1999), and the mechanism that controls cellular P_{O_2} (Del Corso et al. 2002). High GSH levels have been observed in tissues of terrestrial animals that are routinely exposed to changes in oxygen availability by hibernation, estivation, desiccation, environmental hypoxia, or diving (Hermes-Lima and Zenteno-Savín 2002).

To our knowledge, no studies examining the role of GSH in diving birds have been conducted to date. Our results (Zenteno-Savín et al. 2010), however, suggest an active participation of the GSH system in the antioxidant protection of diving birds similar to diving mammals (Wilhelm-Filho et al. 2002; Vázquez-Medina et al. 2006, 2007a, 2010, 2011b). Maintaining high GSH levels in preparation for post-diving ROS production seems to be an important mechanism for counteracting diving-induced ischemia/reperfusion, because GSH recycling is quick and less expensive than enzyme synthesis (Kosower and Kosower 1978; Meister and Anderson 1983). Thus, maintaining high levels of GSH and a reduced cell environment may allow the activation of redox-dependent protective mechanisms in response to diving-induced ischemia, such as the hypoxia inducible factor 1 (HIF-1; Johnson et al. 2004, 2005; Tajima et al. 2009).

The role of other low-molecular antioxidants in the protection against diving-induced ischemia/reperfusion has also been studied in diving birds and mammals. Higher total oxyradical scavenging capacity (TOSC) for ROO• (see Chapter 26) was found in the plasma of Adélie and emperor penguins when compared to plasma of nondiving south polar skuas (*Stercorarius maccormicki*) and snow petrels (*Pagodroma nivea*) (Corsolini et al. 2001). Total antioxidant status is also higher in the heart, kidney, and muscle of the ringed seal than in the heart, kidney, and muscle of the domestic pig (Zenteno-Savín et al. 2002).

Exogenous low-weight antioxidants, such as vitamins, in marine birds and mammals have been examined extensively from nutritional and toxicological perspectives. Kasamatsu et al. (2009) showed that plasma α-tocopherol (vitamin E) content is higher in bottlenose dolphins (*Tursiops truncatus*) and spotted seals (*Phoca largha*) than in dogs or cows (Kasamatsu et al. 2009). However, vitamin content and distribution is highly influenced by the feeding habits of a given species (Williams 1989; Bollengier-Lee et al. 1998). In that regard, studies in several marine mammal species show that α-tocopherol content is higher in fish-eating than in nonfish-eating diving mammals (Kasamatsu et al. 2009). In the same way, vitamin dynamics in Humboldt penguins (*Spheniscus humboldti*) and bottlenose dolphins are largely influenced by their diet, since retinol, retinyl palmitate, and vitamin E storage and utilization are different between captive and free-ranging animals (Crissey et al. 1998; Crissey and Wells 1999), and retinoid levels vary significantly depending on the animals' lipid content (Tornero et al. 2004). Moreover, contaminant exposure also seems to have an effect on vitamin concentration and distribution in marine birds and mammals (Mos and Ross 2002; Jensen et al. 2003; Nyman et al. 2003; Routti et al. 2005; Rosa et al. 2007). More focused studies are needed to elucidate the role of vitamins and other exogenous low-molecular antioxidants on the cellular responses to diving-induced ischemia/reperfusion in marine birds and mammals. However, since it is well established that vitamins have a protective role against ischemia/reperfusion injury and reduce the risk of heart disease in humans (Gey et al. 1991, 1993; Stampfer et al. 1993; Bostick et al. 1999), we hypothesize that vitamins could be important players in the synergistic antioxidant protection of marine birds and mammals in tolerating dive-induced ischemia/reperfusion.

HYPOXIA-INDUCIBLE FACTOR IN DIVING VERTEBRATES

HIF-1α is a key protein in coordinating the adaptive homeostatic responses to hypoxia by regulating the expression of vascular, glycolytic, and cell-cycle regulatory genes (see Chapter 12). HIF-1α also regulates expression of the heat shock factor (HSF), the main factor in the activation of heat shock proteins (HSPs), and a critical element in the adaptation to reoxygenation-derived ROS production (Date et al. 2005; Baird et al.

2006). Recent studies from our laboratories have established that the genome of the ringed seal contains a HIF-1α gene that is similar to the HIF-1α gene of terrestrial mammals, and that the HIF-1α and HIF-1β proteins are constitutively expressed in the heart, kidney, liver, and lungs of the ringed seal. Moreover, our studies also suggest that, in some cases, HIF-1α protein expression in tissues of the ringed seal correlate negatively with the level of protein oxidation (i.e. protein carbonyls content) (Johnson et al. 2004). These results suggest involvement of the HIF-1 system in the responses against dive-induced ischemia in seal tissues.

In contrast to what has been observed in tissues of terrestrial mammals, the constitutive expression of HIF-1 proteins in tissues of the ringed seal cannot be explained by the presence of multiple copies of the HIF-1α gene, the absence of the regulatory von Hippel-Lindau tumor suppressor protein (pVHL), or major differences in the partial amino acid sequence of HIF-1α proteins between tissues of seals and tissues of terrestrial mammals (Johnson et al. 2005). Moreover, the constitutive expression of the HIF-1α protein in tissues of ringed seals correlates with reports of higher hematocrit from more active erythropoiesis and a greater brain capillary density (from increased vascular endothelial growth factor (VEGF) and angiogenesis), as well as more intensive oxidative and glycolytic metabolism in marine compared to terrestrial mammals (Kooyman and Ponganis 1998; Elsner 1999). Thus, it is possible that the constitutive expression of HIF-1α in tissues of the ringed seal is part of the physiological adaptation to frequent progressive asphyxia associated with diving (Elsner and Gooden 1983). However, the mechanism that promotes HIF-1α stabilization in tissues of diving seals remains uninvestigated. It is believed that $HO^{\bullet}$ contributes to controlling the activity of HIF-1α, and that H_2O_2 is required for the binding of HIF-1 to DNA (Soitamo et al. 2001). Although $HO^{\bullet}$ production has not been measured in the ringed seal, the elevated $O_2^{\bullet -}$ production and SOD activity in these diving mammals suggest that H_2O_2 may play a role, via HIF-1, in the physiological adaptation to repetitive diving.

Recent studies also suggest that GSH has a crucial role in controlling the nonhypoxic activation of HIF-1α since a highly reducing cell environment appears to be necessary for HIF-1 stabilization (Salceda and Caro 1997; Haddad 2002a–c; Haddad and Harb 2005). In the North Sea eelpout (*Zoarces viviparus*) acute cold-exposure (which seems to induce similar thermal hypoxia in fish with similar effects as diving ischemia in mammals) promotes a more reduced cellular redox state and high GSH levels that support HIF-1 DNA binding ability previous to reoxygenation (Heise et al. 2006a,b). Further, the high levels of GSH found in tissues of diving mammals probably promote a highly reduced redox status that fosters the redox-dependent activation of HIF-1 genes.

Buetler et al. (2004) identified $O_2^{\bullet -}$ as an important molecule for cellular signaling processes. Haddad and Land (2001), Haddad (2002a–d) and Haddad and Harb (2005) proposed that the nonhypoxic activation of HIF-1 in mammalian lungs is mediated by ROS and GSH, the concentration of which is higher in the lungs of the ringed seal than in pig lungs. Ringed seal lungs have higher levels of HIF-1α than the heart, kidney, liver, or muscle (Johnson et al. 2004, 2005). pVHL is a protein that by binding to HIF-1α under normoxic conditions promotes its ubiquitination and subsequent degradation by the proteasome (Salceda and Caro 1997). During hypoxia, pVHL does not interact with HIF-1α and spares it from being degraded. Therefore, hypoxia promotes the interaction of HIF-1α with HIF-1β, the formation of the HIF-1 complex and its translocation to the nucleus (Maxwell et al. 1999; Jaakkola et al. 2001). High accumulation of HIF-1α, and HIF-1β, despite the presence of high levels of pVHL in ringed seal lungs (Johnson et al. 2005), as well as high levels of GSH (Vázquez-Medina et al. 2007a, 2011b) and $O_2^{\bullet -}$ production (Zenteno-Savín et al. 2002), suggest that there is a nonhypoxic activation of HIF-1 in the lungs of the ringed seal, probably supported by ROS (which inhibit the PHDs) and GSH. This could result in activation of other protective mechanisms against dive-associated ischemia/reperfusion and in recovery of post-dive pulmonary function (Johnson et al. 2005).

CONCLUSIONS AND PERSPECTIVES

It is well established that the tissues of diving vertebrates are capable of producing ROS in response to diving-induced ischemia/reperfusion, and that ROS production is higher in long-duration divers than in short-duration divers. Thus, avoidance of the production of these potentially damaging molecular species is not the mechanism by which diving vertebrates avoid reperfusion-derived oxidative damage. Factors such as the degree of ischemia during dives appear to influence the tissue capacity to produce ROS in diving birds and mammals. Although more evidence is needed to corroborate the potential role of HX salvage in the modulation of ROS production in tissues of diving vertebrates, it is likely that this pathway can maintain the rate of ROS production at the levels needed to activate protective pathways without promoting oxidative damage. Despite presenting higher rates of ROS production, the tissues of diving vertebrates do not have higher oxidative damage than the tissues of nondiving vertebrates. In the same way, despite having higher capacity to produce ROS, the tissues of long deep divers do not have higher oxidative damage than the tissues of short shallow divers.

The latter can be explained by the synergetic effect of constitutively higher antioxidant enzyme activities and GSH content in tissues of diving birds and mammals than in tissues of nondiving birds and mammals. Although the antioxidant patterns are different between seals and penguins, the GSH system seems to play a key role in avoiding reperfusion-derived oxidative damage, as well as a potential activator of redox-dependent mechanisms such as the nuclear accumulation of HIF-1, which is constitutively expressed in the heart, kidney, liver, lungs, and skeletal muscle of the ringed seal during nonhypoxic conditions, and can potentially contribute to seal's tolerance to diving-induced ischemia/reperfusion (Fig. 13.2).

More studies are needed to elucidate the potential role of HX salvage in the modulation of ROS production in tissues of diving birds and mammals. In the same way, the role of other nonenzymatic antioxidants in the tolerance of diving birds and mammals against diving-induced ischemia/reperfusion remains to be investigated. Furthermore, the oxidative stress responses before, during, and after diving in marine birds and mammals, as well as the molecular and cellular mechanisms driving their natural tolerance

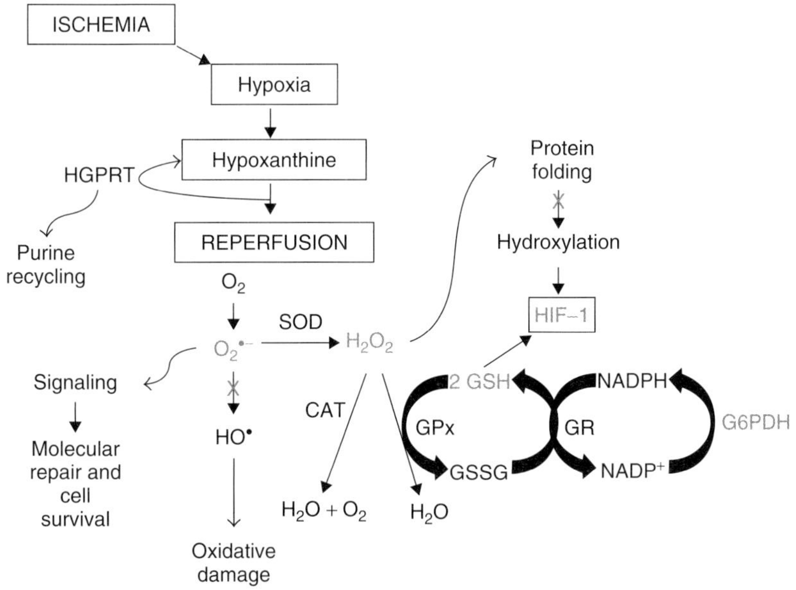

Fig. 13.2 Schematic representation of the contribution of reactive oxygen species and antioxidant defenses in protection of tissues from diving birds and mammals against dive-derived ischemia/reperfusion. $O_2^{\bullet-}$ and H_2O_2 produced in response to dive-induced ischemia/reperfusion act as signaling molecules to induce molecular and cellular protective mechanisms. An enhanced antioxidant system protects diving vertebrates from the deleterious effects of $HO^{\bullet}$. Reactive oxygen species affect protein folding and, therefore, the tridimensional structure of HIF−1α, which precludes it from hydroxylation by prolyl hydroxylases, resulting in its increased constitutive expression in tissues from diving vertebrates. HGPRT = hypoxanthine-guanine phosphoribosyltransferase; O_2 = molecular oxygen; $O_2^{\bullet-}$ = superoxide radical; H_2O_2 = hydrogen peroxide; $HO^{\bullet}$ = hydroxyl radical; H_2O = water; SOD = superoxide dismutase; CAT = catalase; GPx = Glutathione peroxidase; GR = glutathione reductase; G6PDH = glucose 6-phosphate dehydrogenase; GSH = glutathione, reduced; GSSG = Glutathione, oxidized; NADPH = nicotinamide adenine dinucleotide phosphate, reduced; NADP+ = nicotinamide adenine dinucleotide phosphate, oxidized; HIF-1 = hypoxia-inducible factor. See plate section for a color version of this image.

against diving-induced ischemia/reperfusion injury remain unclear but are currently being investigated in our laboratories.

REFERENCES

Ahmad, S., White, C.W., Chang, L.Y., Schneider, B.K., Allen, C.B. (2001) Glutamine protects mitochondrial structure and function in oxygen toxicity. *American Journal of Physiology* 280, L779–L791.

Baird, N.A., Turnbull, D.W., Johnson, E.A. (2006) Induction of the heat shock pathway during hypoxia requires regulation of heat shock factor by hypoxia-inducible factor-1. *Journal of Biological Chemistry* 281, 38675–38681.

Bevan, R.M., Butler, P.J. (1992) The effects of temperature on the oxygen consumption, heart rate and deep body temperature during diving in the tufted duck, *Aythya fuligula*. *Journal of Experimental Biology* 163, 139–151.

Bollengier-Lee, S., Mitchell, M.A., Utomo. D.B., Williams, P. E. V., Whitehead, C.C. (1998) Influence of high dietary vitamin E supplementation on egg production and plasma characteristics in hens subjected to heat stress. *British Poultry Science* 39(1), 106–112.

Bostick, R.M., Kushi, L.M., Wu, Y., Meyer, K.A., Sellers, T.A., Folsom, A.R. (1999) Relation of calcium, vitamin D, and dairy food intake to ischemic heart disease mortality among postmenopausal women. *American Journal of Epidemiology* 149, 151–161.

Buetler, T.M., Krauskopf, A., Ruegg, U.T. (2004) Role of superoxide as a signaling molecule. *News in Physiological Sciences* 19, 120–123.

Cardona-Maldonado, M.A., Mignucci-Giannoni, A.A. (1999) Pygmy and dwarf sperm whales in Puerto Rico and the Virgin Islands, with a review of Kogia in the Caribbean. *Caribbean Journal of Science* 35, 29–37.

Cantú-Medellín, N., Byrd, B., Hohn, A., Vázquez-Medina, J.P., Zenteno-Savín, T. (2011) Differential antioxidant protection in tissues from marine mammals with distinct diving capacities. Shallow/short vs. deep/long divers. *Comparative Biochemistry and Physiology A* 158, 438–443.

Clarke, M.R. (2003) Production and control of sound by the small sperm whales, *Kogia breviceps* and *K. sima* and their implications for other Cetacea. *Journal of the Marine Biology Association UK* 83, 241–263.

Corsolini, S., Nigro, M., Olmastroni, S., Focardi, S., Regoli, F. (2001) Susceptibility to oxidative stress in Adélie and emperor penguin. *Polar Biology* 24, 365–368.

Crissey, S.D., Wells, R. (1999) Serum a- and g-tocopherols, retinol, retinyl palmitate, and carotenoid concentrations in captive and free-ranging bottlenose dolphins (*Tursiops truncatus*). *Comparative Biochemistry and Physiology B* 124, 391–396.

Crissey, S.D., McGill, P., Simeone, A.M. (1998) Influence of dietary vitamins A and E on serum α- and γ-tocopherols, retinol, retinyl palmitate and carotenoid concentrations in Humboldt penguins (*Spheniscus humboldti*). *Comparative Biochemistry and Physiology* 121, 333–339.

Croll, D.A., Acevedo-Gutiérrez, A., Tershy, B.R., Urbán-Ramírez, J. (2001) The diving behavior of blue and fin whales: is dive duration shorter than expected based on oxygen stores? *Comparative Biochemistry and Physiology A* 129, 797–809.

Croxall, J.P., Davis, R.W., O'Connel, M.J. (1988) Diving patterns in relation to diet of Gentoo and Macaroni penguins at South Georgia. *The Condor* 90, 157–167.

Date, T., Mochizuki, S., Belanger, A.J. et al. (2005) Expression of constitutively stable hybrid hypoxia-inducible factor-1α protects cultured rat cardiomyocytes against simulated ischemia-reperfusion injury. *American Journal of Physiology* 288, C314–C320.

Davis, R.W., Fargion, G.S., May, N. et al. (1998) Physical habitat of cetaceans along the continental slope in the northcentral and western Gulf of Mexico. *Marine Mammal Science* 14, 490–507.

Del Corso, A., Vilardo, P.G., Cappiello, M. et al. (2002) Physiological thiols as promoters of glutathione oxidation and modifying agents in protein S-thiolation. *Archives of Biochemistry and Biophysics* 397, 392–398.

Elsner, R. (1999) Living in water: solutions to physiological problems. *In* Reynolds III, J.E., Rommel, S.A. (eds). *Biology of Marine Mammals*. Smithsonian Institution Press, Washington, DC, pp. 73–116.

Elsner, R., Gooden, B.A. (1983) *Diving and Asphyxia. A Comparative Study of Animals and Man*. Monographs of the Physiological Society, 140, Cambridger University Press, Cambridge, 168 pp.

Elsner, R., Franklin, D.L., Van Citters, R.L., Kenney, D.W. (1966) Cardiovascular defense against asphyxia. *Science* 153, 941–949.

Elsner, R., Millard, R.W., Kjekshus, J., White, F.C., Blix, A.S., Kemper, S. (1985) Coronary circulation and myocardial segment dimensions in diving seals, *American Journal of Physiology* 249, H1119–H1126.

Elsner, R., Øyasaeter, S., Saugstad, O.L., Scytte-Blix, A. (1995) Seal adaptations for long dives: recent studies of ischemia and oxygen radicals. *In* Blix, A., Walloe, S.L., Ultang, O. (eds). *Developments in Marine Biology, Whales, Seals, Fish and Man. International Symposium on the Biology or Marine Mammals in the North East Atlantic, Tromso, Norway, 1994*, Vol. 4. Elsevier Science, Amsterdam, pp. 371–376.

Elsner, R., Øyasaeter, S., Almaas, R., Saugstad, O.L. (1998) Diving seals, ischemia-reperfusion and oxygen radicals. *Comparative Biochemistry and Physiology A* 119, 975–980.

Folkow, L.P., Blix, A.S. (1999) Diving behaviour of hooded seals (*Cystophora cristata*) in the Greenland and Norwegian seas. *Polar Biology* 22, 61–74.

Forman, H.J., Zhang, H., Rinna, A. (2009) Glutathione: overview of its protective roles, measurement, and biosynthesis. *Molecular Aspects of Medicine* 30, 1–12.

Gey, K.F., Puska, P., Jordan, P., Moser, U.K. (1991) Inverse correlation between plasma vitamin E and mortality from ischemic heart disease in cross-cultural epidemiology. *American Journal of Clinical Nutrition* 53, 326S–334S.

Gey, K.F., Stähelin, H.B., Eichholzer, M. (1993) Poor plasma status of carotene and vitamin C is associated with higher mortality from ischemic heart disease and stroke Basel prospective study. *Clinical Investigator* 71, 3–6.

Haddad, J.J. (2002a) Science review: redox and oxygen-sensitive transcription factors in the regulation of oxidant-mediated lung injury: role for nuclear factor-ß.B. *Critical Care* 6, 481–490.

Haddad, J.J. (2002b) Antioxidant and prooxidant mechanisms in the regulation of redox(y)-sensitive transcription factors. *Cellular Signalling* 14, 879–897.

Haddad, J.J. (2002c) Oxygen homeostasis, thiol equilibrium and redox regulation of signalling transcription factors in the alveolar epithelium. *Cellular Signalling* 14, 799–810.

Haddad, J.J. (2002d) Oxygen homeostasis, thiol equilibrium and redox regulation of signalling transcription factors in the alveolar epithelium. *Cellular Signalling* 14, 799–810.

Haddad, J.J., Harb, H.L. (2005) L-gamma-glutamyl-l-cysteinyl-glycine (glutathione; GSH) and GSH-related enzymes in the regulation of pro- and anti-inflammatory cytokines: a signaling transcriptional scenario for redox(y) immunologic sensor(s)? *Molecular Immunology* 42, 987–1014.

Haddad, J.J., Land, S.C. (2001) A non-hypoxic, ROS-sensitive pathway mediates TNF-α-dependent regulation of HIF-1α. *FEBS Letters* 505, 269–274.

Halliwell, B., Gutteridge, J. M. C. (1999) *Free Radicals in Biology and Medicine*. Oxford University Press. Oxford.

Heise, K., Puntarulo, S., Nikinmaa, M., Lucassen, M., Portner, H.O., Abele, D. (2006a) Oxidative stress and HIF-1 DNA binding during stressful cold exposure and recovery in the North Sea eelpout (*Zoarces viviparus*). *Comparative Biochemistry and Physiology A* 143, 494–503.

Heise, K., Puntarulo, S., Nikinma, M., Abele, D., Portner, H.O. (2006b) Oxidative stress during stressful heat exposure and recovery in the North Sea eelpout *Zoarces viviparus*. *Journal of Experimental Biology* 209, 353–363.

Hermes-Lima, M., Zenteno-Savín, T. (2002) Animal response to drastic changes in oxygen availability and physiological oxidative stress. *Comparative Biochemistry and Physiology C* 133, 537–556.

Jaakkola, P., Mole, D.R., Tian, Y.M. et al. (2001) Targeting of HIF-alpha to the von Hippel-Lindau ubiquitylation complex by O_2-regulated prolyl hydroxylation. *Science* 292, 468–472.

Jensen, B.J., Haugen, O., Sørmo, E.G., Skaare, J.U. (2003) Negative relationship between PCBs and plasma retinol in low-contaminated free-ranging gray seal pups (*Halichoerus grypus*). *Environmental Research* 93, 79–87.

Johnson, P., Elsner, R., Zenteno-Savín, T. (2004) Hypoxia-inducible factor in ringed seal (*Phoca hispida*) tissues. *Free Radical Research* 38, 847–854.

Johnson, P., Elsner, R., Zenteno-Savín, T. (2005) Hypoxia-inducible factor proteomics and diving adaptations in ringed seal. *Free Radical Biology and Medicine* 39, 205–212.

Kasamatsu, M., Kawauchia, R., Tsunokawab, M. et al. (2009) Comparison of serum lipid compositions, lipid peroxide, α-tocopherol and lipoproteins in captive marine mammals (bottlenose dolphins, spotted seals and West Indian manatees) and terrestrial mammals. *Research in Vetinary Science* 86, 216–222.

Kelly, B.P., Wartzok, D. (1996) Ringed seal diving behavior in the breeding season. *Canadian Journal of Zoology* 74, 1547–1555.

Kirlin, W.G., Cai, J., Thompson, S.A., Diaz, D., Kavanagh, T.J., Jones, D.P. (1999) Glutathione redox potential in response to differentiation and enzyme inducers. *Free Radical Biology and Medicine* 27, 1208–1218.

Kjekshus, J., Blix, A.S., Elsner, R., Hol, R., Amundsen, E. (1982) Myocardial blood flow and metabolism in the diving seal. *American Journal of Physiology* 242, R97–R104.

Kooyman, G.L. (1966). Maximum diving capacities of the Weddell seal, *Leptonychotes weddelli*. *Science* 151, 1553–1554.

Kooyman, G.L. (1989) *Diverse Divers*. Springer-Verlag. Heildelberg.

Kooyman, G.L. (2002) Diving Physiology. *In* Perrin, W.F., Würsig, B., Thewissen, J. G. M. (eds) *Encyclopedia of Marine Mammals*. Academic Press, San Diego, pp. 339–344.

Kooyman, G.L., Ponganis, P.J. (1998) The physiological basis of diving to depth: birds and mammals. *Annual Review of Physiology* 60, 19–32.

Kosower, N.S., Kosower, E.M. (1978) The glutathione status of cells. *International Review of Cytology* 54, 109–160.

Kuppusamy, P., Zweier, J.L. (1989) Characterization of free radical generation by xanthine oxidase. Evidence for hydroxyl radical generation. *Journal of Biological Chemistry* 264, 9880–9884.

Maxwell, P.H., Wiesener, M.S., Chang, G.W. et al. (1999) The tumour suppressor protein VHL targets hypoxia-inducible factors for oxygen- dependent proteolysis. *Nature* 399, 271–275.

Meir, J.U., Ponganis, P.J. (2009) High-affinity hemoglobin and blood oxygen saturation in diving emperor penguins. *Journal of Experimental Biology* 212, 3330–3338.

Meir, J.U., Champagne, C.D., Costa, D.P., Williams, C.L., Ponganis, P.J. (2009) Extreme hypoxemic tolerance and blood oxygen depletion in diving elephant seals. *American Journal of Physiology* 297, R927–R939.

Meister, A., Anderson, M.E. (1983). Glutathione. *Annual Review of Biochemistry* 52, 711–760.

Mohazzab-H, K.M., Agarwal, R., Wolin, M.S. (1999) Influence of glutathione peroxidase on coronary artery responses to alterations in P_{O_2} and H_2O_2. *American Journal of Physiology* 276, H235–H241.

Mos, L., Ross, P.S. (2002) Vitamin A physiology in the precocious harbor seal (*Phoca vitulina*): a tissue-based biomarker approach. *Canadian Journal of Zoology* 80, 1511–1519.

Murphy, B.J., Hochachka, P.W. (1981) Free amino acid profiles in blood during diving and recovery in the Antarctic Weddell seal. *Canadian Journal of Zoology* 59, 455–459.

Naito, Y., Asaga, T., Ohyama, Y. (1990) Diving behavior of Adélie penguins determined by time-depth recorder. *The Condor* 92, 582–586.

Nyman, M., Bergknut, M., Fant, M.L. et al. (2003) Contaminant exposure and effects in Baltic ringed and grey seals as assessed by biomarkers. *Marine Environmental Research* 55, 73–99.

Ponganis, P.J., Kreuter, U., Stockard, T.K. et al. (2008) Blood flow and metabolic regulation in seal muscle during apnea. *Journal of Experimental Biology* 211, 3323–3332.

Ramires, P.R., Ji, L.L. (2001) Glutathione supplementation and training increases myocardial resistance to ischemia-reperfusion in vivo. *American Journal of Physiology* 281, H679–688.

Ramírez, J.M., Folkow, L.P., Blix, A.S. (2007) Hypoxia tolerance in mammals and birds: From the wilderness to the clinic. *Annual Review of Physiology* 69, 113–143.

Routti, H., Nyman, M., Bäckman, C., Koistinen, J., Helle, E. (2005) Accumulation of dietary organochlorines and vitamins in Baltic seals. *Marine Environmental Research* 60, 267–287.

Rosa, C., Blake, J.E., Mazzaro, L., Hoekstra, P., Ylitalo, G.M., O'Hara, T.M. (2007) Vitamin A and E tissue distribution with comparisons to organochlorine concentrations in the serum, blubber and liver of the bowhead whale (*Balaena mysticetus*). *Comparative Biochemistry and Physiology B* 148, 454–462.

Salceda, S., Caro, J. (1997) Hypoxia-inducible factor 1α (HIF-1α) protein is rapidly degraded by the ubiquitin-proteasome system under normoxic conditions. Its stabilization by hypoxia depends on redox-induced changes. *Journal of Biological Chemistry* 272, 22642–22647.

Schafer, F.Q., Qian, S.Y., Buettner, G.R. (2000) Iron and free radical oxidations in cell membranes. *Cell and Molecular Biology* 46, 657–662.

Soitamo, A.J., Rabergh, C.M., Gassmann, M., Sistonen, L., Nikinmaa, M. (2001) Characterization of a hypoxia-inducible factor (HIF-1alpha) from rainbow trout: accumulation of protein occurs at normal venous oxygen tension. *Journal of Biological Chemistry* 276, 19699–19705.

Stampfer, M.J., Hennekens, C.H., Manson, J.E., Colditz, G.A., Rosner, B., Willett, W.C. (1993) Vitamin E consumption and the risk of coronary disease in women. *New England Journal of Medicine* 328, 1444–1449.

Stewart, B.S. (2002) Diving behavior. *In* Perrin, W.F., Würsig, B., Thewissen, J. G. M. (eds). *Encyclopedia of Marine Mammals*. Academic Press, San Diego, pp. 333–339.

Tajima, M., Kurashimab, Y., Sugiyamac, K., Ogurab, T., Sakagamia, H. (2009) The redox state of glutathione regulates the hypoxic induction of HIF-1. *European Journal of Pharmacology* 606, 45–49.

Thompson-Gorman, S.L., Zweier, J.L. (1990) Evaluation of the role of xanthine oxidase in myocardial reperfusion injury. *Journal of Biological Chemistry* 265, 6656–6663.

Tornero, T., Borrella, A., Forcad, J., Pubilla, E., Aquilar, A. (2004) Retinoid and lipid patterns in the blubber of common dolphins (*Delphinus delphis*): implications for monitoring vitamin A status. *Comparative Biochemistry and Physiology B* 137, 391–400.

Vázquez-Medina, J.P., Zenteno-Savín, T., Elsner, R. (2006) Antioxidant enzymes in ringed seal tissues: Potential protection against dive-associated ischemia/reperfusion. *Comparative Biochemistry and Physiology C* 142, 198–204.

Vázquez-Medina, J.P., Zenteno-Savín, T., Elsner, R. (2007a) Glutathione protection against dive-associated ischemia/reperfusion in ringed seal tissues. *Journal of Experimental Marine Biology and Ecology* 345, 110–1118.

Vázquez-Medina, J.P., Zenteno-Savín, T., Burns, J.M. (2007b) Superoxide radical production, oxidative damage and antioxidant defenses in seal and sea lion tissues *Free Radical Biology and Medicine* 43, S72–73.

Vázquez-Medina, J.P., Crocker, D.E., Forman, H.J., Ortiz, R.M. (2010) Prolonged fasting does not increase oxidative damage or inflammation in postweaning northern elephant seal pups. *Journal of Experimental Biology* 213, 2524–2530.

Vázquez-Medina, J.P., Olguín-Monroy, N.O., Maldonado, P.D. et al. (2011a) Maturation increases superoxide radical production without increasing oxidative damage in the skeletal muscle of hooded seals (*Cystophora cristata*) . *Canadian Journal of Zoology* 89, 206–212.

Vázquez-Medina, J.P., Zenteno-Savín, T., Forman, H.J., Crocker, D.E., Ortiz, R.M. (2011b) Prolonged fasting increases glutathione biosynthesis in postweaned northern elephant seals. *Journal of Experimental Biology* 214, 1294–1299.

Whitehead, H. (2002) Sperm whale. *In* Perrin, W.F., Würsig, B., Thewissen J. G. M. (eds). *Encyclopedia of Marine Mammals*. Academic Press, San Diego, pp. 333–339.

Wilhelm-Filho, D., Sell, F., Ghislandi, M. et al. (2002) Comparison between the antioxidant status of terrestrial and diving mammals. *Comparative Biochemistry and Physiology A* 133, 885–892.

Williams, M.H. (1989) Vitamin supplementation and athletic performance. *International Journal of Vitamin and Nutrition Research Supplement* 30, 163–191.

Zenteno-Savín, T., Elsner, R. (1998) Seals and oxidative stress. *Free Radical Biology and Medicine* 25, S42.

Zenteno-Savín, T., Elsner, R. (2000) Differential oxidative stress in ringed seal tissues. *Free Radical Biology and Medicine* 29, S139.

Zenteno-Savín, T., Clayton-Hernández, E., Elsner, R. (2002) Diving seals: are they a model for copying with oxidative stress? *Comparative Biochemistry and Physiology C* 133, 527–536.

Zenteno-Savín, T., St. Leger, J., Ponganis, P.J. (2010) Hypoxemic and ischemic tolerance in emperor penguins. *Comparative Biochemistry and Physiology C* 152, 18–23.

Part III

Marine Animal Models for Aging, Development, and Disease

Chapter 14

AGING IN MARINE ANIMALS

Eva E. R. Philipp[1], Julia Strahl[2], and Alexey A. Sukhotin[3]

[1]Institute of Clinical Molecular Biology, Christian-Albrechts-University Kiel, Kiel, Germany
[2]Alfred Wegener Institute for Polar and Marine Research, Bremerhaven, Germany
[3]White Sea Biological Station, Zoological Institute of Russian Academy of Sciences, St Petersburg, Russia

A GENERAL OVERVIEW OF REACTIVE OXYGEN SPECIES AND AGING

Aging is defined as the progressive deterioration of cells, tissues, and organs associated with a decline in physiological function over time. Organisms show different maximum life spans (Table 14.1) and the question why and how we age has always fascinated researchers. Up to now over 300 aging theories have been formulated (Medvedev 1990). Some of the older theories have now been disproved while others are still valid: the ancient Greeks stated that the secret of extending human life was an ascetic lifestyle, to maintain the flame of life without letting it flame too high. To date this would be called the "caloric restriction theory" and the underlying rationale has changed.

Aging theories can be categorized into evolutionary and mechanistic theories. Evolutionary aging theories ask the question why aging occurs at all. The currently most popular evolutionary aging theories are the "mutation accumulation" theory and the "antagonistic pleiotropy or trade-off" theory. Both theories predict that genes exist that have a negative influence on fitness and performance only in old but not in young animals, or, like a Janus head, even have a beneficial effect (pleiotropy) at young ages. Mechanistic aging theories investigate the systemic and cellular changes leading to the observed decline in performance and fitness with age, i.e., what are the underlying mechanisms of the aging process. Current investigations indicate that stress-resistance mechanisms (such as resistance against cellular damage and increased repair capacities), cellular signaling pathways (such as the insulin pathway), and the level of nutrition in terms of calorie uptake per time in an individual (caloric restriction) are main players in the determination of an individual's life span (Gems and Partridge 2008). Some interactions between these players seem to exist, which are still not resolved.

AGING AND FREE RADICALS: A GENERAL OVERVIEW

The high potential of reactive oxygen species (ROS) to cause oxidative damage led Denham Harman to suspect ROS of being involved in the aging process

Oxidative Stress in Aquatic Ecosystems, First Edition. Edited by Doris Abele, José Pablo Vázquez-Medina, and Tania Zenteno-Savín.

Table 14.1 Maximum life span (MLSP) variations throughout the animal kingdom. Bivalve life span range is 1–407 years, fish life span range is 0.25–205 years, and the oldest animals known so far are sponges with maximum ages up to several thousand years. Life spans data sources: for *Argopecten irradians* (Estabrooks 2007), *Nacella concinna* (Picken 1980), *Sebastes diploproa* (Bennett et al. 1982), *Tindaria callistiformis* (Turekian et al. 1975), *Balaena mysticetus* (George et al. 1999), *Hoplostethus atlanticus* (Fenton et al. 1991), *Panopea abrupta* (Strom et al. 2004), *Margaritifera margaritifera* (Ziuganov et al. 2000), *Arctica islandica* (Wanamaker et al. 2008), *Hydra* (Martinez 1998); other life spans are from AnAge database.

Species	MLSP (yr)
Yeast (*Saccharomyces cerevisiae*)	0.04
Nematoda (*Caenorhabditis elegans*)	0.16
Turquoise killifish (*Nothobranchius furzeri*)[1]	0.25
Transparent goby (*Aphia minuta*)[1]	1
Fruit fly (*Drosophila melanogaster*)	0.3
Bay scallop (*Argopecten irradians*)[1]	1–2
Honey bee (*Apis mellifera*)	8
Queen scallop (*Aequipecten opercularis*)[1]	8–10
Antarctic mud clam (*Laternula elliptica*)[1]	36
Gray seal (*Halichoerus grypus*)	43
Herring gull (*Larus argentatus*)	49
Manx shearwater (*Puffinus puffinus*)[1]	50
Gorilla (*Gorilla gorilla*)	55
Royal albatross (*Diomedea epomophora*)[1]	58
Common limpet (*Nacella concinna*)[1]	>60
Black-headed gull (*Larus ridibundus*)[1]	63
Raven (*Corvus corax*)	69
Splitnose rockfish (*Sebastes diploproa*)[1]	80
American lobster (*Homarus americanus*)[1]	100
Deep-sea clam (*Tindaria callistiformis*) [1]	>100
Bowhead whale (*Balaena mysticetus*)	>100
Human (*Homo sapiens*)	123
Warty oreo (*Allocyttus verrucosus*)[1]	140
Orange roughy (*Hoplostethus atlanticus*)[1]	149
Lake sturgeon (*Acipenser fulvescens*)[1]	152
Geodruck clam (*Panopea abrupta*)[1]	163
Tortoise (*Geochelone elephantopus*)[1]	177
Freshwater pearl shell (*Margaritifera margaritifera*)[1]	190
Red sea urchin (*Strongylocentrotus fransciscanus*)[1]	200
Rockfish (*Sebastes aleutianus*)[1]	205
Ocean quahog (*Arctica islandica*)[1]	407
Epibenthic sponge (*Cinachyra antarctica*)[1]	1550
Hexactinellid sponge (*Scolymastra joubini*)[1]	15,000
Hydra (*Hydra* spp.)[1]	Non-aging?

[1]Aquatic ectothermal species.

(1956). In 1956 he formulated one of the most well-known but also heavily criticized aging theories, the "free radical theory of aging." The theory has since been modified into the "oxidative stress theory of aging" because some ROS-like peroxides, which are formally not free radicals, also play a role in the oxidative damage of cells. The theory states that ROS are produced by all aerobic organisms as by-products of normal metabolism. Due to their destructive action, damage of cellular structures gradually accumulates over the life span of an organism, leading to mitochondrial and cellular degradation and disturbances of physiological processes during aging. Oxidative damage can directly interfere with molecular function and hence impair important cellular processes (gene transcription, metabolism, signal transduction), but can also lead to a nonspecific decline in cellular function caused by the accumulation of damaged molecules (macroaggregrates, lipofuscin) (Brunk and Terman 2002).

ROS are primarily produced in mitochondria; hence these organelles may play a major role in aging. Mitochondrial structures are especially prone to oxidation due to their close vicinity to the major generation site of free radicals, the respiratory chain, located in the inner mitochondrial membrane. Further, the mitochondrial DNA (mtDNA) is located close to the inner mitochondrial membrane and lacks histones, which can protect the DNA against oxidative damage, thus mitochondrial DNA experiences higher oxidative damage compared to nuclear DNA (Richter et al. 1988; Yakes and Van Houten 1997). Mitochondrial damage can drag the cell into a "vicious cycle" with a positive feedback when mitochondria, already damaged by ROS, produce free radicals with greater intensity. The "mitochondrial theory of aging" is a refined version of the "oxidative stress theory of aging" and puts the mitochondria into focus of the aging process (Harman 1972). The generation of these ROS is balanced by the activity of antioxidant enzymes and small molecular antioxidants (e.g., glutathione, vitamins E, C, and coenzyme Q), which remove ROS, stop chain reactions and protect cellular organelles, membranes, and macromolecules from oxidation.

According to the theory, aging in animals is usually associated with a progressive misbalance between ROS production and antioxidant defenses causing "oxidative stress." This process is accompanied by the accumulation of various intermediate and end-products of ROS-induced chain reactions. Some of these products are used for assessment of oxidative damage in cells: ketones and aldehydes, including malondialdehyde (MDA), and "fluorescent age pigments (FAPs)", including lipofuscin.

Mitochondrial ROS generation, cellular redox parameters, and the change of antioxidant capacity and oxidative damage with age, i.e. the relationship between chronological and physiological age, have been investigated to a great extent in terrestrial mammals, birds, and invertebrates (Sestini et al. 1991; Leeuwenburgh et al. 1994; Goodell and Cortopassi 1998; Lopez-Torres et al. 2002). Several observations support the "oxidative stress theory of aging." It is generally found that lipid, protein and DNA damage increase with age (see Aging Markers below) (Shigenaga et al. 1994; Perez-Campo et al. 1998; Barja 1999, 2004). Many long-lived organisms show reduced oxidative damage or increased resistance to oxidative stress compared to shorter-lived comparable species. In mutant mice (*Mus musculus*), flies (*Drosophila melanogaster*) or worms (*Caenorhabditis elegans*) an increased life span correlates with higher resistance to oxidative stress or lower oxidative damage. Prominent mutants are the Snell and Ames dwarf mice (Liang et al. 2003), the age-1 and daf-2 *Caenorhabditis* mutants (Ruvkun et al. 2010), and mth (methuselah) mutant of *Drosophila* (Lin et al. 1998).

While there is good evidence for an age-related increase in oxidative damage of lipids, proteins, and DNA, the relationship of mitochondrial ROS generation or antioxidant capacity and chronological age does not seem to exist. Instead, age-related changes in antioxidant capacity can vary between antioxidants and between tissues of an organism, and also depend on species-specific lifestyle (Buttemer et al. 2010). Further, in transgenic animals, overexpression or knock-out of the classical antioxidant enzymes, such as Mn-SOD (superoxide dismutase), Cu,Zn-SOD, catalase (CAT), or glutathione peroxidase (GPx) gave no consistent results with respect to life expectancy; although, in most cases, animals overexpressing these genes showed a higher resistance to oxidative stress, whereas knock-out animals were more vulnerable to oxidative stress (Jang et al. 2009; Pérez et al. 2009).

The correlation of ROS production in mitochondria and life span was investigated in diverse mammalian and avian species. Mitochondrial ROS generation was lower in long-lived compared to short-lived species, thus tending to inversely correlate with species life span (Barja 1999; Lambert et al. 2007). The existing data are still under debate. One argument is that most measurements on mitochondrial isolates were undertaken under normoxic (21 kPa) conditions, whereas the actual mitochondrial environment, the cytosol, is normally characterized by much lower oxygen concentration (of about 1 kPa). Thus these measurements may not show the true picture.

AGING IN MARINE ANIMALS

Although studies of aging in marine vertebrates and invertebrates have more than 50 years of history (Haranghy et al. 1964; Woodhead 1974) compared to mammals, flies (*Drosophila*) and worms (*Caenorhabditis*), they received far less attention until recent decades

(Reznick et al. 2004; Genade et al. 2005; Gerhard 2007; Philipp and Abele 2010).

As in all aerobic organisms, ROS are produced in marine organisms in the course of cellular respiration. ROS generation may be significantly enhanced by exposure to various adverse environmental factors, such as temperature extremes, hypoxia, ultraviolet radiation (UVR), xenobiotics, etc. In aquatic organisms the majority of studies investigating oxidative stress are focused on the effects of environmental factors (Viarengo et al. 1989, 1995; Buchner et al. 1996; Regoli et al. 1997; Estevez et al. 2002; Keller et al. 2004). Aging studies are mostly focused on growth, reproductive potential and the identification of the chronological age (Brey et al. 1995; Cailliet et al. 2001; Sukhotin and Flyachinskaya 2009; Ziuganov et al. 2000). Only a few studies investigated the relationship between chronological age and oxidative stress parameters, as a measure for physiological fitness, i.e. addressed the relationship between chronological and physiological age (Ivanina et al. 2008; Philipp and Abele 2010; Philipp et al. 2008; Sukhotin et al. 2002; Sukhotin and Pörtner 2001).

Aging and ROS in Fish

Fish represent the largest and most diverse class of vertebrates comprising more than 28,000 marine and freshwater species (Nelson 2006). The class displays a wide variety of life histories, reproductive strategies, age-related mortality rates and, therefore, aging rates, and longevity ranges from several months to more than 200 yr (Table 14.1). Some species, such as the Pacific salmon, *Oncorhynchus* spp., eel, *Anguilla* spp., or capelin, *Mallotus villosus*, are semelparous with rapid aging and high mortality rates after their first spawning. Other species of teleost fish demonstrate slow albeit steady increase of mortality with age, usually after reaching maturity when somatic growth becomes negligible (Woodhead 1979). Among them are the most popular piscine models for aging research – guppies, killifish, and zebrafish. Finally, there are fish with extremely long life spans, associated with persistent slow growth over their lifetime, and very low mortality rates at old age, i.e., showing negligible senescence. These include sturgeons (*Acipenser* spp. and *Huso huso*), warty oreo (*Allocyttus verrucosus*), and rockfish (*Sebastes* spp.).

Historically, fish were the first organisms with asymptotic ("infinite") growth included in aging research, which lead G.P. Bidder (1932) to the erroneous conclusion that fish lack senescence and are potentially immortal. He hypothesized that due to their unlimited growth, fish were able to constantly repair and replace damaged cells and tissues, thereby escaping senescence and achieving immortality. This concept has been thoroughly disproved by aging studies in the laboratory (e.g. Comfort 1960, 1961), as well as in wild fish populations (for review see Woodhead 1998).

Finch and Austad (2001) suggested two criteria for the absence of senescence: (i) no observable decline in reproduction rate or increased mortality after maturation; and (ii) no age-related decline in physiological performance or disease resistance. Like other vertebrates, fish do not meet these criteria and instead show features of aging across all species studied.

The use of fish as models in oxidative stress studies is relatively recent (Kelly et al. 1998; Chapters 12, 20, 22, this volume). Accumulated data on age-related changes in lipid peroxidation in fish are still controversial. In larvae and fry of several fish species, tissue concentrations of extractable FAPs decreased with age (Hammer 1988; Mullin and Brooks 1988; Hill and Womersley 1991; Mourente and Diaz-Salvago 1999). In brain of Dover sole (*Microstomus pacificus*) an initial age-related increase of lipofuscin concentration reached a plateau after 15 years, whereas in the short-lived and fast growing rainbow trout (*Salmo gairdneri*) brain and liver lipofuscin concentrations initially declined to stabilize in individuals older than 1 year (Vernet et al. 1988). In contrast, Passi et al. (2004) observed age-related exponential accumulation of lipid and protein oxidation products in adult rainbow trout and slower but significant accumulation of both substances in sea bass (*Dicentrarchus labrax*). Lipofuscin has been reported to accumulate in liver of annual killifish, *Nothobranchius furzeri* (Genade et al. 2005), while in the other short-lived piscine gerontological model object, the zebrafish (*Danio rerio*), no lipofuscin fluorescence could be detected (Kishi et al. 2003). Zebrafish appear to undergo "very gradual senescence" based on molecular changes including senescence-associated β-galactosidase activity and accumulation of oxidized protein with age in the face of continuously proliferating myocytes and constitutive telomerase activity in adult individuals (Keller and Murtha 2004; Kishi et al. 2003). The pattern of age-associated oxidative damage accumulation reflects the complex interplay of functional change in aging animals (e.g., metabolic alterations, fluctuations in antioxidant defenses, or cellular repair mechanisms), including the level of pro-oxidants and their target molecules such as polyunsaturated fatty acids (PUFA), which are readily oxidized by ROS, or changes in environmental factors. A linear increase of lipofuscin concentration with age suggests a constant ROS production rate, constant oxidative damage balanced by antioxidant protection, and constant aging rate. Acceleration of lipofuscin accumulation may indicate an increase of oxidative stress at advanced age, possibly caused by age-related weakening of antioxidant defenses. Slowing lipofuscin accumulation during aging may denote attenuation of oxidative stress by increased production of antioxidants or by metabolic decline at advanced age, or it could be the consequence of the "dilution" of lipofuscin granules with age in fast-growing young tissue/organism (see also Sukhotin et al. 2002). This phenomenon is well-known from human fibroblast cell cultures (Sitte et al. 2001). Heat shock proteins (HSPs) play a key role in reparation, refolding, or elimination of damaged or denaturated proteins and could be important antagonists of aging. It is interesting that aging modulates heat shock responsiveness and HSP70 expression in zebrafish (*Danio rerio*) (Keller and Murtha 2004). Specifically, both constitutive and induced levels of HSP70 were reduced in mature versus young zebrafish, indicating age-associated decline of repair capacity.

Aging and ROS in Bivalves

Bivalves were recently appreciated as potent models of aging. Some species have a longevity of >400 yr (Table 14.1), and the bivalve shell permits the determination of individual chronological age from annually forming rings, at least in polar and temperate species (Fig. 14.1, see below).

Bivalves belong to different ecotypes, which may influence the aging process. Infaunal bivalves burrow in the sediment and usually have low metabolic rates and low scope for activity. These bivalves often show declining mitochondrial function (respiration, citrate synthase activity) as well as increased accumulation of cellular damage (protein carbonyls, malondialdehyde, lipofuscin) with age (Philipp et al. 2005a,b; Strahl et al. 2007; Abele et al. 2008). Pectinids or scallops are epibenthic (living above the sediment), active swimmers with comparatively higher scope for activity and metabolic rate. In these animals mitochondrial function seems to be conserved, whereas mitochondrial volume density and citrate synthase activity diminish over lifetime (Philipp et al. 2006, 2008).

Decline in some antioxidants (CAT, glutathione (GSH)) and increase in oxidative damage with age was detected in several mussels, clams, and scallops (Canesi and Viarengo 1997; Lomovasky et al. 2002; Sukhotin et al. 2002; Philipp et al. 2006), as well as in other mollusks such as gastropods and cephalopods (Clarke et al. 1990; Zielinski and Pörtner 2000), but these findings cannot be generalized. Shorter-lived species show faster increase in damage parameters and decrease in antioxidants compared to longer-lived species of similar lifestyle (Philipp et al. 2005a,b 2006). In the temperate clam *Mya arenaria* and the Antarctic clam *Laternula elliptica*, shorter life span correlated with higher mitochondrial ROS production in *M. arenaria*, where ROS formation increased with age. Longer-lived bivalve models apparently maintain mitochondrial ROS production low and stable over much of their lifetime (Philipp et al. 2005b). As in fish, changes of cellular aging parameters in mollusks are mostly nonlinear. For *Mya arenaria* and *Mytilus edulis* a decline in lipofuscin at younger ages was followed by an increase at later age (Hole et al. 1995; Philipp et al. 2005a). In the long-lived Iceland clam *Arctica islandica*, which lives >400 yr (Wanamaker et al. 2008), physiological aging appears to be close to negligible (Abele et al. 2008). The observed

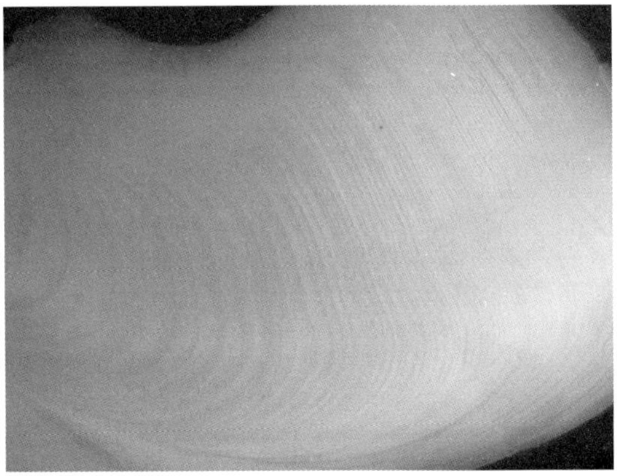

Fig. 14.1 Polished shell cut (umbo) of the marine bivalve *Arctica islandica* with visible year-rings. Magnification × 7. (Photograph J. Strahl, Alfred Wegener Institute.)

decline in antioxidants (CAT, GSH) and the mitochondrial marker citrate synthase was observable only during the first 30 yr; thereafter, values remained constant for another 150 yr. The early decline in antioxidant activities reflects physiological changes during maturation and not senescence. Oxidative damage parameters (lipofuscin, protein carbonyls) were very low compared to shorter-lived bivalves, which appears to be essential for reaching a long lifespan (Strahl et al. 2007).

Age-related molecular changes in bivalves go hand in hand with shifts in cellular function and stress response over age. A swimming experiment with queen scallops demonstrated differences in the capacity for exercise and stress response in young and old specimens. In young animals, exhaustive swimming resulted in a faster decrease of muscle pH and GSH levels, as well as in a reduction of muscle glycogen compared to older scallops going through the same exercise. Young animals closed their shells more often and for longer times upon predator attack, whereas older animals kept the shells open, which matched their lower aerobic (mitochondrial volume density and citrate synthase activity) and anaerobic (glycogen concentration) energetic capacity (Philipp et al. 2008; Schmidt et al. 2008). Hypoxia and temperature experiments with *M. edulis* indicate that younger mussels may be more stress-resistant than older individuals. Younger animals showed faster recuperation of lysosomal integrity destabilized by hypoxia and hyperthermia than older individuals (Hole et al. 1995). In an anoxia–reperfusion experiment older animals showed a marked increase in oxidative damage following reperfusion, whereas in younger animals lipid peroxidation remained low (Viarengo et al. 1989).

The *Mytilus* studies, however, must be taken with a grain of salt, as in these studies age was deduced from size and not from individual age determination through shell ring counts. As shown in several studies (Sukhotin and Pörtner 2001; Sukhotin et al. 2002, 2003), size and age are not necessarily coupled in *Mytilus*, which is also the case in many other bivalve species, especially after reaching asymptotic growth. Indeed, the change with age in the parameters investigated can be quite different when plotted against age or size. Catalase, for example, decreased significantly when plotted against size in *M. edulis*, whereas no significant change was observed when data were related to age (Sukhotin et al. 2002). This example underlines the importance of the individual age determination in bivalve aging studies.

Aging in Crustaceans

ROS-related aging in crustaceans grew from two main standpoints: one relates from the practical need of separating age cohorts in population studies, especially of commercially important species. Crustaceans lack permanent calcified hard structures that could potentially bear age markers; therefore, population age structure is usually determined through the analysis of size-frequency distribution, which identifies modal size classes and considers them to reflect age-cohorts (Sheehy 2002, 2008; Harvey et al. 2008). Although widely used, this procedure is rather imprecise. A second alternative method for age estimation was proposed by Etterhank (1983, 1984) and is based on the quantification of lipofuscin in post-mitotic cells (mainly nervous tissue). Lipofuscin accumulates as a nondegradable end-product of lipid peroxidation and protein damage due to oxidative stress. Since oxidative stress in animals depends on a combination of external and internal factors, including metabolic activity, accumulated lipofuscin reflects "physiological" rather than "chronological" age (Bluhm et al. 2001). Studies so far have demonstrated an age-related increase of lipofuscin or other FAP concentrations in crustacean tissues. FAP accumulation rate is variant – it can be linear (lobsters *Hommarus gammarus*, *Panulirus cygnus*, and *Panulirus argus* (O'Donovan and Tully 1996; Sheehy et al. 1998; Maxwell et al. 2007), crayfish *Cherax quadricarinatus* and *Pacifastacus leniusculus* (Belchier et al. 1998; Sheehy 1996), amphipod *Waldeckia obesa*, (Bluhm et al. 2001) and crab *Carcinus pagurus* (Sheehy and Prior 2008), accelerating with age in an exponential manner or as power function (crab *Callinectes sapidus*; Ju et al. 1999, 2001), lobster *Homarus americanus* (Wahle et al. 1996), and shrimp *Aristeus antennatus* (Mourente and Diaz-Salvago 1999), or decelerating with age (crayfish *Cherax quadricarinatus* (Sheehy 1992) and prawn *Penaeus japonicus* (Vila et al. 2000).

Trends in antioxidant activity and pro-oxidant levels (e.g., PUFA content) in aging crustaceans are controversial. Activities of some antioxidant enzymes (CAT, γ-glutamyl transpeptidase (GTT)) were reported to decline, while others (SOD and GPx) feature an upward trend with age, e.g. in the shrimp *Aristeus antennatus* (Mourente and Diaz-Salvago 1999). In the planktonic cladoceran *Daphnia magna* CAT, SOD, and GPx activities mildly declined over age and significantly increased in the specimens of the oldest age group (Barata et al. 2005). No difference in activities of the same enzymes could be observed between subadult and adult *Gammarus locusta*, but were higher in juveniles of the amphipod (Correia et al. 2003). Overall, crustaceans demonstrate a progressive increase of oxidative damage with age, detectable as accumulation of lipofuscin and other FAPs in cells.

Temperature and Aging in Marine Animal Ectotherms

Most cellular processes, as well as almost all physiological rates, such as rates of respiration, feeding, growth, and locomotion, strongly depend on body temperature, which in ectotherms depends on the environmental temperature. The complex biology of aging, with its dependence on metabolic rate, is likely influenced by environmental temperature in aquatic ectotherms. Comparative studies on wild populations of fish and invertebrates suggest that life span in congeneric species, at least in those with similar lifestyle, increases towards the poles (e.g. Bluhm et al. 2001; Philipp et al. 2005a; Ziuganov et al. 2000; Sukhotin et al. 2007) and towards the deep ocean (Koslow et al. 2000; Cailliet et al. 2001). These observations can be explained by the prolonged life span with retardation of aging by the decelerating effect of low temperatures on ectotherm metabolic rate (e.g. Cailliet et al. 2001). The first experimental confirmation of temperature affecting aging in aquatic vertebrates was obtained by Walford and Liu (1965) studying small short-lived killifish *Cynolebias adolffi*. Rearing fish at lower temperature increased life span due to a fundamental change of the aging process. In fish maintained at low temperatures, morphological signs of aging, such as decreased spinal curvature, eye cataracts, and scale alterations, and reproductive senescence were greatly delayed (review by Yen et al. 2004). Retardation of age-related collagen cross-linking was recorded in *Cynolebias bellottii* at low temperature (Walford et al. 1969), confirming the temperature effect on aging in fish.

The most widely accepted explanation of the temperature-related modulation of the aging rate lies within the framework of the metabolic and oxidative stress theories of aging. In ectotherms environmental temperature directly affects metabolic rate and ROS production rate. The destructive action of ROS in cells is believed to be the primary cause of aging (Sohal 2002; this chapter) and the aging rate is thought to be positively correlated with temperature through metabolic rate. Valenzano et al. (2006) showed increased longevity in the fish species *Nothobranchius furzeri* reared at 22°C compared to individuals reared at 25°C and a decreased accumulation of lipofuscin indicating lower oxidative stress in cold-acclimated fish. There is, however, evidence that aging retardation at low temperatures occurs due to mechanisms other than metabolic slowdown, although these mechanisms are not yet fully understood and require further investigation (Yen et al. 2004).

Temperature effects on age-related oxidative damage determined as lipofuscin accumulation in crustaceans were extensively reviewed by Sheehy (2002). The lipofuscin accumulation rate in brain of the crustacean studied increased significantly with temperature. In the freshwater crayfish, *Cherax quadricarinatus*, lipofuscin accumulated in a temperature-dependent manner with maximal rate at 28°C and minimal at 13°C (Sheehy et al. 1995). Exponential increase of lipofuscin levels at elevated rearing temperature was also found in juvenile lobster *Homarus gammarus* (O'Donovan and Tully 1996). Lipofuscin accumulation rate and its relation to temperature can, however, change with increasing age. In the crayfish *Cherax quadricarinatus* lipofuscin accumulation declined in older individuals at three different temperatures (Sheehy 1992), caused either by the age-related decline of metabolic rate or some kind of compensation or protection against oxidative damage activated at advanced age. The temperature dependence of lipofuscin accumulation rate is often nonlinear. According to the model of Sheehy (2002) an increase of only 0.5°C in average annual seawater temperature from 10 to 10.5°C would lead to more than a doubling of the lipofuscin deposition rate and greatly accelerated physiological aging in cold-water populations of *Homarus gammarus* from northern Scotland. This highlights the fact that current patterns of global warming could lead to severe reductions in life span and age of maturation in many marine ectotherm populations.

An important indirect evidence of the temperature effect on aging in aquatic animals is provided in a number of publications on interspecific comparisons of metabolism and aging in latitudinal separated bivalve mollusks of similar lifestyles (Philipp et al. 2006;

Abele et al. 2009; Philipp and Abele 2010). Actively swimming scallops from temperate (*Aequipecten opercularis*) and Antarctic (*Adamussium colbecki*) regions differ significantly in their maximum life span from 8–10 years to >18 yr, respectively. Oxidative stress was higher in the temperate short-lived species, reflected by more pronounced age-related lipofuscin accumulation and rapid decline in antioxidant activity compared to its cold-adapted counterpart (Philipp et al. 2006). A comparison of two sessile infaunal clams – a cold-water stenotherm *Laternula elliptica* from Antarctica and a eurythermic temperate *M. arenaria* – demonstrated fundamental differences in metabolic strategies that allow the Antarctic species to live three times longer than *M. arenaria* (36 yr vs. 13 yr respectively; Philipp et al. 2005a,b). The Antarctic clam has lower whole animal metabolic rates, showed lower mitochondrial H_2O_2 generation rates, and maintained high GSH levels compared to *M. arenaria*. The long-lived *L. elliptica* was better able to conserve mitochondrial integrity and functional efficiency (i.e., high aerobic capacity, high respiratory control ratio) and featured a relatively high mitochondrial proton leak – the futile proton penetration pore across the inner mitochondrial membrane (Philipp et al. 2005b). Both "mild uncoupling" of mitochondria due to increased proton leak and high GSH concentration theoretically inhibit and counteract ROS production in the Antarctic species. Two hypotheses were derived concerning the prime cause of this strategy (Abele et al. 2009). Firstly, high proton leak and GSH levels in *L. elliptica* tissues could be a mechanism of ROS reduction and possible life extension; this would be strictly species specific and has nothing to do with cold adaptation. The second hypothesis is based on the fact that metabolic cold adaptation in animals often requires increased numbers of mitochondria or elevated inner mitochondrial membrane area (cristae density) to compensate for low aerobic scope and deferred oxygen diffusion in the cold (Hochachka and Somero 2002; Pörtner 2002). Proton leak has been shown to positively correlate with the number of mitochondria, inner mitochondrial membrane area, and the unsaturation ratio of mitochondrial membranes (Porter et al. 1996). Increased cristae density and/or proton permeability of membranes increase proton leak, which in turn mitigates deleterious ROS production, prolonging the life span of cold-adapted animals. Longevity may be a "by-product" of metabolic cold adaptation and generally applicable to polar species.

RECORD HOLDERS OF EXTREMELY SHORT AND LONG LIFE SPANS IN MARINE ORGANISMS

Extremely short- and long-lived multicellular animal species can be found in aquatic ecosystems and are coming more and more into focus as model organisms for comparative aging research. One of the shortest-lived fish species is the coral reef pygmy gobi *Eviota sigillata* with a maximum life span of 56 days (Depczynski and Bellwood 2005). Other short-lived examples are the freshwater killifish *Nothobranchius furzeri* and the transparent goby *Aphia minuta* (Caputo et al. 2002). *Nothobranchius furzeri* has a maximum lifespan of 13 weeks and lives in the dry lowveld of Zimbabwe. It can survive the long (>10 months) dry season as embryos encased in the dry mud (Valdesalici and Cellerino 2003). This species shows explosive growth, early sexual maturation and age-dependent physiological decline, and expresses age-related biomarkers, such as lipofuscin and β-galactosidase (Terzibasi et al. 2007). Current efforts are under way to use this species as a new vertebrate model of aging, and first genome sequences have been obtained (Terzibasi et al. 2007; Reichwald et al. 2009). The pelagic planktophagous marine goby *A. minuta* lives <1 yr. In this species programmed cell death (apoptosis) leads to immediate death after breeding, hence the fish is used as a model to understand general mechanisms of apoptosis and its involvement in the aging process (Caputo et al. 2002). At the other end of the age spectrum, supercentenarians, such as the rockfish *Sebastes aleutianus* with a testified life span of 205 yr, are found within the group of fish (Cailliet et al. 2001).

Another group with very short-lived but also very long-lived species comprises the bivalve mollusks. *Pisidium* spp., *Donax* spp. and *Argopecten irradians irradians* live only 1–2 years, whereas *Arctica islandica* probably is the longest-lived noncolonial animal on Earth. Around Iceland individuals of this species have extremely long life spans of >400 yr (Table 14.1) with slow cellular senescence rates (Abele et al. 2008; Wanamaker et al. 2008). Different populations of *A. islandica* with different maximum life span from diverse marine environments in the northern hemisphere are known, and this species represents a valuable model to study the evolution of life span in different climates and also how longevity is shaped under different environmental and genetic settings.

AGE ESTIMATION IN MARINE ORGANISMS

Reliable age determination methods form the basis of investigations of animal growth, development, and physiological aging. They are also essential for ecological management of commercially important marine species, as they enable determination of life-time growth history, maximum life spans, and the age of reaching sexual maturity in a given species. Physiological parameters, such as metabolic rate, antioxidant defense, and the accumulation of oxidative damage, can be related to chronological age to describe the dynamic aging process in long- and short-lived marine ectotherms (Kishi 2004; Passi et al. 2004; Abele et al. 2008; Philipp and Abele 2010).

Age Estimation Using Hard Structures

In aquatic animals, periodic growth increments of hard structures are often used to estimate age and growth rates, such as teeth of elephant seals and dolphins (Laws 1952; Hohn 1980), turtle bones (Zug et al. 1986; Parham and Zug 1997), fish vertebrae, scales, and otoliths (Chilton and Beamish 1982; Penttila and Dery 1988; Cailliet et al. 2001; Yilmazy and Polat 2002), brittle star skeletons (Gage 1990), statoliths in squids or cubomedusae (Ueno et al. 1995; Arkhipkin 1997), coral skeletons (Dodge and Thomson 1974), and bivalve shells (Hendelberg 1960; Heilmayer et al. 2003; Schöne 2003; Begum et al. 2009). All these organisms exhibit growth-band patterns of different density or structure, triggered either by environmental factors, such as changes in temperature and food supply, or by internal factors, such as reproduction.

The most common method of age determination in fish is to count age rings on the otolith cross-sections (Chilton and Beamish 1982; Christensen 1984). In bivalves, growth lines of younger individuals are mostly visible on the external shell surface, but age determination of older animals has to be conducted on shell cross-sections. The surface of the shell cross-sections is ground and polished to visualize age rings, which can be counted either along the outer shell line or inside the umbo/hinge using a stereomicroscope (Fig. 14.1). Growth increments appear as wide and light bands of homogenous structure and the growth lines as narrow and irregular dark rings (Jones 1980; Ropes et al. 1984). The growth ring pattern is not necessarily formed at regular intervals in time, and more than one growth increment per year can be deposited. Disturbance rings can be produced due to environmental conditions, such as salinity or temperature changes, or to reproductive events.

To determine whether growth layers are formed annually, or how many layers are built per year or day, age determination methods using otoliths, bivalve shells, or other hard structures have to be validated. Tag-recapture or radiometric approaches based on known radioactive decay series are useful to validate the assumed annual growth band formation in fish otoliths and bivalve shells. The analysis of different isotope pairs from the uranium and thorium decay series has been used in aging studies, with each isotope pair being applied in a corresponding time range. Thus, $^{210}Po/^{210}Pb$ is used in the range 0–2 yr in *Nautilus* (Cochran et al. 1981), $^{228}Th/^{228}Ra$ is applied in the 0–10 yr range in corals and clams (Dodge and Thomson 1974; Turekian et al. 1975), while $^{210}Pb/^{226}Ra$ is used in the range 0–100 yr in fishes (Fenton et al. 1991).

Isotope profiles of shell material taken over the growth line also give further information on present-day and former environmental conditions (Turekian et al. 1975; Brey and Mackensen 1997; Schöne et al. 2004). Bivalve shells are increasingly used not only for individual age determination, but also as paleoclimate archives (Schöne et al. 2004, 2005). Analogously to trees, growth rings of the long-lived bivalve *Arctica islandica* are investigated to establish climate models covering the past hundreds of years (Schöne et al. 2005). In each growth increment, the ratio of oxygen $^{18}O/^{16}O$ or carbon $^{13}C/^{12}C$ isotopes mirrors environmental factors, such as temperature and the content of seawater carbonate during the time of increment formation (Brey and Mackensen 1997).

Growth increments have also been found in skeletal structures of deep-sea brittle stars as fine, translucent rings, separated by wider, more opaque zones in the calcitic arm ossicles (Gage 1990), in statoliths of squids and jellyfish (Arkhipkin 1997; Ueno et al. 1995), or even in amphibian and sea turtle bones (Smirina 1994; Parham and Zug 1997). Without an ample validation of data sets, age determinations via growth rings, however, have to be regarded with caution. Coral growth rings, visible as bands of variable skeletal density in the growth direction of the corals, were for example validated to be formed annually; specimens from Eniwetok in the Pacific Ocean had radioactive nucleotides incorporated from USA fission device testing in the years 1948 to 1958, and the radioactive marker layers were in agreement with visible annual growth rings (Knutson et al. 1972).

In marine mammals different age determination methods exist, including the count of growth layer groups in teeth, which are considered to form annually in many odontocetes (Hohn 1980), or the analysis of tissues in mysticetes, which form horny growth layers called ear plugs in the external auditory meatus (Slijper 1962), or the aspartic acid racemization technique in the eye lens of bowhead whales (George et al. 1999), which measures the D/L ratio of the two isomeric forms of the aspartic acid (Bada et al. 1980).

Biochemical and Physiological Proxies of Aging

Alternative methods for age determination in animals lacking permanent hard structures are based on biochemical or physiological proxies in cells and tissues, such as concentrations of lipofuscin. Studies on crustaceans and bivalves show a relationship between age and lipofuscin concentration, and support the idea that lipofuscin is a suitable marker for both physiological and chronological age (Bluhm et al. 2001; Lomovasky et al. 2002; Sheehy 2002). The use of lipofuscin as an age marker has, however, to be calibrated for each species or population, as pigment formation can be strongly affected by temperature or heavy metal contamination. Telomere length is also discussed as a proxy for age. Telomeres are highly conserved repetitive DNA sequences and associated proteins that cap the end of eukaryotic chromosomes, and provide chromosomal stability or promote replication processes and chromosome segregation. In many vertebrate species telomeres shorten during each cell division cycle, because the DNA polymerase is unable to replicate the ends of DNA molecules. At a critical length of the telomeres, cell division stops and replicative senescence starts. The enzyme telomerase can maintain or even re-elongate telomeres through telomeric DNA synthesis. In several aquatic vertebrate and invertebrate species the relationship between telomere length or telomerase activity and individual age was analyzed to test whether or not telomere length could be used as a reliable marker for age determination. The picture is not consistent across species and not even between different tissues in one species. A positive correlation of telomere length with life span was found for short-lived and long-lived scallop species (Estabrooks 2007). In contrast, telomere length in two long-lived birds both increased and declined throughout life (Hall et al. 2004). Lack of age-associated telomere shortening was reported for zebrafish (Lund et al. 2009) and for short- and long-lived sea urchin *Lytechinus variegates* and *Strongylocentrotus franciscanus* (Francis et al. 2006). Therefore, telomere length cannot be considered a reliable age marker.

In summary, common biochemical or physiological proxies for age are difficult to find in marine animals, whereas periodic growth increments and isotope incorporation in hard structures represent an applicable tool for age determination in many species.

REFERENCES

Abele, D., Strahl, J., Brey, T., Philipp, E.E.R. (2008) Imperceptible senescence: ageing in the ocean quahog *Arctica islandica*. *Free Radical Research* 42, 474–480.

Abele, D., Brey, T., Philipp, E.E.R. (2009) Bivalve models of aging and the determination of molluscan lifespans. *Experimental Gerontology* 44, 307–315.

Arkhipkin, A.I. (1997) Age and growth of the mesopelagic squid *Ancistrocheirus lesueurii* (Oegopsida: Ancistrocheiridae) from the central-east Atlantic based on statolith microstructure. *Marine Biology* 129, 103–11.

Bada, J.L., Brown, S.E., Masters, P.M. (1980) Age determination of marine mammals based on aspartic acid racemization in the teeth and lens nucleus. *Reports of the International Whaling Commission Special Issue* 3, 113–118.

Barata, C., Carlos Navarro, J., Varo, I., Carmen Riva, M., Arun, S., Porte, C. (2005) Changes in antioxidant enzyme activities, fatty acid composition and lipid peroxidation in *Daphnia magna* during the aging process. *Comparative Biochemistry and Physiology B* 140, 81–90.

Barja, G. (1999) Mitochondrial oxygen radical generation and leak: sites of production in states 4 and 3, organ specificity, and relation to aging and longevity. *Journal of Bioenergetics and Biomembranes* 31, 347–366.

Barja, G. (2004) Aging in vertebrates, and the effect of caloric restriction: a mitochondrial free radical production-DNA damage mechanism? *Biological Reviews of the Cambridge Philosophical Society* 79, 235–51.

Begum, S., Basova, L., Strahl, J. et al. (2009) A metabolic model for the ocean quahog *Arctica islandica* – effects of animal mass and age, temperature, salinity, and geography on respiration rate. *Journal of Shellfish Research* 28, 533–539.

Belchier, M., Edsman, L., Sheehy, M.R.J., Shelton, P.M.J. (1998) Estimating age and growth in long-lived temperate

freshwater crayfish using lipofuscin. *Freshwater Biology* 39, 439–446.
Bennett, J.T., Boehlert, G. W, Turekian, K.K. (1982) Confirmation of longevity in *Sebastes diploproa* (Pisces: Scorpaenidae) from $^{210}Pb/^{226}Ra$ measurements in otoliths. *Marine Biology* 71, 209–215.
Bidder, G.P. (1932) Senescence. *The British Medical Journal* 24, 583–585.
Bluhm, B.A., Brey, T., Klages, M. (2001) The autofluorescent age pigment lipofuscin: key to age, growth and productivity of the Antarctic amphipod *Waldeckia obesa* (Chevreux, 1905). *Journal of Experimental Marine Biology and Ecology* 258, 215–235.
Brey, T., Mackensen, A. (1997) Stable isotopes prove shell growth bands in the Antarctic bivalve *Laternula elliptica* to be formed annually. *Polar Biology* 17, 465–468.
Brey, T., Pearse, J., Basch, L., McClintock, J., Slattery, M. (1995) Growth and production of *Sterechinus neumayeri* (Echinoidae: Echinodermata) in McMurdo Sound, Antarctica. *Marine Biology* 124, 279–292.
Brunk, U.T., Terman, A. (2002) Lipofuscin: Mechanisms of age-related accumulation and influence on cell function. *Free Radical Biology and Medicine* 33, 611–619.
Buchner, T., Abele-Oeschger, D., Theede, H. (1996) Aspects of antioxidant status in the polychaete *Arenicola marina*: Tissue and subcellular distribution, and reaction to environmental hydrogen peroxide and elevated temperatures. *Marine Ecology Progress Series* 143, 141–150.
Buttemer, W.A., Abele, D., Costantini, D. (2010) From bivalves to birds: oxidative stress and longevity. *Functional Ecology* 24, 971–983.
Cailliet, G.M., Andrews, A.H., Burton, E.J. et al. (2001) Age determination and validation studies of marine fishes: do deep-dwellers live longer?. *Experimental Gerontology* 36, 739–764.
Canesi, L., Viarengo, A. (1997) Age-related differences in glutathione metabolism in mussel tissues (*Mytilus edulis* L.). *Comparative Biochemistry and Physiology B* 116, 217–221.
Caputo, V., Candi, G., Arneri, E. et al. (2002) Short lifespan and apoptosis in *Aphia minuta*. *Journal of Fish Biology* 60, 775–779.
Chilton, D.E., Beamish, R.J. (1982) Age determination methods of fishes studied by the Groundfish Program at the Pacific Biological Station. *Canadian Special Publication of Fisheries and Aquatic Sciences* 60, 120 pp.
Christensen, J.M. (1984) Burning otolith, a technique for age determination of Soles and other fish. *ICES Journal of Marine Science* 29, 73–81.
Clarke, A., Kendall, M.A., Gore, D.J. (1990) The accumulation of fluorescent age pigments in the trochid gastropod *Monodonta lineata*. *Journal of Experimental Marine Biology and Ecology* 144, 185–204.
Cochran, J.K., Rye, D.M., Landman, N.H. (1981) Growth rate and habitat of *Nautilus pompilius* inferred from radioactive and stable isotope studies. *Paleobiology* 7, 469–480.
Comfort, A. (1960) The effect of age on growth resumption in fish (*Lebistes*) checked by food restriction. *Gerontologia* 4, 177–186.
Comfort, A. (1961) Age and reproduction in female *Lebistes*. *Gerontologia* 8, 150–155.
Correia, A.D., Costa, M.H., Luis, O.J., Livingstone, D.R. (2003) Age-related changes in antioxidant enzyme activities, fatty acid composition and lipid peroxidation in whole body *Gammarus locusta* (Crustacea: Amphipoda). *Journal of Experimental Marine Biology and Ecology* 289, 83–101.
Depczynski, M., Bellwood, D.R. (2005) Shortest recorded vertebrate lifespan found in a coral reef fish. *Current Biology* 15, R288–R289.
Dodge, R.E., Thomson, J. (1974) The natural radiochemical and growth records in contemporary hermatypic corals from the Atlantic and Carribean. *Earth and Planetary Science Letters* 23, 313–322.
Estabrooks, S.L. (2007) The possible role of telomeres in the short life span of the bay scallop, *Argopecten irradians irradians* (Lamarck 1819). *Journal of Shellfish Research* 26, 307–13.
Estevez, S.M., Abele, D., Puntarulo, S. (2002) Lipid radical generation in polar (*Laternula elliptica*) and temperate (*Mya arenaria*) bivalves. *Comparative Biochemistry and Physiology B* 132, 729–37.
Etterhank, G. (1983) Age structure and cyclical annual size change in the Antarctic krill *Euphausia superba*. *Polar Biology* 2, 189–93.
Etterhank, G. (1984) A new approach to the assessment of longevity in the antarctic krill *Euphausia superba*. *Journal of Crustacean Biology* 4, 295–305.
Fenton, G.E., Short, S.A., Ritz, D.A. (1991) Age determination of orange roughy, *Hoplostethus atlanticus* (Pisces: Trachichthyidae) using $^{210}Pb/^{226}Ra$ disequilibria. *Marine Biology* 109, 197–202.
Finch, C.E., Austad, S.N. (2001) History and prospects: symposium on organisms with slow aging. *Experimental Gerontology* 36, 593–7.
Francis, N., Gregg, T., Owen, R., Ebert, T., Bodnar, A. (2006) Lack of age-associated telomere shortening in long-lived and short-lived species of sea urchins. *Federation of European Biochemical Societies Letters* 580, 4713–4717.
Gage, J.D. (1990) Skeletal growth markers in the deep-sea brittle stars *Ophiura ljungmani* and *Ophiomusium lymani*. *Marine Biology* 104, 427–35.
Gems, D., Partridge, L. (2008) Stress-response hormesis and aging: "that which does not kill us makes us stronger". *Cell Metabolism* 7, 200–3.
Genade, T., Benedetti, M., Terzibasi, E. et al. (2005) Annual fishes of the genus *Nothobranchius* as a model system for aging research. *Aging Cell* 4, 223–233.
George, J.C., Bada, J., Zeh, J. et al. (1999) Age and growth estimates of Bowhead whales (*Balaena mysticetus*) via aspartic acid racemization. *Canadian Journal of Zoology* 77, 571–580.
Gerhard, G.S. (2007) Small laboratory fish as models for aging research. *Ageing Research Reviews* 6, 64–72.

Goodell, S., Cortopassi, G. (1998) Analysis of oxygen consumption and mitochondrial permeability with age in mice. *Mechanisms of Ageing and Development* 101, 245–256.

Hall, M.E., Nasir, L., Daunt, F. et al. (2004) Telomere loss in relation to age and early environment in long-lived birds. *Proceedings of the Royal Society of London B* 271, 1571–1576.

Hammer, C. (1988) Accumulation of fluorescent age pigments in embryos and larvae of pike, *Esox lucius* (Esocidae, Teleostei) at different incubation temperatures. *Environmental Biology of Fishes* 22, 91–99.

Haranghy, L., Balazs, A., Burg, M. (1964) Investigation on ageing and duration of life of mussels. *Acta Biochimica Hungarica* 16, 57–67.

Harman, D. (1956) Aging: a theory based on free radical and radiation biology. *Journal of Gerontology* 11, 298–300.

Harman, D. (1972) The biological clock: the mitochondria?. *Journal of the American Geriatrics Society* 20, 145–147.

Harvey, H.R., Secor, D.H., Ju, S. (2008) The use of extractable lipofuscin for age determination of crustaceans: Reply to Sheehy (2008). *Marine Ecology Progress Series* 353, 307–311.

Heilmayer, O., Brey, T., Chiantore, M., Cattaneo-Vietti, R., Arntz, W.E. (2003) Age and productivity of the Antarctic scallop, *Adamussium colbecki*, in Terra Nova Bay (Ross Sea, Antarctica). *Journal of Experimental Marine Biology and Ecology* 288, 239–256.

Hendelberg, J. (1960) The freshwater pearl mussel, *Margaritifera margaritifera* (L.). *Report of the Institute of Freshwater Research (Drottningholm)* 41, 149–171.

Hill, K.T., Womersley, C. (1991) Critical aspects of fluorescent age-pigment methologies: modification for accurate analysis and age assessments in aquatic organisms. *Marine Biology* 109, 1–11.

Hochachka, P.W., Somero, G.N. (2002) *Biochemical adaptation Mechanism and Process in Physiological Evolution*. Oxford University Press, New York, 466 pp.

Hohn, A.A. (1980) Age determination and age related factors in the teeth of western north Atlantic bottlenose dolphins. *Scientific Reports of the Whales Research Institute* 32, 33–66.

Hole, L.M., Moore, M.N., Bellamy, D. (1995) Age-related cellular and physiological reactions to hypoxia and hyperthermia in marine mussels. *Marine Ecology Progress Series* 122, 173–178.

Ivanina, A.V., Sokolova, I.M., Sukhotin, A.A. (2008) Oxidative stress and expression of chaperones in aging mollusks. *Comparative Biochemistry and Physiology B* 150, 53–61.

Jang, Y.C., Pérez, V.I., Song, W. et al. (2009) Overexpression of Mn superoxide dismutase does not increase life span in mice. *The Journals of Gerontology A* 64, 1114–1125.

Jones, D.S. (1980) Annual cycle of shell growth increment formation in two continental shelf bivalves and its paleoecologic significance. *Paleobiology* 6, 331–340.

Ju, S.J., Secor, D.H., Harvey, H.R. (1999) Use of extractable lipofuscin for age determination of blue crab *Callinectes sapidus*. *Marine Ecology Progress Series* 185, 171–179.

Ju, S.J., Secor, D.H., Harvey, H.R. (2001) Growth rate variability and lipofuscin accumulation rates in the blue crab *Callinectes sapidus*. *Marine Ecology Progress Series* 224, 197–205.

Keller, E.T., Murtha, J.M. (2004) The use of mature zebrafish (*Danio rerio*) as a model for human aging and disease: Aquatic animal models of human disease. *Comparative Biochemistry and Physiology C* 138, 335–341.

Keller, M., Sommer, A.M., Pörtner, H.-O., Abele, D. (2004) Seasonality of energetic functioning and production of reactive oxygen species by lugworm (*Arenicola marina*) mitochondria exposed to acute temperature changes. *Journal of Experimental Biology* 207, 2529–2538.

Kelly, S.A., Havrilla, C.M., Brady, T.C., Abramo, K.H., Levin, E.D. (1998) Oxidative stress in toxicology: Established mammalian and emerging piscine model systems. *Environmental Health Perspectives* 106, 375–384.

Kishi, S. (2004) Functional aging and gradual senescence in zebrafish. *Annals of the New York Academy of Science* 1019, 521–526.

Kishi, S., Uchiyama, J., Baughman, A.M., Goto, T., Lin, M.C., Tsai, S.B. (2003) The zebrafish as a vertebrate model of functional aging and very gradual senescence. *Experimental Gerontology* 38, 777–86.

Knutson, D.W., Buddemeier, R.W., Smith, S.V. (1972) Coral chronometers: seasonal growth bands in reef corals. *Science* 177, 270.

Koslow, J.A., Boehlert, G.W., Gordon, J.D.M., Haedrich, R.L., Lorancek, P., Parin, N. (2000) Continental slope and deep-sea fisheries: implications for a fragile eco-system. *ICES Journal Marine Science* 57, 548–557.

Lambert, A.J., Boysen, H.M., Buckingham, J.A. et al. (2007) Low rates of hydrogen peroxide production by isolated heart mitochondria associate with long maximum lifespan in vertebrate homeotherms. *Aging Cell* 6, 607–618.

Laws, R.M. (1952) A new method of age determination for mammals. *Nature* 169, 972–973.

Leeuwenburgh, C., Fiebig, R., Chandwaney R., Ji, L.L. (1994) Aging and exercise training in skeletal muscle: responses of glutathione and antioxidant enzyme systems. *American Journal of Physiology* 267, R439–R445.

Liang, H., Masoro, E.J., Nelson, J.F., Strong, R., McMahan, C.A., Richardson, A. (2003) Genetic mouse models of extended lifespan. *Experimental Gerontology* 38, 1353–1364.

Lin, Y.-J., Seroude, L., Benzer, S. (1998) Extended life-span and stress resistance in the *Drosophila* mutant *methuselah*. *Science* 282, 943–946.

Lomovasky, B.J., Morriconi, E., Brey, T., Calvo, J. (2002) Individual age and connective tissue lipofuscin in the hard clam *Eurhomalea exalbida*. *Journal of Experimental Marine Biology and Ecology* 276, 83–94.

Lopez-Torres, M., Gredilla, R., Sanz, A., Barja, G. (2002) Influence of aging and long-term caloric restriction on oxygen radical generation and oxidative DNA damage in rat liver mitochondria. *Free Radical Biology and Medicine* 32, 882–889.

Lund, T.C., Glass, T.J., Tolar, J., Blazar, B.R. (2009) Expression of telomerase and telomere length are unaffected by either age or limb regeneration in *Danio rerio*. *PLoS ONE* 4, e7688.

Martinez, D.E. (1998) Mortality patterns suggest lack of senescence in *Hydra*. *Experimental Gerontology* 33, 217–225.

Maxwell, K.E., Matthews, T.R., Sheehy, M.R.J., Bertelsen, R.D., Derby, C.D. (2007) Neurolipofuscin is a measure of age in *Panulirus argus*, the Caribbean spiny lobster, in Florida. *Biological Bulletin* 213, 55–66.

Medvedev, Z.A. (1990) An attempt at a rational classification of theories of ageing.. *Biological Reviews of the Cambridge Philosophical Society* 65, 375–98.

Mourente, G., Diaz-Salvago, E. (1999) Characterization of antioxidant systems, oxidation status and lipids in brain of wild-caught size-class distributed *Aristeus antennatus* (Risso, 1816) Crustacea, Decapoda. *Comparative Biochemistry and Physiology B* 124, 405–416.

Mullin, M.M., Brooks, E.R. (1988) Extractable lipofuscin in larval marine fish. *Fishery Bulletin* 86, 407–415.

Nelson, J.S. (2006) *Fishes of the World*. John Wiley & Sons, New York.

O'Donovan, V., Tully, O. (1996) Lipofuscin (age pigment) as an index of crustacean age: correlation with age, temperature and body size in cultured juvenile *Homarus gammarus* L. *Journal of Experimental Marine Biology and Ecology* 207, 1–14.

Parham, J.F., Zug, G.R. (1997) Age and growth of loggerehead sea turtles (*Caretta caretta*) of coastal Georgia: an assessment of skleletochronological age-estimates. *Bulletin of Marine Science* 61, 287–304.

Passi, S., Ricci, R., Cataudella, S., Ferrante, I., De Simone, F., Rastrelli, L. (2004) Fatty acid pattern, oxidation product development, and antioxidant loss in muscle tissue of rainbow trout and *Dicentrarchus labrax* during growth. *Journal of Agricultural and Food Chemistry* 52, 2587–2592.

Penttila, J.D., Dery, L.M. (1988) *Age Determination Methods for Northwest Atlantic Species*. NOAA/National Marine Fisheries Service, < http://aquacomm.fcla.edu/2749/ >.

Perez-Campo, R., Lopez-Torres, M., Cadenas, S., Rojas, C., Barja, G. (1998) The rate of free radical production as a determinant of the rate of aging: evidence from the comparative approach. *Journal of Comparative Physiology B* 168, 149–158.

Pérez, V.I., Remmen, H.V., Bokov, A., Epstein, C.J., Vijg, J., Richardson, A. (2009) The overexpression of major antioxidant enzymes does not extend the lifespan of mice. *Aging Cell* 8, 73–75.

Philipp, E., Abele, D. (2010) Masters of longevity: lessons from long-lived bivalves – a mini-review. *Gerontology* 56, 55–65.

Philipp, E., Brey, T., Pörtner, H.-O., Abele, D. (2005a) Chronological and physiological ageing in a polar and a temperate mud clam. *Mechanisms of Ageing and Development* 126, 589–609.

Philipp, E., Pörtner, H.-O., Abele, D. (2005b) Mitochondrial ageing of a polar and a temperate mud clam. *Mechanisms of Ageing and Development* 126, 610–619.

Philipp, E., Heilmayer, O., Brey, T., Abele, D., Pörtner, H.-O. (2006) Physiological ageing in a polar and a temperate swimming scallop. *Marine Ecology Progress Series* 307, 187–198.

Philipp, E.E.R., Schmidt, M., Gsottbauer, C., Saenger, A.M., Abele, D. (2008) Size- and age-dependent changes in adductor muscle swimming physiology of the scallop *Aequipecten opercularis*. *The Journal of Experimental Biology* 211, 2492–2501.

Picken, G.B. (1980) The distribution, growth, and reproduction of the Antarctic limpet *Nacella (Patinigera) concinna* (Strebel, 1908). *Journal of Experimental Marine Biology and Ecology* 42, 71–85.

Porter, R.K., Hulbert, A.J., Brand, M.D. (1996) Allometry of mitochondrial proton leak: influence of membrane surface area and fatty acid composition. *American Journal of Physiology* 271, R1550–R1560.

Pörtner, H.-O. (2002) Climate variations and the physiological basis of themperature dependent biogeography: systemic to molecular hierarchy of thermal tolerance in animals. *Comparative Biochemistry and Physiology* 132, 739–761.

Regoli, F., Principato, G.B., Bertoli, E., Nigro, M., Orlando, E. (1997) Biochemical characterization of the antioxidant system in the scallop *Adamussium colbecki*, a sentinel organism for monitoring the Antarctic environment. *Polar Biology* 17, 251–258.

Reichwald, K., Lauber, C., Nanda, I. et al. (2009) High tandem repeat content in the genome of the short-lived annual fish *Nothobranchius furzeri*: a new vertebrate model for aging research. *Genome Biology* 10, R16.

Reznick, D.N., Bryant, M.J., Roff, D., Ghalambor, C.K., Ghalambor, D.E. (2004) Effect of extrinsic mortality on the evolution of senescence in guppies. *Nature* 431, 1095–9.

Richter, C., Park, J.-W., Ames, B.N. (1988) Normal oxidative damage to mitochondrial and nuclear DNA is extensive. *Proceedings of the National Academy of Sciences USA* 85, 6465–6467.

Ropes, J.W., Jones, D.S., Murawski, S.A., Serchuk, F.A., Jerald, A. (1984) Documentation of annual growth lines in ocean quahog *Arctica islandica* Linné.. *Fisheries Bulletin* 82, 1–19.

Ruvkun, G., Samuelson, A.V., Carr, C.E., Curran, S.P., Shore, D.E. (2010) Signaling pathways that regulate *C. elegans* life span. *In* Clemmons, D., Christen, Y. (eds). *IGFs:Local Repair and Survival Factor Throughout Life Span*. Research

and Perspectives in Endocrine Interactions Series, Springer-Verlag, pp. 69–84.
Schmidt, M., Philipp, E.E.R., Abele, D. (2008) Size and age-dependent changes of escape response to predator attack in the Queen scallop *Aequipecten opercularis*. *Marine Biology Research* 4, 442–450.
Schöne, B.R. (2003) A "clam-ring" master-chronology constructed from a short-lived bivalve mollusc from the nothern Gulf of California, USA. *The Holocene* 13, 39–49.
Schöne, B.R., Freyre Castro, A.D., Fiebig, J., Houk, S.D., Oschman, W., Kröncke, I. (2004) Sea surface water temperatures over the period 1884–1983 reconstructed from oxygen isotope ratios of a bivalve mollusk shell (*Arctica islandica*, southern North Sea). *Palaeogeography, Palaeoclimatology, Palaeoecology* 212, 215–232.
Schöne, B.R., Fiebig, J., Pfeiffer, M. et al. (2005) Climate records from a bivalved Methuselah (*Arctica islandica*, Mollusca; Iceland). *Palaeogeography, Palaeoclimatology, Palaeoecology* 228, 130–148.
Sestini, E.A., Carlson, J.C., Allsopp, R. (1991) The effects of ambient temperature on life span, lipid peroxidation, superoxide dismutase, and phospholipase A2 activity in *Drosophila melanogaster*. *Experimental Gerontology* 26, 385–395.
Sheehy, M.R.J. (1992) Lipofuscin age-pigment accumulation in the brains of ageing field- and laboratory-reared crayfish *Cherax quadricarinatus* (von Martens) (Decapoda: Parastacidae). *Journal of Experimental Marine Biology and Ecology* 161, 79–89.
Sheehy, M.R.J. (1996) Quantitative comparison of *in situ* lipofuscin concentration with soluble autofluorescence intensity in the crustacean brain. *Experimental Gerontology* 31, 421–32.
Sheehy, M.R.J. (2002) Role of environmental temperature in aging and longevity: insights from neurolipofuscin. *Archives of Gerontology and Geriatrics* 34, 287–310.
Sheehy, M.R.J. (2008) Questioning the use of biochemical extraction to measure lipofuscin for age determination of crabs: Comment on Ju et al. (1999, 2001). *Marine Ecology Progress Series* 353, 303–306.
Sheehy, M.R.J., Prior, A.E. (2008) Progress on an old question for stock assessment of the edible crab *Cancer pagurus*. *Marine Ecology Progress Series* 353, 191–202.
Sheehy, M.R.J., Greenwood, J.G., Fielder, D.R. (1995) Lipofuscin as a record of "rate of living" in an aquatic poikilotherm. *Journal of Gerontology A* 50, B327–B336.
Sheehy, M.R.J., Caputi, N., Chubb, C., Belchier, M. (1998) Use of lipofuscin for resolving cohorts of western rock lobster (*Panulirus cygnus*). *Canadian Journal of Fisheries and Aquatic Science* 55, 925–936.
Shigenaga, M.K., Hagen, T.M., Ames, B.N. (1994) Oxidative damage and mitochondrial decay in aging. *Proceedings of the National Academy of Sciences USA* 91, 10771–1078.
Sitte, N., Merker, K., Grune, T., von Zglinicki, T. (2001) Lipofuscin accumulation in proliferating fibroblasts *in vitro*: an indicator of oxidative stress. *Experimental Gerontology* 36, 475–486.
Slijper, E.J. (1962) *Whales*. Hutchingson, New York.
Smirina, E.M. (1994) Age determination and longevity in amphibians. *Gerontology* 40, 133–146.
Sohal, R.S. (2002) Role of oxidative stress and protein oxidation in the aging process, *Free Radical Biology and Medicine* 33, 37–44.
Strahl, J., Philipp, E., Brey, T., Broeg, K., Abele, D. (2007) Physiological aging in the Icelandic population of the ocean quahog *Arctica islandica*. *Aquatic Biology* 1, 77–83.
Strom, A., Francis, R.C., Mantua, N.J., Miles, E.L., Peterson, D.L. (2004) North Pacific climate recorded in growth rings of geoduck clams: A new tool for paleoenvironmental reconstruction. *Geophysical Research Letters* 31, L06206.
Sukhotin, A.A., Flyachinskaya, L.P. (2009) Aging reduces reproductive success in mussels *Mytilus edulis*. *Mechanisms of Ageing and Development* 130, 754–761.
Sukhotin, A.A., Pörtner, H.-O. (2001) Age-dependence of metabolism in mussels *Mytilus edulis* (L.) from the White Sea. *Journal of Experimental Marine Biology and Eology* 257, 53–72.
Sukhotin, A.A., Abele, D., Pörtner, H.-O. (2002) Growth, metabolism and lipid peroxidation in Mytilus edulis: age and size effects. *Marine Ecology Progress Series* 226, 223–234.
Sukhotin, A.A., Lajus, D.L., Lesin, P.A. (2003) Influence of age and size on pumping activity and stress resistance in the marine bivalve *Mytilus edulis* L. *Journal of Experimental Marine Biology and Ecology* 284, 129–144.
Sukhotin, A.A., Strelkov, P.P., Maximovich, N.V., Hummel, H. (2007) Growth and longevity of *Mytilus edulis* (L.) from northeast Europe. *Marine Biology* Research 3, 155.
Terzibasi, E., Valenzano, D.R., Cellerino, A. (2007) The short-lived fish *Nothobranchius furzeri* as a new model system for aging studies. *Experimental Gerontology* 42, 81–89.
Turekian, K.K., Cochran, J.K., Kharkar, D.P. et al. (1975) Slow growth rate of a deep-sea clam determined by 228Ra chronology. *Proceedings of the National Academy of Sciences USA* 72, 2829–2832.
Ueno, S., Imai, C., Mitsutani, A. (1995) Fine growth rings found in statolith of a cubomedusa *Carybdea rastoni*. *Journal of Plankton Research* 17, 1381–1384.
Valdesalici, S., Cellerino, A. (2003) Extremely short lifespan in the annual fish *Nothobranchius furzeri*. *Proceedings of the Royal Society of London B* 270 (Suppl. 2), S189–S191.
Valenzano, D.R., Terzibasi, E., Cattaneo, A., Domenici, L., Cellerino, A. (2006) Temperature affects longevity and age-related locomotor and cognitive decay in the short-lived fish *Nothobranchius furzeri*. *Aging Cell* 5, 275–278.
Vernet, M., Hunter, J.R., Vetter, R.D. (1988) Accumulation of age-pigments (lipofuscin) in two cold-water fishes. *Fisheries Bulletin* 86, 401–407.
Viarengo, A., Pertica, M., Canesi, L., Accomando, R., Mancinelli, G., Orunesu, M. (1989) Lipid peroxidation and

level of antioxidant compounds (GSH, vitamin E) in the digestive glands of mussels of three different age groups exposed to anaerobic and aerobic conditions. *Marine Environmental Research* 28, 291–295.

Viarengo, A., Canesi, L., Martinez, P.G., Peters, L.D., Livingstone, D.R. (1995) Pro-oxidant processes and antioxidant defence systems in the tissues of the Antarctic scallop (*Adamussium colbecki*) compared with the Mediterranean scallop (*Pecten jacobaeus*). *Comparative Biochemistry and Physiology B* 111, 119–126.

Vila, Y., Medina, A., Megina, C., Ramos, F., Sobrino, I. (2000) Quantification of the age-pigment lipofuscin in brains of known-age, pond-reared prawns *Penaeus japonicus* (Crustacea, Decapoda). *Journal of Experimental Zoology* 286, 120–130.

Wahle, R.A., Tully, O., O'Donovan, V. (1996) Lipofuscin as an indicator of age in crustaceans: analysis of the pigment in the American lobster *Homarus americanus*. *Marine Ecology Progress Series* 138, 117–123.

Walford, R.L., Liu, R.K. (1965) Husbandry, life span, and growth rate of the annual fish, *Cynolebias adloffi* E. Ahl. *Experimental Gerontology* 1, 161–168.

Walford, R.L., Liu, R.K., Troup, G.M., Hsiu, J. (1969) Alterations in soluble/insoluble collagen ratios in the annual fish, *Cynolebias bellottii*, in relation to age and environmental temperature. *Experimental Gerontology* 4, 103–109.

Wanamaker, A.D., Heinemeier, J., Scourse, J.D. et al. (2008) Very long-lived mollusks confirm 17th century AD tephra-based radiocarbon reservoir ages for north Icelandic shelf waters. *Radiocarbon* 50, 399–412.

Woodhead, A.D. (1974) Ageing changes in the siamese fighting fish, *Betta splendens* II. The ovary. *Experimental Gerontology* 9, 131–139.

Woodhead, A.D. (1979) Senescence in fishes. *Symposium of the Zoological Society of London* 44, 179–205.

Woodhead, A.D. (1998) Aging, the fishy side: An appreciation of Alex Comfort's studies. *Experimental Gerontology* 33, 39–51.

Yakes, F.M., Van Houten, B. (1997) Mitochondrial DNA damage is more extensive and persists longer than nuclear DNA damage in human cells following oxidative stress. *Proceedings of the National Academy of Sciences USA* 94, 514–519.

Yen, K., Mastitis, J.W., Mobbs, C.V. (2004) Lifespan is not determined by metabolic rate: evidence from fishes and *C. elegans*. *Experimental Gerontology* 39, 947–949.

Yilmazy, S., Polat, N. (2002) Age Determination of Shad (*Alosa pontica* Eichwald, 1838) inhabiting the Black Sea. *Turkish Journal of Zoology* 26, 393–398.

Zielinski, S., Pörtner, H.-O. (2000) Oxidative stress and antioxidative defense in cephalopods: a function of metabolic rate or age?. *Comparative Biochemistry and Physiology B* 125, 147–160.

Ziuganov, V., Miguel, E.S., Neves, R.J. et al. (2000) Life span variation of the freshwater pearl shell: A model species for testing longevity mechanisms in animals. *AMBIO* 29, 102–105.

Zug, G.R., Wynn, A.H., Ruckdeschel, C. (1986) Age determination of Loggerhead Sea Turtle, *Caretta caretta*, by incremental growth marks in the skeleton. *Smithsonian Contributions to Zoology* 427, 1–34.

Chapter 15

OXIDATIVE STRESS AND ANTIOXIDANT SYSTEMS IN CRUSTACEAN LIFE CYCLES

María Luisa Fanjul-Moles[1] *and María E. Gonsebatt*[2]

[1]Facultad de Ciencias, Universidad Nacional Autónoma de México, México, D. F.
[2]Instituto de Investigaciones Biomédicas, Universidad Nacional Autónoma de México, Ciudad Universitaria, México, D.F.

Crustaceans are excellent models to understand how environmental and endogenous factors shape complex life cycles. Factors such as ultraviolet (UV) light, salinity temperature, pollution, and infections alter respiration and metabolism directly or indirectly through alteration of the molt cycle, as well as through endocrine changes that regulate crustacean metabolism, molt, and reproduction.

Molting is an essential and repeated physiological process for somatic growth in arthropods. Ecdysozoans proceed to the next life stage via a molt cycle that consists of intermolt, premolt, ecdysis (shedding of the old cuticle), and postmolt (Skinner 1985). Each species experiences a species-specific fixed number of juvenile molts to reach adulthood. Some crustaceans, particularly female crabs, cease molting at the adult stage, and are terminally anecdysial during their reproductive stage. During molting, concomitant with the increased metabolic activity in the pre-molt phase, oxygen consumption rises, reaching a peak shortly before ecdysis, and then declines rapidly after molting to increase again to basal levels in the intermolt stage (Alcaraz and Sardá 1981; Penkoff and Thurberg 1982; Cockcroft and Wooldridge 1985). This represents an additional burden on the antioxidant system (see below) and increases the risk of oxidative stress (Table 15.1). Interestingly, this increase in oxygen demand is coincident with an increment in hemolymph glucose concentration (Telford 1968). Thus, molting can trigger a series of biochemical changes that may include switching of metabolic processes (Chang and O'Connor 1983).

The ontogeny of crustaceans is complex due to the intricate regulation of their endocrine system (Box 15.1 and Fig. 15.1). Generally, the embryonic development pattern in the eggs of different species is related to the quantity and disposition of the yolk. In some decapods, such as crayfish and crabs, the postembryonic development is short and direct without a free-living larval stage (Helluy et al. 1993). Other species, such as the spiny lobsters, have extensive planktonic larval phases, a short puerulus nektonic phase

Oxidative Stress in Aquatic Ecosystems, First Edition. Edited by Doris Abele, José Pablo Vázquez-Medina, and Tania Zenteno-Savín.

Box 15.1 Crustacean endocrine system

Various aspects of crustacean metabolism require regulation and coordination comparable to that found in vertebrates. The steering mechanisms are controlled by hormones and neurohormones synthesized both in endocrine glands *sensu stricto* and in neuroendocrine organs (Fig. 15.1). Crustacean endocrine systems consist of aggregations of neuroendocrine cells that produce neurohormones released from their axon terminals into neurohemal organs for storage or modification. Some of the neuroendocrine areas are formed by scattered cells with projections within the neuropil. Others form conspicuous clusters whose axons leave the neuropil and project into neurohemal-releasing sites and into non-neuroendocrine glands (Hartenstein 2006). There are also true non-neuroendocrine glands that secrete their products into the circulatory system. The principal endocrine glands in crustaceans are the Y-organ (YO), the mandibular organ (MO), the androgenic gland, the antennary glands, and the ovaries. The main neuroendocrine organs are the X-organ–sinus-gland complex (XO–SG), the post-commissural organ (PCO), and the pericardial organ (PO). The hormones released from endocrine and neuroendocrine organs include peptides, steroids, amines, terpenoids, and prostaglandins (Vogt 2002).

and a long juvenile-to-adult benthic phase. Some of these stages are nonfeeding and the animals rely exclusively on stored energy reserves, especially lipids, as they undergo a molt before feeding. The synthesis of the molting hormone ecdysone (Box 15.2), which controls crustacean growth and development and inhibits the juvenile hormone, is modulated by an insulin-like factor (Riehle and Brown 1999).

Cellular energy metabolism and oxygen consumption are coupled to the generation of reactive oxygen species (ROS). Along their life cycle, larval, juvenile, and adult crustaceans undergo variations in respiratory and metabolic requirements that are also dependent on environmental conditions. Biomass allocated to egg production varies in females in response to environmental conditions such as nutrition. These variations influence biomass at hatching, subsequent developmental pathways, and survival during periods of starvation (Giménez 2006). As it may be inferred from their life cycle, crustaceans represent one of the most complex groups of invertebrates, with various aspects of their metabolism requiring regulatory processes comparable to those found in vertebrates. Studies on antioxidant defenses and oxidative stress in crustacean life cycles, although very important, are relatively new and information is still limited.

CRUSTACEAN METABOLISM AND REACTIVE OXYGEN SPECIES

The principal metabolic routes by which crustaceans derive adenosin triphosphate (ATP) are similar to those in vertebrates and are critical for the production of ROS (Chang and O'Connor 1983). The production of reduced nicotinamide adenine dinucleotide phosphate (NADPH) by glucose-6 phosphate metabolism and the synthesis of ribose by the pentose phosphate pathway are critically important in crustacean energy metabolism.

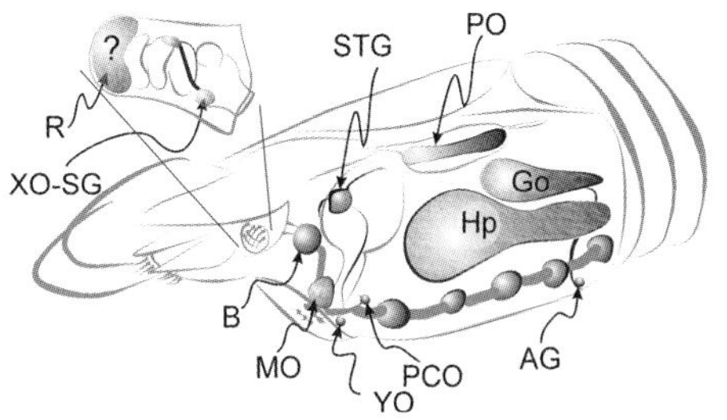

Fig. 15.1 Anatomic position of different gland- and neuropeptide-producing tissues in the cephalotorax of a decapod crustacean. Endocrine organs include androgenic gland (AG), gonads (Go, female ovary or male testis), hepatopancreas (Hp), mandibular organ (MO), pericardial organ (PO), and Y-organ (YO). Neuroendocrine organs include X-organ-sinus-gland complex (XO-SG) and the following neuroendocrine producing sites: brain (B), post-commisural organ (PCO), and stomatogastric ganglion (STG). Thoracic (drawn) and abdominal ganglia (not shown) are also neuropeptide-producing sites. Retina (R) is labeled with a question mark, indicating the possibility of producing peptides with local neural action (see text). (With permission of *Comparative Aspects of Circadian Rhythms* 2008, pp. 41–73, Transworld Research Network, Kerala, India.)

Table 15.1 Antioxidants and antioxidant enzymes described and characterized in crustaceans.

Antioxidants	Antioxidant enzyme(s)	Organism	Observations	Reference
	SOD	*Barytelphusa guerini*	Increased activity levels due to trivalent and hexavalent chromium exposure	Sridevi et al. 1998
	CAT, SOD, GPx	*Macrobrachium malcolmsonii*	Measured during embryonic and larval development	Arun and Subramanian 1998a
GSH, MTT	CAT, SOD,GPx, GRx	*Callinectes sapidus*	Modulation by copper-induced oxidative damage	Brouwer and Brouwer 1998
	CAT, SOD, GPx	*M. malcolmsonii* and *M. lamarrei lamarrei*	Specific activities of the antioxidant enzymes in the subcellular fractions (mitochondrial, cytosolic, and microsomal)	Arun and Subramanian 1998b
GSH, carotenoids, Vitamins A, E, and K	CAT, SOD, Px, GRx	*Palaemon adspersus, Artemia* sp.	Enzyme activities in developing eggs and larvae	Rudneva 1999
	CAT, SOD, GPx, GRx, GST	*Aristeus antennatus*	In neural tissues from males and females wild-caught and size-class distributed from the south coast of Spain	Mourente and Diaz-Salvago 1999
Eye pigments		*Mysis relicta*	Levels in eyes of two different populations with different light damage resistance	Dontsov et al. 1999
	GST	*Hyalella azteca*	Exposure to chlorpyrifos but not to methyl mercury reduced GST activity, no significant changes in GSH levels in adult whole organism homogenates	Steevens and Benson 1999
	CAT, SOD, GPx	*Palaemonetes argentinus Nobili*	Infection by *Probopyrus ringueleti* decreased SOD activity	Neves et al. 2000
	CAT, SOD, GPx	*Carcinus maenas*	Immunolocalization in hepatopancreas	Orbea et al. 2002
Carotenoids		*Penaeu esculentus, Emerita asiatica, Cherax quadricarinatus, Penaeus orientalis*	Content of carotenoids in crutacean ovaries	Liñán-Cabello et al. 2002
Vitamin C, GSH, GSSG	SOD, GPx	*Squilla mantis*	Levels in muscle tissue	Passi et al. 2002
Carotenoids	CAT, SOD, GST	*Daphnia sp*	Intraspecies variation in antioxidant activities was studied in relation to pigmentation and pond characteristics	Borgeras and Hessen 2002

	CAT, SOD, GPx	*Gammarus locusta*	Decreased activity and increased lipid peroxidation with age	Correia et al. 2003
Carotenoids, retinal	SOD	*Litopenaeus vannamei*	Levels in captive and wild shrimp	Liñán-Cabello et al. 2003
Vitamin C, GSH, GSSG	CAT, SOD, GPx	*Macrobrachium rosenbergii*	CAT, SOD, GPx activity changes in the post-larvae receiving vitamin-E, supplementation. Vitamin C and GSH content were determined	Dandapat et al. 2003
Coenzyme Q, α tocopherol GSH, ascorbate, astaxanthin		*Palaemonetes pugio*	Antioxidants and total oxyradical scavenging capacity during embryogenesis	Winston et al. 2004
p26, trehalose		*Artemia embrios*	Protection against oxidant stress	Collins and Klegg 2004
	CAT, GST	*Chasmagnathus granulata*	Daily variations in gill and hepatopancreas	Maciel et al. 2004
Total glutathione	CAT, SOD, GPx	*Farfantepenaeus paulensis*	Higher CAT and GPx activities and GSH levels in shrimp kept in captivity for 45 days with or without eyestalks	Almeida et al. 2004
	SOD	*Callinectes sapidus*	Mitochondrial and cytosolic isoform Mn-SOD modulation by hypoxia	Brower et al. 2004
	SOD	*Macrobrachium nipponense*	Purification and partial characterization of Mn-SOD from muscle tissue	Yao et al. 2004
Vitamin C, E, GSH	CAT, SOD, GPx	*Scylla serrata*	Naphthalene down-regulation of antioxidants and antioxidant enzymes in hepatopancreas, hemolymph and ovary in active vitellogenic stage	Vijayavel et al. 2004
	CAT, SOD, GST	*Chasmagnathus granulata*	Increased CAT and GST and decreased SOD activities during anoxia	De Oliveira et al. 2005
	CAT, GST	*Chasmagnathus granulata*	No change after UV-A and UV-B irradiation	Gouvela et al. 2005
	CAT, SOD, GPx,	*Daphnia magna*	Aging (juvenile to senescent adults) was accompanied by selective loss of antioxidant enzymes	Barata et al. 2005a
	CAT, SOD, GPx, GST	*Daphnia magna*	Differential patterns of antioxidant ezymes responses to redox cycling compounds	Barata et al. 2005b.

(continued)

Table 15.1 *(Continued)*

Antioxidants	Antioxidant enzyme(s)	Organism	Observations	Reference
Astaxanthin		marine pelagic copepods (*Acartia, Paracalanus, Pseudocalanus, Temora, Oithona*)	Production changes due to nutrient dynamics	Van Nieuwerburgh et al. 2005.
GSH	CAT, SOD, GPx; glutaredoxin (Grx), GST	*Penaeus monodon*	Decrease after infection with white spot syndrome virus (WSSV)	Rameshthangam and Ramasamy 2006.
	CAT, GPx, GST	*Gmelinoides fasciatus, Pallasea cancelloides, Ommatogammarus flavus*	Decreased GPx activity and increased CAT activity due to natural organic matter, no change in GST activity	Timofeyev 2006.
GSH	GPx	*Parasesarma erythodactyla*	Modulation by heavy metal contamination	MacFarlane et al. 2006.
	GST, GPx	*Eriocheir sinensis*	Up-regulation by acute cadmiun exposure	Silvestre et al. 2006.
	CAT, SOD, GPx	*Penaeus vannamei*	Decreased activity by nitrite and high selenium dietary supplement	Wang et al. 2006a.
Ascorbic acid	CAT, SOD, GPx, Grx	*Penaeus vannamei*	Increased activity due to ammonia stress in large individuals fed with enriched Artemia diet	Wang et al. 2006b
MTT	CAT, SOD, GPx	*Charybdis japonica*	Cadmiun exposure induced MTT in gill and hepatopancreas CAT, SOD and GPx activities were modulated	Pan and Zhang 2006.
	SOD	*Litopenaeus vannamei*	Molecular cloning and expression of cytosolic Mn-SOD. Transcripts were detected in hemocytes, heart, hepatopancreas, intestine, nervous system, muscle, pleopods and gills. Modulation by WSSV infection in hemocytes	Gómez-Anduro et al. 2006.
GSH	SOD	*Chasmagnathus granulatus*	SOD activity and GSH levels fell after feeding with hexachlorobenzene-treated microalgae	Chaufan et al. 2006.
GSH	CAT, SOD, GST, GPx, Grx	*Fenneropenaeus indicus*	Enzyme activities and antioxidant levels in the white spot syndrome virus (WSSV) infected tissues (hemolymph, hepatopancreas, gills and muscle)	Mohankumar and Ramasamy 2006.

	CAT, SOD, GPx	*Macrobrachium malcolmsonii*	Modulation by exposure to oil effluent	Arun and Subramanian (2007)
	CAT, SOD, GPx	*Litopenaeus vannamei*	Increased activities in shrimp fed with vitamin E diet, which also protected against acute salinity changes	Liu et al. 2007
MTT	SOD	*Hemidiaptomus roubaui*	Immunohistochemistry localization of MTT-like protein in brain and nerve cord after cadmiun exposure. SOD in shell glands after heat stress	Liberge and Barthélémy, 2007.
	Prx	*Fenneropenaeus chinensis*	Cloning and expression in several organs	Zhang et al. 2007
	CAT	*Fenneropenaeus chinensis*	Cloning, and expression in several organs	Zhang et al. 2008
	CAT, SOD	*Litopenaeus vannamei*	Level comparison at different salinities in hepatopancreas and muscle	Li et al. 2008
	CAT, SOD, GST	*Calanus finmarchicus*	Time and concentration modulation of GST, CAT and SOD gene transcription after naphtalene exposure	Hansen et al. 2008
	Trx	*Litopenaeus vannamei*	LvTrX cDNA was sequenced, cloned and overexpressed in bacteria for characterization. Trx is expressed in gill and pleopods, and is not modulated by hypoxia or reoxygenation	Aispuru et al. 2008.
	SOD, CAT, GPx, Trx	*Litopenaeus vannamei*	Acute pH stress (alkaline or acidic) increases mRNA expression	Wang et al. 2009
	Trx	*Fenneropenaeus chinensis*	mRNA detected in hemocytes, heart, hepatopancreas, gills, stomach, and intestine and is first down- and later up-regulated by WSSV virus infection	Ren et al. 2010

Box 15.2 Hormonal control of molting

The molting process in crustaceans involves several steps, each one controlled by physiological changes regulated by XO–SG and YO hormones. The X-organ complex secretes a neuropeptide, the molt-inhibiting hormone (MIH), that suppresses the production of molting hormone (ecdysone), an ecdysteroid secreted by the YO. Other inhibitory factors, as well as changes in YO sensitivity to those factors, are involved in regulating YO activity. Inhibition of molting by environmental stressors (e.g. hypoxia, extreme salinity and temperature) appears to be mediated by crustacean hyperglycemic hormone (CHH). During intermolt, ecdysteroid production is under negative feedback control by circulating ecdysteroids and the MO hormone, methyl farnesoate (MF), which is equivalent to the juvenile hormone of insects. Molting and reproduction are antagonistic processes, and breeding normally occurs during the intermolt stage. Endocrine control of reproduction is complex. The gonad inhibiting hormone (GIH) inhibits the oocytes in females by regulating the uptake of vitellogenin from the hemolymph (Charniaux-Cotton and Payen 1985). In females, the development and activity of the ovary appears to be stimulated by the vitellogenesis-stimulating hormone (VSH). VSH stimulates the synthesis of vitellogenin (VTG), which is probably secreted by thoracic and protocerebral centers in decapods (Martin 1981) and perhaps by the ovary. VTG may be synthesized in the ovary, the hepatopancreas (Quackenbush 1989; Rani and Subramonian 1997), or the adipose tissues (Legrand and Juchault 2006). Both male and female gonads are under neuroendocrine control. GIH, a peptide belonging to the CHH family, is distinct from MIH and inhibits vitellogenesis and gonad maturation. Vitellogenesis requires the synchronization of a circadian cycle both within the molting and seasonal cycles. As a result of this synchronization, juveniles emerge in a favorable season.

One of the most important families of neuropeptides participating in the regulation of mechanisms underlying all these processes is the CHH/MIH/GIH family. Functions attributed to this family are diverse and include regulation of the concentration of hemolymph glucose, molting, reproduction, regulation of other glands, and osmoregulation. The axons of CHH-producing cells form part of the sinus gland (SG) tract, which transverses the medulla terminalis neuropil and ends at the SG. CHH acts on the hepatopancreas to increase hemolymph glucose levels (Komali et al. 2005) and participates in regulation of ecdysis (Webster et al. 2000), reproduction (De Kleijn and Van Herp 1995), carbohydrate and lipid metabolism (Santos et al. 1997, 2001), hydromineral balance (Serrano et al., 2003), and stress response (Lorenzon et al. 1997; Durand et al. 2000; Chang et al. 2001; Santos et al. 2001).

The discontinuous growth of crustaceans, manifested in the molting cycles, results in periods during which the exoskeleton is too soft to allow active feeding. During this period the utilization of energy is under control of hormonal mechanisms. In females the energy demand of the molt cycle is additionally compounded by those needed for egg maturation. Hence crustacean tissues, organs, and life-cycle stages, have different metabolic rates which differentially shape the basal pro-oxidant status. The ratio of glycolytic to nonglycolytic catabolism is always lower in the hepatopancreas and in the gills than in other tissues. During the growth period after ecdysis, the animals have an increased energetic requirement that is predominantly covered via the pentose phosphate pathways (Chang and O' Connors 1983). Both, the synthesis and degradation of glycogen vary throughout the molt cycle and are hormonally regulated (see below). Glucose levels are regulated by the crustacean hyperglycemic hormone (CHH) (see Box 15.2) and glucose is the main circulating carbohydrate in crustacean hemolymph (Hohnke and Sheer 1970). The predominance of catabolism in some points of the crustacean life cycle results in redox changes and oxidative stress (Dandapat et al. 2003; Backup et al. 2008). As a result of hormonal control of metabolism and due to the energy-demanding activities of molt and reproduction, there is an interrelation between metabolism, endocrine regulation, and oxidative status through the animal's life cycle.

Metabolic depression, estivation, and cryptobiosis (diapause) have been reported for various crustacean orders. Embryonic, larval, and adult diapause have been described in brachaipod, ostracod, copepod, isopod, and decapod crustaceans. In copepods, dormancy has been described for egg, nauplius, and copepodid larvae (Guppy and Withers 1999). During this state crustaceans shift from aerobic to anaerobic metabolism, the cell cycle reversibly arrests and stress tolerance is enhanced to protect cellular components for resumption of growth and development when diapause terminates. Internally derived signals including ROS such as H_2O_2 and reactive nitrogen species (RNS) such as $NO^{\bullet}$ might regulate diapause-specific developmental pathways. Ferritin scavenges oxygen radicals and by sequestering iron may curb the Fenton reaction, preventing $HO^{\bullet}$ production and thus oxidative damage. In the rotifer *Brachionus plicatilis* dormancy is accompanied by increased transcription of antioxidant proteins such as γ-glutamyl transpeptidase (GTT), peroxiredoxins (Prx), and thioredoxin (Trx), which control intracellular redox potential (Denekamp et al. 2009).

ROS have a role as signaling molecules during cell growth, division, and development in crustaceans (MacRae et al. 2010). Some NADPH oxidase enzymes generate low concentrations of H_2O_2 which acts as a second messenger that seems to be involved in regulating growth and development in crustacean. In these signaling pathways, H_2O_2 oxidizes protein thiol-residues, perhaps via peroxidase-catalyzed reactions with the target proteins, ultimately promoting tyrosine phosphorylation and dephosphorylation. Protein tyrosine phosphatases, which are active in cell metabolism, growth, and differentiation, mitogen-activated protein kinases (MAPKs), and transcription factors are targeted by H_2O_2, thereby controlling molecular processes ranging from gene expression to protein post-translational modifications (MacRae 2010).

Some xenobiotics and pollutants, such as endosulfan and bisphenol, generate ROS that inhibit or induce molting and reproduction in aquatic decapods and isopods. Interestingly, ROS produced by light in seawater may also promote the development of crustaceans such as *Brachionus plicatilis* after a profound dormancy. This effect may be the result of prostaglandin synthesis due to lipid oxidation (Hagiwara et al. 1995). Exposure to H_2O_2 has also been shown to terminate *Artemia* cyst diapause (Chen et al. 2003). To summarize, changes in metabolic demands during life-cycle stages cause differential ROS production and potentially influence the susceptibility to oxidative stress. Paradoxically, ROS functioning as signaling molecules are required for cell growth, division, and development of crustaceans.

OXIDATIVE STRESS AND HORMONES IN CRUSTACEANS

Hyperglycemic responses have been reported in several crustacean species exposed to different kinds of pollutants and parasites (Fingerman et al. 1998). A wide range of pollutants have endocrine-disrupting functions for invertebrate species. In particular, organic contaminants, such as alkyl phenols, can act as estrogen mimetic compounds. Endocrine disruptors, such as several heavy metals, produce hyperglycemia in crustaceans. Those endocrine disruptors also cause oxidative stress through Fenton reactions (metals) or glutathione (GSH) depletion. Reactive electrophiles (electron absorbing compounds) are potential lipid hydroperoxide generators. Lipid hydroperoxide removal through cellular detoxification pathways causes GSH depletion since GSH is used in GST reactions, yielding glutathione disulphide (GSSG), which if not recycled is excreted from the cell, resulting in a loss of reduced GSH and thus a redox state alteration.

Hyperglycemia and oxidative stress derived from the exposure of aquatic animals to different kinds of pollutants promote a redox signal which, in turn, induces protective responses against oxidative damage and resets the original state of redox homeostasis after temporary exposure to ROS (Rodríguez et al. 2007). Regulation and control produced by the neuroendocrine system (Boxes 15.1 and 15.2) are part of a metabolic feedback loop that may be disrupted by environmental changes and/or pollutants.

Similar to cortisol in humans or corticosterone in rodents, the CHH family of hormones is an important mediator of the stress response in crustaceans. In addition to the above mentioned environmental changes, hyperglycemia and increased CHH levels in circulating hemolymph occur following exposure to hypoxia, anoxia, and reoxygenation, as well as in response to thermal and osmotic stress in several crustaceans (Lorenzon et al. 2002, 2007; Rodriguez et al. 2007). It has been proposed that the regulation of crustacean hemolymph glucose levels is largely due to the secretion of the CHH peptide (the "starved" signal) and insulin-like peptides (the "fed" signal), which control the intracellular metabolism within the hepatopancreas (Verri et al. 2001; Gutierrez et al. 2007). Stress hormones, such as CHH, may operate in concert with other hormones including octopamine, L-encephalin, and serotonin (Basu and Kravitz 2003; Lorenzon et al. 2005). Brief episodes of hyperglycemia as a consequence of starved and fed conditions, dormancy/arousal, or transition between molting and intermolt development stages, could cause tissue damage through mechanisms involving repeated acute changes in cellular metabolism and, thus, in ROS production (de Oliveira et al. 2005).

If CHH regulation is insufficient, biochemical and/or behavioral stress reactions, such as changes in activity or quiescence, occur (Prieto-Sagredo et al. 2000; Fanjul-Moles et al. 1998). Alternatively, if adaptation to stress fails and circulating CHH levels remain chronically elevated, as occurs in polluted or otherwise extremely stressful environmental conditions, the resulting hyperglycemia induces cell damage. Increased glucose levels in vertebrates elicit a rise in ROS production due to the higher flux of reducing equivalents (reduced nicotinamide adenine dinucleotide – NADH) into the mitochondrial electron transport chain (Nishikawa et al. 2000; Brownlee 2001). The resulting loss in GSH produces enhanced sensitivity to oxidative stress. It has also been proposed that the reduction of glucose to sorbitol consumes NADPH. As NADPH is required for regenerating GSH from GSSG, this consumption of NADPH can exacerbate intracellular oxidative stress (Fig. 15.2). A resemblance to carbohydrate metabolism in vertebrates suggests that similar mechanisms exist in crustaceans.

As stated above, in crustaceans the life history seems to be responsible for the differences in the resting levels of glucose. CHH may have a modulating effect through increases in glucose availability, by acting upon glycolysis through the regulation of the Embden–Meyerhof pathway during premolt and through the Krebs cycle and the pentose phosphate pathway in the intermolt stage (Santos and Keller 1993). The pentose phosphate pathway is active in some tissues, including hepatopancreas and gills. This pathway metabolizes glucose, especially in the intermolt period, yielding C5 sugars for the synthesis of nucleic acids and NADPH for synthetic processes. In growing animals, CHH appears in the hemolymph only when the rate of glucose mobilization exceeds its biosynthetic utilization.

Metabolism, behavior, reproduction, and sensory processes are all affected directly and indirectly by periodic replacements of the integument and the underlying cycles of ecdysteroid metabolites during the molting cycle, implying variations in ROS production. The molting cycle is regulated by steroid hormones synthesized and secreted by the Y-organ (YO; Box 15.2, Fig. 15.1). The principal regulator of steroidogenesis in the YO is the molt-inhibiting hormone (MIH), which exerts an inhibitory effect on the YO by regulating the intracellular levels of cAMP or cGMP (Covi et al. 2009). Hemolymph MIH levels and YO responses to changes in MIH depend on the molting cycle stage. For crabs, MIH is structurally distinct from CHH, yet crab CHH also represses ecdysteroid synthesis, albeit to lower extent (Chung and Webster 2003). In contrast, MIH in lobsters has an inhibitory effect on molting and also alters hyperglycemic activity (reviewed in Chang et al. 2001). Thus, CHH and MIH appear to be multifunctional and are involved in the complex regulation of molting and hemolymph glucose regulation.

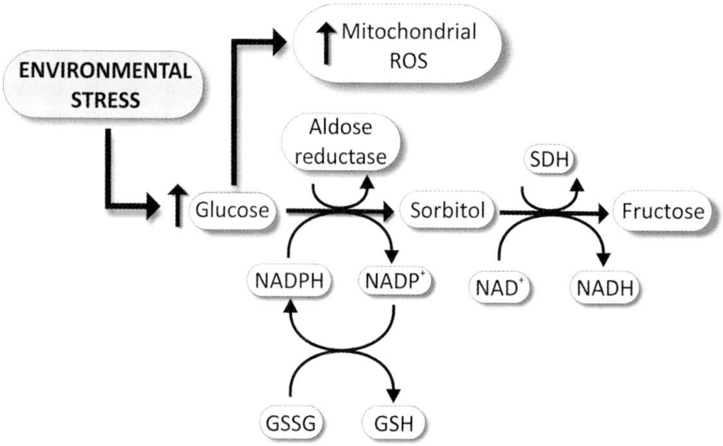

Fig. 15.2 Hyperglycemia induced by environmental stress causes overproduction of superoxide by mitochondria and increased glucose flux through the aldose reductase pathway. This drives the enzymatic conversion of glucose to the polyalcohol sorbitol, with concomitant decreases in NADPH and glutathione.

Most of the knowledge of changes in CHH and MIH hormones in the molting cycle is based on eyestalk ablation studies. Bilateral eyestalk ablation (see Boxes 15.1 and 15.2) induces molting. The second molt after ablation results in a low level or absence of MIH (Chang et al. 1990) which, in turn, stimulates the synthesis and release of ecdysteroids from the YO molting gland in crustaceans. The expression of both MIH (MIH mRNA) in the X-organ and MIH in the sinus gland is 5–10 times lower than that of CHH (Chung and Webster 2003). It has been proposed that CHH is secreted in the endocrine cells of the foregut and hindgut during premolt, implying that its expression may be induced by the elevated ecdysteroid levels in hemolymph. The release of CHH from these cells initiates the ecdysial process with the bursting of the ecdyseal line, which is caused by rapid iso-osmotic water uptake. Halfway to completion of ecdysis (when animals escape from the old cuticle), the level of CHH in hemolymph and, as a consequence, the hemolymph glucose levels reach their maximal concentration, 100 times higher than the level during intermolt. This prolongs the iso-osmotic water uptake for 1–2 h after ecdysis, resulting in a total increment of 20–50% of animal water content. Hence, at molt, CHH has multiple functions in glucose mobilization and ion regulation (Chung et al. 2010). The increment of glycemia as a consequence of stress and its relationship with lipid oxidation and antioxidants, such as GSH, suggest a strong variation in ROS production and oxidative stress that merit further research (Fanjul-Moles et al. 2009).

ENVIRONMENTAL CONDITIONS THAT MODULATE ANTIOXIDANT RESPONSES DURING THE LIFE CYCLE

In Table 15.1 we summarize data on antioxidant enzyme activities in different tissues and life stages of several crustacean species. In addition, modulation of enzyme activity and protein expression has been studied. Recently, antioxidant enzyme genes from different organisms were sequenced, cloned, and characterized.

In some cases, larvae respire through permeable egg membranes. Oxidative stress occurs during larval development due to direct exposure of the larvae to oxygen in water (Peters and Livingstone 1996). In the prawn *Macrobrachium rosenbergii* levels of thiobarbituric acid reactive substances (TBARS) increased in the first larval stages, and superoxide dismutase (SOD) activity and GSH concentrations (in whole tissue homogenates) increased gradually in subsequent larval stages until metamorphosis (Dandapat et al. 2003). Activities of SOD, catalase (CAT), glutathione peroxidase (GPx), and GST in the freshwater prawn *Macrobrachium malcolmsonii* are elevated during the larval stage when compared to the preceding embryonic stages (Arun and Subramanian 1998a). During larval development and metamorphosis of the giant prawn, *Macrobrachium rosenbergii*, lipid peroxidation is higher and SOD activity lower in earlier larval stages (Dandapat et al. 2003). CAT and GPx do not change during development, and GSH increases during larval progression until metamorphosis. The activity of SOD is also significantly lower in *Artemia* dormant cysts than in early development shrimp eggs (Rudneva 1999), probably because cysts are covered by compact shells, whereas shrimp eggs are covered by thinner shells. CAT activity is significantly enhanced in *Artemia nauplii* compared to eggs. Both, higher environmental oxygen levels and elevated metabolic rates after hatching increase the amount of ROS-mediated reactions in *Artemia*. The concentrations of low molecular weight scavengers, such as GSH and vitamins, decrease during brine shrimp development (Rudneva 1999).

The activities of SOD and GPx decrease during the transition from juvenile to adult in *Gammarus locusta*. Lipid peroxidation is 40% higher in older individuals than in juveniles (Correia et al. 2003). Similar patterns are found in *Daphnia magna* (Barata et al. 2005a), although the enzyme most sensitive to aging is CAT. Lower CAT activity may account for the increased lipid peroxidation observed in older individuals.

In *Daphnia magna*, an age-related decline in survival is accompanied by increased oxidative stress and oxidative damage, as demonstrated by changes in the balance between endogenous pro- and antioxidant processes over lifetime. There is a significant increase in the formation of lipid peroxides and loss of antioxidant enzymes in aging *D. magna*. The breakdown of antioxidant defenses may contribute to oxidative stress and compromise survival (Borgeraas and Hessen 2002).

In adult crustaceans circadian rhythms of antioxidant enzymes and low molecular weight scavengers, as well as diurnal rhythms of antioxidant enzymes, melatonin, and oxidative damage have been described. Daily and circadian patterns in GSH levels and enzyme activities occur in the midgut gland and hemolymph of two species of freshwater crayfish with different latitudinal distribution (Duran-Lizarrága et al. 2001; Fanjul-Moles and Prieto-Sagredo 2003), as well as in the nervous system of *P. clarkii* where the levels of antioxidants change with light intensity (Fanjul-Moles et al. 2009). Melatonin N-acetyl-5-methoxytryptamine, a signal that influences the circadian entrainment, is a ROS scavenger that contributes to the antioxidant protection of the cell. Melatonin has been reported in crustaceans such as the prawn *Macrobrachium rosembergii* (Withyachumnarnkul 1992) and crayfish (Agapito et al. 1995; Baltzer 1996) as having a circadian modulatory effect on hemolymph metabolites and neurotransmitter release (Tilden et al. 2003), which can affect GSH rhythms (Durán-Lizárraga et al. 2001).

GENDER DIFFERENCES IN ANTIOXIDANT DEFENSES

The overall activity and production of ecdysteroids is controlled by hormones produced and secreted in the eyestalk by the X-organ-sinus gland, a hormonal axis that parallels the vertebrate system where luteinizing hormone and follicle stimulating hormone regulate gonads, testosterone, and progesterone production. Disruption of these hormonal and metabolic systems by environmental disturbance (e.g., contaminants) has different effects in males and females. A decrement in CAT, SOD, and GPx activities has been reported to occur in adult decapods and marine and estuarine amphipods over their lifetime. Basal antioxidant activities are always lower in amphipods than in decapods, perhaps due to the higher specific metabolic rate of the smaller amphipods. The extent of the decrement seems to be gender specific as SOD activity was reported to be 50% lower in females than in males. Increased polyunsaturated fatty acid (PUFA) levels, lower antioxidant enzyme activities, and decreased metabolic rates in adult amphipods as compared to juveniles, render older animals more susceptible to peroxidation of PUFAs. This was reported for male but not female adults, indicating differential sensitivity to oxidative stress between genders (Correia et al. 2003). Gender-specific sensitivity of the antioxidant system to contaminants was also recently reported for the shore crab, *Carcinus maenas*. The activities of CAT, GPx, and GST recorded in female *C. maenas* hepatopancreas were significantly higher than in males (Pereira et al. 2009), which possibly explains the higher tolerance to a pathological environment in females. Evidence suggests gender differences occur in crustaceans regarding uptake, elimination, and metabolism of toxicants. In crustaceans, gender-specific differences occur in the immune system and in their response to natural toxins. Analysis of the hemolymph from female fairy shrimp revealed the presence of two to three different isozymes that exhibit diphenoloxidase activity, whereas hemolymph from males of the same species featured a single isozyme (Radhika et al. 1998).

ULTRAVIOLET EXPOSURE

Marine and freshwater crustaceans are negatively affected by current levels of UV-A (320–400 nm) and UV-B (290–320 nm) radiation in different life stages (Gouvela et al. 2005). The major effects are observed in planktonic early life stages. In some copepods, UV-B induces naupliar mortality, reduced survival, and fecundity in female larvae so that the sex ratio shifts in the population (Karanas et al. 1979; Chalker-Scott 1995; Naganuma et al. 1997).

Exposure to UV-A or UV-B induces the formation of $O_2^{\bullet-}$, H_2O_2, and $HO^{\bullet}$ (Limón-Pacheco and Gonsebatt 2009), which cause damage in crustacean tissues and can also induce defense mechanisms. For example, GST activity in the pereiopod epidermis is significantly induced by UV-A, but not by UV-B in the eyestalk-less crab *Chasmagnathus granulate* (Gouvela et al. 2005). Mycosporine-like amino acids provide protection against damaging UV-B radiation, which elevates the survival of the *Amphitoe valida* and in the progeny of *Idothea baltica* (Helbling et al. 2002). This protection is provided by the effective absorption of short wavelengths by mycosporine-like amino acids assumed to act as UV-quenchers and antioxidants. In *Cerodaphnia dubia*, the combination of high UV and arsenic ($1.5\ mg\ L^{-1}$) adversely and consistently impacted the survival of the organisms over three generations. This synergism can produce genotoxicity and oxidative stress (Hansen et al. 2002).

Photoprotection of CAT, SOD, GST, and carotenoids occurs in four *Daphnia* species inhabiting shallow ponds, and consequently exposed to UV radiation during the summer. Activity of CAT is elevated in the *D. magna* lab-clone, GST activity is low in melanic *D. pulex*, and SOD activity is high in *D. longispina* from lowland humic ponds. There are no differences in antioxidant capacity between melanic and nonpigmented alpine *D. longispina*. Among the alpine populations of *D. longispina* there is a significant positive correlation between absorbance (300 nm) of the pond water and CAT activity, which could be related to ambient levels of photo-induced H_2O_2 in small water bodies (Borgeraas and Hessen, 2002).

CONCLUSION

Environmental and metabolic factors elicit oxidative stress throughout the life cycle of crustaceans. Endocrine, behavioral, and antioxidant mechanisms allow crustaceans to cope with environmental and oxidative challenges throughout their ontogeny. Moreover, crustaceans elicit differential antioxidant responses in critical points of the life cycle, such as eggs, dormancy, the hatching of larvae, metamorphosis from the larval to the juvenile phase, and as adults. Juvenile-to-adult transitions and aging are associated with a decline in the antioxidant response, while the protection of shells has been observed in eggs and up-regulation of antioxidant enzymes has been documented in larval stages. Therefore, in the face of oxidative stress, larval, juvenile, and adult crustaceans have evolved sophisticated and redundant responses to ensure survival, some of which may have biomass consequences compromising the individual life cycle or the population size.

REFERENCES

Agapito, M.T., Herrero, B., Pablos, M.I. et al. (1995) Circadian rhythms of melatonin and serotonin-N-acetyltransferase activity in *Procambarus clarkii*. *Comparative Biochemistry and Physiology A* 112, 179–185.

Aispuro-Hernandez, E., Garcia-Orozco, K.D., Muhlia-Almazan, A. et al. (2008) Shrimp thioredoxin is a potent antioxidant protein. *Comparative Biochemistry and Physiology C* 148, 94–99.

Alcaraz, M, Sardá, F. (1981) Oxygen consumption by *Nephrops norvegicus* (L.) (Crustacea: Decapoda) in relationship with its moulting stage. *Journal of Experimental Marine Biology and Ecology* 55, 113–118.

Almeida, E.A., Petersen, R.L., Andreatta, E.R. et al. (2004) Effects of captivity and eyestalk ablation on antioxidant status of shrimps (*Farfantepenaeus paulensis*). *Aquaculture* 238, 523–528.

Arun, S., Subramanian, P. (1998a) Antioxidant enzymes in freshwater prawn *Macrobrachium malcolmsonii* during embryonic and larval development. *Comparative Biochemistry and Physiology B* 121, 273–277.

Arun, S., Subramanian, P. (1998b) Antioxidant enzymes activity in subcellular fraction of freshwater prawns *M. malcolmsonii* and *M. lamarrei lamarrei*. *Applied Biochemistry and Biotechnology A* 75, 187–192.

Arun, S., Subramanian, P. (2007) Cytochrome P450-dependent monooxygenase system mediated hydrocarbon metabolism and antioxidant enzyme responses in prawn, *Macrobrachium malcolmsonii*. *Comparative Biochemistry and Physiology C* 145, 610–616.

Baltzer, I. (1996) Encystment of *Gonyaulax polyedra*, dependence on light. *Biological Rhythm Research* 27, 386–389.

Buckup, L., Dutra, B.K., Ribarcki, F.P. et al. (2008) Seasonal variations in the biochemical composition of the crayfish *Parastacus defossus* (Crustacea) *Comparative Biochemistry and Physiology A* 49, 59–67.

Barata, C., Navarro, J.C., Varo, I. et al. (2005a) Changes in antioxidant enzyme activities, fatty acid composition and lipid peroxidation in Daphnia magna during the aging process. *Comparative Biochemistry and Physiology B* 140, 81–90.

Barata, C., Varo, I., Navarro, J.C., Arun, S. et al. (2005b) Antioxidant enzyme activities and lipid peroxidation in the freshwater cladoceran *Daphnia magna* exposed to redox cycling compounds. *Comparative Biochemistry and Physiology C* 140, 175–186.

Basu, A.C., Kravitz, E.A. (2003) Morphology and monoaminergic modulation of Crustacean hyperglycemic hormone-like immunoreactive neurons in the lobster nervous system. *Journal of Neurocytology* 32, 253–263.

Borgeraas, J., Hessen, D.O. (2002) Variations of antioxidant enzymes in *Daphnia* species and populations as related to ambient UV exposure. *Hydrobiologia* 477, 15–30.

Brouwer, M., Brouwer, T.H. (1998) Biochemical defense mechanisms against copper-induced oxidative damage in the blue crab, *Callinectes sapidus*. *Archives of Biochemistry and Biophysics* 351, 257–264.

Brouwer, M., Larkinb, P., Brown-Petersona, N. et al. (2004) Effects of hypoxia on gene and protein expression in the blue crab, *Callinectes sapidus*. *Marine Environmental Research* 58, 787–792.

Brownlee, M. (2001) Biochemistry and molecular cell biology of diabetic complications. *Nature* 414, 813–820.

Chalker-Scott, L. (1995) Survival and sex ratios of the intertidal copepod, *Tigriopus californicus*, following ultraviolet-B (290–320nm) radiation exposure. *Marine Biology* 123, 799–804.

Chang, E.S., O'Connor, J.D. (1983) Metabolism and transport of carbohydrates and lipids. *In* Mantel, L.H. (ed.). *The Biology of Crustacea*, Vol. 5, *Internal Anatomy and Physiological Regulation*. Academic Press, New York, pp. 263–287.

Chang, E.S., Prestwich G.D, Bruce, M. (1990) Amino acid sequence of a peptide with both moult-inhibiting and hyperglycemic activities in the lobster, *Homarus americanus*. *Biochemical and Biophysical Research Communications* 171, 818–826.

Chang, E.S., Chang, S.A., Mulder, E.P. (2001) Hormones in the lives of Crustaceans: An overview. *American Zoologist* 41, 380–388.

Charniaux-Cotton, H., Payen, G.G. (1985) Sexual differentiation. *In* Bliss, D.E., Mantel, L.H. (eds). *The Biology of Crustacea*, Vol. 9. Academic Press, New York, pp. 217–299.

Chaufan, G., Jua'rez, A., Basack, S. et al. (2006) Toxicity of hexachlorobenzene and its transference from microalgae (*Chlorella kessleri*) to crabs (*Chasmagnathus granulatus*). *Toxicology* 227, 262–270.

Chen, C.H., Lehninger, A.L. (1973) Respiration and phosphorilation by mitochondria from the hepatopancreas of the blue crab *Callicnetes sapidus*. *Archives of Biochemistry and Biophysics* 154, 445–447.

Chen T., Reith, M.E., Ross N.W. et al. (2003) Expressed sequence tag (EST)-based characterization of gene regulation in *Artemia* larvae. *Invertebrate Reproduction and Development* 44, 33–44.

Chung, J.S., Webster, S.G. (2003) Moult cycle-related changes in biological activity of moult-inhibiting hormone (MIH) and crustacean hyperglycaemic hormone (CHH) in the crab, *Carcinus maenas*. *European Journal of Biochemistry* 270, 3280–3288.

Chung, J.S., Zmora, N., Katayama, H. et al. (2010) Crustacean hyperglycemic hormone (CHH) neuropeptides family: Functions, titer, and binding to target tissues. *General and Comparative Endocrinology* 166, 447–454.

Collins, C.H., Clegg, J.S. (2004) A small heat-shock protein, p 26, from the crustacean *Artemia* protects mammalian cells (Cos–1) against oxidative damage. *Cell Biology International* 28, 449–455.

Cockcroft, A.C., Wooldridge, T. (1985). The effect of mass, temperature and molting on the respiration of *Macropetasma Africanus* balss (Decapoda: Penaeoidea). *Comparative Biochemistry and Physiology A* 81, 143–148.

Correia, A.D., Costa, M.H., Luis, O.J. et al. (2003) Age-related changes in antioxidant enzyme activities, fatty acid composition and lipid peroxidation in whole body *Gammarus locusta* (crustacea: Amphipoda). *Journal of Experimental Marine Biology and Ecology* 289, 83–101.

Covi, J.A., Chang, E.S., Mykles, D.L. et al. (2009) Conserved role of cyclic nucleotides in the regulation of ecdysteroidogenesis by the crustacean molting gland. *Comparative Biochemistry and Physiology A* 152, 470–477.

Dandapat, J., Chainy, G.B.N., Rao, K.J. (2003) Improved post-larval production in giant prawn, *Macrobrachium rosenbergii*, through modulation of antioxidant defence system by dietary vitamin-E. *Indian Journal of Biotechnology* 2, 195–202.

De Kleijn, D.P., Van Herp, F. (1995) Molecular biology of neurohormone precursors in the eyestalk of Crustacea. *Comparative Biochemistry and Physiology B* 112, 573–579.

Denekamp, NY, Thorne MAS, Clark MS et al. (2009) Discovering genes associated with dormancy in the monogonont rotifer *Brachionus plicatilis*. *BMC Genomics* 10, 108.

De Oliveira, U.O., da Rosa Araújo, A.S., Belló-Klein, A. et al. (2005) Effects of environmental anoxia and different periods of reoxygenation on oxidative balance in gills of the estuarine crab *Chasmagnathus granulata*. *Comparative Biochemistry and Physiology B* 140, 51–57.

Dontsov, A.E., Fedorovich, I.B., Lindstroö, M. et al. (1999) Comparative study of spectral and antioxidant properties of pigments from the eyes of two *Mysis relicta* (crustacea, mysidacea) populations, with different light damage resistance. *Journal of Comparative Physiology B* 169, 157–164.

Durand, F., Devillers, N., Lallier, F. et al. (2000) Nitrogen excretion and change in blood components during emersion of the subtidal spider crab *Maia squinado* (L.). *Comparative Biochemistry and Physiology A* 127, 259–271.

Duran-Lizarraga, M.E., Prieto-Sagredo, J., Gonsebatt, M.E. et al. (2001) The crayfish *Procambarus clarkii* shows circadian variations in different parameters of the GSH cycle. *Photochemistry and Photobiology* 74, 350–355.

Fanjul-Moles, M.L., Prieto-Sagredo, J. (2003) The circadian system of crayfish: a developmental approach. *Microscopy Research and Technique* 60, 291–301.

Fanjul-Moles, M.L., Bosques-Tistler, T., Prieto-Sagredo, J. et al. (1998) Effect of variation in photoperiod and light intensity on oxygen consumption, lactate concentration and behavior in crayfish *Procambarus clarkii* and *Procambarus digueti*. *Comparative Biochemistry and Physiology A* 119, 263–269.

Fanjul-Moles, M.L., Prieto-Sagredo, J., López, D.S. et al. (2009) Crayfish *Procambarus clarkii* retina and nervous system exhibit antioxidant circadian rhythms coupled with metabolic and luminous daily cycles. *Photochemistry and Photobiology* 85, 78–87.

Fingerman, M., Jackson, N., Nagabhushanam, R. (1998) Hormonally-regulated functions in crustaceans as biomarkers of environmental pollution. *Comparative Biochemistry and Physiology C* 120, 343–350.

Giménez, L. (2006) Phenotypic links in complex life cycles: conclusions from studies with decapod crustaceans. *Integrative and Comparative Biology* 46, 615–622.

Gómez-Anduro, G.A., Barillas-Mury, C.V., Peregrino-Uriartea, A.B., et al. (2006) The cytosolic manganese superoxide dismutase from the shrimp *Litopenaeus vannamei*: Molecular cloning and expression. *Developmental and Comparative Immunology* 30, 893–900.

Gouvela, G.R., Marques, D.S., Cruz, B.P. et al. (2005) Antioxidant defenses and DNA damage induced by UV-A and UV-B radiation in the crab *Chasmagnathus granulata* (decapoda, brachyura). *Photochemistry and Photobiology* 81, 398–403.

Guppy, M., Withers, P. (1999) Metabolic depression in animals: physiological perspectives and biochemical generalizations. *Biological Reviews of the Cambridge Philosophy Society* 74, 1–40.

Gutiérrez, A., Nieto, J., Pozo, F. et al. (2007) Effect of insulin/IGF-I like peptides on glucose metabolism in the white shrimp *Penaeus vannamei*. *General and Comparative Endocrinology* 153, 170–175.

Hagiwara, A., Hoshi, N., Kawahara, F., et al. (1995) Resting eggs of the marine rotifer *Brachionus plicatilis* Müller: development, and effect of irradiation on hatching. *Hydrobiologia* 313, 223–229.

Hansen, L.J., Whitehead, J.K., Anderson, S.L. (2002) Solar UV Radiation Enhances the Toxicity of Arsenic in *Ceriodaphnia dubia*. *Ecotoxicology* 1, 279–287.

Hansen, B.H., Altin, D., Vang, S- et al. (2008) Effects of naphthalene on gene transcription in *Calanus finmarchicus* (Crustacea, Copepoda). *Aquatic Toxicology* 86, 157–165.

Hartenstein, V. (2006) The neuroendocrine system of invertebrates: a developmental and evolutionary perspective. *Journal of Endrocrinology* 190, 555–570.

Helbling, W.E., Menchi, C.F., Villafañe, V.E. (2002) Bioaccumulation and role of UV-absorbing compounds in two marine crustacean species from Patagonia, Argentina. *Photochemistry and Photobiology Science* 1, 820–825.

Helluy, S., Sandeman, R., Beltz, B.S. et al. (1993) Comparative brain ontogeny of the crayfish and clawed lobster: implications of direct and larval development. *Journal of Comparative Neurology* 335, 343–354.

Hohnke, L., Sheer, B.J. (1970) Carbohydrate metabolism in crustaceans. *In* Florkin, M., Sheer, B.T. (eds). *Chemical Zoology*, Vol. 5. Academic Press, New York, pp. 147–166.

Karanas, J.J., Van Dyke, H., Worrest, R.C. (1979) Midultraviolet (UV-B) sensitivity of *Acartia clausii* Giesbrecht (Copepoda). *Limnology and Oceanography* 24, 1104–1116.

Komali, M., Kalarani, V., Venkatrayulu, C et al. (2005). Hyperglycaemic effects of 5-hydroxytryptamine and dopamine in the freshwater prawn, *Macrobrachium malcolmsonii*. *Journal of Experimental Zoology A* 303, 448–455.

Legrand, J.J., Juchault, P. (2006) The ontogeny of sex and sexual physiology. *In* Forest, J., von Vaupel Klein, J.C. (eds). *The Crustacea*. Brill Academic Publishers, Leiden, pp. 353–460.

Li, E., Chen, L., Zeng, C. et al. (2008) Comparison of digestive and antioxidant enzymes activities, haemolymph oxyhemocyanin contents and hepatopancreas histology of white shrimp, *Litopenaeus vannamei*, at various salinities. *Aquaculture* 274, 80–86.

Liberge, M., Barthélémy, R-. (2007) Localization of metallothionein, heat shock protein (Hsp70), and superoxide dismutase expression in *Hemidiaptomus roubaui* (Copepoda, Crustacea) exposed to cadmium and heat stress. *Canadian Journal of Zoology* 85, 362–371.

Limón-Pacheco, J., Gonsebatt, M.E. (2009) The role of antioxidants and antioxidant-related enzymes in protective responses to environmentally induced oxidative stress. *Mutation Research* 674, 137–147.

Liñán-Cabello, M.A., Paniagua-Michel, J., Hopkins, P.M. (2002) Bioactive roles of carotenoids and retinoids in crustaceans. *Aquaculture Nutrition* 8, 299–309.

Liñán-Cabello, M.A., Paniagua-Michel, J., Zenteno-Savín T. (2003) Carotenoids and retinal levels in captive and wild shrimp, *Litopenaeus vannamei*. *Aquaculture Nutrition* 9, 383–389.

Liu, Y., Wang, W.-N., Wang, A.-L. et al. (2007) Effects of dietary vitamin E supplementation on antioxidant enzyme activities in *Litopenaeus vannamei* (Boone, 1931) exposed to acute salinity changes. *Aquaculture* 265, 351–358.

Lorenzon, S., Giulianini, P.G., Ferrero, E.A. (1997) Lipopolysaccharide-induced hyperglycemia is mediated by CHH release in crustaceans. *General and Comparative Endocrinology* 108, 395–405.

Lorenzon, S., Pasqual, P., Ferrero, E.A. (2002) Different bacterial lipopolysaccharides as toxicants and stressors in the shrimp *Palaemon elegans*. *Fish and Shellfish Immunology* 13, 27–45.

Lorenzon, S., Edomi, P., Giulianini, P.G. et al. (2005) Role of biogenic amines and cHH in the crustacean hyperglycemic stress response. *Journal of Experimental Biology* 208, 3341–3347.

Lorenzon, S., Giulianini, P.G., Martinis, M. et al. (2007) Stress effect of different temperatures and air exposure during transport on physiological profiles in the American lobster *Homarus americanus*. *Comparative Biochemistry and Physiology A* 147, 94–102.

MacFarlane, G.R., Schreider, M., McLennan, B. (2006) Biomarkers of heavy metal contamination in the red fingered marsh crab, *Parasesarma erythodactyla*. *Archives of Environmental Contamination and Toxicology* 51, 584–593.

Maciel, F.E., Rosa, C.E., Santos, E.A. et al. (2004) Daily variations in oxygen consumption, antioxidant defenses, and lipid peroxidation in the gills and hepatopancreas of an estuarine crab. *Canadian Journal of Zoology* 82, 1871–1877.

MacRae, H.T. (2010) Gene expression, metabolic regulation and stress tolerance during diapause. *Cellular and Molecular Life Sciences* 67, 2405–2424.

Martin, G. (1981) *Contribution à l'étude cytologique et fonctionelle des systèmes de neurosécrétion des Crustacés Isopodes*. Thèse Doctorat d'état, p. 347.

Mohankumar. K., Ramasamy, P. (2006) White spot syndrome virus infection decreases the activity of antioxidant enzymes in *Fenneropenaeus indicus*. *Virus Research* 115, 69–75.

Mourente, G., Diaz-Salvago, E. (1999) Characterization of antioxidant systems, oxidation status and lipids in brain of wild-caught size-class distributed *Aristeus antennatus* (Risso, 1816) Crustacea, Decapoda. *Comparative Biochemistry and Physiology B* 124, 405–416.

Naganuma, T., Inoue, T., Uye, S. (1997) Photoreactivation of UV induced damage to embryos of a planktonic copepod. *Journal of Plankton Research* 19, 783–787.

Neves, C.A, Santos, E.A., Bainy, A.C.D. (2000) Reduced superoxide dismutase activity in *Palaemonetes argentinus* (decapoda, palemonidae) infected by *Probopyrus ringueleti* (isopoda, bopyridae). *Diseases of Aquatic Organisms* 39, 155–158.

Nishikawa, T., Edelstein, D., Du, X.L. et al. (2000) Normalizing mitochondrial superoxide production blocks three pathways of hyperglycaemic damage. *Nature* 404, 787–790.

Orbea, A., Ortiz-Zarragoitia, M., Sole', M. et al. (2002) Antioxidant enzymes and peroxisome proliferation in relation to contaminant body burdens of PAHs and PCBs in bivalve molluscs, crabs and fish from the Urdaibai and Plentzia estuaries (Bay of Biscay). *Aquatic Toxicology* 58, 75–98.

Pan, L., Zhang, H. (2006) Metallothionein, antioxidant enzymes and DNA strand breaks as biomarkers of Cd exposure in a marine crab, *Charybdis japonica*. *Comparative Biochemistry and Physiology C* 144, 67–75.

Passi, S., Cataudella, S., Di Marco, P. et al. (2002) Fatty acid composition and antioxidant levels in muscle tissue of different mediterranean marine species of fish and shellfish. *Journal of Agriculture and Food Chemistry* 50, 7314–7322.

Penkoff, S.J., Thurberg, F.P. (1982) Changes in oxygen consumption of the american lobster, *Homarus americanus*, during the molt cycle. *Comparative Biochemistry and Physiology A* 2, 621–622.

Pereira, P., de Pablo, H. Subida, M.D. et al. (2009) Biochemical responses of the shore crab(*Carcinus maenas*) in a eutrophic and metal-contaminated coastal system (Óbidos lagoon, Portugal). *Ecotoxicology and Environmental Safety* 72, 1471–1480.

Peters, L.D., Livingstone, D.R. (1996) Antioxidant enzyme activities in embryologic and early larval stages of turbot. *Journal of Fish Biology* 49, 986–997.

Prieto-Sagredo, J., Ricalde-Recchia, I., Durán-Lizarraga, M.E. et al. (2000) Changes in hemolymph glutathione status after variation in photoperiod and light irradiance in crayfish *Procambarus clarkii* and *Procambarus digueti*. *Photochemistry and Photobiology* 71, 487–492.

Quackenbush, L.S. (1989) Vitellogenesis in the shrimp, *Penaeus vannamei*: in vitro studies of the isolated hepatopancreas and ovary. *Comparative Biochemistry and Physiology B*, 94, 253–261.

Radhika, M., Abdul Nazar, A.K., Munuswamy, N., Nellaiappan, K., (1998). Sex-linked differences in phenol oxidase in the fairy shrimp *Streptocephalus dichotomus* Baird and their possible role (Crustacea: Anostraca). *Hydrobiology* 377, 161–164.

Rameshthangam, P., Ramasamy, P. (2006) Antioxidant and membrane bound enzymes activity in WSSV-infected *Penaeus monodon* Fabricius. *Aquaculture* 254, 32–39.

Rani, K., Subramonian, T. (1997) Vitellogenesis in the mud crab *Scylla serrata*–an in vivo isotope study. *Journal of Crustacean Biology* 17, 659–665.

Ren, Q., Zhang, R-R., Zhao, X-F. (2010) A thioredoxin response to the WSSV challenge on the Chinese white shrimp, *Fenneropenaeus chinensis*. *Comparative Biochemistry and Physiology C* 151, 92–98.

Riehle, M.A., Brown, M.R. (1999) Insect biochemistry. *Molecular Biology* 29, 855–861.

Rodríguez, E.M., Medesani, A.D., Finguerman, M. (2007) Endocrine disruption in crustaceans due to pollutants: A review. *Comparative Biochemistry and Physiology A* 146, 661–671.

Rudneva, I.I. (1999) Antioxidant system of black sea animals in early development. *Comparative Biochemistry and Physiology C* 122, 265–71.

Santos, E., Keller, R. (1993) Crustacean hyperglycaemic hormone (CHH) and the regulation of carbohydrate metabolism. *Comparative Biochemistry and Physiology A* 106, 405–411.

Santos, E.A., Nery, L.E.M., Keller, R. et al. (1997) Evidence for the involvement of crustacean hyperglycemic hormone in the regulation of lipid metabolism. *Physiological Zoology* 70, 415–420.

Santos, E.A., Keller, R., Rodriguez, E. et al. (2001) Effects of serotonin and fluoxetine on blood glucose regulation in two decapod species. *Brazilian Journal of Medical and Biological Research* 34, 75–80.

Serrano, L., Blanvillain, G., Soyez, D. et al. (2003) Putative involvement of crustacean hyperglycemic hormone isoforms in the neuroendocrine mediation of osmoregulation in the crayfish *Astacus leptodactylus*. *Journal of Experimental Biology* 206, 979–988.

Silvestre, F., Dierick, J., Dumont, V. et al. (2006) Differential protein expression profiles in anterior gills of *Eriocheir sinensis* during acclimation to cadmium. *Aquatic Toxicology* 76, 46–58.

Skinner, D.M., 1985. Molting and regeneration. *In* Bliss, D.E., Mantel, L.H. (eds). *The Biology of the Crustacea*, Vol.9. Academic Press, New York, pp. 43–146.

Sridevi, B., Reddy, K.V., Reddy, S.L.N. (1998) Effect of trivalent and hexavalent chromium on antioxidant enzyme activities and lipid peroxidation in a freshwater field crab, *Barytelphusa guerini*. *Bulletin of Environmental Contamination and Toxicology* 61, 384–390.

Steevens, J.A., Benson, W.H. (1999) Toxicological interactions of chlorpyriphos and methyl mercury in the amphipod *Hyalella azteca*. *Toxicological Sciences* 52, 168–177.

Telford M. (1968) Changes in blood sugar composition during the molt cycle of the lobster, *Homurus umericunus*. *Comparative Biochemistry and Physiology* 26, 917–926.

Tilden, A.R., Shanahan, J.K., Khilji, Z.S. et al. (2003) Melatonin and locomotor activity in the fiddler crab *Uca pugilator*. *Journal of Experimental Zoology A* 297, 80–87.

Timofeyev, M.A. (2006) Antioxidant enzyme activity in endemic baikalean versus palaearctic amphipods, Tagma- and size-related changes. *Comparative Biochemistry and Physiology B* 143, 302–308.

Van Nieuwerburgh, L., Wänstrand, I., Liu, J. et al. (2005) Astaxanthin production in marine pelagic copepods grazing on two different phytoplankton diets. *Journal of Sea Research* 53, 147–160.

Verri, T., Mandal, A., Zilli, L. et al. (2001) D-glucose transport in decapod crustacean hepatopancreas. *Comparative Biochemistry and Physiology A* 130, 585–606.

Vijayavel, K., Gomathi, R.D., Durgabhavani, K. et al. (2004) Sublethal effect of naphthalene on lipid peroxidation and antioxidant status in the edible marine crab *Scylla serrata*. *Marine Pollution Bulletin* 48, 429–433.

Vogt, G. (2002) Functional anatomy. *In* Holdich, D.M. (ed.). *Biology of Freshwater Crayfish*. Blackwell Science, Oxford, pp. 53–151.

Wang, W.-N., Wang, A.-L., Zhang, Y.-J. (2006a) Effect of dietary higher level of selenium and nitrite concentration on the cellular defense response of *Penaeus vannamei*. *Aquaculture* 256, 558–563.

Wang, W.-N., Wang, Y.-J., Wang, A.-L. (2006b) Effect of supplemental l-ascorbyl-2-polyphosphate (APP) in enriched live food on the immune response of *Penaeus vannamei* exposed to ammonia-N. *Aquaculture* 256, 552–557.

Wang, W.-N., Zhou, J., Wang, P. et al. (2009) Oxidative stress, DNA damage and antioxidant enzyme gene expression in the Pacific white shrimp, *Litopenaeus vannamei* when exposed to acute pH stress. *Comparative Biochemistry and Physiology C* 150, 428–435.

Webster, S.G., Dircksen, H., Chung, J.S. (2000) Endocrine cells in the gut of the shore crab Carcinus maenas immunoreactive to crustacean hyperglycaemic hormone and its precursor-related peptide. *Cell and Tissue Research* 300, 193–205.

Winston, G.W., Lemaire, D.G.E., Lee, R.F. (2004) Antioxidants and total oxyradical scavenging capacity during grass shrimp, *Palaemonetes pugio*, embryogenesis. *Comparative Biochemistry and Physiology C* 139, 281–288.

Withyachumnarnkul, B., Pongsa-Asawapaiboon, A., Ajpru, S. et al. (1992) Continuous light increases N-acetyltransferase activity in the optic lobe of the giant freshwater prawn Macrobrachium rosenbergii de Man (Crustacea: Decapoda). *Life Sciences* 51, 1479–1484.

Yao, C.L., Wang, A.L., Wang, W.N. et al. (2004) Purification and partial characterization of Mn superoxide dismutase from muscle tissue of the shrimp, *Macrobrachium nipponense*. *Aquaculture* 241, 621–631.

Zhang,Q., Li, F., Zhang, J., et al. (2007) Molecular cloning, expression of a peroxiredoxin gene in chinese shrimp *Fenneropenaeus chinensis* and the antioxidant activity of its recombinant protein. *Molecular Immunology* 44, 3501–3509.

Zhang, Q., Li, F., Zhang, X. et al. (2008) cDNA cloning, characterization and expression analysis of the antioxidant enzyme gene, catalase, of chinese shrimp *Fenneropenaeus chinensis*. *Fish and Shellfish Immunology* 24, 584–591.

Chapter 16

TRANSFER OF FREE RADICALS BETWEEN PROTEINS AND MEMBRANE LIPIDS: IMPLICATIONS FOR AQUATIC BIOLOGY

Brenda Valderrama[1], *Gustavo Rodríguez-Alonso*[1,2], *and Rebecca Pogni*[3]

[1]Departamento de Medicina Molecular y Bioprocesos, Instituto de Biotecnología, Universidad Nacional Autónoma de México, México
[2]Facultad de Ciencias, Universidad Autónoma del Estado de Morelos, México
[3]Chemistry Department, University of Siena, Italy

The thermodynamic potential of many reaction intermediates of redox enzymes allows the transfer of their electrons to oxygen as an alternative substrate. Membranes act as a physical barrier against the diffusion of oxygen and its reactive species derivatives (ROS); the intensity of the effect is, however, highly specific. For O_2, the barrier effect is so low that it can be considered nonexistent. If at all, O_2 movement is controlled at the lipid–water interface and this control is enhanced at high temperature and, in more rigid membranes, due to accumulation of cholesterol (Moller et al. 2005). For practical purposes, biological membranes do not represent any barrier for oxygen.

Transfer of $O_2^{\bullet-}$ is energetically unfavorable, since it is a charged species and needs to lose its hydration shell to cross membranes. However, cells have specific channels to transport biologically required ions ($O_2^{\bullet-}$ included) across membranes (Lynch and Fridovich 1978). H_2O_2 is a small noncharged hydrophilic molecule which is transported across membranes through specific aquaporins (Bienert et al. 2007). $HO^{\bullet}$ is the most reactive of the partially reduced oxygen species and reacts with phospholipids in the lipid bilayer, first with the polar heads which are inert to other ROS and, afterwards, with the acyl chains. Thus, membranes pose a physical barrier for the diffusion

Oxidative Stress in Aquatic Ecosystems, First Edition. Edited by Doris Abele, José Pablo Vázquez-Medina, and Tania Zenteno-Savín.

of $HO^{\bullet}$ through reaction with the lipid or protein components of the bilayer. Organisms may alter the composition of their membranes in a controlled fashion to reduce permeability as observed in *Saccharomyces cerevisiae* (Folmer et al. 2008).

MEMBRANE LIPIDS

The present view of biomembrane structure is that of an array of lipids and proteins capable of relocating by their slow movement within the leaflet of the membrane (Singer and Nicolson 1972). Some of the naturally occurring lipids in membranes have the capability of self-assembling into one or more of a wide variety of structural arrangements depending on physical factors such as pH, temperature, ionic strength, and hydration (Luzzati et al. 1966). This variety of structures includes the well-known lipid bilayer, micelles, and hexagonal tubes (Seddon 1990). Lipid aggregation in membranes is largely driven by van der Waals interactions. Long and saturated lipids tend to aggregate into gel-like states. Lipids with short acyl chains have a smaller surface area for molecular interactions and enhance membrane fluidity (Fig. 16.1). A useful approach for the biochemical study of membranes is the artificial assembly of lipid bilayers using defined mixtures of natural or chemically modified lipids and phospholipids (Box 16.1).

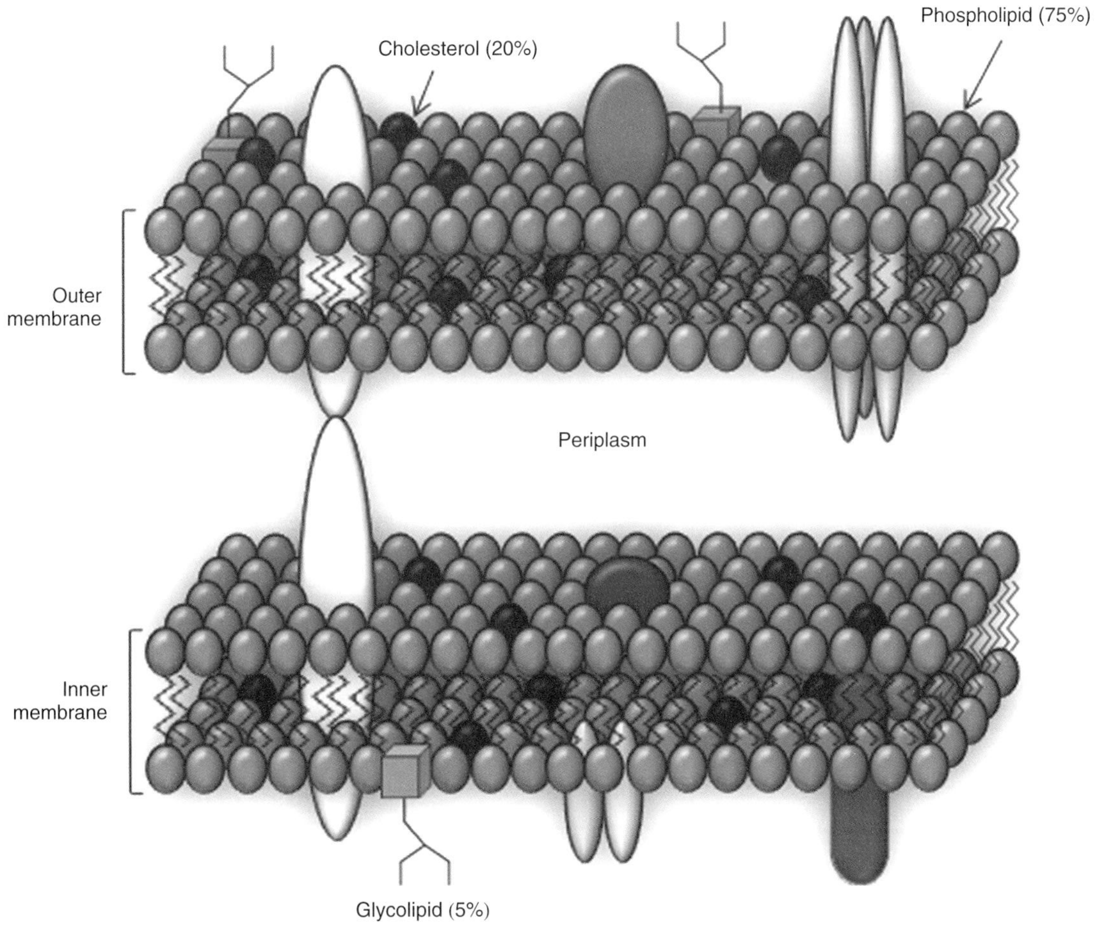

Fig. 16.1 Fluid mosaic model of the mitochondrial membrane.

Box 16.1 A general method for liposome preparation

1. Dissolve each of the phospholipids or other compounds (labels) to be intercalated in the membrane in chloroform at the required molar ratio in new, clean vials.
2. Evaporate the chloroform off the mixture under a stream of nitrogen gas overnight. Alternatively, samples can be dried under vacuum or roto-evaporated.
3. Hydrate the resulting lipid film by adding phosphate buffer saline or other aqueous buffer with vigorous vortexing. Phospholipid concentration at this stage is recommended to be approximately 50 mM.
4. For a more efficient removal of large multilamellar vesicles, the lipid suspension may be subjected to an additional cycle of several freeze–thawing steps.
5. Large unilamellar vesicles (LUVs) are produced by extrusion through 100 nm or 200 nm polycarbonate membrane filters using a miniextruder syringe device (Avanti Polar Lipids) for nine times. Alternatively, LUVs can be obtained by sonication of the lipidic suspension using a titanium-probed ultrasonic disintegrator at 20 kHz output frequencies.
6. To estimate the final lipid concentration, concentrate aliquots of the LUVs mixture to complete dryness in a stream of air at 50°C. Dissolve the dried sample in 2.0 mL chloroform and mix thoroughly for 1 min. Following phase separation, remove the lower chloroform phase and measure Abs at 488 nm. Compare the resulting Abs with a reference curve performed under the same conditions.
7. Liposomes are best if used within 1 week of storage at 4°C.

Membrane composition regulates membrane fluidity and sensitivity to oxidation at warm temperatures in ectotherms through the relative abundance of polyunsaturated fatty acids (PUFA) and a high phosphatidylethanolamine to phosphatidylcholine ratio (Crockett 2008). Membranes experience this modification in fish adapted to cold/warm environmental shifts such as the common carp, goldfish and rainbow trout (Cossins et al. 1978; Tiku et al. 1996; Zehmer and Hazel 2004). *In vitro* studies using fish membranes also suggest that the enrichment of monoenic fatty acids in position sn-1 and a polyenic fatty acid in position sn-2 of the lipid molecule renders the membranes more fluid during adaptation to low temperatures (Farkas et al. 2001).

Cell membranes perform a wide variety of functions, each of them closely related to the type of organelle or cell they belong to. In eukaryotic cells, three organelles are of particular relevance regarding oxidative damage and membrane dysfunction: mitochondria, peroxisomes, and lysosomes.

Mitochondria detoxify endogenous and exogenous ROS with enzymes such as Mn superoxide dismutase (SOD) or Cu,Zn-SOD. As some ROS (such as H_2O_2 or $NO^{\bullet}$) can travel across membranes, there are also enzymatic and nonenzymatic antioxidants outside mitochondria. Glutathione S-transferase (GST) and thioredoxin (Trx), among other ROS scavengers, perform this critical role (Dean et al. 1997; Droge 2002; Finkel and Holbrook 2000; Lenaz 2001).

Peroxisomes are membrane organelles involved in metabolizing oxygen derivatives. One of their multiple functions is the β-oxidation of fatty acids. As for all cellular processes, β-oxidation is highly regulated. However, when PPARS (originally peroxisomes proliferator-activated receptors, now known to be transcription factors responsive to fatty acid-like ligands) overload the system, they promote oxidative stress induced by peroxisomes. Transition metals such as Fe and Cu, usually complexed in peroxisomes can be released and produce $HO^{\bullet}$ via the Fenton reaction, leading to lipid peroxidation, peroxisome membrane damage, and peroxisomal dysfunction (del Río et al. 1992). This was observed after the exposure of sea bass red blood cells to Fe and Cu (Labieniec et al. 2009).

Lysosomes are small organelles containing a large number of hydrolytic enzymes. Lysosomal dysfunction leads to impairment of the recycling capability of the cell. Oxidative stress affects lysosomes, compromising their functionality. One of the most important consequences of the inability to recycle cell components is the accumulation of damaged mitochondria, which, in turn, can act as a positive feedback loop. The functionally deficient mitochondria produce increased amounts of ROS, leading to a growing number of lysosomes undergoing oxidative damage and of abnormal mitochondria as a result of deficiencies in autophagy (Brunk and Terman 2002; Terman et al. 2010).

Lysosomal integrity is widely used in marine environmental monitoring (Moore 1990, Chapter 39). In certain marine organisms, the toxic effect of different pollutants correlates with lysosome functionality and with the occurrence of degenerative, preneoplasic, and neoplasic liver lesions, as in organochlorine-exposed flatfish (Köhler and Pluta 1995; Köhler et al. 2002), as well as with digestive tissue damage and immune inhibition in marine mussels exposed to polycyclic aromatic compounds (Lowe and Pipe 1994; Grundy et al. 1996).

PROTEIN–MEMBRANE ASSOCIATIONS

Proteins, whether attached to the outside, embedded within, or spanning across the bilayer, interact closely with membrane lipids (Fig. 16.1). The negatively charged environment provided by phospholipid polar heads is important in establishing electrochemical interactions with peripheral membrane proteins, whereas the hydrophobic environment at the core of the membrane allows the allocation of proteins spanning across the bilayer. The bonding of phospholipids and proteins is further stabilized by covalent, electrostatic, hydrogen bonds and Van der Waals interactions (Karel 1973).

Polar heads of phospholipids interact with the active site of enzymes affecting their activity by dissociation and reassociation of lipids with membrane-bound enzymes and by modifying the enzymatic activity through changes in the structure of the lipid phase of the membrane (Gazzotti and Peterson 1977). The existence of such an intimate interaction between membrane-associated/membrane-bound proteins and membrane phospholipids has regulatory implications as described above, but also allows the physical communication between functional entities (Chapman et al. 1979). Membrane phospholipids are prone to oxidation and peroxidation chemically (by ROS) or enzymatically. Lipid peroxidation is initiated by the abstraction of a hydrogen atom from a methylene carbon in the side chain of a fatty acid or fatty acyl side chain with sufficient reactivity, and the fate of the unpaired electron is depicted in Fig. 16.2. The higher the number of double bonds in a fatty acid side chain, the easier the removal of the hydrogen atom (Porter 1986; Halliwell and Chirico 1993; Eritsland 2000). Additionally, peroxide radicals generated in the extraliposomal solution

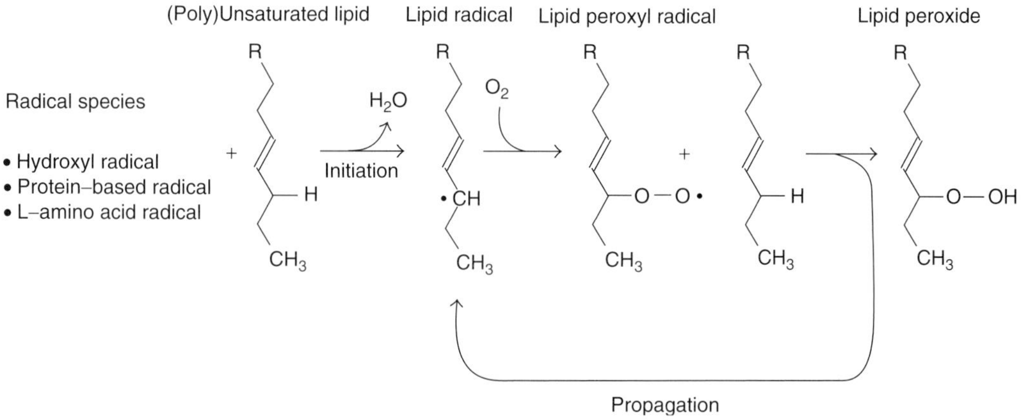

Fig. 16.2 Initiation and propagation reactions leading to lipid peroxidation. Abstraction of a hydrogen atom leaves an unpaired electron on the carbon atom to which it was originally attached. The resulting carbon-centered lipid radical can have several fates: the most likely in aerobic cells is to undergo molecular rearrangement, which by a subsequent reaction with molecular oxygen yields a peroxyl radical. Peroxyl radicals can combine with each other, they can transfer electrons to membrane proteins, and are capable of abstracting hydrogen atoms from adjacent fatty acid side chains propagating the chain reaction of lipid peroxidation.

saponify esters within the liposomal bilayer (Frimer et al. 1996).

In order to fully understand the reactivity of radicals with lipid membranes, it is necessary to determine the depth of radical penetration into a lipid membrane. Frimer et al. (1996) applied the "spin trapping" technique (Buettner 1987). In this procedure, the radical species is immobilized by a diamagnetic nitroso or nitrone compound and detected by electron paramagnetic resonance (EPR, Chapter 35). In the nitroso spin traps, the nitrogen atom of the trap reacts directly with the free radical species, resulting in characteristic EPR spectra. Two nitroso compounds are currently used in biological investigations: 2-methyl-2-nitroso propane (MNP) and 3,5-dibromo-4-nitrosobenzensulfonate (DBNBS). The oxygen-centered spin adducts in nitroso spin traps are, however, quite unstable. Thus, the nitrone spin traps are by far the most popular. Three commonly used nitrone spin traps are: phenyl-t-butyl-nitrone (PBN), α(4-pyridyl-1-oxide)-N-t-butyl nitrone (POBN) and 5,5-dimethyl-1-pyrroline-N-oxide (DMPO) (Tomasi and Iannone 1993). DMPO has received particular attention because it yields distinct and characteristic spin adducts with $O_2^{\bullet-}$ and $HO^\bullet$. The use of DMPO as a probe for ROS generation in biological milieu has, however, some limitations. In recent years a new spin trap, 5-(diethoxyphosphoryl)-5-methyl-1-pyrroline-1-oxide (DEPMPO), has been synthesized (Frejaville et al. 1995). DEPMPO gives more stable adducts with $O_2^{\bullet-}$ and $HO^\bullet$ and has been shown to be a useful probe for free radical formation during myocardial reperfusion injury (Frejaville et al. 1995).

FORMATION OF PROTEIN-BASED FREE RADICALS

Protein-based free radicals occur in proteins belonging to two classes: those containing heme prosthetic groups and others which contain nonheme iron centers or prosthetic groups with other metals. Heme-proteins use H_2O_2 as electron acceptor to catalyze a variety of oxidative reactions through a catalytic cycle with two intermediates (Dunford 1999). A third species (compound III) is produced when ferric peroxidases are exposed to an excess of H_2O_2 (Valderrama et al. 2002). The consensus catalytic mechanism of heme-proteins as peroxidases involves at least five different iron species: ferrous, ferric, compound I, compound II, and compound III (Keilin and Hartree 1951; Dunford 1999). After the reaction of ferric (Fe^{III}) porphyrin with peroxides, a transient sixth-coordination bond is formed with the heme iron, yielding compound I, a cationic oxoFe^{IV} free radical. EPR studies established that in plant heme-peroxidases the second oxidation equivalent in compound I is present as a porphyrin-based π-free radical (Roberts et al. 1981; Patterson et al. 1995). In peroxidases from fungal sources and in other heme-proteins, hydrogen atom abstraction from the protein results in formation of a species where the free radical is based on a residue close to the porphyrin (Valderrama et al. 2002). In some of these enzymes the radical participates in substrate oxidation (DeGray et al. 1992; Blodig et al. 1998; Ruíz-Dueñas et al. 2009).

Protein-based free radicals have also been detected in enzymes using nonheme cofactors, such as class I ribonucleotide reductase (di-Fe center) (Nordlund et al. 1990), class II ribonucleotide reductase (adenosylcobalamine) (Licht et al. 1996), pyruvate formate lyase (FeS cluster) (Wagner et al. 1992), galactose oxidase (mononuclear Cu) (Whittaker and Whittaker 1990), and others (Stubbe and van der Donks 1998). In these enzymes, amino acid-based radicals are involved in catalysis as site for substrate oxidation.

The most common sinks for free radicals in proteins are tyrosine (Nordlund et al. 1990; Whittaker and Whittaker 1990; Wilks and Ortiz de Montellano 1992; Davies and Puppo 1992; DeGray et al. 1992; Barr et al. 1996; Giulivi and Cadenas 1998; Shi et al. 2000), tryptophan (Blodig et al. 1998; Hiner et al. 2001; Ruíz-Dueñas et al. 2009), and cysteine (Licht et al. 1996) side chains. This is consistent with previous observations using oligopeptides (Hawkins and Davies 2001), although at least in one case, a stable glycyl radical has been identified (Wagner et al. 1992). Protein-based radicals are sometimes involved in substrate oxidation; in those cases where the substrate is a membrane lipid, the free radical is transferred to the lipid by a hydrogen abstraction reaction.

OXIDATION OF ARACHIDONIC ACID BY PROSTAGLANDIN H SYNTHASE

Prostaglandin H synthase (PGHS), also known as cyclooxygenase (COX, E.C. 1.14.99.1), is a bifunctional hemeperoxidase, and is highly conserved among vertebrates (including marine mammals and fishes, both cartilaginous and bony), in sequence and folding (Garavito et al. 1994; Chandrasekharan and Simmons 2004). PGHS catalyzes the first committed step in prostaglandin biosynthesis. Prostaglandins are important signaling molecules involved in inflammation, ovulation, modulation of immune responses, and mitogenesis in multicellular organisms (Simmons et al. 2004). PGHS has two activities, COX and peroxidase. In resting cells, long chain PUFA arachidonic acid (AA) is stored within the cell membrane esterified to glycerol in phospholipids. A receptor-dependent event initiates phospholipid hydrolysis and releases the fatty acid. The COX activity catalyzes the conversion of AA and two molecules of oxygen to prostaglandin endoperoxide G_2 (PGG_2), which, in the presence of a reductant, is reduced to prostaglandin endoperoxide H_2 (PGH_2) (Fig. 16.3).

The COX and peroxidase sites in PGHS are located nearby, within a narrow cavity that connects the transmembranal section of the protein with the external side of the nuclear or endoplasmic reticulum membranes. Heme is the only prosthetic group present and is essential for both PGHS activities. Peroxidase activity of PGHS has three functions: activation of the COX reaction, production of a tyrosine-based radical form of compound I and reduction of PGG_2. In the COX reaction, the tyrosyl radical generated from the peroxidase reaction serves as a catalytic residue, first abstracting a hydrogen atom from AA and later (as a tyrosine residue) transferring a hydrogen atom to the PGG_2 radical to form PGG_2. The crystallographic structure of PGHS obtained in the presence of bound AA, demonstrates that the substrate locates in a ~25 Å L-shaped hydrophobic channel that leads to the distal-side cavity of the porphyrin with the fatty acid carboxyl

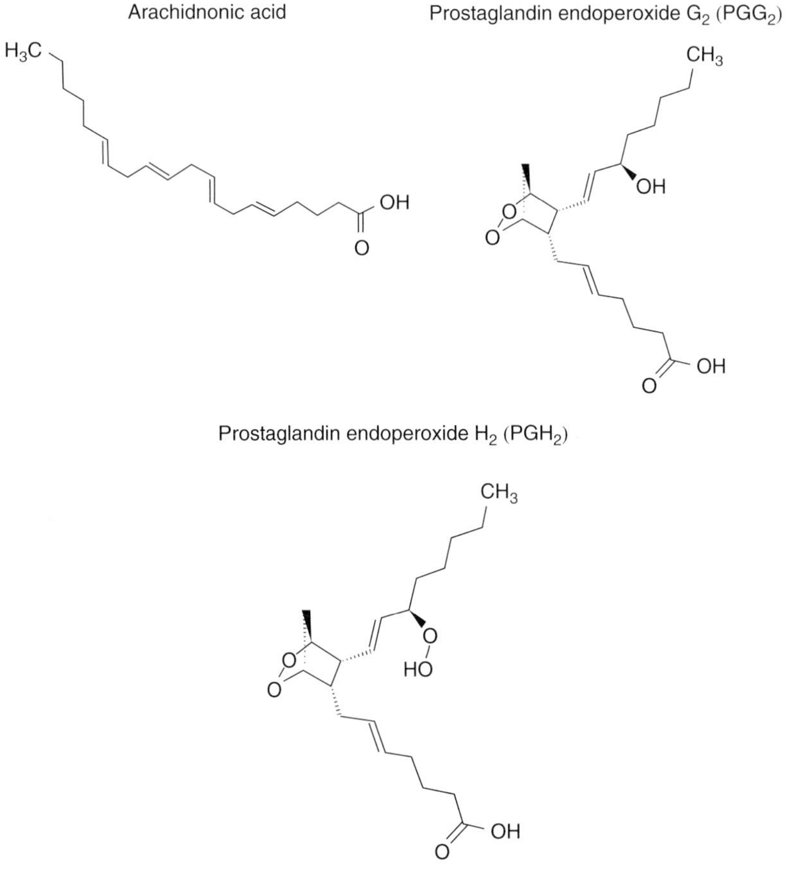

Fig. 16.3 The cyclooxygenase activity of prostaglandin H synthase catalyzes the conversion of arachidonic acid and two molecules of oxygen to prostaglandin endoperoxide G_2 (PGG_2), which in the presence of a reductant is reduced to prostaglandin endoperoxide H_2 (PGH_2). After oxidation to PGG_2, the lipid molecule suffers a dramatic conformational change imposing a tight bend between C8 and C9 with the potential consequence of releasing the stabilizing electrostatic interaction with Arg-120 and displacing the peroxyl moiety at C-15 towards the heme iron.

salt-bridged to Arg-120 and with C-13 of AA adjacent to Tyr-385 (Thuresson et al. 2001). Different studies have converged in the designation of Tyr-385 as the stabilization site for the second oxidation equivalent of compound I and the catalytic residue for the COX reaction (Fig. 16.4).

The evolutionary relevance of the PGHS pathway for prostaglandin formation from marine invertebrates to mammals was elucidated by the cloning and characterization of PGHS homologs in the corals *Gersemia fruticosa* (Koljak et al. 2001) and *Plexaura homomalla* (Valmsen et al. 2001), in the dogfish shark (Yang et al. 2002), in zebrafish (Grosser et al. 2002), and in many terrestrial mammals with the conserved presence of two isoforms (Smith and Langenbach 2001).

Recently, both PGHS isoforms were cloned from brook trout (Liu et al. 2006). Recombinant tPGHS-1 and -2 enzymes have the same cyclooxygenase catalytic chemistry as their mammalian homologs. However, both trout isoforms display substrate specificity with higher affinity toward arachidonate than to eicosapentaenoate or docosahexaenoate, the latter two being the dominant forms in trout tissues. In mammals, aspirin acetylates the conserved Ser-530 residue in PGHS abating oxygenase activity in isoform 1 and shifting the cyclooxygenase activity of isoform 2 to lipid oxygenase activity, which has been related to the pharmacological effect of the drug (Serhan and Oliw 2001). Aspirin treatment did not increase lipoxygenase-type catalysis in either trout enzyme, despite of the sequence conservation. Both trout enzymes have higher requirements for peroxide activator than their mammalian counterparts, although the preferential peroxide activation of PGHS-2 over PGHS-1 seen in mammals was already present in the fish enzymes. This result suggests that mammalian enzymes evolved to operate at lower peroxide levels, correlating with the 3–60 times reduced phospholipid peroxide content of human plasma compared to that of fish (Kaewsrithong et al. 2000). The divergence in cyclooxygenase characteristics between the trout and mammalian PGHS proteins may reflect accommodations to differences in lipid composition and general redox state (Liu et al. 2006).

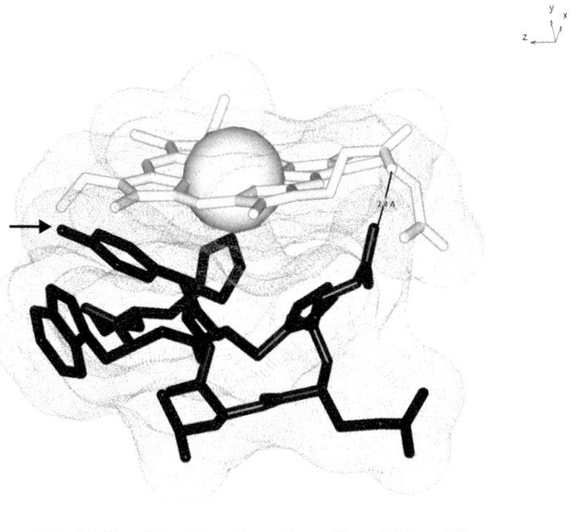

Fig. 16.4 Identification of catalytic Tyr-385 residue (marked with arrow) in the active site of prostaglandin H synthase. Tyr-385 does not interact with the porphyrin directly but through an Asn-382. The amide nitrogen of the Asn-382 residue is hydrogen-bonded to one of the propionate groups, enabling what seems to be the most probable free-radical transfer pathway from the porphyrin to Tyr-385.

OXIDATION OF ARACHIDONIC ACID BY MYOGLOBIN AND HEMOGLOBIN

Several recent studies suggest that fish may become important new models for studying environmental risk factors and the pathology associated with degenerative central nervous system disorders. In kokanee salmon brains, β-amyloid precursor protein and β-amyloid peptide aggregates have been identified by immunohistochemical methods in regions involved in gustation (the sense of taste), olfaction (the sense of smell), vision, the stress response, reproductive behavior, and coordination (Maldonado et al. 2000). Similarly, the rainbow trout (*Oncorhynchus mykiss*) has been proposed to be a useful model for the identification of neurotoxic compounds that lead to degeneration of catecholaminergic neurons (Ryan et al. 2002). The synuclein (Syn) family is exclusive to vertebrates and includes three known proteins: α-synuclein, β-synuclein, and γ-synuclein. α-Synuclein (α-Syn) has been found to be mutated in families with autosomal dominant Parkinson's disease (Ulmer et al. 2005). All synucleins present a common highly conserved α-helical lipid-binding motif with similarity to the class-A2 lipid-binding domains of the exchangeable apolipoproteins. Fish were among the first species in which synuclein proteins were identified, and fish seem to possess high levels of these proteins within their central nervous system (Maroteaux et al. 1988). Recently, α-Syn immunoreactive aggregates resembling Lewy bodies were identified in the brain of the white sucker fish, *Catostomus commersoni*, collected at a site highly contaminated with metal ions from mining activities (Boudreau et al. 2009).

Under oxidative conditions, respiratory proteins such as Mb and Hb can form ferryl heme iron and protein-based free radicals. The ferryl ($Fe^{IV} = O$) derivatives of Mb and Hb are known to be formed whenever these heme proteins escape their normal cellular environment and the protective medium of the plasma during injury, and have also been implicated in oxidative stress in various pathologies (Moore et al. 1998; Holt et al. 1999; Reeder et al. 2002). Mb-induced lipid peroxidation might explain the occurrence of renal tubular necrosis in myoglobinuria, but it has not been demonstrated how this could be causally linked to the reduction in renal blood flow that also characterizes the nephropathy of rhabdomyolysis (rapid breakdown of skeletal muscle tissues). One hypothesis suggests involvement of vasoactive products of lipid peroxidation that cause renal vasoconstriction. Likely candidates are isoprostanes, a group of prostaglandin-like compounds, formed nonenzymatically *in vivo* as products of free radical-induced peroxidation of AA (Morrow et al. 1990) (Fig. 16.5). The environment of the inflammatory sites is acidified, with the consequence of more rapid accumulation of oxidative damage given the low-pH accelerated rate of Mb-induced lipid peroxidation (Rodríguez-Malaver et al. 1997), of the pro-oxidant activity of Hb and cytochrome *c* towards AA, and that of Mb towards microsomal membranes. Furthermore, $O_2^{\bullet -}$ is also more reactive with PUFAs at lower pH (Baker and Gebicki 1984).

High mortality rates of stranded cetaceans have been related to cardiac tissue damage caused by myofibrillar degeneration or contraction band necrosis of the myocardium, characterized by the focal hypercontraction of sarcomeres (a localized atrophia of sarcomeres that hampers the coordinated contraction/expansion of the cardiac tissue) and lysis of small groups of myocardial cells (Turnbull and Cowan 1998). Frequently, contraction band necrosis is preceded by significant kidney damage in the form of rhabdomyolysis and myoglobinuria (excretion of myoglobin associated with lysis of muscle cells) due to severe muscle exertion. It has been observed that acute rhabdomyolysis affects both cardiac and skeletal muscles and may also prompt the occurrence of contraction band necrosis (Turnbull and Cowan 1998; Herraez et al. 2007).

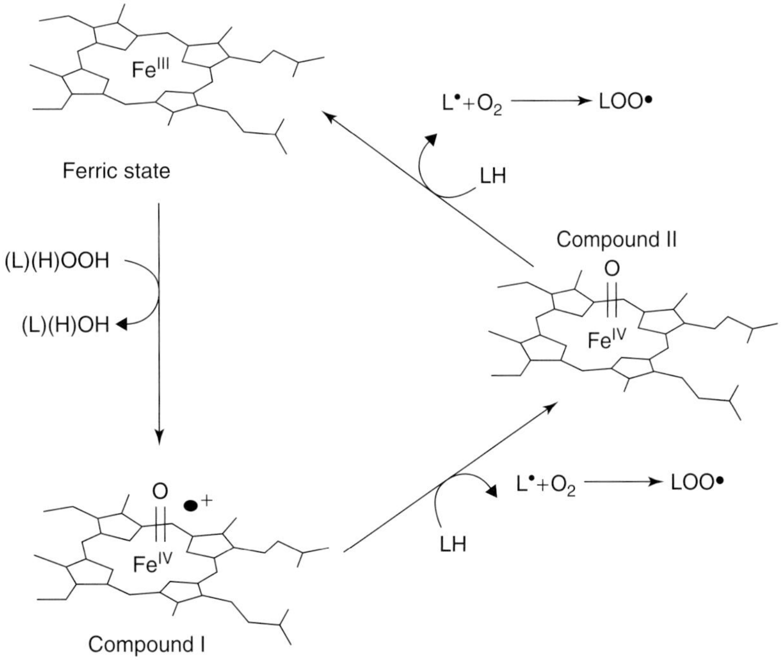

Fig. 16.5 General mechanism for hemeperoxidase-induced lipid peroxidation. The ferric form of the heme group in a protein reacts with H_2O_2 or with organic hydroperoxides, including lipid hydroperoxides, to produce the highly reactive intermediate Compound I. The radical species located in Compound I is frequently allocated to a protein residue, usually a tyrosine moiety, where the oxidation of other compounds is achieved. Unsaturated lipids may be oxidized by Compound I and also by Compound II, triggering the propagation of the lipid-based radical as described in Figure 16.4.

OXIDATION OF CARDIOLIPIN BY α-SYNUCLEIN/CYTOCHROME C COMPLEXES

A handful of proteins have been determined to precede and probably trigger neurodegenerative disorders such as amyloid-β (Alzheimer's disease), α-synuclein (Parkinson's disease), tau (Tauopathies), amyloid A (AA amyloidosis), and transmissible spongiform encephalopathies (TSEs). Most of these studies have been developed using mammalian model systems. However, it has recently been demonstrated that fish models are prone to at least one of these disorders (Málaga-Trillo et al. 2010), and that there is a strong possibility of developing a fish model for synucleinopathies such as Parkinson's disease based on the conservation of the mesostriatocortical system (Vernier et al. 2004).

So far, it is known that the presence of Lewy bodies, mitochondrial impairment, and oxidative stress are cardinal features of neurodegenerative disorders characterized by the oxidative aggregation of α-Syn, collectively known as synucleinopathies (Galvin et al. 2001). The N-terminal fragment of α-Syn participates in binding of different lipids, particularly anionic phospholipids and is important for α-Syn function in regulation of neuronal lipid metabolism, particularly the turnover of cardiolipin (Ellis et al. 2005). In 2006, Trostchansky et al. studied the effect of unsaturated lipids on α-Syn aggregation. Their results show that unsaturated fatty acid, but not saturated acids, modify aggregation patterns of α-Syn by forming adducts with the protein (Trostchansky et al. 2006). The experimental approach used in their work included incubation of α-Syn and unsaturated fatty acids followed by controlled exposure to $ONOO^-$ and determination of the accumulation of lipid-protein adducts. Formation of these adducts may arise from two possible pathways, (i) unsaturated lipids compete with α-Syn for $ONOO^-$, become oxidized, and the lipid oxidation products form adducts with the protein; (ii) α-Syn is oxidized by $ONOO^-$, and the radical formed in the backbone of the protein can be transferred to lipid components in a salvage pathway aimed to restore the protein. The oxidative aggregation of α-Syn has been demonstrated even in the absence of unsaturated lipids, and tyrosine residues in the protein have been associated with the stabilization of the radical formed, playing a key role in the oxidative aggregation of the protein when exposed (Ruf et al. 2008). Bayir et al. (2009) reported that α-Syn forms a triple complex with anionic lipids (such as cardiolipin) and cytochrome *c*. Apparently, α-Syn co-localizes with cytochrome *c* in Lewy bodies, indicating a potential interaction between the two proteins (Hashimoto et al. 1999). The results of Bayir et al. (2009) demonstrate that the covalent conjugation of α-Syn with cytochrome *c* in aggregates prevents the signaling effect of the cytochrome *c* as a death signal in the cytosol, possibly by diverting the flow of free radicals from membranes to α-Syn.

CONCLUSIONS

The formation of free radicals in lipids and membrane lipids is an important mechanism in cell homeostasis finely regulated by specific enzymes. When this regulation is disturbed by chemical or environmental reasons, the accumulation of oxidative damage in membranes may eventually lead to irreversible damage and cell death. Recently, the occurrence of membrane lipid oxidation has been correlated with a number of pathologies in marine organisms. Comparative studies between aquatic and terrestrial organisms, including humans, could lead to novel research in biomedicine and, potentially, to the development of preventive, diagnostic, and/or therapeutic tools to be used in a clinical setting against diseases associated with oxidative stress. The study of the transfer of free radicals between proteins and membrane lipids warrants future research on the regulation of cellular homeostasis in aquatic organisms.

REFERENCES

Baker, M.S., Gebicki, J.M. (1984) The effect of pH on the conversion of superoxide to hydroxyl free radicals. *Archives of Biochemistry and Biophysics* 234, 258–264.

Barr, D.P., Gunther, M.R., Deterding, L.J., Tomer, K.B., Mason, R.P. (1996) ESR spin-trapping of a protein-derived tyrosyl radical from the reaction of Cytochrome *c* with hydrogen peroxide. *Journal of Biological Chemistry* 271, 15498–15503.

Bayir, H., Kapralov, A.A., Jiang, J. et al. (2009) Peroxidase mechanism of lipid-dependent cross-linking of synuclein with cytochrome C: protection against apoptosis versus delayed oxidative stress in Parkinson disease. *Journal of Biological Chemistry* 284, 15951–15969.

Bienert, G.P., Moller, A.L., Kristiansen, K.A. et al. (2007) Specific aquaporins facilitate the diffusion of hydrogen peroxide across membranes. *Journal of Biological Chemistry* 282, 1183–1192.

Blodig, W., Doyle, W.A., Smith, A.T., Winterhalter, K., Choinowski, T., Piontek, K. (1998) Autocatalytic formation of a hydroxy group at C beta of Trp171 in lignin peroxidase. *Biochemistry* 37, 8832–8838.

Boudreau, H.S., Krol, K.M., Eibl, J.K. et al. (2009) The association of metal ion exposure with α-synuclein-like immunoreactivity in the central nervous system of fish, *Catostomus commersoni*. *Aquatic Toxicology* 92, 258–263.

Brunk, U.T., Terman, A. (2002) The mitochondrial-lysosomal axis theory of aging: accumulation of damaged mitochondria as a result of imperfect autophagocytosis. *European Journal of Biochemistry* 269, 1996–2002.

Buettner, G.R. (1987) Spin trapping: ESR parameters of spin adducts. *Free Radical Biology and Medicine* 3, 259–303.

Chandrasekharan, N.V., Simmons, D.L. (2004) The cyclooxygenases. *Genome Biology* 5, 241.

Chapman, D., Gomez-Fernandez, J.C., Goni, F.M. (1979) Intrinsic protein–lipid interactions. Physical and biochemical evidence. *FEBS Letters* 98, 211.

Cossins, A.R., Christiansen, J., Prosser, C.L. (1978) Adaptation of biological membranes to temperature. The lack of homeoviscous adaptation in the sarcoplasmic reticulum. *Biochimica et Biophysica Acta* 511, 442–454.

Crockett, E.L. (2008) The cold but not hard fats in ectotherms: consequences of lipid restructuring on susceptibility of biological membranes to peroxidation, a review. *Journal of Comparative Physiology B* 178, 795–809.

Davies, M.J., Puppo, A. (1992) Direct detection of a globin-derived radical in leghaemoglobin treated with peroxides. *Biochemical Journal* 281, 197–201.

Dean, R.T., Fu, S.L., Stocker, R., Davies, M.J. (1997) Biochemistry and pathology of radical-mediated protein oxidation. *Biochemical Journal* 324, 1–18.

DeGray, J.A., Lassmann, G., Curtis, J.F. et al. (1992) Spectral analysis of the protein-derived tyrosyl radicals from prostaglandin H synthase. *Journal of Biological Chemistry* 267, 23583–23588.

Del Río, L.A., Sandalio, L.M., Palma, J.M., Bueno, P., Corpas, F.J. (1992) Metabolism of oxygen radicals in peroxisomes and cellular implications. *Free Radical Biology and Medicine* 13, 557–580.

Droge, W. (2002) Free radicals in the physiological control of cell function. *Physiological Reviews* 82, 47–95.

Dunford, H.B. (1999) *Heme Peroxidases*. John Wiley & Sons, New York.

Ellis, C.E., Murphy, E.J., Mitchell, D.C. et al. (2005) Mitochondrial lipid abnormality and electron transport chain impairment in mice lacking α-synuclein. *Molecular and Cellular Biology* 25, 10190.

Eritsland, J. (2000) Safety considerations of polyunsaturated fatty acids. *American Journal of Clinical Nutrition* 71, 197S–201S.

Farkas, T., Fodor, E., Kitajka, K., Halver, J.E. (2001) Response of fish membranes to environmental temperature. *Aquaculture Research* 32, 645–655.

Finkel, T., Holbrook, N.J. (2000) Oxidants, oxidative stress and the biology of ageing. *Nature* 408, 239–247.

Folmer, V., Pedroso, N., Matias, A.C. et al. (2008) H_2O_2 induces rapid biophysical and permeability changes in the plasma membrane of Saccharomyces cerevisiae. *Biochimica et Biophysica Acta* 1778, 1141–1147.

Frejaville, C., Karoui, H., Tuccio, B. et al. (1995) 5-(Diethoxyphosphoryl)-5-methyl-1-pyrroline N-oxide: a new efficient phosphorylated nitrone for the in vitro and in vivo spin trapping of oxygen-centered radicals. *Journal of Medicinal Chemistry* 38, 258–265.

Frimer, A.A., Strul, G., Buch, J., Gottlieb, H.E. (1996) Can superoxide organic chemistry be observed within the liposomal bilayer?. *Free Radical Biology and Medicine* 20, 843–852.

Galvin, J.E., Lee, V.M., Trojanowski, J.Q. (2001) Synucleinopathies: clinical and pathological implications. *Archives of Neurology* 58, 186–190.

Garavito, R.M., Picot, D., Loll, P.J. (1994) Prostaglandin H synthase. *Current Opinion in Structural Biology* 4, 529–535.

Gazzotti, P., Peterson, S.W. (1977) Lipid requirement of membrane-bound enzymes. *Journal of Bioenergetics and Biomembranes* 9, 373–386.

Giulivi, C., Cadenas, E. (1998) Heme protein radicals: Formation, fate and biological consequences. *Free Radical Biology and Medicine* 24, 269–279.

Grosser, T., Yusuff, S., Cheskis, E., Pack, M.A., FitzGerald, G.A. (2002) Developmental expression of functional cyclooxygenases in zebrafish. *Proceedings of the National Academy of Sciences USA* 99, 8418–8423.

Grundy, M.M., Ratcliffe, N.A., Moore, M.N. (1996) Immune inhibition in marine mussels by polycyclic aromatic hydrocarbons. *Marine Environmental Research* 42, 187–190.

Halliwell, B., Chirico, S. (1993) Lipid peroxidation: its mechanism, measurement, and significance. *American Journal of Clinical Nutrition* 57, 715S–725S.

Hashimoto, M., Takeda, A., Hsu, L.J., Takenouchi, T., Masliah, E. (1999) Role of cytochrome c as a stimulator of α-synuclein aggregation in Lewy body disease. *Journal of Biological Chemistry* 274, 28849.

Hawkins, C.L., Davies, M.J. (2001) Generation and propagation of radical reactions on proteins. *Biochimica et Biophysica Acta-Bioenergetics* 1504, 196–219.

Herraez, P., Sierra, E., Arbelo, M., Jaber, J.R., De Los Monteros, A.E., Fernandez, A. (2007) Rhabdomyolysis and myoglobinuric nephrosis (capture myopathy) in a striped dolphin. *Journal of Wildlife Diseases* 43, 770–774.

Hiner, A.N.P., Martinez, J.I., Arnao, M.B. et al. (2001) Detection of a tryptophan radical in the reaction of ascorbate peroxidase with hydrogen peroxide. *European Journal of Biochemistry* 268, 3091–3098.

Holt, S., Reeder, B., Wilson, M., Harvey, S., Morrow, J.D., Roberts, I.I. (1999) Increased lipid peroxidation in patients with rhabdomyolysis. *The Lancet* 353, 1241.

Kaewsrithong, J., Qiau, D.F., Ohshima, T., Ushio, H., Yamanaka, H., Koizumi, C. (2000) Unusual levels of phosphatidylcholine hydroperoxide in plasma, red blood cell, and livers of aromatic fish. *Fisheries Science* 66, 768–775.

Karel, M. (1973) Protein-lipid interactions. *Journal of Food Science* 38, 756–763.

Keilin, D., Hartree, E.F. (1951) Purification of horse-radish peroxidase and comparison of its properties with those of catalase and methaemoglobin. *Biochemistry Journal* 49, 88–106.

Köhler, A., Pluta, H.J. (1995) Lysosomal injury and MFO activity in the liver of flounder (*Platichthys flesus* L.) in relation to histopathology of hepatic degeneration and carcinogenesis. *Marine Environmental Research* 39, 255–260.

Köhler, A., Wahl, E., Söffker, K. (2002) Functional and morphological changes of lysosomes as prognostic biomarkers of toxic liver injury in a marine flatfish (*Platichthys flesus* (L.)). *Environmental Toxicology and Chemistry* 21, 2434–2444.

Koljak, R., Järving, I., Kurg, R. et al. (2001) The basis of prostaglandin synthesis in coral. Molecular cloning and expression of a cyclooxygenase from the Arctic soft coral *Gersemia fruticosa*. *Journal of Biological Chemistry* 276, 7033–7040.

Labieniec, M., Milowska, K., Balcerczyk, A. et al. (2009) Interactions of free copper (II) ions alone or in complex with iron (III) ions with erythrocytes of marine fish Dicentrarchus labrax. *Cell Biology International* 33, 941–948.

Lenaz, G. (2001) The mitochondrial production of reactive oxygen species: mechanisms and implications in human pathology. *IUBMB Life* 52, 159–164.

Licht, S., Gerfen, G.J., Stubbe, J.A. (1996) Thiyl radicals in ribonucleotide reductases. *Science* 271, 477–481.

Liu, W., Cao, D., Oh, S.F., Serhan, C.N., Kulmacz, R.J. (2006) Divergent cyclooxygenase responses to fatty acid structure and peroxide level in fish and mammalian prostaglandin H synthases. *The FASEB Journal* 20, 1097.

Lowe, D.M., Pipe, R.K. (1994) Contaminant induced lysosomal membrane damage in marine mussel digestive cells: an in vitro study. *Aquatic Toxicology* 30, 357–365.

Luzzati, V., Reiss-Husson, F., Rivas, E., Gulik-Krzywicki, T. (1966) Structure and polymorphism in lipid-water systems, and their possible biological implications. *Annals of the New York Academy of Science* 137, 409–413.

Lynch, R.E., Fridovich, I. (1978) Permeation of the erythrocyte stroma by superoxide radical. *Journal of Biological Chemistry* 253, 4697–4699.

Málaga-Trillo, E., Salta, E., Figueras, A., Panagiotidis, C., Sklaviadis, T. (2010) Fish models in prion biology: Underwater issues. *Biochimica et Biophysica Acta* 1812, 402–414.

Maldonado, T.A., Jones, R.E., Norris, D.O. (2000) Distribution of b-amyloid and amyloid precursor protein in the brain of spawning (senescent) salmon: a natural, brain-aging model. *Brain Research* 858, 237–251.

Maroteaux, L., Campanelli, J.T., Scheller, R.H. (1988) Synuclein: a neuron-specific protein localized to the nucleus and presynaptic nerve terminal. *Journal of Neuroscience* 8, 2804–2815.

Moller, M., Botti, H., Batthyany, C., Rubbo, H., Radi, R., Denicola, A. (2005) Direct measurement of nitric oxide and oxygen partitioning into liposomes and low density lipoprotein. *Journal of Biological Chemistry* 280, 8850–8854.

Moore, K.P., Holt, S.G., Patel, R.P. et al. (1998) A causative role for redox cycling of Myoglobin and its inhibition by alkalinization in the pathogenesis and treatment of Rhabdomyolysis-induced renal failure. *Journal of Biological Chemistry* 273, 31731–31737.

Moore, M.N. (1990) Lysosomal cytochemistry in marine environmental monitoring. *The Histochemical Journal* 22, 187–191.

Morrow, J.D., Hill, K.E., Burk, R.F., Nammour, T.M., Badr, K.F., Roberts, L.J. (1990) A series of prostaglandin F2-like compounds are produced in vivo in humans by a non-cyclooxygenase, free radical-catalyzed mechanism. *Proceedings of the National Academy of Sciences USA* 87, 9383–9387.

Nordlund, P., Sjöberg, B.M., Eklund, H. (1990) Three-dimensional structure of the free radical protein of ribonucleotide reductase. *Nature* 345, 593–598.

Patterson, W.R., Poulos, T.L., Goodin, D.B. (1995) Identification of a porphyrin pi cation radical in ascorbate peroxidase compound I. *Biochemistry* 34, 4342–4345.

Porter, N.A. (1986) Mechanisms for the autoxidation of polyunsaturated lipids. *Accounts of Chemical Research* 19, 262–268.

Reeder, B., Sharpe, M., Kay, A., Kerr, M., Moore, K., Wilson, M. (2002) Toxicity of myoglobin and haemoglobin: oxidative stress in patients with rhabdomyolysis and subarachnoid haemorrhage. *Biochemical Society Transactions*, 745–748.

Roberts, J.E., Hoffman, B.M., Rutter, R., Hager, L.P. (1981) Electron-nuclear double resonance of horseradish peroxidase compound I. Detection of the porphyrin pi-cation radical. *Journal of Biological Chemistry* 256, 2118–2121.

Rodríguez-Malaver, A.J., Leake, D.S., Rice-Evans, C.A. (1997) The effects of pH on the oxidation of low-density lipoprotein by copper and metmyoglobin are different. *FEBS Letters* 406, 37–41.

Ruf, R.A.S., Lutz, E.A., Zigoneanu, I.G., Pielak, G.J. (2008) α-Synuclein conformation affects its tyrosine-dependent oxidative aggregation. *Biochemistry* 47, 13604–13609.

Ruíz-Dueñas, F.J., Pogni, R., Morales, M. et al. (2009) Protein Radicals in Fungal Versatile Peroxidase. *Journal of Biological Chemistry* 284, 7986–7994.

Ryan, R.W.J., Post, J.I., Solc, M., Hodson, P.V., Ross, G.M. (2002) Catecholaminergic neuronal degeneration in rainbow trout assessed by skin color change: a model system for identification of environmental risk factors. *Neurotoxicology* 23, 545–551.

Seddon, J.M. (1990) Structure of the inverted hexagonal (HII) phase, and non-lamellar phase transitions of lipids. *Biochimica et Biophysica Acta* 1031, 1–69.

Serhan, C.N., Oliw, E. (2001) Unorthodox routes to prostanoid formation: new twists in cyclooxygenase-initiated pathways. *Journal of Clinical Investigation* 107, 1481–1489.

Shi, W., Hoganson, C.W., Espe, M. et al. (2000) Electron paramagnetic resonance and electron nuclear double resonance spectroscopic identifiation and characterizaton of the tyrosyl radicals in prostaglandin H synthase 1. *Biochemistry* 39, 4112–4121.

Simmons, D.L., Botting, R.M., Hla, T. (2004) Cyclooxygenase isozymes: The biology of prostaglandin synthesis and inhibition. *Pharmacological Reviews* 56, 387–437.

Singer, S.J., Nicolson, G.L. (1972) The fluid mosaic model of the structure of cell membranes. *Science* 175, 720–731.

Smith, W.L., Langenbach, R. (2001) Why there are two cyclooxygenase isozymes. *Journal of Clinical Investigation* 107, 1491–1495.

Stubbe, J.A., van der Donks, W.A. (1998) Protein radicals in enzyme catalysis. *Chemistry Reviews* 98, 705–762.

Terman, A., Kurz, T., Navratil, M., Arriaga, E.A., Brunk, U.T. (2010) Mitochondrial turnover and aging of long-lived postmitotic cells: the mitochondrial-lysosomal axis theory of aging. *Antioxidants and Redox Signaling* 12, 503–535.

Thuresson, E.D., Lakkides, K.M., Rieke, C.J. et al. (2001) Prostaglandin endoperoxide H synthase-1. The functions of cyclooxygenase active site residues in the binding, positioning, and oxygenation of arachidonic acid. *Journal of Biological Chemistry* 276, 10347–10357.

Tiku, P.E., Gracey, A.Y., Macartney, A.I., Beynon, R.J., Cossins, A.R. (1996) Cold-induced expression of delta9-desaturase in carp by transcriptional and posttranslational mechanisms. *Science* 271, 815.

Tomasi, A., Iannone, A. (1993) ESR spin trapping artifacts in biological systems," *In* Berlinen, L.J., Reubens, J. (eds). *EMR of Paramagnetic Molecules*, Vol. 13. Plenum Press, New York, pp. 353–384.

Trostchansky, A., Lind, S., Hodara, R. et al. (2006) Interaction with phospholipids modulates α-synuclein nitration and lipidûprotein adduct formation. *Biochemical Journal* 393, 343.

Turnbull, B.S., Cowan, D.F. (1998) Myocardial contraction band necrosis in stranded cetaceans. *Journal of Comparative Pathology* 118, 317–327.

Ulmer, T.S., Bax, A., Cole, N.B., Nussbaum, R.L. (2005) Structure and dynamics of micelle-bound human α-synuclein. *Journal of Biological Chemistry* 280, 9595.

Valderrama, B., Ayala, M., Vazquez-Duhalt, R. (2002) Suicide inactivaction of peroxidases and the challenge of engineering more robust enzymes. *Chemistry and Biology* 9, 555–565.

Valmsen, K., Järving, I., Boeglin, W.E. et al. (2001) The origin of 15R-prostaglandins in the Caribbean coral *Plexaura homomalla*: Molecular cloning and expression of a novel cyclooxygenase. *Proceedings of the National Academy of Sciences USA* 98, 7700–7705.

Vernier, P., Moret, F., Callier, S., Snapyan, M., Wersinger, C., Sidhu, A. (2004) The degeneration of dopamine neurons in Parkinson's disease: insights from embryology and evolution of the mesostriatocortical system. *Annals of the New York Academy of Sciences* 1035, 231–249.

Wagner, A.F., Frey, M., Neugebauer, F.A., Schäfer, W., Knappe, J. (1992) The free radical in pyruvate formate-lyase is located on glycine-734. *Proceedings of the National Academy of Sciences USA* 89, 996–1000.

Whittaker, M.M., Whittaker, J.W. (1990) A tyrosine-derived free radical in apogalactose oxidase. *Journal of Biological Chemistry* 265, 9610–9613.

Wilks, A., Ortiz de Montellano, P.R. (1992) Intramolecular translocation of the protein radical formed in the reaction of recombinant sperm whale myoglobin with H_2O_2. *Journal of Biological Chemistry* 267, 8827–8833.

Yang, T., Forrest, S.J., Stine, N. et al. (2002) Cyclooxygenase cloning in dogfish shark, *Squalus acanthias*, and its role in rectal gland Cl secretion. *American Journal of Physiology* 283, R631–R637.

Zehmer, J.K., Hazel, J.R. (2004) Membrane order conservation in raft and non-raft regions of hepatocyte plasma membranes from thermally acclimated rainbow trout. *Biochimica et Biophysica Acta* 1664, 108–116.

Chapter 17

IMMUNE DEFENSE OF MARINE INVERTEBRATES: THE ROLE OF REACTIVE OXYGEN AND NITROGEN SPECIES

Eva E. R. Philipp[1], *Simone Lipinski*[1], *Jonathan Rast*[2], *and Philip Rosenstiel*[1]

[1]Institute of Clinical Molecular Biology, Cell Biology Department, Christian-Albrechts University Kiel, Kiel, Germany
[2]Sunnybrook Health Sciences Centre, Department of Medical Biophysics and Department of Immunology, University of Toronto, Toronto, ON, Canada

Life in diverse environments and complex communities is coupled with the interaction of the individuals and species living within. Recognition of self and nonself and the maintenance of self-integrity are pivotal for survival and proper physiological function. To consistently achieve and maintain integrity, organisms have evolved an effective immune system. This surveillance system on the one hand recognizes nonself structures like microorganisms, viruses, and other foreign particles, but on the other hand it is important to detect and eliminate modified self-cells (e.g. cancer cells). Invertebrates are a very diverse and successful group of animals with over 2 million species representing more than 20 animal phyla. To cope with pathogenic microorganisms, invertebrates rely exclusively on the germline encoded innate immune system and lack the somatically rearranged adaptive immune system that is found in vertebrates. The adaptive vertebrate immune system recognizes and memorizes newly invading pathogens, targets the immune response, and enhances protection under reinfection.

Several genes and principles of the innate immune system are highly conserved throughout the animal kingdom and also play an important role in vertebrates (for review see Irazoqui et al. 2009; Rosenstiel et al. 2009; Chapter 18). Invertebrates can therefore be regarded as important models to identify and understand evolutionary conserved mechanisms of pathogen detection and immune responses. Many marine invertebrates are filter and deposit feeders which constantly engulf bacteria and other microorganisms.

Oxidative Stress in Aquatic Ecosystems, First Edition. Edited by Doris Abele, José Pablo Vázquez-Medina, and Tania Zenteno-Savín.

In bivalves, bacteria constitute an important food source, and these as well as other marine invertebrates live with a biofilm of bacterial symbionts (Dubilier et al. 2008; Ruby and McFall-Ngai 1999). Contrary to these mostly beneficial bacteria, which sometimes are even necessary for survival, pathogenic bacteria can cause mass mortalities of marine invertebrates (Le Gall et al. 1988; Gómez-León et al. 2005). Thus the control and elimination of these bacteria by the host is of vital importance to ensure optimal physiological function.

The major tasks of the innate immune system include a controlled host/microbial cross talk at epithelial barriers, the recognition of danger signals, the recruitment of mesoderm-derived professional immune cells, clearance of intracellular pathogens by phagocytosis with the aid of reactive oxygen and nitrogen species (ROS and RNS), and the secretion of local or circulating effector-molecules such as antimicrobial peptides and simple opsonic forms of complement.

Pattern recognition receptors (PRRs) sense invariant molecular signatures that are either present in microorganisms (microorganism-associated molecular patterns, MAMPs) or pathogens (pathogen-associated molecular patterns, PAMPs). For example, lipopolysaccharides or unmethylated CpG DNA can signal the presence of nonself. Alternatively they can recognize signals that are derived from endogenous sources and attest profound cellular damage (e.g. extracellular heat shock proteins (HSP), oxidatively modified proteins). The major types of PRRs in the innate immune system are: the Toll-like receptors (TLRs), which are composed of a ligand binding ectodomain and an intracellular adaptor domain, where downstream signaling molecules are recruited upon activation (Takeda et al. 2003); the intracellular nucleotide-binding oligomerization domain (NOD)-like receptor (NLR) family, which plays a pivotal role for the recognition of intracellular PAMPs (Inohara et al. 2005; Rosenstiel et al. 2008) and is related to the apoptosis-inducing apoptosis protease-activating factor (APAF)-like molecules. The third group of PRRs is the family of scavenger receptor cysteine-rich (SRCR) domain-containing proteins (Sarrias et al. 2004). A complex set of transcripts featuring SRCR repeats, which may contribute to the diversity of innate immune responses, can be found from sea urchin to humans. Engagement of these receptors leads to a fast induction of protective programs, e.g. the induction of antimicrobial peptides or the elimination of the infected cell through apoptosis.

In most bilaterian invertebrates, hemocytes/coelomocytes represent the major migratory cell type of the innate immune system. Different hemocyte subpopulations are differentially involved in functions of the host's biology and physiology, including nutrient transport and digestion, tissue and shell formation, maintenance of homeostasis, and immune response (Canesi et al. 2006; Mount et al. 2004). The main function of marine invertebrate hemocytes studied so far is, however, immune response.

Free-floating immune cells can be found in the coelom and hemolymph of nematodes, arthropods, annelids, echinoderms, mollusks, and other marine phyla (Stein and Cooper 1983; Roch 1999; Iwanaga 2002; Smith et al. 2006; Tahseen 2009; Xylander 2009). In echinoderms and annelids these cells are called coelomocytes, whereas in arthropods and mollusks they are referred to as hemocytes. Different types of hemocytes/coelomocytes can be found between and within different invertebrate phyla. In sea urchins, coelomocytes are differentiated in phagocytes, red and colorless spherules, and vibratile cells (Smith et al. 2006, Fig.17.1). In bivalves the hemocytes are mainly divided into agranular (hyaline) and granular hemocytes and the latter further differentiated in eosinophilic and basophilic cells (Canesi et al. 2002; Wootton et al. 2003). These cell types react in number and response to different stress scenarios (like temperature and salinity changes, wound healing), and are most important for the recognition, elimination and control of microorganisms and other foreign particles. A major mechanism of the hemocyte/coelomocyte immune response is phagocytosis, which is linked to the secretion of humoral defense factors such as agglutinins (e.g lectins), lysosomal enzymes, antimicrobial peptides, and complement factors, as well as to the release of ROS and RNS (Roch 1999; Canesi et al. 2002; Tahseen 2009). In certain ways hemocytes/coelomocytes resemble the vertebrate monocyte/macrophage lineage for which the cellular response of ROS generation (oxidative burst) has been described.

We will elaborate on selected aspects of the oxidative burst reaction and the role of ROS and RNS as elicitors, signaling intermediates, and effectors of the immune system, with a focus on knowledge derived from marine invertebrates (see Fig. 17.2 for overview of ROS in immune system functions).

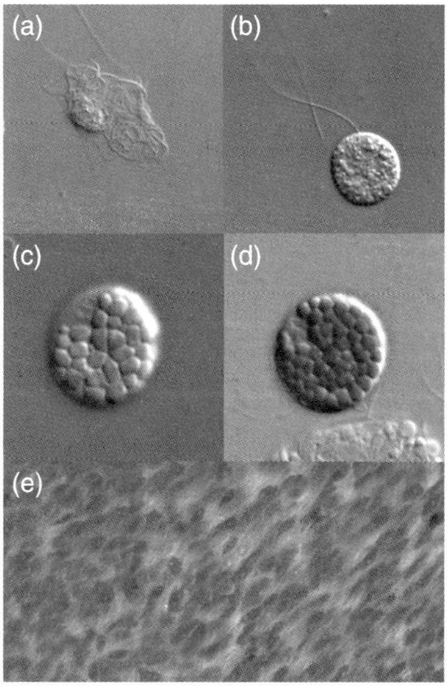

Fig. 17.1 Coelomocytes of the sea urchin *Strongylocentrotus purpuratus*. (a) Phagocytic cells, which diversify into several types, are capable of producing ROS as part of the phagocytosis/killing process. (b) Vibratile cells are flagellated granular cells that participate in clotting and wound healing. (c) Colorless spherule cells are amoeboid cells of unknown function. (d) Red spherule cells are highly motile amoeboid cells that have large granules containing the napthoquinone echinochrome. These cells are active in wound healing and are capable of producing hydrogen peroxide. (e) Intestinal wall with a barrier-like distribution of tissue resident cells with red spherule cell-like structure. (Photographs by Jonathan Rast.) See plate section for a color version of this image.

OXIDATIVE BURST AND REACTIVE OXYGEN SPECIES FORMATION IN MARINE INVERTEBRATES

Immunological functions of hemocytes include recognition, phagocytosis, and elimination of invading microorganisms, such as potentially infective bacteria and parasites.

The major cellular mechanism of immune response in hemocytes is phagocytosis, which includes the action of ROS. Phagocytosis occurs by the invagination of the cell membrane and the formation of an endocytic vacuole, called the primary phagosome. Cytoplasmic granules containing phosphatases, esterases, and amidases, as well oxidative enzymes such as peroxidase and cytochrome *c* oxidase, then migrate and fuse with the phagosome to form a secondary phagosome that will degrade and eliminate the foreign particles (Donaghy et al. 2009). Lysosmal enzymes such as lysozyme can be found within the cell but are also present in extracellular forms in the hemolymph and participate in the killing and degradation of foreign organisms with the aid of ROS generated by the immunocyte, a process called oxidative burst (Lopez et al. 1997).

The term "oxidative" or "respiratory burst" has been coined for humans and other vertebrates and describes the massive release of ROS over a short time period, coupled with a high oxygen uptake (see Chapter 18). It is a key reaction during phagocytosis, carried out by macrophages and neutrophils to degrade bacteria or internalized harmful substances. The cascade of enzymatic and ROS reactions starts upon activation of reduced nicotinamide adenine dinucleotide phosphate (NADPH) oxidase (NOX) complexes (see below) with the production of $O_2^{\bullet-}$. H_2O_2 can then be produced from $O_2^{\bullet-}$ either spontaneously or catalysed by superoxide dismutase (SOD). In the presence of reduced transition metals, H_2O_2 can be further converted into the highly reactive $HO^{\bullet}$. H_2O_2 may also give rise to highly toxic hypochlorous acid (HOCl), a potent antimicrobial oxidant (Klebanoff 2005), catalysed by myeloperoxidase (MPO) in the presence of chloride ions, as described for mammals (Klebanoff 2005) and fish (Castro et al. 2008; Chapter 18), or other halide peroxidases (see below, *Euprymna scolopes*).

In the marine environment, ROS generation has been investigated extensively in bivalve hemocytes and is often also termed "oxidative burst." In bivalves ROS generation is generally less intensive and proceeds more slowly. Peak ROS generation was found to take place 15–70 min after infection and the intensity is between two and ten times higher compared to basal levels, whereas in vertebrates ROS generation reaches values up to 1000–1500 times higher compared to basal levels and lasts only a few minutes (Donaghy et al. 2009). Thus, the term "oxidative burst" should be used with caution for mollusk hemocyte immune response.

The amount of hemocyte-derived ROS varies among different marine invertebrate species and according to the stimulant used to elicit a response (Moss and Allam 2006; Wang et al. 2008b; Donaghy et al. 2009). Common stimulants are lipopolysaccharides (LPS) and peptidoglycans, which are important structural constituents of the outer membrane of gram-negative

(LPS, PGN) and gram-positive (PGN) bacteria. Flagellin is a main component of bacterial flagellar filament. Zymosan, a protein–carbohydrate complex derived from yeast-cell walls, is widely used as an experimental inducer of inflammation or phagocytosis. In most bivalve species, such as oysters (*Crassostrea virginica*, *C. gigas*, *C. ariakensis* and *Ostrea edulis*), mussels (*Mytilus edulis*, *M. galloprovincialis*) and scallops (*Pecten maximus* and *Chlamys farreri*), hemocytes were found to generate ROS when activated by one of these stimulants (for review see Donaghy et al. 2009). Zymosan seems to be a potent stimulator of ROS generation and, in most studies of different species and phylogenetic groups (bivalvia, crustacea), hemocyte activation could be elicited with this stimulant.

Contradictory results with respect to the ability to generate ROS are reported in different studies for similar species. This may mainly arise from the techniques used for ROS or RNS detection. In the past ROS were primarily measured by luminol-dependent chemiluminescence (CL) and the reduction of nitroblue tetrazolium (NBT) to formazan by $O_2^{\bullet-}$. The use of flow cytometry and redox-sensitive fluorescent dyes, like 2′, 7′-dichlorofluorescin-diacetate (DCFH-DA), which detects intracellular ROS, or Amplex red, a fluorescent dye for extracellular ROS detection, allowed more sensitive ROS measurements. Species formerly regarded as unable to generate ROS, for example the clam *Mercenaria mercenaria*, were shown to indeed produce ROS when these new methods were applied (Anderson 1994; Buggé et al. 2007). Further, with these dyes ROS generation was found to occur both inside and outside the hemocytes (Wootton et al. 2003). Different types of hemocytes (e.g. agranular and granular) have a different potential to produce ROS. In *Mytilus edulis* higher capacity for respiratory burst was found for eosinophilic compared to basophilic granulocytes (Pipe et al. 1997). In *C. virginica*, all types of hemocytes, i.e. granulocytes, hyalinocytes, and small granulocytes showed increased production of ROS in the presence of zymosan, with granulocytes showing the highest increase (Hégaret et al. 2003). Generally granulocytes exhibit higher ROS generation capacity than hyalinocytes (Hégaret et al. 2003; Labreuche et al. 2006a,b).

The respiratory burst in bivalve hemocytes is further modulated by factors in the hemolymph. In *Mytilus* 17β-estradiol stimulated oxyradical production (Canesi et al. 2006), and ROS generation of *Mytilus* hemocytes activated with *Vibrio parahemolyticus* and *Escherichia coli* was enhanced in the presence of plasma (Kumazawa et al. 1993); the mechanism underlying these modifications, however, remains to be elucidated. Background ROS generation can be detected in nonstimulated hemocytes, which may demonstrate low ROS production during normal metabolism and serve as signaling molecules also under nonstimulated conditions (Pourova et al. 2010). The isolation procedure or the attachment of the hemocytes to the experimental surface (glass, plastic) might itself lead to activation and ROS generation (Anderson 1994; Moss and Allam 2006).

In the sea urchin, larval pigment cells and adult red spherule cells are loaded with granules that contain the napthoquinone echinochrome (Service and Wardlaw 1984). Red spherule cells are known to accumulate at wound sites in the adult (Johnson 1969) and their pigment cell counterparts display surveillance-like behavior and respond to bacterial exposure in larvae (Hibino et al. 2006). The echinochrome has been described to have antibacterial activity (Service and Wardlaw 1984) and forms H_2O_2 in the presence of extracellular levels of calcium (Perry and Epel 1981). Thus, these cells, which are highly mobile and abundant throughout the tissues of the adult (see Fig. 17.2), may serve as a source of H_2O_2 that can be directed to sites of infection or wounds. This is, in addition to the ROS activity, associated with engulfment of nonself by phagocytic cells of the echinoderms (Ito et al. 1992).

In addition to immunocytes, the generation of ROS can also result from the prophenoloxidase system (proPO) in the hemolymph, which has been observed in several invertebrates, including crustaceans and bivalves (Cerenius and Söderhäll 2004; Cerenius et al. 2008). The main function of the proPO system is the synthesis of melanin, which physically shields the host from invading pathogens. Quinone intermediates, which are generated during melanin synthesis, can enter enzymatic and nonenzymatic redox cycling, producing the corresponding semiquinone radicals and, as a result, generate $O_2^{\bullet-}$ (Bogdan 2007).

The massive release of ROS due to infection has two sides. On the one hand, ROS will help in the degradation of foreign particles; on the other hand, high amounts of ROS can lead to an imbalance, with antioxidant defense producing oxidative stress. Antioxidants such as the enzymes SOD and catalase (CAT) as well as enzymes of the glutathione system, glutathione peroxidase and reductase (GPx, GR), are detected in the hemolymph and hemocytes of bivalves and sea cucumbers (Li et al.

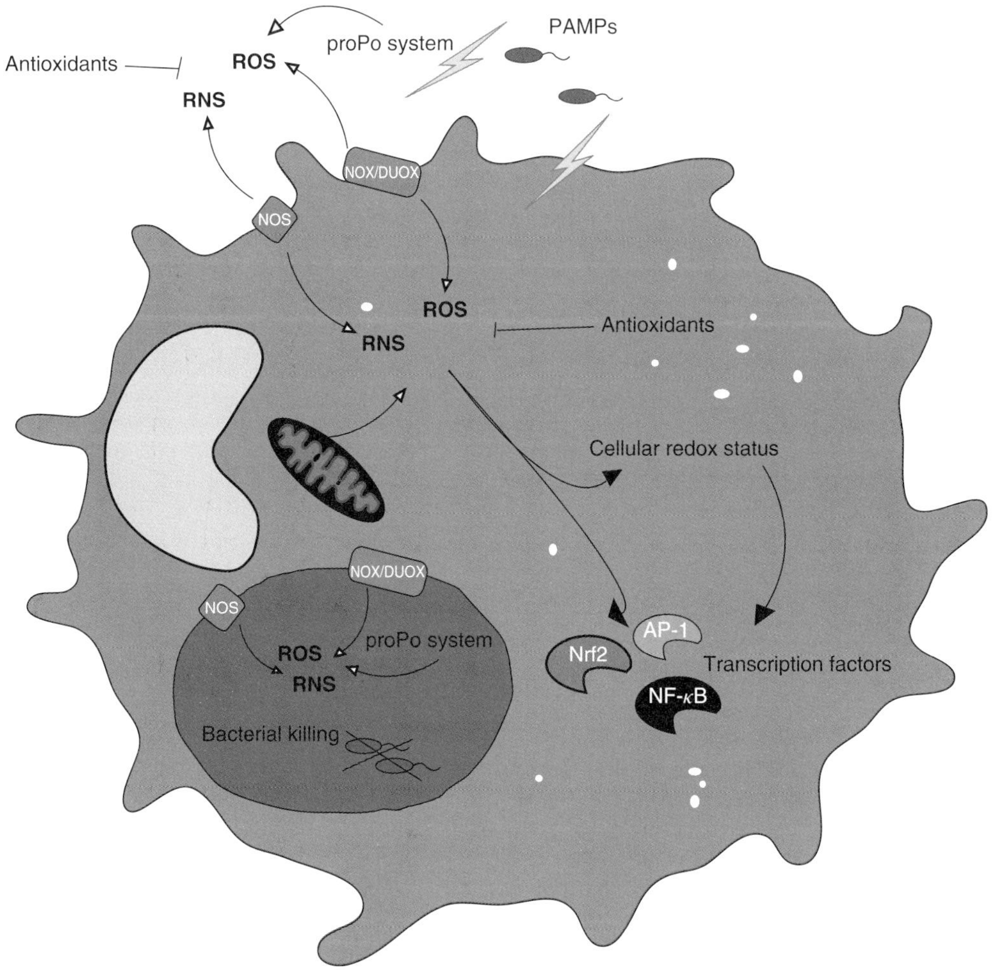

Fig. 17.2 Reactive oxygen species (ROS) in immune system functioning: pathogen-associated molecular patterns (PAMPs) can activate immune cells via extra- and intracellular recognition. The activation leads to the generation of ROS and RNS inside and outside the immunocytes and phagosome (dark gray) by NOX/DUOX, NOS, and the prophenoloxidase system (proPo). ROS and RNS can modulate the cellular redox system and transcription factors, which will in turn modulate downstream signaling pathways (details in the text).

2008; Wang et al. 2008a; Angeles et al. 2009), and may serve as a protective mechanism to counteract ROS production and to avoid oxidative damage.

REACTIVE NITROGEN SPECIES IN THE INVERTEBRATE IMMUNE RESPONSE

Various cells involved in the general immune response are known to release NO•, including monocytes, macrophages, neutrophils, and eosinophils, as well as epithelial cells, fibroblasts, and hepatocytes. RNS production is catalysed by the enzyme NO synthase (endothelial NOS (eNOS), inducible NOS (iNOS), neuronal NOS (nNOS)) and starts with the production of the NO radical (NO•) via the transformation of L-arginine. iNOS is particularly important in immune system function (Bogdan 2001). In vertebrates, NO• itself can inhibit the adhesion of platelets and leukocytes to the endothelium and also influences leucocyte recruitment and adhesion as well as chemotactic response and the production of chemokines (Oliveira et al. 1999). NO• can be converted by reaction with $O_2^{\bullet-}$, RO•, and ROO• to various other RNS such as nitrosonium cation (NO^+), nitroxyl anion (NO^-), or peroxynitrite ($ONOO^-$). $ONOO^-$ is a very powerful oxidant, 2000-fold more potent than H_2O_2 in oxidizing thiols and can also decompose to HO• (Freeman 1994; Victor et al. 2005).

Hemocytes of marine bivalves release NO• in addition to ROS, as described in studies of *Mytilus edulis* (Canesi et al. 2006; Ottaviani et al. 1993), *Mytilus galloprovincialis* (Arumugam et al. 2000; Gourdon et al. 2001; Tafalla et al. 2002), *Viviparus ater* (Conte and Ottaviani 1995), and *Ruditapes decussatus* (Tafalla et al. 2003). In *M. mercenaria* the NOS inhibitors, NMMA and L-NIO, did not suppress DCFH-DA oxidation in hemocytes, indicating that NO• production by NOS was not significant (Buggé et al. 2007). In *Mytilus* hemocytes, however, ROS generation was prevented by L-NMMA and by SOD, which confirms the general involvement of both NO• and $O_2^{\bullet-}$ in the molluscan immune response (Canesi et al. 2006).

Different bacterial strains have different effects on the activation of hemocytes. Lambert and Nicolas (1998) exposed hemocytes from two adult bivalves, *Pecten maximus* and *Crassostrea gigas*, to 12 different bacterial strains and found strains with a range of effectiveness for activating hemocytes. NO• and other RNS have important microbicidal effects against intracellular parasites, bacteria, fungi, and protozoa (Arumugam et al. 2000; Reeves et al. 2003). Whereas $O_2^{\bullet-}$ has in itself low bactericidal activity, once it is dismutated to H_2O_2 it can generate highly bactericidal NOO• from reaction with NO• (Reeves et al. 2003).

In line with the Red Queen Principle, several pathogens have found a way to avoid killing by the oxidative burst system. Taken from the Red Queen's race in Lewis Carroll's *Through the Looking-Glass* the theory states that in an evolutionary system, continuing development is needed just in order to maintain its fitness relative to the systems with which it is coevolving. Bacteria such as *Staphylococcus aureus* (Mandell 1975) and parasites like *Plasmodium berghei* (Fairfield et al. 1986) can avoid killing by the oxidative burst through the action of antioxidant enzymes like CAT. Other parasites like *Perkinsus marinus*, the facultative intracellular parasite of the eastern oyster *C. virginica*, prevent the accumulation of ROS by abrogating the synthesis and/or scavenging the products of the host's phagocytic respiratory burst (by its two SOD forms PmSOD1 and PmSOD2; Fernández-Robledo et al. 2008). In bivalves different pathogenic bacterial strains were found to suppress ROS generation of hemocytes (Labreuche et al. 2006a; Lambert and Nicolas 1998). In symbiotic relationships, quenching of ROS might even be necessary to allow the bacterial symbiont to permanently colonize the host. Such a relationship was studied intensively in the symbiosis between the serpiolid squid *Euprymna scolopes* and the luminous bacterium *Vibrio fisheri* (Ruby and McFall-Ngai 1999). In this special relationship, *V. fisheri* colonizes the crypts of juvenile *E. scoplopes* where it serves as a light organ. To accomplish this relationship and avoid bacterial killing by ROS produced by the host, *V. fisheri* apparently employs a sophisticated strategy. The high oxygen affinity of the bacterial luciferase and the bacterium's respiratory activity keeps the P_{O_2} in the crypts on a low level. This, in turn, reduces the availability of oxygen for the oxidative burst reaction of the host's phagocytes and limits the generation of H_2O_2, which otherwise reacts to bactericidal hypochlorous acid (HOCl) through catalysis of halide peroxidase secreted from the crypts of epithelial cells. Excess H_2O_2 may be further removed by CAT located in the periplasm of *V. fisheri* (McFall-Ngai 1999; Ruby and McFall-Ngai 1999; Visick and McFall-Ngai 2000).

THE NOX/DUOX FAMILY OF NADPH OXIDASES

The major ROS generation sites of hemocytes and other immunocytes are membrane bound NADPH and dual oxidases (NOX, DUOX). The first functional component of the "oxidative burst" in mammalian phagocytes, a heme-containing cytochrome *b* subunit, gp91phox/NOX2, was identified in 1978 and since then ROS have no longer been regarded as a simple metabolic by-product (Segal et al. 1978). Today, in the mammalian system, six homologues of gp91phox/NOX2 are identified by their shared sequence homology: comprising the NADPH oxidases NOX1, NOX3, NOX4 and NOX5, and the dual oxidases DUOX1 and DUOX2 (Dupuy et al. 1999; De Deken et al. 2000; Lambeth et al. 2000). All NOX enzymes transfer electrons across membranes to reduce oxygen to $O_2^{\bullet-}$ and exhibit four conserved domains: a C-terminal NADPH-binding site, a lavin-adenine dinucleotide (FAD) binding site, six transmembrane domains, and heme-binding histidines (Lambeth et al. 2000; Lambeth 2002). NOX5, DUOX1, and DUOX2 additionally possess a calcium-binding EF motive and DUOX1 and DUOX2 further exhibit a peroxidase domain, which gives them the name "dual oxidase" (Banfi et al. 2001; Edens et al. 2001).

Most knowledge about the composition and activation of NADPH oxidases is gained from the well studied phagocyte family member NOX2. In the plasma membrane of resting phagocytes the NADPH complex consists of at least six single proteins: four regulatory components ($p47^{phox}$, $p67^{phox}$, $p40^{phox}$, and Rac1 or Rac2) and two membrane-associated catalytic flavocytochrome *b* components (NOX2 and $p22^{phox}$). Upon activation, a complex series of protein–protein interactions recruits the cytosolic subunits $p47^{phox}$, $p67^{phox}$, and $p40^{phox}$, as well as the GTPase rac2 to the $p22^{phox}$ subunit in the plasma membrane. $p22^{phox}$ serves as a docking site that allows binding of the cytosolic subunits and further contributes to the stabilization of NOX2 in the membrane (Nauseef 2008). Within the assembled active enzyme, electrons are transferred in a stepwise manner from NADPH to FAD, from the proximal heme to the distal heme and finally to molecular oxygen (Cross et al. 1995). The importance of the correct assembling and activation of the phagocyte NADPH complex is emphasized by the finding that mutations in any of the involved genes cause chronic granulomatous disease (CGD) in humans. In this disease phagocytes fail to execute the oxidative burst and patients suffer from a greatly increased susceptibility to bacterial and fungal infections (Heyworth et al. 2003).

NOX2, DUOX1, and DUOX2 are involved in innate immunity, while other NOX family members are implicated in various biological functions. In the DUOX family activation differs from NOX proteins: DUOX1 and DUOX2 require post-translational modifications via their activator proteins DUOXA1 and DUOXA2 and are activated upon calcium stimulation (Ameziane-El-Hassani et al. 2005; Grasberger and Refetoff 2006). Moreover, an additional N-terminal extracellular peroxidase-homology domain and the subsequent production of H_2O_2 instead of $O_2^{\bullet -}$ discriminates this group of proteins from the NOX family (Lambeth et al. 2007). ROS-generating NOX/DUOX orthologs are conserved in vertebrates, urochordates, echinoderms, nematodes, insects, fungi, plants amoeba, and red alga, but not in prokaryotes. Orthologs have been found also in marine invertebrates (Table 17.1). The most basal marine eumetazoan *Nematostella vectensis* solely contains NOX2-like enzymes, and not EF-hand-containing oxidases such as NOX5 and DUOX, although these two families are found in a variety of species of protostomes and deuterostomes. Thus, NOX5 and DUOX may have evolved from NOX2-like prototype oxidases (Sumimoto 2008).

The first functional description of a marine invertebrate NOX/DUOX homolog came from the sea urchin dual oxidase Udx1, which catalyzes the rapid formation of H_2O_2 during fertilization, thus preventing polyspermy (Wong et al. 2004). This is achieved by the activity of ovoperoxidase that uses H_2O_2 as a substrate to cross-link the envelope into a hardened matrix (LaFleur et al. 1998). The expression of Udx1 is maintained during development at the surface of all nonmesenchymal blastomeres and is suggested to have a role in intracellular signaling as well as the additional function of an antimicrobial barrier (Wong and Wessel 2005). Proteins with perforin-like structure (Hibino et al. 2006) are also known to be secreted into the extracellular space in the early embryo (Haag

Table 17.1 Orthologs of NOX/DUOX pathway components in marine invertebrates.

Species	Common name	Gene	Reference
Strongylocentrotus purpuratus	Purple sea urchin	NOX2, Udx1	Kawahara et al. (2007); Maru et al. (2005); Wong et al. (2004)
Lytechinus variegates	Variegated sea urchin	Udx1	Wong et al. (2004)
Ciona intestinalis	Vase tunicate	NOX2, $p22^{phox}$, $p47^{phox}$, $p67^{phox}$, NOX4, DUOX	Hiruta et al. (2006); Inoue et al. (2005)
Nematostella vectensis	Starlet sea anemone	NOX2-like	Sumimoto (2008)

et al. 1999) and, along with ROS, this may be part of a system, or have evolved from a system, to sterilize the exterior of the developing embryo.

Little is known regarding the possible link between NOX/DUOX homologs and immune function in marine invertebrates. In hemocytes of the planorbis snail *Biomphalaria glabrata* there is evidence of regulated H_2O_2 release with regard to immune function, but no NADPH oxidase has been identified at the molecular level (Humphries and Yoshino 2008).

CELLULAR REDOX STATUS, SIGNALING PATHWAYS, AND TRANSCRIPTION FACTORS

In immune function ROS and RNS do not only act as direct microbicidal agents. By operating as intracellular second messengers and modulators of the cellular redox status, they modulate innate immune signaling. Their local production is a crucial factor in ROS-mediated signal transduction. Low production induce the expression of antioxidant enzymes via the redox-sensitive transcription factor Nrf2 (NF-ES-related factor 2). Intermediate ROS production induce an immune response via signaling cascades of the transcription factors NF-κB and AP-1. Finally, high intracellular H_2O_2 amounts induce apoptosis or necrosis (Tonks 2005). From the mechanistic point of view, ROS influence immune signaling cascades by oxidative modifications (e.g. protein-dimerisation, intra- or intermolecular disulfide bonds, protein cross-linking). An additional indirect mechanism by which ROS influence signal transduction is the activation of transcription factors, such as NF-κB, either directly or by alteration of the intracellular redox state that is defined by the ratio of reduced to oxidized glutathione (Meyer et al. 1993; Thannickal and Fanburg 2000; Victor et al. 2005; Oliveira-Marques et al. 2009). In the marine invertebrate system studies on the interplay between cellular redox status and immune system functioning are still lacking.

FUTURE PERSPECTIVES

With the vast diversity of invertebrates, finding diverse immune responses and their many presently unexplored immune mechanisms is not surprising. The recent development of high throughput sequencing technology and whole genome and transcriptome analysis of nonmodel organisms is promising for the potential identification of orthologs for known immune genes (e.g. NOX, DUOX) and the discovery of a whole suite of new immune and defense responses (Rast et al. 2006; Buckley and Smith 2007; Rosenstiel et al. 2009).

REFERENCES

Ameziane-El-Hassani, R., Morand, S., Boucher, J.L. et al. (2005) Dual oxidase-2 has an intrinsic Ca_2^+-dependent H_2O_2-generating activity. *Journal of Biology and Chemistry* 280, 30046–30054.

Anderson, R.S. (1994) Hemocyte-derived reactive oxygen intermediate production in four bivalve mollusks. *Developmental and Comparative Immunology* 18, 89–96.

Angeles, I.P., Chien, Y.-H., Yambot, A.V. (2009) Effect of injected Astaxanthin on survival, antioxidant capacity, and immune response of the giant freshwater prawn *Macrobrachium rosenberhii* (De Man, 1879) challengend with *Lactococcus garvieae*. *Journal of Shellfish Research* 28, 931–937.

Arumugam, M., Romestand, B., Torreilles, J., Roch, P. (2000) In vitro production of superoxide and nitric oxide (as nitrite and nitrate) by*Mytilus galloprovincialis* haemocytes upon incubation with PMA or laminarin or during yeast phagocytosis. *European Journal of Cell Biology* 79, 513–519.

Banfi, B., Molnar, G., Maturana, A. et al. (2001) A Ca(2+)-activated NADPH oxidase in testis, spleen, and lymph nodes. *Journal of Biology and Chemistry* 276, 37594–37601.

Bogdan, C. (2001) Nitric oxide and the immune response. *Nature Immunology* 2, 907–16.

Bogdan, C. (2007) Oxidative burst without phagocytes: the role of respiratory proteins. *Nature Immunology* 8, 1029–1031.

Buckley, K.M., Smith, L.C. (2007) Extraordinary diversity among members of the large gene family, 185/333, from the purple sea urchin, *Strongylocentrotus purpuratus*. *BMC Molecular Biology* 8, 68.

Buggé, D.M., Hégaret, H., Wikfors, G.H., Allam, B. (2007) Oxidative burst in hard clam (*Mercenaria mercenaria*) haemocytes. *Fish and Shellfish Immunology* 23, 188–196.

Canesi, L., Gallo, G., Gavioli, M., Pruzzo, C. (2002) Bacteria–hemocyte interactions and phagocytosis in marine bivalves. *Microscopy Research and Technique* 57, 469–476.

Canesi, L., Betti, M., Ciacci, C., Lorusso, L.C., Pruzzo, C., Gallo, G. (2006) Cell signalling in the immune response of mussel hemocytes. *Invertebrate Survival Journal* 3, 40–49.

Castro, R., Piazzon, M.C., Noya, M., Leiro, J.M., Lamas, J. (2008) Isolation and molecular cloning of a fish myeloperoxidase. *Molecular Immunology* 45, 428–437.

Cerenius, L., Söderhäll, K. (2004) The prophenoloxidase-activating system in invertebrates. *Immunological Reviews* 198, 116–126.

Cerenius, L., Lee, B.L., Söderhäll, K. (2008) The proPO-system: pros and cons for its role in invertebrate immunity. *Trends in Immunology* 29, 263–271.

Conte, A., Ottaviani, E. (1995) Nitric oxide synthase activity in molluscan hemocytes. *FEBS Letters* 365, 120–124.

Cross, A.R., Rae, J., Curnutte, J.T. (1995) Cytochrome b-245 of the neutrophil superoxide-generating system contains two nonidentical hemes. Potentiometric studies of a mutant form of gp91phox. *Journal of Biology and Chemistry* 270, 17075–17077.

De Deken, X., Wang, D., Many, M.C. et al. (2000) Cloning of two human thyroid cDNAs encoding new members of the NADPH oxidase family. *Journal of Biology and Chemistry* 275, 23227–23233.

Donaghy, L., Lambert, C., Choi, K.-S., Soudant, P. (2009) Hemocytes of the carpet shell clam (*Ruditapes decussatus*) and the Manila clam (*Ruditapes philippinarum*): Current knowledge and future prospects. *Aquaculture* 297, 10–24.

Dubilier, N., Bergin, C., Lott, C. (2008) Symbiotic diversity in marine animals: the art of harnessing chemosynthesis. *Nature Reviews Microbiology* 6, 725–740.

Dupuy, C., Ohayon, R., Valent, A., Noel-Hudson, M.S., Deme, D., Virion, A. (1999) Purification of a novel flavoprotein involved in the thyroid NADPH oxidase. Cloning of the porcine and human cdnas. *Journal of Biology and Chemistry* 274, 37265–37269.

Edens, W.A., Sharling, L., Cheng, G. et al. (2001) Tyrosine cross-linking of extracellular matrix is catalyzed by Duox, a multidomain oxidase/peroxidase with homology to the phagocyte oxidase subunit gp91phox. *Journal of Cell Biology* 154, 879–891.

Fairfield, A.S., Eaton, J.W., Meshnick, S.R. (1986) Superoxide dismutase and catalase in the murine malaria, *Plasmodium berghei*: content and subcellular distribution. *Archives of Biochemistry and Biophysics* 250, 526–529.

Fernández-Robledo, J.A., Schott, E.J., Vasta, G.R. (2008) *Perkinsus marinus* superoxide dismutase 2 (PmSOD2) localizes to single-membrane subcellular compartments. *Biochemical and Biophysical Research Communications* 375, 215–219.

Freeman, B. (1994) Free radical chemistry of nitric oxide. Looking at the dark side. *Chest* 105, 79S–84S.

Gómez-León, J., Villamil, L., Lemos, M.L., Novoa, B., Figueras, A. (2005) Isolation of *Vibrio alginolyticus* and *Vibrio splendidus* from *Aquacultured* carpet shell clam (*Ruditapes decussatus*) larvae associated with mass mortalities. *Applied and Environmental Microbiology* 71, 98–104.

Gourdon, I., Guérin, M.C., Torreilles, J., Roch, P. (2001) Nitric oxide generation by hemocytes of the mussel *Mytilus galloprovincialis*. *Nitric Oxide* 5, 1–6.

Grasberger, H., Refetoff, S. (2006) Identification of the maturation factor for dual oxidase. Evolution of an eukaryotic operon equivalent. *Journal of Biology and Chemistry* 281, 18269–18272.

Haag, E.S., Sly, B.J., Andrews, M.E., Raff, R.A. (1999) Apextrin, a novel extracellular protein associated with larval ectoderm evolution in Heliocidaris erythrogramma. *Developmental Biology* 211, 77–87.

Hégaret, H., Wikfors, G.H., Soudant, P. (2003) Flow cytometric analysis of haemocytes from eastern oysters, *Crassostrea virginica*, subjected to a sudden temperature elevation: II. Haemocyte functions: aggregation, viability, phagocytosis, and respiratory burst. *Journal of Experimental Marine Biology and Eology* 293, 249–265.

Heyworth, P.G., Cross, A.R., Curnutte, J.T. (2003) Chronic granulomatous disease. *Current Opinion in Immunology* 15, 578–584.

Hibino, T., Loza-Coll, M., Messier, C. et al. (2006) The immune gene repertoire encoded in the purple sea urchin genome. *Developmental Biology* 300, 349–365.

Hiruta, J., Mazet, F., Ogasawara, M. (2006) Restricted expression of NADPH oxidase/peroxidase gene (Duox) in zone VII of the ascidian endostyle. *Cell Tissue Research* 326, 835–541.

Humphries, J.E., Yoshino, T.P. (2008) Regulation of hydrogen peroxide release in circulating hemocytes of the planorbid snail *Biomphalaria glabrata*. *Developmental and Comparative Immunology* 32, 554–562.

Inohara, N., Chamaillard, M., McDonald, C., Nunez, G. (2005) NOD-LRR proteins: role in host-microbial interactions and inflammatory disease. *Annual Reviews in Biochemistry* 74, 355–383.

Inoue, Y., Ogasawara, M., Moroi, T. et al. (2005) Characteristics of NADPH oxidase genes (Nox2, p22, p47, and p67) and Nox4 gene expressed in blood cells of juvenile *Ciona intestinalis*. *Immunogenetics* 57, 520–534.

Irazoqui, J.R., Urbach, J.M., Ausubel, F.M. (2009) Evolution of host innate defence: insights from *Caenorhabditis elegans* and primitive invertebrates. *Nature Reviews Immunology* 10, 47–58.

Ito, T., Matsutani, T., Mori, K., Nomura, T. (1992) Phagocytosis and hydrogen peroxide production by phagocytes of the sea urchin *Strongylocentrotus nudus*. *Developmental and Comparative Immunology* 16, 287–294.

Iwanaga, S. (2002) The molecular basis of innate immunity in the horseshoe crab. *Current Opinion in Immunology* 14, 87–95.

Johnson, P.T. (1969) The coelomic elements of sea urchins (*Strongylocentrotus*). I. The normal coelomocytes; their morphology and dynamics in hanging drops. *Journal of Invertebrate Pathology* 13, 25–41.

Kawahara, T., Quinn, M.T., Lambeth, J.D. (2007) Molecular evolution of the reactive oxygen-generating NADPH oxidase (Nox/Duox) family of enzymes. *BMC Evolutionary Biology* 7, 109.

Klebanoff, S.J. (2005) Myeloperoxidase: friend and foe. *Journal of Leukocyte Biology* 77, 598–625.

Kumazawa, N.H., Morimoto, N., Okamoto, Y. (1993) Luminol-dependent chemioluminescence in hemocytes derived from marine and estuarine mollusks. *Journal of Veterinary Medical Science* 55, 287–290.

Labreuche, Y., Lambert, C., Soudant, P., Boulo, V., Huvet, A., Nicolas, J.-L. (2006a) Cellular and molecular hemocyte responses of the Pacific oyster, *Crassostrea gigas*, following bacterial infection with *Vibrio aestuarianus* strain 01/32. *Microbes and Infection* 8, 2715–2724.

Labreuche, Y., Soudant, P., Gonçalves, M., Lambert, C., Nicolas, J.-L. (2006b) Effects of extracellular products from the pathogenic *Vibrio aestuarianus* strain 01/32 on lethality and cellular immune responses of the oyster *Crassostrea gigas*. *Developmental and Comparative Immunology* 30, 367–379.

LaFleur, G.J., Horiuchi, Y., Wessel, G.M. (1998) Sea urchin ovoperoxidase: oocyte-specific member of a heme-dependent peroxidase superfamily that functions in the block to polyspermy. *Mechanisms of Development* 70, 77–89.

Lambert, C., Nicolas, J.L. (1998) Specific inhibition of chemiluminescent activity by pathogenic vibrios in hemocytes of two marine bivalves: *Pecten maximus* and *Crassostrea gigas*. *Journal of Invertebrate Pathology* 71, 53–63.

Lambeth, J.D. (2002) Nox/Duox family of nicotinamide adenine dinucleotide (phosphate) oxidases. *Current Opinion in Hematology* 9, 11–17.

Lambeth, J.D., Cheng, G., Arnold, R.S., Edens, W.A. (2000) Novel homologs of gp91phox. *Trends in Biochemical Sciences* 25, 459–461.

Lambeth, J.D., Kawahara, T., Diebold, B. (2007) Regulation of Nox and Duox enzymatic activity and expression. *Free Radical Biology and Medicine* 43, 319–331.

Le Gall, G., Chagot, D., Mialhe, E., Grizel, H. (1988) Branchial Rickettsiales-like infection associated with a mass mortality of sea scallop *Pecten maximus*. *Diseases of Aquatic Organisms* 4, 229–232.

Li, C., Ni, D., Song, L., Zhao, J., Zhang, H., Li, L. (2008) Molecular cloning and characterization of a catalase gene from Zhikong scallop *Chlamys farreri*. *Fish and Shellfish Immunology* 24, 26–34.

Lopez, C., Carballal, M.J., Azevedo, C., Villalba, A. (1997) Enzyme characterisation of the circulating haemocytes of the carpet shell clam, *Ruditapes decussatus* (Mollusca:bivalvia). *Fish and Shellfish Immunology* 7, 595–608.

Mandell, G.L. (1975) Catalase, Superoxide Dismutase, and Virulence of *Staphylococcus aureus*. In vitro and in vivo studies with emphasis on Staphylococcal-leukocyte interaction. *Journal of Clinical Investigation* 55, 561–566.

Maru, Y., Nishino, T., Kakinuma, K. (2005) Expression of Nox genes in rat organs, mouse oocytes, and sea urchin eggs. *DNA Sequence* 16, 83–88.

McFall-Ngai, M.J. (1999) Consequences of evolving with bacterial symbionts: Insights from the squid-*Vibrio* associations. *Annual Review of Ecology and Systematics* 30, 235–256

Meyer, M., Schreck, R., Baeuerle, P.A. (1993) H_2O_2 and antioxidants have opposite effects on activation of NF-xB and AP-1 in intact cells: AP-1 as secondary antioxidant-responsive factor. *The EMBO Journal* 12, 2005–2015.

Moss, B., Allam, B. (2006) Fluorometric measurement of oxidative burst in lobster hemocytes and inhibiting effect of pathogenic bacteria and hypoxia. *Journal of Shellfish Research* 25, 1051–1057.

Mount, A.S., Wheeler, A.P., Paradkar, R.P., Snider, D. (2004) Hemocyte-mediated shell mineralization in the eastern oyster. *Science* 304, 297–300.

Nauseef, W.M. (2008) Biological roles for the NOX family NADPH oxidases. *Journal of Biology and Chemistry* 283, 16961–16965.

Oliveira, D.M., Silva-Teixeira, D.N., Goes, A.M. (1999) Evidence for nitric oxide action on in vitro granuloma formation through pivotal changes in MIP-1a and IL-10 release in human schistosomiasis. *Nitric Oxide* 3, 162–171.

Oliveira-Marques, V., Marinho, H.S., Cyrne, L., Antunes, F. (2009) Role of hydrogen peroxide in NF-κB Activation: From inducer to modulator. *Antioxidants and Redox Signaling* 11, 2223–2243.

Ottaviani, E., Paeman, L.R., Cadet, P., Stefano, G.B. (1993) Evidence for nitric oxide production and utilization as a bacteriocidal agent by invertebrate immunocytes. *European Journal of Pharmacology* 248, 319–324.

Perry, G., Epel, D. (1981) Ca^{2+}-stimulated production of H_2O_2 from naphthoquinone oxidation in *Arbacia* eggs. *Experimental Cell Research* 134, 65–72.

Pipe, R.K., Farley, S.R., Coles, J.A. (1997) The separation and characterisation of haemocytes from the mussel *Mytilus edulis*. *Cell and Tissue Research* 289, 537–545.

Pourova, J., Kottova, M., Voprsalova, M., Pour, M. (2010) Reactive oxygen and nitrogen species in normal physiological processes. *Acta Physiologica* 198, 15–35.

Rast, J.P., Smith, L.C., Loza-Coll, M., Hibino, T., Litman, G.W. (2006) Genomic Insights into the Immune System of the Sea Urchin. *Science* 314, 952–956.

Reeves, E.P., Nagl, M., Godovac-Zimmermann, J., Segal, A.W. (2003) Reassessment of the microbicidal activity of reactive oxygen species and hypochlorous acid with reference to the phagocytic vacuole of the neutrophil granulocyte. *Journal of Medical Microbiology* 52, 643–651.

Roch, P. (1999) Defense mechanisms and disease prevention in farmed marine invertebrates. *Aquaculture* 172, 125–145.

Rosenstiel, P., Jacobs, G., Till, A., Schreiber, S. (2008) NOD-like receptors: Ancient sentinels of the innate immune system. *Cellular and Molecular Life Sciences* 65, 1361–1377.

Rosenstiel, P., Philipp, E.E.R., Schreiber, S., Bosch, T.C.G. (2009) Evolution and function of innate immune receptors – insights from marine invertebrates. *Journal of Innate Immunity* 1, 291–300.

Ruby, E.G., McFall-Ngai, M.J. (1999) Oxygen-utilizing reactions and symbiotic colonization of the squid light organ by *Vibrio fischeri*. *Trends in Microbiology* 7, 414–420.

Sarrias, M.R., Gronlund, J., Padilla, O., Madsen, J., Holmskov, U., Lozano, F. (2004) The Scavenger Receptor Cysteine-Rich (SRCR) domain: an ancient and highly conserved protein module of the innate immune system. *Critical Reviews in Immunology* 24, 1–37.

Segal, A.W., Jones, O.T., Webster, D., Allison, A.C. (1978) Absence of a newly described cytochrome b from neutrophils of patients with chronic granulomatous disease. *Lancet* 2, 446–449.

Service, M., Wardlaw, A.C. (1984) Echinochrome-A as a bactericidal substance in the coelomic fluid of *Echinus esculentus* (L). *Comparative Biochemistry and Physiology B* 79, 161–165.

Smith, L.C., Rast, J.P., Brockton, V. et al. (2006) The sea urchin immune system. *Invertebrate Survival Journal* 3, 25–39.

Stein, E.A., Cooper, E.L. (1983) Inflammatory Responses in Annelids. *American Journal of Zoology* 23, 145–156.

Sumimoto, H. (2008) Structure, regulation and evolution of Nox-family NADPH oxidases that produce reactive oxygen species. *FEBS Journal* 275, 3249–3277.

Tafalla, C., Novoa, B., Figueras, A. (2002) Production of nitric oxide by mussel (*Mytilus galloprovincialis*) hemocytes and effect of exogenous nitric oxide on phagocytic functions. *Comparative Biochemistry and Physiology B* 132, 423–431.

Tafalla, C., Gómez-León, J., Novoa, B., Figueras, A. (2003) Nitric oxide production by carpet shell clam (*Ruditapes decussatus*) hemocytes. *Developmental and Comparative Immunology* 27, 197–205.

Tahseen, Q. (2009) Coelomocytes: Biology and possible immune functions in invertebrates with special remarks on nematodes. *International Journal of Zoology* 2009, 1–13.

Takeda, K., Kaisho, T., Akira, S. (2003) Toll-like receptors. *Annual Review of Immunology* 21, 335–376.

Thannickal, V.J., Fanburg, B.L. (2000) Reactive oxygen species in cell signaling. *American Journal of Physiology* 279; L1005–L1028.

Tonks, N.K. (2005) Redox redux: revisiting PTPs and the control of cell signaling. *Cell* 121, 667–670.

Victor, V.M., Rocha, M., Esplugues, J.V., Fuente, M.D.L. (2005) Role of Free Radicals in Sepsis: Antioxidant Therapy. *Current Pharmaceutical Design* 11, 3141–3158.

Visick, K.L., McFall-Ngai, M.J. (2000) An exclusive contract: Specificity in the *Vibrio fischeri–Euprymna scolopes* partnership. *Journal of Bacteriology* 182, 1779–87.

Wang, F., Yang, H., Gabr, H.R., Gao, F. (2008a) Immune condition of *Apostichopus japonicus* during aestivation. *Aquaculture* 285, 238–243.

Wang, F., Yang, H., Gao, F., Liu, G. (2008b) Effects of acute temperature or salinity stress on the immune response in sea cucumber, *Apostichopus japonicus*. *Comparative Biochemistry and Physiology A* 151, 491–498.

Wong, J.L., Wessel, G.M. (2005) Reactive oxygen species and Udx1 during early sea urchin development. *Developmental Biology* 288, 317–333.

Wong, J.L., Créton, R., Wessel, G.M. (2004) The oxidative burst at fertilization is dependent upon activation of the dual oxidase Udx1. *Developmental Cell* 7, 801–814.

Wootton, E.C., Dyrynda, E.A., Ratcliffe, N.A. (2003) Bivalve immunity: comparisons between the marine mussel (*Mytilus edulis*), the edible cockle (*Cerastoderma edule*) and the razor-shell (*Ensis siliqua*). *Fish and Shellfish Immunology* 15, 195–210.

Xylander, W.E.R. (2009) Hemocytes in Myriapoda (Arthropoda): a review. *Invertebrate Survival Journal* 6, 114–124.

Chapter 18

ATTACK AND DEFENSE: REACTIVE OXYGEN AND NITROGEN SPECIES IN TELEOST FISH IMMUNE RESPONSE AND THE COEVOLVED EVASION OF MICROBES AND PARASITES

Katja Broeg[1] *and Dieter Steinhagen*[2]

[1]Alfred Wegener Institute for Polar and Marine Research, Bremerhaven, Germany
[2]University of Veterinary Medicine Hannover, Centre for Infection Medicine, Fish Disease Research Unit, Hannover, Germany

Fish deal with a broad range of pathogens in their aquatic environment. Evolving with a permanent risk of infection by viruses, bacteria, or parasites from nearly all phyla, fish developed an efficient immune system with multiple humoral and cellular components conferring both innate as well as acquired immunity (Yoder et al. 2002; Levraud and Boudinot 2009). Besides numerous other mechanisms, phagocytic immune cells induce an oxidative burst response to defend the organism against intruders. When activated by invading pathogens or injuries, phagocytes produce reactive oxygen species (ROS) and reactive nitrogen species (RNS). These mechanisms involve two different types of enzymes: NAPDH (reduced nicotinamide adenine dinucleotide phosphate) oxidases (NOX) and nitric oxide ($NO^{\bullet}$) synthases (NOS) (see also Chapter 14). The molecules generated by NOX and NOS also play a pivotal role in cell signaling and in the process of wound healing. While evolution of the present immunological defense system of fish took its time, pathogens have

Oxidative Stress in Aquatic Ecosystems, First Edition. Edited by Doris Abele, José Pablo Vázquez-Medina, and Tania Zenteno-Savín.

taken up the arms race and developed strategies to deal with oxidative attacks of their vertebrate hosts.

PHAGOCYTES GENERATING REACTIVE OXYGEN AND NITROGEN SPECIES

Several leukocyte cell types with phagocytic activity exist in fish. All produce oxygen and nitrogen radicals, but differ in their function for innate and acquired immunity. In contrast to higher vertebrates, fish leukocytes develop in hematopoietic tissues of the kidney and not in the bone marrow, which in fishbone does not exist. In teleosts, the main functional and structural homologue is considered to be the head kidney (pronephros), the primordial kidney of higher vertebrate embryos (Zapata 1981; Zapata and Amemiya 2000).

Macrophages

No single characteristic can be used to define a macrophage. Macrophages that develop from monocytes are ubiquitously distributed in the vertebrate body. Each subpopulation of macrophages is suited explicitly for its specific organ or tissue microenvironment. Cell-type-specific functional differences further relate to microenvironmental stimuli, macrophage developmental status, and the immunoregulatory state of the macrophage (Neumann et al. 2001). Thus, macrophages are "in place," distributed in the vertebrate body to act as the first line of defense against pathogens by rapid phagocytosis of pathogen invaders, but also as potent effector cells.

Macrophages occur in a resting, primed, or activated state. As known for mammals and, since recently, for fish (Joerink et al. 2006), their activation can be subdivided into a classic and an alternative state. Classically activated macrophages (caMF) mainly take part in type-I immune responses against intracellular pathogens by the production of ROS and NO•. Alternatively activated macrophages (aaMF) play an important role in type-II immune responses against extracellular pathogens and are characterized by increased phagocytic activity and enhanced gene expression of major histocompatibility complex (MHC) class II molecules. Arginase plays a key role in the actions of aaMF and is the characteristic enzyme for this type of response (Gordon 2003; Mantovani et al. 2004). Arginase catalyses the hydrolysis of l-arginine to l-ornithine and urea. Ornithine is the precursor of polyamines, which are essential for initiation and progression of the macrophageal cell cycle, and at the same time is the precursor of collagen. Therefore, ornithine promotes cell replication and wound healing.

There are a number of important functions of arginase that relate this enzyme to the function of macrophages. Arginase activity can deplete extracellular and intracellular l-arginine, necessary for proliferation and survival of both pathogen and host cells (Vincendeau et al. 2003). Further, intracellular l-arginine is also the substrate of inducible NOS (iNOS) (Vincendeau et al. 2003). Arginase and iNOS are co-expressed in lipopolysaccharide (LPS) stimulated macrophages (Mori and Gotoh 2000), which enhances NO• production when arginase gene expression is inhibited (Chang et al. 1998) so that arginase modulates NO• production by caMF.

Cultured macrophages derived from carp pronephros could be differentially stimulated to increase either arginase activity or NO• production. Arginase activity was induced by cyclic adenosine monophosphate (cAMP) stimulation; NO• production could be induced by LPS, bacterial mimetics, and well known inducers of iNOS gene expression (Joerink et al. 2006).

Granulocytes (Neutrophils and Acidophils)

Granulocytes represent leukocytes that are rapidly recruited from blood circulation and hematopoietic tissue to areas of inflammation. They take up pathogens and foreign particles and release antimicrobial molecules like ROS and RNS directly at the site of infection and inflammation. Chemotaxis, adherence, and phagocytosis are prerequisites for the successful destruction of pathogens. Extracellular killing plays a minor role (Neumann et al. 2001). Granulocytes are the first to arrive at the site of infection to prevent spreading of pathogenic organisms in the body of the host. In a second phase, granulocytes with high killing capacity migrate to the site of infection, followed by the arrival of monocytes and macrophages. This suite of events has been confirmed for many fish species (Afonso et al. 1998; Waterstrat et al. 1991). Oxidants produced by granulocytes act as potent antimicrobial agents and directly kill microbial pathogens. In addition, in granulocytes ROS and RNS serve as modulators of protein and lipid kinases and phosphatases, membrane receptors, ion channels, and transcription factors, including NF-κB. The latter regulates expression of key cytokines and chemokines that further modulate the inflammatory response. During the inflammatory response, ROS and RNS modulate phagocytosis, secretion, gene expression, and apoptosis (Fialkow et al. 2007).

Taken together, white blood cells, able to generate ROS, can engulf bacteria and kill them principally through oxidative burst, often in concerted action with RNS formation, especially of peroxynitrite. Moreover, neutrophil granulocytes contain myeloperoxidase (MPO) in their cytoplasmic granules (Afonso et al. 1998). In the presence of halide ions and H_2O_2, MPO is able to kill bacteria by halogenations of the bacterial cell walls as well as by production of hypohalite ions (Klebanoff and Clark 1978; cf. Ellis 1999) (Fig. 18.1).

In many parasitic infections both caMF and aaMF play a central role. Carp infected with *Trypanoplasma borreli* responded with an increased $NO^{\bullet}$ production by caMF and died from the infection, whereas *Trypanosoma-carassii*-infected carp responded with increased arginase activation by aaMF and survived the infection (Joerink et al. 2006).

There is also a concerted response of neutrophils and macrophages to bacterial infection (Afonso et al. 1998). Following intraperitoneal injection of sublethal doses of *Yersinia ruckeri* in rainbow trout a rapid

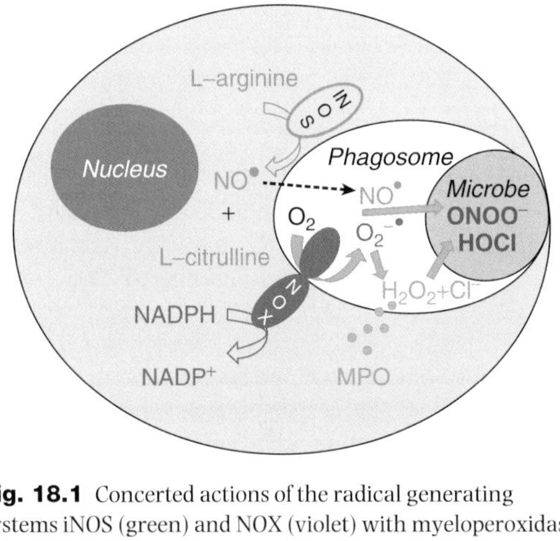

Fig. 18.1 Concerted actions of the radical generating systems iNOS (green) and NOX (violet) with myeloperoxidase (blue, MPO) during phagocytosis leading to the formation of the potent microbicidic products peroxynitrite (red, $ONOO^-$) and hypochloric acid (red, HOCl). See plate section for a color version of this image.

inflammatory reaction ensues with influx of large numbers of neutrophils that phagocytose *Y. ruckeri*. Resident macrophages, which also phagocytosed the bacteria, have been observed to phagocytose neutrophil granulocytes containing bacteria. Furthermore, neutrophils come into close contact with macrophages to which they appear to transfer MPO-containing granules.

In fish, activated macrophages play an important role in the defense against many bacterial pathogens, such as *Aeromonas salmonicida* causing furunculosis, or *Renibacterium salmoninarum* causing bacterial kidney disease (BKD). In an activated status, macrophages exhibit higher production of ROS and increased bactericidal activity.

REACTIVE OXYGEN SPECIES—THE "RESPIRATORY BURST"

The term "respiratory burst" originated from early observations that phagocytes, like neutrophils and macrophages, consume high amounts of oxygen when exposed to bacteria. This consumption was not cyanide-inhibitable and, therefore, not caused by accelerated mitochondrial respiration. Further research investigating the origin of these ROS led to the discovery and the molecular characterization of the phagocytic NADPH oxidase, or "respiratory burst oxidase", abbreviated as "NOX" in the recent literature (Lambeth et al. 2007).

Oxidative killing, which means killing by oxygen, is mediated by a variety of ROS generated during the "respiratory burst" of phagocytes, often in combination with $NO^{\bullet}$ generated by an iNOS (see next paragraph). This burst release of ROS from leukocytes is important for killing microbes but can also cause tissue damage and contribute to the development of chronic inflammation (Nathan 2006). In addition, H_2O_2 has since long been identified as a physiological signaling molecule. In plants H_2O_2 release is one of the earliest responses to infection and helps to activate the host immune responses against pathogens. It also contributes to plant morphogenesis and cell wall lignification (Neill et al. 2002; Apel and Hirt 2004), which wards off pathogen attack. Animal cells use H_2O_2 as a second messenger to regulate transcription, proliferation, or enzyme activity (Bedard and Krause 2007). Exogenous addition of H_2O_2 is able to activate leukocytes, but the physiological effect of H_2O_2 on leukocyte function is still not well understood (Klyubin et al. 1996; Reth 2002).

It has been shown that the reduction of oxygen requires NADPH as reducing equivalent, but in fish NADH is also used, although with a lower efficacy (Shiibashi and Iida, 2001). The reduction is catalyzed by the membrane bound NOX enzyme (Fig. 18.1).

The NOX is dormant in resting cells and can be rapidly activated by a variety of soluble mediators (e.g. chemoattractant peptides or chemokines) and particulate stimuli (e.g. bacteria or immune complexes) that interact with cell-surface receptors. A detailed description of the activation process is found in Chapter 17.

The activity of the respiratory burst and the amount of ROS generated by fish phagocytes is differentially affected by temperature. Cod exposed to an increase of temperature (1°C every 5 days from 10°C up to 19°C) showed no differences in respiratory burst activity during the temperature increase (Perez-Casanova et al. 2008). In rainbow trout exposed to temperatures between 5 and 20°C, respiratory burst activity was higher in fish acclimatized at higher temperature (Nikoskelainen et al. 2004). Moreover, fish kept at 5–10°C had significantly delayed peak of respiratory burst compared to fish kept at 15–20°C. Tilapia (*Oreochromis mossambicus*) acclimated to 27°C exhibited decreased ROS generation of leukocytes when transferred to 19, 31 or 35°C over 24–96 h (Ndong et al. 2006). Even the disease susceptibility increased. The mortality of *Streptococcus iniae*-injected fish held at 19 and 35°C was significantly higher than that of injected-fish held at 23, 27 and 31°C. These findings indicate that different fish species have their individual optimum temperature range with respect to the generation of ROS, and that outside this range disease susceptibility might increase.

Chronic sublethal hypoxia is also an environmental condition that might influence ROS production of phagocytes in fish (Boleza et al. 2001). Decreased oxygen and pH or increased carbon dioxide in blood and tissues of the mummichog, *Fundulus heteroclitus*, ($P_{O_2} = 15$ torr, $P_{CO_2} = 8.0$ torr, pH = 7.0) suppressed ROS production by 58.5–76.0% and reduced bactericidal activity.

Many components of vertebrate immune systems exhibit circadian fluctuations. The zebrafish was used as a model to study molecular pacemakers that may control circadian rhythms of leukocytes (Kaplan et al. 2008). Phagocytosis and the production of ROS by zebrafish leukocytes were found to vary significantly within diurnal cycles. A distinct peak in cellular ROS levels occurred before dawn, whereas the kinetics of respiratory burst responses (i.e., the period of time after stimulation until the maximum response is recorded) was least rapid during this time of the day.

REACTIVE NITROGEN SPECIES AND INDUCIBLE NITRIC OXIDE SYNTHASE

$NO^{\bullet}$ is involved in diverse biological processes as a regulator and effector molecule (Nathan 1992). It may inhibit cell proliferation and mediate antitumor or nonspecific antimicrobial activities by mediating resistance against various pathogens. In addition, it is a neuronal messenger, causes smooth muscle relaxation, and alters platelet function (Nathan 1992).

As part of the immune response, $NO^{\bullet}$ operates through oxidation of thiols, hemes, Fe-S clusters, and other nonheme iron prosthetic groups, and has been proposed to mediate inflammation caused by infection, to be immunosuppressive (Eisenstein et al. 1994), and to have anti-inflammatory effects, e.g. by inhibition of leukocyte adhesion (Gaboury et al. 1993). Similar to ROS, $NO^{\bullet}$ participates in diverse cellular signaling pathways, including regulation of plasma membrane receptors, endocytic pathways, guanosine triphosphate (GTP) binding proteins, ion channels, transcription factors, and tyrosine kinase-mediated signaling (Demple 2002). There are physiologically important interactions between $NO^{\bullet}$ and ROS such as $O_2^{\bullet-}$ (Beckman et al. 1990), which react to form peroxynitrite ($ONOO^-$), a potentially cytotoxic compound.

$NO^{\bullet}$ is synthesized by a series of reactions catalyzed by a group of NOS of which there are three main isoforms: the inducible type (iNOS) first identified in macrophages, and two constitutive types termed neuronal and endothelial NOS (nNOS and eNOS, respectively). The NOS protein contains putative binding sites for heme, tetrahydrobiopterin, calmodulin, flavine mononucleotide, flavine adenine dinucleotide, and NADPH. Phylogenetic analysis, using neighbor joining, showed that the carp iNOS protein clusters together with other vertebrate iNOS proteins. Inducibility of carp iNOS was confirmed by RT-PCR (reverse transcription-polymerase chain reaction) after stimulation of carp phagocytes with lipopolysaccharide or the protozoan blood flagellate *Trypanoplasma borreli*. These stimulators produced

Fig. 1.2 Underwater photograph of shallow coral reef in the Indonesian archipelago near Komodo Island in the year 2000 (Photograph by Michael Lesser.)

Fig. 1.3 Bleached coral (*Montastraea faveolata*) from the shallow coral reefs of Puerto Rico during 2005 bleaching event. (Photograph by Ernesto Weil.)

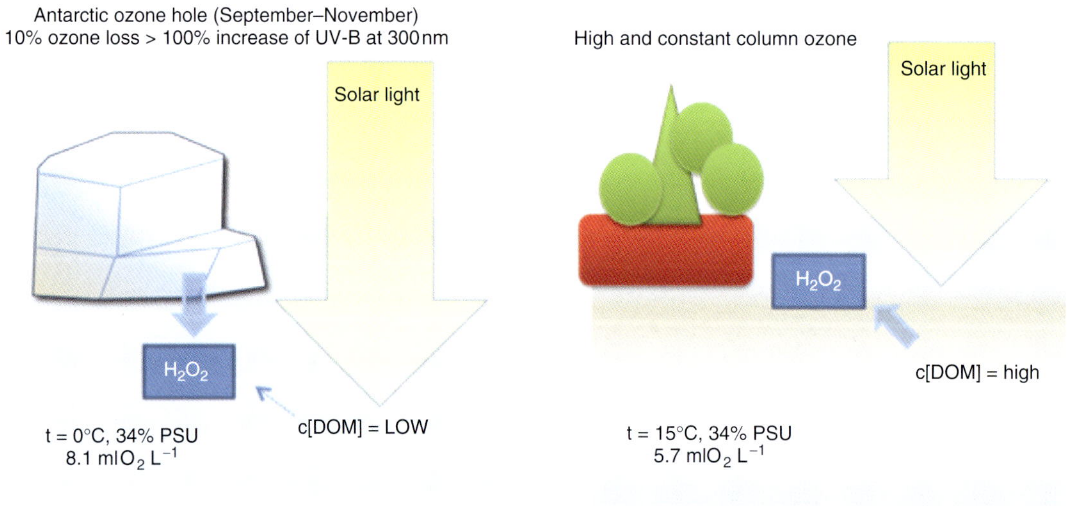

Fig. 2.1 Radiation scenario in polar (left) and temperate regions (right). In oligotrophic polar seas solar light and especially high energy UV-B radiation penetrates up to 60 m deep into the water column. Low DOM (dissolved organic matter) concentrations limit the photochemical formation of H_2O_2. In temperate waters DOM is high and limits light penetration, but also intensifies photochemical H_2O_2 formation in surface waters. In polar regions especially, melting of accumulated snow during sea-ice break-up releases substantial amounts of H_2O_2 into the water column.

Fig. 2.5 Bottom section of a sea-ice core is colored brownish due to the high number of microalgae. (courtesy M. Nicolaus.)

Fig. 2.7 Polar and temperate scallop species.

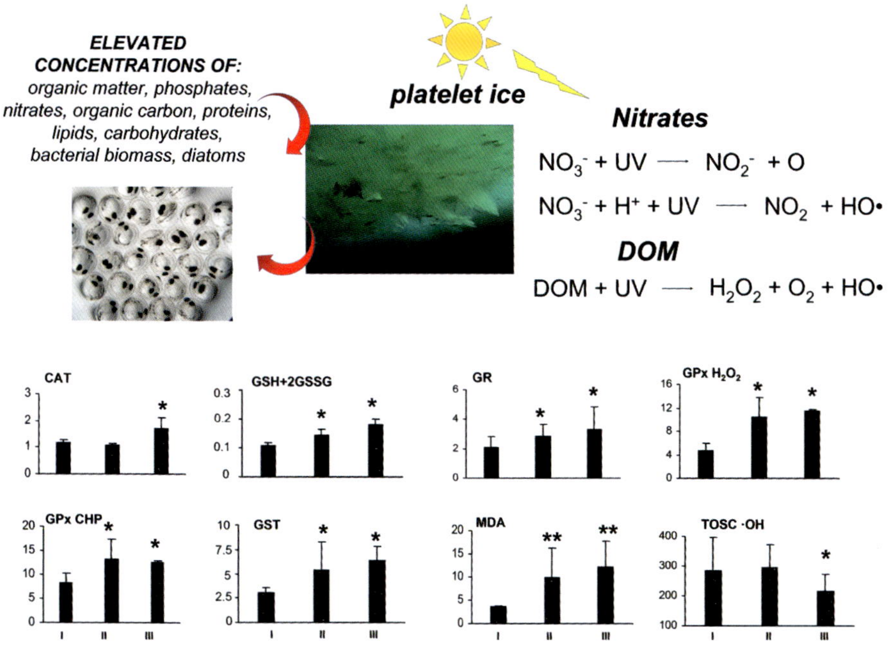

Fig. 2.8 Embryonated eggs of *P. antarcticum* are exposed to increasing pro-oxidant challenge in platelet ice and consequently enhance their antioxidant defenses: CAT (catalase, μmol/min/mg prot), GSH+2GSSG (total glutathione, μmol/g tiss), GR (glutathione reductase, nmol/min/mg prot), $GP \times H_2O_2$ (Se-dependent glutathione peroxidases, nmol/min/mg prot), GP*times*CHP (Se-dependent and Se-independent glutathione peroxidases, nmol/min/mg prot), GST (glutathione S-transferases, nmol/min/mg prot), MDA (malondialdheyde, nmol/g tiss), TOSC •OH (total oxyradical scavenging capacity, UTOSC/mg prot). I, II, III: sampling dates on 2, 9, 16 November respectively.

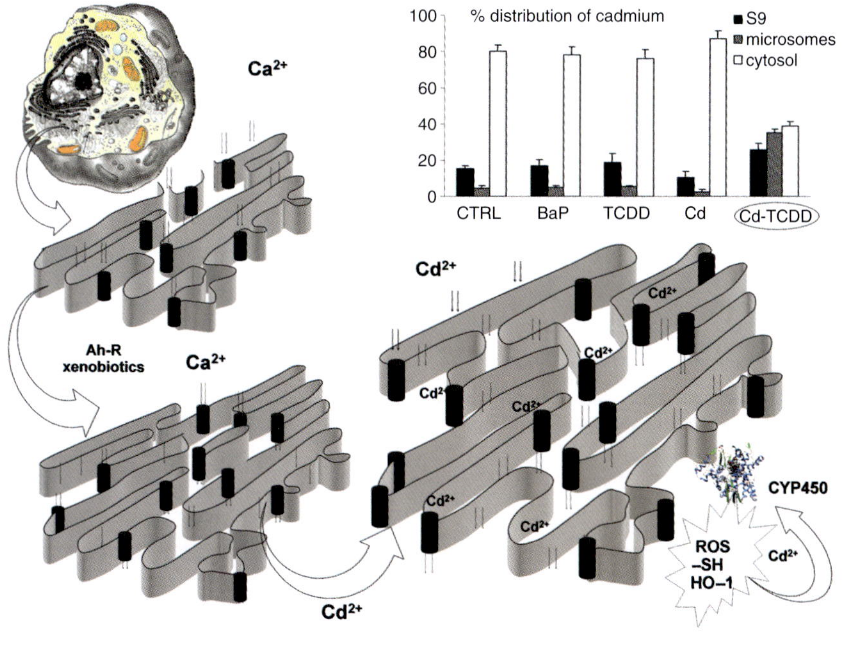

Fig. 2.9 Ah-R xenobiotics induce proliferation of endoplasmic reticulum where Cd can accumulate through Ca^{2+} channels, as confirmed by elevated content of the metal in the microsomal fraction of co-exposed fish. Through post-transcriptional mechanisms, Cd suppresses the induction of cytochrome P450 at protein and catalytic levels.

Fig. 3.2 Example of an intertidal environment at a rocky beach in Pacific Grove, California. Observe the water pools during a low tide (a). Many of these pools are rich with life, such as algae, bivalves, and marine gastropods (b and c). Progressive evaporation of these pools exposes marine invertebrates to increased salinity and temperature, as well as a decrease in dissolved oxygen. (Photographs by Marcelo Hermes-Lima, November 2010.)

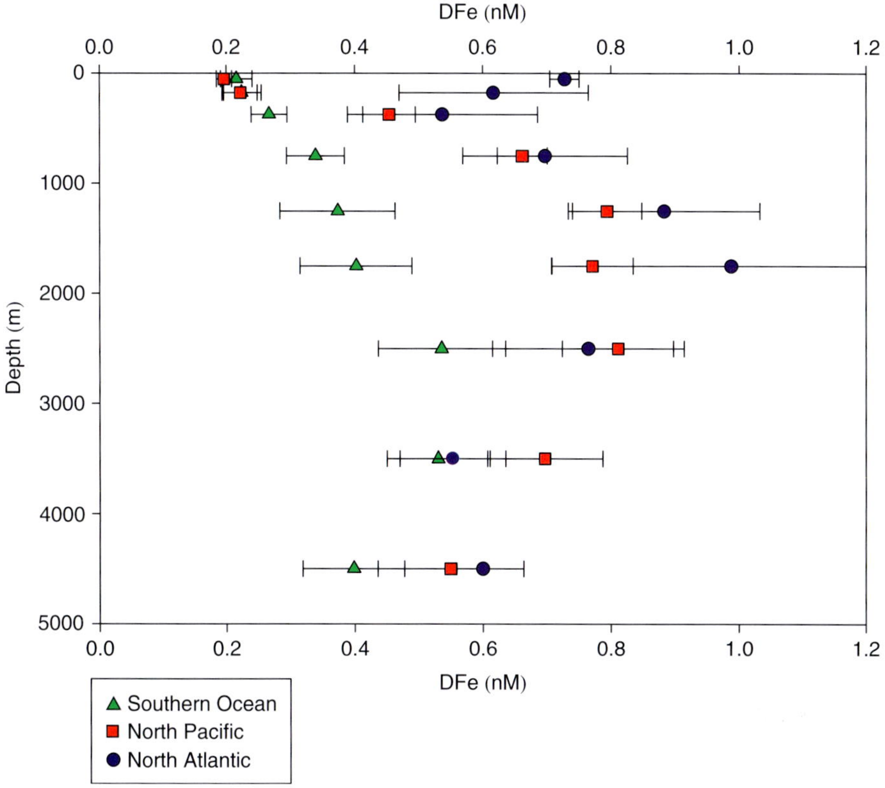

Fig. 8.3 Mean profiles of dissolved iron in the North Atlantic, North Pacific, and Southern Ocean averaged over depth intervals: 0–100 m, 100–250 m, 250–500 m, 500–1000 m, 1000–1500 m, 1500–2000 m, 2000–3000 m, 3000–4000 m, 4000–5000 m. Error bars reflect the 95% confidence interval. (Moore and Braucher 2008 with courtesy of M.B. Klunder.)

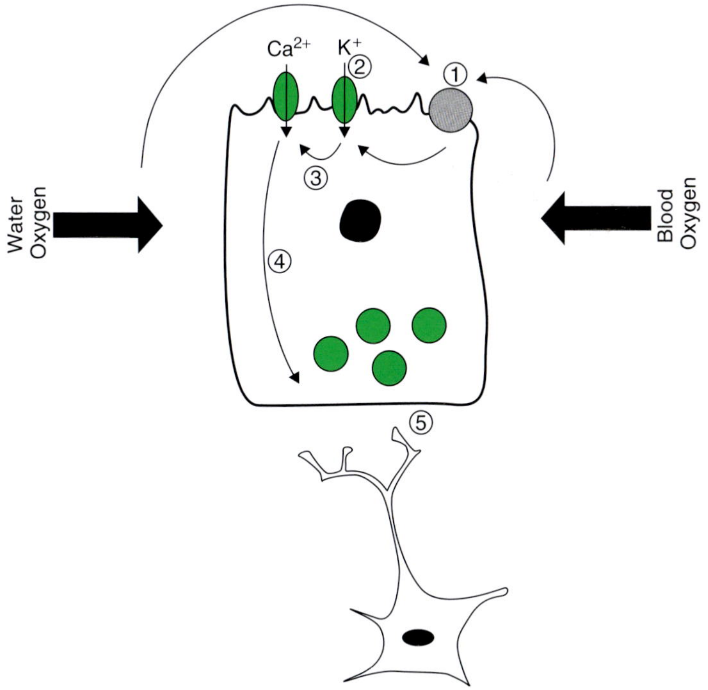

Fig. 12.4 Schematic representation of the function of gill neuroendocrine cells sensing water and blood oxygenation. Oxygen affects, via a primary oxygen sensor – the nature of which is currently unknown (1), the activity of the potassium channel (2). The efflux of potassium via the channel affects the membrane potential (3), which influences calcium flux via a calcium channel (4). The change in intracellular calcium level affects the neurotransmitter release (5), and influences neural activities.

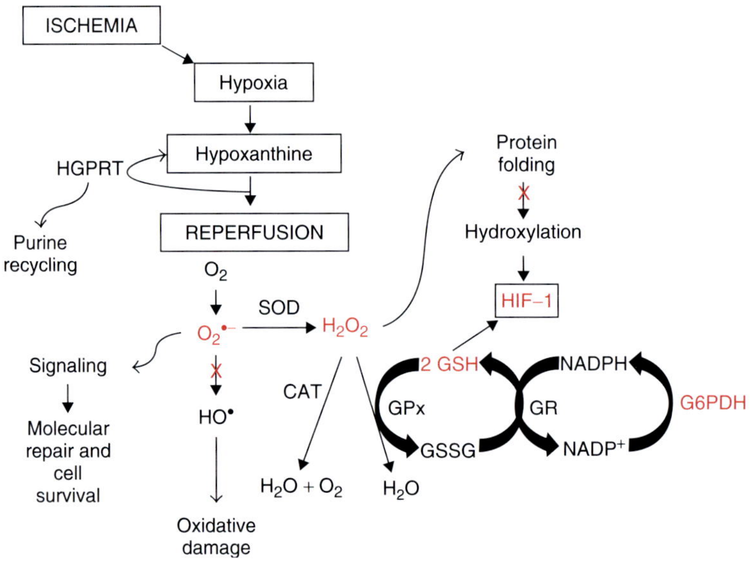

Fig. 13.2 Schematic representation of the contribution of reactive oxygen species and antioxidant defenses in protection of tissues from diving birds and mammals against dive-derived ischemia/reperfusion. $O_2^{\bullet-}$ and H_2O_2 produced in response to dive-induced ischemia/reperfusion act as signaling molecules to induce molecular and cellular protective mechanisms. An enhanced antioxidant system protects diving vertebrates from the deleterious effects of $HO^{\bullet}$. Reactive oxygen species affect protein folding and, therefore, the tridimensional structure of HIF-1α, which precludes it from hydroxylation by prolyl hydroxylases, resulting in its increased constitutive expression in tissues from diving vertebrates. HGPRT = hypoxanthine-guanine phosphoribosyltransferase; O_2 = molecular oxygen; $O_2^{\bullet-}$ = superoxide radical; H_2O_2 = hydrogen peroxide; $HO^{\bullet}$ = hydroxyl radical; H_2O = water; SOD = superoxide dismutase; CAT = catalase; GPx = Glutathione peroxidase; GR = glutathione reductase; G6PDH = glucose 6-phosphate dehydrogenase; GSH = glutathione, reduced; GSSG = Glutathione, oxidized; NADPH = nicotinamide adenine dinucleotide phosphate, reduced; NADP+ = nicotinamide adenine dinucleotide phosphate, oxidized; HIF-1 = hypoxia-inducible factor.

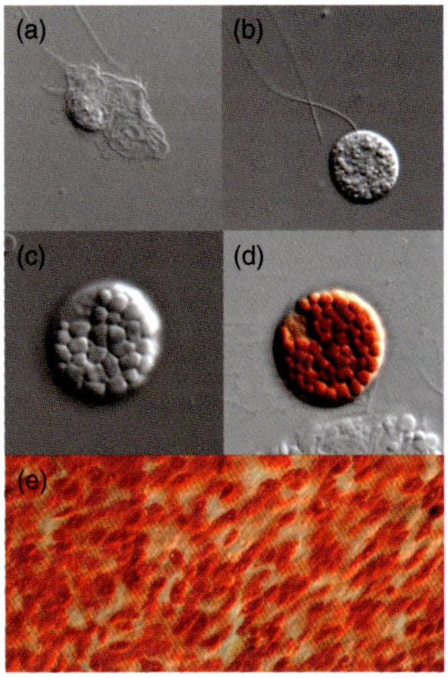

Fig. 17.1 Coelomocytes of the sea urchin *Strongylocentrotus purpuratus*. (a) Phagocytic cells, which diversify into several types, are capable of producing ROS as part of the phagocytosis/killing process. (b) Vibratile cells are flagellated granular cells that participate in clotting and wound healing. (c) Colorless spherule cells are amoeboid cells of unknown function. (d) Red spherule cells are highly motile amoeboid cells that have large granules containing the napthoquinone echinochrome. These cells are active in wound healing and are capable of producing hydrogen peroxide. (e) Intestinal wall with a barrier-like distribution of tissue resident cells with red spherule cell-like structure. (Photographs by Jonathan Rast.)

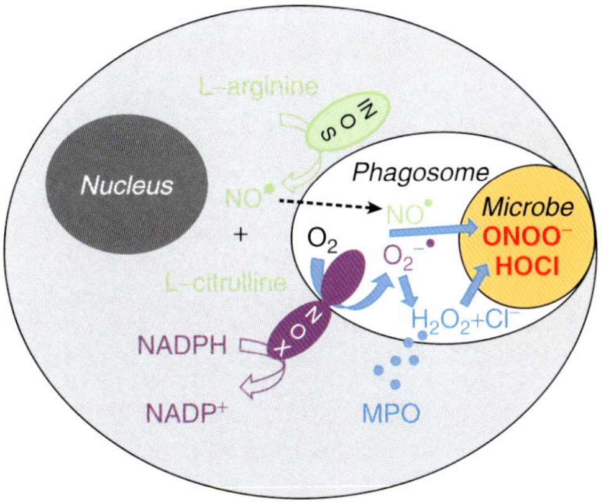

Fig. 18.1 Concerted actions of the radical generating systems iNOS (green) and NOX (violet) with myeloperoxidase (blue, MPO) during phagocytosis leading to the formation of the potent microbicidic products peroxynitrite (red, $ONOO^-$) and hypochloric acid (red, HOCl).

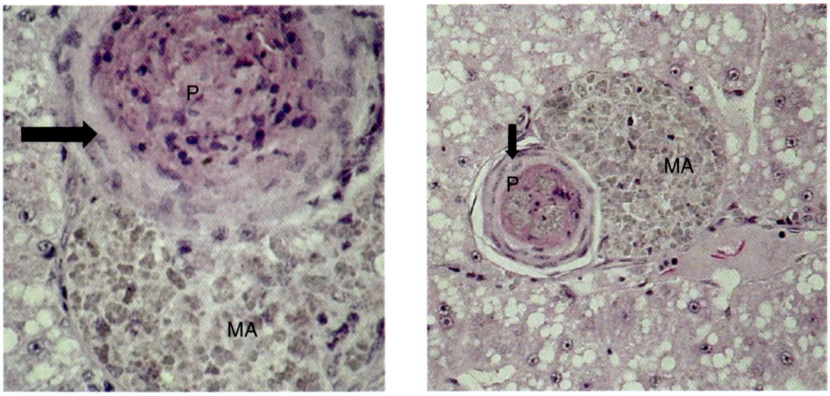

Fig. 18.3 Methacrylate sections of the liver of *Liza aurata* infected with the myxosporean parasite *Myxobolus* sp. MA, macrophage aggregate; P, plasmodium. The border of the plasmodium is formed by epithelioid cells (arrows), which are activated macrophages (see Fig. 18.2; H and E staining). (photographs by Katja Broeg.)

Fig. 19.3 Intertidal mollusks such as the blue mussel (*Mytilus edulis*) routinely experiences transition from (a) active, fully aerobic state during high tides to (b) metabolic rate depression mostly supported by anaerobic metabolism during low tides to avoid desiccation. Other extreme stressors that prompt shell closure and limit gas exchange, such as extremely low or high salinities, hypoxia/anoxia, acute exposure to pollutants (both anthropogenic and natural such as H_2S), or freezing, similarly lead to metabolic arrest. In contrast, some of the strictly subtidal mollusks (such as (c) a nudibranch *Coryphella*) or pelagic species (such as (d) sea angels *Cliona*) are incapable of the metabolic arrest and rely on other, energy-extensive mechanisms of stress protection. (Photographs by Mikhail Fedyuk.)

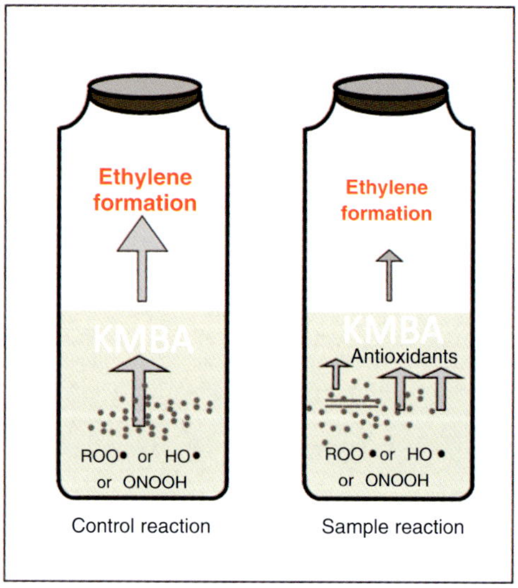

Fig. 26.1 Schematic representation of ethylene formation in the "Control vial", where KMBA is oxidized by ROS, and in the "Sample vial", where cellular antioxidants limit the reaction of ROS with KMBA and thus, ethylene formation.

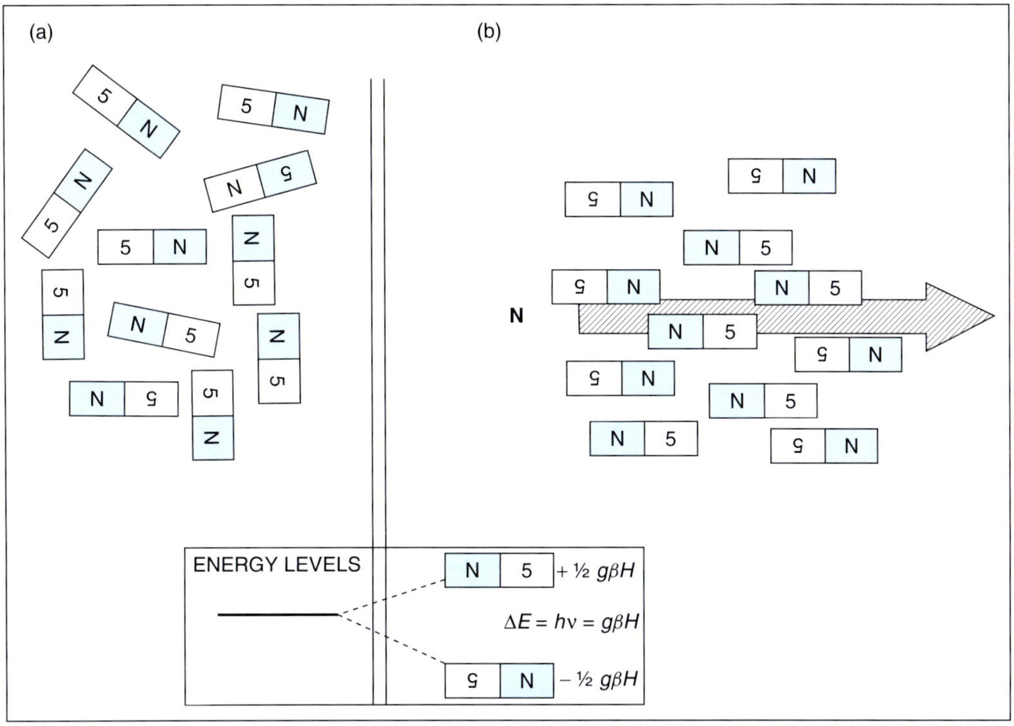

Fig. 35.1 Schematic picture of "free" electrons, in the absence and presence of an external magnetic field. In this simple approximation the spin magnetic moment of a free radical molecular fragment can be represented by that of the unpaired electron alone. (a) In the absence of an external field the spin magnetic moments of free radicals are randomly oriented and are in the same average energy state. (b) In the presence of an external magnetic field the electronic magnets become aligned, according to the laws of space quantization, into one of two allowed orientations: either parallel to the external field, or opposed to it, antiparallel. Inset: Diagram of the energy levels either in the presence or absence of the magnetic field. (Modified from Borg 1976.)

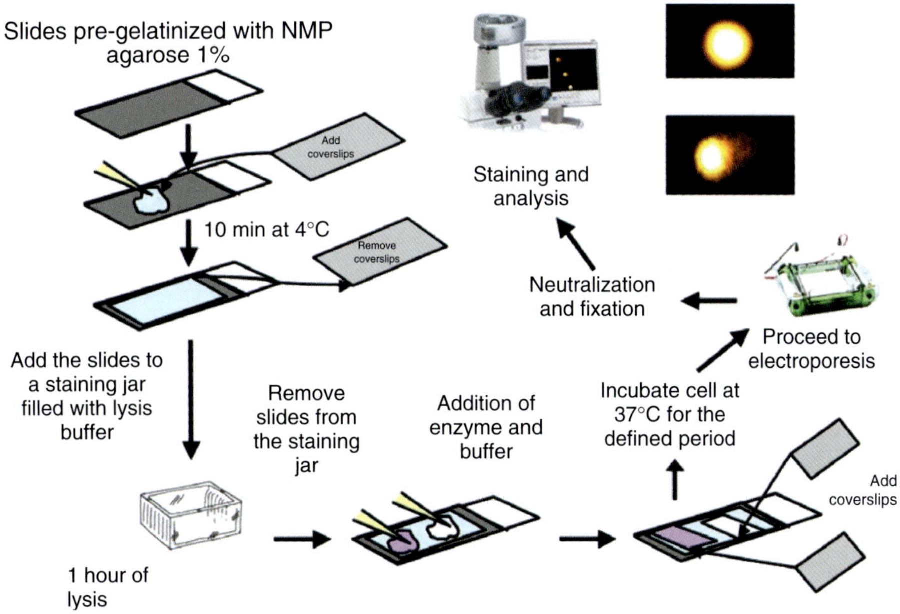

Fig. 37.2 Schematic protocol for comet assay procedure.

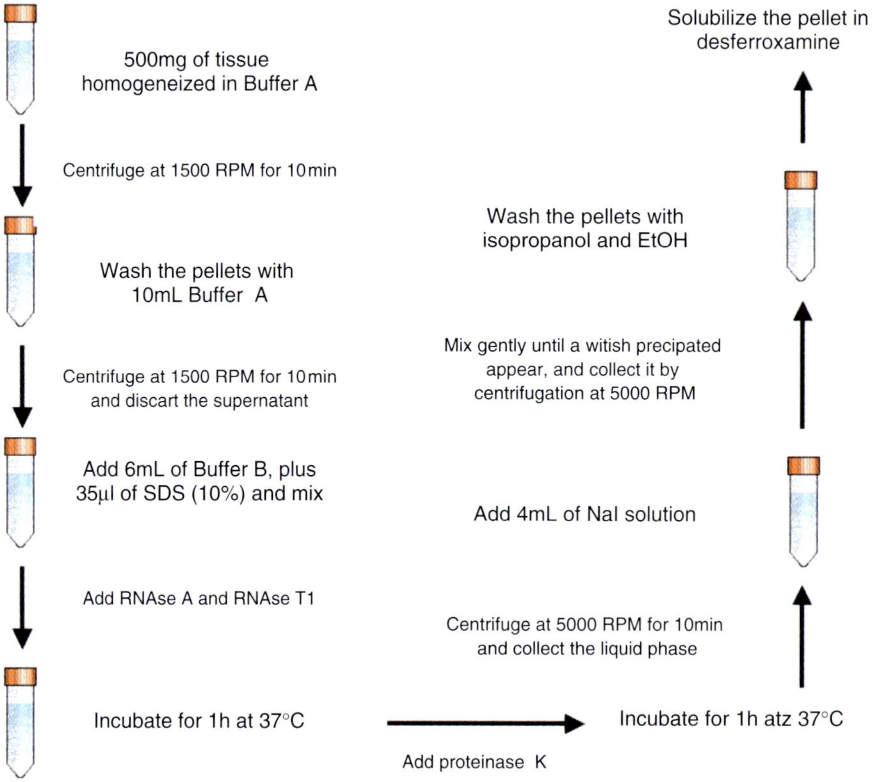
500mg of tissue homogeneized in Buffer A
Centrifuge at 1500 RPM for 10 min
Wash the pellets with 10mL Buffer A
Centrifuge at 1500 RPM for 10 min and discart the supernatant
Add 6mL of Buffer B, plus 35µl of SDS (10%) and mix
Add RNAse A and RNAse T1
Incubate for 1h at 37°C
Add proteinase K
Incubate for 1h atz 37°C
Centrifuge at 5000 RPM for 10min and collect the liquid phase
Add 4mL of NaI solution
Mix gently until a witish precipated appear, and collect it by centrifugation at 5000 RPM
Wash the pellets with isopropanol and EtOH
Solubilize the pellet in desferroxamine

Fig. 37.3 Schematic protocol for DNA extraction.

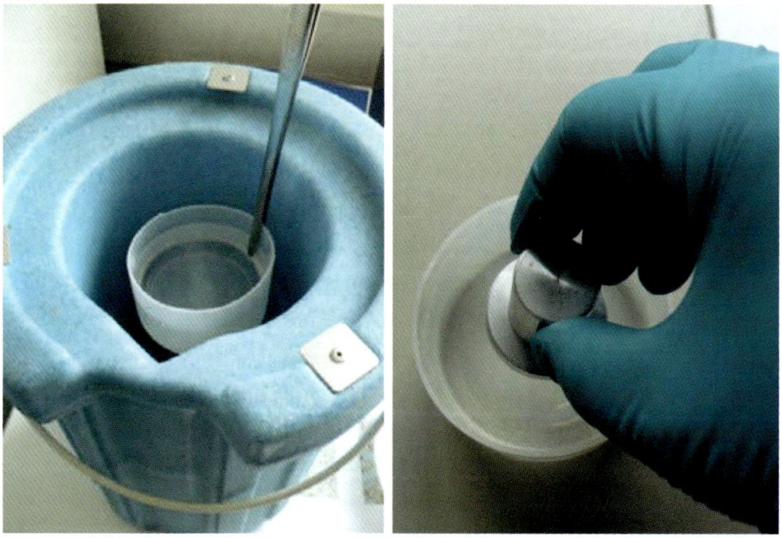

Fig. 39.5 Precooling of hexane followed by the freezing of the tissue on the aluminium chuck. (Photograph by Katja Broeg.)

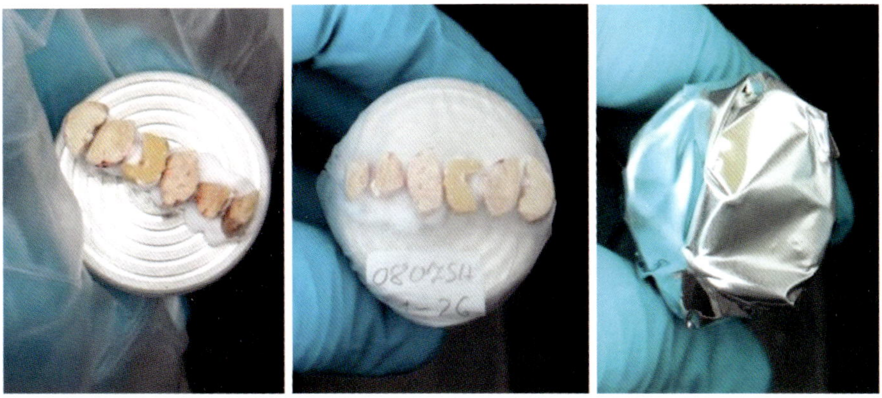

Fig. 39.6 Wrapping of frozen tissue on the chucks in parafilm and aluminium foil for storage in a −80°C freezer. (Photograph by Katja Broeg.)

Fig. 39.7 Working with an automated cryostat. (Photograph by Katja Broeg.)

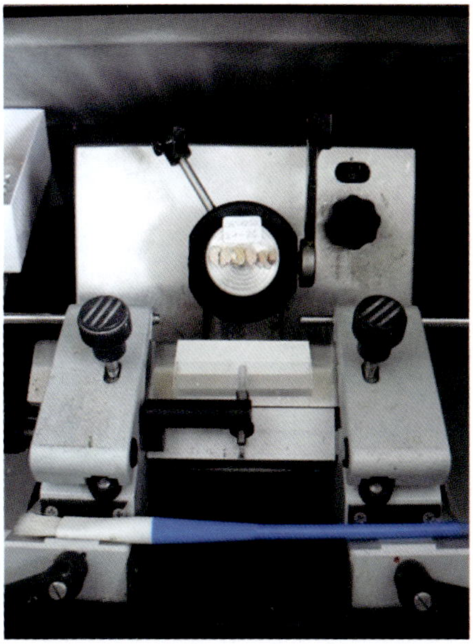

Fig. 39.8 A view into a cryostat cabin with a clumped aluminum chuck. (Photograph by Katja Broeg.)

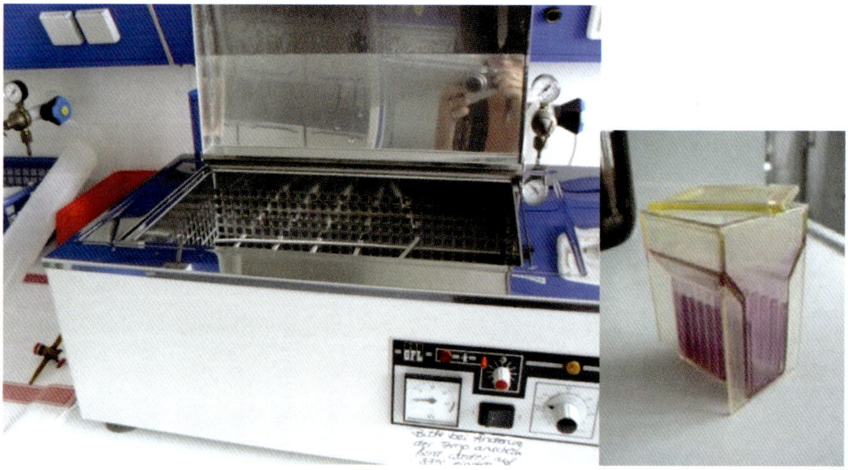

Fig. 39.11 Heated water bath with automated shaking, and a plastic Hellendahl jar. (Photograph by Katja Broeg.)

Fig. 39.12 Positioning of tissue sections in the jars within the water bath at 37°C. (Photograph by Katja Broeg.)

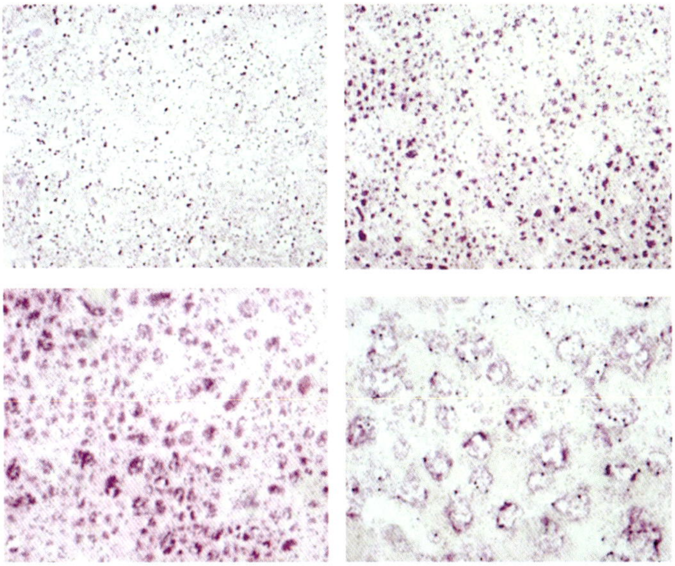

Fig. 39.13 Different sizes of lysosomes (pink) in fish liver (cryostat sections, acid phosphatase, 400×). (Photograph by Katja Broeg.)

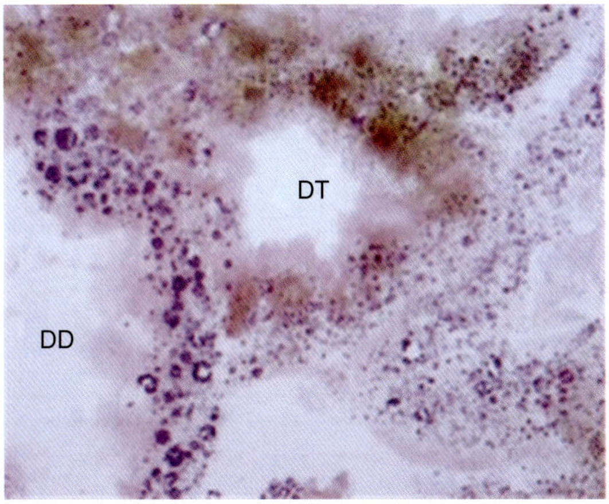

Fig. 39.14 Lysosomes in digestive gland of blue mussel (cryostat section, β-N-acetylhexosaminidase, 400×): DD = digestive duct, DT = digestive tubule.

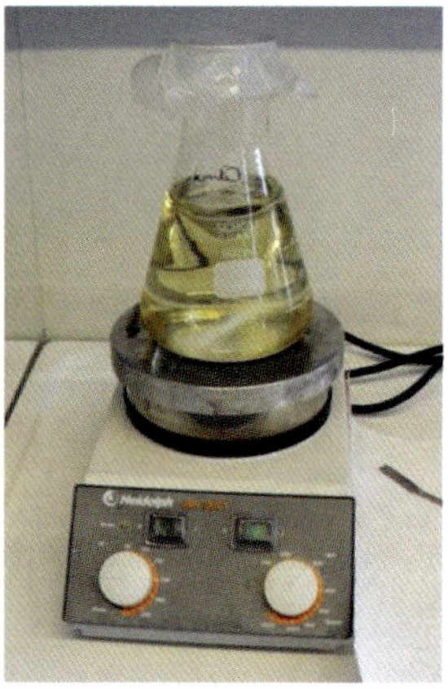

Fig. 39.15 Fast Violet B has to turn from light green/yellow to orange before use. (Photograph by Katja Broeg.)

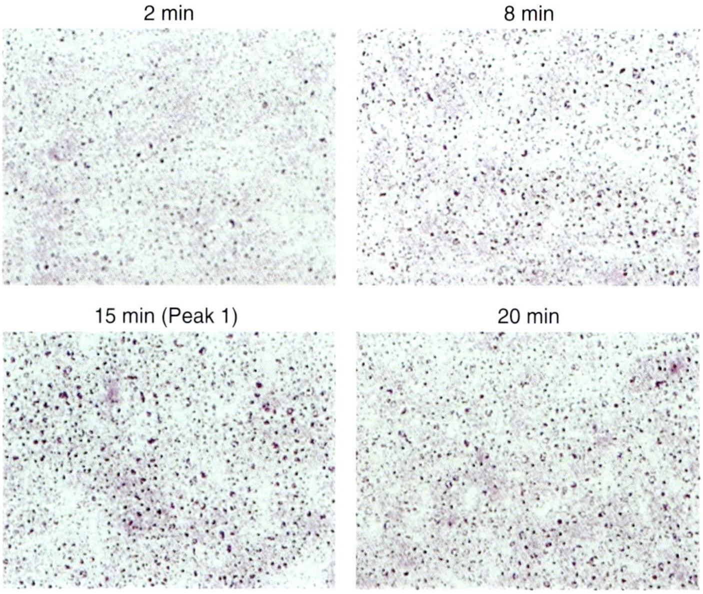

Fig. 39.16 Identification of highest staining intensity in lysosomes (destabilization period/peak) of flounder liver (cryostat sections, AcP, 400×).

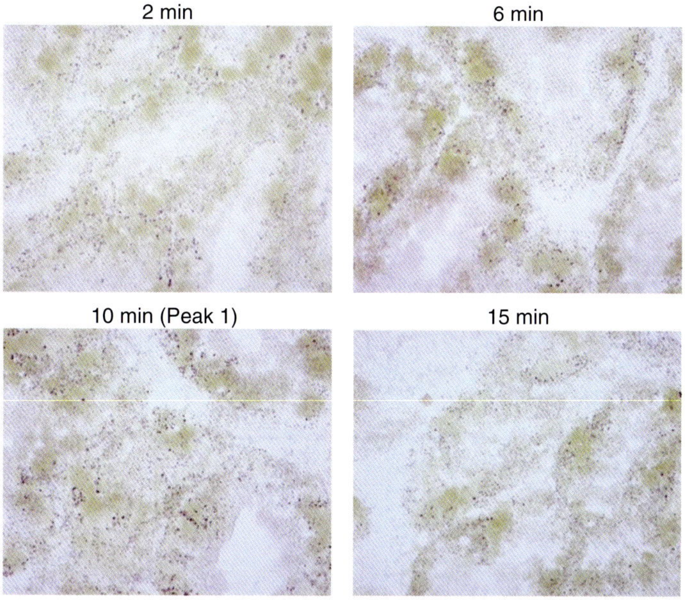

Fig. 39.17 Identification of highest staining intensity in lysosomes (destabilization period/peak) of mussel digestive gland (cryostat sections, Hex, 400×).

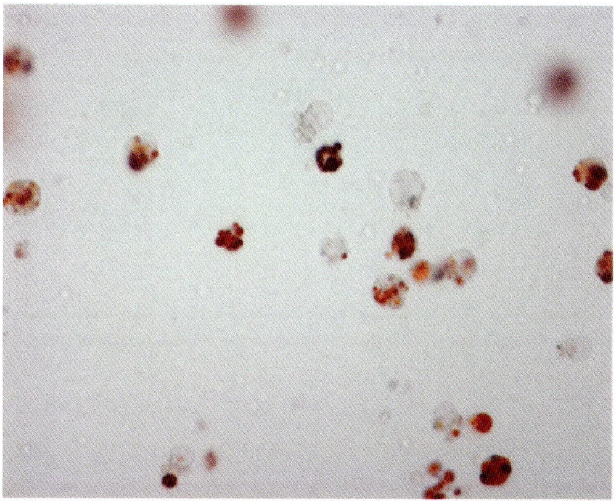

Fig. 39.19 Intact lysosomes in oyster haemocytes. The lysosomes are plainly visible inside the cells. Light microscopy, 400×. (Photograph by Katja Broeg.)

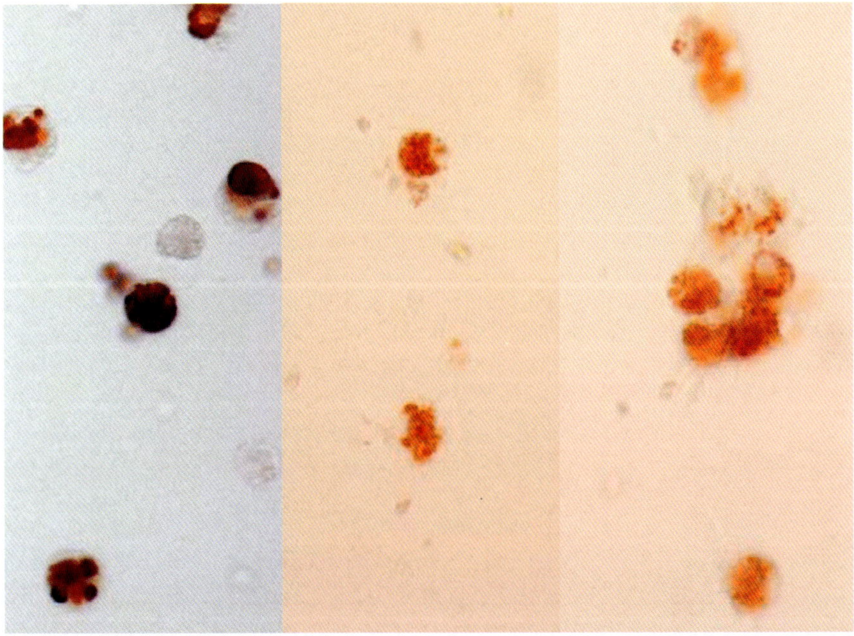

Fig. 39.20 Proceeding breakage of lysosomal membranes and increase of the size of mussel haemocytes. The destabilization is reflected by an even red color of the haemocytes. Lysosomes cannot be identified properly. Light microscopy, 400×. (Photograph by Katja Broeg.)

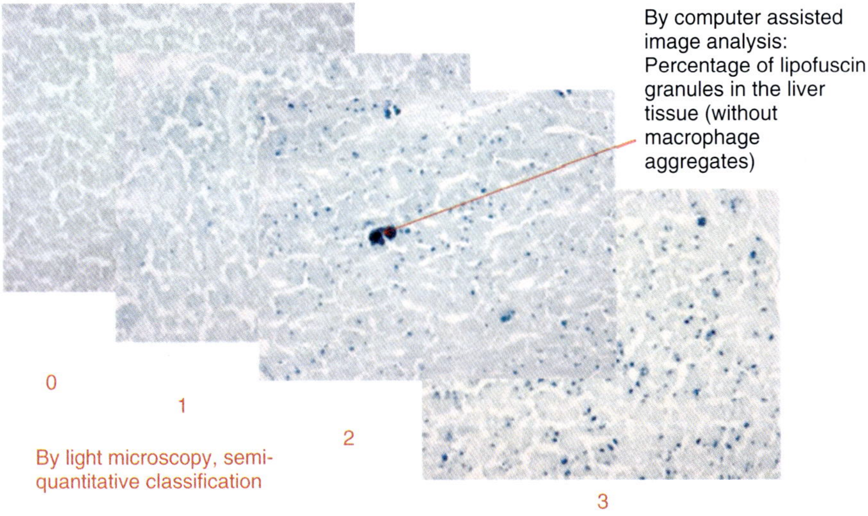

Fig. 39.21 Histochemical technique for the detection of lipofuscin and different methods for its assessment.

high amounts of NO• that were toxic *in vitro* for *T. borreli*. The nuclear transcription factor NF-κB was shown to play a role in the induction of iNOS transcription (Saeij et al. 2000).

Inducible NOS is expressed in cells with phagocytic activity, such as macrophages, Kupffer cells, neutrophils, fibroblasts, and endothelial cells in response to pathological stimuli. Synthesis of NO• is based on the same principle for all isoforms of NOS. In response to various triggers, NOS catalyses production of NO• from l-arginine and molecular oxygen, with l-citrulline as by-product. Oxidation is carried out by the heme moiety, while reduction is achieved using electrons from NADPH, probably due to the action of tetrahydrobiopterin, which is thought to couple the reduction of NADPH to the synthesis of NO•. Electrons are then transferred via FAD and FMN to the heme group, which oxidizes the terminal guanidine nitrogen atoms of l-arginine to NO• and l-citrulline.

As in higher vertebrates, NO• plays a central role in the immune responses of fish. Full-length iNOS genes were sequenced from several fish species (Cox et al. 2001; Wang et al. 2001). Even though there are differences compared to the human NOS gene, the exons show remarkable conservation in sequence and organization.

During ontogenetic development of carp, the first appearance of myeloid cells with an immune response has been demonstrated 2 days postfertilization. Embryos responded to LPS injection with an up-regulation of iNOS (Huttenhuis et al. 2006).

Viral infections (viral haemorrhagic septicemia virus, VHSV) of rainbow trout alter the levels of expression of iNOS genes, indicating a role for NO• production in antiviral defense (Tafalla et al. 2005). Nodaviruses are the etiological agents of one of the most serious viral diseases affecting marine fish aquaculture. In turbot (*Scophthalmus maximus*), Nodavirus infection leads to a significant increase in the production of RNS (Montes et al. 2009).

In mammalian systems, bacterial LPS are stimulating the release of inflammatory factors including tumor necrosis factor alpha (TNF-α), interferon gamma (IFN-γ) and interleukins 1 and 6 (IL-1 and -6), as well as increased production of NO• (Hewett and Roth 1993; Kim and Stadtman 1997; Jungi et al. 1999). At the cell plasma membrane LPS binds to Toll-like receptor TLR4, which results in the initiation of the signal transduction pathway involving the interleukin-1R pathway (Imler and Hoffman 2000; Raetz and Whitfield 2002). This ultimately activates the transcription factor NF-κβ followed by the transcription and translation of the iNOS gene (Monick and Hunnunghake 2003).

The capacity of activated macrophages to produce ROS and RNS is dependent on potassium channel activity (Stafford et al. 2002). In macrophages of goldfish, NO• generation was attenuated when potassium channel activity was blocked with substances such as quinine.

Cyclic AMP is an important second messenger that mediates diverse physiological processes. The function of intracellular cAMP in the regulation of NO• production in fish has been studied in leukocytes derived from head and trunk kidneys (Pietsch et al. 2008). It could be shown that bacteria-stimulated NO• synthesis is dependent on heterotrimeric G protein signal transduction. The elevation of cAMP resulted in a marked decrease of induced NO• production. The duration of cAMP inhibits both *Aeromonas-hydrophila*-stimulated and basal NO• production, suggesting the importance of cAMP-dependent regulation in leukocytes for a fine-tuning of immune response.

The regulation of NO• production in phagocytes of fish seems to be as complex as in higher vertebrates. NO• production of macrophages obtained from different individual turbot differs in its response to LPS. Macrophage populations that do not respond to LPS alone, respond to a combination of LPS and cytokines, with TNF-α being the most effective (Tafalla and Novoa 2000).

While NO• production is important for the defense against intracellular pathogens (Brunet 2001), overproduction may lead to tissue damage in the host and other conditions that are caused by nitrosative stress, and end in nitrosative damage. In mammalian systems, NO• is well documented as a cytostatic agent (Bohle 1998). NO• can inhibit ribonucleotide reductase preventing DNA synthesis (Kwon et al. 1991) and can regulate expression of numerous genes (Bogdan 2001).

High production of NO• is able to suppress immune functions in fish (Saeij et al. 2003). Whereas peripheral blood leukocytes are highly susceptible to NO• stress, pronephros phagocytes are relatively resistant to the immunosuppressive effects of NO•. Glutathione (GSH) has been found to play an important role in the protection against nitrosative stress as pronephros-derived phagocytes, especially the neutrophilic granulocytes, contain higher levels of GSH than peripheral blood leukocytes (Saeij et al. 2003). In addition, neutrophilic granulocytes have higher mRNA levels than peripheral blood lymphocytes (PBL) of glucose-6-phosphate dehydrogenase (G6PDH), manganese superoxide dismutase (Mn-SOD) and γ-glutamylcysteine synthetase (γ-GCS). All enzymes take part in the GSH redox cycle. Thus, neutrophilic granulocytes have a higher capacity than PBL to maintain GSH in a reduced state following nitrosative stress. When stimulated with LPS, neutrophilic granulocytes up-regulate the expression of G6PDH, Mn-SOD and γ-GCS.

REACTIVE OXYGEN AND NITROGEN SPECIES

When $NO^{\bullet}$ and $O_2^{\bullet-}$ are produced concomitantly within the same cellular environment, they react extremely fast to form $ONOO^-$, a potent deleterious reagent towards tyrosine residues. Figure 18.1 summarizes the concerted actions of $NO^{\bullet}$ and $O_2^{\bullet-}$ during the process of phagocytosis. The spontaneous decomposition of $ONOO^-$ generates stable nitrite and nitrate ions (Verchier et al. 2007).

Factors Influencing the ROS and RNS Production of Fish Phagocytes

Different signaling proteins, cytokines, are released by immune cells and are able to modulate ROS and RNS production of phagocytes. The treatment of goldfish macrophages with colony-stimulating factor (CSF-1) enhances their respiratory burst and $NO^{\bullet}$ responses (Grayfer et al. 2009). Macrophages exhibit a concentration-dependent chemotactic response toward CSF-1 as well as an increase in phagocytic activity, indicating that in addition to being an important growth factor of goldfish macrophages, CSF-1 also plays a central role in the regulation of pro-inflammatory responses. CSF-1 administration *in vivo* increases the amount of circulating monocytes in the blood of carp (Hanington et al. 2009). Teleost fibroblasts are important producers of CSF-1. The continuous addition of CSF-1 to primary goldfish macrophage cultures stabilized and extended their longevity and resulted in a long-term culture of functional macrophages capable of mounting a potent $NO^{\bullet}$ response upon activation with goldfish TNF-α.

TNF-α is a pro-inflammatory cytokine that is produced by different immune cells and mediates a wide variety of activities. Turbot TNF-α expression has been studied *in vivo* using different pathogens as stimulus (Ordas et al. 2007). The expression of the cytokine happened early after injection, and was dependent on the pathogen injected and organ analyzed. Virus induced a higher TNF-α expression, but this response was shorter in time than the response induced by bacteria. In addition, TNF-α expression was in general higher in kidney than in liver, as expected since the former is the hematopoietic organ of fish. No effect of recombinant TNF-α alone, nor in combination with LPS, was observed on the respiratory burst activity of turbot macrophages. $NO^{\bullet}$ production was, however, enhanced by the recombinant protein alone and in combination with LPS.

TRIGGERING THE DEFENSE ATTACK

The generation of ROS and $NO^{\bullet}$ requires the activation of cell receptors. Phagocytes identify pathogens and operate on the basis of pattern-recognition receptors (PRRs). These receptors recognize pathogens by binding conserved microbial motifs known as pathogen associated molecular patterns (PAMPs) (Akira and Takeda 2004). These motifs are essential for the physiology of the respective pathogens.

The recognition of these specific motifs induces cellular signals that result in the production of pro-inflammatory cytokines and ROS by the activation of the NOX. Toll-like receptors (TLRs) are the major PRR in the recognition of microbes and the following induction of immune responses. Due to their location at the cell membrane, TLRs recognize pathogens at the cell surface or on the luminal side of vesicles (Akira and Takeda 2004; Latz et al. 2004). Thus, they are presumably not involved in the detection of pathogens that have invaded the cytosol. These pathogens are detected by various cytoplasmic PRRs that have been called NOD–LRR proteins because they contain a nucleotide-binding oligomerization domain (NOD) and several leucine-rich repeats (LRRs) (Inohara et al. 2005).

TLRs have been described in different fish species, such as pufferfish *Fugu rubripes* and zebrafish *Danio rerio* (Oshiumi et al. 2003; Jault et al. 2004; Meijer et al. 2004). The expression profiles of fish PRRs in innate immune cells and their contribution to the activation of downstream signaling pathways are, however, largely unknown. Thus, the ability to recognize and respond to different pathogen-associated molecular patterns has been studied on different phagocyte types of gilthead sea bream (*Sparus aurata* L.) (Sepulcre et al. 2007). Cytokine expression after activation by different PAMPs was shown to be similar but not identical, likely reflecting the differential expression of TLRs by each phagocytic cell type.

Each TLR binds a specific ligand. The binding triggers acute inflammation through production of ROS and pro-inflammatory cytokines. TNF-α, β-glucan, Concanavalin-A (Con-A), and LPS are able to up-regulate dab (*Limanda limanda*) macrophage respiratory burst activity at particular concentrations; LPS, and Con-A induced the largest relative increases in activity. At high concentrations all factors had inhibitory activity (Tahir and Secombes 1996).

In vitro, ROS production is stimulated more rapidly by phorbal myristate acetate (PMA) than zymosan (Kaplan et al. 2008). PMA directly activates the phagocyte NOX system (independent of phagocytosis) to generate a rapid response, whereas zymosan initiates ROS production as a consequence of phagocytosis. This was shown in turbot neutrophils and gilthead sea bream acidophils, which responded in a similar way when incubated with PMA or with particulate glucans (Cousoa et al. 2001). Whereas cells stimulated with PMA released high amounts of intra- and extracellular ROS, ROS release was mainly intracellular when cells were incubated with particulate glucans. Small glucan particles were quickly phagocytosed and ROS were initially produced in intracellular vesicles and tubular structures that later fused with the phagosome or with the cell membrane. Large glucan filaments that were not phagocytosed also induced cell stimulation, and ROS were also produced in intracellular vesicles.

IMMUNE RESPONSE TO INJURY AND WOUND HEALING

The innate immune system not only mediates the initial inflammatory response to infection, but also to injury. Here, the extracellular matrix (ECM) plays an important role in the regulation of innate immunity. Collagen and gelatin have been implicated to prime the respiratory burst of phagocytes from the marine teleost gilthead sea bream *in vitro* (Castillo-Briceno et al. 2009). In addition, collagen and gelatin induced a specific set of immune-related and ECM remodeling enzymes that differed from those induced by pathogens. Thus, besides the well-established response to infection, the innate immune system of fish is able to respond to tissue injury.

The pathway taken by macrophages to metabolize arginine (classic or alternative, see section above) is therefore able to influence the final outcome of inflammatory and injury processes. If the macrophage uses arginine to produce $NO^{\bullet}$ (classic activation), this could lead to inhibited cell proliferation, while the use of arginine to produce ornithine (alternative activation) would promote cell replication/healing.

The question of by which mechanisms the immune cells are directed to a wound has guided Niethammer et al. (2009) during their experiments on zebrafish larvae. They observed that H_2O_2 provides a key signal for leukocyte recruitment to wounds. However, H_2O_2 was measured before the first leukocytes were attracted to the wound at the tailfin of the zebrafish. Niethammer et al. (2009) showed that tissue wounding induces a rapid concentration gradient of H_2O_2 that provides an essential first step in leukocyte recruitment to the injury. Before these observations, the primary source of ROS, including H_2O_2, in tissues has been thought to be due to the respiratory burst activity of leukocytes.

The tissue-scale gradient of H_2O_2 was generated at the wound margin without any help of leukocytes, and Duox, a NOX (Bedard and Krause 2007), was identified as the factor responsible for the oxidative burst of the epithelial cells. The inhibition of H_2O_2 production was sufficient to impair leukocyte recruitment to the wound. Further, when Duox was inhibited, directionality of leukocyte migration towards the wound was impaired, suggesting that the H_2O_2 burst by epithelial cells is necessary to guide leukocytes to wounds. The authors hypothesize that H_2O_2 generation by epithelial cells has two aims: initial killing of bacteria and subsequent recruitment of leukocytes to amplify the immune response and mediate host defense.

PROTECTION STRATEGIES OF BACTERIA

Coevolving with the immune responses of their hosts, bacteria have developed numerous mechanisms and strategies to overcome the aerobic attacks and thus inhibit, and in some cases finally prevent, oxidative killing. Enzymes like catalase and SOD are produced to scavenge phagocyte-derived ROS, bacterial capsules impede their recognition by phagocytes, and the invasion of host cells camouflages the pathogens. A summary of strategies used by fish pathogenic bacteria is given in Table 18.1.

Table 18.1 Protection strategies of bacteria.

Bacterium	Evasion strategy	Reference
Aeromonas salmonicida subsp. *salmonicida*	ROS scavenging enzymes: Manganese superoxide dismutase (MnSOD) Ferric superoxide dismutase (FeSOD) Inducible cytoplasmic catalase	Overview by Bernoth et al. (1997), Ellis (1999)
Renibacterium salmoninarum	Invasion of macrophages and other cell types as possible reservoirs	Gutenberger et al. (1997); Flano et al. (1996)
	Escape of the bacterium from the phagosome to cytosol	Grayson et al. (2002)
	Phagocytosis triggers ROS and NO• production to exhaustion	Hardie et al. (1996)
	Live bacteria induce lower ROS production compared to heat killed	Campos-Pérez et al. (1997)
	Suppression of ROS production by major soluble bacterial antigen p57	Turaga et al. (1987)
	Resistance to NO•, not to peroxynitrite	Campos-Perez et al. (2000a,b)
	Successful killing of bacteria by cytokine activated macrophages, most likely by TNFά	Hardie et al.(1996); Campos-Pérez et al. (1997)
	Induction of a prolonged RNS phase	Vaszques-Torres et al. (2000); Mastroeni et al. (1991)
Mycobacteria	Prevention of phagosome-lysosome fusion ROS scavaging enzymes superoxide dismutase and catalase	Speert (1992)
	TNFα mediates successful killing of mycobacteria	Bekker et al. (2001)
Edwardsiella ictaluri	Invasion of and multiplication in macrophages	Thune et al. (1997)
	Resistance to oxidative killing	Waterstraat et al. (1991)
	Resistance to phagocytosis	Ainsworth and Dexiang (1990)
Edwardsiella tarda	Virulent strains: survive and multiply in macrophages Inhibition of ROS production Modulation of the kinetic of NO• and TNFα production	Ishibe et al. (2009)
Photobacterium damsela subsp. *piscida*	Virulent strains possess a polysaccharide capsule	Margarinos et al. (1996a,b)
	Adhere to intestinal epithelium Are readily killed by macrophages	Skarmeta et al. (1995)
	Avoid phagocyte killing mechanisms by invasion of cells	Ellis (1999)
Vibrio anguillarum	Inhibition of respiratory burst Down-regulation of NOX components p22phox and p40phox Down-regulation of apoptotic caspases 3 and 9	Sepulcre et al. (2007)
Francisella sp.	Entering macrophages via phagocytosis Inhibition of phagosome maturation, escape to the cytosol Intracellular multiplication Trend towards down regulation of iNOS	Vojitech et al. (2009)

PARASITE MECHANISMS TO EVADE IMMUNE RESPONSES OF FISH

Like bacteria, many parasites are able to induce ROS and RNS generation by phagocytes of fish. One example is the myxosporean parasite *Enteromyxum leei*, which invades the intestine of gilthead sea bream and induces a disease which is slow-progressing but may end in the death of fish. The respiratory burst of parasitized fish has been shown to be significantly higher compared to nonparasitized fish (Sitja-Bobadilla et al. 2007). Figure 18.2 illustrates the activity of iNOS in epitheloid cells in the liver of mullet (*Liza aurata*) in response to a *Myxobolus* infection. Epitheloid cells are activated macrophages, in this case surrounding plasmodia of *Myxobolus* (Fig. 18.3; Broeg 2002).

The main mechanisms employed by fish parasites to evade the immune responses of their hosts have been

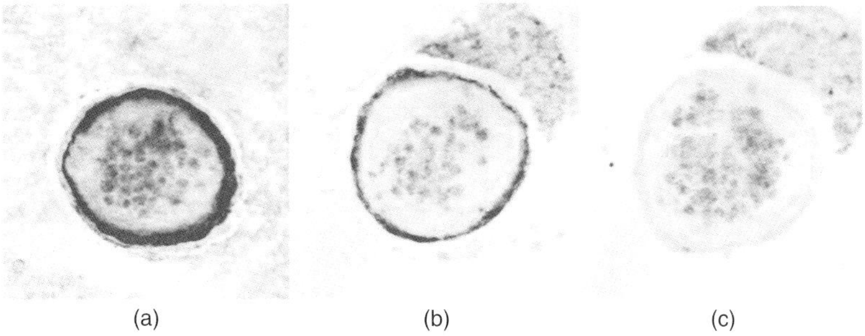

Fig. 18.2 NOS activity in epitheloid cells surrounding plasmodia of the myxosporean parasite *Myxobolus* in the liver of *Liza aurata*. Effect of different NOS inhibitors: (a) without inhibitor, (b) with N^G-methyl-L-arginine (L-NMMA), (c) with diphenyleniodonium. Cryostat sections, ×400. (photographs by Katja Broeg.)

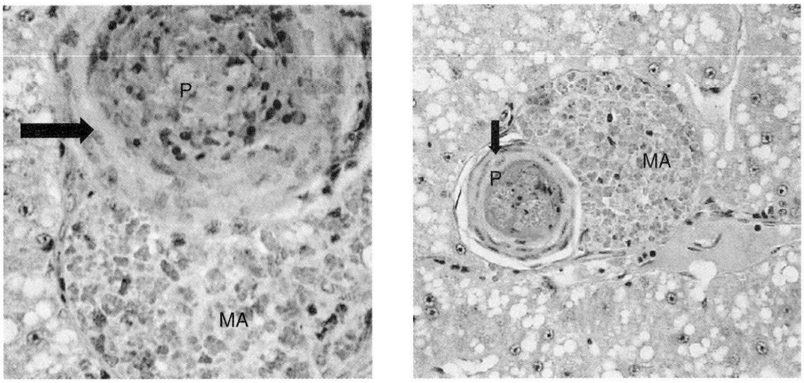

Fig. 18.3 Methacrylate sections of the liver of *Liza aurata* infected with the myxosporean parasite *Myxobolus* sp. MA, macrophage aggregate; P, plasmodium. The border of the plasmodium is formed by epithelioid cells (arrows), which are activated macrophages (see Fig. 18.2; H and E staining). (photographs by Katja Broeg.) See plate section for a color version of this image.

reviewed by Sitjà-Bobadilla (2008). Involved are both direct immunomodulatory impacts of the parasites as well as supportive factors based on specific physiologic conditions of the host, including immune responses. Some parasites chose tissues of the host, such as brain, gonads, or eyes for their development where host barriers prevent or limit the immune response. Other parasites are isolated by the cellular immune response of the host and kept in a dormant stage without being able to kill the host. Intracellular camouflaging is a strategy, typical for intracellular microsporidians, coccidians, and some myxosporeans, and has already been reported for bacteria (Table 18.1).

Another strategy of immune evasion is the parasite's migration to host sites that the immune response has not yet reached or where it is not strong enough to kill them, or the parasites accommodate their life cycles to a season or age in which the host's immune system is down-regulated. As stated above, this is of particular interest in poikilothermic hosts such as fish.

Some parasites have developed anti-immune mechanisms effective towards particular immune responses of the host. These allow parasites to resist innate humoral factors to neutralize host antibodies or to scavenge ROS released by phagocytes. Immunomodulation is further involved in the resistance of parasites against hosts' immune responses. Either the fish immune system is suppressed by reducing the proliferative capacity of lymphocytes and/or the phagocytic activity of the macrophages, or they induce apoptosis of host leukocytes. Parasites also secrete or excrete substances that modulate the secretion of host immune factors such as cytokines to their own benefit. Even the induction of high $NO^{\bullet}$ response

Table 18.2 Modulation of ROS and RNS release by piscine phagocytes under parasite infection.

Parasite	Modulatory effect	Reference
Trypanoplasma borreli	Prolonged survival of neutrophils Up-regulation of ROS and NO• secretion	Scharsack et al. (2003a,b)
	No killing or phagocytosis of parasites Suppression of blood leukocyte proliferation Induction of NO• secretion by cleavage of transferrin	Jurecka et al. (2009); Stafford et al. (2001)
	Induction of nitrosative stress indicated by tyrosine nitration, however nitration of the parasite was nearly absent	Forlenza et al. (2008)
Trypanosoma carassii	Inhibition of LPS-induced NO• production	Saeij et al. (2002)
	T. carassii hsp70 up regulates NO• production by macrophages	Oladiran and Belosevic (2008)
Tetramicra brevifilum	Suppression of ROS production by phagocytes	Leiro et al. (2001)
Schistocephalus solidus	Down-regulation of granulocyte responses to *S. solidus* antigens	Scharsack et al. (2007)

by fish phagocytes is used by parasites to their own advantage, as shown in the case of the *Trypanoplasma borrelli*. A summary of immunomodulatory strategies of piscine parasites is given in Table 18.2.

REFERENCES

Afonso, A., Silva, J., Lousanda, S., Ellis, A.E., Silva, M.T. (1998) Uptake of neutrophils and neutrophilic components by macrophages in the inflamed peritoneal cavity of rainbow trout (*Oncorhynchus mykiss*). *Fish and Shellfish Immunology* 8, 319–338.

Ainsworth, A.J., Dexiang, C. (1990) Differences in the phagocytosis of four bacteria by channel catfish neutrophils. *Development and Comparative Immunology* 14, 201–210.

Akira, S., Takeda, K. (2004) Toll-like receptors signaling. *Nature Reviews Immunology* 4, 499–511.

Apel, K., Hirt, H. (2004) Reactive oxygen species: metabolism, oxidative stress, and signal transduction. *Annual Reviews in Plant Biology* 55, 373–399.

Beckman, J.S., Beckman, T.W., Chen, J., Marshall, P.A., Freeman B.A. (1990) Apparent hydroxyl radical production by peroxynitrite: implications for endothelial injury from nitric oxide and superoxide. *Proceedings of the National Academy of Sciences USA* 87, 1620–1624.

Bedard, K., Krause, K.H. (2007) The NOX family of ROS-generating NADPH oxidases: physiology and pathophysiology. *Physiological Reviews* 87, 245–313.

Bekker L.-G., Freeman S., Murray P.J., Ryffel, B., Kaplan, G. (2001) TNF-α) controls intracellular mycobacterial growth by both inducible nitric oxide synthase-dependent and inducible nitric oxide synthase-independent pathways. *Journal of Immunology* 166, 6728–34.

Bernoth, E.-M., Ellis, A.E., Midtlying, P.J., Olivier, G., Smith, P. (1997) *Furunculosis: Multidisciplinary Fish Disease Research.* Academic Press, London.

Bogdan, C. (2001) Nitric oxide and the regulation of gene expression. *Trends in Cell Biology* 11, 66–75.

Bohle, D.S. (1998) Pathophysiological chemistry of nitric oxide and its oxygenation by products. *Current Opinion in Chemical Biology* 2, 194–200.

Boleza, K.A., Burnett, L.E., Burnett, K.G. (2001) Hypercapnic hypoxia compromises bactericidal activity of fish anterior kidney cells against opportunistic environmental pathogens. *Fish and Shellfish Immunology* 11, 593–610.

Broeg, K. (2002). *Funktionen von Makrophagen und ihren Aggregaten als histochemische Biomarker für Immunmodulation in Fischen verschiedener Klimazonen.* PhD dissertation, Universität Hannover, 213 pp.

Brunet, L.R. (2001) Nitric oxide in parasitic infections. *International Immunopharmacology* 1, 1457–1467.

Campos-Pérez, J.J., Ellis, A.E., Secombes, C.J. (1997) Investigation of factors influencing the ability of *Renibacterium salmoninarum* to stimulate rainbow trout macrophage respiratory burst activity. *Fish and Shellfish Immunology* 7, 555–566.

Campos-Perez, J.J., Ellis, A.E., Secombes, C.J. (2000a) Toxicity of nitric oxide and peroxynitrite to bacterial pathogens of fish. *Diseases of Aquatic Organisms* 43, 109–15.

Campos-Perez, J.J., Ward, M., Grabowski, P.S., Ellis, A.E., Secombes, C.J. (2000b) The gills are an important site of iNOS expression in rainbow trout *Oncorhynchus mykiss* after challenge with the Gram-positive pathogen *Renibacterium salmoninarum*. *Immunology* 99, 153–61.

Castillo-Briceno, P., Sepulcre, M.P., Chaves-Pozo, E., Meseguer, J., García-Ayala, A., Mulero, V. (2009) Collagen regulates the activation of professional phagocytes of the teleost fish gilthead sea bream. *Molecular Immunology* 46, 1409–1415.

Chang, C.I., Liao, J.C., Kuo, L. (1998) Arginase modulates nitric oxide production in activated macrophages. *American Journal of Physiology* 274, H342–H348.

Cousoa, N., Castroa, R., Noyaa, M., Obach, A., Lamas, J. (2001) Location of superoxide production sites in turbot neutrophils and gilthead seabream acidophilic granulocytes during phagocytosis of glucan particles. *Developmental and Comparative Immunology* 25, 607–618.

Cox, R.L., Mariano, T., Heck, D.E., Laskin, J.D., Stegeman, J.J. (2001) Nitric oxide synthase sequences in the marine fish *Stenotomus chrysops* and the sea urchin *Arbacia punctulata*, and phylogenetic analysis of nitric oxide synthase calmodulin-binding domains. *Comparative Biochemistry and Physiology B* 130, 479–491.

Demple, B. (2002) Signal transduction by nitric oxide in cellular stress responses. *Molecular and Cellular Biochemistry* 234–235, 11–18.

Eisenstein, T.K., Huang, D., Meissler Jr., J.J., Al-Ramadi, B. (1994) Macrophage nitric oxide mediates immunosuppression in infectious inflammation. *Immunobiology* 191, 493–502.

Ellis, A.E. (1999) Immunity to bacteria in fish. *Fish and Shellfish Immunology* 9, 291–308.

Fialkow, L., Wang, Y., Downey, G.P. (2007) Reactive oxygen and nitrogen species as signaling molecules regulating neutrophil function. *Free Radical Biology and Medicine* 42, 153–164.

Flano, E., López-Fierro, P., Razquin, B., Kaattari, S.L., Villena, A. (1996) Histopathology of the renal and splenic haemopoietic tissues of coho salmon *Oncorhynchus kisutch* experimentally infected with *Renibacterium salmoninarum*. *Diseases of Aquatic Organisms* 24, 107–115.

Forlenza, M., Scharsack, J.P., Kachamakova, N.M., Taverne-Thiele, A.J., Rombout, J.H.W.M., Wiegertjes, G.F. (2008) Differential contribution of neutrophilic granulocytes and macrophages to nitrosative stress in a host–parasite animal model. *Molecular Immunology* 45, 3178–3189.

Gaboury, J., Woodman, R.C., Granger, D.N., Reinhardt, P., Kubes, P. (1993) Nitric oxide prevents leukocyte adherence: role of superoxide. *American Journal of Physiology* 265, H862–H867.

Gordon, S., (2003) Alternative activation of macrophages. *Nature Reviews Immunology* 3, 23–35.

Grayfer, L., Hanington, P.C., Belosevic, M. (2009) Macrophage colony-stimulating factor (CSF-1) induces proinflammatory gene expression and enhances antimicrobial responses of goldfish (*Carassius auratus* L.) macrophages. *Fish Shellfish Immunology* 26, 406–13.

Grayson, T.H., Cooper, L.F., Wrathmell, A.B., Roper, J., Evenden, A.J., Gilpin, M.L. (2002) Host responses to *Renibacterium salmoninarum* and specific components of the pathogen reveal the mechanisms of immune suppression and activation. *Immunology* 106, 273–283.

Gutenberger, S.K., Duimstra, J.R., Rohovec, J.S., Fryer, J.C. (1997) Intracellular survival of *Renibacterium salmoninarum* in trout mononuclear phagocytes. *Diseases of Aquatic Organisms* 28, 93–106.

Hanington, P.C., Hitchen, S.J., Beamish, L.A., Belosevic, M. (2008) Macrophage colony stimulating factor (CSF-1) is a central growth factor of goldfish macrophages. *Fish and Shellfish Immunology* 26, 1–9.

Hardie, L.J., Ellis, A.E., Secombes, C.J. (1996) *In Vitro* activation of rainbow trout macrophages stimulates killing of *Renibacterium salmoninarum* concomitant with augmented generation of respiratory burst products. *Diseases of Aquatic Organisms* 25, 175–183.

Hewett, J.A., Roth, R.A. (1993) Hepatic and extrahepatic pathobiology of bacterial lipopolysaccharides. *Pharmacological Reviews* 45, 381–411.

Huttenhuis, H.B., Taverne-Thiele, A.J., Grou, C.P. et al. (2006) Ontogeny of the common carp (*Cyprinus carpio* L.) innate immune system. *Developmental and Comparative Immunology* 30, 557–74.

Imler, J.L., Hoffman, J.A. (2000) Toll and Toll-like proteins: an ancient family of receptors signaling infection. *Reviews in Immunogenetics* 2, 294–304.

Inohara, N., Chamaillard, M., McDonald, C., Nunez, G. (2005) NOD–LRR proteins: role in host-microbial interactions and inflammatory disease. *Annual Reviews of Biochemistry* 74, 355–383.

Ishibe, K., Yamanishi, T., Wang, Y. et al. (2009) Comparative analysis of the production of nitric oxide (NO) and tumor necrosis factor-alpha (TNF-alpha) from macrophages exposed to high virulent and low virulent strains of *Edwardsiella tarda*. *Fish and Shellfish Immunology* 27, 386–389.

Jault, C., Pichon, L., Chluba, J. (2004) Toll-like receptor gene family and TIRdomain adapters in *Danio rerio*. *Molecular Immunology* 40, 759–771.

Joerink, M., Forlenza, M., Ribeiro, C.M.S., de Vries, B.J. Savelkoul, H.F.J., Wiegertjes, G.F. (2006) Differential macrophage polarisation during parasitic infections in common carp (*Cyprinus carpio* L.). *Fish and Shellfish Immunology* 21, 561–571.

Jungi, T.W., Valentin-Weigand, P., Brcic, M. (1999) Differential induction of nitric oxide synthesis by Gram-positive and Gram-negative bacteria and their components in bovine monocyte-derived macrophages. *Microbial Pathogenesis* 27, 43–53.

Jurecka, P., Irnazarow, I., Stafford, J.L. et al. (2009) The induction of nitric oxide response of carp macrophages by transferrin is influenced by the allelic diversity of the molecule. *Fish and Shellfish Immunology* 26, 632–638.

Kaplan, J.E., Chrenek, R.D., Morash, J.G., Ruksznis, C.M., Hannum, L.G. (2008) Rhythmic patterns in phagocytosis and the production of reactive oxygen species by zebrafish

leukocytes. *Comparative Biochemistry and Physiology A* 151, 726–730.
Kim, I.Y., Stadtman, T.C. (1997) Inhibition of NK-kB binding and nitric oxide induction in human T cells and lung adenocarcinoma cells by salenite treatment. *Proceedings of the National Academy of Sciences USA* 94, 12904–12907.
Klebanoff, S.J., Clark, R.D. (ed.) (1978) *The Neutrophil: Function and Clinical Disorders*. Amsterdam: North Holland.
Klyubin, I.V., Kirpichnikova, K.M., Gamaley, I.A. (1996) Hydrogen peroxide-induced chemotaxis of mouse peritoneal neutrophils. *European Journal of Cell Biology* 70, 347–351.
Kwon, N.S., Stuehr, D.J., Nathan, C.F. (1991) Inhibition of tumor cell ribonucleotide reductase by macrophage-derived nitric oxide. *Journal of Experimental Medicine* 174, 761–767.
Lambeth, D.L., Kawahara, T., Diebold, B. (2007) Regulation of Nox and Duox enzymatic activity and expression. *Free Radical Biology and Medicine* 43, 319–331.
Latz, E., Schoenemeyer, A., Visintin, A. et al. (2004) TLR9 signals after translocating from the ER to CpG in the lisosome. *Nature Immunology* 5, 190–198.
Leiro, J., Iglesias, R., Paramá, A., Sanmartín, M.L.& Ubeira, F.M. (2001) Effect of *Tetramicra brevifilum* (Microspora) infection on respiratory-burst responses of turbot (*Scophthalmus maximus* L.) phagocytes. *Fish and Shellfish Immunology* 11, 639–52.
Levraud, J.P., Boudinot, P. (2009) The immune system of teleost fish. *Médecine Science* 25, 405–411.
Magarinos, B., Bonet, R., Romalde, J.L., Martinez, M.J., Congregado, F., Toranzo, A.E. (1996a) Influence of the capsular layer on the virulence of *Pasturella piscicida* for fish. *Microbial Pathogenesis* 21, 289–297.
Magarinos, B., Romalde, J.L., Noya, M., Barja, J.L., Toranzo, A.E. (1996b) Adherence and invasive capacities of the fish pathogen *Pasteurella piscicida*. *FEMS Microbiology Letters* 138, 29–34.
Mantovani, A., Sica, A., Sozzani, S., Allavena, P., Vecchi, A., Locati, M. (2004) The chemokine system in diverse forms of macrophage activation and polarization. *Trends in Immunology* 25, 677–686.
Mastroeni, P., Vazquez-Torres, A., Fang, F.C. et al. (2000) Antimicrobial actions of the NADPH phagocyte oxidase and inducible nitric oxide. *Journal of Experimental Medicine* 192, 237–247.
Meijer, A., Gabby-Krens, S., Medina-Rodriguez, I. et al. (2004) Expression analysis of the Toll-like receptor and TIR domain adaptor families of zebrafish. *Molecular Immunology* 40, 773–783.
Monick, M.M., Hunnunghake, G.W. (2003) Second messenger pathways in pulmonary host defense. *Annual Review of Physiology* 65, 643–647.
Montes, A., Figueras, A., Novoa, B. (2009) Nodavirus encephalopathy in turbot (*Scophthalmus maximus*): Inflammation, nitric oxide production and effect of anti-inflammatory compounds. *Fish and Shellfish Immunology* 28, 281–288.
Mori, M., Gotoh, T. (2000) Regulation of nitric oxide production by arginine metabolic enzymes. *Biochemical and Biophysical Research Communications* 275, 715–719.
Nathan, C. (1992) Nitric oxide as a secretory product of mammalian cells. *FASEB Journal* 6, 3051–3064.
Nathan, C. (2006) Neutrophils and immunity: challenges and opportunities. *Nature Reviews: Immunology* 6, 173–182.
Ndong, D., Chen, Y.-Y., Lin, Y.H., Vaseeharan, B., Chen, J.C. (2006) The immune response of tilapia *Oreochromis mossambicus* and its susceptibility to *Streptococcus iniae* under stress in low and high temperatures. *Fish and Shellfish Immunology* 22, 686–94.
Neill, S., Desikan, R., Hancock, J. (2002) Hydrogen peroxide signalling. *Current Opinion in Plant Biology* 5, 388–395.
Neumann, N.F., Stafford, J.L., Barreda, D., Ainsworth, A.J., Belosevic, M. (2001) Antimicrobial mechanisms of fish phagocytes and their role in host defense. *Developmental and Comparative Immunology* 25, 807–825.
Niethammer, P., Grabher, C., Look, A.T., Mitchison, T.J. (2009) A tissue-scale gradient of hydrogen peroxide mediates rapid wound detection in zebrafish. *Nature* 459, 996–999.
Nikoskelainen, S., Bylund, G., Lilius, E.-M. (2004) Effect of environmental temperature on rainbow trout (*Oncorhynchus mykiss*) innate immunity. *Developmental and Comparative Immunology* 28, 581–92.
Oladiran, A., Belosevic, M. (2008) *Trypanosoma carassii* hsp70 increases expression of inflammatory cytokines and chemokines in macrophages of the goldfish (*Carassius auratus* L.). *Veterinary Immunology and Immunopathology* 126(4), 171–198.
Ordas, M.C., Costa, M.M., Roca, F.J., Lopez-Castejon, G., Mulero V., Meseguer, J., Figueras, A., Novoa, B. (2007) Turbot TNF gene: Molecular characterization and biological activity of the recombinant protein. *Molecular Immunology* 44, 389–400.
Oshiumi, H., Tsujita, T., Shida, K., Matsumoto, M., Ikeo, K., Seya, T. (2003) Prediction of the prototype of the human Toll-like receptor gene family from the pufferfish, *Fugu rubripes*, genome. *Immunogenetics* 54, 791–800.
Pérez-Casanova, J.C., Rise, M.L., Dixon, B. et al. (2008) The immune and stress responses of Atlantic cod to long-term increases in water temperature. *Fish and Shellfish Immunology* 24, 600–609.
Pietsch, C., Vogt, R., Neumann, N., Kloas, W. (2008) Production of nitric oxide by carp (*Cyprinus carpio* L.) kidney leukocytes is regulated by cyclic 3',5'-adenosine monophosphate. *Comparative Biochemistry and Physiology A* 150, 58–65.

Raetz, C.R.H., Whitfield, C. (2002) Lipopolysaccharide endotoxins. *Annual Review of Biochemistry* 71, 635–700.

Reth, M. (2002) Hydrogen peroxide as second messenger in lymphocyte activation. *Nature Immunology* 3, 1129–1134.

Saeij, J.P.J., Stet, R.J.M., Groeneveld, A., Verburg-van Kemenade, L.B.M., van Muiswinkel, W.B., Wiegertjes, G.F. (2000) Molecular and functional characterization of a fish inducible-type nitric oxide synthase. *Immunogenetics* 51, 339–346.

Saeij, J.P.J., Van Muiswinkel, W.B., Groeneveld, A., Wiegertjes, G.F. (2002) Immune modulation by fish kinetoplastid parasites: a role for nitric oxide. *Parasitology* 124, 77–86.

Saeij, J.P.J., van Muiswinkel, W.B., van de Meent, M., Amaral, C., Wiegertjes, G.F. (2003) Different capacities of carp leukocytes to encounter nitric oxide-mediated stress: a role for the intracellular reduced glutathione pool. *Developmental and Comparative Immunology* 27, 555–568.

Scharsack, J.P., Steinhagen, D., Kleczka, C. et al. (2003a) Head kidney neutrophils of carp (*Cyprinus carpio* L.) are functionally modulated by the haemoflagellate *Trypanoplasma borreli*. *Fish and Shellfish Immunology* 14: 389–403.

Scharsack, J.P., Steinhagen, D., Kleczka, C. et al. (2003b) The haemoflagellate *Trypanoplasma borreli* induces the production of nitric oxide, which is associated with modulation of carp (*Cyprinus carpio* L.) leucocyte functions. *Fish and Shellfish Immunology* 14, 207–22.

Scharsack, J.P., Koch, K., Hammerschmidt, K. (2007) Who is in control of the stickleback immune system: interactions between *Schistocephalus solidus* and its specific vertebrate host. *Proceedings of the Royal Society of London* 274, 3151–3158.

Sepulcre, M.P., Lopez-Castejon, G., Meseguer, J., Mulero, V. (2007) The activation of gilthead seabream professional phagocytes by different PAMPs underlines the behavioural diversity of the main innate immune cells of bony fish. *Molecular Immunology* 44, 2009–2016.

Shiibashi, T., Iida, T. (2001) NADPH and NADH serve as electron donor for the superoxide-generating enzyme in tilapia (*Oreochromis niloticus*) neutrophils. *Developmental and Comparative Immunology* 25, 461–465.

Sitjà-Bobadilla, A. (2008) Living off a fish: A trade-off between parasites and the immune system. *Fish and Shellfish Immunology* 25, 358–372.

Sitjà-Bobadilla, A., Calduch-Giner, J., Saera-Vila, A. et al. (2007) Chronic exposure to the parasite *Enteromyxum leei* (Myxozoa: Myxosporea) modulates the immune response and the expression of growth, redox and immune relevant genes in gilthead sea bream, *Sparus aurata* L. *Fish and Shellfish Immunology*, 22, 686–694.

Skarmeta, A.M., Bandin, I., Santos, Y., Toranzo, A.E. (1995) *In Vitro* killing of *Pasteurella piscicida* by fish macrophages. *Diseases of Aquatic Organisms* 23, 51–57.

Speert, D.P. (1992) Macrophages in bacterial infection. *In* Lewis, C.E., McGee, J.O'D. (eds). *The Macrophage*. IRL Press, Oxford, pp. 215–263.

Stafford, J.L., Neumann, N.F., Belosevic, M. (2001) Products of proteolytic cleavage of transferrin induce nitric oxide response of goldfish macrophages. *Developmental and Comparative Immunology* 25, 101–115.

Stafford, J.L., Galvez, F., Goss, G.G., Belosevic, M. (2002) Induction of nitric oxide and respiratory burst response in activated goldfish macrophages requires potassium channel activity. *Developmental and Comparative Immunology* 26, 445–459.

Tafalla, C., Novoa, B. (2000) Requirements for nitric oxide production by turbot (*Scophthalmus maximus*) head kidney macrophages. *Developmental and Comparative Immunology* 24, 623–631.

Tafalla, C., Coll, J., Secombes, C.J. (2005) Expression of genes related to the early immune response in rainbow trout (*Oncorhynchus mykiss*) after viral haemorrhagic septicemia virus (VHSV) infection. *Developmental and Comparative Immunology* 29, 615–626.

Tahir, A., Secombes, C.J. (1996) Modulation of dab (*Limanda limanda*, L.) macrophage respiratory burst activity. *Fish and Shellfish Immunology* 6, 135–146.

Thune, R.L., Hawke, J.P., Fernandez, D.H., Lawrence, M.L., Moore, M.M. (1997) Immunisation with bacterial antigens: Edwardsiellosis. *In* Gudding, R., Lillehaug, A., Midtlying P., Brown, F. (eds). *Fish Vaccinology*. Developments in Biological Standardisation 90, Karger, Basel, pp. 125–134.

Turaga, P.S.D., Wiens, G.D., Kaattari, S.L. (1987) Bacterial kidney disease: the potential role of soluble protein antigen. *Journal of Fish Biology A* 31, 191–194.

Vazquez-Torres, A., Jones-Carson, J., Mastroeni, P., Ischiropoulos, H., Fang, F.C. (2000) Antimicrobial actions of the NADPH phagocyte oxidase and inducible nitric oxide synthase in experimental salmonellosis. I. Effects on microbial killing by activated peritoneal macrophages *in vitro*. *Journal of Experimental Medicine* 192, 227–36.

Verchier, Y., Lardy, B., Chuong Nguyen, M.V., Morel, F., Arbault, S., Amatore, C. (2007) Concerted activities of nitric oxide synthases and NADPH oxidases in PLB–985 cells *Biochemical and Biophysical Research Communications* 361, 493–498.

Vincendeau, P., Gobert, A.P., Daulouede, S., Moynet, D., Mossalayi, M.D. (2003) Arginases in parasitic diseases. *Trends in Parasitology* 19, 9–12.

Vojtech, L.N., Sanders, G.E., Conway, C., Ostland, V., Hansen, J.D. (2009) Host immune response and acute disease in a zebrafish model of *Francisella* pathogenesis. *Infection and Immunity* 77, 914–925.

Wang, T.H., Ward, M., Grabowski, P., Secombes, C.J. (2001) Molecular cloning, gene organization and expression of rainbow trout (*Oncorhynchus mykiss*) inducible nitric oxide synthase (iNOS) gene. *Biochemical Journal* 358, 747–755.

Waterstrat, P.R., Ainsworth, A.J., Capley, G. (1991) *In Vitro* responses of channel catfish, *Ictalurus punctatus*, neutrophils to *Edwardsiella ictaluri*. *Developmental and Comparative Immunology* 15, 53–63.

Yoder, J.A., Nielsen, M.E., Amemiya, C.T., Litman, G.W. (2002) Zebrafish as an immunological model system. *Microbes and Infection* 4, 1469–1478.

Zapata, A. (1981) Lymphoid organs of teleost fist. II. Ultrastructure of renal lymphoid tissue of *Rutilus rutilus* and *Gobio gobio*. *Developmental and Comparative Immunology* 5, 685–690.

Zapata, A., Amemiya, C.T. (2000) Phylogeny of lower vertebrates and their immunological structures. *In* Du Pasquier, L., Litman, G.W. (eds). *Origin and Evolution of the Vertebrate Immune System*. Springer-Verlag, Berlin, pp. 67–107.

Part IV

Marine Animal Stress Response and Biomonitoring

Chapter 19

STRESS EFFECTS ON METABOLISM AND ENERGY BUDGETS IN MOLLUSKS

Inna M. Sokolova[1], *Alexey A. Sukhotin*[2], *and Gisela Lannig*[3]

[1]Department of Biology, University of North Carolina at Charlotte, Charlotte, NC, USA
[2]White Sea Biological Station, Zoological Institute of Russian Academy of Sciences, St. Petersburg, Russia
[3]Alfred Wegener Institute for Polar and Marine Research in the Hermann von Helmholtz Association of National Research Centres e.V., Integrative Ecophysiology, Bremerhaven, Germany

Adaptations of energy metabolism play a key role in the suite of the adaptive traits that allow mollusks to thrive in a variety of environments including cold Arctic and Antarctic waters, intertidal zones, estuaries and coastal waters with high anthropogenic impact, temporary freshwater bodies, and deserts. Indeed, meeting energy demands with sufficient energy supply is a prerequisite not only for immediate survival, but for all other energy-demanding functions that ensure evolutionary success such as growth, reproduction, and escape from predators.

Stress displaces an animal from its energetic homeostasis and can impair its performance (Hochachka and Somero 2002; Pörtner and Knust 2007). Recently, studies of the metabolic aspects of adaptations to environmental stress have gained momentum due to concerns of climate change and other anthropogenically induced impacts on animal performance (Somero 2005; Pörtner and Farrell 2008; Sokolova and Lannig 2008). In this chapter we address how marine mollusks cope with the environmental stress in their habitats, focusing on the common stressors and emphasizing the role of energy metabolism in setting the limits for stress tolerance. We also discuss the role of energy homeostasis in preventing or limiting oxidative stress in marine mollusks which often is a common denominator for the effects of environmental stressors at the cellular level.

THE ROLE OF ENERGY HOMEOSTASIS AND ASSOCIATED TRADE-OFFS IN SURVIVAL AND STRESS TOLERANCE

Stress affects the structure and functioning of an organism in such a way that its Darwinian fitness is reduced relative to the optimal conditions (Selye

Oxidative Stress in Aquatic Ecosystems, First Edition. Edited by Doris Abele, José Pablo Vázquez-Medina, and Tania Zenteno-Savín.

1973; Calow 1989a). Overall, animals adapted to less variable environments tend to become specialists accompanied by a narrowing of the stress tolerance windows and, thus, are much more sensitive to small environmental alterations compared to their generalist counterparts. The tolerance limits are closely related to the metabolic performance of an organism so that the environmentally relevant ranges for both short-term and long-term survival can be determined by measuring key metabolic parameters such as aerobic scope, transition to partial anaerobiosis, and/or onset of the metabolic arrest (Pörtner 2002; Pörtner and Knust 2007; Kassahn et al. 2009; and discussion below). Environmental change is considered as stress if an organism needs to increase expenditure on maintenance, defense, or repair in response to it, and as a result, suffers a reduction of net energy gain. The main goal of metabolic stress adaptation is therefore to reinstate the metabolic balance to maximize the organism's fitness. Energy homeostasis implies that the energy demand is covered by sufficient energy supply and that there is a net energy gain to invest into production (somatic growth and reproduction). Thus, any environmental disturbance that reduces this energy investment into production will have direct consequences for the organism's fitness and can be classified as stress.

In any attempt to measure an organism's response to environmental stress, three general considerations should be borne in mind. First, the effect of stress is an integrated response involving all levels of the functional complexity within the organism (molecular, cellular, and physiological). Second, the stress response is dynamic and involves alterations in functional or structural properties over time. Last, potential stress may be compensated at least partly by the homeostatic mechanisms, although these processes can be metabolically costly. It is when compensation for the adverse stress effects is incomplete or impossible that lasting effects of stress become measurable as a decline in the organism's fitness or, ultimately, as death.

Energy can be conserved in the form of adenosin triphosphate (ATP) aerobically and anaerobically. During aerobic metabolism the ability to provide sufficient oxygen supply via ventilation and circulation is a critical physiological bottleneck. Marine mollusks are predominantly water breathers. Depending on lifestyle, oxygen supply mechanisms vary, with higher capacities in more active mollusks than in more hypometabolic benthic ones (Gilbert et al. 1990; Jørgensen 1990; Pörtner 1997). If the oxygen supply mechanisms fail to deliver a sufficient amount of oxygen, mollusks can shift to anaerobic metabolism and sustain themselves by anaerobic ATP production. However, transition to anaerobiosis heralds a time-limited situation because mollusks cannot survive without oxygen indefinitely.

The aerobic scope (i.e. the amount of aerobically produced ATP that is left over after the maintenance costs are covered) can be subdivided into scope for activity, scope for growth, and scope for reproduction. Among these, scope for growth has been studied most extensively in marine mollusks and used to provide a rapid and quantitative assessment of the energy status (Riisgård 2001; Widdows et al. 2002; Smolders et al. 2004; Mubiana and Blust 2007). The overall aerobic scope is dependent not only on food intake and oxygen supply but also on the energetic efficiency of the organism (Beiras et al. 1995, and references therein). In mollusks, a reduction of the baseline costs is found in individuals with multilocus heterozygosity (MLH) and correlates with increased aerobic scope, and higher fitness and stress tolerance (Hawkins 1991; Myrand et al. 2002). Individual variation in energy allocation between the costs for maintenance and other functions may affect both growth rates and balance between growth and reproduction (Bayne 2004). Allocation of the energy remaining after the organism's basal maintenance costs are covered is also subject to trade-offs depending on how the organism prioritizes its different energy-consuming functions to maximize fitness. Phenotypic plasticity in energy allocation to different functions such as somatic growth or reproduction and the associated trade-offs in energy budget are normal and adaptive for mollusks, allowing for a flexible response to the changing environmental conditions (Bayne and Worrall 1980; Lee 1985; Cheung 1993; Schöne et al. 2003, and references therein). Overall, under stress the flexible energy allocation ensures survival of the individual with often negative consequences at the population level, as more energy is allocated for maintenance leaving less energy for growth and/or reproductive effort. This is especially true for long-lived organisms such as many mollusks in which postponing growth and reproduction until the environmental conditions become more favorable can maximize the lifetime fitness in stressful environments. Furthermore, the baseline energy costs can also be temporarily reduced in response to extreme stress exposures that require metabolic rate depression to ensure survival (see the following page).

ENERGY BALANCE DURING STRESS EXPOSURES: MODERATE AND EXTREME STRESS

The role of energy metabolism in stress tolerance and survival of marine mollusks can be generalized using the theory of the dynamic energy budget (DEB; Kooijman 2000; Fig. 19.1). We have modified this model to account for the storage of the surplus energy in the form of lipid, carbohydrate, and/or protein reserves, which can occur when food intake exceeds energy requirements; these reserves play an important role during periods of elevated energy requirements, such as reproduction or stress. Using these changes in energy

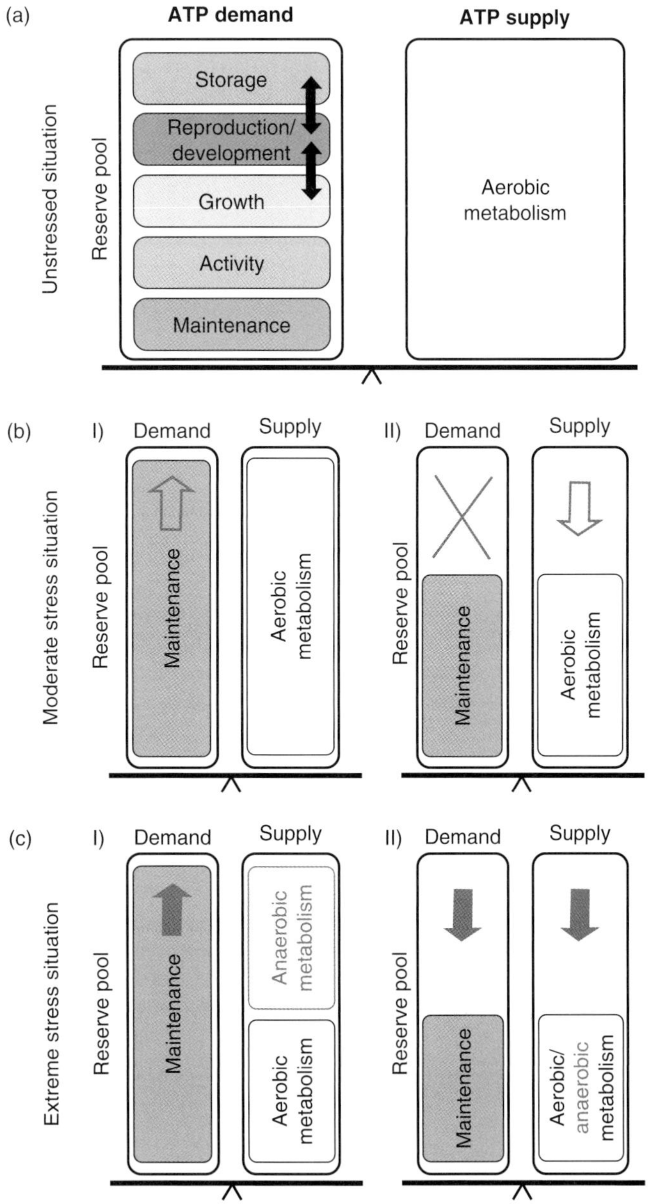

Fig. 19.1 The role of energy metabolism in stress response based on the dynamic energy budget (DEB) model. (a) Under normal conditions (unstressed situation) ATP supply via aerobic metabolism is sufficiently high to cover the costs for activity, growth, and reproduction/development, and usually allows the support of energy storages in addition to the vital maintenance costs. "Normal" trade-offs exist in energy allocation between reserves, reproduction, and growth (indicated by black arrows). (b) During moderate stress, the costs for maintenance increase and/or aerobic metabolism is impaired, depending on the nature of the stressor. In either case, energy homeostasis is achieved at the expense of other processes (e.g. growth, reproduction, or energy storage) via the energy trade-offs to cover maintenance costs and ensure survival. (c) During severe stress situations, the progressive rise in ATP demand for maintenance overrides ATP supply capacities via aerobic metabolism, anaerobic metabolism and, eventually, metabolic depression sets in to ensure energy homeostasis at the reduced energy turnover. This represents a time-limited situation, although metabolic depression can extend survival time from several months up to several years.

allocation and metabolic responses to stress (Fig. 19.1), a useful classification of the degree of environmental stress that allows distinguishing between moderate and extreme stress can be derived. Such classification was first proposed for environmental temperature (Pörtner 2001, 2002) but can be extended to any environmental stressor.

As was shown for temperature, aerobic scope can be maintained only within the defined range, with highest aerobic scope in the optimal range and a decreasing scope in the pejus range with, finally, zero aerobic scope in the critical range when anaerobic metabolism sets in and all ATP is utilized for maintenance (Fig. 19.2). Such temperature-induced limitation of the aerobic scope and transition to partial anaerobiosis at critical temperatures was demonstrated in marine mollusks (Sokolova and Pörtner 2003; Pörtner et al. 2006). The disappearance of the aerobic scope and transition to partial or complete anaerobiosis inevitably heralds a transition from the moderate-stress situation where long-term survival of the population is possible to the time-limited situation where survival is impossible unless the environmental conditions return to the optimum or at least the pejus range. In stress tolerant organisms such as many intertidal mollusks, this transition is often accompanied by metabolic rate depression that reduces basal metabolic costs and allows avoiding negative energy scope (i.e. situation when ATP

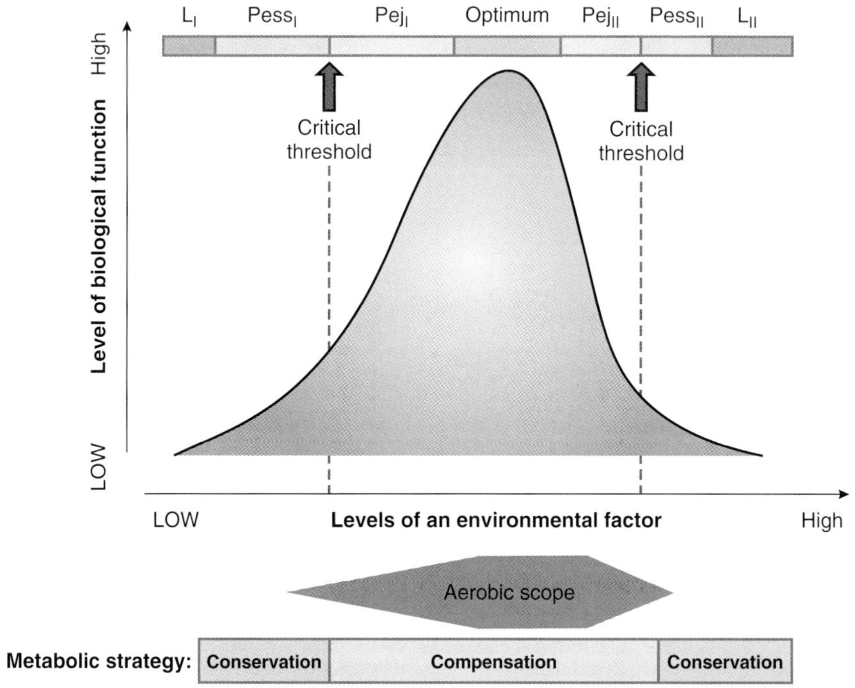

Fig. 19.2 A generalized scheme of the effects of environmental variables (temperature, salinity, oxygen availability, etc.) on metabolic function and adaptation strategies of mollusks. Aerobic scope is maximum when the levels of the environmental variable are close to optimum and decreases in the suboptimal (pejus) range (Pej_I and Pej_{II}). The aerobic scope eventually disappears when the environmental variable reaches the critical threshold levels and crosses into the pessimum range ($Pess_I$ and $Pess_{II}$), where the organism becomes partially or completely sustained by anaerobiosis. In the lethal range (L_I and L_{II}), only short-term passive survival is possible. In the suboptimal and pejus ranges, the main adaptive metabolic strategy is compensation for the increased energy demand for elevated costs of homeostasis, stress protection, and damage repair, whereas energy conservation strategies (i.e. metabolic rate depression) predominate in the pessimum range. Due to the fact that metabolic processes are down-regulated and many of the essential cellular and organismal functions are placed on hold during the metabolic rate depression, survival in the pessimum range is time limited. (Modified from Pörtner and Farrell 2008.)

consumption for basal maintenance outstrips ATP production; see below).

The same principles can be applied to any environmental stressor to determine the ecologically relevant limits of the tolerance window using a transition from energy-sustainable (i.e. optimum and pejus) to non-sustainable (i.e. pessimum) range (Fig. 19.2). Such transition occurs much earlier than the onset of lethality, and in the case of temperature, was shown to be a better predictor of the species' distribution limits in the field than the lethal thresholds (Pörtner et al. 2001; Pörtner and Knust 2007). The advantage of this metabolically based approach is providing the common basis for the analysis of the ecologically relevant tolerance limits for a variety of environmental factors irrespective of their nature.

STRATEGIES OF METABOLIC STRESS RESPONSE: COMPENSATION VERSUS CONSERVATION

In mollusks, strategies of metabolic adaptation to environmental stress depend both on the nature and the degree of environmental stress (Fig. 19.2). Exposure to either moderate or extreme stress elicits distinct gene expression profiles and strikingly different metabolic responses indicating different mechanisms of stress tolerance (Hochachka and Guppy 1987; Hochachka and Somero 2002; Gracey et al. 2008). Generally, moderate stress (when environmental conditions slightly deviate from the optima) results in elevated metabolic rates and ATP turnover that allows an organism to cover additional energy expenses for cellular maintenance and damage repair due to the stress exposure. In contrast, extreme stress often results in metabolic rate depression that slows down the biological clock of an organism. Metabolic depression allows conservation of metabolizable resources and decelerates the unfavorable change in intracellular milieu due to accumulation of metabolic waste and end products. This general rule broadly applies to a wide range of environmental stressors including low oxygen availability, desiccation stress, fluctuations in pH or salinity, exposure to pollutants, and others. Irrespective of the degree of the stressor, stress exposure often leads to oxidative stress and other types of cellular damage and requires the cell to re-direct its resources towards maintenance. This may pose a significant challenge to a stress-exposed organism, especially under conditions of metabolic rate depression when ATP production is severely limited.

COMPENSATORY METABOLIC RESPONSES TO MODERATE STRESSORS

Exposure to low levels of environmental stress can be compensated by physiological mechanisms that alleviate the negative effects of stress on the organism's performance, viability, and production. Generally, sublethal stress of any kind induces a set of common responses. Many of these common stress responses are evolutionarily strongly conserved and referred to as the minimal stress proteome of eukaryotes (Kültz 2005). However, such compensation comes at a cost leading to an increased energy demand due to the additional costs of, for example, synthesis of stress proteins or damage repair as well as increased activity associated with the physiological compensatory responses (e.g. hyperventilation to compensate for low oxygen supply) or behavioral escape mechanisms. Even though each of these mechanisms usually contributes less than 5–10% to the overall ATP demand under the optimal conditions (Calow 1989a; Kültz 2003), their joint activation during stress exposures may significantly increase basal metabolic cost of the maintenance.

Protein Homeostasis

In mollusks, heat shock proteins (HSP) are ubiquitously expressed in response to heat (Clegg et al. 1998; Piano et al. 2004; Ioannou et al. 2009; Ivanina et al. 2009), cold (Lesser and Kruse 2004; Brun et al. 2008), and chemical stress including metal and organic pollutants (Cruz-Rodríguez and Chu 2002; Piano et al. 2004; Franzellitti and Fabbri 2005; Choi et al. 2008). HSPs play a key role in the folding of *de novo* synthesized proteins as well as in repair, refolding, and elimination of damaged or denatured proteins (Nagao et al. 1990; Hartl 1996; Feder and Hofmann 1999). HSPs act as molecular chaperons stabilizing non-native conformation and facilitating correct folding of protein subunits, often coupling ATP binding/hydrolysis to the folding process. HSPs are highly expressed in the cell and may have rapid turnover, especially during stress exposures constituting up to 10% of the total synthesized proteins under control conditions and an even larger fraction during stress (Hofmann and Somero 1996; Tomanek and Somero 2000; Kültz 2003). Therefore, both the synthesis and chaperoning action of HSPs are energy-requiring. Such energy expense was shown to contribute to reduced somatic growth during heat stress in mollusks (Hofmann and Somero 1995). Additional energy allocation to HSPs synthesis was also shown to contribute to increased summer mortality of mussels *Mytilus galloprovincialis* (Anestis et al. 2007) and decline in reproduction in *Mytilus californianus* (Petes et al. 2008).

Irreparably damaged proteins are subject to proteolysis and must be removed from the cell. One of the major proteolytic mechanisms induced by stress starts with ubiquitination of damaged proteins (Schlesinger 1990). Ubiquitination is an ATP-dependent process that requires the support of high levels of aerobic metabolism (Hofmann and Somero 1995, 1996). Other protein degradation mechanisms can also be activated by environmental stressors, as shown by up-regulation of mitochondrial proteases and lysosomal proliferation during exposure to pollutants in mollusks and fish (Rangsayatorn et al. 2004; Sanni et al. 2008). These mechanisms are also likely to be ATP-dependent, although their contribution to cellular and whole-organism energy requirements in mollusks remains to be investigated. Given that protein synthesis in mollusks comprises about 20% of the energy costs for maintenance (Hawkins 1991; Cherkasov et al. 2006), protein homeostasis maintenance during stress exposure can be energetically costly and take up a substantial chunk of the overall organismal energy budget, diverting energy from growth and reproduction (Feder and Hofmann 1999; Kassahn et al. 2009).

Redox Status and Oxidative Stress

Reactive oxygen species (ROS) are continuously generated by a variety of cellular processes in aerobic organisms, and unless rapidly detoxified by enzymatic and nonenzymatic antioxidants, can cause oxidative stress, damage of vital macromolecules, and eventually cell death. Here, we briefly list the major components of the antioxidant defense and highlight their role in the cellular energy budget. A major role in cellular antioxidant defense belongs to enzymes including superoxide dismutase (SOD), catalase (CAT), glutathione peroxidase (GPx), glutathione *S*-transferase (GST), and glutathione reductase (GR). There are also several low molecular weight nonenzymatic antioxidants including carotenoids, flavonoids, vitamins C, A and E, as well glutathione (GSH). A variety of environmental stressors such as heat stress, ultraviolet radiation (UVR), metals, and xenobiotics increase ROS production and thus require up-regulation of antioxidant defenses (Winston and Giulio 1991; Valko et al. 2005; Kakkar and Singh 2007). This will inevitably lead to elevated energy costs for the synthesis of antioxidant enzymes, as well as repair, degradation, and resynthesis of oxidatively damaged cellular components. Despite the fact that the contribution of antioxidants to the energy costs of stress responses has been widely recognized (Calow 1989b; Sibly and Calow 1989; Calow and Forbes 1998), the quantitative aspect of this contribution has not been studied extensively and must be addressed in future research.

Biotransformation and Detoxification of Xenobiotics

As the first line of defense against xenobiotics, aquatic organisms including mollusks employ multixenobiotic resistance proteins (MXR) belonging to the superfamily of ATP-binding cassette (ABC) transporters (Kurelec 1995; Smital and Kurelec 1998; Bard 2000). These proteins pump xenobiotics out of the cell either in their native form, or conjugated to small organic ligands (Litman et al. 2001; Wortelboer et al. 2008). MXR transporters have a broad specificity towards a range of organic xenobiotics and catalyze their ATP-dependent efflux from the cell, preventing accumulation of toxins and their interactions with critical cellular components (Kurelec 1995; Smital and Kurelec 1998; Bard 2000). MXR and related ABC transporters also play an important role in resistance to toxic metals such as Cd and Hg (Huynh-Delerme et al. 2005; Kimura et al. 2005; Ivanina and Sokolova 2008; Bošnjak et al. 2009). Although divalent metals are usually not considered classic MXR substrates, studies suggest MXR and its homologs can extrude Cd conjugated to GSH, ATP or other organic ligands (Li et al. 1997; Borst et al. 2000; Thévenod et al. 2000). Despite the high efficiency of ABC transporters including MXR in keeping intracellular concentrations of xenobiotics at bay, their ATP requirement for pumping are relatively low and account for 0.02–0.04% of the basal metabolic rate (Ivanina and Sokolova

2008). These are conservative calculations that do not take into account the basal ATP hydrolysis of idling MXR and their synthesis. They, however, suggest that the contribution of the stress-induced induction of MXR to the basal metabolic costs of an organism are likely to be due to the costs of protein synthesis and maintenance rather than the pumping cost itself.

If this first line of defense represented by MXR fails, cellular mechanisms of detoxification of xenobiotics become activated. Organic pollutants are metabolized with a wide range of biotransformation enzymes (Livingstone 1993; Stegeman and Hahn 1994). Many of the biotransformation enzymes are also involved in the normal metabolism and can alter their substrates in the presence of xenobiotics. Furthermore, all enzymes require oxygen thus contributing to the nonmitochondrial respiration, which can comprise up to 10% in molluscan cells (Cherkasov et al. 2006). Moreover, their induction by exposure to xenobiotics will also incur in protein synthesis costs similar to other cellular protection proteins described above. In contrast to organic xenobiotics, toxic metals in the cell are mostly handled by a family of small cysteine-rich proteins called metallothioneins (MTs) that play an important role in metal homeostasis and maintenance of the cellular redox status (Palmiter 1998; Haq et al. 2003). Impairment or knockout of MTs strongly sensitizes the organism to metal toxicity and these proteins may be the single most important defense system against the toxic metal insult (Klaassen et al. 1999, 2009; Dallinger et al. 2009). MTs play a role in ROS scavenging (Viarengo et al. 1999) and can be induced by metals, and thermal and osmotic stress (Viarengo et al. 1999; Piano et al. 2004; Ivanina et al. 2009; Gourgou et al. 2010). Elevated MTs and HSPs expression was correlated with elevated basal metabolic rate in mollusks (Ivanina et al. 2008) suggesting their contribution to the rising energy cost during stress response.

As evident from the examples discussed above, cellular protection systems are expensive in terms of the energy required for their production, maintenance, and function. This energy requirement is universal for all general cellular defense system and may explain why elevated metabolic costs are often a common feature of (moderate) stress response irrespective of the nature of stressor (Cherkasov et al. 2006; Jeong and Cho 2007; Lannig et al. 2006a, 2008; Walther et al. 2009).

METABOLIC RESPONSE TO EXTREME STRESS: ENERGY CONSERVATION

When the levels of an environmental variable strongly deviates from the optimum, the degree of disturbance of metabolic functions and the additional energy expense required to counteract the stressor may become incompatible with the normal functioning and maintenance of homeostasis. When this is the case, mollusks resort to metabolic arrest. Transition into the pessimum range is often heralded by the onset of partial anaerobiosis as the organism attempts to supplement the failing aerobic metabolism and increased energy costs of homeostasis by anaerobic ATP production (see above; Figs 19.1 and 19.2). Such critical metabolic limits characterized by the disappearance of the aerobic scope and aerobic/anaerobic transition were described for temperature, salinity, and oxygen concentrations (Matsushima et al. 1984; Sokolova et al. 2000; David et al. 2005; Le Moullac et al. 2007) and are also likely to be found for other environmental factors, such as pH, CO_2 concentration, pollutants, and their combinations. Thus, stress-induced transition to partial or complete anaerobiosis may represent a universal physiological indicator that allows distinguishing between moderate and extreme stressors, and delineation of the threshold between the pejus and pessimum ranges for a given environmental factor.

Metabolic rate depression, or metabolic arrest, is a common response to extreme stress incompatible with the normal physiological functioning in a variety of organisms, including mollusks. Abilities to temporarily and reversibly suppress metabolic rates well below standard metabolic rate (SMR) are especially well developed in mollusks from extreme environments such as intertidal zones of the oceans, ephemeral freshwater bodies, and cold or arid terrestrial biotopes (Fig. 19.3). During exposure to extreme conditions mollusks can reduce their metabolism down to 1–10% of the standard metabolic rates (Hochachka and Guppy 1987; Storey and Storey 1988; Sokolova et al. 2000).

From a bioenergetic perspective, metabolic adaptations to extreme stress typically revolve around the energy balance, particularly that between the consumption and production of ATP. Metabolic arrest involves coordinated suppression of anaerobic and aerobic ATP production with the commensurate reduction of the energy demand. During metabolic arrest, an organism usually seeks out a shelter and becomes inactive, and all available energy is devoted to the

Fig. 19.3 Intertidal mollusks such as the blue mussel (*Mytilus edulis*) routinely experiences transition from (a) active, fully aerobic state during high tides to (b) metabolic rate depression mostly supported by anaerobic metabolism during low tides to avoid desiccation. Other extreme stressors that prompt shell closure and limit gas exchange, such as extremely low or high salinities, hypoxia/anoxia, acute exposure to pollutants (both anthropogenic and natural such as H_2S), or freezing, similarly lead to metabolic arrest. In contrast, some of the strictly subtidal mollusks (such as (c) a nudibranch *Coryphella*) or pelagic species (such as (d) sea angels *Cliona*) are incapable of the metabolic arrest and rely on other, energy-extensive mechanisms of stress protection. (Photographs by Mikhail Fedyuk.) See plate section for a color version of this image.

maintenance of homeostasis with no investment into somatic or gonad growth, reproduction, or locomotion. As a result of the coordinated decrease in energy metabolism, mollusks generally maintain tissue energy balance as indicated by the minimal depletion of tissue ATP and phosphagens, even during the prolonged stress exposure (Isani et al. 1989; Churchill and Storey 1996; Sokolova et al. 2000; Kurochkin et al. 2008). Physiological and molecular mechanisms of metabolic arrest have been discussed previously in several excellent reviews (Hochachka and Somero 1984; Hochachka and Guppy 1987; Larade and Storey 2002, 2007).

OXIDATIVE STRESS AND CELLULAR PROTECTION DURING STRESS-INDUCED METABOLIC MODULATION

Stress-induced interference with metabolic functions strongly affects cellular homeostasis leading to shifts in acid-base, ion, and energy balance, and cellular redox status. Among these consequences of metabolic disruption, disturbance of the redox status and oxidative stress play an important role. Indeed, effects of many common environmental stressors including elevated temperature, hypoxia/anoxia, reoxygenation, and exposure to pollutants converge on oxidative stress in mollusks. The percentage of oxygen converted to ROS is significantly higher in mollusks than in endotherms but the overall ROS generation rates are similar (0.01–0.13 nmol H_2O_2 mg^{-1} protein min^{-1} in mollusks versus 0.01–0.7 nmol H_2O_2 mg^{-1} protein min^{-1} in birds and mammals) due to the slower respiration rate of molluskan mitochondria (Abele et al. 2002; Heise et al. 2003; Abele and Puntarulo 2004; Cherkasov et al. 2007). In addition to mitochondrial ROS sources, other enzymes may contribute to ROS generation in mollusks, especially under stress conditions. These include reduced nicotinamide adenine dinucleotide phosphate (NADPH) oxidase, which produces $O_2^{\bullet -}$ during immune challenge in molluscan hemocytes, nitric oxide synthase (NOS), fatty acid β-oxidation enzymes in peroxisomes, and xenobiotic-metabolizing enzymes such as cytochrome

450 monooxygenase that can produce $O_2^{\bullet-}$ and H_2O_2 during metabolism of organic pollutants (Winston and Giulio 1991; Stegeman and Hahn 1994; Boyd and Burnett 1999; Sokolova 2009). Xanthine oxidase (XO), which plays an important role in cytosolic ROS generation in mammals, has low activity in mollusks and is unlikely to significantly contribute to ROS generation in their tissues (Cancio and Cajaraville 1997; Osman et al. 2004). Some pollutants (especially transition metals or redox-cycling organic pollutants) can also directly induce ROS production via nonenzymatic (Fenton-type) reactions (Di Giulio et al. 1989; Winston and Giulio 1991; Livingstone 2001). All ROS-producing mechanisms are oxygen-dependent and can generate ROS in normoxia or hypoxia, but not in anoxia.

Environmental stressors that disturb energy metabolism in mollusks are also often associated with oxidative stress reflecting elevated ROS production, reduced antioxidant capacity, or some combination of both. This may occur either via direct mechanisms – by the stressors interfering with the function of the mitochondrial electron transport chain or stimulating extramitochondrial ROS-producing enzymes – or indirectly – due to the stress-induced energy deficiency and insufficient ATP generation to sustain production and/or regeneration of antioxidants. Typically, during exposures to moderate stress, antioxidants and other cellular protectors (e.g. HSPs or detoxification enzymes) are induced in order to maintain the normal cellular functions at the expense of elevated ATP turnover and energy expenditure. In contrast, during exposure to extreme stressors, which leads to metabolic rate depression, many of the mechanisms that the cells normally harness to protect intracellular structures are maintained in nonfunctional state in order to conserve energy stores and to ensure survival. The profile of cellular proteins shows an overall depression of the *de novo* protein synthesis and a heavy bias towards expression of the ultimate defense systems (such as HSPs) preventing protein denaturation and irreversible cellular damage (Tomanek and Somero 2000; Larade and Storey, 2002, 2007). Antioxidant levels are often either suppressed or unchanged during metabolic arrest, paralleled by a decreased ROS production due to the overall reduction of metabolism (Abele and Puntarulo 2004). An exception to this pattern is shown by several anoxia- and hypoxia-tolerant intertidal mollusks, which experience cycles of oxygen deprivation and recovery on a daily basis. In these organisms, antioxidants are often induced during hypoxia or anoxia, a likely adaptation to protect against a ROS surge during reoxygenation (Pannunzio and Storey 1998; Hermes-Lima et al. 2001; Abele and Puntarulo 2004). This induction of antioxidants may be partially mediated by elevated expression of the *cfos* and *cjun* transcription factors, which in turn up-regulate transcription of several key antioxidant enzymes (such as SOD and GST) setting the stage for successful post-hypoxic/-anoxic recovery (Hochachka and Lutz 2001; see Chapters 1, 6, 9, 12, 14, 22).

Temperature Stress

Environmental temperature has a direct effect on all metabolic processes in mollusks as well as in other ectotherms (Hochachka and Somero 2002). Mitochondrial oxygen consumption increases with increasing temperature leading to elevated rates of one-electron reduction of oxygen yielding $O_2^{\bullet-}$. At elevated temperatures activities of cytosolic and microsomal enzymes that generate ROS in extramitochondrial compartments also increase and contribute to both elevated rates of metabolism and increased ROS production. However, increases in ROS production are not directly proportional to the temperature-induced changes in metabolism. Typically, oxidative stress is minimal at and around optimal temperature, due to the relatively low rates of ROS generation and optimal efficiency of antioxidant defenses, and may increase at high and low temperature extremes (Abele and Puntarulo 2004).

High temperatures (particularly in pejus and critical ranges) increase mitochondrial ROS production in mollusks (Abele et al. 2002; Heise et al. 2003). This heat-induced increase in ROS production goes hand-by-hand with the heat and free-radicals-mediated membrane damage in mitochondria, their progressive uncoupling, and decrease in phosphorylation efficiency (Abele et al. 2002; Abele and Puntarulo 2004). At critically high temperatures the inner mitochondrial membrane becomes leakier and more oxygen is univalently reduced to $O_2^{\bullet-}$, which may eventually lead to extensive intracellular damage and cell death if not counterbalanced by antioxidant defenses. A moderate temperature increase stimulates antioxidant defenses inducing expression of antioxidant enzymes and/or production of nonenzymatic antioxidants such as GSH or MTs (Lannig et al. 2006b; Ivanina et al. 2009; Gourgou et al. 2010). In an Antarctic clam, *Laternula elliptica*, extreme thermal stress (10°C) led to a significant increase in GPx and GST activities paralleled by an elevation of GSH tissue levels. Acute heat was also shown to induce MTs production in temperate bivalves *M. galloprovincialis* and *C. virginica* (Ivanina et al. 2009; Gourgou et al. 2010).

Cold acclimation and adaptation can also enhance oxidative stress in marine mollusks. Although metabolic activity and therefore ROS generation is suppressed by low temperatures, certain aspects of cold adaptation can make mollusks more prone to oxidative stress. It has been proposed that higher oxygen solubility in cold water than in the cytosol may create an intracellular environment with elevated P_{O_2} favoring ROS formation (Abele and Puntarulo 2004). To some extent, these effects are counteracted by the reduced oxygen diffusion rate at low temperatures and high degree of molecular crowding in the cytosol, which decreases the oxygen solubility and makes it less temperature-dependent (Sidell 1998). More importantly, restructuring of cellular membranes in the cold (specifically an increase in the content of polyunsaturated fatty acids (PUFAS) and phosphatidylethanolamine that are particularly vulnerable to oxidation) makes membranes more susceptible to lipid radical chain propagation and oxidative damage (Crockett 2008). Thus, it is perhaps not surprising that polar mollusks are characterized by higher level of antioxidants compared to temperate species (Regoli et al. 1997a,b; Heise et al. 2003; Camus et al. 2005). In addition to the higher risk of oxidative damage due to the membrane properties and oxygen concentration, low RNA and protein turnover rates, as well as low food availability, may limit the capacity for repair and replacement of damaged biomolecules in polar mollusks, making them susceptible to the deleterious consequences of oxidative stress (Fraser et al. 2002; Camus et al. 2005). Extremely low (subfreezing) temperatures result in tissue anoxia and metabolic arrest in mollusks due to the cessation of ventilation and circulation; therefore, the effects of freezing on ROS formation and antioxidant defenses are similar to those of environmental anoxia and will be reviewed below.

Oxygen Deficiency

Unless the tissue is completely anoxic, even trace amounts of oxygen (~5–7 torr) are sufficient for continuing ROS generation in mitochondria (Guzy et al. 2005; Guzy and Schumacker 2006). In fact, ROS production often increases in hypoxic and hypoxemic states due to the increased electron slip from the highly reduced electron carriers in the mitochondrial electron transport chain (ETC). Anoxia also leads to the highly reduced state of ETC and although ROS are not generated in anoxia due to the absence of oxygen, reoxygenation after anoxia and hypoxia results in a burst of ROS production. Thus, it is not surprising that up-regulation of antioxidant defense systems is an important adaptation to cope with intermittent hypoxia and anoxia in mollusks. In *M. edulis*, intertidal populations that frequently experience anoxia/reoxygenation stress during the air–water transitions showed higher antioxidant defense capacities than their subtidal continuosly submerged counterparts (Letendre et al. 2009). Antioxidants were also upregulated during estivation and recovery in a land snail *Otala lactea* (Hermes-Lima et al. 1998). In *O. lactea*, estivation stimulated activity and substrate affinity of glucose-6-phosphate dehydrogenase (G6PDH), a major enzyme in the NADPH-producing pentose phosphate pathway (Ramnanan and Storey 2006). This activation was mediated by phosphorylation of G6PDH and was proposed to redirect carbon flow through the pentose phosphate cycle away from the glycolysis, to help maintain NADPH production for use in antioxidant defense (Ramnanan and Storey 2006). This boost of antioxidant protection is considered to be an adaptation to prevent oxidative damage during the ROS burst in response to reoxygenation.

Oxygen deficiency during freezing or anoxia also stimulates MT expression. Thus, in a marine gastropod *Littorina littorea*, MT transcripts were 2.5–6-fold induced within 1 h of anoxia or freezing stress and remained elevated throughout the stress exposure, returning back to the basal levels after 24 h of recovery (English and Storey 2003). An increase in MT transcripts may be mediated either via the same signaling pathways as other antioxidants (due to the presence of an antioxidant response element in MT promoter) or through the redox-mediated changes in intracellular free metal pools that activate metal transcription factor 1 (MTF-1) (Andrews 2000; Saydam et al. 2003), and may represent an anticipatory response preparing mollusks to cope with the ROS burst during reoxygenation. Antioxidant roles of MTs have been demonstrated in vertebrate models as well as in intertidal bivalve *M. edulis* (English and Storey 2003). MTs may either directly scavenge ROS via thiolate oxidation of the cysteine residues, and/or assist in binding excess metals released by ROS from intracellular storage sites (Andrews 2000).

Metabolic acidosis that develops during prolonged anaerobiosis in hypoxia and anoxia may also contribute to ROS generation. Low intracellular pH due to the anaerobic accumulation of protons may enhance oxidative stress due to the release of chelated Fe^{2+} from intracellular stores, which can catalyze Fenton reactions and generate $HO^{\bullet}$. Indeed, intracellular acidosis results in elevated cellular injury and lipid peroxidation in mammalian cells (Bronk and Gores 1991; Levraut et al. 2003), and these effects can be prevented by iron chelators (Bronk and Gores 1991). In mollusks, the effects of intracellular pH on oxidative damage and ROS generation are unknown and warrant further investigation. Interestingly, anoxic exposure led to a rapid increase in mRNA and protein levels of an iron-chelating protein, ferritin, in *L. littorea* (English and Storey 2003), suggesting that iron release contributes to the cellular stress during anoxia and may be triggered by intracellular acidosis. Besides the intertidal mollusks, up-regulation of ferritin as well as the transferrin receptor protein needed for the iron uptake into the cell have been also demonstrated in several other anoxia- and hypoxia-tolerant organisms (Storey and Storey 2007; Larade and Storey 2009), suggesting that this adaptation may be common in anoxia and hypoxia tolerance. Future studies are needed to determine the degree to which iron chelation contributes to survival of intermittent oxygen deficiency in mollusks and whether buffering of metabolic protons by shell and connective tissue calcium carbonate, which stabilizes intracellular pH during prolonged anaerobiosis, also attenuates iron release and/or ROS production in hypoxia/anoxia and subsequent recovery.

It is also worth noting that many mollusks are routinely exposed to environmental acidosis during seasonal and diurnal hypercapnia in estuaries and coastal areas with limited water exchange, where pH may drop down to 6–7 (Hubertz and Cahoon 1999; Keppler and Ringwood 2002). These effects of seasonal and diurnal hypercapnia may become exacerbated in the coming years due to the ocean acidification driven by anthropogenically released CO_2 (Caldeira and Wickett 2003, 2005). Given that most mollusks have a limited capability to regulate extra- and intracellular pH during such events (Burnett 1997), it would be interesting to determine whether ROS generation increases during the environmental acidosis in mollusks, whether or not it is associated with the release of iron from intracellular stores, and what is the threshold pH (if any) during metabolic and/or environmental acidosis that triggers elevated oxidative stress. These important questions require further investigation and will provide new insights into the relationship between metabolic function and cellular redox status in mollusks.

Pollution

Oxidative stress is a hallmark of exposure to a variety of organic and inorganic pollutants as evidenced by a wide use of the oxidative stress biomarkers in pollution monitoring programs (Kelly et al. 1998; Livingstone 2001). Toxic metals can either directly catalyze ROS production via Fenton-like chemistry or indirectly induce elevated ROS generation via the interference with MTCs (Stohs et al. 2000; Valko et al. 2005). Many organic xenobiotics are capable of modulating oxidative stress either by acting directly as redox cycling compounds or as a consequence of biotransformation to quinones (Livingstone et al. 1990; Sheehan and Power 1999). Some forms of CYP450 monoxygenases involved in Phase I biotransformation preferentially yield ROS in place of oxygenated metabolites, and their induction results in oxidative stress (Symons and King 2003). In mollusks, oxidative stress has been documented extensively, both during chronic and acute laboratory exposure as well as during field exposure in polluted environments (Niyogi et al. 2001; Funes et al. 2006; Pytharopoulou et al. 2008; Kurochkin et al. 2009). Effects of pollutant exposure on antioxidants in mollusks strongly depends on the concentration of the pollutant. In agreement with the metabolic model proposed in Figs 19.1 and 19.2, exposure to moderate sublethal levels of metals and organic pollutants typically results in energy-dependent up-regulation of the antioxidant defenses, although the degree of induction and the suite of antioxidants involved in this up-regulation varies between species and populations, and depends on the nature of the pollutant (Niyogi et al. 2001; Orbea et al. 2002; Funes et al. 2006; Gagné et al. 2008). During acute exposure, antioxidant defenses are often suppressed, reflecting the general toxicity of the pollutant (Orbea et al. 2002; Chandran et al. 2005). During exposures to both moderate and acutely toxic concentrations, pollutant-induced oxidative stress results in a reduction of the aerobic scope in marine mollusks due to either elevated costs of antioxidant protection and damage repair or damage to the metabolic machinery, or combination thereof (Sokolova and Lannig 2008).

CONCLUSIONS

The various aspects of stress tolerance reviewed in this chapter emphasize the role of a balanced energy metabolism via compensation and conservation in determining an organism's ability to cope with environmental stressors. The essential activities of survival, reproduction, and growth are governed by the rate at which organisms acquire, process, and transform energy. As summarized in this chapter, cellular stress response is energetically costly; however, we are yet to gain a full understanding of how different cellular mechanisms contribute to energy requirements and thus set the limits in environmental stress tolerance. Furthermore, both stress tolerance and energy allocation vary with life stage (Kikkawa et al. 2003; Dupont et al. 2008; Pörtner and Farrell 2008), adding another level of complexity to the analysis of stress responses in nature. As was proposed in Pörtner's concept of oxygen- and capacity-limited thermal tolerance in aquatic ectotherms (Pörtner 2001, 2002) and extended in this chapter to other environmental stressors, thresholds in stress tolerance (pejus, critical, and lethal ranges) depend on metabolic adaptation and vary within and between species (Gazeau et al. 2007; CIESM 2008; Kassahn et al. 2009; Melzner et al. 2009). We showed that in analyzing stress tolerance limits, moderate and extreme stresses have to be distinguished because they elicit different metabolic responses: compensation (mainly associated with a rise in energetic needs) versus conservation (mainly associated with suppressed energy turnover).

The present chapter and recent reviews (Drinkwater et al. 2009; Polechová et al. 2009; Pörtner and Lannig 2009; Melzner et al. 2009) show that evaluating general predictions for marine ecosystems in face of global environmental change has a long way to go and further intensive investigations are required in order to gain a more complete picture of the ocean of tomorrow. To make it even more complicated, it has long been recognized that the impact of one stressor can be modified by others (Kinne 1964) and that there is a need to study their actions concurrently, in particular with respect to temperature interactions (Pörtner et al. 2005; Sokolova and Lannig 2008; Dupont and Thorndyke 2008, Lannig et al. 2010). Analyses of energy metabolism and energy allocation provide a useful and efficient tool for such studies, allowing integration of stress responses irrespective of the nature of interacting stressors.

Assessment of the impacts of environmental stressors associated with global climate change and marine pollution on mollusks is important not only from the viewpoint of fundamental ecology and biogeography, but also for economical reasons. Mollusks are key components of marine ecosystems: affecting energy and nutrient flow in estuarine and coastal systems; acting as ecosystem engineers in these habitats; providing niches for other species; and serving as food sources for top-level predators, including birds and humans (Jackson et al. 2001; Kimbro et al. 2009). Shellfish represent a large part of the world's aquaculture and fisheries. Mollusks contributed to 19% of the USA commercial fishery revenues in 2007 (Cooley and Doney 2009). Accordingly, any stressor that reduces welfare and fitness of marine mollusks will have major consequences on natural populations and aquaculture output, with important economic losses. Improvement of our knowledge about how and why marine ectotherms, including mollusks, react to environmental stressors will help to improve human effort in nature and resource conservation and thus help to retain diversity – in the marine ecosystems as well as on our menus.

REFERENCES

Abele, D., Puntarulo, S. (2004) Formation of reactive species and induction of antioxidant defence systems in polar and temperate marine invertebrates and fish. *Comparative Biochemistry and Physiology A* 138, 405–415.

Abele, D., Heise, K., Pörtner, H.O., Puntarulo, S. (2002) Temperature-dependence of mitochondrial function and production of reactive oxygen species in the intertidal mud clam *Mya arenaria*. *Journal of Experimental Biology* 205, 1831–1841.

Andrews, G.K. (2000) Regulation of metallothionein gene expression by oxidative stress and metal ions. *Biochemical Pharmacology* 59, 95–104.

Anestis, A., Lazou, A., Pörtner, H.O., Michaelidis, B. (2007) Behavioral, metabolic, and molecular stress responses of marine bivalve *Mytilus galloprovincialis* during long-term acclimation at increasing ambient temperature. *American Journal of Physiology* 293, R911–R921.

Bard, S.M. (2000) Multixenobiotic resistance as a cellular defense mechanism in aquatic organisms. *Aquatic Toxicology* 48, 357–389.

Bayne, B.L. (2004) Phenotypic flexibility and physiological tradeoffs in the feeding and growth of marine bivalve molluscs. *Integrative and Comparative Biology* 44, 425–432.

Bayne, B.L., Worrall, C.M. (1980) Growth and production of mussels *Mytilus edulis* from two populations. *Marine Ecolology Progress Series* 3, 317–328.

Beiras, R., Camacho, A.P., Albentosa, M. (1995) Short-term and long-term alterations in the energy budget of young oyster *Ostrea edulis* L. in response to temperature change. *Journal of Experimental Marine Biology and Ecology* 186, 221–236.

Borst, P., Evers, R., Kool, M., Wijnholds, J. (2000) A family of drug transporters: the multidrug resistance-associated proteins. *Journal of the National Cancer Institute* 92, 1295–1302.

Bošnjak, I., Uhlinger, K.R., Heim, W. et al. (2009) Multidrug efflux transporters limit accumulation of inorganic, but not organic, mercury in sea urchin embryos. *Environmental Science and Technology* 43, 8374–8380.

Boyd, J.N., Burnett, L.E. (1999) Reactive oxygen intermediate production by oyster hemocytes exposed to hypoxia. *Journal of Experimental Biology* 202, 3135–3143.

Bronk, S.F., Gores, G.J. (1991) Acidosis protects against lethal oxidative injury of liver sinusoidal endothelial cells. *Hepatology* 14, 150–157.

Brun, N.T., Bricelj, V.M., MacRae, T.H., Ross, N.W. (2008) Heat shock protein responses in thermally stressed bay scallops, *Argopecten irradians*, and sea scallops, *Placopecten magellanicus*. *Journal of Experimental Marine Biology and Ecology* 358, 151–162.

Burnett, L.E. (1997) The challenges of living in hypoxic and hypercapnic aquatic enviroments. *American Zoologist* 37, 633–640.

Caldeira, K., Wickett, M.E. (2003) Oceanography: Anthropogenic carbon and ocean pH. *Nature* 425, 365–365.

Caldeira, K., Wickett, M.E. (2005) Ocean model predictions of chemistry changes from carbon dioxide emissions to the atmosphere and ocean. *Journal of Geophysical Research–Oceans* 110, C09S04.

Calow, P. (1989a) Proximate and ultimate responses to stress in biological systems. *Biological Journal of Linnean Society* 37, 173–181.

Calow, P. (1989b) Ecotoxicology? *Journal of Zoology* 218, 701–704.

Calow, P., Forbes, V.E. (1998) How do physiological responses to stress translate into ecological and evolutionary processes? *Comparative Biochemistry and Physiology A* 120, 11–16.

Camus, L., Gulliksen, B., Depledge, M.H., Jones, M.B. (2005) Polar bivalves are characterized by high antioxidant defences. *Polar Research* 24, 111–118.

Cancio, I., Cajaraville, M.P. (1997) Histochemistry of oxidases in several tissues of bivalve molluscs. *Cell Biology International* 21, 575–584.

Chandran, R., Sivakumar, A.A., Mohandass, S., Aruchami, M. (2005) Effect of cadmium and zinc on antioxidant enzyme activity in the gastropod, *Achatina fulica*. *Comparative Biochemistry and Physiology C* 140, 422–426.

Cherkasov, A.S., Biswas, P.K., Ridings, D.M., Ringwood, A.H., Sokolova, I.M. (2006) Effects of acclimation temperature and cadmium exposure on cellular energy budgets in a marine mollusk *Crassostrea virginica*: Linking cellular and mitochondrial responses. *Journal of Experimental Biology* 209, 1274–1284.

Cherkasov, A.S., Overton, R.A., Jr., Sokolov, E.P., Sokolova, I.M. (2007) Temperature dependent effects of cadmium and purine nucleotides on mitochondrial aconitase from a marine ectotherm, *Crassostrea virginica*: a role of temperature in oxidative stress and allosteric enzyme regulation. *Journal of Experimental Biology* 210, 46–55.

Cheung, S.G. (1993) Population dynamics and energy budgets of green-lipped mussel *Perna viridis* (Linnaeus) in a polluted harbour. *Journal of Experimental Marine Biology and Ecology* 168, 1–24.

Choi, Y.K., Jo, P.G., Choi, C.Y. (2008) Cadmium affects the expression of heat shock protein 90 and metallothionein mRNA in the Pacific oyster, *Crassostrea gigas*. *Comparative Biochemistry and Physiology C* 147, 286–293.

Churchill, T.A., Storey, K.B. (1996) Metabolic responses to freezing and anoxia by the periwinkle*Littorina littorea*. *Journal of Thermal Biology* 21, 57–63.

CIESM (2008) *Impacts of Acidification on Biological, Chemical and Physical Systems in the Mediterranean and Black Seas*, Briand, F. (ed.). CIESM Monography 36, International Commission for Scientific Exploration of the Mediterranean Sea, Monaco.

Clegg, J.S., Uhlinger, K.R., Jackson, S.A., Cherr, G.N., Rifkin, E., Friedman, C.S. (1998) Induced thermotolerance and the heat shock protein 70 family in the Pacific oyster *Crassostrea gigas*. *Molecular Marine Biology and Biotechnology* 7, 21–30.

Cooley, S.R., Doney, S.C. (2009) Anticipating ocean acidification's economic consequences for commercial fisheries. *Environmental Research Letters* 4. DOI: 10.1088/1748-9326/4/2/024007.

Crockett, E. (2008) The cold but not hard fats in ectotherms: consequences of lipid restructuring on susceptibility of biological membranes to peroxidation, a review. *Journal of Comparative Physiology B* 178, 795–809.

Cruz-Rodríguez, L.A., Chu, F.-L.E. (2002) Heat-shock protein (HSP70) response in the eastern oyster, *Crassostrea virginica*, exposed to PAHs sorbed to suspended artificial clay particles and to suspended field contaminated sediments. *Aquatic Toxicology* 60, 157–168.

Dallinger, R., Egg, M., Höckner, M., Schuler, D., Atrian, S., Capdevila, M. (2009) Molecular adaptation to cope with environmental stress of terrestrial gastropod cadmium-metallothioneins: High metal specificity at the protein level combines with multifunctionality of genes. *Comparative Biochemistry and Physiology A* 154, S33.

David, E., Tanguy, A., Pichavant, K., Moraga, D. (2005) Response of the Pacific oyster *Crassostrea gigas* to hypoxia exposure under experimental conditions. *FEBS Journal* 272, 5635–5652.

Di Giulio, R.T., Washburn, P.C., Wenning, R.J., Winston, G.W., Jewell, C.S. (1989) Biochemical responses in aquatic animals: a review of determinants of oxidative stress. *Environmental Toxicology and Chemistry* 8, 1103–1123.

Drinkwater, K.F., Beaugrand, G., Kaeriyama, M. et al. (2009) On the processes linking climate to ecosystem changes. *Journal of Marine Systems* 79, 374–388.

Dupont, S., Thorndyke, M. (2008) Ocean acidification and its impact on the early life-history stages of marine animals. *In* Briand, F. (ed.) *Impacts of Acidification on Biological, Chemical and Physical Systems in the Mediterranean and Black Seas*. CIESM Monographs 36, International Commission for Scientific Exploration of the Mediterranean Sea, Monaco, pp. 89–97.

Dupont, S., Havenhand, J., Thorndyke, W., Peck, L., and Thorndyke, M. (2008) Near-future level of CO_2-driven ocean acidification radically affects larval survival and development in the brittlestar *Ophiothrix fragilis*. *Marine Ecology Progress Series* 373, 285–294.

English, T.E., Storey, K.B. (2003) Freezing and anoxia stresses induce expression of metallothionein in the foot muscle and hepatopancreas of the marine gastropod *Littorina littorea*. *Journal of Experimental Biology* 206, 2517–2524.

Feder, M.E., Hofmann, G.E. (1999) Heat-shock proteins, molecular chaperons, and stress response: Evolutionary and Ecological Physiology. *Annual Review of Physiology* 61, 243–282.

Franzellitti, S., Fabbri, E. (2005) Differential HSP70 gene expression in the Mediterranean mussel exposed to various stressors. *Biochemical and Biophysical Research Communications* 336, 1157–1163.

Fraser, K.P.P., Clarke, A., Peck, L.S. (2002) Low-temperature protein metabolism: seasonal changes in protein synthesis and RNA dynamics in the Antarctic limpet *Nacella concinna* Strebel 1908. *Journal of Experimental Biology* 205, 3077–3086.

Funes, V., Alhama, J., Navas, J.I., Lopez-Barea, J., Peinado, J. (2006) Ecotoxicological effects of metal pollution in two mollusc species from the Spanish South Atlantic littoral. *Environmental Pollution* 139, 214–223.

Gagné, F., Burgeot, T., Hellou, J., St-Jean, S., Farcy, É., Blaise, C. (2008) Spatial variations in biomarkers of *Mytilus edulis* mussels at four polluted regions spanning the Northern Hemisphere. *Environmental Research* 107, 201–217.

Gazeau, F., Quiblier, C., Jansen, J.M., Gattuso, J.P., Middelburg, J.J., Heip, C.H.R. (2007) Impact of elevated CO_2 on shellfish calcification. *Geophysical Research Letters* 34, L07603.

Gilbert, D.L., Adelman, W.J. Jr., Arnold, J.M. (1990) *Squid as Experimental Animals*. Springer-Verlag, New York.

Gourgou, E., Aggeli, I.K., Beis, I., Gaitanaki, C. (2010) Hyperthermia-induced Hsp70 and MT20 transcriptional upregulation are mediated by p38-MAPK and JNKs in *Mytilus galloprovincialis* (Lamarck); a pro-survival response. *Journal of Experimental Biology* 213, 347–357.

Gracey, A.Y., Chaney, M.L., Boomhower, J.P., Tyburczy, W.R., Connor, K., Somero, G.N. (2008) Rhythms of gene expression in a fluctuating intertidal environment. *Current Biology* 18, 1501–1507.

Guzy, R.D., Schumacker, P.T. (2006) Oxygen sensing by mitochondria at complex III: the paradox of increased reactive oxygen species during hypoxia. *Experimental Physiology* 91, 807–819.

Guzy, R.D., Hoyos, B., Robin, E., Chen, H., Liu, L., Mansfield, K.D. et al. (2005) Mitochondrial complex III is required for hypoxia-induced ROS production and cellular oxygen sensing. *Cell Metabolism* 1, 401–408.

Haq, F., Mahoney, M., Koropatnick, J. (2003) Signaling events for metallothionein induction. *Mutation Research* 533, 211–226.

Hartl, F.U. (1996) Molecular chaperones in cellular protein folding. *Nature* 381, 571–580.

Hawkins, A.J.S. (1991) Protein turnover: a functional appraisal. *Functional Ecology* 5, 222–223.

Heise, K., Puntarulo, S., Pörtner, H.O., Abele, D. (2003) Production of reactive oxygen species by isolated mitochondria of the Antarctic bivalve *Laternula elliptica* (King and Broderip) under heat stress. *Comparative Biochemistry and Physiology C* 134, 79–90.

Hermes-Lima, M., Storey, J.M., Storey, K.B. (1998) Antioxidant defenses and metabolic depression. The hypothesis of preparation for oxidative stress in land snails. *Comparative Biochemistry and Physiology B* 120, 437–448.

Hermes-Lima, M., Storey, J.M., Storey, K.B. (2001) Antioxidant defenses and animal adaptation to oxygen availability during environmental stress. *In* Storey, K.B., Storey, J.M. (eds) *Cell and Molecular Response to Stress*. Elsevier Press, Amsterdam, pp. 263–287.

Hochachka, P.W., Guppy, M. (1987) *Metabolic Arrest and the Control of Biological Time*. Harvard University Press, Cambridge, MA, 227 pp.

Hochachka, P.W., Lutz, P.L. (2001) Mechanism, origin, and evolution of anoxia tolerance in animals. *Comparative Biochemistry and Physiology B* 130, 435–459.

Hochachka, P.W., Somero, G.N. (1984) *Biochemical Adaptation*. Princeton University Press, Princeton, NJ.

Hochachka, P.W., Somero, G.N. (2002) *Biochemical Adaptation: Mechanism and Process in Physiological Evolution*. Oxford University Press, Oxford, 466 pp.

Hofmann, G.E., Somero, G.N. (1995) Evidence for protein damage at environmental temperatures: seasonal changes in levels of ubiquitin conjugates and hsp70 in the intertidal mussel *Mytilus trossulus*. *Journal of Experimental Biology* 198, 1509–1518.

Hofman, G.E., Somero, G.N. (1996) Protein ubiquitination and stress protein synthesis in *Mytilus trossulus* occurs during recovery from tidal emersion. *Molecular Marine Biology and Biotechnology* 5, 175–184.

Hubertz, E., Cahoon, L. (1999) Short-term variability of water quality parameters in two shallow estuaries of North Carolina. *Estuaries and Coasts* 22, 814–823.

Huynh-Delerme, C., Huet, H., Noel, L., Frigieri, A., Kolf-Clauw, M. (2005) Increased functional expression of P-glycoprotein in Caco-2 TC7 cells exposed long-term to cadmium. *Toxicology in Vitro* 19, 439–447.

Ioannou, S., Anestis, A., Pörtner, H.O., Michaelidis, B. (2009) Seasonal patterns of metabolism and the heat shock response (HSR) in farmed mussels *Mytilus galloprovincialis*. *Journal of Experimental Marine Biology and Ecology* 381, 136–144.

Isani, G., Cattani, O., Carpene, E., Tacconi, S., Cortesi, P. (1989) Energy metabolism during anaerobiosis and recovery in the posterior adductor muscle of the bivalve *Scapharca inaequivalvis* (Bruguiere). *Comparative Biochemistry and Physiology B* 93, 193–200.

Ivanina, A.V., Sokolova, I.M. (2008) Effects of cadmium exposure on expression and activity of P-glycoprotein in eastern oysters, *Crassostrea virginica* Gmelin. *Aquatic Toxicology* 88, 19–28.

Ivanina, A.V., Cherkasov, A., Sokolova, I.M. (2008) Effect of cadmium on cellular protein and glutathione synthesis and expression of stress proteins in eastern oysters, *Crassostrea virginica* Gmelin. *Journal of Experimental Biology* 211, 577–586.

Ivanina, A.I., Taylor, C., Sokolova, I.M. (2009) Effects of elevated temperature and cadmium exposure on stress protein response in eastern oysters *Crassostrea virginica* (Gmelin). *Aquatic Toxicology* 91, 245–254.

Jackson, J.B.C., Kirby, M.X., Berger, W.H. et al. (2001) Historical overfishing and the recent collapse of coastal ecosystems. *Science* 293, 629–637.

Jeong, W.G., Cho, S.M. (2007) Long-term effect of polycyclic aromatic hydrocarbon on physiological metabolisms of the Pacific oyster, *Crassostrea gigas*. *Aquaculture* 265, 343–350.

Jørgensen, C.B. (1990) *Bivalve Filter Feeding: Hydrodynamics, Bioenergetics, Physiology and Ecology*. Olsen & Olsen, Fredensborg, Denmark, 140 pp.

Kakkar, P., Singh, B.K. (2007) Mitochondria: a hub of redox activities and cellular distress control. *Molecular and Cellular Biochemistry* 305, 235–253.

Kassahn, K.S., Crozier, R.H., Pörtner, H.O., Caley, M.J. (2009) Animal performance and stress: responses and tolerance limits at different levels of biological organisation. *Biological Reviews* 84, 277–292.

Kelly, S.A., Havrilla, C.M., Brady, T.C., Abramo, K.H., Levin, E.D. (1998) Oxidative Stress in Toxicology: Established Mammalian and Emerging Piscine Model Systems. *Environmental Health Perspectives* 106, 375–384.

Keppler, C.J., Ringwood, A.H. (2002) Effects of metal exposures on juvenile clams, *Mercenaria mercenaria*. *Bulletin of Environmental Contamination and Toxicology* 68, 43–48.

Kikkawa, T., Ishimatsu, A., Kita, J. (2003) Acute CO_2 tolerance during the early developmental stages of four marine teleosts. *Environmental Toxicology* 18, 375–382.

Kimbro, D., Grosholz, E., Baukus, A. et al. (2009) Invasive species cause large-scale loss of native California oyster habitat by disrupting trophic cascades. *Oecologia* 160, 563–575.

Kimura, O., Endo, T., Hotta, Y., Sakata, M. (2005) Effects of P-glycoprotein inhibitors on transepithelial transport of cadmium in cultured renal epithelial cells, LLC-PK1 and LLC-GA5-COL 150. *Toxicology* 208, 123–132.

Kinne, O. (1964) The effects of temperature and salinity on marine and brackish water animals: II. Salinity and temperature–salinity combinations. *Oceanography and Marine Biology* 2, 281–339.

Klaassen, C.D., Liu, J., Choudhuri, S. (1999) Metallothionein: An intracellular protein to protect against cadmium toxicity. *Annual Review of Pharmacology and Toxicology* 39, 267–294.

Klaassen, C.D., Liu, J., Diwan, B.A. (2009) Metallothionein protection of cadmium toxicity. *Toxicology and Applied Pharmacology* 238, 215–220.

Kooijman, S. (2000) *Dynamic Energy and Mass Budgets in Biological Systems*. Cambridge University Press, Cambridge, UK, 144p.

Kültz, D. (2003) Evolution of the cellular stress proteome: from monophyletic origin to ubiquitous function. *Journal of Experimental Biology* 206, 3119–3124.

Kültz, D. (2005) Molecular and evolutionary basis of the cellular stress response. *Annual Review of Physiology* 67, 225–257.

Kurelec, B. (1995) Inhibition of multixenobiotic resistance mechanism in aquatic organisms: ecotoxic consequences: Environmental toxicology: hazards to the environment and man in the Mediterranean region. *Science of the Total Environment* 171, 197–204.

Kurochkin, I., Ivanina, A., Eilers, S., Sokolova, I. (2008) Effects of environmental anoxia and re-oxygenation on mitochondrial function and metabolism of eastern oysters (*Crassostrea virginica*). *Comparative Biochemistry and Physiology A* 150, S161.

Kurochkin, I.O., Ivanina, A.V., Eilers, S., Downs, C.A., May, L.A., Sokolova, I.M. (2009) Cadmium affects metabolic responses to prolonged anoxia and reoxygenation in eastern oysters Crassostrea virginica. *American Journal of Physiology* 297, R1262–R1272.

Lannig, G., Cherkasov, A.S., Sokolova, I.M. (2006a) Temperature-dependent effects of cadmium on mitochondrial and whole-organism bioenergetics of oysters (*Crassostrea virginica*). *Marine Environmental Research* 62, S79–S82.

Lannig, G., Flores, J.F., Sokolova, I.M. (2006b) Temperature-dependent stress response in oysters, *Crassostrea virginica*: Pollution reduces temperature tolerance in oysters. *Aquatic Toxicology* 79, 278–287.

Lannig, G., Bock, C., Cherkasov, A., Pörtner, H.O., Sokolova, I.M. (2008) Cadmium-dependent oxygen limitation affects temperature tolerance in eastern oysters (*Crassostrea virginica* Gmelin). *American Journal of Physiology* 294, R1338–R1346.

Lannig, G., Eilers, S., Pörtner, H.O., Sokolova, I.M., Bock, C. (2010) Impact of ocean acidification on energy metabolism of oyster, *Crassostrea gigas* – changes in metabolic pathways and thermal response. *Marine Drugs* 8, 2318–2339.

Larade, K., Storey, K.B. (2002) A profile of metabolic responses to anoxia in marine invertebrates. *In* Storey, K.B., Storey, J.M. (eds). *Sensing, Signaling and Cell Adaptation*. Elsevier Science, Amsterdam, pp. 27–46.

Larade, K., Storey, K.B. (2007) Arrest of transcription following anoxic exposure in a marine mollusc. *Molecular and Cellular Biochemistry* 303, 243–249.

Larade, K., Storey, K.B. (2009) Living without oxygen: Anoxia-responsive gene expression and regulation. *Current Genomics* 10, 76–85.

Le Moullac, G., Queau, I., Le Souchu, P. et al. (2007) Metabolic adjustments in the oyster *Crassostrea gigas* according to oxygen level and temperature. *Marine Biology Research* 3, 357–366.

Lee, S.Y. (1985) The population dynamics of the green mussel, *Perna viridis* (L.) in Victoria Harbour, Hong Kong – dominance in a polluted environment. *Asian Marine Biology* 2, 107–118.

Lesser, M.P., Kruse, V.A. (2004) Seasonal temperature compensation in the horse mussel, *Modiolus modiolus*: metabolic enzymes, oxidative stress and heat shock proteins. *Comparative Biochemistry and Physiology A* 137, 495–504.

Letendre, J., Chouquet, B., Manduzio, H. et al. (2009) Tidal height influences the levels of enzymatic antioxidant defences in *Mytilus edulis*. *Marine Environmental Research* 67, 69–74.

Levraut, J., Iwase, H., Shao, Z.H., Vanden Hoek, T.L., Schumacker, P.T. (2003) Cell death during ischemia: relationship to mitochondrial depolarization and ROS generation. *American Journal of Physiology* 284, H549–H558.

Li, Z.S., Lu, Y.P., Shen, R.G., Szczypka, M., Thiele, D.J., Rea, P.A. (1997) A new pathway for vacuolar cadmium sequestration in *Saccharomyces cerevisiae*: YCF1-catalyzed transport of bis(glutathionato)cadmium. *Proceedings of the National Academy of Sciences USA* 94, 42–47.

Litman, T., Druley, T.E.S.W.D., Bates, S.E. (2001) From MDR to MXR: new undersanding of multidrug resistance systems, their properties and clinical significance. *Cellular and Molecular Life Sciences* 58, 931–959.

Livingstone, D.R. (1993) Biotechnology and pollution monitoring: Use of molecular *Biomarkers* in the aquatic environment. *Journal of Chemical Technology and Biotechnology* 57, 195–211.

Livingstone, D.R. (2001) Contaminant-stimulated reactive oxygen species production and oxidative damage in aquatic organisms. *Marine Pollution Bulletin* 42, 656–666.

Livingstone, D.R., Garcia Martinez, P., Michel, X. et al. (1990) Oxyradical production as a pollution-mediated mechanism of toxicity in the common mussel, *Mytilus edulis* L., and other molluscs. *Functional Ecology* 4, 415–424.

Matsushima, O., Katayama, H., Yamada, K., Kado, Y. (1984) Effect of external salinity change on the adenylate energy charge in the brackish bivalve *Corbicula japonica*. *Comparative Biochemistry and Physiology A* 77, 57–61.

Melzner, F., Gutowska, M.A., Langenbuch, M. et al. (2009) Physiological basis for high CO_2 tolerance in marine ectothermic animals: pre-adaptation through lifestyle and ontogeny? *Biogeosciences* 6, 1–19.

Mubiana, V.K., Blust, R. (2007) Effects of temperature on scope for growth and accumulation of Cd, Co, Cu and Pb by the marine bivalve *Mytilus edulis*. *Marine Environmental Research* 63, 219–235.

Myrand, B., Tremblay, R., Sevigny, J.-M. (2002) Selection against blue mussels (*Mytilus edulis* L.) homozygotes under various stressful conditions. *Journal of Heredity* 93, 238–248.

Nagao, R.T., Kimpel, J.A., Key, J.L. (1990) Molecular and cellular biology of the heat-shock response. *Advances in Genetics* 28, 235–273.

Niyogi, S., Biswas, S., Sarker, S., Datta, A.G. (2001) Antioxidant enzymes in brackishwater oyster, *Saccostrea cucullata* as potential *Biomarkers* of polyaromatic hydrocarbon pollution in Hooghly estuary (india): seasonality and its consequences. *Science of the Total Environment* 281, 237–246.

Orbea, A., Ortiz-Zarragoitia, M., Cajaraville, M.P. (2002) Interactive effects of benzo(a) pyrene and cadmium and effects of di(2-ethylhexyl) phthalate on antioxidant and peroxisomal enzymes and peroxisomal volume density in the digestive gland of mussel *Mytilus galloprovincialis* Lmk. *Biomarkers* 7, 33–48.

Osman, A.M., Rotteveel, S., den Besten, P.J., van Noort, P.C.M. (2004) *In vivo* exposure of *Dreissena polymorpha* mussels to the quinones menadione and lawsone: menadione is more toxic to mussels than lawsone. *Journal of Applied Toxicology* 24, 135–141.

Palmiter, R.D. (1998) The elusive function of metallothioneins. *Proceedings of the National Academy of Sciences USA* 95, 8428–8430.

Pannunzio, T.M., Storey, K.B. (1998) Antioxidant defenses and lipid peroxidation during anoxia stress and aerobic recovery in the marine gastropod *Littorina littorea*. *Journal of Experimental Marine Biology and Ecology* 221, 277–292.

Petes, L., Mouchka, M., Milston-Clements, R., Momoda, T., Menge, B. (2008) Effects of environmental stress on intertidal mussels and their sea star predators. *Oecologia* 156, 671–680.

Piano, A., Valbonesi, P., Fabbri, E. (2004) Expression of cytoprotective proteins, heat shock protein 70 and metallothioneins, in tissues of *Ostrea edulis* exposed to heat and heavy metals. *Cell Stress Chaperones* 9, 134–142.

Polechová, J., Barton, N., Marion, G. (2009) Species' range: Adaptation in space and time. *The American Naturalist* 174, E186–E204.

Pörtner, H.O. (1997) Oxygen limitation of metabolism and performance in pelagic squid. *In* Hawkins, L.E., Hutchinson, S., Jensen, A.C., Sheader, M., Williams, J.A. (eds). *The Responses of Marine Organisms to their Environments. Proceeding of the 30th European Marine Biology Symposium*, University of Southampton, pp. 45–56.

Pörtner, H.O. (2001) Climate change and temperature dependant biogeography: oxygen limitation of thermal tolerancein animals. *Naturwissenschaften* 88, 137–146.

Pörtner, H.O. (2002) Climate variations and the physiological basis of temperature dependent biogeography: systemic to molecular hierarchy of thermal tolerance in animals. *Comparative Biochemistry and Physiology A* 132, 739–761.

Pörtner, H.O., Farrell, A.P. (2008) Physiology and climate change. *Science* 322, 690–692.

Pörtner, H.O., Knust, R. (2007) Climate change affects marine fishes through the oxygen limitation of thermal tolerance. *Science* 315, 95–97.

Pörtner, H.O., Lannig, G. (2009) Oxygen and capacity limited thermal tolerance. *In* Richards, J.G., Farrell, A.P., Brauner, C.J. (eds). *Hypoxia. Fish Physiology*, Vol. 27. Elsevier, Academic Press, pp. 143–191.

Pörtner, H., Berdal, B., Blust, R., et al. (2001) Climate induced temperature effects on growth performance, fecundity and recruitment in marine fish: developing a hypothesis for cause and effect relationships in Atlantic cod (*Gadus morhua*) and common eelpout (*Zoarces viviparus*). *Continental Shelf Research* 21, 1975–1997.

Pörtner, H.O., Langenbuch, M., Michaelidis, B. (2005) Synergistic effects of temperature extremes, hypoxia, and increases in CO_2 on marine animals: From earth history to global change. *Journal of Geophysical Research* 110, C09S10.

Pörtner, H.O., Peck, L., Hirse, T. (2006) Hyperoxia alleviates thermal stress in the Antarctic bivalve, *Laternula elliptica*: evidence for oxygen limited thermal tolerance. *Polar Biology* 29, 688–693.

Pytharopoulou, S., Sazakli, E., Grintzalis, K., Georgiou, C.D., Leotsinidis, M., Kalpaxis, D.L. (2008) Translational responses of *Mytilus galloprovincialis* to environmental pollution: Integrating the responses to oxidative stress and other biomarker responses into a general stress index. *Aquatic Toxicology* 89, 18–27.

Ramnanan, C.J., Storey, K.B. (2006) Glucose-6-phosphate dehydrogenase regulation during hypometabolism. *Biochemical and Biophysical Research Communications* 339, 7–16.

Rangsayatorn, N., Kruatrachue, M., Pokethitiyook, P., Upatham, E.S., Lanza, G.R., Singhakaew, S. (2004) Ultrastructural changes in various organs of the fish *Puntius gonionotus* fed cadmium-enriched cyanobacteria. *Environmental Toxicology* 19, 585–593.

Regoli, F., Nigro, M., Bertoli, E., Principato, G., Orlando, E. (1997a) Defenses against oxidative stress in the Antarctic scallop *Adamussium colbecki* and effects of acute exposure to metals. *Hydrobiologia* 355, 139–144.

Regoli, F., Principato, G.B., Bertoli, E., Nigro, M., Orlando, E. (1997b) Biochemical characterization of the antioxidant system in the scallop *Adamussium colbecki*, a sentinel organism for monitoring the Antarctic environment. *Polar Biology* 17, 251–258.

Riisgård, H.U. (2001) On measurement of filtration rates in bivalves – the stony road to reliable data: review and interpretation. *Marine Ecolology Progress Series* 211, 275–291.

Sanni, B., Williams, K., Sokolov, E.P., Sokolova, I.M. (2008) Effects of acclimation temperature and cadmium exposure on mitochondrial aconitase and LON protease from a model marine ectotherm, *Crassostrea virginica*. *Comparative Biochemistry and Physiology C* 147, 101–112.

Saydam, N., Steiner, F., Georgiev, O., Schaffner, W. (2003) Heat and Heavy Metal Stress Synergize to Mediate Transcriptional Hyperactivation by Metal-responsive Transcription Factor MTF-1. *Journal of Biological Chemistry* 278, 31879–31883.

Schlesinger, M.J. (1990) The ubiquitin system and the heat shock response. *In* Schlesinger, M.J., Santoro, M.G., Garaci, E. (eds). *Stress Proteins: Induction and Function*. Springer-Verlag, Berlin, pp. 80–88.

Schöne, B.R., Flessa, K.W., Dettman, D.L., Goodwin, D.H. (2003) Upstream dams and downstream clams: growth rates of bivalve mollusks unveil impact of river management on estuarine ecosystems (Colorado River Delat, Mexico). *Estuarine, Coastal and Shelf Science* 58, 715–726.

Selye, H. (1973) The evolution of stress concept. *American Scientist* 61, 629–699.

Sheehan, D., Power, A. (1999) Effects of seasonality on xenobiotic and antioxidant defence mechanisms of bivalve molluscs. *Comparative Biochemistry and Physiology C* 123, 193–199.

Sibly, R.M., Calow, P. (1989) A life-cycle theory of responses to stress. *Biological Journal of the Linnean Society* 37, 101–116.

Sidell, B.D. (1998) Intracellular oxygen diffusion: the roles of myoglobin and lipid at cold body temperature. *Journal of Experimental Biology* 201, 1119–1128.

Smital, T., Kurelec, B. (1998) The chemosensitizers of multixenobiotic resistance mechanism in aquatic invertebrates: a new class of pollutants. *Mutation Research/Fundamental and Molecular Mechanisms of Mutagenesis* 399, 43–53.

Smolders, R., Bervoets, L., De Coen, W., Blust, R. (2004) Cellular energy allocation in zebra mussels exposed along a pollution gradient: linking cellular effects to higher levels of biological organization. *Environmental Pollution* 129, 99–112.

Sokolova, I.M. (2009) Apoptosis in molluscan immune defense. *Invertebrate Survival Journal* 6, 49–58.

Sokolova, I.M., Lannig, G. (2008) Interactive effects of metal pollution and temperature on metabolism in aquatic ectotherms: Implications of global climate change. *Climate Research* 37, 181–201.

Sokolova, I.M., Pörtner, H.O. (2003) Metabolic plasticity and critical temperatures for aerobic scope in a eurythermal marine invertebrate (*Littorina saxatilis*, Gastropoda: Littorinidae) from different latitudes. *Journal of Experimental Biology* 206, 195–207.

Sokolova, I.M., Bock, C., Pörtner, H.O. (2000) Resistance to freshwater exposure in White Sea *Littorina spp.* Anaerobic metabolism and energetics. *Journal of Comparative Physiology B* 170, 91–103.

Somero, G.N. (2005) Linking biogeography to physiology: evolutionary and acclimatory adjustments of thermal limits. *Frontiers in Zoology* 2, 1–9.

Stegeman, J.J., Hahn, M.E. (1994) Biochemistry and molecular biology of monooxygenases: Current perspectives on forms, functions, and regulation of cytochrome P450 in aquatic species. *In* Malins, D.C., Ostrander, G.K. (eds), *Aquatic Toxicology*: Molecular, Biochemical and Cellular Perspectives. Lewis Publishers, Boca Raton, Ann Arbor, London, Tokyo, pp. 87–206.

Stohs, S.J., Bagchi, D., Hassoun, E., Bagchi, M. (2000) Oxidative mechanisms in the toxicity of chromium and cadmium ions. *Journal of Environmental Pathology, Toxicology and Oncology* 19, 201–213.

Storey, K., Storey, J.M. (1988) Freeze tolerance: constraining forces, adaptive mechanisms. *Canadian Journal of Zoology* 66, 1122–1127.

Storey, K.B., Storey, J.M. (2007) Tribute to P.L. Lutz: putting life on 'pause' – molecular regulation of hypometabolism. *Journal of Experimental Biology* 210, 1700–1714.

Symons, A.M., King, L.J. (2003) Inflammation, reactive oxygen species and cytochrome P450. *Inflammopharmacology* 11, 75–86.

Thévenod, F., Friedmann, J.M., Katsen, A.D. and Hauser, I.A. (2000) Up-regulation of multidrug resistance P-glycoprotein via nuclear factor-κB activation protects kidney proximal tubule cells from cadmium- and reactive oxygen species-induced apoptosis. *Journal of Biological Chemistry* 275, 1887–1896.

Tomanek, L., Somero, G.N. (2000) Time course and magnitude of synthesis of heat-shock proteins in congeneric marine snails (genus *Tegula*) from different tidal heights. *Physiological and Biochemical Zoology* 73, 249–256.

Valko, M., Morris, H., Cronin, M.T. (2005) Metals, toxicity and oxidative stress. *Current Medicinal Chemistry* 12, 1161–1208.

Viarengo, A., Burlando, B., Cavaletto, M., Marchi, B., Ponzano, E., Blasco, J. (1999) Role of metallothionein against oxidative stress in the mussel (*Mytilus galloprovincialis*). *American Journal of Physiology* 277, R1612–R1619.

Walther, K., Sartoris, F.J., Bock, C., Pörtner, H.O. (2009) Impact of anthropogenic ocean acidification on thermal tolerance of the spider crab *Hyas araneus*. *Biogeosciences Discussion* 6, 2837–2861.

Widdows, J., Donkin, P., Staff, F.J. et al. (2002) Measurement of stress effects (scope for growth) and contaminant levels in mussels (*Mytilus edulis*) collected from the Irish Sea. *Marine Environmental Research* 53, 327–356.

Winston, G.W., Giulio, R.T.D. (1991) Prooxidant and antioxidant mechanism in aquatic organisms. *Aquatic Toxicology* 19, 137–161.

Wortelboer, H.M., Balvers, M.G.J., Usta, M., van Bladeren, P.J., Cnubben, N.H.P. (2008) Glutathione-dependent interaction of heavy metal compounds with multidrug resistance proteins MRP1 and MRP2. *Environmental Toxicology and Pharmacology* 26, 102–108.

Chapter 20

STARVATION, ENERGETICS, AND ANTIOXIDANT DEFENSES

Amalia E. Morales[1], *Amalia Pérez-Jiménez*[1], *Miriam Furné*[1], *and Helga Guderley*[2]

[1]Department of Animal Biology, University of Granada, Granada, Spain
[2]Department of Biology, University of Laval, Québec, Canada

Marine and freshwater ectotherms can experience marked seasonal periods with low food availability. Seasonal changes of temperature or day length and irradiation intensity are the main factors determining primary productivity in temperate and polar zones. However, environmental factors, such as the annual periodicity in strong winds or coastal upwelling, can affect productivity in tropical systems.

Starvation has been suggested to cause reallocation of energy resources towards the most necessary processes for tissue maintenance and survival, often at the cost of metabolic and locomotory scopes. In contrast, to terrestrial animals that simply lose mass when starved, many aquatic organisms, in particular fish, increase tissue hydration to counteract loss in body dry mass and conserve their original body shape by this low cost strategy. Thus, nesting male three-spine stickleback (*Gasterosteus aculeatus* L) reduce their food intake, but maintain higher muscle hydration levels than non-nesting males, apparently to give the illusion of greater size to potential intruders (Fitzgerald et al. 1989). During prolonged starvation of cod (*Gadus morhua*), the breakdown of muscle protein causes an accumulation of solutes and ensuing retention of water that can cause muscle hydration to increase from a normal value of 80% to 90% (Lambert and Dutil 1997).

Unlike mammals, fish can survive extended periods of starvation by mobilizing energy reserves stored during times of favorable feeding conditions. This mobilization of stored energy during starvation leads to significant weight loss (Paul et al. 1995; Collins and Anderson 1997; Power et al. 2000; Guderley et al. 2003). The switch from metabolizing ingested food to mobilization of body deposits causes metabolic adaptations that involve up- and down-regulation of the major pathways of intermediary metabolism, particularly in the liver (see below). Thus, many investigators studying the effect of prolonged starvation in fish report depression of the activity of glucose-degrading and lipogenic enzymes, whereas β-oxidation of fatty acids

Oxidative Stress in Aquatic Ecosystems, First Edition. Edited by Doris Abele, José Pablo Vázquez-Medina, and Tania Zenteno-Savín.

is enhanced to provide energy for most tissues. Amino acids and glycerol are also used as gluconeogenic precursors to maintain glycemia (Navarro and Gutiérrez 1995; Sánchez-Muros et al. 1998; Dou et al. 2002; Guderley et al. 2003; Pérez-Jiménez 2008).

Starvation has been reported to have pro-oxidant effects due to both the inadequate neutralization of reactive oxygen species (ROS) generated by sustained aerobic metabolism and the reduced level of antioxidant defenses. The depression in protein synthesis induced under unfavorable feeding conditions reduces the levels of enzymes involved in the direct neutralization of ROS and in the recycling of important antioxidant molecules, such as glutathione (GSH). Also, food-derived nonenzymatic antioxidants, such as the vitamins C and E, are diminished in aquatic organisms during starvation. When the rate of ROS generation exceeds the capacity of ROS neutralization, oxidative stress occurs and causes reversible and irreversible cellular damage, and finally cell death.

METABOLIC MODIFICATIONS DURING STARVATION

To survive periods of starvation, fish mobilize their endogenous reserves and reduce their energy expenditures, which in a high percentage are derived from protein turnover (Salem et al. 2007). This imposes metabolic adjustments that are species-dependent. Fish are the most diverse group of vertebrates, and many species differ in susceptibility and adaptive responses to fasting (Navarro and Gutiérrez 1995; Wang et al. 2006). Fish age or nutritional status will also affect the response to food limitation.

Navarro and Gutiérrez (1995) provided an excellent and comprehensive review on the metabolic consequences of starvation in fish. In the present chapter we provide only a brief overview of such metabolic adjustements, emphasizing the studies appearing since then.

Plasma glucose levels decline in starved fish (Gillis and Ballantyne 1996; Soengas et al. 1996; Figueiredo-Garutti et al. 2002; Pérez-Jiménez et al. 2007, Pérez-Jiménez 2008). However, most studies report that glycemia never falls below the basal levels established for most fish, 65–70 mg 100 mL^{-1} (Echevarría et al. 1997; Rios et al. 2006; Pérez-Jiménez et al. 2007, Pérez-Jiménez 2008). The maintenance of plasma glucose levels during food deprivation is attributed to three processes: glycogen mobilization, decreased glucose consumption, and gluconeogenesis. A large reduction of liver glycogen has been reported in many starved fish, at least during the initial stages of starvation (Navarro and Gutiérrez 1995; Metón et al. 2003; Rios et al. 2006; Pérez-Jiménez et al. 2007, Pérez-Jiménez 2008), but in most fish glycogen deposits were not completely exhausted, suggesting that a strategy is operating to preserve liver reserves (Collins and Anderson 1995; Rios et al. 2006; Pérez-Jiménez et al. 2007, Pérez-Jiménez 2008). The reduced activity of hepatic enzymes from the glycolytic and the pentose phosphate pathways (Collins and Anderson 1997; Metón et al. 2003; Kirchner et al. 2005; Pérez-Jiménez et al. 2007; Polakof et al. 2007) as well as enhanced or sustained hepatic gluconeogenesis from amino acids and glycerol (Navarro and Gutiérrez 1995; Metón et al. 2003; Sangiao-Alvarellos et al. 2005; Pérez-Jiménez et al. 2007; Pérez-Jiménez 2008) are other strategies operating in fish to maintain glycemia during starvation.

A reduction in glycogen deposits in white muscle occurs in most starved fish (Collins and Anderson 1995; Rios et al. 2006; Pérez-Jiménez et al. 2007; Pérez-Jiménez 2008). However, as in the liver, glycogen deposits in white muscle also seem to be preserved at a minimal level during starvation. The conservation of a minimal quantity of muscle glycogen may reflect its utility as a fuel during predator avoidance.

During starvation, fatty acids derived from triglyceride hydrolysis are preferentially used through β-oxidation as fuels for most fish tissues. Triglycerides are the main energy stores in fish, and may be deposited as perivisceral fat, in liver, or in muscle. The quantitative importance of each tissue for storage and, thus, for mobilization during starvation varies among species. Liver is the main site for lipid storage in gadids; salmonids accumulate perivisceral fat, whereas clupeiformes generally rely on the deposition of lipid in muscle. Independently of the storage site, a marked reduction in fat stores occurs in starved fish (Grigorakis and Alexis 2005; Rios et al. 2006; Furné 2008; Pérez-Jiménez 2008). Regarding the enzymes involved in fatty acid catabolism, Morales et al. (2004) reported a 123% increase in hepatic β-hydroxyacyl CoA dehydrogenase (HOAD) activity in starved common dentex (*Dentex dentex*), whereas in zebrafish (*Danio rerio*) the hepatic expression of genes associated with β-oxidation was depressed (Drew et al. 2008).

Although the role of ketone bodies as an integral part of fish intermediary metabolism has been debated for many years, many studies have already demonstrated β-hydroxybutyrate dehydrogenase (β-OHBDH) activity in several teleost species (Navarro and Gutiérrez 1995; Pérez-Jiménez 2008; Furné et al. 2009). Further, an increase in plasma levels of ketone bodies (Soengas et al. 1996) and in the activity of enzymes involved in ketone body synthesis (Furné 2008; Pérez-Jiménez 2008) has been reported in fish deprived of food.

STARVATION, PHYSICAL ACTIVITY, AND MUSCLE CAPACITIES

Generally in fish, as in other vertebrates, the metabolic capacities of muscle in a given species reflect the requirements of its type of activity. Sprint specialists undertake brief bursts of activity, powered by anaerobic glycolysis in the fast glycolytic (white) fibers. Accordingly, their musculature has primarily fast glycolytic fibers. Endurance activities require muscles with high oxidative capacities, attained both by a high proportion of oxidative (red) muscle and by a greater aerobic capacity per gram of muscle, as well as an appropriate cardiovascular capacity. Interspecific differences in muscle capacities are such that the metabolic capacities of same fiber type will vary markedly according to the species in which it is examined.

According to Méndez and Wieser (1993), the general response of fish to food deprivation consists of three phases: stress, transition, and adaptation, all involving changes in physical activity. The stress phase is short (approximately 24 h) and reflects a state of hyperactivity in which the food searching activity of fish increases. In the transition phase fish gradually reduce swimming activity and thus energy expenditure. Finally, the adaptation phase is characterized by low activity and metabolism and lasts until the food is again available (or to the demise of the fish). These changes in energetic status modify many aspects of swimming performance, since the shift of energy metabolism parallels the reduction of carbohydrate-fuelled locomotor activity in the transiton phase. Changes in food availability strongly affect lactate dehydrogenase (LDH) activity in white muscle. LDH is central to burst-swimming performance because its activity allows glycolytic ATP production critical for muscle contraction during functional hypoxia (Willmer et al. 2000). A decrease in LDH activity because of low food availability should directly affect swimming performance, causing a decline in the ability of an individual to escape from predators or capture prey.

The metabolic sensitivity of muscle to food limitation varies both with fiber type and with position in the body. Generally, white fibers are more susceptible than red fibers (Greer-Walker 1971; Beardall and Johnston 1985; Lowery and Somero 1990; Martínez et al. 2002, 2003). In white muscle of cod (*Gadus morhua*), glycolytic enzyme activities seem to rise continuously with increases in condition (Pelletier et al. 1993; Dutil et al. 1998). In both red and white fibers of cod, starvation decreases the activities of glycolytic enzymes more than those of mitochondrial enzymes (Martínez et al. 2003). Longitudinal differences in muscle metabolic capacities are marked in fed cod, but become less pronounced with starvation (Martínez et al. 2003). In fed cod, the highest activities of glycolytic and mitochondrial enzymes occur in caudal samples of red and white muscle, whereas the activity of nucleotide diphosphokinase is highest in the rostral muscles. These differences may be related to longitudinal differences in activity during swimming (Davies et al. 1995).

Endurance swimming, given its name, would seem to be primarily aerobic. Nonetheless, virtually all sustained and prolonged swimming protocols lead fish to recruit their glycolytic muscle (Burgetz et al. 1998; Richards et al. 2002) and this muscle is the most affected by starvation. Swimming endurance was 70% lower in starved than fed cod (*Gadus morhua*), both in time and distance swam. Of a variety of metabolic, behavioural, and anatomic variables, the number of burst–coast movements (consisting of an acceleration or "burst" phase and a phase of gliding or "coasting") and the activity of LDH in caudal white muscle were the best predictors of swimming endurance (Martínez et al. 2003). Fed cod performed 2.5-fold as many burst–coasts as starved cod. Thus, in both fed and starved cod, an "aerobic" activity receives considerable support from glycolytic muscle.

Sprint swimming or startle responses are very rapid, brief movements powered by white muscle. As these movements last less than 5 s, energetic support is provided by anaerobic metabolism, most likely the generation of ATP from creatine phosphate. Martínez et al. (2002) followed individual cod throughout a starvation–feeding–starvation cycle. Sprint swimming was measured in a computerized "drag strip" at each point in the cycle and biopsies of glycolytic muscle were taken to follow how individuals changed their muscle metabolic capacities with the feeding cycle. While enzyme activities increased with feeding and decreased with starvation, sprint swimming speed increased with feeding, but remained unchanged during the subsequent starvation. A performance hierarchy was maintained throughout the feeding cycle, with fast cod remaining fast under all feeding states. This hierarchy did not reflect the differences in muscle metabolic capacities. Despite the sparing of muscle aerobic capacity during food limitation, prolonged and sprint swimming of cod are affected to a similar extent (Martínez et al. 2004). White muscle lactate levels and lactate accumulation per burst–coast movement were considerably higher during critical swimming tests by fed than starved cod, indicating more intensive use of fast muscle in cod in good condition. The loss of fast muscle therefore reduced both prolonged and sprint swimming. The speed at which burst–coast movements began (the end of sustained swimming) was strongly correlated with the activity of pyruvate dehydrogenase in the caudal and central red muscle samples. These patterns indicate that food limitation will modify all levels of swimming performance in fish, leading to a spiralling decline in condition and in the capacity to recuperate from starvation.

STARVATION AND ANTIOXIDANT DEFENSES: ANTIOXIDANT ENZYMES

Food deprivation leads to depletion of antioxidant stores and increased levels of ROS in organs of living organisms. Starvation has been reported to have pro-oxidant effects in mammals because ROS generation is not adequately neutralized by antioxidant systems (Robinson et al. 1997; Domenicali et al. 2001). Studies regarding the influence of food deprivation on antioxidant defenses in fish are scarce and a general or similar trend cannot be deduced from available data, which seems to indicate that the antioxidant response to cope with starvation is species-dependent.

Liver

Most available studies on the impact of starvation on antioxidant defenses of fish are focused on liver, since it is the main organ for metabolic control and has a key role in ROS generation. Most of these studies show that starvation increases lipid peroxidation in liver (Pascual et al. 2003; Morales et al. 2004; Zhang et al. 2008) but, as reported above, results on antioxidant enzymes are not unanimous (Table 20.1). Thus, rainbow trout (*Oncorynchus mykiss*) subjected to three weeks of starvation showed a decrease of roughly 35% in hepatic glutathione reductase (GR) activity, maintaining this decreased activity at 7 weeks of food restriction (Blom et al. 2000). In the sparid common dentex (*Dentex dentex*) starved for 5 weeks, a reduced GR activity in liver was reported (Morales et al. 2004). Prolonged starvation induced an enhancement in GR activity in the liver of sea bream (*Sparus aurata*) (Pascual et al. 2003), which reflects the different responses to starvation, even between these two closely related species. Similarly, liver catalase (CAT) activity increased in common dentex and decreased in sea

Table 20.1 Relative changes[1] in antioxidant enzymes and lipid peroxidation (TBARS) in liver of starved fish.

Fish species	CAT	SOD	GPx	GR	GST	TBARS	Reference
Atlantic cod (*Gadus morhua*)	↑		↑		↑		Guderley et al. (2003)
Common dentex (*Dentex dentex*)	↑	↑	↑	↓		↑	Morales et al. (2004)
Gilthead sea bream (*Sparus aurata*)	↓	↑	↑	↑		↑	Pascual et al. (2003)
Large yellow croaker (*Pseudosciaena crocea*)		↑	↑			↑	Zang et al. (2008)
Rainbow trout (*Oncorhynchus mykiss*)				↓	↓		Blom et al. (2000)
			↓		↓		Salem et al. (2007)
	↓	=	↓	=		=	Furné et al. (2009)
Rock bream (*Oplegnatus fasciatus*)	↑	↑	↑		↑		Nam et al. (2005)
Adriatic sturgeon (*Acipenser naccarii*)	↓	↓	↓	↓		=	Furné et al. (2009)
Zebrafish (*Danio rerio*)		↓	↓				Drew et al. (2008)

[1]Relative changes with respect to fed fish: ↑, increased; ↓, decreased; =, unchanged.
CAT, catalase; SOD, superoxide dismutase; GPx, glutathione peroxidase; GR, glutathione reductase; GST, glutathione-*S*-transferase; TBARS; thiobarbituric acid reacting substances.

bream. Although the specific activity of hepatic superoxide dismutase (SOD) and glutathione peroxidase (GPx) increased in both species, starvation changed the SOD isozyme pattern in the liver of sea bream, with starved fish displaying two new Mn-SOD isozymes (Pascual et al. 2003; Fig. 20.1), whereas in common dentex no changes in the hepatic SOD isozyme pattern were found, although both Cu,Zn-SOD bands were enhanced in food-deprived fish (Fig. 20.2). Studies in Atlantic cod (*Gadus morhua*) indicate increased activity of liver CAT, GPx, and glutathione *S*-transferase (GST) after 12 weeks of starvation (Guderley et al. 2003). In large yellow croaker (*Pseudosciaena crocea*) starved for 28 days, hepatic SOD and GPx activities and lipid

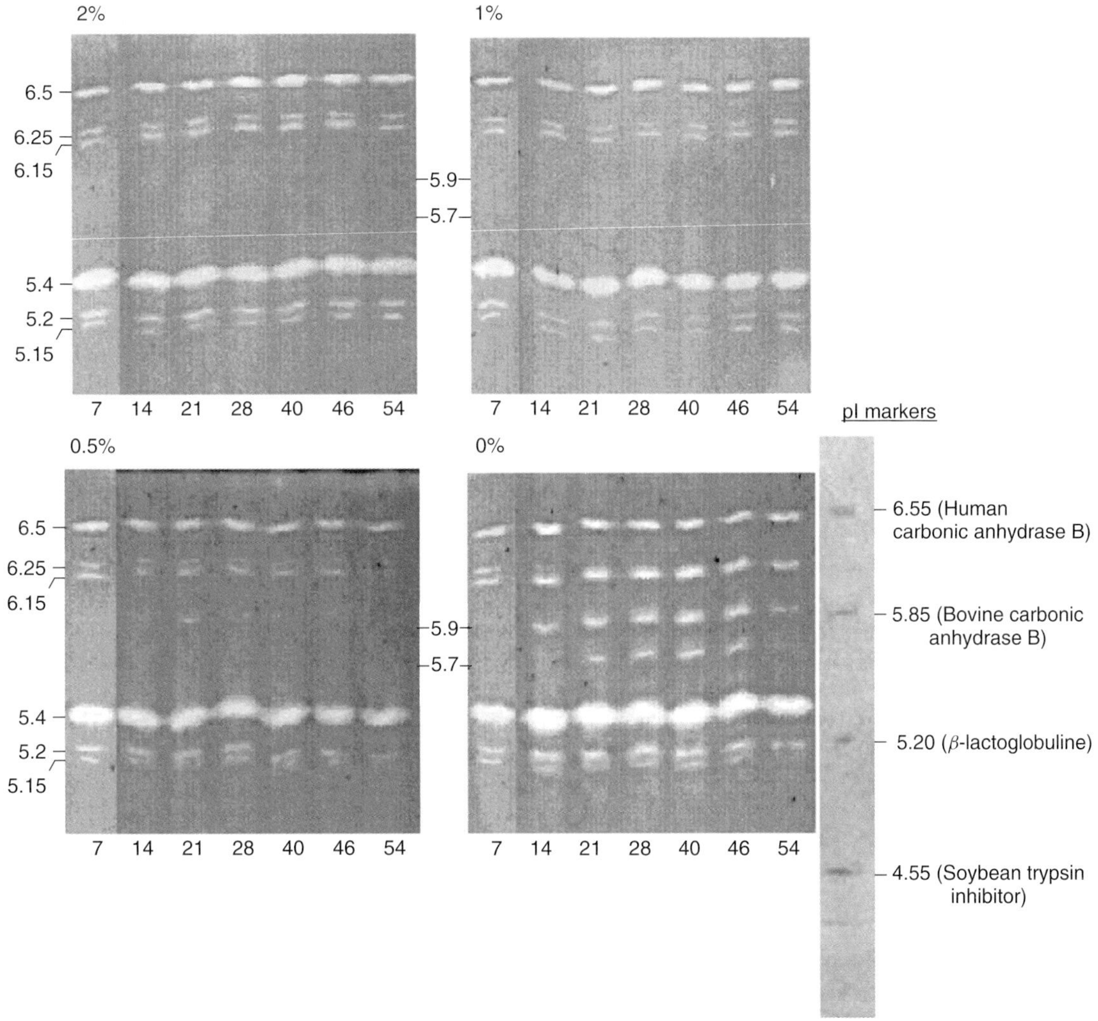

Fig. 20.1 Time-course of liver SOD isoenzymes in sea bream (*Sparus aurata*) subjected to food deprivation or starvation. Feeding ratio (percentage of total weight of the animals in the tank): 2% (used as control), 1, 0.5, and 0% (starved). *X* axis corresponds to the different sampling days (7, 14, 21, 28, 40, 46), and the last sampling (54) was carried out 1 week later, during that week all fish were fed 2% diet. (Reprinted from *Chemico-Biological Interactions*, 145, Pascual, P., Pedrajas, J.R., Toribio, F., López-Barea, J. and Peinado, J., Effect of food deprivation on oxidative stress biomarkers in fish (*Sparus aurata*), 191–199, 2003, with permission from Elsevier.)

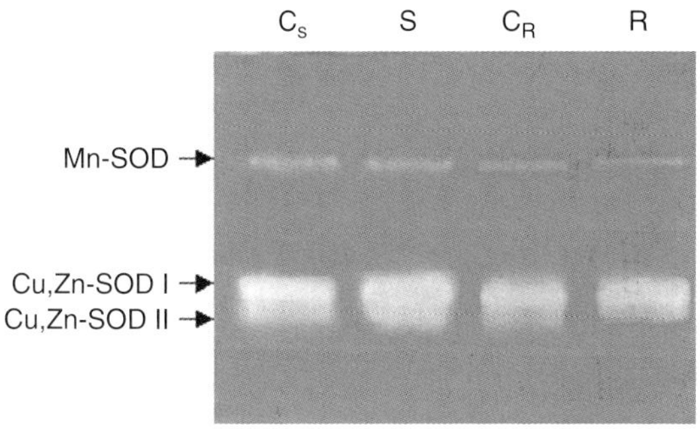

Fig. 20.2 SOD isozyme pattern in liver of common dentex (*Dentex dentex*) starved for 5 weeks and refed for 3 weeks. Control-starvation, starved, control-refeeding, and refed values are represented as C_S, S, C_R, and R, respectively (Morales et al. 2004).

peroxidation were significantly increased with respect to fed fish (Zhang et al. 2008).

Recent studies also evaluate the transcriptional responses to starvation of major antioxidant enzyme genes in the liver of some fish species. In rock bream (*Oplegnatus fasciatus*) significant alterations were apparent, with increased SOD, CAT, GPx, and GST mRNA levels in liver from fish subjected to short-term starvation (Nam et al. 2005). In rainbow trout (*Oncorhynchus mykiss*) starved for 3 weeks, expression of GST and GPx was reduced (Salem et al. 2007). Also in the zebrafish (*Danio rerio*), genes involved with neutralizing ROS, such as SOD, GPx, and several other selenium-binding proteins, were down-regulated in livers of fish starved for 21 days (Drew et al. 2008). These results seem to indicate that prolonged starvation decreases the capacity of fish liver to ameliorate oxidative stress.

Heart

Studies analyzing the antioxidant defenses of heart in starved fish are scarce and the results contradictory. In Adriatic sturgeon (*Acipenser naccarii*) and rainbow trout (*Oncorhynchus mykiss*) a starvation-period of 72 days did not induce lipid peroxidation in heart. The activity of antioxidant enzymes in heart of starved sturgeon remained unchanged whereas cardiac CAT and SOD activities were enhanced by starvation in rainbow trout (Furné et al. 2009). These results might indicate that the heart, being a crucial organ for life, would be in some way protected from stressful circumstances, such as starvation. However, in common dentex (*Dentex dentex*) 5 weeks of starvation depressed cardiac activities of SOD, GPx, and GR. Although CAT activity increased under similar circumstances, it was not sufficient to avoid an increase in lipid peroxidation (Pérez-Jiménez 2008).

White Muscle

The few studies in which the impact of starvation on antioxidant defenses of fish white muscle has been analyzed unanimously report that antioxidant enzymes are spared during food deprivation, presumably to neutralize ROS. Further, an enhanced ROS generation due to increased oxidative metabolism in white muscle should not take place, as the specific activity of mitochondrial enzymes decreases with starvation in cod (*Gadus morhua*) (Pelletier et al. 1993; Guderley et al. 2003; Martínez et al. 2003) and common dentex (*Dentex dentex*) (Pérez-Jiménez 2008). Thus, endogenous rates of ROS production are likely decreased. Probably, cellular disruption during macromolecular mobilization may increase the sensitivity to damage by ROS, explaining why maintenance of antioxidant defenses would be useful during starvation. Muscle-specific activity of the antioxidant enzymes CAT, GPx, and GST did not differ between fed and starved cod (Guderley et al. 2003). In common dentex the specific activity of SOD, CAT, GPx, and GR, as well as the levels of lipid peroxidation in white muscle remained unchanged by starvation (Pérez-Jiménez 2008). In Adriatic sturgeon (*Acipenser naccarii*) and rainbow trout (*Oncorhynchus mykiss*) starved for 72 days, an enhanced activity of SOD, CAT, GPx, and GR in white muscle was found whereas lipid peroxidation remained at the same levels as in fed fish (Furné et al. 2009).

Independently of the tissue considered, a reduced rate of nicotinamide adenine dinucleotide phosphate (NADPH) generation and decreased production of antioxidants are thought to be responsible for the impact of starvation on antioxidant capacities. NADPH is generated by several enzymes that are sensitive to nutritional status. The pentose phosphate pathway seems to be the primary source of NADPH during the response to oxidative stress (Pandolfi et al. 1995; Tian et al. 1999; Leopold and Loscalzo 2000). In higher vertebrates, nutritional status markedly affects the activity of the dehydrogenases in the pentose phosphate pathway (Protsko et al. 1989; Goodridge 1992;

Amir-Ahmady and Salati 2001). The crucial role of glucose-6-phosphate dehydrogenase (G6PDH) in modulating the cellular response to oxidative stress and in maintaining intracellular GSH levels has been confirmed in mammals (Vulliamy et al. 1992; Pandolfi et al. 1995; Salvemini et al. 1999; Tian et al. 1999; Leopold and Loscalzo 2000). NADPH protects CAT from inactivation (Gaetani et al. 1994; Kirkman et al. 1999) because each of the four monomers of CAT contains a NADPH-binding site, necessary for enzymatic activity (Kirkman and Gaetani 1984; Kirkman et al. 1999). The lower availability of NADPH is responsible for the reduced GR activity reported above, and as this enzyme regenerates GSH from glutathione disulphide (GSSG) a depressed activity will reduce antioxidant capacity. This reduction in the recycling GSH rate, along with the depletion of the endogenous GSH pool, has frequently been observed during food deprivation (Papadopoulos et al. 1997; Robinson et al. 1997; Vendemiale et al. 2001; Pascual et al. 2003). A significant depression of the G6PDH activity by food deprivation occurs in many fish species (Viganò et al. 1993; Barroso et al. 1998; Caseras et al. 2002; Metón et al. 2003; Morales et al. 2004), suggesting that these mechanisms also apply to fish.

STARVATION AND ANTIOXIDANT DEFENSES: EXOGENOUS ANTIOXIDANTS

During starvation, certain metabolic pathways are protected more than others, leading, as described above, to greater conservation of aerobic than glycolytic capacity in muscle, to maintain proteolytic capacity and, in some cases, to conserve antioxidant enzyme activities. As animals produce their own GSH, starvation may inhibit its synthesis by reducing availability of precursors through the impact of lower NADPH levels on the reduction of GSSG to GSH. On the other hand, many antioxidant molecules must be supplied by food. Their levels, therefore, are certain to decrease as fasting progresses.

Low molecular weight compounds, such as ascorbic acid (vitamin C), α-tocopherol (vitamin E), or β-carotenes, and some minerals, such as Se, Zn, Cu, and Mn, which play both catalytic and structural roles in antioxidant enzymes, are among the most important exogenous antioxidants. Unfortunately, few studies have focused on how starvation affects their levels (Hamre and Lie 1995; Welker and Congleton 2005). Thus, most available studies investigate how dietary deficiencies of such micronutrients affect antioxidant enzymes and organismal oxidative status (Martínez-Álvarez et al. 2005; Table 20.2).

Besides the direct action of vitamin C as a ROS scavenger, it may indirectly affect the activity of the antioxidant enzymes SOD, CAT, and GPx. The functional sites of these enzymes contain trace metals (Co, Fe, Mn, Se, Zn) of which the absorption, metabolism, and excretion are modified by vitamin C (Jacob 1995; Hidalgo and Morales 2009). Dietary vitamin C deficiencies lead to oxidative stress in most fish. Elbaraasi et al. (2004) reported an enhanced GPx activity and increased levels of GSSG in liver of African catfish (*Clarias gariepinus*) fed a diet deficient in vitamin C. Similar results were observed by Hamre et al. (1997) in Atlantic salmon (*Salmo salar*). In liver of common dentex (*Dentex dentex*), vitamin C deficiency induced an increase in GPx and CAT activities, whereas SOD and GR remained unchanged and liver lipids and proteins showed oxidative damage (Pérez-Jiménez 2008). Lipid peroxidation levels increased in the hepatopancreas of thornfish (*Terapon jarbua*) fed on diets without vitamin C (Chien and Hwang 2001). Similar results have been reported for common carp (*Cyprinus carpio*) (Hwang and Lin 2002), hybrid tilapia (*Oreochromis niloticus* × *O. aureus*) (Shiau and Hsu 2002), Black Sea bream (*Acanthopagrus schlegeli*), and Red Sea bream (*Pagrus major*) (Ji et al. 2003). Trenzado et al. (2009) reported an increased SOD activity in erythrocytes of rainbow trout (*Oncorhynchus mykiss*) fed on a diet without supplementary vitamin C, whereas the activity of CAT, GPx, and GR did not change with respect to control fish.

Vitamin E also plays a central role in the antioxidant defenses of fish. Tocher et al. (2002) reported that the level of dietary vitamin E affected the activities of antioxidant enzymes in liver of sea bream (*Sparus aurata*), turbot (*Scophthalmus maximus*), and halibut (*Hippoglossus hippoglossus*). Thus, in turbot fed on diets without vitamin E, hepatic GPx activity was higher than in the animals fed on the supplemented diet. In the case of halibut, the highest levels of CAT and GPx were observed in fish fed the lowest dietary vitamin E level. Finally, sea bream showed higher CAT and SOD activities when fed a vitamin E deficient diet. Notwithstanding, the GR and GST activities were not related with the dietary or tissue levels of vitamin E in turbot or sea bream. In rainbow

Table 20.2 Effects of deficiencies in several exogenous antioxidants on antioxidant enzymes and oxidative status of fish liver.

Antioxidant	Species	Hepatic Effect	Reference
Vitamin C	*Clarias gariepinus*	Decreased GPx activity and levels of oxidized glutathione	Elbaraasi et al. (2004)
	Salmo salar	Decreased GPx activity	Hamre et al. (1997)
	Dentex dentex	Increased GPx and CAT activities, unchanged SOD and GR activities, increased oxidative damage to lipid and protein	Pérez-Jiménez (2008)
	Terapon jarbua	Increased lipid peroxidation	Chien and Hwang (2001)
	Cyprinus carpio	Increased lipid peroxidation	Hwang and Lin (2002)
	Oreochromis niloticus × *O. aureus*	Increased lipid peroxidation	Shiau and Hsu (2002)
	Acanthopagrus schlegeli	Increased lipid peroxidation	Ji et al. (2003)
	Pagrus major	Increased lipid peroxidation	Ji et al. (2003)
Vitamin E	*Sparus aurata*	Increased CAT and SOD activities, unchanged GR and GST activities, Increased TBARS	Tocher et al. (2002)
	Scophthalmus maximus	Increased GPx activity, unchanged GR and GST activity, Increased TBARS	Tocher et al. (2002)
	Hippoglossus hippoglossus	Increased CAT and GPx activities and TBARS	Tocher et al. (2002)
	Oncorhynchus mykiss	Increased CAT, GPx, and GR activities and TBARS	Trenzado (2008)
	Dicentrarchus labrax	Increased oxidative processes	Frigg et al. (1990); Gatta et al. (2000)
	Clarius gariepinus	Increased TBARS	Baker and Davies (1996, 1997)
	Oncorhynchus tshawytscha	Increased TBARS	Welker and Congleton (2005)
	Salvelinus alpines	Indicators of lipid oxidative damage unchanged	Olsen et al. (1999)
Selenium	*Brycon cephalus*	Increased oxidative damage and GPx, SOD, and CAT activities	Monteiro et al. (2009)
	Oncorhynchus mykiss	Increased oxidative damage to lipid	Ates et al. (2008)
	Salmo salar	Increased oxidative damage to lipid	Bell and Cowey (1987)
Zinc	*Oncorhynchus mykiss*	Unchanged oxidative status	Ogino and Yang (1978)
	Oncorhynchus mykiss	Increased CAT activity and lipid peroxidation, changes in the SOD isozyme pattern	Hidalgo et al. (2002)
Manganese	*Oncorhynchus mykiss*	Depressed MnSOD and CuZnSOD activities	Knox et al. (1981)

CAT, catalase; SOD, superoxide dismutase; GPx, glutathione peroxidase; GR, glutathione reductase; GST, glutathione-*S*-transferase; TBARS, thiobarbituric acid reacting substances.

trout (*Oncorhynchus mykiss*), Trenzado (2008) reported that vitamin E deficient diets induced an increase in CAT, GPx, and GR activities in liver and in CAT and SOD activities in kidney. However, in the erythrocytes of these fish CAT and GPx activities seemed not to be influenced by the lack of vitamin E, whereas SOD activity increased (Trenzado et al. 2009). Interestingly, in the latter study, the SOD isozyme pattern showed a clear interaction with dietary nutrient availability and rearing density of the fish. Thus, in uncrowded groups only one Cu,Zn-SOD isozyme was detected independently of diet supplied, whereas in crowded fish isoenzyme variability was linked with dietary composition, with up to five isozymes becoming apparent (Fig. 20.3). In contrast, no interactions were observed between dietary vitamin E and antioxidant enzyme

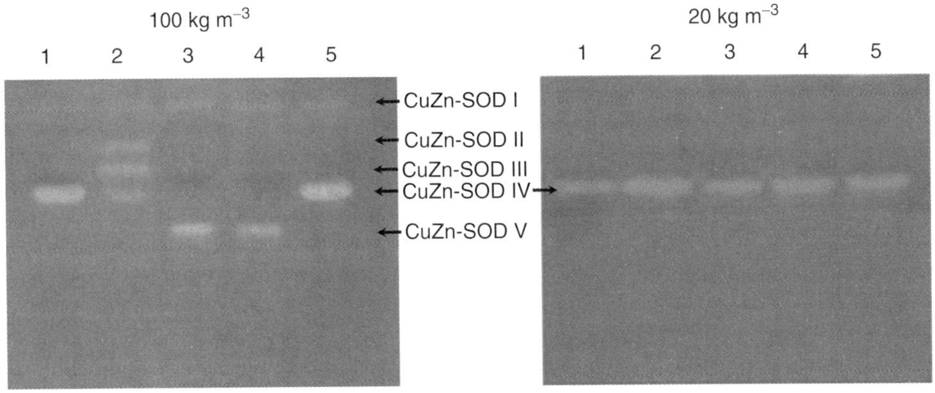

Fig. 20.3 SOD isozyme pattern in erythrocytes of rainbow trout (*Oncorhynchus mykiss*) maintained under two stocking densities (high: 100 kg m^{-3}; low: 20 kg m^{-3}) and fed on five experimental diets (1, -E-HUFA; 2, -E+HUFA; 3, +E-HUFA; 4, +E+HUFA; 5, -C+E+HUFA) (Trenzado et al. 2009).

activities in Atlantic salmon (*Salmo salar* L.) (Lygren et al. 2000).

Regarding the oxidative damage due to vitamin E deficiency, in African catfish (*Clarius gariepinus*), sea bream (*Sparus aurata*), turbot (*Scophthalmus maximus*), and halibut (*Hippoglossus hippoglossus*), the highest liver thiobarbituric acid reactive substances (TBARS) and isoprostane levels occurred when fish were fed diets without vitamin E (Baker and Davies 1996, 1997; Tocher et al. 2002). Similarly, in chinook salmon (*Oncorhynchus tshawytscha*), an inverse correlation between polar lipid peroxidation and vitamin E levels in liver was observed (Welker and Congleton 2005). This trend has also been found in liver and erythrocytes of rainbow trout (*Oncorhynchus mykiss*) (Trenzado 2008; Trenzado et al. 2009). On the contrary, lipid oxidative damage in liver of arctic charr (*Salvelinus alpinus*) seems independent of dietary vitamin E levels (Olsen et al. 1999). Finally, in seabass (*Dicentrarchus labrax*) larvae, inadequate dietary vitamin E levels compromised survival (Guerriero et al. 2004).

Of the limited information available concerning the role of dietary minerals in antioxidant defense mechanisms in fish, most studies focus on selenium (Se). Se is an essential trace mineral obtained partly from the water (Lall and Bishop 1977) but mostly from the diet, in both inorganic and organic forms (Arteel and Sies 2001; Hidalgo and Morales 2009). This element constitutes part of the active site of antioxidant enzymes, such as Se-dependent GPx, necessary for the removal of free radicals and the maintenance of organismal health (Halliwell and Gutteridge 2000; Köhrle et al. 2005). Monteiro et al. (2009) reported that dietary Se supplementation protected against lipid peroxidation in liver, gills, and white muscle of matrinxã (*Brycon cephalus*) exposed to stress, reducing ROS generation and consequently maintaining or reducing the GPx, SOD, and CAT activities, as well as increasing GSH availability in all tissues. In the case of black bullhead catfish (*Ameiurus melas*), Heisinger and Dawson (1983) observed that both Se-dependent GPx and Se-independent GPx activities decreased in liver and erythrocytes of animals fed on Se-deficient diets, although Se-dependent GPx activity was more affected. Similarly, Se is fundamental in the prevention of oxidative damage to lipids and in the maintenance of Se-dependent GPx activity in Atlantic salmon (*Salmo salar*) (Bell and Cowey 1987) or rainbow trout (*Oncorhynchus mykiss*) (Ates et al. 2008).

Nutritional studies have also revealed the crucial role of Cu, Zn, and Mn in preventing oxidative stress, but studies analyzing the effect of deficiencies of these elements in fish are scarce. Mn is commonly present in mitochondrial SOD, whereas cytosolic SODs contain Cu and Zn. Zn plays both catalytic and structural roles in enzymes, and its antioxidant properties are widely recognized (Tate et al. 1999; Maret 2000; Powell 2000). Although it was reported that Zn deficiency did not induce hepatic degeneration in trout liver (Ogino and Yang 1978), Hidalgo et al. (2002) have demonstrated that dietary Zn deficiency induced oxidative stress in rainbow trout liver, with greater lipid peroxidation and changes in the SOD isozyme pattern, suggesting that a compensation in the isoenzymatic profile could take place with an induction of the Mn-SOD isoenzyme in those treatments where the Cu,Zn-SOD activity decreased. With respect to Cu availability, Knox et al. (1984) reported that in rainbow trout (*Oncorhynchus mykiss*) fed a high Cu diet the relative proportions of the Cu,Zn-SOD and the Mn-SOD were correlated with the dietary Zn intake. Increasing dietary Zn reduced the activity of the Mn enzyme and increased that of the Cu,Zn metalloenzyme. Regarding Mn, Knox et al. (1981) reported that rainbow trout fed Mn-deficient diets reduced not only the Mn but also the Zn and Cu levels in liver and heart. This led to a simultaneous decrease in the hepatic Mn-SOD and Cu,Zn-SOD activities and a decrease of Mn-SOD activity in heart. While beyond the scope of this chapter, the combined impact of heavy metal pollution and starvation on mechanisms protecting against oxidative stress provides an important direction of research in ecotoxicology of fish.

PERSPECTIVES

Fluctuations in food availability are natural in most aquatic systems, and fish show an impressive capacity to withstand prolonged periods of food limitation, particularly in comparison with mammals. Part of this stems from their lower metabolic requirements. As food availability often falls with environmental temperature, reduced metabolic demands at low temperature extend the period that fish can survive with little food. As reduced food availability leads to major metabolic modifications, with mitochondrial capacities better conserved than glycolytic capacities, rates of ROS production may be more intense in starved than in fed fish. Effectively, decreased food availability does not reduce routine metabolic rates (expressed per kg body mass), but does decrease the active metabolic rate in parallel with reductions in prolonged and sprint swimming performance (Lapointe et al. 2005). Well-fed cod (*Gadus morhua*) also have a lower cost of transport than cod fed maintenance rations. If ROS production is proportional to routine metabolic rates, starved fish would face unchanged rates of ROS production with an impaired defensive system and a higher cost of displacing a kilogram of body mass. This points to a major conflict. As outlined above, dietary antioxidants are crucial elements protecting against oxidative stress. Antioxidant enzymes and GSH levels are not consistently conserved with starvation. These conclusions support the idea that oxidative stress is a mechanism whereby the negative impacts of starvation occur.

Despite the many ways in which food limitation affects the metabolic capacities of fish, perhaps even more striking is their capacity to rebound from these difficulties. In fact, after starvation, fish enter into a period of compensatory growth during which they rapidly accumulate reserves and reinstate their metabolic capacities. For example, when cod (*Gadus morhua*) are offered more food after being fed maintenance rations, macromolecules are deposited as glycogen and protein in white muscle and as glycogen and lipid in liver (Black and Love 1986). As neither white muscle nor hepatic lipids contribute much to routine metabolic expenditures, the increased synthetic costs in fed animals would be offset by their increased mass of metabolically undemanding structures, keeping routine metabolic rates constant. This compensatory growth occurs despite the previous atrophy of the digestive system. The malleability of the piscine digestive system and its capacity to rise to renewed feeding opportunities is impressive. Although our chapter has revealed several areas in which information about antioxidant defense mechanisms is scant, new investigations during renewed tissue growth after starvation should help elucidate how and why fish are so apt to survive prolonged periods of food limitation.

REFERENCES

Amir-Ahmady, B., Salati, L.M. (2001) Regulation of the processing of glucose–6-phosphate dehydrogenase mRNA by nutritional status. *Journal of Biological Chemistry* 276, 10514–10523.

Arteel, G.E., Sies, H. (2001) The biochemistry of selenium and glutathione system. *Environmental Toxicology and Pharmacology* 10, 153–158.

Ates, B., Orun, I., Talas, Z.S., Durmaz, G., Yilmaz, I. (2008) Effects of sodium selenite on some biochemical and hematological parameters of rainbow trout (*Oncorhynchus mykiss* Walbaum, 1792) exposed to Pb^{2+} and Cu^{2+}. *Fish Physiology and Biochemistry* 34, 53–59.

Baker, R.T.M., Davies, S.J. (1996) Changes in tissue α-tocopherol status and degree of lipid peroxidation with varying α-tocopherylacetate inclusion in diets for African catfish. *Aquaculture Nutrition* 2, 71–79.

Baker, R.T.M., Davies, S.J. (1997) Muscle and hepatic fatty acid profiles and α-tocopherol status in African catfish

(*Clarius gariepinus*) given diets varying in oxidative state and vitamin E inclusion. *Animal Science* 64, 187–195.

Barroso, J.B., Peragón, J., Contreras-Jurado, C. et al. (1998) Impact of starvation-refeeding on kinetics and protein expression of trout liver NADPH-production systems. *American Journal of Physiology* 274, R1578–R1587.

Beardall, C.H., Johnston, I.A. (1985) The ultrastructure of myotomal muscles in the saithe (*Pollachius virens* L.) following starvation and refeeding. *European Journal of Cell Biology* 39, 105–111.

Bell, J.G., Cowey, C.B. (1987) Some effects of selenium deficiency on enzyme activities and indices of tissue peroxidation in Atlantic salmon parr (*Salmo salar*). *Aquaculture* 65, 43–54.

Black, D., Love, R.M. (1986) The sequential mobilization and restoration of energy reserves in tissues of Atlantic cod during starvation and refeeding. *Journal of Comparative Physiology B* 156, 469–479.

Blom, S., Andersson, T.B., Förlin, L. (2000) Effects of food deprivation and handling stress on head kidney 17α-hydroxyprogesterone 21-hydroxylase activity, plasma cortisol and the activities of liver detoxification enzymes in rainbow trout. *Aquatic Toxicology* 48, 265–274.

Burgetz, I.J., Rojas-Vargas, A., Hinch, S.G., Randall, D.J. (1998) Initial recruitment of anaerobic metabolism during sub-maximal swimming in rainbow trout (*Oncorhynchus mykiss*). *Journal of Experimental Biology* 201, 2711–2721.

Caseras, A., Metón, I., Vives, C., Egea, M. Fernández, F., Baanante, I.V. (2002) Nutritional regulation of glucose-6-phosphate gene expression in liver of *Sparus aurata*. *British Journal of Nutrition* 88, 607–614.

Chien, L.T., Hwang, D.F. (2001) Effects of thermal stress and vitamin C on lipid peroxidation and fatty acid composition in the liver of thornfish *Terapon jarbua*. *Comparative Biochemistry and Physiology B* 128, 91–97.

Collins, A.L., Anderson, T.A. (1995) The regulation of endogenous energy stores during starvation and refeeding in the somatic tissues of the golden perch. *Journal of Fish Biology* 47, 1004–1015.

Collins, A.L., Anderson, T.A. (1997) The influence of changes in food availability on the activities of key degradative and metabolic enzymes in the liver and epaxial muscle of the golden perch. *Journal of Fish Biology* 50, 1158–1165.

Davies, M.L.F., Johnston, I.A., van der Wal, J.W. (1995) Muscle fibers in rostral and caudal myotomes of the Atlantic cod (*Gadus morhua* L.) have different mechanical properties. *Physiological Zoology* 68, 673–697.

Domenicali, M., Caraceni, P., Vendemiale, G. et al. (2001) Food deprivation exacerbates mitochondrial oxidative stress in rat liver exposed to ischemia-reperfusion injury. *Journal of Nutrition* 131, 105–110.

Dou, S., Masuda, R., Tanaka, M., Tsukamoto, K. (2002) Feeding resumption, morphological changes and mortality during starvation in Japanese flounder larvae. *Journal of Fish Biology* 60, 1363–1380.

Drew, R.E., Rodnick, K.J., Settles, M. et al. (2008) Effect of starvation on transcriptomes of brain and liver in adult female zebrafish (*Danio rerio*). *Physiological Genomics* 35, 283–295.

Dutil, J-D., Lambert, Y., Guderley, H., Blier, P.U., Pelletier, D., Desroches, M. (1998) Nucleic acids and enzymes in Atlantic cod (*Gadus morhua*) differing in condition and growth rate trajectories. *Canadian Journal of Fisheries and Aquatic Sciences* 55, 788–795.

Echevarría, G., Martínez-Bebiá, M., Zamora, S. (1997) Evolution of biometric indices and plasma metabolites during prolonged starvation in European sea bass (*Dicentrarchus labrax*, L). *Comparative Biochemistry and Physiology A* 118, 111–123.

Elbaraasi, H., Mezes, M., Balogh, K., Horvath, L., Csengeri, I. (2004) Effects of dietary ascorbic acid/iron ratio on some production traits, lipid peroxide state and amount/activity of the glutathione redox system in African catfish *Clarias gariepinus* (Burchell) fingerlings. *Aquaculture Research* 35, 256–262.

Figueiredo-Garutti, M.L., Navarro, I., Capilla, E. et al. (2002). Metabolic changes in *Brycon cephalus* (Teleostei, Characidae) during post-feeding and fasting. *Comparative Biochemistry and Physiology A* 132, 467–476.

FitzGerald, G.J., Guderley, H., Picard, P. (1989) Hidden reproductive cost in the three-spined stickleback (*Gasterosteus aculeatus* L). *Experimental Biology* 48, 295–300.

Frigg, M., Prabucki, A.L., Ruhdel, E.U. (1990) Effect of dietary vitamin E levels on oxidative stability of trout fillets. *Aquaculture* 84, 145–158.

Furné, M. (2008) *Study of different physiological aspects in the sturgeon Acipenser naccarii. A comparative study with rainbow trout Oncorhynchus mykiss*. PhD Thesis, University of Granada, Granada.

Furné, M., García-Gallego, M., Hidalgo, M.C. et al. (2009) Oxidative stress parameters during starvation and refeeding periods in adriatic sturgeon (*Acipenser naccarii*) and rainbow trout (*Oncorhynchus mykiss*). *Aquaculture Nutrition* 15, 587–595.

Gaetani, G.F., Kirkman, H.N., Mangerini, R., Ferraris, A.M. (1994) Importance of catalase in the disposal of hydrogen peroxide within human erythrocytes. *Blood* 84, 325–330.

Gatta, P.P., Pirini, M., Testi, S., Vignola, G., Monetti, P.G. (2000) The influence of different levels of dietary vitamin E on sea bass *Dicentrarchus labrax* flesh quality. *Aquaculture Nutrition* 6, 47–52.

Gillis, T.E., Ballantyne, J.S. (1996) The effects of starvation on plasma free amino acid and glucose concentrations in Lake Sturgeon. *Journal of Fish Biology* 49, 1306–1316.

Goodridge, A.G. (1992) Fatty acids synthesis in eucaryotes. *In* Vance, D.E., Vance, J. (eds). *Biochemistry of Lipids, Lipoproteins and Membranes*. Elsevier, Amsterdam, pp. 111–139.

Greer-Walker, M. (1971) Effects of starvation and exercise on skeletal muscle fibres of the cod (*Gadus morhua* L.) and the coalfish (*Gadus virens* L.). *ICES Journal of Marine Science* 33, 421–426.

Grigorakis, K., Alexis, M.N. (2005) Effects of fasting on the meat quality and fat deposition of commercial-size farmed gilthead sea bream (*Sparus aurata*, L.) fed different dietary regimes. *Aquaculture Nutrition* 11, 341–344.

Guderley, H., Lapointe, D., Bédard, M., Dutil, J.-D. (2003) Metabolic priorities during starvation: enzyme sparing in liver and white muscle of Atlantic cod, *Gadus morhua* L. *Comparative Biochemistry and Physiology A* 135, 347–356.

Guerriero, G., Ferro, R., Russo, G.L., Ciarcia, G. (2004) Vitamin E in early stages of sea bass (*Dicentrarchus labrax*) development. *Comparative Biochemistry and Physiology A* 138, 435–439.

Halliwell, B., Gutteridge, J.M.C. (2000) *Free Radicals in Biology and Medicine*, 3rd edn. Oxford University Press, New York.

Hamre, K., Lie, Ø. (1995) α-Tocopherol levels in different organs of Atlantic salmon (*Salmo salar* L.): effect of smoltification, dietary levels of n−3 polyunsaturated fatty acids, and vitamin E. *Comparative Biochemistry and Physiology A* 111, 547–554.

Hamre, K., Waagbø, R., Berge, R.K., Lie, Ø. (1997) Vitamins C and E interact in juvenile Atlantic salmon (*Salmo salar*, L.). *Free Radical Biology and Medicine* 22, 137–149.

Heisinger, J.F., Dawson, S.M. (1983) Effect of selenium deficiency on liver and blood glutathione peroxidase activity in the black bullhead. *Journal of Experimental Zoology* 225, 325–328.

Hidalgo, M.C., Morales, A.E. (2009) Vitaminas y minerales. *In* Sanz, F. (ed.). *La Nutrición y Alimentación en Piscicultura*. Fundación Observatorio Español de Acuicultura, Madrid, pp. 329–406.

Hidalgo, M.C., Expósito, A., Palma, J.M., de la Higuera, M. (2002) Oxidative stress generated by dietary Zn-deficiency: studies in rainbow trout (*Oncorhynchus mykiss*). *International Journal of Biochemistry and Cell Biology* 34, 183–193.

Hwang, D.F., Lin, T.K. (2002) Effect of temperature on dietary vitamin C requirement and lipid in common carp. *Comparative Biochemistry and Physiology B* 131, 1–7.

Jacob, R.A. (1995) The integrated antioxidant system. *Nutrition Research* 15, 755–766.

Ji, H., Om, A.D., Yoshimatsu, T. et al. (2003) Effect of dietary vitamins C and E fortification on lipid metabolism in red sea bream *Pagrus major* and black sea bream *Acanthopagrus schlegeli*. *Fisheries Science* 69, 1001–1009.

Kirchner, S., Seixas, P., Kaushik, S., Panserat, S. (2005) Effects of low protein intake on extra-hepatic gluconeogenic enzyme expression and peripheral glucose phosphorylation in rainbow trout (*Oncorhynchus mykiss*). *Comparative Biochemistry and Physiology B* 140, 333–340.

Kirkman, N.H., Gaetani, G.F. (1984) Catalase: A tetrameric enzyme with four tightly bound molecules of NADPH. *Proceedings of the National Academy Sciences of USA* 81, 4343–4347.

Kirkman, H.N., Rolfo, M., Ferraris, A.M., Gaetani, G.F. (1999) Mechanisms of protection of catalase by NADPH. Kinetics and stoichiometry. *Journal of Biological Chemistry* 274, 13908–13914.

Knox, D., Cowcy C.B., Adron, J.W. (1981) The effect of low dietary manganese intake on rainbow trout (*Salmo gairdneri*). *British Journal of Nutrition* 46, 495–501.

Knox, D., Cowey, C.B., Adron, J.W. (1984) Effects of dietary zinc intake upon copper metabolism in rainbow trout (*Salmo gairdneri*). *Aquaculture* 40, 199–207.

Köhrle, J., Jakob, F., Contempré, B., Dumont, J.E. (2005) Selenium, the thyroid, and the endocrine system endocrine. *Endocrine Reviews* 26, 944–984.

Lall, S.P., Bishop, F.J. (1977) *Studies on Mineral and Protein Utilization by Atlantic Salmon (Salmo salar) Grown in Sea Water*. Fisheries and Marine Service, Environment Canada, Ottawa, ON, Technical Report No. 688, 16 pp.

Lambert, Y., Dutil, J.D. (1997) Can simple condition indices be used to monitor and quantify seasonal changes in the energy reserves of Atlantic cod (*Gadus morhua*). *Canadian Journal of Fisheries and Aquatic Sciences* 54, 104–112.

Lapointe, D., Guderley, H., Dutil, J.D. (2005) Changes in the condition factor have an impact on metabolic rate and swimming performance relationships in Atlantic cod (*Gadus morhua* L.). *Physiological and Biochemical Zoology* 79, 109–119.

Leopold, J.A., Loscalzo, J. (2000) Cyclic strain modulates resistance to oxidant stress by increasing G6PDH expression in smooth muscle cells. *American Journal of Physiology* 279, H2477–H2485.

Lowery, M.S., Somero, G.N. (1990) Starvation effects on protein synthesis in red and white muscle of the barred sand bass *Palabrax nebulifer*. *Physiological Zoology* 63, 630–648.

Lygren, B., Hamre, K., Waagbo, R. (2000) Effect of induced hyperoxia on the antioxidant status of Atlantic salmon *Salmo salar* L. fed three different levels of dietary vitamin E. *Aquaculture Research* 31, 401–407.

Maret, W. (2000) The function of zinc metallothionein: a link between cellular zinc and redox site. *Journal of Nutrition* 130, 1455S–1458S.

Martínez, M., Guderley, H., Nelson, J.A., Webber, D., Dutil, J.D. (2002) Once a fast cod, always a fast cod: Maintenance of performance hierarchies despite changing food availability in cod. *Physiological and Biochemical Zoology* 75, 90–100.

Martínez, M., Guderley, H., Dutil, J.-D., Winger, P.D., Walsh, S.J. (2003) Condition, prolonged swimming performance and muscle metabolic capacities of cod (*Gadus morhua*). *Journal of Experimental Biology* 206, 503–511.

Martínez, M., Bédard, M., Dutil, J.-D., Guderley, H. (2004) Does condition of Atlantic cod (*Gadus morhua*) have a greater impact upon swimming performance at Ucrit or sprint speeds?. *Journal of Experimental Biology* 207, 2979–2990.

Martínez-Álvarez, R.M., Morales, A.E., Sanz, A. (2005) Antioxidant defenses in fish: Biotic and abiotic factors. *Reviews in Fish Biology and Fisheries* 15, 75–88.

Mendez, G., Wieser, W. (1993) Metabolic responses to food deprivation and refeeding in juveniles of *Rutilus rutilus* (Teleostei: Cyprinidae). *Environmental Biology of Fishes* 36, 73–81.

Metón, I., Fernández, F., Baanante, I.V. (2003) Short- and long-term effects of refeeding on key enzyme activities in glycolysis-gluconeogenesis in the liver of gilthead seabream (*Sparus aurata*). *Aquaculture* 225, 99–107.

Monteiro, D.A., Rantin, F.T., Kalinin, A.L. (2009) The effects of selenium on oxidative stress biomarkers in the freshwater characid fish matrinxã (*Brycon cephalus*) exposed to organophosphate insecticide Folisuper 600 BR® (methyl parathion). *Comparative Biochemistry and Physiology C* 149, 40–49.

Morales, A.E., Pérez-Jiménez, A., Hidalgo, M.C., Abellán, E., Cardenete, G. (2004) Oxidative stress and antioxidant defenses after prolonged starvation in *Dentex dentex* liver. *Comparative Biochemistry and Physiology C* 139, 153–161.

Nam, Y.K., Cho, Y.S., Choi, B.N., Kim, K.H., Kim, S.K., Kim, D.S. (2005) Alteration of antioxidant enzymes at the mRNA level during short-term starvation of rockbream *Oplegnathus fasciatus*. *Fisheries Science* 71, 1385–1387.

Navarro, I., Gutiérrez, J. (1995) Fasting and starvation. *In* Hochachka, P.W., Mommsen, T.P. (eds). *Biochemistry and Molecular Biology of Fishes*. Elsevier, Amsterdam, pp. 393–434.

Ogino, C., Yang, G.Y. (1978) Requirement for rainbow trout for dietary zinc. *Bulletin of the Japanese Society of Fisheries Science* 44, 1015–1018.

Olsen, R.E., Lovaas, E., Lie, O. (1999) The influence of temperature, dietary poly-unsaturated fatty acids, α-tocopherol and spermine on fatty acid composition and indices of oxidative stress in juvenile Arctic char, *Salvelinus alpinus* (L.). *Fish Physiology and Biochemistry* 20, 13–29.

Pandolfi, P., Sonatí, F., Rivi, R., Mason, P., Grosveld, F., Luzzatto, J. (1995) Targeted disruption of the housekeeping gene encoding glucose–6-phosphate dehydrogenase (G6PD): G6PD is dispensable for pentose synthesis but essential for defense against oxidative stress. *EMBO Journal* 14, 5209–5215.

Papadopoulos, M.C., Koumenis, I.L., Dugan, L.L., Giffard, R.G. (1997) Vulnerability to glucose deprivation injury correlates with glutathione levels in astrocytes. *Brain Research* 748, 151–156.

Pascual, P., Pedrajas, J.R., Toribio, F., López-Barea, J., Peinado, J. (2003) Effect of food deprivation on oxidative stress biomarkers in fish (*Sparus aurata*). *Chemico-Biological Interactions* 145, 191–199.

Paul, A.J., Paul, J.M., Smith, R.L. (1995) Compensatory growth in Alaska yellowfin sole, *Pleuronectes asper*, following food deprivation. *Journal of Fish Biology* 46, 442–448.

Pelletier, D., Guderley, H., Dutil, J.D. (1993). Effects of growth rate, temperature, season, body size on glycolytic enzyme activities in the white muscle of Atlantic cod (*Gadus morhua*). *Journal of Experimental Zoology* 265, 477–87.

Pérez-Jiménez, A. (2008) *Nutritive and metabolic response and redox balance of common dentex (Dentex dentex) under different nutritional conditions*. PhD Thesis, University of Granada, Granada.

Pérez-Jiménez, A., Guedes, M.J., Morales, A.E., Oliva-Teles, A. (2007) Metabolic responses to short starvation and refeeding in *Dicentrarchus labrax*. Effect of dietary composition. *Aquaculture* 265, 325–335.

Polakof, S., Míguez, J.M., Soengas, J.L. (2007) Daily changes in parameters of energy metabolism in liver, white muscle, and gills of rainbow trout: Dependence on feeding. *Comparative Biochemistry and Physiology A* 147, 363–374.

Powell, S.R. (2000) The antioxidant properties of zinc. *Journal of Nutrition* 130, 1447S–1454S.

Power, D.M., Melo, J., Santos, C.R.A. (2000) The effect of food deprivation and refeeding on the liver, thyroid hormones and transthyretin in sea bream. *Journal of Fish Biology* 56, 374–387.

Protsko, C.R., Fritz, R.S., Kletzien, R. (1989) Nutritional regulation of hepatic glucose–6-phosphate dehydrogenase. *Biochemical Journal* 258, 295–299.

Richards, J.G., Mercado, A.J., Clayton, C.A., Heigenhauser, G.J.F., Wood, C.M. (2002) Substrate utilization during graded aerobic exercise in rainbow trout. *Journal of Experimental Biology* 205, 2067–2077.

Rios, F.S., Moraes, G., Oba, E.T. et al. (2006) Mobilization and recovery of energy stores in traíra, *Hoplias malabaricus* Bloch (Teleostei, Erythrinidae) during long-term starvation and alter re-feeding. *Journal of Comparative Physiology B* 176, 721–728.

Robinson, M.K., Rustum, R.R., Chambers, E.A., Rounds, J.D., Wilmore, D.W., Jacobs D.O. (1997) Starvation enhances hepatic free radical release following endotoxemia. *Journal of Surgical Research* 69, 325–330.

Salem, M., Silverstein, J., Rexroad, C.E., Yao, J. (2007) Effect of starvation on global gene expression and proteolysis in rainbow trout (*Oncorhynchus mykiss*). *BMC Genomics* 8, 328–343.

Salvemini, F., Franzé, A., Iervolino, A., Filosa, S., Salzano, S., Ursini, M.V. (1999) Enhanced glutathione levels and oxidoresistance mediated by increased glucose–6-phosphate dehydrogenase expression. *Journal of Biological Chemistry* 274, 2750–2757.

Sánchez-Muros, M.J., García-Rejón, L., García-Salguero, L., de la Higuera, M., Lupiáñez, J.A. (1998) Long-term nutritional effects on the primary liver and kidney metabolism in rainbow trout. Adaptative response to starvation and a high-protein, carbohydrate-free diet on glutamate dehydrogenase and alanine aminotransferase kinetics. *International Journal of Biochemistry and Cell Biology* 30, 55–63.

Sangiao-Alvarellos, S., Guzmán, J.M., Láiz-Carrión, R. et al. (2005) Interactive effects of high stocking density and food deprivation on carbohydrate metabolism in several tissues of gilthead sea bream (*Sparus auratus*). *Journal of Experimental Zoology A* 303, 761–775.

Shiau, S.Y., Hsu, C.Y. (2002) Vitamin E sparing effect by dietary vitamin C in juvenile hybrid tilapia, *Oreochromis niloticus* × *O. aureus*. *Aquaculture* 210, 335–342.

Soengas, J.L., Strong, E.F., Fuentes, J., Veira, J.A.R., Andres, M.D. (1996) Food deprivation and refeeding in Atlantic salmon, *Salmo salar*: effects on brain and liver carbohydrate and ketone bodies metabolism. *Fish Physiology and Biochemistry* 15, 491–511.

Tate, D.J., Miceli, M.V., Newsome, D.A. (1999) Zinc protects against oxidative damage in cultured human retinal pigment epithelial cells. *Free Radical Biology and Medicine* 26, 704–713.

Tian, W.-N., Braunstein, L.D., Apse, K. et al. (1999) Importance of glucose–6-phosphate dehydrogenase activity in cell death. *American Journal of Physiology* 276, C1121–C1131.

Tocher, D.R., Mourente, G., Van der Eecken, A. et al. (2002) Effects of dietary vitamin E on antioxidant defence mechanisms of juvenile turbot (*Scophthalmus maximus* L.), halibut (*Hippoglossus hippoglossus* L.) and sea bream (*Sparus aurata* L.). *Aquaculture Nutrition* 8, 195–207.

Trenzado, C.E. (2008) *Selective breeding and diet as strategies for chronic stress tolerance in rainbow trout Oncorhynchus mykiss (Walbaum, 1972)*. PhD Thesis. University of Granada, Granada.

Trenzado, C.E., Morales, A.E., Palma, J.M., de la Higuera, M. (2009) Blood antioxidant defenses and hematological adjustments in crowded/uncrowded rainbow trout (*Oncorhynchus mykiss*) fed on diets with different levels of antioxidant vitamins and HUFA. *Comparative Biochemistry and Physiology C* 149, 440–447.

Vendemiale, G., Grattagliano, I., Caraceni, P. et al. (2001) Mitochondrial oxidative injury and energy metabolism alteration in rat fatty liver: Effect of the nutritional status. *Hepatology* 33, 808–815.

Viganò, L., Arillo, A., Bagnasco, M., Bennicelli, C., Melodia, F. (1993) Xenobiotic metabolizing enzymes in uninduced and induced rainbow trout (*Oncorhynchus mykiss*): effects of diets and food deprivation. *Comparative Biochemistry and Physiology C* 104, 51–55.

Vulliamy, T., Mason, P., Luzzatto, L. (1992) The molecular basis of glucose–6-phosphate dehydrogenase deficiency. *Trends in Genetics* 4, 138–143.

Wang, T., Hung, C.C.Y., Randall, D.J. (2006) The comparative physiology of food deprivation: From feast to famine. *Annual Review of Physiology* 68, 223–251.

Welker, T.L., Congleton, J.L. (2005) Oxidative stress in migrating spring Chinook salmon smolts of hatchery origin: changes in vitamin E and lipid peroxidation. *Transactions of the American Fisheries Society* 134, 1499–1508.

Willmer, P., Stone, G., Johnston, I. (2000) *Environmental Physiology of Animals*. Blackwell Science, Oxford.

Zhang, X.D., Zhu, Y.F., Cai, L.S., Wu, T.X. (2008) Effects of fasting on the meat quality and antioxidant defenses of market-size farmed large yellow croaker (*Pseudosciaena crocea*). *Aquaculture* 280, 136–139.

Chapter 21

ENVIRONMENTALLY INDUCED OXIDATIVE STRESS IN FISH

Volodymyr I. Lushchak

Department of Biochemistry and Biotechnology, Vassyl Stefanyk Precarpathian National University, Ukraine

Every severe stress in living organisms is accompanied by oxidative stress. Interestingly, some reactive oxygen species (ROS) also have a signaling function and support adaptation to environmental stress.

TEMPERATURE CHANGE

Both the increase and decrease of environmental temperature induces oxidative stress in fish. Thermal increase stimulates metabolism and accelerates ROS production and may induce oxidative stress (Parihar and Dubey 1995; Heise et al. 2006a,b; Lushchak and Bagnyukova 2006a,b; Bagnyukova et al. 2007a). Under some circumstances, temperature decrease can also cause oxidative stress in fish (Malek et al. 2004) and two mechanisms could be responsible: (i) decreased efficiency of ROS elimination systems, and (ii) enhanced ROS production in the cold. Unfortunately, we cannot discriminate between these two mechanisms.

OXYGEN LEVEL

In concert with temperature, oxygen is a critical environmental parameter for fish distribution in aquatic environments. Intuitively, one would suppose that enhanced environmental oxygen levels would stimulate ROS production by enhancing the probability that electrons escape from the mitochondrial respiratory chain to interact with oxygen molecules in hyperoxic tissues. The case is, however, not that straightforward, as fish limit oxygen consumption by adjusting ventilation and blood circulation to decrease oxygen extraction from water in order to reduce the risk of oxidative stress under environmental hyperoxia (Lushchak et al. 2005a; Olsvik et al. 2005, 2006; Lushchak and Bagnyukova 2006c; see Chapter 10).

Environmental hypoxia reduces ROS formation in ectotherms. In the European glass eel *Anguilla anguilla*, hypoxia decreased expression of respiratory chain components and antioxidant defenses (Pierron et al. 2007), a phenomenon observed in many organisms. In contrast, exposure of goldfish to anoxia enhanced superoxide dismutase (SOD) and catalase (CAT) activities (Lushchak et al. 2001), which we interpreted as "preparation for oxidative stress" (Hermes-Lima et al. 1998; see Chapter 3). These changes in antioxidant enzyme activities may form part of a transient oxidative stress signal during transition from normoxia via hypoxia to anoxia. Transition to hypoxia also increased SOD and CAT activities in liver of common carp *Cyprinus carpio* (Lushchak et al. 2005b). These results agree with the "preparation for oxidative

Oxidative Stress in Aquatic Ecosystems, First Edition. Edited by Doris Abele, José Pablo Vázquez-Medina, and Tania Zenteno-Savín.

stress" idea, but also suggest mechanisms related with enhanced ROS generation during certain steps in the normoxic–hypoxic transition. Finally, our third fish model, rotan *Percottus glenii*, clearly demonstrated the induction of oxidative stress in hypoxia (Lushchak and Bagnyukova 2007). Hypoxia-induced oxidative stress was also described in the context of high-altitude hypoxia (Møller et al. 2008) and enhanced ROS generation under hypoxia was documented *in vitro* by Schumacker and colleagues (Becker et al. 1999; Bell et al. 2007). In our opinion, at least two mechanisms responsible for hypoxia-induced oxidative stress should be discussed: (i) enhanced numbers of electrons from reduced electron transport chains (ETC) of the mitochondria may escape and univalently reduce molecular O_2; (ii) proteolytic transformation of xanthine reductase to xanthine oxidase (XO) and ROS formation by XO. Conclusive experimental evidence in support of either of these mechanisms is still missing, and future investigation should focus on this process.

OZONE

Ozone has positive and negative effects on biological systems, involving ROS formation. Low ozone concentrations can be beneficial for organisms, whereas higher concentrations are usually deleterious (Bocci et al. 2009). High ozone concentrations are not to be expected in natural aquatic systems and occur only during technological water purification using ultraviolet radiation. In experiments with red blood cells (RBC) of rainbow trout (*Oncorhynchus mykiss*), ozone induced hemolysis, formation of methemoglobin, and RBC membrane lipid peroxidation (Fukunaga et al. 1999). Incubation with ozone enhanced the generation of H_2O_2, which was supposed to mediate RBC damage. The authors concluded that neither ozone nor its derivatives directly attacked the cell from the outside but penetrated the membrane to form ROS with hemoglobin. The effects of ozone on rainbow trout could also be connected with the stimulation of ROS production (Hébert et al. 2008).

CONTAMINANT-INDUCED OXIDATIVE STRESS

In my opinion the classification of environmental stressors as natural or anthropogenic is rather artificial. Human-born pollution of aquatic ecosystems is of significant importance; and in many cases involves contaminant-induced generation of ROS and RNS. The most important contaminant groups are metal ions, pesticides, oil products, and chlorinated hydrocarbons.

Metal Ions

Heavy metals in aquatic systems originate from natural processes such as rock erosion and from human activity. Depending on chemical properties, they can induce oxidative stress via different mechanisms. In this respect, metal ions can be divided into two groups: (i) ions with variable valence (transition metals), and (ii) ions with fixed valence. The ions with a fixed valence can interfere with metabolic pathways. For example, bivalent Mg^{2+}, Sr^{2+}, and Ba^{2+} can either interfere with Ca^{2+} and Zn^{2+} involving processes or substitute these ions in enzymes or regulatory proteins. Substitution of ions with changeable valence such as Fe and Cu in certain proteins/enzymes can be even more critical. These metals interact with protein thiol groups, causing protein inactivation and metabolic disturbance, which also enhances ROS formation. The main way by which transition metals (Fe, Cu, Mn, Cr) stimulate ROS production is via the Fenton reaction. Oxidized metal ions may be reduced via a reaction with $O_2^{\bullet-}$ to form molecular oxygen. The net of both reactions gives the Haber–Weiss reaction.

There are several ways to prevent oxidative stress caused by exposure to ROS produced in metal-dependent situations: to prevent their entrance into the organism; to bind them tightly by specific and nonspecific proteins (see Chapter 8) or include them into cellular vesicles such as lysosomes; to enhance the antioxidant defense; and, finally, to release absorbed ions into the environment. The toxicity of metal ions connected with ROS processes for animals and humans was recently reviewed by Valko and colleagues (2005, 2007), and for fish by Lushchak (2008). General mechanisms of metal involvement in ROS processes are summarized in Fig. 21.1.

Iron

Although Fe is an essential element undergoing redox cycling, information on ROS processes induced by this element in fish is very limited. Fish absorb Fe from the

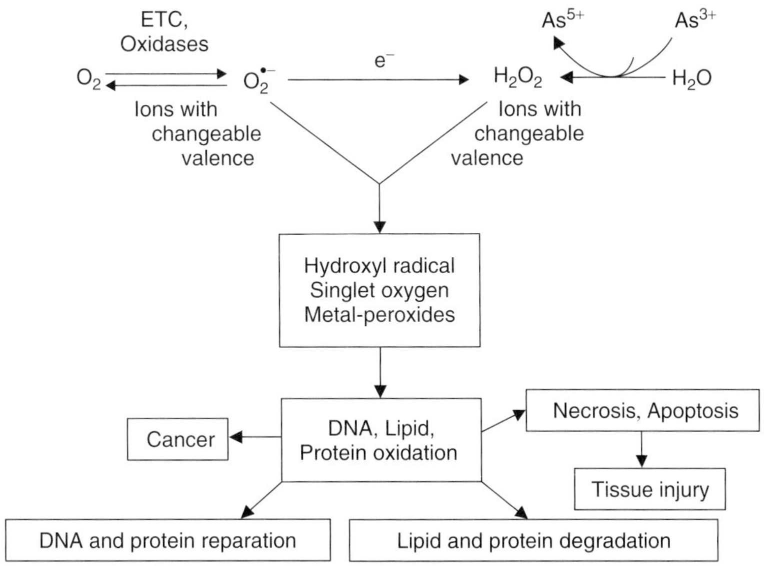

Fig. 21.1 Involvement of metal ions with changeable valence in the metabolism of free radicals in biological systems: their generation, interconversion, and the functional consequences. Superoxide anion radicals may be produced as a side-product by electron transport chains and some oxidases. They can enter three different processes: first, to be reduced, for example, by accepting electrons from ions of metals with changeable valence; second, by receiving one more electron and combining with two protons to give hydrogen peroxide; and third, by accepting an electron from different compounds leading to formation of a hydroxyl radical, etc. Hydrogen peroxide may be produced by some oxidases, but also from oxidation of arsenic(III) to arsenic(V) in the presence of oxygen. The hydroxyl radical, singlet oxygen, and peroxides of metal ions formed are capable of interacting with virtually any constituent of living organisms. Damage to DNA, proteins, and lipids is critically important for cell survival and proper functioning. Oxidative modification of these components may cause cancer, and other diseases, necrosis, and apoptosis leading to cell injury. However, even if severe damage takes place, the cell may recover from oxidative insult. The systems for reparation of damaged DNA are well characterized and are critically important for cell survival. During the past decade, information on reparation of oxidatively modified proteins has accumulated also. However, most modified proteins, as well as oxidized lipids, are degraded.

water across the gill epithelium and through intestinal uptake from food (Bury et al. 2003). Fe metabolism in rainbow trout at normal and Fe-deficient diets has been studied in detail (Walker and Fromm 1976; Carriquiriborde et al. 2004). In a series of experiments, African catfish *Clarias gariepinus* were fed normal and Fe-enriched diets during 5 weeks (Baker et al. 1997). Exposure to the high dietary Fe-ration suppressed fish growth, indicating that this kind of dietary Fe administration is toxic although sublethal. Tissue Fe concentrations were unaffected by the dietary regimen within 5 weeks. However, the concentration of malondialdehyde (MDA) in liver and heart increased with dietary Fe dosage. Simultaneously, fat-soluble antioxidant α-tocopherol (vitamin E) was significantly depleted in liver of fish fed the high Fe diet, illustrating the oxidative stress induced by high Fe diet in fish (Baker et al. 1997).

The effects of waterborn Fe on ROS-related processes in goldfish *Carassius auratus* liver and kidney were studied in our laboratory (Bagnyukova et al. 2006). Fish were exposed to 20 and 500 μM $FeSO_4$ during 7 days. The treatment increased protein carbonyl levels, but reduced lipid peroxide concentration. The concentration of thiobarbituric acid reactive substances (TBARS) was increased in liver and kidney of fish treated with 500 μM $FeSO_4$, indicating onset of oxidative stress. Exposure to high $FeSO_4$ concentrations did not affect SOD activity, but reduced CAT activity in goldfish. Glutathione *S*-transferase (GST) activity was decreased at the high $FeSO_4$ concentration, whereas glutathione reductase (GR) activity was suppressed in kidney of goldfish treated with 20 μM $FeSO_4$. A strong correlation between lipid peroxidation products and CAT activities in liver and GR in kidney indicates a possible up-regulation of these enzymes by $FeSO_4$, and led us to propose that high environmental Fe concentrations elicit a coordinated response of the antioxidant systems in goldfish (Bagnyukova et al. 2006).

Copper

Exposure of fish to dietary and waterborne copper, as well as peritoneal Cu injection can induce oxidative stress. Injection of gilthead sea bream *Sparus aurata* with $CuCl_2$ increased TBARS concentration (Pedrajas et al. 1995). The treatment increased SOD specific activity and resulted in detection of two new Cu,Zn-SOD isoforms. These SOD isoforms were also generated *in vitro* by incubation of cell-free extracts with Cu solutions containing $O_2^{\bullet-}$ or H_2O_2. The authors suggested that the new SOD isoforms in Cu-treated fish resulted from the oxidative modification of original enzymes (Pedrajas et al. 1995).

Forty days of exposure of goldfish *C. auratus* to 0.005–0.025 mg Cu^{2+} L^{-1} decreased catalase and Se-dependent GPx activities, whereas concentrations below 0.0025 mg Cu^{2+} L^{-1} increased GST activity (Liu et al. 2006). Thus, Cu modifies antioxidant enzyme activities in goldfish. In warm adapted African walking catfish *Clarias gariepinus*, dietary Cu administration caused elevated Cu concentrations in intestine, liver, and gills (Hoyle et al. 2007). Cu treatment significantly increased TBARS concentrations in gills and intestine, and total GSH content in intestinal tissue doubled. Liver was depleted from glycogen consistent with the reduced food intake; no histological changes were found in gills, liver, or intestine (Hoyle et al. 2007).

Hansen et al. (2006a,b) investigated the effects of heavy metal contamination on the activity of antioxidant enzymes and on mRNA expression levels in brown trout (*Salmo trutta*) and found: (i) fish exposure to metal ions, particularly to Cu, enhanced the activities of primary (SOD, CAT, GPx) and secondary (GR, metallothionein) antioxidant enzymes/proteins; (ii) the level of mRNA did not always correspond to the respective enzyme activity level; and (iii) metallothioneins were not necessarily up-regulated by the addition of metal ions such as Cu.

Because Cu is contained in several antioxidants, such as Cu,Zn-SOD, and ETS components, such as cytochrome oxidase, Cu deficiency may also be expected to induce oxidative stress. In such a situation, Cu ions can clearly be regarded as antioxidants. In most Cu-related pathologies, however, we deal with an overload of the intracellular Cu pool and saturation of chelators, newly synthesized during Cu intoxication to prevent the appearance of free copper ions in the cells. Metallothioneins and other chaperones play a critical role in binding excess Cu, keeping ROS in balance so that only negligible oxidative stress occurs. The situation tips when the cell is not able to bind the excess of Cu ions, which are then involved in Fenton chemistry that produce highly reactive $HO^{\bullet}$. The bioavailability of Cu ions may be dramatically altered by certain pesticides, such as thiocarbamates in fish, which can induce oxidative stress via abstraction of Cu from the active centers of Cu,Zn-SOD (Lushchak et al. 2007). This example clearly demonstrates that a multifaceted approach to Cu toxicity may show the real response of fish to toxic insult.

Chromium

Cr occurs predominantly in two valence states – hexavalent Cr^{6+} and trivalent Cr^{3+}. Cr^{6+} compounds are used in diverse industries, whereas Cr^{3+} salts, such as chromium picolinate ($Cr(C_6H_4NO_2)_3$), $CrCl_3$, and niacin-bound Cr are used as micronutrients and dietary supplements. Like other metal ions, Cr is toxic and, once released to the ecosystem, it is persistent. Direct and indirect bioaccumulation have been observed in polluted waters so that Cr does not only contaminate individuals, but accumulates in food chains. Two aspects should be noted: (i) cellular reduction of Cr is needed before $HO^{\bullet}$ generation appears; and (ii) Cr salts can play a role of catalyst entering reversible oxidation cycles. It is commonly accepted that biological effects of Cr are at least partially connected to ROS generation. Cr is especially involved in cellular glucose metabolism. In experiments with guppies *Poecilia reticulata*, Perez-Benito (2006) found that low concentrations $<10^{-4}$ M Cr^{6+} increased the maximum lifespan in both males and females. A toxic effect of Cr salts was decreased by antioxidant D-mannitol, which showed potential ROS involvement in Cr toxicity (Perez-Benito 2006).

Potassium dichromate ($K_2Cr_2O_7$) induced oxidative stress in European eel *A. anguilla* L. (Ahmad et al. 2006). In gills, 1 mM dichromate increased GPx activity and decreased GSH concentration. Lipid peroxidation, assessed as TBARS, was intensified in kidney. DNA integrity, evaluated as DNA strand breaks, was higher in both tissues of dichromate-treated animals (Ahmad et al. 2006).

Oxidative stress resulting from Cr exposure seems to be involved in several physiological pathologies in fish. For example, Cr exposure induced lipid peroxidation in tissues of Chinook salmon, *Oncorhynchus tshawytscha*, and high Cr concentrations significantly impaired salmon health (Farag et al. 2006). The kidney of salmons showed gross and microscopic lesions (e.g. necrosis of cells lining kidney tubules); Cr concentrations were elevated in kidneys, and fish growth and survival reduced. The authors suggested that Cr-induced lipid peroxidation and DNA damage caused the observed necrosis and tissue damage (Farag et al. 2006). Another approach to resolve Cr toxicity was chosen by Kuykendall et al. (2006) who studied formation of DNA-protein cross-links (DPXs) in erythrocytes of largemouth bass, *Microproterus salmonoides*, and fathead minnows, *Pimephales promelas*, exposed to waterborne and dietary Cr^{6+}. Fathead minnow exposure to 2 ppm Cr^{6+} led to a significant DPX formation in erythrocytes to more than 140–200% above background levels after 3–4 days. Similarly exposed largemouth basses had 62% elevation of DPX levels after 4 days. Feeding largemouth basses with meat of Cr-injected minnows over 5 days also resulted in significant increase of DPXs in erythrocytes, indicating waterborne and high dose dietary exposure to Cr^{6+} could result in DPX formation in erythrocytes of predatory fish species such as bass (Kuykendall et al. 2006).

There is considerable information on the involvement of Cr-containing compounds in toxicity of wastewaters, but not with respect to oxidative stress involvement. Parvez et al. (2006) investigated oxidative stress in liver, kidney and gills of Indian freshwater fish *Wallago attu* during and after a fish mortality at Panipat, India, and suggested Cr and Cu ions to be responsible. Both are Fenton and Haber–Weiss reactants, suggesting involvement of oxidative stress. Tissue levels of GSH and nonprotein thiols were higher after the fish-mortality episode, whereas TBARS and protein carbonyl levels were lower, leaving involvement of oxidative stress in the fish mortality an open question (Parvez et al. 2006). The effects of Cr^{6+} and Cr^{3+} on ROS processes in goldfish have been summarized in recent papers (Lushchak et al. 2008; Lushchak et al. 2009a,c). In addition to induction of oxidative stress (Shi and Dalal 1990; Valko et al. 2005, 2007; Ahmad et al. 2006; Lushchak 2008), Cr ions can disrupt metal ion-regulated processes, especially in carbohydrate metabolism (Opperman et al. 2008; Stout et al. 2009). Note that in fish, skin mucus can reduce Cr^{6+} and Cr^{3+}, a mechanism implicated in Cr^{6+} detoxification (Arillo and Melodia 1990). Further, the GSH system is likely to be involved in Cr^{6+} reduction (Lushchak et al. 2008).

Mercury

Mercury exists in the oxidation state Hg^{+} (mercuron) and Hg^{2+} (mercuric). In the environment, Hg may be found as methylmercury, produced mainly by Hg^{2+} methylation by microorganisms in soil and water (Valko et al. 2005, 2007). Environmental Hg is ubiquitous and consequently it is practically impossible to avoid exposure. Elemental inorganic and organic Hg is neurotoxic and nephrotoxic, and causes gastrointestinal ulceration and hemorrhage. The biological effects of inorganic or organic Hg are related to their interaction with sulfhydryl-containing residues (Rooney 2007). Hg^{2+} conjugates of cysteine and GSH are transported by organic anion transporters. Fish can accumulate high Hg levels (Salonen et al. 1995; Guallar et al. 2002). Therefore, it should be noted that although the usage of fish-oil-derived fatty acids in human diet reduces the risk of acute coronary events, high Hg content in fish could counteract this protective effect (Valko et al. 2007). Even high content of polyunsaturated fatty acid (PUFA), including docosahexoenoic acid (C 22:6n-3) may not compensate the hazardous Hg effect.

Most studies of Hg effects on fish health report brain pathology and behavioral changes (Baatrup 1991; Ribeiro et al. 1995). Atlantic salmon (*Salmo salar parr*) exposed during 4 months to $HgCl_2$ accumulated significant methylmercury levels in brain that did not cause mortality or growth reduction (Berntssen et al. 2003). However, it resulted in a significant increase in TBARS concentrations and a concomitant decrease in SOD and GPx activities. High Hg concentrations induced pathological damage, visible as vacuole formation and necrosis, and also decreased monoamine oxidase activity. Compared with other organs, the brain was particularly vulnerable to dietary Hg, whereas kidney and liver appeared less sensitive. It should be noted also that low dietary Hg concentrations induced redox protection and SOD activity in brain (Berntssen et al. 2003).

Arsenic

The most common oxidation states of As are +5, +3, and −3. It can form both inorganic and organic compounds in the environment and in cells. Under physiological conditions, the oxidation of As^{3+} to As^{5+} results in H_2O_2 formation inducing oxidative stress (Kalia et al. 2007; Valko et al. 2007; Hébert et al. 2008):

$$H_3AsO_3 + H_2O + O_2 \rightarrow H_3AsO_4 + H_2O_2 \qquad (21.1)$$

The few studies of As toxicity in fish indicate induction of apoptosis in exposed whole animals and cell cultures (Wang et al. 2004; Datta et al. 2007). In two cell lines, TF (fin cells of *Therapon jarbua*) and TO-2 cells (ovary cells of tilapia), sodium arsenite ($NaAsO_2$) produced time- and concentration-dependent apoptotic cell death. Since the antioxidants *N*-acetyl-cysteine and dithiothreitol mitigated arsenite-induced apoptosis in TF cells, ROS involvement in cell death was suggested. However, Schlenk et al. (1997), investigating the effects of different arsenic-compounds on liver of channel catfish, found increased metallothionein levels, but no evidence of alteration in either TBARS or GSH concentration, which casts some doubts concerning the participation of ROS and suggests metallothionein expression and oxidative stress to occur independently of each other. Likewise, treatment of goldfish with arsenate during 1–4 days did not affect TBARS and protein carbonyl levels in the liver, but increased the oxidation of GSH, total GSH concentration, and lipid peroxides, indicating some oxidative stress to be occurring (Bagnyukova et al. 2007b). The activities of the main antioxidants, SOD, CAT, and GPx were correspondingly enhanced. Thus, As compounds apparently affect cellular metabolism in fish, in some cases by enhancing ROS production.

Other Metals

Several other ions of transition metals have biological and/or toxic significance. Cobalt, vanadium, and nickel are found in organisms and food chains, but information on their effects on fish is scarce. Soares et al. (2007) found that vanadium affected the concentrations of lipid peroxide products and activities of certain antioxidants in fish *S. aurata* heart. Vanadates accumulated in the mitochondria, impairing respiration by depolarization of the mitochondrial membrane, which altered the redox state of complex III. Oligomeric vanadium species were much more effective than monomeric ones. The authors concluded that mitochondria are targets of vanadate toxicity (Soares et al. 2007). Changes in mitochondrial functioning might provoke ROS formation, but further investigations are needed to confirm this.

Pesticides

Pesticides are physical, chemical, or biological agents intended to kill undesirable plant and animal pests. Major classes of pesticides are: insecticides, herbicides, and fungicides. Most pesticides are synthetic agents, new to the environment and humans and, therefore, their effects on biological systems are poorly predictable. According to their use, they are described in the following three subsections, and since there are many hundreds of different pesticides, the focus is on only the most commonly used or those known to be oxidative stress inducers.

Insecticides

This group includes pesticides and acaricides. I will concentrate on organophosphate (OP) compounds used in agriculture and fish farming. For example, dichlorvos has been used to treat infestations by copepod parasites in Atlantic salmon cultures (Bron et al. 2006). Potential mechanisms of oxidative stress induction by organophosphate insecticides are shown in Fig. 21.2.

Dichlorvos was found to induce oxidative stress in carp, catfish (*Ictalurus nebulosus*) (Hai et al. 1997a), and European eel (Peca-Llopis et al. 2003a); trichlorfon in Nile tilapia (Thomaz et al. 2009); Folisuper 600 BR in characid fish matrinxã (*Brycon cephalus*) (Monteiro et al. 2009); fenthion in *Oreochromis niloticus* (Piner et al. 2007); and malathion in gilthead sea bream (Rosety et al. 2005). The mechanisms of interference of OP insecticides and nontarget pesticides are under investigation, and of growing concern especially in tropical regions, where increasing demand arises

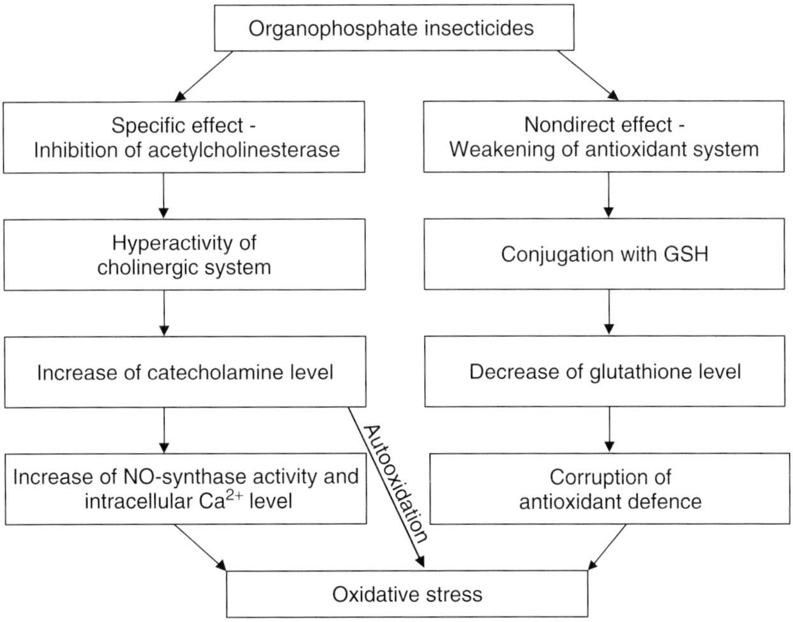

Fig. 21.2 Potential pathways leading to induction of oxidative stress by organophosphate insecticides. These pesticides may affect oxidative processes as least in two ways. The first one is specific and connected with the inhibition of acetylcholinesterase (AChE) and follows a sequence of reactions, and the second is a nonspecific effect related with a decrease in antioxidant potential. The inhibition of the enzyme AChE, which is responsible for terminating the transmission of the nerve impulse, are the primary effects of OPs in vertebrates and invertebrates. They block the hydrolysis of the neurotransmitter acetylcholine (ACh) at the central and peripheral neuronal synapses, leading to excessive accumulation of ACh and activation of ACh receptors. The overstimulation of cholinergic neurones initiates a process of hyperexcitation and convulsive activity that progresses rapidly to *status epilecticus*, leading to profound structural brain damage, respiratory distress, coma, and ultimately the death of the organism if the muscarinic ACh receptor antagonist atropine is not rapidly administered. The toxic effects of OPs are believed to be largely due to the hyperactivity of the cholinergic system, as a result of the accumulation of ACh at the synaptic cleft. It also affects other neurotransmitter systems becoming progressively more disrupted, releasing initially catecholamines and afterwards excitatory amino acids (EAAs), such as glutamate and aspartate, which prolongs the convulsive activity. The break of neuronal and endocrine signaling leads to intracellular influx of Ca^{2+}, triggering the activation of proteolytic enzymes, nitric oxide synthase, and the generation of free radicals resulting in oxidative stress. Since the metabolism of OPs (particularly, dichlorvos) in the liver is connected with glutathione consumption, this also may initiate oxidative stress. Glutathione consumption decreases its concentration, which reduces the power of the antioxidant system disturbing the balance between ROS generation and elimination, and may result in oxidative stress.

due to the economic importance of crops in many countries. Toxic effects of other insecticides, including lindane and DDT, seem to involve the induction of oxidative stress through activation of cytochrome P450, e.g. in skin tumor cell lines of carp (Ruiz-Leal and George 2004) and hepatocytes from *Hoplias malabaricus* (Harada et al. 2003; Filipak Neto et al. 2008). The induction of oxidative stress by DDT may play an important role in tumor promotion and progression (malignant transformation) in fish liver.

Herbicides

I will mainly concentrate on two types of herbicides: the first type are redox cycling compounds, such as paraquat (N,N′-dimethyl-4,4′-bipyridinium dichloride), which constantly generate ROS. Paraquat was shown to induce oxidative stress in *Channa punctata* (Parvez and Raisuddin 2006), zebrafish (Bretaud et al. 2004), and rainbow trout (Stephensen et al. 2002). Its close relative diquat also stimulated ROS production in carp cell lines (Wright et al. 2000) and rainbow trout (Hook et al. 2006). The second group of herbicides consists of several classes of compounds including aminotriazole and dithiocarbamates mainly known as inhibitors of antioxidant enzymes such as CAT and SOD (Fig. 21.3).

To investigate their effects we used a "classic" fish model – goldfish, whereas information for other fish species is from other laboratories. Babich et al. (1993) demonstrated that diethyldithiocarbamate (DDC), an inhibitor of Cu,Zn-SOD, slightly increased the sensitivity of bluegill sunfish BF-2 fibroblasts to H_2O_2. Hai et al. (1997b) found that, although DDC modified pro- and antioxidant systems in tissues of common carp, it could, probably, also serve as an antioxidant due to its abundant thiol-groups. We (Lushchak et al. 2007) clearly demonstrated that DDC induces oxidative stress in goldfish presumably through extraction of Cu from the active center of Cu,Zn-SOD. Aminotriazole (AMT, 3-amino-1,2,4-triazole), a heterocyclic organic compound, is the second most broadly used nonselective herbicide. It is a competitive inhibitor of imidazoleglycerol-phosphate dehydratase, but it also inhibits CAT by binding to Fe in its active site (Lushchak et al. 2003). The exposure to aminotriazole induced oxidative stress in rainbow trout (Dorval and Hontela 2003) and goldfish (Bagnyukova et al. 2005).

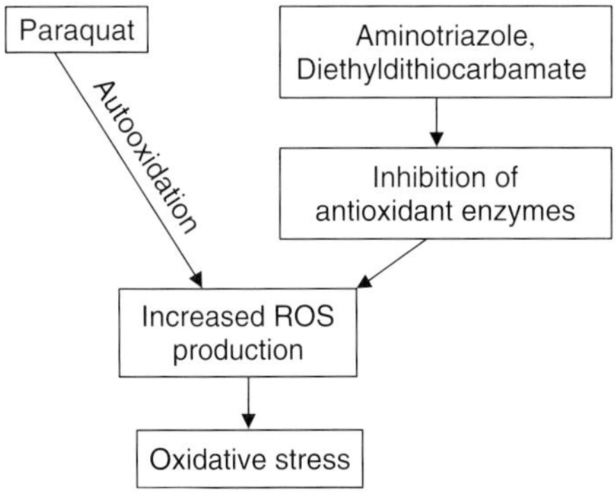

Fig. 21.3 Two suggested mechanisms of herbicide-induced oxidative stress. The first group of herbicides, such as paraquat, is capable of entering redox cycles in living organisms. This can lead to an enhanced level of reactive species and cause oxidative stress. The second group of herbicides does not directly enter redox cycles; rather they compromise the antioxidant system. For example, aminotriazole interacts with the active center of catalase and inactivates the enzyme. Similarly, diethyldithiocarbamate inactivates Cu,Zn-containing superoxide dismutase via extraction of copper ions from the active site of the enzyme. The inactivation of both enzymes enhances the steady-state level of ROS and, if the antioxidant potential is overwhelmed, oxidative stress may be induced.

Herbicides can also exert their biological activity via induction of oxidative stress. For example, glyphosate (N-phosphoromethyl glycine) is a nonselective herbicide that inhibits plant growth through interference with the production of essential aromatic amino acids, by inhibiting the enzyme enolpyruvylshikimate phosphate synthase. This enzyme is responsible for amino acid biosynthesis. When we exposed goldfish to Roundup®, a glyphosate-based herbicide, the induction of oxidative stress was observed. Although the effect was rather weak (Lushchak et al. 2009b), it confirmed findings of induction of oxidative stress by this herbicide in other fish species (Glusczak et al. 2006, 2007). Since the toxicity of herbicides for organisms depends, at least partly, on the induction of oxidative stress, certain antioxidants may be protective. For example, it was found that antioxidant *N*-acetylcysteine reduced the toxic effects of OP dichlorvos on European eel (Peca-Llopis et al. 2003b).

Fungicides

Little work is available on induction of oxidative stress by fungicides in fish. Copper is a frequently used fungicide and was discussed above. Here we will discuss only the organic fungicides such as hexachlorobenzene (HCB), or perchlorobenzene, a chlorocarbon, which induced oxidative stress in liver and brain of common carp (Song et al. 2006). This compound was used as a fungicide formerly for seed treatment, especially on wheat to control the fungal disease bunt. It has been banned globally under the Stockholm Convention on persistent organic pollutants. However, up to now it is one of the most widespread persistent organic pollutants, because it is still released into the environment as a by-product in several industrial processes. In inland waters, HCB is frequently detected, with concentration generally below 1 ng L^{-1}, but higher values have been reported in aquatic systems that receive industrial discharges and surface runoff. In sediments of Lake Ya-Er near a chemical plant in Hubei Province, China, the reported concentrations of HCB reached up to 57 mg kg^{-1} in sediments (Song et al. 2006).

Lipid-soluble HCB was suggested to stimulate ROS production by binding to cytochromes and uncoupling the ETC from monooxygenase activity. Pentachlorophenol, one of HCB major metabolites, is a potent source of ROS during its metabolism (Song et al. 2006).

The group of azole fungicides is widely used in agriculture and its representative, prochloraz (*N*-propyl-*N*-(2-(2,4,6-trichlorophenoxy)ethyl)imidazole-1-carboxamide) is a broad spectrum contact fungicide. Its fungicidal mechanisms are similar to AMT and other imidazole compounds. It inhibits cytochrome P450-dependent 14α-demethylase activity, required for the conversion of lanosterol to ergosterol, an essential component of fungal biological membranes, and disturbs cell membrane assembly. Agricultural application of prochloraz contributes to diffuse freshwater pollution. Prochloraz particularly modulates fish cytochrome P450 enzyme activity. In rainbow trout, prochloraz can also modulate the phase II biotransformation enzymatic activities such as GST (Sanchez et al. 2008).

Oil and Accompanying Pollutants

A substantial part of the oil consumed globally derives from off-shore oil fields and, as recently seen in the Gulf of Mexico, oil spill has become a major threat in marine environments. The main pollutants from this activity include polycyclic aromatic hydrocarbons (PAH), alkylphenols, hydrocarbons, etc. (Sturve et al. 2006). They may affect aquatic organisms in many ways, but oxidative stress is one of the key elements of their toxicity. Several PAHs have cytotoxic, immunotoxic, mutagenic, and/or carcinogenic effects in aquatic organisms (Vogelbein 2003), related to their capacity to enhance ROS production. PAHs are primarily metabolized via hydroxylation and detoxified by enzymes of the cytochrome P450 system (Goksøyr and Förlin 1992), producing redox cycling compounds. CYP1A activity is inhibited by alkylphenols in several fish (Arukwe et al. 1997; Hasselberg et al. 2004a,b), which can affect clearance of xenobiotics causing accumulation of harmful compounds, such as PAHs, in fish tissues.

At least two potential mechanisms of ROS generation by PAHs should be mentioned. The first one includes their conjugation with GSH. This results in the consumption of GSH and a decrease in oxidative stress defense potential. The second one consists in the formation of metabolites capable of entering redox cycles such as metabolites of benzo[a]pyrene (den Besten et al. 1994). Exposure to some PAHs is found to impact oxidative stress parameters in aquatic organisms (Winston and Di Giulio 1991; Livingstone 2001). Exposure to alkylphenols resulted in elevated GR activity and GSH levels in Atlantic cod (Hasselberg et al. 2004a,b), whereas nonylphenol depleted GSH in cod (Hasselberg et al. 2004a), and increased GST activity in largemouth bass *Micropterus salmoides* and in rainbow trout (Uguz et al. 2003; Hughes and Gallagher 2004). Depletion of GSH could also indicate excretion of its oxidized form resulting from GR inactivation (DeLeve and Kaplowitz 1991).

PERSPECTIVES IN FISH OXIDATIVE STRESS RESEARCH

1. In addition to "classic" fish models of pollution, such as zebrafish and fugu (*Fugu rubripes*), goldfish and rainbow trout are promising candidates. The classic models are easy to maintain and their genomes have been sequenced, which helps to clarify their genetic response to chemical insult. However, they are rather small in size and, therefore, conventional methods to study biochemical parameters are hardly applicable. This problem can be solved with goldfish and trout, and genome sequencing is technically now so far advanced that it seems only a question of time before this gap is closed.
2. Standardization of methods to measure oxidative stress is needed. In many papers, authors measure one or two potential markers, and then they elaborate on the development of oxidative stress. Often this leads to erroneous results, and a defined suit of markers should be analyzed which allows clear conclusions.
3. In addition to whole organism experiments, isolated organs, primary cell isolates, and permanent cell cultures of fish should be more broadly established. This may extend the possibilities not only to record the development of oxidative stress, but also to disclose mechanisms and pathways within the adaptive responses.
4. It would be necessary to apply a complex approach to reveal the mechanisms involved in induction of oxidative stress, clearly identifying not only major but also minor ROS sources, and to evaluate their relative significance.

5. The system approach should be applied to evaluate the complex fish response to factors including oxidative stress. This means that reliable markers demonstrating the development of oxidative stress should be measured concomitantly with the activities of antioxidant enzymes, levels of low molecular mass antioxidants, steady-state levels of corresponding mRNA and proteins, etc. Since the response to oxidative challenges is a dynamic process, the inhibitors/activators of transcription, translation, and degradation of RNA and proteins may be useful additional tools.
6. The investigation of mechanisms involved in adaptive response to oxidative stress in fish needs to be developed. Analysis of antioxidant enzymes should involve gene transcription and protein activities. For example, HIF-1α in fish is sensitive to ROS, and its transactivation activity for hypoxic genes may well turn out to be modified by oxidative stress. Little is known about the operation of "classic" gene activators such as AP1, Nrf2/Keap1, NF-κB, etc., within the response to oxidative stress in fish (see Chapter 18), which opens a field for future investigators.

REFERENCES

Ahmad, I., Maria, V.L., Oliveira, M. et al. (2006) Oxidative stress and genotoxic effects in gill and kidney of *Anguilla anguilla* L. exposed to chromium with or without pre-exposure to β-naphthoflavone. *Mutation Research* 608, 16–28.

Arillo, A., Melodia, F. (1990) Protective effect of fish mucus against Cr(VI) pollution. *Chemosphere* 20, 397–402.

Arukwe, A., Förlin, L., Goksøyr, A. (1997) Xenobiotic and steroid biotransformation enzymes in Atlantic salmon (*Salmo salar*) liver treated with an estrogenic compound, 4-nonylphenol. *Environmental and Toxicological Chemistry* 16, 2576–2583.

Baatrup, E. (1991) Structural and functional effects of heavy metals on the nervous system, including sense organs, of fish. *Comparative Biochemistry and Physiology* 100, 253–257.

Babich, H., Palace, M.R., Stern, A. (1993) Oxidative stress in fish cells: *in vitro* studies. *Archives of Environmental Contamination and Toxicology* 24, 173–178.

Bagnyukova, T.V., Storey, K.B., Lushchak, V.I. (2005) Adaptive response of antioxidant enzymes to catalase inhibition by aminotriazole in goldfish liver and kidney. *Comparative Biochemistry and Physiology B* 142, 335–341.

Bagnyukova, T.V., Chahrak, O.I., Lushchak, V.I. (2006) Coordinated response of goldfish antioxidant defenses to environmental stress. *Aquatic Toxicology* 78, 325–331.

Bagnyukova, T.V., Danyliv, S.I., Zin'ko, O.S. et al. (2007a) Heat shock induces oxidative stress in rotan *Perccottus glenii* tissues. *Journal of Thermal Biology* 32, 255–260.

Bagnyukova, T.V., Luzhna, L.I., Pogribny, I.P. et al. (2007b) Oxidative stress and antioxidant defenses in goldfish liver in response to short-term exposure to arsenite. *Environmental and Molecular Mutagenesis* 48, 658–665.

Baker, R.T.M., Martin, P., Davies, S.J. (1997) Ingestion of sub-lethal levels of iron sulphate by African catfish affects growth and tissue lipid peroxidation. *Aquatic Toxicology* 40, 51–61.

Becker, L.B., van den Hoek, T.L., Shao, Z.H. et al. (1999) Generation of superoxide in cardiomyocytes during ischemia before reperfusion. *American Journal of Physiology* 277, H2240–H2246.

Bell, E.L., Klimova, T.A., Eisenbart, J. et al. (2007) Mitochondrial reactive oxygen species trigger hypoxia-inducible factor-dependent extension of the replicative life span during hypoxia. *Molecular and Cellular Biology* 27, 5737–5745.

Berntssen, M.H.G., Aatland, A., Handy, R.D. (2003) Chronic dietary mercury exposure causes oxidative stress, brain lesions, and altered behaviour in Atlantic salmon (*Salmo salar*) parr. *Aquatic Toxicology* 65, 55–72.

Bocci, V., Borrelli, E., Travagli, V. et al. (2009) The ozone paradox: Ozone is a strong oxidant as well as a medical drug. *Medical Research Reviews* 29, 646–682.

Bretaud, S., Lee, S., Guo, S. (2004) Sensitivity of zebrafish to environmental toxins implicated in Parkinson's disease. *Neurotoxicology and Teratology* 26, 857–864.

Bron, J.E., Sommerville, C., Wootten, R. et al. (2006) Fallowing of marine Atlantic salmon, Salmo salar L., farms as a method for the control of sea lice, *Lepeophtheirus salmonis* (Kroyer, 1837). *Journal of Fish Diseases* 16, 487–493.

Bury, N.R., Walker, P.A., Glover, C.N. (2003) Nutritive metal uptake in teleost fish. *Journal of Experimental Biology* 206, 11–23.

Carriquiriborde, P., Handy, R.D., Davies, S.J. (2004) Physiological modulation of iron metabolism in rainbow trout (*Oncorhynchus mykiss*) fed low and high iron diets. *Journal of Experimental Biology* 207, 75–86.

Datta, S., Saha, D.R., Ghosh, D. et al. (2007) Sub-lethal concentration of arsenic interferes with the proliferation of hepatocytes and induces *in vivo* apoptosis in *Clarias batrachus* L. *Comparative Biochemistry and Physiology* 145, 339–349.

DeLeve, L.D., Kaplowitz, N. (1991) Glutathione metabolism and its role in hepatotoxicity. *Pharmacology and Therapeutics* 52, 287–305.

den Besten, P.J., Lemaire, P., Livingstone, D.R. (1994) NADPH-and cumene hydroperoxide-dependent metabolism of benzo[*a*]pyrene by pyloric caeca microsomes of the sea star *Asterias rubens* L. (Echinodermata: Asteroidea). *Xenobiotica* 24, 989–1001.

Dorval, J., Hontela, A. (2003) Role of glutathione redox cycle and catalase in defense against oxidative stress induced by endosulfan in adrenocortical cells of rainbow trout (*Oncorhynchus mykiss*). *Toxicology and Applied Pharmacology* 192, 191–200.

Farag, A.M., May, T., Marty, G.D. et al. (2006) The effect of chronic chromium exposure on the health of Chinook salmon (*Oncorhynchus tshawytscha*). *Aquatic Toxicology* 76, 246–257.

Filipak Neto, F., Zanata, S.M., Silva de Assis, H.C. et al. (2008) Toxic effects of DDT and methyl mercury on the hepatocytes from *Hoplias malabaricus*. *Toxicology In Vitro* 22, 1705–1713.

Fukunaga, K., Nakazono, N., Suzuki, T. et al. (1999) Mechanism of oxidative damage to fish red blood cells by ozone. *IUBMB Life* 48, 631–634.

Glusczak, L., dos Santos Miron, D., Crestani, M. et al. (2006) Effect of glyphosate herbicide on acetylcholinesterase activity and metabolic and hematological parameters in piava (*Leporinus obtusidens*). *Ecotoxicology and Environmental Safety* 65, 237–241.

Glusczak, L., dos Santos Miron, D., Moraes, B.S. et al. (2007) Acute effects of glyphosate herbicide on metabolic and enzymatic parameters of silver catfish (*Rhamdia quelen*). *Comparative Biochemistry and Physiology* 146, 519–524.

Goksøyr, A., Förlin, L. (1992) The cytochrome P450 system in fish, aquatic toxicology and environmental monitoring. *Aquatic Toxicology* 22, 287–312.

Guallar, E., Sanz-Gallardo, M.I., van't Veer, P. et al. (2002) Heavy metals and myocardial infarction study group. Mercury, fish oils, and the risk of myocardial infarction. *New England Journal of Medicine* 347, 1747–1754.

Hai, D.Q., Varga, S.I., Matkovics, B. (1997a) Organophosphate effects on antioxidant system of carp (*Cyprinus carpio*) and catfish (*Ictalurus nebulosus*). *Comparative Biochemistry and Physiology* 117, 83–88.

Hai, D.Q., Varga, S.I., Matkovics, B. (1997b) Effects of diethyldithiocarbamate on antioxidant system in carp tissue. *Acta Biologica Hungarica* 48, 1–8.

Hansen, B.H., Rømma, S., Garmo, Ø.A. et al. (2006a) Antioxidative stress proteins and their gene expression in brown trout (*Salmo trutta*) from three rivers with different heavy metal levels. *Comparative Biochemistry and Physiology C* 143, 263–274.

Hansen, B.H., Rømma, S., Søfteland, L.I. et al. (2006b) Induction and activity of oxidative stress-related proteins during waterborne Cu-exposure in brown trout (*Salmo trutta*). *Chemosphere* 65, 1707–1714.

Harada, T., Yamaguchi, S., Ohtsuka, R. et al. (2003) Mechanisms of promotion and progression of preneoplastic lesions in hepatocarcinogenesis by DDT in F344 rats. *Toxicology and Pathology* 31, 87–98.

Hasselberg, L., Meier, S., Svardal, A. (2004a) Effects of alkylphenols on redox status in first spawning Atlantic cod (*Gadus morhua*). *Aquatic Toxicology* 69, 95–105.

Hasselberg, L., Meier, S., Svardal, A. et al. (2004b) Effects of alkylphenols on CYP1A and CYP3A expression in first spawning Atlantic cod (*Gadus morhua*). *Aquatic Toxicology* 67, 303–313.

Hébert, N., Gagné, F., Cejka, P. et al. (2008) Effects of ozone, ultraviolet and peracetic acid disinfection of a primary-treated municipal effluent on the immune system of rainbow trout (*Oncorhynchus mykiss*). *Comparative Biochemistry and Physiology C* 148, 122–127.

Heise, K., Puntarulo, S., Nikinmaa, M. et al. (2006a) Oxidative stress during stressful heat exposure and recovery in the North Sea eelpout *Zoarces viviparus* L. *Journal of Experimental Biology* 209, 353–363.

Heise, K., Puntarulo, S., Nikinmaa, M. et al. (2006b) Oxidative stress and HIF-1 DNA binding during stressful cold exposure and recovery in the North Sea eelpout (*Zoarces viviparus*). *Comparative Biochemistry and Physiology A* 143, 494–503.

Hermes-Lima, M., Storey, J.M., Storey, K.B. (1998) Antioxidant defenses and metabolic depression. The hypothesis of preparation for oxidative stress in land snails. *Comparative Biochemistry and Physiology B* 120, 437–448.

Hook, S.E., Skillman, A.D., Small, J.A. et al. (2006) Gene expression patterns in rainbow trout, *Oncorhynchus mykiss*, exposed to a suite of model toxicants. *Aquatic Toxicology* 77, 372–785.

Hoyle, I., Shaw, B.J., Handy, R.D. (2007) Dietary copper exposure in the African walking catfish, *Clarias gariepinus*: Transient osmoregulatory disturbances and oxidative stress. *Aquatic Toxicology* 83, 62–72.

Hughes, E.M., Gallagher, E.P. (2004) Effects of 17-[beta] estradiol and 4-nonylphenol on phase II electrophilic detoxification pathways in largemouth bass (*Micropterus salmoides*) liver. *Comparative Biochemistry and Physiology* 137, 237–247.

Kalia, K., Narulaa, G.D., Kannan, G.M. et al. (2007) Effects of combined administration of captopril and DMSA on arsenite induced oxidative stress and blood and tissue arsenic concentration in rats. *Comparative Biochemistry and Physiology* 144, 372–379.

Kuykendall, J.R., Miller, K.L., Mellinger, K.N. et al. (2006) Waterborne and dietary hexavalent chromium exposure causes DNA-protein crosslink (DPX) formation in erythrocytes of largemouth bass (*Micropterus salmoides*). *Aquatic Toxicology* 78, 27–31.

Liu, H., Wang, W., Zhang, J.-F. et al. (2006) Effects of copper and its ethylenediaminetetraacetate complex on the antioxidant defenses of the goldfish, *Carassius auratus*. *Ecotoxicology and Environmental Safety* 65, 350–354.

Livingstone, D.R. (2001) Contaminant-stimulated reactive oxygen species production and oxidative damage in aquatic organisms. *Marine Pollution Bulletin* 42, 656–666.

Lushchak, V.I. (2008) Oxidative stress as a component of transition metal toxicity in fish. *In* Svensson, E.P. (eds). *Aquatic Toxicology Research Focus*. Nova Science Publishers, Hauppaug, NY, pp. 1–29.

Lushchak, V.I., Bagnyukova, T.V. (2006a) Temperature increase results in oxidative stress in goldfish tissues. 1. Indices of oxidative stress. *Comparative Biochemistry and Physiology C* 143, 30–35.

Lushchak, V.I., Bagnyukova, T.V. (2006b) Temperature increase results in oxidative stress in goldfish tissues. 2. Antioxidant and associated enzymes. *Comparative Biochemistry and Physiology C* 143, 36–41.

Lushchak, V.I., Bagnyukova, T.V. (2006c) Effects of different environmental oxygen levels on free radical processes in fish. *Comparative Biochemistry and Physiology B* 144, 283–289.

Lushchak, V.I., Bagnyukova, T.V. (2007) Hypoxia induces oxidative stress in tissues of a goby, the rotan *Perccottus glenii*. *Comparative Biochemistry and Physiology B* 148, 390–397.

Lushchak, V.I., Lushchak, L.P., Mota, A.A. et al. (2001) Oxidative stress and antioxidant defenses in goldfish *Carassius auratus* during anoxia and reoxygenation. *American Journal of Physiology* 280, R100–R107.

Lushchak, O.V., Bagnyukova, T.V., Lushchak, V.I. (2003) Effect of aminotriazole on the activity of catalase and glucose–6-phosphate dehydrogenase of two frog species, *Rana ridibunda* and *R. esculenta*. *Ukranian Biochemical Journal* 75, 45–50.

Lushchak, V.I., Bagnyukova, T.V., Husak, V.V. et al. (2005a) Hyperoxia results in transient oxidative stress and an adaptive response by antioxidant enzymes in goldfish tissues. *International Journal of Biochemistry and Cell Biology* 37, 1670–1680.

Lushchak, V.I., Bagnyukova, T.V., Lushchak, O.V. et al. (2005b) Hypoxia and recovery perturb free radical processes and antioxidant potential in common carp (*Cyprinus carpio*) tissues. *International Journal of Biochemistry and Cell Biology* 37, 1319–1330.

Lushchak, V.I., Bagnyukova, T.V., Lushchak, O.V. et al. (2007) Diethyldithiocarbamate injection induces transient oxidative stress in goldfish tissues. *Chemico-Biological Interactions* 170, 1–8.

Lushchak, O.V., Kubrak, O.I., Nykorak, M.Z. et al. (2008) The effect of potassium dichromate on free radical processes in goldfish: Possible protective role of glutathione. *Aquatic Toxicology* 87, 108–114.

Lushchak, O.V., Kubrak, O.I., Lozinsky, O.V. et al. (2009a) Chromium(III) induces oxidative stress in goldfish liver and kidney. *Aquatic Toxicology* 93, 45–52.

Lushchak, O.V., Kubrak, O.I., Storey, J.M. et al. (2009b) Low toxic herbicide Roundup induces mild oxidative stress in goldfish tissues. *Chemosphere* 76, 932–937.

Lushchak, O.V., Kubrak, O.I., Torous, I.M. et al. (2009c) Trivalent chromium induces oxidative stress in goldfish brain. *Chemosphere* 75, 56–62.

Malek, R.L., Sajadi, H., Abraham, J. et al. (2004) The effects of temperature reduction on gene expression and oxidative stress in skeletal muscle from adult zebrafish. *Comparative Biochemistry and Physiology C* 138, 363–373.

Møller, P., Risom, L., Lundby, C. et al. (2008) Hypoxia and oxidation levels of DNA and lipids in humans and animal experimental models. *IUBMB Life*. 60, 707–723.

Monteiro, D.A., Rantin, F.T., Kalinin, A.L. (2009) The effects of selenium on oxidative stress biomarkers in the freshwater characid fish matrinxã, *Brycon cephalus* (Günther, 1869) exposed to organophosphate insecticide Folisuper 600 BR (methyl parathion). *Comparative Biochemistry and Physiology C* 149, 40–49.

Olsvik, P.A., Kristensen, T., Waagbø, R. et al. (2005) mRNA expression of antioxidant enzymes (SOD, CAT and GSH-Px) and lipid peroxidative stress in liver of Atlantic salmon (*Salmo salar*) exposed to hyperoxic water during smoltification. *Comparative Biochemistry and Physiology C* 141, 314–323.

Olsvik, P.A., Kristensen, T., Waagbø, R. et al. (2006) Effects of hypo- and hyperoxia on transcription levels of five stress genes and the glutathione system in liver of Atlantic cod *Gadus morhua*. *Journal of Experimental Biology* 209, 2893–2901.

Opperman, D.J., Piater, L.A., van Heerden, E. (2008) A novel chromate reductase from *Thermus scotoductus* SA-01 related to old yellow enzyme. *Journal of Bacteriology* 190, 3076–3082.

Parihar, M.S., Dubey, A.K. (1995) Lipid peroxidation and ascorbic acid status in respiratory organs of male and female freshwater catfish *Heteropneustes* fossilis exposed to temperature increase. *Comparative Biochemistry and Physiology C* 112, 309–313.

Parvez, S., Raisuddin, S. (2006) Effects of paraquat on the freshwater fish *Channa punctata* (Bloch): non-enzymatic antioxidants as biomarkers of exposure. *Archives of Environmental Contamination and Toxicology* 50, 392–397.

Parvez, S., Pandey, S., Ali, M. et al. (2006) Biomarkers of oxidative stress in *Wallago attu* (Bl. and Sch.) during and after a fish-kill episode at Panipat, India. *Science of the Total Environment* 368, 627–636.

Pedrajas, J.R., Peinado, J., Lypez-Barea, J. et al. (1995) Oxidative stress in fish exposed to model xenobiotics. Oxidatively modified forms of Cu,Zn-superoxide dismutase as potential biomarkers. *Chemico-Biological Interactions* 98, 267–282.

Perez-Benito, J.F. (2006) Effects of chromium (VI) and vanadium (V) on the lifespan of fish. *Journal of Trace Elements in Medicine and Biology* 20, 161–170.

Peca-Llopis, S., Ferrando, M.D., Peca, J.B. (2003a) Increased recovery of brain acetylcholinesterase activity in dichlorvos-intoxicated European eels *Anguilla anguilla* by bath treatment with N-acetylcysteine. *Diseases of Aquatic Organisms* 55, 237–245.

Peca-Llopis, S., Ferrando, M.D., Peca, J.B. (2003b) Fish tolerance to organophosphate-induced oxidative stress is dependent on the glutathione metabolism and enhanced by N-acetylcysteine. *Aquatic Toxicology* 65, 337–360.

Pierron, F., Baudrimont, M., Gonzalez, P. et al. (2007) Common pattern of gene expression in response to hypoxia or cadmium in the gills of the European glass eel (*Anguilla anguilla*). *Environmental Science and Technology* 41, 3005–3011.

Piner, P., Sevgiler, Y., Uner, N. (2007) *In vivo* effects of fenthion on oxidative processes by the modulation of glutathione metabolism in the brain of *Oreochromis niloticus*. *Environmental Toxicology* 22, 605–612.

Ribeiro, C., Fernandes, L., Carvlaho, C. et al. (1995) Acute effects of mercuric chloride on the olfactory epithelium of *Trichomycterus brasiliensis*. *Ecotoxicology and Environmental Safety* 31, 104–109.

Rooney, J.P.K. (2007) The role of thiols, dithiols, nutritional factors and interacting ligands in the toxicology of mercury. *Toxicology* 234, 145–156.

Rosety, M., Rosety-Rodríguez, M., Ordonez, F.J. et al. (2005) Time course variations of antioxidant enzyme activities and histopathology of gilthead seabream gills exposed to malathion. *Histology and Histopathology* 20, 1017–1020.

Ruiz-Leal, M., George, S. (2004) An *in vitro* procedure for evaluation of early stage oxidative stress in an established fish cell line applied to investigation of PHAH and pesticide toxicity. *Marine Environmental Research* 58, 631–635.

Salonen, J.T., Seppanen, K., Nyyssonen, K. et al. (1995) Intake of mercury from fish, lipid peroxidation, and the risk of myocardial infarction and coronary, cardiovascular, and any death in eastern Finnish men. *Circulation* 91, 645–655.

Sanchez, W., Piccini, B., Porcher, J.M. (2008) Effect of prochloraz fungicide on biotransformation enzymes and oxidative stress parameters in three-spined stickleback (*Gasterosteus aculeatus* L.). *Journal of Environmental Science and Health B* 43, 65–70.

Schlenk, D., Wolford, L., Chelius, M. et al. (1997) Effect of arsenite, arsenate, and the herbicide monosodium methyl arsonate (MSMA) on hepatic metallothionein expression and lipid peroxidation in channel catfish. *Comparative Biochemistry and Physiology C* 118, 177–183.

Shi, X., Dalal, N.S. (1990) On the hydroxyl radical formation in the reaction between hydrogen peroxide and biologically generated chromium(V) species. *Archives of Biochemistry and Biophysics* 277, 342–350.

Soares, S.S., Gutiérrez-Merino, C., Aureliano, M. (2007) Mitochondria as a target for decavanadate toxicity in Sparus aurata heart. *Aquatic Toxicology* 83, 1–9.

Song, S.B., Xu, Y., Zhou, B.S. (2006) Effects of hexachlorobenzene on antioxidant status of liver and brain of common carp (*Cyprinus carpio*). *Chemosphere* 65, 699–706.

Stephensen, E., Sturve, J., Förlin, L. (2002) Effects of redox cycling compounds on glutathione content and activity of glutathione-related enzymes in rainbow trout liver. *Comparative Biochemistry and Physiology C* 133, 435–442.

Stout, M.D., Herbert, R.A., Kissling, G.E. et al. (2009) Hexavalent chromium is carcinogenic to F344/N rats and B6C3F1 mice after chronic oral exposure. *Environmental Health Perspectives* 117, 716–722.

Sturve, J., Hasselberg, L., Fälth, H. et al. (2006) Effects of North Sea oil and alkylphenols on biomarker responses in juvenile Atlantic cod (*Gadus morhua*). *Aquatic Toxicology* 78, 73–78.

Thomaz, J.M., Martins, N.D., Monteiro, D.A. et al. (2009) Cardio-respiratory function and oxidative stress biomarkers in Nile tilapia exposed to the organophosphate insecticide trichlorfon (NEGUVON). *Ecotoxicology and Environmental Safety* 72, 1413–1424.

Uguz, C., Iscan, M., Erguven, A. et al. (2003) The bioaccumulation of nonyphenol and its adverse effect on the liver of rainbow trout (*Onchorynchus mykiss*). *Environmental Research* 92, 262–270.

Valko, M., Morris, H., Cronin, M.T. (2005) Metals, toxicity and oxidative stress. *Current Medicinal Chemistry* 12, 1161–1208.

Valko, M., Leibfritz, D., Moncol, J. et al. (2007) Free radicals and antioxidants in normal physiological functions and human disease. *International Journal of Biochemistry and Cellular Biology* 39, 44–84.

Vogelbein, W.K. (2003) Polycyclic aromatic hydrocarbons and liver carcinogenesis in fish. *In* Newman, M.C., Unger, M.A. (eds). *Fundamentals of Ecotoxicology*, CRC Press LLC, Boca Raton, FL, pp. 143–145.

Walker, R.L., Fromm, P.O. (1976) Metabolism of iron by normal and iron deficient rainbow trout. *Comparative Biochemistry and Physiology* 55, 311–318.

Wang, Y.-C., Chaung R.-H., Tung, L.-C. (2004) Comparison of the cytotoxicity induced by different exposure to sodium arsenite in two fish cell lines. *Aquatic Toxicology* 69, 67–79.

Winston, G.W., Di Giulio, R.T. (1991) Prooxidant and antioxidant mechanisms in aquatic organisms. *Aquatic Toxicology* 19, 137–161.

Wright, J., George, S., Martinez-Lara, E. et al. (2000) Levels of cellular glutathione and metallothionein affect the toxicity of oxidative stressors in an established carp cell line. *Marine Environmental Research* 50, 503–508.

Chapter 22

CHEMICAL POLLUTANTS AND THE MECHANISMS OF REACTIVE OXYGEN SPECIES GENERATION IN AQUATIC ORGANISMS

Francesco Regoli

Dipartimento di Biochimica, Biologia e Genetica, Università Politecnica delle Marche, Ancona, Italy

CHEMICAL POLLUTANTS IN MARINE ORGANISMS

Marine organisms can be exposed to a large variety of chemical compounds released into the environment due to the continuous expansion of human activities. Pollutants can enter marine systems via a large number of sources and routes: with direct and indirect pathways; differences in affinity and distribution within environmental matrices; and variations in bioavailability and mechanisms of toxicity. Uptake of contaminants in target tissues occurs from sediments, suspended particulate matter, the water column, or through the diet, depending on trophic level and ecological lifestyle of the organisms. Trace metals and organic xenobiotics are environmentally relevant classes of pro-oxidant chemicals which can exhibit different pathways of oxidative challenge at the molecular and cellular level, and several mechanisms may exist for a single compound, determining a very complex network of oxidative interactions and cascade effects.

MAIN INTRACELLULAR OXIDATIVE PATHWAYS FOR TRACE METALS

The most important mechanisms for reactive oxygen species (ROS) generation by trace metals are summarized in Fig. 22.1. These elements enhance the generation of ROS as a result of their ability to lose electrons and catalyze Haber–Weiss and Fenton reactions (Halliwell and Gutteridge 1984, 2007). Haber–Weiss reactions generate $HO^{\bullet}$ after the reduction of an oxidized metal by $O_2^{\bullet -}$ and its reaction with H_2O_2 (Halliwell and Gutteridge 1984):

$$\text{metal}^{n+1} + O_2^{\bullet -} \rightarrow \text{metal}^{n+} + O_2$$
$$\text{metal}^{n+} + H_2O_2 \rightarrow \text{metal}^{n+1} + HO^{\bullet} + HO^{-}$$
$$\text{overall:}\quad O_2^{\bullet -} + H_2O_2 \rightarrow +O_{2+}HO^{\bullet} + HO^{-} \quad \text{with } (\text{metal}^{n+1}/\text{metal}^{n+})$$

Haber–Weiss reactions are of particular importance during activated phagocytosis when a large amount of $O_2^{\bullet -}$ is generated and a limited amount of metals is

Oxidative Stress in Aquatic Ecosystems, First Edition. Edited by Doris Abele, José Pablo Vázquez-Medina, and Tania Zenteno-Savín.

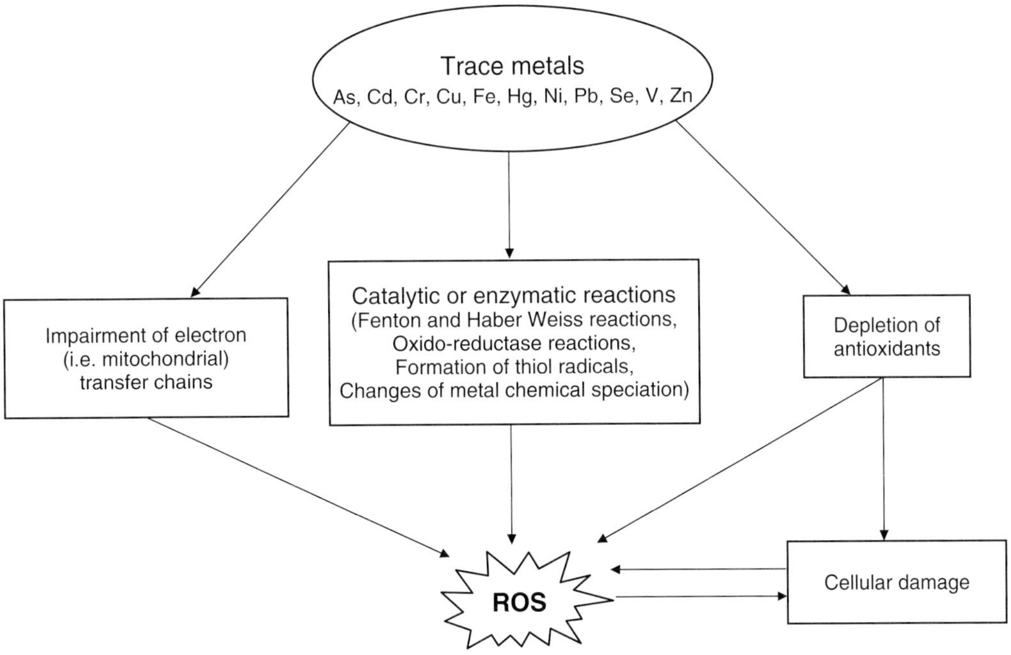

Fig. 22.1 Main mechanisms of reactive oxygen species (ROS) formation by trace metals.

thus needed as catalyst (Freeman and Crap 1982). This reaction can represent a source of cellular oxidative stress; it is generally slow, unless a transition metal ion reacts with H_2O_2 to produce $HO^•$ and the oxidized metal (Fenton reaction):

$$\text{metal}^{n+} + H_2O_2 \rightarrow \text{metal}^{n+1} + HO^• + HO^-$$

Addition of a reducing agent, such as ascorbate, leads to a cyclic generation of $HO^•$ with a marked increase of damage to biological molecules.

Several elements can catalyze these reactions with different efficiencies, depending on redox potential. Typical inducers are Fe(II), Cu(I), Cr(III), (IV), (V), (VI), and V(V), while in the presence of chelating agents, such as Gly-Gly-His and thiol-containing agents, Co(II) and Ni(II) also react with H_2O_2 and lipid peroxides to generate $HO^•$ and lipid radicals (Inoue and Kawanishi 1989; Shi et al. 2004).

Cr exists in several oxidation states (0, III, IV, V, VI). While Cr(III) is a widely distributed form in the environment and an essential nutrient, chromate (Cr(VI)) is typically associated with industrial processes and induces toxicological and oxidative effects. Cr(VI) is reduced to an intermediate Cr(V) before being further reduced to the stable Cr(III) (Shi et al. 2004). Cr(VI) is reduced by several flavoenzymes, like ferrodoxin- nicotinamide adenine dinucleotide phosphate $(NADP)^+$ oxidoreductase, glutathione reductase (GR), and lipoil dehydrogenase, with formation of a relatively long-lived Cr(V)–NADPH complex. During this reduction process, molecular oxygen is simultaneously reduced to H_2O_2, which can react with the Cr(V)–NADPH complex to generate $HO^•$, and regenerate Cr(VI). Cr(VI) acts as a catalyst and, thus, trace amounts of this compound can determine a cyclic generation of $HO^•$.

The same flavoenzymes are responsible for the reduction of vanadate leading to active intermediates which react with oxygen to form $O_2^{•-}$, while arsenite activates reduced nicotinamide adenine dinucleotide (NADH) oxidase and produces $O_2^{•-}$ (Halliwell and Gutteridge 1984; Halliwell and Gutteridge 2007). For arsenite, oxidative mechanisms include effects on mitochondrial electron transfer chain and increased ROS formation at the ubiquinone site of the respiratory chain (Corsini et al. 1999; Shi et al. 2004). In addition, intermediary arsine species, like $(CH_3)_2AsOO^•$, may be formed from the reaction of dimethylarsine and oxygen, with concomitant formation of $O_2^{\bullet-}$; this peroxyl radical has been suggested to play a more active role in DNA damage than ROS (Yamanaka et al. 1990). When present as methylated compounds, As can release redox-active Fe from ferritin and generate ROS by promoting the Fenton reaction, in synergy with ascorbic acid.

Some trace metals react with cysteine leading to formation of thiol radicals ($GS^{\bullet}$), which cause direct damage or react with other thiols to generate ROS (Shi and Dalal 1988).

$$GS^{\bullet} + RSH \rightarrow RSSR^{\bullet-} + H^{+}$$
$$RSSR^{\bullet-} + O_2 \rightarrow RSSR + O_2^{\bullet-}$$

The inbalance or depletion of antioxidant defenses can also indirectly increase ROS formation by trace metals, even for those elements with only one oxidation state, like Cd. Due to their elevated affinity for -SH groups, trace metals react with glutathione (GSH), where a sulfhydryl, an amino and two carboxylic acid groups, as well as two peptide linkages, represent reactive sites for metals. The reactions of metals with GSH may result in either the formation of complexes, or in its oxidation to glutathione disulphide (GSSG) (Christie and Costa 1984; Meister 1989). Metals promoting GSH oxidation include Cu(II), Co(II), Mn(II), Fe(II), and Cr(VI), whereas more stable coordination complexes are formed with GSH by Zn(II), Cd(II), Hg(II), Pb(II), and Ni(II). Depletion of antioxidant capacity enhances the onset of oxidative damage, altered redox enzyme activity, and structural and functional organelle dysfunctions. Accumulation of oxidatively damaged cellular components may further exacerbate the risk of oxidative stress through decomposition of peroxides, release of metal ions from storage sites, heme protein release, conversion of xanthine dehydrogenase to xanthine oxidase, impaired mitochondrial function, and increased intracellular Ca^{2+} levels (Halliwell and Gutteridge 2007).

OXIDATIVE METABOLISM OF ORGANIC XENOBIOTICS

Aromatic xenobiotics, like polycyclic aromatic hydrocarbons (PAHs), polychlorinated biphenyls (PCBs), halogenated hydrocarbons, dioxin (TCDD), and dioxin-like chemicals, enhance intracellular ROS production through induction of the cytochrome P450 system (Stegeman and Lech 1991). Cytochrome P450 is a multigene family of heme-containing enzymes, which catalyze a variety of oxidative reactions, including hydroxylation, epoxidation, dealkylation, deamination, sulfoxidation, and desulfuration, termed phase I reactions (Di Giulio et al. 1995). During these reactions, the lipophilicity of the xenobiotic is lowered by the addition of a polar functional moiety (like a hydroxyl group), which often makes the metabolite more hydrophilic and, in some cases, ready for further conjugation reactions and/or excretion (Fig. 22.2). After binding of the xenobiotic to cytochrome P450, the heme group is reduced by NADPH-cytochrome P450 reductase; an oxygen molecule is incorporated in the heme pocket forming a peroxide, and when a second electron cleaves the O–O bond, the substrate is oxygenated at the R–H bond and released, while Fe in the heme group is reconverted to the ferric state. These reactions generally proceed quickly; however, the slow oxidation of some chemicals can result in uncoupling of the electron transfer and oxygen reduction, thus causing the release of ROS (Schlezinger et al. 2006). For some xenobiotics, ROS are the final metabolites of the biotransformation process. Metabolites produced in phase I reactions can be eliminated or covalently conjugated to various endogenous compounds, such as GSH, glucuronic acid, and sulfate, to further decrease their lipophilic properties. These phase II reactions, catalyzed respectively by glutathione *S*-transferase (GST), uridine diphosphate-glucoronosyl transferases (UDP-GTs), and sulfotransferases (STs), result in water-soluble nontoxic products readily excreted through bile or urine.

Some PAH metabolites may have pro-oxidant effects through the "redox cycle", a well-known source of chemically-mediated ROS generation activated by quinones, diols, nitroaromatics, hydroxylated nitroamines, and transition metal chelates (Livingstone 2001). In this cycle (Fig. 22.3) the compound is initially reduced to a radical intermediate with reducing equivalents provided by the activity of NAD(P)H reductases. The radical intermediates can be toxic themselves or, by reacting with oxygen, produce $O_2^{\bullet-}$ regenerating the parent compound, which can go through another reaction cycle. The net results of these cycles are generation of $O_2^{\bullet-}$ and consumption of NADPH. Metabolism of quinones also includes the activity of NADPH-quinone reductase 1 (NQO1 or DT-diaphorase), which catalyzes their two-electron reduction to hydroquinones (Burczynski and Penning 2000), while univalent reductions yield ROS, such as semiquinone radicals and $O_2^{\bullet-}$ (Fig. 22.4).

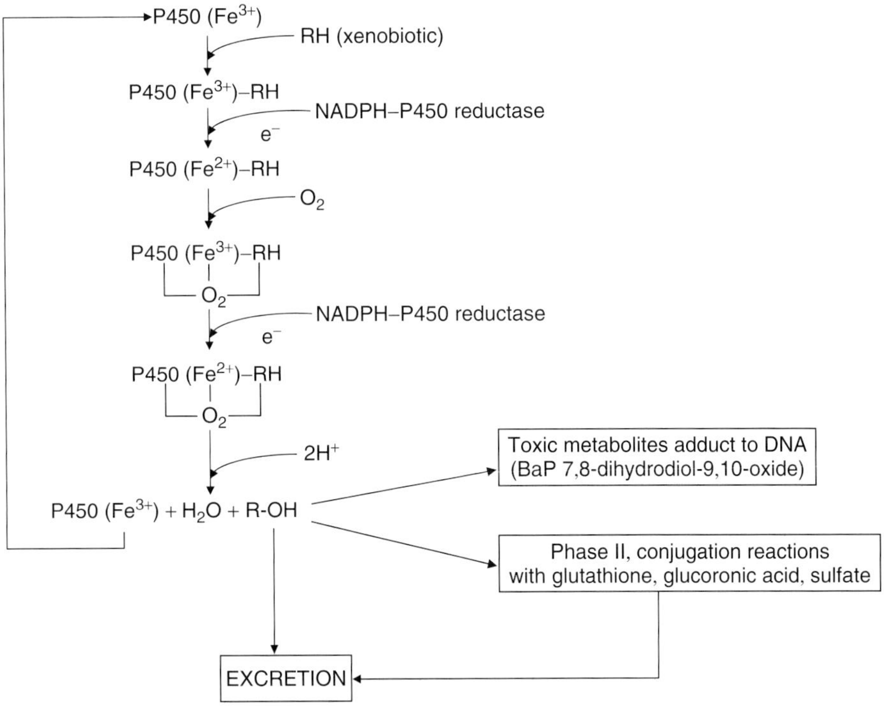

Fig. 22.2 Phase I reaction catalyzed by cytochrome P450.

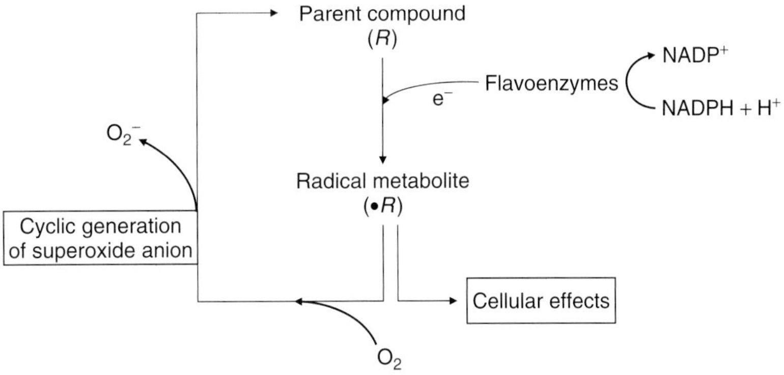

Fig. 22.3 Cyclic generation of reactive oxygen species through the "redox cycle".

OXIDATIVE INTERACTIONS BETWEEN VARIOUS CLASSES OF CHEMICALS

Pollutant-mediated generation of ROS is further modulated through transcriptional and post-transcriptional mechanisms with several indirect and cascade effects contributing to the overall oxidative pressure (Fig. 22.5). Some classes of nonplanar PCBs and pesticides induce transcription of cytochrome P450, eliciting a response similar to that described for phenobarbitol (PB) and pregnenolone-16α-carbonitrile (PCN). Two different receptors of the nuclear receptor family, pregnane X receptor (PXR) and constitutively active receptor (CAR), form a heterodimer with the retinoid X receptor (RXR) which then binds to DNA PB-responsive elements to activate gene transcription

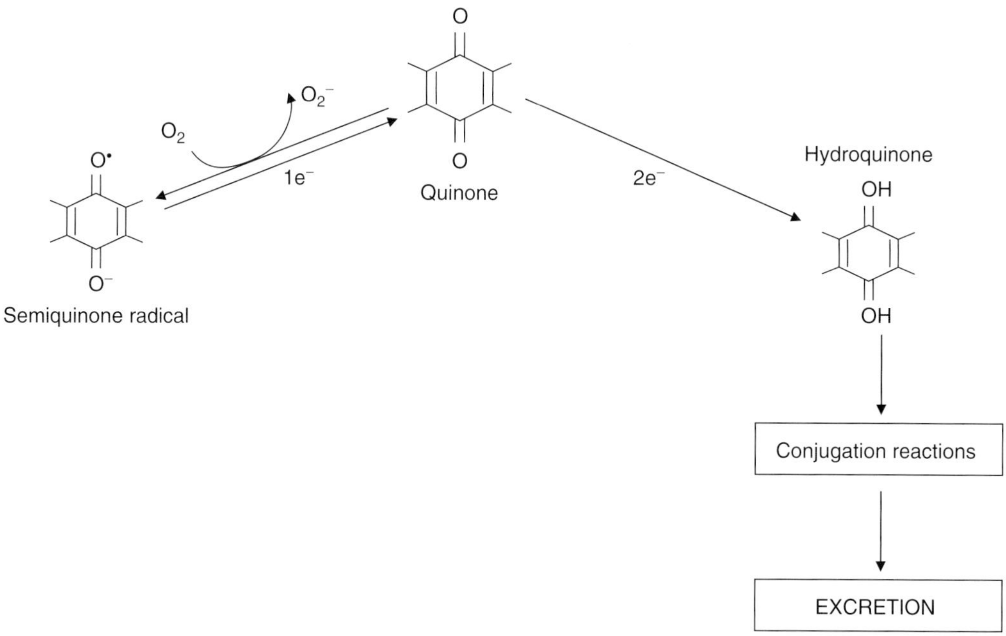

Fig. 22.4 Univalent and divalent reduction of quinones.

of the cytochrome P450, specifically CYP2B and CYP3A subfamilies, enhancing ROS formation during their catalytic activities (Honkakoski and Negishi 2000; Wei et al. 2000; Xie et al. 2000).

In a similar mechanism, crude oil, lubricants, PAHs, PCBs, phthalate ester plasticizers, and alkylphenols induce peroxisomal gene transcription responsible for ROS formation (Cajaraville et al. 2003; Schrader and Fahimi 2006). These effects are mediated by peroxisomal proliferator activated receptors (PPAR α, β, or γ) which, like many members of the nuclear hormone receptor superfamily, heterodimerize with RXR and bind to specific DNA regions of target genes, peroxisome proliferator responsive elements (PPREs). As an effect of xenobiotics, transition metals like Fe and Cu, abundant in peroxisomes, are released, and catalyze $HO^\bullet$ generation through Fenton reactions (Bacon and Britton 1989).

Aromatic chemicals, e.g. TCDD, PAHs, and PCBs, induce cytochrome P450 through interaction with the aryl hydrocarbon (Ah) receptor, a cytosolic xenobiotic binding protein that is activated after dissociation from a heat shock 90 protein. The subunit AhR ligand is then transported into the nucleus by the AhR nuclear translocator (ARNT), where it interacts with specific DNA regions (dioxin responsive elements, DRE) and initiates the transcription of Ah genes (Whitlock Jr et al. 1996; Rowland and Gustafsson 1997). These genes include the cytochrome P450s, CYP1A and CYP1B, UDPGT 1AG, GST Ya, class 3 aldehyde dehydrogenase, and NQO1.

Induction of cytochrome P450 and xenobiotic metabolism enzymes has both pro-oxidant and antioxidant effects. Enhanced ROS formation can be caused by the induction of cytochrome P450; but, excessive levels of ROS induce an inhibitory effect on the biotransformation pathway. Cytochrome P450 is sensitive to ROS, particularly to H_2O_2 at pre-transcriptional (binding with Ah receptor), transcriptional, protein, and catalytic levels (Morel and Barouki 1998, 1999; Barouki and Morel 2001). Similar down-regulation of cytochrome P450 by ROS has environmental implications since other pro-oxidant chemicals and stressors can modulate the metabolism of AhR agonists (Barker et al. 1994; Morel and Barouki 1998). Trace metals frequently occur as co-contaminants with organic xenobiotics, and increase ROS formation through several direct and indirect mechanisms, potentially reducing cytochrome P450 biotransformation efficiency (Vakharia et al. 2001).

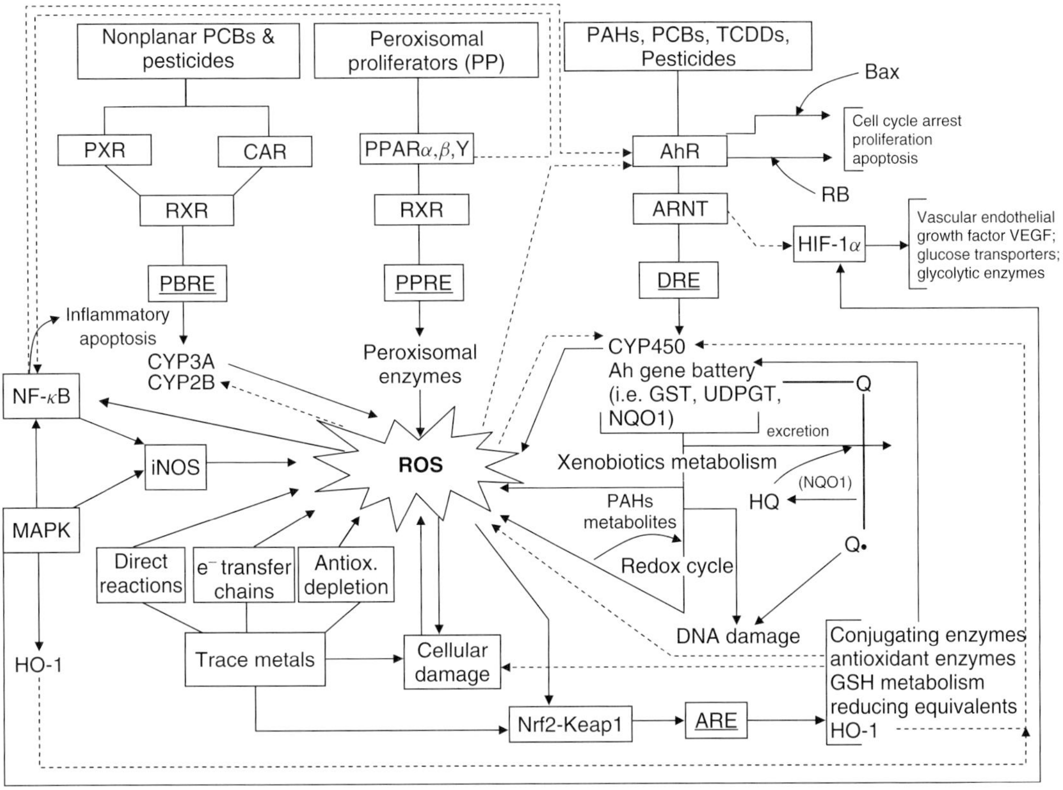

Fig. 22.5 Overall schematic presentation of mechanisms, interactions, and signaling pathways involved in reactive oxygen species (ROS) generation by the main classes of environmental chemical pollutants.

Interactions between different classes of chemicals, generation of ROS, and onset of oxidative stress are also modulated through indirect mechanisms and cascade effects reflecting changes in levels and functions of redox-sensitive signaling proteins and transcription factors. Among these, Nrf2 regulates the expression of cytoprotective enzymes after exposure to levels of ROS, RNS, lipid aldehydes, trace metals, electrophilic xenobiotics, and their metabolites (Dinkova-Kostova et al. 2005; Osburn and Kensler 2008). Nrf2-regulated genes include a wide array of enzymes catalyzing electrophilic conjugating reactions, antioxidant mechanisms, metabolism of GSH, and production of reducing equivalents, including GST, epoxide hydrolase, NQO1, UDP-GT, aldehyde dehydrogenase 1A1, aldo-keto reductase, GR, peroxiredoxin, thioredoxin, thioredoxin reductase, CAT, Cu, Zn-superoxide dismutase (SOD), glutathione peroxidase (GPx), glutamate cysteine ligase, UDP-glucose dehydrogenase, glucose-6-phosphate dehydrogenase (G6PDH), and heme oxygenase-1 (HO1) (Thimmulappa et al. 2002; Kwak et al. 2003; Hu et al. 2006; Leonard et al. 2006; Osburn and Kensler 2008). HO1 transforms heme (pro-oxidant) to bilirubin (antioxidant) as a response to oxidant challenge and oxidative damage; however, this mechanism also reduces the availability of heme groups for the cytochrome P450 proteins (Carpenter et al. 2002). Trace metals, which induce HO1 through antioxidant response elements (ARE), can, in this way, affect the induction capability of cytochrome P450 at the post-transcriptional level (Jacobs et al. 1999; Vakharia et al. 2001). A similar hypothesis was formulated for fish, in which Cd and Cu were shown to depress dioxin- and PAH-mediated induction of cytochrome P450 through different oxidative effects acting at transcriptional, translational and catalytic levels (Regoli et al. 2003, 2005; Benedetti et al. 2007, 2009). Under nonstressed conditions, the role of Nrf2 in counteracting basal ROS formation is minimal; enhanced transcription of cytoprotective genes occurs only when oxidative challenge is higher, thus energetically favourable. Up-regulation of Nrf2 has potential to reduce oxidative toxicity of environmental pollutants, targeting multiple steps along the pathway of induced cellular damage.

The AhR-mediated responses are not limited to the activation of Ah metabolic enzymes, but also affect other cellular processes such as development, cell proliferation, regulation of cell cycle, and apoptosis (Ma and Whitlock Jr 1996; Kolluri et al. 1999). Interaction of AhR with retinoblastoma (RB) protein delays cell cycle progression, which would give a cell more time to repair oxidative adduct damage caused by elevated rates of cytochrome P450-catalyzed metabolism (Puga et al. 1999). AhR has also been shown to modulate apoptosis by interacting with the Bax gene and cross-talk between pathways mediated by AhR and by HIF-1α. HIF-1α regulates the expression of vascular endothelial growth factor, glucose transporters, and glycolytic enzymes, which are down-regulated by AhR agonists due to a competition for ARNT and a direct interaction between AhR and HIF-1 (Chan et al. 1999; Matikainen et al. 2001; Nie et al. 2001). Deficits in both glucose transport and the activity of several glycolytic enzymes would be involved in some of the toxic effects of dioxins, such as vascularization and the wasting syndrome (rapid loss of body mass) after acute exposure (Liu and Matsumura 1995). The pathway of HIF-1 genes, down-regulated by AhR agonists, is induced by carcinogenic metals through the mitogen-activated protein kinase (MAPK) signaling pathway causing cell proliferation and cancer (Drevs et al. 2003; Leonard et al. 2004).

Another signaling pathway modulating metabolism of xenobiotics involves the stress–response transcription factor nuclear factor-κB (NF-κB), which is activated by a number of stimuli, including ROS and trace metals, through MAPK signaling (Barnes and Karin 1997; Li and Karin 1999). After its activation in the cytoplasm, NF-κB is translocated to the nucleus, binds to specific DNA sequences (κB elements), and regulates the transcriptional induction of inducible $NO^{\bullet}$ synthase (iNOS), as well as a number of genes mediating inflammatory responses and apoptosis. Redox regulation of NF-κB occurs in the nucleus as a direct modification caused by ROS on specific cysteine residues of the DNA domain, which inhibits binding of NF-κB to the nucleus (Toledano and Leonard 1991; Tsukamoto et al. 2002). A direct binding interaction of the AhR and the NF-κB subunit, RelA, has been demonstrated to suppress AhR-mediated gene transcription (Tian et al. 1999). This role of NF-κB can represent an additional mechanism to explain the negative effect of oxidative stress on cytochrome P450 1A1 inducibility and metabolism of organic xenobiotics. Activated PPARγ antagonizes the activity of NF-κB, thus down-regulating iNOS (Li et al. 2000).

Metals modulate cellular redox status and cytochrome P450 metabolism also through activation of MAPK, a group of proteins collectively termed mitogen-activated protein kinases, which control intracellular signaling via a cascade of protein phosphorylation along a given pathway. Activation of HIF-1 and NF-κB, and the enhanced transcription of HO1 and of iNOS, are among the numerous oxidative effects caused by trace metals through the MAPK system (Leonard et al. 2004).

The examples discussed address a small subset of the possible pro-oxidant mechanisms, interactions at the molecular level, and signaling pathways that may be affected by pollutants. They also provide evidence of the potential for additive, synergistic, and antagonistic effects of different classes of chemicals. The elevated complexity of interactions between pro-oxidant mechanisms, signaling pathways, detoxification, and biotransformation processes, influences the formation and transformation rate of ROS and their potential to cause oxidative damage. Overall, evidence suggests that oxidative stress plays an important role in the molecular mechanism of chemically-induced toxicity.

REFERENCES

Bacon, B.R., Britton, R.S. (1989) Hepatic injury in chronic iron overload. Role of lipid peroxidation. *Chemico-Biological Interactions* 70, 183–226.

Barker, C.W., Fagan, J.B., Fasco, D.S. (1994) Down-regulation of P4501A1 and P4501A2 in isolated hepatocytes by oxidative stress. *Journal of Biological Chemistry* 269, 3985–3990.

Barnes, P.J., Karin, M. (1997) Nuclear factor-kappa B: a pivotal transcription factor in chronic inflammatory diseases. *New England Journal of Medicine* 336, 1066–1071.

Barouki, R., Morel, Y. (2001) Repression of cytochrome P4501A1 gene expression by oxidative stress: mechanisms and biological implications. *Biochemistry and Pharmacology* 61, 511–516.

Benedetti, M., Martuccio, G., Fattorini, D. et al. (2007) Oxidative and modulatory effects of trace metals on metabolism of polycyclic aromatic hydrocarbons in the Antarctic fish *Trematomus bernacchii*. *Aquatic Toxicology* 85, 167–175.

Benedetti, M., Fattorini, D., Martuccio, G. et al. (2009). Interactions between trace metals (Cu, Hg, Ni, Pb) and 2,3,7,8-tetrachlorodibenzo-p-dioxin in the Antarctic fish *Trematomus bernacchii*: oxidative effects on biotransformation pathway. *Environmental Toxicology and Chemistry* 28, 818–825.

Burczynski, M.E., Penning, T.M. (2000) Genotoxic polycyclic aromatic hydrocarbon ortho-quinones generated by aldoketo reductases induce CYP1A1 via nuclear translocation of the aryl hydrocarbon receptor. *Cancer Research* 60, 908–915.

Cajaraville, M.P., Cancio, I., Ibabe, A., Orbea, A. (2003) Peroxisome proliferation as a biomarker in environmental pollution assessment. *Microscopy Research and Technique* 61, 191–202.

Carpenter, D.O., Arcaro, K., Spink, D.C. (2002) Understanding the human health effects of chemical mixtures. *Environmental Health Perspectives* 110, 25–42.

Chan, W.K., Yao, G., Gu, Y.Z., Bradfield, C.A. (1999) Cross-talk between the aryl hydrocarbon receptor and hypoxia inducible factor signaling pathways-demonstration of competition and compensation. *Journal of Biological Chemistry* 274, 12115–12123.

Christie, N.T., Costa, M. (1984) *In vitro* assessment of the toxicity of metal compounds. IV. Disposition of metals in cells: Interactions with membranes, glutathione, metallothionein, and DNA. *Biological Trace Element Research* 6, 139–158.

Corsini, E., Asti, L., Viviani, B. et al. (1999) Sodium arsenate induces overproduction of interleukin-1 alpha in murine keratinocytes: role of mitochondria. *Journal of Investigative Dermatology* 113, 769–765.

Di Giulio, R.T., Benson, W.H., Sanders, B.M., Van Veld, P.A. (1995) Biochemical mechanisms: metabolism, adaptation, and toxicity. *In* Rand, G.M. (ed.) *Fundamentals of Aquatic Toxicology: Effects, Environmental Fate and Risk Assessment*. Taylor & Francis, Washington, pp. 523–561.

Dinkova-Kostova, A.T., Holtzclaw, W.D., Kensler, T.W. (2005) The role of keap1 in cellular protective responses. *Chemical Research in Toxicology* 18, 1779–1791.

Drevs, J., Medinger, M., Schimdt-Gersbach, C. et al. (2003) Receptor tyrosine kinases: the main targets for new anticancer therapy. *Current Drug Targets* 4, 113–121.

Freeman, B.A., Crap, J.D. (1982) Biology and disease: free radicals and disease injury. *Laboratory Investigation* 47, 412–426.

Halliwell, B., Gutteridge, J.M.C. (1984) Oxygen toxicity, oxygen radicals, transition metals, and disease. *Journal of Biochemistry* 219, 1–14.

Halliwell, B., Gutteridge, J.M.C. (2007) *Free Radicals in Biology and Medicine*, 4rd edn. Oxford University Press, New York.

Honkakoski, P., Negishi, M. (2000) Regulation of cytochrome P450 (CYP) genes by nuclear receptors. *Biochemical Journal* 347, 321–337.

Hu, R., Xu, C., Shen, G. et al. (2006) Gene expression profiles induced by cancer chemopreventive isothiocyanate sulforaphane in the liver of C57BL/6J mice and C57BL/6J/Nrf2 (-/-) mice. *Cancer Letters* 243, 170–192.

Inoue, S., Kawanishi, S. (1989) ESR evidence for superoxide, hydroxyl radicals and singlet oxygen produced from hydrogen peroxide and nickel(II) complex of glycylglycyl-L-histidine. *Biochemical and Biophysical Research Communications* 159, 445–451.

Jacobs, J.M., Nichols, C.E., Andrew, A.S. et al. (1999) Effect of arsenite on induction of CYP1A 1, CYP 2B, and CYP3A in primary cultures of rat hepatocytes. *Toxicology and Applied Pharmacology* 157, 51–59.

Kolluri, S.K., Weiss, C., Koff, A., Göttlicher, M. (1999) p27Kip1 induction and inhibition of proliferation by intracellular Ah receptor in developing thymus and hepatoma cells. *Genes and Development* 13, 1742–1753.

Kwak, M.K., Wakabayashi, N., Itoh, K. et al. (2003) Modulation of gene expression by cancer chemopreventive dithiolethiones through the Keap1-Nrf2 pathway. Identification of novel gene clusters for cell survival. *Journal of Biological Chemistry* 278, 8135–8145.

Leonard, M., Kieran, N., Howell, K. et al. (2006) Reoxygenation-specific activation of the antioxidant transcription factor Nrf2 mediates cytoprotective gene expression in ischemia-reperfusion injury. *FASEB Journal* 20, 2624–2626.

Leonard, S.S., Harris, G.K., Shi, X., (2004) Metal-induced oxidative stress and signal transduction. *Free Radical Biology and Medicine* 37, 1921–1942.

Li, N.X., Karin, M. (1999) Is NF-kappa B the sensor of oxidative stress? *FASEB Journal* 13, 1137–1143.

Li, M., Pascual, G., Glass, C.K. (2000) Peroxisome proliferator-activated receptor γ-dependent repression of the inducible nitric oxide synthase gene. *Molecular and Cellular Biology* 20, 4699–4707.

Liu, P.C.C., Matsumura, F. (1995) Differential effects of 2,3,7,8-tetrachlorodibenzo-p-dioxin on the adipose-type and brain-type glucose transporters in mice. *Molecular Pharmacology* 47, 65–73.

Livingstone, D.R. (2001) Contaminant-stimulated reactive oxygen species production and oxidative damage in aquatic organisms. *Marine Pollution Bulletin* 42, 656–666.

Ma, Q., Whitlock Jr., J.P. (1996) The aromatic hydrocarbon receptor modulates the Hepa 1c1c7 cell cycle and differentiated state independently of dioxin. *Molecular and Cellular Biology* 16, 2144–2150.

Matikainen, T., Perez, G.I., Jurisicova, A. et al. (2001) Aromatic hydrocarbon receptor-driven Bax gene expression is required for premature ovarian failure caused by biohazardous environmental chemicals. *Nature Genetics* 28, 355–360.

Meister, A. (1989) On the biochemistry of glutathione. *In* Taniguchi, N., Higashi, T., Sakamoto S., Meister, A. (eds). *Glutathione Centennial. Molecular Perspectives and Clinical Implications*. Academic Press, San Diego, CA, pp. 3–21.

Morel, Y., Barouki, R. (1998) Down-regulation of cytochrome P4501A1 gene promoter by oxidative stress-critical contribution of nuclear factor 1. *Journal of Biological Chemistry* 273, 26969–26976.

Morel, Y., Barouki, R. (1999) Repression of gene expression by oxidative stress. *Biochemical Journal* 342, 481–496.

Nie, M., Blankenship, A.L., Giesy, J.P. (2001) Interaction between aryl hydrocarbon receptor (AhR) and hypoxia signaling pathways. *Environmental Toxicology and Pharmacology* 10, 17–27.

Osburn, W.O., Kensler, T.W. (2008) Nrf2 signaling: an adaptive response pathway for protection against environmental insults. *Mutation Research* 659, 31–39.

Puga, A., Barnes, S.J., Dalton, T.P. et al. (1999) Aromatic hydrocarbon receptor interaction with the retinoblastoma protein potentiates repression of E2F-dependent transcription and cell cycle arrest. *Journal of Biological Chemistry* 275, 2943–2950.

Regoli, F., Winston, G.W., Gorbi, S. et al (2003) Integrating enzymatic responses to organic chemical exposure with total oxyradical absorbing capacity and DNA damage in the European eel *Anguilla anguilla*. *Environmental Toxicology and Chemistry* 22, 2120–2129.

Regoli, F., Nigro, M., Benedetti, M. et al. (2005) Interactions between metabolism of trace metals and xenobiotic agonists of the aryl hydrocarbon receptor in the Antarctic fish Trematomus bernacchii: environmental perspectives. *Environmental Toxicology and Chemistry* 24, 1475–1482.

Rowland, J.C., Gustafsson, J.A. (1997) Aryl hydrocarbon receptor mediated signal transduction. *Critical Reviews in Toxicology* 27, 109–134.

Schlezinger, J.J., Struntz, W.D.J., Goldstone, J.V., Stegeman, J.J. (2006) Uncoupling of cytochrome P450 1A and stimulation of reactive oxygen species production by co-planar polychlorinated biphenyl congeners. *Aquatic Toxicology* 77, 422–432.

Schrader, M., Fahimi, H.D. (2006) Peroxisomes and oxidative stress. *Biochimica et Biophysica Acta* 1763, 1755–1766.

Shi, H., Hudson, L.G., Liu, K.J. (2004) Oxidative stress and apoptosis in metal ion-induced carcinogenesis. *Free Radical Biology and Medicine* 37, 582–593.

Shi, X.L., Dalal, N.S. (1988) On the mechanism of the chromate reduction by glutathione: ESR evidence for the glutathionyl radical and an isolable Cr(V) intermediate. *Biochemical and Biophysical Research Communications* 156, 137–142.

Stegeman, J.J., Lech, J.J. (1991) Cytochrome P-450 monooxygenase systems in aquatic species: carcinogen metabolism and biomarkers for carcinogen and pollutant exposure. *Environmental Health Perspectives* 90, 101–109.

Thimmulappa, R.K., Mai, K.H., Srisuma, S. et al. (2002) Identification of Nrf2-regulated genes induced by the chemopreventive agent sulforaphane by oligonucleotide microarray. *Cancer Research* 62, 5196–5203.

Tian, Y., Ke, S., Denison, M.S. et al. (1999) Ah receptor and NF-kappa B interactions, a potential mechanism for dioxin toxicity. *Journal of Biological Chemistry* 274, 510–515.

Toledano, M.B., Leonard, W.J. (1991) Modulation of transcription factor NF-kappa B binding activity by oxidation-reduction *In vitro*. *Proceedings of Natural Academy of Sciences USA* 88, 4328–4332.

Tsukamoto, H. (2002) Redox regulation of cytokine expression in Kupffer cells. *Antioxidants and Redox Signaling* 4, 405–414.

Vakharia, D.D., Liu, N., Pause, R. et al. (2001) Effect of metals on polycyclic aromatic hydrocarbon induction of CYP1A1 and CYP1A2 in human hepatocyte cultures. *Toxicology and Applied Pharmacology* 170, 93–103.

Wei, P., Zhang, J., Egan-Hafley, M. et al. (2000) The nuclear receptor CAR mediates specific xenobiotic induction of drug metabolism. *Nature* 407, 920–923.

Whitlock Jr, J.P., Okino, S.T., Dong, L. et al. (1996) Induction of cytochrome P4501A1: a model for analyzing mammalian gene transcription. *FASEB Journal* 10, 809–818.

Xie, W., Barwick, J.L., Simon, C.M. et al. (2000) Reciprocal activation of xenobiotic response genes by nuclear receptors SXR/PXR and CAR. *Genes and Development* 14, 3014–3023.

Yamanaka, K., Hoshino, M., Okamoto, M. et al. (1990) Induction of DNA damage by dimethylarsine, a metabolite of inorganic arsenics, is for the major part likely due to its peroxyl radical. *Biochemical and Biophysical Research Communications* 168, 58–64.

Chapter 23

BIOMARKERS OF OXIDATIVE STRESS: BENEFITS AND DRAWBACKS FOR THEIR APPLICATION IN BIOMONITORING OF AQUATIC ENVIRONMENTS

José María Monserrat[1,2], *Rafaela Elias Letts*[2,3], *Josencler L. Ribas Ferreira*[2], *Juliane Ventura-Lima*[1,2], *Lílian L. Amado*[1,2], *Alessandra M. Rocha*[2], *Stefania Gorbi*[3], *Raffaella Bocchetti*[3], *Maura Benedetti*[3], *and Francesco Regoli*[3]

[1]Biological Sciences Institute, Federal University of Rio Grande – FURG, Rio Grande, RS, Brazil
[2]Post-graduation program in Physiological Sciences, Comparative Animal Physiology, Federal University of Rio Grande – FURG, Rio Grande, RS, Brazil
[3]Department of Biochemistry, Biology and Genetic, Polytechnic University of Marches, Ancona, Italy

Oxidative Stress in Aquatic Ecosystems, First Edition. Edited by Doris Abele, José Pablo Vázquez-Medina, and Tania Zenteno-Savín.

Aquatic environments are potentially vulnerable to pollution, since almost all kinds of chemicals used in anthropogenic activities, directly or indirectly, reach the water bodies (Islam and Tanaka 2004). The high diversity of contaminants is responsible for several deleterious effects in organisms inhabiting these ecosystems, including direct damage caused by xenobiotics and bioaccumulation of these compounds along the food chain (Rodríguez-Ariza et al. 1999). Consequences at the ecosystem level take too long to become evident and when this happens, it is usually too late to take corrective actions (Goksoyr et al. 1996).

Chemical analyses can reveal the presence of xenobiotics in the aquatic environment, but do not provide information on bioavailability and, therefore, on potentially harmful effects for biological systems (Monserrat et al. 2007). As the ultimate goal of environmental monitoring is to protect biological and ecological systems, the main focus in pollution studies has shifted from contaminants (chemical-based) to effects (biological-based) monitoring (Lan 2009; Box 23.1). Thus the measurement of biomarkers in appropriate sentinel organisms has been widely used to evaluate both the exposure to chemical pollutants and the onset of biological effects in aquatic organisms (Newman 1998). The term biomarker has been defined by the National Academy of Sciences (USA) as "a xenobiotically induced variation in cellular or biochemical components or processes, structures or functions that is measured in a biological system or sample" (NRC 1987). This definition was later extended in order to include behavioral parameters, as they were recognized as an important link between alterations in biochemical/physiological responses and ecological effects (Monserrat et al. 2007).

A number of biochemical and physiological parameters have been used both in laboratory conditions (bioassays) and in field studies as early warning signals of chemical toxic effects (Monserrat et al. 2007). Various biomarkers can respond to toxic stress with different degrees of specificity (Lan 2009) and are, therefore, divided into specific and nonspecific responses (NRC 1987). Specific biomarkers include those parameters responding to a select group of chemicals, such as metallothionein for trace metals (Linde-Arias et al. 2008), acetylcholinesterase (AChE) activity for organophosphorus and carbamate pesticides as well as cyanotoxins like anatoxin-a(s) (Monserrat et al. 2001, 2007), δ-aminolevulinic acid dehydratase (ALAD) activity for lead contamination (Company et al. 2008; Hylland et al. 2009), and vitellogenin for estrogenic chemicals (Porte et al. 2006). Among nonspecific biomarkers, oxidative stress responses can reveal pro-oxidant challenges caused by several forms of chemical, environmental, and biological disturbance (Regoli et al. 2003, 2004a; Benedetti et al. 2007, 2009).

THE APPLICATION OF OXIDATIVE STRESS PARAMETERS AS POLLUTION BIOMARKERS

Antioxidant defenses are not 100% effective and any stressor that overwhelms cellular capability in counteracting these reactive molecules, both through an increase in reactive oxygen species (ROS) formation and/or through the depletion of antioxidant defenses, can lead to toxic effects (Regoli 2000). Natural events that enhance ROS formation in aquatic animals include the exposure to some cyanobacterial toxins such as microcystin (Amado et al. 2009; Leão et al. 2008), hypoxia–hyperoxia cycles induced by phytoplanktonic blooms through photosynthetic and respiratory processes (Rosa et al. 2005), ultraviolet radiation (UVR) (Lesser 2006), fluctuations in oxygen availability (Gorbi et al. 2005), presence of symbionts (Regoli et al. 2000a; Downs et al. 2002), specific physiological features (Corsolini et al. 2001), and adaptation to extreme environmental conditions (Regoli et al. 2000b; Corsolini et al. 2001;

Box 23.1 Pollution *versus* contamination

Pollution can be defined as contamination of air, food, or water in such a manner as to cause a potential harm to human health and other organisms, or to damage nature without any justification. Whereas contamination is defined as any perturbation of the quality of the water or environment in general, by sewage, industrial waste, toxins, and metals, to a degree that creates a hazard to humans or aquatic organisms through poisoning or disease.

Camus et al. 2003). Chemical-related sources of enhanced ROS formation include exposure to organic contaminants, such as cycling redox compounds (quinones, nitroaromatics, nitroamines, bipyridil herbicides), polycyclic aromatic hydrocarbons (PAHs), halogenated hydrocarbons (pesticides, polychlorinated biphenyls (PCBs), polybrominated diphenyl ethers (PBDE)), dioxins, and trace metals (As, Al, Cd, Cr, Cu, Hg, Ni, Pb, Se, V) (Livingstone 2001; Regoli et al. 2005a; Halliwell and Gutteridge 2007).

Because different classes of contaminants can interfere with the normal functioning of antioxidant defenses and also contribute to intracellular ROS formation, several studies measured variations of antioxidant defense and onset of oxidative damage as useful biomarkers to evaluate the pro-oxidant effects of chemical pollutants in aquatic environments (Regoli and Principato 1995; Regoli et al. 2004a; Bocchetti et al. 2008a). Determination of these biomarkers has been successfully employed in field studies aimed to characterize impacted areas, where complex mixtures of contaminants are present (Geracitano et al. 2004a,b; Amado et al. 2006a,b; Bocchetti et al. 2008b; Gorbi et al. 2008).

ENVIRONMENTAL AND BIOLOGICAL FACTORS THAT MODULATE ANTIOXIDANT RESPONSES

Besides pollution, many environmental and biological factors can naturally influence the response of oxidative stress biomarkers, potentially leading to misinterpretation of data (Minguez et al. 2009). Factors that influence baseline antioxidant enzyme activities in various seasons include variations in food availability, increase of water temperature in summer, and spawning processes (Bocchetti and Regoli 2006).

Activities of antioxidant enzymes frequently follow site- and species-specific seasonal cycles. Mediterranean mussels, *Mytillus galloprovincialis*, sampled from an unpolluted area of the Western Adriatic coast, exhibited natural fluctuations of antioxidants, especially during spring and summer (Bocchetti and Regoli 2006). Catalase (CAT) and glutathione reductase (GR) showed higher activities in spring than in the warmer period when high values of glutathione peroxidase (GPx) and a generally low content of total glutathione (GSH) were also detected (Bocchetti and Regoli 2006). Mussels from the Tyrrhenian or eastern Adriatic revealed some differences in seasonal cycle of antioxidants (Regoli 1998; Petrovic et al. 2004), suggesting that the same environmental and biological factors can cause pro-oxidant effects temporally shifted in different geographical areas (Bocchetti and Regoli 2006).

The scallop, *Adamussium colbecki*, is a key species for monitoring the Antarctic environment and its antioxidant efficiency has recently been characterized (Regoli et al. 1998, 2002). Seasonal fluctuations were also highlighted in the susceptibility of this species toward oxidative stress, with an increased capability to neutralize $ROO^{\bullet}$ and $HO^{\bullet}$ at the end of December, and a quite constant resistance towards $ONOO^{-}$ during the Antarctic summer (Regoli et al. 2002). Despite the limited time-frame analyzed, these results suggest the occurrence of metabolic changes mainly influencing intracellular generation of specific ROS, probably related to the period of highest feeding activity and, to a lesser extent, to the phase of reproductive cycle (Regoli et al. 2002).

The Antarctic silverfish, *Pleuragramma antarcticum*, is another key organism in the ecology and food webs of the Southern Ocean. Eggs with fully developed yolk-sac embryos and newly hatched larvae occur in the platelet ice accumulating below the sea-ice layer. This environment has strong pro-oxidant characteristics at the beginning of the austral spring, when the rapid growth of algal ice communities, the massive release of nutrients, and the photoactivation of dissolved organic carbon and nitrates, all represent important sources for ROS formation. Such processes are concentrated in a short period of a few weeks, which overlaps with the final development of *P. antarcticum* in platelet ice. A marked temporal increase of antioxidants in embryos of *P. antarcticum* has been demonstrated to occur as an adaptive counteracting response to oxidative conditions of platelet ice (Regoli et al. 2005b). Particularly prompt responses were observed for GSH metabolism which, however, did not prevent formation of increasing levels of lipid peroxidation products; from the analysis of total oxyradical scavenging capacity (TOSC), the overall efficiency to neutralize $ROO^{\bullet}$ remained almost constant while slightly lower TOSC values were obtained toward $HO^{\bullet}$ at the end of the sampling period. Laboratory exposures to 0.5–5 $\mu g\ L^{-1}$ benzo[*a*]pyrene (B[*a*]P) caused significant accumulation of this PAH but limited variations of antioxidants in exposed embryos; these results suggest that the elevated pro-oxidant challenge, to which these organisms are naturally adapted, may be responsible for the moderate responsiveness to pro-oxidant chemicals (Regoli et al. 2005a).

Seasonal variations of pro-oxidant challenge have been investigated in the Antarctic sponge *Haliclona dancoi* exposed to hyperoxic conditions for the photosynthetically produced oxygen by endosymbiotic diatoms (Regoli et al. 2004b). However, the presence of diatoms is typically characterized by marked seasonal differences during the austral summer. The algae are practically absent in sponge tissues in early November at the beginning of summer and increase again in number from the middle of December. The efficiency of antioxidant defenses in the sponge showed a marked response to symbionts, with clearly enhanced values corresponding to the peak in diatom abundance (Regoli et al. 2004b).

The effect of season was also observed in the cichlid fish acara (*Geophagus brasiliensis*) when healthy individuals collected in a nonpolluted site during the high temperature period showed higher oxygen consumption, SOD, and GST activities, and both total-GSH and GSSG concentrations in the liver, compared to fish collected during autumn (Wilhelm Filho et al. 2001). The barnacle, *Balanus balonoides*, also seems to possess an antioxidant system which follows a seasonal pattern. The decreasing activity of antioxidant enzymes during the post-monsoon or winter period depends on physiological factors, such as gonad maturation and food availability (Niyogi et al. 2001).

Besides the direct increase of metabolic rate and oxygen consumption during the summer, the enhanced activity of ROS generating systems, such as cytochrome P450, can further increase the flow of intracellular ROS in these periods (Gorbi et al. 2005). Mediterranean symbioses between the demosponge *Petrosia ficiformis* and the cyanobacterium *Aphanocapsa feldmanni* naturally experience marked seasonal variations in symbiont content, light intensity, and seawater temperature, fluctuations that modify the pro-oxidant challenge for the sponge (Regoli et al. 2004c). Antioxidant efficiency of symbiotic sponges showed significant seasonal changes in antioxidant capacity, with more marked variations observed during the summer months (Regoli et al. 2004c). These results indicate a greater production of H_2O_2 in the symbiotic sponges during this period, supporting the hypothesis that seawater temperature can significantly modulate the pro-oxidant challenge in Mediterranean symbioses. The results also suggest that species with lower antioxidant efficiency may indeed be less tolerant of conditions effecting oxidative damage; e.g. increases in temperature during the summer months. Based on such considerations, increasing temperature and solar radiation can be considered pro-oxidant factors that favor the generation of ROS during the summer months. The polychaete *Laeonereis acuta* also showed higher activity of antioxidant and phase II enzymes (CAT and GST) when collected in a polluted site during autumn and summer (Geracitano et al. 2004b).

Also, the phase of life cycle can be important as demonstrated in the shrimp *Palaemonetes pugio holthiuswas*. Exposure of shrimp to coal combustion residues significantly decreased survival in larvae but not in juveniles and adults (Palomero et al. 2001). The understanding of factors that modulate age-associated differences in sensitivity to toxic agents might be important for predicting the effects of xenobiotic compounds on aged populations (Palomero et al. 2001). This aspect is often neglected since limited information is available on the age structure of aquatic animals used in biomonitoring studies, where only gross age estimates such as length are often measured.

Oxidative biomarkers can also be affected by changes during the reproductive period, resulting in increased oxygen demand and activation of various biochemical pathways. In fish, a retinoid reserve is deposited in the eggs as a source of vitamin A, fundamental for reproductive and proliferative processes, and as a protective element against oxidative stress for the future embryo (Irie and Seki 2002). Female zebrafish, *Danio rerio*, fed with Cu or B[*a*]P supplemented diets showed lower retinoid levels at the end of the experiment (260 days), a result that would affect reproduction considering the importance of retinoids in fish gonadogenesis (Alsop et al. 2007).

Many studies demonstrated that the type of nutrition can favor growth and enhance fish health also through modulation of the immune system. Juvenile fish supplemented with yeast revealed lower mortality associated with enhanced humoral and antioxidant immune responses, like SOD activity (Reyes-Becerril et al. 2008). Due to the elevated fluctuations of food availability in natural aquatic systems, this factor should also be considered when interpreting antioxidant results in environmental monitoring. Fish can be adapted to survive extended periods of starvation with a dietary regime low in calories without undernutrition, and increased levels of TBARS were shown as useful indicators of oxidative stress under such food deprivation conditions (Abele et al. 2007). Augmented autophagy induced by nutritional deprivation has been shown to act as a cellular defense against environmentally induced oxidative stress in marine mollusks (Moore et al. 2007; Moore 2008). The above examples point out the importance of considering several environmental and biological factors beyond pollution when using oxidative stress biomarkers in environmental monitoring.

PARASITES AS MODULATORS OF ANTIOXIDANT RESPONSES

The presence of parasites is another factor potentially modulating the responses of oxidative stress biomarkers. Parasites can induce various physiological changes in the host, such as an energetic reallocation for their own development or a weakening of the host immune system (Minguez et al. 2009). The energetic cost of parasitic infection is evidenced by the variety of nutrients that parasites directly consume from their host. Cestodes, for instance, actively take amino acids, fatty acids, glucose, and whole proteins from their host's alimentary tract (Kelly and Janz 2008).

Parasitic infections can also cause an energetic burden associated with the repair of damaged tissues and the activation of immune responses against the infection (Mackenzie et al. 1995). When a fish is injured by a parasite, for example, an inflammatory process will initiate to arrest, reduce, and repair the damage; such physiological processes can reduce resistance of fish to other stressing factors (Dautremepuits et al. 2003). Carp infected with *Ptychobothrium* sp. exhibited a significant increase of antioxidant enzymes in liver and head kidney (also known as forekidney or primordial kidney) compared with healthy fish. Parasitic infections in mammals generally lead to the inhibition of antioxidant activities associated with the production of ROS by macrophages at the infection site to eliminate the parasite (Dautremepuits et al. 2003). In this multistress context, parasitism could thus represent a confounding factor interacting with other stress factors, like pollution, and such relationships could have serious implications for environmental risk assessment. Some trematodes, like *Bunodera luciopercae* and *Schistosoma mansoni*, can influence immune and endocrine systems of their mollusk host, leading to false-positive or false-negative results (Minguez et al. 2009). Parasitized clams, *Pisidium amnicum*, demonstrated an increased tolerance towards organic xenobiotics (Heinonen et al. 2000), whereas the bivalve *Cerastoderma edule* infected by the trematode *Labratrema minimus* showed a decreased synthesis of metallothioneins involved in the homeostasis and detoxification of trace metals (Baudrimont et al. 2005). The activities of antioxidant enzymes, such as CAT, GR, GST and the content of lipid peroxides, have been shown to increase in infected fish (Dautremepuits et al. 2003). Other studies suggested that the antioxidant activities of the host can modulate parasite life cycles, like in mammals where the *Schistosome* parasite was most susceptible to immune elimination during the early life stages and least as an adult form. Such a difference is explained by the greater resistance to oxidative killing in the adult worm.

OXIDATIVE STRESS AND CELL SIGNALING

It is well accepted that ROS are not only mere by-products of cellular metabolism, but also important participants in cell signaling and regulation (Thannickal and Fanburg 2000). Therefore, ROS have been considered as secondary messengers for other signals, such as receptor-binding-dependent phosphorylation (Halliwell and Gutteridge 2007). Low doses of ROS, particularly H_2O_2, are typically mitogenic and promote cell proliferation, while intermediate doses result in either temporary or permanent growth arrest, such as replicative senescence. Furthermore, high levels of ROS cause cell death via apoptotic or necrotic mechanisms. A large number of signaling pathways are involved in coordinating the response to enhanced ROS production and influence the cellular fate (Martindale and Holbrook 2002).

Although the presence of antioxidant molecules and antioxidant enzymes in all cellular compartments could be considered an obstacle to ROS signaling, it has been shown that activities of antioxidant proteins are dynamically regulated through phosphorylation/dephosphorylation reactions (Bartosz 2009), and this kind of regulation can modulate the half-life of pro-oxidants, such as H_2O_2.

Experimental evidence demonstrated that Nrf2, an important transcription factor involved in the expression of antioxidant and detoxification enzymes, can be modulated by phosphorylation of Ser 40 by protein kinase C (PKC, EC 2.7.11.13), which increases Nrf2 half-life (Huang et al. 2002; Kwak et al. 2004;

Kobayashi et al. 2009). Phosphorylation of this transcription factor would lead to activated expression of different antioxidant enzymes, which, however, is not directly related to an increase of catalytic activity since these enzymes are also a target for phosphorylation. For example, phosphorylation of the catalytic subunit of the enzyme glutamate-cysteine ligase (GCL, EC 6.3.2.2), triggered by protein kinase A (PKA, also known as cAMP-dependent protein kinase EC 2.7.11.11) and PKC, affects the enzymatic activity, mainly by lowering the enzyme's maximum velocity (V_{max}) (Sun et al. 1996; Toroser et al. 2006). GCL activity is the rate-limiting step in the *de novo* synthesis of GSH (Dickinson and Forman 2002; Maher 2005).

Phosphorylation of peroxiredoxin 1 (Prx1) on Thr90 by cyclin-dependent kinases (CDKs) reduced the peroxidase activity of this protein by 80% in HeLa cells (Chang et al., 2002). Peroxiredoxins are a family of peroxidases that reduce H_2O_2 and alkyl hydroperoxides to water and alcohol, respectively, with the use of reducing equivalents provided by thiol-containing proteins (Hoffman et al. 2002).

CAT and GPx, both involved in removal of H_2O_2, are modulated at catalytic level by phosphorylation, performed by c-Abl and Arg nonreceptor tyrosine kinases. The cytoplasmic forms of both kinases, activated by oxidative stress, bind to CAT and GPx1, phosphorylating the enzymes and thus increasing their activities. However, at higher levels of ROS, c-Abl and Arg dissociate from CAT, which is dephosphorylated by protein tyrosine phosphatases, ubiquitinylated, and degraded, thus potentiating the increase in ROS levels (Cao et al. 2003a,b; Rhee et al. 2005).

GST, phase II detoxification enzymes involved in conjugation reactions between GSH and xenobiotics or cellular metabolites (including by-products of oxidative stress, such as lipid peroxides), are also subjected to post-transcriptional regulation via phosphorylation. Okamura et al. (2009) showed that activated epidermal growth factor receptor (EGFR), a receptor-type tyrosine kinase, binds to and phosphorylates GSTP1 in a cell-free system and in human tumor cells, increasing the catalytic efficiency of the enzyme. Therefore, any alteration in normal phosphorylation/dephosphorylation reactions caused by pollutants can lead to increased/diminished activity of antioxidant enzymes, contributing to abnormal ROS levels, and altered cell redox status. However, it is important to consider that other factors, including cyanotoxins like microcystins and nodularins, are phosphatase inhibitors (De Figueiredo et al. 2004) and, in this way, can modulate, alter, and/or disrupt the integrated antioxidant responses at the cellular level. Some of the toxic effects elicited by microcystins can, thus, be considered as consequences of phosphatases inhibition and post-transcriptional regulation of antioxidant enzymes (Amado and Monserrat 2010).

OXIDATIVE STRESS RESPONSES INDUCED BY NANOMATERIALS

Nanoparticles are natural or engineered compounds considered as potential pollutants in aquatic environments, due to their massive production in recent years. These particles possess at least one dimension in the nanoscale (less than 100 nm), with a high surface/volume ratio greatly enhancing their reactivity with other molecules, including biomolecules (Colvin 2003). Moreover, at this supramolecular size, compounds acquire novel physicochemical properties that are not present in the bulk state, such as photo-excitability, conductivity, optical properties, catalytic capability, adsorption behavior, and many others (Limbach et al. 2007). Besides the size of the particles, surface area, shape, lattice structure, surface chemistry, surface charge, and aggregation state are other key factors that influence reactivity of nanomaterials (Maynard 2007). Because of the exponential growth of nanomaterial production and resulting release in the environment, Donaldson and co-workers (2004) proposed a new branch of toxicology termed Nanotoxicology, to "study the effects of nanodevices and nanostructures in living organisms".

Nanoparticles in the aquatic environment can acquire different behaviors influenced by particle dimensions, solubility, reactivity, aggregation state, and chemical composition (Thomas et al. 2006). For example, titanium dioxide (TiO_2) nanoparticles can aggregate more than nanotubes and colloidal fullerenes (C_{60}) in a medium with weak ionic strength; but with increasing salt concentrations, carbon nanoparticles rapidly aggregate (Brant et al. 2005). Sedimentation and deposition reduce particle mobility but increase concentration in the sediments, enhancing potential exposure of benthic organisms. In a modeling study with sand soil matrix, Lecoanet and Wiesner (2004) showed that single-walled nanotubes (SWNT) are more mobile than fullerene C_{60}, as well as TiO_2 and ferroxane (nano-sized iron). The potential sublethal impacts of nano-TiO_2 and SWNT to the infaunal species have been investigated recently in the lugworm *Arenicola marina* exposed to natural sediments (Galloway et al. 2010). A significant decrease in casting rate and a significant increase in cellular and DNA damage was measured for nano-TiO_2. Aggregates of TiO_2 were also observed within the lumen of the gut and adhered to the outer epithelium of the worms.

Several lines of evidence indicate that nanomaterials are frequently associated with oxidative stress generation. DNA damage can be induced by generation of $^1O_2$ by C_{60} or even via C_{60} binding to DNA forming DNA adducts (Nakanishi 2002; Zhao et al. 2005). Also, light exposure can substantially affect toxicity of nanoparticles, with significantly higher levels of lipid peroxidation in brain homogenates of the fish *Cyprinus carpio* exposed to C_{60} with fluorescent light than without illumination (Shinohara et al., 2009). These results are in agreement with previous findings of Kamat et al. (2000), who proposed the generation of ROS, including $^1O_2$ and $HO^\bullet$, by photoactivated C_{60}. In goldfish skin cells exposed to TiO_2, UV incidence lowered cell viability and increased oxidative DNA damage due to generation of $HO^\bullet$ (Reeves et al. 2007). From an ecotoxicological point of view, organisms exposed to nanocompounds near the water surface might be more susceptible to oxidative impairment due to natural UV light incidence.

Nanoparticles have also been hypothesized to act as carriers for other environmental toxicants, reaching organs/cells/organelles that would not be reached by the xenobiotics alone (Limbach et al. 2007). When algae *Pseudokirchneriella subcapitata* were exposed to atrazine, methyl parathion, pentachlorophenol, and phenanthrene concomitantly with C_{60} nanoparticles, phenanthrene showed a higher effect in terms of inhibition of algal growth when co-exposed with C_{60} (Baun et al. 2008).

REFERENCES

Abele, D., Roecken, D., Grave, M., Buck, B.H. (2007) Body growth, mitochondrial enzymatic capacities and aspects of the antioxidant system and redox balance under calorie restriction in young turbot (*Scophthalmus maximus*, L.). *Aquaculture Research* 38, 467–477.

Alsop, D., Brown, S., Van der Kraak, G. (2007) The effects of copper and benzo(*a*)pyrene on retinoids and reproduction in zebrafish. *Aquatic Toxicology* 82, 281–295.

Amado, L.L., Monserrat, J.M. (2010) Oxidative stress generation by microcystins in aquatic animals: why and how. *Environment International* 36, 226–235.

Amado, L.L., da Rosa, C.E., Meirelles Leite, A. et al., (2006a) Biomarkers in croakers *Micropogonias furnieri* (Teleostei, Scianidae) from polluted and non-polluted areas from the Patos Lagoon estuary (Southern Brazil): evidences of genotoxic and immunological effects. *Marine Pollution Bulletin* 52, 199–206.

Amado, L.L., Robaldo, R.B., Geracitano, L., Monserrat, J.M., Bianchini, A. (2006b) Biomarkers of exposure and effect in the Brazilian flounder *Paralichthys orbignyanus* (Teleostei: Paralichthyidae) from the Patos Lagoon estuary (Southern Brazil). *Marine Pollution Bulletin* 52, 207–213.

Amado, L.L., Garcia, M.L., Ramos, P.B. et al. (2009) A method to measure total antioxidant capacity against ROO in aquatic organisms: Application to evaluate microcystins toxicity. *Science of The Total Environment* 407, 2115–2123.

Bartosz, G. (2009) Reactive oxygen species: Destroyers or messengers? *Biochemical Pharmacology* 77, 1303–1315.

Baudrimont, M., de Montaudouin, X., Palvadeau, A. (2005) Impact of digenean parasite infection on metallothionein synthesis by the cockle (*Cerastoderma edule*): A multivariate field monitoring. *Marine Pollution Bulletin* 52, 494–502.

Baun, A., Sørensen, S.N., Rasmussen, R.F., Hartmann, N.B., Koch, C.B. (2008) Toxicity and bioaccumulation of xenobiotic organic compounds in the presence of aqueous suspensions of aggregates of nano-C60. *Aquatic Toxicology* 86, 379–387.

Benedetti, M., Martuccio, G., Fattorini, D. et al. (2007) Oxidative and modulatory effects of trace metals on metabolism of polycyclic aromatic hydrocarbons in the Antarctic fish *Trematomus bernacchii*. *Aquatic Toxicology* 85, 167–175.

Benedetti, M., Fattorini, D., Martuccio, G., Nigro, M., Regoli, F. (2009) Interactions between trace metals (Cu, Hg, Ni, Pb) and 2,3,7,8-tetrachlorodibenzo-p-dioxin in the Antarctic fish *Trematomus bernacchii*: oxidative effects on biotransformation pathway. *Environmental Toxicology and Chemistry* 28, 818–825.

Bocchetti, R., Regoli, F. (2006) Seasonal variability of oxidative biomarkers, lysosomal parameters, metallothioneins and peroxisomal enzymes in the Mediterranean mussel *Mytilus galloprovencialis* from Adriatic Sea. *Chemosphere* 65, 913–921.

Bocchetti, R., Virno Lamberti, C., Pisanelli, B. et al. (2008a) Seasonal variations of exposure biomarkers, oxidative stress responses and cell damages in the clams, Tapes philippinarum, and mussels, *Mytilus galloprovincialis*, from Adriatic Sea. *Marine Environmental Research* 66, 24–26.

Bocchetti, R., Fattorini, D., Pisanelli, B. et al. (2008b) Contaminant accumulation and biomarkers responses in caged mussels, *Mytilus galloprovincialis*, to evaluate bioavailability and toxicological effects of remobilized chemicals during

dredging and disposal operations in harbour areas. *Aquatic Toxicology* 89, 257–266.
Brant, J., Lecoanet, H., Wiesner, M.R. (2005) Aggregation and deposition characteristics of fullerene nanoparticles in aqueous systems. *Journal of Nanoparticle Research* 7, 545–53.
Camus, L, Birkely, S.R., Jones, M.B. et al. (2003) Biomarker responses and PAH uptake in *Mya truncata* following exposure to oil-contaminated sediment in an Arctic fjord (Svalbard). *Science of the Total Environment* 308, 221–234.
Cao, C., Leng, Y., Huang, W., Liu, X., Kufe, D. (2003a) Glutathione peroxidase 1 is regulated by the c-abl and arg tyrosine kinases. *Journal of Biological Chemistry* 278, 39609–39614.
Cao, C., Leng, Y., Kufe, D. (2003b) Catalase activity is regulated by c-abl and arg in the oxidative stress response. *Journal of Biological Chemistry* 278, 29667–29675.
Chang, T.-S., Jeong, W., Choi, S.Y., Yu, S., Kang, S.W., Rhee, S.G. (2002) Regulation of peroxiredoxin 1 activity by cdc2-mediated phosphorylation. *Journal of Biological Chemistry* 277, 25370–25376.
Colvin, V.K. (2003) The potential environmental impact of engineered nanomaterials. *Nature Biotechnology* 21, 1166–1170.
Company, R., Serafim, A., Lopes, B. et al. (2008) Using biochemical and isotope geochemistry to understand the environmental and public health implications of lead pollution in the lower Guadiana River, Iberia: A freshwater bivalve study. *Science of the Total Environment* 405, 109–199.
Corsolini, S., Nigro, M., Olmastroni, S., Focardi, S., Regoli, F. (2001) Susceptibility to oxidative stress in Adélie and Emperor penguin. *Polar Biology* 24, 365–368.
Dautremepuits, C., Betoulle, S., Vernet, G. (2003) Stimulation of antioxidant enzymes levels in carp (*Cyprinus carpio*) infected by *Ptychobothrium* sp. (Cestoda). *Fish and Shellfish Immunology* 15, 467–471.
De Figueiredo, D.R., Azeiteiro, U.M., Esteves, S.M., Gonçalves, F.J.M., Pereira, M.J. (2004) Microcystin-producing blooms – A serious global public health issue. *Ecotoxicology and Environmental Safety* 59, 151–163.
Dickinson, D.A., Forman, H.J. (2002) Cellular glutathione and thiols metabolism. *Biochemical Pharmacology* 64, 1019–1026.
Donaldson, K., Stone, V., Tran, C.L., Kreyling, W., Borm, P.J.A. (2004) Nanotoxicology. *Occupational Environmental Medicine* 61, 727–728.
Downs, C.A., Fauth, J.E., Halas, J.C., Dustan, P., Bemiss, J., Woodley, C.M. (2002) Oxidative stress and seasonal coral bleaching. *Free Radical Biology and Medicine* 33, 533–543.
Galloway, T., Lewis, C., Dolciotti, I., Johnston, B.D., Moger, J., Regoli, F. (2010) Sublethal toxicity of nano-titanium dioxide and carbon nanotubes in a sediment dwelling marine polychaete. *Environmental Pollution* 158, 1748–1755.
Geracitano, L.A., Bocchetti, R., Monserrat, J.M., Regoli, F., Bianchini, A. (2004a) Oxidative stress responses in two populations of *Laeonereis acuta* (Polychaeta, Nereididae) after acute and chronic exposure to copper. *Marine Environmental Research* 58, 1–17.
Geracitano, L.A., Monserrat, J.M., Bianchini, A. (2004b) Oxidative stress in *Laeonereis acuta* (Polychaeta, Nereididae): environmental and seasonal effects. *Marine Environmental Research* 58, 625–630.
Goksoyr, A., Beyer, J., Egaas, E. et al. (1996) Biomarker responses in flounder (*Platichthys flesus*) and their use in pollution monitoring. *Marine Pollution Bulletin* 33, 36–45.
Gorbi, S., Baldini, C., Regoli, F. (2005) Seasonal variability of metallothioneins, cytochrome P450, bile metabolites and oxyradical metabolism in the European eel *Anguilla anguilla* L. (Anguillidae) and striped mullet *Mugil cephalus* L. (Mugilidae). *Archives of Environmental Contamination and Toxicology* 49, 62–70.
Gorbi, S., Virno Lamberti, C., Notti, A., Benedetti, M., Fattorini, D., Moltedo, G., Regoli, F. (2008) An ecotoxicological protocol with caged mussels *Mytilus galloprovincialis*, for monitoring the impact of an offshore platform in the Adriatic sea. *Marine Environmental Research* 65: 34–49.
Halliwell, B., Gutteridge, J.M.C. (eds). (2007) *Free Radicals in Biology and Medicine*, 4th edn. Oxford University Press, New York.
Heinonen, J., Kukkonen, J.V., Holopainen, I.J. (2000) Toxicokinetics of 2,4,5-trichlrophenol and benzo(a)pyrene in the clam *Pisidium amnicum*: effects of seasonal temperatures and trematode parasites. *Archives of Environmental Contamination and Toxicology* 39, 352–359.
Hoffmann, B., Hecht, H.J. Flohé, L. (2002) Peroxiredoxins. *Biological Chemistry* 383, 347–364.
Huang, H.-C., Nguyen, T. and Pickett, C.B. (2002) Phosphorylation of Nrf2 at Ser-40 by protein kinase C regulates antioxidant response element-mediated transcription. *Journal of Biological Chemistry* 277, 42769–42774.
Hylland, K., Ruus, A., Grung, M., Green, N. (2009) Relationships between physiology, tissue contaminants, and biomarker responses in Atlantic cod (*Gadus morhua* L.). *Journal of Toxicology and Environmental Health A* 72, 226–233.
Irie, T., Seki, T. (2002) Retinoid composition and retinol localization in the eggs of teleost fishes. *Comparative Biochemistry and Physiology B* 131, 209–219.
Islam, S.M., Tanaka, M. (2004) Impacts of pollution on coastal and marine ecosystems including coastal and marine fisheries and approach for management: a review and synthesis. *Marine Pollution Bulletin* 48, 624–649.
Kamat, J.P., Devasagayam, T.P.A., Proyadarsini, K.I., Mohan, H. (2000) Reactive oxygen species mediated membrane damage induced by fullerene derivatives and its possible biological implications. *Toxicology* 155, 55–61.
Kelly, J.M., Janz, D.M. (2008) Altered energetics and parasitism in juvenile northern pike (*Esox lucius*) inhabiting

metal-mining contamined lakes. *Ecotoxicology and Environmental Safety* 70, 357–369.
Kobayashi, M., Li, L., Iwamoto, N. et al. (2009) The antioxidant defense system Keap1-Nrf2 comprises a multiple sensing mechanism for responding to a wide range of chemical compounds. *Molecular and Cellular Biology* 29, 493–502.
Kwak, M.-K., Wakabayashi, N., Kensler, T.W. (2004) Chemoprevention through the Keap1–Nrf2 signaling pathway by phase 2 enzyme inducers. *Mutation Research* 555, 133–148.
Lan, P.K.S. (2009) Use of biomarkers in environmental monitoring. *Ocean and Coastal Management* 52, 348–354.
Lecoanet, H.F., Wiesner, M.R. (2004) Velocity effects on fullerene and oxide nano-particle deposition in porous media. *Environmental Science and Technology* 38, 4377–82.
Leão, J.C., Geracitano, L.A., Monserrat, J.M., Amado, L.L., Yunes, J.S. (2008) Microcystin-induced oxidative stress in *Laeonereis acuta* (Polychaeta, Nereididae). *Marine Environmental Research* 66, 92–94.
Lesser, M.P. (2006) Oxidative stress in marine environments: biochemistry and physiological ecology. *Annual Reviews of Physiology* 68, 253–278.
Limbach, L.K, Wick, P., Manser, P., Grass, R.N., Bruinink, A., Stark, W.J. (2007) Exposure of engineered nanoparticles to human lung epithelial cells: influence of chemical composition and catalytic activity on oxidative stress. *Environmental Science and Technology* 41, 4158–4163.
Linde-Arias, A.R., Inácio, A.F., Novo, L.A., Alburquerque, C.d., Moreira, J.C. (2008) Multibiomarker approach in fish to assess the impact of pollution in a large Brazilian river, Paraiba do Sul. *Environmental Pollution* 156, 974–979.
Livingstone, D.R. (2001) Contaminant-stimulated reactive oxygen species production and oxidative damage in aquatic organisms. *Marine Pollution Bulletin* 42, 656–666.
Mackenzie, K., Williams, H.H., Williams, B., McVicar, A.H., Siddall, R. (1995) Parasites as indicators of water quality and the potencial use of helminth transmission in marine pollution studies. *Advances in Parasitology* 35, 85–144.
Maher, P. (2005) The effects of stress and aging on glutathione metabolism. *Ageing Research Review* 4, 288–314.
Martindale, J.L., Holbrook, N.J. (2002) Cellular response to oxidative stress: signaling for suicide and survival. *Journal of Cellular Physiology* 192, 1–15.
Maynard, A.D. (2007) Nanotoxicology: laying a firm foundation for sustainable nanotechnologies. *In*: Monteiro-Riviere, N.A., Tran, C.L. (eds). *Nanotoxicology – Characterization, Dosing and Health Effects*. Informa Healthcare USA, New York, pp. 1–6.
Minguez, L., Meyer, A., Molloy, D.P., Giambérini, L. (2009) Interactions between parasitism and biological responses in zebra mussels (*Dreissena polymporpha*): importance in ecotoxicological studies. *Environmental Research* 109, 843–850.
Monserrat, J.M., Yunes, J.S., Bianchini, A. (2001) Effects of *Anabaena spiroides* (Cyanobacteria) aqueous extracts on the acetylcholinesterase activity of aquatic species. *Environmental Toxicology and Chemistry* 20, 1228–1235.
Monserrat, J.M., Martínez, P.E., Geracitano, L.A. et al. (2007) Pollution biomarkers in estuarine animals: Critical review and new perspectives. *Comparative Biochemistry and Physiology C* 146, 221–234.
Moore, M.N. (2008) Autophagy as a second level protective process in conferring resistance to environmentally-induced oxidative stress. *Autophagy* 4, 254–256.
Moore, M.N., Viarengo, A., Donkin, P., Hawkins, A.J.S. (2007) Autophagic and lysosomal reactions to stress in the hepatopancreas of blue mussels. *Aquatic Toxicology* 84, 80–91.
Nakanishi, I. (2002) DNA cleavage via superoxide anion formed in photoinduced electron transfer from NADH to gamma-cyclodextrin-bicapped C_{60} in an oxygen-saturated aqueous solution. *Journal of Physical Chemistry B* 106, 2372–2380.
Newman, M.C. (1998) *Fundamentals of Ecotoxicology*. Ann Arbor Press, Chelsea, 402 pp.
Niyogi, S., Biswas, S., Sarker, S., Datta, A.G. (2001) Seasonal variation of antioxidant and biotransformation enzymes in barnacle, *Balanus balanoides*, and their relation with polyaromatic hydrocarbons. *Marine Environmental Research* 52, 13–26.
NRC (1987) Biological markers in environmental health research. *Environmental Health Perspectives* 74, 3–9.
Okamura, T., Singh, S., Buolamwini, J. et al. (2009) Tyrosine phosphorylation of the human glutathione S-transferase p1 by epidermal growth factor receptor. *Journal of Biological Chemistry* 284, 16979–16989.
Palomero, J., Galán, A.I., Muñoz, M.R., Tuñon, M.J., González-Galego, J., Jiménez, R. (2001) Effects of aging on the susceptibility to the toxic effects of cyclosporin A in rats. Changes in liver glutathione and antioxidant enzymes. *Free Radical Biology and Medicine* 8, 836–845.
Petrovic, S., Semencic, L., Ozretic, B., Ozretic, M. (2004) Seasonal variations of physiological and cellular biomarkers and their use in the biomonitoring of north Adriatic coastal waters (Croatia). *Marine Pollution Bulletin* 49, 713–720.
Porte, C., Janer, G., Lorusso, L.C. et al. (2006) Endocrine disruptors in marine organisms: approaches and perspectives. *Comparative Biochemistry and Physiology C* 143, 303–315.
Reeves, J.F., Davies, S.J., Dodd, N.J.F., Jha, A.N. (2007) Hydroxyl radicals ($^{\bullet}OH$) are associated with titanium dioxide (TiO_2) nanoparticle-induced cytotoxicity and oxidative DNA damage in fish cells. *Mutation Research* 640, 113–122.
Regoli F. (1998) Trace metal and antioxidant mechanisms seasonal variations in gills and digestive gland of the Mediterranean mussel *Mytilus galloprovincialis*. *Archives of Environmental Contamination Toxicology* 34, 48–63.
Regoli, F. (2000) Total oxyradical scavenging capacity (TOSC) in polluted and translocated mussels: a predictive biomarker of oxidative stress. *Aquatic Toxicology* 50, 351–361.
Regoli, F., Principato, G. (1995) Glutathione, glutathione-dependent and antioxidant enzymes in mussel, *Mytilus*

galloprovincialis, exposed to metals in different field and laboratory conditions: implications for a proper use of biochemical biomarkers. *Aquatic Toxicology* 31, 143–164.

Regoli, F., Nigro, M., Orlando, E. (1998) Lysosomal and antioxidant responses to metals in the antarctic scallop *Adamussium colbecki*. *Aquatic Toxicology* 40, 375–392.

Regoli, F., Cerrano, C., Chierici, E., Bompadre, S., Bavestrello, G. (2000a) Susceptibility to oxidative stress of the Mediterranean demosponge *Petrosia ficiformis*: role of endosymbionts and solar irradiance. *Marine Biology* 137, 453–461.

Regoli, F., Nigro, M., Bompadre, S., Winston, G.W. (2000b) Total oxidant scavenging capacity (TOSC) of microsomal and cytosolic fractions from Antarctic, Arctic and Mediterranean scallops: differentiation between three potent oxidants. *Aquatic Toxicology* 49, 13–25.

Regoli, F., Nigro, M., Chiantore, M., Winston, G.W. (2002) Seasonal variations of susceptibility to oxidative stress in *Adamussium colbecki*, a key bioindicator species for the Antarctic marine environment. *Science of the Total Environment* 289, 205–211.

Regoli, F., Winston, G.W., Gorbi, S. et al. (2003) Integrating enzymatic responses to organic chemical exposure with total oxyradical absorbing capacity and DNA damage in the European eel *Anguilla anguilla*. *Environmental Toxicology and Chemistry* 22, 2120–2129.

Regoli, F., Frenzilli, G., Bocchetti, R. et al. (2004a) Time-course variation in oxyradical metabolism, DNA integrity and lysosomal stability in mussels, *Mytilus galloprovincialis*, during a field translocation experiment. *Aquatic Toxicology* 68, 167–178.

Regoli, F., Nigro, M., Chierici, E., Cerrano, C., Schiapparelli, S., Totti, C., Bavestrello, G. (2004b) Variations of antioxidant efficiency and presence of endosymbiotic diatoms in the Antarctic porifera *Haliclona dancoi*. *Marine Environmental Research* 58, 637–640.

Regoli, F., Cerrano, C., Chierici, E., Chiantore, M.C., Bavestrello, G. (2004c) Seasonal variability of prooxidant pressure and antioxidant adaptation to symbiosis in the Mediterranean demosponge *Petrosia ficiformis*. *Marine Ecology Progress Series* 275, 129–137.

Regoli, F., Nigro, M, Benedetti, M. et al. (2005a) Interactions between metabolism of trace metals and xenobiotic agonists of the Ah receptor in the Antarctic fish *Trematomus bernacchii*: environmental perspectives. *Environmental Toxicology and Chemistry* 24, 1475–1482.

Regoli, F., Nigro, M, Benedetti, M., Fattorini, D., Gorbi, S. (2005b) Antioxidant efficiency in early life stages of the Antarctic silverfish, *Pleuragramma antarcticum*: responsiveness to prooxidant conditions of platelet ice and chemical exposure. *Aquatic Toxicology* 75, 43–52.

Reyes-Becerril, M., Tovar-Ramírez, D., Ascencio-Valle, F. et al. (2008) Effects of dietary live yeast Debaryomyces hansenii on the immune and antioxidant system in juvenile leopard grouper *Mycteroperca rosacea* exposed to stress. *Aquaculture* 280, 39–44.

Rhee, S.G., Yang, K.-S., Kang, S.W., Woo, H.A., Chang, T.-S. (2005) Controlled elimination of intracellular H_2O_2: regulation of peroxiredoxin, catalase, and glutathione peroxidase via post-translational modification. *Antioxidants & Redox Signaling* 7, 619–626.

Rodríguez-Ariza, A., Alhama, J., Díazz-Méndez, F.M., López-Barea, J. (1999) Content of 8-Oxo-dG in chromosomal DNA of *Sparus aurata* fish as biomarker of oxidative stress and environmental pollution. *Mutation Research* 438, 97–107.

Rosa, C.E., de Souza, M.S., Yunes, J.S., Proença, L.A.O., M. Nery, L.E., Monserrat, J.M. (2005) Cyanobacterial blooms in estuarine ecosystems: Characteristics and effects on *Laeonereis acuta* (Polychaeta, Nereididae). *Marine Pollution Bulletin* 50, 956–964.

Shinohara, N., Matsumoto, T., Gamo, M. et al. (2009) Is lipid peroxidation induced by the aqueous suspension of fullerene C_{60} nanoparticles in the brains of *Cyprinus carpio*? *Environmental Science and Technology* 43, 948–953.

Sun, W.-M., Huang, Z.-Z., Lu, S.C. (1996) Regulation of γ-glutamylcysteine synthetase by protein phosphorylation. *Biochemical Journal* 320, 321–328.

Thannickal, V.J., Fanburg, B.L. (2000) Reactive oxygen species in cell signaling. *American Journal of Physiology* 279, L1005–L1028.

Thomas, T., Thomas, K., Sadrieh, N., Savage, N., Adair, P., Bronaug, R. (2006) Research strategies for safety evaluation of nanomaterials. Part VII: evaluating consumer exposure to nanoscale materials. *Toxicological Sciences* 91, 14–19.

Toroser, D., Yarian, C.S., Orr, W.C., Sohal, R.S. (2006) Mechanisms of γ-glutamylcysteine ligase regulation. *Biochimica et Biophysica Acta* 1760, 233–244.

Wilhelm Filho, D., Torres, M.A., Tribess, T.B., Pedrosa, R.C., Soares, C.H.I. (2001) Influence of season and pollution on the antioxidant defenses of the cichlid fish acará (*Geophagus brasiliensis*). *Brazilian Journal of Medical and Biological Research* 34, 719–726.

Zhao, X., Striolo, A., Cummings, P.T. (2005) C_{60} binds to and deforms nucleotides. *Biophysics Journal* 89, 3856–3862.

Part V

Methods of Oxidative Stress Detection

Chapter 24

DETECTION OF REACTIVE METABOLITES OF OXYGEN AND NITROGEN

Matthew B. Grisham

Immunology and Inflammation Research Group, Department of Molecular and Cellular Physiology, Louisiana State University Health Sciences Center, Shreveport, LA, USA

All forms of marine life are exposed to exogenous and endogenous sources of reactive oxygen species (ROS) and oxidized metabolites of nitrogen (NO_x) (Palumbo 2005; Lesser 2006; al Housari et al. 2010). Visible and UV light absorption by chromophoric dissolved organic matter in seawater produces large amounts of ROS via the photochemical generation of $O_2^{\bullet-}$ and H_2O_2. Hydrothermal vents also generate large amounts of $O_2^{\bullet-}$, H_2O_2, and sulfur-centered radicals via the oxidation of H_2S by O_2 (Lesser 2006). In order to protect themselves from these potentially noxious oxidants, vent worms, clams, and bacteria express a number of different oxidant defense mechanisms.

The propensity for environmentally generated ROS, organic ROS, and NO_x to damage such a diverse ecosystem required that all marine organisms evolved a varied set of oxidant defenses. Understanding how ROS and NO_x affect the delicate balance of the marine environment necessitates the ability to accurately quantify these metabolites. The complex and overlapping nature of these reactive species makes their detection and quantification possible in theory but difficult in practice.

DETECTION AND QUANTIFICATION OF $O_2^{\bullet-}$

Cytochrome *c* Reduction

The superoxide dismutase (SOD)-inhibitable reduction of ferricytochrome *c* (Fe^{+3}*cyt c*) to ferrocytochrome *c* (Fe^{+2}*cyt c*) has been used to quantify the rate of $O_2^{\bullet-}$ formation by a variety of different enzymes, cells, and vascular tissue (Azzi et al. 1975; Massey 1959; Tarpey et al. 2004).

$$Fe^{3+}\,\text{cyt}\,c + O_2^{\bullet-} \rightarrow Fe^{2+}\text{cyt}\,c + O_2 \qquad (24.1)$$

The reaction is followed kinetically by continuously recording the reduction of Fe^{3+} cyt *c* at 550 nm. This determination can be used to quantify $O_2^{\bullet-}$ production as the molar extinction coefficient for Fe^{2+} cyt *c* at 550 nm is $2.1 \times 10^4\ M^{-1}\,cm^{-1}$ (Massey 1959). As with any spectrophotometric assay, there are several precautions and limitations that need to be recognized:

1. Reduction of Fe^{3+} cyt *c* is not absolutely specific for $O_2^{\bullet-}$. Cellular reductants such as ascorbate

Oxidative Stress in Aquatic Ecosystems, First Edition. Edited by Doris Abele, José Pablo Vázquez-Medina, and Tania Zenteno-Savín.

and glutathione (GSH) and certain cellular reductases can reduce Fe^{3+} cyt. The rate of Fe^{3+} cyt *c* reduction must always be performed in the presence of SOD such that the SOD-inhibitable rate may be determined.

2. Enzymes such as xanthine oxidase (XO) are capable of reducing quinones or redox-active phenolic compounds that may also be present in cell or tissue extracts and whose reduced forms are capable of directly reducing Fe^{3+} cyt *c* (Fridovich 1985).
3. Fe^{2+} cyt *c* can be reoxidized by cytochrome oxidases, cellular peroxidases, and oxidants including H_2O_2 and $ONOO^-$ (Thomson et al. 1995). Because the apparent rate of Fe^{3+} cyt *c* reduction is decreased by these reoxidation reactions one will underestimate the rate of $O_2^{\bullet -}$ formation. Enzyme inhibitors (10 μM CN for cytochrome oxidase) or ROS scavengers (100 U mL^{-1} catalase (CAT) for H_2O_2, 10 mM urate for $ONOO^-$) can be added to the reaction mixture to avoid this shortcoming. To enhance the specificity of the cytochrome *c* assay for $O_2^{\bullet -}$, Fe^{3+} cyt *c* can be acetylated (Azzi et al. 1975; O'Brien 1984). Succinoylation of cytochrome *c* is very effective in decreasing the reduction of cytochrome *c* by nicotinamide adenine dinucleotide phosphate hydrogen (NADPH)-cytochrome P-450 reductase or cytochrome b_5 and oxidation by cytochrome *c* oxidase (Kuthan et al. 1982).
4. Because of its size and charge, cytochrome *c* has limited ability to gain access to the intracellular space, making this assay more appropriate for detecting extracellular $O_2^{\bullet -}$ production.
5. The cytochrome *c* assay represents the "gold standard" for quantifying $O_2^{\bullet -}$ *in vitro*; however, its relative insensitivity to detect low rates of $O_2^{\bullet -}$ production, as well as interference by a variety of endogenous reductants, limits its applicability for *in vivo* detection of this reactive species (Landmesser et al. 2003; Markert et al. 1984).

Inhibition of Aconitase Activity

Intracellular fluxes of $O_2^{\bullet -}$ can be estimated by quantifying the $O_2^{\bullet -}$-dependent inactivation of aconitase (Castro et al. 1994; Gardner et al. 1995; Hausladen and Fridovich 1996). Aconitase is a cytosolic and mitochondrial enzyme that catalyzes the conversion of citrate to isocitrate. $O_2^{\bullet -}$ inactivates aconitase due to oxidation and loss of Fe from its nonheme iron cluster [4Fe-4S] within the active site, with a rate constant of 10^6–10^7 M^{-1} s^{-1} (Flint et al. 1993; Castro et al. 1994; Hausladen and Fridovich 1996). Because of low but continuous $O_2^{\bullet -}$ production during normal metabolism, a small but significant fraction of aconitase activity may, at any one time, be inactive (but capable of reactivation). As the rate of $O_2^{\bullet -}$ production increases, the ratio of inactive/active aconitase will increase (Gardner and Fridovich 1992; Gardner et al. 1995, 1996). Using the aconitase assay, relative rates of $O_2^{\bullet -}$ production within the cytosol and/or mitochondria can be determined. Aconitase activity has been used to determine relative rates of $O_2^{\bullet -}$ formation in a variety of cell types as well as in liver, heart, lung, and brain (Gardner and Fridovich 1992; Gardner et al. 1995; Melov et al. 1999). As with the cytochrome *c* assay, there are precautions and limitations that need to be considered:

1. In addition to $O_2^{\bullet -}$, other oxidants may inhibit aconitase activity (Castro et al. 1994; Hausladen and Fridovich 1994, 1996).
2. It has been reported that $NO^\bullet$ itself can inhibit aconitase activity; however, it is doubtful that $NO^\bullet$ is capable of directly interacting with and inactivating the iron sulfur cluster of the enzyme. The small amount of aconitase inactivation that appears to be due to $NO^\bullet$ is most probably mediated by $ONOO^-$. The effect of NOS inhibition and/or SOD administration will allow one to determine the role(s) of $O_2^{\bullet -}$ and $NO^\bullet$ derived species in aconitase inactivation.

$O_2^{\bullet -}$-Dependent Oxidation of Hydroethidine

Hydroethidine (dihydroethidium; HE) is a cell permeant compound that can undergo a two-electron oxidation to form the DNA-binding fluorophore ethidium (Fig. 24.1; Benov et al. 1998; Zhao et al. 2002). The observation that $O_2^{\bullet -}$ oxidizes HE to ethidium created a great deal of interest since neither H_2O_2 nor HOCl are capable of oxidizing HE. The intracellular oxidation of HE to ethidium by $O_2^{\bullet -}$ has been analyzed using flow cytometry in cell suspensions (Carter et al. 1994) and by visualization of cellular and anatomic regions displaying increased rates of $O_2^{\bullet -}$ production

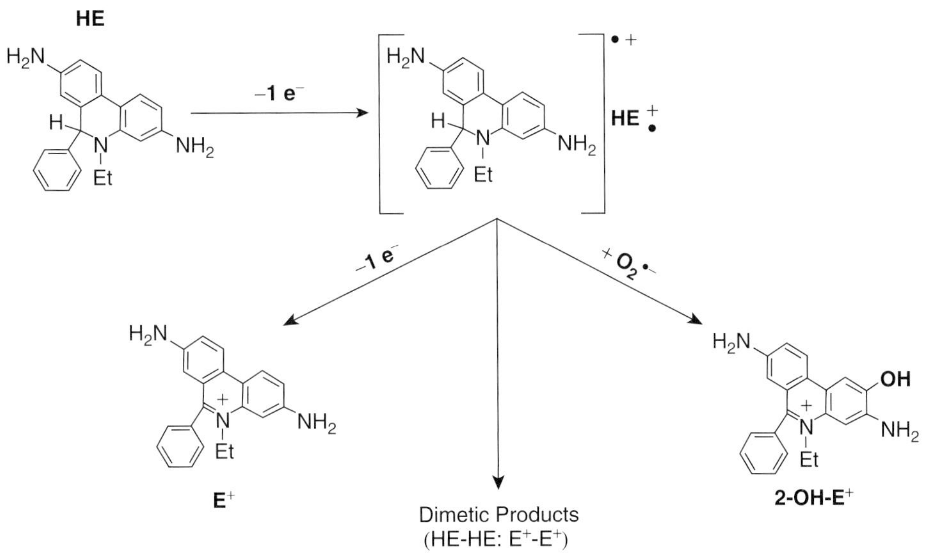

Fig. 24.1 Formation of 2-hydroxyethdium (2-OH-E^+) from the interaction of superoxide ($O_2^{\bullet-}$) with hydroethidine (HE). A one electron oxidation of HE leads to formation of the HE cation radical ($HE^{\bullet+}$) intermediate that may be oxidized further to yield ethidium (E^+) or it may react with $O_2^{\bullet-}$ to yield 2-OH-E^+. Different dimeric products may also be formed during the oxidation of HE. (Adapted from Zielonka et al. 2008, 2009.)

using digital imaging microfluorometry *in vivo* (Suzuki et al. 1995) and in tissue slices/isolated blood vessels (Bindokas et al. 1996; Miller et al. 1998). Unlike other "intracellular" probes for $O_2^{\bullet-}$ (e.g. luminol, lucigenin), there is little artifactual formation of $O_2^{\bullet-}$ by HE due to redox cycling. There are a few critical reactions that may limit the use of HE conversion to ethidium as a quantitative marker for $O_2^{\bullet-}$ production:

1. Fe^{+3} cyt *c* is also capable of oxidizing HE. This may be important when working with mitochondria or when Fe^{+3} cyt *c* is released into the cytosol during apoptosis (Green and Reed 1998).
2. Use of high concentrations of HE can result in $O_2^{\bullet-}$-independent fluorescence as a result of ethidium formation that exceeds the binding capacity of mitochondrial nucleic acids, allowing the E^+ to bind to nuclear DNA with marked enhancement of fluorescence (Budd et al. 1997).
3. The HE assay may underestimate $O_2^{\bullet-}$ production as HE may enhance the rate of $O_2^{\bullet-}$ dismutation to H_2O_2 (Benov et al. 1998).
4. Although the HE assay is reasonable for qualitative determinations of intracellular $O_2^{\bullet-}$ production, it is not a quantitative assay. The interaction of HE with $O_2^{\bullet-}$ produces 2-hydroxyethidium rather than ethidium (Fig. 24.1). Because of their spectral overlap, 2-hydroxyethidium and ethidium cannot be distinguished by simple fluorescence spectroscopy (Zielonka et al. 2008, 2009). All of the oxidation products of HE including 2-hydroxyethidium and ethidium can be separated and quantified using HPLC providing a **quantitative** method to measure intracellular formation of $O_2^{\bullet-}$ (Zielonka et al. 2008, 2009).

Chemiluminescence Probes

Chemiluminescent methods for $O_2^{\bullet-}$ detection have been used in more than 1000 published studies because of the ability of the different probes to gain access to the intracellular space, the professed specificity of reaction of the probe with $O_2^{\bullet-}$, minimal cellular toxicity, and increased sensitivity when compared with other biochemical methods. The most widely used chemiluminescent compound for intracellular $O_2^{\bullet-}$ detection is lucigenin (bis-N-methylacridinium dintrate). The use of this probe has raised questions regarding its ability to faithfully measure $O_2^{\bullet-}$ generation in cells and tissue (Tarpey et al. 1999, 2004; Wardman 2007).

It was shown more than 10 years ago that lucigenin will undergo redox cycling within the cell generating $O_2^{\bullet-}$, the very metabolite it is supposed to measure! (Liochev and Fridovich 1997; Tarpey and Fridovich 2001). It was suggested that reducing the concentration of lucigenin (below 5 μM) eliminated the potential for redox cycling (Skatchkov et al. 1999; Munzel et al. 2002); however, other studies demonstrated that even at this low concentration of probe, redox cycling with enhanced $O_2^{\bullet-}$ formation occurred (Barbacanne et al. 2000; Janiszewski et al. 2002). Despite these limitations in estimating rates of $O_2^{\bullet-}$ formation, the use of low concentrations of lucigenin may provide qualitative measures of $O_2^{\bullet-}$ production in well-controlled and well-defined *in vitro* systems.

In an attempt to take advantage of the sensitivity of chemiluminescent techniques, other nonredox-cycling compounds have been studied for their utility to specifically react with $O_2^{\bullet-}$. Most of these probes are similar in structure to the luciferins derived from bioluminescent marine life. For example, coelenterazine (2-(4-hydroxybenzyl)-6-(4-hydroxyphenyl)-8-benzyl-3,7-dihydroimidazo[1,2-α] pyrazin-3-one) is a lipophilic luminophore originally isolated from the coelenterate *Aequorea* and is the light-producing chromophore in the aequorin complex (Tarpey et al. 1999; Barbacanne et al. 2000; Janiszewski et al. 2002). $O_2^{\bullet-}$-stimulated chemiluminescence occurs following direct oxidation of coelenterazine (Goto et al. 1968; Teranishi and Shimomura 1997), thereby eliminating the redox-cycling-dependent artifactual generation of $O_2^{\bullet-}$ associated with lucigenin (Liochev and Fridovich 1997; Teranishi and Shimomura 1997; Vasquez-Vivar et al. 1997). Although the intensity of light emitted from the interaction of coelenterazine with $O_2^{\bullet-}$ is greater than lucigenin-dependent chemiluminescence, coelenterazine-dependent chemiluminescence is not entirely specific for $O_2^{\bullet-}$, as $ONOO^-$ will also result in luminescence in the presence of coelenterazine (Tarpey et al. 1999). The selective use of nitric oxide synthase (NOS) inhibitors and/or $ONOO^-$ scavengers may help to differentiate between the contributions made by $O_2^{\bullet-}$- and $ONOO^-$-mediated chemiluminescence. Coelenterazine-dependent chemiluminescence has been used to detect $O_2^{\bullet-}$ in cultured cells (Duerrschmidt et al. 2000), neutrophils (Lucas and Solano 1992), vascular tissue (Hink et al. 2001), and isolated mitochondria (Raha et al. 2000). Structural analogs, CLA (2-methyl-6-phenyl-3,7-dihydroimidazo[1,2-α] pyrazin-3-one) and MCLA (2-methyl-6-(4-methoxyphenyl)-3,7-dihydroimidazo [1,2-α] pyrazin-3-one) have also been used to assess $O_2^{\bullet-}$ formation in several systems (Nakano 1990a,b; Ishii et al. 1997; Skatchkov et al. 1998). The MCLA assay will detect $O_2^{\bullet-}$ formation on the surface of liver (Uehara et al. 1993), within the intestine (Saitoh et al. 1995), heart (Ushiroda et al. 1997), and lung (Lucas and Solano 1992) *in vivo*. More work is needed to confirm the specificity of these newer probes as well as potential interfering reactions that may limit their use under certain conditions.

DETECTION AND QUANTIFICATION OF H_2O_2

Peroxidase-based Assays

A variety of peroxidase-based assays have been developed to specifically quantify H_2O_2 generation in cells and tissues. These assays are based upon the horseradish peroxidase (HRP)-catalyzed, H_2O_2-dependent oxidation of an electron donating detector compound (AH) (Boveris et al. 1977; Boveris 1984):

$$\text{HRP-Fe}^{+3} + H_2O_2 \rightarrow {}^{+\bullet}\text{HRP-Fe}^{+4} \tag{24.2}$$

$$^{+\bullet}\text{HRP-Fe}^{+4} + \text{AH} \rightarrow \text{HRP-Fe}^{+3} + \text{A}^{+} \tag{24.3}$$

where $^{+\bullet}$HRP-Fe^{+4} represents the ferryl porphyrin cation radical intermediate of HRP, which is called Compound I. The amount of H_2O_2 present is estimated by following the decrease in fluorescence of initially fluorescent electron donors such as scopoletin (7-hydroxy-6-methoxy-coumarin), or by monitoring the increase in fluorescence from nonfluorescent electron donors such as diacetyldichlorofluorescin (Hinkle et al. 1967), p-hydroxyphenylacetate (Hyslop and Sklar 1984), homovanillic acid (3-methoxy-4-hydroxyphenylacetic acid) (Ruch et al. 1983), or Amplex Red (N-acetyl-3,7-dihydroxyphenoxazine) (Cohen et al. 1996; Zhou et al. 1997). Oxidation of nonfluorescent dyes such as tetramethylbenzidine (Staniek and Nohl 1999) or phenol red (Pick and Keisari 1980) may be followed spectrophotometrically. Several precautions are required for accurate interpretation using these types of assays to quantify H_2O_2 generation:

1. Numerous reductants can serve as alternative electron donating substrates for HRP leading to underestimation of H_2O_2 formation.

2. Competition with HRP by endogenous CAT for H_2O_2 can also lead to underestimation of H_2O_2.
3. Quenching of fluorescent signals by cell and tissue components can lead to overestimation or underestimation of H_2O_2 depending on the specific electron donor used. When measuring cellular or subcellular H_2O_2 production, accuracy of the assay can be improved by separation of the incubation media from cellular components before addition of the detection system, limiting confounding interactions of the detection system with cellular elements (Staniek and Nohl 1999).
4. Direct reduction of the oxidized detector molecule by electron transport components can limit the utility of the HRP assay for H_2O_2 determinations in mitochondria (Staniek and Nohl 1999). While spectrophotometric detection of H_2O_2 with phenol red has been used in purified enzyme-substrate mixtures and to detect H_2O_2 release from activated leukocytes (Pick and Keisari 1980), the phenol red-HRP assay is exquisitely pH-dependent and less sensitive than fluorescent methods, reducing its utility when attempting to detect H_2O_2 from cell types with low endogenous rates of oxidant formation. The HRP-linked assays represent a particularly useful method for the quantification of H_2O_2 in cultured cells, organ cultures, and isolated buffer-perfused tissue preparations. However, these methods are not suitable for determinations of H_2O_2 in the presence of plasma or serum since many reducing agents present in extracellular fluid will interfere with the assay.

Dichlorofluorescein Fluorescence

The oxidation of 2′-7′-dichlorofluorescin (DCFH) to yield the fluorescent compound 2′-7′-dichlorofluorescein (DCF) was initially proposed to be a relatively specific indicator for H_2O_2 formation (Keston and Brandt 1965) and has been used in hundreds of studies to detect extracellular and/or intracellular ROS production by a variety of different cells (reviewed in Wardman 2007). With the development and widespread use of flow cytometry, DCFH oxidation is now routinely used to quantify ROS generation within individual cells or populations of cells. The diacetate form of DCFH (DCFH-DA) is taken up by cells, where intracellular esterases hydrolyze the diacetate moiety, thereby "trapping" DCFH within the intracellular space. It was assumed that intracellular H_2O_2 oxidizes DCFH to DCF and the fluorescence is measured with excitation at 498 nm and emission at 522 nm. Quantitative and detailed mechanistic studies demonstrate that the use of DCFH as a specific probe for H_2O_2 is fraught with spurious and confounding side reactions that make this probe virtually unusable for the specific detection of H_2O_2 for the following reasons (Bonini et al. 2006; Wardman 2007):

1. The H_2O_2-dependent oxidation of DCFH to DCF occurs very slowly, if at all, in the absence of trace/contaminating amounts of redox active metals such iron or copper (Bonini et al. 2006; Wardman 2007).
2. The H_2O_2-dependent oxidation of DCFH is greatly enhanced in the presence of heme-containing substances (Cathcart et al. 1983; LeBel et al. 1992). Lipid peroxides are also capable of inducing DCF fluorescence in the presence of heme-containing compounds.
3. Peroxidases are capable of inducing DCFH oxidation in the absence of H_2O_2 (LeBel et al. 1992; Larsen et al. 1996; Rota et al. 1999).
4. DCFH does not necessarily remain within the intracellular space as it may leak out and accumulate in extracellular media where it would be available for reaction with extracellular metals and oxidants (Royall and Ischiropoulos 1993).
5. The peroxidase-dependent formation of DCF from DCFH, both in the presence and absence of H_2O_2, has been found to be a complex reaction generating both H_2O_2 and $O_2^{\bullet-}$ via the DCF semiquinone free radical $DCF^{\bullet-}$ intermediate (Rota et al. 1999; Bonini et al. 2006; Wardman 2007).
6. Other potent oxidizing agents are capable of directly oxidizing DCFH to DCF in the absence of H_2O_2, including $ONOO^-$ and HOCl (Crow 1997; Kooy et al. 1997).

A structurally related analog of DCFH is dihydrorhodamine 123 (DHR), a cell-permeant, mitochondrial-avid compound that is oxidized to the fluorophore rhodamine 123. Again, H_2O_2 is incapable of directly oxidizing DHR, however, heme-containing proteins will catalyze the H_2O_2-dependent oxidation of DHR to rhodamine (LeBel et al. 1992; Royall and Ischiropoulos 1993). Several other cell-derived oxidants such as

ONOO$^-$ and HOCl are also capable of directly oxidizing DHR to rhodamine (Crow 1997). Because of the various biologic substances that can lead to DCF and rhodamine fluorescence coupled to the inherent uncertainty relating to endogenous versus artifactual oxidant generation, these fluorescent assays are best applied as qualitative markers of cellular oxidant stress, rather than as precise indicators of rates of H_2O_2 formation.

Aminotriazole Inhibition of Catalase

A major pathway for the decomposition of H_2O_2 is through its reaction with CAT. A relatively stable intermediate called Compound I ($^{+\bullet}$CAT-Fe^{+4}) is rapidly formed during the CAT-dependent metabolism of H_2O_2. Aminotriazole is a cell permeant compound that reacts with the Compound I intermediate causing irreversible inhibition of the enzyme. Catalase is inhibited by aminotriazole only in the presence of H_2O_2 (Margoliash et al. 1960). Intracellular rates of H_2O_2 formation can be estimated by determining the half-time of aminotriazole-dependent inactivation of CAT. Steady-state H_2O_2 concentration can be approximated according to (Kinnula et al. 1993; Paler-Martinez et al. 1994):

$$[H_2O_2] = \frac{k_{cat}}{k_1} \quad \text{where} \quad \frac{0.5}{t_{1/2}} = -k_{cat} \quad \text{and} \quad k_1 = 1.7 \times 10^7 M^{-1}s^{-1} \tag{24.4}$$

This method has been used to assess the rates of H_2O_2 formation in numerous isolated cell types as well as different tissues *in vivo* (Yusa et al. 1987; Guidet and Shah 1989; Piantadosi and Tatro 1990).

QUANTIFICATION OF NO$^\bullet$ AND ITS OXIDIZED METABOLITES

It is now well-established that the endogenous production of NO$^\bullet$ by different NOS isozymes is important in modulating vascular homeostasis, neurotransmission, and host defense mechanisms in terrestrial and marine organisms (Palumbo 2005; Bryan and Grisham 2007). The major pathway for NO$^\bullet$ metabolism is its stepwise oxidation to nitrite and nitrate. When introduced into oxygenated blood plasma or physiological fluids/buffers, NO$^\bullet$ will spontaneously auto-oxidize to yield exclusively nitrite (NO_2^-). The oxidation of NO by molecular oxygen is second order with respect to NO$^\bullet$:

$$2NO + O_2 \rightarrow 2NO_2 \tag{24.5}$$

$$2NO + 2NO_2 \rightarrow 2N_2O_3 \tag{24.6}$$

$$2N_2O_3 + 2H_2O \rightarrow 4NO_2^- + 4H^+ \tag{24.7}$$

whereby NO_2, N_2O_3, and NO_2^- represent nitrogen dioxide, dinitrogen trioxide, and nitrite, respectively. It should be noted that N_2O_3 is a potent nitrosating agent by its ability to generate the nitrosonium ion (NO^+). Introduction of NO$^\bullet$ and NO_2^- into whole blood results in the rapid oxidation of both metabolites to yield almost exclusively NO_3^-. Although the mechanisms by which NO$^\bullet$ and NO_2^- are converted to NO_3^- *in vivo* are not entirely clear, there are several possibilities. One mechanism suggests that the NO_2^- derived from NO$^\bullet$ autoxidation is rapidly converted to NO_3^- via its oxidation by certain oxyhemoproteins (P-Fe^{2+} O_2) such as oxyhemoglobin or oxymyoglobin (Ignarro et al. 1993):

$$2P\text{-}Fe^{2+}O_2 + 3NO_2^- + 2H^+ \rightarrow 2P\text{-}Fe^{3+} + 3NO_3^- + H_2O \tag{24.8}$$

or

$$4P\text{-}Fe^{2+}O_2 + 4NO_2^- + 4H^+ \rightarrow 4P\text{-}Fe^{3+} + 4NO_3^- + O_2 + 2H_2O \tag{24.9}$$

These investigators used large concentrations of NO$^\bullet$ (300 μM), which will rapidly autoxidize to NO_2^-. Although the authors suggested that the NO_2^- would in turn react with the hemoproteins, this reaction requires 2–3 h. Another, possibly more reasonable explanation for the predominant presence of NO_3^- *in vivo* may be due to the fact that the levels of NO$^\bullet$ produced by NOS *in vivo* are much smaller and, thus, the half-life of NO$^\bullet$ would be much longer. In this case, NO$^\bullet$ would react directly and very rapidly with oxyhemoproteins (P-Fe$^{+2}O_2$) to yield NO_3^- before it has an opportunity to autoxidize to NO_2^-:

$$P\text{-}Fe^{2+}O_2 + NO^\bullet \rightarrow P\text{-}Fe^{3+} + NO_3^- \tag{24.10}$$

These mechanisms of autoxidation of NO$^\bullet$ would also be important in tissues and cell cultures where NO$^\bullet$ may interact with a variety of different hemoproteins.

In addition to its reaction with oxygen, NO$^\bullet$ may rapidly interact with $O_2^{\bullet-}$ to yield ONOO$^-$ and its conjugate acid ONOOH (Beckman and Koppenol 1996):

$$O_2^{\bullet -} + NO^{\bullet} \rightarrow ONOO^{-} + H^{+} \leftrightarrow ONOOH \rightarrow NO_3^{-} + H^{+} \quad (24.11)$$

$ONOO^-/ONOOH$ may play an important role in several pathophysiological situations. The question of whether $ONOO^-/ONOOH$ is actually formed *in vivo* and exerts significant physiologic and/or pathophysiologic activity remains the subject of active debate. $ONOO^-$ promotes nitration and hydroxylation of bioorganic molecules. ONOOH exists as pair of the caged radicals $NO_2^{\bullet}$ and $HO^{\bullet}$. Because both $O_2^{\bullet -}$ and $NO^{\bullet}$ are produced in large amounts during active inflammation, the ability to specifically detect $ONOO^-/ONOOH$ *in vitro* and *in vivo* would represent an advancement in understanding the pathophysiological role of these oxidants and nitrating agents. It was originally proposed that $ONOO^-/ONOOH$ could be specifically detected *in vitro* and *in vivo* via its ability to nitrate tyrosine residues to yield 3-nitrotyrosine (3NT). We now know that 3NT may be generated by multiple pathways suggesting that the presence of 3NT is not a specific "footprint" for $ONOO^-/ONOOH$ formation *in vivo* (see below).

Measurement of $NO^{\bullet}$-Derived Nitrate and Nitrite

Griess Reaction

One method that has been used to quantify NO_2^- and NO_3^- in freshwater or saltwater, as well as $NO^{\bullet}$-derived NO_2^- and NO_3^- in physiological fluids, is called the Griess reaction. This method requires that NO_3^- first be reduced to NO_2^- and then NO_2^- is directly determined using the Griess reagent (Fig. 24.2). The Griess reaction is a two-step diazotization reaction in which N_2O_3, generated from the acid-catalyzed formation of nitrous acid (HNO_2) from NO_2^-, reacts with sulfanilic acid to produce a diazonium ion which is then coupled to *N*-(1-napthyl)ethylenediamine to form a chromophoric azo product that absorbs at 543 nm (Grisham et al. 1996; Nims et al. 1996; Bryan and Grisham 2007). For quantification of NO_3^- and NO_2^- in extracellular fluids, we have found enzymatic reduction of NO_3^- to NO_2^- using a commercially available preparation of nitrate reductase (NR) to be the most satisfactory method:

$$NO_3^- + NADPH \xrightarrow{NR} NO_2^- + NADP^+ \quad (24.12)$$

Aspergillus NR (Boehringer Mannheim, Indianapolis, IN) is highly efficient at reducing very small amounts

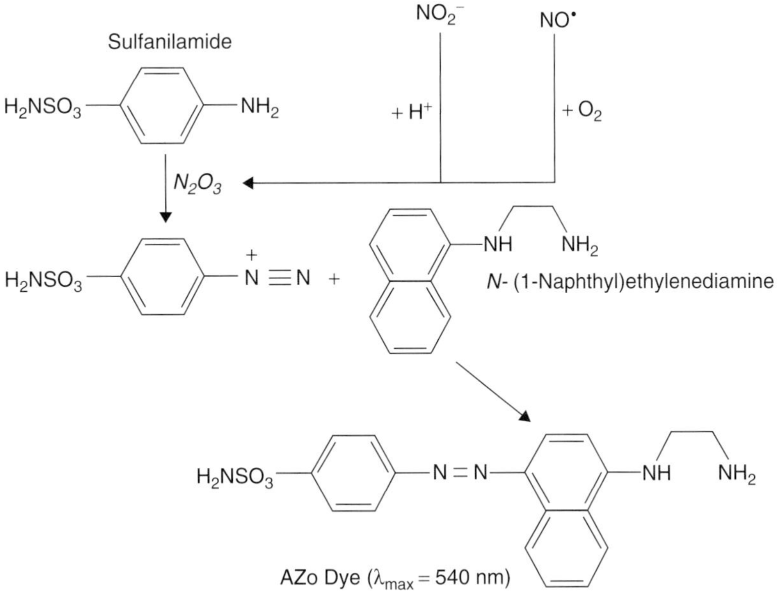

Fig. 24.2 The Griess reaction. The nitrosating agent dinitrogen trioxide (N_2O_3) generated from the autoxidation of $NO^{\bullet}$ or from the acidification of nitrite (NO_2^-) reacts with sulfanilamide to yield a diazonium derivative. This reactive intermediate will interact with *N*-1-(naphthyl)ethelenediamine to yield a colored diazo product that absorbs strongly at 540 nm. (Reproduced with permission from Bryan and Grisham 2007.)

of NO_3^- to NO_2^-. Following the incubation, any unreacted NADPH is oxidized by addition of lactate dehydrogenase and pyruvic acid because NADPH and NADH strongly inhibit the Griess reaction. An alternative method for oxidizing any unreacted NADPH is to replace the lactate dehydrogenase (LDH)/pyruvate system with 1 mM potassium ferricyanide (Grisham et al. 1996). A known volume of premixed Griess reagent is then added to each incubation mixture, incubated for 10 min, and the absorbance of each sample determined at 543 nm. The Griess reaction may also be used to estimate $NO^•$ fluxes under more physiologocial conditions (e.g. neutral pH) by taking advantage of the fact that at neutral pH and ambient O_2 tension (100 mm Hg) the rapid autoxidation of $NO^•$ produces N_2O_3 (Nims et al. 1996). This N-nitrosating agent can then react with sulfanilic acid to produce a diazonium ion, which then interacts with *N*-(1-napthyl)ethylenediamine to form a chromophoric azo product. $NO^•$-derived NO_2^- and NO_3^- may be quantified in tissue culture media and organ culture supernatants.

Diaminonaphthalene Assay

Although the Griess reaction is a simple, rapid, and inexpensive assay for quantifying NO_2^- and NO_3^-, it has a practical sensitivity limit of only 2–3 μM. In attempts to enhance its sensitivity, different fluorimetric methods have been developed. One method to quantify NO_2^- and NO_3^- in polluted waters, as well as to measure $NO^•$-derived NO_2^- and NO_3^- in physiological fluids, uses the aromatic diamino compound 2,3-diaminonaphthalene (DAN) (Miles et al. 1996; Bryan and Grisham 2007). The relatively nonfluorescent DAN reacts rapidly with NO_2^- or NO-derived N_2O_3 generated from the acid-catalyzed formation of HNO_2 or the autoxidation of $NO^•$ to yield the highly fluorescent product 2,3-naphthotriazole (NAT; Fig. 24.3). This assay offers the additional advantages of specificity, sensitivity, and versatility. This assay is capable of detecting as little as 10–30 nM (i.e. 10–30 pmol mL^{-1}) naphthotriazole and may be used to quantify $NO^•$ generated under biologically relevant conditions (e.g. neutral pH) with minimal interference by nitrite decomposition (Miles et al. 1996; Nims et al. 1996). As with the Griess reaction, the DAN assay can be used to continually quantify $NO^•$ production under physiological conditions in biological fluids, tissue culture media, and organ culture supernatants.

Diaminofluoroscein-2 Assay

In addition to the DAN assay, diaminofluoroscein-2 (DAF-2) may be used to determine the presence of NO *in vitro* and *in situ* (Kojima et al. 1998a,b). Like

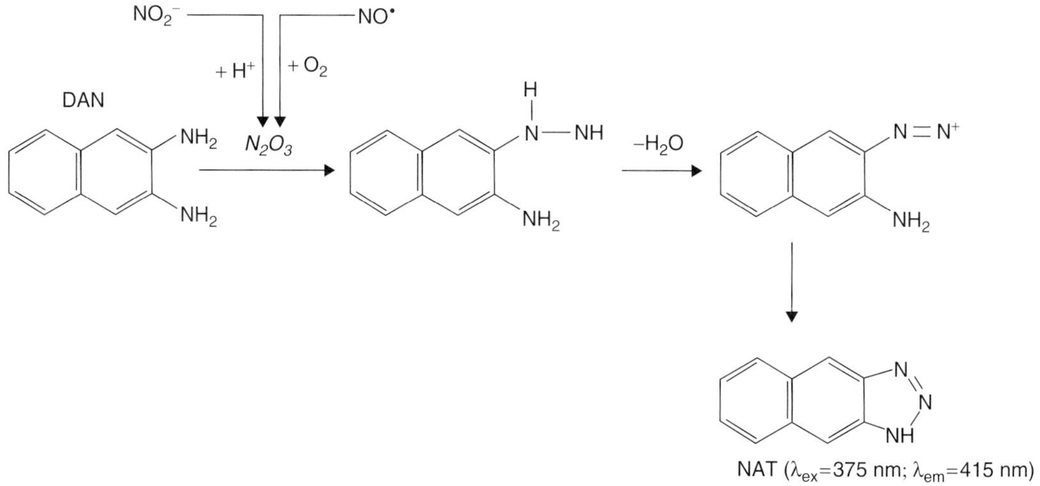

Fig. 24.3 Fluorometric detection of $NO^•$ or nitrite using diaminonaphthalene (DAN). The nitrosating agent dinitrogen trioxide (N_2O_3) generated from the autoxidation of $NO^•$ or from the acidification of nitrite (NO_2^-) reacts with DAN to yield the highly fluorescent product naphthotriazole (NAT). (Reproduced with permission from Bryan and Grisham 2007.)

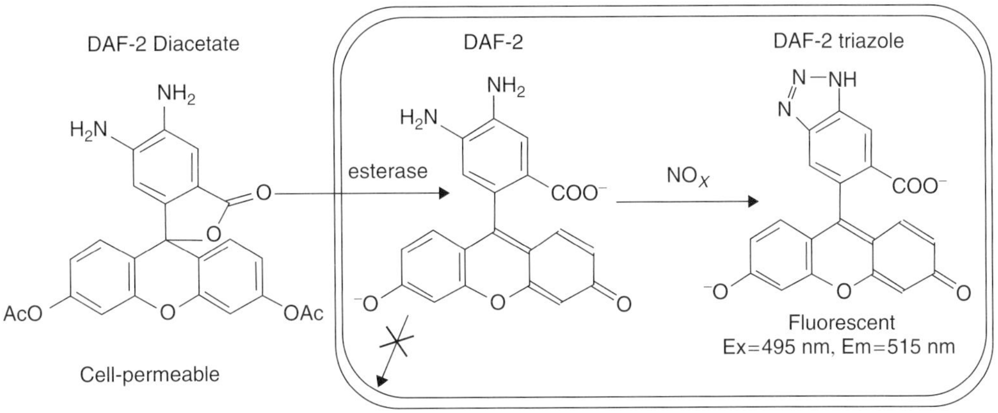

Fig. 24.4 Fluorometric detection of NO$^{\bullet}$ using diaminofluroscein-2 diacetate (DAF-2 DA). DAF-2DA diffuses into cells and tissue where nonspecific esterases hydrolyze the diacetate residues thereby trapping DAF-2 within the intracellular space. NO-derived nitrosating agents such as N_2O_3 nitrosate DAF-2 to yield its highly fluorescent product DAF-2 triazole (DAF-2T). (Reproduced with permission from Bryan and Grisham 2007.)

DAN, the N-nitrosation of the diamino groups results in a nitrosamine, which through an internal rearrangement forms the fluorescent triazole (Fig. 24.4). The advantages of this compound are that wavelength associated with fluorescein can be used, making equipment currently used for other bioassays as well as cell and tissue imaging easily adapted to detect nitric oxide *in vitro* and *in vivo*. Although NO$^{\bullet}$-derived N_2O_3 is thought to be the primary mechanism by which cells and tissues generate the triazole derivative, recent investigations suggest that oxidative nitrosylation may represent an alternative pathway for triazole generation (Espey et al. 2001, 2002a,b). Since DAF-2 can be oxidized by one electron by species such as NO_2, it may first oxidize the diamino complex to an aromatic radical that then undergoes radical–radical coupling to form the nitrosamine, which subsequently rearranges to the fluorescent triazole. This implies under biological conditions that the triazole can be formed from either nitrosative or oxidative chemistry.

In buffered salt solutions or tissue culture media these two pathways can be differentiated. The use of nitrosative scavengers such as azide will prevent nitrosative but not oxidative stress (Espey et al. 2001, 2002a,b). Urate will quench oxidative nitrosylation but not nitrosative stress. Using different scavengers, the chemistry of DAF-2 formation can be teased out in different experiments. Though DAF-2 has been thought to be an indicator of NO$^{\bullet}$, it recently has been shown that nitroxyl (HNO) reacts with DAF-2 giving even higher yields of triazole than NO$^{\bullet}$. In light of the potential for HNO in biological systems, it could be interesting to entertain the possibility that some of the NO$^{\bullet}$ detected by DAF-2 is in fact HNO. One cautionary note is that using powerful light sources such as lasers can result in photochemistry that can lead to false positives. Controls need to be done to ensure that it is the chemistry from NO$^{\bullet}$ and not some artifact of the photochemistry.

Quantification of S-Nitrosothiols

The formation and biological properties of NO$^{\bullet}$-derived S-nitrosothiols (RSNOs) play an important part of the biology of NO$^{\bullet}$ (Foster et al. 2003). Autoxidation of NO$^{\bullet}$ in the presence of thiols (RSH) generates RSNOs via the following mechanism:

$$2NO + O_2 \rightarrow 2NO_2^{\bullet} \quad (24.13)$$

$$2NO + 2NO_2^{\bullet} \rightarrow 2N_2O_3 \quad (24.14)$$

$$2N_2O_3 + 2RSH \rightarrow 2RSNO + 2NO_2^{-} \quad (24.15)$$

Different RSNOs are known to stimulate guanylate cyclase, thereby promoting vasorelaxation (Allen et al. 2009; Foster et al. 2009). Here we describe methods for the colorimetric and fluorometric detection of RSNO (Foster et al. 2003, 2009).

The detection of RSNO has often employed the Saville reaction (Saville 1958), which involves the displacement of the nitrosonium ion (NO^+) by mercury salts (Fig. 24.5). The colorimetric techniques described

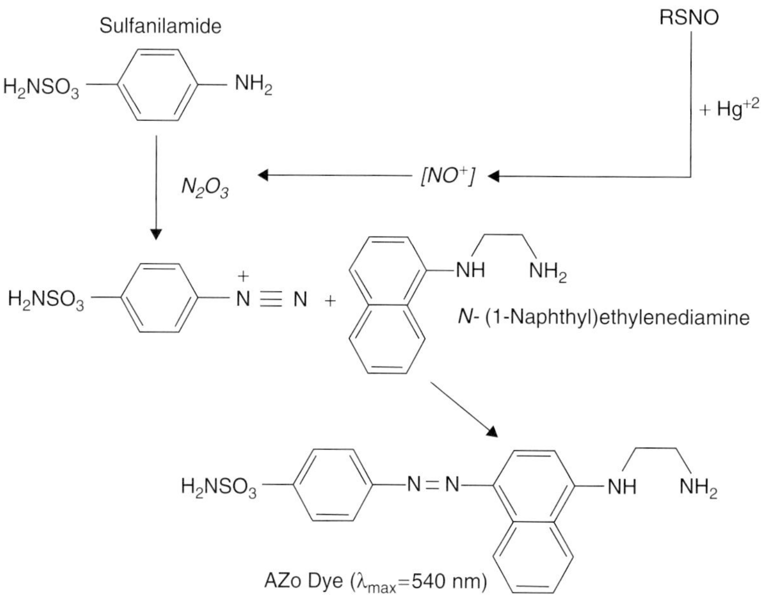

Fig. 24.5 Detection of S-nitrosothiols (RSNO) by the Saville reaction. Liberation of the nitrosonium (NO^+) -derived nitrosating agent (N_2O_3) from the interaction between RSNO with $HgCl_2$ in the presence of the Griess reagents results in the formation of the same diazo product as described in Fig. 24.1. (Reproduced with permission from Bryan and Grisham 2007.)

above to detect RSNO by utilizing the Griess reaction to quantify the NO_2^- formed from the hydrolysis of NO^+ liberated following the treatment of RSNO with mercuric chloride can be used. Samples that contain large amounts of NO_2^- can interfere with and limit the detection range of these methods under acidic conditions. Two methods have been devised to detect RSNO-derived nitrosating species at neutral pH (Wink et al. 1999). The colorimetric method uses the components of the Griess reaction while the fluorimetric method utilizes the conversion of DAN to its fluorescent triazole derivative. These methods may be conducted at neutral rather than acidic pH, which eliminates the interference of contaminating NO_2^- and allows the detection of the RSNO-derived nitrosating agent.

The colorimetric reaction utilizes the same chemistry as described previously in which NO^+-derived N_2O_3 generated by the interaction between RSNO and mercuric chloride (at neutral pH) reacts with sulfanilamide to form a diazonium ion (Fig. 24.5). The resulting diazonium salt then reacts with naphthylethylenediamine to form the colored azo complex. The fluorometric assay is based on the reaction of DAN with NO^+-derived N_2O_3 liberated from RSNO following mercuric chloride addition to yield a primary nitrosamine which is converted rapidly to a fluorescent triazole (Fig. 24.6). The colorimetric assay has a detection range of 0.5–100 μM, while the fluorometric assay is effective in the range of 50–1000 nM RSNO (Wink et al. 1999). The combination of the two assays provides a detection range from 50 nM to 100 μM, RSNO which generally is more than adequate for most biological experiments. Variations of these methods have been used to quantify high and low molecular weight RSNOs in human and rat plasma as well as the s-nitrosated derivatives of human and rat hemoglobin (Jourd'Heuil et al. 2000a,b).

Measurement of 3-Nitrotyrosine

The vast majority of studies implicating $ONOO^-$ as an important cytotoxic species have used immunohistochemical or HPLC detection of 3NT as proof of $ONOO^-$ (or ONOOH) formation *in vivo*. In reality, it is not a specific "footprint" for $ONOO^-$ formation *in vivo* because heme and a variety of heme-containing proteins may catalyze the H_2O_2-dependent formation of 3NT (Fig. 24.7) (Eiserich et al. 1996, 2002):

$$H_2O_2 + \text{P-Fe}^{+3} \rightarrow {}^{+\bullet}\text{P-Fe}^{+4} \text{ (porphyrin cation radical)} \quad (24.16)$$

$$^{+\bullet}\text{P-Fe}^{+4} + NO_2^- \rightarrow NO_2^\bullet \quad (24.17)$$

$$2\,NO_2^\bullet + \text{Tyrosine} \rightarrow \text{3-Nitrotyrosine} \quad (24.18)$$

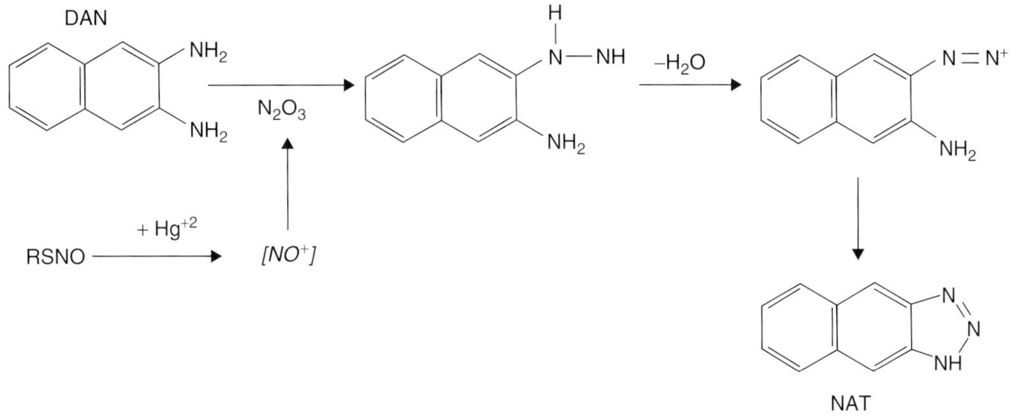

Fig. 24.6 Fluorometric detection of RSNO. Liberation of the nitrosonium (NO^+) -derived nitrosating agent (N_2O_3) from the interaction between RSNO and $HgCl_2$ in the presence of the diaminonaphthlene (DAN) results in the formation of the same fluorometric triazole derivative described in Fig. 24.3. (Reproduced with permission from Bryan and Grisham 2007.)

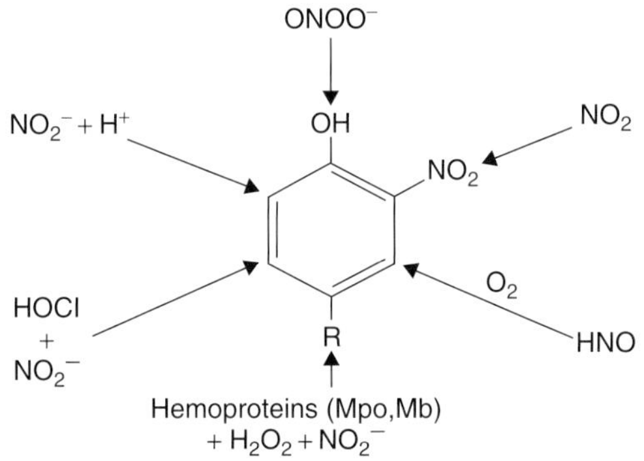

Fig. 24.7 Multiple pathways for the formation of 3-nitrotyrosine. 3-nitrotyrosine may be generated by peroxynitrite ($ONOO^-$), autoxidation of $NO^\bullet$, nitroxyl (HNO) in the presence of oxygen, heme or hemoprotein catalyzed, H_2O_2-dependent oxidation of nitrite (NO_2^-), NO_2^- interaction with HOCl, and acidified NO_2^-. (Reproduced with permission from Bryan and Grisham 2007.)

Myeloperoxidase (MPO) as well as other peroxidases (and heme/hemoproteins) will mediate this reaction. Co-localization of MPO (or eosinophil peroxidase) with 3NT during inflammation strongly indicates this reaction is the primary source of tyrosine nitration *in vivo*. Although some studies have indicated the possibility of peroxidase-independent sources of 3NT formation (Brennan et al. 2002), it should be remembered that virtually any hemoprotein possess the potential to catalyze the H_2O_2-dependent oxidation of nitrite to yield 3NT (Kilinc et al. 2001; Thomas et al. 2002). In addition to heme or hemoprotein-mediated NT formation, the MPO-derived oxidant hypochlorous acid (HOCl) will interact with NO_2^- to yield nitryl choride ($ClNO_2$), which will nitrate tyrosine to form 3NT (Eiserich et al. 1996). Thus, the presence of 3NT in inflammatory foci most probably represents the sum total of all nitration reactions mediated by $ONOO^-$, $NO_2^\bullet$, and $ClNO_2$.

Another example of $ONOO^-$ independent generation of 3NT is the reaction of tyrosine with acidified NO_2^-:

$$HNO_2 + H^+ \leftrightarrow H_2NO_2^+ + NO_2^- \leftrightarrow N_2O_3 + H_2O \quad (24.19)$$

$$N_2O_3 \leftrightarrow NO_2 + NO \quad (24.20)$$

$$2NO_2 + \text{Tyrsosine} \rightarrow 3NT + NO_2^- \quad (24.21)$$

When the pH approaches 4–5, NO_2^- will generate NT. Whether acidified nitrite can produce NT *in vivo* remains the subject of active investigation. This reaction may be important in acidic microenvironments such as the lipid bilayer or phagolysosomes where NO accumulation and oxidation to nitrite would readily occur. The gastric lumen would provide a potentially important environment where NO_2^- could easily participate in acid catalyzed nitration chemistry.

CONCLUSIONS

It is becoming increasingly appreciated that in addition to land-based organisms, both freshwater and marine life are exposed to large amounts of exogenous and endogenous sources of ROS and NO_x. Understanding how these microorganisms, plants and animals are affected by environmental and metabolic sources of these reactive species requires the use of analytical techniques that are capable of accurately measuring these species *in vitro* and *in vivo*. The objective of this chapter was to present a critical evaluation of several methods used to quantify the major ROS and NO_x found in biological systems, with special emphasis on methods that use instrumentation commonly found in most research laboratories.

REFERENCES

Al Housari, F., Vione, D., Chiron, S., Barbati, S. (2010) Reactive photoinduced species in estuarine waters. Characterization of hydroxyl radical, singlet oxygen and dissolved organic matter triplet state in natural oxidation processes. *Photochemistry and Photobiology Science* 9, 78–86.

Allen, B.W., Stamler, J.S., Piantadosi, C.A. (2009) Hemoglobin, nitric oxide and molecular mechanisms of hypoxic vasodilation. *Trends in Molecular Medicine* 15, 452–460.

Azzi, A., Montecucco, C., Richter, C. (1975) The use of acetylated ferricytochrome c for the detection of superoxide radicals produced in biological membranes. *Biochemical and Biophysical Research Communications* 65, 597–603.

Barbacanne, M.A., Souchard, J.P., Darblade, B. et al. (2000) Detection of superoxide anion released extracellularly by endothelial cells using cytochrome c reduction, ESR, fluorescence and lucigenin- enhanced chemiluminescence techniques. *Free Radical Biology and Medicine* 29, 388–396.

Beckman, J.S., Koppenol, W.H. (1996) Nitric oxide, superoxide, and peroxynitrite: the good, the bad, and ugly. *American Journal of Physiology* 271, C1424–C1437.

Benov, L., Sztejnberg, L., Fridovich, I. (1998) Critical evaluation of the use of hydroethidine as a measure of superoxide anion radical. *Free Radical Biology and Medicine* 25, 826–831.

Bindokas, V.P., Jordan, J., Lee, C.C., Miller, R.J. (1996) Superoxide production in rat hippocampal neurons: selective imaging with hydroethidine. *Journal of Neuroscience* 16, 1324–1336.

Bonini, M.G., Rota, C., Tomasi, A., Mason, R.P. (2006) The oxidation of 2′,7′-dichlorofluorescin to reactive oxygen species: a self-fulfilling prophesy? *Free Radical Biology and Medicine* 40, 968–975.

Boveris, A. (1984) Determination of the production of superoxide radicals and hydrogen peroxide in mitochondria. *Methods in Enzymology* 105, 429–435.

Boveris, A., Martino, E., Stoppani, A.O. (1977) Evaluation of the horseradish peroxidase-scopoletin method for the measurement of hydrogen peroxide formation in biological systems. *Annals of Biochemistry* 80, 145–158.

Brennan, M.L., Wu, W., Fu, X. et al. (2002) A tale of two controversies: defining both the role of peroxidases in nitrotyrosine formation *in vivo* using eosinophil peroxidase and myeloperoxidase-deficient mice, and the nature of peroxidase-generated reactive nitrogen species. *Journal of Biological Chemistry*, 277, 17415–17427.

Bryan, N.S., Grisham, M.B. (2007) Methods to detect nitric oxide and its metabolites in biological samples. *Free Radical Biology and Medicine* 43, 645–657.

Budd, S.L., Castilho, R.F., Nicholls, D.G. (1997) Mitochondrial membrane potential and hydroethidine-monitored superoxide generation in cultured cerebellar granule cells. *FEBS Letters* 415, 21–24.

Carter, W.O., Narayanan, P.K., Robinson, J.P. (1994) Intracellular hydrogen peroxide and superoxide anion detection in endothelial cells. *Journal of Leukocyte Biology* 55, 253–258.

Castro, L., Rodriguez, M., Radi, R. (1994) Aconitase is readily inactivated by peroxynitrite, but not by its precursor, nitric oxide. *Journal of Biological Chemistry* 269, 29409–29415.

Cathcart, R., Schwiers, E., Ames, B.N. (1983) Detection of picomole levels of hydroperoxides using a fluorescent dichlorofluorescein assay. *Analytical Biochemistry* 134, 111–116.

Cohen, G., Kim, M., Ogwu, V. (1996) A modified catalase assay suitable for a plate reader and for the analysis of brain cell cultures. *Journal of Neuroscience Methods* 67, 53–56.

Crow, J.P. (1997) Dichlorodihydrofluorescein and dihydrorhodamine 123 are sensitive indicators of peroxynitrite *in vitro*: implications for intracellular measurement of reactive nitrogen and oxygen species. *Nitric Oxide* 1, 145–157.

Duerrschmidt, N., Wippich, N., Goettsch, W., Broemme, H.J., Morawietz, H. (2000) Endothelin-1 induces NAD(P)H oxidase in human endothelial cells. *Biochemical and Biophysical Research Communications* 269, 713–717.

Eiserich, J.P., Cross, C.E., Jones, A.D., Halliwell, B., van der Vliet, A. (1996) Formation of nitrating and chlorinating species by reaction of nitrite with hypochlorous acid. A novel mechanism for nitric oxide-mediated protein modification. *Journal of Biological Chemistry*, 271, 19199–19208.

Eiserich, J.P., Baldus, S., Brennan, M.L. et al. (2002) Myeloperoxidase, a leukocyte-derived vascular NO oxidase. *Science* 296, 2391–2394.

Espey, M.G., Miranda, K.M., Thomas, D.D., Wink, D.A. (2001) Distinction between nitrosating mechanisms within human cells and aqueous solution. *Journal of Biological Chemistry* 276, 30085–30091.

Espey, M.G., Miranda, K.M., Thomas, D.D., Wink, D.A. (2002a) Ingress and reactive chemistry of nitroxyl-derived species within human cells. *Free Radical Biology and Medicine* 33, 827–834.

Espey, M.G., Thomas, D.D., Miranda, K.M., Wink, D.A. (2002b) Focusing of nitric oxide mediated nitrosation and oxidative nitrosylation as a consequence of reaction with superoxide. *Proceedings of the National Academy of Sciences USA* 99, 11127–11132.

Flint, D.H., Tuminello, J.F., Emptage, M.H. (1993) The inactivation of Fe-S cluster containing hydro-lyases by superoxide. *Journal of Biological Chemistry* 268, 22369–22376.

Foster, M.W., McMahon, T.J., Stamler, J.S. (2003) S-nitrosylation in health and disease. *Trends in Molecular Medicine* 9, 160–168.

Foster, M.W., Forrester, M.T., Stamler, J.S. (2009) A protein microarray-based analysis of S-nitrosylation. *Proceedings of the National Academy of Sciences USA* 106, 18948–18953.

Fridovich, I. (1985) Cytochrome *c*. *In* Greenwald, R.A. (ed.). *Handbook of Methods for Oxygen Radical Research*. Raton, FL, CRC Press, pp. 121–122.

Gardner, P.R., Fridovich, I. (1992) Inactivation-reactivation of aconitase in Escherichia coli. A sensitive measure of superoxide radical. *Journal of Biological Chemistry* 267, 8757–8763.

Gardner, P.R., Raineri, I., Epstein, L.B., White, C.W. (1995) Superoxide radical and iron modulate aconitase activity in mammalian cells. *Journal of Biological Chemistry* 270, 13399–13405.

Gardner, P.R., Nguyen, D.D., White, C.W. (1996) Superoxide scavenging by Mn(II/III) tetrakis (1-methyl–4-pyridyl) porphyrin in mammalian cells. *Archives of Biochemistry and Biophysics* 325, 20–28.

Goto, T., Inoue, S., Sugiura, S., Nishikawa, K., Isobe, M., Abe, Y. (1968) Cypridina bioluminescence V. structure of emitting species in the luminescence of cypridina luciferin and its related compounds. *Tetrahedron Letters* 37, 4035–4038.

Green, D.R., Reed, J.C. (1998) Mitochondria and apoptosis. *Science* 281, 1309–1312.

Grisham, M.B., Johnson, G.G., Lancaster, J.R., Jr. (1996) Quantitation of nitrate and nitrite in extracellular fluids. *Methods in Enzymology* 268, 237–246.

Guidet, B., Shah, S.V. (1989) Enhanced *in vivo* H2O2 generation by rat kidney in glycerol-induced renal failure. *American Journal of Physiology* 257, F440–F445.

Hausladen, A., Fridovich, I. (1994) Superoxide and peroxynitrite inactivate aconitases, but nitric oxide does not. *Journal of Biological Chemistry* 269, 29405–29408.

Hausladen, A., Fridovich, I. (1996) Measuring nitric oxide and superoxide: rate constants for aconitase reactivity. [Review] [19 refs]. *Methods in Enzymology* 269, 37–41.

Hink, U., Li, H., Mollnau, H. et al. (2001) Mechanisms underlying endothelial dysfunction in diabetes mellitus. *Circulation Research*, 88, E14–E22.

Hinkle, P.C., Butow, R.A., Racker, E., Chance, B. (1967) Partial resolution of the enzymes catalyzing oxidative phosphorylation. XV. Reverse electron transfer in the flavin-cytochrome beta region of the respiratory chain of beef heart submitochondrial particles. *Journal of Biological Chemistry* 242, 5169–5173.

Hyslop, P.A., Sklar, L.A. (1984) A quantitative fluorimetric assay for the determination of oxidant production by polymorphonuclear leukocytes: its use in the simultaneous fluorimetric assay of cellular activation processes. *Analytical Biochemistry* 141, 280–286.

Ignarro, L.J., Fukuto, J.M., Griscavage, J.M., Rogers, N.E., Byrns, R.E. (1993) Oxidation of nitric oxide in aqueous solution to nitrite but not nitrate: comparison with enzymatically formed nitric oxide from L-arginine. *Proceedings of the National Academy of Sciences USA* 90, 8103–8107.

Ishii, M., Shimizu, S., Yamamoto, T., Momose, K., Kuroiwa, Y. (1997) Acceleration of oxidative stress-induced endothelial cell death by nitric oxide synthase dysfunction accompanied with decrease in tetrahydrobiopterin content. *Life Sciences* 61, 739–747.

Janiszewski, M., Souza, H.P., Liu, X., Pedro, M.A., Zweier, J.L., Laurindo, F.R. (2002) Overestimation of NADH-driven vascular oxidase activity due to lucigenin artifacts. *Free Radical Biology and Medicine* 32, 446–453.

Jourd'Heuil, D., Gray, L., Grisham, M.B. (2000a) S-nitrosothiol formation in blood of lipopolysaccharide-treated rats. *Biochemical and Biophysical Research Communications* 273, 22–26.

Jourd'Heuil, D., Hallen, K., Feelisch, M., Grisham, M.B. (2000b) Dynamic state of S-nitrosothiols in human plasma and whole blood. *Free Radical Biology and Medicine* 28, 409–417.

Keston, A.S., Brandt, R. (1965) The fluorometric analysis of ultramicro quantities of hydrogen peroxide. *Annals of Biochemistry* 11, 1–5.

Kilinc, K., Kilinc, A., Wolf, R.E., Grisham, M.B. (2001) Myoglobin-catalyzed tyrosine nitration: no need for peroxynitrite. *Biochemical and Biophysical Research Communications* 285, 273–276.

Kinnula, V.L., Mirza, Z., Crapo, J.D., Whorton, A.R. (1993) Modulation of hydrogen peroxide release from vascular endothelial cells by oxygen. *American Journal of Respiratory Cell and Molecular Biology* 9, 603–609.

Kojima, H., Nakatsubo, N., Kikuchi, K. et al. (1998a) Detection and imaging of nitric oxide with novel fluorescent indicators: diaminofluoresceins. *Analytical Chemistry* 70, 2446–2453.

Kojima, H., Sakurai, K., Kikuchi, K. et al. (1998b) Development of a fluorescent indicator for nitric oxide based on the fluorescein chromophore. *Chemical and Pharmaceutical Bulletin* 46, 373–375.

Kooy, N.W., Royall, J.A., Ischiropoulos, H. (1997) Oxidation of 2′,7′-dichlorofluorescin by peroxynitrite. *Free Radical Research* 27, 245–254.

Kuthan, H., Ullrich, V., Estabrook, R.W. (1982) A quantitative test for superoxide radicals produced in biological systems. *Biochemical Journal* 203, 551–558.

Landmesser, U., Dikalov, S., Price, S.R. et al. (2003) Oxidation of tetrahydrobiopterin leads to uncoupling of endothelial cell nitric oxide synthase in hypertension. *Journal of Clinical Investigation* 111, 1201–1209.

Larsen, L.N., Dahl, E., Bremer, J. (1996) Peroxidative oxidation of leuco-dichlorofluorescein by prostaglandin H synthase in prostaglandin biosynthesis from polyunsaturated fatty acids. *Biochimica et Biophysica Acta* 1299, 47–53.

LeBel, C.P., Ischiropoulos, H., Bondy, S.C. (1992) Evaluation of the probe 2′,7′-dichlorofluorescin as an indicator of reactive oxygen species formation and oxidative stress. *Chemistry Research in Toxicology* 5, 227–231.

Lesser, M.P. (2006) Oxidative stress in marine environments: biochemistry and physiological ecology. *Annual Review of Physiology* 68, 253–278.

Liochev, S.I., Fridovich, I. (1997) Lucigenin (bis-N-methylacridinium) as a mediator of superoxide anion production. *Archives of Biochemistry and Biophysics* 337, 115–120.

Lucas, M., Solano, F. (1992) Coelenterazine is a superoxide anion-sensitive chemiluminescent probe: its usefulness in the assay of respiratory burst in neutrophils. *Analytical Biochemistry* 206, 273–277.

Margoliash, E., Novogrodsky, A., Schejter, A. (1960) Irreversible reaction of 3-amino 1, 2,4-triazole and related inhibitors with the protein catalase. *Biochemical Journal* 74, 339–348.

Markert, M., Andrews, P.C., Babior, B.M. (1984) Measurement of O_2^- production by human neutrophils. The preparation and assay of NADPH oxidase-containing particles from human neutrophils. *Methods in Enzymology* 105, 358–365.

Massey, V. (1959) The microestimation of succinate and the extinction coefficient of cytochrome *c*. *Biochimica et Biophysica Acta* 34, 255–256.

Melov, S., Coskun, P., Patel, M. et al. (1999) Mitochondrial disease in superoxide dismutase 2 mutant mice. *Proceedings of the National Academy of Sciences USA* 96, 846–851.

Miles, A.M., Wink, D.A., Cook, J.C., Grisham, M.B. (1996) Determination of nitric oxide using fluorescence spectroscopy. *Methods in Enzymology* 268, 105–120.

Miller, F.J., Gutterman, D.D., Rios, C.D., Heistad, D.D., Davidson, B.L. (1998) Superoxide production in vascular smooth muscle contributes to oxidative stress and impaired relaxation in atherosclerosis. *Circulation Research* 82, 1298–1305.

Munzel, T., Afanas'ev, I.B., Kleschyov, A.L., Harrison, D.G. (2002) Detection of superoxide in vascular tissue. *Arteriosclerosis, Thrombosis, and Vascular Biology* 22, 1761–1768.

Nakano, M. (1990a) Assay for superoxide dismutase based on chemiluminescence of luciferin analog. *Methods in Enzymology* 186, 227–232.

Nakano, M. (1990b) Determination of superoxide radical and singlet oxygen based on chemiluminescence of luciferin analogs. *Methods in Enzymology* 186, 585–591.

Nims, R.W., Cook, J.C., Krishna, M.C. et al. (1996) Colorimetric assays for nitric oxide and nitrogen oxide species formed from nitric oxide stock solutions and donor compounds. *Methods in Enzymology* 268, 93–105.

O'Brien, P.J. (1984) Superoxide production. *Methods in Enzymology* 105, 370–378.

Paler-Martinez, A., Panus, P.C., Chumley, P.H., Ryan, U., Hardy, M.M., Freeman, B.A. (1994) Endogenous xanthine oxidase does not significantly contribute to vascular endothelial production of reactive oxygen species. *Archives of Biochemistry and Biophysics* 311, 79–85.

Palumbo, A. (2005) Nitric oxide in marine invertebrates: a comparative perspective. *Comparative Biochemistry and Physiology A* 142, 241–248.

Piantadosi, C.A., Tatro, L.G. (1990) Regional H_2O_2 concentration in rat brain after hyperoxic convulsions. *Journal of Applied Physiology* 69, 1761–1766.

Pick, E., Keisari, Y. (1980) A simple colorimetric method for the measurement of hydrogen peroxide produced by cells in culture. *Journal of Immunological Methods* 38, 161–170.

Raha, S., McEachern, G.E., Myint, A.T., Robinson, B.H. (2000) Superoxides from mitochondrial complex III: the role of manganese superoxide dismutase. *Free Radical Biology and Medicine* 29, 170–180.

Rota, C., Chignell, C.F., Mason, R.P. (1999) Evidence for free radical formation during the oxidation of 2′–7′-dichlorofluorescin to the fluorescent dye 2′–7′-dichlorofluorescein by horseradish peroxidase: possible implications for oxidative stress measurements. *Free Radical Biology and Medicine* 27, 873–881.

Royall, J.A., Ischiropoulos, H. (1993) Evaluation of 2′,7′-dichlorofluorescin and dihydrorhodamine 123 as fluorescent probes for intracellular H_2O_2 in cultured endothelial cells. *Archives of Biochemistry and Biophysics* 302, 348–355.

Ruch, W., Cooper, P.H., Baggiolini, M. (1983) Assay of H2O2 production by macrophages and neutrophils with homovanillic acid and horse-radish peroxidase. *Journal of Immunological Methods* 63, 347–357.

Saitoh, D., Kadota, T., Okada, Y., Masuda, Y., Ohno, H., Inoue, M. (1995) Direct evidence for the occurrence of superoxide radicals in the small intestine of the burned rat. *American Journal of Emergency Medicine* 13, 37–40.

Saville, B. (1958) A scheme for the colorimetric determination of microgram amounts of thiols. *Analyst* 83, 670–672.

Skatchkov, M.P., Sperling, D., Hink, U., Anggard, E., Munzel, T. (1998) Quantification of superoxide radical formation in intact vascular tissue using a cypridina luciferin analog as an alternative to lucigenin. *Biochemical and Biophysical Research Communications* 248, 382–386.

Skatchkov, M.P., Sperling, D., Hink, U. et al. (1999) Validation of lucigenin as a chemiluminescent probe to monitor vascular superoxide as well as basal vascular nitric oxide production. *Biochemical and Biophysical Research Communications* 254, 319–324.

Staniek, K., Nohl, H. (1999) H(2)O(2) detection from intact mitochondria as a measure for one-electron reduction of dioxygen requires a non-invasive assay system. *Biochimica et Biophysica Acta* 1413, 70–80.

Suzuki, H., Swei, A., Zweifach, B.W., Schmid-Schonbein, G.W. (1995) *In Vivo* evidence for microvascular oxidative stress in spontaneously hypertensive rats. Hydroethidine microfluorography. *Hypertension* 25, 1083–1089.

Tarpey, M.M., Fridovich, I. (2001) Methods of detection of vascular reactive species: nitric oxide, superoxide, hydrogen peroxide, and peroxynitrite. *Circulation Research* 89, 224–236.

Tarpey, M.M., White, C.R., Suarez, E., Richardson, G., Radi, R., Freeman, B.A. (1999) Chemiluminescent detection of oxidants in vascular tissue: lucigenin but not coelenterazine enhances superoxide formation. *Circulation Research* 84, 1203–1211.

Tarpey, M.M., Wink, D.A., Grisham, M.B. (2004) Methods for detection of reactive metabolites of oxygen and nitrogen: *in vitro* and *in vivo* considerations. *American Journal of Physiology* 286, R431–R444.

Teranishi, K., Shimomura, O. (1997) Coelenterazine analogs as chemiluminescent probe for superoxide anion. *Analytical Biochemistry* 249, 37–43.

Thomas, D.D., Espey, M.G., Vitek, M.P., Miranda, K.M., Wink, D.A. (2002) Protein nitration is mediated by heme and free metals through Fenton-type chemistry: an alternative to the NO/O_2^- reaction. *Proceedings of the National Academy of Sciences USA* 99, 12691–12696.

Thomson, L., Trujillo, M., Telleri, R., Radi, R. (1995) Kinetics of cytochrome *c*2+ oxidation by peroxynitrite: implications for superoxide measurements in nitric oxide-producing biological systems. *Archives of Biochemistry and Biophysics* 319, 491–497.

Uehara, K., Maruyama, N., Huang, C.K., Nakano, M. (1993) The first application of a chemiluminescence probe, 2-methyl-6-[p-methoxyphenyl]-3,7-dihydroimidazo[1,2-a]pyrazin-3-one (MCLA), for detecting O_2^- production, *in vitro*, from Kupffer cells stimulated by phorbol myristate acetate. *FEBS Letters* 335, 167–170.

Ushiroda, S., Maruyama, Y., Nakano, M. (1997) Continuous detection of superoxide *in situ* during ischemia and reperfusion in the rabbit heart. *Japanese Heart Journal* 38, 91–105.

Vasquez-Vivar, J., Hogg, N., Pritchard, K.A., Jr., Martasek, P., Kalyanaraman, B. (1997) Superoxide anion formation from lucigenin: an electron spin resonance spin-trapping study. *FEBS Letters* 403, 127–130.

Wardman, P. (2007) Fluorescent and luminescent probes for measurement of oxidative and nitrosative species in cells and tissues: progress, pitfalls, and prospects. *Free Radical Biology and Medicine* 43, 995–1022.

Wink, D.A., Kim, S., Coffin, D. et al. (1999) Detection of S-nitrosothiols by fluorometric and colorimetric methods. *Methods in Enzymology* 301, 201–211.

Yusa, T., Beckman, J.S., Crapo, J.D., Freeman, B.A. (1987) Hyperoxia increases H2O2 production by brain *in vivo*. *Journal of Applied Physiology* 63, 353–358.

Zhao, H., Kalivendi, S.V., Joseph, J., Zhang, H., Kalyanaraman, B. (2002) Superoxide reacts with hydroethidine but forms a fluorescent product that is distinctly different from ethidium. *Free Radical Biology and Medicine* 33, S425.

Zhou, M., Diwu, Z., Panchuk-Voloshina, N., Haugland, R.P. (1997) A stable nonfluorescent derivative of resorufin for the fluorometric determination of trace hydrogen peroxide: applications in detecting the activity of phagocyte NADPH oxidase and other oxidases. *Analytical Biochemistry* 253, 162–168.

Zielonka, J., Srinivasan, S., Hardy, M., Ouari, O., Lopez, M., Vasquez-Vivar, J., Avadhani, N.G., Kalyanaraman, B. (2008) Cytochrome c-mediated oxidation of hydroethidine and mito-hydroethidine in mitochondria: identification of homo- and heterodimers. *Free Radical Biology and Medicine* 44, 835–846.

Zielonka, J., Hardy, M., Kalyanaraman, B. (2009) HPLC study of oxidation products of hydroethidine in chemical and biological systems: ramifications in superoxide measurements. *Free Radical Biology and Medicine* 46, 329–338.

Chapter 25

ROLE OF SINGLET MOLECULAR OXYGEN IN THE OXIDATIVE DAMAGE TO BIOMOLECULES

Graziella Eliza Ronsein[1], *Glaucia Regina Martinez*[2], *Eduardo Alves de Almeida*[3], *Sayuri Miyamoto*[1], *Marisa Helena Gennari de Medeiros*[1], *and Paolo Di Mascio*[1]

[1]Departamento de Bioquímica, Instituto de Química, Universidade de São Paulo, São Paulo, SP, Brazil

[2]Biologia Molecular, Setor de Ciências Biológicas, Universidade Federal do Paraná, Curitiba, PR, Brazil

[3]Departament of Chemistry and Environmental Sciences, IBILCE-UNESP, São José do Rio Preto, SP, Brazil

In its ground state, molecular oxygen is a poorly reactive triplet state molecule ($^3O_2$), which carries two electrons of parallel spin in its binding orbital (Frimer 1985). In contrast, singlet molecular oxygen ($^1O_2$) is regarded as a ROS formed by an input of energy (Foote et al. 1968; Frimer 1985) (See Box 25.1). In biological systems it can be produced by the action of peroxidases (Kanofsky 1983, 1984), UV-A light in the presence of a sensitizer (Kielbassa et al. 1997; Baier et al. 2006; Box 25.1), and also during the dimerization reactions of peroxyl radicals in the process of lipid peroxidation (Miyamoto et al. 2003a,b, 2006). Electron-rich biomolecules, such as lipids, proteins, and DNA, can be important targets for $^1O_2$ oxidation. Reactions of $^1O_2$ with biomolecules will be reviewed in this chapter.

Generation of $^1O_2$ through photosensitization plays a role in the toxic reactions of dissolved environmental pollutants in aquatic ecosystems. Environmental pollutants may act as photosensitizers (see Box 25.1) by absorbing energy from sunlight and transferring this energy to molecular oxygen, generating $^1O_2$.

Oxidative Stress in Aquatic Ecosystems, First Edition. Edited by Doris Abele, José Pablo Vázquez-Medina, and Tania Zenteno-Savín.

Box 25.1 Jablonski diagram

Simplified Jablonski diagram illustrating energy transfer from a photosensitizer to molecular oxygen, generating $^1O_2$. The ground state of the photosensitizer (and of most organic molecules) is a singlet state (S_0), which has all electrons spin-paired. The absorption of one photon leads to the change of an electron of the photosensitizer to the excited singlet state (S_1). When electron de-excitation occurs, the excess energy is released as thermal or radiation energy. The emitted radiation is called fluorescence. Excited photosensitizers can also undergo rapid intersystem crossing (ISC) from S_1 to an excited triplet state (T_1), by inverting the spin of the excited electron. Once formed, the T_1 state may emit phosphorescence returning to the S_0 state. Alternatively, the energy can be transferred to an oxygen molecule converting it from the triplet ground state (T_0) to the excited singlet state ($^1O_2$). Ab, absorption; Fl, fluorescence; Ph, phosphorescence; ET, energy transfer.

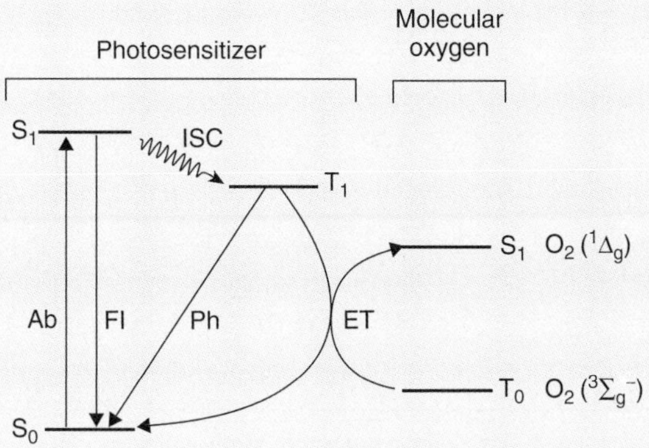

As an example, the decline in frog populations and the concomitant increase in malformations in these animals were suggested to result from the photosensitized generation of $^1O_2$, since $^1O_2$ phosphorescent light emission from natural water has been observed (Bilski et al. 2003). $^1O_2$ toxicity is also technically used to control parasite densities in aquaculture. Photochemical activation represents a well controllable and environmentally less obtrusive alternative to toxic pesticides (see Chapter 21). A recent work by Robertson et al. (2009) showed that methylene blue and a mixture of methylene blue and nuclear fast red reagents (well known photosensitizers) are an effective treatment against a marine copepod, a model of parasitic sea lice. The authors pointed out that significant research has to be done to develop a marketable product, including: efficacy, fish safety, and environmental safety. However, growing resistance to the few existing products for sea lice treatment may encourage development. Singlet molecular oxygen can also be used for photodegradation of organic contaminants in aquatic environments. Nevertheless, questions concerning the mechanisms of reaction during photodegradation, as well as the efficiency of the process warrant further investigation.

OXIDATIVE DAMAGE GENERATED BY SINGLET MOLECULAR OXYGEN

Among the different DNA constituents, guanine is the only normal nucleic acid component that significantly reacts with $^1O_2$ at neutral pH (Steenken and Jovanovic 1997). The reaction of $^1O_2$ and guanine in aqueous solution generates two diastereoisomers of spiroiminodihydantoin nucleosides as the main products and also minor amounts of 8-oxo-7,8-dihydro-2′-deoxyguanosine (8-oxodGuo) (Buchko et al. 1992; Ronsein et al. 2007; Fig. 25.1). 8-OxodGuo is the unique product detected in double stranded DNA oxidation by $^1O_2$ (Ravanat et al. 2001). 8-oxodGuo is, however, an ubiquitous marker of DNA damage by oxidizing agents (Cadet et al. 1997). UV-A causes DNA damage acting indirectly via photosensitization reactions that require the presence of chromophores

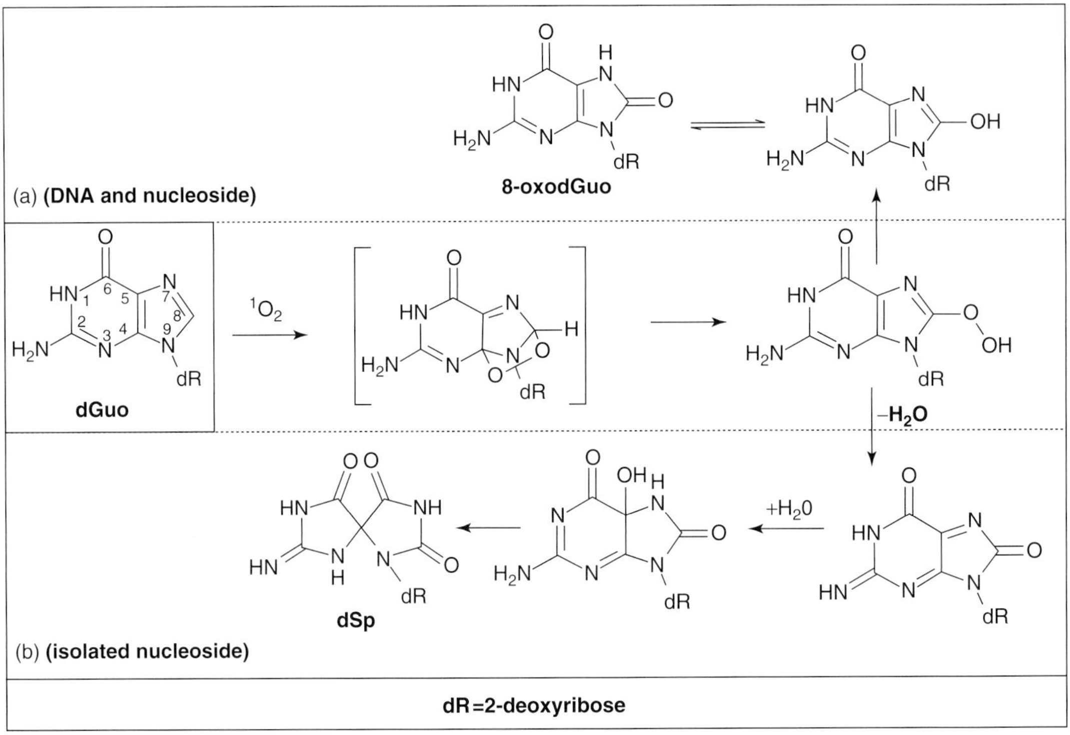

Fig. 25.1 Main oxidation products generated by the reaction of $^1O_2$ with DNA: (a) dGuo oxidation in DNA and nucleoside; (b) dGuo oxidation in isolated nucleoside. (Adapted from Ronsein et al. 2007.)

(Cadet et al. 2009). Thus, UV-A-induced oxidative DNA damage is mainly caused by $^1O_2$ produced through photosensitization and to a lesser extent through direct formation of hydroxyl radicals (HO•). As a result, major amounts of 8-oxodGuo were formed along with smaller amounts of oxidized pyrimidine bases and DNA strand breaks ascribed to the oxidizing action of HO• (Pflaum et al. 1994; Kielbassa et al. 1997).

Unsaturated lipids and cholesterol (Ch) in cell membranes are prominent targets of $^1O_2$. Lipid hydroperoxides (LOOHs) are the primary products of $^1O_2$ attack, with double bonds shifted to the allylic position (Terao and Matsushita 1977; Frankel et al. 1979). LOOH causes changes in the structural organization and packing of membrane lipids, leading to alteration in membrane fluidity (Girotti 1998). Free radical mediated oxidation of unsaturated lipids yields conjugated diene hydroperoxides. In contrast, the reaction of $^1O_2$ with unsaturated lipids yields not only the conjugated diene hydroperoxides, but also non-conjugated hydroperoxides (Terao and Matsushita 1977; Frankel 1984). Free radical and $^1O_2$-mediated oxidation of lipids are illustrated in Fig. 25.2A, using linoleic acid as the example. Thus, the identification of the nonconjugated hydroperoxides is suggested to serve as signature products (fingerprint) of $^1O_2$-mediated oxidation. $^1O_2$ was recently found to be the major ROS responsible for photo-oxidative damage in plants. The authors established a method that allows discrimination between free radicals and $^1O_2$-mediated oxidation of unsaturated lipids, by using nonconjugated hydroperoxides as specific reporters of $^1O_2$-dependent lipid oxidation. They were able to show that, whereas in *Arabidopsis* roots nonenzymatic lipid oxidation depends on free radicals, in leaves it is almost exclusively mediated by $^1O_2$ (Triantaphylides et al. 2008).

Some cholesterol hydroperoxides (ChOOHs) are especially well suited molecules to be used as signature products of $^1O_2$ (Girotti and Korytowski 2000).

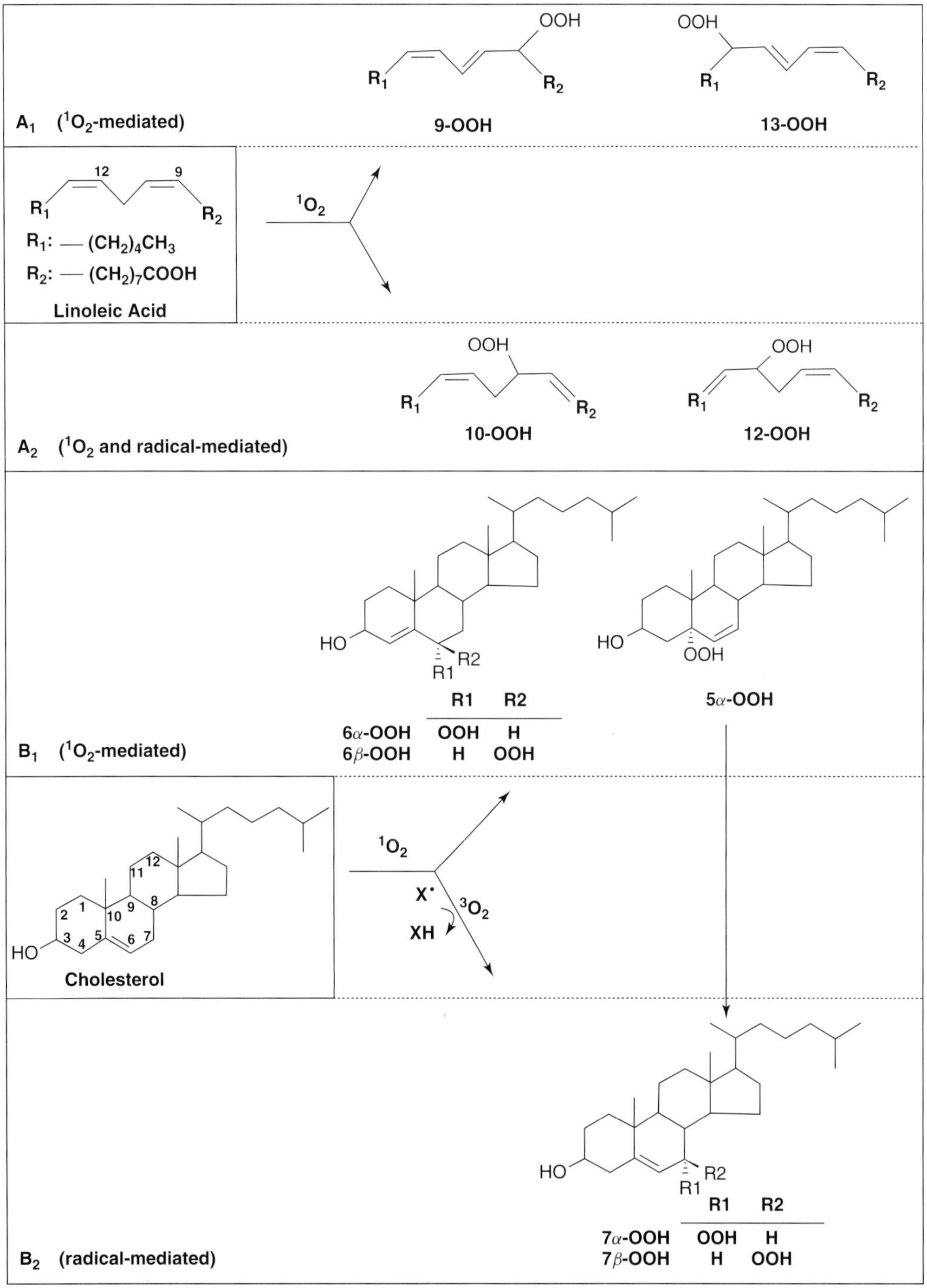

Fig. 25.2 Main oxidation products generated in the reaction of $^1O_2$ with lipids. (A) Linoleic acid oxidation: (A_1) $^1O_2$-mediated oxidation; (A_2) $^1O_2$ and radical-mediated oxidations. (B) Cholesterol oxidation: (B_1) $^1O_2$-mediated oxidation; (B_2) radical-mediated oxidation. $X^\bullet$ = a radical. (Adapted from Ronsein et al 2007).

In $^1O_2$-mediated reactions, only three primary ChOOH are produced: 3β-hydroxy-5α-cholest-6-ene-5-hydroperoxide (5α-OOH), 3β-hydroxycholest-4-ene-6α-hydroperoxide (6α-OOH), and 3β-hydroxycholest-4-ene-6β-hydroperoxide (6β-OOH; Fig. 25.2B). Although $^1O_2$ does not produce 3β-hydroxycholest-5-ene-7α-hydroperoxide (7α-OOH) or 3β-hydroxycholest-5-ene-7β-hydroperoxide (7β-OOH), this pair can arise via allylic rearrangement of 5α-OOH, especially in low polarity solvents (Beckwith et al. 1989). This contrasts with free-radical-mediated reactions, which give rise to the epimeric pair 7α-OOH and 7β-OOH (Fig. 25.2B). Thus, identification of 5α-OOH, 6α-OOH or 6β-OOH unambiguously indicates $^1O_2$ involvement. Given the importance of a correct differentiation among the isomers, a new method for detection and characterization of Ch-oxidized products was recently developed. This method discriminates isomers not only by retention times, but also by differences in fragmentation pattern (Ronsein et al. 2010). However, special care should be taken when using signature products of a specific oxidant. A few years ago, Wentworth et al. (2003) reported the generation of O_3 in biological systems based on the oxidation of cholesterol yielding 3-hydroxy-5-oxo-5,6-secocholestan-6-al (CSec) and 3-hydroxy-5-hydroxy-β-norcholestane-6-carboxaldehyde (ChAld). However, as recently shown, this product can also be generated by $^1O_2$, either through catalyzed decomposition of 5α-OOH, the product of $^1O_2$ oxidation of cholesterol, or through decomposition of a 1,2-dioxetane formed by $^1O_2$ at the Δ^5 bond of cholesterol (Brinkhorst et al. 2008; Uemi et al. 2009).

At neutral pH, the chemical reaction of $^1O_2$ with proteins is limited to tryptophan (Trp), tyrosine (Tyr), histidine (His), methionine (Met), and cysteine (Cys) residues (Wilkinson et al. 1995). The products depend on which amino acid residue is oxidized. Hydroperoxides and/or endoperoxides were found as primary products in the case of Trp, Tyr, and His oxidation (Fig. 25.3; Ronsein et al. 2007). For example, the main Trp-oxidation products are two isomeric hydroperoxides. Further, these hydroperoxides generate radicals in the presence of metal ions, or decompose to the corresponding alcohols or *N*-formylkynurenine (*N*-FMK) (Ronsein et al. 2008, 2009; Fig. 25.3a). The reaction of $^1O_2$ with Tyr residues leads to the formation of long-lived hydroperoxides, which decay slowly at room temperature to the corresponding alcohols (Wright et al. 2002; Fig. 25.3b). Histidine (His) oxidation is proposed to occur via the formation of a His endoperoxide, followed by rearrangement to His hydroperoxide. The decomposition of this hydroperoxide intermediate to a carbonyl intermediate is followed by addition of a water molecule, yielding hydroxylated imidazolone. Aspartic acid, asparagines, and urea were detected as final products of His photooxidation (Tomita et al. 1969; Agon et al. 2006; Fig. 25.3c). The reaction of $^1O_2$ with sulphur containing amino acids Met and Cys still needs further clarification. In the case of Met, the product distribution is pH-dependent and involves a persulfoxide as a common intermediate at all pH values (Ando and Takata 1985; Straight and Spikes 1985). Cysteine is predominantly photo-oxidized to the corresponding disulfide (Ando and Takata 1985).

STRATEGIES FOR SINGLET MOLECULAR OXYGEN GENERATION

In order to confirm if a biological effect results from $^1O_2$ oxidation it is important to examine whether other processes known to generate $^1O_2$ can mimic the effect.

Singlet molecular oxygen can be generated by either physical or chemical processes. The former involves the use of a photosensitizer, a compound that absorbs the energy of a UV or visible light photon, to become an excited singlet state (Box 25.1 and Fig. 25.4). This state is rapidly converted to an excited triplet state, which can transfer energy directly to dissolved molecular oxygen generating $^1O_2$. This process is called the type II photosensitization. Common photosensitizers used in biological systems are methylene blue and Rose Bengal (Foote and Denny 1968; Foote 1991; Kochevar and Redmond 2000). The problem with this form of $^1O_2$ generation is the so-called type I process, which can lead to the formation of charged radicals derived from the photosensitizer (from the target molecules), but it can also result in the formation of $O_2^{\bullet-}$ (Foote 1991; Cadet et al. 2009; Fig. 25.4). Chemical sources of $^1O_2$ include the decomposition of dioxetanes (Briviba et al. 1996) and endoperoxides (Aubry 1991; Di Mascio and Sies 1989), and the reactions of H_2O_2 with either hypochlorite (Held et al. 1978) or peroxynitrite (Di Mascio et al. 1994). Recently, the generation of $^1O_2$ was also shown to occur during the decomposition of peroxynitrite, via formation of labile peroxynitrate (Gupta et al. 2009; Miyamoto et al. 2009; Fig. 25.4).

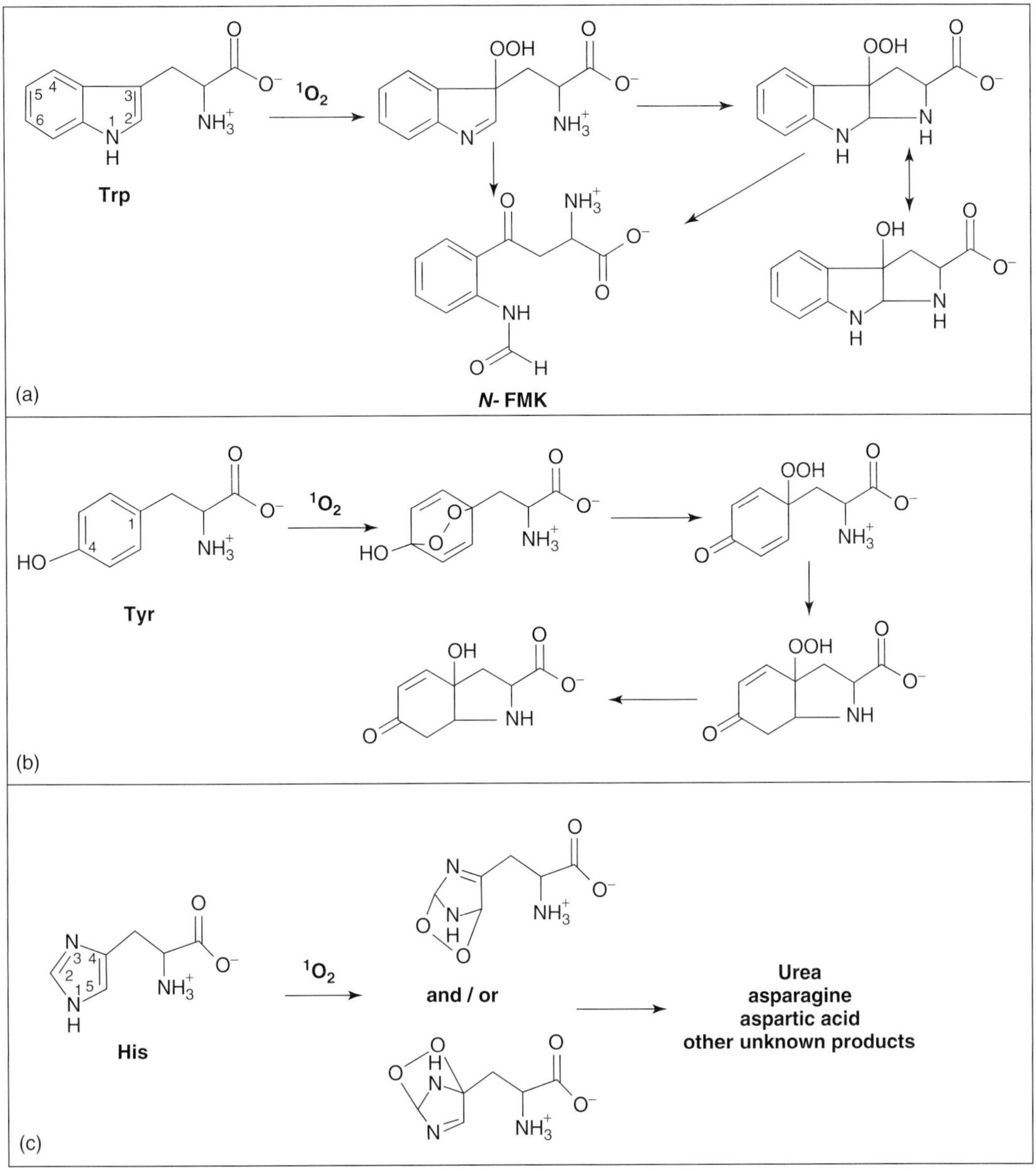

Fig. 25.3 Main oxidation products generated by the reaction of $^1O_2$ with amino acids: (a) Trp oxidation; (b) Tyr oxidation; (c) Hys oxidation. (Adapted from Ronsein et al. 2007.)

Considering the complexity of biological systems and the great variety of ROS generated in photochemistry, it is difficult to unambiguously assess the specific role of $^1O_2$ in the resulting biological effects. Additionally, most chemical reactions involve reactive molecules, such as H_2O_2 and hypochlorite, which can react with biological targets. Furthermore, conditions for the majority of the chemical reactions are incompatible with biological systems (Pierlot et al. 2000). Therefore, efforts have been made to develop

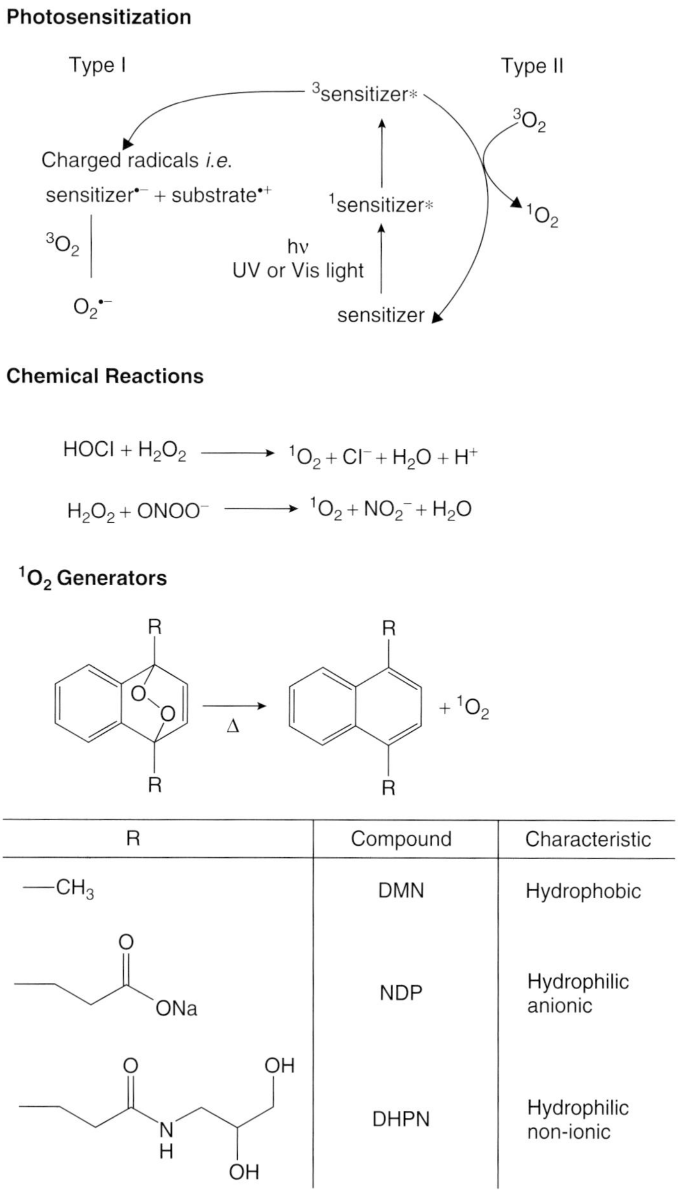

Fig. 25.4 Sources of singlet molecular oxygen.

suitable $^1O_2$ generators based on the thermolysis of endoperoxides. These compounds are chemically inert and capable of reversibly binding oxygen. For example, the endoperoxide of 1,4-dimethylnaphthalene (DMN) releases $^1O_2$ at 37°C with a high yield. This specific endoperoxide is a good $^1O_2$ generator in lipophilic media. Water-soluble and nontoxic derivatives of DMN have been used as carriers of $^1O_2$ for aqueous environments, since they trap $^1O_2$ at low temperatures (0–5°C), can be stored for months at −80°C, and release a defined amount of $^1O_2$ on warming to 37°C. Examples of these hydrophilic generators are the anionic disodium 1,4-naphthalenedipropanoate (NDP) and non-ionic *N*, *N*′-di(2,3-dihydroxypropyl)-1,

4-naphthalenedipropanamide (DHPN) (Pierlot et al. 1996, 2000; Dewilde et al. 1998; Fig. 25.4).

Singlet molecular oxygen labeled with $[^{18}O]$ was generated using oxygen-labeled endoperoxide of DHPN ($DHPN^{18}O_2$). This approach was used to assess the reactivity of $^1O_2$ towards cellular DNA, distinguishing this injury from the damage caused by other oxidants or by secondary reactive species, generated in the reaction of $^1O_2$ with biomolecules. In this study, the detection of ^{18}O-labeled 8-oxodGuo confirmed that intracellular formation of $^1O_2$ is able to induce direct DNA damage (Ravanat et al. 2000).

PROCEDURES FOR SINGLET MOLECULAR OXYGEN DETECTION

Methods Based on the use of Quenchers and Lifetime Enhancers

The lifetime of $^1O_2$ can be 10-fold increased when the species is generated in deuterium oxide (D_2O) instead of H_2O (Rodgers and Snowden 1982). This change is expected to stimulate $O_2(^1\Delta_g)$-mediated reactions with an increase in product yields. It should be pointed out that the biological enhancement factors (around two) are much lower than would be predicted from the *in vitro* lifetime experiments, because components of the cellular milieu will quench $O_2(^1\Delta_g)$ (Tyrrell 2000).

Carotenoids and tocopherols are efficient $^1O_2$ quenchers, and the carotenoid lycopene is the most effective of all. The quenching abilities of carotenoids and tocopherols are mainly due to physical quenching (Di Mascio et al. 1989, 1990). Nevertheless, these molecules have not been widely used since they are extremely lipophilic and require appropriate solubilization in water. However, they are well suited molecules to test $^1O_2$ effects in lipophilic environments, i.e. lipoproteins and membranes. For example, tocopherols protect against photoinhibition and photo-oxidative stress in *Arabidopsis thaliana*. The protective mechanism is related to the capacity of tocopherols to quench $^1O_2$ in its production sites within the chloroplast (Havaux et al. 2005).

Sodium azide is another $^1O_2$ quencher (Li et al. 2001) and has been used to verify the participation of this species in biological processes, although it is also known to have many side effects, such as inhibition of a number of enzymes (e.g. cytochrome *c* oxidase, adenosin triphosphate (ATP) synthase). Unfortunately, all these methods are indirect and of low specificity (Girotti 2001) and must be used with caution, preferably coupled with more specific methodologies.

Chemical Trapping of Singlet Molecular Oxygen

Chemical trapping is the most widely used and sensitive method to detect and quantify $^1O_2$ (Nardello and Aubry 2000). This technique is based on the formation of an endoperoxide, which is a specific product of the reaction of $^1O_2$ with a chemical trap (Nardello and Aubry 2000). The endoperoxide can be detected and identified using appropriate techniques, such as HPLC coupled to mass spectrometry in tandem. The most suitable chemical traps are anthracene derivatives, since the endoperoxides formed are stable at room temperature and their decomposition occurs only at elevated temperatures (100°C). In recent years, many efforts have been made to obtain chemical traps compatible with different environments. An important illustration of the usefulness of anthracene derivatives was the demonstration that $^1O_2$ is generated from lipid hydroperoxides, treated with metals or peroxynitrite, by using the hydrophobic chemical trap 9,10-diphenylanthracene (DPA) and ^{18}O-labeled lipid hydroperoxides. The detection of the resulting endoperoxides was performed by HPLC-MS analysis. Formation of ^{18}O-labeled endoperoxide of DPA confirmed the reaction between two lipid peroxyl radicals by the Russell mechanism (Miyamoto et al. 2003a,b).

The biological environment is not restricted to the hydrophobic moiety of membranes and lipoproteins. Since DPA is hydrophobic, it is not appropriate for aqueous systems. Contrary to DPA, anthracene-9,10-diyldiethyl disulfate (EAS) is an interesting chemical trap for aqueous systems. EAS reacts with $^1O_2$ generating the corresponding endoperoxide; its solubility is not pH dependent. Moreover, it is stable up to 120°C and it can be detected in small amounts by HPLC (Di Mascio and Sies 1989). Recently, EAS was employed to show the generation of $^1O_2$ in the reaction of hypochlorous acid and lipid hydroperoxides (Miyamoto et al. 2006). A compound similar to EAS is the anthracene-9,10-bisethanesulphonic acid (AES), which is soluble in water and buffer solutions and stable over a wide range of pH (Botsivali and Evans 1979; Nardello et al. 2005). Interestingly, it was shown that anthracene-9,10-divinylsulfonate (AVS), an intermediate in AES synthesis, is also a suitable $^1O_2$ chemical trap. AVS was recently employed to demonstrate that thymine hydroperoxides can be a potential source of $^1O_2$ in DNA. However, EAS, AES, and AVS are ionic compounds. The synthesis of a hydrophilic and nonionic anthracene derivative, the *N*,*N'*-di-(2,3-dihydroxypropyl)-9,10-anthracenedipropanamide (DHPA), might be of fundamental importance for biological investigations (Martinez et al. 2006). Figure 25.5 presents structures of some $^1O_2$ chemical traps.

Furthermore, the anthracene moiety has been associated with the fluorescein derivative (9-[2-(3-carboxy-9,10-diphenyl)antryl]-6-hydroxy-3H-xanten-3-one). This compound has a weak fluorescence, and the formation of anthracene endoperoxide through reaction with $^1O_2$ promotes a structural change that renders the product highly fluorescent (Tanaka et al. 2001).

The drawback of anthracene derivatives is that they absorb light and may act as photosensitizers, generating $^1O_2$. Naphthalene derivatives do not feature this problem, however, these compounds react slowly, and products are unstable at room temperature. For example, disodium 1,3-cyclohexadiene-1,4-dietanoate does not absorb visible light, is water soluble, and highly reactive towards $^1O_2$ (kr = 2.6×10^7 $M^{-1}s^{-1}$). Also, its product is a stable and specific endoperoxide (Nardello and Aubry 2000).

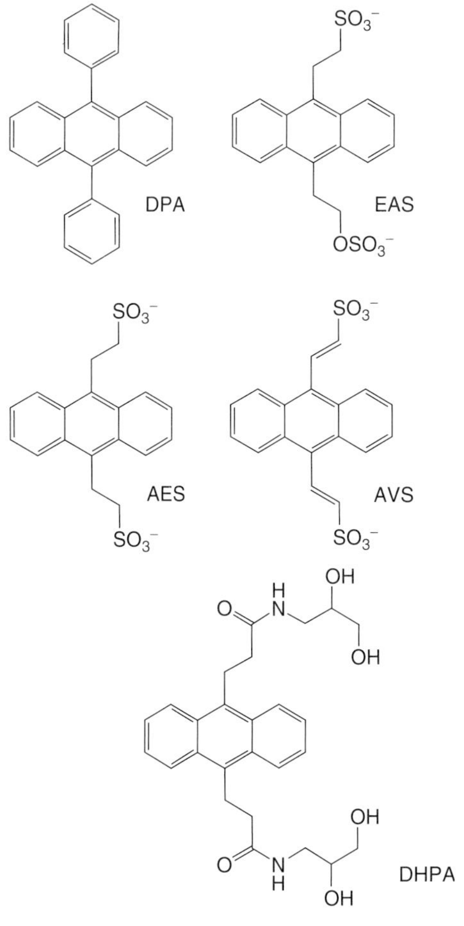

Fig. 25.5 Chemical traps of singlet molecular oxygen.

Description of a Method for Detection and Quantification of Singlet Molecular Oxygen

The following method describes the detection and quantification of AVS endoperoxide ($AVSO_2$). This chemical trap can be added to the reaction system at a final concentration of 8 mM. This is the β-value obtained for the EAS molecule (Di Mascio and Sies 1989). The β-value is the concentration needed to trap 50% of the $^1O_2$ generated in the system (Monroe 1985). Unfortunately, the β-value for AVS has not yet been determined, but considering the structural similarities between AVS and EAS, we used AVS at the same final concentration recommended for EAS. The incubation of the investigated system and AVS can be performed in 50 mM phosphate buffer at pH 7.4 prepared in D_2O. Deuterium oxide should be used as the solvent because it increases $^1O_2$ lifetime (Rodgers and Snowden 1982). The incubation can be performed for 5 min, with stirring and in the dark. Light should be excluded, since exposure of anthracene derivatives to light can generate the corresponding endoperoxides. After the incubation period, the product can be injected into a HPLC or, even better, into a HPLC/MS/MS system. A good chromatographic separation can be achieved using a Phenomenex Gemini C-18 column (250 mm × 4.6 ID mm, particle size 5 μm) and a gradient of methanol/acetonitrile (7:3 v/v) (solvent B) and 25 mM ammonium formate (solvent A) at a flow rate of 1 mL min^{-1}. In the case of mass spectrometry analyses, the selected reaction monitoring mode (SRM) in the negative mode is especially suited to monitor the specific mass transition of $AVSO_2$ (m/z 210 to 194). Standard $AVSO_2$ can be obtained by the exposure of AVS to light in the presence or even in the

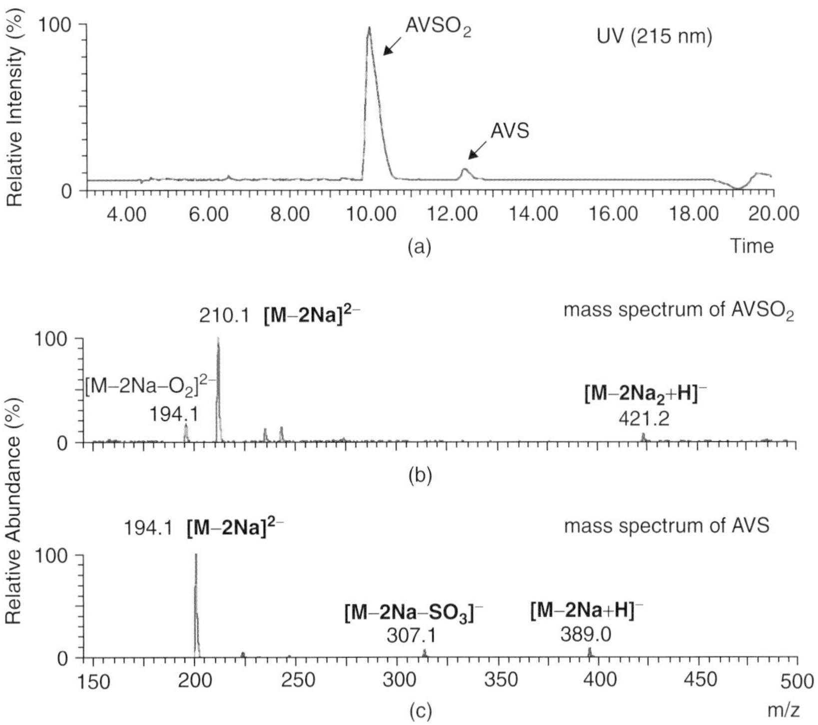

Fig. 25.6 Detection of $AVSO_2$ and AVS by HPLC coupled to UV and MS detections. (a) Detection of $AVSO_2$ and AVS by HPLC/UV (215 nm). (b) Mass spectrum of $AVSO_2$. (c) Mass spectrum of AVS.

absence of a sensitizer. The method described above was adapted from Prado et al. (2009). Figure 25.6 shows an example of $AVSO_2$ analysis by HPLC coupled to UV (215 nm) and MS detections. The $AVSO_2$ was generated by incubation of 8 mM AVS with 30 mM $DMNO_2$ (a clean source of $^1O_2$, as discussed above). However, the choice of a chemical trap will depend on the characteristics of the system that is under investigation (i.e. aqueous or hydrophobic environment).

LIGHT EMISSION FROM SINGLET MOLECULAR OXYGEN

Singlet molecular oxygen is an excited species that decays to ground state emitting light. Relaxation of $^1O_2$ to the ground state produces emission at 1270 nm in a process called monomol light emission (equation 25.1; Browne and Ogryzlo 1964). Additionally, two molecules of $^1O_2$ can collide, resulting in the simultaneous relaxation of both molecules with light emission at 634 and 703 nm. This process is called dimol light emission (equation 25.2; Khan and Kasha 1963). Taking advantage of $^1O_2$ light emissions, common approaches used for its detection are measurements at the near infrared (1270 nm) and visible (634 and 703 nm) regions.

$$O_2(^1\Delta_g) \rightarrow O_2(^3\Sigma_g^-) + \lambda\nu \quad (\lambda = 1270) \qquad (25.1)$$

$$O_2(^1\Delta_g) + O_2(^1\Delta_g) \rightarrow 2O_2(^3\Sigma_g^-) + \lambda\nu \quad (\lambda = 634 \text{ and } 703 \text{ mm}) \qquad (25.2)$$

Dimol Light Emission Measurement

Light emission by $^1O_2$ in the visible region can be measured with a photon-counter apparatus, following a previously described method (Cadenas and Sies 1984; Miyamoto et al. 2003a; Murphy and Sies 1990). The photon counting device consists of a red-sensitive photomultiplier tube (9203BM Thor EMI Electron Tubes, UK) cooled to −20°C by a thermoelectric cooler (FACT 50 MKIII, EMI Gencom, Plainview, NY). The potential applied to the photomultiplier can vary between −900 and −1200 V. The phototube output is connected to an amplifier discriminator (model 1121, Princeton Instruments, NJ), which transmits the signal to a computer. Selective light emission at wavelengths

$\lambda > 570$ nm can be obtained with a long pass cut-off filter (03IFS006, Melles Griot visible filters) placed between the cuvette and the photomultiplier tube. It is extremely important to use this cut-off filter in order to discriminate $^1O_2$ dimol emission from excited carbonyls, which emit light between 380 and 460 nm. Sample solutions should be poured into a thermostated glass cuvette (35 mm × 7 mm × 55 mm) with reflexion coated walls, and the reactant is injected through a small polyethylene tube under continuous stirring.

Monomol Light Emission Measurement

Infrared light emission of $^1O_2$ at 1270 nm can be monitored with a liquid nitrogen cooled photodiode germanium detector (Ge-Diode, model EI-L, Edinburgh Instruments Ltd, Livingston, UK), sensitive in the spectral region of 800–1700 nm. A silicon filter and a band-pass filter at 1270 nm with a 10 nm half-bandwidth should be used (Spectrogon UK Ltd, Glenrothes, UK). The Ge-Diode system is equipped with a bias power supply (model PS-1) set to provide 160 V to the detector, a muon filter (model MF-1) to filter cosmic and ionizing radiation, an optical chopper (Bentham 218, Bentham Instruments, UK) set at a frequency of 125 Hz, and a lock-in amplifier (Bentham 225, Bentham Instruments, UK) are required to processes the signal from the detector. The detector signals can be registered by the F-900 version 6.22 software program (Edinburgh Analytical Instruments, Livingston, UK) (Fig. 25.7).

The $^1O_2$ monomol light emission spectrum can be measured with a special photocounting apparatus

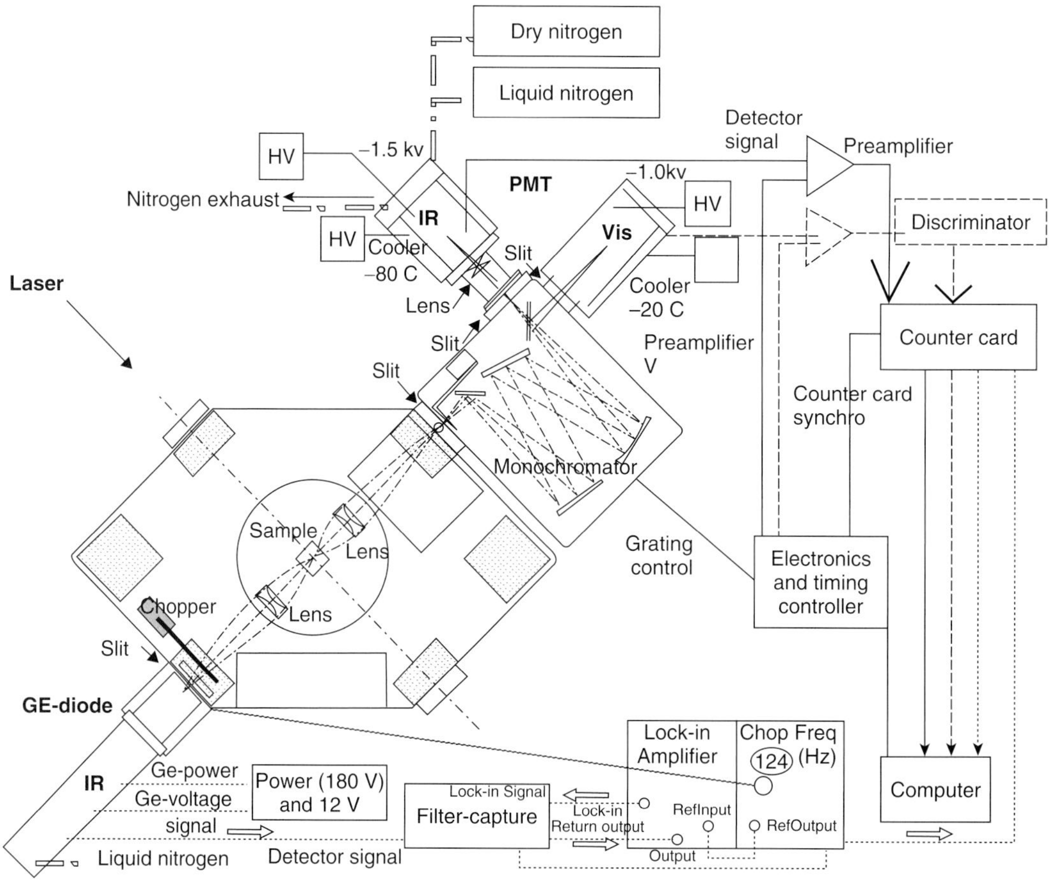

Fig. 25.7 Singlet molecular oxygen monomol light emission photocounting apparatus.

equipped with a monochromator, capable of selecting emissions in the near-infrared region (800–1400 nm; Fig. 25.7). The apparatus consists of a photomultiplier tube (R5509 PMT, Hamamatsu Photoniks KK, Shizuoka, Japan) cooled to −80°C with liquid nitrogen (S600 PHOTOCOOLTM, PC176TSCE005 cooler, Products for Research Inc., MA) to reduce the dark current. The power is provided by a high voltage DC power supply (Model C3360, Hamamatsu Photoniks KK, Shizuoka, Japan) and the applied potential is set at −1.5 kV. The light emitted from the sample passes through a monochromator (M300, Bentham Instruments, UK) equipped with a diffraction grating (Type G306R1u0, Bentham Instruments, UK) capable of selecting wavelengths in the infrared region. The control of the monochromator and the acquisition of data can be carried out using the F900 software described earlier (Miyamoto et al. 2003a,b).

Description of a Method for Detection of Singlet Molecular Oxygen by Monomol Light Emission

The optimal condition to detect the $^1O_2$ monomol light emission can be adjusted with the endoperoxide of 1,4-dimethylnaphthalene ($DMNO_2$, 15 mM in chloroform), or with the endoperoxide of *N*, *N'*-di(2,3-dihydroxypropyl)-1, 4-naphthalenedipropanamide ($DHPNO_2$, in D_2O) heated to 37°C. For the experiments, the photodiode detector must be filled with liquid nitrogen and left to equilibrate for 2 h before use. The assay can be conducted at 37°C in a thermostated quartz cuvette (10 mm × 10 mm × 30 mm) under continuous stirring with a small magnetic stirrer (Cuv-o-stirr, model 333, Hellma, Mülheim, Germany). After recording the baseline with the assay solvent, the reactant is injected into the cuvette with a syringe injection pump (Syringe Pump Model 22, Harvard Apparatus, MA) (Miyamoto et al. 2003a,b).

In order to obtain unambiguous evidence of $^1O_2$ in the system, a spectrum in the infrared region can be recorded. As mentioned above, relaxation of $^1O_2$ to the ground state produces emission at 1270 nm, which is characteristic for this species. For this reason, experiments should be conducted in quartz cuvettes under continuous stirring, as described above. Typically, 3–5 scans in the range of 1200–1350 nm must be recorded and averaged to yield the spectrum.

CONCLUSION

Growing evidence suggests $^1O_2$ toxicity can be used as an environmentally secure alternative for the control of parasites in aquaculture. Also, $^1O_2$ can be employed as a controlled method to achieve photodegradation of organic contaminants in aquatic environments. Furthermore it is now recognized that $^1O_2$ is involved in oxidative damage in plants and animals. In addition to cellular sources, there are numerous environmental sources of this excited species, important examples are near-ultraviolet radiation (UVA, 320–400 nm) and visible radiation (400–700 nm), which, together with appropriate photoexcitable compounds and molecular oxygen, can produce $^1O_2$ in marine systems. Since under stress conditions different ROS are simultaneously produced, a causal link between the production of a specific ROS and its damaging effects has always been difficult to establish. In this chapter we have described methodologies to evaluate the role of $^1O_2$ in biological systems. Looking for such a process in a specific system, the first approach can be to detect the increase or decrease of a specific marker of damage (i.e. 8-oxodGuo) in the presence of D_2O and known $^1O_2$ quenchers (i.e. azide, carotenoids). Moreover, one can detect well-defined signature products of $^1O_2$-mediated oxidation (i.e. cholesterol 5α-OOH). Another helpful approach is to use a clean source of $^1O_2$ ($^1O_2$ generators, the choice of which depends on the polarity of the system) to cross check whether the generator causes the same alterations observed in the investigated system. Chemical trapping of $^1O_2$ is another appropriate and sensitive method to detect $^1O_2$. Again, the trap should be chosen based on the characteristics of the system. Light emission measurements in the visible, and especially in the near-infrared region, are also valuable tools to detect $^1O_2$. Appropriated methodologies should be chosen according to the characteristics of the investigated systems, and the concomitant use of two methodologies in order to obtain reliable results is recommended.

REFERENCES

Agon, V.V., Bubb, W.A., Wright, A., Hawkins, C.L., Davies, M.J. (2006) Sensitizer-mediated photo-oxidation of histidine residues: Evidence for the formation of reactive side-chain peroxides. *Free Radical Biology and Medicine* 40, 698–710.

Ando, W., Takata, T. (1985) Photooxidation of sulfur compounds. *In* Frimer, A.A. (ed.). *Singlet O_2*. CRC Press; Boca Raton, FL, pp. 1–117.

Aubry, J.M. (1991) New chemical sources of singlet oxygen. *In* Vigo-Pelfrey, C. (ed.). *Membrane Lipid Oxidation*. CRC Press; Boca Raton, FL, pp. 65–102.

Baier, J., Maisch, T., Maier, M., Engel, E., Landthaler, M., Baumler, W. (2006) Singlet oxygen generation by UVA light exposure of endogenous photosensitizers. *Biophysical Journal* 91, 1452–1459.

Beckwith, A.L.J., Davies, A.G., Davison, I.G.E., Maccoll, A., Mruzek, M.H. (1989) The mechanisms of the rearrangements of allylic hydroperoxides–5-α-hydroperoxy–3-β-hydroxycholest–6-ene and 7-α-hydroperoxy–3-β-hydroxycholest–5-ene. *Journal of the Chemical Society, Perkin Transactions* 2 815–824.

Bilski, P., Burkhart, J.G., Chignell, C.F. (2003) Photochemical characterization of water samples from Minnesota and Vermont sites with malformed frogs: potential influence of photosensitization by singlet molecular oxygen ($^1O_2$) and free radicals on aquatic toxicity. *Aquatic Toxicology* 65, 229–241.

Botsivali, M., Evans, D.F. (1979) New trap for singlet oxygen in aqueous solution. *Journal of the Chemical Society, Chemical Communications* 1114–1116.

Brinkhorst, J., Nara, S.J., Pratt, D.A. (2008) Hock cleavage of cholesterol 5ähydroperoxide: An ozone-free pathway to the cholesterol ozonolysis products identified in arterial plaque and brain tissue. *Journal of the American Chemical Society* 130, 12224–12225.

Briviba, K., Saha-Moller, C., Adam, W., Sies, H. (1996) Formation of singlet oxygen in the thermal decomposition of 3-hydroxymethyl–3,4,4-trimethyl–1,2-dioxetane, a chemical source of triplet-excited ketones. *Biochemistry and Molecular Biology International* 38, 647–651.

Browne, R., Ogryzlo, E. (1964) Chemiluminescence from reaction of chlorine with aqueous hydrogen peroxide. *Proceedings of the Chemical Society of London* 117.

Buchko, G.W., Cadet, J., Berger, M., Ravanat, J.-L. (1992) Photooxidation of d(TpG) by phthalocyanines and riboflavin. Isolation and characterization of dinucleoside monophosphates containing the 4R* and 4S* diastereoisomers of 4,8-dihydro–4-hydroxy–8-oxo–2′-deoxyguanosine. *Nucleic Acids Research* 20, 4847–4851.

Cadenas, E., Sies, H. (1984) Low-level chemiluminescence as an indicator of singlet molecular oxygen in biological systems. *Methods in Enzymology* 105, 221–231.

Cadet, J., Berger, M., Douki, T., Ravanat, J.L. (1997) Oxidative damage to DNA: formation, measurement, and biological significance. *Reviews of Physiology Biochemistry and Pharmacology* 131, 1–87.

Cadet, J., Douki, T., Ravanat, J.L., Di Mascio, P. (2009) Sensitized formation of oxidatively generated damage to cellular DNA by UVA radiation. *Photochemical and Photobiological Sciences* 8, 903–911.

Dewilde, A., Pellieux, C., Pierlot, C., Wattre, P., Aubry, J.M. (1998) Inactivation of intracellular and non-enveloped viruses by a non-ionic naphthalene endoperoxide. *Biological Chemistry* 379, 1377–1379.

Di Mascio, P., Sies, H. (1989) Quantification of singlet oxygen generated by thermolysis of 3,3′-(1,4-naphthylene)dipropionate endoperoxide. Monomol and dimol photoemission and the effects of 1,4-diazabicyclo[2.2.2]octane. *Journal of the American Chemical Society* 111, 2909–2914.

Di Mascio, P., Kaiser, S., Sies, H. (1989) Lycopene as the most efficient biological carotenoid singlet oxygen quencher. *Archives of Biochemistry and Biophysics* 274, 532–538.

Di Mascio, P., Devasagayam, T.P., Kaiser, S., Sies, H. (1990) Carotenoids, tocopherols and thiols as biological singlet molecular oxygen quenchers. *Biochemical Society Transactions* 18, 1054–1056.

Di Mascio, P., Bechara, E., Medeiros, M.H.G., Briviba, K., Sies, H. (1994) Singlet molecular oxygen production in the reaction of peroxynitrite with hydrogen peroxide. *FEBS Letters* 355, 287–289.

Foote, C.S. (1991) Definition of type I and type II photosensitized oxidation. *Photochemistry and Photobiology* 54, 659.

Foote, C.S., Denny, R.W. (1968) Chemistry of singlet oxygen. VII. Quenching by.beta.-carotene. *Journal of the American Chemical Society* 90, 6233–6235.

Foote, C.S., Wexler, S., Ando, W., Higgins, R. (1968) Chemistry of singlet oxygen. IV. Oxygenations with hypochlorite-hydrogen peroxide. *Journal of the American Chemical Society* 90, 975–981.

Frankel, E.N. (1984) Chemistry of free radical and singlet oxygen of lipids. *Progress in Lipid Research* 23, 197–221.

Frankel, E.N., Neff, W.E., Bessler, T.R. (1979) Analysis of autoxidized fats by gas chromatography-mass spectrometry.5. Photosensitized oxidation. *Lipids* 14, 961–967.

Frimer, A. (ed.) (1985) *Singlet O_2*, Vols I–III. Boca Raton: CRC Press.

Girotti, A.W. (1998) Lipid hydroperoxide generation, turnover, and effector action in biological systems. *Journal of Lipid Research* 39, 1529–1542.

Girotti, A.W. (2001) Photosensitized oxidation of membrane lipids: reaction pathways, cytotoxic effects, and cytoprotective mechanisms. *Journal of Photochemistry and Photobiology B* 63, 103–113.

Girotti, A.W., Korytowski, W. (2000) Cholesterol as a singlet oxygen detector in biological systems. *Methods in Enzymology* 319, 85–100.

Gupta, D., Harish, B., Kissner, R., Koppenol, W.H. (2009) Peroxynitrate is formed rapidly during decomposition of peroxynitrite at neutral pH. *Dalton Transactions* 5730–5736.

Havaux, M., Eymery, F., Porfirova, S., Rey, P., Dormann, P. (2005) Vitamin E protects against photoinhibition and photooxidative stress in *Arabidopsis thaliana*. *Plant Cell* 17, 3451–3469.

Held, A.M., Halko, D.J., Hurst, J.K. (1978) Mechanisms of chlorine oxidation of hydrogen peroxide. *Journal of the American Chemical Society* 100, 5732–5740.

Kanofsky, J. (1983) Singlet oxygen production by lactoperoxidase. *Journal of Biological Chemistry* 258, 5991–5993.

Kanofsky, J. (1984) Singlet oxygen production by chloroperoxidase-hydrogen peroxide-halide systems. *Journal of Biological Chemistry* 259, 5596–5600.

Khan, A., Kasha, M. (1963) Red chemiluminescence of molecular oxygen in aqueous solution. *Journal of Chemical Physics* 39, 2105–2106.

Kielbassa, C., Roza, L., Epe, B. (1997) Wavelength dependence of oxidative DNA damage induced by UV and visible light. *Carcinogenesis* 18, 811–816.

Kochevar, I.E., Redmond, R.W. (2000) Photosensitized production of singlet oxygen. *Methods in Enzymology* 319, 20–28.

Li, M.Y., Cline, C.S., Koker, E.B., Carmichael, H.H., Chignell, C.F., Bilski, P. (2001) Quenching of singlet molecular oxygen ($^1O_2$) by azide anion in solvent mixtures. *Photochemistry and Photobiology* 74, 760–764.

Martinez, G.R., Garcia, F., Catalani, L.H. et al. (2006) Synthesis of a hydrophilic and non-ionic anthracene derivative, the N,N′-di-(2,3-dihydroxypropyl)–9,10-anthracenedipropanamide as a chemical trap for singlet molecular oxygen detection in biological systems. *Tetrahedron* 62, 10762–10770.

Miyamoto, S., Martinez, G., Martins, A., Medeiros, M.H.G., Di Mascio, P. (2003a) Direct evidence of singlet molecular oxygen [$O_2(^1\Delta_g)$] production in the reaction of linoleic acid hydroperoxide with peroxynitrite. *Journal of the American Chemical Society* 125, 4510–4517.

Miyamoto, S., Martinez, G., Medeiros, M.H.G., Di Mascio, P. (2003b) Singlet molecular oxygen generated from lipid hydroperoxides by the Russell mechanism: studies using ^{18}(O)-labeled linoleic acid hydroperoxide and monomol light emission measurements. *Journal of the American Chemical Society* 125, 6172–6179.

Miyamoto, S., Martinez, G.R., Rettori, D., Augusto, O., Medeiros, M.H.G., Di Mascio, P. (2006) Linoleic acid hydroperoxide reacts with hypochlorous acid, generating peroxyl radical intermediates and singlet molecular oxygen. *Proceedings of the National Academy of Sciences USA* 103, 293–298.

Miyamoto, S., Ronsein, G.E., Corrêa, T.C., Martinez, G.R., Medeiros, M.H.G., Mascio, P.D. (2009) Direct evidence of singlet molecular oxygen generation from peroxynitrate, a decomposition product of peroxynitrite. *Dalton Transactions* 5720–5729.

Monroe, B.M. (1985) Singlet oxygen in solution: lifetimes and reaction rate constants. *In* Frimer, A.A. (ed.). *Singlet O_2*, CRC Press, Boca Raton, FL, pp. 177–224.

Murphy, M.E., Sies, H. (1990) Visible-range low-level chemiluminescence in biological systems *Methods in Enzymology* 186, 595–610.

Nardello, V., Aubry, J.M. (2000) Measurement of photogenerated singles oxygen in aqueous media. *Methods in Enzymology* 319, 50–58.

Nardello, V., Aubry, J.-M., Johnston, P., Bulduk, I., Vries, A.H.M., Alsters, P.L. (2005) Facile preparation of the water-soluble singlet oxygen traps anthracene–9, 10-divinylsulfonate (AVS) and anthracene–9,10-diethylsulfonate (AES) via a Heck reaction with vinylsulfonate. *Synlett* 17, 2667–2669.

Pflaum, M., Boiteux, S., Epe, B. (1994) Visible light generates oxidative DNA base modifications in high excess of strand breaks in mammalian cells. *Carcinogenesis* 15, 297–300.

Pierlot, C., Hajjam, S., Bathelemy, C., Aubry, J.-M. (1996) Water-soluble naphthalene derivatives as singlet oxygen ($^1O_2$, 1[Delta]g) carriers for biological media. *Journal of Photochemistry and Photobiology B* 36, 31–39.

Pierlot, C., Aubry, J., Briviba, K., Sies, H., Di Mascio, P. (2000) Naphthalene endoperoxides as generators of singlet oxygen in biological media. *Methods in Enzymology* 319, 3–20.

Prado, F.M., Oliveira, M.C., Miyamoto, S. et al. (2009) Thymine hydroperoxide as a potential source of singlet molecular oxygen in DNA. *Free Radical Biology and Medicine* 47, 401–409.

Ravanat, J.-L., Di Mascio, P., Martinez, G.R., Medeiros, M.H.G., Cadet, J. (2000) Singlet oxygen induces oxidation of cellular DNA. *Journal of Biological Chemistry* 275, 40601–40604.

Ravanat, J.L., Saint-Pierre, C., Di Mascio, P., Martinez, G.R., Medeiros, M.H.G., Cadet, J. (2001) Damage to isolated DNA mediated by singlet oxygen. *Helvetica Chimica Acta* 84, 3702–3709.

Robertson, P.K.J., Black, K.D., Adams, M. et al. (2009) A new generation of biocides for control of crustacea in fish farms. Journal of *Photochemistry and Photobiology B* 95, 58–63.

Rodgers, M.A.J., Snowden, P.T. (1982) Lifetime Of O–2(1delta-G) In Liquid Water As Determined By Time-Resolved Infrared Luminescence Measurements. *Journal of the American Chemical Society* 104, 5541–5543.

Ronsein, G.E., Miyamoto, S., Medeiros, M.H.G., Di Mascio, P. (2007) Singlet oxygen generates biological peroxides. *Current Topics in Biochemical Research* 9, 79–91.

Ronsein, G.E., Oliveira, M.C.B., Miyamoto, S., Medeiros, M.H.G., Di Mascio, P. (2008) Tryptophan oxidation by singlet molecular oxygen [O_2 ($^1\Delta_g$)]: mechanistic studies using ^{18}O-labeled hydroperoxides, mass spectrometry, and light emission measurements. *Chemical Research in Toxicology* 21, 1271–1283.

Ronsein, G.E., Oliveira, M.C.B., Medeiros, M.H.G., Di Mascio, P. (2009) Characterization of $O_2(^1\Delta_g)$-derived oxidation products of tryptophan: a combination of tandem mass spectrometry analyses and isotopic labeling studies. *Journal of the American Society for Mass Spectrometry* 20, 188–197.

Ronsein, G.E., Prado, F.M., Mansano, F.V. et al. (2010) Detection and characterization of cholesterol-oxidized products using HPLC coupled to dopant assisted atmospheric pressure photoionization tandem mass spectrometry. *Analytical Chemistry* 82, 7293–7301.

Steenken, S., Jovanovic, S.V. (1997) How easily oxidizable is DNA? One-electron reduction potentials of adenosine and guanosine radicals in aqueous solution. *Journal of the American Chemical Society* 119, 617–618.

Straight, R.C., Spikes, J.D. (1985) Photosensitized oxidation of biomolecules. *In Singlet O_2*, Frimer, A.A. (ed.), CRC Press; Boca Raton.P. 91–143.

Tanaka, K., Miura, T., Umezawa, N. et al. (2001) Rational design of fluorescein-based fluorescence probes, mechanism-based design of a maximum fluorescence probe for singlet oxygen. *Journal of the American Chemical Society* 123, 2530–2536.

Terao, J., Matsushita, S. (1977) Products formed by photosensitized oxidation of unsaturated fatty-acid esters. *Journal of the American Oil Chemists Society* 54, 234–238.

Tomita, M., Irie, M., Ukita, T. (1969) Sensitized photooxidation of histidine and its derivatives. Products and mechanism of the reaction. *Biochemistry* 8, 5149–5160.

Triantaphylides, C., Krischke, M., Hoeberichts, F.A. et al. (2008) Singlet oxygen is the major reactive oxygen species involved in photooxidative damage to plants. *Plant Physiology* 148, 960–968.

Tyrrell, R.M. (2000) Role for singlet oxygen in biological effects of ultraviolet A radiation. *Methods in Enzymology* 319, 290–295.

Uemi, M., Ronsein, G.E., Miyamoto, S., Medeiros, M.H.G., Di Mascio, P. (2009) Generation of cholesterol carboxyaldehyde by the reaction of singlet molecular oxygen [O_2 ($1\Delta_g$)] as well as ozone with cholesterol. *Chemical Research in Toxicology* 22, 875–884.

Wentworth, P., Jr., Nieva, J., Takeuchi, C., et al. (2003) Evidence for ozone formation in human atherosclerotic arteries. *Science* 302, 1053–1056.

Wilkinson, F., Helman, W.P., Ross, A.B. (1995) Rate constants for the decay and reactions of the lowest electronically excited singlet state of molecular oxygen in solution. An expanded and revised compilation. *Journal of Physical and Chemical Reference Data* 24, 663–1021.

Wright, A., Bubb, W.A., Hawkins, C.L., Davies, M.J. (2002) Singlet oxygen-mediated protein oxidation: evidence for the formation of reactive side chain peroxides on tyrosine residues. *Photochemistry and Photobiology* 76, 35–46.

Chapter 26

TOTAL OXYRADICAL SCAVENGING CAPACITY ASSAY

Stefania Gorbi and Francesco Regoli

Dipartimento di Biochimica, Biologia e Genetica, Università Politecnica delle Marche, Ancona, Italy

The total oxyradical scavenging capacity assay (TOSC assay) is a reliable approach to quantify the overall capability of cellular antioxidants to neutralize different forms of reactive oxygen species (ROS), such as ROO•, HO•, and ONOOH (Di Giulio et al. 1995; Winston et al. 1998; Regoli and Winston 1999; Halliwell and Gutteridge 2007).

The TOSC assay is based on the reaction between these oxyradicals, which are artificially generated at constant rate, with α-keto-γ-methiolbutyric acid (KMBA), which is oxidized to ethylene. Antioxidants present in cellular compartments will efficiently compete with KMBA in neutralizing ROS and thus reduce ethylene formation compared to a control reaction (Fig. 26.1). The time course of reactions is followed by gas chromatographic (GC) analyses of ethylene formation in the headspace of vials, and the antioxidant capacity of the sample is calculated from the quantitative inhibition of KMBA oxidation.

Appropriate assay conditions are described for three different oxyradical generating systems, ROO•, HO•, and ONOOH. Each oxidant is produced at a rate giving an equivalent yield of ethylene production in control reactions throughout the whole duration of the assay. Thus, the relative efficiency of antioxidants in cellular extracts can be compared from their ability to counteract a pro-oxidant challenge, quantitatively similar in terms of KMBA oxidation, but qualitatively different because the oxidation is due to different ROS, to which the antioxidant responds in a different manner.

The TOSC assay has been widely applied to characterize the total oxyradical scavenging capacity of marine organisms from different environments or exposed to chemical pollutants under field or laboratory conditions (Corsolini et al. 2001; Frenzilli et al. 2001; Camus et al. 2002, 2003; Gorbi and Regoli 2003; Regoli et al. 2003, 2004; Vinagre et al. 2003; Bocchetti and Regoli 2006; Gorbi et al. 2008, 2009; Krapp et al. 2009). Compared to variations in individual antioxidants, TOSC is less sensitive in revealing a pro-oxidant challenge but has a greater biological relevance, because it indicates onset of several forms of cellular toxicity, like lysosomal destabilization, increase of lipid peroxidation and onset of genotoxic damage (Regoli et al. 2002; Gorbi and Regoli 2003).

SAMPLE PREPARATION

Procedures for sample preparation and storage are essentially the same as described in Chapter 27, except that bacitracin and PMSF are not present in the homogenization buffer. Briefly, samples are homogenized with

Oxidative Stress in Aquatic Ecosystems, First Edition. Edited by Doris Abele, José Pablo Vázquez-Medina, and Tania Zenteno-Savín.

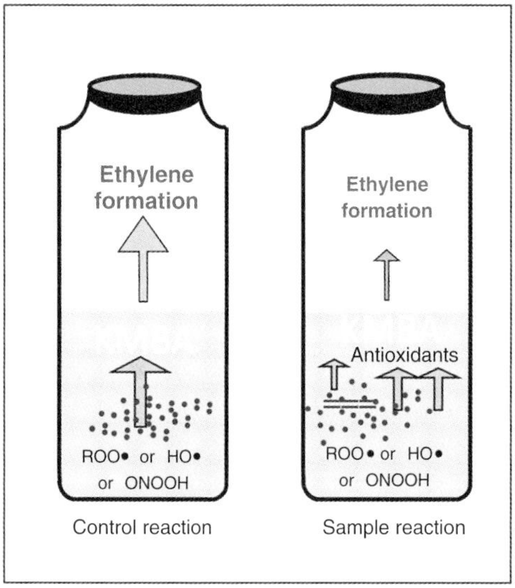

Fig. 26.1 Schematic representation of ethylene formation in the "Control vial", where KMBA is oxidized by ROS, and in the "Sample vial", where cellular antioxidants limit the reaction of ROS with KMBA and thus, ethylene formation. See plate section for a color version of this image.

a hand-held glass potter on ice, using 100 mM KPi buffer (pH 7.5) with 2.5% NaCl (v:w) for marine invertebrates and 1.8% NaCl (v:w) for vertebrates, respectively, and protease inhibitors (see Chapter 27). A homogenization ratio 1:5 tissue wet weight : buffer volume is appropriated for most tissues, but different dilutions should be tested for specific organisms. Homogenized samples are centrifuged at 100,000 *g* for 1 h 10 min at 4°C to obtain cytosolic fractions. Supernatants are removed, subdivided in small aliquots (100 μL) and stored at −80°C until analyses. For mussel digestive gland or fish liver cytosolic fractions, a final dilution of 1:100 is usually applied. Detailed protocols for solution preparation are given in Chapter 27.

VIALS FOR OXYRADICAL GENERATING SYSTEMS AND TOSC REACTIONS

The reactions for oxyradical generation ($ROO^\bullet$, $HO^\bullet$, and ONOOH) are carried out in 10 mL vials sealed with gas-tight Mininert valves, in a final volume of 1 mL (Fig. 26.2). During the whole reaction time (94–106 min) the vials are maintained at 35°C in a shaking water bath.

GENERATION OF ROO•, REQUIRED SOLUTIONS, AND PREPARATION OF VIALS FOR MEASURING TOSC-ROO•

$ROO^\bullet$ is artificially produced by the thermal homolysis of 20 mM ABAP (2-2′-azo-bis-(2 methylpropionamidine)-dihydrochloride) at 35°C in 100 mM K Pi buffer (pH 7.4). Reaction of $ROO^\bullet$ with 0.2 mM KMBA will produce a different rate of ethylene generation in control and sample reactions:

$$\text{ABAP} \xrightarrow[\text{ROO}^\bullet \text{ formation at } 35^\circ\text{C}]{} \text{ROO}^\bullet + \text{KMBA}$$

→ ethylene formation in the **Control Reaction**

Fig. 26.2 (a) 10 mL vials used for the TOSC-analysis and (b) gas-tight Mininert valves used for aspiring the ethylene from the head space with a 200 μL GC syringe.

$$ABAP \xrightarrow[\text{ROO}^\bullet \text{ formation at } 35^\circ C]{} ROO^\bullet + KMBA + sample$$

$\rightarrow$ reduced ethylene formation in the

Sample Reaction

The following solutions must be prepared for the assay (see Chapter 27 for details):

- 100 mM K Pi buffer pH 7.4.
- 200 mM ABAP working solution: dissolve 0.0542 g mL^{-1} of distilled water.
- 100 mM KMBA stock solution: dissolve 0.017 g mL^{-1} of distilled water; dilute to prepare a 2 mM KMBA working solution.

Solutions should be prepared daily.

Prepare eight 10 mL vials, two for the control reactions and six for samples. In each vial, 1 mL of reaction mixture is prepared by adding the solutions in the order and volumes indicated in the following table.

Reactions start with the addition of ABAP. In a final volume of 1 mL different aliquots of sample are compensated by different volumes of K-Pi buffer to maintain standardized assay concentrations of KMBA (0.2 mM) and ABAP (20 mM). Appropriate sample aliquots should give an experimental TOSC value ranging between 25 and 45 (see below), which corresponds to the maximum linearity interval (Regoli and Winston 1999).

GENERATION OF HO•, REQUIRED SOLUTIONS, AND VIAL PREPARATION FOR MEASURING TOSC-HO•

$HO^\bullet$ is generated at 35°C from a Fenton driven reaction of Fe-EDTA (1.8 μM Fe^{3+} – 3.6 μM EDTA) plus 180 μM ascorbate (ASC) as reducing agent in 100 mM KPi buffer (pH 7.4). Reaction of $HO^\bullet$ with 0.2 mM KMBA will produce a different rate of ethylene generation in the control and sample reactions:

$$Fe^{3+}\text{-(EDTA)} + ASC \rightarrow Fe^{2+}\text{-(EDTA)} + ASC^\bullet + H^+$$
$$ASC^\bullet + Fe^{3+}\text{-(EDTA)} \rightarrow \text{DHA (dehydroascorbate)} + Fe^{2+}\text{-(EDTA)}$$

and

$$2ASC^\bullet + 2H^+ \rightarrow ASC + DHA$$
$$ASC + O_2 \rightarrow DHA + H_2O_2$$
$$Fe^{2+}\text{-(EDTA)} + H_2O_2 \rightarrow HO^\bullet + OH^- + Fe^{3+}\text{-(EDTA)}$$

$HO^\bullet$ + KMBA

$\rightarrow$ ethylene formation in the **Control Reaction**

$HO^\bullet \times$ KMBA $\times$ sample

$\rightarrow$ reduced ethylene formation in the

Sample Reaction

The following solutions must be prepared for the assay:

- 100 mM KPi buffer pH 7.4.
- 50 mM ascorbic acid stock solution: dissolve 0.0088 g mL^{-1} of distilled water. Dilute to

Reaction	KPi buffer (μL)	Sample (μL)	KMBA working solution (μL)	ABAP working solution (μL)
Control 1	800	0	100 (to obtain 0.2 mM assay concentration)	100 (to obtain 20 mM assay concentration)
Control 2	800	0	100 (to obtain 0.2 mM assay concentration)	100 (to obtain 20 mM assay concentration)
Sample 1	800–x	x sample 1	100 (to obtain 0.2 mM assay concentration)	100 (to obtain 20 mM assay concentration)
Sample 2	800–x	x sample 2	100 (to obtain 0.2 mM assay concentration)	100 (to obtain 20 mM assay concentration
Sample . . .	800–x	x sample . . .	100 (to obtain 0.2 mM assay concentration)	100 (to obtain 20 mM assay concentration

prepare a fresh 1.8 mM ascorbic acid working solution.

- 2 mM KMBA working solution (prepare freshly).
- 20 mM Fe^{3+} stock solution: dissolve 0.0054 g mL^{-1} of distilled water (from $FeCl_3 \times 6H_2O$, FW 270.3). Dilute to prepare a 4 mM Fe^{3+} **solution A**.
- 50 mM EDTA stock solution: dissolve 0.186 g in 10 mL of distilled water. Dilute to prepare a 8 mM EDTA **solution B**.
- Working solution: 2 mM Fe^{3+}–4 mM EDTA. Mix equal volumes of **solution A** and **solution B** (1 : 1). Wait 15–20 min for a complete chelating reaction between Fe^{3+} and EDTA. Then dilute to prepare 18 μM Fe^{3+}–36 μM EDTA working solution.

Solutions should be prepared daily.

Prepare eight 10 mL vials, two for the control reactions and six for the samples. It is very important to clean the vials with nitric acid before assaying for TOSC-HO$^{\bullet}$. In each vial, 1 mL of reaction mixture is prepared by adding the solutions in the order and volumes indicated in the following table.

Reactions start with the addition of the ascorbic acid solution. In a final volume of 1 mL, different aliquots of sample are compensated for by different volumes of KPi buffer to maintain constant assay concentrations of KMBA (0.2 mM), Fe^{3+}-EDTA (1.8 μM–3.6 μM), and ascorbic acid (180 μM). Experimental TOSC values for appropriate sample aliquots should range between 25 and 45 (see below), which correspond to the maximum linearity interval (Regoli and Winston 1999).

GENERATION OF ONOOH, REQUIRED SOLUTIONS, AND VIAL PREPARATION FOR MEASURING TOSC-ONOOH

ONOOH is generated at 35°C from the decomposition of SIN-1 (3-morpholinosydnonimine *N*-ethylcarbamide) (Lomonsova et al. 1998) in 100 mM KPi buffer (pH 7.4) with 0.1 mM DTPA (diethylene-triamine-pentaacetic acid). The reaction of ONOOH with 0.2 mM KMBA produces a different rate of ethylene generation in control and sample reactions:

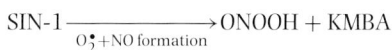

$$\text{SIN-1} \xrightarrow[O_2^{\bullet}+\text{NO formation}]{} \text{ONOOH} + \text{KMBA}$$

→ ethylene formation in the **Control Reaction**

$$\text{SIN-1} \xrightarrow[O_2^{\bullet}+\text{NO formation}]{} \text{ONOOH} + \text{KMBA} \times \text{sample}$$

→ reduced ethylene formation in the **Sample Reaction**

The following solutions must be prepared for the assay:

- 100 mM KPi buffer pH 7.4.
- 1 mM DTPA working solution: dissolve 0.0039 of DTPA in 10 mL of distilled water.
- 10 mM SIN-1 stock solution: dissolve 0.0021 g mL^{-1} of distilled water. Dilute to prepare a SIN-1 1 mM working solution.
- 100 mM KMBA stock solution and 2 mM KMBA working solution.

Solutions must be prepared daily.

Reaction	KPi buffer (μL)	Sample (μL)	Fe^{3+}–EDTA working solution (μL)	KMBA working solution (μL)	Ascorbic acid working solution (μL)
Control	700	0	100 (to obtain 1.8 μM–3.6 μM assay concentration)	100 (to obtain 0.2 mM assay concentration)	100 (to obtain 180 μM assay concentration)
Control 2	700	0	100 (to obtain 1.8 μM–3.6 μM assay concentration)	100 (to obtain 0.2 mM assay concentration)	100 (to obtain 180 μM assay concentration)
Sample 1	700–x	x sample 1	100 (to obtain 1.8 μM–3.6 μM assay concentration)	100 (to obtain 0.2 mM assay concentration)	100 (to obtain 180 μM assay concentration)
Sample 2	700–x	x sample 2	100 (to obtain 1.8 μM–3.6 μM assay concentration)	100 (to obtain 0.2 mM assay concentration)	100 (to obtain 180 μM assay concentration)
Sample . . .	700–x	x sample . . .	100 (to obtain 1.8 μM–3.6 μM assay concentration)	100 (to obtain 0.2 mM assay concentration)	100 (to obtain 180 μM assay concentration)

Reaction	KPi buffer (μL)	Sample (μL)	DTPA working solution (μL)	KMBA working solution (μL)	SIN-1 working solution (μL)
Control	720	0	100 (to obtain 0.1 mM assay concentration)	100 (to obtain 0.2 mM assay concentration)	80 (to obtain 80 μM assay concentration)
Control 2	720	0	100 (to obtain 0.1 mM assay concentration)	100 (to obtain 0.2 mM assay concentration)	80 (to obtain 80 μM assay concentration)
Sample 1	720–x	x sample 1	100 (to obtain 0.1 mM assay concentration)	100 (to obtain 0.2 mM assay concentration)	80 (to obtain 80 μM assay concentration)
Sample 2	720–x	x sample 2	100 (to obtain 0.1 mM assay concentration)	100 (to obtain 0.2 mM assay concentration)	80 (to obtain 80 μM assay concentration)
Sample . . .	720–x	x sample . . .	100 (to obtain 0.1 mM assay concentration)	100 (to obtain 0.2 mM assay concentration)	80 (to obtain 80 μM assay concentration)

Prepare eight 10 mL vials, two for the control reactions and six for the samples. In each vial, 1 mL of reaction mixture is prepared by adding the solutions in the order and volume indicated in the above table.

Reactions start with the addition of SIN-1 solution. In a final volume of 1 mL different aliquots of sample are compensated by different volumes of KPi buffer to maintain constant assay concentrations of KMBA (0.2 mM), DTPA (0.1 mM) and SIN-1 (1 mM). An appropriate aliquot of sample should give an experimental TOSC value ranging between 25 and 45 (see below), which corresponds to the maximum linearity interval (Regoli and Winston 1999).

TOSC ASSAY: GC ANALYSIS OF ETHYLENE FROM HEAD SPACE OF VIALS

Reactions between the ROS produced in the assay mixtures and KMBA (and thus ethylene formation) start in each vial when the last component is added (200 mM ABAP for ROO$^{\bullet}$; 1.8 mM ascorbic acid for HO$^{\bullet}$; 1 mM SIN-1 for ONOOH). It is most applicable to start reactions in different vials at specific time intervals and monitor the ethylene production at 12-min intervals in each vial, according to the following time scheme.

	Times[1] (min)						
	Start	Measurements of ethylene formation					
Vial 1 – Control Reaction	0	12	24	36	48	60	72
Vial 2 – Control Reaction	1	13	25	37	49	61	73
Vial 3 – Sample 1	3	15	27	39	51	63	75
Vial 4 – Sample 1 (replicate)	4	16	28	40	52	64	76
Vial 5 – Sample 2	6	18	30	42	54	66	78
Vial 6 – Sample 2 (replicate)	7	19	31	43	55	67	79
Vial 7 – Sample 3	9	21	33	45	57	69	81
Vial 8 – Sample 3 (replicate)	10	22	34	46	58	70	82

[1]Time schedule indicating, for each vial, the minutes at which reaction starts, and those at which ethylene is measured.

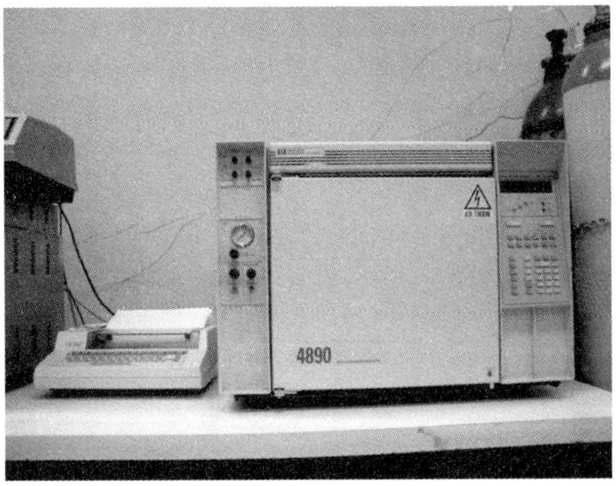

Fig. 26.3 GC and integrator to analyze ethylene formation in the reaction vials.

It is recommended to analyze controls and samples in duplicate. Eight serial vials can be monitored in sequence at 12-min intervals for a total assay duration of 106 min. Ethylene production in vials is measured by GC analysis of 200 μL aliquots directly taken from the head space of the vials and injected into a gas GC equipped with a capillary column (30 m × 0.32 mm × 0.25 mm) and a flame ionization detector (FID). The oven, injection and FID temperatures are set to 35 (oven), 160 (injection) and 220°C (FID). Helium can be used as carrier gas (1 mL min^{-1} flow rate). A split ratio of 20:1 should be used (Fig. 26.3).

QUANTIFICATION OF TOTAL OXYRADICAL SCAVENGING CAPACITY

For each control and sample reaction, the time course of ethylene formation is monitored every 12 min. From the resulting plots exemplarily shown in the accompanying graph (Fig. 26.4) the area under the curves is mathematically calculated from the integral of the equation that defines the experimental time points for both control and sample reactions (usually a second order equation is calculated) (see Excel sheet in Fig. 26.5):

The integral of the area below each curve is calculated using the following equation:

$$(A/3 \times TIME1^3 \times B/2 \times TIME1^2 \times C \times TIME1) - (A/3 \times TIME0^3 \times B/2 \times TIME02 \times C \times TIME0)$$

where A, B, C are the coefficients obtained from the equation $y = Ax^2 + Bx + C$; TIME 0 is the first measurement time in min (i.e. 12); and TIME 1 is the last measurement time in min (i.e. 96). For each sample, TOSC is then quantified according to the equation:

$$TOSC = 100 - (\int SA/\int CA \times 100)$$

where ∫SA and ∫CA are the integrated areas calculated under the least squares kinetic curve produced during the reaction for sample (SA) and control (CA) reactions (Winston et al. 1998). Thus, a TOSC value of 0 ($\int SA/\int CA = 1$) indicates a sample with no scavenging capacity (no inhibition of ethylene formation), while a maximum theoretical TOSC value of 100 would correspond to the total inhibition of ethylene formation in the assay ($\int SA = 0$).

Good experimental TOSC values generally range between 25 and 45. This corresponds to the best interval in terms of linearity of ethylene production during the reaction time (12–96 min). If the resulting TOSC values are not within this range (i.e. TOSC value <25 or >45), the assay should be repeated using

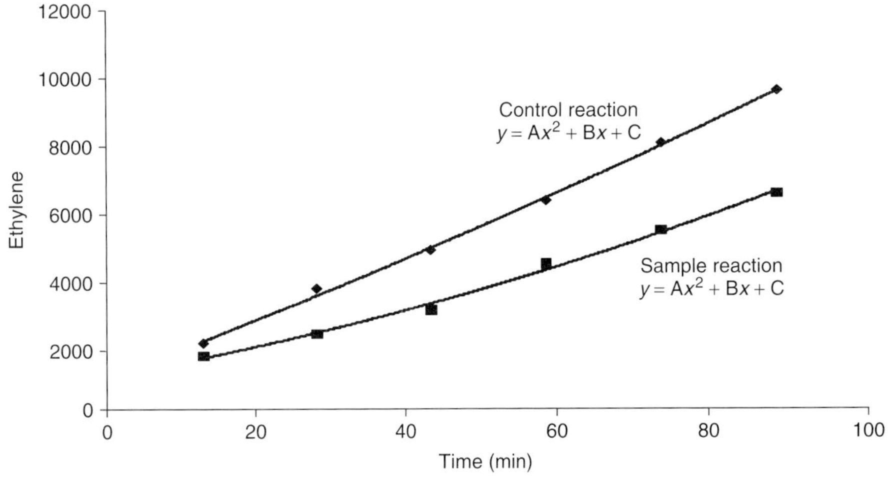

Fig. 26.4 Time course of ethylene formation.

1. Ethylene area calculation at various times for both control and samples (in duplicate)

time	Control	Control	Sample 1	Sample 1 (replicate)
12	1000	954	800	851
24	2041	1873	1606	1586
36	3856	3357	2264	2234
48	4677	4835	2977	2945
60	6258	6230	4378	4344
72	8021	7944	5377	5341
84	9890	9230	6490	6452
96	11100	11097	7300	7350

2. Average values are obtained from replicates of control and sample reactions, and equations of kinetic curves calculated (generally second order is better)

MEDIA

time	Control	sample
12	977	825,5
24	1957	1596
36	3606,5	2249
48	4756	2961
60	6244	4361
72	7982,5	5359
84	9560	6471
96	11098,5	7325

y = 0,2595x2 + 94,472x - 281,81 Control reaction

y = 0,2351x2 + 54,576x + 83,223 Sample Reaction

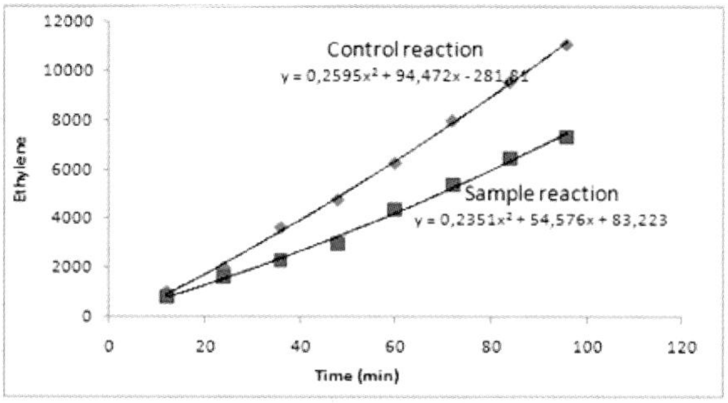

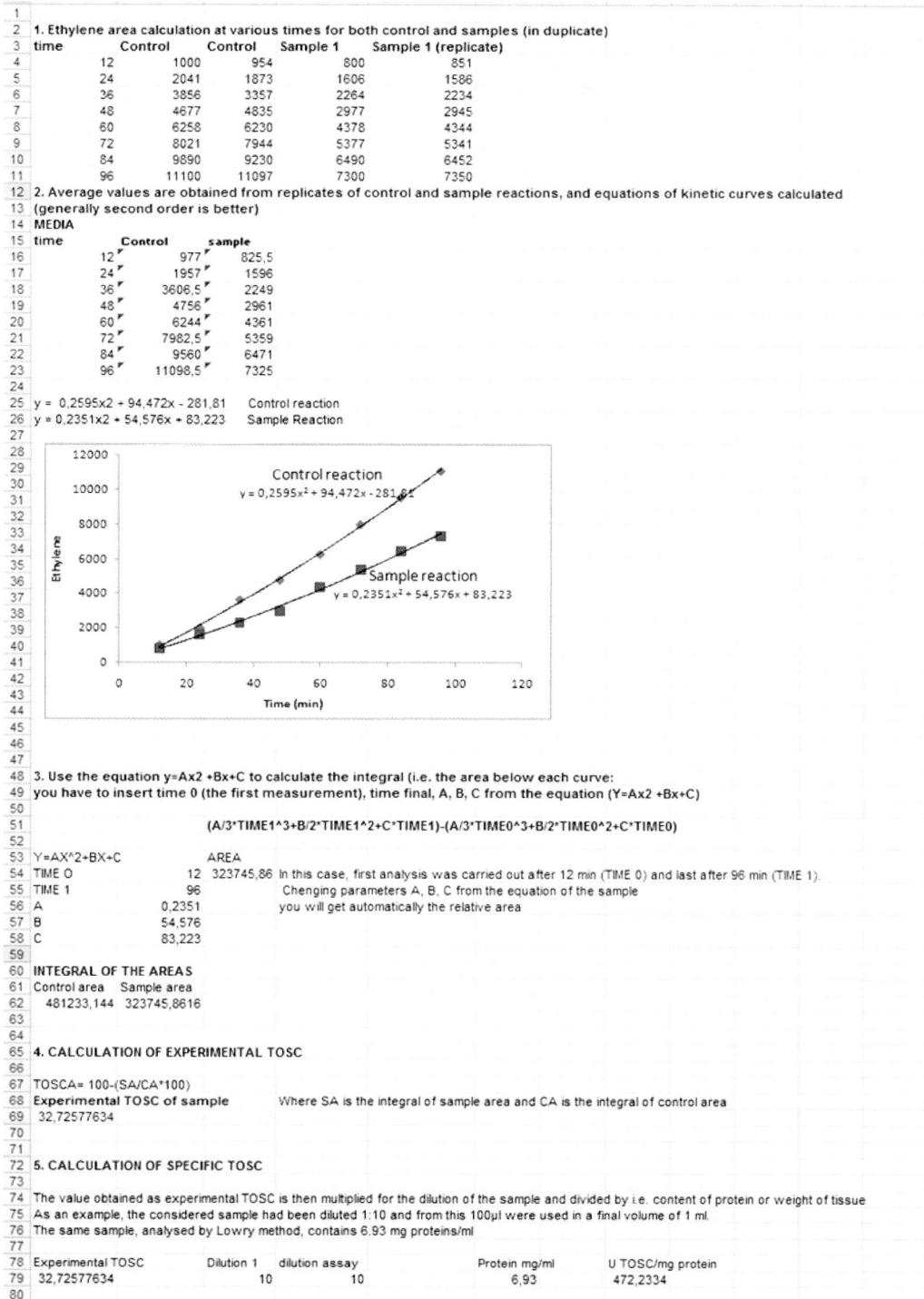

3. Use the equation y=Ax2 +Bx+C to calculate the integral (i.e. the area below each curve: you have to insert time 0 (the first measurement), time final, A, B, C from the equation (Y=Ax2 +Bx+C)

(A/3*TIME1^3+B/2*TIME1^2+C*TIME1)-(A/3*TIME0^3+B/2*TIME0^2+C*TIME0)

Y=AX^2+BX+C		AREA
TIME O	12	323745,86
TIME 1	96	
A	0,2351	
B	54,576	
C	83,223	

In this case, first analysis was carried out after 12 min (TIME 0) and last after 96 min (TIME 1).
Chenging parameters A, B, C from the equation of the sample
you will get automatically the relative area

INTEGRAL OF THE AREAS

Control area	Sample area
481233,144	323745,8616

4. CALCULATION OF EXPERIMENTAL TOSC

TOSCA= 100-(SA/CA*100)

Experimental TOSC of sample Where SA is the integral of sample area and CA is the integral of control area

32,72577634

5. CALCULATION OF SPECIFIC TOSC

The value obtained as experimental TOSC is then multiplied for the dilution of the sample and divided by i.e. content of protein or weight of tissue
As an example, the considered sample had been diluted 1:10 and from this 100μl were used in a final volume of 1 ml.
The same sample, analysed by Lowry method, contains 6.93 mg proteins/ml

Experimental TOSC	Dilution 1	dilution assay	Protein mg/ml	U TOSC/mg protein
32,72577634	10	10	6,93	472,2334

Fig. 26.5 Excel sheet with the procedures for TOSC values calculation.

another (higher or lower) sample dilution. Generally for invertebrate digestive gland and fish liver (homogenized 1:4), a final dilution of 1:100 is appropriate.

To obtain the specific TOSC of a sample, the experimental value is normalized to protein content:

$$\text{TOSC (Utosc mg of protein}^{-1}) = \text{Experimental TOSC value} \times \text{sample dilution/mg of protein}$$

The specific TOSC values have been reported for GSH towards ROO$^{\bullet}$ (0.84 ± 0.01), HO$^{\bullet}$ (0.11 ± 0.02) and ONOOH (0.17 ± 0.02) (Regoli and Winston 1999). This enables us to use GSH as a standard to verify whether the analysis is correctly performed. In the assay towards ROO$^{\bullet}$, vials containing GSH at concentrations of 15 μM, 30 μM and 45 μM can be tested as standards. In Fig. 26.5 an example excel sheet for TOSC value calculations is shown.

REFERENCES

Bocchetti, R., Regoli, F. (2006) Seasonal variability of oxidative biomarkers, lysosomal parameters, metallothioneins and peroxisomal enzymes in the Mediterranean mussel *Mytilus galloprovincialis* from Adriatic Sea. *Chemosphere* 65, 913–921.

Camus, L., Jones, M.B., Børseth, J.F. et al. (2002) Total oxyradical scavenging capacity and cell membrane stability of haemocytes of the Arctic scallop, *Chlamys islandicus*, following benzo(a)pyrene exposure. *Marine Environmental Research* 54, 425–430.

Camus, L., Birkely, S.R., Jones, M.B. et al. (2003) Biomarker responses and PAH uptake in *Mya truncata* following exposure to oil-contaminated sediment in an Arctic fjord (Svalbard). *The Science of the Total Environment* 308, 221–234.

Corsolini, S., Nigro, M., Olmastroni, S. et al. (2001) Susceptibility to oxidative stress in Adélie and emperor penguin. *Polar Biology* 24, 365–368.

Di Giulio, R.T., Benson, W.H., Sanders, B.M., Van Veld, P.A. (1995) Biochemical mechanisms: metabolism, adaptation and toxicity. *In* Rand, G. (ed.) *Fundamentals of Aquatic Toxicology: Effects, Environmental Fate and Risk Assessment*. Taylor and Francis, London, pp. 523–561.

Frenzilli, G., Nigro, M., Scarcelli, V. et al. (2001) DNA integrity and total oxyradical scavenging capacity in the Mediterranean mussel, *Mytilus galloprovincialis*: a field study in a highly eutrophicated coastal lagoon. *Aquatic Toxicology* 53, 19–32.

Gorbi, S., Regoli, F. (2003) Total oxyradical scavenging capacity as an index of susceptibility to oxidative stress in marine organisms. *Comments on Toxicology* 9, 303–322.

Gorbi, S., Virno Lamberti, C., Notti, A. et al. (2008) An ecotoxicological protocol with caged mussels, *Mytilus galloprovincialis*, for monitoring the impact of an offshore platform in the Adriatic sea. *Marine Environmental Research* 65, 34–49.

Gorbi, S., Benedetti, M., Virno Lamberti, C. et al. (2009) Biological effects of diethylene glycol (DEG) and produced waters (PWs) released from offshore activities: A multi-biomarker approach with the sea bass *Dicentrarchus labrax*. *Environmental Pollution* 157, 3166–3173.

Halliwell, B., Gutteridge, J.M.C. (2007) *Free Radicals in Biology and Medicine*, 4rd edn. Oxford University Press, New York.

Krapp, R.H., Bassinet, T., Berge, J. et al. (2009) Antioxidant responses in the polar marine sea-ice amphipod *Gammarus wilkitzkii* to natural and experimentally increased UV levels. *Aquatic Toxicology* 94, 1–7.

Lomonsova, E.E., Kirsch, M., Rauen, U., De Groot, H. (1998) The critical role of HEPES in SIN–1 cytotoxicity, peroxynitrite versus hydrogen peroxide. *Free Radical Biology and Medicine* 24, 522–528.

Regoli, F., Winston, G.W. (1999) Quantification of total oxidant scavenging capacity (TOSC) of antioxidants for peroxynitrite, peroxyl radicals and hydroxyl radicals. *Toxicology and Applied Pharmacology* 156, 96–105.

Regoli, F., Gorbi, S., Frenzilli, G. et al. (2002) Oxidative stress in ecotoxicology: from the analysis of individual antioxidants to a more integrated approach. *Marine Environmental Research* 54, 419–423.

Regoli, F., Winston, G.W., Gorbi, S. et al. (2003) Integrating enzymatic responses to organic chemical exposure with total oxyradical absorbing capacity and DNA damage in the European eel *Anguilla anguilla*. *Environmental Toxicology and Chemistry* 22, 2120–2129.

Regoli, F., Frenzilli, G., Bocchetti, R. et al. (2004) Time-course variations of oxyradical metabolism, DNA integrity and lysosomal stability in mussels, *Mytilus galloprovincialis*, during a field translocation experiment. *Aquatic Toxicology* 68, 167–178.

Vinagre, T.M., Alciati, J.C., Regoli, F. et al. (2003) Effect of microcystin on ion regulation and antioxidant system in gills of the estuarine crab *Chasmagnathus granulatus* (Decapoda, Grapsidae). *Comparative Biochemistry and Physiology C* 135, 67–75.

Winston, G.W., Regoli, F., Dugas, A.J. et al. (1998) A rapid gas chromatographic assay for determining oxyradical scavenging capacity of antioxidants and biological fluids. *Free Radical Biology and Medicine* 24, 480–493.

Chapter 27

SPECTROPHOTOMETRIC ASSAYS OF ANTIOXIDANTS

Francesco Regoli[1], *Raffaella Bocchetti*[1], *and Danilo Wilhelm Filho*[2]

[1]Dipartimento di Biochimica, Biologia e Genetica, Università Politecnica delle Marche, Ancona, Italy
[2]Departamento de Ecologia e Zoologia, Centro de Ciencias Biologicas, Universidade Federal de Santa Catarina, Florianópolis, Brazil

This chapter provides technical protocols for measuring the most common antioxidants in tissues of marine organisms by means of simple spectrophotometric assays. After a brief description of sample preparation procedures, methods for the determination of superoxide dismutase (SOD), catalase (CAT), glutathione peroxidase (GPx; both Se-dependent and the sum of Se-dependent and Se-independent forms), glutathione *S*-transferase (GST), glutathione reductase (GR), glyoxalase I (GI), and glyoxalase II (GII) activities, as well as glutathione (GSH) content, will be presented. Dilutions and volumes of samples in assay conditions are often suggested, based on our experience with temperate invertebrates and fish. However, it should always be considered that marine organisms might present huge variations in the levels or activities of antioxidant defenses. Thus, although the principle and reagents of various assays are of general applicability, appropriate dilutions of samples should always be checked before starting a working session with new species and/or tissues.

SAMPLE STORAGE AND PREPARATION

It is highly recommended that tissue samples for analysis of antioxidants are frozen in liquid N_2 immediately after animal dissection, and stored at $-80^\circ C$ until homogenate preparation. A suitable homogenization buffer is 100 mM K-phosphate buffer (KH_2PO_4), pH 7.5, with 2.5% and 1.8% NaCl for marine invertebrates and marine vertebrates, respectively, and 0.1 mM phenylmethylsulphonyl fluoride (PMSF). The addition of protease inhibitors is required when homogenizing tissues with elevated levels of these enzymes (typically digestive glands). An appropriate mix of inhibitors is 0.008 trypsin inhibitor unit (TIU) mL^{-1} aprotinin, 1 ng mL^{-1} leupeptin, 0.5 ng mL^{-1} pepstatin, and 0.1 mg mL^{-1} bacitracin.

Samples are homogenized on ice, preferably using a hand-held potter with Teflon or glass pestle. A common homogenization ratio of 1:5 weight : volume (w : v) is recommended. Higher volumes of buffer facilitate tissue disintegration, while diluting the proteins. Conversely, lower buffer volumes make samples

Oxidative Stress in Aquatic Ecosystems, First Edition. Edited by Doris Abele, José Pablo Vázquez-Medina, and Tania Zenteno-Savín.

more concentrated but may reduce the efficiency of homogenization. Homogenization of each sample should take approximately 30–60 s. To obtain the post-mitochondrial S9 fraction, containing microsomes and cytosol, homogenates are centrifuged at 12,000 *g* for 15 min at 4°C. Supernatants are collected without the lipid phase, immediately assayed, or subdivided into small aliquots and stored at −80°C. For most antioxidants, however, it is recommended to use a more purified, cytosolic fraction due to the presence of potentially interfering molecules in the post-mitochondrial S9 fraction.

If microsomes should also be collected, the supernatants obtained after the 12,000 *g* centrifugation are further centrifuged at 100,000 *g* for 1 h 10 min at 4°C, to obtain the cytosolic fraction whereas the pellets represent the microsomes. Supernatants are removed, immediately assayed, or subdivided in small aliquots (100–200 μL), and stored at −80°C. Before freezing, add 1 μL of dithiothreitol (DTT; from a 100 mM stock solution in methanol stored at 4°C) to 100 μL of the cytosolic fractions designated for GPx analyses.

HOMOGENIZATION REAGENTS AND SOLUTIONS

- 100 mM potassium phosphate (KPi) buffer, pH 7.5 (homogenization buffer). Dissolve 0.680 g of KH_2PO_4 in 50 mL of distilled water; adjust pH to 7.5 with concentrated KOH.
- 1.8% or 2.5% NaCl. Dissolve 1.8 g or 2.5 g of NaCl in 100 mL of homogenization buffer.
- 100 mM PMSF. Dissolve 0.174 g of PMSF in 10 mL of methanol. This stock solution is stable for 6 months at 4°C. Add 100 μL of stock solution to 100 mL of homogenization buffer to obtain a 0.1 mM concentration.
- Aprotinin (0.008 TIU mL^{-1}). This trypsin inhibitor can be purchased as a powder or solution. Its activity is expressed as TIU and reported in the product specification as total activity (for powders) or activity mL^{-1} (for solutions). Aprotinin concentration in homogenization buffer is 0.008 TIU mL^{-1}. Depending on the aprotinin stock solution, calculate the volume containing 0.8 TIU and add to 100 mL of homogenization buffer.
- Leupeptin (1 ng mL^{-1}). Dissolve 1 mg in 1 mL of distilled water to obtain a stock solution. Add 100 μL to 100 mL of homogenization buffer.
- Pepstatin (5 ng mL^{-1}). Dissolve 1 mg in 1 mL of methanol 100% to obtain a stock solution. Add 50 μL to 100 mL of homogenization buffer.
- Bacitracin (0.1 mg mL^{-1} solution). Dissolve 100 mg in 1 mL of distilled water to obtain a stock solution. Add 100 μL to 100 mL of homogenization buffer.
- 100 mM DTT. Dissolve 0.01542 g of DTT in 1 mL of methanol to obtain a stock solution.

Stock solutions of PMSF, leupeptin, pepstatin, bacitracin, and DTT are stable for 6 months at 4°C.

DETERMINATION OF SUPEROXIDE DISMUTASES ACTIVITY (EC1.15.1.1)

SOD catalyzes the dismutation of $O_2^{\bullet-}$ into O_2 and H_2O_2 (Fridovich 1986):

$$2O_2^{\bullet-} + 2H^+ \rightarrow H_2O_2 + O_2$$

SOD activity is determined in an indirect spectrophotometric assay by monitoring the reduction of cytochrome *c* by $O_2^{\bullet-}$ at 550 nm. $O_2^{\bullet-}$ is generated by a HX/XO system (McCord and Fridovich 1976). One unit of SOD is defined as the amount of enzyme needed to inhibit the reduction of cytochrome *c* by 50%. Different volumes of sample are used to determine 50% inhibition of the reaction rate.

Reagents and Solutions

- 100 mM KPi buffer, pH 7.8. Dissolve 0.680 g of KH_2PO_4 in 50 mL of distilled water; adjust pH to 7.8 with concentrated KOH.
- 100 mM EDTA stock solution. Dissolve 3.7224 g of EDTA (disodium salt, dehydrated) in 100 mL of distilled water. This solution is stable for 3 months at 4°C.
- 300 mU mL^{-1} xanthine oxidase (XO, prepare immediately before the use). Depending on the product, dilute the XO to 300 mU mL^{-1} with cold distilled water.
- Working buffer: 100 mM KPi buffer, pH 7.8; 0.2 mM EDTA; 100 μM hypoxanthine (HX); 20 μM cytochrome *c*. Prepare immediately before use:
  - 50 mL of 100 mM KPi buffer
  - + 100 μL of 100 mM EDTA
  - + 0.68 mg of HX
  - + 12.3 mg of cytochrome *c*.

Procedure

Spectrophotometric analyses are carried out at $\lambda = 550$ nm (light path 1 cm) at constant temperature ($18 \pm 1°C$ for temperate species). A reference (uninhibited) reaction is carried out to calculate the change in absorbance per minute (ΔAbs), which reflects the reduction of cytochrome *c* by $O_2^{\bullet-}$ in the absence of SOD. For each sample at least three different sample volumes should be tested in the reaction, to calculate the amount needed to inhibit the reference reaction by 50%.

Reference Reaction

Add to a plastic cuvette:

- 500 μL working buffer
- 480 μL of 100 mM KPi buffer
- + 20 μL of XO.

Mix vigorously after the addition of XO. Read the increase in absorbance for at least 1 min. The reaction should be linear. ΔAbs should be approximately 0.1.

Sample Reactions

Three readings (R-I to R-III) with three different volumes for each sample are performed. Note that the final volume in the assay is always 1000 μL.

Add to 3 plastic cuvettes:

(R-I)	(R-II)	(R-III)
500 μL working buffer	500 μL working buffer	500 μL working buffer
470 μL KPi buffer	460 μL KPi buffer	440 μL KPi buffer
10 μL sample	20 μL sample	40 μL sample
+ 20 μL XO	+ 20 μL XO	+ 20 μL XO

Start reaction by adding XO and mix vigorously. Read the increase in absorbance for at least 1 min. Reactions should also be linear. Increasing the amount of sample in the assay should decrease the ΔAbs compared to the reference reaction (reduction of cytochrome *c* is lowered by a higher content of SOD in the samples). The three volumes of sample should give an inhibition of between 20 and 70% compared to the

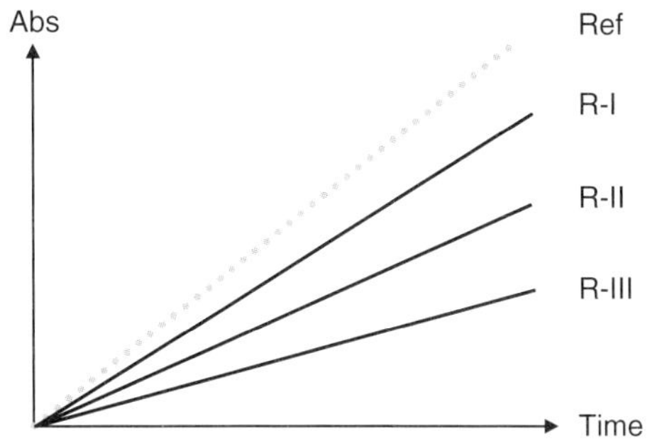

Fig. 27.1 The reaction starts by adding XO. Absorbance increase must be linear for 1 min.

reference reaction range (Fig. 27.1). It is important to note that the three volumes reported in the example are suggestive and should be adjusted according to the observed inhibition of the reference reaction for each type of sample. However, in a final volume of 1 mL, different aliquots of samples will be compensated by different volumes of KPi buffer. In this way, the volume of working buffer does not change in different readings, so that the applied concentrations of HX, cytochrome *c*, and XO in the assay remain constant.

Calculation of the Volume Necessary to Reduce the Reference Reaction by 50%

The relationship between added volume of sample and observed ΔAbs is often described as linear in a semilogarithmic scale (Fig. 27.2).

Plotting the relationship between added volume of sample (μL) and observed ΔAbs (as shown in Fig. 27.2) it is possible in the majority of conventional worksheets

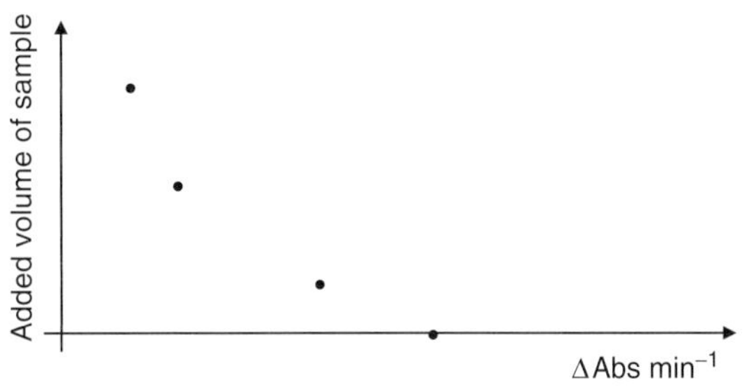

Fig. 27.2 Relationship between added volume of sample and observed ΔAbs.

to interpolate these points choosing among different equations (i.e linear, polynomial, exponential). By selecting a second-order regression on a linear scale, the worksheets also provide the coefficients for this equation, which relates absorbance to added volume of sample (μL):

$$y = ax^2 + bx + c$$

Using the coefficients a, b, c from the equation and setting x to the value corresponding to 50% of the ΔAbs measured in the reference reaction, y will yield the volume (μL) of sample needed to reduce the reference reaction by 50%, e.g. the volume of sample (Vol) that contains 1 Unit of SOD. According to the present protocol, the total volume in the assay is 1000 μL (see table for RI–RIII above), in which the number of contained SOD Units will thus correspond to 1000/Vol.

Calculations

SOD units are typically normalized to protein content, tissue weight, or mL (for blood, plasma or hemolymph). Normalization to tissue weight often precludes comparisons between different species or organs with different relative water content. When normalizing the activity to mg protein, comparability of the values from different tissues may be influenced by the basal content of total proteins, which can be highly variable between organs (Vlahogianni et al. 2007; Fernandez et al. 2009).

When normalizing to protein content:

$$\text{SOD (U mg protein}^{-1}) = (1000/\text{Vol}) \times (\text{sample dilution})/\text{proteins}$$

where "1000/Vol" corresponds to the number of SOD units contained in 1 mL (final assay volume, see above), "sample dilution" is the dilution factor applied (if any) to the cytosolic fractions before the assay, and "proteins" is the protein content in the cytosolic fractions (e.g. undiluted sample extract expressed as mg mL^{-1}).

DETERMINATION OF CATALASE ACTIVITY (EC1.11.1.6)

CAT is a heme containing enzyme found in peroxisomes of all aerobic organisms (Mueller et al. 1997). This enzyme catalyzes the decomposition of H_2O_2 to H_2O and O_2.

$$2H_2O_2 \longrightarrow 2H_2O + O_2$$

The physiological role of CAT is fundamental in counteracting the production of H_2O_2, which can react with transition metals to form $HO^{\bullet}$. By removing one of the main precursors of these highly toxic reactive oxygen species (ROS), CAT has been indicated as one of the main antioxidant defenses towards $HO^{\bullet}$ in marine organisms (Halliwell and Aruoma 1991; Halliwell and Gutteridge 2001; Regoli et al. 2002). The spectrophotometric assay described by Aebi (1984) is one of the most used methods to measure CAT activity in vertebrate and invertebrate tissues. Other methods, such as the polarographic determination, are less common. The spectrophotometric assay quantifies the loss of absorbance at 240 nm due to the decomposition of 5–12 mM H_2O_2. The rate of H_2O_2 decay is proportional to the amount of CAT contained in the sample, and is calculated using an extinction coefficient $\varepsilon = 0.04$ mM^{-1} cm^{-1}.

Since H_2O_2 is not stable, the stock solution needs to be titrated before performing the assay to test its effective concentration. The titration consists in reading the absorbance of different dilutions of the H_2O_2 stock, and calculating the concentrations with the extinction coefficient:

Dilution factor	Abs reading	Concentration
df_1	Abs_1	$c_1 = Abs_1/\varepsilon$ mM
df_2	Abs_2	$c_2 = Abs_2/\varepsilon$ mM
df_3	Abs_3	$c_3 = Abs_3/\varepsilon$ mM

H_2O_2 concentration in the stock solution is calculated by multiplying the concentrations obtained in diluted solutions with their respective dilution factors:

$$c_1 * X_1 = C_1$$
$$c_2 * X_2 = C_2$$
$$c_3 * X_3 = C_3$$

The mean of these values is the effective H_2O_2 concentration in the stock solution.

Importantly: to increase lifetime of the originally purchased stock solution, do not dip the pipette-tip into the stock bottle. Instead pour an aliquot into a small, clean beaker, use it to make your solutions, and discharge the rest of the aliquot.

Reagents and Solutions

- 100 mM KPi buffer, pH 7.0 (adjust pH with concentrated KOH).
- 1.2 M H_2O_2. Prepare fresh by adding 100 μL of the aliquot taken from the 12 M H_2O_2 stock to 900 μL of distilled water.

Procedure

Spectrophotometric analyses are carried out at $\lambda = 240$ nm (light path 1 cm) and at constant temperature (18± 1°C for temperate species).

Add to a quartz cuvette (final volume 1 mL):

- 980 μL of KPi buffer

INSTRUMENT AUTOZERO

- 10 μL of 1.2 M H_2O_2

Read the absorbance and add:

- 10 μL of sample.

Mix vigorously after the addition of sample. The instrument autozero before the addition of H_2O_2 allows a check of the proper assay concentration of this substrate (12 mM), which should give an absorbance value of 0.48. Other protocols "autozero" with buffer and sample, and start the reaction by adding a known volume of a previously prepared 12 mM H_2O_2 dilution (35 μL of commercial 30% H_2O_2 solution is added to 10 mL buffer, absorbance ≈ 1.400–1.500 AU).

Register the decrease in the absorbance for at least 1 min. Reactions should be linear for at least 30 s; otherwise a different dilution or sample volume should be used. Each sample is measured twice. Maintain a final volume of 1 mL even if different sample dilutions are needed. Compensate by changing the volume of KPi buffer. Appropriate dilutions for measuring CAT in cytosolic fractions of marine organisms can be highly variable ranging for samples homogenized at (w : v) 1:5 to 1:10 to 1:100 in digestive gland of invertebrates, and between 1:100 and 1:500 in fish liver. Proper dilutions to obtain reliable readings need to be predetermined for each batch of samples.

Calculations

Results are usually expressed in μmol or nmol H_2O_2 min^{-1} mg $protein^{-1}$, g of wet tissues, or mL of blood, plasma or hemolymph (see Box 27.1). When evaluating blood, hemolysates must be diluted at least 500 times in the cuvette, to avoid interference of peroxidase activity from hemoglobin (Aebi 1984). Normalization to tissue weight often precludes comparisons between different species or organs with different relative water content (see Box 27.2). When normalizing to mg of protein, the values might also be distorted by different protein content of specific tissues.

When the enzymatic activity is normalized to protein content:

$$\text{CAT activity } (\mu\text{mol min}^{-1}\text{ mg}^{-1}\text{ proteins}) = (\Delta\text{Abs}/-0.04) \times (\text{sample dilution})/\text{proteins}$$

where "ΔAbs" is the change in absorbance per minute, "−0.04" is the extinction coefficient (mM^{-1} cm^{-1}), "sample dilution" is the dilution factor (if any) applied to the cytosolic fractions before the assay,

Box 27.1 Less common units for expressing CAT activity

Other units are less common for expressing CAT activity and might confound or make comparisons more difficult. Among these, the Bergmeyer Unit (UB) corresponds to the amount of CAT that converts half of the O_2 from a H_2O_2 solution of any concentration in 100 s at 25°C (Bergmeyer 1965). International Units (UI) correspond to μmoles of substrate converted in 1 min at 25°C per mg of protein (Bergmeyer 1965). Finally, one Sigma Unit (SU) corresponds to the amount of CAT that converts 1 mol of H_2O_2 per minute at 25°C, pH 7.0, while the decay in H_2O_2 concentration falls in the interval of 10.3–9.2 mM. Therefore, the interconversions are: (i) for pure enzyme concentration (considering the MW of CAT as 30,000; Chance at al. 1973) – 1 pmol of CAT = 0.033 UB = 0.2 UI; or 1 UB = 13 UI; (ii) for CAT activity, multiply the value of CAT concentration in pmoles for 28.2 to obtain the amount of H_2O_2 min^{-1} (μmoles min^{-1} g^{-1}), or for 0.47 if the activity is recorded per second (μmoles s^{-1} g^{-1} or mg $protein^{-1}$).

Box 27.2 Cautionary notes

Some organisms, especially invertebrates, might have some peculiarities which complicate determination of CAT activity. For instance, fresh homogenates from two mollusks cultivated in South Brazil, *Perna perna* and *Crassostea gigas*, revealed a sort of lag phase, contrasting with the "normal" rapid and sharp profile from different tissues of many aquatic species so far studied (De Almeida et al. 2007). In short, there is a relatively long delay of approximately 1 min in *P. perna* and of about 3 min in *C. gigas* before a decay in H_2O_2 can be detected. If the researcher is not aware about this delay and intends to record only the first seconds or minute, no apparent CAT activity is detected.

If the researcher is interested in the calculation of the V_{max} of CAT activity, then a cautious procedure should be considered. With long monitoring periods, e.g. 1–3 min, the profile would not reflect the V_{max} value of CAT activity (better measured in the first seconds of reaction), and a relatively low or underestimated value is obtained. CAT activity is characterized as a pseudo-first-order category, which makes this enzyme unique compared to the other antioxidant enzymes (Chance et al. 1973). As a consequence, the comparison of results on CAT activities coming from different laboratories can be more difficult.

and "proteins" is the protein content in the cytosolic fractions expressed as mg mL^{-1}.

DETERMINATION OF GLUTATHIONE PEROXIDASES ACTIVITY (EC1.11.1.9 AND EC2.5.1.18)

GPx are a large family of selenium-dependent and selenium-independent enzymes with peroxidase activities. The main biological role of GPx is to protect cells from oxidative damage by reducing both organic hydroperoxides to their corresponding alcohols and free H_2O_2 to water. The Se-dependent GPx react with a wide variety of hydroperoxides, including both H_2O_2 and organic peroxides, whereas the Se-independent forms reduce only organic hydroperoxides. GPx have various cellular localizations including cytosol, mitochondrial matrix, and membranes, thus combining with CAT in removing H_2O_2 from different cellular compartments (Halliwell and Gutteridge 2007).

$$2GSH + H_2O_2 \rightarrow GSSG + H_2O$$

$$2GSH + ROOH \rightarrow GSSG + ROH + H_2O$$

GPx activities are measured by a coupled spectrophotometric assay (Lawrence and Burk 1976). GPx reduce H_2O_2 or organic hydroperoxides using GSH as cofactor. Variation in absorbance is thus measured due to consumption of reduced nicotinamide adenine dinucleotide phosphate (NADPH), used by GR to reconvert glutathione disulphide (GSSG) to GSH. The decrease in NADPH concentration is proportional to the GPx activity in the sample and it is followed at 340 nm ($\varepsilon = 6.22\ mM^{-1}\ cm^{-1}$).

$$2GSH + H_2O_2 \rightarrow GSSG + H_2O$$
(Se-dependent forms)

$$2GSH + ROOH \rightarrow GSSG + ROH + H_2O$$
(both Se-dependent and Se-independent forms)

$$GSSG + NADPH + H^+ \rightarrow 2GSH + NADP^+ \quad \text{(GR)}$$

Reagents and Solutions

- 100 mM KPi buffer (adjust pH with concentrated KOH).
- 100 mM EDTA.
- 100 mM GSH working solution (freshly prepared). Dissolve 0.0307 g of GSH (MW 307.32) in 1 mL of distilled water.
- NADPH (20 mg mL^{-1}) working solution (freshly prepared). Dissolve 2 mg of NADPH in 100 μL of distilled water.
- GR (100 U mL^{-1}) working solution (prepare immediately before use). Depending on the product, dilute GR to 100 U mL^{-1} in cold distilled water.
- 100 mM sodium azide (NaN_3) working solution (freshly prepared). Dissolve 6.5 mg in 1 mL of distilled water. NaN_3 is an inhibitor of CAT and is added only when measuring GPx activity towards H_2O_2.

- 200 mM cumene hydroperoxide (CHP) working solution (freshly prepared). Add 38 μL of a 5.2 M CHP stock solution to 962 μL of methanol.
- 100 mM H_2O_2 working solution (freshly prepared). Add 83 μL of the 12 M H_2O_2 stock solution to 9917 μL of distilled water. Since H_2O_2 is not stable, the stock solution needs to be titrated before performing the assay to test the effective concentration (see above).

Again: take care not to contaminate the original H_2O_2 stock solution!

Procedure

Spectrophotometric analyses are carried out at $\lambda = 340$ nm (light pass 1 cm) at constant temperature ($18 \pm 1°C$ for temperate species). Before measuring GPx activities, the rate of the blank reaction has to be carefully determined. At least 10 readings without sample should be performed at the beginning of the session to evaluate the decrease in absorbance due to the oxidation of NADPH by H_2O_2 or CHP. This value will be subtracted from the total rate of sample reaction ($\Delta Abs_{sample} - \Delta Abs_{blank} = \Delta Abs_{final\ sample}$). Run a new blank reaction following every 20 samples.

Since GPx activities can vary greatly in different tissues and species, substrate concentration in the assay might become limiting for the enzymatic reaction. Although assay concentrations are suggested in the following protocols for both H_2O_2 and CHP, these should be tested when analyzing new organisms, verifying the linear relationship between the final rate of sample reaction and the amount of sample added in the assay.

In the assay for measuring the activity of Se-dependent forms, add to a plastic or glass cuvette (final volume 1 mL):

- 835 μL of KPi buffer
- 10 μL of NaN_3 working solution
- 10 μL of EDTA
- 20 μL of GSH working solution
- 10 μL of GR working solution
- 100 μL of blank (KPi buffer) or sample

INSTRUMENT AUTOZERO

- 10 μL of NADPH working solution

Read the absorbance (should be 0.9–1.2)

- 5 μL of H_2O_2 working solution.

For measuring the sum activity of Se-dependent and Se-independent forms add to a plastic/glass cuvette (final volume 1 mL):

- 846 μL of KPi buffer
- 10 μL of EDTA
- 20 μL of GSH working solution
- 10 μL of GR working solution
- 100 μL of blank (KPi) or sample

INSTRUMENT AUTOZERO

- 10 μL of NADPH working solution

Read the absorbance (should be 0.9–1.2)

- 4 μL of CHP working solution.

In both assays, mix vigorously after the addition of substrates (H_2O_2 or CHP). Read decrease in the absorbance for at least 1 min. Reactions should be linear, otherwise a different sample volume should be used. Each sample is measured twice. Maintain a final volume of 1 mL, even if different aliquots of samples are needed. Compensate by changing the volume of added KPi buffer. Appropriate dilutions for measuring GPx activities in cytosolic fractions of marine organisms can vary, e.g. for samples homogenized 1:5, ranging between 1:10 and 1:50 in digestive gland of invertebrates and between 1:10 and 1:100 in fish liver. Proper dilutions need to be determined.

Calculations

Results are expressed in nmol min^{-1} mg^{-1} protein, or g (tissues), or mL (blood, plasma, or hemolymph). Normalization to tissue weight often precludes comparisons between different species or organs with different relative water content. When normalizing to mg of protein, comparability of values may be compromised by different tissue-specific protein content.

When the enzymatic activity is normalized to protein content:

$$\text{GPx activity (nmol min}^{-1}\text{mg}^{-1}\text{proteins)} = (\Delta Abs_{final\ sample}/-6.22) \times (\text{sample dilution}) \times 1000/\text{proteins}$$

where "$\Delta Abs_{final\ sample}$" is the change in absorbance per minute subtracted from the rate of blank reaction ($\Delta Abs_{final\ sample} = \Delta Abs_{sample} - \Delta Abs_{blank}$), "6.22" is the extinction coefficient (mM^{-1} cm^{-1}), "sample dilution" is the dilution factor (if any) applied to cytosolic fractions before the assay and "proteins" is the protein content in cytosolic fractions expressed as mg mL^{-1} (e.g. not diluted samples).

DETERMINATION OF GLUTATHIONE *S*-TRANSFERASES ACTIVITY (EC 2.5.1.18)

GST are a family of multifunctional enzymes with cytosolic, mitochondrial, and microsomal localization. They catalyze the conjugation of GSH to electrophilic centers of a wide variety of endogenous (e.g. peroxidized lipids) and exogenous substrates (e.g. organic xenobiotics). The latter reactions, known as phase II of biotransformations, facilitate the dissolution of lipophilic chemicals in the aqueous cellular and extracellular fluids, and thus their excretion (George 1994).

The spectrophotometric assay for GST activity is based on the GST-catalyzed reaction between GSH and a substrate, among which the 1-chloro-2, 4-dinitrobenzene (CDNB) has the broadest range of isozyme detectability (e.g. alpha-, mu-, pi- and other GST isoforms). The GST-catalyzed formation of GS-DNB produces a dinitrophenyl thioether that can be detected at 340 nm, $\varepsilon = -9.6$ mM^{-1} cm^{-1} (Habig and Jacoby 1981).

Reagents and Solutions

- 100 mM KPi buffer, pH 6.5 (adjust pH with concentrated KOH).
- 50 mM CDNB stock solution. Dissolve 50.6 mg of 1-chloro-2,4-dinitrobenzene in 5 mL of methanol. The stock solution is stable in the dark and at 4°C for 1 month.
- 100 mM GSH working solution (freshly prepared). Dissolve 0.0307 g of GSH (MW 307.32) in 1 mL of distilled water.
- Working buffer (100 mM KPi buffer with 1.5 mM CDNB). Prepare immediately before use by adding 1.5 mL of CDNB stock solution to 50 mL of 100 mM KPi buffer, pH 6.5.

Procedure

Spectrophotometric analyses are carried out at $\lambda = 340$ nm (light path 1 cm) at constant temperature (18 ± 1°C for temperate species).

Add to a plastic cuvette (final volume 1 mL):

- 965 μL of working buffer
- 15 μL of GSH working solution
- 20 μL of sample.

Mix vigorously after the addition of sample. Read increase in absorbance for at least 1 min. Reactions should be linear, otherwise a different dilution or sample volume should be used. Maintain a final volume of 1 mL even if different aliquots of samples are needed. Compensate by changing the volume of working buffer. Appropriated dilutions for measuring GST in cytosolic fractions of marine organisms can be variable, e.g. for samples homogenized 1:5, normally ranging between 1:10 and 1:100 in digestive gland of invertebrates and between 1:100 and 1:500 in fish liver. Proper dilutions must be tested for each new batch of samples.

Calculations

Results are usually expressed in nmol min^{-1} mg^{-1} protein or g of tissues. Normalization to tissue weight often precludes comparisons between different species or organs with different water content. When normalizing to mg of protein, comparability of values may be compromised by different tissue-specific protein contents.

When the enzymatic activity is normalized to protein content:

$$\text{GST activity (nmol min}^{-1}\text{mg}^{-1}\text{proteins)} = (\Delta\text{Abs}/9.6) \times (\text{sample dilution}) \times 1000/\text{proteins}$$

where "ΔAbs" is the variation of absorbance per minute, "9.6" is the extinction coefficient (mM^{-1} cm^{-1}), "sample dilution" is the dilution factor (if any) applied to the cytosolic fractions before the assay, and "proteins" is the protein content in cytosolic fractions expressed as mg mL^{-1} (e.g. not diluted samples).

DETERMINATION OF GLUTATHIONE REDUCTASE ACTIVITY (EC 1.6.4.2)

GR is the enzyme responsible for the reduction of GSSG to GSH, using NADPH as an electron donor. Although GR does not directly act as an antioxidant, it plays an indirect but nevertheless essential role in protecting cells from oxidative damage and in maintaining the proper redox status of GSH. The spectrophotometric assay for measuring GR activity is based on the absorbance decrease caused by the consumption of NADPH during the conversion of GSSG to GSH ($\lambda = 340$ nm, $\varepsilon = -9.6$ mM^{-1} cm^{-1}) (Meister 1989).

$$GSSG + NADPH \rightarrow 2GSH + NADP^{+}$$

Reagents and Solutions

- 100 mM KPi buffer, pH 7.0 (adjust pH to 7.0 with concentrated KOH).
- 100 mM EDTA stock solution.
- 10 mM GSSG working solution (freshly prepared). Dissolve 0.0061 g of GSSG (MW 612.63) in 1 mL of distilled water.
- NADPH (1 mg mL^{-1}) working solution (freshly prepared). Dissolve 5 mg of NADPH in 5 mL of distilled water.

Procedure

Spectrophotometric analyses are carried out at $\lambda =$ 340 nm (light path 1 cm) at constant temperature ($18 \pm 1^{\circ}$C for temperate species).

Add to a plastic cuvette (final volume 1 mL):

- 750 μL of KPi buffer
- 10 μL of EDTA stock solution
- 100 μL of GSSG working solution

INSTRUMENT AUTOZERO

- 100 μL of NADPH working solution

Read the absorbance, which should be approximately 0.6

- 40 μL of sample.

Mix vigorously after the addition of sample. Read the decrease in absorbance for at least 1 min. Reactions should be linear, otherwise a different dilution or sample volume should be used. Each sample is measured twice. Maintain a final volume of 1 mL even if different aliquots of samples are needed. Compensate by changing the volume of working buffer. Appropriate dilutions for measuring GR in cytosolic fractions of marine organisms usually vary, e.g. for samples homogenized 1:5, ranging between 1:10 and 1:100 in digestive gland of invertebrates and in fish liver. Proper dilutions should always be tested prior to new batch measurements.

Calculations

Results are usually expressed in nmol min^{-1} mg^{-1} protein or g of tissue. Normalization to the tissue weight often precludes comparisons between different species or organs with different relative water content. When normalizing to mg of protein, comparability of values may be compromised by different tissue-specific protein contents.

When the enzymatic activity is normalized to protein content:

$$\text{GR activity (nmol min}^{-1}\text{mg}^{-1}\text{proteins)} = (\Delta\text{Abs}/-6.22) \times \text{(sample dilution)} \times 1000/\text{proteins}$$

where "ΔAbs" is the variation of absorbance per minute, "−6.22" is the extinction coefficient (mM^{-1} cm^{-1}), "sample dilution" is the dilution factor (if any) applied to the cytosolic fractions before the assay, and "proteins" is the protein content in cytosolic fractions expressed as mg mL^{-1}.

DETERMINATION OF GLYOXALASE I (EC 4.4.1.5) AND II (EC3.1.2.6) ACTIVITIES

The glyoxalase system catalyses the detoxification of reactive α-ketoaldehydes formed in cellular oxidative processes. Using GSH as cofactor, GI forms an intermediate thiolester, which is subsequently hydrolysed by GII to the corresponding D-hydroxy acid, with GSH regeneration.

$$\text{R–CO–CHO} + \text{GSH} \rightarrow \text{R–CHOH–COSG} \quad \text{(GI)}$$

$$\text{R–CHOH–COSG} \rightarrow \text{GSH} + \text{R–CHOH–COOH} \quad \text{(GII)}$$

Increased activities of glyoxalase enzymes have been observed in marine organisms exposed to chemical pollutants, probably reflecting an efficient detoxification mechanism against enhanced levels of α-ketoaldehydes, extremely toxic and reactive compounds formed in cellular oxidative processes (Mannervik et al. 1989; Viarengo 1989; Viarengo et al. 1990; Regoli 1992; Regoli and Principato 1995).

The spectrophotometric assay for GI activity is based on the formation of S-D-lactoyl-glutathione from the hemimercaptal adduct of methylglyoxal (MG) and GSH (Principato et al. 1983). GII activity is followed by the reaction of 5,5′-dithiobis(2-nitrobenzoic acid) (DTNB)

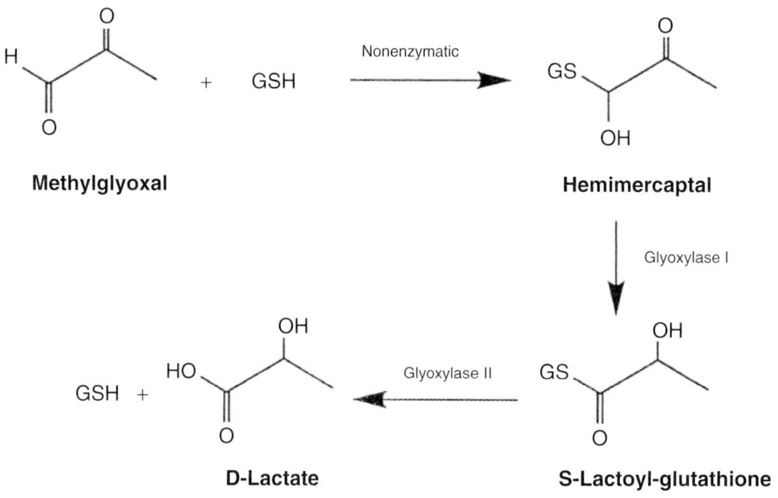

Fig. 27.3 Glyoxalase I generates an intermediate thioester and S-Lactoyl-glutathione from the adduct of methylglyoxal and GSH; glyoxalase II catalyzes the formation of D-lactate with recycle of GSH.

with GSH formed from S-D-lactoylglutathione (Principato et al. 1987) (Fig. 27.3).

Reagents and Solutions

- 100 mM KPi buffer, pH 6.8 (adjust pH with concentrated KOH).
- 100 mM GSH working solution (freshly prepared). Dissolve 0.0307 g of GSH (MW 307.32) in 1 mL of distilled water.
- Methylglyoxal 2% (v/v) working solution. Dilute 50 μL of a stock methylglyoxal solution (40%) into 950 μL of distilled water.
- 100 mM 3-(N-morpholino)propanesulfonic acid (MOPS), pH 7.2. Dissolve 2.09 g of MOPS in 100 mL of distilled water.
- 53 mM S-D- lactoylglutathione. Dissolve 0.02 g in 1 mL of distilled water.
- 20 mM DTNB stock solution (freshly prepared). Dissolve 0.07926 g of DTNB in 10 mL of methanol. This solution is stable for 3 months in the dark at 4°C.

Procedure

Spectrophotometric analyses are carried out at $\lambda = 240$ nm and $\lambda = 412$ nm (light path 1 cm), for GI and GII respectively, at constant temperature ($18 \pm 1°C$).

For GI activity, add to a quartz cuvette (final volume 1 mL):

- 840 μL of 100 mM KPi buffer
- 100 μL of methylglyoxal working solution
- 10 μL of GSH working solution
- 50 μL of blank (KPi buffer) or sample.

For GII activity, add to a plastic cuvette (final volume 1 mL):

- 930 μL of MOPS buffer
- 10 μL of S-D- lactoylglutathione working solution*
- 10 μL of DTNB stock solution
- 50 μL of blank (KPi buffer) or sample.

*After adding S-D-lactoylglutathione a spontaneous hydrolysis is observed; wait for the stabilization of absorbance before adding blank or sample.

In both GI and GII assays, mix vigorously after the addition of blank/samples. Register the increase in the absorbance for at least 1 min. Reactions should be linear, otherwise different dilutions or sample volumes should be used in the assay. Each sample is measured twice. Maintain a final volume of 1 mL even if different aliquots of sample are needed. Compensate by changing the volume of KPi buffer for GI or MOPS for GII. Appropriate dilutions for measuring glyoxalase activities in cytosolic fractions of marine organisms can vary, e.g. for samples homogenized 1:5, ranging between 1:10 and 1:50 in digestive gland of invertebrates and between 1:100 and 1:500 in fish liver (Regoli and Principato 1995; Romero-Ruiz et al. 2003). Proper dilutions should always be tested prior to new batch measurements.

Calculations

Results are usually expressed in μmol min^{-1} mg^{-1} protein or g of tissue. Normalization to tissue weight often precludes comparisons between different species or organs with different relative water content. When normalizing to mg of protein, comparability of values may be compromised by different tissue-specific protein contents.

When the enzymatic activity is normalized to protein content:

$$\text{GI activity}(\mu\text{mol min}^{-1}\text{mg}^{-1}\text{proteins}) = (\Delta\text{Abs}_{\text{sample}}/3.37) \times (\text{sample dilution})/\text{proteins}$$

$$\text{GII activity}(\mu\text{mol min}^{-1}\text{mg}^{-1}\text{proteins}) = (\Delta\text{Abs}_{\text{sample}}/13.6) \times (\text{sample dilution})/\text{proteins}$$

where "ΔAbs_{sample}" is the variation of absorbance per minute, "3.37" and "13.6" are the extinction coefficients (mM^{-1} cm^{-1}) for GI and GII respectively, "sample dilution" is the dilution factor (if any) applied to the cytosolic fractions before the assay, and "proteins" is the protein content in cytosolic fractions expressed as mg mL^{-1} (e.g. not diluted samples).

SPECTROPHOTOMETRIC DETERMINATION OF TOTAL GLUTATHIONE

GSH is the most abundant cellular thiol involved in metabolic and transport processes, and in cell protection against the toxic effects of a variety of endogenous and exogenous compounds, including trace metals and ROS (Meister and Anderson 1983). GSH plays multiple protective roles against oxidative stress, acting as a direct scavenger of ROS, and also as cofactor of several antioxidant enzymes. GSSG is reconverted to GSH by GR, but when oxidative processes exceed the reducing capacity of GR, the excess of GSSG is excreted resulting in a net loss of total GSH from the tissue (Meister 1989).

Total GSH concentration (GSH + 2GSSG) is measured in a colorimetric reaction using DTNB, also known as Ellman's Reagent, which reacts with thiolic compounds. 2 GSH react with DTNB to generate GSSG and 2-nitro-5-thiobenzoic acid (Akerboom and Sies 1981). GSSG in the sample is reconverted by GR to GSH, which undergoes another reaction, with a continuous formation of 2-nitro-5-thiobenzoic acid, proportional to total GSH content. The formation of 2-nitro-5-thiobenzoic acid is monitored at $\lambda = 412$ nm (Fig. 27.4).

Tissue Homogenates

To eliminate the interference of thiolic groups of proteins, tissues are homogenized in 5% sulphosalicylic acid with 4 mM EDTA. The homogenization buffer is prepared by dissolving 5 g of sulfosulfosalicylic acid in 100 mL of distilled water and adding 4 mL of 100 mM EDTA. This solution is stable for 1 month at room temperature. Samples are homogenized (1:5 w : v) at 4°C using a hand-held glass potter with a teflon or glass pestle, maintained on ice for 45 min for de-proteinization and centrifuged at 37,000 *g* for 15 min at 4°C. Supernatants are collected, assayed or subdivided into aliquots (100–200 μL) and stored at

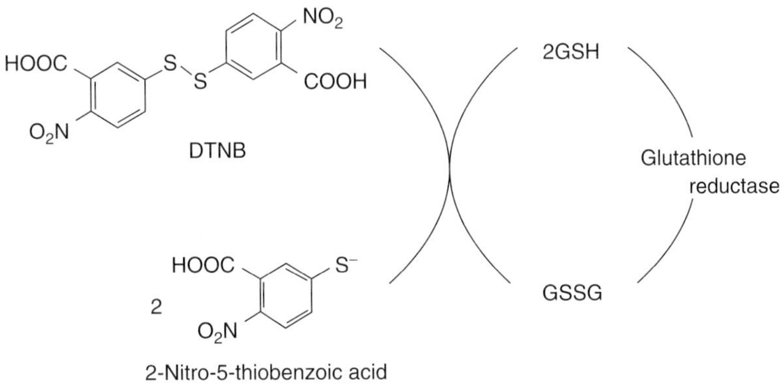

Fig. 27.4 Glutathione (GSH) reacts with DTNB to generate 2-Nitro-5-thiobenzoic acid and GSSG that is reconverted to GSH by glutathione reductase.

−80°C. Total GSH content can be normalized to tissue weight or protein content (see below). In the latter case, pellets are re-suspended in 5 mL of 1 M NaOH for protein measurement.

Reagents and solutions

- 100 mM KPi buffer, pH 7.0 (adjust pH with concentrated KOH).
- 100 mM EDTA stock solution.
- 20 mM DTNB stock solution. Dissolve 0.07927 g of DTNB in 10 mL of methanol. This solution is stable for 3 months in the dark at 4°C.
- NADPH ($4\ mg\ mL^{-1}$) working solution (freshly prepared). Dissolve 4 mg of NADPH in 1 mL of distilled water.
- GR ($100\ U\ mL^{-1}$) working solution (prepare immediately before use). Depending on the product, dilute GR to $100\ U\ mL^{-1}$ in cold distilled water.
- Working buffer. Add 1 mL of 100 mM EDTA to 100 mL of KPi buffer.

Standards

Both GSH or GSSG can be used as standards.

1. When using GSH standards, prepare a fresh stock solution of 100 mM GSH by dissolving 0.0307 g of GSH in 1 mL of distilled water. Obtain at least three standards by diluting the stock solution: 10 μM GSH, 20 μM GSH and 30 μM GSH. These standards are diluted 10-fold in the assay, giving 1 μM, 2 μM, 3 μM final GSH concentrations in the calibration curve.
2. When using GSSG standards, prepare a fresh stock solution of 1 mM GSSG by dissolving 0.00325 g of GSSG in 5 mL of distilled water. Obtain at least three standards by diluting the stock solution: 5 μM GSSG, 10 μM GSSG and 15 μM GSSG. These standards will be diluted 10-fold in the assay, giving 0.5 μM, 1 μM, 1.5 μM final GSSG concentrations.

Procedure

Spectrophotometric analyses are carried out at $\lambda =$ 412 nm (light path 1 cm) and at constant temperature (18 ± 1°C for temperate species). Blank and standard reactions generate a calibration curve to relate absorbance to glutathione concentration, which will be used to quantify GSH content in the samples.

For blank and standard reactions, add to a plastic or glass cuvette (final volume 1 mL):

- 835 μL of working buffer
- 100 μL of blank (KPi buffer) or standards
- 5 μL of DTNB
- 50 μL of NADPH
- 10 μL of GR.

For sample reactions, add to a plastic cuvette (final volume 1 mL):

- 835 μL of working buffer
- 100 μL of sample
- 5 μL of DTNB
- 50 μL of NADPH
- 10 μL of GR.

Mix vigorously after the addition of GR. Read increase in absorbance for at least 1 min. Reactions should be linear, otherwise a different dilution or sample volume should be used. Maintain a final volume of 1 mL, even if different aliquots of samples are needed. Compensate by changing the volume of working buffer. Appropriate dilutions for measuring GSH content in tissues of marine organisms usually vary, e.g. for samples homogenized 1:5, ranging between 1:10 and 1:100 in digestive gland of invertebrates and in fish liver. Proper dilutions should always be tested.

Calculations

The ΔAbs of blank and standards are used to obtain the linear calibration curve between absorbance and GSH or GSSG concentration ($y = ax + b$). This equation allows calculation of the GSH concentration in samples (in $\mu M = \mu mol\ L^{-1}$). Results are expressed in GSH equivalents and normalized to tissue weight (or protein content).

To normalize to tissue weight when GSH has been used as standard:

$$\mathrm{GSH} + 2\mathrm{GSSG}\ (\mu\mathrm{mol\ g^{-1}tissue}) = (\mathrm{concentration}/1000) \times (\mathrm{sample\ dilution}) \times (\mathrm{w/v\ ratio}) \times 1$$

To normalize GSH concentration to tissue weight when GSSG has been used as standard:

$$\text{GSH} + 2\text{GSSG}\ (\mu\text{mol g}^{-1}\text{tissue}) = (\text{concentration}/1000) \times (\text{sample dilution}) \times (\text{w/v ratio}) \times 2$$

where, "concentration" is glutathione concentration (as GSH or GSSG depending on the standard used) derived from the calibration curve, "sample dilution" is the dilution factor (if any) applied to homogenized fractions before the assay, and "w/v ratio" is the homogenization ratio.

To normalize glutathione concentration to protein content:

$$\text{GSH} + 2\text{GSSG}(\text{nmol mg}^{-1}\text{protein}) = (\text{GSH} + 2\text{GSSG})/\text{protein}$$

where "GSH + 2GSSG" is GSH concentration previously calculated and expressed as μmol g^{-1} of tissue, and "protein" is the protein content measured in the pellet obtained after centrifugation of homogenates and expressed as mg g^{-1} of tissue.

REFERENCES

Aebi, H. (1984) Catalase *in vitro*. *Methods in Enzymolology* 105, 121–126.

Akerboom, T.P.M., Sies, H. (1981) Assay of glutathione, glutathione disulfide and glutathione mixed disulfides in biological samples. *Methods in Enzymology* 71, 373–382.

Bergmeyer, H.U. (ed.) (1965) *Methods of Enzymatic Analysis*. Academic Press, New York, pp. 885–894.

Chance, B., Boveris, A., Oshino, N., Loschen, G. (1973) The nature of the catalase intermediate in the biological function. *In* King, T., Mason, H., Morrison, M. (eds). *Oxidases and Related Redox Systems*, Vol. I. University Park Press, Baltimore.

De Almeida, E.A., Bainy, A.C.D., de Melo Loureiro, A.P. et al. (2007) Oxidative stress in *Perna perna* and other bivalves as indicators of environmental stress in the Brazilian marine environment: antioxidants, lipid peroxidation and DNA damage. *Comparative Biochemistry and Physiology A* 146, 588–600.

Fernandez, C., San Miguel, E., Fernandez-Briera, A. (2009) Superoxide dismutase and catalase: tissue activities and relation with age in the long-lived species *Margaritifera margaritifera*. *Biological Research* 42, 57–68.

Fridovich, I. (1986) Superoxide dismutases. *Advances in Enzymology and Related Areas of Molecular Biology* 58, 61–97.

George, S.G. (1994) Enzymology and molecular biology of phase II xenobiotic conjugation enzymes in fish. *In* Malins, D.C., Ostrander, G.K. (eds). *Molecular Biological Approaches to Aquatic Toxicology*. Lewis Publisher, Boca Raton, FL, pp. 37–85.

Habig, W.H., Jacoby, W.B. (1981) Assays for differentiation of glutathione S-transferases. *Methods in Enzymology* 77, 398–405.

Halliwell, B., Aruoma, O.I. (1991) DNA damage by oxygen derived species. Its mechanism and measurement in mammalian system. *FEBS Letters* 281, 9–19.

Halliwell, B., Gutteridge, J.M.C. (2001) *Free Radicals in Biology and Medicine*. University Press, Oxford, UK, pp. 936.

Halliwell, B., Gutteridge, J.M.C. (2007) *Free Radicals in Biology and Medicine*, 4rd edn. Oxford University Press, New York.

Lawrence, R.A., Burk, R.F. (1976) Glutathione peroxidase activity in selenium-deficient rat liver. *Biochemical and Biophysical Research Communications* 71, 952–958.

Mannervik, B.I., Carlberg, I., Larson, K. (1989) Glutathione. General review of mechanisms of action. *In* Dolphin, D., Poulsen, R., Wiley A.O. (eds) *Glutathione: Chemical, Biochemical and Medical aspects*, Part A. John Wiley & Sons, New York, pp. 475–516.

McCord, J.M., Fridovich, I. (1976) Superoxide dismutase: an enzymatic function for erythrocuprein (hemocuprein). *Journal of Biological Chemistry* 244, 6049–6055.

Meister, A. (1989) On the biochemistry of glutathione. *In* Taniruchi, N., Higashi, T., Sakamoto, S., Meyster, A. (eds) *Glutathione Centennial: Molecular Prospectives and Clinical implications*. Academic Press, San Diego, CA, pp. 3–22.

Meister, A., Anderson, M.E. (1983). Glutathione. *Annual Review of Biochemistry* 52, 711–760.

Mueller, S., Riedel, H.D., Stemmel, W. (1997) Direct evidence for catalase as the predominant H_2O_2-removing enzyme in human erythrocytes. *Blood* 90, 4973–4978.

Principato, G.B., Locci, P., Rosi, G. et al. (1983) Activity changes of glyoxalase I–II and glutathione reductase in regenerating rat liver. *Biochemical International* 6, 249–255.

Principato, G.B., Rosi, G., Talesa, V. et al. (1987) Purification and characterization of two forms of glyoxalase II from the liver and brain of Wistar rats. *Biochimica et Biophysica Acta* 911, 349–355.

Regoli, F. (1992) Lysosomal responses as a sensitive stress index in biomonitoring heavy metal pollution. *Marine Ecology Progress Series* 84, 63–69.

Regoli, F., Principato, G.B. (1995) Glutathione, glutathione-dependent and antioxidant enzymes in mussels, *Mytilus galloprovincialis*, exposed to metals under field and laboratory conditions: implications for the use of chemical biomarkers. *Aquatic Toxicology* 31, 143–164.

Regoli, F., Pellegrini, D., Winston, G.W. et al. (2002) Application of biomarkers for assessing the biological impact of dredged materials in the Mediterranean: The relationship between antioxidant responses and susceptibility to oxidative stress in the red mullet (*Mullus barbatus*). *Marine Pollution Bulletin* 44, 912–922.

Romero-Ruiz, A., Amezcua, O., Rodríguez-Ortega, M.J., Muñoz, J.L. et al. (2003) Oxidative stress biomarkers in bivalves transplanted to the Guadalquivir estuary after Aznalcollar spill. *Environmental Toxicology and Chemistry* 22, 92–100.

Viarengo, A. (1989) Heavy metals in marine invertebrates: mechanisms of regulation and toxicity at the cellular levels. *Critical Review in Aquatic Science* 1, 295–317.

Viarengo, A., Canesi, L., Pertica, M. et al. (1990) Heavy metals effects on lipid peroxidation in the tissues of *Mytilus galloprovincialis* Lam. *Comparative Biochemistry and Physiology C* 97, 37–42.

Vlahogianni, T., Dassenakis, M., Scoullos, M.J., Valavanidis, A. (2007) Integrated use of biomarkers (superoxide dismutase, catalase and lipid peroxidation) in mussel *Mytilus galloprovincialis* for assessing heavy metals' pollution in coastal areas from the Saronikos Gulf of Greece. *Marine Pollution Bulletin* 54, 1361–1371.

Chapter 28

EVALUATION OF GLUTATHIONE STATUS IN AQUATIC ORGANISMS

Eduardo Alves de Almeida[1], *Danilo Grunig Humberto Silva*[1], *Afonso Celso Dias Bainy*[2], *Florêncio Porto Freitas*[3], *Flávia Daniela Motta*[3], *Osmar Francisco Gomes*[3], *Marisa Helena Gennari de Medeiros*[3], *and Paolo Di Mascio*[3]

[1]Department of Chemistry and Environmental Sciences, IBILCE-UNESP, São José do Rio Preto, São Paulo, Brazil
[2]Department of Biochemistry, Federal University of Santa Catarina, UFSC, Florianópolis, Santa Catarina, Brazil
[3]Department of Biochemistry, IQ-USP, São Paulo, Brazil

Glutathione (δ-glutamylcysteinylglycine, GSH, Fig. 28.1) is the predominant nonprotein sulfhydryl in cells which plays important roles in antioxidant defense, xenobiotic, eicosanoid and estrogen metabolism, and regulation of the cell cycle and gene expression (Meister and Anderson 1983; Sies and Ketterer 1988; Meister 1989, 1992). GSH is used as electron donor in reductive processes essential for synthesis and degradation of proteins, formation of deoxyribonucleotides, and for the reduction of H_2O_2 and organic peroxides by glutathione peroxidases (GPx). Oxidized glutathione is generally bound to another oxidized molecule to form glutathione disulfide (GSSG).

Cellular GSH levels vary between tissues and species from mM concentrations in the cytosol of mammalian hepatocyte to μM levels in blood plasma and about 1–2 mM in most cells. GSSG is generally maintained at 1–10% of total glutathione through reduction by glutathione disulphide reductase (GR), an enzyme which uses electrons from nicotinamide adenine dinucleotide phosphate hydrogen (NADPH) derived from pentose phosphate shunt (Akerboom and Sies 1981; Meister 1988; Forman et al. 2009). Examples of glutathione concentration in fish liver can be found in Chapter 2. When cells are exposed to oxidative stress, GSSG levels can increase significantly with a concomitant decrease in GSH (Baskin and Salan 1997). Therefore, the GSH/GSSG ratio is frequently used to evaluate oxidative stress status in biological systems (Araujo et al. 2008).

Oxidative Stress in Aquatic Ecosystems, First Edition. Edited by Doris Abele, José Pablo Vázquez-Medina, and Tania Zenteno-Savín.

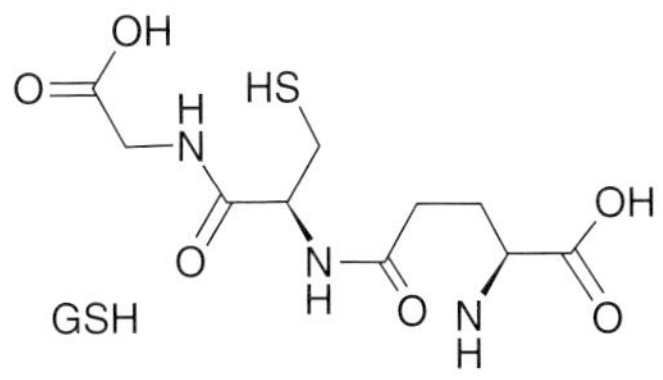

Fig. 28.1 Structure of the tripeptide glutathione.

GSH is also conjugated with a great variety of both endogenous (estrogens, prostaglandins, leucotrienes) and exogenous (drugs, organic xenobiotic pollutants) compounds through catalysis by glutathione *S*-transferases (GST; Boyland and Chasseaud 1969). Conjugation with GSH is an essential aspect of the metabolism of xenobiotics, and increased rates of conjugation reactions can also lead to GSH depletion in cells, causing increased susceptibility to oxidative stress. Toxic metals and carbon-centered radicals are able to bind GSH causing its depletion in cells (Viarengo 1989; Mason and Jenkins 1996; Dickinson and Forman 2002). Thus, the evaluation of the glutathione status can also be useful to assess the exposure of animals to environmental pollutants like metals and organic xenobiotics (Conners and Ringwood 2000; Wu et al. 2007).

Due to the relatively low concentrations of GSH and especially GSSG in cells, their measurement in biological samples requires very sensitive and accurate methods. Several methods have been established for glutathione measurement in organisms, and in this chapter we will describe three of the most frequently applied methods for GSH and GSSG measurements: (i) spectrophotometric measurement of continuous GSSG reduction promoted by glutathione reductase in the presence of NADPH; (ii) HPLC coupled to electrochemical detection; and (iii) HPLC coupled to fluorescence detection.

SPECTROPHOTOMETRIC MEASUREMENTS

The total amount of glutathione (GSH + GSSG) can be measured in tissues or body fluids based on the ability of GSH to react with 5,5′-dithio-bis(2-nitrobenzoic acid) (DTNB, Ellmann's Reagent; Fig. 28.2) (Akerboom and Sies 1981), producing the conjugate GS-TNB and the yellow TNB (5′-thio-2-nitrobenzoic acid) anion that can be detected at 412 nm. The rate of TNB production is proportional to the GSH concentration in the extract.

The conjugate GS-TNB can further react with another GSH molecule producing GSSG and another free TNB, so the more GSH in the sample the more TNB is formed, which is monitored at 412 nm during 1 or 2 min. In order to improve the spectrophotometric signal, a recycling reaction catalyzed by GR in the

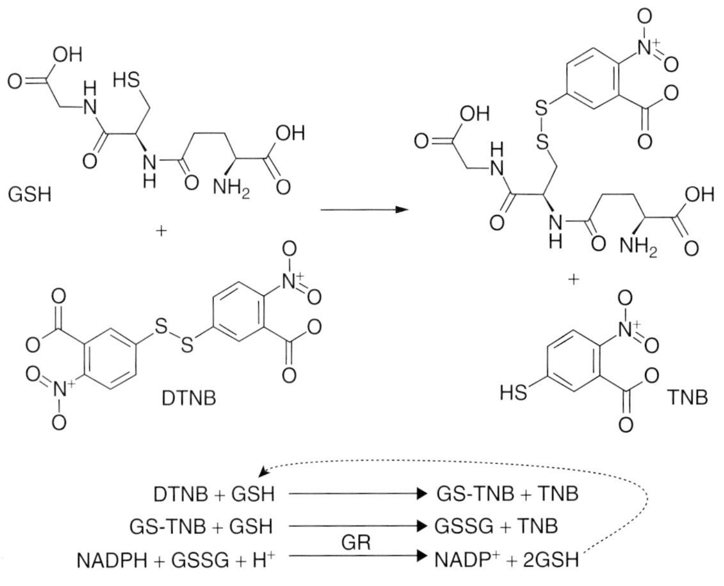

Fig. 28.2 A general scheme for total GSH measurement through spectrophotometry.

Box 28.1 General protocol for total GSH measurement by spectrophotometry

This protocol has been proposed by Tietze (1969) and was modified by Akerboom and Sies (1981).

Sample processing

Tissues

Tissues are homogenized in 10 volumes (w/v) of 0.5 M perchloric acid and centrifuged at 4°C for 10 min at 3020 *g*. The supernatant must be neutralized using 1.75 M K_3PO_4 and immediately centrifuged at 15,600 *g* for 1 min.

Blood

500 μL of total blood are precipitated with perchloric acid 2 M, containing 4 mM EDTA and centrifuged for 2 min at 15,600 *g*. The supernatant is neutralized with 2 M KOH, containing 0.3 M MOPS (morpholino propanesulphonic) and centrifuged again at 15,600 *g*.

Assay and calculation

- The assay mixture for regular 1 cm cuvettes (1 mL final volume) contains:
  0.1 M potassium phosphate buffer with 1 mM EDTA, pH 7.0,
  240 μM NADPH,
  76 μM DTNB,
  0.12 units mL^{-1} GR.
- Start assay by adding 10–20 μL of sample extract and read absorbance ΔA_{412} at 25°C.
- The increase of the absorbance rate per min is recorded during 2 min at 412 nm, 25°C, measuring the formation of 5-thio-2-nitrobenzoate (TNB).
- The GSH concentration is then calculated comparing the result to a 10 μM GSSG standard and expressed as μmoles of total glutathione per mg wet tissue or per mL total blood.

presence of NADPH should be coupled to the assay, regenerating 2GSH from the GSSG previously formed by the reaction with DTNB. This allows a continuous production of TNB, which is proportional to the amount of total GSH (GSSG + GSH) in the sample. The change in absorbance per minute (ΔA) obtained in the assay is then used to calculate the μmol GSH g^{-1} of tissue in the sample based on a previously run standard curve with known amounts of GSH.

With the same technique it is also possible to measure only oxidized GSSG levels. Briefly, all free GSH is first removed from the sample by incubating the homogenate with *N*-ethylmaleimine (NEM) or 4-vinylpyridine (4-VP), both of which form a stable complex with GSH at room temperature, and remove it from the reaction with DTNB. NEM has an inhibiting effect on GR activity which is not the case with 4-VP. Either of these scavengers can be used after testing the minimal amount that needs to be added and the incubation time necessary to remove all GSH from the sample. The concentration of the scavenger should optimally be 10 times greater than the expected GSH concentration in the sample. The GSH content can then be calculated as the difference between total GSH and GSSG. It should be noted that in many samples the GSSG concentration will be below detection limit of this method. A general protocol for GSH measurement through this method is given in Box 28.1.

HIGH PERFORMANCE LIQUID CHROMATOGRAPHY

Fluorescence Detection

Different HPLC-fluorescence methods can be found in the literature to measure GSH and GSSG levels in animal samples. They provide robust and specific analytical methodology for the quantification of intracellular thiols, but sample derivatization is always needed. Several compounds are available for derivatization, such as dinitrobenzyl derivatives, 7-fluoro-2,1,3-benzoxadiazole-4-sulfonate (SBD-F), monobromobimane, dansyl chloride, 5-methyl-(2-(*m*-iodoacetylaminophenyl)-benzoxazole) (MIPBO), and *orto*-phthaldialdehyde (OPA) (Kand'ár et al. 2007).

The sensitivity of the procedure in which dinitrophenyl derivatives are generated in the range of 500 pmol per sample is adequate for tissues with high GSH and GSSG concentrations such as the liver, but requires large sample sizes in extrahepatic tissues. Moreover, since GSSG is normally present at concentrations at least 10-fold less than reduced GSH, GSSG often cannot be measured. The fluorescent bimane adducts provide a methodology with excellent sensitivity (of the order of 0.5 pmol) but the method is not capable of monitoring disulfides directly. Two-step derivatization of glutathione and glutathione disulfide with iodoacetate and dansyl chloride yields derivatives that can be monitored with high sensitivity (1–2 pmol) (Lakritz et al. 1997).

A simple and sensitive protocol to measure both GSH and GSSG in different biological samples like plasma and liver by previous derivatization with OPA was developed by Kand'ár et al. (2007), based on the previous method described by Hissin and Hilf (1976) (see Box 28.2). Reduced glutathione reacts with *o*-phthaldialdehyde (OPA) to form a stable, highly fluorescent tricyclic derivate at pH 8, whereas GSSG reacts with OPA at pH 12. For the measurement of GSSG, GSH is complexed to *N*-ethylmaleimide. This method provides excellent sensitivity, precision, and accuracy in follow-up samples, with a limit of detection of 14.0 fmol for GSH, and 5.6 fmol for GSSG.

Electrochemical Detection

Another useful and much employed method to measure GSH and GSSG in biological samples is the use of HPLC coupled to electrochemical detection (ECD).

Box 28.2 General protocol for GSH and GSSG measurement in biological samples with HPLC and fluorescence detection (Hissin and Hilf 1976)

Sample preparation

Samples should be diluted or homogenized (1:10 w/v) in cold 10% metaphosphoric acid, incubated at 4°C for 10 min, and centrifuged at 22,000 *g* for 15 min at 4°C. Supernatants are collected and stored at −80°C. Glass homogenizers must be pre-cooled prior to use and homogenization should be carried out on ice.

GSH assay

Add 1 mL of 0.1% EDTA in 0.1 M NaH_2PO_4, pH 8.0 to 50 μL of the sample supernatant and mix well. To 20 μL of this mixture, add 300 mL of 0.1% EDTA in 0.1 M NaH_2PO_4 and 20 μL of 0.1% OPA in methanol and mix well. The mixtures are incubated at 25°C (room temperature) for 15 min in the dark. Filter the mixture through a 0.2 μM nylon filter (Nalgene) and store at 4°C, or inject (20 μL) into the HPLC system.

GSSG assay

Add 200 μL of 40 mM N-ethylmaleimine to 200 μL of sample supernatant and incubate for 25 min in the dark at 25°C. Then, add 750 μL of 0.1 M NaOH and store at 4°C, or inject (20 μL) into the HPLC system.

HPLC conditions

Column: C18, 150 mm × 4 mm i.d., 5 μM, at 37°C.
Mobile phase: 15% methanol in 25 mM NaH_2PO_4 (v/v), pH 6.0, pumped isocratically at a flow rate of 0.5 mL min^{-1}.
Peak detection at λ_{ex}: 350, λ_{em}: 420 nm.

Earlier methods were based on the reduction of the disulfide at a first Au/Hg amperometric electrode followed by its oxidation at a second downstream electrode (Kleinman and Richie 1995). Samples are isocratically separated at 1 mL min^{-1} on a C18 column with a mobile phase consisting of 93:25% (v/v) 0.1 M monochloroacetic acid, 5% methanol, 1.75% *N,N*-dimethylformamide, and 2.25 mM heptanesulfonic acid adjusted to a final pH of 2.8. With this method, GSH and GSSG signals elute after ~6 and 12 min, respectively.

Caution must be taken to rigorously exclude oxygen from the system by degassing all solvents. Another complication is that the Au/Hg electrodes lose sensitivity after a few hundred injections and must be reconditioned. Further, a high potential (see below) should be applied in electrochemical cells to obtain a good GSSG signal in the chromatogram, since many thiol disulfides are found in very low concentrations in tissues and body fluids and produce a high baseline noise.

More recent studies have indicated porous graphite coulometric electrodes to be more robust and sensitive for GSH and GSSG quantification in biological samples (Rodríguez-Ariza et al. 1994). Although here high oxidation potentials are also required, at least for GSSG detection, peaks from both GSH and GSSG can be observed easily, even in those tissues where concentrations are low.

This procedure represents a robust and simple technique where the sample passes through an electrochemical cell in which the electrochemical potential reduces or oxidizes the chemical compound to be analyzed according to its redox potential. Porous graphite electrodes increase sensitivity because the sample passes directly through the electrode. In contrast, in amperometric detectors in which the electrode is parallel to the sample/solvent flow, only compounds close to the electrode are oxidized/reduced, whereas compounds flowing at greater distance from the electrode are not properly captured by the electrochemical potential (Fig. 28.3).

The patented ESA® (Bedford, MA) coulometric detectors are usually the most used and have the advantage to possess two working electrodes in the electrochemical cell, allowing for parallel detection of compounds with different electrochemical potentials. The detector can be adjusted to lower potential (i.e. +0.65 V) at electrode 1 for GSH monitoring, while electrode 2 is adjusted to a higher potential

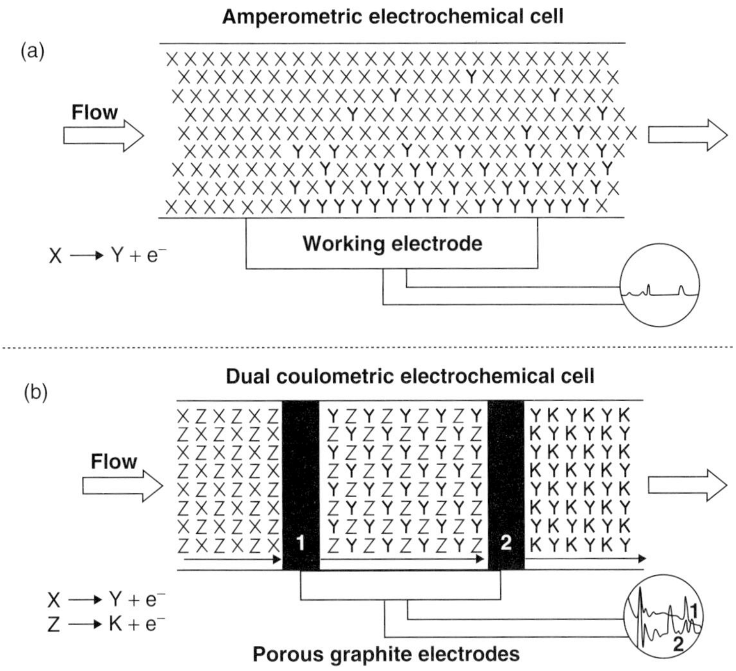

Fig. 28.3 Schemes showing main differences between (a) amperometric electrochemical detectors and (b) the dual coulometric electrochemical cell. In the amperometric detectors, the working electrode is generally parallel with the solute flow, and just those samples close to the electrode suffer its potential. In coulometric detectors, the porous graphite electrodes are placed perpendicular to the solute flow, so that all samples pass through the electrode, being fully oxidized or reduced.

(i.e. +0.90 V) for GSSG monitoring. A guard cell set at +0.9 V oxidizes the mobile phase before it passes through the analytical electrochemical cell, decreasing mobile phase noise in the chromatogram.

The novel boron-doped diamond (BDD) thin-film electrode offers advantages over other carbon-based working electrodes in the detection of GSH and GSSG (Fig. 28.3). This electrode works in a wide potential window in aqueous solutions. It is possible to apply potentials over +1.2 V, obtaining maximum response for GSH and GSSG (Rodríguez-Ariza et al. 1994), achieving a good signal-to-noise ratio. Another advantage of this system is that the potential does not cause mobile phase oxidation, a problem generally found with glassy carbon or graphitic electrodes, and resulting in a decrease in the signal to noise ratio. Other attractive features of BDD electrodes are: low background currents that do not reduce electrode lifetime; long-term stability of response, low sensitivity to dissolved oxygen, inertness and lack of fouling.

Box 28.3 Protocol for GSH and GSSG measurement with HPLC coupled to a dual coulometric electrochemical detector (Rodríguez-Ariza et al. 1994)

Sample preparation

- Biological fluids: add 5–10% of trichloroacetic acid to the fluid to precipitate proteins. Centrifuge samples at 5000 *g* for 5 min, collect the supernatant, filter through a syringe filter (0.22 μM) and inject directly into the HPLC system.
- Tissues: can be homogenized in 5% metaphosphoric acid or 10% perchloric acid. After centrifugation of the sample at 5000 *g* for 5 min, collect the supernatant, filter through a syringe filter (0.22 μM) and inject directly into the HPLC system.
- Once prepared, samples can be stored at −80°C until analysis, but it is recommended to perform the analysis immediately after sample preparation to avoid GSH oxidation during the freezing process. Also, it is strongly recommended that all procedures are carried out on ice and cooled, also to avoid GSH oxidation and artifactual results.
- The injection volume for the HPLC system depends on the GSH concentration in your sample. To inject larger volumes, you have to change the sample loop. Remember that the injected volume should be no less than half the loop volume, and for accurate reproducibility it is best to "overfill" the loop. For most tissues, a volume of 20 μL or 50 μL will be sufficient, and you have to test this for your sample material. Sample dilution can be done using the mobile phase.

HPLC conditions

- Mobile phase: 50 mM NaH_2PO_4
  0.25 mM octanesulfate acid
  pH 2.5
  2% acetonitrile.

*Important: adjust pH with orthophosphoric acid, because HCl interferes with the electrochemical signal.

- Column: C18 (250 × 4.6 mm, 5 μM).
- Electrochemical potentials (for ESA5011 cell) [for BDD cell]:

  Guard cell: (+0.90 V) [+0.90 V] (mobile phase)
  Electrode 1: (+0.65 V) [+0.90 V] (GSH)
  Electrode 2: (+0.90 V) [+1.50 V] (GSSG).
- Flow rate: 1 mL min^{-1}, isocratic.

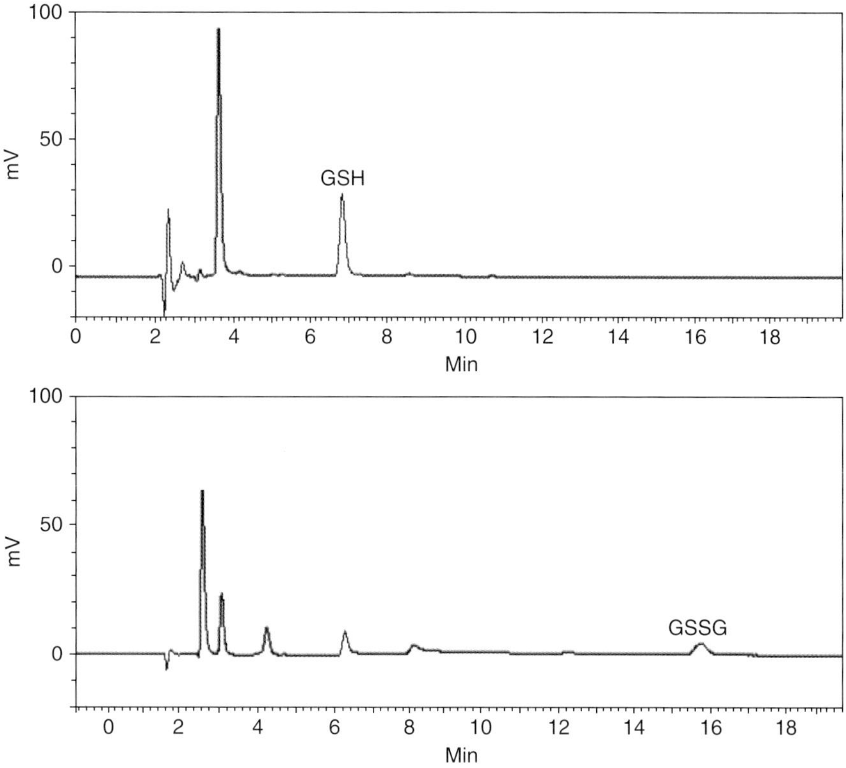

Fig. 28.4 Chromatogram of a fish liver sample, showing the peaks of GSH (channel 1) and GSSG (channel 2), using the ESA5011 model electrochemical cell.

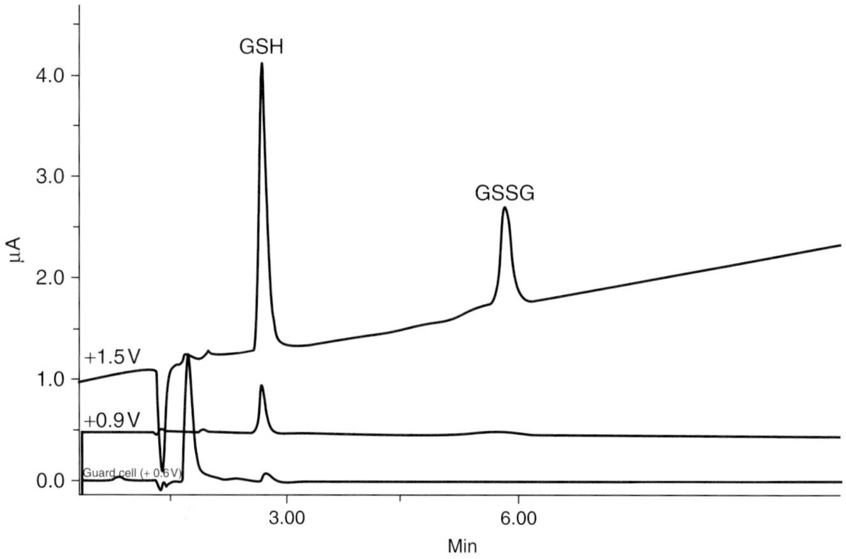

Fig. 28.5 Standard chromatogram of GSH and GSSG (100 pmol) using the BDD electrode. Mobile phase: 25 mM sodium phosphate, 50 µM octanesulfonic acid, 1% acetonitrile; flow: 0.7 mL min^{-1}, isocratic; column Kinetex 2.6µ C18 100 Å (100 × 4.6 mm, Phenomenex).

Generally, ion-pair chromatography (IPC) is the most widely applied method in combination with ECD. Octanesulfate acid is the most routinely used ion-pair reagent. In IPC, column equilibration is generally slower, which can generate problems when gradient elution is applied. Moreover, when many samples are sequentially run for an extended time (over night or during several days), a slight change in retention time can occur. In this case, it is necessary to remove the ion-pairing reagents from the column with sequential washings of the column, whereafter re-equilibration is necessary.

Box 28.3 contains a general procedure to measure GSH and GSSG levels through HPLC coupled to a coulometric ECD, using the ESA5011 and the BDD electrochemical cells. The chromatogram for a fish liver sample analyzed using these conditions is shown in Figs 28.4 and 28.5, respectively. Good GSH and GSSG resolution can be achieved using this method, and the concentration of the thiols in samples can be calculated from separate calibration curves for GSH and GSSG. Reported detection limits for GSH and GSSG are in the range of 0.5 and 1.0 pmol, respectively.

REFERENCES

Akerboom, T.P.M., Sies, H. (1981) Assay of glutathione, glutathione disulfide and glutathione mixed disulfides in biological samples. *Methods in Enzymology* 77, 373–382.

Araujo, A.R.T.S., Saraiva, M.L.M.F.S., Lima, J.L.F.C. (2008) Determination of total and oxidized glutathione in human whole blood with a sequential injection analysis system. *Talanta* 74, 1511–1519.

Baskin, S.I., Salem, H. (1997) *Oxidants, Antioxidants and Free Radicals*. Taylor & Francis, Washington, pp. 173–174.

Boyland, E., Chasseaud, L.F. (1969) The role of glutathione and glutathione S-transferases in mercapturic acid biosynthesis. *Advances in Enzymology and Related Areas of Molecular Biology* 32, 173–219.

Conners, D.E., Ringwood, A.H. (2000). Effects of glutathione depletion on copper cytotoxicity in oysters (*Crassostrea virginica*). *Aquatic Toxicology* 50, 341–349.

Dickinson, D.A., Forman, H.J. (2002) Cellular glutathione and thiols metabolism. *Biochemical Pharmacology* 64, 1019–1026.

Forman, H.J., Zhang, H., Rinna, A. (2009) Glutathione: overview of its protective roles, measurement, and biosynthesis. *Molecular Aspects of Medicine* 30, 1–12.

Hissin, P.J., Hilf, R. (1976) A fluorometric method for determination of oxidized and reduced glutathione in tissues. *Analytical Biochemistry* 74, 214–226.

Kand'ár, R., Žáková, P., Lotková, H., Kučera, O., Červinková, Z. (2007) Determination of reduced and oxidized glutathione in biological samples using liquid chromatography with fluorimetric detection. *Journal of Pharmaceutical and Biomedical Analysis* 43, 1382–1387.

Kleinman, W.A., Richie J.P. (1995) Determination of thiols and disulfides using high-performance liquid chromatography with electrochemical detection. *Journal of Chromatography B* 672, 73–80.

Lakritz, J., Plopper, C.G., Buckpitt, A.R. (1997) Validated high-performance liquid chromatography-electrochemical method for determination of glutathione and glutathione disulfide in small tissue samples. *Analytical Biochemistry* 247, 63–68.

Mason, A.Z., Jenkins, K.D. (1996). Metal detoxification in aquatic organisms. *In* Tessier, A., Turner, D.R. (eds), *Metal Speciation and Bioavailability in Aquatic Systems*. IUPAC Press, London, pp. 479–608.

Meister, A. (1988) Glutathione metabolism and its selective modification. *Journal of Biological Chemistry* 263, 17205–17208.

Meister, A. (1989) Metabolism and function of glutathione. *In* Dolphin D, Avramovich A, Poulson R (eds). *Glutathione: Chemical, Biochemical and Medical Aspects*. New York, Willey, pp. 423–442.

Meister, A. (1992) Biosynthesis and function of glutathione, an esssential biofactor. *Journal of Nutritional Science and Vitaminology* Special No. 1–6.

Meister, A., Anderson, M.E. (1983) Glutathione. *Annual Review in Biochemistry* 52, 711–760.

Rodriguez-Ariza, A., Toribio, F., López-Barea, J. (1994) Rapid determination of glutathione status in fish liver using high-performance liquid chromatography and electrochemical detection. *Journal of Chromatography B* 656, 311–318.

Sies, H., Ketterer, B. (1988) *Glutathione Conjugation: Mechanisms and Biological Significance*. Academic Press, London, 496 pp.

Tietze, F. (1969). Enzymic method for quantitative determination of nanogram amounts of total and oxidized glutathione: applications to mammalian blood and other tissues. *Analytical Biochemistry* 27, 502–22.

Viarengo, A. (1989) Heavy metals in marine invertebrates: mechanisms of regulation and toxicity at the cellular level. *Reviews in Aquatic Sciences* 1, 295–317.

Wu, Y.Q., Wang, C.G., Wang, Y., Zhao, Y., Chen, Y.X., Zuo, Z.H. (2007). Antioxidant responses to benzo[*a*]pyrene, tributyltin and their mixture in the spleen of *Sebasticus marmoratus*. *Journal of Environmental Science* 19, 1129–35.

Chapter 29

MEASUREMENT OF ANTIOXIDANT PIGMENTS AND VITAMINS IN PHYTOPLANKTON, ZOOPLANKTON, AND FISH

Pauline Snoeijs[1], *Norbert Häubner*[2], *Peter Sylvander*[1], *and Xiang-Ping Nie*[1,3]

[1]Department of Systems Ecology, Stockholm University, Stockholm, Sweden
[2]Department of Ecology and Evolution, Uppsala University, Uppsala, Sweden
[3]Department of Ecology, Jian University, Guangzhou, China

A major pitfall for ecological studies on the dynamics of essential compounds, such as antioxidant pigments and vitamins, in aquatic food webs is that chemical analysis of these compounds can be complicated. It is often necessary to adapt methods for use with aquatic organisms because existing methods are developed for other systems, e.g. for human blood or vegetables. It may be necessary to develop methods with lower detection limits because plankton samples are usually very low in biomass, and antioxidants mostly occur in small quantities. It may also be necessary to simplify extraction procedures to allow for the large sample throughput required in ecology. However, during the past decades method development has been significant (Tsao and Deng 2004).

In this chapter we focus on practical aspects of relatively simple analytical methods, mainly employing high performance liquid chromatography (HPLC), which are applicable to aquatic food-web studies. The ecological significance of the pigments and vitamins treated are discussed in Chapter 5 and recommended methods are summarized in Table 29.1. Comprehensive overviews of the analysis of pigments

Oxidative Stress in Aquatic Ecosystems, First Edition. Edited by Doris Abele, José Pablo Vázquez-Medina, and Tania Zenteno-Savín.

Table 29.1 Basic extraction and HPLC conditions for recommended methods. For references, see text.

Analyte	Extraction	Separation	Mobile phase (v:v)	Detection
Algal carotenoids (e.g. β-carotene)	Methanol	Reversed-phase C18 column	Gradient A: methanol, 0.5 M ammonium acetate 80 : 20 B: acetonitrile : ddH_2O 90 : 10 C: ethyl acetate	UV 436 nm
Animal carotenoids (e.g. astaxanthin)	Zooplankton: Acetone Animal tissue: With or without saponification Double-phase acetone + hexane	Reversed-phase C18 column	Gradient A: methanol, 0.5 M ammonium acetate 80 : 20 B: acetonitrile : ddH_2O 90 : 10 C: ethyl acetate	UV 470 nm
Phycobiliproteins	Buffer	–	–	Scanning spectrophotometer or fluorometer Cyanobacteria: 565, 620, 650 nm Red algae: 455, 564, 592, 618, 645 nm
Vitamin A (Retinol)	Plankton: Saponification Petroleum spirit Animal tissue 1: Hexane Animal tissue 2: Saponification Hexane	Plankton: Reversed-phase C18 column Animal tissue 1: Normal-phase silica column Animal tissue 2: Normal-phase silica column	Plankton: isocratic Aqueous methanol, 5% Animal tissue 1: gradient A: *n*-heptane B: 1,4-dioxane Animal tissue 2: isocratic *n*-heptane : tert-butyl methyl ether 92 : 8	Retinol Retinyl ester: UV 325 nm Retinal: UV 368 nm
Vitamin B_1 (Thiamine)	Plankton: HCl Thiochrome conversion Animal tissue: TCA, boiling Ethyl acetate/hexane Thiochrome conversion	Reversed-phase NH_2 column	Plankton: isocratic Methanol : 0.1 M phosphate buffer 43 : 57 Animal tissue: isocratic Acetonitrile : 85 mM potassium phosphate buffer 35 : 65	Plankton: Fluorescence excitation 375 nm emission 450 nm Animal tissue: Fluorescence excitation 375 nm emission 433 nm

Vitamin C (Ascorbic acid)	Plankton: Metaphosphoric acid + acetic acid + EGTA Enzymatic oxidation Fluorescent reagent OPD	Reversed-phase C18 column	Plankton: isocratic 80 mM potassium dihydrogen phosphate : methanol 80 : 20	Plankton: Fluorescence excitation 355 nm emission 425 nm
	Animal tissue: Metaphosphoric acid Iodine oxidation Fluorescent reagent DMPD		Animal tissue: isocratic Methanol : 0.1% metaphosphoric acid 80 : 20	Animal tissue: Fluorescence excitation 362 nm emission 444 nm
Vitamin D (Calciferol)	Plankton: Saponification Petroleum spirit HPLC purification Methanol	Reversed-phase C18 column	Plankton: isocratic Aqueous methanol, 3%	Plankton: UV 280 nm
	Animal tissue: Saponification Double-phase n-hexane + ethyl-acetate HPLC purification Methanol		Animal tissue: isocratic Methanol : ddH_2O 93 : 7	Animal tissue: UV 265 nm
Vitamin E (Tocopherol)	Plankton 1: Saponification Petroleum spirit	Plankton 1: Reversed-phase C18 column	Plankton 1: isocratic Aqueous methanol, 3%	Plankton 1: Fluorescence excitation 292 nm emission 350 nm
	Plankton 2: Double-phase *n*-hexane + DMF + SDS	Plankton 2: Reversed-phase C18 column	Plankton 2: isocratic Methanol : 1-propanol : 0.5 M ammonium acetate 75 : 20 : 5	Plankton 2: Electrochemical (coulometric) two-channel analytical cell (250 and 800 mV)
	Animal tissue: Saponification Double-phase *n*-hexane + ethanol or ethyl-acetate	Animal tissue: Normal-phase silica column	Animal tissue: isocratic Hexane-based	Animal tissue: Fluorescence excitation 295 nm emission 325 nm

and antioxidants in general can be found in Jeffrey et al. (1997) and Ball (2006), respectively. All compounds discussed here have documented antioxidant capacity (Wiseman 1993; Combs 1998; Gliszczyńska-Świgło 2006).

GENERAL NOTES ON HPLC ANALYSIS OF ANTIOXIDANT PIGMENTS AND VITAMINS

Conditions

It should always be kept in mind that antioxidants are sensitive to light, high temperature, atmospheric oxygen, and changes in pH. This implies that samples must be taken fast and be stored dark, cold, and well packed. Extractions and HPLC analyses must be carried out under dim light conditions and, while handling the samples, they should always be kept on ice and often also under nitrogen gas.

Calibration

Each HPLC column must be carefully calibrated for all analytes and over the whole range of concentrations measured in the samples. Calibration factors for each analyte are obtained by plotting HPLC peak area against the amount of analyte standard injected into the HPLC system. The regressions for each analyte should be linear over the whole measurement range with $R^2 > 0.95$. Triplicate injections of five different standard concentrations are recommended for column calibration. It is also possible to use an internal standard, a substance that is added to all standard solutions and samples. The ratio of the peak areas of the internal standard and the peak area of the analyte is then used as a correction factor for the quantification of the analyte. Standards and blanks (extraction solvent treated in the same way as the samples) should be run routinely, at least once per day.

Extraction

The analytes are extracted from the sample with a suitable solvent, with or without saponification and/or homogenization, before injection into the HPLC system. The completeness of extraction is verified by re-extracting pellets until the analyte is fully extracted. Saponification is the hydrolysis of esters by heating under basic conditions and is used to remove (co-extracted) neutral lipids prior to analysis because fats can obscure HPLC peaks of analytes or clog the HPLC column. Animal fats are fatty esters in the form of triglycerides. The alkali breaks the ester bonds and releases the fatty acid salts (crude soap) and glycerol. Soaps can be precipitated by adding NaCl. Saponification is often recommended prior to HPLC analysis of animal tissues, despite the fact that it can cause extensive loss of analytes (Kramer et al. 1997, and references therein).

Injection

The rate by which the sample solvent passes through the column (the elution strength) should not be higher than that of the HPLC mobile phase, otherwise peaks will be deformed (Patil et al. 2008). Dissolving the analyte in the mobile phase, or the first mobile phase when a gradient is used, will prevent peak deformations that can occur when solvent and mobile phase are incompatible. To remove the last possible particulates from an extract before injection into the HPLC system, the extract should always be filtered through a solvent-resistant syringe filter with a maximum pore size of 0.5 μm. The HPLC column should also preferably be protected with a guard column. Particles will destroy or shorten the lifetime of the HPLC column. Injection volumes are usually between 10 and 150 μL. If the concentrations of all target analytes in a sample are sufficient, smaller injection volumes will produce sharper peaks and better separation.

SAMPLING AND SAMPLE TREATMENT

Sampling and Storage

Plankton samples are usually collected on glass fiber filters, which are inert to extraction chemicals, but have less well-defined pore-sizes than polycarbonate filters. For phytoplankton, Whatman® GF/F filters with a pore size of ~0.7 μm are suitable and for zooplankton Whatman® GF/C filters with a pore size of ~1.2 μm. These pore sizes are a highly approximate measure; organisms larger than 0.7 or 1.2 μm are most likely to be caught in the filter, while smaller organisms may pass but can also get stuck in small crevices between the glass fibers. To avoid exposure to oxygen, it is recommended to filtrate as fast as possible and to not let the filter become dry and exposed to air for longer than a few seconds. The filter is folded with forceps, tightly wrapped in aluminum foil, and immediately frozen in liquid nitrogen. Also for macroalgae it is recommended to freeze the samples immediately in the field. For example, concentrations of carotenoids involved in xanthophyll cycles change within minutes when light conditions change. Likewise, animal tissue samples should be taken from freshly killed animals and immediately frozen in the field to avoid decomposition or transformation of the analytes. For pigment analysis it is recommended to store the samples directly in methanol before freezing, which stabilizes pigments for longer storage times. Sample storage until analysis is best in liquid nitrogen and second best in a -80°C freezer.

Sample Treatment

Homogenization of small plankton organisms on filters is best performed with a sonicator. The ultrasound breaks the cell walls and the analytes are released into the solvent. An easy way to get rid of glass fiber filters or other debris is to use a double centrifuge tube system as described by Wright and Mantoura (1997, p. 439, fig. 17.4). The upper tube (extraction tube) contains the filter and the solvent and has a small hole in the bottom covered with Parafilm®. The lower tube is the collection tube. The tubes are kept on ice while sonicating in the extraction tube. To avoid overheating of the sample, sonication should be performed in intervals, if longer than 30 s of sonication is necessary. After centrifugation in a cooled centrifuge, the Parafilm has broken and the extract is in the collection tube while the debris is left in the extraction tube. For some extractions, sonication can be replaced by addition of the strong solvent N,N-dimethylformamide (DMF) with only vortex mixing (Porra et al. 1989; Schumann et al. 2005). Larger samples of macroalgae can be homogenized with a bead beater or mortar in liquid nitrogen, or directly in the solvent. Animal tissue samples are usually homogenized using a mortar or an overhead stirrer (e.g. Ultra Turrax®) directly in the extraction solvent, but it can be worthwhile testing sonication or bead beating. An overhead stirrer is more difficult to clean; it should be rinsed with the solvent several times and the different fractions are combined as the final extract. To dispose of animal cell debris, the supernatant can be filtered through a glass funnel with fine glass wool before centrifugation, or the same system with double tubes as described above can be applied.

ANALYSIS OF ALGAL CAROTENOIDS

Extraction

The suitability of different solvents for the extraction of fat-soluble pigments from phytoplankton (carotenoids and chlorophylls) was evaluated by Wright et al. (1997). They found that sonication in DMF gave the best general pigment extraction of all protocols tested and methanol the second-best. DMF is a highly toxic substance, and therefore sonication in the less toxic solvent methanol is recommended for routine analyses. However, a large number of pigments are extracted simultaneously and it is impossible to find the ideal solvent for all pigments, e.g. the most nonpolar carotenoid β-carotene is slightly better extracted in acetone than in methanol. We tested four extraction solvents for nine sets of four replicate samples of natural phytoplankton communities and found that the extraction results were not significantly different for DMF, acetone, and methanol for the dominant pigments. However, a solvent mixture with a composition similar to the mobile phase (methanol : 1.8 M ammonium acetate : ethyl acetate = 8 : 1 : 1.2) gave significantly lower extraction yields for chlorophyll-*a* and β-carotene than the other three solvents (Fig. 29.1). Addition of 20% ammonium acetate buffer to a methanol extract immediately before injection as proposed by Wright and Mantoura (1997) slightly improves chromatographic resolution of the early eluting peaks over that obtained with 100% methanol. However, unless the autosampler can be programmed to automatically mix sample and buffer immediately prior to injection, this cannot be applied when samples are waiting in an autosampler for several hours because pigments, especially chlorophylls, will degrade faster with this addition.

Detection

The goal of finding one HPLC method capable of separating all significant fat-soluble phytoplankton

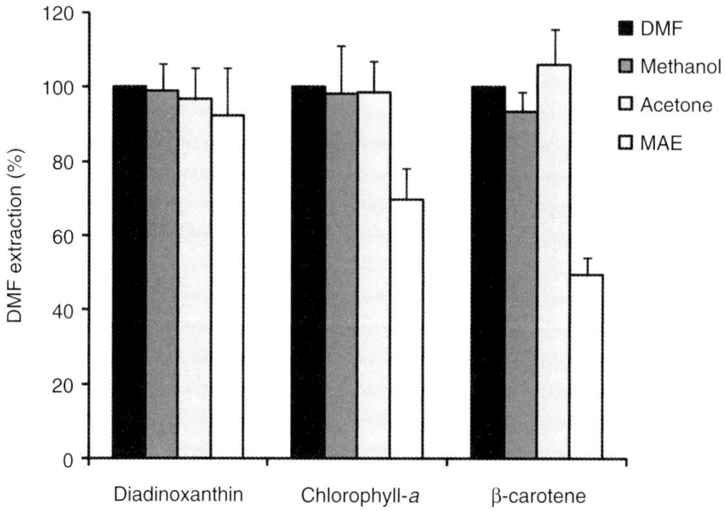

Fig. 29.1 Comparison of the extraction yields of four different extraction solvents, expressed as percentage of the extraction yield of DMF. The graph shows means of nine different natural phytoplankton communities of which four replicate samples were each extracted with one of the four solvents DMF, methanol, acetone, and a solvent mixture with a composition resembling the HPLC mobile phase (MAE, methanol : 1.8 M ammonium acetate : ethyl acetate = 8 : 1 : 1.2). Altogether 36 samples were analyzed. For chlorophyll-*a* and β-carotene the yield in MAE was significantly ($p < 0.05$) lower than in the other three solvents in repeated ANOVA measures, with the phytoplankton community as the repeated measure. No further differences were found between the solvents. Error bars represent 95% confidence limit.

pigments – from the polar acidic chlorophylls, xanthophylls of intermediate polarities and esterified chlorophylls, to the nonpolar carotene hydrocarbons – has not yet been achieved (Jeffrey et al. 1997). Although method refinements are still made (e.g. Van Heukelem and Thomas 2001), the 1980s was the most intense period of HPLC method development. Great progress was made with the use of reversed-phase C18 columns (Mantoura and Llewellyn 1983) and most phytoplankton researchers now follow the reversed-phase HPLC method of Wright and Jeffrey (1997) with no or only slight modifications. This method uses a reversed-phase C18 column (e.g. Sphereclone, Phenomenex™) with a three-solvent gradient mobile phase and UV detection at 436 nm for chlorophylls and carotenoids and at 405 nm for pheophytin *a* and pheophorbide *a*. Over 40 fat-soluble algal pigments can be separated with this method. To simplify peak identification, absorption can be measured over a range of wavelengths with a diode-array UV detector to get the absorption spectra of individual peaks, which later can be compared to a library with absorption spectra of known compounds. Some C18 columns can even separate between lutein and zeaxanthin, which are very similar molecules but represent different metabolic pathways (Britton 1993). Single-species macroalgal samples usually have a lower carotenoid diversity than phytoplankton communities, but the same method (Wright and Jeffrey 1997) can be applied successfully (Ursi et al. 2003; Choo et al. 2005; Andersson et al. 2006). Calibration of HPLC columns has in recent years become facilitated by the commercial availability of algal pigment standards, e.g. some pigments isolated from higher plants are available from Sigma-Aldrich™, and DHI Lab™ in Denmark produces about 30 algal carotenoid standards by HPLC separation.

ANALYSIS OF ANIMAL CAROTENOIDS

Extraction

A large variety of carotenoids occurs in the different tissues of aquatic animals, among which astaxanthin is a major representative (Matsuno 2001). For simple extraction of astaxanthin and other animal carotenoids from zooplankton samples on glass fiber filters, sonication in ice-cold acetone can be used (Andersson et al. 2003). Extraction from fish eggs and fry tissues as described by Pettersson and Lignell (1999) includes a two-phase extraction. After homogenization in ice-cold acetone with an overhead stirrer and centrifugation, the supernatant is vigorously shaken with cyclohexane and distilled water to obtain phase separation. The pigment-containing cyclohexane fraction is N_2-evaporated to dryness and resolved in mobile phase A before injection. With this method carotenoid esters stay esterified.

Saponification

Samples with high amounts of esterified carotenoids can first be homogenized and extracted, followed by saponification. The rather complicated saponification method used by Bjerkeng et al. (1990, 2000) for fish tissues includes thin-layer chromatography (TLC). Less elaborate methods were described for cysts of the green alga *Haematococcus lacustris* (Yuan and Chen 1997) and shells of the shrimp *Parapenaeopsis hardwickii* (Lin et al. 2005), in which astaxanthin esters are hydrolyzed by the addition of NaOH to the extracts.

Detection

HPLC detection of astaxanthin and other animal carotenoids is mostly performed with UV detection at 470–480 nm. Different carotenoid isomers occur in natural samples and those of astaxanthin are most studied. Major geometric isomers are all-*trans*-, 9-*cis*- and 13-*cis*-astaxanthin (synonyms are all-*E*-, 9*Z*- and 13*Z*-astaxanthin, respectively). The geometric isomers can be separated by a reversed phase C18 column (e.g. Gemini, Phenomenex™). Commercial astaxanthin standards (e.g. from Sigma-Aldrich™) consist of nearly 100% all-*trans*-astaxanthin. The *cis*-isomers can be produced from all-*trans*-astaxanthin by treatment with strong organic solvents and temperature (Yuan and Chen 1999), which is a reversible process (Yuan and Chen 2001). We tested incubations in different organic solvents and found that isomerization from all-*trans*-astaxanthin into 13-*cis*-astaxanthin was highest (ca. 50%) after 18 h incubation in dichloromethane (DCM) at 4°C or 110 h incubation in methanol at 35°C while 9-*cis*-astaxanthin remained below 1%. Highest isomerization from all-*trans*-astaxanthin into 9-*cis*-astaxanthin (3–4%) was found in DCM at 70°C for 18 h and in chloroform at 35°C for 110 h (Fig. 29.2).

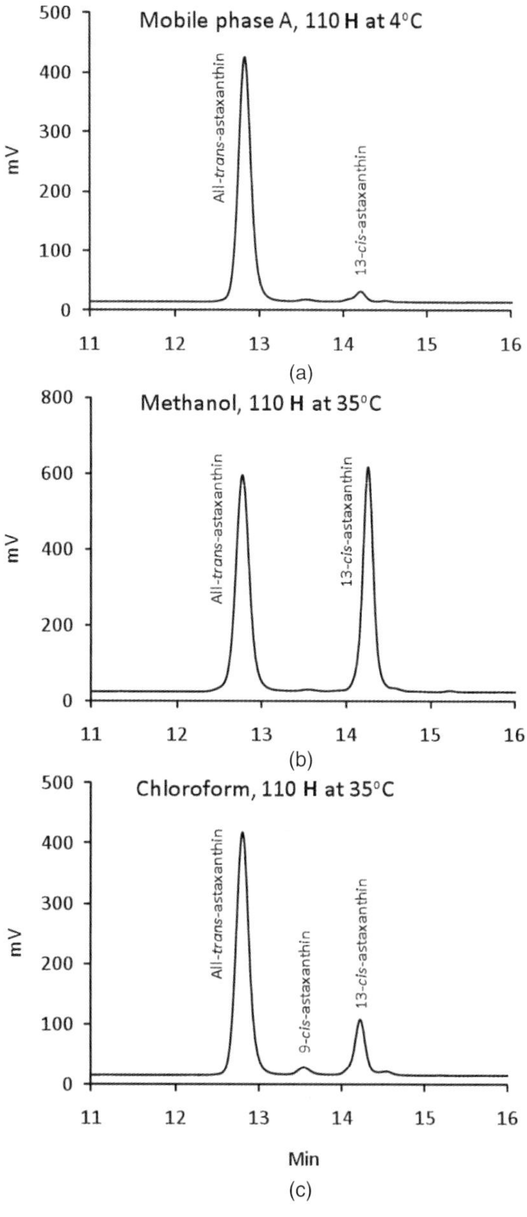

Fig. 29.2 Chromatograms of *cis–trans* mixtures of astaxanthins produced by 110 h dark incubations of artificial all-*trans*-astaxanthin standard powder purchased from Sigma-Aldrich™ in different organic solvents. (a) Reference: Mobile phase A = methanol : 0.5 M ammonium acetate buffer 80 : 20 at 4°C. (b) Methanol at 35°C. (c) Chloroform at 35°C. After 110 h the solvents were evaporated with nitrogen gas and the pellets were dissolved in mobile phase A and injected into the HPLC system. Ultaviolet detection was at 470 nm.

In addition to geometric isomers, and considering that each molecule has two chiral centers in C-3 and C-3′, astaxanthin may have three optical (configurational) isomers: two enantiomers (3R,3′R and 3S,3′S) and a meso form (3R,3′S) (Turujman et al. 1997). The analysis of optical isomers requires chiral HPLC columns and more complicated extraction procedures (Vecchi and Müller 1979; Abu-Lafi and Turujman 1999). Østerlie et al. (1999) showed that geometrical and optical isomers of astaxanthin are distributed selectively in different tissues of rainbow trout. Distinct species utilize these carotenoid isomers differently and the biochemical basis for this is still unclear (Bjerkeng et al. 1997, Østerlie et al. 2000).

ANALYSIS OF PHYCOBILIPROTEINS

Phycobiliproteins are usually quantified spectrophotometrically. Plankton samples collected on glass fiber filters (Rapala et al. 2002), pellets from cell suspensions (Hong and Lee 2008) or macroalgal fragments (Godínez-Ortega 2008) are sonicated in a buffer solution. The extracts are incubated in darkness at 4°C and centrifuged before measurement in a scanning spectrophotometer (400–700 nm). Phycobiliprotein concentrations are calculated using specific formulae, e.g. those proposed by Beer and Eshel (1985), Tandeau de Marsac (1988) or Sampath-Wiley and Neefus (2007). If a higher sensitivity is necessary, fluorometry can be used for the analysis of phycobilin pigments in the same type of extracts (Downes and Hall 1998). Separation of phycobiliprotein subunits is achieved by removal of chlorophyll components and protein purification followed by HPLC (Swanson and Glazer, 1990; Swanson et al. 1991; Glazer et al. 1997).

ANALYSIS OF VITAMIN A

Algal samples

For the analysis of retinol in algae, following the method used by Brown et al. (1999), the samples are first saponified with ethanolic KOH, after which the vitamin is extracted from the algal cells with petroleum spirit. The petroleum spirit is evaporated and the residue is re-dissolved in methanol. Detection of retinol is with UV at 325 nm using a reversed-phase C18 column and an isocratic mobile phase of 5% aqueous methanol.

Animal Tissues

Moren et al. (2004b) analyzed retinal, retinol, and retinyl esters in fish larvae with a simple method based on Bankson et al. (1986). The sample is homogenized and shaken in distilled water containing acetic acid, ascorbic acid, EDTA, and ethanol. The retinoids are then extracted three times with hexane. After N_2 evaporation of the solvent, the dry samples are redissolved in hexane. Detection is with UV at 325 nm for retinol and retinyl esters and at 368 nm for retinal using a normal-phase silica column and a mobile phase gradient of *n*-heptane and 1,4-dioxane. For samples with a high degree of esterified vitamin A, Moren et al. (2004a) modified the HPLC method for analysis of retinol and didehydro-retinol of Nöll (1996) for fish samples. Fish tissue is saponified in ethanolic KOH, heated for 20 min at 100°C and cooled. The samples are then extracted three times with hexane. After N_2 evaporation of the solvent, the dry samples are redissolved in hexane. HPLC detection is with UV as above, using a normal-phase silica column and an isocratic mobile phase of *n*-heptane : tert-butyl methyl ether 92 : 8.

ANALYSIS OF VITAMIN B_1

Lynch and Young (2000) reviewed the published methods for HPLC analysis of thiamine. Some methods use UV detection, but the more sensitive methods require fluorescence detection. For fluorescence detection the conversion of thiamine to thiochrome is necessary.

Plankton Samples

For the analysis of free (nonphosphorylated) thiamine (TF), thiamine monophosphate (TMP), and thiamine diphosphate (TDP) in small phyto- and zooplankton samples, a highly sensitive method was developed by Pinto et al. (2002). Glass fiber filters with plankton samples are added to ice-cold HCl, sonicated and centrifuged. Thiochrome conversion is achieved by adding a freshly made potassium ferricyanide solution and NaOH. A final pH >8 is necessary for maximal fluorescence of thiochrome and its phosphates (Kawasaki and Sanemori 1985). Detection is with a fluorescence detector using a reversed-phase NH_2 column and an isocratic mobile phase of methanol and phosphate buffer. The sensitivity of this method is 15 fmol of TF, TMP, and TDP, calculated by the method proposed by Knoll (1985).

Animal Tissues

For analysis of TF, TMP, and TDP in tissue samples of fish, birds, and mammals, the protocol of Brown et al. (1998b) with modifications by Kankaanpää et al. (2002), Mancinelli et al. (2003) and Balk et al. (2009) can be used. The tissue is carefully homogenized in ice-cold 2% trichloroacetic acid (TCA), boiled for 10 min in a water bath and homogenized again in ice-cold 10% TCA. The supernatant is washed four times with an equal volume of ethyl acetate/hexane (3 : 2), which removes most TCA and raises the pH to 5. Thiochrome conversion is achieved by adding a freshly made potassium ferricyanide solution and NaOH. Detection is with a fluorescence detector using a reversed-phase NH_2 column and an isocratic mobile phase of methanol and acetonitrile and phosphate buffer. Poel et al. (2009) recently described a method that detects TF, TMP, TDP, thiamine triphosphate (TTP), and an unidentified fifth substance, which may be related to thiamine tetraphosphate or adenosine thiamine triphosphate.

ANALYSIS OF VITAMIN C

Plankton Samples

Following the method used for microalgae and rotifers by Brown and Miller (1992) and Brown et al. (1998a), ascorbic acid is extracted from the samples by sonication in a mixture of 3% metaphosphoric acid, 8% acetic acid, and an EGTA-glutathione solution. Some algae require overnight incubation, and for wild phytoplankton communities this is also recommended. Sodium acetate buffer is added to the supernatant, and the tube is equilibrated at 37°C prior to incubation with ascorbic acid oxidase. The basis of the assay is the enzymatic oxidation of ascorbic acid to dehydroascorbic acid and the condensation of the latter with *o*-phenylenediamine (OPD) to form 3-(1,2-dihydroxyethyl) furo[3,4-b]-quinoxaline-1-one. This fluorescent derivative is then chromatographed and analyzed by HPLC. Detection of ascorbic acid is with fluorescence using a reversed-phase C18 column and an isocratic mobile phase made up of potassium dihydrogen phosphate and methanol. Hapette and Poulet (1990) described a different method for the determination of ascorbic acid in marine plankton using a polystyrene divinyl benzene column.

Animal Tissues

For the analysis of ascorbic acid and dehydroascorbic acid in fish tissues a large number of methods have been published. Following Ito et al. (1995), fish tissue samples are homogenized in 5% metaphosphoric acid, centrifuged, and the supernatant is filtered for deproteination. Detection is with UV at 300 nm using an ion-exclusion column and an isocratic mobile phase of 5 mM oxalic acid. Following Maeland et al. (1999), fish tissue samples are homogenized in ice-cold 0.5% metaphosphoric acid with EDTA and DL-dithiothreitol. After centrifugation, the supernatant is mixed with a chloroform–methanol mixture and centrifuged again. Aliquots of the water phase are mixed with the mobile-phase buffer with added metaphosphoric acid and DL-dithiothreitol. Detection is with an amperometric electrochemical detector and a reversed-phase C18 column. Following Iglesias et al. (2006), fish tissue samples are homogenized in 4.5% metaphosphoric acid. All ascorbic acid is oxidized to dehydroascorbic acid by mixing the sample with an iodine solution, decomposing the excess of iodine with $Na_2S_2O_3$, and adjusting the pH with NaOH. A fluorescent-derivative reagent solution of 4,5-dimethylo-phenilenediamine (DMPD) is added and the DMPD derivatives are re-extracted from the aqueous phase with isobutanol. Detection is with fluorescence using a reversed-phase C18 column and an isocratic mobile phase of methanol and 0.1% metaphosphoric acid. Nelis et al. (1997) evaluated different extraction methods for vitamin C. They found that high ionic strength acidic mixtures (e.g. 5% metaphosphoric acid) may be associated with a negative drift in HPLC retention times. Their recommendation is to use a mixture of 1% acetic acid, 0.1% metaphosphoric acid, and 1 mM EDTA for extracting ascorbic acid and dehydroascorbic acid from aquatic organisms (brine shrimp *Artemia*, *Brachionus*, fish eggs, fish larvae, shrimp larvae). HPLC analysis used by Nelis et al. (1997) is with a reversed-phase ion-pair liquid chromatographic method using electrochemical detection.

ANALYSIS OF VITAMIN D

The analysis of vitamin D in biological samples typically requires saponification, extraction, and one or two HPLC purification steps prior to final quantification by HPLC (Mattila et al. 1996).

Algal Samples

For the analysis of ergocalciferol (vitamin D_2) and cholecalciferol (vitamin D_3) in algae, following the method of Brown et al. (1999), the samples are first saponified with ethanolic KOH, after which the vitamins are extracted from the algal cells with petroleum spirit. Vitamin D is isolated from the petroleum spirit extract by normal-phase HPLC before being chromatographed with UV detection at 280 nm using a reversed-phase C18 column and an isocratic mobile phase of 3% aqueous methanol.

Animal Tissues

For the analysis of vitamin D in fish, following the method used by Salo-Väänänen et al. (2000), the samples are first saponified with KOH and precipitated by NaCl, after which the vitamin is extracted in a double-phase extraction with *n*-hexane : ethyl acetate 8 : 2, N_2 evaporated and redissolved in *n*-hexane. The extract is purified by normal-phase HPLC method with a UV detector set at 265 nm and a mobile phase consisting of *n*-hexane : tetrahydrofuran : 2-propanol 98 : 1 : 1. The Vitamin D containing fraction is collected and N_2-evaporated. The residue is redissolved in methanol and detection is with reversed-phase HPLC, with a UV detector set at 265 nm and a mobile phase of methanol : water 93 : 7. The method described by Lu et al. (2007) is similar, but includes two HPLC purification steps. Ostenmeyer et al. (2006) developed a highly selective and sensitive detection method for vitamin D and provitamin D in fish samples with (Coulochem) electrochemical detection.

ANALYSIS OF VITAMIN E

Algal Samples

For the analysis of α-tocopherol in algae, following the method of Brown et al. (1999), the samples are first saponified with ethanolic KOH, after which the vitamin is extracted from the algal cells with petroleum spirit. The petroleum spirit is evaporated and the residue is redissolved in methanol. Detection of α-tocopherol is with fluorescence, excitation at 292 nm and emission at 350 nm, using a reversed-phase C18 column and an isocratic mobile phase of 2% aqueous methanol.

Plankton Samples

For the simultaneous extraction and detection of α-, γ-, and δ-tocopherol in plankton samples Coulochem electrochemical detection can be used (Häubner, Lewander and Snoeijs, unpublished). A two-phase extraction with DMF and *n*-hexane as extraction solvents and sodium dodecyl sulphate (SDS) as additive simplifies the extraction procedure because elaborate homogenization and saponification steps (Huo et al. 1996, 1997; Sanchez-Machado 2002) are unnecessary.

Animal Tissues

While Kalogeropoulos et al. (2007) used a simple extraction in hexane with butylated hydroxytoluene for the analysis of α-tocopherol in fish, normally saponification of animal tissue samples is carried out before extraction. Kramer et al. (1997) evaluated the use of saponification for α-, β-, γ-, and δ-tocopherol analysis in heart and liver samples of pigs and concluded that normally there was no need to remove the co-eluting lipids by saponification, since they did not interfere with the quantification of the tocopherols. Saponification was only necessary when tocopherol esters were present, and should be carried out after an initial hexane extraction to remove the free tocopherols in order to avoid their loss by saponification, particularly non α-tocopherol and tocotrienols. Salo-Väänänen et al. (2000) and Mestre Prates et al. (2006) described relatively simple methods to analyze α-, β-, γ-, and δ-tocopherols with saponification. The suitability of silica columns and hexane-based mobile phases for the separation of tocopherols and tocotrienols by normal-phase HPLC with fluorescence detection has been demonstrated previously by Eldin et al. (2000) and is commonly used.

REFERENCES

Abu-Lafi, S., Turujman, S.A. (1999) Reproducibility of the separation of astaxanthin stereoisomers on Pirkle covalent L-leucine and D-phenylglycine columns. *Journal of Chromatography A* 855, 157–170.

Andersson, M., Van Nieuwerburgh, L., Snoeijs, P. (2003) Pigment transfer from phytoplankton to zooplankton with emphasis on astaxanthin production in the Baltic Sea food web. *Marine Ecology Progress Series* 254, 213–224.

Andersson, M., Schubert, H., Pedersén, M., Snoeijs, P. (2006) Different patterns of carotenoid composition and photosynthesis acclimation in two tropical red algae. *Marine Biology* 149, 653–665.

Balk, L., Hägerroth, P.Å., Åkerman, G. et al. (2009) Wild birds of declining European species are dying from a thiamine deficiency syndrome. *Proceedings of the National Academy of Sciences USA* 106, 12001–12006.

Ball, G.F.M. (2006) *Vitamins in Foods. Analysis, Bioavailability, and Stability*. Taylor & Francis, Boca Raton, FL.

Bankson, D.D., Russell, R.M., Sadowski, J.A. (1986) Determination of retinyl esters and retinol in serum or plasma by normal phase liquid chromatography: method and applications. *Clinical Chemistry* 32, 35–40.

Beer, S., Eshel, A. (1985) Determining phycoerythrin and phycocyanin concentrations in aqueous crude extracts of red algae. *Australian Journal of Marine and Freshwater Research* 36, 785–792.

Bjerkeng, B., Storebakken, T., Liaaen-Jensen, S. (1990) Response to carotenoids by rainbow trout in the sea: resorption and metabolism of dietary astaxanthin and canthaxanthin. *Aquaculture* 91, 153–162.

Bjerkeng, B., Følling, M., Lagocki, S., Storebakken, T., Olli, J.J., Alsted, N. (1997) Bioavailability of all-*E*-astaxanthin and Z-isomers of astaxanthin in rainbow trout (*Oncorhynchus mykiss*). *Aquaculture* 157, 63–82.

Bjerkeng, B., Hatlen, B., Jobling, M. (2000) Astaxanthin and its metabolites idoxanthin and crustaxanthin in flesh, skin, and gonads of sexually immature and maturing Arctic charr (*Salvelinus alpinus* (L.)). *Comparative Biochemistry and Physiology B* 125, 395–404.

Britton, G. (1993) Biosynthesis of carotenoids. *In* Young, A., Britton G. (eds). *Carotenoids in Photosynthesis*. Chapman & Hall, London.

Brown, M.R., Miller, K.A. (1992) The ascorbic acid content of eleven species of microalgae used in mariculture. *Journal of Applied Phycology* 4, 205–215.

Brown, M.R., Skabo, S., Wilkinsson, B. (1998a) The enrichment and retention of ascorbic acid in rotifers fed microalgal diets. *Aquaculture Nutrition* 4, 151–156.

Brown, S.B., Honeyfield, D.C., Vandenbyllaerdt, L. (1998b) Thiamine analysis in fish tissues. *American Fisheries Society Symposium* 21, 73–91.

Brown, M.R., Mular, M., Miller, I., Farmer, C., Trenerry, C. (1999) The vitamin content of microalgae used in aquaculture. *Journal of Applied Phycology* 11, 247–255.

Choo, K.S., Nilsson, J., Pedersén, M., Snoeijs, P. (2005) Photosynthesis, carbon uptake and antioxidant defense in two coexisting filamentous green algae under different stress conditions. *Marine Ecology Progress Series* 292, 127–138.

Combs, G.F. (1998) *The Vitamins, Fundamental Aspects in Nutrition and Health*, 2nd edn. Academic Press, San Diego, CA.

Downes, M.T., Hall, J. (1998) A sensitive fluorometric technique for the measurement of phycobilin pigments and its application to the study of marine and freshwater picophytoplankton in oligotrophic environments. *Journal of Applied Phycology* 10, 357–363.

Eldin, A.K., Gorgen, S., Petterson, J., Lampi, A.M. (2000) Normal-phase high-performance liquid chromatography of tocopherols and tocotrienols: comparison of different chromatographic columns. *Journal of Chromatography A* 881, 217–227.

Glazer, A.N., Chan, C.F., West, J.A. (1997) An unusual phycocyanobilin-containing phycoerythrin of several bluish-colored acrochaetioid, freshwater red algal species. *Journal of Phycology* 33, 617–624.

Gliszczyńska-Świgło, A. (2006) Antioxidant activity of water soluble vitamins in the TEAC (trolox equivalent antioxidant capacity) and the FRAP (ferric reducing antioxidant power) assays. *Food Chemistry* 96, 131–136.

Godínez-Ortega, J.L., Snoeijs, P., Robledo, D., Freile-Pelegrín, Y., Pedersén, M. (2008) Growth and pigment composition in the red alga *Halymenia floresii* cultured under different light qualities. *Journal of Applied Phycology* 20, 253–260.

Hapette, A.M., Poulet, S.A. (1990) Application of high-performance liquid chromatography to the determination of ascorbic acid in marine plankton. *Journal of Liquid Chromatography and Related Technologies* 13, 357–370.

Hong, S.J., Lee, C.G. (2008) Statistical optimization of culture media for production of phycobiliprotein by *Synechocystis* sp. PCC 6701. *Biotechnology and Bioprocess Engineering* 13, 491–498.

Huo, J.Z., Nelis, H.J., Lavens, P., Sorgeloos, P., De Leenheer, A.P. (1996) Determination of vitamin E in aquatic organisms by high-performance liquid chromatography with fluorescence detection. *Analytical Biochemistry* 242, 123–128.

Huo, J.Z., Nelis, H.J., Lavens, P., Sorgeloos, P., De Leenheer, A.P. (1997) Determination of E vitamers in microalgae using high-performance liquid chromatography with fluorescence detection. *Journal of Chromatography A*, 782, 63–68.

Iglesias, J., González, M.J., Medina, I. (2006) Determination of ascorbic and dehydroascorbic acid in lean and fatty fish species by high-performance liquid chromatography with fluorometric detection. *European Food Research and Technology* 223, 781–786.

Ito, T., Murata, H., Yasui, Y., Matsui, M., Sakai, T., Yamauchi, K. (1995) Simultaneous determination of ascorbic acid and dehydroascorbic acid in fish tissues by high-performance liquid chromatography. *Journal of Chromatography B* 667, 355–357.

Jeffrey, S.W., Mantoura, R.F.C., Wright, S.W. (eds). (1997) *Phytoplankton Pigments in Oceanography: Guidelines to Modern Methods*. Unesco, Paris.

Kalogeropoulos, N., Chiou, A., Mylona, A., Ioannou, M.S., Andrikopoulos, N.K. (2007) Recovery and distribution of natural antioxidants (α-tocopherol, polyphenols and terpenic acids) after pan-frying of Mediterranean finfish in virgin olive oil. *Food Chemistry* 100, 509–517.

Kankaanpää, H., Vuorinen, P.J., Sipiä, V., Keinänen, M. (2002) Acute effects and bioaccumulation of nodularin in sea trout (*Salmo trutta* m. *trutta* L.) exposed orally to *Nodularia spumigena* under laboratory conditions. *Aquatic Toxicology* 61, 155–168.

Kawasaki, T., Sanemori, H. (1985) Vitamin B_1: thiamines. *Modern Chromatographic Science Series* 30, 385–412.

Knoll, J.E. (1985) Estimation of the limit of detection in chromatography. *Journal of Chromatographic Science* 23, 422–423.

Kramer, J.K.G., Blais, L., Fouchard, R.C., Melnyk, R.A., Kallury, K.M.R. (1997) A rapid method for the determination of vitamin E forms in tissues and diet by high-performance liquid chromatography using a normal-phase diol column. *Lipids* 32, 323–330.

Lin, W.C., Chien, J.T., Chen, B.H. (2005) Determination of carotenoids in spear shrimp shells (*Parapenaeopsis hardwickii*) by liquid chromatography. *Journal of Agricultural and Food Chemistry* 53, 5144–5149.

Lu, Z., Chena, T.C., Zhang, A. et al. (2007) An evaluation of the vitamin D_3 content in fish: Is the vitamin D content adequate to satisfy the dietary requirement for vitamin D? *Journal of Steroid Biochemistry and Molecular Biology* 103, 642–644.

Lynch, P.L.M., Young, I.S. (2000) Determination of thiamine by high-performance liquid chromatography. *Journal of Chromatography A*, 881, 267–284.

Maeland, A., Rosenlund, G., Stoss, J., Waagbø, R. (1999) Weaning of Atlantic halibut (*Hippoglossus hippoglossus L.*) using formulated diets with various levels of ascorbic acid. *Aquaculture Nutrition* 5, 211–219.

Mancinelli, R., Ceccanti, M., Guiduccia, M.S. et al. (2003) Simultaneous liquid chromatographic assessment of thiamine, thiamine monophosphate and thiamine diphosphate in human erythrocytes: a study on alcoholics. *Journal of Chromatography B* 789, 355–363.

Mantoura, R.F.C., Llewellyn, C.A. (1983) The rapid determination of algal chlorophyll and carotenoid pigments and their breakdown products in natural waters by reversed-phase high-performance liquid chromatography. *Analytica Chimica Acta* 151, 297–314.

Matsuno, T. (2001) Aquatic animal carotenoids. *Fisheries Science* 67, 771–783.

Mattila, P.H., Piironen, V.I., Uusi-Rauvab, E.J., Koivistoinen, P.E. (1996) New analytical aspects of vitamin D in foods. *Food Chemistry* 57, 95–99.

Mestre Prates, J.A., Gonçalves Quaresma, M.G., Branquinho Bessa, R.J., Andrade Fontes, C.M.G., Mateus Alfaia, C.M.P. (2006) Simultaneous HPLC quantification of total cholesterol, tocopherols and β-carotene in Barrosã-PDO veal. *Food Chemistry* 94, 469–477.

Moren, M., Næss, T., Hamre, K. (2004a) Conversion of β-carotene, canthaxanthin and astaxanthin to vitamin A in Atlantic halibut (*Hippoglossus hippoglossus* L.) juveniles. *Fish Physiology and Biochemistry* 27, 71–80.

Moren, M., Opstad, I., Hamre, K. (2004b) A comparison of retinol, retinal and retinyl ester concentrations in larvae of Atlantic halibut (*Hippoglossus hippoglossus* L.) fed *Artemia* or zooplankton. *Aquaculture Nutrition* 10, 253–259.

Nelis, H.J., De Leenheer, A.P., Merchie, G., Lavens, P., Sorgeloos, P. (1997) Liquid chromatographic determination of vitamin C in aquatic organisms. *Journal of Chromatographic Science* 35, 333–336.

Nöll, G.N. (1996) High-performance liquid chromatographic analysis of retinal and retinol isomers. *Journal of Chromatography A* 721, 247–259.

Østerlie, M., Bjerkeng, B., Liaaen-Jensen, S. (1999) Accumulation of astaxanthin all-*E*, 9*Z* and 13*Z* geometrical isomers and 3 and 3′ RS optical isomers in rainbow trout (*Oncorhynchus mykiss*) is selective. *Journal of Nutrition* 129, 391–398.

Østerlie, M., Bjerkeng, B., Liaaen-Jensen, S. (2000) Plasma appearance and distribution of astaxanthin E/Z and R/S isomers in plasma lipoproteins of men after single dose administration of astaxanthin. *Journal of Nutritional Biochemistry* 11, 482–490.

Ostermeyer, U., Schmidt, T. (2006) Vitamin D and provitamin D in fish – determination by HPLC with electrochemical detection. *European Food Research and Technology* 222, 403–413.

Patil, N.S., Mendhe, R.B., Sankar, A.A., Iyer, H. (2008) Procedure for chromatography involving sample solvent with higher elution strength than the mobile phase. *Journal of Chromatography A*, 1177, 234–242.

Pettersson, A., Lignell, Å. (1999) Astaxanthin deficiency in eggs and fry of Baltic salmon (*Salmon salar*) with the M74 syndrome. *Ambio* 28, 43–47.

Pinto, E., Pedersén, M., Snoeijs, P., Van Nieuwerburgh, L., Colepicolo, P. (2002) Simultaneous detection of thiamine and its phosphate esters from microalgae by HPLC. *Biochemical and Biophysical Research Communications* 291, 344–348.

Poel, C., Bäckermann, S., Ternes, W. (2009) Degradation and conversion of thiamin and thiamin phosphate esters in fresh stored pork and in raw sausages. *Meat Science* 83, 506–510.

Porra, R.J., Thompson, W.A., Kriedmann, P.E. (1989) Determination of accurate extinction coefficients and simultaneous equations for assaying chlorophylls *a* and *b* extracted with four different solvents: verification of the concentration of chlorophyll standards by atomic absorption spectroscopy. *Biochimica et Biophysica Acta* 975, 384–394.

Rapala, J., Lahti, K., Räsänen, A.L., Esala, A.L., Niemelä, S.I., Sivonen, K. (2002) Endotoxins associated with cyanobacteria and their removal during drinking water treatment. *Water Research* 36, 2627–2635.

Salo-Väänänen, P., Ollilainen, V., Mattila, P., Lehikoinen, K., Salmela-Mölsä, E., Piironen, V. (2000) Simultaneous HPLC analysis of fat-soluble vitamins in selected animal products after small-scale extraction. *Food Chemistry* 71, 535–543.

Sampath-Wiley, P., Neefus, C.D. (2007) An improved method for estimating R-phycoerythrin and R-phycocyanin contents from crude aqueous extracts of *Porphyra* (Bangiales, Rhodophyta). *Journal of Applied Phycology* 19, 123–129.

Sanchez-Machado, D.I., Lopez-Hernandez, J., Paseiro-Losada, P. (2002) High-performance liquid chromatographic determination of alpha-tocopherol in macroalgae. *Journal of Chromatography A*, 976, 277–284.

Schumann, R., Häubner, N., Klausch, S., Karsten, U. (2005) Chlorophyll extraction methods for the quantification of green microalgae colonizing building facades. *International Biodeterioration and Biodegradation* 55, 213–222.

Swanson, R.V., Glazer, A.N. (1990) Separation of phycobiliprotein subunits by reversed-phase high-pressure liquid chromatography. *Analytical Biochemistry* 188, 295–299.

Swanson, R.V., Ong, L.J., Wilbanks, S.M., Glazer, A.N. (1991) Characterization of phycobiliproteins with unusually high phycourobilin content. *Journal of Biological Chemistry* 266, 9528–9534.

Tandeau de Marsac, N., Houmard, J. (1988) Complementary chromatic adaptation: physiological conditions and action spectra. *Methods in Enzymology* 167, 318–328.

Tsao, R., Zeyuan Deng, Z. (2004) Separation procedures for naturally occurring antioxidant phytochemicals. *Journal of Chromatography B* 812, 85–99.

Turujman, S.A., Wamer, W.G., Wei, R.R., Albert, R.H. (1997) Rapid liquid chromatographic method to distinguish wild salmon from aquacultured salmon fed synthetic astaxanthin. *Journal of AOAC International* 3, 622–632.

Ursi, S., Pedersén, M., Plastino, E., Snoeijs, P. (2003) Intraspecific variation of photosynthesis, respiration and photoprotective carotenoids in *Gracilaria birdiae* (Gracilariales: Rhodophyta). *Marine Biology* 142, 997–1007.

Van Heukelem, L., Thomas, C.S. (2001) Computer-assisted high-performance liquid chromatography method development with applications to the isolation and analysis of phytoplankton pigments. *Journal of Chromatography A* 910, 31–49.

Vecchi, M., Müller R.K. (1979). Separation of (3S,3′S)-(3R,3′R)- and (3S,3′R)-astaxanthin via -(-)camphanic acid esters. *Journal of High Resolution Chromatography* 2: 195–196.

Wiseman, H. (1993) Vitamin D is a membrane antioxidant – ability to inhibit iron-dependent lipid peroxidation in liposomes compared to cholesterol, ergosterol and tamoxifen and relevance to anticancer action. *FEBS Letters* 326, 285–288.

Wright, S.W., Jeffrey, S.W. (1997) High-resolution HPLC system for chlorophylls and carotenoids of marine phytoplankton. *In* Jeffrey, S.W., Mantoura, R.F.C., Wright S.W. (eds). *Phytoplankton Pigments in Oceanography: Guidelines to Modern Methods*. Unesco, Paris, pp. 327–341.

Wright, S.W., Mantoura, R.F.C. (1997) Guidelines for collection and pigment analysis of field samples. *In* Jeffrey, S.W., Mantoura, R.F.C., Wright, S.W. (eds). *Phytoplankton Pigments in Oceanography: Guidelines to Modern Methods*. Unesco, Paris, pp. 261–282.

Wright, S.W., Jeffrey, S.W., Mantoura, R.F.C. (1997) Evaluation of methods and solvents for pigment extraction. *In* Jeffrey, S.W., Mantoura, R.F.C., Wright S.W. (eds). *Phytoplankton Pigments in Oceanography: Guidelines to Modern Methods*. Unesco, Paris, pp. 261–282.

Yuan, J.P., Chen, F. (1997) Identification of astaxanthin isomers in *Haematococcus lacustris* by HPLC-photodiode array detection. *Biotechnology Techniques* 11, 455–459.

Yuan, J.P., Chen, F. (1999) Isomerization of *trans*-astaxanthin to *cis*-isomers in organic solvents. *Journal of Agricultural Food Chemistry* 47, 3656–3660.

Yuan, J.P., Chen, F. (2001) Kinetics for the reversible isomerization reaction of *trans*-astaxanthin. *Food Chemistry* 73, 131–137.

Chapter 30

CAROTENOID ANALYSIS AND IDENTIFICATION IN MARINE ANIMALS

Eduardo Alves de Almeida[1], *Glaucia Regina Martinez*[2], *and Paolo Di Mascio*[3]

[1]Department of Chemistry and Environmental Sciences, IBILCE-UNESP, São José do Rio Preto, SP, Brazil
[2]Department of Biochemistry, Federal University of Paraná, Curitiba, PR, Brazil
[3]Department of Biochemistry, IQ-USP, São Paulo, SP, Brazil

Carotenoids are usually C_{40} tetraterpenoids built from eight isoprenoid units, joined so that the sequence is reversed at the center (O'Neil and Shwartz 1992). They can be totally linear or cyclic at one or both ends and oxygen atoms may be present. The oxygenated derivatives are known as xanthophylls, while the hydrocarbon carotenoids are referred to as carotenes (Careri et al. 1999). The structures of some carotenoids are shown in Fig. 30.1. The great frequency of conjugated double bounds serves as the light-harvesting chromophore, responsible for the typical yellow, orange, or red color of carotenoids. In nature, they are generally found in the all-*trans* isomeric form, but they may also be present in *cis* configuration (O'Neil and Shwartz 1992).

Carotenoids have many important physiological functions at all trophic levels in the aquatic environment (Krinsky 1994; Olson and Owens 1998; Fraser and Bramley 2004; Škerget et al. 2005; Carrol and Berenbaum 2006). Plants are able to synthesize carotenoids in chloroplasts and thereby adjust the amounts of synthesized pigment to meet the prevailing light conditions, whereas marine animals usually obtain carotenoid precursors from their diet (McGraw and Hill 2001) and thus, depend on environmental availability.

Carotenoids, particularly vitamin A precursors, have received increasing interest in recent years because of the reported health benefits for humans (Krinsky 1991). In aquatic animals, the effects of carotenoids are multifaceted: they enhance larval growth and survival (Torrissen 1984), improve the performance of broodstock (Watanabe et al. 1991; Verakunpiriya et al. 1997) and nauplii quality (Wyban et al. 1997), and increase resistance to diseases (Tachubana et al. 1997).

The most outstanding quality of carotenoids is, however, their antioxidant property. They neutralize radical electrons by absorbing them into their conjugated double bond system. Thus, elevated amounts of carotenoids in mantle tissue of female *Perna perna* mussel as compared to males have been suggested to provide an antioxidant effect that protects developing eggs from oxidative damage (Louda et al. 2008). Further, red yeast (*Phaffia rhodozyma*) diet, rich in astaxanthin, effectively suppresses lipid peroxidation

Oxidative Stress in Aquatic Ecosystems, First Edition. Edited by Doris Abele, José Pablo Vázquez-Medina, and Tania Zenteno-Savín.

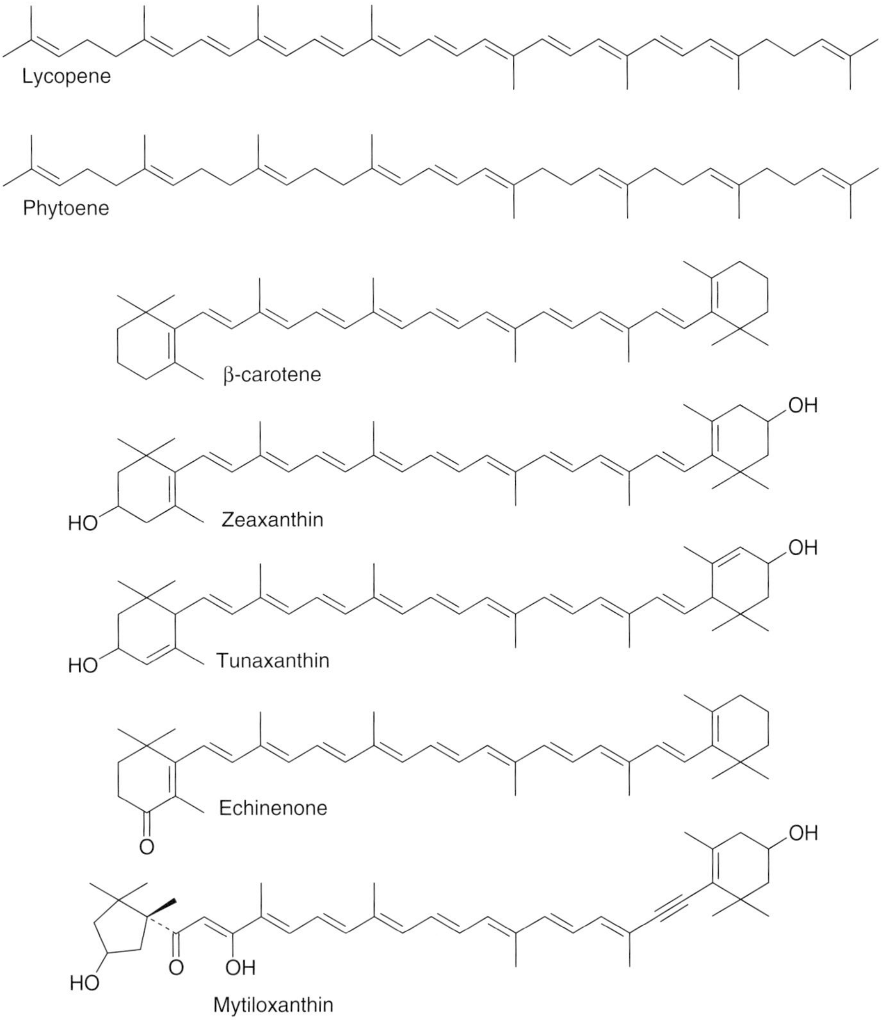

Fig. 30.1 Structure of some carotenoids.

in rainbow trout (*Oncorhynchus mykiss*) that were fed oxidized oil (Nakano et al. 1999).

The antioxidant mechanism of carotenoids, including those from aquatic organisms, is generally attributed to their ability to quench singlet oxygen ($^1O_2$) (Di Mascio et al. 1990, 1991a,b) and scavenge free radicals (Mayne and Parker 1989; Greenberg et al. 1990; Edge et al. 1997). In marine photosynthetic organisms, such as the marine algae *Gonyaulax polyedra*, carotenoids like peridinin, diadinoxanthin, dinoxanthin, diatoxanthin, and β-carotene protect the light-harvesting pigments (LHC) in the antenna complexes against photochemical damage caused by reactive oxygen species (ROS), including $^1O_2$ which was evidenced by increases in carotenoid levels during the light phase of the day, compared to the dark phase (Di Mascio et al. 1995; Hollnagel et al. 1996; Pinto et al. 2000).

Foote et al. (1970) described the mechanism by which carotenoids quench $^1O_2$. Energy transfer between $^1O_2$ and the carotenoid happens in a first step and generates a triplet state carotenoid ($^3CAR^*$) which, when returning to the ground state, dissipates energy as heat (Fig. 30.2a). Chemical quenching (Fig. 30.2b) can also occur, leading to the destruction of the carotenoid

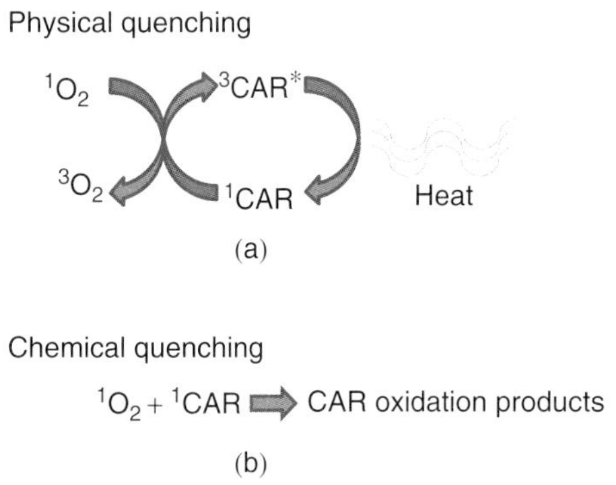

Fig. 30.2 (a) Physical and (b) chemical quenching of $^1O_2$ by carotenoids (CAR).

(Farnillo and Wilkinson 1973). For example, oxidation products of β-Carotene generated by $^1O_2$ were identified as β-ionone, β-apo-14′-carotenal, β-apo-10′-carotenal, β-apo-8′-carotenal, and β-carotene-5,8-endoperoxide (Stratton et al. 1993).

The antioxidant activity of carotenoids is related to their structure (Krinsky and Deneke 1982). The number of double bonds and the presence of functional groups affect the interaction of carotenoids with different radicals (Wörner et al. 1979). It is also observed that the carotenoid–radical reactions depend not only on the carotenoids, but also on the nature of the radical and oxygen concentration (Telfer et al. 1994). Depending on the nature of the radical, the mechanism of scavenging may be by electron abstraction or donation, resulting in the formation of a carotenoid or a cation radicals, respectively ($CAR^{\bullet}$ or $CAR^{\bullet+}$), or by generation of a carotenoid–radical adduct (Telfer et al. 1994). Pro-oxidant activity of carotenoids is observed at high oxygen concentration, due to the reaction of the carotenoid radical with molecular oxygen, generating deleterious peroxyl radicals.

Sachindra et al. (2007) studied the radical scavenging and $^1O_2$ quenching capacity of marine carotenoid fucoxanthin and its metabolites and showed that all compounds were able to efficiently scavenge $HO^{\bullet}$ and $O_2^{\bullet-}$, as well as to quench $^1O_2$. Indeed, the fish carotenoid astaxanthin was proved to be a better antioxidant than α-tocopherol under several experimental conditions. Thus, the evaluation of carotenoid levels in aquatic organisms can provide important information on antioxidant capacity in these organisms.

EVALUATION OF CAROTENOIDS IN AQUATIC ORGANISMS

Sample Preparation

Carotenoids are generally lipid soluble and can be extracted using organic solvents. The extraction procedure involves macerating tissue samples or direct addition of fluid sample into aqueous–organic mixtures like water–ethanol, water–dichloromethane or water–acetone (Krinsky and Welankiwar 1984; Vershinin 1996; Lowe et al. 1999). These solvents denature proteins and extract the carotenoids into the organic phase, which can be dissolved in petroleum ether or diethyl ether for further extraction. Addition of relatively strong salt solutions to the extract (i.e. 10% NaCl) prevents emulsion formation (Krinsky and Welankiwar 1984; Hertzberg et al. 1988). The organic phase can be directly analyzed or dried under nitrogen to concentrate the pigments.

Care must be taken during extraction, since many carotenoids can be easily oxidized or undergo isomerization in the presence of light. Protection from direct light and cooling of extraction solutions will improve the results of the carotenoid analysis. Conservation under nitrogen can also avoid oxidation during sample freezing. It has also been proposed that the presence of trace amounts of transition metals in extraction media can promote *trans*-to-*cis* isomerization of carotenoids (Zhao et al. 2005). Thus, we recommend the use of ultrapure water to avoid trace-metal contamination of the aqueous phase. Indeed, it is also recommended to use a fume hood for extraction, as many of the organic solvents, especially dichloromethane or acetone, are toxic and cancerogenic.

Carotenoid analysis with high performance liquid chromatography

Different chromatographic methods for carotenoid analysis have been described, including open-column methods and thin-layer chromatography (TLC) (Guiochon et al. 1983). High performance liquid chromatography (HPLC) has been employed as a powerful technique to quantify low levels and various forms of carotenoids in biological samples, offering several advantages over TLC and open-column methods. Normal-phase columns can be employed for carotenoid analysis by HPLC, but better carotenoid separations are achieved on reversed-phase C18 columns, eluted with isocratic (i.e., a single solvent) mixtures of organic solvents such as, e.g. 70% acetonitrile, 20% dichloromethane, and 10% methanol, or binary systems (i.e., mobile phase gradients). As the composition of carotenoids can vary among different species, optimization of the solvents used in the mobile phase should be carried out. Because of this, numerous different HPLC conditions are described in the literature for carotenoid analysis in different species (Ohkubo et al. 1999).

Most of the normal-phase separations of different carotenoids including their isomers use $Ca(OH)_2$ columns, and the mobile phase usually contains small amounts of acetone in hexane or pentane (O'Neil and Schwartz 1992). However, $Ca(OH)_2$ columns are generally not commercially available, and reverse-phase C18 columns are therefore more applicable for carotenoid analyses. Further, a C30 column was recently shown to achieve better resolution of carotenoid stereo-isomers than a C18 column (Lacker et al. 1999).

An example of the analysis of mussel (*Perna perna*) carotenoids through HPLC coupled to UV-Vis detection is shown in Fig. 30.3. (These procedures were especially developed for the present chapter by the authors). Mantle tissue of female *P. perna* mussels were homogenized (1 : 10 w/v) in hexane : methanol : water (5 : 4 : 1) and centrifuged at 5000 *g* for 5 min. The hexane fraction was directly injected (100 μL) into the HPLC system, and separated on a C18 column (250 × 4.6 mm, 5 μm), using a binary mobile phase gradient of 80 to 100% of acetonitrile : methanol (75:25) in water during 30 min, at 1 mL min^{-1}.

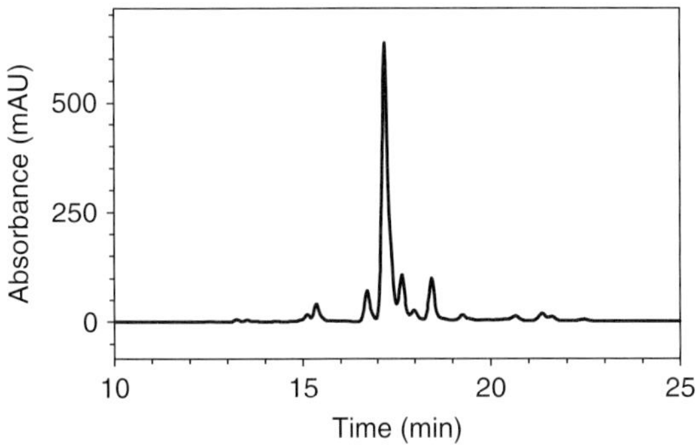

Fig. 30.3 Chromatogram of mussel (*Perna perna*) carotenoids.

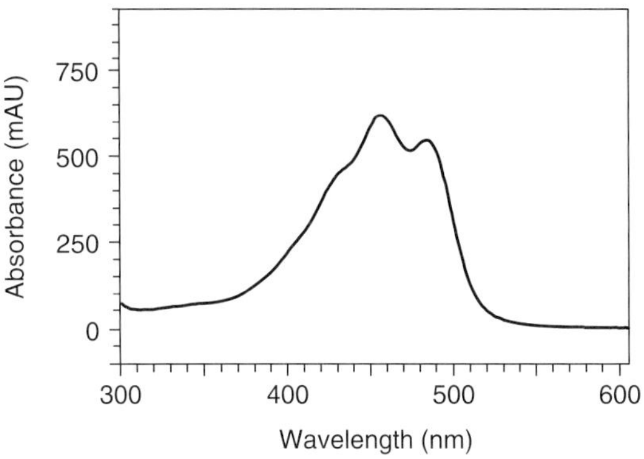

Fig. 30.4 UV-Vis spectrum of a carotenoid.

The UV-Vis detector used for peak detection in Fig. 30.3 was set at 470 nm. Most carotenoids absorb light at maximum of three wavelengths, resulting in three-peak spectra in the 400–500 nm region of the visible spectrum, with the middle peak having the highest intensity (Zechmeister 1962), as shown in Fig. 30.4. For this reason, the UV-Vis detector is generally set at 450–470 nm to monitor the carotenoid peaks. Diodearray-based detectors are useful tools that allow viewing of the spectral characteristics of a carotenoid peak and, thereby, identify the compound.

Another common method for carotenoid analysis is the estimation of total carotenoids, based on the maximum absorbance of most carotenoids in 450–470 nm. For example, the reported $E^{1\%} = 2592$ for β-carotene at 453 nm can be used to determine the total carotenoid amount in aquatic species (Vershinin 1996; Tsushima et al. 1997).

Application of combined techniques for the characterization of carotenoids in aquatic species

Due to the great variety and the presence of species-specific carotenoids in different organisms, previous characterization and identification of the pigments is needed. Commercially available carotenoid standards (e.g. astaxanthin, β-carotene) are useful in the identification and quantification of carotenoids in aquatic species, and a standard calibration curve is needed for each individual compound. However, many species-specific main carotenoid compounds are not commercially available as standards, which

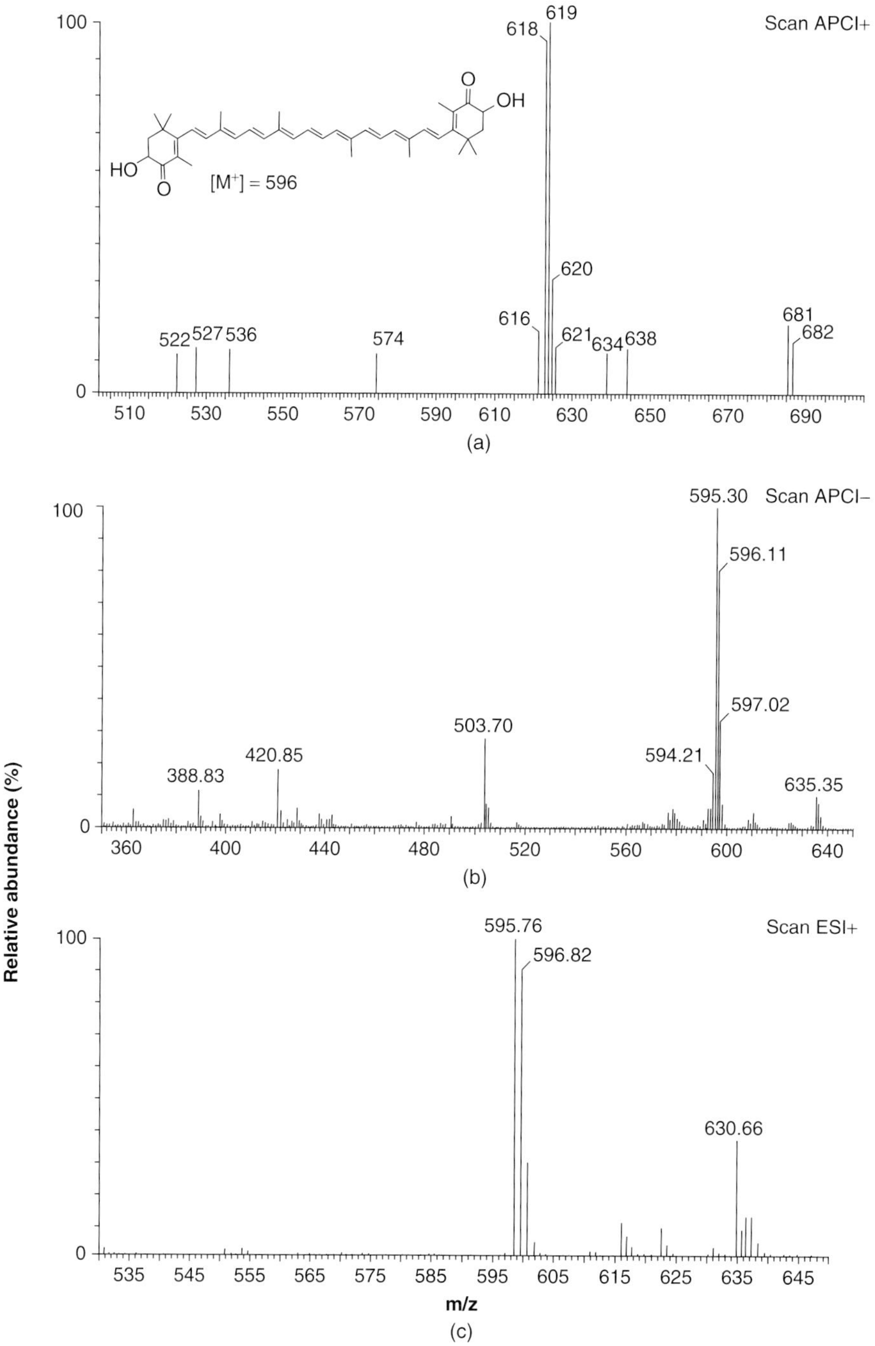

Fig. 30.5 Mass spectra (m/z 400–800) of astaxanthin in the (a) APCI+, (b) APCI−, and (c) ESI+ modes.

renders both identification and quantification very difficult. As an example, mussels contain highly specific carotenoids in their tissues, including mytiloxanthin and isomytiloxanthin, with names referring to the species *Mytillus edulis*, in which these carotenoids were first isolated and identified (Hertzberg et al. 1988). In this case, it can be suggested that researchers exchange their identified carotenoid samples between their laboratories for a better attribution of carotenoids when studying different marine species.

In a classic work, Hertzberg et al. (1988) characterized a total of 19 carotenoids from *M. edulis*, including ten C_{40} and two C_{37} acetylenic carotenoids, four C_{40} and two C_{37} skeletal allenic carotenoids, and occasionally β,β-carotene. As expected, none of these compounds were commercially available, creating the need for chemical characterization to establish them as carotenoid standards. The application of combined techniques like mass spectrometry (MS), nuclear magnetic resonance (NMR), infrared (IR) spectra or even UV-Vis spectra can serve to provide the structural elucidation of newly identified carotenoid compounds in chemical laboratories.

Moreover, the combination of different HPLC conditions is needed to better separate different carotenoids, allowing more precise collection for further characterization by other methods. Ohkubo et al. (1999) for example, used seven different HPLC conditions to characterize carotenoids from goldfish (*Carassius auratus*).

Mass spectrometry can be used to analyze samples containing very small amounts of carotenoids (fentomol levels) and is also useful in the identification of specific carotenoid structures. As aquatic species sometimes possess unusual or species-specific carotenoids, confirmation of the carotenoid structure is facilitated when the mass of the carotenoid is known.

Mass spectrometry techniques are based on the ionization of the sample by an ionizing system and its further detection in a mass detector, which detects the ions in the mass/charge ratio (m/z), so that the mass of the compound is obtained. Ionization of the samples is a limiting step, because different numbers of ions can be generated, resulting in complex spectra. The ionization source should be finely adjusted in order to obtain preferentially the molecular ion of the compound (i.e. $[M + H]^+$).

Samples can be injected directly into the mass spectrometer, or previously separated by coupling an HPLC or a GC system to the mass spectrometer. Different kinds of ionization systems are applied in mass spectrometry, and some of them have been described for carotenoid

Box 30.1 Example for carotenoid separation by HPLC and mass spectrometry (MS) for analysis in ESI+, APCI+, and APCI− ionization modes

HPLC conditions:

- Mobile phase: binary gradient of acetonitrole (solvent A) and methanol (solvent B) from 80A : 20B% to 100A : 0B% within 20 min at a flow rate of 150 μL min^{-1}.
- Column: Luna (Phenomenex) C18 (150 mm × 4.6 mm, 5 μm) with its own guard column.

For the ESI+ mode the conditions are:

- Cone voltage: 20 V
- Probe temperature: 100°C
- Drying gas flow: 500 L h^{-1}
- Capillary voltage: 3.5 kV.

For the APCI+ and APCI− the conditions are:

- Cone voltage: 25 V
- Probe temperature: 400°C
- Drying gas flow: 200 L h^{-1}
- Capillary voltage: 3 kV.

analyses, including moving belt interface (MBI) combined with electron impact, fast atom bombardment (FAB), and most recently, electrospray ionization (ESI), and atmospheric pressure chemical ionization (APCI) (van Breemen 1997).

The moving belt interface suffers from poor sensitivity for carotenoids because of incomplete analyte volatilization, pyrolysis, and excessive fragmentation. FAB techniques are preferred over MBI due to the formation of much more abundant molecular ions, lack of fragmentation, and high sensitivity. However, it is limited to very low flow rates from the HPLC system and the need to clean the ion source every 3 h (Dembitsky and Rezanka 1996; van Breemen 1997).

The ESI and APCI ionization modes are the main techniques used for carotenoid analysis in recent years, with great sensitivity and specificity (van Breemen 1997; Careri et al. 1999; Lacker et al. 1999). Figure 30.5 shows the mass spectra (m/z 400–800) of astaxanthin in the ESI+, APCI+ and APCI− modes. HPLC and MS conditions are described in Box 30.1.

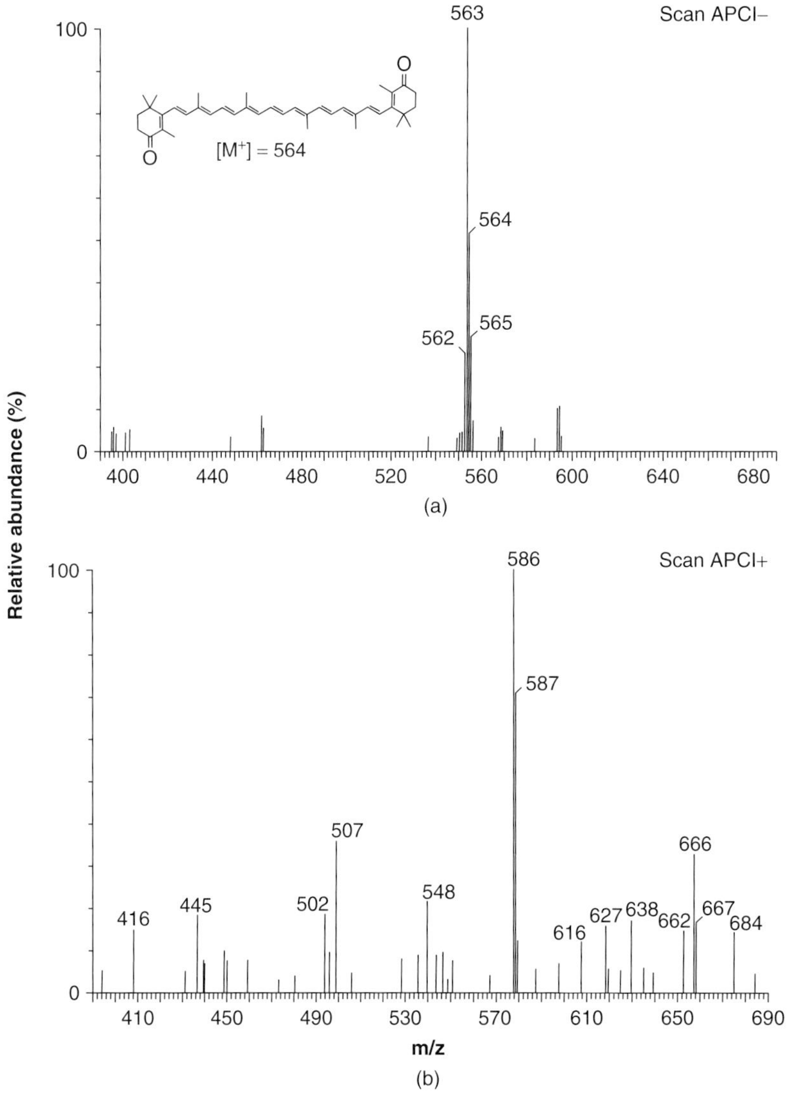

Fig. 30.6 Cantaxanthin mass spectra in the (a) APCI− and (b) APCI+ modes.

The carotenoid astaxanthin has a mass of 596, as confirmed in the spectra B and C of Fig. 30.4. In the APCI+ mode, a signal at m/z 619 was obtained, corresponding to a sodium adduct of the carotenoid (sodium mass = 23; [M + 23] = 619). Water, sodium, ammonium, and other adducts are sometimes observed in mass spectra, depending on the solution used to prepare samples, or the solvent used in the mass spectrometer. The same conditions were also applied for cantaxanthin analysis by the APCI+ and APCI− modes, as show in Fig. 30.6.

Using the same procedure for astaxanthin and cantaxanthin standards as described in Box 30.1, we can analyze and confirm the structure of unknown carotenoids from specific species. For example, the analyses of the main carotenoid of the mussel *P. perna* shown in Fig. 30.3 (most intense peak in the chromatogram) using the MS techniques described above, revealed a compound with a signal at $m/z = 580$, for both APCI− and ESI+ modes, as shown in Fig. 30.7.

Figure 30.8 shows the structure of two possible carotenoids, alloxanthin and phoenicoxanthin, both with a mass of 580 Da. Further studies by subjecting the purified carotenoid to IR techniques can reveal the absence of triple bonds in carotenoid composition, thus in this case eliminating alloxanthin and suggesting phoenicoxanthin as the probable carotenoid.

Since many other possibilities exist because different carotenoids have the same mass, further methods are needed to confirm carotenoid identity. Isomytiloxanthin, for example, has a m/z signal at 598 for the molecular ion, but a $m/z = 580$ value was also found for the [M − 18] ion (Hertzberg et al. 1988).

Infrared and UV-Vis spectra can also be useful in carotenoid identification due to the presence of characteristic bands for functional groups (Lowe et al. 1999). These approaches, however, can be inadequate for unambiguous identification of carotenoids in animal samples, owing to the risk of spectral interferences. Thus, if available, further characterization by NMR

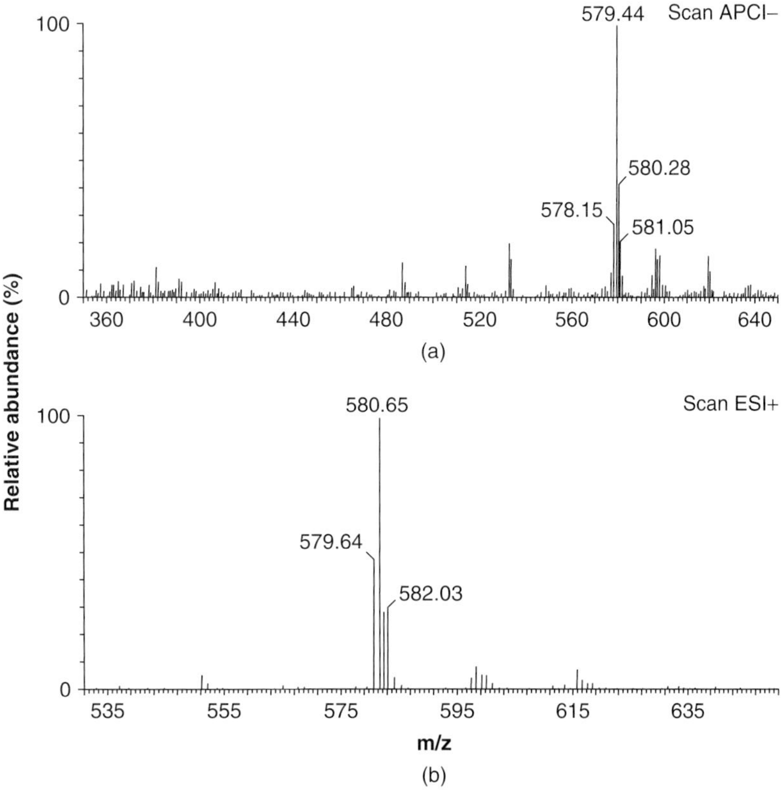

Fig. 30.7 (a) APCI− and (b) ESI+ mass spectrum of the most intense peak from the chromatogram of Fig. 30.3.

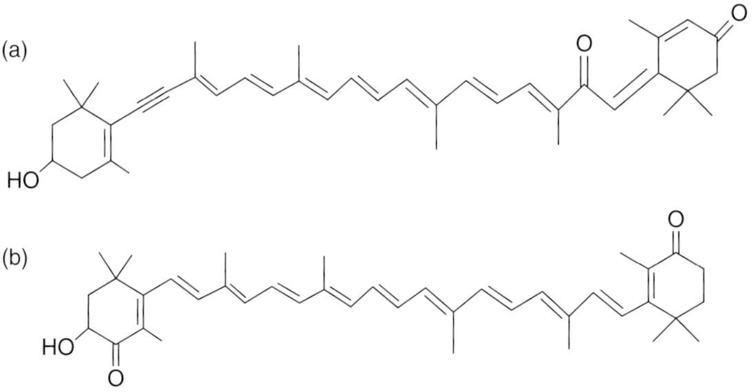

Fig. 30.8 Two examples of possible structures for the *P. perna* carotenoid found at $m/z = 580$: (a) alloxanthin; (b) phoenicoxanthin.

spectroscopy represents a better technique to confirm the structure of the compound. ^{1}H NMR spectra and chemical shift attributions for several carotenoids are currently described in the literature, allowing comparisons and facilitating carotenoid identification in appropriately understood species.

REFERENCES

Careri, M., Corradini, C., Elviri, L., Nicoletti, I., Zagnoni, I. (1999) Liquid chromatography-electrospray mass spectrometry of beta-carotene and xanthophylls. Validation of the analytical method. *Journal of Chromatography A* 854, 233–244.

Carrol, M.J., Berenbaum, M.R. (2006) Lutein sequestration and furanocoumarin metabolism in Parsnip webworms under different ultraviolet light regimes in the montane west. *Journal of Chemical Ecology* 32, 277–305.

Dembitsky, V.M., Řezanka, T. (1996) Comparative study of the endemic freshwater fauna of Lake Baikal – VII. Carotenoid composition of the deept-water amphipod crustacean *Acanthogammarus (Brachyuropus) grewingkii*. *Comparative Biochemistry and Physiology B* 114, 383–387.

Di Mascio, P., Devasagayam, T.P., Kaiser, S., Sies, H. (1990) Carotenoids, tocopherols and thiols as biological singlet molecular oxygen quenchers. *Biochemical Society Transactions* 18, 1054–1056.

Di Mascio, P., Kaiser, S.P., Devasagayam, T.P., Sies, H. (1991a) Biological significance of active oxygen species: *in vitro* studies on singlet oxygen-induced DNA damage and on the singlet oxygen quenching ability of carotenoids, tocopherols and thiols. *Advances in Experimental Medicine and Biology* 283, 71–77.

Di Mascio, P., Murphy, M.E., Sies, H. (1991b) Antioxidant defense systems: the role of carotenoids, tocopherols, and thiols. *American Journal of Clinical Nutrition* 53, 194–200.

Di Mascio, P., Hollnagel, H.C., Sperança, M., Colepicolo, P. (1995) Diurnal rhythm of beta-carotene in photosynthetic alga *Gonyaulax polyedra*. *Biological Chemistry Hoppe-Seyler* 376, 297–301.

Edge, R., McGarvey, D.J., Truscott, T.G. (1997) The carotenoids as anti-oxidants: a review. *Journal of Photochemistry and Photobiology B* 41, 189–200.

Farnillo, A., Wilkinson, F. (1973) On the mechanism of quenching of singlet oxygen in solution. *Photochemistry and Photobiology* 18, 447–450.

Foote, C.S., Chang, Y.C., Denny, R.W. (1970) Chemistry of singlet oxygen. X. Carotenoid quenching parallels biological protection. *Journal of American Chemical Society* 92, 5216–5218.

Fraser, P.D., Bramley, P.M. (2004) The biosynthesis and nutritional uses of carotenoids. *Progress in Lipid Research* 43, 228–265.

Greenberg, E.R., Baron, J.A., Stukel, T.A. et al. (1990) A clinical trial of beta carotene to prevent basal-cell and squamous-cell cancers of the skin. The Skin Cancer Prevention Study Group. *New England Journal of Medicine* 323, 789–795.

Guiochon, G., Gonnord, M.F., Zakaria, M., Beaver, L.A., Siouffi, A.M. (1983) Chromatography with a two-dimensional column. *Chromatographia* 17, 121–124.

Hertzberg, S., Partali, V., Liaaen-Jensen, S. (1988) Animal carotenoids. 32. Carotenoids of *Mytilus edulis* (edible mussel). *Acta Chemica Scandinavica B* 42, 495–503.

Hollnagel, H.C., Di Mascio, P., Asano, C.S. et al. (1996) The effect of light on the biosynthesis of beta-carotene and superoxide dismutase activity in the photosynthetic alga *Gonyaulax polyedra*. *Brazilian Journal of Medical and Biological Research* 29, 105–110.

Krinsky, N. (1991) Effects of carotenoids in cellular and animal systems. *American Journal of Clinical Nutrition* 53, 238–246.

Krinsky, N.I. (1994) The biological properties of carotenoids. *Pure and Applied Chemistry* 66, 1003–1010.

Krinsky, N.I., Deneke, S.M. (1982) Interaction of oxygen and oxy-radicals with carotenoids. *Journal of the National Cancer Institute* 69, 205–210.

Krinsky, N.I., Welankiwar, S. (1984) Assay of carotenoids. *Methods in Enzymology* 105, 155–162.

Lacker, T., Strohschein, S., Albert, K. (1999) Separation and identification of various carotenoids by C30 reversed-phase high-performance liquid chromatography coupled to UV and atmospheric pressure chemical ionization mass spectrometric detection. *Journal of Chromatography A* 854, 37–44.

Louda, J.W., Neto, R.R., Magalhaes, A.R.M., Schneider, V.F. (2008) Pigment alterations in the brown mussel *Perna perna*. *Comparative Biochemistry and Physiology* B 150, 385–394.

Lowe, G.M., Booth, L.A., Young, A.J., Bilton, R.F. (1999) Lycopene and β-carotene against oxidative damage in HT29 cells at low concentrations but rapidly lose this capacity at higher doses. *Free Radical Research* 30, 141–151.

Mayne, S.T., Parker, R.S. (1989) Antioxidant activity of dietary canthaxanthin. *Nutrition and Cancer* 12, 225–36.

McGraw, K.J., Hill, G.E. (2001) Carotenoid access and intraspecific variation in plumage pigmentation in male American goldfinches (Carduelis trsitis) and northern cardinals (*Cardinalis cardinalis*). *Functional Ecology* 15, 732–739.

Nakano, T., Kanmuri, T., Sato, M., Takeuchi, M. (1999) Effect of astaxanthin rich red yeast (*Phaffia rhodozyma*) on oxidative stress in rainbow trout. *Biochimica et Biophysica Acta* 1426, 119–125.

Ohkubo, M., Tsushima, M., Maoka, T., Matsuno, T. (1999) Carotenoids and their metabolism in the goldfish *Carassius auratus* (Hibuna). *Comparative Biochemistry and Physiology B* 124, 333–340.

Olson, V.A., Owens, I.P.F. (1998) Costly sexual signals: are carotenoids rare, risky or required. *Trends in Ecology and Evolution* 13, 510–514.

O'Neil, C.A., Schwartz, S.J. (1992) Chromatographic analysis of cis/trans carotenoid isomers. *Journal of Chromatography* 624, 235–252.

Pinto, E., Catalani, L.H., Lopes, N.P., Di Mascio, P., Colepicolo, P. (2000) Peridinin as the major biological carotenoid quencher of singlet oxygen in marine algae *Gonyaulax polyedra*. *Biochemical and Biophysical Research Communications* 268, 496–500.

Sachindra, N.M., Sato, E., Maeda, H. et al. (2007) Radical scavenging and singlet oxygen quenching activity of marine carotenoid fucoxanthin and its metabolites. *Journal of Agricultural and Food Chemistry* 55, 8516–8522.

Škerget, M., Kotnik, P., Handolin, M., Hraš, A.R., Simonič, M., Knez, Ž. (2005) Phenols, proanthocyanidins, flavones in some plant materials and their antioxidant activities. *Food Chemistry* 89, 191–198.

Stratton, S.P., Schaefer, W.H., Liebler, D.C. (1993) Isolation and identification of singlet oxygen oxidation products of β-carotene. *Chemical Research in Toxicology* 6, 542–547.

Tachibana, K., Yagi, M., Hara, K., Mishima, T., Tsuchimoto, M. (1997) Effects of feeding b-carotene supplemented rotifers on survival and lymphocyte proliferation reaction of fish larvae of Japanese parrotfish (*Oplegnathus fasciatus*) and Spotted parrotfish (*Oplegnathus punctatus*): preliminary trials. *Hydrobiologia* 358, 313–316.

Telfer, A., Dhami, S., Bishop, S.M., Phillips, D., Barber, J. (1994) Beta-carotene quenches singlet oxygen formed by isolated photosystem II reaction centers. *Biochemistry* 33, 14469–14474.

Torrissen, O.J. (1984) Pigmentation of salmonids: effects of carotenoids in eggs and start-feeding diet on survival and growth rate. *Aquaculture* 43, 185–193.

Tsushima, M., Katsuyama, M., Matsuno, T. (1997) Metabolism of carotenoids in the Apple snail, *Pomacea canaliculata*. *Comparative Biochemistry and Physiology B* 118, 431–436.

Van Breemen, R.B. (1997) Liquid chromatography/mass spectrometry of carotenoids. *Pure and Applied Chemistry* 69, 2061–2066.

Verakunpiriya, V., Watanabe, K., Mushiake, K. et al. (1997) Effect of krill meal supplementation in soft-dry pellets on spawning and quality of egg of yellowtail. *Fisheries Science* 63, 433–439.

Vershinin, A. (1996) Carotenoids in mollusca: approaching the functions. *Comparative Biochemistry and Physiology B* 113, 63–71.

Watanabe, T., Lee, M.J., Mizutani, J. et al. (1991) Effective component of cuttlefish meal and raw krill for improvement of quality of red sea bream *Pagrus major* eggs. *Nippon Suisan Gakkaishi* 57, 681–694.

Wörner, P., Patscheke, H., Paschen, W. (1979) Response of platelets exposed to potassium tetraperoxochromate, an extracellular source of singlet oxygen, hydroxyl radicals, superoxide anions and hydrogen-peroxide. *Hoppe-Seyler's Zeitschrift für Physiologische Chemie* 360, 559–570.

Wyban, J., Martinez, G., Sweeney, J. (1997) Adding paprika to Penaeus vannamei maturation diet improves nauplii quality. *World Aquaculture* 28, 59–62.

Zechmeister, L. (1962) *Cis-Trans Isomeric Carotenoids Vitamin A and Arylpolyenes*, Academic Press, New York, 251 pp.

Zhao, L., Chen, F., Zhao, G., Wang, Z., LIAO, X., Hu, X. (2005) Isomerization of *trans*-astaxanthin induced by copper(II) ion in ethanol. *Journal of Agricultural and Food Chemistry* 53, 9620–9623.

Chapter 31

LINOLEIC ACID OXIDATION PRODUCTS AS BIOMARKERS OF OXIDATIVE STRESS *IN VIVO*

Etsuo Niki and Yasukazu Yoshida
National Institute of Advanced Industrial Science and Technology, Health Research Institute, Osaka, Japan

Many toxic compounds induce the oxidation of biologically important molecules, which eventually lead to various disorders and diseases (Kelly et al. 1998; Slaninova et al. 2009). Recently, the effect of nanomaterials such as C60 fullerenes as inducers of oxidative stress has received increasing attention (Oberdörster 2004). It has been shown that nanomaterials produce reactive oxygen species (ROS) within cells and induce oxidation of biological molecules (Horie et al. 2010, and references therein). Lipids, especially polyunsaturated fatty acids (PUFA) and their esters, are susceptible to oxidation. Lipid peroxidation induces disturbance of cellular structures and functional loss of biomembranes, modifies lipoproteins, and generates potentially toxic products (Niki 2009). Lipid peroxidation has been used as an indicator of xenobiotic-induced oxidative stress and oxidative damage in fish. Toxic compounds, such as cadmium (Cd) (Thomas and Wofford 1993) and aromatic hydrocarbons (Di Giulio et al. 1993) enhance the levels of the lipid peroxidation biomarker malondialdehyde (MDA) in hepatic tissue of fish (see Chapters 32 and 34). Lipid peroxidation products induce muscular dystrophy, known as "Sekoke disease" in carp and other fish, a disease that is prevented by vitamin E administration (Watanabe and Hashimoto 1968).

Lipid peroxidation proceeds by three distinct mechanisms: (i) free-radical-mediated chain oxidation, (ii) free-radical-independent nonenzymatic oxidation, and (iii) enzymatic oxidation (Niki 2009). Specific products are formed by each pathway and specific antioxidants are required to inhibit each mechanism. The characteristics of each type of oxidation are summarized in Table 31.1, including the products formed when linoleic acid (18:2) is used as substrate.

FREE-RADICAL-MEDIATED LIPID PEROXIDATION

The free-radical-mediated lipid peroxidation proceeds by a chain reaction, where one initiating free radical can oxidize many lipid molecules. The mechanisms

Oxidative Stress in Aquatic Ecosystems, First Edition. Edited by Doris Abele, José Pablo Vázquez-Medina, and Tania Zenteno-Savín.

Table 31.1 Mechanisms of lipid peroxidation and characteristics of HPODE produced from linoleate.

Type	Characteristics	Isomers of HPODE		
		region	stereo	enantio
1. Enzymatic (15LOX)	Specific catalytic	13	*cis,trans*	*S*
2. Nonenzymatic, free-radical-mediated (LOO•)	Random chain oxidation	9,13	*cis,trans* *trans,trans* 9-*ZE* = 13-*ZE* 9-*EE* = 13-*EE*	*R* = *S* (racemic)
3. Nonenzymatic, nonradial ($^1O_2$)	Random stoichiometric	9,10,12,13	*cis,trans*	

and kinetics of this type of oxidation have been studied extensively and are now fully understood (Yin and Porter 2005). The major reactions include: (1) abstraction of *bis*-allylic hydrogen from PUFAs yielding a carbon-centered radical, which rearranges to *cis,trans*-pentadienyl radical; (2) addition of oxygen to the pentadienyl radical to form lipid peroxyl radical; (3) release of oxygen from the peroxyl radical to produce oxygen and pentadienyl radical (this pentadienyl radical rapidly reacts with oxygen to the more thermodynamically stable *trans,trans*-peroxyl radical); and (4) intramolecular addition to the double bond to yield bicyclic prostaglandin type products (Fig. 31.1). The latter proceeds when lipids have three or more double bonds, that is, linolenates (18:3), arachidonates (20:4), eicosapentaenoates (20:5, EPA), and docosahexaenoates (22:6, DHA). The fatty acid and its moieties of phospholipids, glycerides, and cholesteryl esters are oxidized by the same mechanisms. Numerous products are formed, and the product distribution is determined by the relative importance of the competing reactions.

The lipid peroxidation of linoleic acid and its esters has been studied extensively, since they are the most abundant PUFA *in vivo*. Their oxidation proceeds by a straightforward mechanism, which yields much simpler products with high selectivity than arachidonates and other highly unsaturated fatty acids. The free-radical-mediated oxidation of linoleates proceeds by steps 1, 2 and 3 (see above) to give

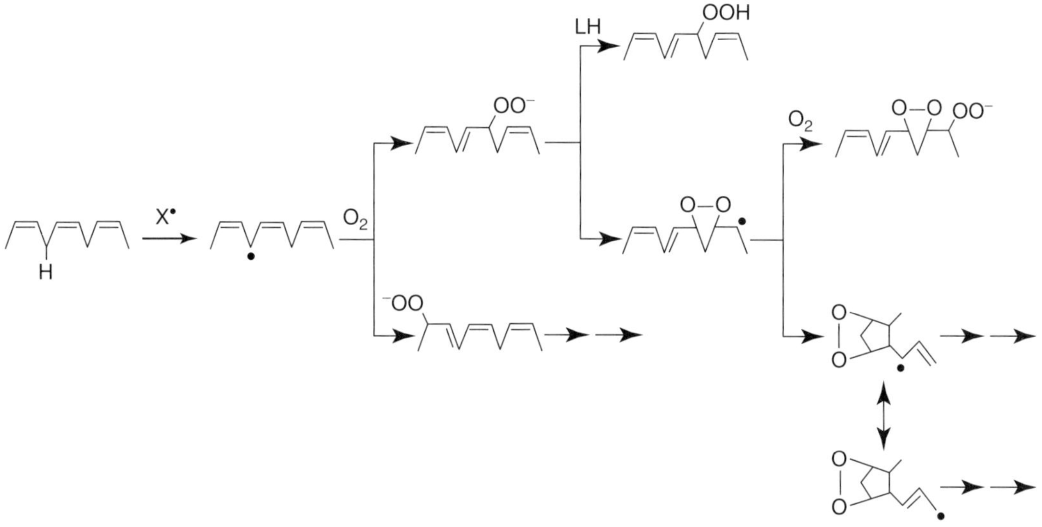

Fig. 31.1 Major reaction pathways of free-radical-mediated lipid peroxidation of polyunsaturated fatty acids.

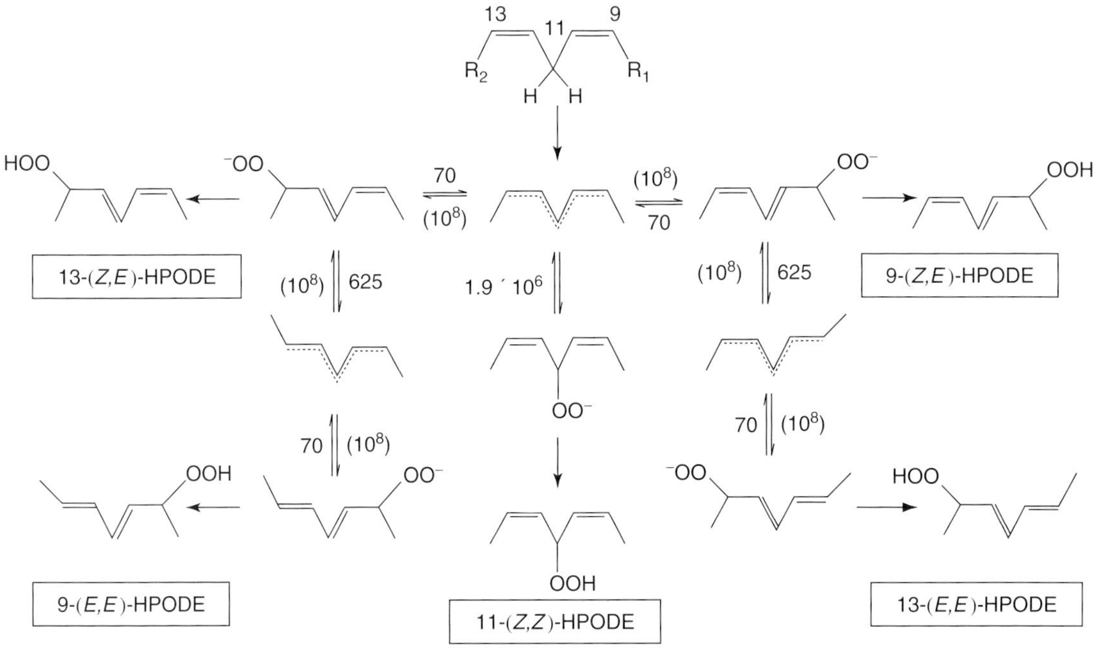

Fig. 31.2 Mechanisms of peroxidation of linoleates.

conjugated diene hydroperoxides. The mechanism of oxidation of linoleates is shown in Fig. 31.2. The initial step is the abstraction of *bis*-allylic hydrogen at carbon 11. The resulting radical rearranges rapidly to the more stable pentadienyl radical, with which oxygen reacts to give 9- or 13-*cis,trans*-peroxyl radicals. These peroxyl radicals attack another lipid by abstraction of *bis*-allylic hydrogen, which yields 9-(10E,12Z)-hydroperxyoctadecadienoic acid (HPODE) or 13-(9Z,11E)-HPODE and continues the propagation reaction. The 9- or 13-*cis,trans*-peroxyl radicals may undergo β-scission to release oxygen and a pentadienyl radical, to which oxygen adds to give more stable 9-(10E,12E)- or 13-(9E,11E)-peroxyl radicals preferentially to *trans,cis*-peroxyl radical formation. Alternatively, the peroxyl radical is scavenged by antioxidants, such as vitamin E, to give the same hydroperoxide. Thus, four kinds of hydroxyoctadecadienoic acids (HODEs) are formed in the oxidation of linoleates as shown in Fig. 31.2. In fact, in a system containing 402 mM methyl linoleate in acetonitrile solution induced by a constant flux of free radicals generated from an azo initiator, the amounts of methyl linoleate and oxygen consumed, as well as the formation of conjugated diene and hydroperoxide agree well, 87, 91, 98, and 90 μmol, respectively (Yamamoto et al. 1982). The stoichiometry shows that methyl linoleate is oxidized by a straightforward mechanism that results quantitatively in methyl ester formation giving four conjugated diene hydroperoxyoctadecadienoic acid (HPODE). Note that the amounts of 9-*cis,trans*- and 13-*cis,trans*-HPODE are equivalent, as are the amounts of 9-*trans,trans*- and 13-*trans,trans*-HPODE, and that they arebreak racemic.

NONRADICAL, NONENZYMATIC LIPID PEROXIDATION

$^1O_2$ and O_3 oxidize lipids by nonradical mechanisms. $^1O_2$ may be formed photochemically and cause deleterious damage. It oxidizes unsaturated lipids mainly by ene-reaction giving hydroperoxide with concomitant double-bond migration. For example, the oxidation of linoleates by $^1O_2$ gives 9, 10, 12, and 13-HPODE, among which 10- and 12-HPODE are specific products for $^1O_2$ oxidation. O_3 oxidizes lipids to the resulting ozonides and cleavage products.

Myeloperoxidase, MPO, a heme protein secreted by activated phagocytes, reacts with H_2O_2 in the presence of chlorine (Cl) and bromine (Br) to give hypochlorous acid (HOCl) and hypobromous acid (HOBr), respectively. The dissociated form hypochlorite and hypobromide oxidize lipids to give chlorohydrine and bromohydrine. Hypochlorite reacts with H_2O_2 to yield $^1O_2$, whereas it reacts with organic hydroperoxides to give alkoxyl and/or peroxyl radicals, with little concomitant formation of $^1O_2$. The oxidation of methyl linoleate by hypochlorite yields racemic 9- and 13-, *cis,trans*-, and *trans,trans*-conjugated diene hydroperoxides, showing that the oxidation proceeded by the free-radical mechanism.

ENZYMATIC LIPID PEROXIDATION

Lipoxygenase (LOX) and cyclooxygenase (COX) oxidize arachidonic acid to hydroperoxyeicosatetraenoic aicd (HPETE), prostaglandins, prostacyclin, thromboxane, and leukotrienes. COX and LOX oxidize lipids regio-, stereo-, and enantio-specifically. For example, 15-lipoxygenase oxidizes linoleates to give 13(S)-9Z, 11E-HPODE exclusively, while the free-radical-mediated oxidation of linoleates gives four racemic products as shown in Table 31.1. LOX oxidizes phospholipids and cholesteryl esters in low density lipoprotein (LDL) particles as well as free fatty acids, although the specificity is lower.

PUFAs can be oxygenated by cytochrome P450 to hydroxy and epoxy fatty acids (Oliw et al. 1996). *Bis*-allylic cytochrome P450-hydroxylases transform linoleic acid to 11-HODE, arachidonic acid to 13-5Z,8Z,11Z,14Z-, 10-5Z,8Z,11Z,14Z-, and 7-5Z, 8Z,11Z,14Z-hydroxytetraenoic acid (HETE), and eicosapentaenoic acid (EPA) to 16-5Z,8Z,11Z,14Z,17Z-, 13-5Z,8Z,11Z,14Z,17Z-, and 10-5Z,8Z,11Z,14Z,17Z-hydroxyeicosapentaenoic acid as major metabolites. The *bis*-allylic hydroxy fatty acids are chemically unstable and decompose rapidly to *cis-trans* conjugated hydroxy fatty acids. The fatty acid epoxides, which are formed by cytochrome P450, are chemically stable but are hydrolyzed to diols by soluble epoxide hydrolases. Epoxidation of PUFAs is a prominent pathway of metabolism in the liver and the renal cortex.

CHOLESTEROL OXIDATION

Cholesterol is also oxidized by three mechanisms (Van Reyk et al. 2006). The oxidation products of cholesterol are termed *oxysterols*. The free-radical-mediated oxidation of cholesterol gives 7-hydroperoxycholesterol (7-OOHCh), 7-hydroxycholesterol (7-OHCh), 7-ketocholesterol (7-KCh), cholesterol-5,6-epoxide (5,6-epoxyCh), and cholestane−3β,5α, 6β-triol as major products. Both, alpha and beta forms of 7-OOHCh, 7-OHCh, and 5,6-epoxyCh are formed. $^1O_2$ oxidizes cholesterol to give 5-OOHCh as the major primary product.

Various enzymes oxidize cholesterol to give specific hydroxycholesterols. Many of the enzymes belong to the cytochrome P450 family. 7α-Hydroxycholesterol, an important intermediate in bile acid biosynthesis, is formed by cholesterol 7α-hydroxylase (CYP7A1) as well as free-radical oxidation. The drug-metabolizing enzyme CYP3A4 oxidizes cholesterol to give 4β-hydroxycholesterol. The mitochondrial enzyme CYP27A1 gives 27-hydroxycholesterol. 25-Hydroxycholesterol is formed by cholesterol 25-hydroxylase which is not a cytochrome P450 enzyme.

ISOPROSTANES

Morrow and Roberts discovered that isoprostanes (IsoPs) are formed by a free-radical-mediated peroxidation of arachidonic acid independent of COX (Morrow et al. 1990). IsoPs are a series of prostaglandin-like compounds of which those containing an F-type prostane ring (F_2-IsoP) have been studied more extensively (Fig. 31.3). The 8-, 9-, 11- and 12-peroxyl radicals, but not 5- and 15-peroxyl radicals, derived from arachidonic acid, undergo two consecutive intramolecular cyclizations, oxygen addition, and hydrogen abstraction to form PGG_2-like compounds, which are reduced by ketoreductase to F_2-IsoP. Compounds are denoted as 5-, 8-, 12-, and 15-series regio-isomers depending on the carbon atom to which the side chain HO is attached. A total of 64 isomers of F_2-IsoPs can be formed from arachidonic acid. An important structural distinction between IsoPs and COX-derived PGs is that the former contain side chains that are predominantly oriented *cis* to the prostane ring, whereas the latter possess exclusively *trans* side chains. The prostaglandin H_2-like bicyclic endoperoxide intermediates isomerizes to D_2- and E_2-IsoP, which are unstable and spontaneously dehydrate to J_2- and A_2-IsoP, chemically reactive electrophilic cyclopentenones. The oxidation of EPA and DHA by similar mechanisms gives F3-IsoPs and

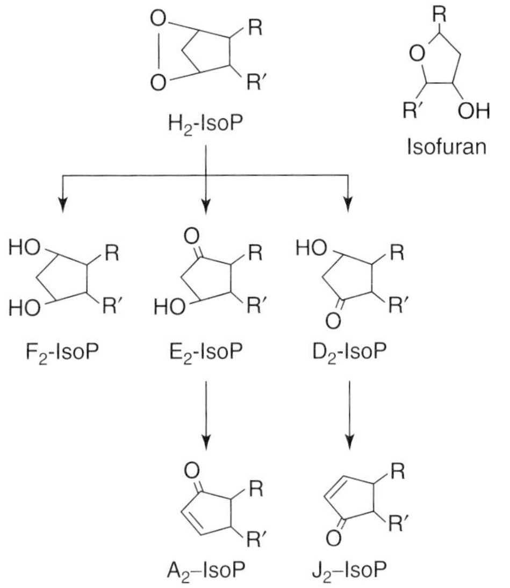

Fig. 31.3 Isoprostanes and isofurans from arachidonates.

F4-neuroprostanes (Roberts et al. 2005). Isofuranes (IsoFs) with a substituted tetrahydrofuran ring structure have also been found.

ALDEHYDES

Various small molecular weight aldehydes such as acrolein, MDA, and 4-hydroxy-2-nonenal (HNE) are formed during lipid peroxidation as secondary or decomposition products (Esterbauer et al. 1991). Among them, HNE has been studied most extensively. The α,β-unsaturated aldehydes are highly reactive and readily react with proteins, DNA (see Chapter 38 for analytical methods of DNA adducts), and phospholipids to cause deleterious effects. Modification of amino acid residues in proteins and peptides by these aldehydes occurs mainly as cysteine, lysine, and histidine stable adducts by Michael addition (Poli et al. 2008). Carbonyl groups of aldehydes may alternatively react with the amino groups to form Schiff bases. HNE can undergo both a Michael addition and a Schiff base formation with phospholipids such as phosphatidylethanolamine and phosphatidylserine. HNE- and acrolein-protein adducts are considered good biomarker of lipid peroxidation *in vivo*, and they are usually detected using antibodies. Aldehydes are not only toxic end-products and remnants of lipid peroxidation, but are also potential signaling messengers. The significance of lipid peroxidation products such as 4-hydroxynonenal as a physiological signaling messenger is a subject of future studies (Forman et al. 2008).

15-Deoxy-delta12,14-prostaglandin J_2 (15d-PGJ_2) is another metabolite of lipid peroxidation which exerts versatile biological effects (Uchida and Shiabata 2008). 15d-PGJ_2 is formed by nonenzymatic two-step dehydration of PGD_2.

BIOMARKERS OF OXIDATIVE STRESS

The oxidation products of biological molecules are used as markers to monitor oxidative stress in marine animal and plant material. Various markers are summarized in Table 31.2 and can be applied in different laboratories after considering the simplicity of the assay, the required instrumentation, the quantitation method, and biological relevance. The identification and quantification of lipid peroxidation products in biological fluids and tissues are difficult because numerous products are present in very low concentrations. Thiobarbituric acid reactive substances (TBARS) have been used frequently, because the assay is simple and can be measured with conventional UV/visible absorption spectrophotometer (see Chapter 32). This assay is, however, not specific nor quantitative. Recently, IsoPs formed by free-radical-mediated oxidation of arachidonates independent of COX have been used and are now accepted as the most reliable biomarker for lipid peroxidation and oxidative stress *in vivo*. In fact, good correlation has been observed between the level of IsoPs in human fluids and progress of various diseases (Yin 2008). Neuroprostanes, the oxidation products of eicosapentaenoic acid (EPA) and docosahexaenoic acid (DHA) formed by similar mechanisms as IsoPs, may also be important biomarkers in fish.

HODE, the major product of linoleate oxidation, is another reliable biomarker for oxidative stress *in vivo*. HPODE, the primary products of linoleates, are enzymatically reduced *in vivo* to form the corresponding HODE. More highly unsaturated fatty acids such as EPA and DHA and their esters are oxidized by more complicated mechanisms to give numerous products with much lower yield. Lipid hydroxides after reduction with triphenylphosphine were estimated as 0.93, 5.7, and 4.5 mg g^{-1} tissue in skeletal muscle, dark muscle, and liver, respectively, of yellowtail. The levels were found to be higher in cultured yellowtail, carp,

Table 31.2 Biomarkers of oxidative stress.

Lipid	Protein	DNA
Ethane and pentane in exhaled gas	Protein carbonyls	Comet assay
TBARS	Hydroperoxides	Thymine glycol
Conjugated dienes	Nitro-, chloro-, bromo-amino acid	5-Hydroxyuracil
Hydroperoxides	Disulfide -SS-	2-, 8-Hydroxyadenine
Aldehydes	-SOH, -SOOH, -SOOOH	8-Hydroxyguanine
Ketones	Aldehyde-modified proteins	8-Nitro-, chloro-, brome-
Isoprostanes	Hydroperoxide-modified proteins	Guanine
Neuroprostanes	Cross-linked proteins	
Isofuranes	Dityrosine	
Neurofuranes	Albumin dimmer	
HODE	Advanced oxidation products	
Lyso PC	Creatol	
Oxidized LDL	Myeloperoxidase	
Oxysterols		

sweet fish, and red sea bream than wild fishes (Tanaka et al. 1999).

As described above, lipid peroxidation proceeds by three distinct mechanisms and different antioxidants are required to counteract each type of oxidation. The decrease in the level of antioxidant capacity will result in an increase in lipid peroxidation. The oxidation products derived from antioxidants, such as tocopherylquinone from vitamin E, may serve as a marker for lipid peroxidation and oxidative stress.

MEASUREMENT OF HYDROXYOCTADECADIENOIC ACIDS

Physiological samples such as plasma and tissue homogenates are reduced at room temperature after addition of internal standards, 8-iso-PGF$_{2\alpha}$-d$_4$ and 13-HODE-d$_4$, and antioxidant, followed by hydrolysis with potassium hydroxide under nitrogen in the dark at 40°C (Fig. 31.4) (Yoshida et al. 2008). The mixture is cooled on ice, acidified with 10% acetic acid in water, and then extracted with chloroform : ethyl acetate (4:1 v/v). The sample is mixed with a vortex mixer for 1 min and centrifuged at 1500 *g* for 5 min at 4°C. The chloroform and ethyl acetate layer is concentrated to approximately 1 mL after removal of the water layer and evaporated to dryness under nitrogen gas. The derived sample is reconstituted with methanol and water (70:30 v/v); a portion of the sample (10 μL)

Reduction (Ph_3P) and saponification (KOH)

↓

pH=3 (HCl)

↓

C18 Sep-Pak (precondition; methanol, H_2O)
Sample load, wash; H_2O, CH_3CN/ H_2O,
elution; Hex/AcOEt/iPA

↓

NH_2 Sep-Pak (precondition; Hex)
Sample load, wash, Hex/AcOEt, CH_3CN,
elution; AcOEt/MeOH/AcOH

↓

Evaporation

↓

LC-MS/MS

Fig. 31.4 Measurement of HODE from biological samples.

is subjected to HPLC–tandem mass spectrometry (LC-MS/MS) analysis. LC is carried out on an ODS column in a column oven maintained at 30°C. The LC apparatus consists of an automatic sample injector, three pumps, and a dynamic mixer. A mixture of solvent A (2 mM ammonium acetate in water) and solvent B (methanol: acetonitrile 5:95 v/v) is used as eluent at a flow rate of 0.2 mL min^{-1}, the initial composition of the gradient was 80% A and 20% B for 2 min, and changed to 50% A and 50% B within 45 min. MS analysis is carried out using a triple-quadrupole mass spectrometer fitted with electrospray ionization (ESI). A specific precursor-to-product ion transition is carried out by selected reaction monitoring (SRM) after collision-induced dissociation in the negative mode. Argon is used as the collision gas. The precursor, product ions, and collision energy are determined after the optimization of MS/MS as follows: $m/z = 353.5$ and 192.6–193.6 at 29 eV for 8-iso-PGF$_{2\alpha}$; $m/z = 357.0$ and 196.5–197.5 at 29 eV for 8-iso-PGF$_{2\alpha}$-d$_4$; $m/z = 319.0$ and 114.5–115.5 at 10 eV for 5-HETE; $m/z = 319.3$ and 162.8–163.8 at 13 eV for 12-HETE; $m/z = 319.3$ and 202.5–203.5 at 10 eV for 15-HETE; $m/z = 295.0$ and 194.6–195.6 at 21 eV for both 13-(Z,E)-HODE and 13-(E,E)-HODE; $m/z = 295.0$ and 170.5–171.5 at 24 eV for both 9-(E,Z)-HODE and 9-(E,E)-HODE; $m/z = 295.0$ and 182.6–183.6 at 22 eV for both 10-(Z,E)-HODE and 12-(Z,E)-HODE; and $m/z = 299.0$ and 197.6–198.6 at 26 eV for 13-HODE-d$_4$. The standard samples of HODE isomers (9- and 13-(Z,E) and (E,E)-HODE), HETE isomers (5-, 12-, and 15-HETE), 8-iso-PGF$_{2\alpha}$, 7α-OHCh, 7β-OHCh, cholesterol, and linoleic acid (18:2) are added to human plasma to determine the factors for quantification. The concentrations of HODE, HETE isomers, and 8-iso-PGF$_{2\alpha}$ are determined using 13-HODE-d$_4$ and 8-iso-PGF$_{2\alpha}$-d$_4$ as internal standards. The factors apparently depend on the recovery efficiency of solvent extraction as well as the sensitivities of the targeted fragment ions.

CONCLUDING REMARKS

The oxidative stress encountered in aquatic environments is increasing and its effects on living organisms are receiving much attention. Various biomarkers such as those summarized in Table 31.2 may be useful for monitoring those effects. Each marker has inherent merits and demerits. HODE is a potentially useful biomarker, since it is a major lipid peroxidation product of a straightforward mechanism and can be quantified by several methods.

MS analysis is a powerful tool due to its high sensitivity and specificity and allows lipid peroxidation products to be identified in their intact forms (Yoshida et al. 2008; Kuksis et al. 2009). For example, the hydroxyoctadecadienoate in different phospholipids or triglycerides can be identified separately. While sample preparation for gas chromatography–mass spectroscopy (GC-MS) requires extensive derivatization, including reduction, hydrolysis, pentafluorobenzyl (PFB) esterification, and trimethylsilyl (TMS) derivatization, LC–MS/MS analysis does not. However, analyses with MS are labor intensive, time consuming, and very expensive. For a high throughput analysis of large-scale samples, enzyme-linked immunosorbent assay (ELISA) is recommended, although the preparation of antibody with high affinity and specificity may not be easy because many similar conjugated diene hydroxides from other PUFAs, as well as regio- and stereo-isomers of HODE, may be present in the samples. Antibodies for several lipid peroxidation products including HODE have been developed (Shibata et al. 2009; Uchida 2010) and are now commercially available:

http://www.usbio.net/item/H9110-18A
http://www.abnova.com/products/products_detail.asp?catalog_id = KA0317
http://www.cellbiolabs.com/lipid-peroxidation
http://www.nof.co.jp/business/life/shindan/english/index.html
http://www.jaica.com/e/selection_guide.html

REFERENCES

Di Giulio, R.T., Habic, C., Gallagher, E.P. (1993) Effects of Black Rock Harbor sediments on indices of biotransformation, oxidative stress, and DNA integrity in channel catfish. *Aquatic Toxicology* 26, 1–22.

Esterbauer, H., Schaur, J.S., Zollner, H. (1991) Chemistry and biochemistry of 4-hydroxynonenal, malonaldehyde and related aldehydes. *Free Radical Biology and Medicine* 11, 81–128.

Forman, H.J., Fukuto, J.M., Miller, T., Zhang H., Rinna A., Levy S. (2008) The chemistry of cell signaling by reactive oxygen and nitrogen species and 4-hydroxynoneanal. *Archives of Biochemistry and Biophysics* 477, 183–195.

Horie, M., Nishio, K., Fujita, K. et al. (2010) *In vitro* evaluation of cellular responses induced by sable fullerene C60 medium dispersions. *Journal of Biochemistry* 148, 289–298.

Kelly, S.A., Havrilla, C.M., Brady, T.C., Abramo, K.H., Levin, E.D. (1998) Oxidative stress in toxicology: Established mammalian and emerging piscine model systems. *Environmental Health Perspectives* 106, 375–384.

Kuksis, A., Suomela, J-P., Tarvainen, M., Kallio, H. (2009) Lipidomic analysis of glycerolipid and cholesteryl ester autooxidation products. *Molecular Biotechnology* 42, 224–268.

Morrow, J.D., Hill, K.E., Burk, R.F., Nammour, T.M., Badr, K.F., Roberts, L.J. 2nd. (1990) A series of prostaglandin F2-like compounds are produced *in vivo* in humans by a non-cyclooxygenase, free radical-catalyzed mechanisms. *Proceedings of National Academy of Sciences USA* 87, 9383–9387.

Niki, E. (2009) Lipid peroxidation: Physiological levels and dual biological effects. *Free Radical Biology and Medicine* 47, 468–484.

Oberdörster, E. (2004) Manufactured nanomaterials (fullerenes, C60) induce oxidative stress in the brain of juvenile largemouth bass. *Environmental Health Perspectives* 112, 1058–1062.

Oliw, E.H., Bylund, J., Herman, C. (1996) Bisallylic hydroxylation and epoxidation of polyunsaturated fatty acids by cytochrome P450. *Lipids* 31, 1003–1021.

Poli, G., Biasi, F., Leonarduzzi, G. (2008) 4-Hydroxynonenal-protein adducts: A reliable biomarker of lipid oxidation in liver disease. *Molecular Aspects of Medicine* 29, 67–71.

Roberts, L.J. 2nd, Fessel, J.P., Davies, S.S. (2005) The biochemistry of the IsoP, neuroprostane, and isofuran. Pathways of lipid peroxidation. *Brain Pathology* 15, 143–148.

Shibata, N., Toi, S., Shibata, T. et al. (2009) Immunohistochemical detection of 13(R)-hydroxyoctadecadienoic acid in atherosclerotic plaques of human carotid arteries using a novel specific antibody. *Acta Histochemica et Cytochemica* 42, 197–203.

Slaninova, A., Smutna, M., Modra, H., Svobodova, Z. (2009) A review: Oxidative stress in fish induced by pesticides. *Neuroendocrinology Letters* 30, 2–12.

Tanaka, R., Higo, Y., Murata, H., Nakamura, T. (1999) Accumulation of hydroxyl lipids in live fish with oxidative stress. *Fisheries Science* 65, 796–797.

Thomas, P., Wofford, H.W. (1993) Effects of cadmium and Arocolor 1254 on lipid peroxidation, glutathione peroxidase activity, and selected antioxidants in Atlantic croaker tissues. *Aquatic Toxicology* 27, 159–178.

Uchida, K. (2010) Immunochemical detection of lipid peroxidation-specific epitopes. *In* Aldini, G., Yeum, K-J., Niki, E., Russell, R.M. (eds). *Biomarkers for Antioxidant Defense and Oxidative Damage. Principles and Practical Applications.* Wiley-Blackwell, New York, pp. 157–171.

Uchida, K., Shiabata, T. (2008) 15-Deoxy-delta(12,14)-prostaglandin J2: An elctrophilic trigger of cellular responses. *Chemical Research in Toxicology* 21, 138–144.

Van Reyk, D.M., Brown, A.J., Hult'en, L.M., Dean, R.T., Jessup, W. (2006) Oxysterols in biological systems: sources, metabolism and pathophysiological relevance. *Redox Report* 11, 255–262.

Watanabe, T., Hashimoto, Y. (1968) Toxic components of oxidized saury oil inducing muscular dystrophy in carp. *Bulletin of the Japanese Society of Scientific Fisheries* 34, 1131–1140.

Yamamoto, Y., Niki, E., Kaimya, Y. (1982) Oxidation of lipids. I. Quantitative determination of the oxidation of methyl linoleate and methyl linolenate. *Bulletin of Chemical Society of Japan* 55, 1548–1550.

Yin, H. (2008) New techniques to detect oxidative stress markers: mass spectrometry-based methods to detect isoprostanes as the gold standard for oxidative stress *in vivo*. *BioFactors* 34, 109–124.

Yin, H., Porter, N.A. (2005) New insights regarding the autoxidation of polyunsaturated fatty acids. *Antioxidants and Redox Signaling* 7, 170–184.

Yoshida, Y., Kodai, S., Takemura, S., Minamiyama, Y., Niki, E. (2008) Simultaneous measurement of F_2-isoprostane, hydroxyoctadecadienoic acid, hydroxyeicosatetraenoic acid, and hydroxycholesterol from physiological samples. *Analytical Biochemistry* 379, 105–115.

Chapter 32

THE CLASSIC METHODS TO MEASURE OXIDATIVE DAMAGE: LIPID PEROXIDES, THIOBARBITURIC-ACID REACTIVE SUBSTANCES, AND PROTEIN CARBONYLS

Volodymyr I. Lushchak, Halyna M. Semchyshyn, and Oleh V. Lushchak

Department of Biochemistry and Biotechnology, Vassyl Stefanyk Precarpathian National University, Ivano-Frankivsk, Ukraine

LIPID PEROXIDATION

Lipid peroxidation can contribute to loss of cell function and viability. For example, (i) peroxidation of mitochondrial membranes can disturb cellular energetics, (ii) peroxidation of endoplasmic reticulum and mitochondrial membranes may trigger Ca^{2+} release and uncontrolled activation of Ca^{2+}-dependent pathways, and (iii) peroxidation of biological membranes can result in their permeabilization and cell lysis (Sies 1993; Lushchak 2007). In addition, accumulation of lipoperoxidation products may lead to propagation of oxidative modifications. Polyunsaturated fatty acids (PUFAs) are highly susceptible to peroxidation by reactive oxygen species (ROS) (Yin and Porter 2003). ROS-induced lipid oxidation yields primary, secondary, and end products (Box 32.1).

Quantitative and qualitative evaluation of lipid peroxidation is not easy because of the complex nature of lipid peroxidation product mixtures.

Oxidative Stress in Aquatic Ecosystems, First Edition. Edited by Doris Abele, José Pablo Vázquez-Medina, and Tania Zenteno-Savín.

Box 32.1 Suggested pathways of lipid peroxidation

Oxidation of arachidonic acid (LH) involves an allylic hydrogen abstraction to form a tetradienyl radical (L•) followed by insertion of molecular oxygen. Addition of oxygen results in peroxyl radical formation (LOO•), which is further transformed to hydroperoxide (LOOH) by hydrogen abstraction from another lipid molecule (LH). The latter gives another free radical (L•) and propagates oxidation. All radical compounds that appear from oxidation of lipids are named as primary products of lipid peroxidation. Having very high reactivity, they can interact with other lipid molecules (LH) yielding lipid hydroperoxides (LOOH). The latter belong to secondary lipid peroxidation products. Reinitiation is one of many complications in lipid peroxidation. In this case, other oxidants such as Fe^{3+} ions are capable of initiating new free-radical chain oxidation of LOOH. In addition, peroxyl radical (LOO•) can undergo further oxidation to form other highly oxidized products such as bicyclic endoperoxides, monocyclic peroxides, serial cyclic peroxides, and other complex peroxides (Yin et al. 2002). Most of them are unstable and can be readily decomposed to a wide array of products, such as reactive aldehydes, alkanes (ethane and pentane), isoprostanes and other compounds. Cytotoxic and mutagenic compounds such as malondialdehyde (MDA) and 4-hydroxynonenal belong to end products of lipid peroxidation (Esterbauer et al. 1991). Secondary and end products of lipid peroxidation can react with other compounds present in biological material, propagating oxidation.

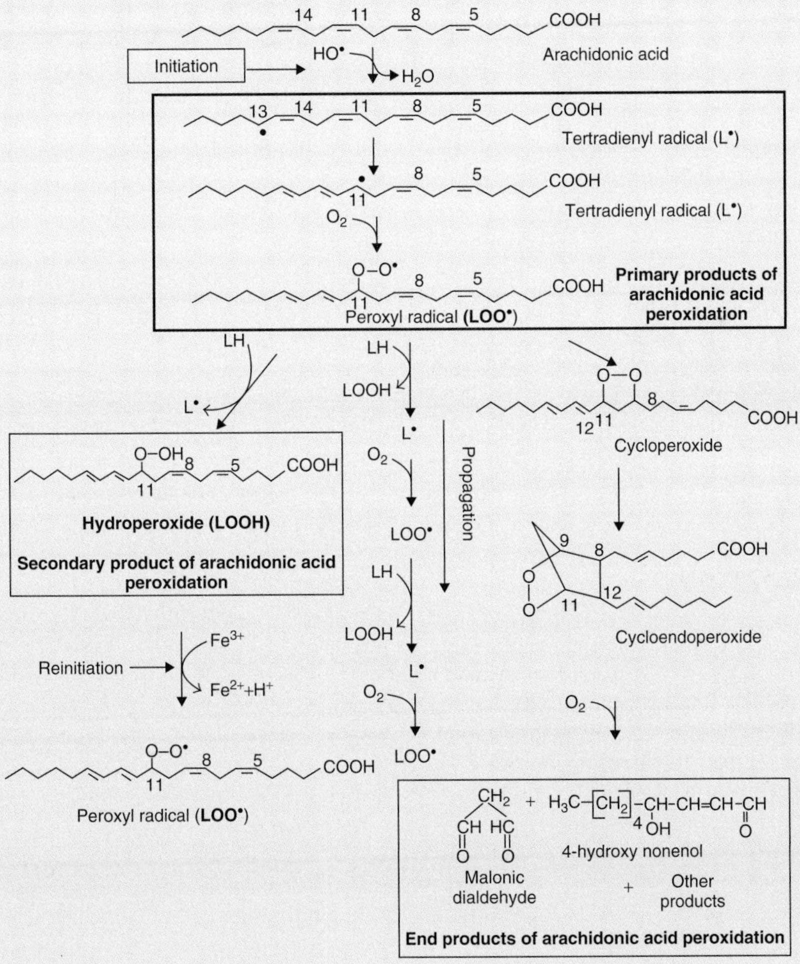

Lipid Peroxide Determination with Xylenol Orange

Assay Principle

Measurement of lipid hydroperoxides is widely used as an indication of oxidative stress and of the oxidative status in biological samples (Box 32.2).

The ferrous oxidation-xylenol orange (FOX) method was first applied to detect H_2O_2 formation in irradiated aqueous solutions containing xylenol orange (Gupta 1973). Later, it was adopted to demonstrate that protein–glucose mixtures generate H_2O_2 under certain conditions (Jiang et al. 1990). Nowadays, the method is widely used to assay for H_2O_2, butyl- and cumyl-hydroperoxides (Jiang et al. 1992; Wolff 1994), linoleic acid hydroperoxides (Jiang et al. 1992), hydroperoxides in liposomes (Jiang et al. 1991), low-density lipoproteins (Nourooz-Zadeh et al. 1996, 1997), DNA and its constituents (Michaels and Hunt 1978), and lysed erythrocytes (Ou and Wolf 1994). Xylenol orange can also be successfully used for the determination of trace metals, including noble metal ions, such as palladium (II), rhodium (III), and ruthenium (III) capable of forming complexes with xylenol orange (Solovey-Vandersteen et al. 2004).

The main advantage of the FOX method is its broad applicability to various biological samples. This method is simple to perform and a spectrophotometer, available in most laboratories, is the only equipment required. Therefore, this method offers the possibility of determining the total content of hydroperoxides rapidly and at low cost, and to assess the susceptibility of biological structures and organic compounds to oxidation (Bou et al. 2008).

As a drawback, determination of lipid hydroperoxides is quite challenging, because different kinds of hydroperoxides are produced during the process of lipid oxidation, and these reactive compounds can react and decompose rapidly, even at moderate temperatures (Box 32.1). Another disadvantage of the FOX method is its small linear range and low reproducibility (Bou et al. 2008). The FOX assay has a low specificity and several substances can interfere with the method. Some of these compounds are oxidizing/reducing agents that can be present in the samples endogenously, whereas others such as chelators are sometimes added. Here we will describe the modification of the FOX method proposed by Hermes-Lima et al. (1995) and used many times by us for different organisms, particularly for fish (Lushchak and Bagnyukova 2006; Bagnyukova et al. 2006).

Box 32.2 Structure of $(Fe^{3+})_2$: xylenol orange complex

The determination of lipid peroxide with xylenol orange (ferrous oxidation-xylenol orange (FOX) method) is based on the ability of hydroperoxides to oxidize Fe^{2+} to Fe^{3+} ions under acidic conditions at room temperature:

$$LOOH + Fe^{2+} \rightarrow LO^{\bullet} + OH^{-} + Fe^{3+}$$

The xylenol orange dye binds Fe^{3+} ion to form a chromophore complex:

$$2Fe^{3+} + XO \rightarrow (Fe^{3+})_2 : XO$$

where XO is xylenol orange. The complex absorbs strongly at 540–600 nm. The development of color at room temperature using xylenol orange and Fe^{3+} or hydroperoxide standards is maximal after 30 min and is stable up to 2 h in most cases.

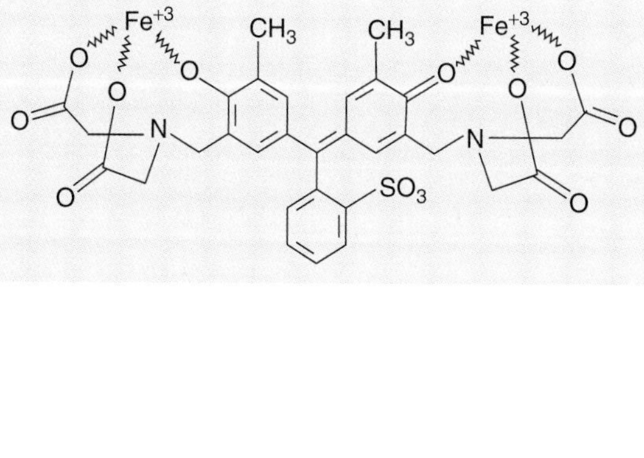

Materials

- Reagents: 96% ethanol, 4 mM 3,3′-bis[N,N-D(carboxymethyl) aminomethyl]-*o*-cresolsulphonephthalein tetrasodium salt (xylenol orange); 1 M sulfuric acid (H_2SO_4); 1 mM cumene hydroperoxide (CHP); 1 mM ferrous sulphate ($FeSO_4 \cdot 7H_2O$); bidistilled water.
- Equipment: Test and centrifuge tubes; Potter-Elvehjem glass homogenizer; pipettes (20–1000 µL); glass or plastic cuvettes; spectrophotometer; refrigerated centrifuge.

Preparation of Samples and Solutions

Sample preparation

Homogenize 30–40 mg of the sample (fish tissue) in five volumes of 96% ice-cold ethanol using a glass pre-cooled Potter-Elvehjem homogenizer. Centrifuge the sample at 10,000 *g* for 10 min at 4°C, transfer supernatants to clean centrifuge tubes and keep on ice.

Preparation of 4 mM xylenol orange, 3 mL

Dissolve 9 mg of xylenol orange powder (FW 760.6) in an exact volume of 3 mL bi-distilled water. Prepare before reaction mixture preparation and keep at room temperature (~20°C).

Preparation of 1 mM cumene hydroperoxide (CHP), 1 mL

Pre-weigh Eppendorf centrifugation tube and add about 10 µL of 1 mM CHP (FW 152.2, 80%). Calculate amount of water needed as $x/7.61$ mL (following the proportion 7.61 mg – 50 mM – 1 mL). Make 50-fold dilution of this stock solution.

Determination of appropriate amount of extract

For each new sample batch it is critically important to predetermine the range of sample volume that should be used in order to achieve a linear relationship between absorbance and reaction, and also the adequate incubation time. Figure 32.1 represents two examples of fish tissues to illustrate this.

Samples of goldfish brain and kidney are weighted and homogenized in five volumes of 96% ice-cold ethanol. Sample amounts from 5 to 50 µL are incubated for up to 75 min with incubation medium (see "Procedure" section below) and optical density read at 10-min time intervals. Optical density depends on incubation time and amount of tissue sample used (Fig. 32.1). Incubation time of 45 min was optimal for both brain and kidney, and the recommended sample volume is 30 µL for brain and 40 µL for kidney. Within this range of sample volume the absorbance increases linearly with sample volume in both tissues (see Fig. 32.1).

Procedure

1. Prepare two glass test tubes for each sample and for at least two blanks. It is strongly recommended to measure each sample at least in duplicate to enhance the accuracy of the method. The volume of all components must be calculated before the experiment (see Table 32.1).
2. Add 37.5 µL of 4 mM xylenol orange.‡
3. Add water and mix gently.
4. Add 37.5 µL of 1 M H_2SO_4 and mix gently with xylenol orange to prevent $FeSO_4$ oxidation.
5. Add 380 µL of 1 mM $FeSO_4$ and mix again.*
6. Add 20–40 µL of sample extract and water to final volume 1.5 mL.†
7. **First incubation**. Incubate the samples for 30–60 min at room temperature (~20°C).§
8. **First measurement**. Measurement of optical density of samples must be performed in glass or plastic cuvettes at 580 nm as fast as possible. Start determining $^1OD_{580}$ in two to three repeat blanks then measure your samples. Completely return sample to test tube after measurement.¶
9. Add 7.5 µL of CHP (1 mM) to each sample test tube to oxidize all Fe^{2+} to Fe^{3+}. Do not add CHP to blanks.
10. **Second incubation**. Incubate samples for 60 min at room temperature (~20°C).
11. **Second measurement**. Re-measure $^2OD_{580}$ of all samples and blanks.

‡Add xylenol orange directly to the bottom of the vial as a single drop.

*Add $FeSO_4$ solution to incubation medium after H_2SO_4, because Fe^{2+} at neutral pH is quickly oxidized to Fe^{3+}.

†Take care that all of the exact volume (see Fig. 32.1) is transferred to the vial. Cover the vials to prevent contamination.

§The incubation time must be established as previously described (see Fig. 32.1). Start timing immediately after addition of the extract into the first vial.

¶Do not leave anything in the cuvette!

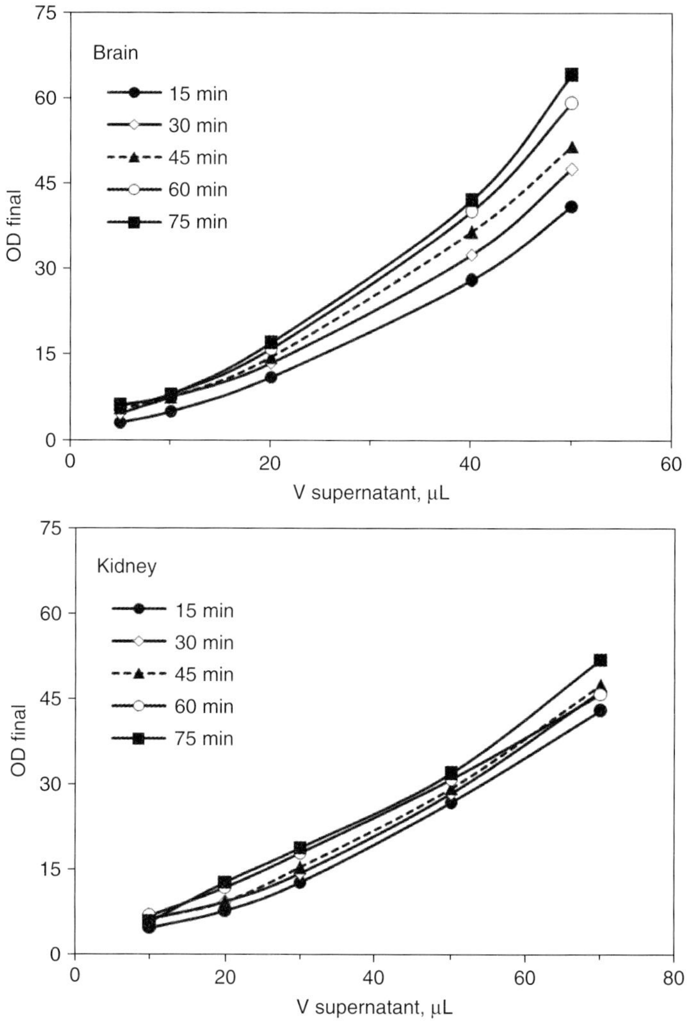

Fig. 32.1 Dependence of optical density OD_{final} (see "Procedure" and "Calculation of results") on supernatant volume in goldfish brain and kidney samples at different incubation periods.

Table 32.1 Calculation of volumes of the components in the incubation medium for peroxide lipid measurement using the FOX assay.

Samples (total volume 1.5 mL)	Blanks (total volume 1.5 mL)	Final concentration (mM)
37.5 μL xylenol orange (4 mM)	37.5 μL xylenol orange (4 mM)	0.1
37.5 μL H_2SO_4 (1 M)	37.5 μL H_2SO_4 (1 M)	25
380 μL $FeSO_4$ (1 mM)	380 μL $FeSO_4$ (1 mM)	0.25
1045 μL (H_2O + sample- supernatant)	1045 μL H_2O	

Calculation of Results

$$CHPeq/gww = \frac{OD_{final}}{(OD_{CHP} - OD_{final})} \cdot 5 \cdot \frac{1500}{V} \cdot 6$$

Where CHPeq/gww, cumene hydroperoxide equivalents (nmol) per gram wet weight represents the concentration of lipid peroxides; OD_{final} is the optical density of the sample obtained at the first measurement ($^{1}OD_{sample} - {}^{1}OD_{blank}$); OD_{CHP} is the optical density of the sample obtained at the second measurement after incubation with CHP ($^{2}OD_{sample} - {}^{2}OD_{blank}$); 5 is the concentration of CHP, nmol; 1500 is the sample volume, μL; V is the volume of supernatant, μL; 6 is the dilution of the tissue during homogenization (1 mg tissue : 5 μL ethanol).

Troubleshooting

The main problems that can affect the measurement of lipid peroxides were considered above in the section of method description. Briefly, the most critical points include:

- Determination of lipid peroxides depends on the kind of samples and their previous extraction and/or purification. The majority of samples are extracts from protein precipitation with alcohols, or from lipid extraction by organic solvents. The samples are then added to the reaction media to measure the peroxides. Therefore, it is important to select an appropriate solvent to extract the lipid peroxides in each case.
- Determine the appropriate amount of extract to provide sufficient sensitivity while remaining within the linear range of the method (see Fig. 32.1). Check the dependence of OD_{580} on incubation time and amount of sample for each tissue studied.
- Fe^{2+} in aerated water is rapidly oxidized to Fe^{3+} at pH above 7 and, therefore, it has to be directly dissolved at acidic pH to make it more stable. Nevertheless, in some cases, even when the Fe-solution is acidic, Fe has a poor stability, so it is recommended to prepare the reactants immediately before the measurement.
- Maintain the proposed concentrations of $FeSO_4$ and xylenol orange, because the latter can bind different amounts of Fe^{3+} ions. As a result, various complexes with different absorption properties can be formed (Babic et al. 2008);
- Contamination with metals is quite common. Therefore, it is recommended that high-purity reagents and clean glassware are used. Glassware and cuvettes should be cleaned with a sulfuric acid–dichromate solution and rinsed with bi-distilled water.
- Always prepare reagents mentioned as "Fresh" immediately prior to use.
- Do not forget to cover the test tubes during the incubation.

Determination of Thiobarbituric Acid Reactive Substances

Assay Principle

The thiobarbituric acid reactive substances (TBARS) assay is one of the most popular methods to measure end products of lipid peroxidation in biological systems – from bacteria to human tissues (Lushchak et al. 2005a,b; Semchyshyn et al. 2005; Lushchak and Bagnyukova 2006; Bagnyukova et al. 2006; Chakraborty et al. 2009). This method is based on the reaction of TBA with malondialdehyde (MDA) and other aldehyde products derived from secondary products of lipid peroxidation. The red compound is assumed to form mainly from reaction of TBA with MDA. In this chapter we will describe the colorimetric method of TBARS detection, but other methods also exist, like fluorescent measurement which increases sensitivity, or different types of high performance liquid chromatography (HPLC), which enhance the specificity of the assay (see Chapter 34; Box 32.3).

Compounds other than MDA such as saturated and unsaturated aldehydes, substituted pyrimidines, biliverdin, sucrose, glucose, fructose, 2-deoxyribose, N-acetylneuraminic acid, and amino acids can also react with TBA, thus interfering with determinations (Rice-Evans et al. 1991). Although many of these potentially disturbing substances do not normally occur in sufficient concentrations in a tissue extract to interfere with the assay, it should be kept in mind that the method is clearly not specific. Therefore, MDA and all other compounds that can interact with TBA are collectively named "thiobarbituric acid reactive substances". Additional formation of TBARS may occur during the heating step of the assay, presumably from decomposition of lipid peroxides by ROS generation. To solve this problem, antioxidants can be added to the incubation mixture. For instance, addition of

Box 32.3 Interaction between malondialdehyde and thiobarbituric acid

The method is based on the reaction of thiobarbituric acid (TBA) with malondialdehyde (MDA) and other aldehyde products derived from secondary lipid peroxidation products. One MDA molecule reacts stoichiometrically with two TBA molecules. The most generally used procedure utilizes biological samples that have been mixed with trichloracetic acid (TCA), a strong acid precipitant; the mixture is then centrifuged and the resulting supernatant is heated with TBA in the acidic medium (Kikugawa 1997). After cooling, absorbance of the red TBARS complex (with maximum at 532–535 nm) is measured.

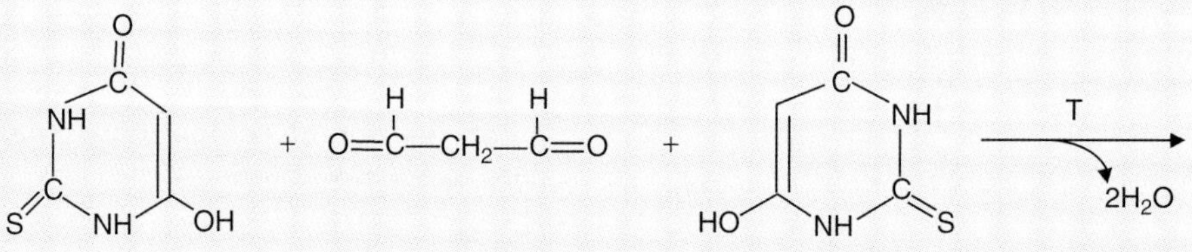

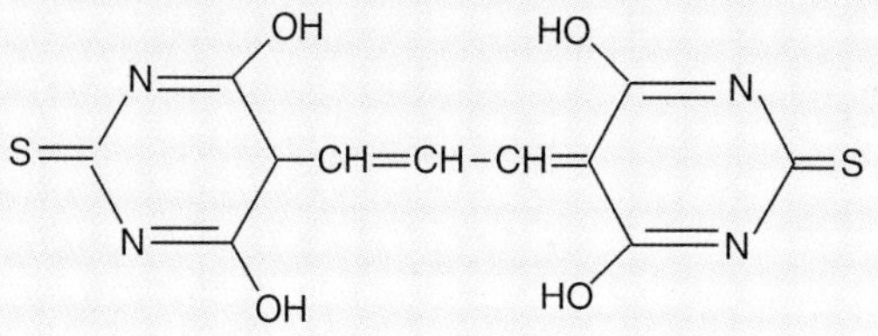

butylated hydroxytoluene (BHT) to the TBA reagents lowers metal-catalyzed oxidation of lipids during heating (Rice-Evans et al. 1991). Overall, the TBARS assay is quick and very easy to use and can be applied with some degree of reliability to evaluate lipid peroxidation in complex biological systems *in vitro* and *in vivo*. We have used this method with these precautions in several of our studies of fish (Lushchak et al. 2005a,b; Lushchak and Bagnyukova 2006; Bagnyukova et al. 2006) and frogs (Bagnyukova et al. 2005).

Materials

- Reagents: 40% w/v trichloracetic acid (TCA); 1% 2-TBA; *n*-butanol; 10 mM BHT; homogenization medium (50 mM potassium phosphate buffer (pH 7.0), 0.5 mM EDTA, 1 mM phenylmethanesulfonylfluoride (PMSF) diluted in ethanol); 0.1 M hydrochloric acid (HCl); distilled water.
- Equipment: Glass centrifuge tubes with screw cap suitable for heating (10–15 mL); centrifuge tubes 2 mL; Potter-Elvehjem glass homogenizer; pipettes (20–1000 μL); heated water bath; glass cuvette; centrifuge; spectrophotometer.

Preparation of Samples and Solutions

Sample preparation

Homogenize 130–150 mg of tissue in 10 volumes of homogenization medium containing 50 mM potassium phosphate buffer (pH 7.0) and 0.5 mM EDTA using a Potter-Elvehjem homogenizer. Centrifuge the sample at 10,000 *g* for 10 min at 4°C. Use supernatants for the assay.

Preparation of the TBA reagent (1% TBA, 10 mM BHT, 0.1 M HCl), 50 mL

Prepare a stock solution of 0.25 M BHT in ethanol (dissolve 110 mg of BHT powder in 2 mL of ethanol). Transfer this solution to a narrow-necked volumetric flask containing 45 mL of 0.1 M HCl. Add 0.5 g TBA and adjust the volume to 50 mL with 0.1 HCl. Adjust pH to 2.5 with KOH.

Procedure

1. Add 1 mL of 40% TCA to each 1 mL of tissue homogenate in 2 mL centrifuge tubes.

2. Centrifuge the samples at 5000 *g* for 5 min at room temperature.
3. Transfer 1.5 mL of supernatants to 10–15 mL heat resistant centrifuge tubes with screw cap.
4. Prepare reagent blank by adding 1.5 mL of bi-distilled water.
5. Add 1.5 mL of TBA reagent to each sample and blank.
6. Cover the tubes losely with screw caps and boil for 60 min in a water bath. You can also use glass marbles instead of screw tops to close vials during heating.
7. Place samples and blanks on ice for cooling.
8. Add 3 mL of *n*-butanol and mix thoroughly.
9. Centrifuge samples and blanks at 5000 *g* for 10 min at room temperature.
10. Gently transfer the upper part of the butanol layer (2 mL) into clean tubes.‡
11. Measure the optical density of the samples at 535 nm using glass cuvettes.

‡Do not shake the samples after centrifugation and transfer to clean tube.

Calculation of Results

$$[\text{TBARS}]/\text{gww} = \frac{\text{OD} \cdot V \cdot P \cdot 1000}{\varepsilon_{535} \cdot n}$$

Where [TBARS] is the concentration of TBARS, mmol per gram wet weigh; OD is the optical density of the sample (against the reagent blank); V is volume of butanol (3 mL); P is dilution (in this case $1 + 10 = 11$); $\varepsilon_{535} = 156{,}000\ \text{M}^{-1}\ \text{cm}^{-1}$ (extinction coefficient for TBARS); n is the homogenate volume (0.75 mL).

Troubleshooting

- The reaction occurs at pH 2–3, but excess acid (pH$<$ 2) can inhibit color development.
- Do not use plastic cuvettes for OD measurement, because butanol etches plastic.

PROTEIN OXIDATION

Proteins can be modified by a large number of reactions involving ROS and lipid oxidation products (Lushchak 2007). Oxidative modification of proteins is often associated with the appearance of additional carbonyl groups (such as aldehyde or ketone groups) in proteins or their derivatives (see Box 32.4). Formation of additional carbonyl groups often leads to loss of protein function. For example, enzyme interaction with ROS frequently results in enzyme inactivation or modification of regulatory properties (Johnson et al. 1985; Friguet et al. 1994; Bagnyukova et al. 2005; Lushchak 2007). Additional carbonyl groups can alter protein folding, increasing hydrophobicity, and often results in formation of toxic aggregates, which can reduce the activity of proteasomes (Squier 2001).

There are several reasons why measurement of oxidized proteins is preferred over oxidized lipids. Since proteins carry out specific biological functions, it is possible to register not only the formation of oxidized products, but also modification of protein function. The end products of protein oxidation are rather stable, and many highly sensitive methods for analysis of products of ROS oxidation of proteins can give clues as to the type of oxidants involved (Shacter 2000).

Determination of Carbonyl Groups of Proteins with 2,4-Dinitrophenylhydrazine

Assay Principle

The most widely used method for determination of protein carbonyls utilizes the reaction of carbonyl groups with 2,4-dinitrophenylhydrazine (DNPH; Box 32.5).

Protein carbonyl content is widely used as both a marker for oxidative stress and a measure of oxidative damage (see Chapter 33). However, the appearance of carbonyl groups is certainly not specific for oxidative modification. For instance, glycation of proteins may result in formation of additional carbonyl groups in amino acid residues (Lushchak 2007). Despite this, an increase in the concentration of protein carbonyls is a standard marker for oxidative stress, and the method is very convenient to detect and quantify oxidative modification of proteins.

Materials

- Reagents: 40% w/v trichloracetic acid (TCA); 10 mM DNPH; ethanol : ethylacetate mixture (1:1); 6 M guanidine hydrochloride; phosphoric acid (H_3PO_4); homogenization medium (50 mM potassium phosphate buffer (pH 7.0), 0.5 mM

Box 32.4 Suggested pathways of peptide chain oxidation

Peptide chain oxidation results in the cleavage of the peptide bond. The abstraction of a hydrogen by $HO^{\bullet}$ from the α-carbon atom of an amino acid residue starts a sequence of reactions leading to the formation of alkyl radical and water. The subsequent addition of oxygen to the alkyl radical yields an alkylperoxyl radical which reacts with protonated superoxide anion ($HO_2^{\bullet}$) or Fe^{2+} and H^+. The alkylperoxide formed may be converted to H_2O_2 and a Schiff base. The latter is rather unstable and can be decomposed to carbonyl compounds. Further, oxidation of some amino acid residues also leads to carbonyl derivatives. Carbonyl groups may be formed during oxidation of the side chains of lysine, arginine, histidine, proline, threonine, glutamic acid, and aspartic acid. The oxidation of the first four amino acid residues converts them directly into aldehyde or ketone derivatives, whereas the oxidation of threonine and glutamic acid results in peptide bond cleavage and formation of carbonyl compounds.

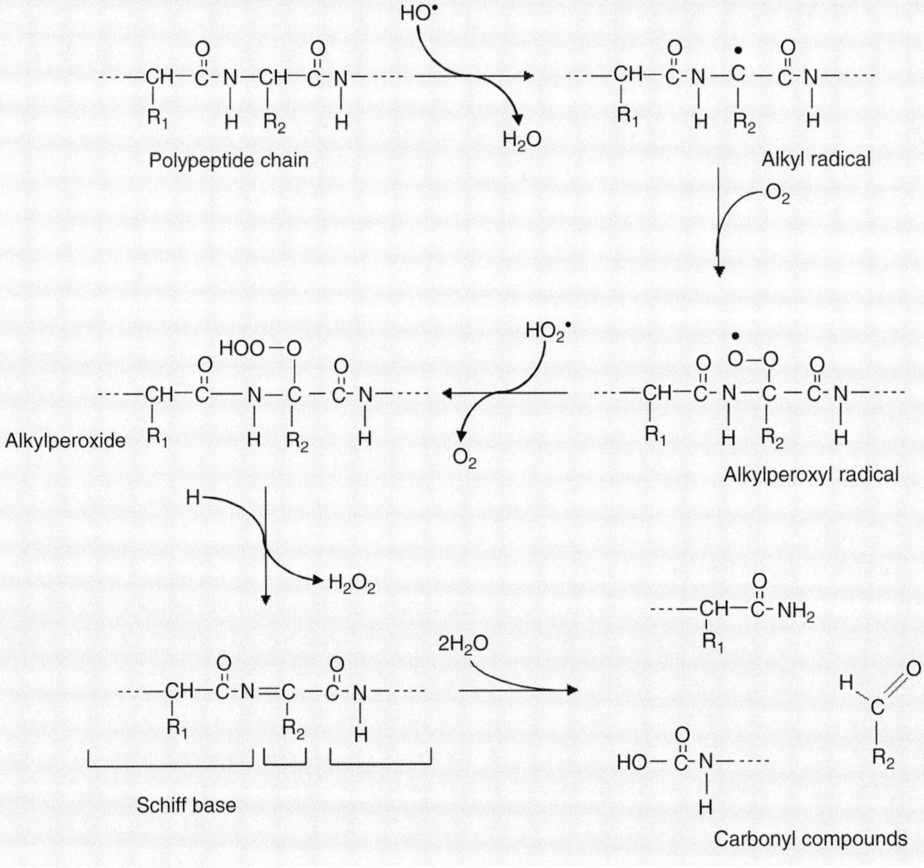

EDTA, 1 mM PMSF in ethanol); 0.1 M hydrochloric acid (HCl) 36.5–38%.

- Equipment: Test tubes; centrifuge tubes (1.5 mL); Potter-Elvehjem glass homogenizer; pipettes (20–1000 μL); microsyringes; spectrophotometer; centrifuge; glass sticks.

Preparation of Samples and Solutions

Preparation of samples

Homogenize 130–150 mg of tissue in 10 volumes of homogenization medium containing 50 mM potassium phosphate buffer (pH 7.0) and 0.5 mM EDTA using the

Box 32.5 Interaction between protein carbonyl groups and 2,4-dinitrophenylhydrazine

The method is based on the reaction of carbonyl groups with 2,4-dinitrophenylhydrazine (DNPH), a classic carbonyl reagent which forms protein-bound 2,4-dinitrophenylhydrazones. Hydrazones can be quantified spectrophotometrically at 370 nm. The development of color at room temperature (~20°C) using DNPH and protein solutions reaches a plateau after 60 min of incubation.

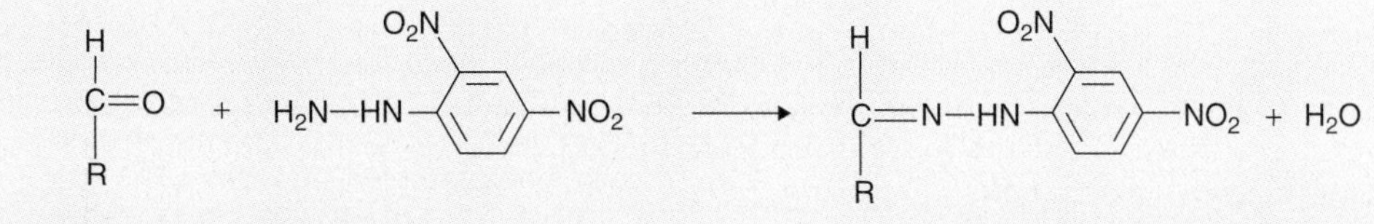

Protein carbonyl 2,4-dinitrophenylhydrezine 2,4-dinitrophenylhydrazone

Potter-Elvehjem homogenizer. Centrifuge samples at 10,000 *g* for 10 min at 4°C. Use the supernatants.

Preparation of 6 M Guanidine hydrochloride, 30 mL

Dissolve 17.19 g of guanidine hydrochloride (FW 95.53) in 12 mL of distilled water. Add 1.5 mL of concentrated H_3PO_4 and mix. Adjust volume to 30 mL with water. Mix vigorously and filtrate over Whatman 1 filter paper.

Preparation of 10 mM DNPH in 0.2 M HCl, 10 mL

Dissolve 28 mg of DNPH (FW 198.1, 98%, contains 30% water) in 10 mL of 2 M HCl. Note that DNPH is hard to dissolve. Filtrate the solution over Whatman 1 filter paper. The solution is stable at 5°C for at least two weeks.

Procedure

1. Prepare two centrifuge tubes for each sample and label them as control and experimental.
2. To each pair of tubes (experimental and control) add 0.25 mL of supernatant.
3. Add 0.5 mL of 2 M HCl to control tubes and 0.5 mL of 10 mM DNPH in 2 M HCl to experimental tubes. Incubate for 1 h at room temperature (~20°C).
4. Add 0.5 mL of 40% TCA.
5. Centrifuge samples for 5 min at 5000 *g* at room temperature.
6. Remove and discard liquid fraction.‡
7. Add 1 mL of ethanol–ethylacetate mixture and break pellets with glass sticks.*
8. Centrifuge samples for 10 min at 10,000 *g* at room temperature.
9. Repeat procedures in steps 6 and 7 twice to remove excess of unbound DNPH.
10. To each sample add 1.5 mL of 6 M guanidine hydrochloride and vortex.
11. Wait for ~30 min for protein to dissolve fully.
12. Centrifuge samples for 10 min at 10,000 *g* at room temperature.
13. Transfer sample supernatants into clean test tubes.
14. Measure optical density of samples at 370 nm.†
15. Measure the concentration of protein in experimental samples; Bradford method may be used.

*Do not use the same stick in control after experimental tube because of the risk of contamination with DNPH.

†Remember that each experimental sample has its own control sample. Measure OD_{370} of controls prior to experimental samples.

Calculations of Results

$$[CP] = \frac{\Delta OD \cdot V \cdot 1000}{\varepsilon_{370} \cdot [\text{protein}]}$$

Where [CP] is the protein carbonyls content, nmol per mg protein; ΔOD is the optical density (OD experimental sample – OD control sample); V is the total volume of sample (3 mL); $\varepsilon_{370} = 22,000\ M^{-1}\ cm^{-1}$ (extinction coefficient for dinitrophenylhydrazones); [protein] is the protein concentration in the experimental samples, $mg\ mL^{-1}$.

Troubleshooting

The method requires excess DNPH to be present in the incubation mixture. DNPH has a significant absorbance at 370 nm. Therefore, the reagent must be removed accurately before spectrophotometric determination of the protein-bound hydrazone. Otherwise, the residual reagent can cause an increase in the apparent carbonyl concentration in the sample.

REFERENCES

Babic, S., Battista, J., Jordan, K. (2008) An apparent threshold dose response in ferrous xylenol-orange gel dosimeters when scanned with a yellow light source. *Physics in Medicine and Biology* 53, 1637–1650.

Bagnyukova, T.V., Vasylkiv, O.Yu., Storey, K.B. et al. (2005) Catalase inhibition by amino triazole induces oxidative stress in goldfish brain. *Brain Research* 1052, 180–186.

Bagnyukova, T.V., Chahrak, O., Lushchak V.I. (2006) Coordinated response of goldfish antioxidant defenses to environmental stresses. *Aquatic Toxicology* 78, 325–331.

Bou, R., Codony, R., Tres, A. et al. (2008) Determination of hydroperoxides in foods and biological samples by the ferrous oxidation-xylenol orange method: A review of the factors that influence the method's performance. *Analytical Biochemistry* 377, 1–15.

Chakraborty, S., Singh, O.P., Dasgupta, A. (2009) Correlation between lipid peroxidation-induced TBARS level and disease severity in obsessive-compulsive disorder. *Progress in Neuropsychopharmacology, Biology and Psychiatry* 33, 363–366.

Esterbauer, H., Schaur, R.J., Zollner, H. (1991) Chemistry and biochemistry of 4-hydroxynonenal, malondialdehyde and related aldehydes. *Free Radical Biology and Medicine* 11, 81–128.

Friguet, B., Szweda, L.I., Stadtman, E.R. (1994) Susceptibility of glucose–6-phosphate dehydrogenase modified by 4-hydroxy–2-nonenal and metal-catalyzed oxidation to proteolysis by the multicatalytic protease. *Archives of Biochemistry and Biophysics* 311, 168–173.

Gupta, B.L. (1973) Microdetermination techniques for H2O2 in irradiated solutions. *Microchemistry Journal* 18, 363–374.

Hermes-Lima, M., Willmore, W.G., Storey, K.B. (1995) Quantification of lipid peroxidation in tissue extracts based on Fe(III) xylenol orange complex formation. *Free Radical Biology and Medicine* 19, 271–280.

Jiang, Z-Y., Woollard, A.C.S., Wolff, S.P. (1990) Hydrogen peroxide production during experimental protein glycation. *FEBS Letters* 268, 69–71.

Jiang, Z-Y., Woollard, A.C.S., Wolff, S.P. (1991) Lipid hydroperoxide measurement by oxidation of Fe2+ in the presence of xylenol orange. Comparison with the TBA assay and an iodometric method. *Lipids* 26, 853–856.

Jiang, Z-Y., Hunt, J.V., Wolff, S.P. (1992) Ferrous ion oxidation in the presence of xylenol orange for detection of lipid hydroperoxide in low density lipoprotein. *Analytical Biochemistry* 202, 384–389.

Johnson, E.A., Levine, R.L., Lin, E.C. (1985) Inactivation of glycerol dehydrogenase of *Klebsiella pneumoniae* and the role of divalent cations. *Journal of Bacteriology* 164, 479–483.

Kikugawa, K. (1997) Use and limitation of thiobarbituric acid (TBA) test for lipid peroxidation. *Recent Research Developments in Lipids Research* 1, 73–96.

Lushchak, V.I. (2007) Free radical oxidation of proteins and its relationship with functional state of organisms. *Biochemistry (Moscow)* 72, 809–827.

Lushchak, V.I., Bagnyukova, T.V. (2006) Temperature increase results in oxidative stress in goldfish tissues. I. Indices of oxidative stress. *Comparative Biochemistry and Physiology* 143, 30–35.

Lushchak, V.I., Bagnyukova, T.V., Husak, V.V. et al. (2005a) Hyperoxia results in transient oxidative stress and an adaptive response by antioxidant enzymes in goldfish tissues. *International Journal of Biochemistry and Cell Biology* 37, 1670–1680.

Lushchak, V.I., Bagnyukova, T.V., Lushchak, O.V. et al. (2005b) Hypoxia and recovery perturb free radical processes and antioxidant potential in common carp (*Cyprinus carpio*) tissues. *International Journal of Biochemistry and Cell Biology* 37, 1319–1330.

Michaels, H.B., Hunt, J.W. (1978) A model for radiation damage in cells by direct effect and by indirect effect: a radiation chemistry approach. *Radiation Research* 74, 23–34.

Nourooz-Zadeh, J., Tajaddini-Sarmadi, J., Ling, K.L.E. et al. (1996) Low-density lipoprotein is the major carrier of lipid hydroperoxides in plasma. Relevance to determination of total plasma lipid hydroperoxide concentrations. *Biochemistry Journal* 313, 781–786.

Nourooz-Zadeh, J., Tajaddini-Sarmadi, J., Tritschler, H. et al. (1997) Relationships between plasma measures of oxidative stress and metabolic control in NIDDM. *Diabetologia* 40, 647–653.

Ou, P., Wolf, S.P. (1994) Erythrocyte catalase inactivation (H2O2 production) by ascorbic acid and glucose in the presence of aminotriazole: role of transition metals and relevance to diabetes. *Biochemistry Journal* 303, 935–940.

Rice-Evans, C.A., Diplock, A.T., Symins, M.C.R. (1991) Methods for TBA-reaction. *In* Burdon, R.H., van Knippenberg, P.H. (eds). *Laboratory Techniques in Biochemistry and Molecular Biology*, Vol. 22. Elsevier, Amsterdam, pp. 147–149.

Semchyshyn, H.I., Bagnyukova, T.V., Storey, K. et al. (2005) Hydrogen peroxide increases the activities of soxRS regulon enzymes and the levels of oxidized proteins and lipids in *Escherichia coli*. *Cell Biology International* 29, 898–902.

Shacter, E.Y. (2000) Quantification and significance of protein oxidation in biological samples. *Drug Metabolism Reviews* 32, 307–326.

Sies, H. (1993) Strategic of antioxidant defense. *European Journal of Biochemistry* 215, 213–295.

Solovey-Vandersteen, O., Vrublevska, T., Lang, H. (2004) UV-visible and IR spectroscopic studies of ruthenium(ii)-xylenol orange complex. *Acta Chimica Slovenica* 51, 95–106.

Squier, T.C. (2001) Oxidative stress and protein aggregation during biological aging. *Experimental Gerontology* 36, 1539–1550.

Wolff, S.P. (1994) Ferrous ion oxidation in presence of ferric ion indicator xylenol orange for measurement of hydroperoxides. *Methods in Enzymology* 233, 182–189.

Yin, H., Porter, N.A. (2003) Specificity of the ferrous oxidation of xylenol orange assay: analysis of autoxidation products of cholesteryl arachidonate. *Analytical Biochemistry* 313, 319–326.

Yin, H., Havrilla, C.M., Morrow, J.D. et al. (2002) Formation of isoprostane bicyclic endoperoxides from the autoxidation of cholesteryl arachidonate. *Journal of the American Chemistry Society* 124, 7745–7754.

Chapter 33

PROTEIN CARBONYL MEASUREMENT BY ENZYME LINKED IMMUNOSORBENT ASSAY

Betul Catalgol[1,2,3], *Stefanie Grimm*[1,2], *and Tilman Grune*[1]

[1]Institute of Nutrition, Friedrich Schiller University, Jena, Germany
[2]Institute of Biological Chemistry and Nutrition, University Hohenheim, Stuttgart, Germany
[3]Department of Biochemistry, Faculty of Medicine, Marmara University, Istanbul, Turkey

Reactive oxygen species (ROS) can cause several modifications in proteins, lipids, and DNA. Following ROS interactions with cellular components, specific products are known to be formed. Since proteins are the most abundant molecules in organisms, assessment of oxidatively modified proteins has become popular. Active ROS that are produced as by-products of cellular metabolism or from environmental sources cause modifications of amino acids within proteins that, in general, result in loss of protein function/enzymatic activity. The degree of protein oxidation caused by a given oxidant depends on many factors, including nature, relative location, flux rate of the oxidant, and available antioxidant systems (Grune et al. 1997).

Oxidative protein modifications occur in several ways, such as direct interaction of ROS with proteins and indirect reactions following the interaction of ROS with other cellular components. In addition, protein modifications that change amino acid side chains can occur. Oxidation of protein backbones is characterized by fragmentation of polypeptide chains, and fragmentation results in peptide fragments with derivatized terminal amino acids. Oxidation of side chains results in the formation of different products (Stadtman and Levine 2000; Davies 2003).

Cells ameliorate the modifications following protein oxidation to restore protein function and maintain cellular integrity. As mentioned above, various protein modifications exist, but repair mechanisms have been evolved only for the frequent and easily repairable products, such as disulfide bonds and methionine sulfoxides (Holmgren 1989; Puig and Gilbert 1994; Vogt 1995). For the most part, oxidatively modified proteins are not repaired and must be removed by proteolytic degradation. Protein degradation is a physiological process required to maintain cellular function. Therefore, cells have developed highly regulated intracellular

Oxidative Stress in Aquatic Ecosystems, First Edition. Edited by Doris Abele, José Pablo Vázquez-Medina, and Tania Zenteno-Savín.

proteolytic systems responsible for the removal of such nonfunctional proteins before they start to aggregate (Mehlhase and Grune 2002; Bader and Grune 2006). It is well known that cytosolic and nuclear oxidized proteins are mainly degraded by the proteasome (Grune et al. 2003).

PROTEIN CARBONYL FORMATION

Protein carbonyl groups, irreversible and nonenzymatic oxidative damage markers, are introduced into proteins by a variety of oxidative pathways. Carbonyl formation occurs as a consequence of protein side chain oxidation or from fragmentation (Buss et al. 1997). Carbonyl derivatives are formed following oxidation of protein side chains of lysine, arginine, proline, histidine, or threonine. The main carbonyl products of metal-catalyzed protein oxidation are glutamic semialdehyde, a product of arginine and proline oxidation, and aminoadipic semialdehyde, a product of lysine oxidation (Stadtman and Oliver 1991; Shringarpure and Davies 2002) (Table 33.1). Additionally, cleavage of peptide bonds by the α-amidation pathway or by oxidation of glutamyl residues, leading to the formation of a fragment in which the N-terminal amino acid is blocked by an α-ketoacyl derivative, can result in the formation of carbonyl groups (Dalle-Donne et al. 2006).

Several reactive species, including $O_2^{\bullet-}$, $^1O_2$, $HO^{\bullet}$, $ROO^{\bullet-}$, hypochlorous acid, $ONOO^-$ and O_3, produce carbonyl groups as a consequence of amino acid residue oxidation (Berlett and Stadtman 1997; Dean et al. 1997). Next to these ROS catalyzed reactions, secondary reactions with products resulting from oxidation of nonprotein cellular constituents may partially be responsible for the introduction of carbonyl groups into the protein pool. Secondary reactions, such as the formation of advanced glycation end products, involve the reaction of reducing sugars with protein residues, particularly lysine residues. Furthermore, the generation of advanced lipoxidation end products is conducted by the reaction of carbonyl-containing oxidized lipids (malondialdehyde or 4-hydroxynonenal) from polyunsaturated fatty acid (PUFA) oxidation with protein residues of lysine, cysteine, and histidine (Berlett and Stadtman 1997; Buss et al. 1997; Alamdari et al. 2005; Dalle-Donne et al. 2006).

Table 33.1 Amino acids contributing to carbonyl modification (Berlett and Stadtman 1997; Voss and Grune 2006).

Native amino acid residues	Carbonyl derivatives formed by oxidation
Alanine	Acetone
Valine	Formaldehyde
Leucine	Isobutyraldehyde
Aspartate	Glyoxylic acid
Proline	2-pyrrolidone, glutamic semialdehyde
Arginine	Glutamic semialdehyde
Lysine	Aminoadipic semialdehyde
Histidine	2-oxo-histidine, asparagine
Tryptophan	*N*-formylkynurenine

Previous studies described a possible role for protein carbonylation in protein quality control (Nystrom 2005). These studies suggest that when the protein is irreparably damaged, carbonylation acts as a tagging system for the degradation pathways (similar to ubiquitination). Some regulated cellular processes utilize the carbonylation of specific proteins as a mechanism for triggering their degradation (e.g. iron regulatory protein-2) (Iwai et al. 1998). As a result of this process, carbonylation is not only an undesirable by-product of aerobic metabolism, but can also function as a regulatory principle.

Protein carbonyls are the most widely measured biomarkers of protein oxidation, as they are, in general, stable products formed relatively early during oxidative stress. Thus, the quantification of protein carbonyls as an indicator of ROS-mediated protein modification is a useful tool in biochemical stress research.

PROTEIN CARBONYL DETECTION BY ENZYME LINKED IMMUNOSORBENT ASSAY

At present, the most widely used assay for protein carbonyl detection involves the derivatization of the carbonyl group with 2,4-dinitrophenylhydrazine (DNPH) (Fig. 33.1). Derivatization with DNPH leads to the formation of a stable protein-conjugated dinitrophenylhydrazone product, which has a peak absorbance of nearly 360 nm. Therefore, DNPH provides a basic agent for the quantification of protein carbonyl content in purified proteins as well as in protein mixtures. This hydrazone derivate can be quantified by different methods (Halliwell and Gutteridge 2007), including spectrophotometric (Reznick and Packer 1994;

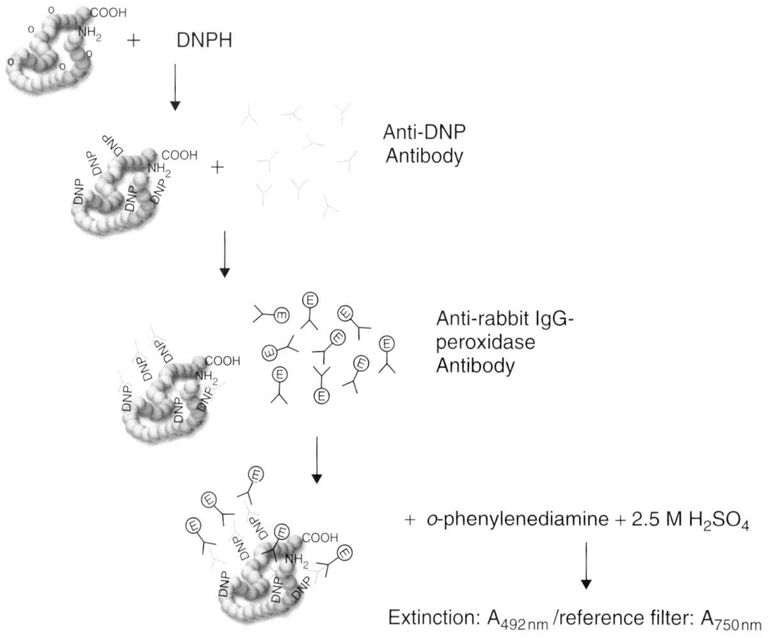

Fig. 33.1 Principle of the carbonyl ELISA method.

Table 33.2 Methods used to measure protein carbonyls (Berlett and Stadtman 1997).

Method	Advantages	Disadvantages
Spectrophotometric	Simple equipment	Requires high amount of protein, not suitable with proteins that have the same absorption peak as protein-DNP
Immunoblotting	Increased sensitivity and selectivity[1], identification of carbonylated proteins	Semiquantitative
HPLC	Increased sensitivity and specificity,[2,3] requires only micrograms of protein. Full spectra of the peaks provide minimum interference of nucleic acid	High-cost equipment
ELISA	Requires only micrograms of protein, sensitive and reproducible, minimal interference	Protein loss during washing Two-day experiment
Immunohistochemistry[4,5]	Requires only a few cells. Localization in tissues and organs is visible.	Exact quantification is difficult

[1]Shacter et al. 1994; [2]Gladstone and Levine 1994; [3]Levine et al. 1994; [4]Smith et al. 1998; [5]Keller et al. 1993.

see Chapter 32), one-dimensional or two-dimensional electrophoresis followed by immunoblotting (Robinson et al. 1999), immunohistochemistry (Keller et al. 1993; Smith et al. 1998), high performance liquid chromatography (HPLC) (Levine et al. 1994), and enzyme linked immunosorbent assay (ELISA) (Buss et al. 1997) (Table 33.2).

The spectrophotometric method is generally not recommendable because of practical problems. The most important disadvantage of the spectrophotometric assay is its high protein requirement, sometimes more than is available in clinical samples (Reznick and Packer 1994). Quantification of protein carbonyls by spectrophotometry following DNPH modification is sometimes not feasible (with the proteins that contain high amounts of chromophore that absorbs at 360 nm such as hemoglobin, myoglobin, retinoids), and in these cases the reaction of carbonyls with tritiated sodium borohydride provides an alternative method for quantitative measurement. Tritiated sodium borohydride transforms protein carbonyls to protein-bound ethanol groups, and tritium is simultaneously incorporated into these proteins (Yan 2009).

A widely used assay is the protein carbonyl ELISA method with different modifications between laboratories. This method can be applied to cells, tissue, and plasma. Its advantages are reliability and sensitivity and, further, the assay can be applied to both experimental studies and clinical samples (Quinlan et al. 1994). Free DNPH and nonprotein constituents are easily washed away during ELISA performance and give minimal interference. This results in higher sensitivity and accuracy at lower protein carbonyl concentrations compared to the spectrophotometric DNPH assay. The first protocol for quantifying protein carbonyls by ELISA was developed by Buss et al. (1997). The protocol described in this chapter includes some modifications from Sitte et al. (1998) and Voss et al. (2006).

In this method, samples can be used fresh or stored at −80°C, but before starting with determination of carbonyls, the protein concentration of the samples should be determined. The commercially available Bradford reagent can be used for determination of protein content. All samples and standards should be adjusted to 4 mg mL^{-1} or the lowest protein concentration, in order to have the same protein concentration (minimum 1 mg mL^{-1}).

If the protein concentration is very low, then the samples can be concentrated. For this purpose, 0.8 vol of 28% cold trichloracetic acid (TCA) is added to 60 μg protein, mixed, left on ice for 10 min, and centrifuged at 10,000 *g* for 10 min. 15 μL of PBS (standard phosphate-buffered saline: 140 mM NaCl, 2.7 mM KCl, 8.1 mM Na_2HPO_4, 1.5 mM KH_2PO_4) is added to the protein precipitate before adding the DNPH reagent as described below. Another possibility for low protein samples is the use of an alternative method (Alamdari et al. 2005) as described below.

Oxidized and reduced bovine serum albumin (BSA) should be prepared in advance. A serial dilution of oxidized and reduced BSA should be prepared for the standard curve. A minimum six-point standard curve of oxidized BSA with reduced BSA should be included in each plate. DNPH reagent in PBS without protein should be used as blank for the calculation of samples.

In the principle of the ELISA microassay, DNPH reacts with free carbonyl groups on proteins, leading to adduction of the dinitrophenyl (DNP) group to the carbonyl group. A primary antibody to DNP and a secondary antibody which is anti-rabbit IgG peroxidase (POD), are used. Peroxidase in the secondary antibody reacts with substrates *o*-phenylenediamine and 3,3′,5,5′-tetramethylbenzidine dihydrochloride together with hydrogen peroxide. The latter is used in the method for minimal yield of proteins instead of *o*-phenylenediamine because of its higher sensitivity.

Measurement Procedure

Derivatization is carried out by incubating 15 μL of samples, standards, and blank with 45 μL of fresh DNPH solution for 45 min at room temperature in the dark, with intervals of vortexing every 10 min. Samples and standards should have the same protein concentration. 5 μL of this incubation solution are added to 1 mL coating buffer. 200 μL of samples are loaded into 96-well Nunc Immuno Plate MaxiSorp (eBioscience, San Diego, CA) and incubated overnight at 4°C. On the next day, samples not absorbed are discarded and plates are washed with PBS to remove excess DNPH. Wells are blocked with 250 μL of blocking solution in the dark for 1.5 h at room temperature. Blocking solution is discarded, 200 μL per well anti-DNP antibody is added and plates are incubated for 1 h at 37°C. Wells are washed three times with PBS-T. 200 μL per well anti-rabbit-IgG-POD antibody is added and incubated in the dark for 1 h at room temperature. Wells are rewashed three times with PBS-T. 200 μL per well of substrate solution containing *o*-phenylendiamine, as described in Table 33.3, is added and incubated for 15 min at 37°C. The reaction is stopped with 100 μL of 2.5 M H_2SO_4. Absorbance is measured at 492 nm in a microplate reader, and the carbonyl content is calculated from its peak absorption using a molar absorption coefficient (ε) of 22,000 M^{-1} cm^{-1}. Carbonyl content (nmol L^{-1}) is calculated using the Lambert-Beer equation:

Abs_{492} (test-blank) × $10^9/\varepsilon$

The final carbonyl content in the proteins is expressed as nmol mg $protein^{-1}$.

For minimal yield of proteins (≥5 μg per 1 mL PBS), protein samples can be first adsorbed to the plate. After overnight adsorbance, samples can be derivatized with DNPH (Alamdari et al. 2005). For this, 200 μL per

Table 33.3 Solutions for protein carbonyl measurement by ELISA.

Name/description	Composition	Note/remark
Lysis buffer	200 μL of 500 μM HEPES + 40 μL of 250 mM DTT + 9.76 mL H_2O	
DNPH buffer	6 M Guanidinehydrochloride + 0.5 M KH_2PO_4	pH 2.5
DNPH solution	10 mM = 2 mg mL^{-1} in DNPH buffer	DNPH is dissolved in the buffer shortly before starting the experiment to prevent DNPH precipitation
Coating buffer	0.71 g Na_2HPO_4 + 0.6 g NaH_2PO4+ 4.1 g NaCl in 500 mL H_2O	pH 7.0
Blocking solution	10 mL 11 mg mL^{-1} BSA + 90 mL PBS-T	
Developer	0.71 g Na_2HPO4 + 0.5 g citric acid in 100 mL H_2O	
Anti-DNP antibody (rabbit) (Sigma D9656)	1 : 1 000 dilution in blocking solution	Several other polyclonal and monoclonal anti-DNP antibodies are commercially available, for example, from Intergen, Oncor, Molecular probes
Anti-rabbit-IgG-POD antibody	1 : 10 000 dilution in blocking solution	Sigma A-1949, monoclonal
Substrate solution (60 mg tablet *o*-phenylenediamine, Sigma P-1 063)	60 mg 5 mL^{-1} in developer → 1 mL + 19 mL developer + 8 μL H_2O_2	H_2O_2 and *o*-phenylenediamine should be added immediately before utilization
Stop solution	2.5 M sulfuric acid	
PBS	0.2 g KCl + 0.2 g KH_2PO_4 + 8 g NaCl + 1.45 g Na_2HPO_4 in 1000 mL H_2O	
PBS-T	0.1 mL Tween 20 in 100 mL PBS	

well DNPH is added, and plates incubated in the dark at room temperature for 45 min, and then washed at least three times with 300 μL PBS : ethanol (v/v 1 : 1), and once with 300 μL PBS. Then plates are blocked and antibodies are added (see above). One tablet of 3,3′,5,5′-tetramethylbenzidine dihydrochloride is dissolved in 40 mL substrate buffer containing 50 mM Na_2HPO_4 and 50 mM citric acid pH 5.0, and 4 μL of 30% H_2O_2 is added. 200 μL substrate solution are added to each well and incubated for 5 min at 37°C. The reaction is stopped with 100 μL of 2.5 N HCl. Absorbance is measured at 450 nm wavelength using a microplate reader. This modification makes it unnecessary to concentrate proteins with TCA.

Frequently Occurring Problems with the ELISA Method

Care should be taken during the washing steps. The aim is to sufficiently remove excess reagents and minimize protein loss.

Protein carbonyl content is generally expressed as nmol mg $protein^{-1}$. Consequently, the only critical parameter for carbonyl measurement is protein concentration. It should be taken into consideration that protein-conjugated DNP interferes with the BCA protein assay based on the reaction between cuprous cation and bicinchoninic acid. Therefore, it is recommendable to use other protein assays such as Bradford or Lowry, or to measure protein concentration before adding DNPH (Yan 2009).

In general, the ELISA measurement of carbonyls takes two days (without counting the preparation of reduced and oxidized BSA). The first day is destined for sample preparation and determination of protein concentration, derivatization of carbonyls with DNPH, and plate coating, whereas the second day is used for blocking, antibody incubation, and data acquisition.

The procedures used in different laboratories are often not precisely specified in the published papers. This point is of crucial importance when comparing data from different working groups because there is a considerable variation in the basal levels of protein carbonyls in certain literature sources, depending on how the carbonyl assay is performed.

Commercial BSA already contains carbonyl groups. Therefore, it has to be reduced with sodium borohydride. Fully reduced BSA contains only small amounts of carbonyls and its unspecific background signal must be subtracted from all samples.

PREPARATION OF OXIDIZED AND REDUCED BSA STANDARDS AND BLOCKING SOLUTIONS

The calibration of the ELISA method should be performed against a standard solution of BSA which is oxidized with hypochlorous acid (HOCl). This oxidation is a simple way to generate carbonyl groups. Subsequently, the oxidized BSA is modified with DNPH and stored in aliquots. The amount of protein carbonyls (as nmol mg protein^{-1}) should be quantified by the spectrophotometric method (Reznick and Packer 1994; see Chapter 32).

Reduced BSA for washing and blocking should be produced more frequently. The reduction is performed with sodium borohydride. Since absolute results from different preparations show high deviations in the absorbance, the assay is thought to be dependent on blank values, which vary with different washing steps and the amount of the retained free DNP. The absorbance of fully reduced BSA is substracted as blank during the standard curve calculation. If the blank values are not subtracted, the absolute results are higher but relative comparison between (comparative) results does not change. Under some conditions, this subtraction gives negative values because of the difference between the amounts of DNP retained by protein precipitates of pure BSA and samples (Buss et al. 1997).

Production of Oxidized and Fully Reduced BSA

Oxidized BSA is prepared by modifying 50 mg BSA in 1 mL PBS with 50 mM hypochlorous acid for 2 h under continuous shaking (90 rpm) at room temperature (Buss et al. 1997; Winterbourn and Buss 1999). Some modified methods use other agents such as an ascorbic acid-ferrous amonium sulfate mixture (Alamdari et al. 2005), H_2O_2–ferrous-sulfate mixture (Robinson et al. 1999) and H_2O_2–vanadyl-sulfate mixture (Keller et al. 1993) for BSA oxidation.

For the preparation of reduced BSA, 1 g BSA is dissolved in 100 mL PBS and 2 g sodium borohydride is added and incubated for 30 min. After incubation, the solution is neutralized with 2 N HCl (warning: H_2 is formed). In order to remove sodium borohydride, the protein has to be dialyzed against PBS overnight and protein concentration is measured at 280 nm. For the standard curve, reduced BSA has to be adjusted to 4 mg mL^{-1} or to 0.1% for the blocking solution.

Following the production, reduced and oxidized BSA should be stored at −80°C to avoid any oxidation by aerial oxygen.

DETECTION IN AQUATIC SYSTEMS

Several studies measured protein carbonyl formation in aquatic systems following various inducing conditions. Following the polycyclic aromatic hydrocarbon-rich oil spill in Göteborg harbor in 2003, eelpout (*Zoarces viviparus*) was used to monitor the impact of this event. Eelpout are known to be bottom dwelling and relatively stationary fish and, therefore, suitable for use in environmental studies. In this study, fish were caged on site 2–3 days before sampling. During sampling, fish were killed, measured and weighed. Livers were excised, divided into pieces, and stored in liquid nitrogen (Almroth et al. 2005). Protein carbonylation was measured using an ELISA method similar to the one we described above in homogenized samples. Results showed differences between reference and polluted field sites. Protein carbonyl levels were 3–5 nmol carbonyl mg^{-1} protein, similar to levels measured in humans and in rats, the organisms in which protein carbonyl formation are best studied (Fagan et al. 1999; Das and Chainy 2004).

In another study performed with *Z. viviparus*, carbonyl groups were determined as indicators of oxidative damage upon cold exposure (Heise et al. 2007). 200 mg of tissue were homogenized in 800 μL of 50 mM *N*-2-hydroxyethylpiperazine-*N'*-2-ethanesulfonic acid (HEPES) buffer, and centrifuged at 100,000 *g* for 15 min. 0.4 mL of each supernatant was incubated at room temperature for 1 h with 1.4 mL 10 mM 2,4-dinitrophenylhydrazine (DNPH) in 2 M HCl. For blanks 0.4 mL of each supernatant was mixed with 1.4 mL 2 M HCl. Subsequently, proteins were precipitated with 200 μL of 100% TCA and centrifuged for 10 min at 10,000 *g*. Protein pellets were washed three times with ethanol : ethylacetate

(1 : 1) and centrifuged for 10 min at 10,000 *g*. Pellets were dried on air for 2 h, resuspended in 0.6 mL 6 M guanidine hydrochloride in 20 mM potassium phosphate (pH 2.3) and incubated at 37°C until complete resuspension. Carbonyl content was measured spectrophotometrically at 360 nm. Results showed that summer *Zoarces viviparus* (2.2 nmol mg^{-1} protein) had reduced carbonyl contents compared to winter fish (3.2 nmol mg^{-1} protein).

Chelated ferric iron, H_2O_2, and ascorbate were used to generate $HO^{\bullet}$ in washed, minced cod muscle. Protein carbonyl content increased between 2 and 24 h of storage at 5°C on treatment with the free-radical-generating system. This increase was less when the storage period included a freeze–thaw process (Srinivasan and Hultin 1997).

REAGENTS REQUIRED FOR THE CARBONYL ELISA METHOD

The reagents needed for the carbonyl measurements are summarized in Table 33.3.

Standard Solutions

Standards should be prepared with oxidized and reduced BSA in various ratios, from 0 to 100% of oxidized BSA, using reduced BSA to adjust the protein concentration to 1 mg mL^{-1}.

Colorimetric Assay for Standards

Carbonyl amounts in the standard solutions are determined together with the samples by ELISA. The absorbance of "fully" reduced BSA indicates a small amount of carbonyls (e.g. Abs_{375} of about 0.13 per 10 mg corresponds to 0.6 nmol carbonyls mg^{-1} protein). This absorbance cannot be reduced totally as free DNP or other unspecific effects are attributed to this background signal. This unspecific signal should be substracted from all standards.

CONCLUSION

The described ELISA method used to determine protein carbonyl amounts is sensitive and reproducible. It requires only a small amount of protein (ca. 60 μg or less). The disadvantage of this method is that absolute values are subjected to some uncertainty; therefore, the method is best for utilization in comparative studies.

In the original method it was recommended that samples containing less than 4 mg protein mL^{-1} must be concentrated by TCA before adding DNPH (Buss et al. 1997). It was argued that TCA modifies the absorption characteristics of the samples and a loss of protein was also assumed (Buss et al. 1997; Winterbourn and Buss 1999; Alamdari et al. 2005). Because of the disadvantages of TCA, another specific ELISA method should be developed to determine carbonyls in samples containing low amounts of protein.

REFERENCES

Alamdari, D.H., Kostidou, E., Paletas, K. et al. (2005) High sensitivity enzyme-linked immunosorbent assay (ELISA) method for measuring protein carbonyl in samples with low amounts of protein. *Free Radical Biology and Medicine* 39, 1362–1367.

Almroth, B.C., Sturve, J., Berglund, A., Förlin, L. (2005) Oxidative damage in eelpout (*Zoarces viviparus*), measured as protein carbonyls and TBARS, as biomarkers. *Aquatic Toxicology* 73, 171–180.

Bader, N., Grune, T. (2006) Protein oxidation and proteolysis. *Biological Chemistry* 387, 1351–1355.

Berlett, B.S., Stadtman, E.R. (1997) Protein oxidation in aging, disease, and oxidative stress. *Journal of Biological Chemistry* 272, 20313–20316.

Buss, H., Chan, T.P., Sluis, K.B., Domigan, N.M., Winterbourn, C.C. (1997) Protein carbonyl measurement by a sensitive ELISA method. *Free Radical Biology and Medicine* 23, 361–366.

Dalle-Donne, I., Aldini, G., Carini, M., Colombo, R., Rossi, R., Milzani, A. (2006) Protein carbonylation, cellular dysfunction, and disease progression. *Journal of Cellular and Molecular Medicine* 10, 389–406.

Das, K., Chainy, G.B.N. (2004) Thyroid hormone influences antioxidant defense system in adult rat brain. *Neurochemistry Research* 29, 1755–1766.

Davies, M.J. (2003) Singlet-oxygen mediated damage to proteins and its consequences. *Biochemical and Biophysical Research Communications* 305, 761–770.

Dean, R.T., Fu, S., Stocker, R., Davies, M.J. (1997) *Biochemistry* and pathology of radical-mediated protein oxidation. *Biochemistry Journal* 324, 1–18.

Fagan, J.M., Sleczka, B.G., Sohar, I. (1999) Quantitation of oxidative damage to tissue proteins. *International Journal of Biochemistry and Cell Biology* 31, 751–757.

Gladstone, I.M.J., Levine, R.L. (1994) Oxidation of proteins in neonatal lungs. *Pediatrics* 93, 764–768.

Grune, T., Reinheckel, T., Davies, K.J.A. (1997) Degradation of oxidized proteins in mammalian cells. *FASEB Journal* 11, 526–534.

Grune, T., Merker, K., Sandig, G., Davies, K.J.A. (2003) Selective degradation of oxidatively modified protein substrates by the proteasome. *Biochemical and Biophysical Research Communications* 305, 709–718.

Halliwell, B., Gutteridge, J.M.C. (2007) *Free Radicals in Biology and Medicine*, 4th edn. Oxford University Press, 261 pp.

Heise, K., Estevez, M.S., Puntarulo, S. et al. (2007) Effects of seasonal and latitudinal cold on oxidative stress parameters and activation of hypoxia inducible factor (HIF–1) in zoarcid fish. *Journal of Comparative Physiology B* 177, 765–777.

Holmgren, A. (1989) Thioredoxin and glutaredoxin systems. *Journal of Biological Chemistry* 264, 13963–13966.

Iwai, K., Drake, S.K., Wehr, N.B. et al. (1998) Iron-dependent oxidation, ubiquitination, and degradation of iron regulatory protein 2: implications for degradation of oxidized proteins. *Proceedings of the National Academy of Sciences USA* 95, 4924–4928.

Keller, R.J., Halmes, N.C., Hinson, J.A., Pumford, N.R. (1993) Immunochemical detection of oxidized proteins. *Chemistry Research in Toxicology* 6, 430–433.

Levine, R.L., Williams, J., Stadtman, E.R., Shacter, E. (1994) Carbonyl assays for determination of oxidatively modified proteins. *Methods in Enzymology* 233, 346–357.

Mehlhase, J., Grune, T. (2002) Proteolytic response to oxidative stress in mammalian cells. *Biological Chemistry* 383, 559–567.

Nystrom, T. (2005) Role of oxidative carbonylation in protein quality control and senescence. *EMBO Journal* 24, 1311–1317.

Puig, A., Gilbert, H.F. (1994) Protein disulfide isomerase exhibits chaperone and anti-chaperone activity in the oxidative refolding of lysozyme. *Journal of Biological Chemistry* 269, 7764–7771.

Quinlan, G.J., Evans, T.W., Gutteridge, J.M. (1994) Oxidative damage to plasma proteins in adult respiratory distress syndrome. *Free Radicals Research* 20, 289–298.

Reznick, A.Z., Packer, L. (1994) Oxidative damage to proteins: spectrophotometric method for carbonyl assay. *Methods in Enzymology* 233, 263–357.

Robinson, C.E., Keshavazian, A., Pasco, D.S., Frommel, T.O., Winshop, D.H., Holmes, E.W. (1999) Determination of protein carbonyl groups by immunoblotting. *Analytical Biochemistry* 266, 48–57.

Shacter, E., Williams, J.A., Lim, M., Levine, R.L. (1994) Differential susceptibility of plasma proteins to oxidative modification: Examination by Western blot immunoassay. *Free Radical Biology and Medicine* 17, 429–437.

Shringarpure, R., Davies, K.J.A. (2002) Protein turnover by the proteasome in aging and disease. *Free Radical Biology and Medicine* 32, 1084–1089.

Sitte, N., Merker, K., Grune, T. (1998) Proteasome-dependent degradation of oxidized proteins in MRC–5 fibroblasts. *FEBS Letters* 440, 399–402.

Smith, M.A., Sayre, L.M., Anderson, V.E. et al. (1998) Cytochemical demonstration of oxidative damage in alzheimer disease by immunochemical enhancement of the carbonyl reaction with 2,4-dinitrophenylhydrazine. *Journal of Histochemistry and Cytochemistry* 46, 731–735.

Stadtman, E.R., Oliver, C.N. (1991) Metal-catalyzed oxidation of proteins: Physiological consequences. *Journal of Biological Chemistry* 266, 2005–2008.

Stadtman, E.R., Levine, R.L. (2000) Protein oxidation. *Annals of the New York Academy of Science* 899, 191–208.

Srinivasan, S., Hultin, H.O. (1997) Chemical, physical, and functional properties of cod proteins modified by a nonenzymic free-radical-generating system. *Journal of Agriculture and Food Chemistry* 45, 310–320.

Vogt, W. (1995) Oxidation of methionyl residues in proteins: tools, targets, and reversal. *Free Radical Biology and Medicine* 18, 93–105.

Voss, P., Grune, T. (2006) *In* Dalle-Donne, I., Scaloni, A., Butterfield, D.A. (eds). *Redox Proteomics: From Protein Modifications to Cellular Dysfunction and Diseases*. John Wiley & Sons, New York, pp. 527–562.

Voss, P., Horáková, L., Jakstadt, M., Kiekebusch, D., Grune, T. (2006) Ferritin oxidation and proteasomal degradation: protection by antioxidants. *Free Radicals Research* 40, 673–683.

Winterbourn, C.C., Buss, I.H. (1999) Protein carbonyl measurement by enzyme-linked immunosorbent assay. *Methods in Enzymology* 300, 106–111.

Yan, L.J. (2009) Analysis of oxidative modification of proteins. *Current Protocols in Protein Science* 56, 1404–1456.

Chapter 34

EVALUATION OF MALONDIALDEHYDE LEVELS

Sayuri Miyamoto[1], Eduardo Alves de Almeida[2], Lílian Nogueira[2], Marisa Helena Gennari de Medeiros[1], and Paolo Di Mascio[1]

[1]Departamento de Bioquímica, Instituto de Química, Universidade de São Paulo, São Paulo, SP, Brazil

[2]Departament of Chemistry and Environmental Sciences, IBILCE-UNESP, São José do Rio Preto, SP, Brazil

Oxidative damage of lipids can lead to the formation of lipid hydroperoxides that can be further decomposed to generate several aldehydes, including malondialdehyde (MDA). This oxidative by-product of lipid peroxidation is used as an index of tissue oxidative stress in many laboratories. This aldehyde has been measured for many years mainly through spectrophotometric detection of thiobarbituric-acid–MDA derivative at 535 nm. This methodology has been criticized due to its low specificity and sensitivity (see also Chapter 32). Several other methodologies to access MDA content in tissues and cells have been developed. Among them, the use of high performance liquid chromatography (HPLC) coupled to UV-Vis or fluorescence detection allows MDA analysis with higher specificity, sensitivity, and better reproducibility than the spectrophotometric detection. Besides thiobarbituric acid, MDA can be directly detected by other derivatizing agents. This chapter will describe some simple and accurate methods for MDA evaluation in animal tissues by spectrophotometric and HPLC-based assays.

MDA is formed from polyunsaturated fatty acids (PUFAs; Fig. 34.1) with more than two methylene-interrupted double bonds. It is generally agreed that major precursors of MDA are arachidonic acid (20:4) and docosahexaenoic acid (22:6), whereas fatty acids with less than two double bonds, such as oleic acid (18:1) and linoleic acid (18:2), are poor precursors for MDA (Esterbauer et al. 1991). Two major possible routes of formation have been reported. Pryor and Stanley (1975) suggested a pathway involving cyclization of peroxyl radicals and formation of bicycle-endoperoxide intermediates, which subsequently decomposes to give free MDA by thermic or acid-catalyzed reactions. Esterbauer and colleagues proposed a pathway involving successive hydroperoxide formation and β-cleavage of the fatty acid chain to a hydroperoxyaldehyde that yields MDA by β-scission (Esterbauer et al. 1991). MDA can also be formed as a by-product of enzymatic pathways involved in eicosanoid metabolism, such as in the biosynthesis of tromboxane A2 (Hecker and Ullrich 1989) and prostaglandins (Pryor and Stanley 1975).

Oxidative Stress in Aquatic Ecosystems, First Edition. Edited by Doris Abele, José Pablo Vázquez-Medina, and Tania Zenteno-Savín.

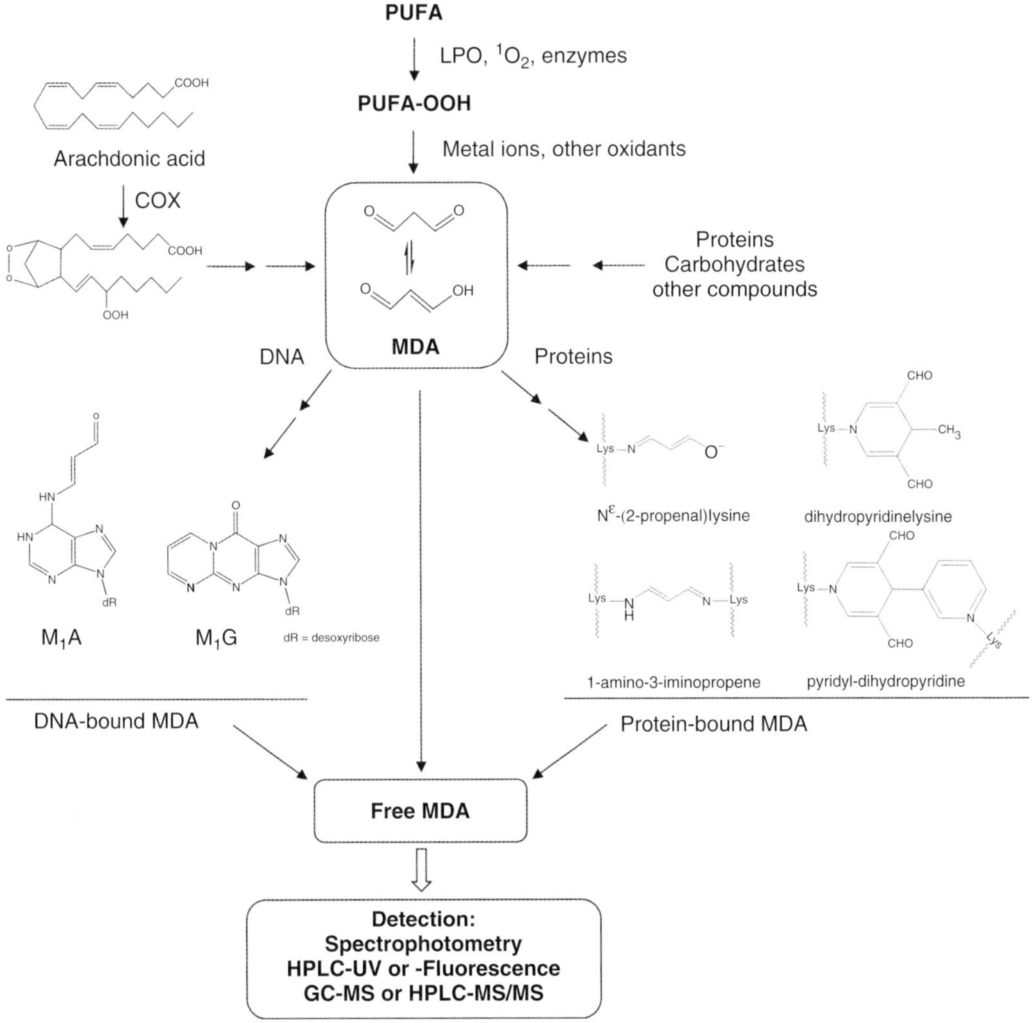

Fig. 34.1 Scheme of MDA formation, reactions with biomolecules and detection. MDA can be formed by lipid peroxidation (LPO) of polyunsaturated fatty acids (PUFA) present in membrane lipids generating hydroperoxides (PUFA-OOH) as primary products. MDA can also be formed as a by-product of eicosanoid metabolism and during the oxidation of proteins and carbohydrates. In biological tissues, MDA can react with DNA and proteins forming covalent adducts. Free and/or bound MDA can be measured by spectrophotometric and chromatographyc techniques (HPLC or GC) coupled to UV-Vis, fluorescence, and mass spectrometry detectors.

Compared to free radicals, aldehydes are relatively stable. Nonetheless, in biological systems aldehydes react readily with thiol and amine groups of proteins and nucleic acids so that only low amounts of MDA exist in free form. MDA reacts with Lys residues in proteins through Schiff's base addition reaction, which forms N^{ε}-(2-propenal)lysine adducts. MDA can further generate lysine–lysine cross-links with 1-amino-3-iminopropene and pyridyl–dihydropyridine type bridges (Uchida 2000). Similar to proteins, MDA reacts with nucleic acid bases in DNA or RNA, to form multiple adducts (Marnett 1999) and can also introduce cross-linkages (Voitkun and Zhitkovich 1999). Cross-links between DNA and histones are mediated by MDA under ionic and physiological pH conditions (Voitkun and Zhitkovich 1999). The major adduct to DNA is a pyrimidopurinone called pyrimido-[1,2α]purin-10(3*H*)-one (M_1G) followed by N_6-(3-oxo-propenyl)deoxyadenosine (M_1A) (Marnett 1999).

MALONDIALDEHYDE DETERMINATION BY 2-THIOBARBITURIC ACID ASSAY

The most common and widely used method to determine MDA levels in biological samples is the 2-thiobarbituric acid (TBA) assay (Uchiyama and Mihara 1978; Ohkawa et al. 1979; see Chapter 32). In this assay, MDA reacts with two molecules of TBA under acidic condition (usually pH between 2 and 3) and heating to form a pink chromogen [$(TBA)_2$-MDA adduct (Esterbauer and Cheeseman 1990), with a maximum absorbance between 530 and 535 nm and a fluorescence emission at 553 nm.

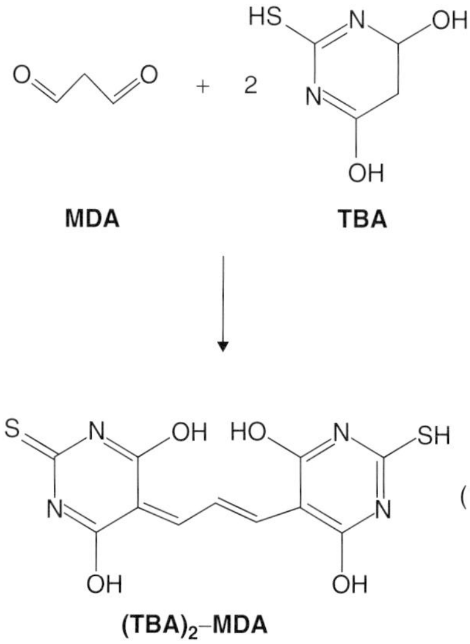

(1)

There are many variations of the TBA assay, differing in sample pretreatment, acid type and concentration, heating temperature and time, and presence or absence of antioxidants. Here we describe two protocols, one general method by Uchiyama and Mihara (1978) (Protocol 1) and a method that has been adapted from Almeida et al. (2004) to measure MDA in tissues from aquatic organisms (Protocol 2). Variations in sample pretreatment include the use of compounds to precipitate proteins and alkaline or acidic treatments aimed to hydrolyze bound MDA. It is generally believed that acid or alkaline treatments hydrolyze the bound fraction of MDA and so allow the determination of total MDA (Cighetti et al. 1999; Del Rio et al. 2005). Free MDA is determined using organic solvents to precipitate proteins.

The spectrophotometric and spectrofluorimetric TBA assays works well with defined *ex vivo* membrane systems. However, these assays lack specificity and does not measure only MDA (discussed in Chapter 32 in detail). For this reason, it is often said that the TBA assay measures "TBA Reactive Substances" (TBARS) rather than MDA. Another critical point of the TBA assay is the artifactual formation of MDA (Janero 1990), especially from lipid hydroperoxides during the acid heating step. This problem is usually solved by sample pretreatment with compounds that reduce hydroperoxides to hydroxides and by the addition of metal chelating agents and antioxidants (e.g., butylated hydroxytoluene, BHT). In spite of all these problems, the TBA assay is still often used and several attempts have been undertaken to improve its specificity. Application of HPLC to measure $(TBA)_2$-MDA adduct has greatly contributed to increase the specificity as well as the sensitivity of the assay.

Various HPLC-based assays have been developed (Esterbauer and Cheeseman 1990; Draper et al. 1993; Agarwal and Chase 2002; Mendes et al. 2009). One of the earliest studies using HPLC for analysis of $(TBA)_2$-MDA adduct was published by Bird et al. (1983). The method was based on the separation of the adduct on a reversed-phase C18 column (3 × 22 mm, Bondapak) isocratically eluted with 15% methanol in water followed by absorbance detection. Depending on the sample, isocratic elution appears to not always effectively separate of the adduct from other TBA-reactive compounds, which causes overestimation of the real MDA concentration. In these cases, gradient elution is recommended (Korchazhkina et al. 2003; Sigolo et al. 2008).

PROTOCOL 1

TBA Assay (Adapted from Uchiyama and Mihara 1978)

Reagents

- 1% phosphoric acid.
- 0.67% TBA in deionized water. Requires heating to dissolve the TBA. Solution cannot be stored, always prepare freshly.

- *n*-Butanol.
- Standard MDA solution: 5 mM stock solution of 1,1,3,3-tetramethoxypropane (TMP, malonaldehyde bisdimethylacetal) or 1,1,3,3-tetraethoxypropane (TEP, malonaldehyde bisdiethylacetal) in ethanol. This standard stock solution should be prepared precisely in a volumetric flask. Store at 4°C. TMP or TEP yields equimolar amounts of MDA under the reaction conditions. Alternatively, MDA can be obtained by preparing the stock solution in 1% (v/v) sulfuric acid and leaving it at room temperature for 2 h. The concentration of MDA is checked by measuring the UV absorbance at 245 nm ($\varepsilon = 13\,700\,M^{-1}\,cm^{-1}$, Esterbauer and Cheeseman 1990). This preparation has been shown to also generate small amounts of highly reactive β-ethoxy and β-methoxy acrolein (Leuratti et al. 1998).

Procedure

Sample preparation

1. To avoid the formation of MDA during the TBA assay, it is recommended to add an antioxidant, such as, butylated hydroxytoluene (0.01 vol% of a 2% BHT solution in ethanol) and a chelating agent (EDTA, DTPA, or desferal at 1 mM final concentration). Transition metals strongly affect the TBA reaction and can lead to misleading values through artifactual lipid peroxidation. Addition of BHT and chelating agents lowers the metal-catalyzed autoxidation of lipids, minimizing erroneous increase in color during the acid heating stages.
2. Deproteinization to determine free MDA can be accomplished by the addition of organic solvents, such as acetonitrile (sample : acetonitrile, 1:1 v/v), followed by vortex mixing and centrifugation. Use the clear supernatant for MDA determination.

Sample incubation with TBA

1. Pipette 50 μL of the sample or the standard into a screw capped tube containing 3 mL of phosphoric acid solution. Make a blank by adding 50 μL of sample buffer.
2. For the standard curve prepare TMP or TEP standards between 100 and 500 μM for the colorimetric assay and between 0.5 and 10 μM for the fluorometric assay.
3. Add 1 mL TBA solution, close the tubes tightly and vortex to mix.
4. Heat the reaction mixture in boiling water for 30 min.
5. Leave it to cool.
6. Extract the pink chromogen by adding 2 mL of buthanol.
7. Mix vigorously for 30 s and centrifuge at 1000 *g* for 5 min.
8. Use the upper layer for measurements.

MDA measurements

MDA levels can be measured by using one of the following techniques:

1. Colorimetric: measure absorbance in the range of 530–535 nm.
2. Fluorometric: measure fluorescence at λ_{ex} 515 nm and λ_{em} 553 nm.

PROTOCOL 2

TBA Assay using HPLC (Adapted from Almeida et al. 2004)

The $(TBA)_2$-MDA adduct is separated from other components by a reversed-phase column (usually C18 column) using isocratic or gradient elution and detected by UV-Vis or fluorescence detection. Isocratic elution is usually carried out with a mixture of 40% methanol in 50 mM phosphate buffer, pH 6.8–7.0. Gradient elution uses a mixture of acetonitrile with water.

Sample Preparation

1. To promote the reaction of 2-thiobarbituric acid (TBA) with MDA, mix 100 mg of tissue in 300 μL of 100 mM Tris, pH 8.0 (ratio 1:3 w/v).
2. Add 300 μL TBA 0.4%. This solution is prepared by diluting 40 mg of TBA in 10 mL of 0.2 M hydrochloric acid.
3. Incubate samples at 90°C for 60 min. Cool samples on ice.
4. Add 1 mL of *n*-butanol for MDA extraction, mix and centrifuge at 1000 *g* or 3 min.

5. Collect supernatant.
6. Inject 20 μL in an HPLC/UV-Vis system.

Standard Curve

To quantify the MDA present in the samples it is necessary to make a curve with known concentrations of authentic MDA standards, also derivatized with TBA. An MDA standard curve can be obtained by incubating increasing concentrations of tetramethoxypropane (TMP) or tetraethoxypropane (TEP) diluted in Tris buffer (300 μL final volume) with 300 μL of the TBA 0.4% solution at 90°C for 60 min and then following the protocol for sample extraction with *n*-butanol and centrifugation. A good start point for HPLC measurements could be the preparation of a standard curve containing 0.5 to 5.0 nmol MDA in 20 μL of *n*-butanol extract.

HPLC with UV-Vis Detector

For the measurement of MDA levels, both samples and standards can be directly injected into the HPLC system (i.e. 20 μL). The general conditions used to obtain the chromatograms shown in Fig. 34.2 are as follows:

- Mobile phase: 0.05 M KH_2PO_4, pH 7.0, with 40% methanol.
- Flow rate: isocratic, 1 mL min^{-1}.
- Column: C18 (150 × 4.6 mm, 5 μm)
- Wavelength: 535 nm

The UV-Vis detector is set to 535 nm. The chromatogram of the TBA-MDA HPLC separation of fish liver samples is very clear (Fig. 34.2) and it can be supposed that the interference of other TBA reactive substances is not so alarming.

MALONDIALDEHYDE DETERMINATION WITHOUT DERIVATIZATION

MDA can also be detected without derivatization. In this case, MDA is separated by an HPLC method using an amino-phase column using a mobile phase consisting of 10% acetonitrile in 30 mM Tris buffer pH 7.4 and detection of MDA at UV 267 nm. This method has, however, low sensitivity (limit of detection: 0.25 μM for a 20 μL injection) and requires a deproteinization step before injection of the sample into the column (Esterbauer and Cheeseman 1990).

OTHER METHODS FOR DETERMINING MALONDIALDEHYDE

MDA measurements can also be done by using derivatizing agents other than TBA, including hydrazines (e.g., 2,4-dinitrophenylhydrazine, phenylhydrazine, Cighetti et al. 1999; 9-fluorenylmethoxycarbonyl

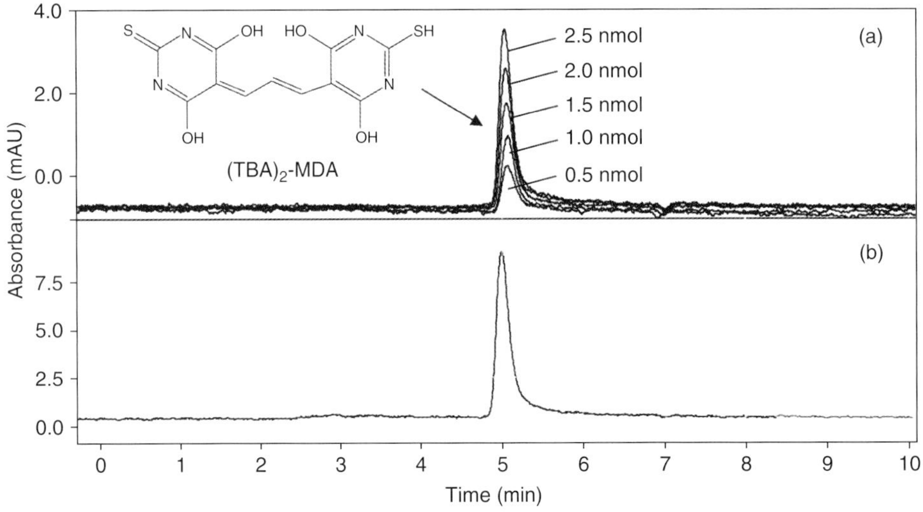

Fig. 34.2 Chromatograms showing increasing concentrations (0.5, 1.0, 1.5, 2.0 and 2.5 nmol) of (a) TBA-MDA standards and (b) a fish liver sample derivatized with TBA, and measured using HPLC-UV/Vis.

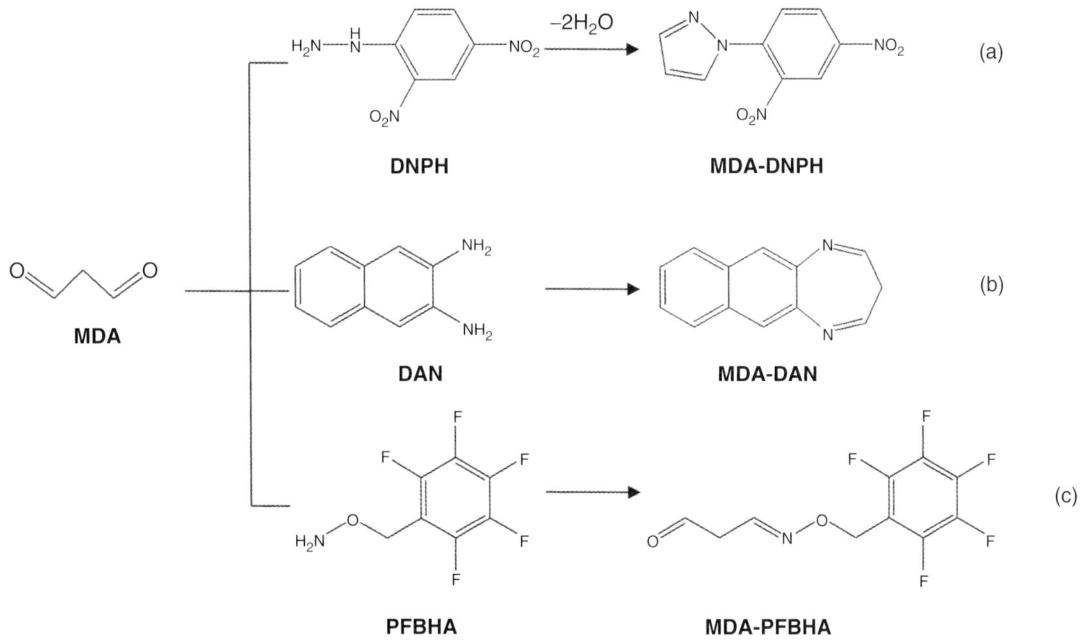

Fig. 34.3 MDA derivatization by reagents other than TBA. Common derivatization reagents are: (a) DNPH, 2,4-dinitrophenylhydrazine; (b) DAN, diaminonaphtalene; and (c) PFBHA, O-pentafluorobenzyl hydroxylamine.

hydrazine, Mao et al. 2006; diaminonaphtalene (DAN), Steghens et al. 2001; O-(2, 3, 4, 5, 6-pentafluorobenzyl) hydroxylamine hydrochloride (PFBHA · HCl), Luo et al., 1995) (Fig. 34.3). The latter is most commonly used for gas chromatography–mass spectrometry (GC–MS) analysis.

The reaction of aldehydes with hydrazines, in particular with 2,4-dinitrophenylhydrazine (DNPH) to give the stable hydrazone derivatives has largely been used to estimate sample carbonyl levels. The advantage of this method over the TBA assay is that it occurs at room temperature under mild pH conditions, avoiding artifactual formation of the aldehyde. The MDA-DNPH derivatives strongly absorb in the region of 300–380 nm ($\varepsilon = 25\,000 - 28\,000\,M^{-1}$ cm^{-1}; Esterbauer and Cheeseman, 1990) and can be determined by HPLC coupled with UV-Vis detection (Luo et al. 1995; Korchazhkina et al. 2003; Sim et al. 2003) and GC–MS (Cighetti et al. 1999).

More accurate ways to measure MDA are based on its detection by GC–MS (Cighetti et al. 1999) or liquid chromatography–mass spectrometry (LC–MS) with the use of stable-isotope-labeled internal standards (Bruenner et al. 1996). These techniques are considered the most reliable and precise. However, they require extensive sample preparation and GC–MS or LC–MS are not readily available in many laboratories.

MALONDIALDEHYDE AS AN INDEX OF THE OXIDANT STATUS IN AQUATIC ORGANISMS

Environmental factors such as temperature, dissolved oxygen and season can be responsible for significant changes in MDA levels, indicating that organisms are facing natural alterations of oxidative stress conditions. MDA levels can further vary as a consequence of seasonal or reproductive cycles in aquatic species (Almeida et al. 2007; Padmini et al. 2008), which are often accompanied by changes in membrane lipid composition, uptake of fatty acids for energy supply, or changes in antioxidant defense systems. Seasonal variations in diet can also contribute to the natural changes of MDA levels in aquatic animals. In larvae of Manchurian trout (*Brachymystax lenok*), MDA levels increase in viscera, muscles, and brain as dietary lipid levels increase (Zhang et al. 2009).

MDA levels also vary between tissues (Almeida et al. 2003; Duran and Talas 2009) or in the same tissue over time of development (Li et al. 2009). A previous characterization of typical MDA levels in new species is therefore pertinent before starting to assess environmental stress conditions. Finally, numerous works relate significant increase in MDA levels in different aquatic organisms to exposure to several classes of pollutants, including metals (Almeida et al. 2004; Verlecar et al. 2007; Bouraoui et al. 2009), organic compounds (Bebianno and Barreira 2009; Koenig et al. 2009; Kopecka-Pilarczyk and Correia 2009), and pesticides (Kavitha and Rao 2009; Lushchak et al. 2007, 2009). For this reason, the measurement of MDA levels in aquatic organisms is considered an excellent pollutant biomarker to indicate the presence of contaminants in the environment.

REFERENCES

Agarwal, R., Chase, S.D. (2002) Rapid, fluorimetric-liquid chromatographic determination of malondialdehyde in biological samples. *Journal of Chromatography B* 775, 121–126.

Almeida, E.A., Marques, S.D., Klitzke, C.F. et al. (2003) DNA damage in digestive gland and mantle tissue of the mussel Perna perna. *Comparative Biochemistry and Physiology C* 135, 295–303.

Almeida, E.A., Miyamoto, S., Bainy, A.C.D., Medeiros, M.H.G., Di Mascio, P. (2004) Protective effect of phospholipid hydroperoxide glutathione peroxidase (PHGPx) against lipid peroxidation in mussels *Perna perna* exposed to different metals. *Marine Pollution Bulletin* 49, 386–392.

Almeida, E.A., Bainy, A.C.D., Loureiro, A.P.M. et al. (2007) Oxidative stress in Perna perna and other bivalves as indicators of environmental stress in the Brazilian marine environment: Antioxidants, lipid peroxidation and DNA damage. *Comparative Biochemistry and Physiology A* 146, 588–600.

Bird, R.P., Hung, S.S.O., Hadley, M., Draper, H.H. (1983) Determination of malonaldehyde in biological-materials by high-pressure liquid-chromatography. *Analytical Biochemistry* 128, 240–244.

Bebianno, M. J., Barreira, L.A. (2009) Polycyclic aromatic hydrocarbons concentrations and biomarker responses in the clam Ruditapes decussatus transplanted in the Ria Formosa lagoon. *Ecotoxicology and Environmental Safety* 72, 1849–1860.

Bouraoui, Z., Banni, M., Ghedira, J., Clerandeau, C., Narbonne, J. F., Boussetta, H. (2009) Evaluation of enzymatic Biomarkers and lipoperoxidation level in Hediste diversicolor exposed to copper and benzo a pyrene. *Ecotoxicology and Environmental Safety* 72, 1893–1898.

Bruenner, B. A., Jones, A. D., German, J.B. (1996) Simultaneous determination of multiple aldehydes in biological tissues and fluids using gas chromatography/stable isotope dilution mass spectrometry. *Analytical Biochemistry* 241, 212–219.

Cighetti, G., Debiasi, S., Paroni, R., Allevi, P. (1999) Free and total malondialdehyde assessment in biological matrices by gas chromatography–mass spectrometry: what is needed for an accurate detection. *Analytical Biochemistry* 266, 222–229.

Del Rio, D., Stewart, A.J., Pellegrini, N. (2005) A review of recent studies on malondialdehyde as toxic molecule and biological marker of oxidative stress. *Nutrition, Metabolism and Cardiovascular Diseases* 15, 316–328.

Draper, H.H., Squires, E.J., Mahmoodi, H., Wu, J., Agarwal, S., Hadley, M. (1993) A comparative-evaluation of thiobarbituric acid methods for the determination of malondialdehyde in biological-materials. *Free Radical Biology and Medicine* 15, 353–363.

Duran, A., Talas, Z.S. (2009) Biochemical changes and sensory assessment on tissues of carp (*Cyprinus carpio*, Linnaeus 1758) during sale conditions. *Fish Physiology and Biochemistry* 35, 709–714.

Esterbauer, H., Cheeseman, K.H. (1990) Determination of aldehydic lipid-peroxidation products – malonaldehyde and 4-hydroxynonenal. *Methods in Enzymology* 186, 407–421.

Esterbauer, H., Schaur, R.J., Zollner, H. (1991) Chemistry and biochemistry of 4-hydroxynonenal, malonaldehyde and related aldehydes. *Free Radical Biology and Medicine* 11, 81–128.

Hecker, M., Ullrich, V. (1989) On the mechanism of prostacyclin and thromboxane A2 biosynthesis. *Journal of Biological Chemistry* 264, 141–150.

Janero, D.R. (1990) Malondialdehyde and thiobarbituric acid-reactivity as diagnostic indexes of lipid-peroxidation and peroxidative tissue-injury. *Free Radical Biology and Medicine* 9, 515–540.

Kavitha, P., Rao, J.V. (2009) Sub-lethal effects of profenofos on tissue-specific antioxidative responses in a Euryhyaline fish, Oreochromis mossambicus. *Ecotoxicology and Environmental Safety* 72, 1727–1733.

Koenig, S., Savage, C., Kim, J.P. (2009) Two novel non-destructive biomarkers to assess PAH-induced oxidative stress and porphyrinogenic effects in crabs. *Biomarkers* 14, 452–464.

Kopecka-Pilarczyk, J., Correia, A.D. (2009) Biochemical response in gilthead seabream (Sparus aurata) to *in vivo* exposure to a mix of selected PAHs. *Ecotoxicology and Environmental Safety* 72, 1296–1302.

Korchazhkina, O., Exley, C., Andrew Spencer, S. (2003) Measurement by reversed-phase high-performance liquid chromatography of malondialdehyde in normal human urine following derivatisation with 2,4-dinitrophenylhydrazine. *Journal of Chromatography B* 794, 353–362.

Li, H.C., Zhou, Q.F., Wu, Y., Fu, J.J., Wang, T., Jiang, G.B. (2009) Effects of waterborne nano-iron on medaka (*Oryzias latipes*): Antioxidant enzymatic activity, lipid peroxidation and histopathology. *Ecotoxicology and Environmental Safety* 72, 684–692.

Luo, X.P., Yazdanpanah, M., Bhooi, N., Lehotay, D.C. (1995) Determination of aldehydes and other lipid-peroxidation products in biological samples by gas-chromatography mass-spectrometry. *Analytical Biochemistry* 228, 294–298.

Lushchak, V.I., Bagnyukova, T.V., Lushchak, O.V., Storey, J.M., Storey, K.B. (2007) Diethyldithiocarbamate injection induces transient oxidative stress in goldfish tissues. *Chemico-Biological Interactions* 170, 1–8.

Lushchak, O.V., Kubrak, O.I., Storey, J. M., Storey, K.B., Lushchak, V.I. (2009) Low toxic herbicide Roundup induces mild oxidative stress in goldfish tissues. *Chemosphere* 76, 932–937.

Leuratti, C., Singh, R., Lagneau, C. et al. (1998) Determination of malondialdehyde-induced DNA damage in human tissues using an immunoslot blot assay. *Carcinogenesis* 19, 1919–1924.

Mao, J., Zhang, H., Luo, J. et al. (2006) New method for HPLC separation and fluorescence detection of malonaldehyde in normal human plasma. *Journal of Chromatography B* 832, 103–108.

Marnett, L.J. (1999) Lipid peroxidation – DNA damage by malondialdehyde. *Mutation Research* 424, 83–95.

Mendes, R., Cardoso, C., Pestana, C. (2009) Measurement of malondialdehyde in fish: A comparison study between HPLC methods and the traditional spectrophotometric test. *Food Chemistry* 112, 1038–1045.

Ohkawa, H., Ohishi, N., Yagi, K. (1979) Assay for lipid peroxides in animal tissues by thiobarbituric acid reaction. *Analytical Biochemistry* 95, 351–358.

Padmini, E., Geetha, B.V., Rani, M.U. (2008) Liver oxidative stress of the grey mullet *Mugil cephalus* presents seasonal variations in Ennore estuary. *Brazilian Journal of Medical and Biological Research* 41, 951–955.

Pryor, W.A., Stanley, J.P. (1975) Suggested mechanism for the production of malonaldehyde during the autoxidation of polyunsaturated fatty acids. Nonenzymic production of prostaglandin endoperoxides during autoxidation. *Journal of Organic Chemistry* 40, 3614–3615.

Sigolo, C.A.O., Di Mascio, P., Kowaltowski, A.J., Garcia, C.C.M., Medeiros, M. H.G. (2008) Trans, trans–2,4-decadienal induces mitochondrial dysfunction and oxidative stress. *Journal of Bioenergetics and Biomembranes* 40, 103–109.

Sim, A. S., Salonikas, C., Naidoo, D., Wilcken, D.E.L. (2003) Improved method for plasma malondialdehyde measurement by high-performance liquid chromatography using methyl malondialdehyde as an internal standard. *Journal of Chromatography B* 785, 337–344.

Steghens, J.-P., van Kappel, A. L., Denis, I., Collombel, C. (2001) Diaminonaphthalene, a new highly specific regent for HPLC-UV measurement of total and free malondialdehyde in human plasma or serum. *Free Radical Biology and Medicine* 31, 242–249.

Uchida, K. (2000) Role of reactive aldehyde in cardiovascular diseases. *Free Radical Biology and Medicine* 28, 1685–1696.

Uchiyama, M., Mihara, M. (1978) Determination of malonaldehyde precursor in tissues by thiobarbituric acid test. Analytical *Biochemistry* 86, 271–278.

Verlecar, X. N., Jena, K. B., Chainy, G.B.N. (2007) Biochemical markers of oxidative stress in Perna viridis exposed to mercury and temperature. *Chemico-Biological Interaction* 167, 219–226.

Voitkun, V., Zhitkovich, A. (1999) Analysis of DNA-protein crosslinking activity of malondialdehyde *in vitro*. *Mutation Research* 424, 97–106.

Zhang, H., Mu, Z. B., Xu, L. M., Xu, G. F., Liu, M., Shan, A.S. (2009) Dietary lipid level induced antioxidant response in Manchurian trout, *Brachymystax lenok* (Pallas) larvae. *Lipids* 44, 643–654.

Chapter 35

THE USE OF ELECTRON PARAMAGNETIC RESONANCE IN STUDIES OF OXIDATIVE DAMAGE TO LIPIDS IN AQUATIC SYSTEMS

Gabriela Malanga and Susana Puntarulo

Physical Chemistry-PRALIB, School of Pharmacy and Biochemistry, University of Buenos Aires, Buenos Aires, Argentina

Oxidative damage to lipids has been studied over several decades and has been characterized in terms of the nature of the oxidant, the type of lipid, and the severity of the oxidation. Many stable products are formed during the process and, accordingly, several procedures and techniques have been developed to assess these products and evaluate lipid peroxidation. Currently the most widely used method is the spectrophotometric assay, which determines malondialdehyde (MDA) formation by its reaction with thiobarbituric acid (TBA, see also Chapters 33 and 34). The assay is easy to perform and therefore frequently applied. However MDA is only one out of the many end products of lipid oxidation, and the reaction with TBA to form thiobarbituric acid reactive substances (TBARS) can be produced by many other compounds in cells. In contrast, electron paramagnetic resonance (EPR) spectroscopy uses exogenous traps to detect lipid radicals as unique and stable products.

EPR, also known as electron spin resonance (ESR), is presently the only analytical approach that allows direct detection of free radicals. This technique reports on the magnetic properties of unpaired electrons and their molecular environment (Tarpey 2004). Biologically important paramagnetic species include free radicals and many transition elements. In addition, there is a substantial and rapidly growing use of synthetic stable free radicals (spin labels) to obtain information on a wide variety of complex biochemical environments such as macromolecules and membranes. Therefore, EPR techniques have become increasingly used in biochemical and biophysical investigations, and most likely their use will increase further as more biologically oriented scientists become aware of the capabilities

Oxidative Stress in Aquatic Ecosystems, First Edition. Edited by Doris Abele, José Pablo Vázquez-Medina, and Tania Zenteno-Savín.

of these techniques. Moreover, EPR spectroscopy has developed into a multifaceted field that employs several different techniques that have the common basis of resonant absorption of microwaves by paramagnetic substances. In this chapter we provide a general description of the EPR theory, followed by illustrations of particular biochemical uses of EPR in aquatic organisms.

GENERAL FEATURES OF THE EPR TECHNIQUE

The phenomenon of EPR occurs because all the electrons have identical values of mass, charge, intrinsic angular moments, and magnetic moments, which allow them to interact with their atomic or molecular surroundings in ways that may reveal chemical structure and electronic or bonding characteristics with potentially high sensitivity and resolution (Borg 1976). The critical electronic property that underlies EPR is the quantized value of the electron's magnetic moment, which, in turn, derives from its spin. The primary physical picture of the electron envisions it as a unit of negative charge that spins without friction on its own axis. Since a moving charge generates a magnetic field, the axis of each spinning electron has an associated magnetic dipole moment not dissimilar to the north and south magnetic poles of the earth (Borg 1976) (Fig. 35.1). However, quantum restrictions fix the spins of all electrons to be the same, and relative to any axis of atomic or molecular reference only two orientations of spin (and hence of electronic magnetic dipoles) are allowed. The two opposite senses of spin

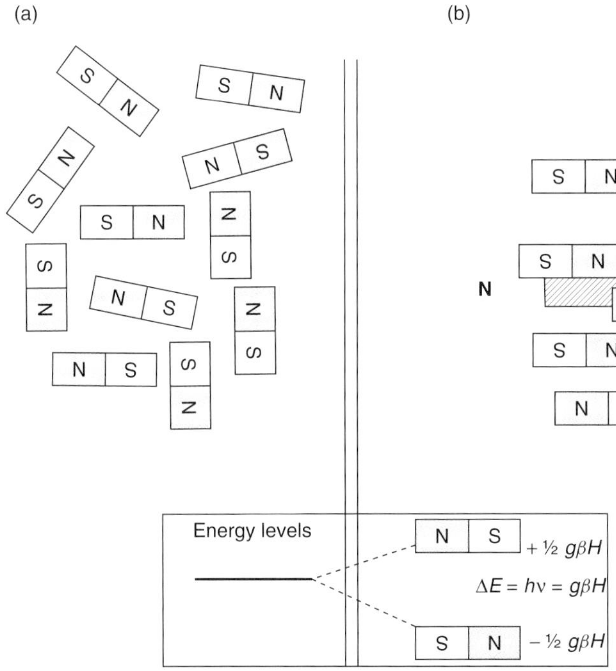

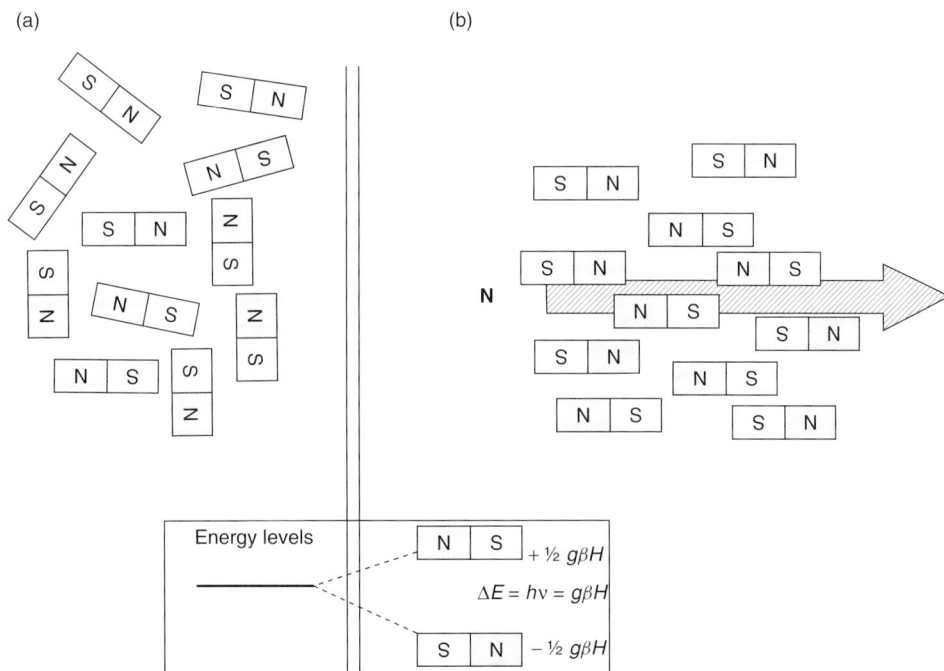

Fig. 35.1 Schematic picture of "free" electrons, in the absence and presence of an external magnetic field. In this simple approximation the spin magnetic moment of a free radical molecular fragment can be represented by that of the unpaired electron alone. (a) In the absence of an external field the spin magnetic moments of free radicals are randomly oriented and are in the same average energy state. (b) In the presence of an external magnetic field the electronic magnets become aligned, according to the laws of space quantization, into one of two allowed orientations: either parallel to the external field, or opposed to it, antiparallel. Inset: Diagram of the energy levels either in the presence or absence of the magnetic field. (Modified from Borg 1976.) See plate section for a color version of this image.

are designated as α and β, and in the absence of an external magnetic field, a free electron has no preference for either spin state. Thus, the two allowed states, designated by the electron spin quantum number, m_s, are of equal energy or doubly degenerate. However, in the presence of an applied magnetic field, the "degeneracy is lifted" because the spin state, with its magnetic moment aligned parallel to the applied field, will be more stable as when antiparallel oriented, e.g., it will be at lower energy (Fig. 35.1). Indeed induced energy separation varies directly with the strength of the applied magnetic field, such that if the two allowed electronic spin states are designated as $m_s = +{}^1\!/_2$ and $m_s = -{}^1\!/_2$, the separation between the two respective energy levels is proportionate to the product of β, the Bohr magneton (the value of the intrinsic magnetic moment associated with the spin of a free electron), times a proportionality constant (the g factor) multiplied by the strength of the field, H. In other words, the energy difference in changing from α to β or from β to α spin (i.e., for $\Delta m_s = \pm 1$) is as shown in equation 35.1:

$$\Delta E = g\beta H \tag{35.1}$$

For the free electron, the value of g resulting from the electron's spin is very close to 2, but because the electron's unit negative charge spins at relativistic velocities, there is a small correction so that the actual g value resulting from electron spin is 2.002319278 (Borg 1976). The key to EPR is that energy transitions are possible when the orientation of electronic spins aligned by a magnetic field will change. If the spin ensemble (e.g. "free" electrons in the sample) is irradiated simultaneously with oscillating electromagnetic radiation whose photon energy, E, equals $h\nu$ (with h being Planck's constant and ν the frequency of the radiation field oscillation) such that $h\nu$ exactly corresponds to the energy difference, ΔE, between the antiparallel and parallel electronic magnetic moments, transitions will occur (equation 35.2). Under these conditions, namely,

$$\Delta \mathrm{E} = \mathrm{h}\nu = g\beta H \tag{35.2}$$

magnetically aligned electrons are in resonance with the radiation field, which means that they can take energy from it or give energy to it. Some parallel magnetic dipoles will absorb a quantum of radiation energy to "jump" from $m_s = -{}^1\!/_2$ to the slightly higher energy state corresponding to $m_s = +{}^1\!/_2$, whereas some electrons with antiparallel spins will flip to the parallel state and thus release the same amount of energy (given in equation 35.2) to the electromagnet field (Fig. 35.1).

However, according to the Boltzmann distribution, quantized particles thermally distributed among different energy levels will preferably reside in the more stable states (Borg 1976). For the two-state electron spin system, if there are N^+ electrons in the antiparallel, or higher, energy level with energy E^+ and N^- parallel electrons with electron spin energy, E^- (equation 35.3)

$$N^+/N^- = \exp(-(E^+ - E^-)/kT) \tag{35.3}$$

where k is Boltzmann's constant and T is the absolute temperature. Since $(E^+ - E^-) = \Delta E$, substitution from equation 35.2 yields the equation 35.4

$$N^+/N^- = \exp(-g\beta H/kT) \tag{35.4}$$

If ΔN is the net fractional difference between the population of each spin, then equation 35.5 results

$$\Delta N = 1 - N^+/N^- \tag{35.5}$$

In fact, it is the presence of the difference, ΔN, that produces the net absorption of electromagnetic energy at resonance, and, in turn, it is this net absorption of energy from the applied electromagnetic field that provides the sample signal in EPR spectroscopy. It is conventional in EPR spectroscopy to present first derivative spectra rather than the integrated absorption or emission spectra characteristic of most other applied spectroscopies used by chemists and life scientists.

Since most radicals contain one or more atoms with magnetic nuclei, an unpaired electron in a radical fragment is apt to interact with internal fields due to nuclear magnetic dipolar moments, as well as with the applied field. This is termed "the nuclear hyperfine interaction" and results in the splitting of resonance lines into two or more components, so-called hyperfine splitting (HFS). Nuclear HFS can enhance the informational content of an EPR spectrum, because many characteristic patterns of lines may result, and these may aid in the identification of free radicals. In addition, their full resolution and analysis may allow accurate determination of the orbital distributions of unpaired electrons, and hence of reactivities, spatial orientations, and structural information (Borg 1976).

The major components of an EPR spectrometer are: scanning electromagnet, microwave source and conductors, sample cavity, and sensitive detection and signal amplification systems (Fig. 35.2). Usual EPR

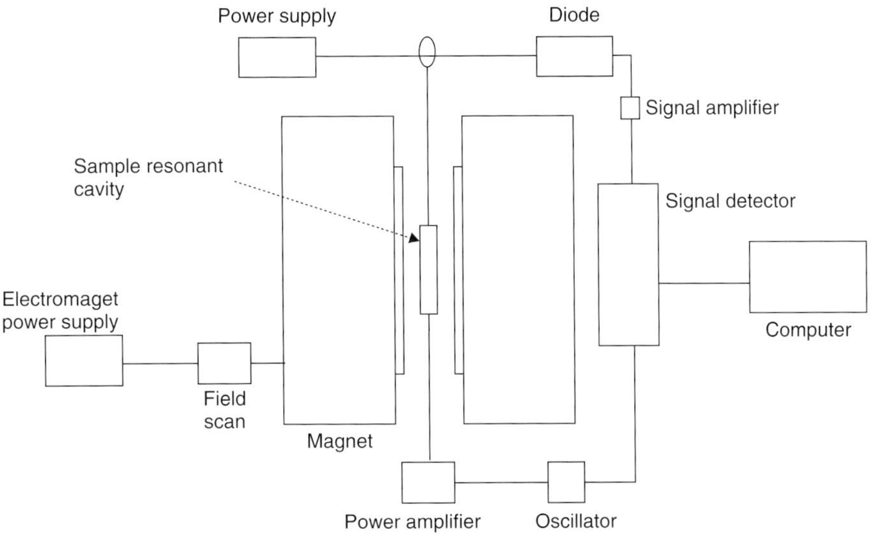

Fig. 35.2 Simplified diagram of a typical EPR spectrometer. (Modified from Borg (1976) and Atkins and de Paula (2006).)

spectra are the first derivatives of microwave power absorbed plotted versus the applied magnetic field strength. The electron spin resonance spectral lines have shape, width, intensity and position (g value), and hyperfine spectral line splitting from the interaction of unpaired electrons with magnetic nuclei that can determine the structure or positions of free radical components, and is a powerful tool in free radical identification. Table 35.1 summarizes the main features of an EPR spectrum.

THE SIGNIFICANCE OF AN EPR SIGNAL IN A BIOLOGICAL SYSTEM

Although it is true that the absence of an EPR signal does not necessarily denote absence of meaningful free radical quantities, it is likewise true that the mere presence of EPR within a chemical or biochemical reaction, or in a biological system, does not suffice to prove that a significant free radical component or pathway has been detected. Further correlations are needed to substantiate the relevance of an EPR signal; for example, quantification of an EPR spectrum may distinguish free radicals formed within the main reaction pathway from those arising from minor side reactions or from adventitious impurities. Even more conclusive, kinetic analysis of a sample system or reaction wherein EPR signal intensities obey the relationships expected can provide especially firm support for the role imputed to the radical.

Because free radicals are highly reactive entities whose lifetimes in chemical or biochemical reactions tend to be very short, a common problem in EPR measurements is to find experimental techniques that will allow time-consuming EPR spectra to be recorded from detectable levels of short-lived paramagnetic intermediates. Free radical detection by EPR requires

Table 35.1 Main features of EPR spectra.

Features	Description
Integrated intensity	The actual magnitude of the microwave absorption is proportional to the total number of unpaired electrons in the sample.
g values	The actual resonance position, 2.002319278 for a free electron.
Line width	Energy spread of the levels occupied by the unpaired electron spin-lattice interaction, spin-spin interaction and line broadening.
Splittings	Interaction of the unpaired electron with the nuclear magnetic moment (hyperfine splitting) or other(s) unpaired electrons (electronic splitting).

concentrations on the order of 10^{-8} M, and the recording of resolved HFS usually is done at 100- to 1000-fold that concentration (Atkins and de Paula 2006). However, second-order decay kinetics are common for free radicals, with decay by dismutation alone occurring in some cases at rates greater than 109 M^{-1} s^{-1}, even where there are no specific radical scavengers present (Bolton et al. 1972). Even assuming a modest rate constant of 108 M^{-1} s^{-1} and taking a starting concentration of free radicals of 10^{-6} M, the first half-life would be only approximately 10 ms in duration. With a starting concentration of 10^{-5} M, the first half-life would be shortened to approximately 1.0 ms. First-order decay constants of 70 s^{-1} and 700 s^{-1} would also yield half-lives of approximately 10 and 1 ms, respectively (Borg 1976).

Any of the approaches listed below will serve to retain intrinsically short-lived intermediates at constant concentrations during the relatively long time span for EPR measurements: (i) trapping, in which reactive species are sequestered into an unreactive matrix; (ii) regenerative procedures, which produce dynamic steady states whose stationary concentrations can be maintained over suitable prolonged intervals; and (iii) fast perturbation methods that can be cycled repeatedly when only kinetic EPR data are sought. Chemically reactive species may be trapped if they are formed by physical means in a dehydrated matrix, in solid samples, in aprotic or apolar solvents, at low temperatures, or under some other conditions that suppress their characteristic reactivity.

Figure 35.3 summarizes the employed methodology to detect free radicals in biological systems. Stable secondary radical species, such as ascorbyl radical, can be detected directly by EPR at room temperature. In contrast, the detection of some other radicals, such as protein radicals and Fe, requires low temperatures (77°K) to stabilize those radicals. More stable radical species are also formed by adding exogenous "spin traps," molecules that react with primary radical species to give more enduring radical adducts with characteristic EPR "signatures". These spin traps, frequently nitroxide and nitrone derivatives, can also be used to label biomolecules and probe basal and oxidative-induced molecular events in protein and lipid environments (Borbat et al. 2001). At room temperature, spin traps have been used successfully to detect HO• and lipid radicals. In some other cases, such as NO• detection, both spin-trapping and low-temperature strategies are employed (Fig. 35.3).

Although probe instability, tissue metabolism, and lack of spin specificity are drawbacks for employing EPR for the *in vivo* determination of free radicals, the dependability of this technique, mostly by combining it with other biochemical strategies, drastically enhances the value of these procedures.

DETECTION OF LIPID RADICALS IN AQUATIC SYSTEMS

As stated above, endogenously produced lipid radicals have extremely short half-lives and are present in low concentrations, rendering their detection difficult. Spin trapping followed by EPR analysis overcomes the limits of sensitivity of endogenous radicals in biological

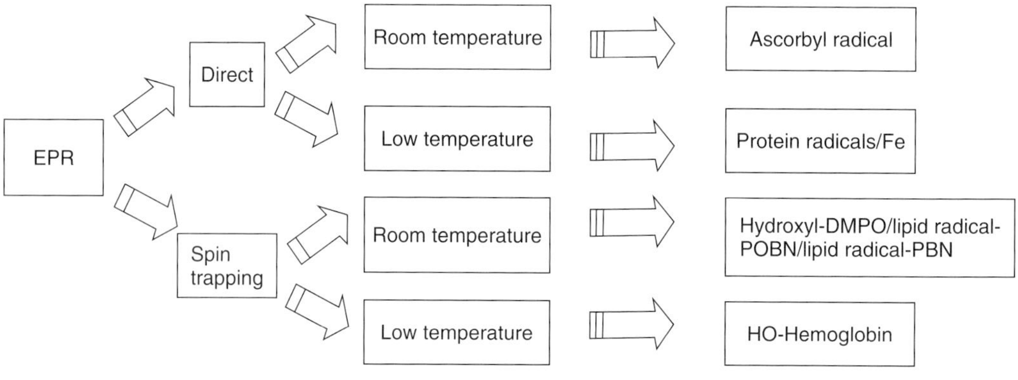

Fig. 35.3 Scheme summarizing the EPR techniques usually applied to biological systems. (Malanga and Puntarulo, unpublished data.)

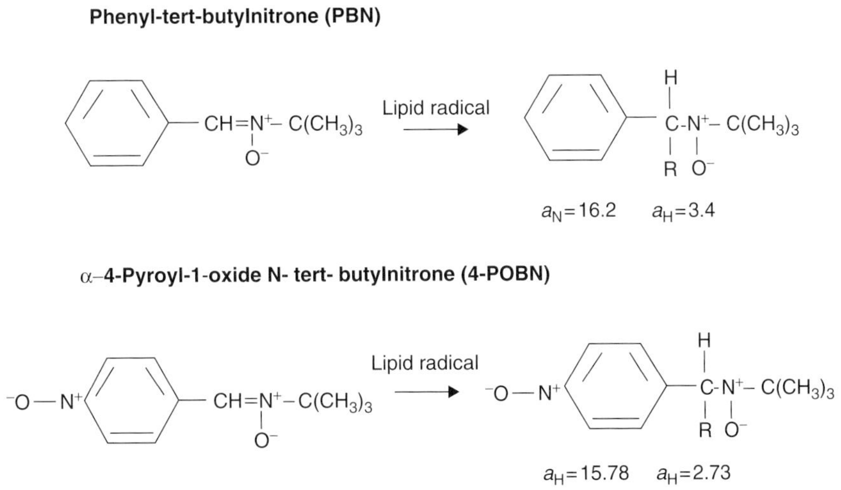

Fig. 35.4 Reactions between the spin traps POBN and PBN with lipid radicals (LR•). (Malanga and Puntarulo, unpublished data.)

systems and it has proved to be the least ambiguous method to detect short-lived reactive free radicals generated in low concentrations in aquatic systems (Luo et al. 2006). Even though EPR detection of lipid radicals can be considered a fingerprint of radical presence, spin-trapping studies cannot really distinguish among ROO•, alcohoxyl RO• and R• adducts owing to the similarity of the corresponding coupling constants (Buettner 1987).

The use of spin traps such as phenyl-tert-butylnitrone (PBN) and α-4-pyroyl-1-oxide-N-tert-butylnitrone (POBN) (Fig. 35.4) have had a memorable history in detecting organic radical products of lipid peroxidation (Detcho 1999).

EPR Measurements of Lipid Radical Content in Algae, Sea Urchin, and Limpets

EPR techniques have been applied successfully to several unrelated aquatic systems, such as algae, sea urchin, and limpets. EPR spectra were obtained from intact *Chlorella vulgaris* algae in log phase growth state by suspending them in phosphate buffer with POBN, and 100 μM desferal (Bold and Wynne 1978; Jurkiewicz and Buettner 1994; Malanga and Puntarulo 1995; Estevez et al. 2001). The spectrum obtained is shown in Fig. 35.5B. The POBN itself was examined and no POBN-spin adduct was observed (Fig. 35.5E). Fe is recognized as a bioactive element (Bruland et al. 1991) and a deficiency in Fe has been suggested to limit primary productivity in some ocean regions (Martin et al. 1990, 1993). Bioassays conducted in high nitrate surface waters in the subarctic North Pacific (Martin et al. 1989; Coale 1991), equatorial Pacific (Price et al. 1991), Southern Ocean (De Baar et al. 1983; Hebling et al. 1991), and North Atlantic (Martin et al. 1993) provide support for the assertion that the phytoplankton community is affected by Fe availability. Data in Fig. 35.6a show the lipid radical spectra recorded from cells grown in the presence

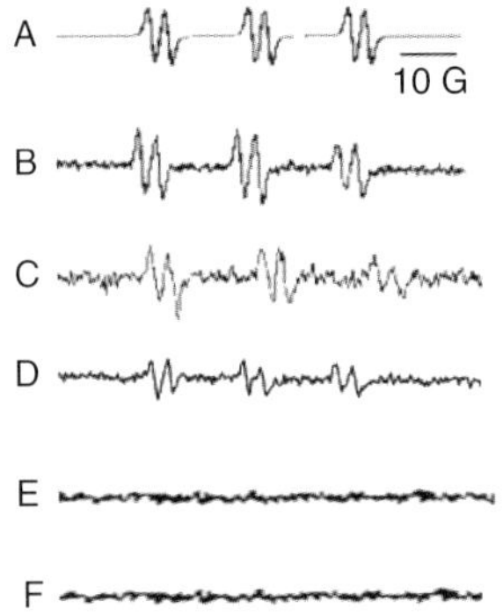

Fig. 35.5 EPR detection of lipid radicals in aquatics systems. (A) Computer-simulated EPR spectra exhibiting hyperfine splittings that are characteristic of POBN/lipid radicals a_N = 15.78 G and a_H = 2.73 G. (B) *Chlorella vulgaris* cells in log phase of growth. (C) *Nacella (Patinigera) magellanica* digestive gland. (D) Gonads from *Loxechinus albus* in the presence of PBN, a_N = 16.2 G and a_H = 3.4 G. (E) Spectra of POBN itself. (F) Spectra of PBN itself. (Malanga and Puntarulo, unpublished data.)

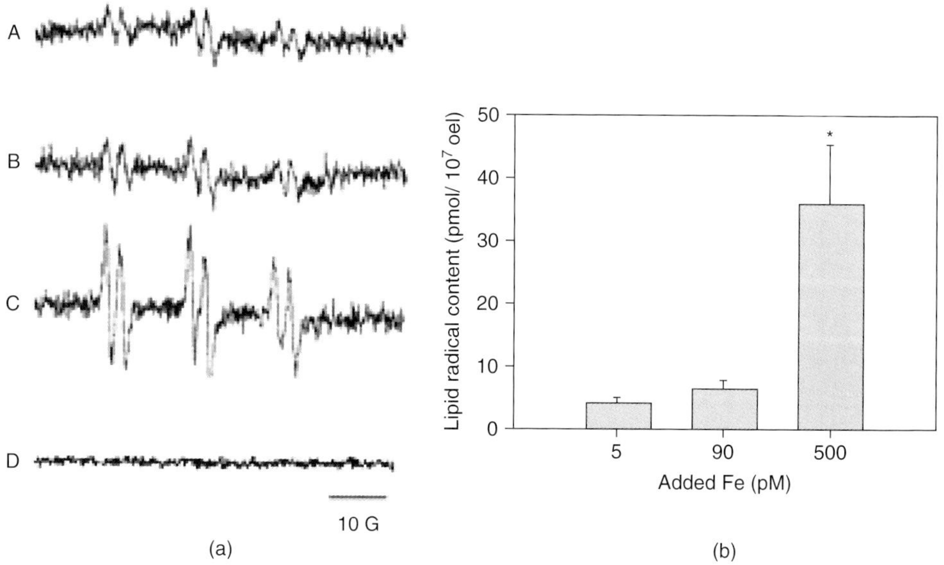

Fig. 35.6 Effect of Fe overload on lipid radical content in cells of *Chlorella vulgaris*. (a) Typical EPR spectra with POBN of *C. vulgaris* cells in log phase of development grown in the presence of (A) 5 μM Fe in the incubation medium, (B) 90 μM Fe, (C) 500 μM Fe, and (D) spectrum of POBN itself. (b) Quantification of the effect of Fe supplementation on lipid radical content in *C. vulgaris*. *Significant difference from lipid radical content in the presence of 5 μM Fe at $p < 0.05$, ANOVA. (Data from Estevez et al. 2001.)

and absence of excess Fe (Estevez et al. 2001), and quantification data in Fig. 35.6b indicate that cellular lipid radical content significantly increased under Fe overload (exposure to 500 μM).

EPR spectra from limpets digestive gland homogenates prepared in PBS containing 50 mM POBN were also obtained using previously published setting parameters (Malanga et al. 2007; Fig. 35.5C). Since metabolic rates, locomotory activity, and growth rates in marine ectotherms are a function of temperature and seasonality of food availability in temperate and Antarctic waters (Kirchin et al. 1992), seasonal differences of lipid peroxidation in the digestive glands of the limpet *Nacella (Patinigera) magellanica* were assessed as the tissue content of lipid radicals (Malanga et al. 2007). Lipid radicals in the digestive gland gave a characteristic EPR spectrum with hyperfine coupling constants of $a_N = 15.78$ G and $a_H = 2.73$ G (Fig. 35.7A). Bulk lipid radical content in limpet digestive glands was significantly lower (67%) in winter than summer indicating that higher metabolic activity is associated with elevated seasonal loads of lipid radical formation (Fig. 35.7B).

Lipid radical content was detected employing spin-trapping techniques using PBN in homogenates prepared from sea urchin (*Loxechinus albus*) gonads (Lai et al. 1986; Malanga et al. 2009). The spectrum obtained is shown in Fig. 35.5D. The PBN itself was examined and no PBN-spin adduct was observed (Fig. 35.5F). In spite of the great seasonal fluctuations in environmental parameters such as productivity, water temperature, and photoperiod, bulk lipid radical content in sea urchin gonads (Fig. 35.8) did not change significantly over the year of study (Malanga et al. 2009).

Quantification of possible spin adducts was performed using an aqueous solution of 2,2,5,5-tetramethyl piperidine 1-oxyl (TEMPO) introduced into the same sample cell used for spin trapping. EPR spectra for both sample and TEMPO solutions were recorded at exactly the same spectrometer settings, and the first derivative EPR spectra were integrated to obtain the area intensity, before calculating the concentration of spin adduct according to Kotake et al. (1996). Data shown in Table 35.2 correspond to lipid radical content in each of the samples tested.

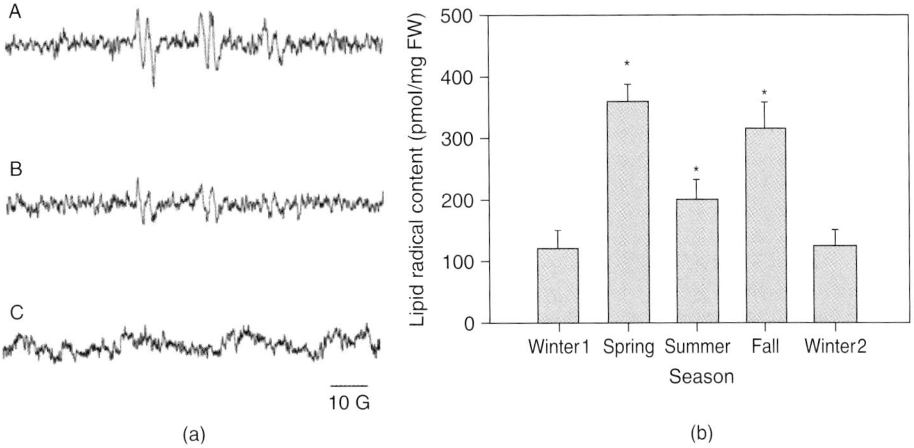

Fig. 35.7 Effect of seasonality on lipid radical content in digestive glands of *N. (P) magellanica*. (a) Typical EPR spectra of digestive glands isolated from animals captured in (A) April (fall) and (B) July (winter), and (C) EPR spectra of POBN itself. (b) Quantification of lipid radical content in the *N. (P) magellanica* digestive gland over the year. *Significant difference from lipid radical content in the samples taken in winter at $p < 0.05$, ANOVA. (Data from Malanga et al. 2007.)

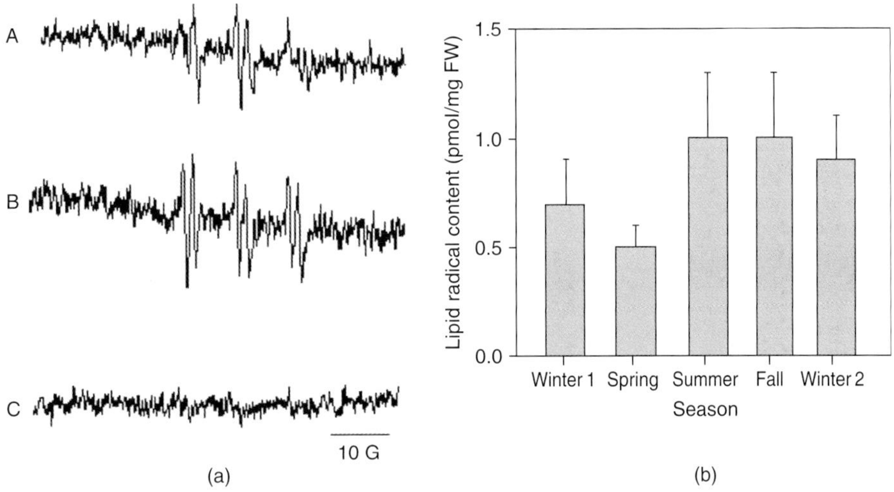

Fig. 35.8 Effect of seasonality on lipid radical content in gonads of *L. albus*. (a) EPR detection of lipid radicals in gonads from animals collected in (A) winter 1 (August) and (B) summer (December), and (C) spectrum of PBN itself. (b) Quantification of lipid radical content in the gonads of *L. albus* over the year. (Data from Malanga et al. 2009.)

EPR Measurements of Lipid Radical Generation in Bivalves

To test for a potential effect of high environmental Fe concentrations, which are naturally occurring in Antarctic coastal environments (see Chapter 2), the rate of lipid radical generation was measured in tissues of marine invertebrates from these areas (Estevez et al. 2002). The bivalve *Laternula elliptica* (Laternulidae) is abundant in nearshore waters around the Antarctic continent and islands, whereas the ecologically similar clam *Mya arenaria* (Myoidea) occurs on intertidal mudflats of the German North Sea coast, where dissolved and particulate Fe concentrations

Table 35.2 Quantification of lipid radical content in aquatic systems.

Aquatic organisms	Condition	Lipid radical content
Chlorella vulgaris[1]	Log phase of growth	6 ± 1 pmol 107 $cell^{-1}$
N. (P) magellanica[2]	Natural habitat in winter	122 ± 29 pmol mg^{-1} FW
Loxechinus albus[3]	Natural habitat in winter	0.7 ± 0.2 pmol mg^{-1} FW

Taken from [1]Estevez et al. (2001), [2]Malanga et al. (2007) and [3]Malanga et al. (2009).

are notoriously lower. Animals of both species were collected, their tissues frozen at -30°C, and their isolated digestive glands suspended in phosphate buffer with POBN and 100 μM desferal. EPR spectra were obtained as described in Estevez et al. (2001). Spin adduct was quantified as previously indicated using an aqueous solution of TEMPO, and the concentration of the spin adduct was calculated according to Kotake et al. (1996). The rate of lipid radicals generation was measured in digestive glands isolated from both animals after *in vivo* incubation at temperatures between 2 and 15°C. Lipid radical content after 10 min at a fixed temperature was significantly higher in the Antarctic bivalve than in the temperate mud clam (50 ± 5 and 10 ± 2 pmol lipid radical mg^{-1} protein min^{-1}, respectively) suggesting that higher levels of lipid peroxidation may be a consequence of transition metals enrichment occurring naturally in Antarctic marine filter feeders (Estevez et al. 2002).

CONCLUDING REMARKS

Spin trapping followed by EPR analysis has proven to be the most direct and sensitive method to detect short-lived ROS generated in low concentrations in biological systems, and also provides a better way of identifying ROS actually present in those systems. Spin-trapping reagents have extremely high rates of reactivity and, thus, have the potential for detection and quantization of radicals that would not be possible by other methods. Regarding EPR detection of lipid radicals, both PBN and POBN have the advantage of forming very stable spin adducts not affected by light, heat, and O_2, and have been widely used to trap radicals generated in biological systems. The data presented here briefly summarize the optimization of the assay performed in our laboratory and successful application of this methodology to marine organisms under both natural environmental conditions and after oxidative stress. Recently, EPR analysis has also been applied to study ROS generation in the fish *Carassius auratus* exposed to several toxic compounds (Luo et al. 2006, 2009; Yin et al. 2007). In summary, the EPR-spin label methods have become a routinely used spectroscopic approach in many biological systems and there are no doubts that in the field of free radicals no other technique allows such a powerful combination of flexibility of approach, sensitivity to concentration, and clear evidence of the chemical conformation of the reaction scheme.

REFERENCES

Atkins, P., de Paula, J. (2006) Molecular spectroscopy 3: magnetic resonance. *In* Atkins, P., de Paula, J. (eds). *Atkins's Physical Chemistry*. Oxford,University Press, pp. 549–555.

Bold, H.C., Wynne, M.J. (1978) Cultivation of Algae in the Laboratory. *In* McElroy, W.D., Swanson, C.P. (eds). *Introduction to the Algae Structure and Reproduction*. Englewood Cliffs, Prentice Hall, pp. 571.

Bolton, J.R., Borg, D.C., Swartz, H.M. (1972) Experimental aspects of biological election spin reasonance. *In* Swartz, H.M., Bolton, J.R., Borg, D.C. (eds). *Applications of Electron Spin Resonance*. Wiley Interscience, New York, p. 63.

Borbat, P.P., Costa-Filho, A.J., Earle, K.A., Moscicki, J.K., Freed, J.H. (2001) Electron spin resonance in studies of membranes and proteins. *Science* 291, 266–269.

Borg, D.C. Applications of electron spin resonance in biology. *In* Pryor, W.A. (ed.). *Free Radicals in Biology*, Vol. 1. New York, Academic Press, pp. 69–147.

Buettner, G.R. (1987) Spin trapping: ESR parameters of spin adducts. *Free Radical Biology and Medicine* 3, 259–303.

Bruland, K.W., Donat, J.R., Hutchins, D.A. (1991) Interactive influences of bioactive metals on biological production in oceanic waters. *Limnology and Oceanography* 36, 1555–1577.

Coale, K.H. (1991) Effects of iron, manganese, copper and zinc enrichments on productivity and biomass in the subArctic Pacific. *Limnology and Oceanography* 36, 1851–1864.

De Baar, H.J.W., Buma, A.G.J., Nolting, R.F., Cadee, G.C., Jacques, G., Treguer, P.J. (1990) On iron limitation of the Southern Ocean: Experimental observations in the Weddell and Scotia Sea. *Marine Ecology Progress Series* 65, 105–122.

Detcho, A., Stoyanovsky, D.A., Melnikov, Z., Cederbaum, A.I. (1999) ESR and HPLC-EC Analysis of the interaction of hydroxyl radical with DMSO: rapid reduction and quantification of POBN and PBN nitroxides. *Analytical Chemistry* 71, 715–721.

Estevez, M.S., Malanga, G., Puntarulo, S. (2001) Iron-dependent oxidative stress in *Chlorella vulgaris*. *Plant Science* 161, 9–17.

Estevez, M.S., Abele, D., Puntarulo, S. (2002) Lipid radical generation in polar (*Laternula elliptica*) and temperate (*Mya arenaria*) bivalves. *Comparative Biochemistry and Physiology B* 132, 729–737.

Jurkiewicz, B.A., Buettner, G.R. (1994) Ultraviolet light-induced free radical formation in skin: an electron paramagnetic resonance study. *Photochemistry and Photobiology* 59, 1–4.

Hebling, E.W., Villafañe, V., Holm-Hansen, O. (1991) Effect of iron on productivity and size distribution of Antarctic phytoplankton. *Limnology and Oceanography* 36, 1879–1885.

Kirchin, M.A., Wiseman, A., Livingstone, D.R. (1992) Seasonal and sex variation in the mixed function oxygenase system of digestive gland microsomes of the common mussel, *Mytilus edulis* (L.). *Comparative Biochemistry and Physiology C* 101, 81–91.

Kotake, Y., Tanigawa, T., Tanigawa, M., Ueno, I., Allen, D.R., Lai, C. (1996) Continuous monitoring of cellular nitric oxide generation by spin trapping with an iron-dithiocarbamate complex. *Biochimica et Biophysica Acta* 1289, 362–368.

Lai, E.K., Crossley, C., Sridhar, R., Misra, H.P., Janzen, E.G., McCay, P.B. (1986) *In vivo* spin trapping of free radicals generated in brain, spleen, and liver during γ radiation of mice. *Archives of Biochemistry and Biophysics* 244, 156–160.

Luo, Y., Su, Y., Lin, R., Shi, H., Wang, X. (2006) 2-Chlorophenol induced ROS generation in fish *Carassius auratus* based on the EPR method. *Chemosphere* 65, 1064–1073.

Luo, Y., Wang, X., Ji, L, Su, Y. (2009) EPR detection of hydroxyl radical generation and its interaction with antioxidant system in Carassius auratus exposed to pentachlorophenol. *Journal of Hazardous Materials* 171, 1096–1102.

Malanga, G., Puntarulo, S. (1995) Oxidative stress and antioxidant content in *Chlorella vulgaris* after exposure to ultraviolet-B radiation. *Physiologia Plantarum* 94, 672–679.

Malanga, G., Estevez, M.S., Calvo, J., Abele, D., Puntarulo, S. (2007) The effect of seasonality on oxidative metabolism in *Nacella (Patinigera) magellanica*. *Comparative Biochemistry and Physiology A* 146, 551–558.

Malanga, G., Perez, A., Calvo, J., Puntarulo, S. (2009) The effect of seasonality on oxidative metabolism in the sea urchin *Loxechinus albus*. *Marine Biology* 156, 763–770.

Martin, J.H., Gordon, R.M., Fitzwater, S.E., Broenkow, W.W. (1989) Vertex: phytoplankton/iron studies in the Gulf of Alaska. *Deep-Sea Research* 36, 649–680.

Martin, J.H., Fitzwater, S.E., Gordon, R.M. (1990) Iron deficiency limits phytoplankton growth in Antarctic waters. *Global Biogeochemistry Cycles* 4, 5–12.

Martin, J.H., Fitzwater, S.E., Gordon, R.M., Hunter, C.N., Tanner, S.J. (1993) Iron, primary productivity and carbon nitrogen flux studies during the JGOFS North Atlantic Bloom Experiment. *Deep-Sea Research* 40, 115–134.

Price, N.M., Andersen, L.F., Morel, F.M.M. (1991) Iron and nitrogen nutrition of equatorial Pacific phytoplankton. *Deep-Sea Research* 37, 295–315.

Tarpey, M.M., Wink, D.A., Grisham, M.B. (2004) Methods for detection of reactive metabolites of oxygen and nitrogen: *in vitro* and *in vivo* considerations. *American Journal of Physiology* 286, R431–R444.

Yin, Y., Jia, H., Sun, Y. et al. (2007) Bioaccumulation and ROS generation in liver of *Carassius auratus*, exposed to phenanthrene. *Comparative Biochemistry and Physiology C* 145, 288–293.

Chapter 36

THE ASCORBYL RADICAL/ASCORBATE RATIO AS AN INDEX OF OXIDATIVE STRESS IN AQUATIC ORGANISMS

Gabriela Malanga, María Belén Aguiar, and Susana Puntarulo
Physical Chemistry-PRALIB, School of Pharmacy and Biochemistry, University of Buenos Aires, Buenos Aires, Argentina

GENERAL FEATURES OF $Asc^{\bullet -}$ IN BIOLOGICAL SYSTEMS

The ascorbyl radical ($Asc^{\bullet -}$) content in biological tissues is a newly discovered, noninvasive and natural indicator of oxidative stress (Roginsky et al. 1994) within the cellular hydrophilic compartment.

Aquatic organisms contain a variety of antioxidants including water-soluble compounds (i.e. ascorbic acid). Ascorbic acid has low redox potential, which allows it to donate one single electron to almost any free radical occurring in a biological system, or to reduce oxidized biological radical scavengers, such as α-tocopherol (Vergely et al. 2003). The $Asc^{\bullet -}$ is the intermediate in the oxidation of ascorbate (Asc) to dehydroascorbate (DHA) (Fig. 36.1) (Hubel et al. 1997). It has an unpaired electron (e^-) in a highly delocalized π-system, giving it stability as the "terminal small-molecule antioxidant" (Buettner 1993).

The concentration of $Asc^{\bullet -}$ is a steady-state value, determined by the rates of generation and decay of this species. $Asc^{\bullet -}$ is formed by Asc oxidation processes mediated or not by metal catalysts such as Fe and Cu (Martell 1982). Reaction (R1) given below leads to $Asc^{\bullet -}$ generation, and the rate of $Asc^{\bullet -}$ formation can be calculated by equation (36.1).

$$Asc + Fe^{3+} - k_1 \longrightarrow Asc^{\bullet -} + Fe^{2+} \quad (R1)$$

$$\frac{d[A^{\bullet -}]}{dt} = k_1 [Fe^{3+}][Asc] \quad (36.1)$$

Self-disproportionation has been postulated as the main, or even the only, way for $Asc^{\bullet -}$ decay in

Oxidative Stress in Aquatic Ecosystems, First Edition. Edited by Doris Abele, José Pablo Vázquez-Medina, and Tania Zenteno-Savín.

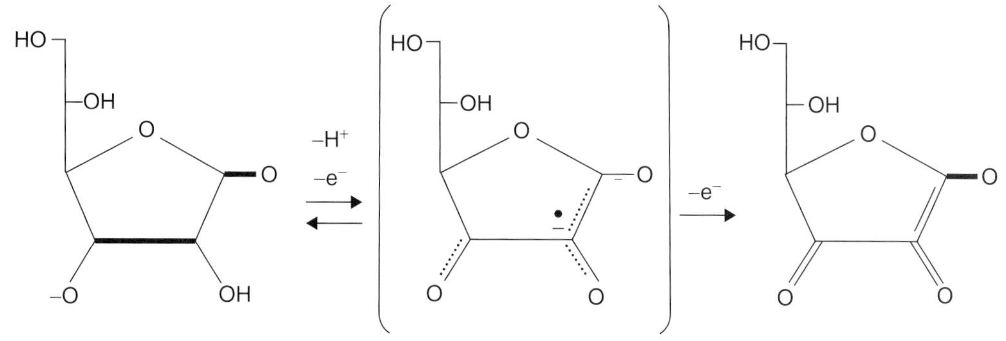

Fig. 36.1 Generation and decay reactions of the radical Asc$^{\bullet -}$ (Malanga et al., unpublished data).

biological systems (reaction R2), and the rate of this reaction is described by equation (36.2).

$$2Asc^{\bullet -} - k_2 \longrightarrow Asc + DHA \quad (R2)$$

$$-\frac{d[Asc^{\bullet -}]}{dt} = 2\,k_2\,[Asc^{\bullet -}]^2 \quad (36.2)$$

The rate constant k_2 has been estimated as 2×10^6 $M^{-1}\,s^{-1}$. Under steady-state conditions, the generation rate of Asc$^{\bullet -}$ should be identical to the decay rate of Asc$^{\bullet -}$, according to equation (36.3).

$$\frac{d[Asc^{\bullet -}]}{dt} = -\frac{d[Asc^{\bullet -}]}{dt} = 2\,k_2[Asc^{\bullet -}]^2 \quad (36.3)$$

DETECTION OF Asc$^{\bullet -}$ BY ELECTRON PARAMAGNETIC RESONANCE

The production of Asc$^{\bullet -}$, a resonance stabilized tricarbonyl species, is easily observable by EPR since it is a relatively stable radical with a biologic half-life of 30–60 min at room temperature, but stable at freezing temperature (Ahola et al. 2004). This half-life is longer compared to other free radicals generated in biological systems (Buettner and Kiminyo 1992). Therefore, Asc$^{\bullet -}$ is directly detectable by EPR in aqueous solutions at room temperature (Buettner and Jurkiewicz 1993).

To determine the optimal experimental conditions for the detection of Asc$^{\bullet -}$ an ascorbic acid solution containing ferrous sulfate in an alkaline pH is prepared. EPR spectra obtained present a duplet with the following spectral parameters: $g = 2.005$ and $a_H = 1.8$ G, consistent with the EPR spectrum of Asc$^{\bullet -}$ (Fig. 36.2; Chapter 35). Pietri et al. (1990) optimized the EPR method for the detection of the Asc$^{\bullet -}$ radical at room temperature by the addition of dimethylsulfoxide (DMSO). The basic properties of DMSO lead to the kinetic stabilization of Asc$^{\bullet -}$, and DMSO has been used now for several years for the EPR analysis of Asc$^{\bullet -}$ in many biological systems (Pietri et al. 1994; Barbehenn et al. 2003).

Asc$^{\bullet -}$ signal intensity is a function of pH, temperature, catalytic metal concentration, oxygen concentration, and Asc concentration. Only when these variables are controlled can the intensity of the EPR signal of the Asc$^{\bullet -}$ in steady-state concentration serve as a marker for the oxidative stress in a system (Buettner and Jurkiewicz 1993). However, in biological systems it is very difficult to control all these variables, especially the total Asc content, which can be modified by the pro-oxidant conditions in a given homogenate or cellular preparation. In such cases, the Asc$^{\bullet -}$ content is not applicable as a sole indicator of oxidative stress.

Asc$^{\bullet -}$/Asc RATIO IN AQUATIC SYSTEMS

Examples from Algae and Marine Invertebrates

Asc$^{\bullet -}$/Asc ratio was established by measuring Asc$^{\bullet -}$ and Asc content, in different systems and under different conditions. For Asc$^{\bullet -}$ quantification a Bruker ECS 106 spectrometer has been used. Homogenates were prepared in pure DMSO (1:3) and the spectra scanned at 20°C under the following conditions: 50 kHz field modulation, microwave power 20 mW, modulation

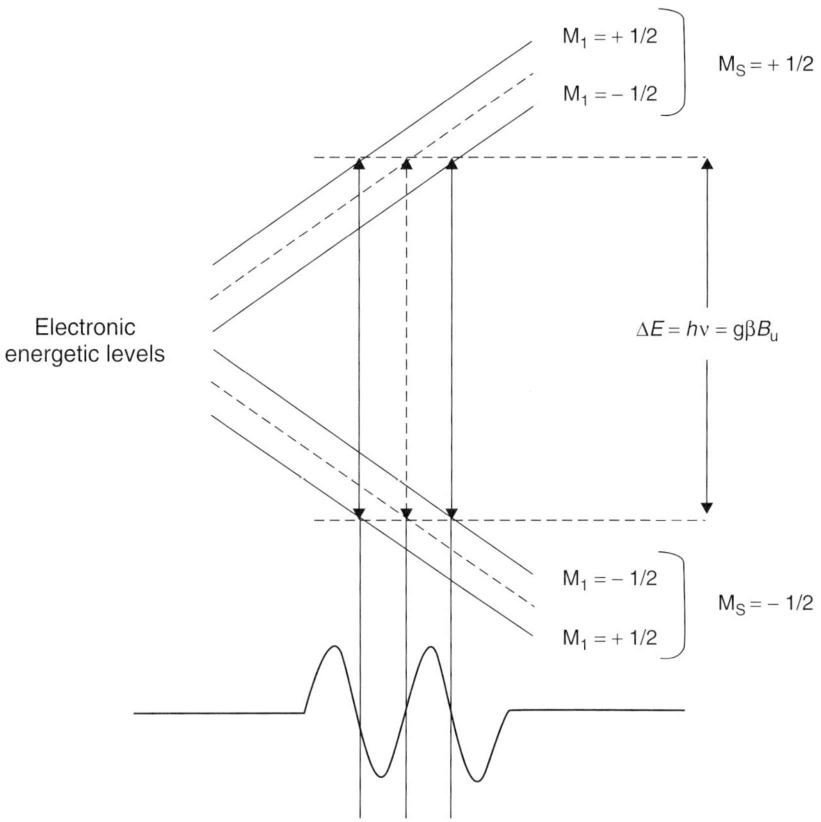

Fig. 36.2 Typical EPR spectrum corresponding to 2 mM solution of ascorbic acid in phosphate buffer solution (pH 7.8) with 100 μM EDTA and 10 μM ferrous sulfate. Spectral parameters, $g = 2.005$ and $a_H = 1.8$ G were observed (Malanga et al., unpublished data).

amplitude 1 G, time constant 655 ms, receiver gain 1×10^5, microwave frequency 9.81 GHz, and scan rate 0.18 G s^{-1} (Giulivi and Cadenas 1993). Quantification was performed according to Kotake et al. (1996). Figure 36.3 shows the EPR signal characteristic of $Asc^{\bullet-}$ for (A) algal cells, (B) isolated digestive glands of the limpet *Nacella Patinigera magellanica*, and (C) gonads from the Chilean sea urchin *Loxechinus albus*. Both animals were sampled from the Beagle Channel. The signals exhibit a strong EPR doublet with the following spectral features, $a_H = 1.8$ G, and $g = 2.005$. When DMSO itself was examined, no DMSO spin adduct was observed (Fig. 36.3D). Asc content was measured by reverse phase HPLC with electrochemical detection. Cells or tissue samples were homogenized in metaphosphoric acid 10% (w/v) (1:9 w/v) according to Kutnink et al. (1987).

In our first example, the $Asc^{\bullet-}$/Asc ratio was used for the assessment of the oxidative stress associated with the growth of intact cells of the green alga *Chlorella vulgaris* under laboratory conditions. Stock cultures of *C. vulgaris* were grown in Bold's basal medium, supplemented with 1.5 g L^{-1} glucose (Bold and Wynne 1978). Cells were grown at 20–25°C under light/dark cycles of 12:12 h. The irradiance at the surface of the culture was approximately 38 W m^{-2} of photosynthetically active radiation (Philips 40-W daylight-fluorescent light). The measurements of $Asc^{\bullet-}$ revealed no difference between log and stationary growth phase. During the physiological development of the cell culture a significant increase in Asc content was observed in the stationary phase as compared to the log growth phase (log phase: 1.4 ± 0.2 and stationary phase: 1.8 ± 0.2 nmol 10^{-7} cell). A very mild and indeed nonsignificant increase in the $Asc^{\bullet-}$/Asc ratio from $(5.7 \pm 0.8) \times 10^{-3}$ in the log phase to $(6.1 \pm 0.5) \times 10^{-3}$ arbitrary units (AU) in the stationary phase indicates no oxidative stress to occur during growth of *C. vulgaris* under the applied culture conditions (Malanga et al. 2001).

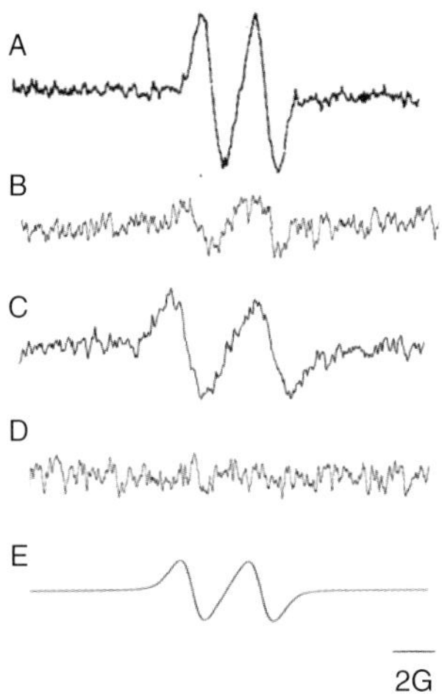

Fig. 36.3 EPR detection of Asc$^{\bullet-}$ in aquatic systems: (A) *Chlorella vulgaris* cells in log phase of growth; (B) *Nacella (Patinigera) magellanica* digestive gland; (C) gonads from *Loxechinus albus*; (D) a spectra of DMSO itself; (E) computer-simulated EPR spectra exhibiting hyperfine splittings that are characteristic for Asc$^{\bullet-}$ (Malanga et al., unpublished data).

Excess Fe (II) leads to an increase in $O_2^{\bullet-}$ formation and to oxidative stress. Measuring the Asc$^{\bullet-}$/Asc ratio, we further assessed the pro-oxidant effect of Fe-supplementation (500 μM) on *C. vulgaris* cultures. Quantification of the EPR signals indicated that Asc$^{\bullet-}$ was significantly increased in cells supplemented with 500 μM Fe as compared to cells grown at 90 μM Fe (Table 36.1). The study confirmed Fe to be a major catalyst of oxidative stress in algal cultures (Estevez et al. 2001). This may have implications in the context of Fe fertilization in oligotropic and "anaemic" open oceanic waters (Marchetti et al. 2009).

Further, two interesting examples of natural oxidative stress conditions in marine invertebrates were investigated in our laboratory using the Asc$^{\bullet-}$/Asc ratio as stress indicator. Digestive glands of intertidal limpets *Nacella (P.) magellanica* from the Beagle Channel were investigated for the effect of the natural oxidative stress in response to seasonal change (Malanga et al. 2007). Digestive glands are directly affected by the animal's feeding status and also a major target for oxidative disruption related to environmental stress and pollution (Livingstone 1991; Malanga et al. 2004). The Asc$^{\bullet-}$/Asc ratio indicated significantly (4.9 fold) lower oxidative stress levels in digestive glands of intertidal limpets during winter as compared to the rest of the year (Table 36.2). The results highlight the need for taking into account the seasonal variations of oxidative stress conditions in invertebrates when analyzing other stress effects, such as contamination.

Another seasonal survey was carried out with the sea urchin *Loxechinus albus* from the Beagle Channel. In this case, the oxidative status of the gonads was analyzed. The Asc$^{\bullet-}$ EPR spectrum was lower in gonads isolated from sea urchins in spring and summer, as compared to winter (Malanga et al. 2009). The Asc$^{\bullet-}$ content, assessed by quantification of EPR signals, decreased by 64% in samples collected in summer as compared to winter values. Asc$^{\bullet-}$ values from two subsequent winter samplings (2005 and 2006) did not differ beyond the levels of interindividual variability. The data shown in Table 36.3 indicate that Asc content remained unchanged in gonads of *L. albus* collected in winter, spring, and summer, but significantly decreased in gonads from sea urchins collected in fall. The low values of the Asc$^{\bullet-}$/Asc ratio in summer compared to winter-collected sea urchin gonads can be related to the low level of maturation in summer

Table 36.1 Effect of Fe supplementation to the incubation medium[1] on cellular Asc$^{\bullet-}$/Asc ratio.

Fe in the medium (μM)	**Asc$^{\bullet-}$ (pmol 10^7 cell^{-1})**	**Asc (nmol 10^7 cell^{-1})**	**Asc$^{\bullet-}$/Asc (10^3 AU)**
5	5 ± 1	1.9 ± 0.3	3 ± 1
90	9 ± 2*	1.3 ± 0.2	5 ± 1
500	20 ± 6*	2.5 ± 0.4*	10 ± 3*

[1]Algal cultures (2×10^5 cells mL$^{\bullet-1}$) were added with Fe starting day 0 of growth, and were assayed on day 12 (log phase of growth). Data are expressed as means ± SE of 4–6 independent experiments, with two replicates in each experiment. (Summarized data from Estevez et al. 2001.)
AU stands for arbitrary units.
*Significantly different, at $p \leq 0.05$, from values for cells with 5 μM Fe added (Statview SE+Graphics, v 1.03, Abacus Concepts Inc, Berkeley, CA).

Table 36.2 $Asc^{\bullet -}$/Asc ratio in digestive glands of *Nacella (P.) magellanica* over the year.[1]

Season	$Asc^{\bullet -}$ (pmol mg^{-1} FW)	Asc (pmol mg^{-1} FW)	$Asc^{\bullet -}$/Asc (10^5 AU)
Winter$_1$	1.1 ± 0.6	30 ± 3	3.7 ± 0.2
Spring	$5 \pm 1^*$	$15 \pm 2^*$	$32 \pm 5^*$
Summer	$3 \pm 1^*$	$17 \pm 2^*$	$18 \pm 5^*$
Fall	$5 \pm 2^*$	$11 \pm 3^*$	$45 \pm 6^*$
Winter$_2$	1.7 ± 0.4	28 ± 4	6 ± 2

[1]Data are expressed as means ± SEM of six independent experiments. (Summarized data from Malanga et al. 2007.)
FW stands for fresh weight.
AU stands for arbitrary units.
*Significantly different at $P \leq 0.05$ from values in winter$_1$ samples (Statview for Windows, ANOVA, SAS Institute Inc., v 5.0).

Table 36.3 $Asc^{\bullet -}$/Asc in *Loxechinus albus* over the year period.[1]

Season	$Asc^{\bullet -}$ (pmol mg^{-1} FW)	Asc (pmol mg^{-1} FW)	$Asc^{\bullet -}$/Asc (10^5 AU)
Winter$_1$	0.3 ± 0.1	16 ± 2	16 ± 1
Spring	0.12 ± 0.04	17 ± 3	13 ± 1
Summer	$0.09 \pm 0.02^*$	17 ± 3	$6.5 \pm 0.4^*$
Fall	0.16 ± 0.05	$4 \pm 1^*$	38 ± 11
Winter$_2$	0.4 ± 0.1	11 ± 2	24 ± 8

[1]Data are expressed as means ± SEM of six independent experiments. (Summarized data from Malanga et al. 2009.)
FW stands for fresh weight.
AU stands for arbitrary units.
*Significantly different at $p \leq 0.05$ (Kruskal–Wallis, non-parametric ANOVA) from values in winter$_1$ samples.

when the sea urchins feed to store reserve intermediates for gametogenesis, which starts in fall. The $Asc^{\bullet -}$/Asc ratio was indeed significantly increased in autumn, indicating that the state of gamete maturation has a significant effect on the natural levels of oxidative stress in gonad tissues.

Examples from Fish

Lattuca et al. (2009) report data on *in-situ* $Asc^{\bullet -}$/Asc ratio in gills and liver of silverside, *Odontesthes nigricans*, from the Beagle Channel. The gill was tested because it is in direct contact with the oxygenated surrounding seawater, and the liver is a strongly vasculated organ with a high organ-specific metabolic rate. Thus, both tissues are prone to suffer some basal oxidative stress. $Asc^{\bullet -}$ content did not differ significantly between gills and liver in *O. nigricans*, but the Asc content was 12 ± 2 nmol mg^{-1} FWT in the gills and 159 ± 28 nmol mg^{-1} FWT in the liver. Thus, the ratio $Asc^{\bullet -}$/Asc differed significantly between organs, $(6 \pm 2) \times 10^{-5}$ and $(5 \pm 2) \times 10^{-6}$ AU for the gills and the liver, respectively. This shows that individual organs within one organism also feature different levels of basal radical production and, therefore, different oxidative stress levels.

CONCLUDING REMARKS

The $Asc^{\bullet -}$/Asc ratio, but not the $Asc^{\bullet -}$-EPR signal alone, is a valid proxy for oxidative stress in aquatic plants and animal tissues. An increase of the $Asc^{\bullet -}$/Asc ratio under natural and experimentally induced stress conditions provides an early and simple diagnostic tool in the investigation of aquatic organisms. EPR measurements with tissue homogenates provide

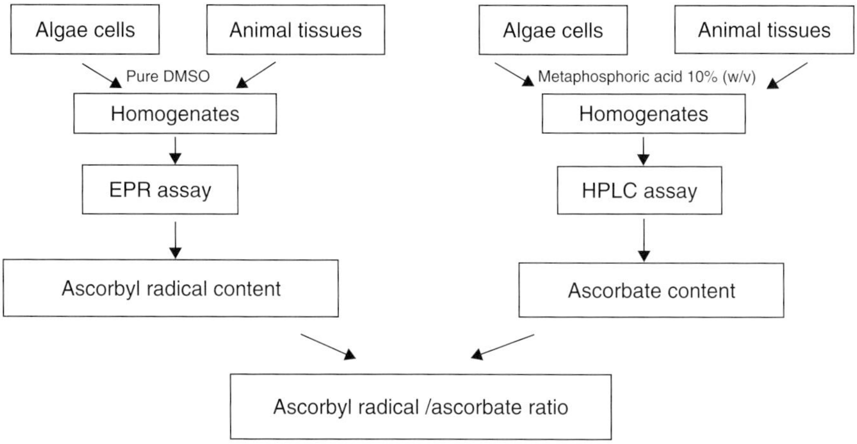

Fig. 36.4 Experimental design to estimate the ascorbyl radical/ascorbate ratio as an oxidative stress indicator.

important advantages over the frequently used in vitro assays, because this limits the formation of ROS artifacts during time-consuming sample preparation. Moreover, the measurement of the Asc$^{\bullet-}$ content is performed at room temperature avoiding more difficult procedures required for low-temperature studies. Thus, EPR spectroscopy in conjunction with HPLC detection of the Asc content can be considered as a powerful tool for detection of the initial stages of oxidative stress in the cellular hydrophilic compartment. Drawbacks of the approach are mainly the high costs and the difficult access to an EPR laboratory not available in many institutes. Furthermore, it is necessary to combine EPR detection of the Asc$^{\bullet-}$ with other biochemical techniques, such as HPLC (Fig. 36.4). However, we enthusiastically encourage and recommend young researchers to use and test this methodology in different types of biological marine materials, such as cell suspensions, tissues, or organs to further develop the method and fully appreciate its values and limitations.

REFERENCES

Ahola, T., Fellman, V., Kjellmer, I., Raivio, K.O., Lapatto, R. (2004) Plasma 8-isoprostane is increased in preterm infants who develop bronchopulmonary dysplasia or periventricular leukomalacia. *Pediatric Research* 56, 88–93.

Barbehenn, V.R., Poopat, U., Spencer, B. (2003) Semiquinone and ascorbyl radicals in the gut fluids of caterpillars measured with EPR spectrometry. *Insect Biochemistry and Molecular Biology* 33, 125–130.

Bold, H.C., Wynne, M.J. (1978) Cultivation of algae in the laboratory. *In* McElroy, W.D., Swanson, C.P. (eds). *Introduction to the Algae Structure and Reproduction*. Prentice Hall, Englewood Cliffs, NJ, 571 pp.

Buettner, G.R. (1993) The pecking order of free radicals and antioxidants: lipid peroxidation, alpha-tocopherol, and ascorbate. *Archives of Biochemistry and Biophysics* 300, 535–543.

Buettner, G.R., Kiminyo, K.P. (1992) Optimal EPR detection of weak nitroxide spin adduct and ascorbyl free radical signals. *Journal of Biochemical and Biophysical Methods* 24, 147–151.

Buettner, G.R., Jurkiewicz, B.A. (1993) Ascorbate free radical as a marker of oxidative stress: an EPR study. *Free Radical Biology and Medicine* 14, 49–55.

Estevez, M.S., Malanga, G., Puntarulo, S. (2001) Iron-dependent oxidative stress in *Chlorella vulgaris*. *Plant Science* 161, 9–17.

Giulivi, C., Cadenas, E. (1993) The reaction of ascorbic acid with different heme iron redox states of myoglobin. *FEBS Letters* 332, 287–290.

Hubel, C. A., Kagan, V.E., Kosin, E.R., McLaughlin, M., Roberts, J.M. (1997) Increased ascorbate radical formation and ascorbate depletion in plasma from women with preeclampsia: implications for oxidative stress. *Free Radical Biology and Medicine* 23, 597–609.

Kotake, Y., Tanigawa, T., Tanigawa, M., Ueno, I., Allen, D. R., Lai, C. (1996) Continuous monitoring of cellular nitric oxide generation by spin trapping with an iron-dithiocarbamate complex. *Biochimica et Biophysica Acta* 1289, 362–368.

Kutnink, M. A., Hawkes, W. C., Schaus, E. E., Omaye, S.T. (1987) An internal standard method for the unattended high performance liquid chromatographic analysis of ascorbic acid in blood components. *Analytical Biochemistry* 166, 424–430.

Lattuca, M. E; Malanga, G., Aguilar Hurtado, C., Pérez, A. F., Calvo, J., Puntarulo, S. (2009) Main features of the oxidative metabolism in gills and liver of *Odontesthes nigricans* Richardson (Pisces, Atherinopsidae). *Comparative Biochemistry and Physiology B.* 154, 406–411.

Livingstone, D.R. (1991) Organic xenobiotic metabolism in marine invertebrates. *Advances in Comparative Environmental Physiology* 7, 45–185.

Malanga, G., Juarez A. B., Albergheria, J. S., Veléz, C. G., Puntarulo S. (2001) Efecto de la radiación UVB sobre el contenido de ascorbato y radical ascorbilo en algas verdes. *In* Alveal, K., Antezana, T. (eds). *Sustentabilidad de la Biodiversidad, Un problema actual, bases científico-técnicas, teorizaciones y proyecciones*. Universidad de Concepción-Chile, pp. 389–398.

Malanga, G., Estevez, M. S., Calvo, J., Puntarulo, S. (2004) Oxidative stress in limpets exposed to different environmental conditions in the Beagle Channel. *Aquatic Toxicology* 69, 299–309.

Malanga, G., Estevez, M. S., Calvo, J., Abele, D., Puntarulo, S. (2007) The effect of seasonality on oxidative metabolism in *Nacella (Patinigera) magellanica*. *Comparative Biochemistry and Physiology A* 146, 551–558.

Malanga, G., Perez, A., Calvo, J., Puntarulo, S. (2009) The effect of seasonality on oxidative metabolism in the sea urchin *Loxechinus albus*. *Marine Biology* 156, 763–770.

Marchetti, A., Parker, M.S., Moccia, L.P. et al. (2009) Ferritin is used for iron storage in bloom-forming marine pennate diatoms. *Nature* 457, 467–471.

Martell, A.E. (1982) Chelates of ascorbic acid formation and catalytic properties. *In* Sieb, P.A., Tolbert, B.M. (eds) *Ascorbic Acid: Chemistry Metabolism and Uses*. American Chemical Society, Washington, DC, pp. 153–178.

Pietri, S., Culcasi, M., Stella, L., Cozzone, P.J. (1990) Ascorbyl free radical as a reliable indicador of free-radical-mediated myocardial ischemic and post-ischemic injury a real-time continuous-flow ESR study. *European Journal of Biochemistry* 193, 845–854.

Pietri, S., Séguin, J. R., D'Arbigny, P., Culcasi, M. (1994) Ascorbyl free radical: A noninvasive marker of oxidative stress in human open-heart surgery. *Free Radical Biology and Medicine* 16, 523–528.

Roginsky, V. A., Stegmann, H.B. (1994) Ascorbyl radical as natural indicator of oxidative stress: quantitative regularities. *Free Radical Biology and Medicine* 17, 93 103.

Vergely, C., Maupoil, V., Clermont, G., Bril, A., Rochette, L. (2003) Identification and quantification of free radicals during myocardial ischemia and reperfusion using electron paramagnetic resonance spectroscopy. *Archives of Biochemistry and Biophysics* 420, 209–216.

Chapter 37

EVALUATION OF OXIDATIVE DNA DAMAGE IN AQUATIC ANIMALS: COMET ASSAYS AND 8-OXO-7,8-DIHIDRO-2′-DEOXYGUANOSINE LEVELS

José Pedro Friedmann Angeli[1], *Glaucia Regina Martinez*[2], *Flávia Daniela Motta*[1], *Eduardo Alves de Almeida*[3], *Marisa Helena Gennari de Medeiros*[1], *and Paolo Di Mascio*[1]

[1]Departamento de Bioquímica, Instituto de Química, Universidade de São Paulo, São Paulo, SP, Brazil
[2]Biologia Molecular, Setor de Ciências Biológicas, Universidade Federal do Paraná, Curitiba, PR, Brazil
[3]Department of Chemistry and Environmental Sciences, IBILCE-UNESP, São José do Rio Preto, SP, Brazil

The analysis of DNA alterations in aquatic organisms is a highly suitable method for evaluating the level of genotoxic contamination of marine and freshwater environments. It enables the detection of exposure to low concentrations of contaminants in all eukaryotic cells. The advantage of using these methods is that genotoxic lesions can be detected and quantified without a detailed knowledge of the identity and the physical/chemical properties of the contaminants present. Analysis of DNA damage helps to assess the actual state of contamination and impact in aquatic (and also terrestrial) environments.

Environmental parameters/agents that increase the levels of cellular reactive oxygen species (ROS) formation are mainly ultraviolet radiation (UVR) and chemical toxic pollutants, capable of redox cycling.

Oxidative Stress in Aquatic Ecosystems, First Edition. Edited by Doris Abele, José Pablo Vázquez-Medina, and Tania Zenteno-Savín.

Transition metals, quinones, dyes, bipyridyl, herbicides, and aromatic nitro compounds are known for their redox-cycling properties and their potential to cause oxidative stress (Kappus and Sies 1981; Valavadinis et al. 2006, see Chapters 21 and 22). ROS play a critical role on the biological effects promoted by these contaminants (Ames et al. 1993) as they interact with cellular biomolecules causing oxidative damage which, especially DNA damage, can have potentially serious consequences for the cell and the organism if not rapidly repaired.

DNA oxidation is the most common (and most studied) form of DNA damage. It is mainly induced directly and indirectly by ROS such as hydroxyl radical ($HO^•$), hydrogen peroxide (H_2O_2), singlet molecular oxygen ($^1O_2$), nitric oxide ($NO^•$), and peroxynitrite ($ONOO^-$), alcoxyl ($RO^•$) and peroxyl ($ROO^•$) radicals. A great variety of oxidized bases have been identified in DNA exposed to an oxidative insult, but 8-oxo-7,8-dihydro-2′-deoxyguanosine (8-oxodGuo) is the most abundant and readily formed, mainly because of the high oxidation potential of guanine. The presence of 8-oxodGuo in DNA may lead to misincorporation of adenine opposite to this lesion unless repair occurs prior to DNA replication (Collins 2009). Figure 37.1 depicts the mechanism of 8-oxodGuo and other lesions formed through oxidation of the guanine moiety by $^1O_2$, $HO^•$, and other one-electron oxidants. $^1O_2$ adds across the 4–8 bond generating an unstable endoperoxide (**3**), which in turn decomposes generating several products, including 8-oxodGuo (**5**) and 4R* and 4S* diastereomers of spiroiminodihydantoin nucleosides (**8**).

The one-electron oxidation process of guanine was divided in two parts for better illustration. After oxidation in aqueous solution, the cation radical of guanine (**2**) may proceed by two competing pathways, the nucleophilic addition and the deprotonation reaction. In the nucleophilic addition, hydration takes place after the initial oxidation generating the reducing 8-hydroxy-7,8-dihydro-7-yl radical (**9**). This radical can also be generated by direct attack of $HO^•$. This radical can then rearrange to form (**5**) in the presence of oxygen, or 2,6-diamino-4-hydroxy-5-formamidopyrimidine (**10**) under anaerobic conditions. In the deprotonation pathway, the radical cation (**2**) undergoes a deprotonation giving rise to the highly oxidizing guanine(−H)$^•$ radical (**11**), the reaction of (**11**) and its tautomer (**12**) proceed very slowly with oxygen ($k < 10^3$), whereas both react promptly with $O_2^{•-}$ ($k \sim 10^9$). Subsequently, the nucleoside undergoes a complicated and not very well understood pathway that generates oxazolone (**17**) (Cadet et al. 2006a, b, 2008). Another important characteristic of (**5**) is that it is more readily oxidized than (**1**). One of the known mechanisms for (**5**) consumption is through the reaction with the guanine radical cation (**11**) generating (**8**), although there is still much debate regarding the importance of this reaction under *in vivo* conditions (Misiaszek et al. 2004).

In the past decade considerable effort has been applied to establish the best method for the detection of oxidative DNA damage. The widespread use of measurements of 8-oxodGuo can be attributed mainly to the easiness with which it can be measured by high performance liquid chromatography (HPLC). Nevertheless, the background level of 8-oxodGuo can vary by two to three orders of magnitude depending on the method used. This is attributed to artifact formation during DNA isolation and hydrolysis (ESCODD 2002). During the ESCODD trial it was also stated that the worst method for 8-oxodGuo measurements was gas chromatograph–mass spectrometry (GC–MS), since the method is prone to generate high amounts of artificial damage. Another widely applied method is the modified version of the comet assay with enzymatic detection (Collins and Dusinska, 2002). This technique uses bacterial DNA repair endonuclease, namely formamidopyrimidine DNA N-glycosylase (FPG) and endonuclease III (ENDO III), two enzymes that recognize oxidized purines and pyrimidines, generating strand breaks that are detected by electrophoresis.

Both the modified version of the comet assay and the chromatographic separation with electrochemical detection are methods of choice for detection of DNA oxidative status. However, both assays also present some drawbacks. On the positive side the modified version of the comet assay produces less background damage and is also of higher sensitivity, but on the negative side there is a lack of dose response at higher levels of damage. This can be explained by the assay characteristics since addition of extra damage to bases already present in the tail will not lead to further increase in DNA migration. The chromatographic method using electrochemical detection (ECD) is less sensitive with a higher background but presents a good dose–response relationship.

The aim of this chapter is to present the two most widely applied methods for DNA oxidative damage detection, the comet assay and HPLC–ECD.

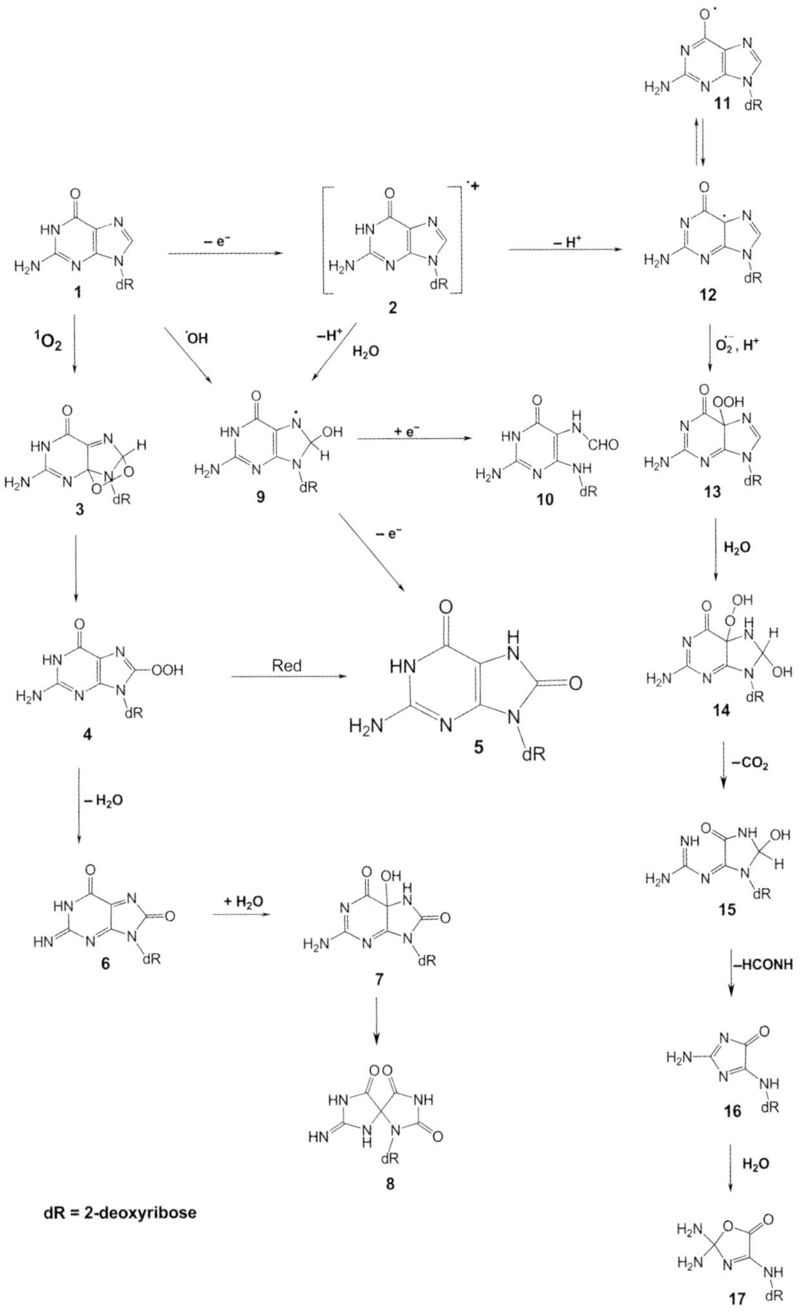

Fig. 37.1 Schematic representation of possible pathways involved in dGuo oxidation.

COMET ASSAY

The single cell gel electrophoresis (SCGE) or comet assay is a versatile and sensitive method for measuring single- (SSB) and double-strand breaks (DSB) in DNA. Comets are formed by the relaxation of DNA supercoiling in a structural loop of the DNA by a single DNA break release, allowing this loop to migrate to the anode under eletrophoretic conditions (Collins, 2009). Besides its ability to detect apurinic sites, SSB, and DSB, the comet assay can also be used to measure oxidized bases. This is achieved with an additional step that relies on the addition of bacterial endonucleases following cell lysis; these enzymes are able to generate a strand break in places where an oxidative modification took place.

The comet assay (Fig. 37.2) was first applied in ecotoxicology about 15 years ago and since then has become one of the most popular tests for detecting DNA strand breaks in aquatic animals following *in vitro*, *in vivo*, and *in situ* exposure to toxicants or natural stress. For a comprehensive review of this method we suggest reading Mitchelmore and Chipman (1998) as well as Lee and Steinert (2003).

Protocol

Reagent Preparation

Lysis solution

Ingredients for 1000 mL (stock solution):

2.5 M NaCl	146.1 g
100 mM EDTA	37.2 g
10 mM Tris-HCl	1.2 g.

Fill up with approximately 890 mL of deionized water (dH_2O); adjust the pH to 10 with approximately 8 g of solid NaOH; complete the volume to 1000 mL with dH_2O; store at room temperature. Solution is stable for over a month.

Working solution (100 mL): Add 1 mL of Triton X-100, 10 mL of DMSO, and 89 mL of lysis stock solution; cool the solution to 4°C before use.

Electrophoresis solution

Preparation of stock solutions:

1. 10 N NaOH* (200 g 500 mL^{-1} dH_2O)
2. 200 mM EDTA (14.89 g 200 mL^{-1} dH_2O)
3. Store at room temperature.
   *Do not use if older than 10 days.

Working solution: To prepare 1000 mL of electrophoresis buffer (300 mM NaOH, 1 mM EDTA) add 5 mL of the 200 mM EDTA and 30 mL of the 10 N NaOH and fill up to 1000 mL with dH_2O, mix properly. Check that pH is >13. The electrophoresis solution must be freshly made, and the total volume depends on the electrophoresis chamber capacity.

Neutralization buffer

0.4 M Tris	48.5 g 950 mL^{-1} dH_2O

Add Tris to 950 mL of dH_2O, adjust pH to 7.5 with concentrated HCl (approximately 30 mL) and store at room temperature. Working solution must be prepared fresh.

Staining solution

- Ethidium bromide stock solution (10×, 200 μg mL^{-1}): 10 mg dissolved in 50 mL of dH_2O. Store this solution at room temperature. Stock solution is indefinitely stable.
- Ethidium bromide working solution: 1 mL of ethidium bromide stock solution and 9 mL of dH_2O. Filtrate this solution prior to use.

Ethidium bromide is extremely toxic because it intercalates with DNA, wear gloves, work under fume hood.

Phosphate buffer saline (PBS)

KCl	0.2 g
KH_2PO_4	0.2 g
NaCl	8.0 g
Na_2HPO_4	1.15 g.

Add weighted salts to 950 mL and adjust pH to 7.4. Complete volume to 1000 mL with dH_2O and store this solution at 4°C. Solution is stable for several months.

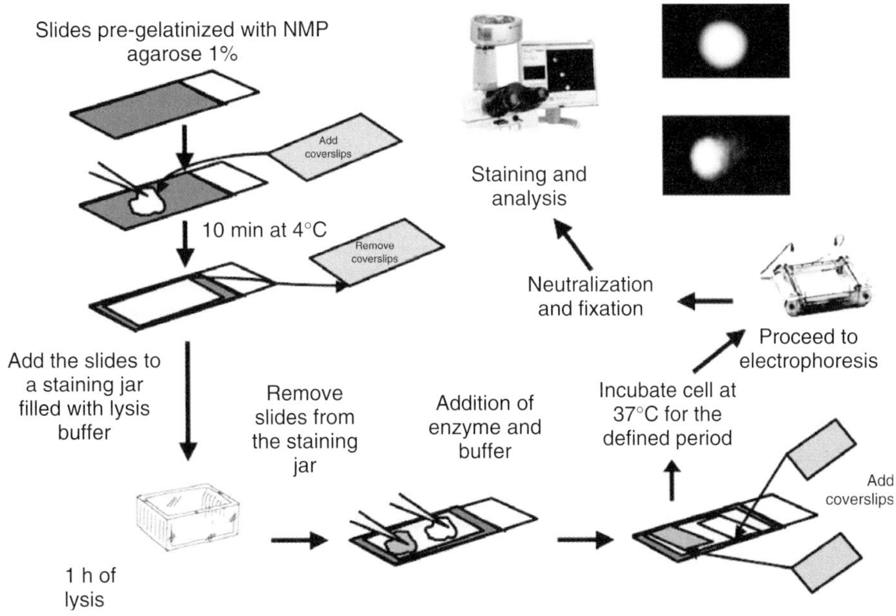

Fig. 37.2 Schematic protocol for comet assay procedure. See plate section for a color version of this image.

Enzyme reaction buffer for ENDO III and FPG

To a total volume of 100 mL dH_2O, add:

40 mM HEPES*	1.132 g
0.1 M KCl	0.74 g
0.5 mM EDTA	0.0146 g
0.02 mg mL^{-1} BSA	

*N-2-Hydroxyethylpiperazine-N′-2-ethanesulfonic acid.

Adjust pH to 8.0 with KOH. This solution can be prepared as a 10× stock solution and frozen at −20°C.

Recommendations for enzyme dilution and storage

The enzymes should be sorted into small aliquots (such as 2 μL) and stored at −80°C. This minimizes repeated freezing and thawing of the enzymes. The final dilution of the working solution may vary from batch to batch. Examples: (i) if ENDO III has a suggested dilution of 1000×, dilute a 2 μL aliquot with buffer to 2 mL and store in suitable aliquots (e.g. 300 μL of 1000× diluted ENDO III) at −80°C; (ii) FPG is less stable than ENDO III, and repeated freezing–thawing must be minimized. If the suggested dilution is 3000×, first dilute a 2 μL aliquot to 200 μL (100×) and store 10 μL aliquots at −80° C. **For enzyme dilutions, use enzyme reaction buffer containing 10% glycerol for ENDO and FPG**. For use in the assay, dilute a 10 μL aliquot to 300 μL of diluted enzyme (ENDO III) with buffer (no glycerol!) and use at once; do not refreeze (Collins and Dusinská 2002).

Slide Preparation

1. Prior to use, clean all slides with ethanol.
2. Prepare normal melting point agarose (NMP) 1.5% (300 mg in 20 mL PBS) and dissolve it by heating to 90°C in a water bath. Dip slides in the hot agarose (~ 60°C), remove excess agarose from one side of the slides and leave to dry at room temperature overnight in a horizontal position. These slides can be stored and used for several weeks.
3. Prepare 0.5% low melting point (LMP) agarose (100 mg in 20 mL PBS) by heating it in a microwave and keep it in a water bath at 37°C until use.
4. Add 120 μL of LMP agarose (37°C) to 5–15 μL of cell suspension. Place the coverslips on top and leave the slides at 4°C for 10 to 20 min. After applying the cells to the slide, avoid light exposure.

5. Coverslips must be removed gently and the slides must be accommodated in the electrophoresis chamber containing the lysis buffer and kept at 4°C for at least 1 h.

Enzyme treatment (endonuclease III, formamidopyrimidine glycosylase)

Prepare 300 mL of enzyme reaction buffer. Put aside 1 mL for enzyme dilutions. Wash slides three times with this buffer (4°C) in a staining jar keeping the solution for 5 min each time. Meanwhile, prepare dilutions of the enzymes. (***Note:*** The buffer in which the enzyme is stored contains β-mercaptoethanol to preserve the enzyme. However, inclusion of sulphydryl reagents in the reaction buffer would significantly increase background of DNA breakage.)

Remove slides from the last wash, and dab-off excess liquid with tissue paper. Place 50 μL of enzyme solution (or buffer alone, as control) onto slides, and cover with a 22 × 22 mm cover slip. Put slides into a moist box to prevent desiccation and incubate at 37°C for 45 min (ENDO III) or 30 min (FPG).

After incubation with the enzyme, the slides may be refrigerated at 4°C for about 10 min, since incubation at 37°C can lead to gel melting.

Electrophoresis and Staining

1. After the lysis, remove the coverslips gently in order to avoid gel detaching. Place the slides into the electrophoresis chamber (the slides can also be washed in electrophoresis solution in order to avoid foaming), and if necessary complete with clean slides.
2. Gently add the electrophoresis solution so that all the slides are covered with it. The electrophoresis chamber must be put in a receptacle filled with ice (this is done in order to avoid heating of the slides during electrophoresis).
3. Leave the slides in the alkaline solution for 20 to 60-min. In general, an incubation time of 30 min is recommended.
4. Start the electrophoresis setting the equipment at 25 V and 300 mA. These parameters can be adjusted by adding or removing electrophoresis solution. The time of electrophoresis may vary, but in general the time recommended is 20 min.
5. After electrophoresis is completed, carefully remove the slides watching out for possible gel detachment. Neutralize the slides with neutralization buffer by washing three times for 5 min (further washing helps to decrease possible backgrounds).
6. After completing neutralization, dry the slides and fix them with ethanol for 5 min. The slides can be stored in the refrigerator until analysis.
7. For staining, add 100 μL of ethidium bromide solution and cover it with a cover slip. After 5 min the slide is ready to be analyzed. The slides must be stained one per turn, and analyzed immediately.

Evaluation of DNA Damage

For visualization of DNA migration, the slides are analyzed under a 40× objective (400×) using a fluorescence microscope equipped with an excitation filter of 515–560 nm and a 590 long-pass filter of 590 nm. Usually 50 cells per replicate are analyzed (giving 100 cells analyzed). The evaluation can be done in three different ways:

1. Visual analysis of 50 cells by allocating them arbitrarily to five different categories: (0, no damage; 1, small; 2, medium damage; 3, high damage; 4, maximum damage). In order to obtain a score, multiply the number of cells in each class by its respective class number, thus obtaining a score ranging from 0 to 400.
2. Analysis of 50 cells using an ocular equipped with a calibration ruler. For each cell, the length of the tail is calculated in microns, calculating the mean tail length for each treatment.
3. Automated analysis equipped with a CCD camera and a specific program designed for comet evaluation.

8-oxodGuo MEASUREMENTS WITH HPLC-ECD

HPLC is widely used for measurements of 8-oxodGuo in biological samples of DNA. Since modifications occur at very low frequencies, the use of sensitive detectors coupled to HPLC is needed. Detectors used routinely for the detection include mass spectrometers (MS) but mainly ECD. In the case of ECD, coulometric detectors are the most widely applied and also the most sensitive, although amperometric detectors are also applied with success (Kaur and Halliwell 1996).

Protocol

Reagent Preparation

Buffer A (pH 7.5)

Ingredients for 500 mL:

320 mM sucrose	54.77 g
5 mM $MgCl_2$	0.508 g
10 mM Tris-HCl	0.605 g
0.1 mM desferroxamine	0.032 g
Triton X-100 (1%)	5 mL.

Buffer B (pH 8)

Ingredients for 1 L:

100 mM Tris-HCl	12 g
5 mM EDTA	1.85 g
0.15 mM desferroxamine	0.1 g.

RNAse T1 buffer (pH 7.4)

Ingredients for 200 mL:

10 mM Tris-HCl	0.24 g
1 mM EDTA	0.075 g
2.5 mM desferroxamine	0.33 g.

Preparation of enzyme solution: 20,000 U of RNAse T1 for 1 mL of RNAse T1 buffer.

RNAse A buffer (pH 5.2)

10 mM sodium acetate.

Preparation of enzyme solution: 10 mg of RNAse A for 1 mL of RNAse A buffer, heat the solution for 15 min at 100°C.

NaI buffer

Ingredients for 200 mL:

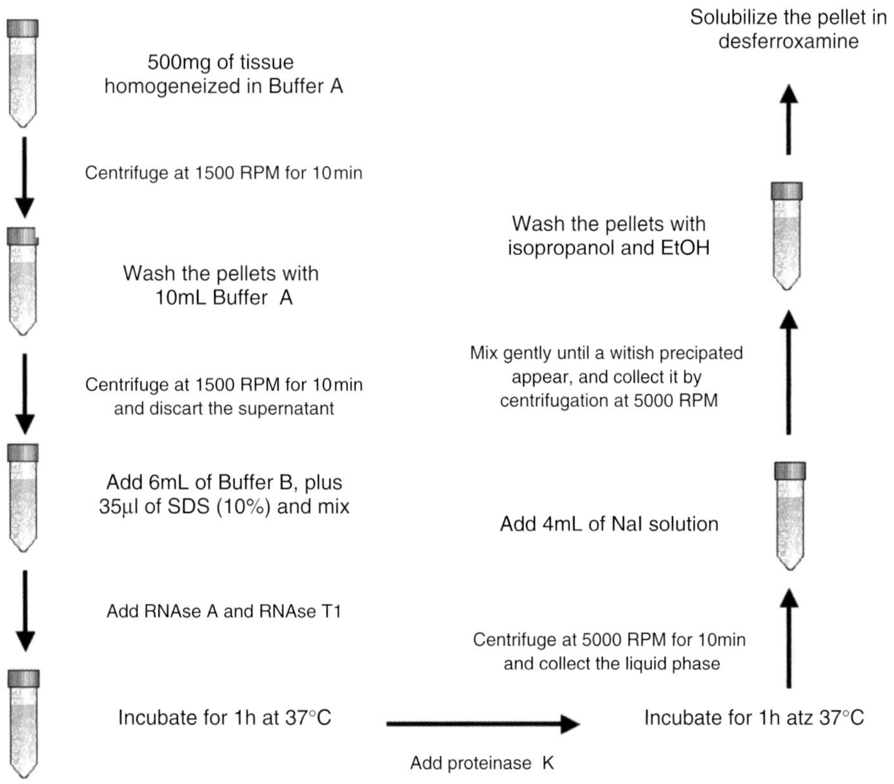
500mg of tissue homogeneized in Buffer A
Centrifuge at 1500 RPM for 10 min
Wash the pellets with 10mL Buffer A
Centrifuge at 1500 RPM for 10 min and discart the supernatant
Add 6mL of Buffer B, plus 35µl of SDS (10%) and mix
Add RNAse A and RNAse T1
Incubate for 1h at 37°C
Add proteinase K
Incubate for 1h atz 37°C
Centrifuge at 5000 RPM for 10min and collect the liquid phase
Add 4mL of NaI solution
Mix gently until a witish precipated appear, and collect it by centrifugation at 5000 RPM
Wash the pellets with isopropanol and EtOH
Solubilize the pellet in desferroxamine

Fig. 37.3 Schematic protocol for DNA extraction. See plate section for a color version of this image.

7.6 M NaI	227.8 g
40 mM Tris-HCl	0.971 g
20 mM EDTA	1.489 g
0.3 mM desferroxamine	0.004 g.

1 M sodium acetate buffer (pH 5)

Ingredients for 100 mL:

1 M CH_3COONa	8.20 g.

Tris-HCl 1M buffer, pH 7.4

Ingredients for 500 mL:

1 M Tris-HCl	60.54 g.

Phosphatase buffer, pH 7.0

Ingredients for 50 mL:

100 μM $ZnCl_2$	0.7 mg
5 mM $MgCl_2$	51 mg
5 mM Tris-HCl	30 mg
50% (v/v) glycerol	25 mL.

Extraction, Hydrolysis, and Quantification of DNA

For DNA isolation (Fig. 37.3), the chaotropic NaI method is the most common. Tissues (approximately 500 mg) are homogenized in 10 mL of a lysis solution (Buffer A). After centrifugation at 1500 *g* for 10 min, the pellets are resuspended in 10 mL of the lysis solution and centrifuged again at 1500 *g* for 10 min. The pellets are then suspended in 6 mL of Buffer B and 35 μL of 10% SDS (sodium dodecyl sulfate) are added. The

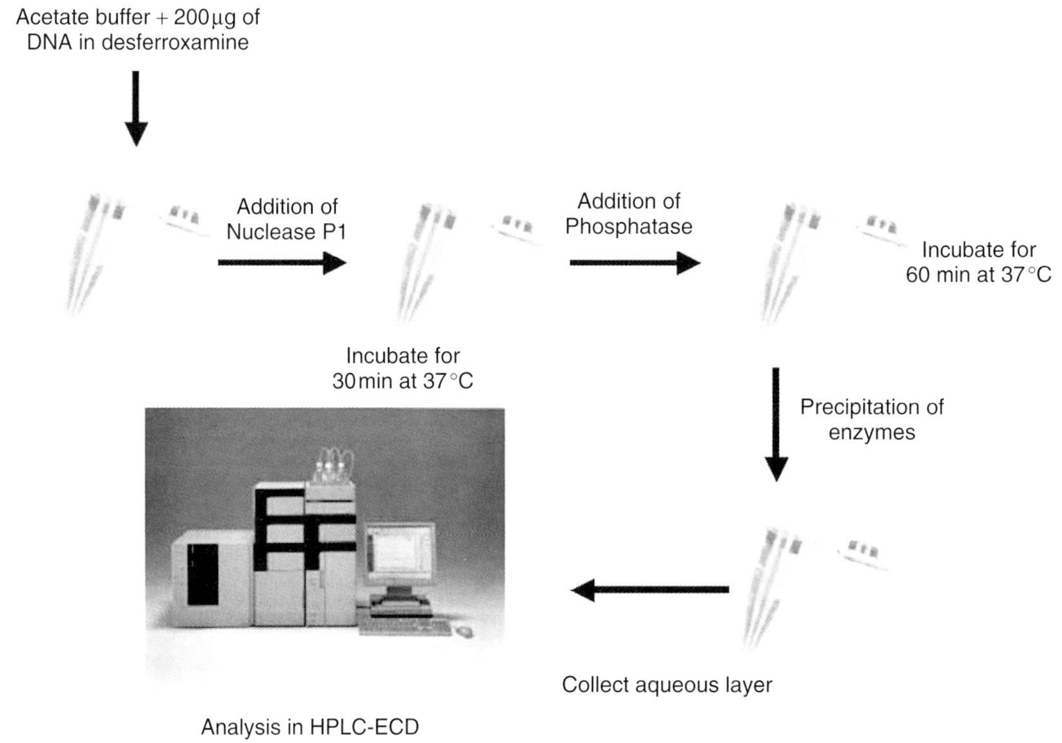
Acetate buffer + 200 μg of DNA in desferroxamine
Addition of Nuclease P1
Addition of Phosphatase
Incubate for 60 min at 37 °C
Incubate for 30 min at 37 °C
Precipitation of enzymes
Collect aqueous layer
Analysis in HPLC-ECD

Fig. 37.4 Schematic protocol for DNA hydrolysis.

enzymes RNase A (30 μL, 10 mg mL^{-1}) and RNase T1 (4 μL, 20 U μL^{-1}) are added, and the reaction mixture is incubated at 37°C. After 1 h, add 300 μL proteinase K (20 mg mL^{-1}) and incubate at 37°C for 1 h. After centrifugation at 5000 *g* for 15 min, the liquid phase is collected and 4 mL of a solution containing 7.6 M NaI buffer is added, followed by the addition of 8 mL of isopropanol. The content in the tube must be well mixed by inversion until a whitish precipitate appears. The precipitate is collected by centrifugation at 5000 *g* for 15 min and washed with 5 mL of isopropanol 60% (5000 *g*, 15 min), followed by 5 mL of ethanol 70% (5000 *g*, 15 min). The DNA pellet is dissolved in 500 μL of desferroxamine (0.1 mM), and the DNA concentration measured spectrophotometrically at 260 nm.

For DNA hydrolysis (Fig. 37.4), add 4 μL of 1 M sodium acetate buffer (pH 5) to an aliquot of 0.1 mM desferroxamine solution containing 200 μg of DNA and digest with two units of nuclease P1 at 37°C for 30 min. Then, add Tris-HCl 1 M buffer, pH 7.4 (12 μL), 12 μL of phosphatase buffer, and six units of alkaline phosphatase, adjust the final volume to 200 μL with dH_2O and incubate for 1 h at 37°C. The enzymes are then precipitated by the addition of one volume of chloroform and, after centrifugation at 1000 *g* for 5 min, the resulting aqueous layer is ready for HPLC–ECD analysis (100 μL of the DNA solution/injection).

Analysis of 8-oxodGuo with HPLC–ECD

The method described below is currently applied in our laboratory, researchers are encouraged to use it as a starting protocol and adjust it, if necessary, for use with their samples and equipment. For analysis of 8-oxodGuo, samples of 35 μg of hydrolyzed DNA are injected into the HPLC–ECD system (Figs 37.5 and 37.6), consisting of a Shimadzu model LC-10AD pump connected to a luna reversed-phase C18 column (250 mm ×4.6 mm i.d., 5 μm) (Phenomenex). The isocratic mobile phase is composed of 8% methanol in 25 mM potassium phosphate buffer pH 5.5. Flow rate is 1 mL min^{-1}. Use UV detection at 254 nm to follow the deoxynucleoside chromatogram, since 2′-deoxyguanosine should also be quantified to enable expression of the level of 8-oxodGuo as the number of lesions per 10^6 dGuo. Coulometric detection of the 8-oxodGuo lesion is achieved with a Coulchem III detector (ESA, Chelmsord, M A, USA). The potential of the two electrodes is set at 130 and 280 mV respectively. This kind of detector is composed of two cells in tandem (Fig. 37.5).

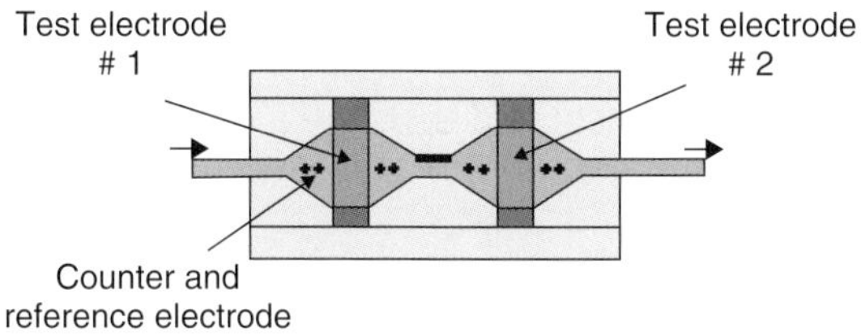

Fig. 37.5 Shematic presentation of the analytical cell of an electrochemical detector. The molar ratio of 8-oxodGuo to dGuo in each DNA sample is determined based on coulometric detection at 280 mV for 8-oxodGuo and absorbance at 254 nm for dGuo in each injection. A standard calibration curve of 8-oxodGuo (0.1–1 pmol) and dGuo (2–20 nmol) must be constructed for quantification. The results are expressed as 8-oxodGuo 10^{-6} dGuo. Figure 37.6 shows the typical chromatogram of a DNA sample.

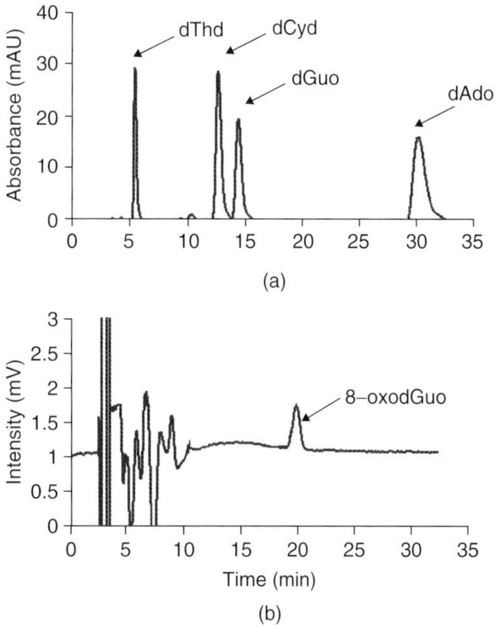

Fig. 37.6 Schematic chromatogram of 35 μg of DNA injected in to HPLC–EC. (a) Chromatogram for UV detection (λ = 254 nm) of DNA samples, showing 2′-deoxytimidine dThd, 2′-deoxycytidine dCyd, 2′-deoxyguanosine dGuo, and 2′-deoxyadenosine dAdo. (b) Chromatogram for electrochemical detection of the same sample, with 8-oxodGuo indicated.

REFERENCES

Ames, B.N., Shigenaga, M.K., Hagen, T.M. (1993) Oxidants, antioxidants and the degenerative diseases of aging. *Proceedings of the National Academy of Sciences* 90, 7915–7922.

Cadet, J., Douki, T., Ravanat, J.L. (2006a) One-electron oxidation of DNA and inflammation processes. *Nature Chemical Biology* 2, 348–349.

Cadet, J., Ravanat, J.L., Martinez, G.R., Medeiros, M.H.G., Di Mascio, P. (2006b) Singlet oxygen oxidation of isolated and cellular DNA: product formation and mechanistic insights. *Photochemistry and Photobiology* 82, 1219–1225.

Cadet, J., Douki, T., Ravanat, J.L. (2008) Oxidatively generated damage to the guanine moiety of DNA: mechanistic aspects and formation in cells. *Accounts in Chemical Research* 41, 1075–1083.

Collins, A.R. (2009) Investigating oxidative DNA damage and its repair using the comet assay. *Mutation Research* 681, 24–32.

Collins, A.R., Dusinská, M. (2002) Oxidation of cellular DNA measured with the comet assay. *Methods in Molecular Biology* 186, 147–159.

ESCODD (European Standards Committee on Oxidative DNA Damage). (2002) Comparative analysis of baseline 8-oxo-7,8-dihydroguanine in mammalian cell DNA, by different methods in different laboratories: an approach to consensus. *Carcinogenesis* 23, 2129–2133.

Kappus, H., Sies, H. (1981) Toxic drug effects associated with oxygen metabolism: redox cycling and lipid peroxidation. *Experientia* 37, 1233–1241.

Kaur, H., Halliwell, B. (1996) Measurment of oxidizes and methylated DNA bases by HPLC with electrochemical detection. *Biochemical Journal* 318, 21–23.

Lee, R.F., Steinert, S. (2003) Use of the single cell gel electrophoresis/comet assay for detecting DNA damage in aquatic (marine and freshwater) animals. *Mutation Research* 544, 43–64.

Misiaszek, R., Crean, C., Joffe, A., Geacintov, N.E., Shafirovich, V. (2004) Oxidative DNA damage associated with combination of guanine and superoxide radicals and repair mechanisms via radical trapping. *Journal of Biological Chemistry* 279, 32106–32115.

Mitchelmore, C.L., Chipman, J.K. (1998) DNA strand breakage in aquatic organisms and the potential value of the comet assay in environmental monitoring. *Mutation Research* 399, 165–147.

Valavanidis, A., Vlahogianni, T., Dassenakis, M., Scoullos, M. (2006) Molecular biomarkers of oxidative stress in aquatic organisms in relation to toxic environmental pollutants. *Ecotoxicology and Environmental Safety* 64, 178–189.

Chapter 38

EVALUATION OF DNA ADDUCTS FORMED BY LIPID PEROXIDATION BY-PRODUCTS

Camila Carrião Machado Garcia[1], *José Pedro Friedmann Angeli*[1], *Eduardo Alves de Almeida*[2], *Marisa Helena Gennari de Medeiros*[1], *and Paolo Di Mascio*[1]

[1]Departamento de Bioquímica, Instituto de Química, Universidade de São Paulo, São Paulo, SP, Brazil

[2]Department of Chemistry and Environmental Sciences, IBILCE-UNESP, São José do Rio Preto, SP, Brazil

Reactive oxygen species (ROS) can be produced by endogenous sources such as aerobic metabolism, inflammation, or exposure to a variety of chemical and physical agents. Cell membranes are potential targets of ROS/RNS and enzymes that initiate an autocatalytic process generating hydroperoxides (LOOH) (Schneider 2009). Not detoxified LOOH can participate in reactions that enhance its toxicity forming LOO• and LO•. These lipid radicals can propagate the oxidation process to other cellular compartments and also undergo fragmentation of the parental molecule generating ketones, epoxides, and aldehydes (Terao and Matsushita 1977; Esterbauer 1993, Almeida et al. 2006; Schneider 2009).

Lipid peroxidation products, such as malonaldehyde (MDA), 4-hydroxy-2-nonenal (HNE), 1,4-hydroperoxy-(2*E*)-nonenal (HPNE), 4-oxo-(2*E*)-nonenal (ONE), 9,12-dioxo-(10*E*)-dodecenoic acid (DODE), 5,8-dioxo-(10*E*)-octenoic acid (DOOE), 2,4-decadienal (DDE), 4,5-epoxy-(2*E*)-decenal (EDE), hexenal (HE), acrolein (ACL), and crotonaldehyde (CRT) (Fig. 38.1), can covalently modify lipids, proteins, and DNA. As for DNA, the nucleophilic groups present in purine and pyrimidine bases are altered by electrophilic sites forming adducts. The four DNA bases, 2′-deoxyguanosine (dGuo), 2′-deoxyadenosine (dAdo), 2′-deoxycytidine (dCyd), and thymidine (dThd) have multiple locations that can be attacked, one

Oxidative Stress in Aquatic Ecosystems, First Edition. Edited by Doris Abele, José Pablo Vázquez-Medina, and Tania Zenteno-Savín.

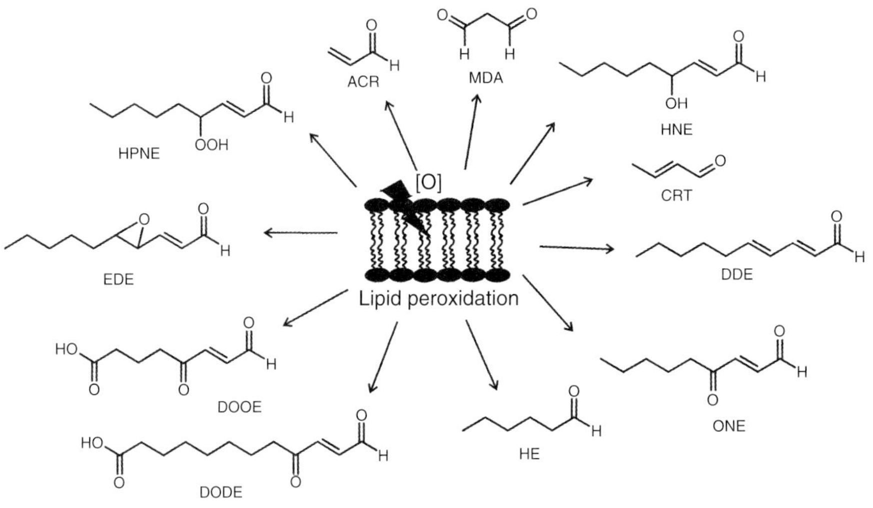

Fig. 38.1 Structure of some lipid peroxidation by-products.

of the most reactive atoms in DNA bases are the nitrogen 7 of dGuo (reviewed by Himmelstein et al. 2009). The reaction of α, β unsaturated aldehydes (acrolein, crotonaldehyde, and HNE) with nucleophilic sites in DNA bases, results in the formation of cyclic propano adducts (Winter et al. 1986; Blair 2008; Medeiros 2009). In addition, epoxidized unsaturated aldehydes can generate ethano or etheno adducts after reacting with DNA (Golding et al. 1996; Blair 2008; Medeiros 2009).

The amount of each product formed during lipid peroxidation can vary with membrane composition and fatty acids content. HNE is one of the best studied lipid peroxidation by-products. It is formed in abundance and presents high genotoxicity. HNE can react with DNA generating 1,N^2-propanodeoxyguanosine (1,N^2-propanodGuo) (Fig. 38.2) as a major product (Minko et al. 2009). Among the aldehydes formed during lipid peroxidation, MDA is the most mutagenic. Its reaction with dGuo, dCyd, and dAado generates promutagenic adducts, such as cyclic pyrimidopurinone 3-(2-deoxy-β-D-erythro-pentofuranosyl)pyrimido[1,2-α]purin-10(3H)-one (M1dGuo), the acyclic N^6-(3-oxo-1-propenyl)-2′-deoxyadenosine (M1dAdo), and N^4-(3-oxo-1-propenyl)-2′-deoxycytidine (M1dCyd) (Marnett and Tuttle 1980; Marnett 1999a, 1999b; Fig. 38.2).

Etheno adducts are well validated biomarkers of DNA damage arising from the reaction of lipid peroxidation end products, and are recognized as a typical signature of DNA damage in white blood cells (Arab et al. 2009; Fig. 38.2). 1,N^2-εdGuo, for example, is derived from the reaction between *trans,trans*-2,4-decadienal (DDE) and dGuo (Loureiro et al. 2000). This adduct can also be formed by reactive aldehydes generated by lipid, amino acid, or sugar oxidation, 2,3-epoxyaldehydes or vinyl chloride (Barbin 2000; Petrova et al. 2007). The biological relevance of these adducts has been extensively demonstrated. For example, 1,N^2-εdGuo has been shown to be mutagenic in both *Escherichia coli* and human cells (Langouet et al. 1998; Akasaka and Guengerich 1999). *In vitro* misincorporation studies found that 1,N^2-εdGuo tends to block the replicative polymerase, misincorporation of dATP and dGTP, and frameshift deletions (Langouet et al. 1997; Zhang et al. 2009). In addition, 1,N^2-εdGuo is mutagenic in *E. coli* with more abundant G→A mutations followed by G→T mutations (Langouet et al. 1998).

Besides endogenous lipid peroxidation by-products, toxic aldehydes can be generated from industrial activities and fuel exhaust. For instance, aldehydes such as crotonaldehyde are produced in large quantities as synthetic intermediates of industrial products. They are also formed by combustion of plant materials including tobacco (IARC 1995; Eder et al. 1996).

The presence of aldehydes in the environment and their increased formation in pathological conditions has raised concern about their high reactivity, toxicity, and mutagenicity. Therefore, DNA adducts from lipid peroxidation end products are useful biomarkers of oxidative stress, as well as predictors of disease

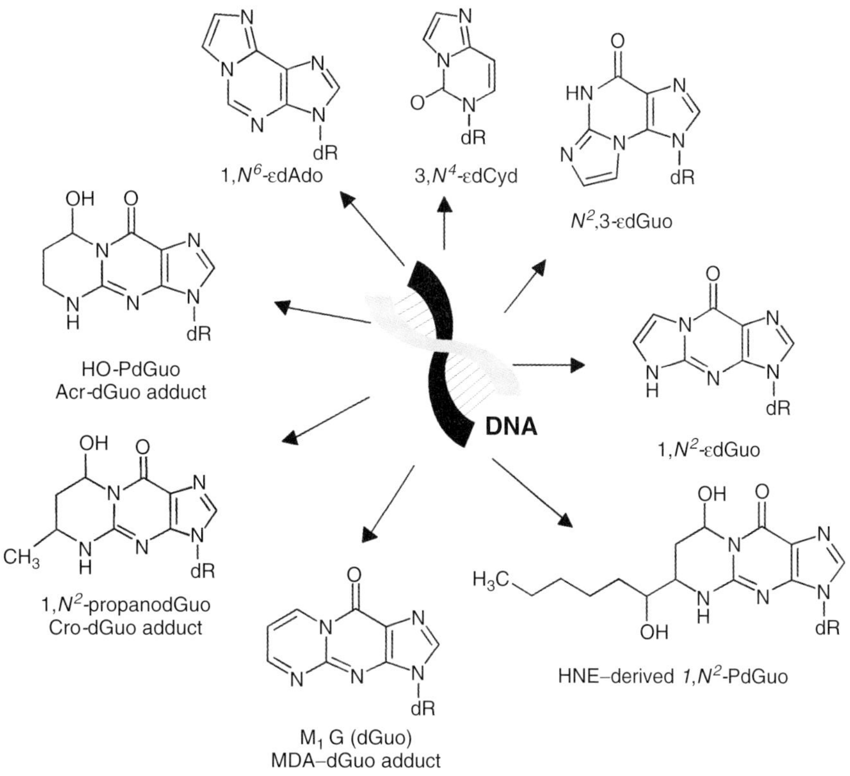

Fig. 38.2 Structure of DNA adducts derived from lipid peroxidation products.

dR =2- deoxyribose

development or environmental biomonitoring (Bartsch and Nair 2004, 2005; Nair et al. 2007). The analysis of DNA alterations in aquatic organisms has been shown to be a highly suitable method for evaluating environmental genotoxic contamination. This chapter describes the analysis of DNA adducts using HPLC–MS (high performance liquid chromatography coupled to mass spectrometry). Detailed protocols are given and discussed below.

GENERAL ANALYSIS

Several methods for separation and quantification of DNA adducts have been developed that quantify extremely low background levels of DNA adducts including immunoassays (Yang et al. 2000; Kawai et al. 2002), ^{32}P and fluorescent postlabeling (Bartsch and Nair 2000; Phillips and Arlt 2007; Emami et al. 2008), ultrasensitive, highly specific gas chromatography/electron capture negative chemical ionization high-resolution mass spectrometry (GC/ECNCI–HRMS) (Rouzer et al. 1997; Chen and Chiu 2003) and HPLC/MS–MS with off-line and on-line reverse-phase extraction to purify the adducts from milligram amounts of DNA (Churchwell et al. 2002; Loureiro et al. 2002; Zhang et al. 2006; Taghizadeh et al. 2008).

Methods involving antibody and ^{32}P-postlabeling require less DNA and are in general more sensitive than other techniques, but lack specificity because they do not use internal standards. MS methods employing a specific isotopically labeled internal standard provide unequivocal identification and rigorous quantification of low levels of DNA adducts (Koc and Swenberg 2002; Medeiros 2009).

MS methods are generally associated with two main chromatographic methods to separate the DNA bases: gas and liquid. GC is a sensible method but can produce

Table 38.1 Summarized methodologies for quantification of various DNA adducts using different approaches.

DNA adduct	Method	Detection Limit	Reference
1,N^2-propanodGuo (Cr-dGuo and Acr-dGuo)	^{32}P-postlabeling method and high performance liquid chromatography for prepurification of nucleotides. The aducts identity were attested by HPLC	Not Reported	Nath and Chung (1994)
εdAdo and εdCyd	Immunoaffinity purification of the etheno adducts and subsequent ^{32}P-postlabelling followed separation as 5′-monophosphates on polyethyleneimine-cellulose-coated thin-layer plates. Unmodified nucleotides in the DNA samples were quantitated by HPLC	25 amol of dAdo and dCyd for a 50 μg DNA sample	Nar, J. et al. (1995)
1,N^2-propanodGuo	3′-5′-biphosphates are two-directionally separated on PEI-cellulose TLC and quantified by autoradiography (^{32}P-postlabeling method)	$<$ 21 adducts 10^{-9} normal nucleotides	Wacker, M. et al. (2000)
1,N^2-εdGuo	Based on on-line reversed-phase high-performance liquid chromatography with electrospray tandem mass spectrometry detection, previous purification of DNA basis.	7.4 adducts 10^{-8} normal nucleotides (350 μg of DNA sample)	Loureiro, A.P.M. et al. (2002)
εdAdo and εdCyd and M1dGuo and other products	On-line sample preparation coupled with liquid chromatography in *tandem* mass spectrometry	$<$ 1 adduct 10^{-8} normal nucleotides (100 μg of DNA sample)	Churchwell, M. I. et al. (2002)
1,N^2-propanodGuo	Capillary liquid chromatography nanoelectrospray isotope dilution/high capacity ion trap mass spectrometer in the MS/MS mode. Nucleotides enzymatically digested were cleaned up by solid phase extraction (SPE)	40 adducts 10^{-9} normal nucleosides	Liu et al. (2006)
Ciclic-1,N^2-propanodGuo (Acr-Cro-, Pen-, Hep-, and HNE-dGuo)	SPE/HPLC-based ^{32}P-postlabeling assay The adducts are converted to ring-open form for structural confirmation and quantification	9 adducts 10^{-9} normal nucleosides (80 μg of DNA sample)	Pan, J., et al. (2006)
1,N^2-εdGuo, εdAdo, εdCyd, and other products	Individual nucleosides are purified by HPLC and quantified by isotope–dilution, electrospray ionization liquid chromatographic tandem mass spectrometry (LC/MS–MS) method	10 fmol in 100 μg to 1,N^2-εdGuo. 1 fmol in 10–20 μg of DNA to dA and 0.5 fmol in 50 DNA to dCyd	Taghizedeh, K. et al. (2008)
dG-HNE (ring open form), 6 and 8-hydroxi-propanodGuo and other products	Two-dimensional linear quadrupole ion trap mass spectrometer (LIT/MS) was employed to screen DNA adducts. This comprised data-dependent neutral loss scanning followed by triple-stage mass spectrometry (CNL–MS3) in the MS/MS/MS scan mode.	1 adduct 10^{-8} normal DNA nucleosides for a 10 μg DNA sample	Bessette, E. E. et al. (2009)

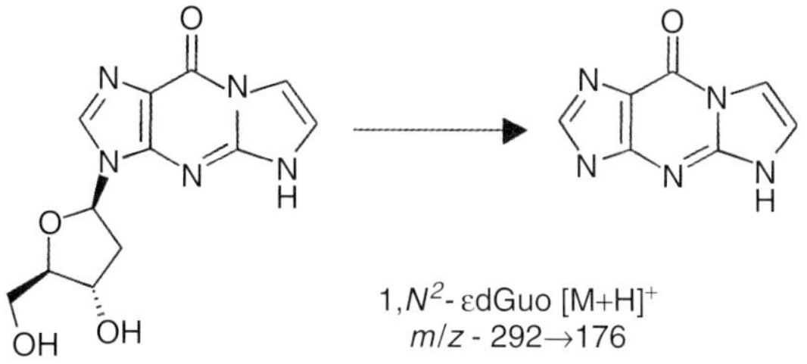

Fig. 38.3 Representative fragmentation and loss of 2-deoxyribose for 1,N^2-$^\varepsilon$dGuo in MS analysis.

artifacts during measurement. HPLC yields better separation of DNA bases and can be employed to analyze a large number of adducts (Koc and Swenberg 2002; Singh 2006). Table 38.1 summarizes some methods that can be used for quantification of various DNA adducts using different approaches.

Technological improvements in HPLC–MS have made this technique the most effective to measure DNA adducts. The high sensitivity and specificity are attributed to excellent separation of unmodified and modified DNA bases, isotopically labeled internal standards, and specific detection of DNA adduct molecule fragments after the loss of 2-deoxyribose and consequently the loss of 116 units of mass during collision-induced dissociation (as exemplified in Fig. 38.3) (Koc and Swenberg 2002; Singh and Farmer 2006).

SAMPLE PREPARATION

Sample preparation involves tissue collection and storage, DNA extraction, hydrolysis isolation, and separation of excess unmodified bases from DNA adducts. All these preparatory steps need to be carried out with utmost care and cleanliness, to minimize DNA contamination, artifactual oxidation, as well as unwanted DNA adduct disappearance or formation. To prevent lipid peroxidation and oxidative DNA damage during sample handling and storage, antioxidants and metal chelating agents, such as butylated hydroxytoluene (BHT) and desferoxamine, are added to all solutions (see concentrations in solutions protocols). Fresh tissues need to be snap-frozen and stored at −80°C. The DNA from live cells, lymphocytes, and hemolymph needs to be isolated immediately after cell treatment and lymphocyte (or hemolymph) separation.

Some specific types of DNA adducts require special care during sample preparation and analysis.

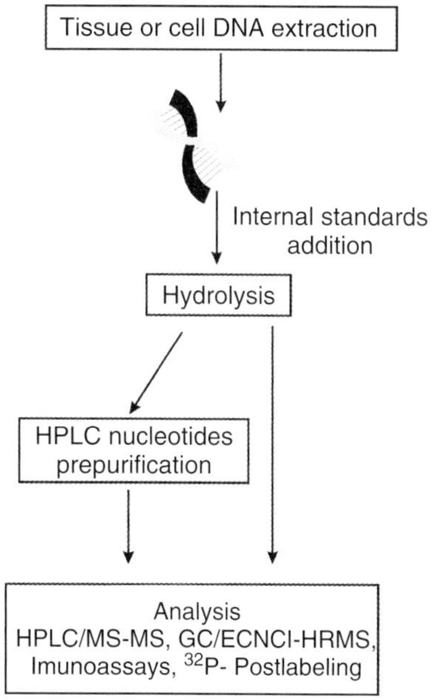

Sch. 38.1 Generalized flow chart of DNA adduct analysis.

For instance, Taghizadeh et al. (2008) observed DNA alteration during extraction and hydrolysis: dI and dU were formed from nucleobase and deaminase activities. This problem is resolved by adding the dAdo and dCyt-deaminase inhibitors coformycin and tetrahydrouridine to all buffers. DNA extraction is one of the most critical steps requiring attention to minimize contamination from RNA, lipids, proteins, and glycogen. These compounds are eliminated by specific enzymes or organic extractions, because they can interfere in the subsequent adduct detection and reduce the sensitivity of the method. Other contaminants (i.e. inorganic salts, polar compounds in DNA matrix) interfere with MS analysis as they can suppress ESI ionization and increase the baseline.

Another problem concerns adduct stability during sample preparation. Here, the structure and reactivity of each molecule needs to be considered. In general, temperature, metal contamination, and pH are the determinants of stability and degradation.

Scheme 38.1 presents a general protocol to analyze different DNA adducts. These analyses consist in multiple steps: sample storage, DNA isolation/hydrolysis, adduct enrichment (depending on methodology), adduct detection/quantification, and unmodified bases quantification.

DNA EXTRACTION

DNA is isolated from tissues or cells by the adapted chaotropic NaI method (Wang et al. 1994; Loureiro et al. 2002). As a general rule for tissue samples, 100 mg of tissue yield 100 μg of DNA; however, in the case of fatty and fibrous tissues higher amounts of tissue fresh mass need to be processed.

Protocol

Solutions

- Lysis buffer: 320 mM sucrose, 5 mM $MgCl_2$, 10 mM Tris-HCl, 0.1 mM desferroxamine, and 1% (v/v) Triton X-100, pH 7.5.
- Tris-HCl buffer: 10 mM Tris-HCl buffer containing 5 mM EDTA, 0.15 mM desferroxamine, pH 8.0.
- RNAse A: 10 mg mL^{-1} re-suspended in water.*
- RNAse T1: 20 U μL^{-1} in 10 mM Tris-HCl buffer containing 1 mM EDTA and 2.5 mM desferroxamine, pH 7.4.
- Proteinase K: 20 mg mL^{-1} resuspended in water.*
- NaI solution: 7.6 M NaI, 40 mM Tris-HCl (pH 8), 20 mM EDTA, 0.3 mM desferroxamine.
- Isopropanol 60%: 60 mL isopropanol in 100 mL water.*
- Ethanol 70%: 70 mL ethanol dilute with water to 100 mL.*
- Desferroxamine mesilate solution: 0.1 mM of desferroxamine mesilate in water.

*MiliQ water is pretreated overnight with Chelex®.

DNA Extraction Procedure

100 mg of tissue or cellular pellet (3×10^8 cells) are homogenized (i.e. using a Thurrax blender) in 10 mL of cold lysis buffer. After centrifugation at 1500 *g* for 10 min, 4°C, pellets are resuspended in 10 mL of cold lysis buffer and centrifuged at 1500 *g* for 10 min, 4°C. Pellets are then suspended in 6 mL of Tris-HCl buffer. 350 μL of 10% SDS, RNase A (30 μL) and RNase T1 (4 μL) are added and the reaction mixture is incubated at 37°C. After 1 h of incubation, 300 μL proteinase K are added and incubation continued for 1 h. After centrifugation at 5000 *g* for 15 min at 15°C, the liquid phase is collected and 600 μL of NaI solution and then 5 mL of isopropanol added. The content in the tube is well mixed by inversion until a whitish precipitate appears. Precipitates are collected by centrifugation at 5000 *g* for 15 min, 4°C and washed with 5 mL of isopropanol 60% (5000 *g*, 15 min, 4°C) followed by 5 mL of ethanol 70% (5000 *g*, 15 min, 4°C). DNA pellets are dissolved in 500 μL of desferroxamine mesilate solution after ethanol evaporation (at room temperature with inverted tubes for approximated 5 min). DNA concentration is measured at 260 nm (concentration in μg mL^{-1} is equal to absorbance × dilution × 50, when using 1 mL cuvettes), and its purity is assessed by ensuring $A_{260}/A_{280} \geq 1.7$.

DNA extraction can also be performed using a DNA extraction kit. DNA isolated with this procedure gives a DNA with a clear appearance. Use the kit according to the manufacturer's instructions.

INTERNAL STANDARDS

A constant amount of synthesized internal standard (see below) is added to the DNA samples prior to enzymatic hydrolysis. The use of an isotopically labeled nucleotide derivative is recommended to insure high specificity and accurate quantification of adducts because it helps to control for losses of analyte during sample preparation (hydrolysis) and analysis (electrospray ionization, ESI). Most frequently applied isotopic internal standards are the change of nitrogen-14 to nitrogen-15, increasing some units of mass compared to the normal nucleotides (i.e. five units of mass for dGuo and dAdo, three for dCyd and two for dThd) and, consequently, represent modified bases, with the same structure. In addition, some groups choose to use the substitution of nitrogen and carbon (12 to 13) in the same molecule, increasing more units of mass in the normal nucleotide, depending of the analyzed nucleotide.

Synthesis is different for each DNA adduct that is analyzed. There are protocols in the literature describing the synthesis and purification methods (in general HPLC) for several types of adducts. In this chapter we will exemplify the detection and quantification methodology of 1,N^2-etheno-2′-deoxyguanosine, (1,N^2-edGuo) using the protocol proposed by Loureiro et al. (2002) with slight modifications.

Synthesis of 1,N^2-Etheno-2′-Deoxyguanosine Unlabeled Standard

The 1,N^2-εdGuo unlabeled standard is obtained by reacting dGuo with chloroacetaldehyde as described by Loureiro et al. (2002). Its identity is confirmed by the following spectroscopic features: UV λ_{max}222 ($\varepsilon = 40,570\ M^{-1}\ cm^{-1}$), 271 ($\varepsilon = 10,971\ M^{-1}\ cm^{-1}$), 295 nm ($\varepsilon = 11,912\ M^{-1}\ cm^{-1}$) at pH 1 (50 mM HCl-KCl); 226 ($\varepsilon = 49,937\ M^{-1}\ cm^{-1}$), 285 nm ($\varepsilon = 16,785\ M^{-1}\ cm^{-1}$) at pH 7 (50 mM phosphate buffer); and 233 ($\varepsilon = 42,764\ M^{-1}\ cm^{-1}$), 280 ($\varepsilon = 8643\ M^{-1}\ cm^{-1}$), 307 nm ($\varepsilon = 11,687\ M^{-1}\ cm^{-1}$) at pH 11 (50 mM carbonate-bicarbonate buffer). ESI–MS m/z176 ($[M + H]^+$ − 2-D-*erythro*-pentose, 65% relative intensity), 292 ($[M + H]^+$, 100% relative intensity). ^1H NMR (DMSO-d_6) δ7.97 (s, 1H, H-2), 7.49 (d, 1H, $J = 2.43$ Hz, H-6), 7.24 (d, 1H, $J = 2.58$ Hz, H-7), 6.22–6.25 (dd, 1H, H-1'), 5.32 (bs, 1H, OH-5'), 5.23 (s, 1H, OH-3'), 4.37 (m, 1H, H-3'), 3.83–3.86 (m, 1H, H-4'), 3.58–3.60 (m, 1H, H-5'), 3.49–3.52 (m, 1H, H-5''), 2.64–2.69 (m, 1H, H-2'), 2.17–2.21 (m, 1H, H-2'').

Synthesis of [$^{15}N_5$]-1,N^2-Etheno-2′-Deoxyguanosine Internal Standard

[$^{15}N_5$]-1,N^2-εdGuo is obtained by reacting [$^{15}N_5$]-dGuo with chloroacetaldehyde, with subsequent purification by HPLC as described by Loureiro et al. (2002). Its identity is confirmed by MS. ESI–MS for [$^{15}N_5$]-1,N^2-εdGuo: m/z 181 ($[M + H]^+$ − 2-D-*erythro*-pentose, 58% relative intensity), 297 ($[M + H]^+$, 100% relative intensity).

DNA DIGESTION

DNA can be hydrolyzed by different methods; here we describe the acid hydrolysis using nuclease P1.

Protocol

Solutions:

- Sodium acetate buffer: 1 M sodium acetate in water* (pH 5).
- Nuclease P1 solution: 0.5 units per μL of buffer containing $ZnCl_2$ (enzyme cofactor) – this buffer is described by each supplier.
- Tris-HCl buffer: Tris-HCl 1 M buffer (pH 7.4).
- Phosphatase buffer: six units of alkaline phosphatase, two units per μL.
- Alkaline phosphatase: comes together with the phosphatase enzyme.

*MiliQ water is pretreated overnight with Chelex®.

DNA digestion

Step 1:

- 200 μg of DNA in a solution of 0.1 mM desferroxamine (the volume depends on the calculated DNA concentration). However, the maximum volume of DNA solution that can be added is 150 μL, if the DNA is highly concentrated, the volume should be completed with MiliQ water to 150 μL.
- 4 μL of sodium acetate buffer 1 M, pH 5.0.
- 20 μL of internal standard (10 fmol μL^{-1}), corresponding to a 200 fmol.
- 4 μL of nuclease P1, corresponding to two units.
- Incubation for 30 min at 37°C.

*MiliQ water is pretreated overnight with Chelex®.

Step 2:

- 10 μL Tris-HCl buffer 1 M, pH 7.0.
- 10 μL of phosphatase buffer.
- 3 μL of alkaline phosphatase, corresponding to six units.
- Incubation for 60 min at 37°C.

Enzymes are precipitated by the addition of 1 volume of chloroform and, after centrifugation at 1000 *g* for 5 min, the resulting aqueous layer is subjected to HPLC–ESI/MS–MS (100 μL of DNA). The amounts of the reagents and labeled internal standard are proportionally adjusted for hydrolysis and analysis of other DNA quantities.

CALIBRATION CURVE

The calibration curve is constructed from 10 different concentration points by plotting the chromatographic peak area ratios of unlabeled adducts and a fixed amount of internal standard, which is added to each injection, in order to avoid discrepancies of ionization between samples versus the amount of the unlabeled adduct injected (0.1–100 fmol per injection). These calibration curve quantities can vary with the MS sensitivity and DNA amount.

The amount of injected [$^{15}N_5$]-labeled internal standard is always 100 fmol. Blank injections (sample solvent plus 100 fmol of [$^{15}N_5$]-1,N^2-εdGuo) in the same conditions show no carryover of the unlabeled 1,N^2-εdGuo from previous analyses and in the [$^{15}N_5$]-1,N^2-εdGuo standards.

UNMODIFIED DNA BASES PREPURIFICATION

Purification of unmodified DNA bases is performed in order to increase adduct amount by minimizing the interference of major products present in the sample and consequently increasing sensitivity. Some methods utilize a first step to separate the adduct from the analyte (digest DNA). This step, the isolation of DNA adducts from normal nucleotides, consists in a solid phase extraction or HPLC separation for collection of the adducts in a specific DNA-bases-free fraction.

In MS analysis there are two primary methods, one uses previously purified nucleotides whereas the other uses an automated switching valve that separates nucleotides on-line into the HPLC–MS in tandem. In this chapter we describe the method using the automated switching valve (Loureiro et al. 2002).

DETECTION

Mass Spectrometry

To prevent interference, loss of sensitivity, and to achieve DNA-sample prepurification, we connect an automated switching valve, programmed to change its position after elution of the last unmodified DNA-base (dThd), allowing only the modified DNA bases to reach the mass spectrometer. Briefly, this valve as depicted in Fig. 38.4a (top) is configured to discard all analytes eluted from column 1 in the first 19 min. After that, the position is changed (Fig. 38.4a, bottom), directing the modified bases eluted from column 1 to column 2 and then into the mass spectrometer. This valve and the two columns allow a better purification of the analyte, leading to a clear improvement in the detection.

The monitored transitions corresponding to the adducts are represented by a predominant fragmentation resulting from a decrease of 116 mass units by the loss of 2-deoxyribose, $[M+H]^+ - $ 2-D-*erythro*-pentose ion formation.

The developed liquid chromatography program allows for adequate separation of the adduct from the last eluted nucleoside, 2′-deoxythymidine (dThd). In addition to the solvent flow and the acetonitrile content, the use of 0.1% formic acid in the mobile phase was particularly important for the elution of the normal nucleosides in the following order: dCyd< dAdo< dGuo< dThd. When pure water is used instead of 0.1% formic acid, the last eluted nucleoside is dAdo, the tail of which is frequently a problem that reduces MS sensitivity. The automated switching valve, programmed to change its position in the period between the end of elution of dThd and the beginning of elution of 1,N^2-εdGuo + [$^{15}N_5$]-1,N^2-εdGuo, ensures that no excess of normal nucleosides enters the MS, preventing loss of sensitivity.

High-Performance Liquid Chromatography–Electrospray Ionization Tandem Mass Spectrometry

On-line high-performance liquid chromatography–electrospray ionization tandem mass spectrometry (HPLC – ESI/MS – MS) analysis in the positive mode is carried out using a API-4000 QTRAP MS (Applied Biosystems, Foster City, CA). The 1,N^2-εdGuo adducts in DNA samples are detected by multiple-reaction monitoring (MRM). An Agilent HPLC system (Kyoto, Japan) consisting of an autosampler (1200 High performance), a column oven (1200 G1216B) set at 18°C, with an automated switching valve, 1200 binary pump SL, 1200 isocratic pump SL, an UV detector (1200 DAD G1315C) is used for sample injection and cleanup of the analytical column (Luna C18(2) 250 mm × 4.6 mm i.d., 5 μm, Phenomenex, Torrance, CA). The adduct is eluted from this first column with a gradient of formic acid (0.1% in water) and acetonitrile (from 0 to 18 min, 8% acetonitrile and 0.4 mL min^{-1}; from 18 to 19 min, 8 to 50% acetonitrile and 0.4 to 0.2 mL min^{-1}; from 19 to 19.5 min, 50% acetonitrile and 0.2 to 0.1 mL min^{-1}; from 19.5 to 32 min, 50% acetonitrile and 0.1 mL min^{-1}; from 32 to 32.5 min, 50% acetonitrile and 0.1 to 0.05 mL min^{-1}; from 32.5 to 35 min, 50% acetonitrile and 0.05 mL min^{-1}; from 35 to 36 min, 50% acetonitrile and 0.05 to 0.8 mL min^{-1}; from 36 to 38 min, 50% acetonitrile and 0.8 mL min^{-1}; from 38 to 39 min, 50 to 8% acetonitrile and 0.8 mL min^{-1}; from 39 to 49 min, 8% acetonitrile and 0.8 to 0.4 mL min^{-1}).

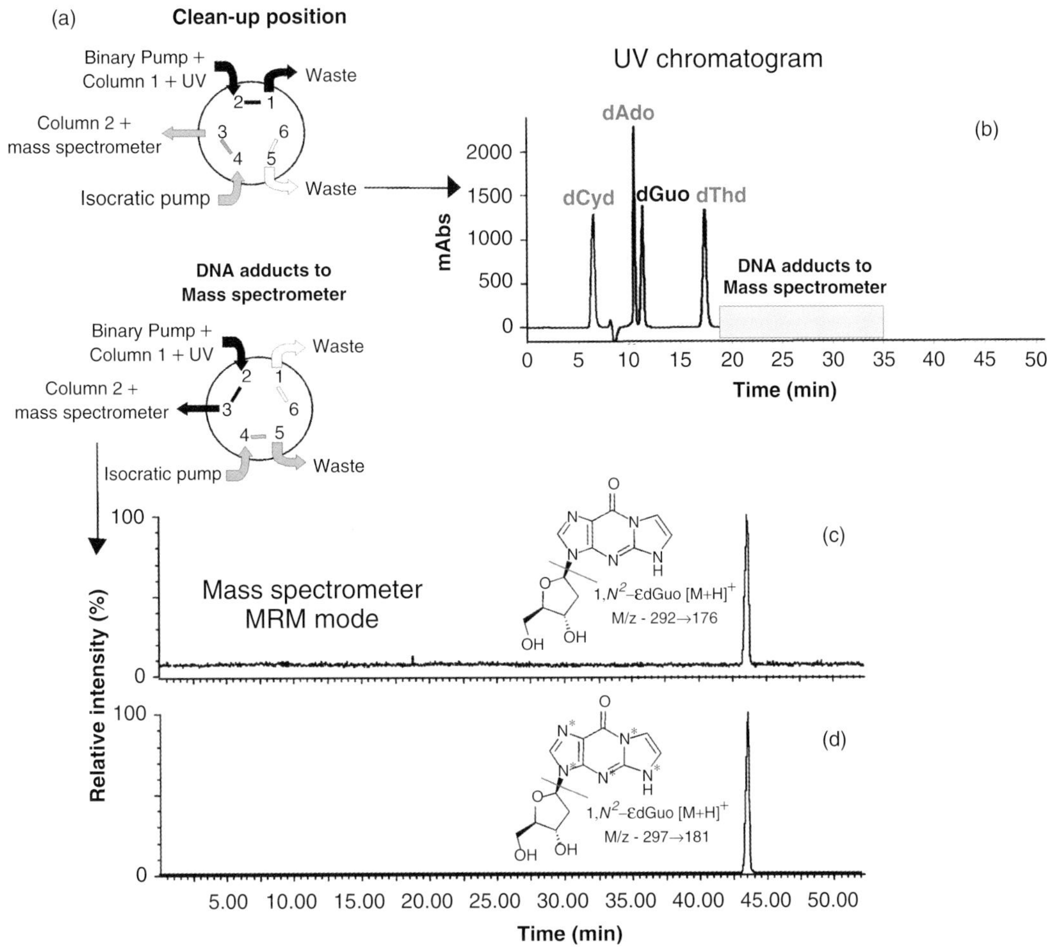

Fig. 38.4 HPLC–ESI/MS–MS detection DNA adducts in calf thymus DNA: (a) positions of the automated switching valve during the HPLC–ESI/MS–MS analysis; (b) enzymatically hydrolyzed DNA, 260 nm; (c) 1,N^2-εdGuo: m/z 292 → 176; and (d) [$^{15}N_5$]-1,N^2-εdGuo: m/z 297 → 181.

An isocratic pump is used to simultaneously load a second column (C18 (2) Phenosphere-Next 150 mm × 2 mm i.d., 3 μm, Phenomenex) with an isocratic flow (0.05 mL min^{-1}) of a solution of formic acid 0.1% in water : acetonitrile (50:50), maintaining a constant flow of the mobile phase to the MS during analysis. The position of the switching valve is changed twice: at 24 min, allowing the first column eluent to enter the second column; and at 35 min, permitting the first column to be washed while the adduct is eluted through the second column into the mass spectrometer. The total time for the analysis is 50 min.

The DNA hydrolysates containing 10 fmol of the [$^{15}N_5$]-1,N^2-εdGuo and [$^{15}N_5$]-1,N^2-propanodGuo internal standard are injected into the system. The m/z 292/176 (1,N^2-εdGuo), 297/181 ([$^{15}N_5$]-1,N^2-εdGuo) $[M + H]^+$ ions are monitored with a dwell time of 200 ms.

All parameters of the MS are adjusted for acquisition of the best $[M + H]^+/[M + H]^+$ − 2-D-*erythro*-pentose transition. The curtain gas is adjusted to 25 psi, source temperature is held at 450°C, nebulizer and auxiliary gas is kept at 60 psi, Turbo Ion Spray voltage on 5500 V, collision gas set on high, interface heater is held at 100°C and the entrance potential at 10 V. To 222/176 and 227/181 transitions are selected 17 V to collision energy, 16 V to collision cell exit and 41 V to declustering potential. Data are processed using Analyst 1.4.2. (Applied Biosystems, Foster City, CA).

Unmodified Nucleotides

2′-Deoxyguanosine Quantification in DNA Sample

The following HPLC system is used to quantitatively determine dGuo in the DNA hydrolysates: a Luna 10 C18 (2) (250 mm × 4.6 mm i.d., 5 μm) semi-preparative column (Phenomenex, Torrance, CA) is eluted with a gradient of water and acetonitrile (from 0 to 5 min, 5% acetonitrile; from 5 to 30 min, 5 to 20% acetonitrile) at a flow rate of 1 mL min^{-1}, and absorbance is monitored at 254 nm. A standard calibration curve prepared within the range of dGuo expected to be present in the DNA hydrolysates is used for this quantification (see also Chapter 37).

Some variations in adduct and DNA bases retention time can occur with the use of different HPLC equipment. However, some adjustment in acetonitrile percentage, flow, and automated switching valve time of change is sufficient for methodology adaptation.

QUANTIFICATION

The quantification of DNA adducts is performed by the ratio of adducts to dGuo, for this the peak area is plotted against concentration yielding a calibration curve (see above), to quantify adducts in fmol and dGuo in nmol. The results are expressed by n adducts per 10^9, 10^8, 10^7 or 10^6 dGuo (n = number of adducts).

DNA ADDUCTS IN AQUATIC ANIMALS

Exposure of aquatic organisms to pollutants can promote an increase in ROS/RNS production. Thus the assessment of oxidative stress in specific organisms provides information about the environmental status (Kelly et al. 1998; Livingstone 2001; Valavadinis et al. 2006). There are few works measuring DNA adducts in aquatic models. Here, we have described the results from Almeida et al. (2003, 2006), in which for the first time the levels of the etheno-DNA adduct 1,N^2-etheno-2′-deoxyguanosine (1,N^2-εdGuo) were analyzed in tissues of the mussel *Perna perna* to check for effects of seasonality. The levels of etheno adducts measured in autumn in digestive gland of *P. perna* (~5 adducts 10^{-7} dGuo) and in mantle tissue (~15 adducts 10^{-7}dGuo) were similar to basal levels reported from mammalian tissues and cells. In addition, these results indicate higher levels in mantle tissue than in digestive gland. In summer, the levels of 1,N^2-εdGuo in mantle were similar to those found in autumn, but in the digestive gland the values increased during summer to ~17 adducts 10^{-7}dGuo, indicating that the level of this lesion in *P. perna* may change with season (unpublished results). Preliminary results from our group's contaminant studies with *P. perna* also indicate higher levels of 1,N^2-εdGuo in mussels transplanted to a polluted site.

The use of DNA adducts as a tool for assessing oxidative stress-derived DNA lesions has opened new perspectives in the field of aquatic toxicology. Nevertheless, more studies are needed to characterize the natural levels of DNA adducts generated by lipid peroxidation by-products in order to obtain a reliable method and standards that can be used in biomarker studies of oxidative stress and pollution effects with aquatic models.

REFERENCES

Akasaka, S., Guengerich, F.P. (1999) Mutagenicity of site-specifically located 1,n^2-ethenoguanine in Chinese hamster ovary cell chromosomal DNA. *Chemical Research in Toxicolology* 12, 501–507.

Almeida, E.A., Marques, S.A., Klitzke, C.F. et al. (2003) DNA damage in digestive gland and mantle tissue of the mussel *Perna perna*. *Comparative Biochemistry and Physiology C* 135, 295–303.

Almeida, E.A., Bainy, A.C.D., Loureiro, A.P.M. et al. (2006) Oxidative stress in Perna perna and other bivalves as indicators of environmental stress in the Brazilian marine environment: Antioxidants, lipid peroxidation and DNA damage. *Comparative Biochemistry and Physiology A* 146, 588–600.

Arab, K., Pedersen, M., Nair, J., Meerang, M., Knudsen, L.E., Bartsch, H. (2009) Typical signature of DNA damage in white blood cells: a pilot study on etheno adducts in Danish mother–newborn child pairs. *Carcinogenesis* 30, 282–285.

Barbin, A. (2000) Etheno-adduct-forming chemicals: from mutagenicity testing to tumor mutation spectra. *Mutation Research* 462, 55–69.

Bartsch, H., Nair, J. (2000) Ultrasensitive and specific detection methods for exocylic DNA adducts: markers for lipid peroxidation and oxidative stress. *Toxicology* 153, 105–114.

Bartsch, H., Nair, J. (2004) Oxidative stress and lipid peroxidation-derived DNA-lesions in inflammation driven carcinogenesis. *Cancer Detection and Prevention* 28, 385–391.

Bartsch, H., Nair, J. (2005) Accumulation of lipid peroxidation-derived DNA lesions: potential lead markers for chemoprevention of inflammation-driven malignancies. *Mutation Research* 591, 34–44.

Bessette, E.E., Goodenough, A.K., Langouet, S. et al. (2009) Screening for DNA adducts by data-dependent constant neutral loss-triple stage mass spectrometry with a linear quadrupole ion trap mass spectrometer. *Analytical Chemistry* 81, 809–819.

Blair, I.A. (2008) DNA adducts with lipid peroxidation products. *Journal of Biological Chemistry* 283, 15545–15549.

Chen, H.J., Chiu, W.L. (2003) Detection and quantification of 1,N^6-ethenoadenine in human urine by stable isotope dilution capillary gas chromatography/negative ion chemical ionization/mass spectrometry. *Chemical Research in Toxicology* 16, 1099–1106.

Churchwell, M.I., Beland, F.A., Doerge, D.R. (2002) Quantification of multiple DNA adducts formed through oxidative stress using liquid chromatography and electrospray tandem mass spectrometry. *Chemical Research in Toxicology* 15, 1295–1301.

Eder, E., Budiawan, Schuler, D. (1996) Crotonaldehyde: a carcinogenic and mutagenic air, water and food pollutant. *Central European Journal of Public Health* 4, 21–22.

Emami, A., Dyba, M., Cheema, A.K., Pan, J., Nath, R.G., Chung, F.L. (2008) Detection of the acrolein-derived cyclic DNA adduct by a quantitative ^{32}P-postlabeling/solid-phase extraction/HPLC method: blocking its artifact formation with glutathione. *Analytical Biochemistry* 374, 163–172.

Esterbauer, H. (1993) Cytotoxicity and genotoxicity of lipid-oxidation products. *American Journal of Clinical Nutrition* 57, 779S–785S.

Golding, B.T., Slaich, P.K., Kennedy, G., Bleasdale, C., Watson, W.P. (1996) Mechanisms of formation of adducts from reactions of glycidaldehyde with 2′-deoxyguanosine and/or guanosine. *Chemical Research in Toxicology* 9, 147–157.

Himmelstein, M.W., Boogaard, P.J., Cadet, J. et al. (2009) Creating context for the use of DNA adduct data in cancer risk assessment: II. Overview of methods of identification and quantitation of DNA damage. *Critical Reviews in Toxicology* 39, 679–694.

IARC. (1995) Crotonaldehyde. In *Monographs on the Evaluation of the Carcinogenic Risk of Chemicals to Humans*, Vol. 63. International Agency for Research on Cancer, Lyon, pp. 373–391.

Kawai, Y., Kato, Y., Nakae, D. et al. (2002) Immunohistochemical detection of a substituted 1,N^2-ethenodeoxyguanosine adduct by omega–6 polyunsaturated fatty acid hydroperoxides in the liver of rats fed a choline-deficient, L-amino acid-defined diet. *Carcinogenesis* 23, 485–489.

Kelly, K.A., Havrilla, C.M., Brady, T.C., Abramo, K.H., Levin, E.D. (1998) Oxidative stress in toxicology: established mammalian and emerging piscine model systems. *Environmental Health Perspectives* 106, 375–384.

Koc, H., Swenberg, J.A. (2002) Applications of mass spectrometry for quantitation of DNA adducts. *Journal of Chromatography B* 778, 323–343.

Langouet, S., Muller, M., Guengerich, F.P. (1997) Misincorporation of dNTPs opposite 1,N^2-ethenoguanine and 5,6,7,9-tetrahydro–7-hydroxy–9-oxoimidazo[1,2-a]purine in oligonucleotides by Escherichia coli polymerases I exo- and II exo-, T7 polymerase exo-, human immunodeficiency virus–1 reverse transcriptase, and rat polymerase beta. *Biochemistry* 36, 6069–6079.

Langouet, S., Mican, A.N., Muller, M. et al. (1998) Misincorporation of nucleotides opposite five-membered exocyclic ring guanine derivatives by escherichia coli polymerases In vitro and in vivo: 1,N^2-ethenoguanine, 5,6,7,9-tetrahydro–9-oxoimidazo[1, 2-a]purine, and 5,6,7,9-tetrahydro–7-hydroxy–9-oxoimidazo[1, 2-a]purine. *Biochemistry* 37, 5184–5193.

Liu, X., Lovell, M.A., Lynn, B.C. (2006) Detection and quantification of endogenous cyclic DNA adducts derived from trans–4-hydroxy–2-nonenal in human brain tissue by isotope dilution capillary liquid chromatography nanoelectrospray tandem mass spectrometry. *Chemical Research in Toxicology* 19, 710–718.

Livingstone, D.R. (2001) Contaminant-stimulated reactive oxygen species production and oxidative damage in aquatic organisms. *Marine Pollution Bulletin* 42, 656–666.

Loureiro, A.P.M., Di Mascio, P., Gomes, O.F., Medeiros, M.H.G. (2000) trans,trans–2,4-decadienal-induced 1,N^2-etheno–2′-deoxyguanosine adduct formation. *Chemical Research in Toxicology* 13, 601–609.

Loureiro, A.P.M., Marques, S.A., Garcia, C.C.M., Di Mascio, P., Medeiros, M.H.G. (2002) Development of an on-line liquid chromatography-electrospray tandem mass spectrometry assay to quantitatively determine 1,N^2-etheno–2′-deoxyguanosine in DNA. *Chemical Research in Toxicology* 15, 1302–1308.

Marnett, L.J. (1999a) Chemistry and biology of DNA damage by malondialdehyde. *IARC Scientific Publications* 150, 17–27.

Marnett, L.J. (1999b) Lipid peroxidation-DNA damage by malondialdehyde. *Mutation Research* 424, 83–95.

Marnett, L.J., Tuttle, M.A. (1980) Comparison of the mutagenicities of malondialdehyde and the side products formed during its chemical synthesis. *Cancer Research* 40, 276–282.

Medeiros, M.H.G. (2009) Exocyclic DNA Adducts as biomarkers of lipid oxidation and predictors of disease. Challenges in developing sensitive and specific methods for clinical studies. *Chemical Research in Toxicology* 22, 419–425.

Minko, I.G., Kozekov, I.D., Harris, T.M., Rizzo, C.J., Lloyd, R.S., Stone, M.P. (2009) Chemistry and biology of DNA containing 1,N^2-deoxyguanosine adducts of the alpha,beta-unsaturated aldehydes acrolein, crotonaldehyde, and 4-hydroxynonenal. *Chemical Research in Toxicology* 22, 759–778.

Nair, J., Barbin, A., Guichard, Y., Bartsch, H. (1995) 1,N^6-ethenodeoxyadenosine and 3,N4-ethenodeoxycytine in liver DNA from humans and untreated rodents detected by immunoaffinity/^{32}P-postlabeling. *Carcinogenesis* 16, 613–617.

Nair, U., Bartsch, H., Nair, J. (2007) Lipid peroxidation-induced DNA damage in cancer-prone inflammatory diseases: a review of published adduct types and levels in humans. *Free Radical Biology and Medicine* 43, 1109–1120.

Nath, R.G., Chung, F.L. (1994) Detection of exocyclic 1,N^2-propanodeoxyguanosine adducts as common DNA lesions in rodents and humans. *Proceedings of the National Academy of Sciences USA* 91, 7491–7495.

Pan, J., Davis, W., Trushin, N. et al. (2006) A solid-phase extraction/high-performance liquid chromatography-based ^{32}P-postlabeling method for detection of cyclic 1,N^2-propanodeoxyguanosine adducts derived from enals. *Analytical Biochemistry* 348, 15–23.

Petrova, K.V., Jalluri, R.S., Kozekov, I.D., Rizzo, C.J. (2007) Mechanism of 1,N^2-etheno-2′-deoxyguanosine formation from epoxyaldehydes. *Chemical Research in Toxicology* 20, 1685–1692.

Phillips, D.H., Arlt, V.M. (2007) The ^{32}P-postlabeling assay for DNA adducts. *Nature Protocols* 2, 2772–2781.

Rouzer, C.A., Chaudhary, A.K., Nokubo, M. et al. (1997) Analysis of the malondialdehyde–2′-deoxyguanosine adduct pyrimidopurinone in human leukocyte DNA by gas chromatography/electron capture/negative chemical ionization/mass spectrometry. *Chemical Research in Toxicology* 10, 181–188.

Schneider, C. (2009) An update on products and mechanisms of lipid peroxidation. *Molecular Nutrition and Food Research* 53, 315–321.

Singh, R., Farmer, P.B. (2006) Liquid chromatography-electrospray ionization-mass spectrometry: the future of DNA adduct detection. *Carcinogenesis* 27, 178–196.

Taghizadeh, K., McFaline, J.L., Pang, B. et al. (2008) Quantification of DNA damage products resulting from deamination, oxidation and reaction with products of lipid peroxidation by liquid chromatography isotope dilution tandem mass spectrometry. *Nature Protocols* 3, 1287–1298.

Terao, J., Matsushita, S. (1977) Products formed by photosensitized oxidation of unsaturated fatty acid esters. *Journal of the American Oil Chemistry Society* 54, 234–238.

Valavanidis, A., Vlahogianni, T., Dassenakis, M., Scoullos, M. (2006) Molecular biomarkers of oxidative stress in aquatic organisms in relation to toxic environmental pollutants. *Ecotoxicology and Environmental Safety* 64, 178–189.

Wacker, M., Schuler, D., Wanek, P., Eder, E. (2000) Development of a ^{32}P-postlabeling method for the detection of 1,N(2)-propanodeoxyguanosine adducts of trans–4-hydroxy–2-nonenal *in vivo*. *Chemical Research in Toxicology* 13, 1165–1173.

Wang, L., Hirayasu, K., Ishizawa, M., Kobayashi, Y. (1994) Purification of genomic DNA from human whole blood by isopropanol-fractionation with concentrated NaI and SDS. *Nucleic Acids Research* 22, 1774–1775.

Winter, C.K., Segall, H.J., Haddon, W.F. (1986) Formation of cyclic adducts of deoxyguanosine with the aldehydes trans–4-hydroxy–2-hexenal and trans–4-hydroxy–2-nonenal *In vitro*. *Cancer Research* 46, 5682–5686.

Yang, Y., Nair, J., Barbin, A., Bartsch, H. (2000) Immunohistochemical detection of 1,N(6)-ethenodeoxyadenosine, a promutagenic DNA adduct, in liver of rats exposed to vinyl chloride or an iron overload. *Carcinogenesis* 21, 777–781.

Zhang, H., Beckman, J.W., Guengerich, F.P. (2009) Frameshift deletion by *Sulfolobus solfataricus* P2 DNA polymerase Dpo4T239W is selective for purines and involves normal conformational change followed by slow phosphodiester bond formation. *Journal of Biological Chemistry* 284, 35144–35153.

Zhang, S., Villalta, P.W., Wang, M., Hecht, S.S. (2006) Analysis of crotonaldehyde- and acetaldehyde-derived 1,N^2-propanodeoxyguanosine adducts in DNA from human tissues using liquid chromatography electrospray ionization tandem mass spectrometry. *Chemical Research in Toxicology* 19, 1386–1392.

Chapter 39

METHODS TO QUANTIFY LYSOSOMAL MEMBRANE STABILITY AND THE ACCUMULATION OF LIPOFUSCIN

Katja Broeg[1] *and Stefania Gorbi*[2]

[1]Alfred Wegener Institute for Polar and Marine Research, Bremerhaven, Germany
[2]Department of Biochemistry, Biology and Genetic, Polytechnic University of Marches, Ancona, Italy

Lysosomes, membrane bound cellular organelles, contain a broad range of different acid hydrolases such as proteases, nucleases, glycosidases, lipases, phospholipases, phosphatases, and sulfatases. These enzymes are able to degrade all cellular components (Fig. 39.1) and, therefore, need to be safely isolated. Lysosomal membranes act as barriers between the lumen of the lysosomes and the cytosol. Their integrity determines, among other factors, the lifespan of a cell. The stability and operational capacity of the lysosomal membrane is of fundamental importance for the viability of the cell because:

- release of hydrolytic enzymes into the cytoplasm would lead to self-digestion of cell components and generation of a death signal that leads to the induction of apoptosis;
- damage of proton-pumps in the lysosomal membranes would lead to suppressed activity of lysosomal enzymes due to a shift in pH to higher than optimum values;
- intact membranes prevent acidification of the cytoplasm, and release of accumulated contaminants into the extralysosomal cytoplasmic compartment.

The stability of lysosomal membranes is a sensitive tool for the assessment of cell damage. Membranes are a prime target for the action of reactive oxygen and nitrogen species (ROS/RNS). ROS target lipids and proteins, which can easily be peroxidized from the inside or the outside of the lysosomes (Fig. 39.2).

Various mechanisms inflict toxic injury on lysosomes. The first one relates to the alkalinization of lysosomes through uncoupling of the lysosomal proton pumps. Another mechanism involves modification of the activity of lysosomal enzymes. Enhanced autophagic activity of lipids and cell organelles might

Oxidative Stress in Aquatic Ecosystems, First Edition. Edited by Doris Abele, José Pablo Vázquez-Medina, and Tania Zenteno-Savín.

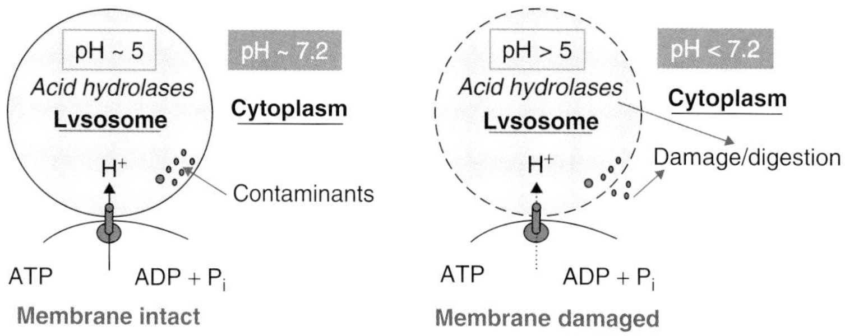

Fig. 39.1 Alterations of the cellular integrity caused by the damage of lysosomal membranes.

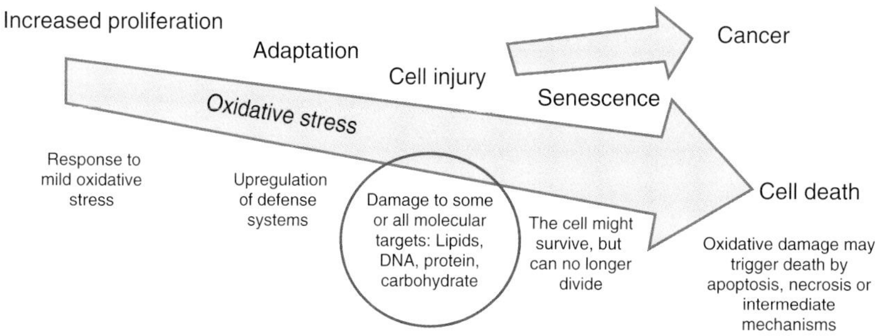

Fig. 39.2 Oxidative stress and the central role of cellular damage during its progression into oxidative damage.

lead to an overload of lipofuscin and increased trapping of metals within the lysosomes, potentiating the formation of ROS. This is also induced by ingested xenobiotics (polycyclic aromatic hydrocarbons, PAHs) and their interaction with heavy metals in redox cycling reactions, such as the Fenton reaction.

The products of lipid peroxidation modify the physical characteristics of biological membranes. Incorporation of lipid peroxides changes the physical structure of the membrane by decreasing its fluidity and increasing its fragility.

Factors affecting lysosomal membrane stability are:

- starvation/food availability
- estradiol
- progesterone
- PAHs
- heavy metals
- organochlorines
- hyperthermia
- hypoxia
- hyposalinity.

Even the increase of cytosolic calcium can destabilize lysosomal membranes through activation of Ca^{2+}-dependent phospholipase A2 (cPLA2) (Marchi et al. 2004). Lysosomal membrane stability (LMS) is a well-established integrative indicator for the assessment of health and fitness of vertebrate and invertebrate organisms exposed to harmful environmental conditions.

Lysosomal membrane destabilization increases linearly with the toxically induced alterations in the central organs for detoxification, the liver of vertebrates, and the digestive gland of invertebrates (Broeg et al. 2005; Viarengo et al. 2007). Especially, contaminant exposure usually provokes alterations of LMS.

Taking into account potential confounding factors during sampling minimizes erroneous determination of stress, i.e.:

- sampling outside the reproductive period of a species;
- sampling of defined size classes;
- sampling of healthy organisms without externally visible pathology;
- consideration of ambient salinity (invertebrates).

Numerous studies have already been conducted on the use of lysosomal membrane stability as a biomarker of stress in aquatic organisms. The list of different indicator organisms is long and contains the mollusks *Mytilus galloprovincialis* (Marigómez and Baybay-Villacorta 2003; Gorbi et al. 2008), *Mytilus edulis* (Schiedeck et al. 2006), *Mytilus trossolus* (Kopecka et al. 2006), *Pecten maximus* (Hauton et al. 2001), *Mercenaria mercenaria* (Zaroogian and Voyer 1995), *Corbicula fluminea* (Champeau and Narbonne 2006), *Ruditapes philippinarum* (Caselli and Fabbri 2005; Da Ros and Nesto 2005; Coughlan et al. 2009), *Radix peregra* (Guerlet et al. 2006), *Dreissena polymorpha* (Guerlet et al. 2007), *Physa* (Zaldibar et al. 2006); the fish *Platichthys flesus* (Köhler 1991; Broeg et al. 1999), *Limanda limanda* (Köhler et al. 1992), *Dicentrachus labrax* (Romeo et al. 2000), *Zoarces viviparus* (Sturve et al. 2005), *Mullus barbatus* (Zorita et al. 2007b), *Pollachius virens* (Bilbao et al. 2006), *Siganus rivulatus, Liza aurata, Diplodus sargus* (Diamant et al. 1999; Broeg 2002); as well as other invertebrate species, such as caddisfly larvae (Werner et al. 2000), earthworm (Lee and Kim 2009), and gastropods *Helix aspersa* (Regoli et al. 2005, 2006).

Historically, LMS has been used as an end point in biomedical research aimed at determining the effects of drugs and other compounds on human cells (Dingle and Dean 1976; de Duve 1983). Assessment of the lysosomal membrane labilization period started by using tissue homogenates and isolated lysosomes. An indicator for a labilized membrane was the lysosomal enzyme activity (mainly acid phosphatase) within the cytosolic fraction or the medium (reviewed by Dingle and Dean 1976).

A histochemical method, based on the demonstration of the latent activity of the lysosomal hydrolase N-acetyl-hexosaminidase (Hex) was developed by Bitensky et al. (1973) for biomedical purposes. It measures the influx of the substrate into lysosomes (the LMS test). Moore (1976, 1988) applied this test to mussel digestive gland. The test was then adopted for application in fish liver by Köhler (1991). To date, it is widely accepted for ecosystem health assessments (UNEP/RAMOGE 1999) and is recommended as a core biomarker in international monitoring programmes on the biological effects of pollution (JAMP, 2003, UNEP/MAP 2005, ICES 2006, HELCOM 2010).

In the decade of the 'omics sciences', methods to assess the functional capability of membranes might fall under the term of "functionomics." These methods are explained in detail within this chapter.

LYSOSOMAL MEMBRANE STABILITY, THE HISTOCHEMICAL APPROACH

See Box 39.1 for the equipment needed for the LMS test and Broeg et al. (1999), BEEP (2002), and Moore et al. (2004) for details of this histochemical approach.

Basic Concept

Cryostat sections of liver or digestive gland tissue are exposed to artificial acid and temperature stress for different time periods from 0 to 50 min to labilize the lysosomal membranes (Fig. 39.3).

The second step is the incubation with a substrate for a lysosomal enzyme, coupled with a naphthol AS-BI group that penetrates the destabilized membranes. The naphthol AS-BI group is a functional group that undergoes electrophilic aromatic substitution with a diazonium salt, finally leading to a color product. When the coupled substrate enters the lysosome it is hydrolyzed by the corresponding lysosomal hydrolase and the AS-BI group is released. The last step is the reaction of a diazonium salt with the hydrolyzed naphthol AS-BI group in the lysosome. The reaction product is a brightly colored diazonium dye. The positive reaction reflects the breakdown of the lysosomal membrane, offering entry to the substrate, followed by the hydrolysis and postcoupling of the naphthol AS-BI group.

The stability of the lysosomal membrane is assessed as the time required for its permeabilization (destabilization period), which is represented by the maximum staining intensity within the lysosomes ("peak"). The longer the time needed to destabilize the membrane by acid labilization and temperature stress, the higher is the stability of the lysosomal membranes. Other terms in the literature are labilization period (LP) and latency of lysosomal enzymes.

Sampling and Preparation

Procedure for Fish

After the measurement of length and weight, the fish is killed, the peritoneal cavity opened, and the liver carefully removed without damaging the gall bladder. A piece of the central part of the liver (5 mm × 5 mm × 5 mm) is placed in a prelabeled cryovial and

Box 39.1 Equipment needed for the assessment of LMS (histochemical approach)

- High quality motorized cryostat microtome
- Cryostat chucks
- Freeze boxes for sections
- Shaking water bath up to 40°C
- Hellendahl histological staining jars
- Cleaned but untreated microscope slides with writing area, coverslips
- Bright-field binocular microscope with 10× and 40× objectives
- A 580 nm green filter to enhance contrast of the purple reaction product.
- Optional computer assisted image analysis
- In addition to standard laboratory equipment

- Cryo-Matrix (Neg -50, Richard Allen Scientific)
- Acetone
- Naphthol AS-BI N-acetyl-β-glucosaminide (Sigma) for mussel
- Naphthol AS-BI phosphate (Sigma) for fish
- Fast Violet B (Fluka)
- Collagen-derived polypeptide (POLYPEP, P5115, Sigma)
- Citrate buffer 0.1 M, pH 4.5, containing 2.5% sodium chloride (w/v)
- Phosphate buffer 0.1 M, pH 7.4
- Aqueous mounting medium (Difco, Kaiser's glycerol-gelatine, Sigma or others)

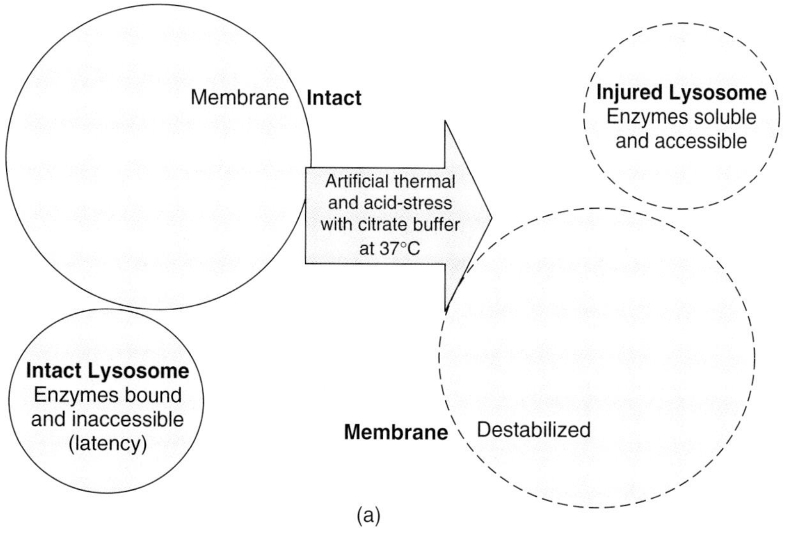

Fig. 39.3 (a)–(e) Scheme of the basic mechanisms of the histochemical approach of the lysosomal membrane stability test.

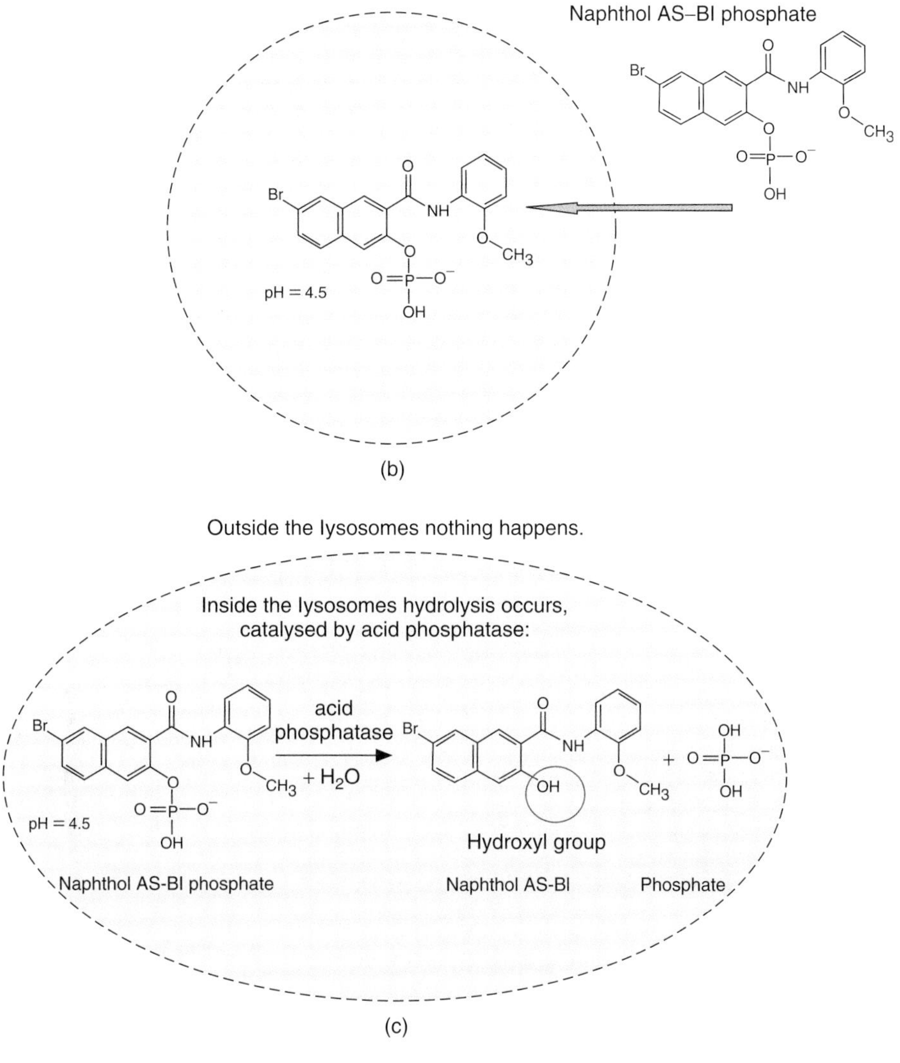

Fig. 39.3 *(continued)*

immediately frozen in liquid nitrogen. Samples can then be stored in a freezer at $-80°C$ for up to 1 yr. Following this procedure eliminates the risk of large ice crystal formation and avoids structural damage of the subcellular components. In the case of macroscopic diagnosis of neoplastic changes, the lesion and the surrounding extralesional parenchyma are dissected. General advantages of the freeze fixation are that the tissue is not chemically modified. Thus, lysosomal membranes are not altered or damaged, which allows testing of their functional capability.

Procedure for Bivalves

Bivalves should be sampled in the sublittoral part of the population, or should be sampled when submerged, since this minimizes the effects of air exposure during low tide on LMS. Collection can be by divers, benthic corers, or drag net. Sampling should be avoided during spawning. The line of maximum growth of the shell is measured. The specimens used for biomarker analyses are selected according to predefined size classes. The bivalves are washed 2–3 times (ca. 15 min) in seawater

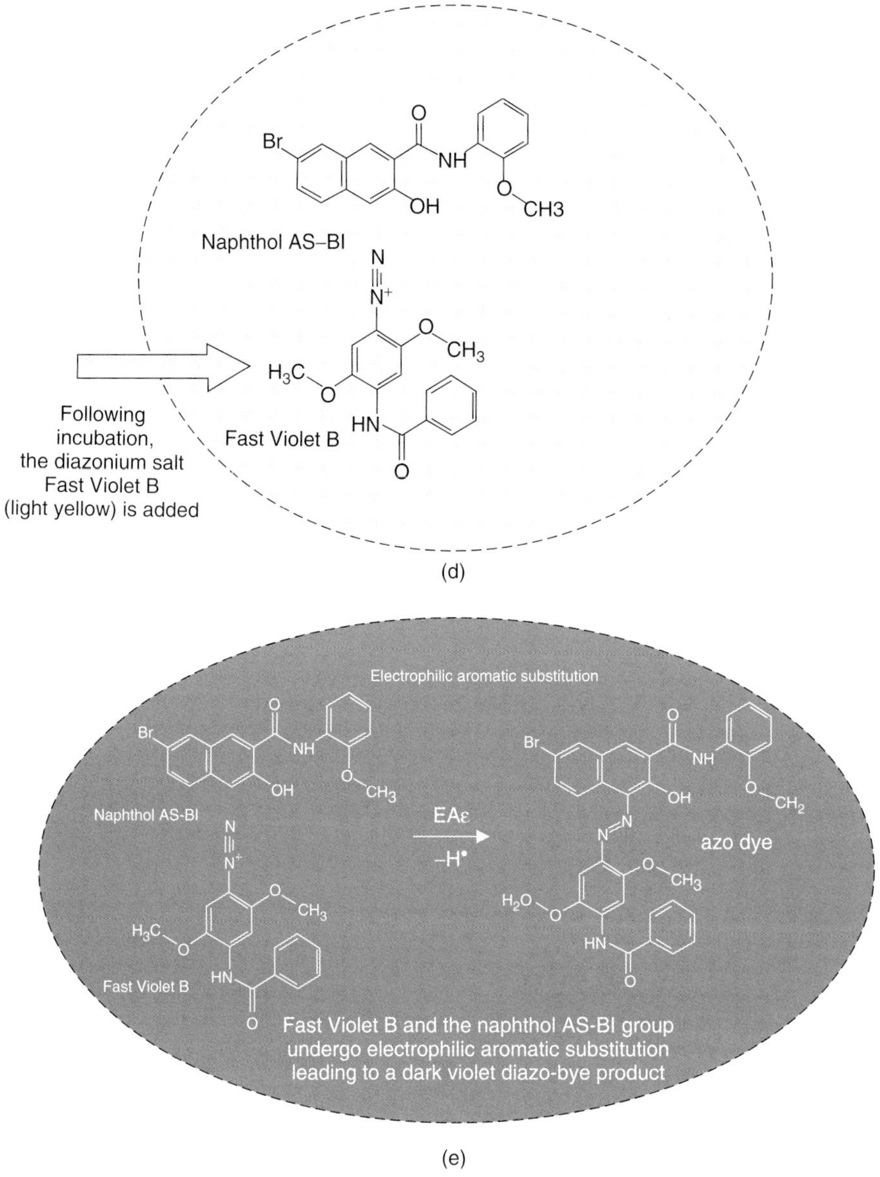

Fig. 39.3 *(continued)*

from the sampling site. Dissection should take place immediately after the washing.

If possible, transport of animals should be avoided, or kept as short as possible. If necessary, animal transport to the laboratory should avoid rough handling and mussels should be transported in containers with paper moistened with seawater. Lysosomal parameters can change within 1 h following air exposure (Brenner et al. (2011) submitted ms). Small pieces (5 mm × 5 mm × 5 mm) of freshly excised digestive gland tissue are placed on metal cryostat chucks (up to five pieces in a row across the centre, Fig. 39.4).

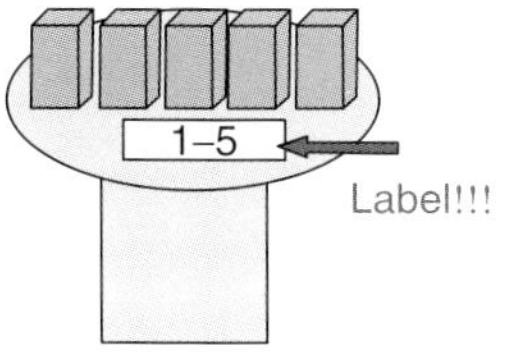

Fig. 39.4 Positioning of the digestive glands of five mussels on an aluminum chuck.

Each chuck is then placed for 1 min in a small bath of *n*-hexane that has been pre-cooled to $-70^{\circ}C$ in liquid nitrogen (Fig. 39.5). The chucks with the supercooled solidified tissues are subsequently sealed by double-wrapping with parafilm and aluminium foil (Fig. 39.6) and stored at $-80^{\circ}C$ until required for sectioning. Tissues may be stored for up to 1 yr.

Zorita et al. (2007a) showed that in mussel, as already described for fish, there are no differences in membrane stability when the tissue is directly frozen in liquid nitrogen. Thus, the procedure presented for fish liver can also be applied to mussel digestive gland.

The digestive gland is dissected, placed in premarked cryovials, and snap-frozen in liquid nitrogen. Samples can be stored at $-80^{\circ}C$ for up to 1 yr. A piece of the mantle is also dissected and fixed in Baker's calcium formalin to identify the reproductive stage and the gender of the specimens. Alternatively, the mantle can also be snap-frozen for these purposes, sectioned and stained for H & E.

A subsample (10–15 individuals) can be taken for the measurement of shell dimensions as well as tissue wet and dry weights. These will be used for the calculation of morphometric indices. The subsample can be used for the assessment of seasonal/nutritional condition of the species at the sampling site as a reference for all the biomarker measurements performed in that site during the sampling period.

Sectioning

Cryovials with tissue samples and aluminium chucks are placed in the cryostat cabin (Figs 39.7 and 39.8). Frozen tissue pieces are taken out of the vials and arranged according to their number. Up to five samples are glued to the precooled aluminium chuck with cryomatrix glue solution. The chuck is then firmly fixed in the cryostat. Section thickness, cutting velocity, and the antiroll guide are adjusted. The thickness of the section should not exceed 10 μm. First the "trimming mode" of the cryostat has to be chosen, to allow the knife to carefully approach the tissue. When the complete tissue section is evenly cut, the automatic sectioning mode is chosen and sectioning is started. The first "normal" section has to be discarded.

Before starting the sectioning, slides must be prelabeled. Cryostat sections are placed in duplicate on one slide. For LMS, prelabeled slides with labilization periods from 0 min to 50 min are prepared (Fig. 39.9).

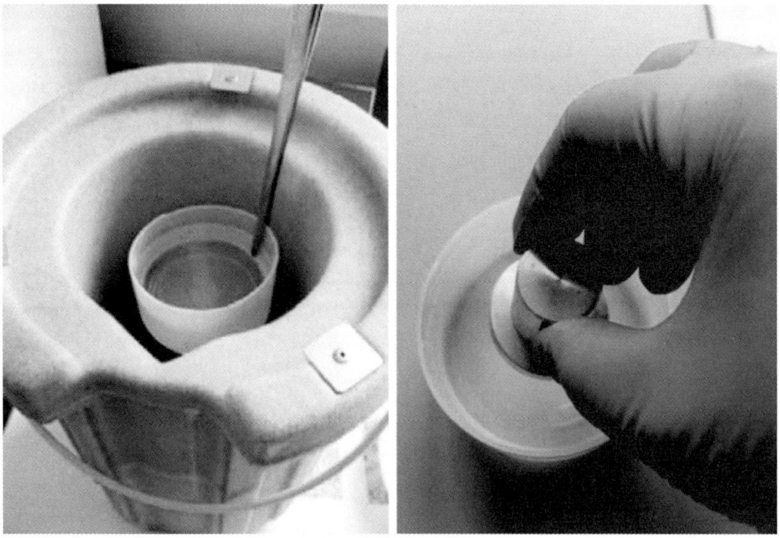

Fig. 39.5 Precooling of hexane followed by the freezing of the tissue on the aluminium chuck. (Photograph by Katja Broeg.) See plate section for a color version of this image.

Fig. 39.6 Wrapping of frozen tissue on the chucks in parafilm and aluminium foil for storage in a −80°C freezer. (Photograph by Katja Broeg.) See plate section for a color version of this image.

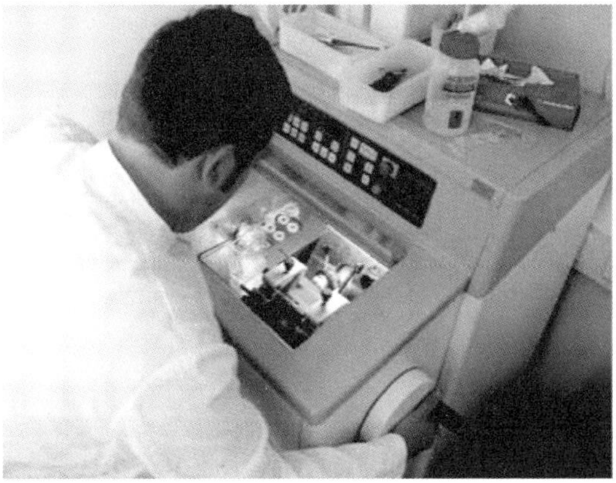

Fig. 39.7 Working with an automated cryostat. (Photograph by Katja Broeg.) See plate section for a color version of this image.

Fig. 39.8 A view into a cryostat cabin with a clumped aluminum chuck. (Photograph by Katja Broeg.) See plate section for a color version of this image.

Recommendations

- Glue the tissue on the chuck with a small space between each sample and with similar height for each of the samples.
- Clean the knife with acetone between two sections.
- Wait until acetone is evaporated before you continue to cut.

Another important factor is the temperature of the cryostat cabin. In a room-cooled cryostat, temperature should be between −25°C and −30°C depending on the fat content of the tissue (the higher the fat content, the lower the temperature). The best temperature for an object- and knife-cooled cryostat is −18°C (object cooling) and −16°C (knife cooling). The sectioning velocity should be kept constant for all sections. The

Fig. 39.9 Prelabeled slides with destabilization periods from 0 to 50 min. (Photograph by Katja Broeg.)

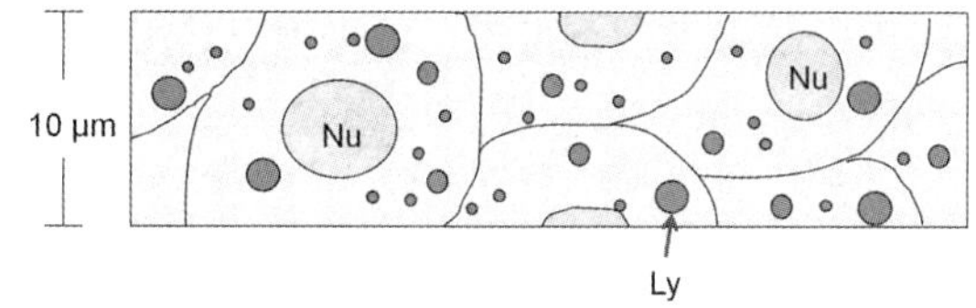

Fig. 39.10 Scheme of a tisssue section: Each cell is cut, but there are plenty of intact lysosomes within the cytosol (Nu = nuclei, Ly = lysosomes).

first section is placed on a slide (ambient temperature) and stored in the cryostat (Fig. 39.10). For the collection of the second section, the slide has to be warmed up slightly with the finger at the back of the slide.

After sectioning is completed, slides are stored frozen at −20°C for up to 1 day until further processing.

Labilization, Incubation, and Histochemistry

The LMS test comprises labilization, incubation, and enzyme histochemistry. The following list summarizes the laboratory procedures "step by step."

1. Plastic Hellendahl jars (as many as there are chucks that have been prepared for incubation) are filled with 50 mL citrate buffer (0.1 M, pH 4.5, containing 2.5% NaCl) each, closed and placed in a shaking water bath (Fig. 39.11) at 37°C until the buffer is warmed to bath temperature.
2. The first slides (50 min) are then placed into the jar (Fig. 39.12). Now the stopwatch is started, running from 50 min back to zero.
3. The next slides (45 min) are placed into the jar when the stopwatch shows 45, and so forth.
4. 2 min is the shortest time period of acid labilization.
5. When the alarm starts, all jars are taken out of the water bath and the citrate buffer is decanted.
6. Then, the "0 min" slides are placed into the jars.
7. The next step starts: Substrate incubation.
8. The two components of the incubation medium (see below) are thoroughly mixed.
9. The jars containing the slides and otherwise empty are filled with the freshly prepared incubation medium and closed.
10. They are placed again into the heated water bath and shaken for 15 min (fish, Köhler et al. 1992), or 20 min (mussel, Moore et al. 2004).
11. After this time period, jars are removed from the bath and the medium decanted.
12. Slides are thoroughly rinsed twice for 3 min with 3% saline to remove all residues of the incubation medium, prior to the next step: Azo-coupling.
13. The saline is decanted.
14. The Fast Violet B solution (see below) is poured into the jars.
15. Azo-coupling is performed for 10 min at room temperature.
16. Then Fast Violet B is decanted, followed by:
17. rinsing for 10 min in tap water;
18. fixation for 15 min in Baker's calcium formalin (see below) at room temperature;
19. rinsing three times in distilled water;
20. Air-drying and mounting in glycerine gelatine.

The lysosomal marker enzymes and respective substrates mainly used for LMS in the different indicator species (e.g. Figs 39.13 and 39.14) are:

For mollusks	For fish
β-N-acetylhexosaminidase (Hex): EC 3.2.1.52	Acid Phosphatase (AcP): EC 3.1.3.2
Reaction: Hydrolysis of terminal nonreducing N-acetyl-D-hexosamine residues in N-acetyl-β-D hexosaminides	Reaction: A phosphate monoester + H_2O = alcohol + phosphate
Systematic name: β-N-acetyl-D-hexosaminide N-acetylhexosaminohydrolase	Systematic name: phosphate-monoester phosphohydrolase (acid optimum)
Substrate: Naphthol AS-BI N-acetyl-β-D glucosaminide	Comments: Wide specificity. Also catalyzes transphosphorylations, marker enzyme for phagocytes.
	Substrate: Naphthol AS-BI phosphate

Fig. 39.11 Heated water bath with automated shaking, and a plastic Hellendahl jar. (Photograph by Katja Broeg.) See plate section for a color version of this image.

Preparation of Incubation Medium, Fast Violet B Solution, Citrate Buffer, and Baker's Formalin

All solutions are prepared before the first slides are placed in the water bath for acid labilization. The two components of the incubation medium (in beakers closed with Parafilm) are placed in the water bath for 15 min before the labilization period ends, to equilibrate it at the same temperature. The Fast Violet solution has to be stirred during the whole time period until use on a magnetic stirrer.

The Incubation Medium

1. Component 1: Dissolve 3.5 g low viscosity polypeptide (Polypep P5115, Sigma, tissue stabilizer) in 50 mL of 0.1 M citrate buffer (pH 4.5) containing 2.5% NaCl.
2. Component 2: Dissolve 20 mg naphthol AS-BI N-Acetyl-β-D glucosaminide (Sigma, substrate) (mussel and fish) or 10 mg naphthol AS-BI phosphate (Sigma, substrate) (fish) in 2.5 mL 2-methoxyethanol. The substrate has to be totally dissolved.
3. 1 min before incubation, both solutions are thoroughly mixed.

Fast Violet B Solution

Dissolve Fast Violet B (Fig. 39.15) in 0.1 M phosphate buffer (pH 7.4) with the concentration 1 mg mL^{-1}.

Citrate Buffer*

1. Make up 0.2 M solutions of trisodium citrate and citric acid.

2. Add citrate solution to citric acid until pH 4.5 is reached.
3. Measure volume of mixture.
4. Take an equal volume of distilled water and add sufficient NaCl to give a 5% solution.
5. Mix the two solutions.

*Remember: If you work with mussels from brackish water you must adjust the buffer to the ambient salinity.

Fig. 39.12 Positioning of tissue sections in the jars within the water bath at 37°C. (Photograph by Katja Broeg.) See plate section for a color version of this image.

Baker's Calcium Formalin

40% Formalin	300 mL
Calcium acetate	60 g

Fill up to 3 L with distilled water

Assessment

The time needed to destabilize the membranes is represented by the maximum staining intensity of the lysosomes, the "peak" (Figs 39.16 and 39.17). In most cases there are at least two peaks of maximum staining intensity, possibly due to differential membrane properties of the subpopulations of lysosomes (Broeg 2010).

The staining intensity is lower between the peaks, since the diazonium product leaks out when the membrane is further damaged and the lysosome is broken. Early first and second peaks indicate low stability of the lysosomal membranes.

Moore et al. (2004) suggested using the first peak of maximum staining intensity to determine the labilization period. Therefore, the first peak has been defined with respect to baseline/reference values (Broeg et al. 2005; Viarengo et al. 2007) as well as values representing reversible and irreversible liver damage (Köhler et al. 2002) and the progression of toxically induced alterations in fish (Broeg et al. 2005).

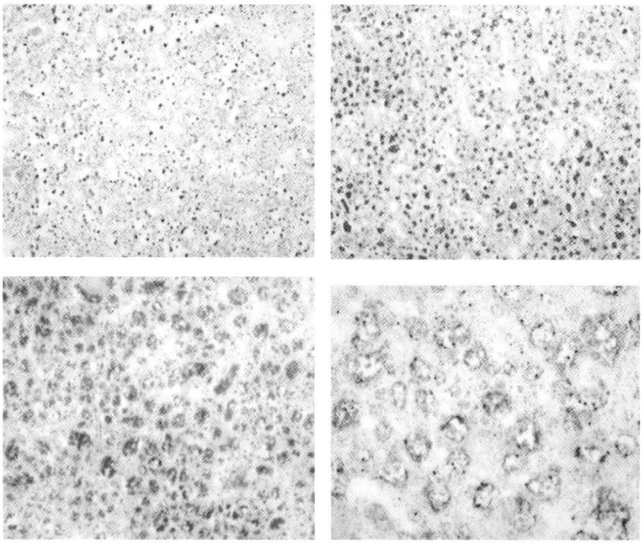

Fig. 39.13 Different sizes of lysosomes (pink) in fish liver (cryostat sections, acid phosphatase, 400×). (Photograph by Katja Broeg.) See plate section for a color version of this image.

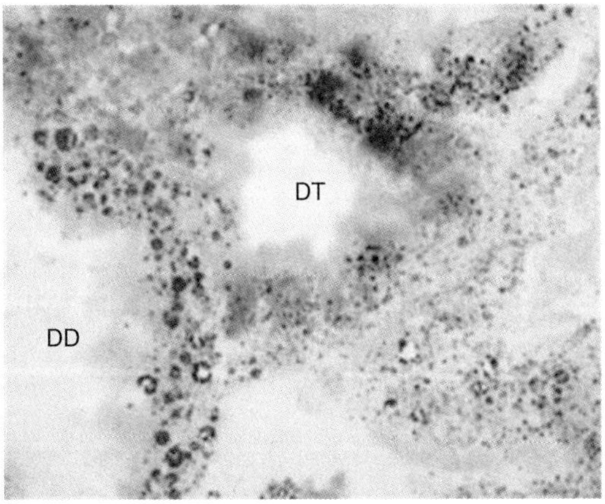

Fig. 39.14 Lysosomes in digestive gland of blue mussel (cryostat section, β-N-acetylhexosaminidase, 400×): DD = digestive duct, DT = digestive tubule. See plate section for a color version of this image.

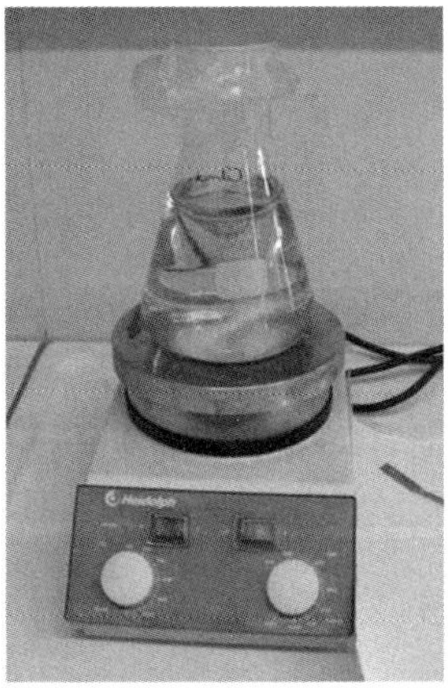

Fig. 39.15 Fast Violet B has to turn from light green/yellow to orange before use. (Photograph by Katja Broeg.) See plate section for a color version of this image.

To avoid incorrect measurements:

- Areas with artifacts like cracks and folding should not be used for the assessment.
- Choose areas with clearly identified digestive alveoli in mussel and homogeneous areas of liver tissue in fish.
- Avoid other areas such as digestive ducts, stomach, and connective tissue in mussel and areas with large vessels, inflammation, or pancreatic tissue in fish.

There are different procedures for identifying the destabilization period.

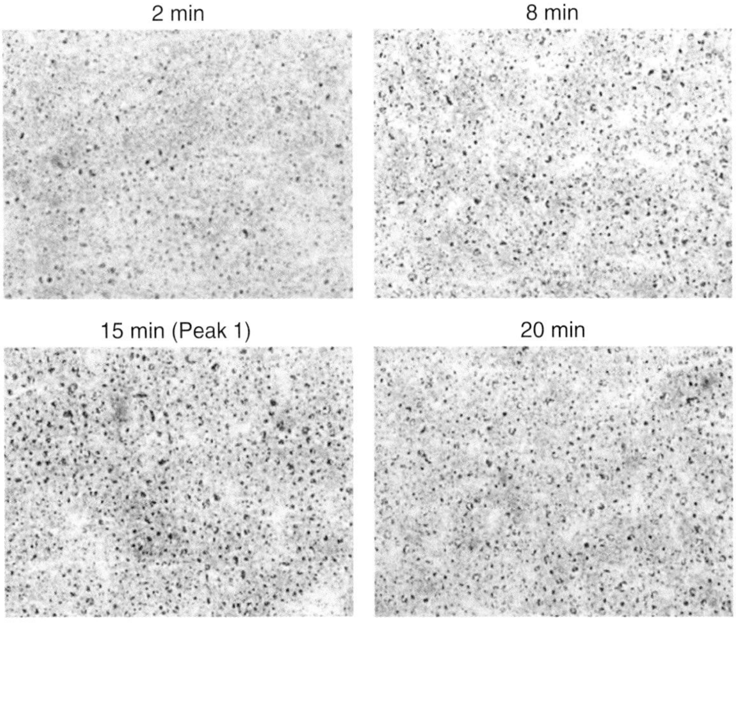

Fig. 39.16 Identification of highest staining intensity in lysosomes (destabilization period/peak) of flounder liver (cryostat sections, AcP, 400×). See plate section for a color version of this image.

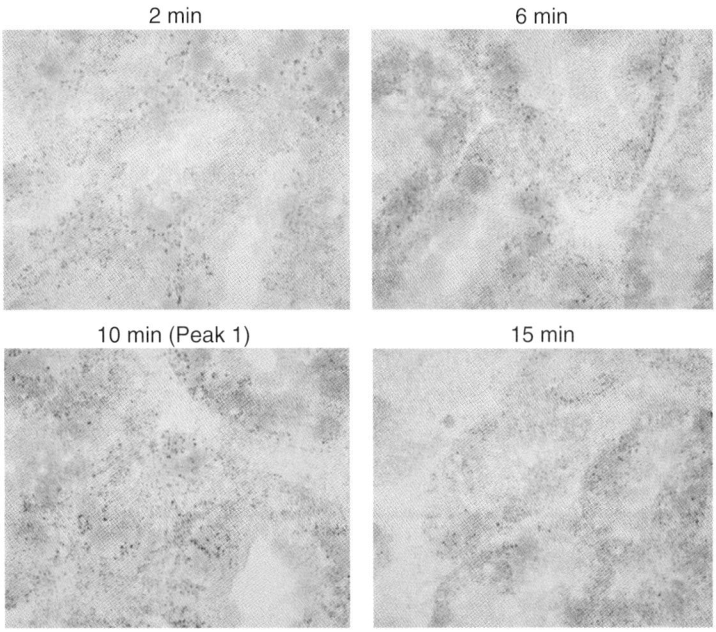

Fig. 39.17 Identification of highest staining intensity in lysosomes (destabilization period/peak) of mussel digestive gland (cryostat sections, Hex, 400×). See plate section for a color version of this image.

Assessment by Light Microscopy Without Image Analysis

The staining intensity can be assessed using microscopic examination (Moore et al. 2004) of the maximum staining intensity in the serial sections. For this procedure:

- each tissue section should be divided into four roughly equal areas for assessment.
- This can be done by means of drawing a cross with a very fine marker pen on the cover slide over each section, compartmentalizing it into quadrants. The positions and orientation of the cross should be nearly the same on all sections.
- Then the staining intensities of the different time intervals are compared stepwise to each other to determine the peaks.
- A mean value is obtained from the average of the assessments obtained in each of the quartiles.

Helpful in this context is the use of a diagram to mark the staining intensity of each section semiquantitatively (Fig. 39.18).

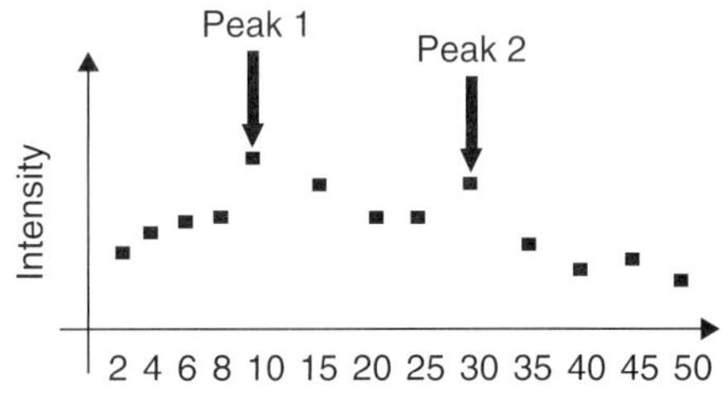

Fig. 39.18 Semiquantitative diagram for the assessment of the highest staining intensities by direct comparison of the different labilization periods.

Scanning of Standardized Images of Each Labilization Period for a Direct Comparison

- Scanning of the images of the different destabilization periods is performed by the use of computer-assisted image analysis.
- A prerequisite is the defined calibration of microscope and camera (constant light intensities).
- Using image analysis software, a "macro" has to be designed to scan images and present them in a "gallery" allowing a direct comparison between the staining intensities of the different labilization periods.

Automatic Measurement of Number and Percentage of Dark Stained Compartments in the Tissue with Computer Assisted Image Analysis

The example presented here uses a Zeiss Axioskop light microscope and the KS300 software. Considering the basic requirements, this can also be done with other microscopes and image-analysis software. The basic adjustments for the Zeiss Axioskop light microscope prerequisites to assure comparability of results are:

- objective 40×
- aperture of the condenser open
- lamp voltage 5 Volt
- aperture of the light source adjusted at 15
- green filter on (= monochromatic light)
- condenser lens pivoted.

To compare different images concerning their staining properties by computer-assisted image analysis, the light intensity must stay constant during the comparison. Assessment of the destabilization period is done by the measurement of the percentage of dark stained lysosomes in the area of measurement. For this purpose, the image is taken from the tissue at 400× magnification under standardized microscope and camera adjustments. The features of the measurement, in this case the "area%" of dark stained lysosomes, have to be defined with the image analyzing software. In the next step, the "threshold" of staining intensity is defined and must also be kept constant during the different measurement labilization periods. The threshold should be defined on the basis that the lysosomes are clearly identified by the chosen threshold and at the same time appear as individual lysosomes distinguishable from each other on the image. This measurement is conducted for all time periods from 2 to 50 min of LMS period. The resulting area% values show at which time period the dark stained lysosomes occupied the largest area in the image frame, representing the destabilization time, i.e. the peak.

Mussel digestive gland is not as homogeneous as fish liver. Therefore, the measurement has to be modified for the assessment of mussel LMS. The areas where the LMS is going to be assessed (regions of interest, ROI) have to be marked interactively prior to the measurement.

LYSOSOMAL MEMBRANE STABILITY—THE NEUTRAL RED RETENTION TEST

The neutral red retention technique for mussel haemocyte lysosomes is a nondestructive test (Moore et al. 2004). Thus, it has to be performed with live haemocyte cells immediately after sampling. Following careful extraction of a haemolymph sample, the animals can be returned to their habitat. Transport of the mussels should again be avoided. In case this is not possible, animals must be maintained in a moist environment at constant temperature (according to the ambient temperature of the sampling location) during the transport to the laboratory (see above, Histochemical Approach).

Stock solutions of physiological saline and neutral red are prepared in advance and stored in a refrigerator. Since the neutral red stock solution will become solid in the refrigerator, it has to be warmed to room temperature for dilution to the working concentration.

Haemolymph Extraction

1. The mussel valves are carefully separated along the ventral side, using a scalpel, which is also used to keep the valves apart during haemolymph extraction.
2. Remaining water is allowed to drain out of the shell cavity.
3. 0.1 mL of haemolymph is withdrawn from the posterior adductor muscle using a 1–2 mL hypodermic syringe with a 25-gauge needle containing 0.1 mL of physiological saline.
4. The needle is removed from the syringe and the contents emptied into a 1.5–2.0 mL Eppendorff tube.
5. The cells should be used within the next 20 min and kept in a refrigerator prior to use.
6. Following gentle mixing of the contents, 50 μL haemolymph and physiological saline mixture are applied to each slide. For each sample, a clean pipette tip is used.
7. Slides are placed in a lightproof cooled humidity chamber. They incubate for 15–20 min. Do not allow the slides to touch ice, this will damage the cells.

8. After this period, excess suspension is drained off and carefully wiped around the area containing adhered cells, to remove any remaining excess fluid.

Neutral Red Incubation

1. 40 μL of the neutral red (NR) working solution are pipetted onto the haemocytes. The solution has to be carefully applied by touching the surface of the slide with the pipette tip and gently purging the dye onto the cells. Wait 15 min to allow neutral red to penetrate the cells (in a dark humidity chamber).
2. A coverslip is gently applied.
3. Slides are then systematically examined under a light microscope after 15 min and then again after another15 min. Subsequent examinations are made at intervals of 30 min up to 120 min.
4. The final examination is made after 180 min of incubation.

If possible, the whole slide is scanned and placed again into the chamber as quickly as possible. Observations should not be longer than1 min per slide. All slides must receive the same exposure to light under the microscope, and the light intensity has to be kept as low as possible since neutral red is sensitive to light exposure and acts as a photosensitizer.

Assessment

Cells are examined for structural abnormalities and NR retention time. The retention time of the NR staining by the lysosomes is recorded by assessing the proportion of cells displaying leakage from the lysosomes into the cytosol (Figs 39.19 and 39.20). This is reflected by a uniform pale red coloration of the cytoplasm in contrast to clearly visible dark red lysosomes. Structural abnormalities comprise abnormalities in lysosomal size and cell shape. Conditions are recorded in a table at each time interval of evaluation.

The end point is reached when 50% or more of the cells, based on a visual or a digital photographic determination (see below), exhibit lysosomal leakage or show abnormalities such as enlargement. For the evaluation of retention time and the recording of structural abnormalities, the use of standardized evaluation sheets is recommended. An example is given in Moore et al. (2004).

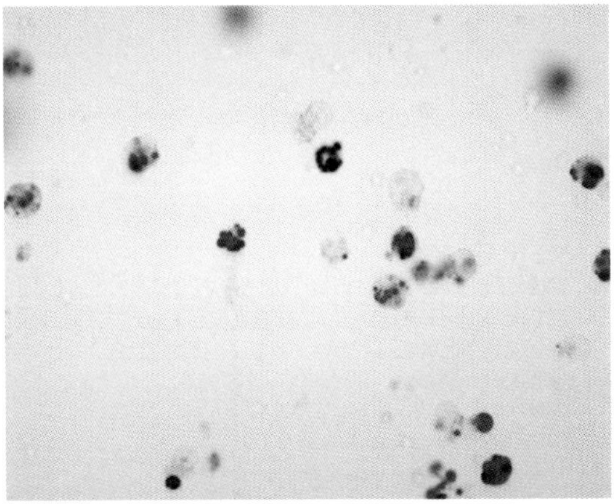

Fig. 39.19 Intact lysosomes in oyster haemocytes. The lysosomes are plainly visible inside the cells. Light microscopy, 400×. (Photograph by Katja Broeg.) See plate section for a color version of this image.

Reagents and Solutions

Mussel Physiological Saline

HEPES*	4.77 g
Sodium chloride	25.48 g
Magnesium sulphate	13.06 g
Potassium chloride	0.75 g
Calcium chloride	1.47 g

*N-2-Hydroxyethylpiperazine-N′-2-ethanesulfonic acid.

The salts are dissolved in approximately 800 mL of distilled water and then filled up to 1 L with more distilled water. The solution has to be stored in a refrigerator and equilibrated at room temperature prior to use, and before pH is checked and adjusted to 7.36 with 1 M NaOH. The saline has always to be adjusted to the salinity of the area in which the mussels have been sampled.

Neutral Red Stock Solution

- The NR stock solution (100 mM) is prepared by dissolving 28.8 mg of dye in 1 mL of DMSO (dimethyl sulfoxide). The mixture has to be stored in the refrigerator and can then be used for up to 2–3 weeks. Since the solution solidifies in the refrigerator, it must be warmed to room temperature before dilution to the working solution.

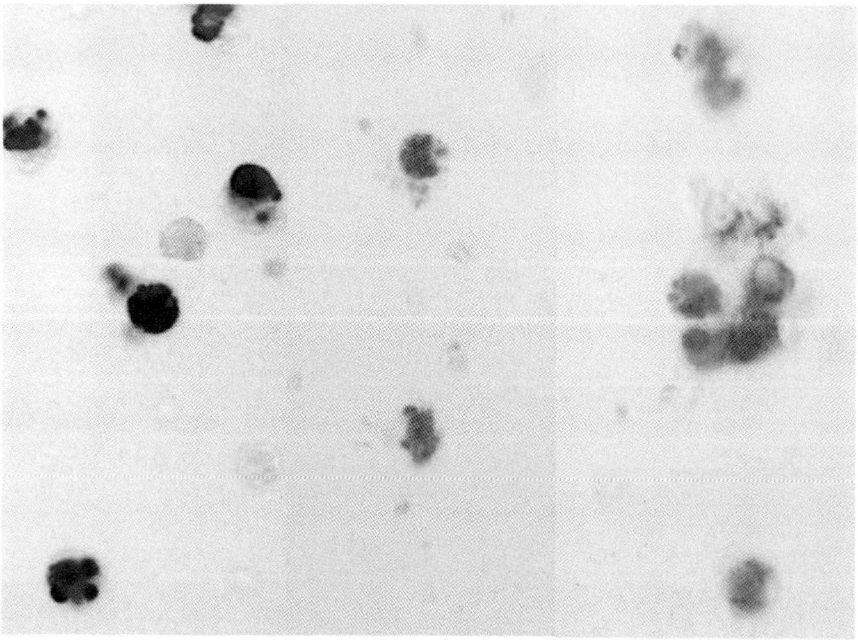

Fig. 39.20 Proceeding breakage of lysosomal membranes and increase of the size of mussel haemocytes. The destabilization is reflected by an even red color of the haemocytes. Lysosomes cannot be identified properly. Light microscopy, 400×. (Photograph by Katja Broeg.) See plate section for a color version of this image.

- To prepare the working solution, 10 μL of stock neutral red is diluted in 5 mL of mussel physiological saline. The working solution will last about 4 h before the dye starts to precipitate.

LIPOFUSCIN–HISTOCHEMICAL APPROACH

Lipofuscin is a permanent and not further degradable end product of lipid and protein oxidation. It is mainly stored in residual bodies in the cells or in late stages of phagocytes.

Unsaturated fatty acids present in membrane phospholipids are a major target for peroxidation. Reactions involving ROS occur in reaction chains: hydrogen is abstracted from the fatty acid by hydroxyl radical, leaving a carbon-centered radical within the fatty acid structure. This radical then reacts with oxygen to yield the peroxy radical, which can then continue to react with other fatty acids or proteins.

This process can have numerous effects: increased membrane rigidity, decreased activity of membrane-bound enzymes and ion pumps, altered activity of membrane receptors, and altered permeability. In addition to effects on phospholipids, radicals can also directly attack membrane proteins and induce lipid–lipid, lipid–protein and protein–protein cross-linking, all of which obviously have effects on membrane function.

Lipofuscin is detected by the ferric(III)-chloride/potassium ferricyanide technique according to Schmorl and modified by Bahns (Pearse, 1960).

- Cryostat sections (7–10 μm) are fixed in Baker's calcium formalin for 15 min.
- Rinsed in distilled water.
- Stained in Schmorl's solution. Maximum staining period 5 min.
- Washed in 1% acetic acid for 2 min.
- Rinsed in tap water for 10 min.
- 3 × rinsed in distilled water.
- Sections are air-dried and mounted in Histomount (Thermo Shandon, USA) or other UV-free mounting media.

The staining period should be kept as short as possible to minimize blue background staining. Schmorls solution can be used only once.

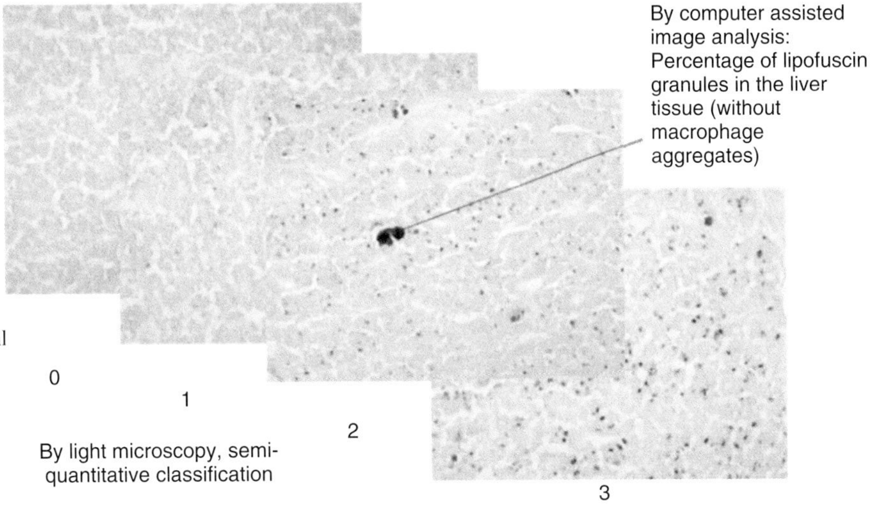

Fig. 39.21 Histochemical technique for the detection of lipofuscin and different methods for its assessment. See plate section for a color version of this image.

Staining Solution

Solution A

Ferric(III)-chloride	500 mg
Distilled water	50 mL

Solution B

Potassium ferricyanide	500 mg
Distilled water	50 mL

Solutions are mixed at a 1:1 ratio and used within 30 min after preparation

Assessment

The quantification of the staining intensity can be performed semiquantitatively (Fig. 36.21) or by image analysis measuring the area% of lipofuscin within the region of interest in digestive tubules or liver. Five measurements of each section are performed in digestive tubules or liver.

REFERENCES

BEEP. (2002) *Standard Operation Procedures of WP2.* EU Project Biological Effects of Environmental Pollution (BEEP) in Marine Coastal Ecosystems, http://www.beep.u-bordeaux1.fr

Bilbao, E., Soto, M., Cajaraville, M.P. et al. (2006) Cell and tissue level *Biomarkers* of pollution in feral pelagic fish, herring *(Clupea harengus)* and cod *(Gadus modua)* caged along a pollution gradient in Statfjord area (North Sea). *In* Hylland, K., Lang, T., Vethaak, D. (eds). ICES *Biological Effects of Contaminants in Pelagic Ecosystems.* BECPELAG Workshop, SETAC Press, Brussels, pp. 121–141.

Bitensky, L., Butcher, R.S., Chayen, J. (1973) Quantitative cytochemistry in the study of lysosomal function. *In* Dingle, J.T. (eds). *Lysosomes in Biology and Pathology*, Vol. 3. Elsevier, Amsterdam, pp. 465–510.

Brenner, M., Broeg, K., Wilhelm, C., Buchholz, C., Koehler, A. (2011) Effect of air exposure on lysosomal tssues of *Mytilus edulis* L. from intertidal wild banks and submerged culture ropes.. Submitted to *Comparative Biochemistry and Physiology* (A).

Broeg, K. (2002) *Funktionen von Makrophagen und ihren Aggregaten als histochemische Biomarker für Immunmodulation in Fischen verschiedener Klimazonen.* Dissertation, Universität Hannover.

Broeg, K. (2010) The activity of macrophage aggregates in the liver of flounder (*Platichthys flesus*) and wrasse (*Symphodus melops*) is associated with tissue damage. *Marine Environmental Research* 69, S14–S16.

Broeg, K., Zander, S., Diamant, A. et al. (1999) The use of fish metabolic, pathological and parasitological indices in pollution monitoring I. North Sea. *Helgoland Marine Research* 53, 171–194.

Broeg, K., Westernhagen, H.v., Zander, S. et al. (2005) The "Bioeffect Assessment Index" (BAI) – A concept for the quantification of effects of marine pollution by an integrated biomarker approach. *Marine Pollution Bulletin* 50, 495–503.

Caselli, F., Fabbri, E. (2005) Use of *Mytilus galloprovincialis* and *Tapes philippinarum* as sentinel organisms for the development of a biosurveillance program in the Piallassa Baiona coastal lagoon (Ravenna, Italy). *Chemistry and Ecology* 21, 465–477.

Champeau, O., Narbonne, J.F. (2006) Effects of tributyltin and 17b-estradiol on immune and lysosomal systems of the Asian clam *Corbicula fluminea* (M.). *Environmental Toxicology and Pharmacology* 21, 323–330.

Coughlan, B.M., Moroney, G.A, van Pelt, F.N.A.M. et al. (2009) The effects of salinity on the Manila clam (*Ruditapes philippinarum*) using the neutral red retention assay with adapted physiological saline solutions. *Marine Pollution Bulletin* 58, 1680–1684.

Da Ros, L., Nesto, N. (2005) Cellular alterations in *Mytilus galloprovincialis* (LMK) and *Tapes philippinarum* (Adams and Reeve, 1850) as biomarkers of environmental stress: field studies in the Lagoon of Venice (Italy). *Environment International* 31, 1078–1088.

De Duve, C. (1983) Lysosomes revisited. *European Journal of Biochemistry* 137, 391–397.

Diamant, A., Banet, A., Paperna, I. et al. (1999) The use of fish metabolic, pathological and parasitological indices in pollution monitoring II. Red Sea and Mediterranean Sea. *Helgoland Marine Research* 53, 195–208.

Dingle, J.T., Dean, R.T. (eds) (1976) *Lysosomes in Biology and Pathology*, Vol. 5. North Holland, Amsterdam, p. 404.

Gorbi, S., Virno Lamberti, C., Notti, A. et al. (2008) An ecotoxicological protocol with caged mussels, Mytilus galloprovincialis, for monitoring the impact of an offshore platform in the Adriatic Sea. *Marine Environmental Research* 65, 34–49.

Guerlet, E., Ledy, K., Giamberini., L (2006) Field application of a set of cellular *Biomarkers* in the digestive gland of the freshwater snail *Radix peregra* (Gastropoda, Pulmonata). *Aquatic Toxicology* 77, 19–32.

Guerlet, E., Ledy, K., Meyer, A., Giamberini, L (2007) Towards a validation of a cellular biomarker suite in native and transplanted zebra mussels: a 2-year integrative field study of seasonal and pollution-induced variations. *Aquatic Toxicology* 81, 377–388.

Hauton, C., Hawkins, L.E., Hutchinson, S. (2001) Response of haemocyte lysosomes to bacterial inoculation in the oysters *Ostrea edulis* L., *Crassostrea gigas* (Thunberg) and the scallop *Pecten maximus* (L). *Fish and Shellfish Immunology* 11, 143–53.

ICES (2006) *Report of the Working Group on Biological Effects of Contaminants (WGBEC)*, 27–31 March, Copenhagen, Denmark. ICES CM 2006/MHC, 04, 79 pp.

HELCOM (2010) Hazardous substances in the Baltic Sea – An integrated thematic assessment of hazardous substances in the Baltic Sea. Helsinki Commission, Baltic Marine Environment Protection. *Baltic Sea Environment Proceedings* 120 B, 156 pp.

JAMP (Joint Assessment and Monitoring Program). (2003) *JAMP Guidelines for Contaminant-Specific Biological Effects Monitoring*. OSPAR Commission, Ref. No: 2003–10, 38 pp.

Köhler, A. (1991) Lysosomal perturbations in fish liver as indicators for toxic effects of environmental pollution. *Comparative Biochemistry and Physiology C* 100, 123–127.

Köhler, A., Deisemann, H., Lauritzen, B. (1992) Histological and cytochemical indices of toxic injury in the liver of dab *Limanda limanda*. *Marine Ecology Progress Series* 91, 141–153.

Köhler, A., Wahl, E., Soffker, K. (2002) Functional and morphological changes of lysosomes as prognostic *Biomarkers* of toxic liver injury in marine flatfish *(Platichthys flesus* (L)). *Environmental Toxicology and Chemistry* 21, 2434–2444.

Kopecka, J., Lehtonen, K.K., Barsiene, J., et al. (2006) Measurements of biomarker levels in flounder (*Platichthys flesus*) and blue mussel (*Mytilus trossulus*) from the Gulf of Gdańsk (southern Baltic). *Marine Pollution Bulletin* 53, 406–421.

Lee, B.T., Kim, K.W. (2009) Lysosomal membrane response of earthworm, Eisenia fetida, to arsenic contamination in soils. *Environmental Toxicology* 24, 369–376.

Marchi, B., Burlando, B., Moore, M.N., Viarengo, A. (2004) Mercury- and copper-induced lysosomal membrane destabilisation depends on $[Ca^{2+}]$ dependent phospholipase A2 activation. *Aquatic Toxicology* 66, 197–204.

Marigómez, I., Baybay-Villacorta, L. (2003) Pollutant specific and general lysosomal responses in digestive cells of mussels exposed to model organic chemicals. *Aquatic Toxicology* 64, 235–257.

Moore, M.N. (1976) Cytochemical demonstration of latency of lysosomal hydrolases in digestive cells of the common mussel *Mytilus edulis*, and changes induced by thermal stress. *Cell and Tissue Research* 175, 279–287.

Moore, M.N. (1988) Cytochemical responses of the lysosomal system and NADPH-ferrihemoprotein reductase in molluscan digestive cells to environmental and experimental exposure to xenobiotics. *Marine Ecology Progress Series* 46, 81–89.

Moore, M.N., Lowe, D., Koehler, A. 2004. Biological effects of contaminants: measurement of lysosomal membrane stability. *ICES Times* 36, 31.

Pearse, A.G.E. (1960) *Histochemistry, Theoretical and Applied*. Churchill, London, 998 pp.

Regoli, F., Gorbi, S., Machella, N. et al. (2005) Pro-oxidant effects of extremely low frequency electromagnetic fields in the land snail *Helix aspersa*. *Free Radical Biology and Medicine* 39, 1620–1628.

Regoli, F., Gorbi, S., Fattorini, D. et al. (2006) Use of the land snail *Helix aspersa* as sentinel organism for monitoring ecotoxicologic effects of urban pollution: an integrated approach. *Environmental Health Perspectives* 114, 63–69.

Romeo, M., Bennani, N., Gnassia-Barelli, M. et al. (2000) Cadmium and copper display different responses towards oxidative stress in the kidney of the sea bass *Dicentrarchus labrax*. *Aquatic Toxicology* 48, 185–194.

Schiedeck, D., Broeg, K., Barsiene, J. et al. (2006) Biomarker responses and indication of contaminant effects in blue mussel (*Mytilus edulis*) and eelpout (*Zoarces viviparus*) from the western Baltic Sea. *Marine Pollution Bulletin* 53, 387–405.

Sturve, J., Balk, L., Berglund, A. et al. (2005) Effects of dredging in Göteborg harbour assessed by biomarkers in eelpout (*Zoarces viviparus*). *Environmental Toxicology and Chemistry* 24, 1951–1961.

UNEP/MAP. (2005) *Facts Sheets on Marine Pollution Indicators*. UNEP (DEC)/MED/WG.264/Inf. 14, Athens, 249 pp.

UNEP/RAMOGE. (1999) *Manual on the Biomarkers Recommended for the MED POL Biomonitoring Programme*. UNEP, Athens, 39 pp.

Viarengo, A., Lowe, D., Bolognesi, C. et al. (2007) The use of biomarkers in biomonitoring: a 2-tier approach assessing the level of pollutant-induced stress syndrome in sentinel organisms. *Comparative Biochemistry and Physiology* 146 C, 281–300.

Werner, I., Broeg, K., Cain, D. et al. (2000) Biomarkers of heavy metal effects in two species of caddisfly larvae from Clark Fork River, Montana: stress proteins (HSP70) and lysosomal membrane integrity. *Clark Fork Symposia Papers and Abstracts on the Upper Clark Fork or Blackfoot Rivers*.

Zaldibar, B., Rodrigues, A., Lopes, M. et al. (2006) Freshwater molluscs from volcanic areas as model organisms to assess adaptation to metal chronic pollution. *Science of Total Environment* 371, 168–175.

Zaroogian, G., Voyer, R.A. (1995) Interactive cytotoxicities of selected organic and inorganic substances to brown cells of *Mercenaria mercenaria*. *Cell Biology and Toxicology* 11, 263–271.

Zorita, I., Apraiz, I., Ortiz-Zarragoitia, M. et al., (2007a) Assessment of biological effects of environmental pollution along the NW Mediterranean Sea using mussels as sentinel organisms. *Environmental Pollution* 148, 236–250.

Zorita, I., Apraiz, I., Ortiz-Zarragoitia, M. et al. (2007b) Assessment of biological effects of environmental pollution along the NW Mediterranean Sea using red mullets as sentinel organisms. *Environmental Pollution* 153, 157–168.

INDEX

Oxidative Stress in Aquatic Ecosystems, First Edition. Edited by Doris Abele, José Pablo Vázquez-Medina, and Tania Zenteno-Savín.

Index compiled by Laurence Errington